About Pearson

Pearson is the world's learning company, with presence across 70 countries worldwide. Our unique insights and world-class expertise comes from a long history of working closely with renowned teachers, authors and thought leaders, as a result of which, we have emerged as the preferred choice for millions of teachers and learners across the world.

We believe learning opens up opportunities, creates fulfilling careers and hence better lives. We hence collaborate with the best of minds to deliver you class-leading products, spread across the Higher Education and K12 spectrum.

Superior learning experience and improved outcomes are at the heart of everything we do. This product is the result of one such effort.

Your feedback plays a critical role in the evolution of our products and you can contact us - reachus@pearson.com. *We look forward to it.*

Engineering Physics

Engineering Physics

Shatendra Sharma
Professor and Director
University Science Instrumentation Centre
Jawaharlal Nehru University
New Delhi

Jyotsna Sharma
Associate Professor
School of Basic and Applied Sciences
K. R. Mangalam University
Gurugram

Pearson

Editor—Acquisitions: Harsha Singh
Editor—Production: M. Balakrishnan

ISBN 978-93-868-7331-6

First Impression, 2018
Second Impression, 2024

Published by Pearson India Education Services Pvt. Ltd, CIN: U72200TN2005PTC057128

Head Office: 1st Floor, Berger Tower, Plot No. C-001A/2, Sector 16B, Noida – 201 301,Uttar Pradesh, India.

Registered Office: Featherlite, 'The Address' 5th Floor, Survey No 203/10B, 200 Ft MMRD Road,Zamin Pallavaram, Chennai - 600044
Website: in.pearson.com; Email:companysecretary.india@pearson.com

Digitally printed in India by Trinity Academy for Corporate Training Ltd, New Delhi in the year of 2024.

Dedicated to
Our Parents and Family Members
who have been
source of motivation and encouragement

—**Shatendra Sharma** and **Jyotsna Sharma**

Brief Contents

Contents

Preface

The book "Engineering Physics" is intended to serve as a text book for B.Tech., B.Sc. and Diploma for Engineering students of Indian universities and technical education boards. The chapters of the book cover the revised and latest syllabi of most of the Indian universities and is written to provide a simple, clear and logical presentation of subject with a large number of diagrams and illustrations. The basic concepts and derivations are included in a simple and easy manner so that the students do not find any difficulty in going through the different chapters of the book. Throughout the book, the emphasis is laid on the physical concepts. The students shall not need to refer the elementary books to supplement their course of study. Great efforts have been made to make the subject matter lucid and comprehensive. To make the book more useful and to clarify the concepts several solved examples are included in each chapter. Brief summary of each chapter is also given at the end as points to remember. A number of unsolved problems, short and long answer questions are also included with the view that the students may practice to solve using incorporated principles in different sections of the chapter. The key terms with definitions and important formulae are given at the end of each chapter. This will help the students to revise the important definitions and formulae used in the chapter.

Four online chapters are available with this book and can be found at www.pearsoned.co.in/shatendrasharma. These can be accessed by the students after registration with the key available with the book. Multiple-choice Questions are also available on the book website. Also, complete solutions manual is also available for the instructors at www.pearsoned.co.in/shatendrasharma.

In spite of the best efforts of the authors, it is possible that some unintentional errors and misprints might have crept in. Hence, any suggestions from the readers for the improvement of articles would be cordially accepted. We hope the book will be useful for the students.

Shatendra Sharma
Jyotsna Sharma

Acknowledgements

We express our deep sense of gratitude to our teachers, students, colleagues and friends for their help and suggestions in refining the contents and presentation of the book.

Jyotsna Sharma would like to express special thanks to Prof. K. K. Agrawal, Chancellor, KRMU for his humane and able guidance during the preparation of the manuscript. She is also grateful to the university management for providing cooperation at all stages.

We are also indebted to our spouses and family members for their patience to bear with us during the preparation of the book.

The authors are thankful to the publisher and composer for bringing out the book on time and in the desired format and Mr. Prabin Kumar for providing secretarial assistance.

We acknowledge the courtesy of M/s. Zeiss, Jeol Ltd., PANalytical B.V., Brucker Inc., GE, Digital Instruments and Varian Inc. for the photographs of some of their instruments that are used as illustrations in this book.

The publishers and the authors would also like to thank the reviewers who took time to go through the book and gave us their valuable feed back. Their names are as follows:

Asit Kumar Gupta
Asansol Engineering College
West Bengal

Manish Kaushik
Shankara Institute of Technology
Rajasthan

Praful D. Shirbhate
G S Gawande College of Engineering
Maharashtra

B. M. Pratima
Formerly with Swarna Bharathi Institute of Science & Technology
Telangana

S. Jayakumar
SNS College of Engineering
Tamil Nadu

Sandeep Chhoker
Jaypee Institute of Information Technology
Uttar Pradesh

Akumalla Sanathkumar
IIIT-RK Valley (Idupulapaya) Institute
Andhra Pradesh

Sanjeev Kumar
PEC University of Technology
Chandigarh

Vipin Gupta
KIET Group of Institutions
Uttar Pradesh

Shatendra Sharma
Jyotsna Sharma

About the Authors

Dr Shatendra Sharma is currently Professor and Director of University Science Instrumentation Centre Jawaharlal Nehru University, New Delhi, India. He has worked as Lecturer in Physics; Scientist with (DRDO) Defence Research and Development Organization, Ministry of Defence; Associate Professor, Jawaharlal Nehru University; Director as well Advisor, All India Council for Technical Education Ministry of HRD; Founder Director, Advanced Instrumentation and Research Facility at JNU, New Delhi. He is working as a Professor since 1999 at JNU, New Delhi, India. The areas of his research interest include atomic and nuclear radiation physics, XRF, electron microscopy and de-excitation of ionized atoms after ion–atom interactions, characterization of materials, etc. He is currently working in the area of characterization and processing of nano-structures. He has authored and edited 12 books in Physics and other engineering subjects and has published more than 50 research papers in referred journals. He has participated and presented research papers in several national and international conferences. He is also a Fellow Member of Institution of Engineers (India) and Institution of Electronics and Telecommunication Engineers.

Dr Jyotsna Sharma is currently working as an Associate Professor in Physics at K. R. Mangalam University, Gurugram. She obtained her Ph.D. in Plasma Physics from Jawaharlal Nehru University, New Delhi. She has over 12 years of teaching experience in Physics to engineering students. In recognition of excellent teaching skills and significant contributions in other professional activities, she was bestowed upon with *"Best Teacher Award"* in 2011 by KIIT College of Engineering. The areas of her research interest are the study of instabilities in plasma & dusty plasmas and interaction of ion beams with plasma. She is actively engaged in research and has been awarded a research project by Department of Science and Technology in the area of plasma physics. She has published more than 20 research papers in international journals. She has also participated in many international conferences for the presentation of her research work. She is also a member of many professional bodies.

At a Glance

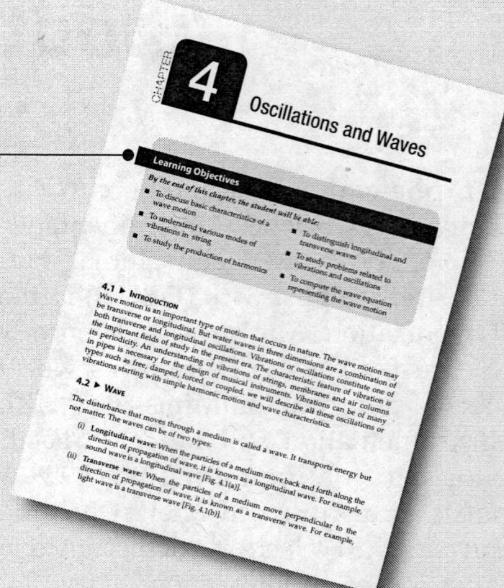

Learning Objectives: These are included in the beginning of each chapter to identify the expected outcomes after completion of the chapter

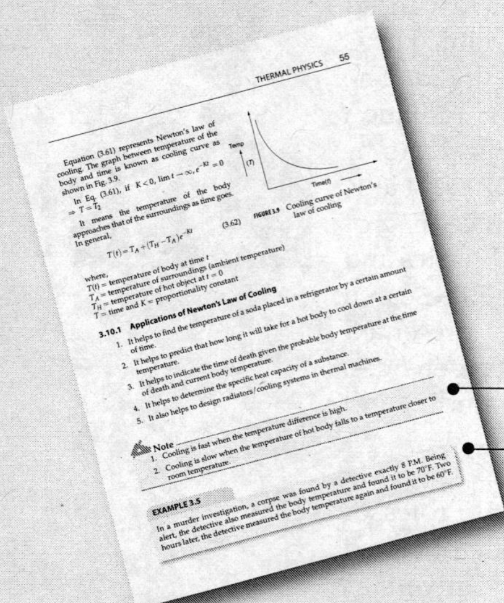

Note: Important and interesting facts are highlighted in each chapter

Solved Examples: Numerous examples solved in a step-by-step manner

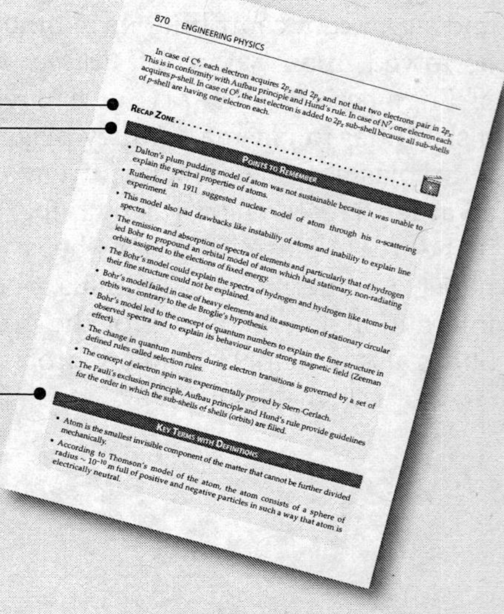

RECAP ZONE is aimed at summarizing the concepts learnt in the chapter and is ideal for revision. It includes

- Points to Remember
- Key Terms with Definitions
- Important Formulae and Equations

Points to Remember summarizes the concepts and topics presented in each chapters

Key Terms with Definitions presents all the important terms with their definitions

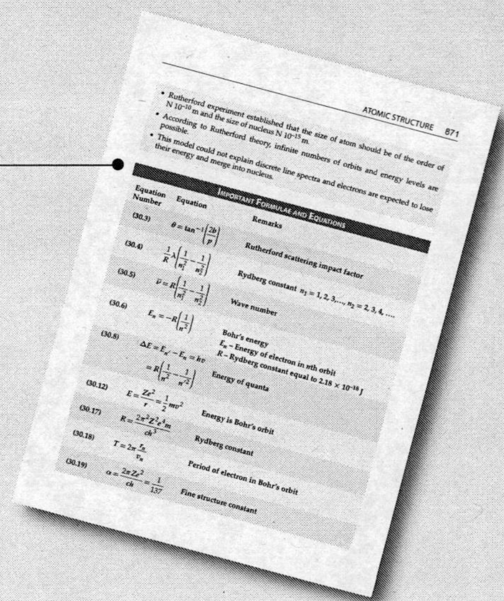

Important Formulae and Equations list downs all the important formulae and equations encountered in each chapter

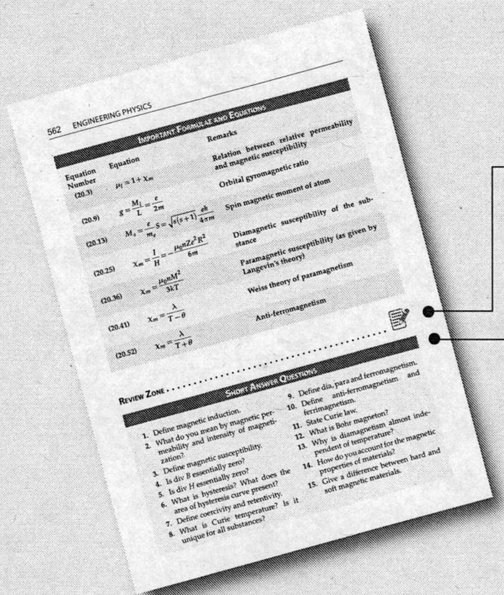

REVIEW ZONE is aimed at assessment. It includes

- Short Answer Questions
- Long Answer Questions
- Numerical Problems

Short Answer Questions prepare students for viva-voce examinations and can be answered in one-word or short sentences

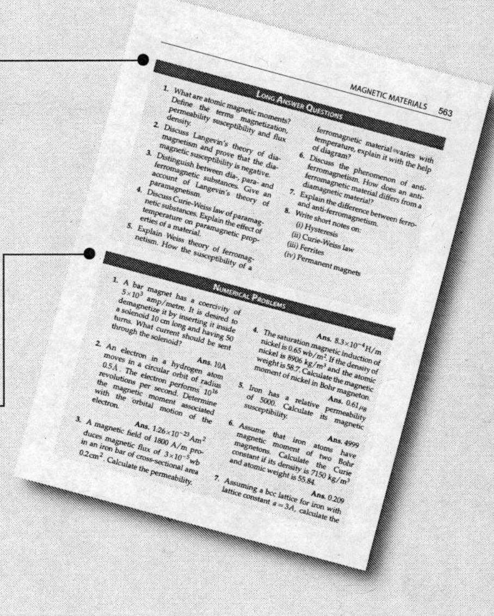

Long Answer Questions provides descriptive type questions and are built around frequently asked questions in examinations

Numerical Problems include problems which are mathematical in nature and are provided with the final answers

CHAPTER 1

Basic Concepts

Learning Objectives

By the end of this chapter, the student will be able:

- To review the concept of electric charge, field and electric force
- To understand the Blackbody radiation and derive the Planck's radiation formula
- To explain molecular model of gas and energy–pressure relation
- To derive Maxwell's velocity distribution, Boltzmann's distribution and law of equi-partition of energy
- To estimate the size of atom and atomic radii
- To define Mole and Avogadro number

1.1 ▶ Electric Charge

The phenomenon of two electrically charged objects attracting each other can be explained by the fact that objects gain a net electric charge, when rubbed with some other material.

These charges are of two types, namely positive (+) and negative (–), with the following basic properties:

- Charges of the same sign repel each other.
- Charges of the opposite sign attract each other.

Detailed experiments have established the following fundamental characteristics of electric charge:

- Charge is neither created nor destroyed, it is always conserved.
- Charge always exists as an integral multiple of a basic unit which means that it is quantized.

This basic unit of charge is conventionally denoted by e:

$$e = 1.602 \times 10^{-19} \text{ Coulombs (C)} \tag{1.1}$$

In an atom, the charge on an electron is e^- and that on a proton is e^+.

Note

The positive and negative sign designation of two opposite types of charges are relative for mathematical purpose only. Physics would have not changed even if these signs were interchanged.

1.2 ▶ Electric Forces and Fields

The amount of attraction or repulsion between charged objects can be put in quantitative terms by the introduction of the electric force. The simplest case is the force between two point charges (charges with a negligible size). Experiments by Coulomb and others uncovered the following formula for the magnitude of force $\vec{F}$ between two such charges Q_1 and Q_2 separated by a distance r, as in Fig. 1.1.

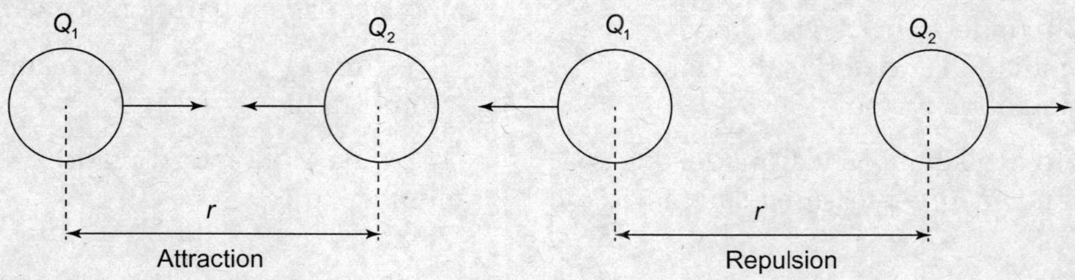

FIGURE 1.1 Force between two point charges Q_1 and Q_2

$$\vec{F} \propto \frac{Q_1 Q_2}{r^2} \text{ if } Q_1 = Q_2 = 1\,\text{C} \text{ and } r = 1\,\text{m} \tag{1.2}$$

$$F = k$$

where k is a constant called Coulomb's constant

$$k = 9.0 \times 10^9 \frac{\text{N}\cdot\text{m}^2}{\text{C}^2} \tag{1.3}$$

The direction of the force is along a line joining the two charges. It is repulsive if the charges have the same sign and is attractive if the charges have the opposite sign.

If there are more than two charges present, then the force on any one charge must be found by adding vectorially the forces found by Coulomb's law [Eq. (1.2)] between each pair of charges.

It is convenient under many circumstances to introduce the concept of the electric field, conventionally denoted by

$$f_x = E_x \qquad f_y = E_y - vB_z \qquad f_z = E_z + vB_y.$$

Suppose we have a "background" distribution of charges Q_1 Q_2, ..., Q_n in some region of space and measure the force $\vec{F}$ of a charge q placed nearby. The electric held $\vec{E}$ associated with this charge distribution is defined through the relation:

$$\vec{F} = q\vec{E} \tag{1.4}$$

Thus, the unit of $\vec{E}$ is N/C (Newton/Coulomb). In a sense, the charge q is a *test charge*, which probes the strength at different points of the potential electric force due to the charges $Q_1, Q_2, \ldots, Q_n$. Note that, for a given electric field $\vec{E}$, the force on a positive charge is opposite in direction to the force on a negative charge.

For a single point charge Q, the electric field at a distance r away is found from Eqs. (1.2) and (1.4) to have the magnitude:

$$E = k\frac{Q}{r^2} \tag{1.5}$$

with a direction equal to the direction of the force on a positive test charge placed at the point of interest. As with the electric force, the electric field due to multiple point charges can be found by adding vectorially the electric field found by Eq. (1.5) for each individual charge.

1.3 ▶ Blackbody Radiation

Until the late eighteenth century, the wave nature of light was the prevalent theory, as it could explain most of the experiments conducted on light. However, there were a few notable exceptions. One such exception was that associated with blackbody (perfect absorber and emitter) radiation—the radiation that a body emits when heated. It was known that the nature of this radiation changes as the temperature changes. Experiments on "blackbodies" show the following typical curves of the intensity of the radiation (energy emitted per unit time per unit area) versus the wavelength at a fixed temperature T, as shown in Fig. 1.2.

However, when the wave nature of light was applied to explain blackbody radiation, it could not explain the phenomena. Rather, it predicted that the intensity, I, for a given temperature should behave as:

$$I \sim \frac{1}{\lambda^4} \tag{1.6}$$

which agrees with the experimental data for long wavelengths but deviates for short wavelengths. This is illustrated in Fig. 1.3.

In 1900, Max Planck devised a theory of blackbody radiation which was in agreement with all wavelengths. In this theory, the molecules of a body cannot have arbitrary energies, but are instead quantized, i.e. the energies can only have discrete values. The magnitude of these energies is given by the formula:

$$E = nh\nu, \tag{1.7}$$

where $n = 0, 1, 2, \ldots$ is an integer, ν is the frequency of vibration of the molecule and h is a constant, called Planck's constant.

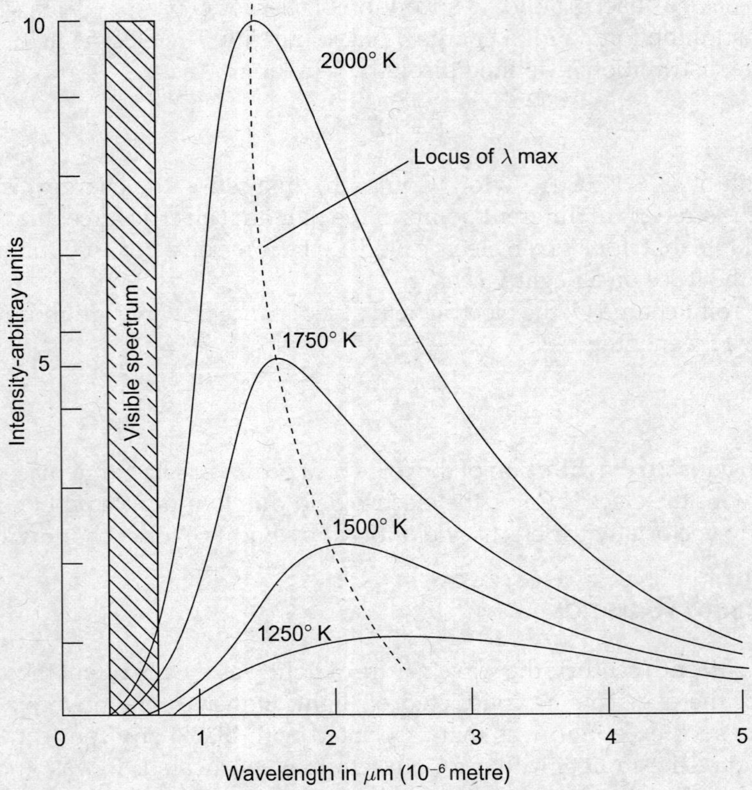

FIGURE 1.2 Energy distribution in spectrum of an incandescent solid at various temperatures

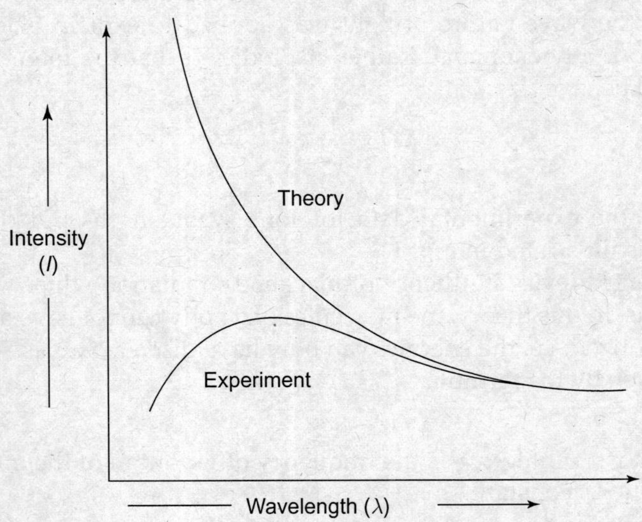

FIGURE 1.3 Classical prediction of blackbody radiation

and, Planck's constant $h = 6.626 \times 10^{-34}$ Joul·sec

$$= 4.136 \times 10^{-15} \text{ eV·sec}$$

$$\nu = \text{frequency of radiation}$$

This quantum idea later explained the photoelectric effect. It became a part of the Bohr's theory of discrete atomic spectra and laid the foundation of modern quantum theory.

1.3.1 Planck's Radiation Formula

With the assumption that the electromagnetic modes in a cavity are quantized with the quantum energy equal to Planck's constant times the frequency, Planck derived a radiation formula. The average energy per "mode" or "quantum" is the energy of the quantum times the probability that it will be occupied (the Einstein-Bose distribution function):

$$\langle E \rangle = e \frac{h\nu}{kV / kT} - 1 \tag{1.8}$$

This average energy times the density of such states, expressed in terms of either frequency or wavelength

$$\rho(\nu) = \frac{dn_s}{dv} = \frac{8\pi}{c^3}\nu^2$$

$$\rho(\lambda) = \frac{dn_s}{d\lambda} = \frac{8\pi}{\lambda^4} \tag{1.9}$$

gives the energy density, the Planck's radiation formula.

Energy per unit volume per unit frequency

$$S_\nu = \frac{8\pi h}{c^3} \frac{\nu^3}{e^{hv/kT} - 1} \tag{1.10}$$

Energy per unit volume per unit wavelength

$$S_\lambda = \frac{8\pi hc}{\lambda^5} \frac{1}{e^{hc/\lambda kT} - 1} \tag{1.11}$$

The Planck's radiation formula is an example of the distribution of energy according to Bose-Einstein statistics as shown in Fig. 1.4. The above expressions are obtained by multiplying the density of the states in terms of frequency or wavelength times the photon energy times the Bose-Einstein distribution function with normalization constant $A = 1$.

In comparison Rayleigh-Jeans law and the quantum Planck's radiation formula, experiments confirm the Planck's relationship (Fig. 1.5).

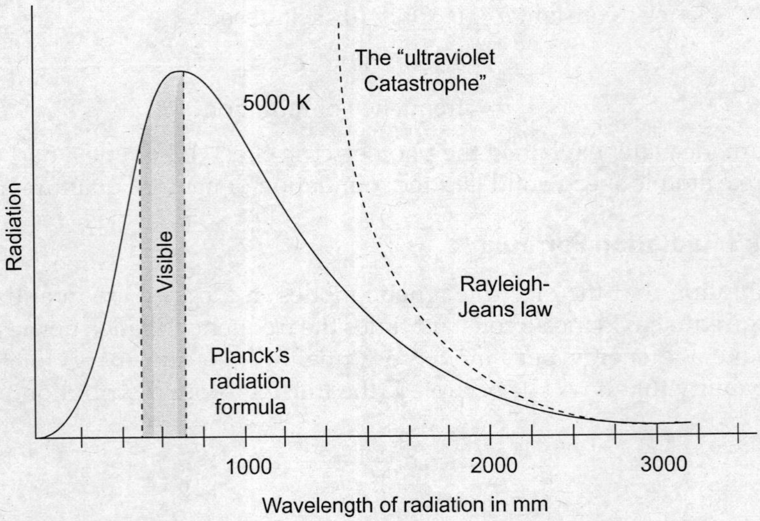

FIGURE 1.4 Blackbody radiation intensity as a function of frequency

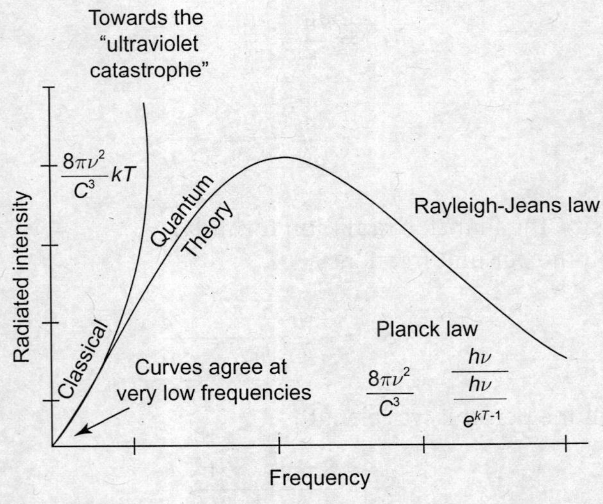

FIGURE 1.5 Rayleigh-Jeans curve agrees with the Planck's radiation formula for long wavelengths (low frequencies) in experiments

1.4 ▶ Energy and Pressure Relationship in a Molecular Model of Gas

It is not difficult to extend Bernoulli's theory to a quantitative description, relating the gas pressure to the molecular velocities.

Let us consider a single perfectly elastic gas particle of mass m, bouncing rapidly back and forth at speed v inside a narrow cylinder of length L with a piston at one end. It is presumed that the particle moves in a straight line, and its back and forth motion is along the same line.

Obviously, non-continuous force is exerted on the piston but at regular intervals. However, if the piston is much heavier than the particle, the effect similar to a smooth force over a period of time compared to the interval between impacts of force. So we can find the value of the equivalent force.

Using Newton's law in the form

force = rate of change of momentum,

we see that the particle's momentum changes by $2mv$ each time it hits the piston. The time between the particles striking the piston is $2L/v$, and so the frequency of the particles hitting the piston at regular intervals is $v/2L$ per second. This means that if there were no balancing force, then by conservation of momentum, the particle would cause the momentum of the piston to change by $2mv \cdot v/2L$ units in each second. This is the rate of change of momentum and must be equal to the balancing force,

$$F = mv^2/L$$

On generalization, consider an infinite number of particles bouncing inside a rectangular box, of length L, in the x-direction (which is along the edge of the box).

The total force on the side of the box of area A perpendicular to the x-direction is the sum of single particle forces, the relevant velocity being the component of the velocity in the x-direction. Pressure is force per unit area, $P = F/A$. Of course, we do not know the velocities of the particles in an actual gas. If we sum N contributions, one from each particle in the box, each contribution proportional to v_x^2 for that particle, we will get the sum as N times the average value of v_x^2.

That is to say,

$$P = \frac{F}{A} = \frac{Nm\bar{v}_x^2}{LA} = \frac{Nm\bar{v}_x^2}{V} \tag{1.12}$$

where N is the number of particles in a box of volume V.

Note that the particles move haphazardly in all directions, so the average value of v_x^2 must be the same as that of v_y^2 or v_z^2; hence, $v^2 = v_x^2 + v_y^2 + v_z^2$.

Finally,
$$P = \frac{Nm\bar{v}^2}{3V} \tag{1.13}$$

The macroscopic pressure of a gas relates directly to average kinetic energy per molecule. In the above description, we have taken into account the possible complications caused by interactions between the particles at room temperature; these interactions are negligible for gases like air. Furthermore, it is experimentally established that most gases satisfy the gas law over a wide temperature range:

$$PV = nRT \tag{1.14}$$

for n moles of gas, i.e. $n = N/N_A$, with N_A is the Avogadro's number and R the gas constant.

Introducing Boltzmann's constant as $k = R/N_A$, the following relationship is derived. The average molecular kinetic energy is proportional to the absolute temperature,

$$E_k = \frac{1}{2}mv^2 = \frac{3}{2}kT \tag{1.15}$$

where Boltzmann's constant $k = 1.38 \times 10^{-23}$ joules/K.

1.4.1 Mole and Avogadro Number

Mole

The number of gas molecule in a quantity that is equal to its molecular weight (in grams) shall contain 10^{23} times of its molecular weight. 1 Mole of a compound (or molecule in case of gases) is the weight equal to its molecular weight in grams.

Avogadro number

In 1811, Avogadro gave a hypothesis that in case of two equal volume samples of ideal gases at the same temperature and pressure conditions shall contain equal number of molecules. It states that as number of units in one mole of any substance (defined as its molecular weight in grams), equal to 6.02252×10^{23}. Avogadro's number is a dimensionless quantity. For example, the molecular weight of NH_4 is 11, if 11 g of gas is taken then it shall contain 6.02252×10^{23} molecules. Similarly, 85 g of $NaNO_3$ (Mol weight 85) shall contain 6.02252×10^{23} molecules, 12.011 g of carbon (one mole of carbon) contains 6.02252×10^{23} carbon atoms, 180.16 g of glucose, $C_6H_{12}O_6$, contains 6.02252×10^{23} molecules of glucose. Avogadro's number is determined by calculating the spacing of the atoms in a crystalline solid through X-ray methods and combining this data with the measured volume of one mole of the solid to obtain the number of molecules per molar volume.

Recap Zone

Points to Remember

- The amount of attraction or repulsion between charged objects can be put in quantitative terms by the introduction of the electric force. The direction of the force is along a line joining the two charges. It is repulsive if the charges have the same sign and is attractive if the charges have opposite sign.
- Planck devised a theory of blackbody radiation which was in agreement with all wavelengths. In this theory, the molecules of a body cannot have arbitrary energies, but are instead quantized—the energies can only have discrete values. The magnitude of these energies is given by the formula: $E = nh\nu$, where $n = 0, 1, 2, \ldots$ is an integer, ν is the frequency of vibration of the molecule and h is a constant, called Planck's constant.

- It is experimentally established that most gases satisfy the gas law over a wide temperature range:

$$PV = nRT$$

REVIEW ZONE

SHORT ANSWER QUESTIONS

1. What is an electric charge and its units?
2. What is Coulomb's law?
3. Discuss the blackbody radiation and its importance.
4. What are the modes of cavity radiation?
5. What is Rayleigh-Jean's formula of radiation?
6. How Planck's radiation formula is derived?
7. Discuss the energy–pressure relationship of a gas.
8. Define mole and Avogadaro number.

CHAPTER

2 Properties of Matter

Learning Objectives

By the end of this chapter, the student will be able:

- To understand the basic properties of matter like elasticity, modulus of elasticity and Hook's law
- To understand the classification of materials
- To demonstrate working of cantilever, depression and finding the Young's modulus of cantilever
- To prepare the design of girders with the applications of elasticity

2.1 ▶ Introduction

The elastic behaviour of materials plays very important role in engineering as the automobiles, bridge construction, structural knowledge of the columns and beams as well as supports need the knowledge of strength of the materials used.

If a weight is hung on the lower end of a vertically suspended steel spring, the spring get stretched. When the weight is removed, the spring regains its original size and shape. It means that the spring has elastic behaviour.

Thus, the property of the body by virtue of which it tends to regain its original shape and size, when the applied force is removed, is called elasticity.

2.2 ▶ Some Important Terms Related to Elasticity

(a) **Elasticity:** The property of a body to attain its original configuration (volume and length) again when the applied deforming forces are removed is called elasticity.

(b) **Perfectly elastic body:** A body that regains its original configuration (length, volume or shape) completely and immediately after the removal of applied deforming force, is called perfectly elastic body.

(c) **Perfectly plastic body:** A body that cannot regain its original configuration after the removal of deforming force from it however small the deforming forces may be is called perfectly plastic body.

(d) **Stress:** When a deforming force is applied on the body, it changes the configuration of the body by changing the positions of the atoms or molecules of the body. As a result of it, the internal restoring force tends to bring the deformed body back to its initial configuration which is called stress,

i.e.,

$$\text{Stress} = \frac{\text{internal restoring force}}{\text{area}}$$

The SI unit of stress is Nm^{-2} or Pascal (Pa).

Types of stress

(i) **Normal stress:** When a deforming force applied on a body acts normally over an area, the internal restoring force acting per unit area of the body is called normal stress.

(ii) **Tangential or shearing stress:** When a deforming force applied on a body acting tangentially to the surface of the body and produces a change in the shape without any change in the volume, then the stress acting on the body is called tangential stress.

Note

Stress that is uniform in all directions is called hydrostatic but the stress that varies with direction is called directed stress.

(e) **Strain:** When a deforming force applied on a body produces a change in its configuration, the body is said to be strained or deformed. Thus, the ratio of change in configuration to the original configuration of the body is called strain, i.e.,

$$\text{Strain} = \frac{\text{change in configuration}}{\text{original configuration}}$$

Strain has no units, i.e. it is dimensionless.

Types of strain:

The change in configuration produces a change in either length or volume or shape of the body; thus, it is of three types:

(i) **Longitudinal strain:** If the deforming force applied on the body produces a change in length alone, the strain acting on the body is called longitudinal or tensile strain. It is defined as

$$\text{Longitudinal strain} = \frac{\text{change in length } (\Delta l)}{\text{original length } (l)}$$

(ii) **Volumetric strain:** If the deforming force applied on the body produces a change in volume alone, the strain acting on the body is called volumetric strain. It is defined as

$$\text{Volumetric strain} = \frac{\text{change in volume } (\Delta V)}{\text{original volume } (V)}$$

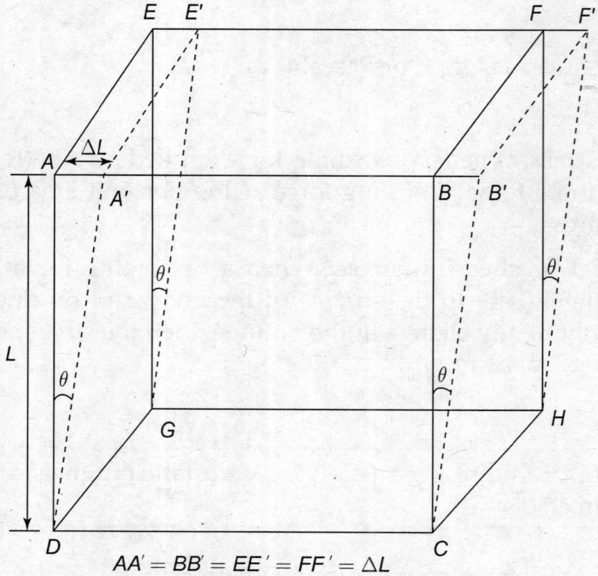

FIGURE 2.1 Shearing strain

(iii) **Shearing strain:** If the applied deforming force produces a change in the shape of the body without any change in volume, the strain acting on the body is called shearing strain (Fig. 2.1). It is defined as the angle in radians through which a plane perpendicular to the fixed surface of the cubical body gets turned under the effect of applied tangential force, i.e,

$$\text{Shearing strain, } \theta = \frac{\Delta L}{L}$$

(f) **Elastic limit:** Elastic limit is the internal property of a body. It is the upper limit up to which, if the deforming force is removed, the body regains its original shape completely and beyond which if the deforming force is increased, the body gets permanently deformed.

Note

The term for uniform stress in the earth is called lithostatic.

2.3 ▶ HOOKE'S LAW

Hooke's law states that the stress developed in the body is directly proportional to the strain produced in it, i.e.,

$$\text{Stress} \propto \text{Strain}$$

or
$$\text{Stress} = \text{E strain}$$

where E is a constant of proportionality and is called Modulus of Elasticity of the material of the body.

Also, load applied on a body is proportional to strain produced in it.

2.3.1 Experimental Verification of Hooke's Law

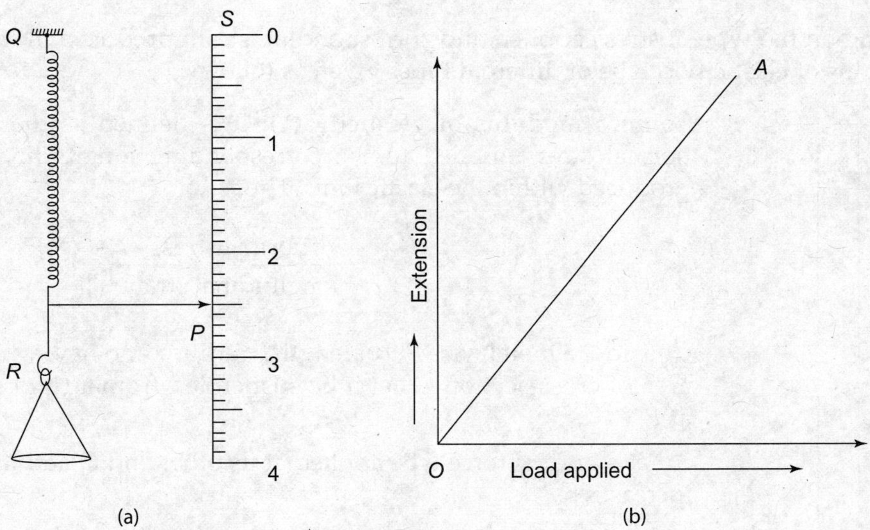

FIGURE 2.2 Experimental verification of Hooke's law

Consider a metallic spring *QR* suspended vertically from a rigid support and a scale pan is attached to the lower end *R* of the spring. A pointer *P* is fixed near end *R* of the spring, which can slide over a fixed linear vertical scale *S* [Fig. 2.2(a)]. Note the reading of pointer when the pan is empty. Add some weight (10 g) in the pan. The spring gets stretched and its length increases. Now, note down the reading of the pointer on the scale again. The difference between these two readings gives the extension in the spring for the weight added in the pan. Add more and more weight in the pan in steps and note the corresponding extensions produced in the spring. On plotting a graph between weight added in the pan and the extension produced in the spring, we get a straight line *OA* [Fig. 2.2(b)]. This experiment shows that the extension produced is directly proportional to the load applied. Thus, Hooke's law is verified.

2.4 ▶ Modulus of Elasticity

According to Hooke's law, within the elastic limit, the stress is directly proportional to strain, i.e.

$$\text{Stress} \propto \text{Strain}$$

$$\Rightarrow \qquad \boxed{\textbf{Stress} = \textbf{E} \times \textbf{Strain}} \tag{2.1}$$

where E is known as the modulus of elasticity and it is defined as the ratio of the stress applied to the corresponding strain produced in the body, within the elastic limit.

Note

E is constant and does not change for a given material.

2.4.1 Types of Modulus of Elasticity

Depending on the type of stress applied and corresponding strain produced in the body, the modulus of elasticity can be of different types given as follows:

FIGURE 2.3

1. **Young's modulus of elasticity (*Y*):** It is defined as the ratio of normal stress applied to the corresponding longitudinal strain produced within the elastic limit. Thus,

$$Y = \frac{\text{normal stress}}{\text{longitudinal strain}}$$

Consider a metal wire *PQ* of length *l*, radius *r* and having uniform area of cross-section *a*. Let it be suspended from a rigid support at *P*.

Let a normal force *F* be applied at its other end *Q* and its length increases by

$$\Delta l \, (= QQ')$$

$$\text{Longitudinal strain} = \frac{\Delta l}{l}$$

$$\text{Normal stress} = \frac{F}{a} = \frac{F}{\pi r^2}$$

and
$$\boxed{Y = \frac{F / \pi r^2}{\Delta l / l} = \frac{Fl}{\pi r^2 (\Delta l)}} \tag{2.2}$$

The SI unit of Young's modulus of elasticity is N/m^2 or Pascal (Pa) and the CGS unit is dyne/cm^2. Young's modulus of elasticity is involved in solids only.

Note

The value of modulus of elasticity, i.e. E, in Pa can turn out to be a very large number. Therefore, sometimes the value of E may be given in N/m^2.

2. **Bulk modulus of elasticity (B):** It is defined as the ratio of normal stress applied to the corresponding volumetric strain produced in the body, within the elastic limit.

   Thus,

$$B = \frac{\text{normal stress}}{\text{volumetric strain}}$$

   Let us consider a spherical solid body, having volume V and surface area a. Let a force F be applied normally on the surface of the body in order to compress it and its volume V decreases by ΔV.

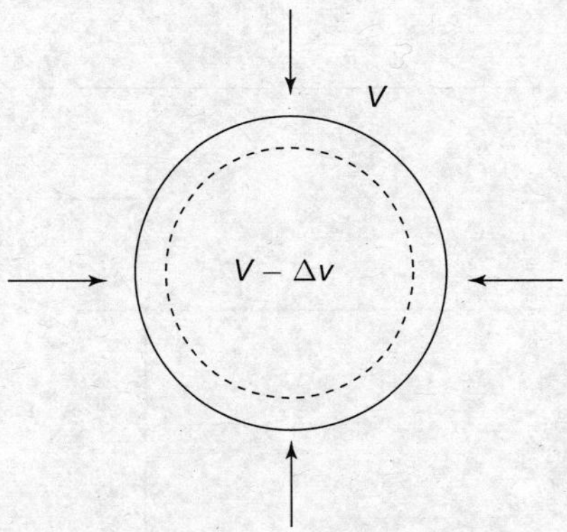

FIGURE 2.4 Volumetric strain

Thus,

$$\text{volumetric strain} = -\Delta V/V$$

and

$$\text{normal stress} = F/a$$

and

$$B = \frac{F/a}{-\Delta V/V} = \frac{-FV}{a\Delta V}$$

If P denotes the increase in pressure applied on the body, then

$$F/a = P$$

$$\therefore \quad \boxed{B = -PV/\Delta V} \tag{2.3}$$

The SI unit of bulk modulus of elasticity is N/m^2 and the CGS unit is $dyne/cm^2$.

Compressibility: Here, $1/B = K$ = compressibility of the material. It is the reciprocal of bulk modulus of elasticity (B).

The SI unit of compressibility is $N^{-1}m^2$ and the CGS unit is $dyne^{-1}cm^2$.
Bulk modulus is involved in solids, liquids and gases. Bulk moduli for solids are larger than liquids, which are larger than gases. For a perfectly rigid body, Young's modulus and bulk modulus of elasticity are infinity.

Note

Stress that acts to shorten an object is called compression and the stress that lengthens an object is called tension.

3. **Modulus of rigidity or shear modulus of elasticity (G):** It is defined as the ratio of tangential stress applied to the corresponding shearing strain produced in the body, within the elastic limit. Thus,

$$G = \frac{\text{tangential stress}}{\text{shearing strain}}$$

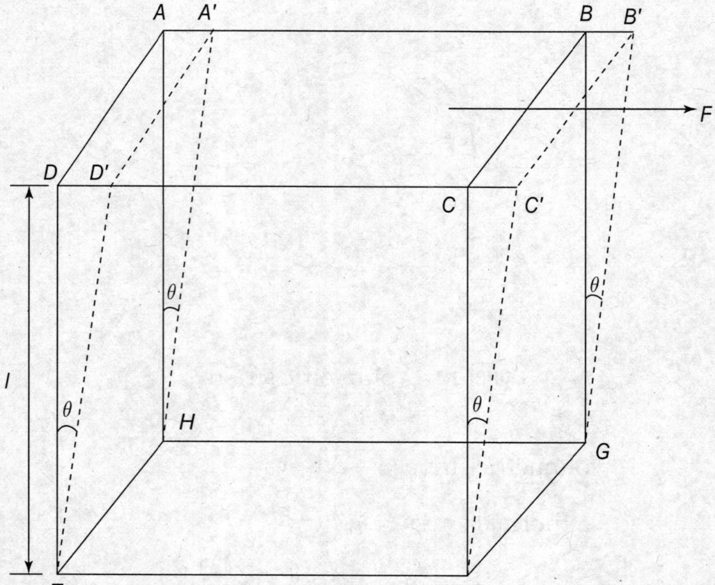

Here, $DD' = \Delta l = CC'$

FIGURE 2.5 Modulus of elasticity

Let us consider a solid metal cube EFGHABCD whose each side is of length l. Let its lower face EFGH be fixed and a tangential face is applied on the top face ABCD and an equal and opposite force F will be set up on the lower face and a couple force is formed by these forces. Thus, the body will not change its volume but suffers a change in its shape. Due to this, the vertical planes EHAD and FGBC be shifted laterally to positions EHA′D′ and FGB′C′, through an angle θ, i.e. $DED' = CFC' = \theta$

and $$DD' = \Delta l$$

$$\text{Shearing strain} = \theta = \frac{DD'}{ED} = \frac{\Delta l}{l}$$

Tangential stress $$= \frac{F}{a}$$

and shear modulus of elasticity (G) $$\boxed{= \frac{F/a}{\theta} = \frac{F}{a\theta} = \frac{Fl}{a\Delta l}} \tag{2.4}$$

The SI unit of G is N/m^2 or Pascal and the CGS unit is $dyne/cm^2$.
The shear modulus of elasticity is involved in solids only. For most of the solids, the value of shear modulus of elasticity is one third of the Young's modulus.

i.e., $$G = \frac{Y}{3}$$

Note

Shear modulus of ideal liquids is zero since the liquids at rest are not able to sustain a tangential force.

2.5 ▶ Stress–Strain Diagram

Suspend a wire AB from a rigid support through its one end A and attach a hanger at the other end B of the wire of uniform area of cross-section on which some different known weights can be placed. Attach a vernier scale V to the lower end B of the wire AB, which can slide over a main scale M, Fig. 2.6(a). Put different known weights in the hanger and note down the corresponding extensions produced in the stretched wire. Calculate the values of stress and strain for various observations. Plot a graph between stress and strain of the wire. We will get a curve of the type as shown in Fig. 2.6(b). From this curve, we note the following:

(i) The portion OP of the graph is a straight line, which shows that stress is directly proportional to the strain. Point P is called point of proportional limit.

(ii) Beyond point P, the variation is not a straight line. The portion between O and Q is called elastic region and point Q is called elastic point.

(iii) Beyond point Q, the strain increases much more rapidly with stress. The slope of the graph between stress and strain from Q to R becomes quite small. In QR, Hooke's law fails and the extension produced in the wire is partly plastic and partly elastic in behaviour.

(iv) Beyond R, the wire starts showing increase in strain without any increase in stress. The point R is called yield point. The behaviour of wire beyond R is perfectly plastic up to S. The stress corresponding to yield point R is called yield strength.

(v) Beyond *S*, the graph is a curved line STU. The point *U* is called breaking point. The stress corresponding to point *T* is called breaking or ultimate stress. The behaviour of the graph from *S* to *U* shows plastic behaviour of the wire.

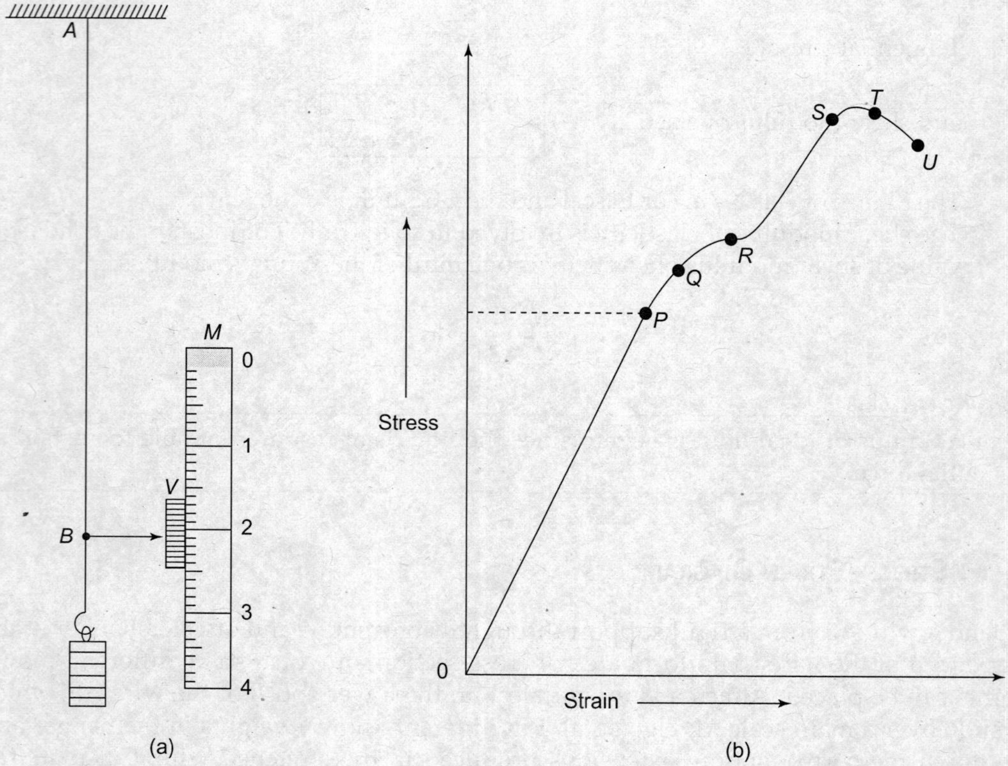

FIGURE 2.6 Stress–strain graph

2.5.1 Classification of Materials from the Study of Stress–Strain Curve

The materials based on the stress–strain variation study can be classified as follows:

1. **Ductile materials:** This type of materials shows large plastic range beyond elastic limit. The breaking point is separated widely from the point of elastic limit for such materials (Fig. 2.7). These materials are used to make springs and sheets. Examples are copper, silver, aluminium and iron.
2. **Brittle materials**: This type of materials shows very small plastic range beyond elastic limit. The breaking point lies near to the elastic limit. Such type of materials cannot be used for making sheets and springs. Examples are glass and cast iron. The stress–strain curve for such materials is shown in Fig. 2.8.

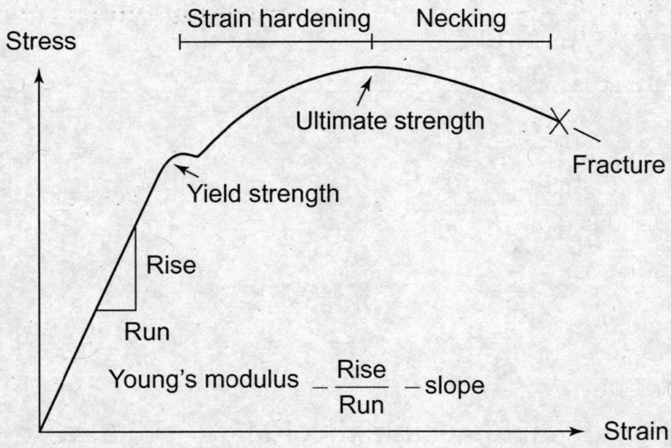

FIGURE 2.7 Stress–strain graph for ductile materials

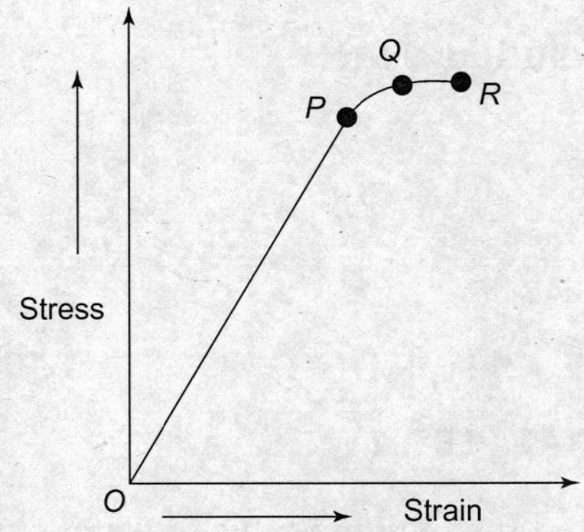

FIGURE 2.8 Stress–strain graph for brittle materials

3. **Elastomers:** In such type of materials, the stress–strain variation is not a straight line within elastic limit. Such materials do not obey Hooke's law in most of the regions. These materials have no plastic range and their elastic region is very large. The breaking point of such materials lies very close to the elastic limit. Examples are rubber, the vessel carrying blood from heart, etc. The stress–strain variation curve is shown in Fig. 2.9.

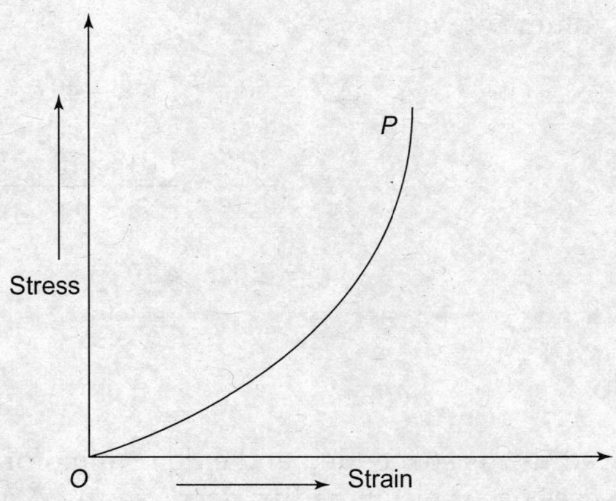

FIGURE 2.9 Stress–strain graph for elastomers

EXAMPLE 2.1

A wire increases by 10^{-3} of its length, when a stress of 10^8 Nm^{-2} is applied on it. What is the Young's modulus of the material of the wire?

Solution: Here,

$$\text{stress} = 10^8\ \text{Nm}^{-2}$$

$$\text{strain} = 10^{-3}$$

$$\therefore \quad y = \frac{\text{stress}}{\text{strain}} = \frac{10^8}{10^{-3}}$$

$$= \mathbf{10^{11}\ Nm^{-2}}$$

EXAMPLE 2.2

Find the bulk modulus of water from the following data: initial volume = 500 litres, pressure increases = 100 atm, final volume = 497.5 litres.

Solution: Given $1\ \text{atm} = 1.013 \times 10^5\ \text{pa}$

Here, $P = 100\ \text{atm} = 100 \times 1.013 \times 10^5$

$$= 1.013 \times 10^7\ \text{pa}$$

Initial volume, $V = 500.0\ \text{litre} = 500 \times 10^{-3}\ \text{m}^3$

Final volume, $V' = 497.5\ \text{litre} = 497.5 \times 10^{-3}\ \text{m}^3$

Therefore, decrease in volume of water

$$\Delta V = 500 \times 10^{-3} - 497.5 \times 10^{-3}\ \text{m}^3$$

Now,

$$K = \frac{PV}{\Delta V} = \frac{1.013 \times 10^7 \times 500 \times 10^{-3}}{2.5 \times 10^{-3}}$$

$$= \mathbf{2.026 \times 10^9\ pa}$$

2.6 ▶ Poisson's Ratio

If a wire of length l and radius r is suspended and a deforming force is applied at its free end, then its length increases by Δl but its radius decreases by Δr. By this single applied force, two types of strains are produced in the wire:

(i) Longitudinal strain $= \dfrac{\Delta l}{l}$

(ii) Lateral strain $= \dfrac{-\Delta r}{r}$

$$\text{The Poisson's ratio is } \sigma = \frac{\text{lateral strain}}{\text{longitudinal strain}}$$

$$\sigma = \frac{-\Delta r / r}{\Delta l / l} = \frac{-\Delta r \times l}{\Delta l \times r} \tag{2.5}$$

The negative sign shows that, if the length increases, the radius of wire will decrease. The Poisson's ratio is a dimensionless and unit less quantity.

Relation between Y, B, G and σ:

(i) $$\boxed{Y = 2G(1+\sigma)} \tag{2.6}$$

(ii) $$\boxed{Y = 3B(1-2\sigma)} \tag{2.7}$$

(iii) $$\boxed{\frac{9}{y} = \frac{1}{B} + \frac{3}{G}} \tag{2.8}$$

(iv) $$\sigma = \frac{3B-2G}{2G+6B} \tag{2.9}$$

Note

Poisson's ratio for all substances lies in between – 1 and + 0.5. It is a measure of Poisson's effect in which a material tends to expand in a direction perpendicular to the direction of compression.

2.7 ▶ Factors Affecting Elasticity

In actual practice, the body loses its elasticity even within elastic limits due to elastic fatigue. The other factors affecting the elastic properties of a material are as follows:

(i) Effect of temperature
(ii) Effect of impurities
(iii) Effect of cyclic stress
(iv) Effect of annealing
(v) Effect of nature of crystals

2.8 ▶ Bending Moment

A structural member acted upon by external loads at right angles to its axis is known as beam, and the bending moment at the cross-section of the beam is defined as the algebraic sum of the moment of the forces, to the left or right of the section.

2.9 ▶ Shear Force

The shear force at the cross-section of the beam is defined as the unbalanced vertical force to the left or right of the section. Whenever a horizontal beam is loaded with the vertical loads, it bends due to the loads. The amount of beam bending depends upon the type of loads, elasticity of the beam, length of the beam, amount of loads and the type of beam. So, we can study the deflection scientifically by analyzing the shear force or bending moment diagrams of a beam.

2.10 ▶ Sign Conventions for Bending Moment and Shear Force

The sign conventions used here are widely followed and internationally recognized. These are given as follows:

(a) **Bending moment:** The bending moment is taken at a section as positive, if it tends to bend the beam to a curvature having concavity at the top, as shown in Fig. 2.10(a).

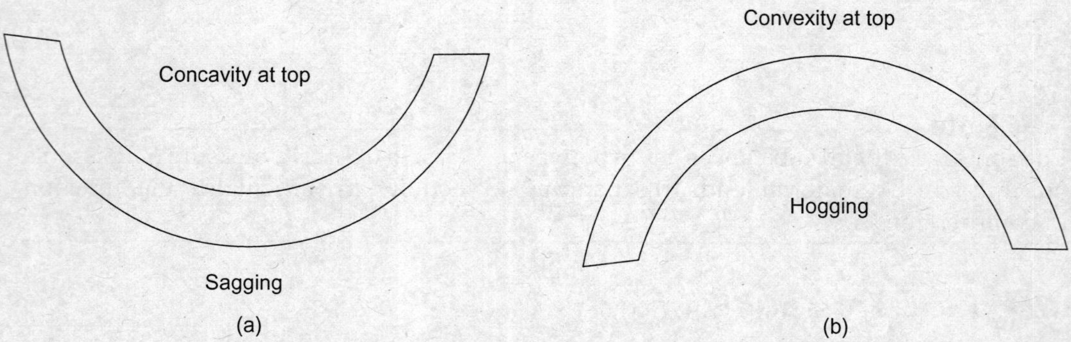

FIGURE 2.10 (a) Positive B.M; (b) negative B.M

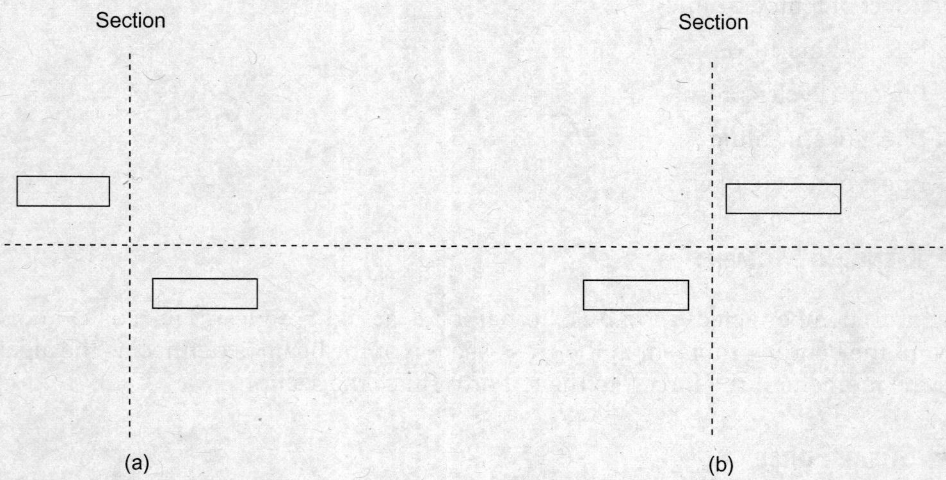

FIGURE 2.11 (a) Positive S.F; (b) negative S.F

On the other hand, bending moment at a section is taken as negative, if it tends to bend the beam to a curvature having convexity at the top as shown in Fig. 2.10(b).

Also, we take the bending moment as positive if it is acting at a section in the clockwise direction to the left. On the other hand, the bending moment is taken as negative if it is acting at a section in the anticlockwise direction to the left.

(b) **Shear force:** Shear force at the cross-section of a beam is the unbalanced vertical force, which tends to slide one portion of the beam with respect to the other portion either in upward or in downward direction.

So, shear force is taken as positive, when the left-hand portion tends to slide upwards or right-hand portion tends to slide downwards as shown in Fig. 2.11(a).

Similarly, shear force is taken as negative, when the left-hand portion tends to slide downwards or right-hand portion tends to slide upwards as shown in Fig. 2.11(b).

2.11 ▶ CANTILEVER

Cantilever is a beam fixed horizontally at one end and its other end is loaded. When subjected to a structural load, the cantilever carries the load to the support where it is forced against by a moment and shear stress. Cantilevers are widely found in cantilever bridges and balconies.

2.12 ▶ DEPRESSION OF A CANTILEVER

Let us consider a beam fixed at one end and loaded at its another end as shown in Fig. 2.12.

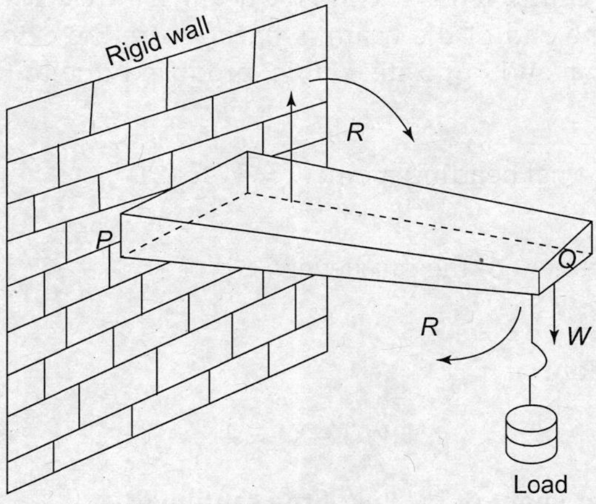

FIGURE 2.12 Depression of a cantilever

Let PQ is the neutral axis of a cantilever of length 'l' fixed at its end P and loaded at free end Q by a weight W. The end Q is depressed to Q'. Here, QQ' represents the vertical depression. Due to the load applied, a couple is created between the two forces (Fig. 2.13). i.e.,

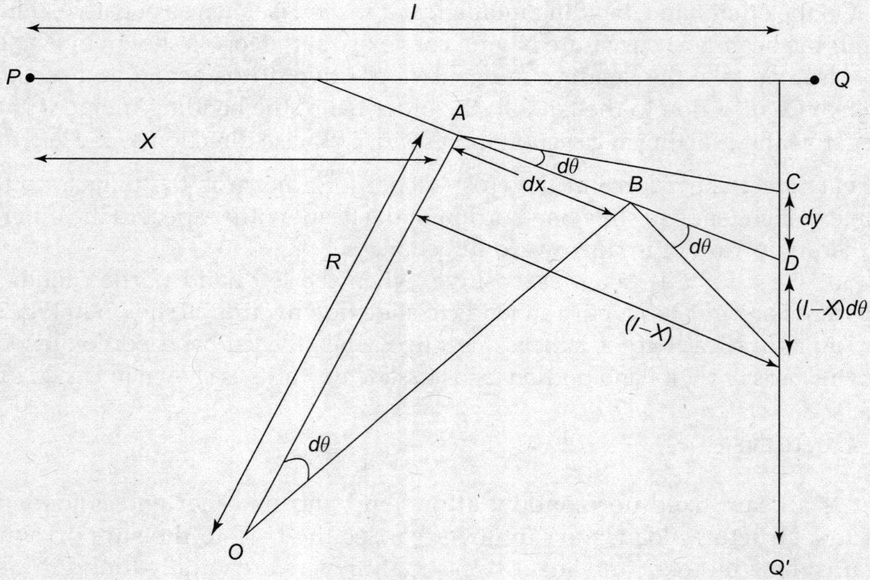

FIGURE 2.13 Cantilever

(i) load W is applied at the free end Q in the downward direction and

(ii) reaction R is acting in the upward direction at the supporting end.

This external bending couple tends to bend the beam in the clockwise direction. The beam cannot rotate because one end of the beam is fixed. Here, the external bending couple is balanced by another equal and opposite couple, produced due to the elastic nature of the material known.

$$\text{As internal bending moment} = W(l - X) \tag{2.10}$$

$$\text{Internal bending moment of the cantilever} = \frac{YI}{R} \tag{2.11}$$

In the equilibrium position,

$$W(l - X) = \frac{YI}{R} \tag{2.12}$$

where I = Geometrical moment of inertia of the cantilever,
R = Radius of curvature of the neutral axis of beam and
Y = Young's modulus of cantilever. Let $d\theta$ be the angle between tangents at A and B.

$$\angle AOB = d\theta = \frac{dX}{R} \tag{2.13}$$

Substituting the volume of R from Eq. (2.13) to Eq. (2.12), we get

$$d\theta = \frac{dX\, W(l-X)}{YI} \tag{2.14}$$

The depression due to curvature at AB is

$$dy = (l - X)\, d\theta \tag{2.15}$$

and

$$dy = (l-X)\left[\frac{dX\, W(l-X)}{YI}\right]$$

(substituting the value of $d\theta$ from Eq. (2.15)

$$dy = \frac{W(l-X)^2\, dX}{YI} \tag{2.16}$$

Total depression at the free end of the cantilever

$$y = \int_0^l dy = \int_0^l \frac{W(l-X)^2\, dX}{YI}$$

$$= \frac{W}{YI}\int_0^l (l-X)^2\, dX$$

$$= \frac{W}{YI}\int_0^l (l^2 + X^2 - 2lX)\, dX$$

$$= \frac{W}{YI}\left[l^2 X + \frac{X^3}{3} - \frac{2X^2 l}{2}\right]_0^l$$

or

$$y = \frac{Wl^3}{3YI}$$

Thus, the depression at the free end of a cantilever is

$$\boxed{y = \frac{\boldsymbol{Wl^3}}{\boldsymbol{3YI}}} \tag{2.17}$$

where Y = Young's modulus of the cantilever. If the depression y is measured, Young's modulus of the material can be calculated.

2.13 ▶ Young's Modulus of a Cantilever

(a) **Rectangular section:** In case of a rectangular section, the moment of inertia is

$$I = \frac{bd^3}{12} \tag{2.18}$$

where b is the breadth and d is the thickness of the section.

Thus, the depression of a cantilever of the rectangular section is

or
$$y = \frac{4Wl^3}{Ybd^3} \tag{2.19}$$

(b) **Circular section:** In case of a circular section, the moment of inertia is

$$I = \frac{\pi r^4}{4}$$

where r is the radius. So, the depression of a cantilever of the circular section is

$$\boxed{y = \frac{Wl^3}{3Y}\frac{4}{\pi r^4}} \tag{2.20}$$

EXAMPLE 2.3

A hollow cylindrical column of steel supports a load of 16,000 kg. The inner and other radii of the column are 40 cm and 50 cm, respectively. Assuming the load distribution to be uniform, calculate the compressional strain of the column. Given the Young's modulus of steel = 2.0×10^{11} Pa.

Solution: Here, $Y = 2.0 \times 10^{11}$ Pa

Inner radius of the column, $r_1 = 40 \text{ cm} = 0.4 \text{ m}$

Outer radius of the column, $r_2 = 50 \text{ cm} = 0.5 \text{ m}$

Therefore, the area of cross-section of the column,

$$a = \pi(r_2^2 - r_1^2) = \pi(0.5^2 - 0.4^2) = 0.09\pi \text{ m}^2$$

The mass supported on the column, $M = 16{,}000$ kg
Therefore, compressional force on the column,

$$F = Mg = 16{,}000 \times 9.8$$

$$= 1{,}56{,}800 \text{ N}$$

Now,

$$Y = \frac{FL}{al}$$

Compressional strain,

$$\frac{l}{L} = \frac{F}{ay} = \frac{1{,}56{,}800}{0.09\pi \times 2.0 \times 10^{11}}$$

$$= \mathbf{2.77 \times 10^{-6}}$$

EXAMPLE 2.4

By how much the pressure on one litre of water be changed to compress it by 0.1%? Given that Bulk modulus of $H_2O = 2.2 \times 10^9$ Pa.

Solution: Here,

$$K = 2.2 \times 10^9 \text{ Pa}$$

$$\frac{\Delta V}{V} = 0.1\% = 0.001$$

Now,

$$K = \frac{\Delta p}{\Delta V / V}$$

or

$$\Delta p = K \times \frac{\Delta V}{V} = 2.2 \times 10^9 \times 0.001 \text{ N}$$

$$= \mathbf{9.2 \times 10^6 N}$$

EXAMPLE 2.5

Draw shear force and bending moment diagrams for a cantilever beam of span 1.5 m carrying point loads as shown in Fig. 2.14.

Solution: Given: span $(l) = 1.5$ m
Point load at B $(w_1) = 1.5$ kN
Point load at C $(w_2) = 2$ kN

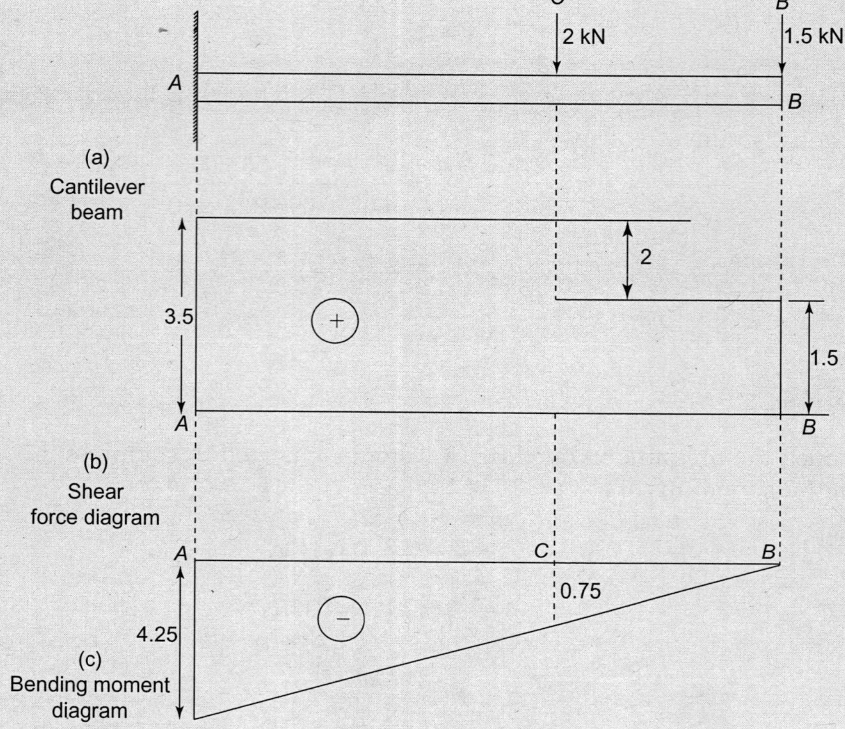

FIGURE 2.14 (a), (b) and (c) Bending moment diagram

Shear force diagram: The values are given as follows:

$$F_B = +w_1 = 1.5 \text{ kN}$$

$$F_C = +(1.5 + w_2) = +(1.5 + 2)$$

$$= 3.5 \text{ kN}$$

$$F_A = +3.5 \text{ kN}$$

Bending moment diagram: The values calculated from Fig. 2.14 are as follows:

$$M_B = 0$$

$$M_C = -[1.5 \times 0.5] = -0.75 \text{ kN-m}$$

$$\mathbf{M_A = -[(1.5 \times 1.5) + (2 \times 1)] = -4.25 \text{ kN-m}}$$

EXAMPLE 2.6

A cantilever beam *AB*, 2 m long carries a uniformly distributed load of 1.5 kN/m over a length of 1.6 m from the free end. Draw shear force and bending moment diagrams for the beam.

Solution: Given: Span (l) = 2 m

Uniformly distributed load (w) = 1.5 kN/m

Length of cantilever *CB* carrying load (a) = 1.6 m

Shear force diagram: The shear force Fig. 2.14 gives the values as follows:

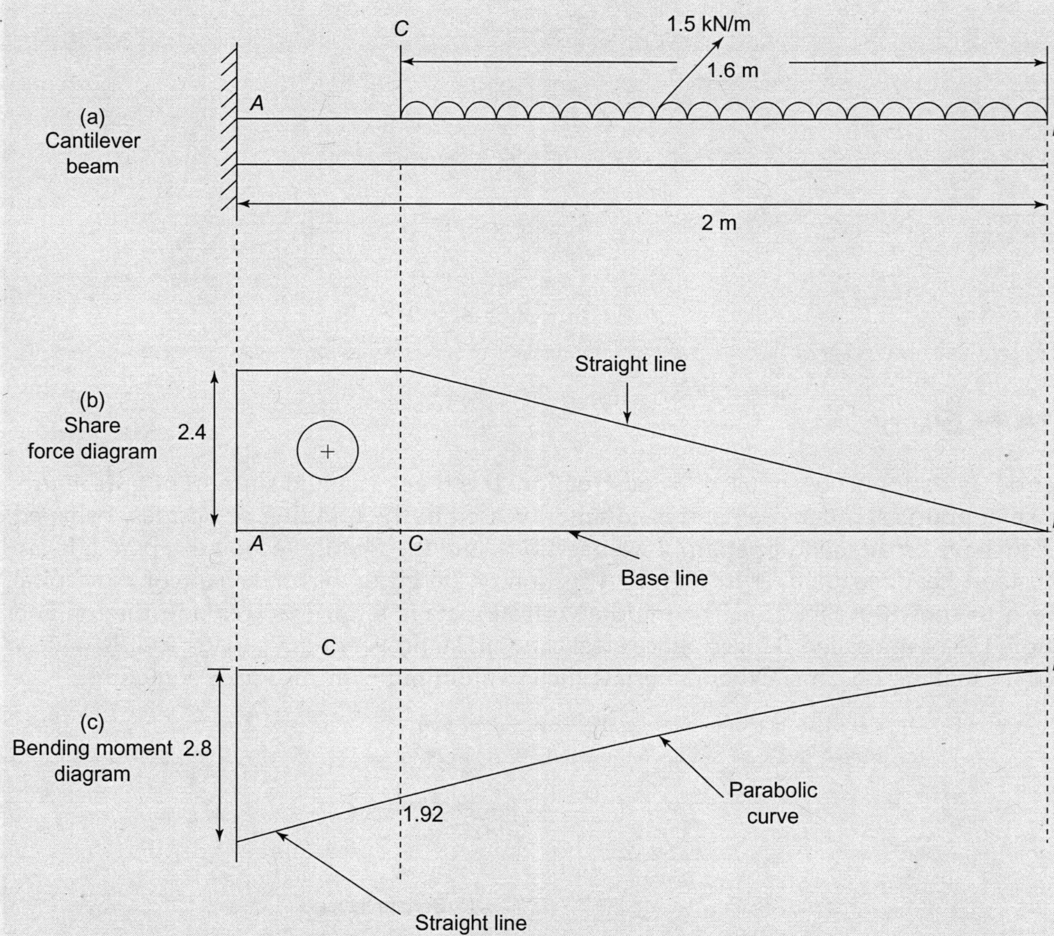

FIGURE 2.15 Bending moment diagram

$$F_B = 0$$

$$F_C = +\,wa = +1.5 \times 1.6$$

$$= +2.4 \text{ kN}$$

$$F_A = +2.4 \text{ kN}$$

Bending moment diagram: The bending moment diagram is shown in Fig. 2.15 and the values are given as follows:

$$M_B = 0$$

$$M_C = \frac{-wa^2}{2}$$

$$= \frac{1.5 \times (1.6)^2}{2}$$

$$= -1.92 \text{ kN-m}$$

$$M_A = -\left[(1.5 \times 1.6)\left(0.4 + \frac{1.6}{2}\right)\right]$$

$$= -\ \mathbf{2.88\ kN\text{-}m}$$

2.14 ▶ GIRDER

A girder is a support beam used in construction. It is the horizontal support of a structure, which supports smaller beams. It is commonly used in the building of bridges. I-shaped girders have broadened upper and lower sections, and the middle section is tapered. It can withstand heavy loads over it. The cross-section of the girder has the shape of the capital letter I as shown in Fig. 2.16. The middle vertical plate is known as web and the top and bottom plates are called flanges. Since steel can withstand very heavy loads so it is widely used for making I-beams. Other materials such as aluminium are also used sometimes.

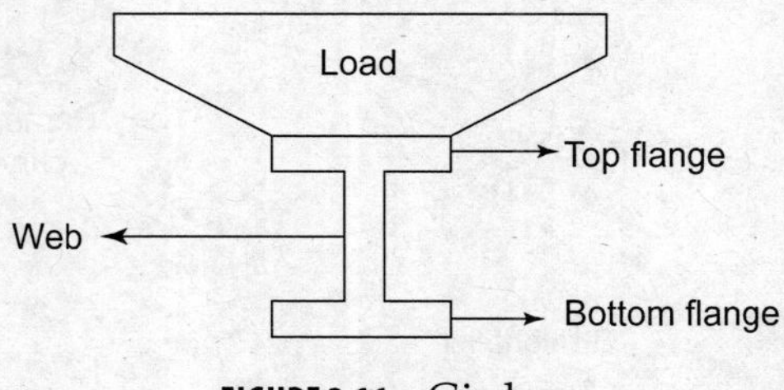

FIGURE 2.16 Girders

The depression in the beam, used as girder for a given load, must be minimum. The depression at the mid-point of a beam is given by

$$y = \frac{4Wl^3}{Ybd^3} \tag{2.21}$$

which can be minimized by

(a) increasing Young's modulus (Y)
(b) decreasing the load (W)
(c) increasing the thickness of the girder (d)
(d) increasing the breadth (b)

Applications of I-shaped girder:

I-shaped girders are used in

(a) commercial and residential construction,
(b) trailer and truck bed framing,
(c) machine bases,
(d) construction of platforms,
(e) support frames and columns for trolley ways, lifts and hoists.

Note

The girders are mainly of three types: (a) rolled steel, (b) plate and (c) box girders. Usually very large beams are called girders.

2.15 ▶ APPLICATIONS OF ELASTICITY

A few of the important applications of the elasticity of materials in daily life are as follows:

1. Any metallic part of a machine is never given to a stress beyond the elastic limits of the material. Otherwise, this metallic part will get permanently deformed and hamper its working.
2. A hollow shaft is always stronger than a solid shaft made up of same and equal amount of a material. Actually, the torque required to twist a hollow shaft is greater than the torque necessary to twist a solid shaft of the same mass, length and material through the same angle. Hence, a hollow shaft is stronger than a solid shaft.
3. The thickness of metallic ropes used in cranes to move heavy weights is decided on the basis of the elasticity of the material of the rope. To give flexibility to the rope, it is made up of thin wires braided together.
4. The bridges are unsafe after long use. During its long use, the bridge undergoes quick alternating strains continually. It results in the loss of elastic strength of the bridge, which may lead to the collapse of the bridge.

5. In designing a beam for using it to support a load (in the construction of bridges and roofs). In this case, the depth of beam is increased rather than the breadth, which is advantageous. Since the value of depression of δ varies inversely as its breadth and as cube of depth, so depression can be reduced by increasing the depth of beam.

Note

Difference between a beam and a girder: Beams span between girders and girders span between columns. Girders are called collector beams. Beam is a short span member (up to 5 m) whereas girder is long span member. Beam is used in context with a residential building whereas girder is used in industrial roofs or in bridges.

Recap Zone

Points to Remember

- Whenever a single force (or a system of forces) is applied on a body, the body undergoes some deformation. This deformation per unit length is known as strain.
- Hooke's law states that when a material is loaded, within its elastic limit, the stress is proportional to the strain.

$$E = \frac{\text{stress}}{\text{strain}}$$

- The ratio of normal stress to volumetric strain is called bulk modulus of elasticity. It is denoted by K.
- Young's modulus is defined as the ratio of normal stress to the longitudinal strain. It is denoted by Y.
- Elastic fatigue is defined as the loss in the strength of a material due to the repeated alternating strains applied on the body.
- The shear force at the cross-section of a beam is defined as the vertical unbalanced force to the left or right of the section.
- The bending moment at the cross-section of a beam is defined as the algebraic sum of the moments of the forces to the left or to the right of the beam section.
- Cantilever is a beam fixed horizontally at one end and load at the other free end.
- Depression of the cantilever at the free end in case of uniform bending is given as

$$y = \frac{W}{YI}\frac{l^3}{3}$$

where I = moment of inertia of the cross-section of cantilever and

Y = Young's modulus of the cantilever.

- A girder is the term used to denote the main horizontal support of a structure which supports the smaller beams. It is commonly used in the building of bridges and planes. It takes the shape of I sometimes; then it is called I-shape girder.

KEY TERMS WITH DEFINITIONS

- **Stress** – The resistance per unit area to applied deformed force is known as stress.
- **Poisson's ratio** – The ratio of lateral strain to the linear strain of a body within its elastic limit.
- **Stiffness** – Elastic moduli measures the stiffness of the material.
- **Yield point** – If the stress is too large, the strain deviates from being proportional to the stress. The point at which this happens is the *yield point*.
- **Yield stress** – Hooke's law is not valid beyond the yield point. The stress at the yield point is called *yield stress*.
- **Plastic deformation** – Deformation caused is caused by the motion of dislocations.
- **Ductility** – The ability to deform before braking. It is the opposite of **brittleness**.
- **Resilience** – Capacity to absorb energy *elastically*.
- **Toughness** – Ability to absorb energy up to fracture.
- **True stress** – The ratio of the force to the initial area.
- **Hardness** – It is the resistance to plastic deformation.
- **Necking** – The sudden decrease in the area of cross-section after ultimate stress.
- **Plastic region** – The region in which the material deforms permanently.

IMPORTANT FORMULAE AND EQUATIONS

Equation Number	Equation	Remarks
(2.1)	**Stress = E × strain**	**Hook's law**
(2.2)	$Y = \dfrac{F/\pi r^2}{\Delta l/l} = \dfrac{Fl}{\pi r^2(\Delta l)}$	**Young's modulus of elasticity**
(2.3)	$B = -PV/\Delta V$	**Bulk's modulus of elasticity**
(2.4)	$\dfrac{F/a}{\theta} = \dfrac{F}{a\theta} = \dfrac{Fl}{a\Delta l}$	**Shear modulus of elasticity (*G*)**
(2.6)	$Y = 2G(1+\sigma)$	**Relation between Young's modulus and shear modulus of elasticity**

(2.7)	$Y = 3B(1 - 2\sigma)$	Relation between Bulk's modulus and Poisson's ratio
(2.8)	$\frac{9}{y} = \frac{1}{B} + \frac{3}{G}$	Relation between three moduli of elasticity
(2.17)	$y = \frac{Wl^3}{YI3}$	The depression at the free end of a cantilever
(2.20)	$y = \frac{Wl^3}{3Y}\frac{4}{\pi r^4}$	Young's modulus of a circular section of a cantilever

REVIEW ZONE

SHORT ANSWER QUESTIONS

1. Define stress, strain and elasticity.
2. State Hooke's law.
3. Define Poisson's ratio.
4. Derive a relation between Bulk's modulus and Young's modulus of elasticity.
5. Define modulus of rigidity.
6. Discuss the stress versus strain graph for a metallic wire.
7. Define elastic limit and elastic fatigue.
8. Name the factors, which affect the elasticity of a solid.
9. Define the terms: shear force, bending moment and I-shaped girders.
10. Derive an expression for depression of a cantilever.

LONG ANSWER QUESTIONS

1. Define normal stress, longitudinal strain and young's modulus of elasticity. Give units of each of them. Also, Derive an expression for the relation between Young's modulus and Bulk's modulus.
2. State and explain Hooke's law. A wire is fixed at one end is subjected to increasing load at the other end. Draw a curve between stress and strain with the help of stress versus strain curve. Also explain the terms elastic limit, yield point, breaking point and permanent set.
3. Derive an expression for the depression of a cantilever with a point load at its free end.
4. Discuss I-shape girders and also explain the applications of I-shape girders and elasticity.

Numerical Problems

1. A steel wire of length 3.6 m and cross-section 2.5×10^{-5} m^2 stretches by the same amount as a copper wire of length 2.4 m and cross-section 3.2×10^{-5} under a given load. What is the ratio of the Young's modulus of steel to that of copper?

   Ans. 1.92

2. The pressure of a medium is changed from 1.01×10^5 Pa to 1.1675×10^5 Pa and change in volume is 10% keeping temperature constant. Find the bulk modulus of the medium.

   Ans. 1.55×10^6 Pa

3. A mass of 5.0 kg is hung from a copper wire of 1 mm diameter and 2 m in length. Calculate the extension produced. What should be the minimum diameter of the wire so that its elastic limit is not exceeded? Elastic limit for copper = 1.5×10^9 dyne cm^{-2}; Y for copper = 1.1×10^{12} dyne/cm^2.

   Ans. ℓ = 0.1134 cm, d' = 0.0645 cm

4. A beam of 6 m long rests on two supports 5 m apart. The right end is overhanging by 1 m. The beam carries a uniformly distributed load of 1.5 kN/m over the entire length of the beam. Draw S.F and B.M diagram and find the amount and position of maximum bending moment.

   Ans. 4.32 kN-m at 2.4 m from left end

5. A cantilever beam of 1.5 m spam is loaded as shown in Fig. 2.17. Draw the shear force and bending moment diagrams.

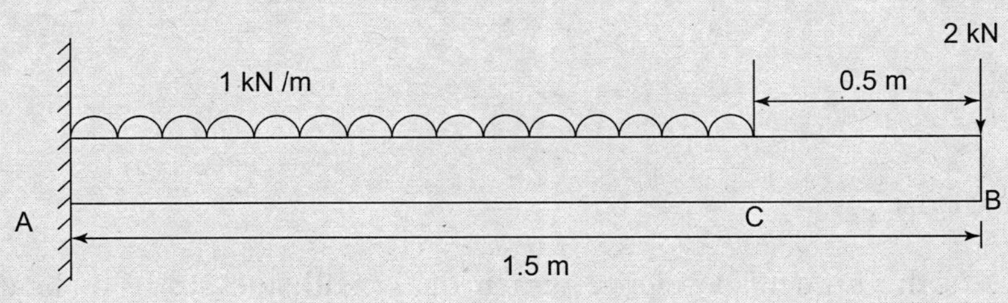

CHAPTER

3 Thermal Physics

Learning Objectives

By the end of this chapter, the student will be able:

- To revise the basic concepts of thermal physics
- To explain the difference between heat and temperature
- To understand the concept of heat conduction and transport phenomenon
- To understand different models of heat conduction
- To analyze the phenomena of viscosity, conduction and diffusion
- To know practical applications of thermal physics

3.1 ▶ Introduction

Thermal physics is the branch of science which deals with the study of heat energy and temperature. Heat energy is the energy of a system in terms of the motion of its molecules. The faster the molecules in a substance move the more heat energy they have. When we talk about the Earth's solid, global warming, liquid cores and the way in which energy flows from the centre of the Sun to its surface, we need to study thermal physics to understand all these concepts. In this chapter, we discuss the thermal properties of solids, modes of heat transfer and many methods to measure the thermal conductivity of different materials.

3.2 ▶ Thermal Properties of Materials

Every material has fundamental physical properties that determine their energy performance like conductivity, resistance, thermal mass, etc. Understanding these intrinsic properties, we can choose the right material to manage heat flows.

3.3 ▶ Specific Heat

Specific heat is the amount of heat energy required to raise the temperature of a mole of a solid by one degree. Different substances need different amounts of heat energy to raise their temperature. For example to raise the temperature of 1 kg of water by 1 K, it requires 4184 joules but the same amount of heat raises the temperature of 1 kg of copper by 11 K. That's why we say that water has a large heat capacity than copper. Moreover, the heat capacity at constant volume C_v is more useful than the heat capacity at constant pressure C_p.

When some heat is supplied to a solid, its temperature rises and its internal energy is increased. The increase in internal energy is due to (a) increase in the vibrations of the atoms about their mean positions and (b) increase in kinetic energy of free electrons.

From the first law of thermodynamics,

$$dQ = dU - dW$$

or

$$dQ = dU - PdV$$

where the specific heat at constant volume is given by

$$C_v = \left[\frac{\partial Q}{\partial T}\right]_v = \left[\frac{\partial U}{\partial T}\right]_v,$$

where U is the internal energy of the system.

And similarly, the specific heat at constant pressure is given by

$$C_p = \left[\frac{\partial Q}{\partial T}\right]_p$$

But the difference between C_p and C_v vanishes at low temperatures. Whereas, we can easily calculate C_v from C_p if the volume expansion coefficient α and the compressibility β of a material are known. In that case,

$$C_p - C_v = \frac{\alpha^2 TV}{\beta},$$

where V is the volume of the solid.

Note

Specific heat per gram for water is much higher than that for a metal.

EXAMPLE 3.1

Calculate the capacity of given copper of mass 75 g and temperature difference 10°C if 200 J of heat is lost?

Solution: Given $m = 75$ g

$$\Delta Q = 200 \text{ J}$$
$$T = 10°\text{C}$$
$$C = \frac{Q}{m\Delta T} = 20 \text{ J/°C}$$

3.4 ▶ Density of States

3.4.1 Frequency Distribution of Normal Modes

Consider a one-dimensional chain of atoms (string) of length L (Fig. 3.1). If the ends of this string are fixed, then stationary waves are set up in it having specific wavelength $n(\lambda/2) = L$, where n is any positive integer and ≥ 1. Suppose $y(x,t)$ represents the displacement of the string at any point x at a time t.

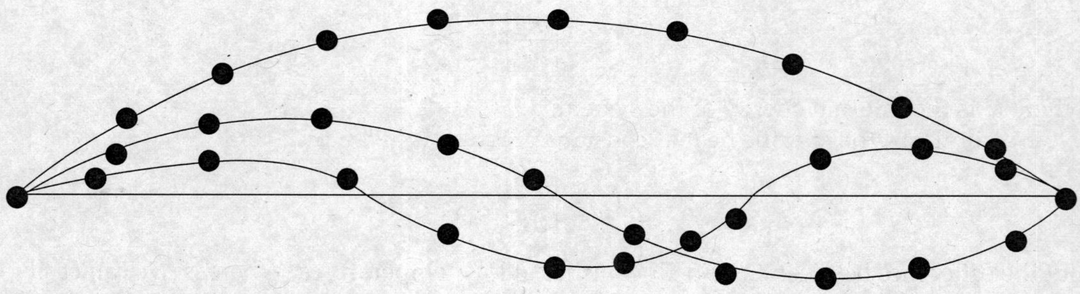

FIGURE 3.1 Normal modes of the chain

The one-dimensional wave equation for this system can be written as

$$\frac{\partial^2 y}{\partial x^2} = \frac{1}{v^2}\frac{\partial^2 y}{\partial t^2} \tag{3.1}$$

where v is the velocity of the wave in the string.

Solving Eq. (3.1) for boundary conditions $y = 0$ at $x = 0$ and $x = L$, the solution of Eq. (3.1) is obtained in the form of standing waves as

$$y(x,t) = A\sin\left(\frac{n\pi x}{L}\right)\cos 2\pi\nu_n t \tag{3.2}$$

Differentiating Eq. (3.2) twice w.r.t x and t, we get

$$\frac{\partial^2 y}{\partial x^2} = -\frac{n^2\pi^2 y}{L^2}$$

and $$\frac{\partial^2 y}{\partial t^2} = -4\pi^2 \nu_n^2 y \tag{3.3}$$

Substituting Eq. (3.3) in Eq. (3.1), we get the frequency as

$$\nu_n^2 = \frac{n^2 v^2}{4L^2} \tag{3.4}$$

The number of possible modes of vibrations is

$$g(\nu) = dn = \frac{2L}{v} d\nu \tag{3.5}$$

The frequency spectrum is discrete. For one-dimensional string, the frequency spectrum corresponds to an infinite number of equidistant lines, as given by Fig. 3.2(a).

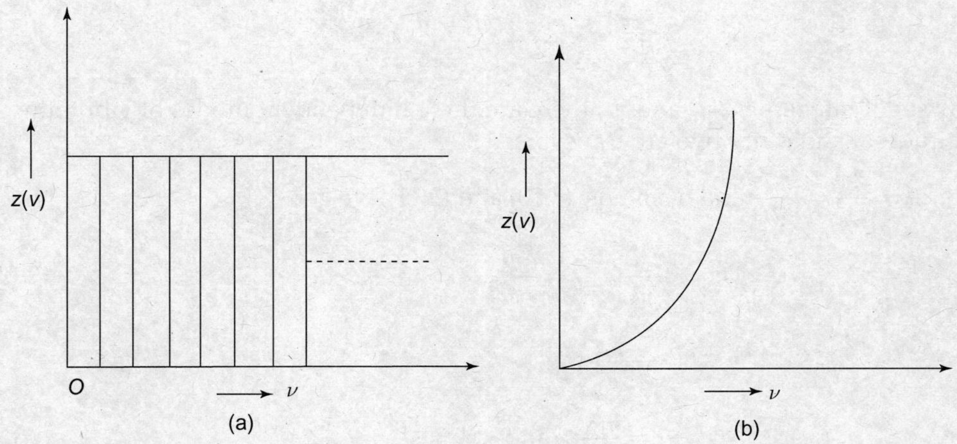

FIGURE 3.2 (a) Frequency spectrum for a finite continuous string according to Eq. (3.5) and (b) three-dimensional continuous frequency spectrum

In three-dimensional cases, the wave equation (3.1) becomes

$$\frac{\partial^2 y_1}{\partial x^2} + \frac{\partial^2 y_1}{\partial y^2} + \frac{\partial^2 y_1}{\partial z^2} = \frac{1}{v_1^2} \frac{\partial^2 y_1}{\partial t^2} \tag{3.6}$$

and $$\frac{\partial^2 y_t}{\partial x^2} + \frac{\partial^2 y_t}{\partial y^2} + \frac{\partial^2 y_t}{\partial z^2} = \frac{1}{v_t^2} \frac{\partial^2 y_t}{\partial t^2} \tag{3.7}$$

Here, we get two velocities of propagation (v_1 for longitudinal wave and v_t for transverse waves).

Again by applying the simplest boundary conditions, we obtain the following standing wave solutions:

$$y_1(x,y,z) = A_1 \sin\left(\frac{n_x \pi x}{L}\right) \sin\left(\frac{n_y \pi y}{L}\right) \sin\left(\frac{n_z \pi z}{L}\right) \cos 2\pi\nu_1 t \tag{3.8}$$

and

$$y_t(x,y,z) = A_t \sin\left(\frac{n_x \pi x}{L}\right) \sin\left(\frac{n_y \pi y}{L}\right) \sin\left(\frac{n_z \pi z}{L}\right) \cos 2\pi\nu_t t \tag{3.9}$$

Again, differentiating Eqs. (3.8) and (3.9) w.r.t x, y, z twice and substituting the resulting values in Eqs. (3.6) and (3.7), we get

$$\nu_1^2 = \frac{v_1^2}{(2L)^2}\left(n_x^2 + n_y^2 + n_z^2\right) \tag{3.10}$$

and

$$\nu_t^2 = \frac{v_t^2}{(2L)^2}\left(n_x^2 + n_y^2 + n_z^2\right) \tag{3.11}$$

For any set of integers n_x, n_y and n_z, there are three independent modes of vibrations, one is longitudinal and other two are transverse.

Using $k = \frac{2\pi}{\lambda} = \frac{2\pi\nu}{V}$ and from Eqs. (3.10) and (3.11), we get

$$n_x = \frac{k_x L}{\pi}, \quad n_y = \frac{k_y L}{\pi}, \quad n_z = \frac{k_z L}{\pi} \tag{3.12}$$

If we take $k^2 = k_x^2 + k_y^2 + k_z^2$, then

$$\nu_1 = \frac{v_1 k}{2\pi} \quad \text{and} \quad \nu_t = \frac{v_t k}{2\pi} \tag{3.13}$$

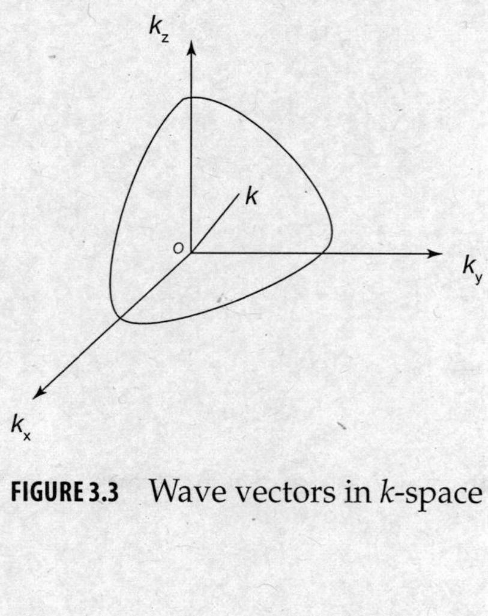

FIGURE 3.3 Wave vectors in k-space

Here, k is a wave vector having components k_x, k_y, k_z. If we plot k in reciprocal space or k-space with r as position vector (Fig. 3.3), then the points defined in Eq. (3.12) can be distributed as lattice points in the positive octant which has the volume

$$\frac{1}{8} \times \frac{4}{3}\pi k^3 = \frac{1}{6}\pi k^3.$$

Volume occupied by each point is $\left(\frac{\pi}{L}\right)^3$ and the total number of lattice points is $\frac{Vk^3}{6\pi^2}$.

For a particular frequency ν, we get two wave vectors k_1 and k_t as

$$k_1 = \frac{2\pi\nu}{v_1} \quad \text{and} \quad k_t = \frac{2\pi\nu}{v_t} \tag{3.14}$$

Now, the total number of lattice points with radii k_1 and k_t will be

$$\frac{V}{6\pi^2}\left(\frac{2\pi\nu}{v_1}\right)^3 \quad \text{and} \quad \frac{V}{6\pi^2}\left(\frac{2\pi\nu}{v_t}\right)^3$$

The total number of modes $z(\nu)$ of frequency less than ν_D will be

$$z(\nu) = \frac{8\pi^3 v}{6\pi^2}\left(\frac{1}{v_1^3} + \frac{2}{v_t^3}\right)\nu^3 \tag{3.15}$$

With mean velocity v as

$$\frac{3}{v^3} = \frac{1}{v_1^3} + \frac{2}{v_t^3} \tag{3.16}$$

Using Eqs. (3.15) and (3.16), we get

$$z(\nu) = \frac{4\pi V \nu^3}{v^3} \tag{3.17}$$

Differentiating Eq. (3.17), we get the required frequency distribution function as

$$z(\nu) = \frac{4\pi V \nu^3}{v^3} \tag{3.18}$$

Also, from Eq. (3.15), we get

$$\boxed{z(\nu)d\nu = 4\pi V\left(\frac{1}{v_1^3} + \frac{2}{v_t^3}\right)\nu^2 d\nu} \tag{3.19}$$

Equation (3.19) gives the number of modes whose frequencies lie between ν and $(\nu + d\nu)$. The function $z(\nu)$ is known as spectral distribution function of normal modes.

Since, in the three-dimensional lattice crystal, the number of normal vibrations is $3N$; thus, $z(\nu)$ should satisfy the normalization condition, i.e.,

$$\int_0^{\nu_D} z(\nu)d\nu = 3N, \tag{3.20}$$

where ν_D is the maximum frequency of the spectrum. It is also known as Debye frequency. Using Eqs. (3.19) and (3.20), we get

$$4\pi V\left(\frac{1}{v_1^3}+\frac{2}{v_t^3}\right)\int_0^{\nu_D}\nu^2 d\nu = 3N$$

or

$$\nu_D^3 = \frac{9N}{4\pi V}\left(\frac{1}{v_1^3}+\frac{2}{v_t^3}\right)^{-1} \tag{3.21}$$

Using Eq. (3.21) into Eq. (3.19), we get

$$f(\nu) = \frac{9N}{\nu_D^3}v^2 \tag{3.22}$$

and it gives

$$\begin{aligned} f(\nu) &= K\nu^2 \qquad (\text{for } \nu \le \nu_D) \\ &= 0 \qquad (\text{for } \nu \ge \nu_D) \end{aligned}$$

where $K = \dfrac{9N}{\nu_D^3}$ is independent of ν and is a constant.

Note

A high density of states at a particular energy level means that there are a number of states available for occupation.

3.5 ▶ Classical Model

The main assumptions of the model are as follows:

(i) Each substance consists of a large number of atomic particles executing harmonic motions. These atomic particles are also known as atomic oscillators.

(ii) The energies of these atomic oscillators are different but they vibrate with the same frequency.

(iii) If N is the total number of atoms in the crystal, then the total energy of the substance at any given temperature is N times the average energy E of the oscillator.

(iv) The internal energy of the substance is mainly contributed by the vibrational energy of all the atomic oscillators.

(v) The atomic oscillators can take up continuum of energy values ranging from 0 to ∞.

Thus, the frequency of such atomic oscillators can be obtained as follows:
We know that the restoring force of a harmonic oscillator is given by

$$\boxed{F \propto x \quad \text{or} \quad F = -fx,} \tag{3.23}$$

where f is the elastic constant.

If m is the mass of the oscillator, then Eq. (3.23) can be written as

$$m\frac{d^2x}{dt^2} = -fx \tag{3.24}$$

or

$$\frac{d^2x}{dt^2} + \left(\frac{f}{m}\right)x = 0$$

where

$$\omega = \sqrt{\frac{f}{m}} \text{ or } 2\pi\nu = \sqrt{\frac{f}{m}} \text{ and } \nu = \frac{1}{2\pi}\sqrt{\frac{f}{m}} \tag{3.25}$$

where ν is the frequency of an atomic oscillator.

For a single one-dimensional isotropic harmonic oscillator, the total energy is given by

$$\boxed{\mathbf{E = \frac{p_x^2}{2m} + P \cdot E}} \tag{3.26}$$

$$P \cdot E = \int F\,dx$$

where

$$P \cdot E = -f\int_x^0 x\,dx$$

$$P \cdot E = m\omega^2\int_0^x x\,dx = \frac{m\omega^2x^2}{2}$$

From Eq. (3.26),

$$E = \frac{p_x^2}{2m} + \frac{m\omega^2x^2}{2} \tag{3.27}$$

where p_x is the momentum along the x-axis.

Let $\overline{E}$ be the mean energy of the oscillator and N_A be the total number of oscillators in a given mass (let us take one kg mole of the substance).

Then, the mean energy of the atomic oscillator can be written as

$$\boxed{\overline{E} = \frac{\sum_0^\infty E\,dN}{\sum_0^\infty dN}} \tag{3.28}$$

Let dN be the number of atomic oscillators having energies in the energy range E and $(E + dE)$ at a given temperature T.

Then,

$$\overline{E} = \frac{\sum_{0}^{\infty} E \exp(-E / K_B T)}{\sum_{0}^{\infty} \exp(-E / K_B T)} \tag{3.29}$$

Substituting the value of E from Eq. (3.27) into Eq. (3.29), we get

$$\overline{E} = \frac{K_B T}{2} + \frac{K_B T}{2} = K_B T \tag{3.30}$$

When $x = a$ (let a be the maximum displacement of oscillator on either side of its equilibrium position)

$$\overline{E} = \frac{1}{2} fa^2 = K_B T$$

or

$$a = \left[\frac{2K_B T}{f}\right]^{1/2} \tag{3.31}$$

The classical value of the vibrational energy of a crystal having N_A atoms is given as

$$\boxed{U = 3N_A K_B T = 3RT} \tag{3.32}$$

Specific heat at constant volume is

$$\boxed{C_v = \left(\frac{dU}{dT}\right)_v = 3R = 5.96 \text{ cal/mole K}} \tag{3.33}$$

Equation (3.33) is given by Dulong and Petit in 1869 and is known as Dulong-Petit's law. This equation is generally found correct at room temperature and above for elements of atomic weight greater than 40. According to this law, the heat capacity is constant for a solid and it is independent of temperature. Thus, Dulong-Petit explained the behaviour of solids at high temperatures but this law fails to explain the variation of heat capacity with temperature.

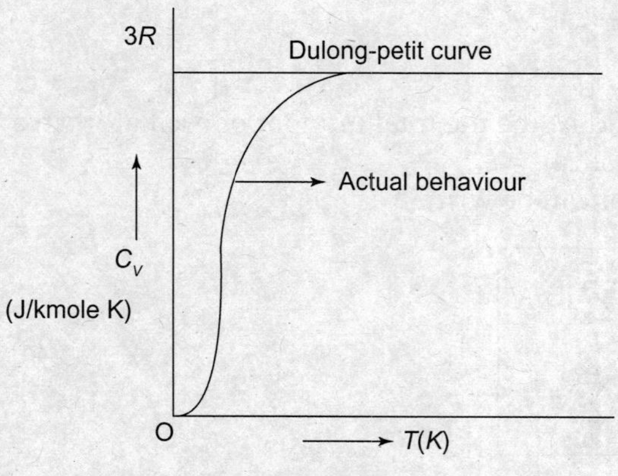

FIGURE 3.4 Specific heat of a solid at low temperature

The above law fails under some other conditions, like as below:

(i) At low temperatures, the specific heat of all solids approaches zero as T^3 in insulators and as T in metals. However, in superconductors, this drop is much faster (Fig. 3.4).

(ii) A number of elements having low atomic numbers and high melting points show low value of specific heats than given by Dulong and Petit.

(iii) The electropositive metals such as Na, Cs, Mg and Ca show an increase in specific heat above the maximum value $3R$ with increase in temperature.

3.6 ▶ Einstein's Model

The behaviour of specific heat of all the solids at low temperatures could not explain satisfactorily on the basis of classical theory. Einstein in 1907 resolved this problem by the application of Planck's quantum theory. He assumed that a crystalline solid is made up of N_A atoms (per mole), which can be regarded as to behave like $3N_A$ independent harmonic oscillators, each has frequency ν. He assumed that the atoms are identical and independent, which vibrate with same natural frequency but all are having discrete energy values. These energy values are given by Planck as

$$E_n = nh\omega = nh\nu \tag{3.34}$$

The mean energy of atomic oscillators is

$$\overline{E} = \frac{\sum E\,dN}{\sum dN} \tag{3.35}$$

where dN denotes the number of oscillators having energy range in between E and $E+dE$. Thus,

$$\overline{E} = \frac{\sum\limits_{n=0}^{\infty} nh\nu \exp(-nh\nu / K_B T)}{\sum\limits_{0}^{\infty} \exp(-nh\nu / K_B T)}$$

$$= \frac{h\nu\left[e^{-h\nu/K_BT} + 2e^{-2h\nu/K_BT} + 3e^{-3h\nu/K_BT} + \cdots\right]}{\left[1 + e^{-h\nu/K_BT} + e^{-2h\nu/K_BT} + e^{-3h\nu/K_BT} + \cdots\right]} \tag{3.36}$$

Let $$x = -\frac{h\nu}{K_B T}$$

Equation (3.36) can be rewritten as

$$\overline{E} = \frac{h\nu\left[e^{x} + 2e^{2x} + 3e^{3x} + \cdots + \infty\right]}{\left[1 + e^{x} + e^{2x} + e^{3x} + \cdots + \infty\right]}$$

$\Rightarrow$ $$\overline{E} = h\nu\left[\frac{d}{dx}\log\left(1 + e^{x} + e^{2x} + e^{3x} + \cdots\right)\right]$$

$$\Rightarrow \qquad \bar{E} = h\nu\left[\frac{d}{dx}\log\frac{1}{\left(1-e^{x}\right)}\right]$$

$$= h\nu\left[\frac{d}{dx}\left\{\log 1 - \log\left(1-e^{x}\right)\right\}\right]$$

$$= \frac{-h\nu\left(-e^{x}\right)}{\left(1-e^{x}\right)} = h\nu\left[\frac{e^{x}}{1-e^{x}}\right]$$

$$= \frac{h\nu\, e^{x}}{\left[1-e^{x}\right]} = \frac{h\nu}{e^{-x}\left(1-e^{x}\right)}$$

$$= \frac{h\nu}{\left(e^{-x}-1\right)}$$

or

$$\bar{E} = \frac{h\nu}{\left(e^{h\nu/K_BT}-1\right)} \tag{3.37}$$

The total energy of the solid is given by

$$U = 3N_AK_BT = 3N_A\left(\bar{E}\right)$$

$$\Rightarrow \qquad U = 3N_A\frac{h\nu}{\left[\exp(h\nu/K_BT)-1\right]}$$

$$\Rightarrow \qquad \frac{dU}{dT} = \frac{-(3N_Ah\nu)\exp[h\nu/K_BT]\left[\dfrac{-h\nu}{K_BT^2}\right]}{\left[\exp(h\nu/K_BT)-1\right]^2} \tag{3.38}$$

Equation (3.38) can be written as

$$C_v = \frac{dU}{dT} = \frac{(3N_AK_B)\exp[(h\nu/K_BT)]\left[\dfrac{h\nu}{K_BT}\right]^2}{\left[\exp(h\nu/K_BT)-1\right]^2}$$

$$\Rightarrow \qquad \frac{C_v}{3R} = \left[\frac{h\nu}{K_BT}\right]^2\left\{\frac{\exp[h\nu/K_BT]}{\left[\exp(h\nu/K_BT)-1\right]^2}\right\} \qquad [\because N_AK_B = R]$$

Put $$h\nu = K_B\theta_E$$

$$\frac{C\nu}{3R} = \left(\frac{\theta_E}{T}\right)^2 \frac{\exp(\theta_E/T)}{(\exp(\theta_E/T)-1)^2}$$

$\Rightarrow$ $$\boxed{\frac{C_v}{3R} = F(E)\left(\frac{\theta_E}{T}\right)} \tag{3.39}$$

where $F(E)$ is known as Einstein's function and θ_E is known as Einsten's temperature.

Case (a): At high temperatures, i.e. when $h\nu << K_BT$. Equation (3.37) can be written as

$$\overline{E} = \frac{h\nu}{e^x - 1},$$

where $$x = \frac{h\nu}{K_BT}$$

Here, $$e^x = 1 + x + \frac{x^2}{2!} + \frac{x^3}{3!} + \cdots \cong 1 + x$$

$\Rightarrow$ $$\overline{E} = \frac{h\nu}{1+x-1} = \frac{h\nu}{x}$$

and $$U = 3N_A\overline{E} = 3N_A 3RT = 3R \tag{3.40}$$

Thus, for large values of temperature, the expression reduces to the classical expression (Fig. 3.5).

Case (b): At low temperatures, i.e., when $h\nu \gg K_BT$.
Equation (3.37) can be written as

$$\overline{E} = \frac{h\nu}{\exp(h\nu/K_BT)}$$

$$U = 3N_A\overline{E} = \frac{3N_A h\nu}{\exp(h\nu/\mathrm{K_BT})}$$

$$C_v = \frac{dU}{dT} = (3N_A h\nu)\exp(-h\nu/K_BT)\left[\frac{h\nu}{K_BT^2}\right]$$

$\Rightarrow$ $$C_v = 3R\left[\frac{h\nu}{K_BT}\right]^2 \exp(-h\nu/K_BT) \tag{3.41}$$

Thus, with decreasing temperature, C_v drops exponentially (Fig. 3.5).

Equation (3.41) is plotted for aluminium with frequency $\nu = 6.4 \times 10^{12}$ Hz . It is clear from Fig. 3.5 that it is in good agreement with data except at very low temperatures.

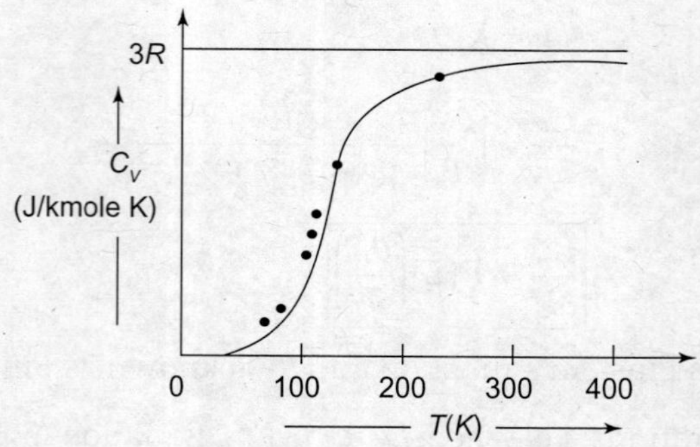

FIGURE 3.5 Comparison of specific heat data for aluminium (dots) and predicted curve from Einstein equation (line)

EXAMPLE 3.2

If the unit cell has 3 atoms, thus how many phonon modes are present?

Solution: 3N modes
i.e. 9 modes

EXAMPLE 3.3

If the velocity of sound in a solid is taken to be 3×10^8 m/s and interatomic distance as 5×10^8 m, calculate the value of cutoff frequency assuming a linear lattice.

Solution: We know that

Critical frequency, $\nu = \dfrac{V}{2a}$

where, $V = 3 \times 10^3$ m/s
and $a = 5 \times 10^{-10}$ m
Thus, $\boldsymbol{\nu = 3 \times 10^{12}}$ **Hz**

Note

Einstein model of the solid predicts the heat capacity accurately at high temperatures; it noticeably deviates from experimental values at low temperatures. At low temperature, the specific heat capacity for the Debye model tends to T^3. At high temperature, it tends to $3R$.

3.7 ▶ Debye's Model

In 1913, Debye improved Einstein's theory by taking a solid as a continuously vibrating medium, which gives a spectrum of frequencies instead of a single frequency as assumed by

Einstein in his model. According to Debye, the atomic oscillator does not vibrate at constant frequency but they are elastically coupled to their neighbours and the motion of such a coupled system can be described in terms of normal modes. Once the frequency distribution is known, we can calculate the total internal energy and hence the specific heat of the solids.

The internal energy of the solid can be written as given by Eq. (3.42) (by taking into account the density of states as discussed in Section 3.4). Using Eq. (3.21), we get

$$U = \int\limits_{\substack{\text{all allowed}\\\text{frequencies}}} \bar{E}(\nu) z(\nu) d\nu \tag{3.42}$$

But we know that

$$\bar{E} = \frac{h\nu}{\exp\left(\frac{h\nu}{K_B T}\right) - 1} \tag{3.43}$$

and

$$z(\nu) d\nu = 4\pi V \left(\frac{1}{v_1^3} + \frac{2}{v_t^3} \right) \nu^2 d\nu \tag{3.44}$$

Using Eqs. (3.43) and (3.44) in Eq. (3.42), we get

$$U = 4\pi V \left(\frac{1}{v_1^3} + \frac{2}{v_t^3} \right) \int_0^{\nu_D} \frac{h\nu \nu^2 d\nu}{\exp\left(\frac{h\nu}{K_B T}\right) - 1}$$

Also, taking Debye frequency,

$$\nu_D^3 = \frac{9N}{4\pi V} \left(\frac{1}{v_1^3} + \frac{2}{v_t^3} \right)^{-1} \tag{3.45}$$

Using Eq. (3.45) into Eq. (3.44), we get

$$U = \frac{9N}{\nu_D^3} \int_0^{\nu_D} \frac{h\nu \nu^2 \, d\nu}{exp\left(\frac{h\nu}{K_B T}\right) - 1} \tag{3.46}$$

Let us suppose that $$x = \frac{h\nu}{K_B T}$$

or $$d\nu = \frac{K_B T dx}{h}$$

and also suppose $$x_D = \frac{h\nu_D}{K_B T}$$

from Eq. (3.46),

$$\boxed{U = 9N\left(\frac{K_B T}{h\nu_D}\right)^3 K_B T \int_0^{\nu_D} \frac{x^3}{e^x - 1} dx} \tag{3.47}$$

Here, we can define θ_D as Debye temperature

where, $$K_B\theta_D = h\nu_D$$

$$\Rightarrow \qquad \theta_D = \frac{h\nu_D}{K_B} \tag{3.48}$$

Equation (3.47) is called Debye expression.

Case (a): At high temperatures, $e^x = 1 + x$
Equation (3.47) becomes

$$U = 9N_A K_B T\left[\frac{T}{\theta_D}\right]^3 \int_0^{\theta_D/T} x^2 dx$$

$$U = 3N_A K_B T$$

$$\Rightarrow \qquad C_v = \frac{dU}{dT} = 3N_A K_B = 3R$$

Thus, at high temperatures, the Debye expression is similar to Einstein's and classical expression.

Case (b): At very low temperatures, $T \to 0, \ \dfrac{\theta_D}{T} \to \infty$

$$\int_0^{\infty} \frac{x^3}{e^x - 1} dx = \int_0^{x_D} \frac{x^3}{e^{x-1}} dx = \frac{\pi^4}{15}$$

and Eq. (3.47) becomes

$$U = 9N_A K_B T\left[\frac{T}{\theta_D}\right]^3 \left[\frac{\pi^4}{15}\right]$$

$$\Rightarrow \qquad U = \frac{9}{15} \frac{N_A K_B T^4 \pi^4}{\theta_D{}^3}$$

and $$C_v = \frac{dV}{dT} = \frac{12}{5}\pi^4 R\left(\frac{T}{\theta_D}\right)^3 \text{ for } T \ll \theta_D \tag{3.49}$$

Thus, according to Debye model, the specific heat at very low temperatures is proportional to T^3, which is in a good agreement with the experimental observations (Fig. 3.6).

Case (c): At intermediate temperatures, the Debye model is not found in a good agreement with Dulong-Petit law.

Note

Treating the solid as a *continuum*, i.e., the atomic structure is ignored. A continuum has vibrational modes of *arbitrary low frequencies*, and at sufficiently low *T* only these low frequency modes are excited. These low frequency normal modes are simply standing sound waves.

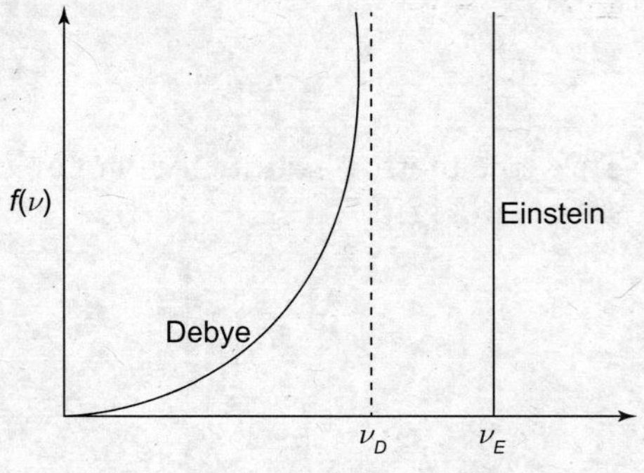

FIGURE 3.6 Frequency spectrum according to Debye's model

EXAMPLE 3.4

Find the Debye temperature for sodium chloride at 10, 15 and 20 K, where its corresponding specific heat values are 0.066, 0.249 and 0.649 KJ/Kmol-k.

Solution: Debye temperature for NaCl is

$$T_D = 3\sqrt{\frac{12}{5} C_\nu \pi^4 N_A K_B T^3}$$

where $\pi = 3.14$.

$$N_A = 6.023 \times 10^{23} \text{ mol}^{-1}$$

$$C_v = 0.066, 0.249, 0.649 \text{ kJ/kmol-k}$$

So,

(i) At 10 K, $T_D = 3.8\text{K}$

(ii) At 15 K, $T_D = 297\text{K}$

(iii) At 20 K, $T_D = 288\text{K}$

3.8 ▶ Thermal Conductivity in Insulators due to Phonons

As we know that there are no mobile electrons in insulators, and so most of the heat energy is carried by phonons. In Eq. (3.49) in last the section (Section 3.7), v is the velocity of sound, λ is the mean free path of phonons and C_v is the lattice specific heat per unit volume and it is given by Debye T^3 law as

$$C_v = \begin{cases} \alpha T^3 & (T < \theta_D) \\ 3N_K & (T > \theta_D) \end{cases}$$

Thus, the thermal conductivity of an insulator, at low temperatures, is proportional to T^3, as shown in Fig. 3.7.

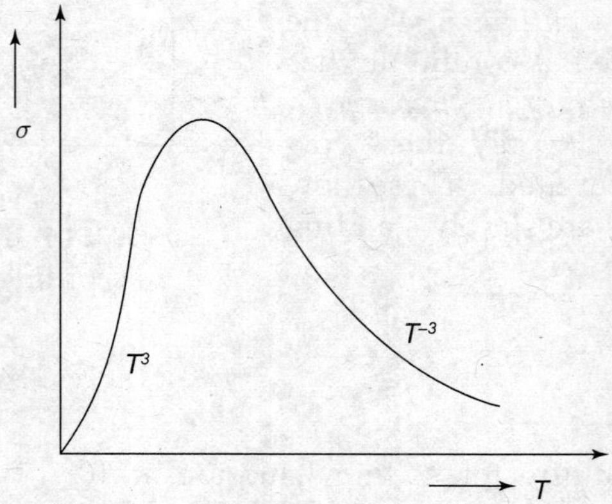

FIGURE 3.7 Variation of conductivity in insulators with temperature

At high temperatures, i.e. for $T \gg \theta_D$, the lattice specific heat is constant and σ dependence on temperature arises due to anharmonicity, i.e. phonon–phonon interaction. However, the collision of phonons with boundaries and impurities also contribute to this variation but phonon–phonon interaction pre-dominates.

In insulators, mean free path of a phonon is inversely proportional to the absolute temperature, i.e.,

$$\lambda \propto \frac{1}{T}.$$

Therefore, the thermal conductivity σ, of an insulator, is proportional to the temperature (at high temperatures).

3.9 ▶ Thermal Conductivity in Solids

In the phenomenon of thermal conductivity, the heat is transferred from one part of the body to the other part. The amount of thermal energy crossing per unit time is proportional to the temperature gradient, i.e.,

$$Q \propto \frac{dT}{dx}$$

$$\Rightarrow \quad \boxed{Q = K\frac{dT}{dx}} \tag{3.50}$$

where K is known as thermal conductivity of the solid.

In conductors, electrons are responsible for the transfer of heat, but in insulators, heat is carried by phonons mainly.

Thus, heat can be transferred by both electrons and phonons, and hence,

$$K_{\text{total}} = K_{\text{electron}} + K_{\text{phonon}} \tag{3.51}$$

To calculate K, let us consider a rod of conductor having uniform area of cross-section (say 1 sq.m) and having uniform temperature gradient.

Let us consider three parallel planes P, Q and R separated by mean free path λ (Fig. 3.8).

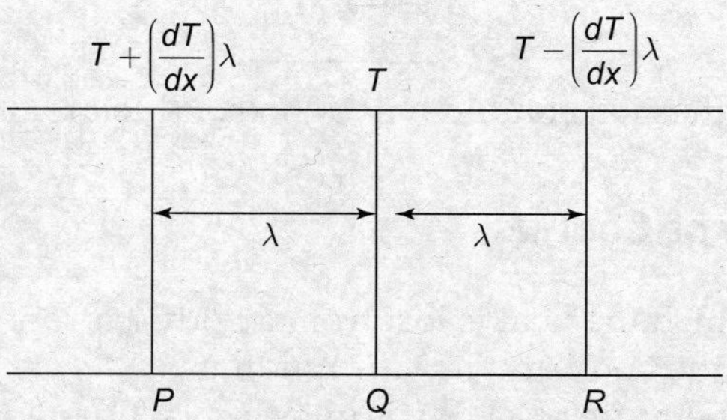

FIGURE 3.8 Thermal conductivity in a conductor

Now, according to kinetic theory, the electrons flowing per unit area per unit time in a given direction are $(1/6)\eta\nu$, where η is the density of electrons moving with average velocity v. The amount of energy carried by an electron from plane P to Q will be $\left(\frac{dU}{dx}\right)\lambda$.

Thus, excess energy flowing through the plane Q to the left is

$$Q_{\text{left}} = -\frac{1}{6}\eta\nu\lambda\left(\frac{dU}{dx}\right) \tag{3.52}$$

Similarly, excess energy flowing through the plane Q to the right is

$$Q_{\text{right}} = \frac{1}{6}\eta\nu\lambda\left(\frac{dU}{dx}\right) \tag{3.53}$$

Therefore, net amount of heat energy flowing through the plane Q is

$$Q = \frac{1}{6}\eta\nu\lambda\left(\frac{dU}{dx}\right) + \frac{1}{6}\eta\nu\lambda\left(\frac{dU}{dx}\right)$$

$$Q = \frac{\eta\nu\lambda}{3}\left(\frac{dU}{dx}\right) \tag{3.54}$$

Here,
$$\eta\left(\frac{dU}{dx}\right) = \eta\left(\frac{dU}{dT}\right)\left(\frac{dT}{dx}\right)$$

$$= \eta C_v \left(\frac{dT}{dx} \right) \tag{3.55}$$

Using Eq. (3.55) in Eq. (3.54), we get

$$Q = \frac{\eta C_v \nu \lambda}{3} \left(\frac{dT}{dx} \right) \tag{3.56}$$

Comparing Eqs. (3.50) and (3.56), we get

$$\boxed{\boldsymbol{K = \frac{1}{3} \nu \lambda C_v}} \tag{3.57}$$

Equation (3.57) gives the thermal conductivity K, in conductors.

3.10 ▶ Newton's Law of Cooling

It states that the rate at which the heat is lost from a body is proportional to the difference in temperatures between the body and its surroundings.

Suppose the initial temperature of a body is given as T_1 °C and it is allowed to cool in air which is maintained at temperature T_2 °C (constant).

Let us suppose that the temperature of the body at time t, be T °C

Then by Newton's law of cooling

$$\frac{dT}{dt} = -K(T - T_2) \tag{3.58}$$

where K is the proportionality constant and its value depends upon the surface properties of the material being cooled.

As the temperature of the body T is greater than the temperature of the surroundings, thus, $\frac{dT}{dt} < 0$.

Initially, at $t = 0$, $T = T_1$

By solving Eq. (3.58), we get

$$-Kt = \log(T - T_2) + \log C$$

$$\Rightarrow \quad T - T_2 = Ce^{-kt} \tag{3.59}$$

Applying initial conditions, from Eq. (3.59), we get

$$C = T_1 - T_2 \tag{3.60}$$

Putting Eq. (3.60) in Eq. (3.59), we get

$$\boxed{\boldsymbol{T = T_2 + \left(T_1 - T_2 e^{-Kt} \right)}} \tag{3.61}$$

Equation (3.61) represents Newton's law of cooling. The graph between temperature of the body and time is known as cooling curve as shown in Fig. 3.9.

In Eq. (3.61), if $K < 0$, $\lim t \to \infty, e^{-Kt} = 0$ $\Rightarrow T = T_2$

It means the temperature of the body approaches that of the surroundings as time goes. In general,

$$T(t) = T_A + (T_H - T_A)e^{-Kt} \qquad (3.62)$$

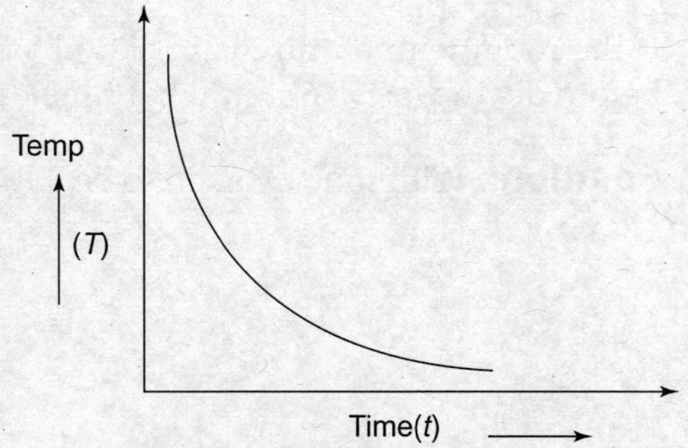

FIGURE 3.9 Cooling curve of Newton's law of cooling

where,
$T(t)$ = temperature of body at time t
T_A = temperature of surroundings (ambient temperature)
T_H = temperature of hot object at $t = 0$
T = time and K = proportionality constant

3.10.1 Applications of Newton's Law of Cooling

1. It helps to find the temperature of a soda placed in a refrigerator by a certain amount of time.
2. It helps to predict that how long it will take for a hot body to cool down at a certain temperature.
3. It helps to indicate the time of death given the probable body temperature at the time of death and current body temperature.
4. It helps to determine the specific heat capacity of a substance.
5. It also helps to design radiators/cooling systems in thermal machines.

Note

1. Cooling is fast when the temperature difference is high.
2. Cooling is slow when the temperature of hot body falls to a temperature closer to room temperature.

EXAMPLE 3.5

In a murder investigation, a corpse was found by a detective exactly 8 P.M. Being alert, the detective also measured the body temperature and found it to be 70°F. Two hours later, the detective measured the body temperature again and found it to be 60°F.

If the room temperature is 50°F and assuming that the body temperature of a person before death was 98.6 F, at what time did the murder occur?

Solution: With time O taken to be 8 P.M., we have
Here,

$$\frac{dT}{dt} = k(50 - T)$$

$$T(0) = 70$$

$$T(2) = 60$$

Thus,

$$T = 50 + 20\left(\frac{1}{2}\right)^{\frac{t}{2}}$$

We have to find the value of t for which $T(t) = 98.6$

Thus,
$$50 + 20\left(\frac{1}{2}\right)^{\frac{t}{2}} = 98.6$$

$$t = 2\frac{\ln\left(\frac{48.6}{20}\right)}{\ln\left(\frac{1}{2}\right)}$$

$$t \cong -2.56$$

Thus, this person was murdered at about 5.30 P.M.

EXAMPLE 3.6

Robin heats the water to 7°F and he waits for 10 minutes. How much would be the temperature if $k = 0.056$ per min and surrounding temperature is 27°C.

Solution: Given

$$T_A = 27°\text{C}$$
$$\text{T}_\text{H} = 70°\text{C}$$
$$t = 10 \text{ min}$$

and
$$k = 0.056$$

The formula is

$$T(t) = T_A + (T_H - T_A)e^{-kt}$$

Thus,

$$T(t) = 27 + (70 - 27)e^{-0.56}$$

or

$$T(t) = 51.56°C$$

Temperature cools down from 70°C to 51.56°C after 10 minutes.

3.11 ▶ Modes of Heat Transfer

Heat transfer means the exchange of thermal energy between two physical systems. The rate at which the heat is transferred depends upon the temperatures of the system and the properties of the medium through which the heat is being transferred.

There are three fundamental modes of heat transfer:

1. **Conduction:** When the two physical systems are in physical contact with each other, then the transfer of energy between them is done by the phenomenon of conduction. As we heat the metal, the particles vibrate, these vibrations make the adjacent particles vibrate, and thus the transmission of heat takes place by molecular vibrations. As it always requires material medium (solid), it takes place in vacuum. Metals are different. The outer electrons of metal atoms drift and are free to move. When the metal is heated, 'electrons' gain kinetic energy and transfer it throughout the metal. Insulators, such as wood and plastic, do not have such electrons which is why they do not conduct heat as well as metals.
2. **Convection:** The transfer of heat energy between an object and its environment due to fluid motion is called convection. The average temperature is taken as reference for measuring or evaluating properties related to convective heat transfer.
3. **Radiation:** The transfer of heat energy by the emission of electromagnetic radiation is known as the phenomenon of heat transfer through radiation. There are no particles between the Sun and the Earth; so it cannot travel by conduction or by convection. Hence, no material medium needed.

EXAMPLE 3.7

Calculate the heat lost by the block when iron block decreases its temperature from 60°C to 40°C if the mass of the body is 2 kg. Specific heat of iron C = 0.45 kJ/kgK.

Solution: Given

$$T_i = 60°C$$

$$T_g = 40°C$$

$$m = 2 \text{ kg}$$

and heat lost is $Q = mC\Delta T$

$$= \mathbf{263.7\ J}$$

3.11.1 Liner Conduction Heat Flow

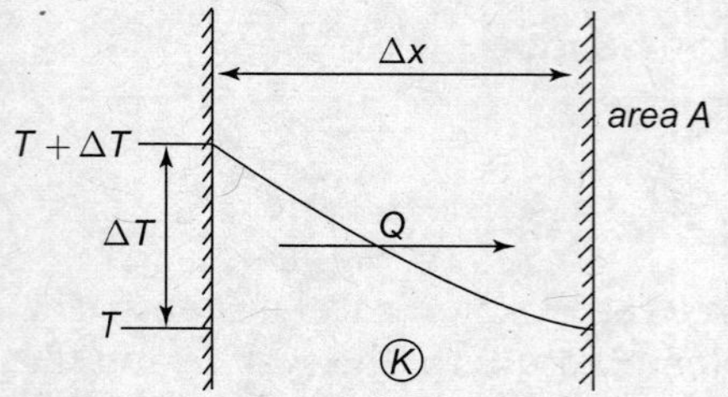

FIGURE 3.10 Linear heat flow

Linear conduction can be easily understood if we imagine two blocks, one very hot and the other is cold. The mode of heat transfer between two solid blocks is known as linear heat flow (Fig. 3.10). If a plane wall of thickness (Δx), area A supports a temperature difference ΔT, heat transfer rate per unit time (Q) by conduction through the wall is given by Fourier's law which states that

$$Q = KA\frac{dT}{dx} \tag{3.63}$$

where,
Q = heat flow rate (in watt)
K = thermal conductivity (watt/km)
A = cross-sectional area of the conduction (m^2)
dT = change of temperatures between 2 points (K)
dx = change of displacements between two points (m).
Here, both Q and K are constant.
Hence,

$$\frac{A_H(dT)}{dx_H} = \frac{A_s(dT)}{dx_s} = \frac{A_c(dT)}{dx_c} \tag{3.64}$$

i.e., the temperature gradient is inversely proportional to the cross-sectional area.

3.11.2 Radial Conduction Heat Flow

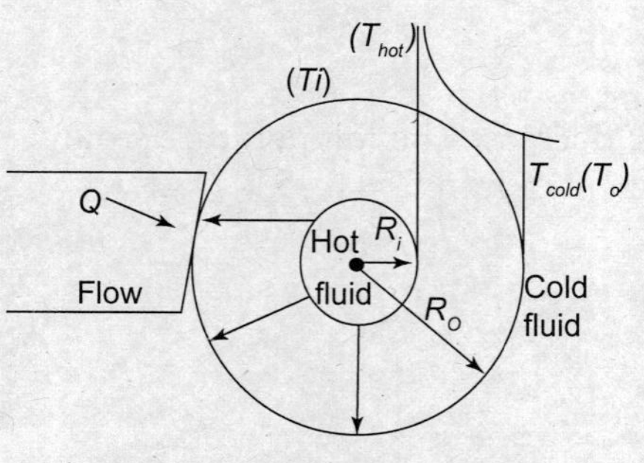

FIGURE 3.11 Radial heat flow

When the inner and outer surfaces of a thick-walled cylinder are each at a uniform temperature, heat flows radially through the walls of the cylinder (Fig. 3.11).

From continuity considerations, the radial heat flow through successive layers in the wall must be constant if the flow is steady but since the area of successive layers increases with radius, the temperature gradient must decrease with radius.

The amount of heat (Q) which is conducted through the wall of the cylinder per unit time is

$$Q = -\frac{2\pi rl K(T_i - T_0)}{\ln \frac{R_0}{R_i}} \tag{3.65}$$

where
Q = radial heat flow rate (W)
l = thickness of material (m)
K = thermal conductivity (w/km)
T_i = inner section temperature (K)
T_o = outer section temperature (K)
R_o = outer radius (m)
R_i = inner radius (m)

3.12 ▶ Rubber Tube Method/Cylindrical Shell Method (Radial Flow of Heat)

We know that in radial flow of heat, heat energy flows from the inner side towards the other side along the radius of the cylindrical shell.

This method is useful in explaining the thermal conduction in bad conductors taken in powder form. Consider a cylindrical tube of length l, inner radius r_i and outer radius r_o (Fig. 3.12).

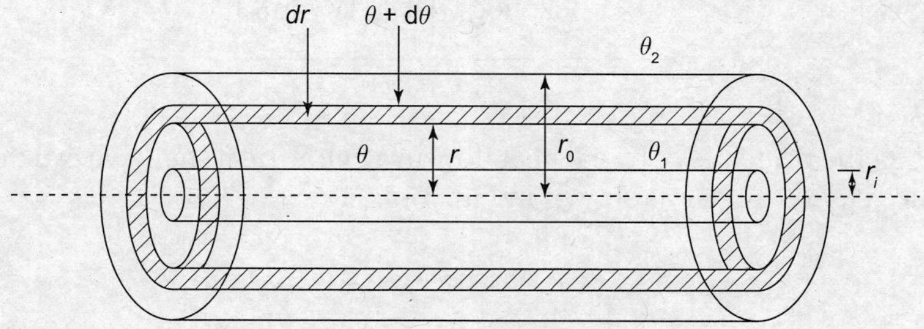

FIGURE 3.12 Cylindrical shell method

Let the tube carries some hot liquid or steam in it. When the steady state is reached, the temperature on the inner surface is θ_1 and the outer surface is θ_2, also $\theta_1 > \theta_2$. Now, the heat is conducted radially across the walls of the tube. Consider an element dr having length l at a distance r from the axis.

The cylinder may be considered to consist of a large number of co-axial cylinders of increasing radii. Let the temperatures of inner and outer surfaces of the elemental shell be θ and $(\theta + d\theta)$. Then, the amount of heat conducted per second is

$$Q = -KA\frac{d\theta}{dr} \tag{3.66}$$

Here, area of cross-section is $A = 2\pi rl$

From Eq. (3.66), we get

$$\therefore \qquad Q = -2\pi K l r \frac{d\theta}{dr} \tag{3.67}$$

Rearranging Eq. (3.67), we get

$$\frac{dr}{r} = -\frac{2\pi K l}{Q} d\theta \tag{3.68}$$

The thermal conductivity of whole cylinder is obtained by

$$\int_{r_i}^{r_o} \frac{dr}{r} = -\frac{2\pi l K}{Q} \int_{\theta_1}^{\theta_2} d\theta$$

$$\Rightarrow \qquad \log_e\left(\frac{r_o}{r_i}\right) = \frac{2\pi l K}{Q}(\theta_1 - \theta_2)$$

$$\Rightarrow \qquad K = \frac{Q \log_e\left(\frac{r_o}{r_i}\right)}{2\pi l(\theta_1 - \theta_2)}$$

$$\Rightarrow \qquad K = \frac{Q \times 2.302 \times \log_{10}\left(\frac{r_o}{r_i}\right) \text{ (W / mK)}}{2\pi l(\theta_1 - \theta_2)} \tag{3.69}$$

Now, based on the principle of radial heat flow in a cylindrical shell, we can determine the thermal conductivity of rubber as shown in Fig. 3.13.

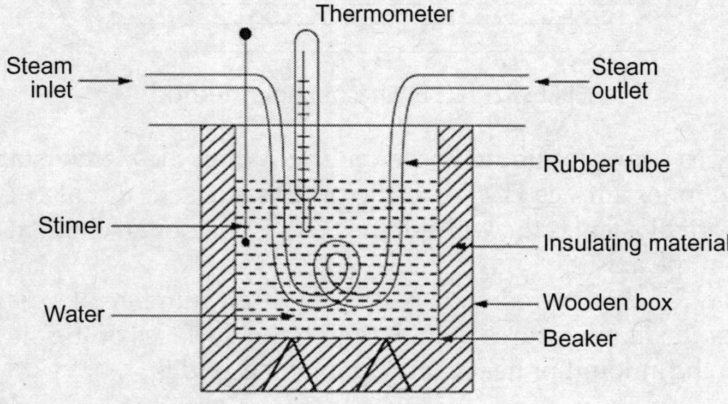

FIGURE 3.13 Rubber tube method

The empty calorimeter is weighed, let it be w_1. Now it is filled with $2/3^{rd}$ of water and weighed again. Let it be w_2. Now, a known length of rubber tube is immersed inside the water contained in the calorimeter. Steam is passed through one end of the rubber tube and let out through the other end of the tube. The heat flows from the inner layer of the tube to the other layer and is radiated. The radiated heat is gained by water in the calorimeter. The time taken for the steam flow to raise the temperature of water about 10°C is noted. Let it be in 't' seconds. Let θ_1 and θ_2 be the initial and final temperature of water, and θ_s is the temperature of steam.

Using the theory of cylindrical shell method, Q can be written as

$$Q = \frac{K\,2\pi l(\theta_s - \theta_3)}{\log_e \frac{r_o}{r_i}} \tag{3.70}$$

where

r_i = inner radius of tube

r_o = outer radius of tube

$$\theta_3 = \frac{\theta_1 + \theta_2}{2} \tag{3.71}$$

$\theta_2 - \theta_1$ = rise in temperature of water

s_1 = specific heat capacity of calorimeter

s_2 = specific heat capacity of water

Thus, the amount of heat gained by calorimeter per second is

$$\frac{w_1 s_1(\theta_2 - \theta_1)}{t} \tag{3.72}$$

And the amount of heat gained by water per second is

$$\frac{(w_2 - w_1)s_2(\theta_2 - \theta_1)}{t} \tag{3.73}$$

The amount of heat gained by water and calorimeter per second is obtained by adding Eqs. (3.72) and (3.73), we get

$$Q = (w_2 - w_1)s_2(\theta_2 - \theta_1) + w_1 s_1(\theta_2 - \theta_1)$$

$$\Rightarrow \qquad Q = \frac{(\theta_2 - \theta_1)[w_1 s_1 + (w_2 - w_1)s_2]}{t} \tag{3.74}$$

but

$$Q = \frac{2\pi l K(\theta_s - \theta_3)}{\log_e\left(\frac{r_o}{r_i}\right)} \tag{3.75}$$

Comparing Eqs. (3.74) and (3.75), we get the thermal conductivity of rubber tube

$$K = \frac{(\theta_2 - \theta_1)\log_e\left(\frac{r_o}{r_i}\right)[w_1 s_1 + (w_2 - w_1)s_2]}{2\pi lt\left[\theta_s - \frac{(\theta_1 + \theta_2)}{2}\right]} \tag{3.76}$$

Equation (3.75) gives the thermal conductivity of rubber (3.76).

3.13 ▶ Lee's Disc Method

The thermal conductivity of a bad conductor like ebonite or cardboard can be determined by this method.

Principle: The quantity of heat conducted in the steady state across any cross-section is equal to the quantity of heat radiated from its surface to the surroundings.

Experimental arrangement: The bad conductor is placed in between the steam chamber (S) and the disc (D), provided the bad conductor, steam chamber and the slab should be of same diameter. The steam chamber and disc are provided with holes where thermometers are inserted as shown in Fig. 3.14.

Working: Steam is passed through the chamber till the steady state is reached. Let the temperature of hot end of chamber and the cold end of disc be θ_s and θ_d, respectively.

Let L be the thickness of the bad conductor, m be the mass, S be the specific heat capacity, r be the radius, and h be the height of the slab.

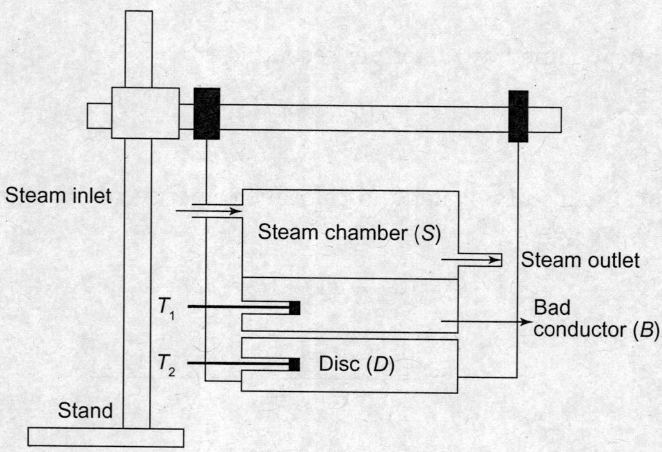

FIGURE 3.14 Lee's disc method

Now, the heat conducted by bad conductor per second is given by $\frac{KA(\theta_s - \theta_d)}{L}$, where the area of cross-section is $A = \pi r^2$.

The amount of heat conducted per second $= \dfrac{K\pi r^2(\theta_s - \theta_d)}{L}$

The amount of heat lost by slab per second $= m \times s \times$ rate of cooling

$$= msR_c$$

But we know that the amount of heat conducted by bad conductor = amount of heat lost by the slab per second. Thus,

$$\frac{K\pi r^2(\theta_s - \theta_d)}{L} = msR_c$$

or

$$\boxed{K = \frac{msR_cL}{\pi r^2(\theta_s - \theta_d)}} \tag{3.77}$$

To find the rate of cooling R_c

To find the rate of cooling for the disc alone, the bad conductor is removed and the steam chamber is placed over the disc and then heated. When the temperature of slab becomes 5°C higher than θ_d, the steam chamber is then removed. The slab is allowed to cool simultaneously a stop watch is switched ON.

A graph is then plotted taking temperature on the y-axis and time along the x-axis, the rate of cooling $\left(\dfrac{d\theta}{dt}\right)$ for the disc alone is found from the graph as shown in Fig. 3.15. The rate of cooling is directly proportional to the surface area exposed.

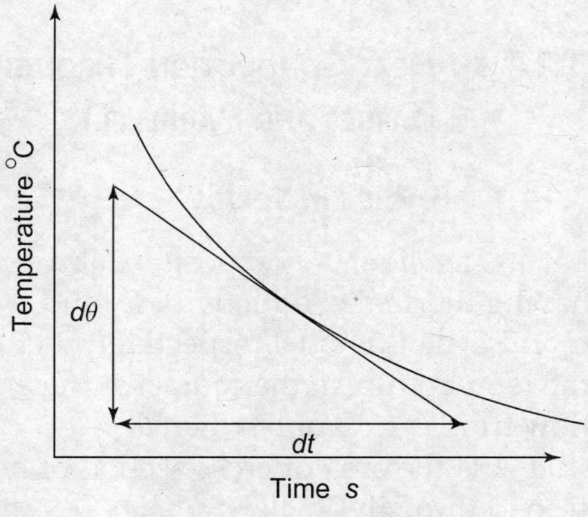

FIGURE 3.15 Rate of cooling for disc

Case (i): Steam chamber and bad conductor are placed over slab, in which radiation takes place from the bottom surface of area (πr^2) of the slab and the sides of the area ($2\pi rh$).

$$\therefore \quad R_c = 2\pi r^2 + 2\pi rh$$

$$R_c = \pi r(r + 2h) \tag{3.78}$$

Case (ii): The heat is radiated by the slab alone, i.e., from the bottom of area (πr^2), top surface of the slab of area (πr^2) and also through the sides of the slab of area $2\pi rh$.

$$\left(\frac{d\theta}{dt}\right)_{\theta_d} = \pi r^2 + \pi r^2 + 2\pi rh$$

$$\left(\frac{d\theta}{dt}\right)_{\theta_d} = 2\pi r(r + h) \tag{3.79}$$

From Eqs. (3.78) and (3.79), we get

$$\frac{R_c}{\left(\frac{d\theta}{dt}\right)_{\theta_d}} = \frac{\pi r(r+2h)}{2\pi r(r+h)}$$

$$\Rightarrow \qquad R_c = \frac{(r+2h)}{2(r+h)}\left(\frac{d\theta}{dt}\right)_{\theta_d} \tag{3.80}$$

By putting Eq. (3.80) into Eq. (3.77), we get

$$K = \frac{msL\left(\frac{d\theta}{dt}\right)_{\theta_d}(r+2h)}{\pi r^2(\theta_s - \theta_d)2(r+h)} \tag{3.81}$$

3.14 ▶ Heat Conduction Through a Compound Media (Series and Parallel)

3.14.1 Bodies in Series

Let us consider a composite slab of two different bodies, C and D, arranged in series having thermal conductivity K_1 and K_2, respectively. Let the thickness of these two layers C and D be t_1 and t_2, respectively. Let the temperatures of the end faces be θ_1 and θ_2, and the temperature at the contact surface be θ, which is unknown. We know that the heat will flow from C to D only when $\theta_1 > \theta_2$.

If A is the area of cross-section of both the layers, then the amount of heat flowing per second through C will be

$$Q = \frac{K_1 A(\theta_1 - \theta)}{t_1} \tag{3.82}$$

Similarly, the amount of heat flowing per second through B will be

$$Q = \frac{K_2 A(\theta_1 - \theta)}{t_2} \tag{3.83}$$

From Eqs. (3.82) and (3.83), we get

$$\frac{K_1 A(\theta_1 - \theta)}{t_1} = \frac{K_2 A(\theta_1 - \theta)}{t_2} \tag{3.84}$$

Also, from Eq. (3.84), we get

$$K_1 A(\theta_1 - \theta)t_2 = K_2 A(\theta_1 - \theta)t_1$$

Rearranging the above equation, we get

$$\theta = \frac{K_1 t_2 \theta_1 + K_2 t_1 \theta_2}{K_1 t_1 + K_2 t_2} \tag{3.85}$$

Equation (3.85) gives the expression for interface temperature of two composite slabs in series (Fig. 3.16).

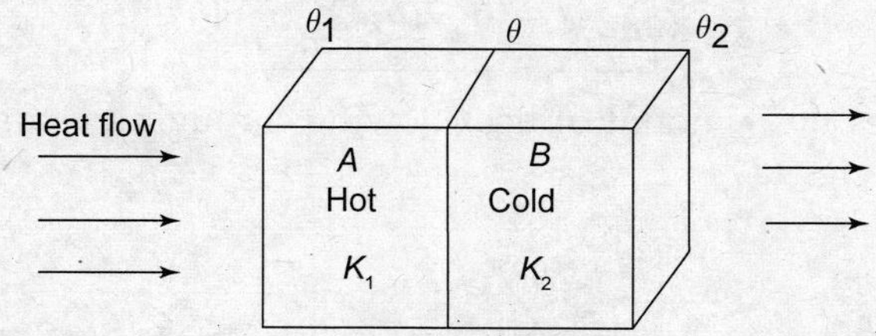

FIGURE 3.16 Heat conduction in series compound media

Substituting Eq. (3.85) into Eq. (3.82) and rearranging this equation, we get

$$Q = \frac{K_1 A\left(\theta_1 - \dfrac{K_1 t_2 \theta_1 + K_2 t_1 \theta_2}{K_1 t_1 + K_2 t_2}\right)}{t_1}$$

$$= \frac{K_1 K_2 A}{t_1}\left(\frac{\theta_1 t_1 - \theta_2 t_1}{K_2 t_1 + K_1 t_2}\right)$$

or

$$\boxed{Q = \frac{A(\theta_1 - \theta_2)}{\dfrac{t_1}{K_1} + \dfrac{t_2}{K_2}}} \tag{3.86}$$

Equation (3.86) gives the amount of heat flowing through the compound wall of two materials (in series).

3.14.2 Bodies in Parallel

Let us consider a compound wall of two different bodies C and D arranged in parallel having thermal conductivities K_1 and K_2 and of thickness t_1 and t_2, respectively. The temperature θ_1 is maintained at one face of the layers C and D and opposite faces of the layers are maintained at temperature θ_2. Let A_1 and A_2 be the areas of cross-section of the bodies.

Amount of heat flowing through the first body C in one second is

$$Q_1 = \frac{K_1 A_1 (\theta_1 - \theta_2)}{t_1} \tag{3.87}$$

Similarly, amount of heat flowing through the second body D in one second is

$$Q_2 = \frac{K_2 A_2(\theta_1 - \theta_2)}{t_2} \tag{3.88}$$

Now, the total heat flowing through the compound body per second will be equal to the sum of the individual bodies, i.e., $Q = Q_1 + Q_2$.

or

$$\boxed{\sum Q = (\theta_1 - \theta_2)\sum \frac{KA}{t}} \tag{3.89}$$

Equation (3.89) gives the amount of heat flowing through the compound wall of two materials (in parallel) (Fig. 3.17).

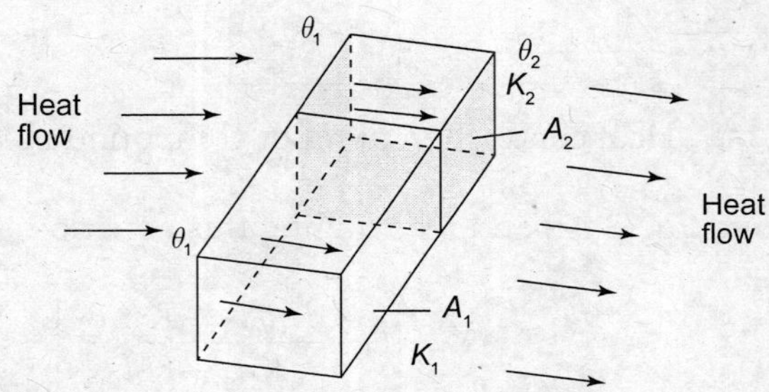

FIGURE 3.17 Heat conduction in parallel compound media

EXAMPLE 3.8

The wall of a bakery oven is built of insulating brick 10 cm thick and thermal conductivity 0.22 $\text{Jm}^{-1}\ \text{s}^{-1}\ ^\circ\text{C}^{-1}$. Steel reinforcing members penetrate the brick, and their total area of cross-section represents 1% of the inside wall area of the oven.

If the thermal conductivity of the steel is 45 $\text{Jm}^{-1}\ \text{s}^{-1}\ ^\circ\text{C}^{-1}$, calculate (a) the relative proportions of the total heat transferred through the wall by the brick and by the steel and (b) the heat loss for each m^2 of oven wall if the inner side of the wall is at 230°C and the outer side is at 25°C.

Solution: Applying $Q = A\Delta Tk/x$, we know that ΔT is the same for the bricks and for the steel. Also x, the thickness, is the same.

(a) Consider the loss through an area of 1 m^2 of wall (0.99 m^2 of brick, and 0.01 m^2 of steel)

For brick $$Q_b = A_b \Delta T k_b / x$$

$$= \frac{0.99(230 - 25)0.22}{0.10}$$

$$= 446\ \text{Js}^{-1}$$

For steel $$Q_s = A_s \Delta T\, k_s / x$$

$$= \frac{0.01(230 - 25)45}{0.10}$$

$$= 923 \text{ Js}^{-1}$$

Therefore, $$Q_b / Q_s = 0.48$$

(b) Total heat loss

$$Q = (Q_b + Q_s) \text{ per m}^2 \text{ of wall}$$

$$= 446 + 923$$

$$= 1369 \text{ J s}^{-1}$$

Therefore, percentage of heat carried by steel

$$= (923/1369) \times 100$$

$$= \mathbf{67\%}$$

EXAMPLE 3.9

A cold store has a wall comprising 11 cm of brick on the outside, then 7.5 cm of concrete and then 10 cm of cork. The mean temperature within the store is maintained at −18°C and the mean temperature of the outside surface of the wall is 18°C. Calculate the rate of heat transfer through the wall. The appropriate thermal conductivities are for brick, concrete and cork, respectively, 0.69, 0.76 and 0.043 J m^{-1} s^{-1} °C^{-1}. Determine also the temperature at the interfaces between the concrete and cork layers, and the brick and concrete layers.

Solution: For brick, $$x_1/k_1 = 0.11/0.69 = 0.16$$

For concrete, $$x_2/k_2 = 0.075/0.76 = 0.10$$

For cork, $$x_3/k_3 = 0.10/0.043 = 2.33$$

But $$\frac{1}{U} = \frac{x_1}{k_1} + \frac{x_2}{k_2} + \frac{x_3}{k_3}$$

$$= 0.16 + 0.10 + 2.33$$

$$= 2.59$$

Therefore, $$U = 0.38 \text{ J m}^{-2} \text{ s}^{-1} {}^\circ\text{C}^{-1}$$

$$\Delta T = 18 - (-18) = 36^\circ\text{C}$$

$$A = 1 \text{ m}^2$$

$$q = UA\Delta T$$

$$= 0.38 \times 1 \times 36$$

$$= 13.7 \text{ J s}^{-1}$$

Further, $$q = \frac{A_3 \Delta T_3 k_3}{x_3}$$

and for the cork wall $A_3 = 1 \text{ m}^2$, $x_3/k_3 = 2.33$ and $q = 13.7 \text{ J s}^{-1}$

Therefore, $13.7 = 1 \times \Delta T_3 \times \frac{1}{2.33}$ from rearranging Eq. (5.2)

$$\Delta T_3 = 32°\text{C}.$$

But ΔT_3 is the difference between the temperature of the cork/concrete surface T_c and the temperature of the cork surface inside the cold store.

Therefore, $$T_c - (-18) = 32$$

where T_c is the temperature at the cork/concrete surface and so $T_c = 14°\text{C}$.

If ΔT_1 is the difference between the temperature of the brick/concrete surface, T_b, and the temperature of the external air.

Then, $$13.7 = 1 \times \Delta T_1 \times \frac{1}{0.16} = 6.25\, \Delta T_1$$

Therefore, $$18 - T_b = \Delta T_1 = \frac{13.7}{6.25} = 2.2$$

So, $$T_b = 15.8°\text{C}$$

Working it through shows approximate boundary temperatures: air/brick 18°C, brick/concrete 16°C, concrete/cork 14°C, cork/air –18°C

This shows that almost all of the temperature differences occur across the insulation (cork), and the actual intermediate temperatures can be significant especially if they lie below the temperature at which the atmospheric air condenses, or freezes.

EXAMPLE 3.10

A cork slab of 10 cm thick having one face at –12°C and the other face at 21°C. If the mean thermal conductivity of the cork is 0.042 $\text{Jm}^{-1}\text{s}^{-1}°\text{C}^{-1}$ in this temperature range, what is the rate of heat transfer through 1 m^2 of wall?

Solution: $$T_1 = 21°\text{C}$$

$$T_2 = -12°\text{C}$$

$$\Delta t = 33°\text{C}$$

$$A = 1\,\text{m}^2$$

$$K = 0.042\,\text{J/ms°C}$$

$$t = 0.1\,\text{m}$$

$$\mathbf{Q = 13.9\,J/s}$$

RECAP ZONE

POINTS TO REMEMBER

- Heat capacity per unit mass is known as specific heat.
- Heat flow is positive in the direction of temperature fall.
- Density of states of a system defines the number of states per interval of energy at each energy level that are available to be occupied.
- The two types of internal energy in most solids are (i) the kinetic energy of the electrons and (ii) vibrational energy of the atoms about their mean positions. As the temperature of solid increases, the important thermal properties such as heat capacity, thermal expansion and thermal conductivity, all are dependent upon energy changes of the atoms and free electrons.
- The contribution to specific heat due to the transition of vibrating atoms in the crystal lattice of vibrational states of higher energy is called lattice specific heat.
- Lattice vibrations or phonon is a quantum of elastic energy and travels with the velocity of sound in the medium.
- Thermal conduction is the transfer of heat energy in a material due to the temperature gradient within it.
- If U is the internal energy of the system, then C_v is the specific heat at constant volume and C_p is the specific heat at constant pressure and

$$C_p - C_v = R\,(\text{Universal constant})$$

$$[C_v]_T = [C_v]_{la} + [C_v]_{el}$$

where,

$$[C_v]_{la} \rightarrow \text{lattic vibrations}$$

and $$[C_v]_{el} \rightarrow \text{electron contribution}$$

- According to classical model, the specific heat is given as

$$C_v = \left(\frac{dU}{dT}\right)_v = 3R,$$

this is known as Dulong-Petit law and it is valid only for high temperatures.

- According to Einstein model, the crystal is made up of N atoms vibrating in three independent directions with frequency v and the specific heat is given by

$$C_v = \left(\frac{dU}{dT}\right)_v = 3NK_B\left(\frac{hv}{K_BT}\right)^2 \frac{\exp\left(\frac{hv}{K_BT}\right)}{\left[\exp\left(\frac{hv}{K_BT}\right)-1\right]^2}$$

- Debye assumes that all the atoms are elastically coupled to its neighbours vibrate collectively with a spectrum of frequencies. The heat capacity at low temperatures is given by

$$C_v = \frac{12}{5}\pi^4 N K_B\left(\frac{T}{\theta_D}\right)^3$$

This is also known as Debye T^3 law. The Debye curve of specific heat capacity is more close to the experimental curve.

- Thermal conductivity of real gases is largely independent of pressure and may be considered a function of temperature alone. For solids and liquids, properties are largely independent of pressure and depend on temperature alone.
- The elastic energy of sound waves in solids is quantized and the quantum unit is called phonon.
- Conduction, convection and radiation are the three main methods of heat flow.

Key Terms with Definitions

- **Density of states** – It gives the number of allowed electron (or hole) states per volume at a given energy.
- **Lattice vibrations** – The oscillations of atoms in a solid about their equilibrium positions.
- **Phonon** – It is a collective excitation in a periodic, elastic arrangement of atoms or molecules in condensed matter, like solids and some liquids.
- **Thermal conductivity** – It is the property of a material to conduct heat.
- **Heat conduction** – It is the transfer of heat by microscopic collisions of particles and movement of electrons within a body.

- **Heat capacity** – It is the amount of heat required to raise the temperature of a system by one kelvin.
- **Temperature** – It is a comparative measurement of hot or cold.
- **Thermal radiation** – Thermal radiation is electromagnetic radiation generated by the thermal motion of charged particle in matter.
- **Einstein's temperature** – It indicates the probability that a molecule has in its degrees of freedom to store energy in its atomic oscillators.
- **Heat transfer** – Heat transfer, the flow of energy in the form of heat, is a process by which a system's internal energy is changed
- **Conduction** – An energy transfer across a system boundary due to a temperature difference by the mechanism of inter-molecular interactions.
- **Convection** – It is the dominant form of heat transfer in liquids and gases.
- **Radiation** – The transfer of energy by the emission of electromagnetic radiation.
- **Advection** – It is the transport phenomenon of a fluid from one location to another, and depends on the motion and momentum of the fluid.

IMPORTANT FORMULAE AND EQUATIONS

Equation Number	Equation	Remarks
(3.19)	$z(\nu)d\nu = 4\pi\nu\left(\frac{1}{v_1^3}+\frac{2}{v_t^3}\right)\nu^2 d\nu$	The number of modes whose frequencies lies between v and $(\nu+d\nu)$, where, v_1 and v_t are two velocities of propagation (v_1 for longitudinal wave and v_t for transverse waves).
(3.23)	$F \propto x$	The restoring force of a harmonic oscillator, where f is the elastic constant and x is the displacement.
(3.26)	$E = \frac{p_x^2}{2m} + P.E$	The total energy of a single one-dimensional isotropic harmonic oscillator, where potential energy is $P.E = \int f dx$, m is the mass of the oscillator and p_x is the momentum along the x-axis.

(3.28)	$\overline{E} = \dfrac{\sum_0^\infty E\,dN}{\sum_0^\infty dN}$	The mean energy of the atomic oscillator, where $\overline{E}$ is the mean energy of the oscillator and dN is the number of atomic oscillators having energies in the energy range E and $(E + dE)$ at a given temperature T.
(3.32)	$U = 3N_A K_B T = 3RT$	The classical value of the vibrational energy of a crystal where N_A be the total number of oscillators in a given mass, R is the universal gas constant.
(3.33)	$C_v = \left(\dfrac{dU}{dT}\right)_v$	Specific heat at constant volume, where U is the internal energy.
(3.39)	$\dfrac{C_v}{3R} = F(E)\left(\dfrac{\theta_E}{T}\right)$	Specific heat as given by Einstein model, where $F(E)$ is known as Einstein's function.
(3.47)	$U = 9N\left(\dfrac{K_B T}{hv_D}\right)^3 K_B T \int_0^{V_D} \dfrac{x^3}{e^x - 1} dx$	Debye expression, where ν_D is Debye frequency.
(3.50)	$Q = K\dfrac{dT}{dx}$	Heat energy of solids, where K is called thermal conductivity of the solids.
(3.57)	$K = \dfrac{1}{3}\nu\lambda C_v$	Thermal conductivity in conductors, where λ is the mean free path of particles.
(3.61)	$T = T_2 + \left(T_1 - T_2 e^{-Kt}\right)$	Newton's law of cooling, where the initial temperature of a body is given as T_1°C and it is allowed to cool in air which is maintained at temperature T_2°C.
(3.65)	$Q = -\dfrac{2\pi rl K(T_i - T_0)}{\ln\dfrac{R_0}{R_i}}$	Amount of heat conducted through the walls of the cylinder per unit time, where Q = radial heat flow rate (W) l = thickness of material (m) K = thermal conductivity (W/km) T_i = inner section temperature (K) T_o = outer section temperature (K) R_o = outer radius (m) R_i = inner radius (m)

(3.76)	$K=\dfrac{(\theta_2-\theta_1)\log_e\left(\dfrac{r_o}{r_i}\right)[w_1s_1+(w_2-w_1)s_2]}{2\pi lt\left[\theta_s-\dfrac{(\theta_1+\theta_2)}{2}\right]}$	Thermal conductivity of rubber tube, where r_i = inner radius of tube, r_o = outer radius of tube, θ_1, θ_2 are the initial, final temperature of water and θ_s is the temperature of steam. $\theta_1-\theta_1$ = rise in temperature of water s_1 = specific heat capacity of calorimeter s_2 = specific heat capacity of water
(3.77)	$K=\dfrac{msR_cL}{\pi r^2(\theta_s-\theta_d)}$	Thermal conductivity of bad conductor where the temperature of hot end (chamber) and the cold end (disc) be θ_s and θ_d, respectively. Let L be the thickness of the bad conductor, m be the mass and S be the specific heat capacity, r be the radius, h be the height of the slab.
(3.86)	$Q=\dfrac{A(\theta_1-\theta_2)}{\dfrac{t_1}{K_1}+\dfrac{t_2}{K_2}}$	Amount of heat flowing through the compound wall of two materials (in series).
(3.89)	$\sum Q=(\theta_1-\theta_2)\sum\dfrac{KA}{t}$	Amount of heat flowing through the compound wall of two materials (in parallel).

Review Zone

Short Answer Questions

1. State Dulong-Petit law.
2. What do you mean by lattice, specific heat, internal energy and heat capacity?
3. Give the expression for Einstein's equation for molar lattice specific heat.
4. Show that at low temperatures C_v varies directly as T^3.
5. Write down the expression for Debye's frequency.
6. What is Debye T^3 law?
7. What do you mean by Phonons?

8. Discuss the merits and demerits of Einstein's quantum theory.
9. What is Newton's law of cooling?
10. What do you mean by linear heat flow?
11. What do you understand by radial heat flow?
12. What is rubber tube method?
13. Why and where we use Lee's disc apparatus?
14. What are the three main modes of heat transfer?
15. Give the expression of heat flowing through the walls of compound media in series and parallel.

Long Answer Questions

1. State and explain Dulong-Petit's law and show how the departure from this law has been explained by Einstein's theory.
2. Obtain an expression for the specific heat capacity of a solid on the basis of
   (a) Einstein's theory and
   (b) Debye's theory.
3. Discuss the salient features of Debye's theory of specific heat and show how far it agrees with the experimental values.
4. Show that the electrons specific heat of a conductor is very small as compared to the lattice specific heat at room temperature.
5. How does the Debye's model differ from the Einstein's model? Mention the essential differences in the model and note the consequences of these differences.
6. Describe Newton's law of cooling. What are the applications of the law?
7. Explain the experimental arrangement and working of Lee's disc method.
8. Explain the experimental arrangement of rubber tube method. Determine its thermal conductivity also.
9. Derive an expression to determine the thermal conductivity of a bad conductor by Lee's disc method.
10. Explain in detail the conduction through compound media (series and parallel both).

Numerical Problems

1. What is the Debye frequency for copper, if it has the Debye temperature of 315 K and find the Debye specific heat at 10 K and 300 K?

   Ans. $-\nu = 6.55 \times 10^{12}\,\text{Hz}$

   $C_v\,(10\text{k}) = 62.67\ \text{J/kmol-k}$

   $C_v\,(300\text{k}) = 1.6 \times 10^{3}\ \text{kJ/kmol-k}$

2. The unit cell parameter of Nacl is 5.65 A° and the modulus of elasticity along [100] direction is $6 \times 10^{10}\,\text{N/m}^2$. Estimate the wavelength at which an electromagnetic radiation is strongly reflected by the crystal. At wt of Na = 23 and of cl = 37.

Hint: Use $w_{max} = \left[2k\left(\frac{1}{M}+\frac{1}{m}\right)\right]^{\frac{1}{2}}$

Ans. $-w_{max} = 5.15\times10^{13}$ rad/sec

and $\lambda = 3.67\times10^{-5}$ m

3. If the velocity of sound in a solid is of the order of 10^{13} m/s, compare the frequency of the sound wave $\lambda = 20\,A°$ for (a) a monoatomic system, (b) acoustic waves and optical waves in a diatomic system containing two identical atoms (M = m) per unit cell of inter-atomic spacing 2.2 A°.

Ans.

(i) $W_{homogenoussystem}$

$= 3.14\times10^{12}$ rad/s

(ii) $W_{acoustic} = 4.54\times10^{12}$ rad/s

(iii) $W_{optical} = 6.42\times10^{12}$ rad/s to

4.54×10^{12} rad/s

4. At midnight, with the temperature inside your house at 70°F and the temperature outside at 20°F, the furnace breaks down. Two hours later, the temperature in your house has fallen to 50°F. Assume that the outside temperature remains 20°F. At what time will the inside temperature of your house reach 40°F?

Hint: Use Newton's law of cooling

Ans. $t = 3.6$, after 3:30 a.m.

5. The oil is heated to 60°C. It cools to 50°C after 6 minutes. Find the time taken by the oil to cool from 50°C to 40°C. (surrounding temperature = 25°C).

Ans. –9.993 min.

6. Calculate the heat capacity if 2000 J of heat is used to heat iron rod of mass 5 kg from 20°C to 30°C.

Ans. 200J/°C

7. A room is maintained at 20°C whereas temperature outside the room is 10°C. There is one of the window in the corner having area 1 m by 2 m. Calculate the heat transfer. (Thermal conductivity of glass is 1. 4 /mk)

Ans. –9333.33 W

8. The opposite faces of a metal plate of 0.2 cm thickness are at difference of temperature 100°C and the area of the plate is 100 sq. m. Find the quantity of heat that will flow through the plate in one minute if $k = 0.2$ CGS units.

Ans. $Q = 10$ W

CHAPTER

4 Oscillations and Waves

Learning Objectives

By the end of this chapter, the student will be able:

- To discuss basic characteristics of a wave motion
- To understand various modes of vibrations in string
- To study the production of harmonics
- To distinguish longitudinal and transverse waves
- To study problems related to vibrations and oscillations
- To compute the wave equation representing the wave motion

4.1 ▶ Introduction

Wave motion is an important type of motion that occurs in nature. The wave motion may be transverse or longitudinal. But water waves in three dimensions are a combination of both transverse and longitudinal oscillations. Vibrations or oscillations constitute one of the important fields of study in the present era. The characteristic feature of vibration is its periodicity. An understanding of vibrations of strings, membranes and air columns in pipes is necessary for the design of musical instruments. Vibrations can be of many types such as free, damped, forced or coupled. we will describe all these oscillations or vibrations starting with simple harmonic motion and wave characteristics.

4.2 ▶ Wave

The disturbance that moves through a medium is called a wave. It transports energy but not matter. The waves can be of two types:

(i) **Longitudinal wave:** When the particles of a medium move back and forth along the direction of propagation of wave, it is known as a longitudinal wave. For example, sound wave is a longitudinal wave [Fig. 4.1(a)].

(ii) **Transverse wave:** When the particles of a medium move perpendicular to the direction of propagation of wave, it is known as a transverse wave. For example, light wave is a transverse wave [Fig. 4.1(b)].

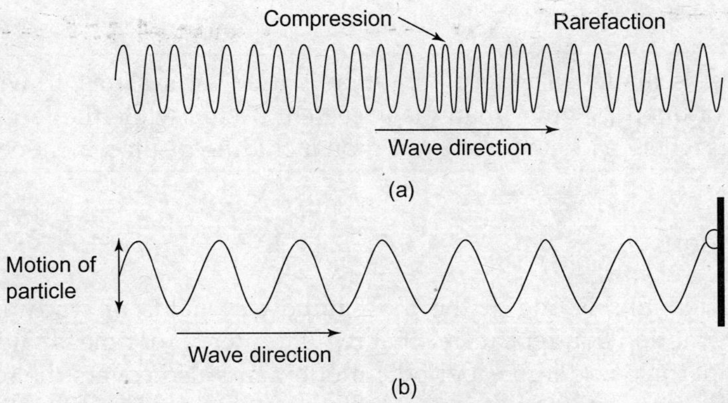

FIGURE 4.1 (a) Longitudinal wave; (b) Transverse wave

4.3 ▶ Wave Properties

Time period T of a wave is defined as the time required to complete one rotation or we can say that the time required for one complete wave to pass a given point.

Frequency, $f = \frac{1}{T}$, where T is the time period.

The unit of frequency is Hz (hertz).

The amplitude A of a wave is defined as the maximum displacement of the particles of the medium about their mean position. In case of a transverse wave, amplitude is half the distance between the top of a crest and the bottom of a trough.

The wavelength λ of a wave is the distance between two successive crests or troughs.

$v = f\lambda$, where f denotes the frequency of the wave and λ represents the wavelength of the wave.

The phase δ of a wave is the physical quantity that expresses the instantaneous position and direction of motion of the wave.

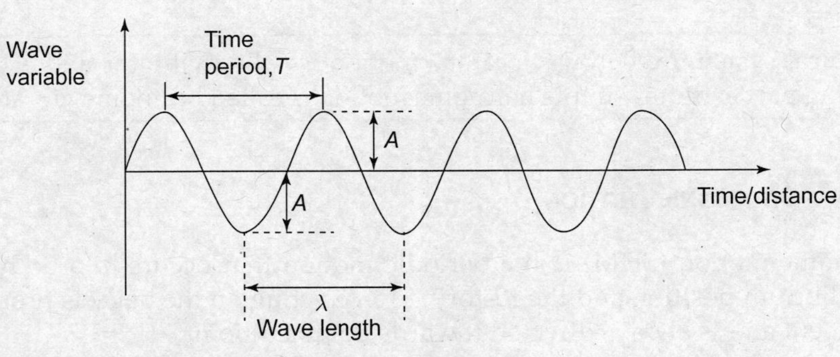

FIGURE 4.2 Representation of a wave

Note

Vibration versus oscillation: The term vibration is used for systems, in which back and forth vibration of quantity other than displacement occurs while the term oscillation is used for systems where back and forth vibration including displacement occurs.

4.4 ▶ Oscillations

Vibrations or oscillations is one of the most important fields of study in physics and engineering. Any motion that repeats itself at regular interval of time is said to be periodic motion. But if a particle undergoes periodic motion and also covers the same path back and forth about its mean position, it is said to be executing oscillatory motion.

Such type of motion is not only periodic but also bounded. The basic characteristic of an oscillatory motion is that it is started by displacing an object from its position of stable equilibrium. But a restoring force is developed which tries to bring the object back to its equilibrium position. This process is repeated and the object oscillates back & forth about its equilibrium position. Vibrations or oscillations are of the following types:

1. **Free oscillations or vibrations:** When the object is allowed to vibrate or oscillate on its own, the types of oscillations are called free oscillations. In this case, the object is not subjected to any external force.
2. **Forced oscillations:** When the object is subjected to a periodic external force, the resulting oscillations are known as forced oscillations.
3. **Undamped oscillations:** If during oscillation, no energy is lost or dissipated in friction, the oscillations are termed as undamped oscillations.
4. **Damped oscillations:** If an object that is set to oscillate comes at rest after some time, i.e. there is some loss of energy during oscillations. These types of oscillations are known as damped oscillations.

Note

The mechanical system executing vibratory motion is called vibrator and a vibrator in which moving part executes simple harmonic motion is called harmonic vibrator.

4.5 ▶ Simple Harmonic Motion

Simple harmonic motion (SHM) is the periodic motion that occurs in a body displaced from its equilibrium position and the restoring force acting on the body is proportional to the displacement and is always directed towards the fixed point.

Let us consider a particle of mass m be executing simple harmonic motion along a straight line. Let 'y' be the displacement of the particle from its mean position at any time t.

Then, the restoring force will be

$$f \propto -y \text{ or } f = -ky \tag{4.1}$$

where k is the proportionality constant and is also known as force constant.

If $m = \frac{d^2y}{dt^2}$ be the acceleration at time 't', then

$$m\frac{d^2y}{dt^2} = -ky$$

$$\Rightarrow \qquad \frac{d^2y}{dt^2} + \frac{ky}{m} = 0 \tag{4.2}$$

Putting $\frac{k}{m} = \omega^2$, we have

$$\frac{d^2y}{dt^2} + \omega^2 y = 0 \tag{4.3}$$

Equations (4.2) and (4.3) are known as differential equations of SHM.

To get the solution of Eq. (4.3), let us assume

$$y = Ce^{\alpha t} \tag{4.4}$$

where both C and α are constants.

$$\frac{dy}{dt} = C\alpha e^{\alpha t} \tag{4.5}$$

and

$$\frac{d^2y}{dt^2} = C\alpha^2 e^{\alpha t} \tag{4.6}$$

Putting Eqs. (4.4) and (4.6) in Eq. (4.3), we get

$$C\alpha^2 e^{\alpha t} + \omega^2 C e^{\alpha t} = 0$$

$$\Rightarrow \qquad Ce^{\alpha t}\left(\alpha^2 + \omega^2\right) = 0$$

But $Ce^{\alpha t} \neq 0$ because $y = Ce^{\alpha t}$ gives the displacement; thus

$$\alpha^2 + \omega^2 = 0 \Rightarrow \alpha = \pm i\omega$$

So,

$$y = Ce^{iwt} \text{ and } y = Ce^{-iwt}$$

Thus, the linear combination, i.e.,

$$y = C_1 e^{i\omega t} + C_2 e^{-i\omega t} \tag{4.7}$$

gives the general solution of Eq. (4.3).
Equation (4.7) can also be rewritten as

$$y = C_1(\cos\omega t + i\sin\omega t) + C_2(\cos\omega t - i\sin\omega t)$$

or
$$y = (C_1 + C_2)\cos\omega t + (iC_1 - iC_2)\sin\omega t \tag{4.8}$$

Let $C_1 + C_2 = A\sin\delta$ and $iC_1 - iC_2 = A\cos\delta$
Equation (4.8) gives

$$y = A\sin\delta\cos\omega t + A\cos\delta\sin\omega t$$

or
$$y = A\sin(\omega t + \delta) \tag{4.9}$$

Equation (4.9) gives the desired solution of Eq. (4.3) and it gives the displacement y of the particle executing SHM.

In Eq. (4.9), if we replace t by $\left(t + \frac{2\pi}{\omega}\right)$, then we have

$$y = A\sin\left[\omega\left(t + \frac{2\pi}{w}\right) + \delta\right] \text{ or } y = A\sin(\omega t + 2\pi + \delta)$$

or
$$y = A\sin(\omega t + \delta) \tag{4.10}$$

Equations (4.9) and (4.10) are same. This shows that the motion is repeated after an interval of $\frac{2\pi}{\omega}$.

Thus, the time period of SHM will be $T = \frac{2\pi}{\omega}$,

but
$$\omega^2 = \frac{k}{m} \Rightarrow \omega = \sqrt{\frac{k}{m}}$$

so,
$$T = \frac{2\pi}{\omega} = 2\pi\sqrt{\frac{m}{k}} \tag{4.11}$$

Similarly, we can calculate the frequency of SHM as given below:

$$\boxed{\boldsymbol{f = \frac{1}{T} = \frac{\omega}{2\pi} = \frac{1}{2\pi}\sqrt{\frac{k}{m}}}} \tag{4.12}$$

In Eq. (4.10), the quantity $(\omega t + \delta)$ is known as the phase of the particle. If $t = 0$, then $(\omega t + \delta) = \delta$. It means that its initial phase of the particle was δ. If a particle starts oscillating from its mean position, its phase will be zero but if it starts oscillating from the extreme position, its phase will be $\frac{\pi}{2}$.

Note

Time period of SHM is given by $T = 2\pi\sqrt{\dfrac{\text{displacement}}{\text{acceleration}}}$.

4.5.1 Harmonic Oscillator

To find the expression of velocity, acceleration and total energy of a harmonic oscillator, let us start with the expression of its displacement. As we know, $y = A\sin(\omega t + \delta)$
Differentiating it with time

$$\frac{dy}{dt} = v = A\omega\cos(\omega t + \delta) \tag{4.13}$$

$$= A\omega\sqrt{1 - \sin^2(\omega t + \delta)}$$

$$= \omega\sqrt{A^2 - A^2\sin^2(\omega t + \delta)}$$

or

$$v = \omega\sqrt{A^2 - y^2} \tag{4.14}$$

The maximum velocity can be calculated by putting $y = 0$ in Eq. (4.14) as

$$\boxed{v_{\max} = \omega A} \tag{4.15}$$

Differentiating Eq. (4.13) again w.r.t y, we get

$$\frac{d^2y}{dt^2} = \frac{dv}{dt} = a = -A\omega^2\sin(\omega t + \delta)$$

$\Rightarrow$

$$a = -\omega^2 y \tag{4.16}$$

Equation (4.16) gives the acceleration of the oscillating particle.
Here, minimum acceleration can be obtained by putting $y = 0$, i.e.,

$$a_{\min} = 0 \text{ (at the mean position)}$$

and the maximum acceleration is

$$a_{\max} = \omega^2 A \text{ (at the extreme position)} \tag{4.17}$$

The total energy of a simple harmonic oscillator is the sum of kinetic and potential energies.

The kinetic energy (KE) of the harmonic oscillator is

$$\text{KE} = \frac{1}{2}mv^2 = \frac{1}{2}m\left(\frac{dy}{dt}\right)^2$$

or

$$\text{KE} = \frac{1}{2}m\omega^2A^2\cos^2(\omega t + \delta) \tag{4.18}$$

and the potential energy (amount of work done in moving the system from 0 to y by applying external force) of the system is

$$\text{PE} = \int F \cdot dy = \int_0^y ky\,dy$$

or $$\text{PE} = \frac{1}{2}ky^2 \tag{4.19}$$

Now the total energy of the oscillator can be obtained as

$$E = \text{PE} + \text{KE}$$

or $$E = \frac{1}{2}m\omega^2 A^2 \sin^2(\omega t + \delta) + \frac{1}{2}m\omega^2 A^2 \cos^2(\omega t + \delta)$$

or $$\boxed{\boldsymbol{E = \frac{1}{2}m\omega^2 A^2}} \tag{4.20}$$

Since the total energy E is constant, it is obvious that the maximum possible value of PE or KE can be $\frac{1}{4}m\omega^2 A^2$ only.

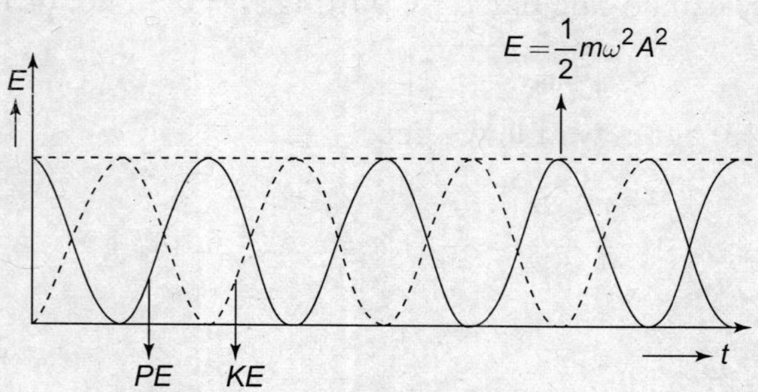

FIGURE 4.3 Total energy of the harmonic oscillator

The average potential energy of the simple harmonic oscillator for one cycle is

$$<\text{PE}> = \frac{1}{T}\int_0^T \frac{1}{2}ky^2 dt = \frac{\int_0^T \frac{1}{2}kA^2 \sin^2(\omega t + \delta)dt}{T}$$

$$= \frac{\frac{1}{2}\int_0^T m\omega^2 A^2 \sin^2(\omega t + \delta)dt}{T}$$

$$= \frac{1}{2}m\omega^2 A^2 \frac{\int_0^T \sin^2(\omega t + \delta)dt}{T}$$

$$\text{or } <\text{PE}>=\frac{1}{4}m\omega^2A^2 \quad \left[\because \frac{\int_0^T \sin^2(\omega t+\delta)dt}{T}=\frac{1}{2}\right] \tag{4.21}$$

Similarly, the average KE for one cycle is

$$<\text{KE}>=\frac{1}{T}\int_0^T \frac{1}{2}m\left(\frac{dy}{dt}\right)^2 dt=\frac{\int_0^T \frac{1}{2}mA^2\omega^2\cos^2(\omega t+\delta)dt}{T}$$

$$=\frac{1}{2}m\omega^2A^2\frac{\int_0^T \cos^2(\omega t+\delta)dt}{T}$$

or

$$<\text{KE}>=\frac{1}{4}m\omega^2A^2 \tag{4.22}$$

$\Rightarrow$

$$<\text{KE}>=<\text{PE}>=\frac{1}{4}m\omega^2A^2 \tag{4.23}$$

Thus, the total average energy for one complete cycle will be

$$<E>=<\text{KE}>+<\text{PE}>$$

or

$$<E>=\frac{1}{2}m\omega^2A^2 \tag{4.24}$$

4.5.2 Types of Simple Harmonic Motion

Simple harmonic motion can be classified into two classes, namely, linear simple harmonic motion and angular simple harmonic motion.

(i) **Linear simple harmonic motion**

The motion is said to be linear simple harmonic motion, if the displacement of a particle executing SHM is linear. Example: simple pendulum.

(ii) **Angular simple harmonic motion**

The motion is said to be angular simple harmonic motion, if the displacement of a particle executing SHM is angular. Examples: torsional oscillations and oscillations of a compound pendulum.

4.6 ▶ Damped Oscillations

When a pendulum is placed from its mean position, its amplitude of oscillations decreases with time, and finally, the pendulum comes to rest after some time. It happens due to

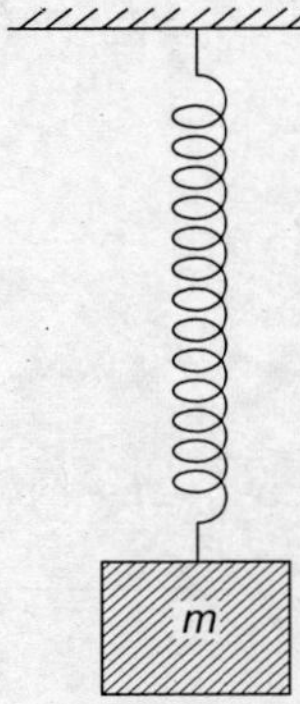

FIGURE 4.4 Damped oscillations

the frictional resistance and such type of oscillations is known as damped oscillations (Fig. 4.4).

Suppose a body of mass m is attached to a spring of force constant k. Let y be the displacement of the body from its equilibrium state. Now, $\frac{dy}{dt}$ will be the instantaneous velocity of the body.

Here, two types of forces acting on the body are as follows:

(i) a restoring force proportional to the displacement y, $f_r = -ky$ and

(ii) a damping force, $-q\frac{dy}{dt}$ proportional to the velocity and oppositely directed.

Thus, the total force acting on the body is

$$F = -ky - q\frac{dy}{dt}$$

or

$$m\frac{d^2y}{dt^2} = -ky - q\frac{dy}{dt}$$

$\Rightarrow$

$$\frac{d^2y}{dt^2} + \frac{q}{m}\frac{dy}{dt} + \frac{k}{m}y = 0$$

Putting $\frac{q}{m} = 2s$ and $\frac{k}{m} = \omega^2$, we get

$$\frac{d^2y}{dt^2} + 2s\frac{dy}{dt} + \omega^2 y = 0 \tag{4.25}$$

Equation (4.25) is the equation of second degree whose solution is of type

$$y = Ae^{\alpha t} \tag{4.26}$$

Differentiating Eq. (4.26) w.r.t. t, we get

$$\frac{dy}{dt} = A\alpha e^{\alpha t} \text{ and } \frac{d^2y}{dt^2} = A\alpha^2 e^{\alpha t}$$

By putting these values in Eq. (4.25), we get

$$A\alpha^2 e^{\alpha t} + 2sA\alpha e^{\alpha t} + \omega^2 Ae^{\alpha t} = 0 \tag{4.27}$$

$$\Rightarrow \qquad Ae^{\alpha t}\left[\alpha^2+2s\alpha+\omega^2\right]=0$$

Here, $Ae^{\alpha t} \neq 0$, so $\alpha^2+2s\alpha+\omega^2=0$

or
$$\alpha=-s\pm\sqrt{s^2-\omega^2}$$

which has two roots:

$$\alpha=-s+\sqrt{s^2-\omega^2} \text{ and } \alpha=-s-\sqrt{s^2-\omega^2}.$$

Thus, the general solution of Eq. (4.25) is

$$\boxed{y=A_1e^{\left(-s+\sqrt{s^2-\omega^2}\right)t}+A_2e^{\left(-s-\sqrt{s^2-\omega^2}\right)t}} \tag{4.28}$$

The solution of Eq. (4.25) depends on the conditions whether $s^2>\omega^2$, $s^2=\omega^2$ or $s^2<\omega^2$. Thus, we will discuss the different cases as given below:

Case (a): When $s^2>\omega^2$, $\sqrt{s^2-\omega^2}$ is real and Eq. (4.28) becomes negative. This shows that displacement y continuously decreases exponentially to zero without performing any oscillation. This type of motion is known as over damped motion.

Case (b): When $s^2=\omega^2$, Eq. (4.28) does not satisfy Eq. (4.25).

Case (c): When $\sqrt{s^2-\omega^2}=\beta$ (very small quantity), Eq. (4.28) gives

$$y=A_1\exp(-s+\beta)t+A_2\exp(-s-\beta)t$$

$$\Rightarrow \qquad y=e^{-st}\left[A_1e^{\beta t}+A_2e^{-\beta t}\right]$$

$$\Rightarrow \qquad y=e^{-st}\left[A_1(1+\beta t+\cdots)+A_2(1-\beta t+\cdots)\right]$$

Here,

$$y=e^{-st}\left[(A_1+A_2)+\beta t(A_1-A_2)+\cdots\right]$$

$$\Rightarrow \qquad y=e^{-st}\left[P+Qt\right] \tag{4.29}$$

where $P=(A_1+A_2)$ and $Q=\beta(A_1-A_2)$. From Eq. (4.29), it is clear that the term $(P+Qt)$ increases with increase in t. But y decreases due to exponential term e^{-st} and finally it becomes zero as t increases. This type of motion is called critical damped motion, e.g. in voltmeters and ammeters, etc.

Case (d): When $s^2<w^2$, then $\sqrt{s^2-w^2}$ is imaginary, and it can be written as

$$\sqrt{s^2-\omega^2}=i\sqrt{\omega^2-s^2}=i\beta'$$

Now, Eq. (4.28) becomes

$$y = A_1 e^{(-s+i\beta')t} + A_2 e^{(-s-i\beta')t}$$

$$\Rightarrow \qquad y = e^{-st}\left[A_1 e^{i\beta' t} + A_2 e^{-i\beta' t}\right]$$

or

$$y = e^{-st}\left[(A_1 + A_2)\cos\beta' t + i(A_1 - A_2)\sin\beta' t\right]$$

$$y = e^{-st}\left(A\sin\delta\cos\beta' t + A\cos\delta\sin\beta' t\right) \tag{4.30}$$

Put $\qquad A\sin\delta = A_1 + A_2$ and $A\cos\delta = i(A_1 - A_2)$

Equation (4.30) becomes

$$y = e^{-st} A\sin(\beta' t + \delta) \tag{4.31}$$

Equation (4.31) shows the oscillatory motion of damped harmonic oscillator, because its amplitude $\left(Ae^{-st}\right)$ decreases with time t. However, the decay of amplitude depends on the damping factor s.

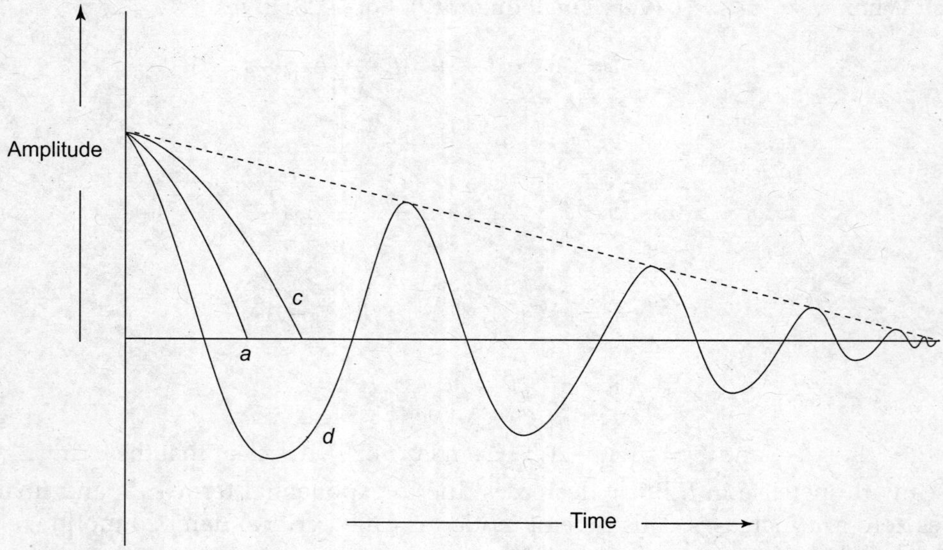

FIGURE 4.5 Case (a) with $s^2 > \omega^2$ (over damped); Case (b) not valid hence not shown; Case (c) $\sqrt{s^2 - \omega^2} = \beta$ (critical damped); Case (d) $s^2 < \omega^2$ (under damped)

EXAMPLE 4.1

The instantaneous displacement of a particle executing SHM is given by $y = A\sin(\omega_0 t + \phi)$. If the displacement of the particle at $t = 0$ is y_0 and its velocity at $t = 0$ is v_0, then find the values of A and ϕ.

Solution: Displacement: $y = A\sin(\omega_0 t + \phi)$

$$\text{Velocity: } \frac{dy}{dt} = A\omega_0 \cos(\omega_0 t + \phi)$$

At $t = 0$, $y = y_0$ and $dy/dt = v_0$, and therefore,

$$y_0 = A\sin\phi$$
$$v_0 = A\omega_0 \cos\phi$$

i.e.,
$$\sin\phi = \frac{y_0}{A} \text{ and } \cos\phi = \frac{v_0}{A\omega_0}$$

Therefore,
$$\tan\phi = \frac{\sin\phi}{\cos\phi} = \frac{y_0/A}{v_0/A\omega_0} = \frac{\omega_0 y_0}{v_0}$$

i.e.,
$$\phi = \tan^{-1}\left(\frac{\omega_0 y_0}{v_0}\right)$$

Also,
$$\sin^2\phi + \cos^2\phi = 1 = \frac{y_0^2}{A^2} + \frac{v_0^2}{A^2\omega_0^2}$$

Hence,
$$1 = \frac{1}{A^2}\left[y_0^2 + \frac{v_0^2}{\omega_0^2}\right]$$

$$A = \left[y_0^2 + \frac{v_0^2}{\omega_0^2}\right]^{\frac{1}{2}}$$

EXAMPLE 4.2

The potential energy of a mass of 1 kg executing SHM is given by $U_p = 2x^2 + 4x + 4$ joule. Find the equilibrium position, force constant and the frequency of oscillations.

Solution: We know that

$$dU_p = -Fdx$$

Therefore,
$$F = -\frac{dU_p}{dx}$$

Here,
$$U_p = 2x^2 + 4x + 4$$

Hence,
$$F = -\frac{dU_p}{dx} = -(4x+4)$$

or
$$F = -4(1+x)$$

If we put $1+x = z$, then

$$F = -4z$$

Taking z as the displacement, we find that $F = -4z$ represents SHM with the force constant $S = 4$.

At the equilibrium position, $F = 0$

Therefore,
$$x + 1 = 0$$

or
$$x = -1$$

The frequency of oscillations is given by

$$f_0 = \frac{1}{2\pi}\sqrt{\frac{S}{m}}$$

Here, $m = 1$ kg, and therefore,

$$f_0 = \frac{1}{2\pi}\sqrt{\frac{4}{1}} = \frac{1}{\pi}$$

Since all quantities are in SI, therefore $S = 4$ Nm^{-1}, equilibrium position is at $x = -1$ m and $f_0 = \frac{1}{\pi}$ Hz.

4.7 ▶ Attenuation Coefficients of an Oscillating System

In case of a harmonic oscillator, logarithmic decrement, relaxation time and quality factor give the attenuation of an oscillating system.

4.7.1 Logarithmic Decrement

The rate at which the amplitude of the oscillation dies is measured by logarithmic decrement. The amplitude of damped harmonic oscillator is given by the factor Ae^{-st}. Let $A_1, A_2, A_3, ...$ be the amplitudes at time $t = T, 2T, 3T, ...,$ respectively, then

$$A_1 = Ae^{-sT}, \; A_2 = Ae^{-s(2T)}, \; A_3 = Ae^{-s(3T)} \text{ and so on.}$$

or
$$\frac{A_0}{A_1} = \frac{A_1}{A_2} = \frac{A_2}{A_3} = \cdots = e^{sT} = e^r \tag{4.32}$$

where r is called logarithmic decrement.
By taking the natural log of Eq. (4.32), we get

$$\ln\frac{A_0}{A_1} = \ln\frac{A_1}{A_2} = \ln\frac{A_2}{A_3} = r \tag{4.33}$$

It is clear from Eq. (4.33) that logarithmic decrement is the natural logarithm of ratio between two successive maximum amplitudes, which are separated by one period.

4.7.2 Relaxation Time

Relaxation time is the time taken by an oscillator for decaying total mechanical energy by the factor $\frac{1}{e}$ times of its initial value.

The mechanical energy of an oscillator is given by

$$E = \frac{1}{2} m A^2 \omega^2 e^{-2st} \tag{4.34}$$

At $t = 0$,
$$E_0 = \frac{1}{2} m A^2 \omega_0^2 \tag{4.35}$$

Total energy of harmonic oscillator will be

$$E = E_0 e^{-2st} \tag{4.36}$$

Suppose τ be the relaxation time, then at time $t = \tau$, $E = \frac{E_0}{e}$
From Eq. (4.36), we get

$$E_0 e^{-2st} = \frac{E_0}{e}$$

or
$$\tau = \frac{1}{2s} \tag{4.37}$$

The dissipated energy can be written as

$$E = E_0 e^{-t/\tau} \tag{4.38}$$

4.7.3 Quality Factor

Quality factor of harmonic oscillator may be defined as 2π times the ratio of energy stored in the system to the energy lost per cycle. This factor of a damped oscillator gives the quality of it. Therefore,

$$Q = 2\pi \frac{E}{P_{\text{diss}}} \tag{4.39}$$

$$\Rightarrow \qquad Q = 2\pi \frac{E}{(E/\tau)P_{\text{diss}}} = \frac{2\pi\tau}{T}$$

$$\Rightarrow \qquad Q = \omega\tau \quad \left[\because P_{\text{diss}} = \frac{E}{\tau}\right] \tag{4.40}$$

From Eq. (4.40), it is clear that Q is higher for higher values of τ.

Here, $$\omega = \sqrt{\frac{k}{m}},\ \tau = \frac{1}{2s} \text{ and } \boxed{Q = \frac{1}{2s}\sqrt{\frac{k}{m}}} \tag{4.41}$$

Equation (4.41) is the required expression for quality factor. Quality factor will be higher, if s is low.

4.8 ▶ Forced Oscillations

The oscillations in which the body oscillates with frequency other than its natural frequency, due to the effect of external applied periodic force are called forced oscillations. A mass-spring system is shown in Fig. 4.6.

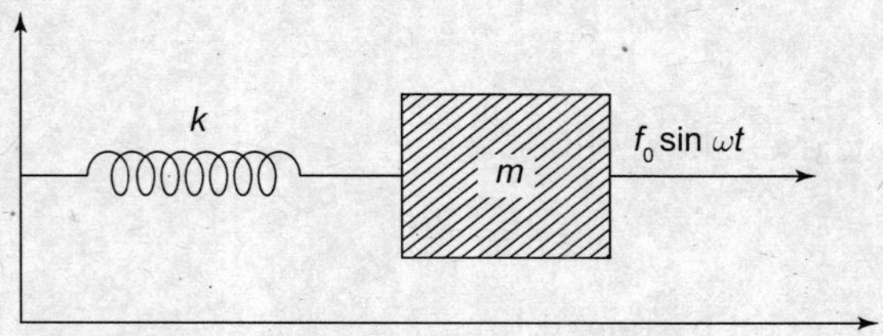

FIGURE 4.6 Forced oscillations

Let the restoring force on the particle of mass m is $-ky$, where y is the displacement of the particle from its mean position, damping force is $-q\frac{dy}{dt}$ and the external periodic force is $f_0 \sin \omega t$. Now, the total force acting on the particle will be

$$F = f_0 \sin \omega t - q\frac{dy}{dt} - ky \tag{4.42}$$

By Newton's second law of motion,

$$F = m\frac{d^2y}{dt^2}$$

Hence, Eq. (4.42) can be written as

$$F = m\frac{d^2y}{dt^2} = f_0 \sin \omega t - q\frac{dy}{dt} - ky$$

$$\Rightarrow \qquad \frac{d^2y}{dt^2} + \frac{q}{m}\frac{dy}{dt} + \frac{k}{m}y = \frac{f_0 \sin \omega t}{m} \tag{4.43}$$

By putting,

$$\frac{q}{m} = 2s, \ \frac{k}{m} = \omega_0^2 \text{ and } \frac{f_0}{m} = f$$

Equation (4.43) becomes

$$\frac{d^2y}{dt^2} + 2s\frac{dy}{dt} + \omega_0^2 y = f \sin \omega t \tag{4.44}$$

The general solution of Eq. (4.44) in steady state will be

$$y = A\sin(\omega t - \delta) \tag{4.45a}$$

where A is the amplitude of the oscillations in the steady state. By differentiating Eq. (4.45a) twice w.r.t. t, we have

$$\frac{dy}{dt} = \omega A \cos(\omega t - \delta)$$

and

$$\frac{d^2y}{dt^2} = -\omega^2 A \sin(\omega t - \delta)$$

By substituting the values of y, $\frac{dy}{dt}$ and $\frac{d^2y}{dt^2}$ in Eq. (4.44), we get

$$-\omega^2 A \sin(\omega t - \delta) + 2s\omega A\cos(\omega t - \delta) + \omega_0^2 A \sin(\omega t - \delta) = f \sin\{(wt - \delta) + \delta\}$$

or

$$A\left(\omega_0^2 - w^2\right)\sin(\omega t - \delta) + 2s\omega A\cos(\omega t - \delta) = f\sin(\omega t - \delta)\cos\delta + f\cos(\omega t - \delta)\sin\delta \tag{4.45b}$$

Comparing the coefficients of $\sin(\omega t - \delta)$ and $\cos(\omega t - \delta)$ on both sides of Eq. (4.45b), we get

$$A\left(\omega_0^2 - \omega^2\right) = f\cos\delta \tag{4.46}$$

and $$2s\omega A = f\sin\delta \tag{4.47}$$

By squaring and adding Eqs. (4.46) and (4.47), we get

$$A^2\left(\omega_0^2-\omega^2\right)^2+4s^2\omega^2A^2=f^2$$

$\Rightarrow$ $$A=\frac{f}{\sqrt{\left(\omega_0^2-\omega^2\right)^2+4s^2\omega^2}} \tag{4.48}$$

Dividing Eq. (4.47) by Eq. (4.46), we get

$$\tan\delta=\frac{2s\omega}{\omega_0^2-\omega^2} \tag{4.49}$$

$\Rightarrow$ $$\delta=\tan^{-1}\left[\frac{2s\omega}{\omega_0^2-\omega^2}\right] \tag{4.50}$$

Equation (4.48) gives the amplitude of forced oscillations and Eq. (4.50) gives the phase of the oscillations. Both quantities depend upon the value of $\left(\omega_0^2-\omega^2\right)$, i.e. the driving frequency ω and the natural frequency ω_0 of the oscillator. The amplitude and the phase can be calculated for different cases as given below:

Case (a): When $\omega << \omega_0$, i.e. for very low driving frequency. The amplitude of the vibrations can be written as

$$A=\frac{f}{\sqrt{\left(\omega_0^2-\omega^2\right)^2+4s^2\omega^2}}\approx\frac{f}{\omega_0^2}$$

or $$A=\frac{f_0}{m\omega_0^2}\quad\left[\because f=\frac{f_0}{m}\right]$$

or $$A=\frac{f_0}{k}\quad\left[\because \omega_0^2=\frac{k}{m}\right] \tag{4.51}$$

Equation (4.51) shows that the amplitude depends on the force constant of the spring and the magnitude of the applied force.

Phase, $$\delta=\tan^{-1}\left[\frac{2sw}{\omega_0^2-\omega^2}\right]=\tan^{-1}\left[\frac{2s\omega}{\omega_0^2}\right] \tag{4.52}$$

Since $\omega_0^2 >> \omega$, $\delta\to 0$ because $\dfrac{2s\omega}{\omega_0}\to 0$.

Thus, the driving force and the displacement are in phase.

Case (b): When $\omega=\omega_0$, the driving and natural frequencies are same; this frequency is called resonant frequency. Under this condition, the amplitude will be

$$A=\frac{f}{\sqrt{\left(\omega_0^2-w^2\right)^2+4s^2\omega^2}}$$

$$A=\frac{f}{2s\omega}=\frac{f_0/m}{(q/m)\omega}=\frac{f_0}{q\omega} \tag{4.53}$$

From Eq. (4.53), it is clear that amplitude of oscillations depends upon the applied force and damping force.

$$\text{Phase, } \delta=\tan^{-1}\left[\frac{2s\omega}{\left(\omega_0^2-\omega^2\right)}\right]$$

or $$\delta=\tan^{-1}\left[\frac{2s\omega}{0}\right] \qquad [\because \omega=\omega_\circ]$$

or $$\delta=\tan^{-1}\infty=\frac{\pi}{2} \tag{4.54}$$

Thus, displacement lags behind the force by a phase of $\frac{\pi}{2}$.

Case (c): When $\omega>>\omega_0$, i.e. when the driving frequency is very large. Then, the amplitude will be

$$A=\frac{f}{\sqrt{\left(\omega_0^2-\omega^2\right)^2+4s^2\omega^2}}$$

Since ω is very large, $\omega^4>>4\omega^2s^2$

$\Rightarrow$ $$A=\frac{f}{\omega^2}=\frac{f_0}{m\omega^2}$$

$$\text{Phase, } \delta=\tan^{-1}\left[\frac{2s\omega}{\left(\omega_0^2-\omega^2\right)}\right]$$

or $$\delta=\tan^{-1}\left[-\frac{2s\omega}{\omega^2}\right]$$

or $$\delta=\tan^{-1}\left[-\frac{2s\omega}{\omega}\right]=\pi$$

In this case, displacement lags behind the force by a phase of π.

4.8.1 Resonance

If an oscillator system has a number of oscillators coupled together, the resultant motion will be complicated. But, if the system oscillates in such a way that every part of the system has same frequency, the oscillations are known as normal oscillations or normal modes and the associated frequencies are called the normal frequencies or natural frequencies. If a mechanical structure is subjected to a driving frequency which is equal to the natural frequencies, the resulting oscillations will have large amplitudes, which can lead to disastrous consequences.

Collapse of roadways and bridges in earthquakes is one of the examples of resonance.

Sharpness of resonance means the rate of fall in amplitude, with the change of forcing frequency on each side of resonance frequency. Smaller is the damping, larger is the resonance.

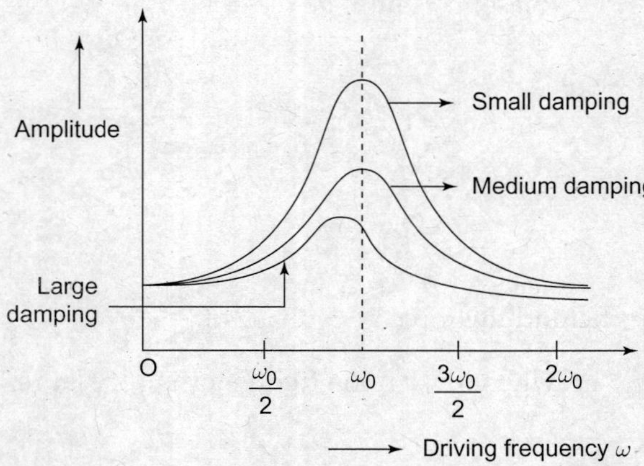

FIGURE 4.7 Variation of amplitude of forced oscillations with the driving force of frequency ω

4.9 ▶ Coupled Oscillations

In case of forced oscillations, the amplitude and the frequency of the driving force remain unaltered. But, when two systems are comparable, motion of each system will affect the other. This type of oscillation/vibration is known as coupled vibration. In the coupled vibrations, two or more oscillators are joined together in such a way that the exchange of energy takes place between them. To understand the coupled vibrations, let us consider two systems of masses m_1 and m_2 coupled together, oscillating in the y-direction. Let their masses be m_1 and m_2 and k_1 and k_2 be their restoring forces. Let k_{12} be the force of restitution on the first or the second for unit displacement of the second or the first, which is called the coupling coefficient.

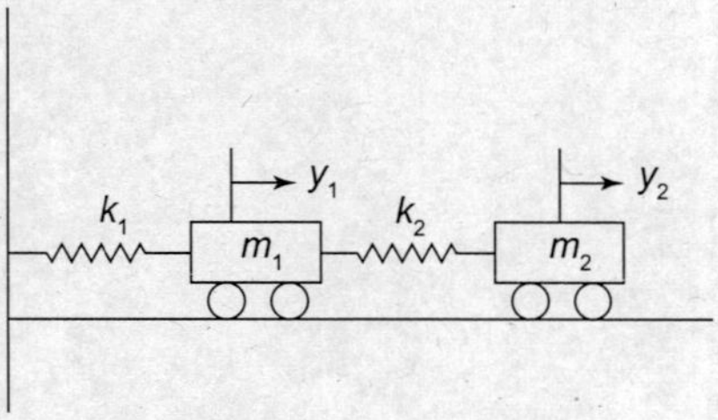

FIGURE 4.8 Coupled oscillators

The equations of motions of these two systems can be written as

$$m_1 \frac{d^2 y_1}{dt^2} + k_1 y_1 + k_{12} y_2 = 0 \tag{4.55}$$

and
$$m_2 \frac{d^2 y_2}{dt^2} + k_2 y_2 + k_{12} y_2 = 0 \tag{4.56}$$

Dividing Eq. (4.55) by $\sqrt{m_1}$ and Eq. (4.56) by $\sqrt{m_2}$, we get

$$\sqrt{m_1} \frac{d^2 y_1}{dt^2} + \frac{k_1}{\sqrt{m_1}} y_1 + \frac{k_{12}}{\sqrt{m_1}} y_2 = 0 \tag{4.57}$$

and

$$\sqrt{m_2} \frac{d^2 y_2}{dt^2} + \frac{k_2}{\sqrt{m_2}} y_2 + \frac{k_{12}}{\sqrt{m_2}} y_1 = 0 \tag{4.58}$$

Put
$$y' = \sqrt{m_1} y_1 \tag{4.59}$$

and
$$y'' = \sqrt{m_2} y_2 \tag{4.60}$$

From Eqs. (4.57) to (4.60), we have

$$\frac{d^2 y'}{dt^2} + \omega_1^2 y' + \omega^2 y'' = 0 \tag{4.61}$$

$$\frac{d^2 y'}{dt^2} + \omega_2^2 y'' + \omega^2 y' = 0 \tag{4.62}$$

where $\omega_1^2 = \dfrac{k_1}{m_1}$, $\omega_2^2 = \dfrac{k_2}{m_2}$ and $\omega^2 = \dfrac{k_{12}}{\sqrt{m_1}\sqrt{m_2}}$.

Both the systems are supposed to oscillate with frequency $f = \dfrac{\alpha}{2\pi}$, and the oscillations are simple harmonic. Now, let us assume

$$y' = Ae^{i\alpha t} \tag{4.63}$$

$$y'' = Be^{i\alpha t} \tag{4.64}$$

The simple harmonic motion associated with each normal coordinate y' or y'' is called normal mode of the coupled system.

Substituting Eqs. (4.63) and (4.64) in Eqs. (4.61) and (4.62), we get

$$\left(\omega_1^2 - \alpha^2\right) A = -\omega^2 B \tag{4.65}$$

and

$$\left(\omega_2^2-\alpha^2\right)B=-\omega^2 A \tag{4.66}$$

Multiplying Eqs. (4.65) and (4.66), we get

$$\left(\omega_1^2-\alpha^2\right)\left(\omega_2^2-\alpha^2\right)=\omega^4$$

$\Rightarrow$

$$\omega_1^2\omega_2^2-\alpha^2\omega_2^2-\alpha^2\omega_1^2+\alpha^4-\omega^4=0 \tag{4.67}$$

Solving Eq. (4.67), we get

$$\alpha^2=\frac{1}{2}\left(\omega_1^2+\omega_2^2\right)\pm\frac{1}{2}\sqrt{\left(\omega_1^2+\omega_2^2\right)^2-4\left(\omega_1^2\omega_2^2-\omega^4\right)}$$

or

$$\alpha^2=\frac{1}{2}\left(\omega_1^2+\omega_2^2\right)\pm\frac{1}{2}\sqrt{\left(\omega_1^2-\omega_2^2\right)^2+4\omega^4} \tag{4.68}$$

Now α has two values:

$$\alpha_1=\left[\frac{1}{2}\left(\omega_1^2+\omega_2^2\right)+\frac{1}{2}\sqrt{\left(\omega_1^2+\omega_2^2\right)^2+4\omega^4}\right]^{1/2} \tag{4.69}$$

and

$$\alpha_2=\left[\frac{1}{2}\left(\omega_1^2+\omega_2^2\right)-\frac{1}{2}\sqrt{\left(\omega_1^2-\omega_2^2\right)^2+4\omega^4}\right]^{1/2} \tag{4.70}$$

But, if the two frequencies of the system are equal, i.e. $\omega_1=\omega_2$, then

$$\alpha^2=\omega_1^2\pm\omega'$$

Hence,

$$\alpha_1=\sqrt{\omega_1^2+\omega^2}=\omega_1\left(1+\frac{\omega^2}{\omega_1^2}\right)^{1/2}$$

$$\alpha_1=\omega_1+\frac{\omega^2}{2\omega_1}(\text{app.}) \tag{4.71}$$

Similarly,

$$\alpha_2=\sqrt{\omega_1^2-\omega^2}=\omega_1-\frac{\omega^2}{2\omega_1}(\text{app.}) \tag{4.72}$$

Here,

$$\alpha_1+\alpha_2=2\omega_1 \tag{4.73}$$

and

$$\alpha_1-\alpha_2=\frac{\omega^2}{\omega_1} \tag{4.74}$$

where $\frac{\alpha_1}{2\pi}$ and $\frac{\alpha_2}{2\pi}$ are two modified frequencies of the coupled system when coupled together. So the general solution can be written as

$$y' = A_1 e^{i\alpha_1 t} + A_2 e^{i\alpha_2 t} \tag{4.75}$$

and
$$y'' = B_1 e^{i\alpha_1 t} + B_2 e^{i\alpha_2 t} \tag{4.76}$$

From Eqs. (4.59) and (4.60),

$$y_1 = \frac{y'}{\sqrt{m_1}} = \frac{1}{\sqrt{m_1}}\left(A_1 e^{i\alpha_1 t} + A_2 e^{i\alpha_2 t}\right)$$

or
$$y_1 = \frac{1}{\sqrt{m_1}}\left[A_1(\cos\alpha_1 t + i\sin\alpha_1 t) + A_2(\cos\alpha_2 t + i\sin\alpha_2 t)\right] \tag{4.77}$$

or
$$y_1 = a_1\cos\alpha_1 t + a_2\sin\alpha_1 t + a_3\cos\alpha_2 t + a_4\sin\alpha_2 t \tag{4.78}$$

Similarly,

$$y_2 = b_1\cos\alpha_1 t + b_2\sin\alpha_1 t + b_3\cos\alpha_2 t + b_4\sin\alpha_2 t \tag{4.79}$$

where

$$a_1 = \frac{A_1}{\sqrt{m_1}},\ a_2 = \frac{iA_1}{\sqrt{m_1}},\ a_3 = \frac{A_2}{\sqrt{m_1}} \text{ and } a_4 = \frac{iA_2}{\sqrt{m_1}}$$

Also,
$$b_1 = \frac{B_1}{\sqrt{m_2}},\ b_2 = \frac{iB_1}{\sqrt{m_2}},\ b_3 = \frac{B_2}{\sqrt{m_2}} \text{ and } b_4 = \frac{iB_2}{\sqrt{m_2}}$$

If the particle m_1 starts at rest initially, then

at $t = 0,\ y_1 = (y_1)_0,\ y_2 = 0,\ \frac{dy_1}{dt} = 0,\ \frac{dy_2}{dt} = 0$

Thus, from Eq. (4.78)

$$a_1 + a_3 = (y_1)_0$$

At the beginning, there is no coupling; thus,

$$a_1 = a_3 = \frac{(y_1)_0}{2} \text{ and } a_2 = a_4 = 0$$

$\Rightarrow$
$$b_1 + b_3 = 0 \Rightarrow b_1 = -b_3 = b \text{ (say)}$$

From Eqs. (4.78) and (4.79), we get

$$y_1 = \frac{(y_1)_0}{2}(\cos\alpha_1 t + \cos\alpha_2 t)$$

$$\Rightarrow \qquad y_2 = b(\cos\alpha_1 t - \cos\alpha_2 t)$$

$$\Rightarrow \qquad y_1 = (y_1)_0 \cos\left(\frac{\alpha_1+\alpha_2}{2}\right)t \cos\left(\frac{\alpha_1-\alpha_2}{2}\right)t \tag{4.80}$$

From Eqs. (4.73), (4.74) and (4.78), (4.79),

$$y_1 = (y_1)_0 \cos\omega_1 t \cos\frac{\omega^2}{2\omega_1}t \tag{4.81}$$

and

$$y_2 = 2b\sin\omega_1 t \sin\frac{\omega_2}{2\omega_1}t \tag{4.82}$$

Substituting the values of α_1^2 and α_2^2 from Eqs. (4.71) and (4.72) in Eqs. (4.65) and (4.66), we get

$$\frac{A}{B} = 1 \text{ or } A = B$$

In other words, we can say that the amplitude of y' is equal to the amplitude of y''. Thus, in equations:

$$y' = \sqrt{m_1}y_1,\ y'' = \sqrt{m_2}y_2 \text{ and } \sqrt{m_1}y_1 = \sqrt{m_2}y_2$$

Hence, the amplitude of $y_1 = \dfrac{(y_1)_0}{2}$ and $y_2 = b$ must be equal.

Thus,

$$\frac{(y_1)_0}{2} = \sqrt{\frac{m_2}{m_1}}b \text{ or } b = \frac{(y_1)_0}{2}\sqrt{\frac{m_1}{m_2}}$$

$$\Rightarrow \qquad y_1 = (y_1)_0 \cos\frac{\omega^2}{2\omega_1}t \cos\omega_1 t \tag{4.83}$$

and

$$y_2 = \sqrt{\frac{m_1}{m_2}}(y_1)_0 \sin\frac{\omega^2}{2\omega_1}t \sin\omega_1 t \tag{4.84}$$

Equations (4.83) and (4.84) represent the oscillations of two systems when they are coupled together.

Note

The driving force and the acceleration differ in phase by $(-\phi + \pi/2)$ rad.

EXAMPLE 4.3

Two identical simple pendulums are connected by a light spring attached to their bobs. The mass of each bob is 10 g and the stiffness constant of the spring is 8×10^3 Nm^{-1}. When one pendulum is clamped, the period of the other is found to be 1.20 s. Find the periods of the normal modes.

Solution: When one pendulum is clamped, the differential equation of motion of the other is given by

$$m\frac{d^2x}{dt^2} = -mg\frac{x}{t} - Sx$$

$$m\frac{d^2x}{dt^2} = -mg\left(\frac{g}{t} + \frac{S}{m}\right)x$$

or
$$m\frac{d^2x}{dt^2} = -m\left(\omega_0^2 + \frac{S}{m}\right)x$$

where ω_0 is the normal mode frequency of the in-phase mode. This gives

$$\omega_1^2 = \omega_0^2 + \frac{S}{m}$$

where ω_1 is the frequency of oscillation of one of pendulums when the other is clamped.

Here, $T = 1.20$ s. Therefore,
$$\omega_1 = \frac{2\pi}{T} = \frac{2\pi}{1.20}\text{s}^{-1}$$

Also $S = 8 \times 10^{-3}$ Nm^{-1} and $m = 10$ g 10^{-2} kg.

Therefore,
$$\left(\frac{2\pi}{1.20}\right)^2 = \omega_0^2 + \frac{8\times10^{-3}}{10^{-2}}$$

or
$$\omega_0^2 = \left(\frac{2\pi}{1.20}\right)^2 + 0.8$$

or
$$\omega_0^2 = 26.616$$

or
$$\omega_0 = 5.159$$

Therefore,
$$T_0 = \frac{2\pi}{\omega_0} = \frac{2\pi}{5.519} = 1.218 \cong 1.22$$

That is, the time period of in-phase mode is 1.22 s (appox.).

Also, the frequency of out of phase mode is

$$\omega_2^2 = \left(\omega_0^2 + \frac{2S}{m}\right)$$

$$= 26.616 + 2\times\frac{8\times10^{-3}}{10^{-2}}$$

$$= 26.616 + 1.6$$

$$\omega_2^2 = 28.216$$

That is, $\omega_2 = 5.312$

Hence, $T_2 = \frac{2\pi}{\omega_2} = \frac{2\pi}{5.312} \cong 1.18 \text{ s}$

That is, the time period of out of phase mode is 1.18 s (approx.).

EXAMPLE 4.4

An inductor, capacitor and resistor of values 0.2 H, 1 μF and 800 Ω are connected in series. Show that the circuit is oscillatory and calculate the frequency of oscillations.

Solution: Given $L = 0.2\text{H}$, $C = 1\ \mu\text{F} = 10^{-6}$ F, $R = 800\ \Omega$

$$\therefore \quad \frac{1}{LC} = \frac{1}{0.2\times10^{-6}} = 5\times10^6$$

And $$\frac{R^2}{4L^2} = \frac{64\times10^4}{4\times4\times10^{-2}} = 4\times10^6$$

$$\therefore \quad \frac{R^2}{4L^2} < \frac{1}{LC}$$

Hence, the circuit is oscillatory.
Frequency of oscillations

$$v = \frac{1}{2\pi}\sqrt{\frac{1}{LC} - \frac{R^2}{4L^2}} = \frac{10^3}{2\times3.142}\text{Hz}$$

$$= \mathbf{159\ Hz}$$

EXAMPLE 4.5

Show that the ratio of the energy lost per cycle to the energy stored in the damped oscillator is $2\pi/Q$ where Q is the quality factor.

Solution: Energy of the damped oscillator is given by

$$U = U_0 e^{-2bt}$$

Rate of loss of energy is given by

$$\frac{dU}{dt} = -2bU_0 e^{-2bt}$$

Loss of energy in one time period is obtained by putting $dt = T$. Therefore,

$$[dU]_T = -2bU_0 e^{-2bt} T$$

Now,
$$\frac{U}{[dU]_T} = \frac{U_0 e^{-2bt}}{-2bU_0 e^{-2bt} T} = -\frac{1}{2bT}$$

Negative sign indicates that the energy decreases.

Now, $\omega_0' = 2\pi/T$ and $b = r/2m$. Therefore,

$$\frac{U}{[dU]_T} = \frac{1}{2 \times \frac{r}{2m} \times \frac{2\pi}{\omega_0'}} = \frac{1}{2\pi} \frac{\omega'0^m}{r} = \frac{Q}{2\pi}$$

or
$$\frac{[dU]_T}{U} = \frac{2\pi}{Q}$$

4.10 ▶ Wave Motion

It is a basic and important method of transferring energy from one place to another place. The important characteristics of a wave motion are given below:

(i) It is a disturbance produced in the medium due to the repeated periodic motion of the particles of the medium.

(ii) There is a regular phase change between the particles of a medium that takes place.

(iii) The velocity of the wave is different from the velocity of the particle. The velocity of particle is different at different positions but the velocity of wave is uniform.

(iv) Wave travels in forward direction but the particles of the medium vibrate about their mean positions.

4.10.1 Wave as Periodic Vibration in Space and Time

The speed of a wave in a medium depends only upon the properties of the medium, and it can be described in terms of the periodicity of the vibrations of medium particles in the space. In Fig. 4.9, an one-dimensional sinusoidal wave is presented, which represents the displacement at different intervals of time:

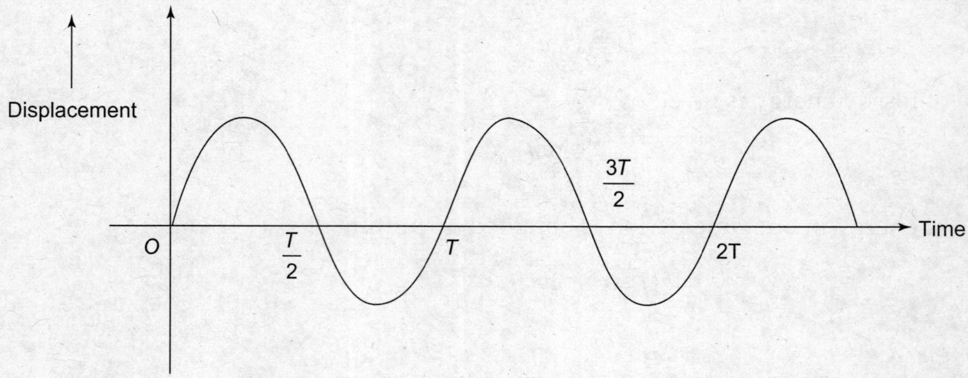

FIGURE 4.9 Wave motion

When $t=0,\ y=0$ (min.)

at $t=\frac{T}{4},\ y=a$ (max.)

When $t=\frac{T}{2},\ y=0$ (min.)

at $t=\frac{3T}{4},\ y=-a$ (max. in $-$ve direction)

When $t=T,\ y=0$ (min.)

The slope of the curve at any point represents the velocity of vibrating particle at that particular instant. The rate of change of slope represents the acceleration at that point.

4.10.2 Differential Equation of Wave Motion

The general equation of wave motion is

$$y=a\sin\frac{2\pi}{\lambda}(vt-x)$$

Differentiating it w.r.t t, we get

$$\frac{dy}{dt}=\frac{2\pi av}{\lambda}\cos\frac{2\pi}{\lambda}(vt-x)$$

Differentiating again w.r.t t, we get

$$\frac{d^2y}{dt^2}=\frac{-4\pi^2av^2}{\lambda^2}\sin\frac{2\pi}{\lambda}(vt-x) \qquad \text{(a)}$$

Similarly differentiating the general equation of wave motion twice w.r.t x, we get

$$\frac{d^2y}{dx^2} = \frac{-4\pi^2 a}{\lambda^2}\sin\frac{2\pi}{\lambda}(vt - x) \qquad \text{(b)}$$

Comparing (a) and (b), we get

$$\frac{d^2y}{dt^2} = v^2\frac{d^2y}{dx^2}$$

This is known as the differential equation of wave motion.

EXAMPLE 4.6

The amplitude of an oscillator of frequency 200 per second falls to 1/10 of its initial value after 2000 cycles. Calculate (i) relaxation time, (ii) quality factor and (iii) damping constant.

Solution: Let the amplitude (instantaneous) of damped oscillator be $a\,e^{-\alpha t}$.
Let at $t = 0$, initial amplitude is a_0.
After 10 seconds, amplitude will be $\frac{a_0}{10}$.

$$\Rightarrow \qquad \frac{a_0}{10} = a_0 e^{-\alpha^{10}}$$

$$\Rightarrow \qquad 10 = e^{10\alpha}$$

Taking log, $$2.3\log_{10} 10 = 10\alpha$$

$$\Rightarrow \qquad \alpha = \frac{2.3}{10} = 0.23$$

(i) Relaxation time

$$\tau = \frac{1}{2}\alpha = \frac{1}{2}\times 0.23$$

$$\Rightarrow \qquad \boldsymbol{\tau = 2.174\ \mathrm{s}}$$

(ii) Quality factor

$$Q = \omega\tau$$

$$\Rightarrow \qquad Q = 2\pi f \times \tau$$

$$\Rightarrow \qquad Q = 2\times 3.14\times 2w\times 2.174$$

$$\Rightarrow \qquad \boldsymbol{Q = 2730}$$

(iii) Damping constant

$$E = E_0 e^{-t/\tau}$$

$$\Rightarrow \quad \frac{E_0}{10} = Ee^{-t/\tau}$$

$$\Rightarrow \quad 10 = e^{t/\tau}$$

$$\Rightarrow \quad \log_e 10 = t/\tau$$

$$\Rightarrow \quad \mathbf{t = \tau \log_e 10 = 2.174 \times 2.3 = 5\ s}$$

4.11 ▶ Reflection and Transmission of Transverse Waves at a Boundary

When a wave reaches at a point where a sudden change of impedance in its path occurs, a part of its amplitude gets reflected and transmitted depending upon the impedance of the medium.

Let us consider two strings of linear mass densities m_1 and m_2 joined together under the tension T along x-axis as shown in Fig. 4.10.

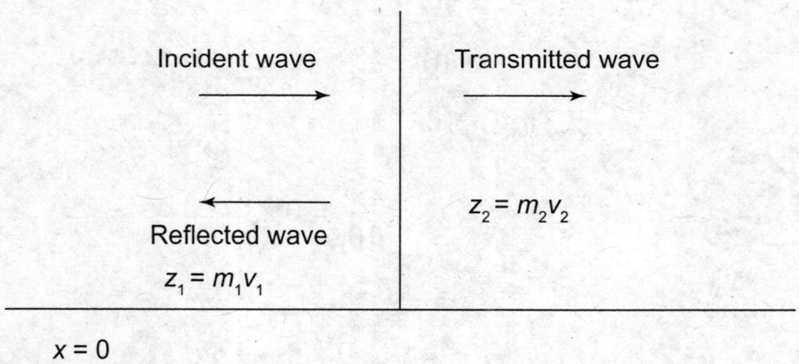

FIGURE 4.10 Reflection and transmission in a transverse wave

Let the wave velocities and impedances at left and right to $x = 0$ be v_1, v_2 and z_1, z_2, respectively.
Here,

$$v_1 = \sqrt{\frac{T}{m_1}},\ v_2 = \sqrt{\frac{T}{m_2}} \text{ and } z_1 = m_1 v_1,\ z_2 = m_2 v_2$$

When a transverse wave is incident on the boundary at $x = 0$, a part of it, is reflected and a part of it, is transmitted into the region of impedance z_2.

The wave equations for incident, reflected and transmitted waves are

$$y_i = a_i \sin\left(\omega t - \frac{x}{v_1}\right) \tag{4.85}$$

$$y_r = a_r \sin\left(\omega t + \frac{x}{v_1}\right) \tag{4.86}$$

and

$$y_t = a_t \sin\left(\omega t - \frac{x}{v_2}\right) \tag{4.87}$$

Consider two boundary conditions:

(i) There should be no discontinuity at $x = 0$ for displacement, so

$$y_i + y_r = y_t \tag{4.88}$$

(ii) The forces on the two sides of boundary must be equal. Thus, the transverse force is

$$T\left[\frac{dy}{dx}\right] \text{ at } x = 0 \tag{4.89}$$

Applying second boundary condition,

$$T\left[\frac{d}{dx}(y_i + y_r)\right] = T\frac{dy_t}{dx} \text{ at } x = 0$$

Putting the values of y_i, y_r and y_t at $x = 0$, from Eqs. (4.85), (4.86), (4.87) and (4.88), we get

$$a_i + a_r = a_t \tag{4.90}$$

Differentiating (4.85), (4.86) and (4.87) w.r.t x, we get

$$\frac{dy_i}{dx} = -\frac{a_i}{v_1}\cos w\left(t - \frac{x}{v_1}\right)$$

$$\frac{dy_r}{dx} = \frac{a_r}{v_1}\cos w\left(t + \frac{x}{v_1}\right)$$

$$\frac{dy_t}{dx} = -\frac{a_t}{v_2}\cos w\left(t - \frac{x}{v_2}\right)$$

Putting above equations in Eqn. (4.90), we get

$$-T\frac{a_i}{v_1}\cos\omega t + T\frac{a_r}{v_1}\cos\omega t = -T\frac{a_t}{v_2}\cos\omega t$$

or $$T\left(\frac{a_i - a_r}{v_1}\right) = T\frac{a_t}{v_2} \tag{4.91}$$

But we know that

$$v_1 = \sqrt{\frac{T}{m_1}} \Rightarrow v_1^2 = \frac{T}{m_1}$$

or $$\frac{T}{v_1} = m_1 v_1 = z_1$$

Similarly, $$\frac{T}{v_2} = m_2 v_2 = z_2$$

Using the above equations in Eq. (4.91), we get

$$z_1(a_i - a_r) = z_2 a_t \text{ and } a_i - a_r = \frac{z_2}{z_1} a_t \tag{4.92}$$

But also, $$a_i + a_r = a_t$$

From the above two equations,

$$2a_i = \left(1 + \frac{z_2}{z_1}\right) a_t$$

$$\Rightarrow \quad \frac{a_t}{a_i} = \frac{2}{1 + \dfrac{z_2}{z_1}}$$

$$\Rightarrow \quad \boxed{\frac{a_t}{a_i} = \frac{2z_1}{z_1 + z_2}} \tag{4.93}$$

Using $\left(\frac{z_1}{z_2}\right) \times (4.92)$ and subtracting from $(a_i + a_r = a_t)$, we get

$$\boxed{a_i\left(1 - \frac{z_1}{z_2}\right) + a_r\left(1 + \frac{z_1}{z_2}\right) = 0 \Rightarrow \frac{a_r}{a_i} = \frac{z_1 - z_2}{z_1 + z_2}} \tag{4.94}$$

where $\frac{a_t}{a_i}$ is known as transmission coefficient and represented by T whereas $\frac{a_r}{a_i}$ is known as reflection coefficient and is represented by R.

Thus, $$R = \frac{z_1 - z_2}{z_1 + z_2} \text{ and } T = \frac{2z_1}{z_1 + z_2}.$$

Both these terms are independent of frequency of wave but depends on the ratio of impedance of two mediums.

Case (a): If $z_1 = z_2$, $R = 0$ and $T = 1$. Entire wave will be transmitted in this case.

Case (b): If $z_2 > z_1$, R is –ve and T is +ve. It means that displacement will be changed by π on reflection. But there is no change in the phase of transmitted wave.

Case (c): If $z_1 > z_2$, reflected wave suffers no change of phase as R is positive.

Case (d): If $z_2 = \infty$, $R = -1$, $T = 0$.

Entire wave will be reflected, no wave will be transmitted but reflected wave also suffers a phase change of π.

Case (e): If $z_2 = 0$, $R = 1$ and $T = 2$. This shows the flick at the end of a whip as free-ended string when a wave reaches it.

4.12 ▶ Reflection and Transmission of a Longitudinal Wave at Boundaries

When a longitudinal or sound wave is incident at the boundary of two media having different acoustic impedances, then a part of it is reflected and a part of it is transmitted. Let us consider a sound wave incident normally at boundary $x = 0$ as shown in Fig. 4.11.

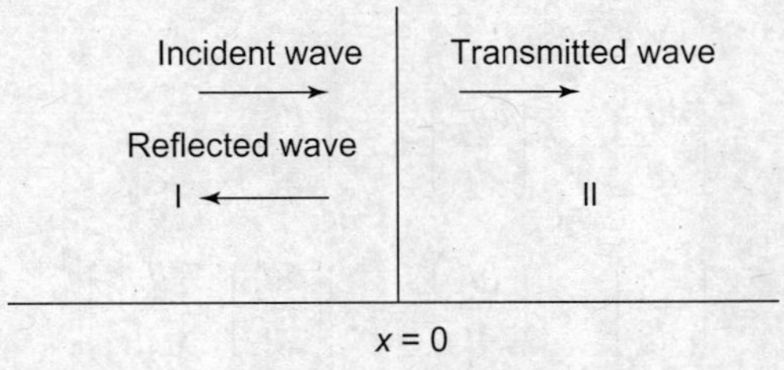

FIGURE 4.11 Reflection and transmission in a longitudinal wave

The equations of incident, reflected and transmitted waves are as follows:

$$y_i = a_i \sin\omega\left(t - \frac{x}{v_1}\right)$$

$$y_r = a_r \sin\omega\left(t + \frac{x}{v_1}\right)$$

$$y_t = a_t \sin\omega\left(t - \frac{x}{v_2}\right) \tag{4.95}$$

Let us consider the two boundary conditions here.

(i) The particle velocity must be continuous at the junction of two media, i.e.,

$$\frac{dy_i}{dt} + \frac{dy_r}{dt} = \frac{dy_t}{dt} \tag{4.96}$$

(ii) The acoustic pressure must be continuous, i.e.,

$$P_i + P_r = P_t \tag{4.97}$$

If K is the Bulk modulus, then

$$K = \frac{\text{stress}}{\text{strain}} = \frac{-P}{(dy/dx)}$$

thus,

$$P_i = -k_1\left(\frac{dy_i}{dx}\right) = p_1 v_2 \left(\frac{dy_i}{dx}\right)$$

$$P_i = p_1 v_1 \left(\frac{dy_i}{dt}\right)$$

or

$$\left.\begin{aligned} P_i &= z_1\left(\frac{dy_i}{dt}\right) \\ P_r &= -z_1\left(\frac{dy_r}{dt}\right) \\ P_t &= z_2\left(\frac{dy_t}{dt}\right) \end{aligned}\right\} \tag{4.98}$$

Similarly

and

Putting Eq. (4.98) in Eq. (4.96), we get

$$z_1\left(\frac{dy_i}{dt}\right) - z_1\left(\frac{dy_r}{dt}\right) = z_2\left(\frac{dy_t}{dt}\right) \tag{4.99}$$

$$\Rightarrow \quad z_1\left(\frac{dy_i}{dt}\right) - z_1\left(\frac{dy_r}{dt}\right) = z_2\left[\left(\frac{dy_i}{dt}\right) + \left(\frac{dy_r}{dt}\right)\right]$$

$$\Rightarrow \quad \frac{dy_r/dt}{dy_i/dt} = \frac{z_1 - z_2}{z_1 + z_2} \tag{4.100}$$

Also, from Eqs. (4.98) and (4.99), we have

$$z_1\left(\frac{dy_i}{dt}\right) - z_1\left(\frac{dy_t}{dt} - \frac{dy_i}{dt}\right)$$

$$= z_2\left(\frac{dy_t}{dt}\right) \tag{4.101}$$

$$\Rightarrow \quad \frac{dy_t/dt}{dy_i/dt} = \frac{2z_1}{z_1 + z_2}$$

We can also write the above equations (4.100) and (4.101) as

$$\frac{P_r}{P_i} = \frac{z_1(dy_r/dt)}{z_i(dy_i/dt)} = \frac{z_2 - z_1}{z_1 + z_2} \tag{4.102}$$

and

$$\frac{P_t}{P_i} = \frac{z_2(dy_t/dt)}{z_1(dy_i/dt)} = \frac{2z_2}{z_1 + z_2} \tag{4.103}$$

The above equations represent the acoustic pressures of reflected and transmitted wave, respectively.

EXAMPLE 4.7

Show that the bandwidth of the resonance absorption curve defines the phase angle range $\tan\phi = \pm 1$.

Solution: We know that

$$\tan\phi = \frac{\omega m - S/\omega}{r} \tag{I}$$

Let ω be the driving frequency on either side of the bandwidth. Then the power supplied at ω is half the maximum value of the power supplied.

Now,
$$P_{av} = \frac{F_0^2}{2Z_m^2} r$$

Also,
$$P_{\max} = \frac{F_0^2}{2r}$$

Therefore,
$$\frac{F_0^2 r}{2Z_m^2} = \frac{1}{2}\frac{F_0^2}{2r}$$

which gives
$$Z_m^2 = 2r^2.$$

or
$$\left[r^2 + (\omega m - S/\omega)^2\right] = 2r^2$$

or
$$(\omega m - S/\omega)^2 = r^2$$

or
$$\omega m - S/\omega = \pm r$$

If ω_1 and ω_2 be the driving frequencies on either side of the bandwidth, then

$$\omega_1 m - S/\omega_1 = -r$$
$$\omega_2 m - S/\omega_2 = +r$$

Let ϕ_1 and ϕ_2 be the phase angles corresponding to the driving frequencies ω_1 and ω_2, respectively. Then from equation (I),

$$\tan\phi_1 = \frac{\omega_1 m - S/\omega_1}{r} = \frac{-r}{r} = -1$$

and
$$\tan\phi_2 = \frac{\omega_2 m - S/\omega_2}{r} = \frac{r}{r} = 1$$

That is, bandwidth of the resonance absorption curve defines the phase angle range $\tan\phi = \pm 1$.

Recap Zone

Points to Remember

- A wave is a disturbance that travels through medium. It carries energy with it but not matter.
- If the resultant force is proportional to the displacement of an object, from its mean position, then the resulting oscillatory motion is called SHM.
- The equation of a damped harmonic oscillator is $y = Ae^{-\delta t}\sin(wt + \delta)$.
- The oscillations in a damped oscillator diminish quickly; to sustain it an external force is required. This is the driven harmonic oscillator.
- Rate of fall of amplitude with the change of forcing frequency on each side of resonance frequency is called sharpness of resonance.
- When coupling is loose, the rate of transfer of energy is small. However, when coupling is tight, the rate of transfer of energy is high.
- The phase of an oscillating wave determines the position and the direction of motion of particle at a particular instant.

Key Terms with Definitions

- **Damping** – The mechanism that results in dissipation of the energy of an oscillator is called damping

- **Quality factor** – Number equal to 2 times the ratio of the instantaneous energy of the oscillator to the energy lost during one time period
- **Logarithmic decrement** – It is the ratio of the time period of the damped oscillator and the relaxation time
- **Relaxation time** – Time interval during which the amplitude of damped oscillator decays to 1/e times its initial value.
- **Bandwidth** – The difference between the frequencies of the driving force on either side of the velocity resonance frequency ω_0 at which the average power supplied to the oscillator reduces to half of its maximum value.
- **Normal modes** – The manner of oscillation of the coupled oscillator is termed as mode.
- **Degrees of freedom** – A number of independent ways in which a system can acquire energy are called the degrees of freedom of the system

IMPORTANT FORMULAE AND EQUATIONS

Equation Number	Equation	Remarks
(4.12)	$T = 2\pi\sqrt{\dfrac{\text{displacement}}{\text{acceleration}}}$	**Frequency of SHM**
(4.15)	$v_{max} = \omega A$	**Maximum velocity of harmonic oscillator**
(4.20)	$E = \dfrac{1}{2} m\omega^2 A^2$	**Total energy of the oscillator**
(4.28)	$y = A_1 e^{\left(-s+\sqrt{s^2-\omega^2}\right)t} + A_2 e^{\left(-s-\sqrt{s^2-\omega^2}\right)t}$	**General solution of damped oscillations**
(4.41)	$Q = \dfrac{1}{2s}\sqrt{\dfrac{k}{m}}$	**Quality factor**
(4.93)	$T = \dfrac{2z_1}{z_1 + z_2}$	**Transmission coefficient of transverse wave**
(4.94)	$R = \dfrac{z_1 - z_2}{z_1 + z_2}$	**Reflection coefficient of transverse wave**

Review Zone

Short Answer Questions

1. Define a wave.
2. What do you mean by SHM?
3. Explain the terms amplitude, frequency and phase of a wave.
4. Differentiate between the damped and undamped vibrations.
5. Distinguish between forced and natural oscillations.
6. Explain coupled oscillations.
7. What do you mean by the terms (a) reflection and (b) transmission of a wave?
8. Define resonance and quality factor.
9. Distinguish between longitudinal and transverse waves.
10. What do you understand by reflection of waves?

Long Answer Questions

1. Define simple harmonic motion. Also explain the terms time period, amplitude and phase of a wave.
2. Derive the differential equation for SHM from energy principle.
3. Write the differential equation for the damped natural oscillation of a particle and solve it. Give the conditions for over damping, critical damping and under damping action.
4. What is sharpness of resonance? Explain the effect of damping on the sharpness of resonance.
5. Give the theory of forced oscillations. Explain it from the physical consideration, when the frequency of the forced vibration is same as the external force.
6. Write the equation of motion of two masses coupled together and connected to two rigid supports by 3 springs of same stiffness constant. Derive the characteristic equation and the frequencies of the normal modes of vibration.

Solve-it-First Problems

1. The acceleration of the particle executing simple harmonic motion is $\frac{\pi^2}{3}$ cm/s^2 when its displacement is 3 cm. Calculate its time period of motion.

   Hint: Refer to Section 4.3

   Ans. 6 s

2. The displacement of the particle executing simple harmonic motion is given by $y = 10\cos\left(4\pi t + \frac{\pi}{3}\right)$ m.

Find out the frequency and displacement after time 1 second.

Hint: Refer to Section 4.5

Ans. 5 m

3. Two vibrations at right angles to each other, described by equations $y_1 = 3\cos 4\pi t$ and $y_2 = 3\cos\left(8\pi t + \frac{\pi}{3}\right)$, where y_1 and y_2 are expressed in centimetres and t in seconds. Write down the equation of motion for the same.

Hint: We have

$$y_1 = 3\cos 4\pi t$$

$$\Rightarrow \frac{y_1}{3} = \cos 4\pi t \qquad \text{(i)}$$

and

$$y_2 = 3\cos\left(8\pi t + \pi/3\right) \qquad \text{(ii)}$$

$$\Rightarrow \frac{y_2}{3} = \cos\left(8\pi t + \pi/3\right)$$

Use (i) and (ii) to solve for equation of motion.

Ans. $y_1^2\left(1 - \frac{y_1^2}{9}\right) = 3\left(\frac{y_1^2}{9} - \frac{1}{2} - \frac{y_2}{3}\right)^3$

4. A mass of 10 g is acted upon by a restoring force 5 dyne/cm and a resistance 2 dyne-s/cm. Find out whether the motion is oscillatory or a periodic. Find the value of resistance for the motion to be critically damped.

Hint:

(i) Use $\frac{k^2}{4} = \frac{(2s)^2}{4} = \frac{(0.2)^2}{4} = 0.01$

(ii) For critical damping, use

$$\frac{k}{2} = \frac{2s}{2} = \omega \Rightarrow \frac{q}{m} = 2\sqrt{\frac{\mu}{m}}$$

Ans. 0.01, 14.14 dyne/cm

5. Two masses m_1 and m_2 are connected by a massless spring of spring constant k. The system is free to oscillate along the length of the spring. Show that the system oscillates with a frequency $f = \frac{1}{2\pi}\sqrt{\frac{k}{\mu}}$, where μ is the reduced mass, which is given as: $\mu = \frac{m_1 m_2}{m_1 + m_2}$.

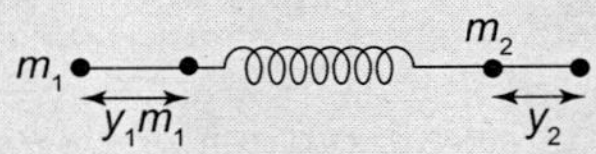

Hint: Use

$$m_1 \frac{d^2 y_1}{dt^2} = k(y_2 - y_1) \qquad \text{(i)}$$

$$m_2 \frac{d^2 y_1}{dt^2} = -k(y_2 - y_1) \qquad \text{(ii)}$$

and $\frac{d^2 y}{dt^2} = -\frac{k}{\mu} y$

Ans. $f = \frac{1}{\sqrt{2}\pi}\sqrt{\frac{k}{\mu}}$

6. The equations of motion of two coupled oscillations are given as $\ddot{q}_1 + \sqrt{2}q_2 + 4q_1 = 0$ and $\ddot{q}_2 + \sqrt{2}q_1 + 5q_2 = 0$.

Find the normal frequencies and the ratio of amplitudes of the two normal modes.

Hint: $\ddot{q}_1 + \sqrt{2}q_2 + 4q_1 = 0 \qquad \text{(i)}$

$$\ddot{q}_2 + \sqrt{2}q_1 + 5q_2 = 0 \qquad \text{(ii)}$$

From (i) and (ii), we get

$$A\alpha^2 e^{\alpha t} + \sqrt{2}Be^{\alpha t} + 4Ae^{\alpha t} = 0 \qquad \text{(iii)}$$

Solving (i), (ii) and (iii),

$\alpha^2 = -3$ or $\alpha^2 = -6$

So the normal frequencies are

$w_1 = \sqrt{3}$ or $w_2 = \sqrt{6}$

$$\frac{A}{B} = \frac{1}{\sqrt{2}}$$

Ans. $-\sqrt{2}$ or $\left(\frac{1}{\sqrt{2}}\right)$

Numerical Problems

1. A body of mass 0.01 kg executes SHM with amplitude 3×10^{-2} m and time period 2 s. Calculate the energy of the particle.

   Ans. 4.43×10^{-5} J

2. Two simple harmonic motions acting on a particle simultaneously are given by $y_1 = 0.003 \sin 2\pi t$ and $y_2 = 0.002 \sin\left(2\pi t - \pi/3\right)$

   Write down the expression for the resultant displacement as a function of time.

   Ans. $y = 0.004347 \sin(2\pi t + \alpha)$, where $\alpha = 23.22$

3. In one-dimensional motion of a mass of 10 g, it is acted upon by a restoring force of 10 dyne per cm and a resisting force of 2 dyne s/cm. Find (i) whether the motion is periodic or oscillatory, (ii) the value of resisting force which will make the motion critically damped and (iii) the value of mass for which given forces will make the motion critically damped.

4. Two particles of the same mass are attached to a weightless spring of length 1 with fixed ends stretched under a tension T. Find the normal frequencies of the vibration of the masses if they are attached at distances $l/3$ and $2l/3$ from one end. Assume that the tension in the string remains constant during the vibration.

CHAPTER 5

Diffraction

Learning Objectives

By the end of this chapter, the student will be able:

- To understand the phenomena of diffraction of light
- To differentiate diffraction and interference
- To understand Fraunhofer's diffraction due to single slit and n slits
- To describe the working of diffraction grating
- To estimate the dispersive power of grating and Rayleigh's criterion
- To describe zone plate and half zone plate
- To compute Fresnel diffraction from a straight edge

5.1 ▶ Introduction

When light falls on an obstacle or a small aperture, there is a departure from straight line propagation.

The light bends round the corners of the obstacle or aperture and spreads to some extent into the region of geometrical shadow. However, the amount of bending depends upon the size of obstacle and the wavelength of light. This bending of light is called diffraction. The phenomenon of diffraction produces bright and dark fringes known as diffraction bands. The light waves diverge from narrow slit S, pass on obstacle AB with a straight edge A parallel to the slit, and the geometrical shadow produces, which can be seen as alternate bright and dark bands on the screen (Fig. 5.1).

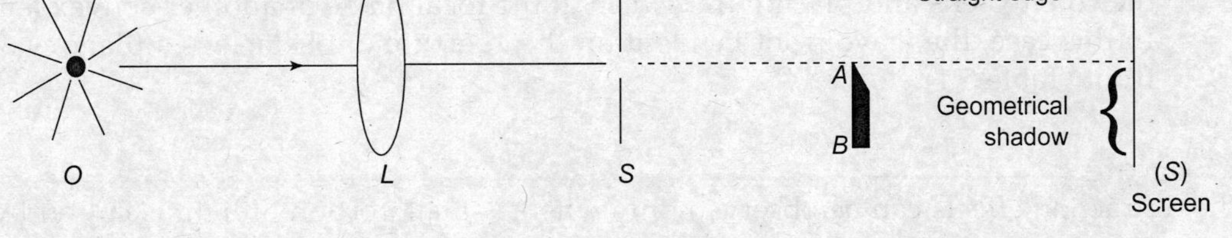

FIGURE 5.1 Diffraction of light from edge of an object

5.2 ▶ Difference Between Diffraction and Interference

The difference between diffraction and interference is given as follows:

(i) In the phenomenon of diffraction, the interaction between the secondary wavelets originating from different points of the same wave front takes place, whereas in the interference, two separate wave fronts originating from two coherent sources superimposed.

(ii) In diffraction, the width of various fringes is never equal, whereas in interference, the width of fringes may or may not be equal.

(iii) All the maxima are of varying intensity in diffraction, whereas in interference, all the maxima are of same intensity.

(iv) In diffraction, the regions of maximum intensity are not perfectly bright and the regions of minimum intensity are not perfectly dark, whereas in interference the regions of minimum intensity are perfectly dark and the regions of maximum intensity are perfectly bright.

Note

The phenomenon of diffraction and interference shows the wave nature of light while photoelectric effect and Compton scattering show the particle nature of light. So light has dual character.

5.3 ▶ Fresnel and Fraunhofer Diffraction

According to Fresnel, diffraction is due to the interference of the various secondary wavelets originated from the wave front, which are not obstructed by the obstacle. The diffraction phenomena are divided into two classes: (i) Fresnel diffraction and (ii) Fraunhofer diffraction.

(i) **Fresnel diffraction:** In this class of diffraction, the source of light and the screen are at finite distance from the diffracting aperture or obstacle having sharp edge. The wave front incident on the aperture or obstacle is either spherical or cylindrical. No lens is used to make the rays parallel.

(ii) **Fraunhofer diffraction:** In this class of diffraction, the source of light and the screen are at infinite distance from the diffraction aperture or obstacle having sharp edge. This can be achieved by placing the light source at the focal plane of the convex lens and placing the screen at the focal plane of another convex lens. In this case, the wave front incident on the aperture or obstacle is a plane wave front (Table 5.1).

Note

The diffraction effects can be observed only when a small portion of light is cut off by an obstacle.

TABLE 5.1 Comparison between Fresnel and Fraunhofer diffraction

Fresnel diffraction	Fraunhofer diffraction
1. Point source of light or an illuminated narrow slit is used as light source.	1. Extended source of light at infinite distance is used as light source.
2. Light incident on the aperture or obstacle is a spherical or cylindrical wave front.	2. Light incident on the aperture or obstacle is a plane wave front.
3. The source and screen are at finite distance from the aperture or obstacle producing diffraction.	3. The source and screen are at infinite distance from the aperture or obstacle producing diffraction.
4. Lenses are not used to focus the rays.	4. Converging lens is used to focus the rays.
5. Single slit or double slit or grating is used to get diffraction pattern.	5. Fresnel's zone plate is used to get diffraction pattern.

5.4 ▶ Fraunhofer's Diffraction at a Single Slit

The single slit is represented as AB in Fig. 5.2. The slit is in the form of a narrow rectangle. The width of the slit AB is equal to 'l' and the plane of the slit is perpendicular to the plane of the paper. A plane wave front WW' of monochromatic light of wave length 'λ' is incident on the slit AB. Every point on the wave front in the slit will act as a source of secondary wavelets. The secondary wavelets travelling in the direction of OA_0 are brought to focus at A_0 on the screen SS' by using a converging lens L. The secondary wavelets from AB that are brought to focus at A_0 have no path difference.

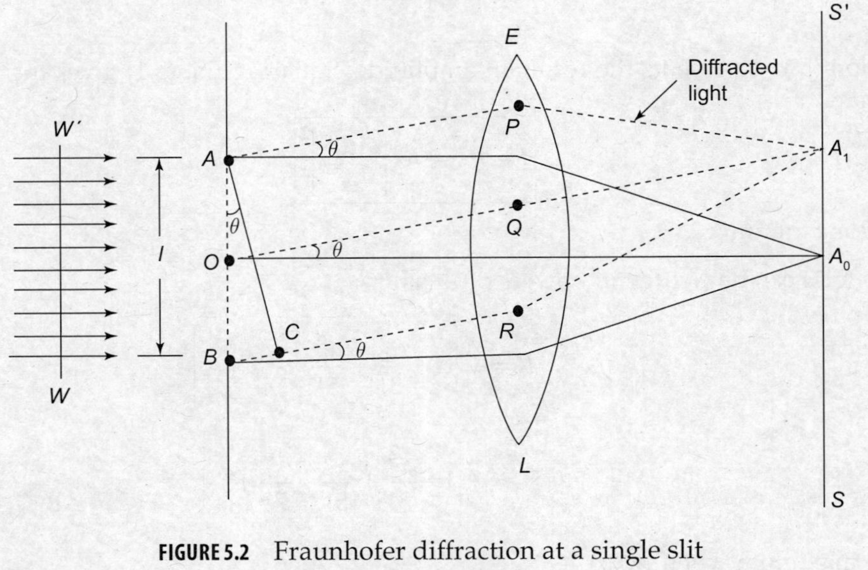

FIGURE 5.2 Fraunhofer diffraction at a single slit

Hence, the intensity at A_0 is high and it is known as central maximum. The secondary wavelets in the slit AB which make an angle 'θ' with OA_0 direction are brought to a point A_1 on the screen.

Intensity at point A_1 depends on the path difference between the wavelets at A and B reaching to point A_1. To find the path difference, a perpendicular AC is drawn to BR from A. Now the path difference between the secondary wavelets from A and B in the direction of OP_1 is BC.

$$BC = AB\sin\theta = l\sin\theta$$

The corresponding phase difference is $(2\pi/\lambda)l\sin\theta$. Let us consider the width of the slit be divided into n equal parts. Then the phase difference between any two successive parts is $(1/n)\times(2\pi/\lambda)l\sin\theta$ and let it be d. Let the amplitude of the wave in each part be a. The resultant amplitude R using the vector addition method is

$$R = a\frac{\sin\frac{nd}{2}}{\sin\frac{d}{2}} = a\frac{\sin\left(\frac{\pi}{\lambda}l\sin\theta\right)}{\sin\left(\frac{\pi}{n\lambda}l\sin\theta\right)}$$

Put

$$\alpha = \frac{\pi}{\lambda}l\sin\theta$$

Then

$$R = a\frac{\sin\alpha}{\sin\left(\frac{\alpha}{n}\right)} \approx a\frac{\sin\alpha}{(\alpha/n)} \quad \left[\text{since } \frac{\alpha}{n} \text{ is very small}\right]$$

$$= an\frac{\sin\alpha}{\alpha} = A\frac{\sin\alpha}{\alpha} \tag{5.1}$$

where $A = na$.

Equation (5.1) represents the resultant amplitude. Squaring Eq. (5.1) gives the intensity (I) of light.

$$\boxed{I = R^2 = A^2\left[\frac{\sin\alpha}{\alpha}\right]^2} \tag{5.2}$$

Principal maximum

Equation (5.1) can be written in powers of α as

$$R = \frac{A}{\alpha}\left[\alpha - \frac{\alpha^3}{3!} + \frac{\alpha^5}{5!} - \frac{\alpha^7}{7!} + \cdots\right]$$

$$= A\left[1 - \frac{\alpha^2}{3!} + \frac{\alpha^4}{5!} - \frac{\alpha^6}{7!} + \cdots\right] \tag{5.3}$$

R will be maximum when $\alpha = 0$

i.e.,
$$\alpha=\frac{\pi}{\lambda}l\sin\theta=0 \Rightarrow \sin\theta=0 \text{ or } \theta=0$$

Hence, the secondary wavelets that travel normal to the slit can produce maximum amplitude known as principal maximum.

Minimum intensity positions

The intensity will be minimum when sin α in Eq. (5.2) is zero, i.e., $\sin\alpha=0$.

The values of α are $\alpha=\pm\pi, \pm 2\pi, \pm 3\pi, 4\pi, \ldots=\pm m\pi$ or $\frac{\pi}{\lambda}l\sin\theta=\pm m\pi$

or
$$l\sin\theta=\pm m\lambda \tag{5.4}$$

where $m=1, 2, 3,\ldots$ and $m=0$ corresponds to principal maximum.

From Eq. (5.4), we know that the positions of minimum intensity are found on both sides of principal maximum.

Secondary maxima

Between equally spaced minima, weak secondary maxima are found. The positions of secondary maxima can be obtained by differentiating the intensity of light (I) with 'α' and equating it to zero. From Eq. (5.2),

$$\frac{dI}{d\alpha}=\frac{d}{d\alpha}\left[A^2\left(\frac{\sin\alpha}{\alpha}\right)^2\right]=0$$

i.e.,
$$A^2\cdot\frac{2\sin\alpha}{\alpha}\left(\frac{\alpha\cos\alpha-\sin\alpha}{\alpha}\right)=0$$

In the above equation, either $\sin\alpha=0$ or $(\alpha\cos\alpha-\sin\alpha)=0$.

We already came across $\sin\alpha=0$ for minimum intensity positions. Hence, $\alpha\cos\alpha-\sin\alpha=0$.

or
$$\alpha=\tan\alpha \tag{5.5}$$

The values of α that satisfy the above equation can be obtained by plotting $y=\alpha$ and $y=\tan\alpha$ on the same graph (Fig. 5.3). The points of intersection of two curves give the value of α which satisfy Eq. (5.5).

The points of intersection are $\alpha=0, \pm\frac{3\pi}{2}, \pm\frac{5\pi}{2}, \ldots$. Substituting these values in Eq. (5.2), we get the intensities of various maxima.

For $\alpha=0$,
$$I_0=A^2 \quad \text{[Principal maximum]}$$

For $\alpha = \frac{3\pi}{2}$, $$I_1 = A^2\left[\frac{\sin\left(3\pi/2\right)}{\left(3\pi/2\right)}\right]^2 = \frac{A^2}{22} \text{ (approximately)} = \frac{I_0}{22}$$

and $$\frac{I_1}{I_0} = \frac{1}{22} = 4.5$$

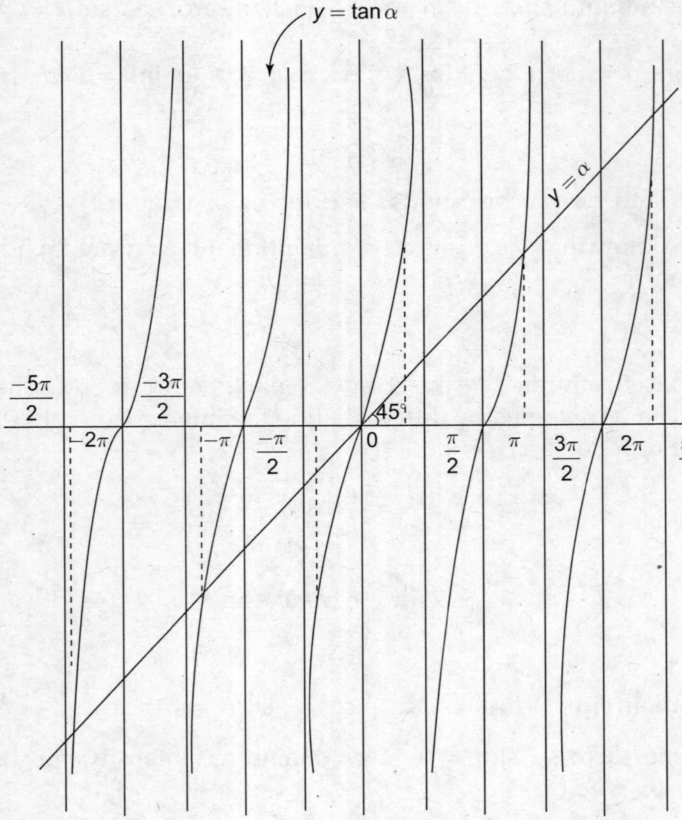

FIGURE 5.3 $y = \alpha$ and $y = \tan\alpha$ graph $\left[\alpha = \frac{\pi}{\lambda} e \sin\theta\right]$ for single slit diffraction

For $\alpha = \frac{5\pi}{2}$, $$I_2 = A^2\left|\frac{\sin 5\pi/2}{5\pi/2}\right|^2 = \frac{A^2}{62} \text{ (approximately)} = \frac{I_0}{62} \text{ and so on.}$$

Also $$\frac{I_1}{I_0} = \frac{1}{62} = 1.61$$

From the above values, we come to know that the most of the incident light is found in the principal maximum.

A graph is drawn between the intensity of light versus α (Fig. 5.4). The principal maximum occurs at the centre of diffraction pattern and the subsidiary maxima of decreasing intensity on both sides of principal maximum. The subsidiary maxima are not exactly at the centres of minima but slightly towards the centre of the diffraction pattern.

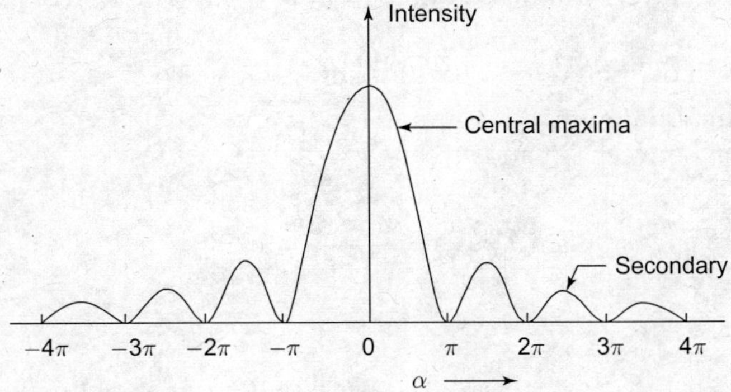

FIGURE 5.4 Intensity distributions due to diffraction at a single slit

Note

The width of central maxima is greater for narrow slits.

EXAMPLE 5.1

A single slit of width 4×10^{-3} mm is illuminated by a monochromatic light of wavelength 6000 Å. Find the angular separation of the diffracted first-order minimum from the central maximum.

Solution: The minimum intensity due to a single slit diffraction is

$$l\sin\theta_m = m\lambda,\ m = 1,2,3,\ldots$$

For the first-order diffraction, $m = 1$ and width $l = 4\times10^{-3}$ mm $= 4\times10^{-6}$ m.
Wavelength,

$$\lambda = 6000 \text{ Å}$$

$$\lambda = 6000\times10^{-10} \text{ m}$$

$$\sin\theta_1 = \frac{\lambda}{l} = \frac{6000\times10^{-10}}{4\times10^{-6}}$$

$$\theta_1 = \sin^{-1}\left(\frac{0.3}{2}\right)$$

$$= 8°37'37''$$

EXAMPLE 5.2

A monochromatic light of wavelength 6000×10^{-8} cm is diffracted by a single slit kept at a distance of 100 cm from the screen. The first diffracted minimum appears at a distance of 1 mm from the central maximum. Find the width of the slit.

Solution: Wavelength of light, $\lambda = 6000\times10^{-10}$ m

The distance to the first minimum, $d = 1$ mm $= 1 \times 10^{-3}$ m

Distance of separation between slit and screen, $D = 100$ cm $= 1$ m

Hence,
$$\sin\theta_1 \approx \frac{d}{D} = \frac{10^{-3}\,\text{m}}{1\,\text{m}} = 10^{-3}$$

We know that $l\sin\theta_1 = \lambda$

$$l = \frac{\lambda}{\sin\theta_1} = \frac{6000\times10^{-10}\,\text{m}}{10^{-3}}$$
$$= \mathbf{6\times10^{-4}\ m}$$

5.5 ▶ Fraunhofer's Diffraction at a Double Slit

The double slits have been represented as A_1B_1 and A_2B_2 in Fig. 5.5. The slits are narrow and rectangular in shape. The planes of the slits are perpendicular to plane of the paper.

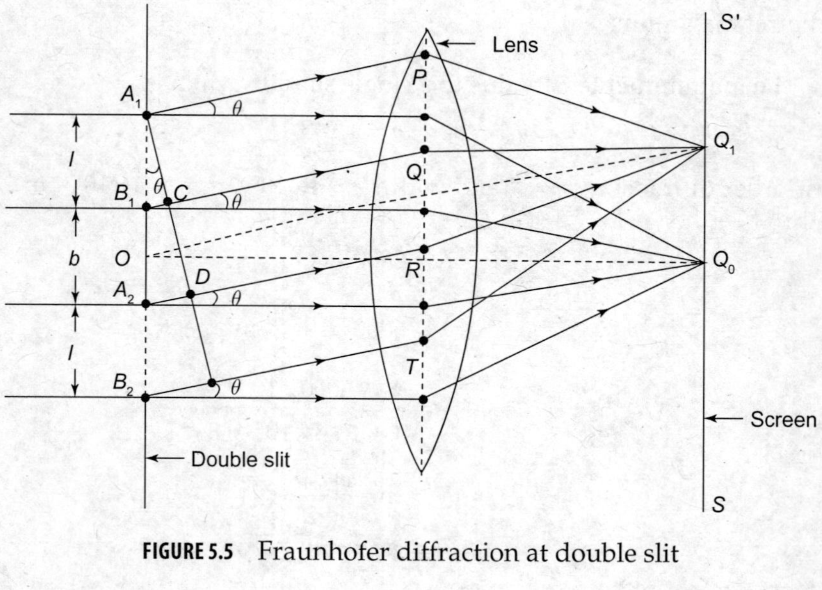

FIGURE 5.5 Fraunhofer diffraction at double slit

Let the width of both the slits be equal and it is 'l' and they are separated by opaque length 'b'. A monochromatic plane wave front of wavelength 'λ' is incident normally on both the slits.

Theory

Every point on the wave front in the slits will act as a source of secondary wavelets. The secondary wavelets travelling in the direction of OQ_0 are brought to focus at Q_0 on the screen SS' by using a converging lens L. Q_0 corresponds to the position of the central bright maximum. The intensity distribution on the screen is the combined effect of interference of diffracted secondary waves from the slits.

The diffracted intensity on the screen is very large along the direction of incident beam (i.e., along OQ_0). Hence, it is maximum at Q_0. This is known as principal maximum of zero order.

The intensity at point Q_1 on the screen is obtained by applying the Fraunhofer diffraction theory at single slit and interference of diffracted waves from the two slits. The diffracted wave amplitude due to single slit at an angle θ with respect to incident beam is $A\sin\alpha/\alpha$, where 2α is the phase difference between the secondary wavelets arising at the end points of a slit. This phase difference can be estimated as follows:

Draw a normal from A_1 to B_1Q. Now, B_1C is the path difference between the diffracted waves at an angle θ at the slit A_1B_1.

From the triangle A_1B_1C,

$$\sin\theta = \frac{B_1C}{A_1B_1} = \frac{B_1C}{l} \quad \text{or} \quad B_1C = l\sin\theta$$

The corresponding phase difference,

$$(2\alpha) = \frac{2\pi}{\lambda} l\sin\theta$$

or

$$\alpha = \frac{\pi}{\lambda} l\sin\theta \tag{5.6}$$

The diffracted wave amplitude, $[A\sin\alpha/\alpha]$, at the two slits combine to produce interference. The path difference between the rays coming from corresponding points in the slits A_1B_1 and A_2B_2 can be found by drawing a normal from A_1 to A_2R. A_2D is the path difference between the waves from the corresponding points of the slits.

In the triangle, A_1A_2D, $\dfrac{A_2D}{A_1A_2} = \sin\theta$ or the path difference $A_2D = A_1A_2\sin\theta = (l+b)\sin\theta$.

The corresponding phase difference

$$(2\beta) = \frac{2\pi}{\lambda}(l+b)\sin\theta \tag{5.7}$$

Applying the theory of interference on the wave amplitudes $[A\sin\alpha/\alpha]$ at the two slits gives the resultant wave amplitude (R).

$$R = 2A\frac{\sin\alpha}{\alpha}\cos\beta \tag{5.8}$$

The intensity at Q_1 is

$$I = R^2 = 4A^2\frac{\sin^2\alpha}{\alpha^2}\cos^2\beta$$

$$\boxed{= 4I_0\frac{\sin^2\alpha}{\alpha^2}\cos^2\beta} \quad [\text{since } I_0 = A^2] \tag{5.9}$$

Equation (5.9) represents the intensity distribution on the screen. The intensity at any point on the screen depends on α and β. The intensity of central maximum is $4I_0$. The intensity distribution at different points on the screen can be explained in terms of path difference between the incident and diffracted rays as follows. In Eq. (5.9), the term $\cos^2\beta$ corresponds to interference and $\sin^2\alpha/\alpha^2$ corresponds to diffraction. Now, we shall look at the conditions for interference and diffraction maxima and minima.

Interference maxima and minima

If the path difference $$A_2D = (l+b)\sin\theta_n = \pm n\lambda \tag{5.10}$$

where $n = 1, 2, 3, \ldots$, then 'θ_n' gives the directions of the maxima due to interference of light waves coming from the two slits. The $\pm$ sign indicates maxima on both sides with respect to the central maximum. On the other hand, if the path difference is odd multiples of $\lambda/2$, i.e.,

$$A_2D = (l + b)\sin\theta_n = \pm(2n-1)\lambda/2 \tag{5.11}$$

then 'θ_n' gives the directions of minima due to interference of the secondary waves from the two slits on both sides with respect to central maximum.

Diffraction maxima and minima

If the path difference $$B_1C = l\sin\theta_n = \pm n\lambda \tag{5.12}$$

where $n = 1, 2, 3, \ldots$, then 'θ_n' gives the directions of diffraction minima. The $\pm$ sign indicates minima on both sides with respect to central maximum. For diffraction maxima,

$$l\sin\theta_n = \pm(2n-1)\lambda/2 \tag{5.13}$$

is the condition. The $\pm$ sign indicates maxima on both sides with respect to central maximum.

The intensity distribution on the screen due to double slit diffraction is shown in Fig. 5.6. Figure 5.6(a) represents the graph for interference term; Fig. 5.6(b) shows the graph for diffraction term; Fig. 5.6(c) represents the resultant distribution.

5.5.1 Missing Orders in a Double Slit Diffraction Pattern

Based on the relative values of l and b, certain orders of interference maxima are missing in the resultant pattern. The direction of interference maxima is given as $(l+b)\sin\theta_n = n\lambda$, where $n = 1, 2, 3, \ldots$, and the directions of diffraction minima are given as $l\sin\theta_m = m\lambda$, where $m = 1, 2, 3, \ldots$.

For some values of 'θ_n', the values of l and b are satisfied such that at these positions the interference maxima and the diffraction minima are formed. The combined effect results in missing of certain orders of interference maxima. Now we see certain values of l and b for which interference maxima are missing.

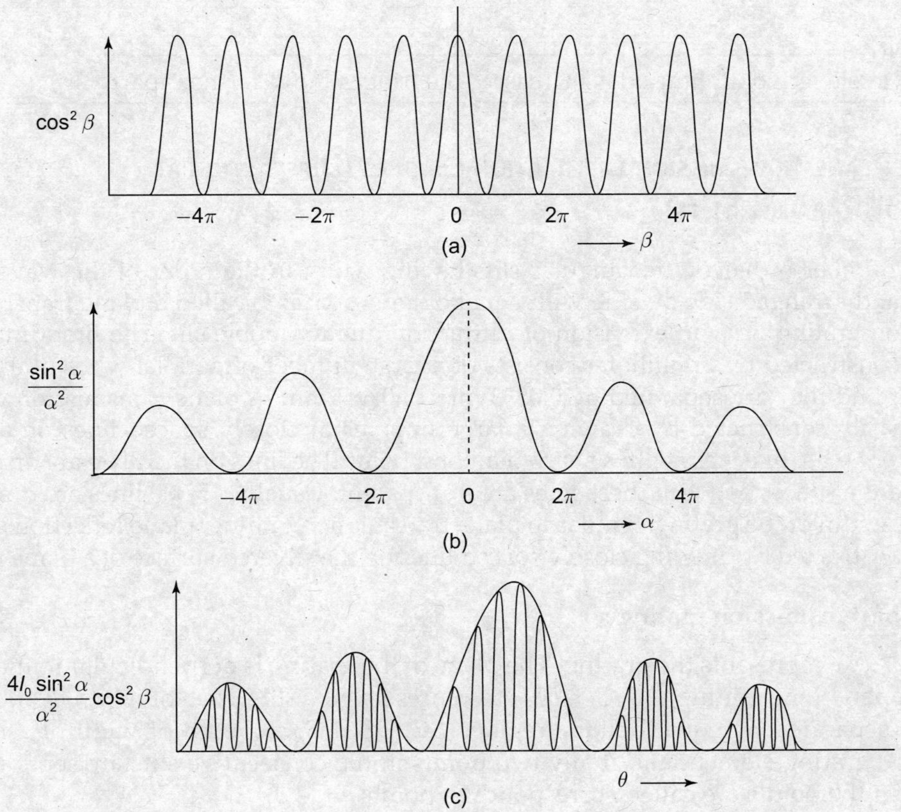

FIGURE 5.6 Intensity distributions due to diffraction at double slit

(i) If $l = b$,

then, $2l \sin\theta_n = n\lambda$ and $l \sin\theta_m = m\lambda$

$$\therefore \qquad \frac{n}{m} = 2 \text{ or } n = 2m \tag{5.14}$$

If $m = 1, 2, 3, \ldots$, then $n = 2, 4, 6, \ldots$, i.e. the interference orders $2, 4, 6, \ldots$ are missing in the diffraction pattern.

(ii) If $2l = b$,

then $3l \sin\theta_n = n\lambda$ and $l \sin\theta_m = m\lambda$

$$\therefore \qquad \frac{n}{m} = 3 \text{ or } n = 3m \tag{5.15}$$

If $m = 1, 2, 3, \ldots$, then $n = 3, 6, 9, \ldots$, i.e.. the interference orders $3, 6, 9, \ldots$ are missing in the diffraction pattern.

(iii) If $l + b = l$, i.e, $b = 0$, the two slits are joined. So, the diffraction pattern is due to a single slit of width $2l$.

Note

If the wavelength of light used is increased, the fringes move farther apart.

5.6 ▶ Plane Transmission Diffraction Grating (Diffraction at *N* Parallel Slits)

A large number of narrow rectangular slits having width in the order of the wavelength of light and arranged side by side with equal opaque spaces is called a plane transmission diffraction grating. It is an extension of diffraction due to double slit. The first diffraction grating constructed by Fraunhofer consists of a large number of parallel wires of 0.05 mm diameter and they are separated by 0.0533 mm to 0.687 mm. A plane transmission grating can be easily constructed by drawing a large number of closely spaced lines on a plane transparent plate like glass with a sharp diamond point. The lines on the plate are opaque to light and the spaces between these lines are transparent. Usually, 15,000 lines are drawn on one inch width of the grating. Similar to plane transmission grating, plane reflection grating can be constructed by drawing closely spaced lines on the silvered surface of a plane mirror.

Theory of transmission grating

In Fig. 5.7, *XY* represents the grating. The plane of the grating is perpendicular to the plane of the paper. In the grating, *AB*, *CD*, *EF*, etc. represents the slits, each of width 'l' and these slits are separated by equal opaque regions *BC*, *DE*, *EG*, etc. each of width 'b'. $l+b$ is called the grating element and if any two points in the consecutive slits are separated by $l + b$, then the points are called corresponding points.

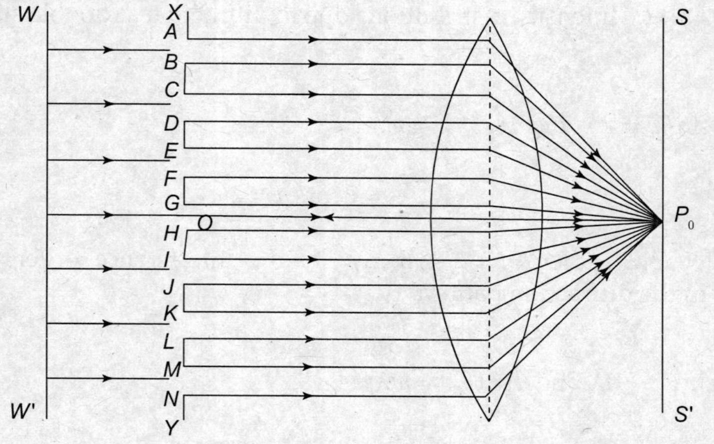

FIGURE 5.7 Diffraction grating

Let a monochromatic plane wave front WW' of wavelength 'λ' be incident on the slits. Each point on the wave front in the slits will act as a source of secondary wavelets. These wavelets will travel in all directions. The secondary wavelets in each slit will produce diffraction. The diffracted secondary wavelets of all the slits combine to produce interference. Now all the secondary wavelets travelling in the direction of the incident beam will get converged at P_0 on the screen SS' by passing through a convex lens L having focal length f.

The screen is at the focal plane of the lens. The intensity at P_0 is very high and it is called central maximum. Next we see the secondary wavelets travelling in a direction that makes an angle 'θ' with the direction of incident beam. As shown in Fig. 5.8, the secondary wavelets that make an angle 'θ' will reach the point P_1 on the screen.

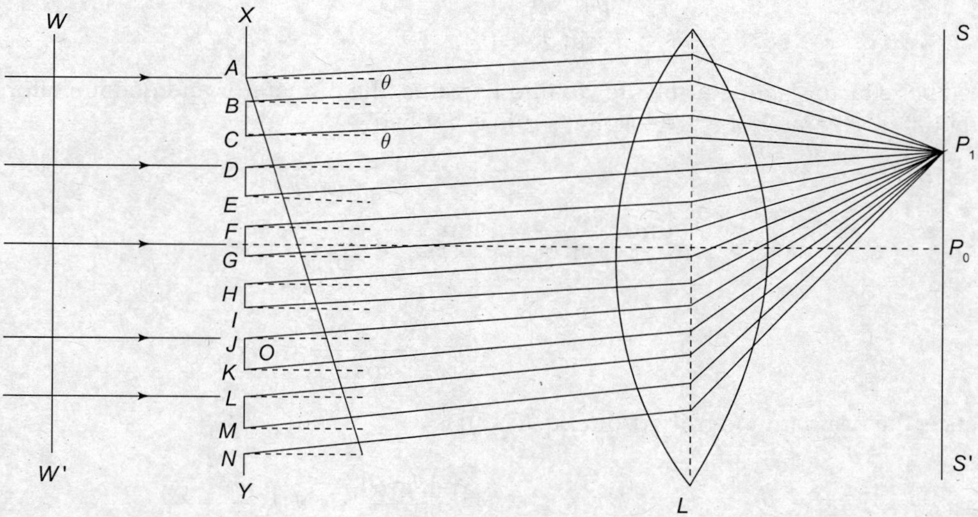

FIGURE 5.8 Diffraction due to grating showing intensity at P_1

The intensity at P_1 can be obtained by considering diffraction at each slit and interference of all these diffracted waves at an angle 'θ' with the incident beam. The resultant amplitude due to diffraction at single slit is $A\sin\alpha/\alpha$, which is now at the middle point of each slit, where $\alpha = \pi/\lambda l \sin\theta$. Let us assume that there are N slits on the grating, the path difference between the corresponding points is $(l+b)\sin\theta$ and in terms of phase difference it is $2\pi/\lambda(l+b)\sin\theta = 2\beta$. The resultant of N amplitudes, each of $A\sin\alpha/\alpha$ with common phase difference (2β) between consecutive diffracted amplitudes can be obtained by the vector addition method.

The resultant amplitude (R) of the N amplitudes each of $A\sin\alpha/\alpha$ is

$$R = \frac{A\sin\alpha}{\alpha}\frac{\sin N\beta}{\beta} \tag{5.16}$$

And intensity

$$I = R^2 = \left(\frac{A\sin\alpha}{\alpha}\right)^2\left(\frac{\sin N\beta}{\sin\beta}\right)^2 \tag{5.17}$$

The value $(A\sin\alpha/\alpha)^2$ shows the intensity distribution due to single slit diffraction and the value $\left(\sin^2 N\beta/\sin^2\beta\right)$ shows the combined intensity distribution due to interference at all the N slits.

Principal maxima

From Eq. (5.17), we know that the intensity would be maximum when $\sin\beta = 0$, where $\beta = 0, \pi, 2\pi, 3\pi, \ldots = n\pi$ and $n = 0, 1, 2, 3, \ldots$ and $n = 1, 2, 3, \ldots$.

Then

$$\frac{\sin N\beta}{\sin\beta} = \frac{0}{0}$$

The above factor is indeterminate. To find its value, the numerator and denominator are differentiated separately, i.e. L' Hospital's rule is applied.

$$\lim_{\beta\to\pm n\pi}\frac{\sin N\beta}{\sin\beta} = \lim_{\beta\to\pm n\pi}\frac{\frac{d}{d\beta}\sin(N\beta)}{\frac{d}{d\beta}\sin\beta}$$

$$= \lim_{\beta\to\pm n\pi}\frac{N\cos N\beta}{\cos\beta} = \pm N$$

Hence, the resultant intensity [from Eq. (5.17)] is

$$I = N^2\left[\frac{A\sin\alpha}{\alpha}\right]^2 = N^2 I_0\frac{\sin^2\alpha}{\alpha^2} \tag{5.18}$$

These maxima are most intense and are called principal maxima. The maxima are obtained for $\beta = \pm n\pi$ or

$$\frac{\pi(l+b)\sin\theta}{\lambda} = \pm n\pi \quad \text{or}$$

$$\boxed{(l+b)\sin\theta = \pm n\lambda} \tag{5.19}$$

$n = 0$ corresponds to zero order (or central) maximum. $n = 1, 2, 3, \ldots$ corresponds to the first, second, third, etc. principal maxima, respectively. The $\pm$ sign shows that the principal maxima are present on both sides of the central maximum.

Minima

From Eq. (5.17), we know a series of minima occur when $\sin N\beta = 0$ but $\sin\beta \neq 0$.

So, $N\beta = \pm m\pi$ or $\frac{N\pi}{\lambda}(l+b)\sin\theta = \pm m\pi$ or $N(l+b)\sin\theta = \pm m\lambda$

Here, m can take integral values except $0, N, 2N, \ldots, nN$. For these values, we get maxima.

Secondary maxima

Between two adjacent principal maxima, there are $(N-1)$ minima and $(N-2)$ other maxima called secondary maxima. To get the positions of secondary maxima, differentiate Eq. (5.17) w.r.t β and then equate it to zero.

Thus,

$$\frac{dI}{d\beta}=\left(\frac{A\sin\alpha}{\alpha}\right)^2\cdot 2\left(\frac{\sin N\beta}{\sin\beta}\right)$$

$$\left[\frac{N\cos N\beta\sin\beta-\sin N\beta\cos\beta}{\sin^2\beta}\right]=0$$

So,

$$N\cos N\beta-\sin N\beta\cos\beta=0$$

or

$$N\tan\beta=\tan N\beta$$

The roots of the above equation (other than $\beta=\pm n\pi$, this corresponds to principal maxima) give the position of secondary maxima. To find the intensity $\left(\sin^2 N\beta/\sin^2\beta\right)$ from $N\tan\beta=\tan N\beta$, we use the triangle shown in Fig. 5.9.

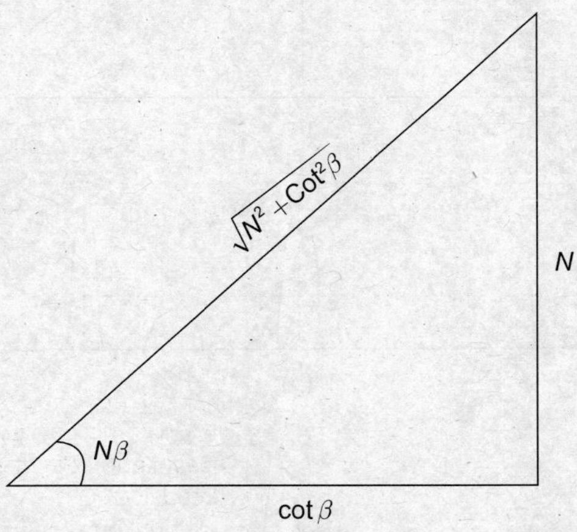

FIGURE 5.9 Triangle for calculation of secondary maxima

$$\sin N\beta=\frac{N}{\sqrt{N^2+\cot^2\beta}}$$

Square the above equation and divide with $\sin^2\beta$, we have

$$\frac{\sin^2 N\beta}{\sin^2\beta}=\frac{N^2}{\left(N^2+\cot^2\beta\right)\times\sin^2\beta}$$

$$= \frac{N^2}{N^2 \sin^2 \beta + \cos^2 \beta}$$

$$= \frac{N^2}{1+\left(N^2-1\right)\sin^2 \beta}$$

Now,

$$\frac{\text{Intensity of secodary maximum}}{\text{Intensity of principal maxima}} = \frac{1}{1+\left(N^2-1\right)\sin^2 \beta}$$

From the above equation, we know that as N increases the intensity of secondary maxima relative to principal maxima decreases. When N is very large, the intensity of secondary maxima is very less as compared to principal maxima.

The resultant intensity distribution on both sides of the central maximum can be represented by the diffraction term $\left(\sin^2 \alpha/\alpha^2\right)$ and by the interference term $\left(\sin^2 N\beta/\sin^2 \beta\right)$. The variation of the above two terms and the resultant from the central maximum is shown in Fig. 5.10.

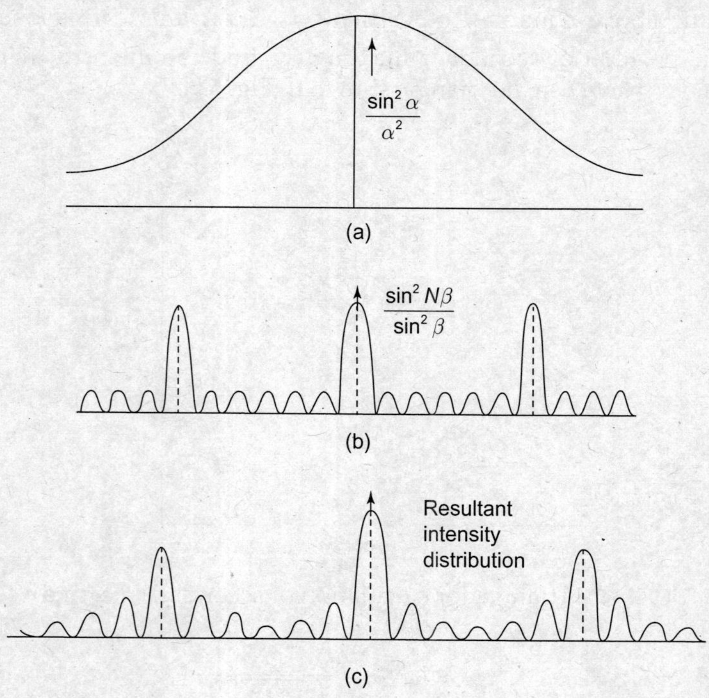

FIGURE 5.10 (a) The variation of intensity due to the factor $\sin^2 \alpha/\alpha^2$, (b) the variation of intensity due to the factor $\sin^2 N\beta/\sin^2 \beta$ and (c) the resultant variation of intensity

Note

The angular separation between two consecutive minima is same.

5.6.1 Grating Spectrum

The diffraction pattern formed with a grating is known as a grating spectrum. From Eq.(5.19), we know the 'θ' values for different principal maxima. If the number of lines in the grating is very large, then the maxima are sharp and bright. Lines parallel to the grating lines are called spectral lines. Different order bright lines are formed on both sides of central maximum. Instead of monochromatic light if white light is used, then light of different wavelengths is diffracted in different directions for each order of diffraction as shown in Fig. 5.11.

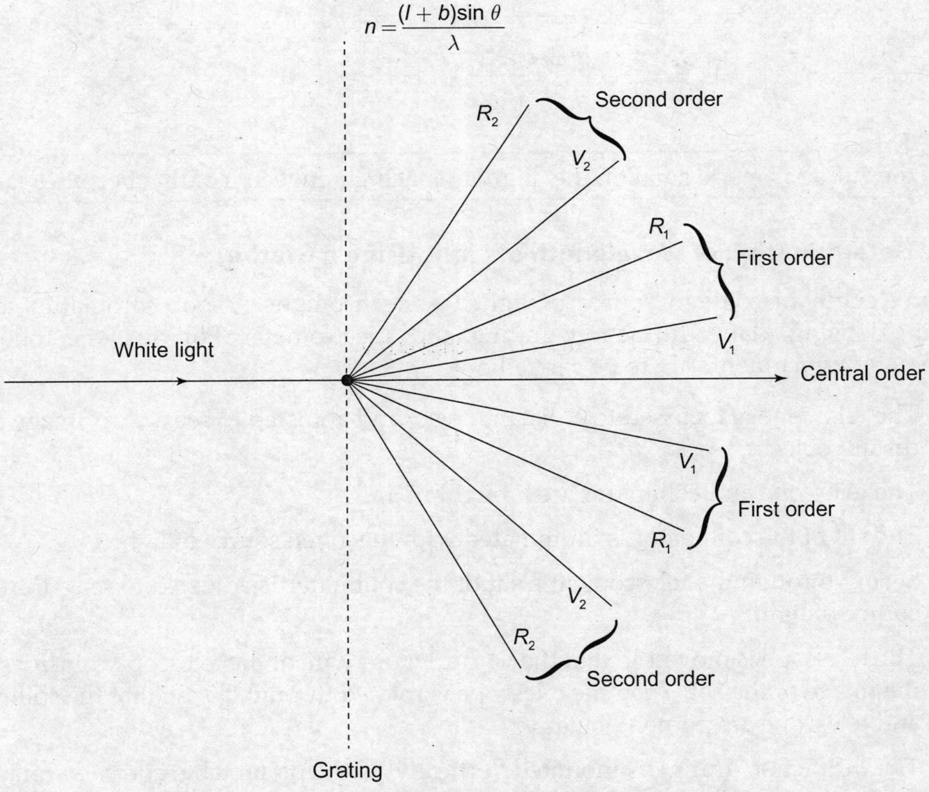

FIGURE 5.11 Grating spectrum for white light source

The central image is white because the central maxima of different wavelength rays coincide at that place. In each order of diffraction, waves having longer wavelengths are diffracted with greater diffraction angle.

5.6.2 Maximum Number of Orders Available with a Grating

From Eq. (5.19), we have

$$n = \frac{(l+b)\sin\theta}{\lambda}$$

The maximum angle of diffraction is 90°; hence, the maximum order of diffraction $n_{\max}$ is

$$n_{\max} \le \frac{(l+b)\sin 90°}{\lambda}$$

$$\le \frac{(l+b)}{\lambda} \le \frac{1}{N\lambda}$$

where $N = 1/(e+b) =$ number of lines per unit distance of grating.

$$\therefore \quad \boxed{n_{\max} \le \frac{1}{N\lambda}} \tag{5.20}$$

Note

When the ruling is made smaller, the angular spacing between maxima becomes larger.

5.6.3 Determination of Wavelength of Light Using a Grating

The wavelength of a given source of light (monochromatic or polychromatic) can be determined using plane diffraction grating and spectrometer. For this, the following spectrometer adjustments are to be carried out:

1. The telescope is focussed to a distant object and adjusted to see a clear image of the distant object.
2. The telescope and collimator are brought in line.
3. The slit of the collimator is illuminated with the given source of light.
4. Seeing through the telescope, the slit of the collimator is adjusted to see a thin clear source of light.
5. The telescope is moved so that the vertical cross-wire of the telescope coincides with the image of the slit. Then the telescope is rotated through 90° so that the collimator and telescope are perpendicular.
6. The diffraction grating is mounted vertically on the prism table and then rotated in the forward and backward directions so that the light from the collimator is reflected on the surface of the grating and the reflected light coincides with the vertical cross-wire of the telescope. Then, fix the grating. Now the plane of the grating makes 45° with the incident beam.
7. Release the base of the spectrometer and rotate the grating through 45° so that the surface of the grating is normal to the incident beam.

Now the telescope is released and brought in line with the collimator to view the direct ray. Seeing through the telescope, it is moved slowly to one side until the first-order diffracted spectral line is seen as shown in Fig. 5.12. The vertical cross-wire of the telescope is made to coincide with the first-order spectral line and note down the spectrometer reading. Again, bring the telescope in line with the collimator and move it to other side to see the first-order

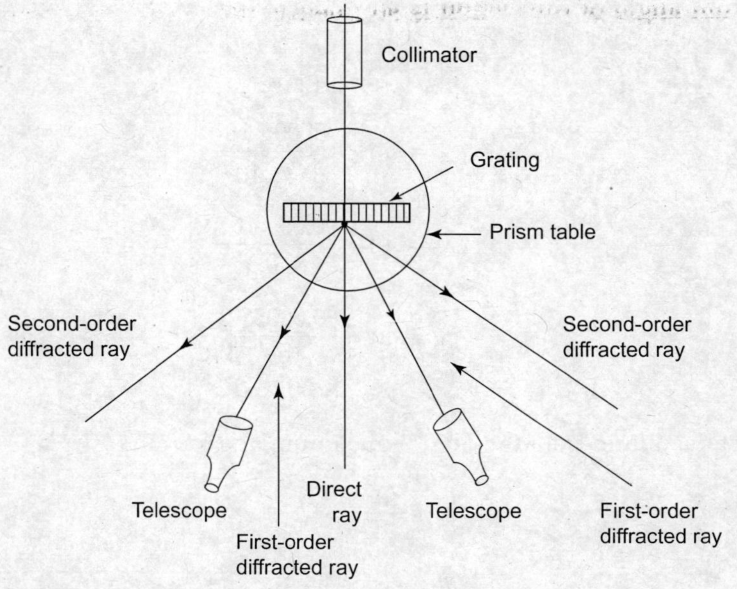

FIGURE 5.12 Determination of wavelength of monochromatic light using a spectrometer

spectral line on the other side. Coincide the vertical cross-wire of telescope with this spectral line and take the reading. The difference between these two readings gives twice the diffraction angle (2θ). Using the formula $\sin\theta = nN\lambda$ (n = order of diffracted ray, N = number of lines per unit width of grating, 'λ' = wavelength of monochromatic light), the wavelength of monochromatic light can be determined. The same process can be repeated for second order and even for higher orders.

In case of white light, the diffraction angle is measured for different colours and hence the different wavelengths are determined.

EXAMPLE 5.3

A monochromatic light of wavelength 5500 Å is incident on a single slit of width 0.3 mm and gets diffracted. Find the diffraction angles for the first minimum and the next maximum.

Solution: Wavelength of light, $\lambda = 5500\ \text{Å} = 5500 \times 10^{-10}$ m

Width of the slit, $l = 0.3\ \text{mm} = 0.3 \times 10^{-3}$ m

The diffraction angle for the first minimum is given using the formula

$$l \sin\theta_n = n\lambda \quad \text{and} \quad \sin\theta_1 = \frac{\lambda}{l}$$

$$\theta_1 = \sin^{-1}\left(\frac{\lambda}{l}\right)$$

$$= \sin^{-1}\left[\frac{5500\times10^{-10}}{0.3\times10^{-10}}\right]$$

$$= \sin^{-1}\left(\frac{0.55\times10^{-3}}{0.3}\right)$$

$$\theta_1 = \sin^{-1}\left(\frac{0.00055}{0.3}\right)$$

$$= 0°\,6'18''$$

For the first maximum from the central maximum, the formula is

$$l\sin\theta_1' = \frac{3\lambda}{2}$$

$$\theta_1' = \sin^{-1}\frac{3\lambda}{2l}$$

$$= \sin^{-1}\left[\frac{3\times5500\times10^{-10}}{2\times0.3\times10^{-3}}\right]$$

$$= \sin^{-1}0.0028$$

$$\mathbf{= 0°9'27''}$$

EXAMPLE 5.4

In an experiment for observing diffraction pattern due to a straight edge, the distance between the slit source $(\lambda = 6000\text{ Å})$ and the straight edge is 6 m. Calculate their separations.

Solution: If l be the distance between straight edge and the slit, and b be the distance between eyepiece and straight edge. Now, the position of the nth maximum (bright band) is

$$x_n = \sqrt{\frac{b(l+b)(2n+1)\lambda}{l}},\ n = 0, 1, 2, 3, \ldots$$

Here, $l = 600$ cm, $b = 400$ cm, $\lambda = 6000\times10^{-8}$ cm

$$x_n = \sqrt{\frac{400\times1000(2n+1)\times6000\times10^{-8}}{600}}$$

$$= \frac{\sqrt{(2n+1)}}{5} \text{cm, where } n = 0, 1, 2, 3, \ldots$$

The position of the first three maxima (putting $n = 0, 1, 2, 3, \ldots$)

$$x_1 = \frac{1}{5} = 0.20 \text{ cm}$$

$$x_2 = \frac{\sqrt{3}}{5} = 0.346 \text{ cm}$$

$$x_3 = \frac{\sqrt{5}}{5} = 0.447 \text{ cm}$$

The separations are as follows:

$$\boldsymbol{x_2 - x_1 = 0.346 - 0.200 = 0.146}$$

and $$\boldsymbol{x_3 - x_1 = 0.447 - 0.346 = 0.101}$$

EXAMPLE 5.5

There are 15,000 lines per inch in a grating. What is the maximum number of order obtained by using light of wavelength 6000 Å?

Solution: Wavelength, $\lambda = 6000$ Å

$$= 6000 \times 10^{-10} \text{ m}$$

Number of lines grating, $N = 15{,}000$ lines/inch

$$= \frac{15{,}000}{2.54} \text{ lines/cm}$$

$$= \frac{15{,}00{,}000}{2.54} \text{ lines/m}$$

We know,

$$nN\lambda = \sin\theta$$

$$n_{max}N\lambda = \sin\theta_{max}$$

$$= \sin 90° = 1$$

$$n_{max} = \frac{1}{N\lambda} = \frac{2.54}{15{,}00{,}000 \times 6000 \times 10^{-10}}$$

$$= \mathbf{2.82}$$

5.7 ▶ Dispersive Power of a Grating

The dispersive power of a grating is defined as the rate of change of the diffraction angle with the wavelength of light. It is expressed as $d\theta/d\lambda$.

The diffraction of the nth order principal maxima for a wavelength is given by

$$(l+b)\sin\theta = n\lambda \tag{5.21}$$

Differentiating it with respect to λ,

$$(l+b)\cos\theta\, d\theta = nd\lambda$$

or

$$\boxed{\frac{d\theta}{d\lambda} = \frac{n}{(l+b)\cos\theta}} \tag{5.22}$$

This is the expression of dispersive power.

From Eq. (5.22), it is clear that

$$\text{Dispersive power} \propto n$$
$$\propto \frac{1}{(l+b)}$$
$$\propto \frac{1}{\cos\theta}$$

Case 1: If θ is large, cos θ will be smaller, and dispersive power of grating will be high.
Also, $\theta_{\text{red}} > \theta_{\text{violet}}$
Thus, dispersion in red region is greater than that in violet region.

Case 2: If θ is small, cos $\theta = 1$ and $d\theta \propto d\lambda$.
This type of spectrum is called normal spectrum.

Note
Higher is the diffraction order, higher is the dispersive power.

5.8 ▶ Rayleigh Criterion for Resolving Power

According to Rayleigh's criterion, two nearby images are said to be resolved if they are separated by at least a certain minimum distance so that the position of diffraction central maximum of the first image coincides with the first diffracted minimum of the second image in any order of diffraction and vice versa.

Rayleigh's criterion can be explained by considering the diffraction patterns due to two spectral lines of wavelengths 'λ_1' and 'λ_2'. Let A and B represent central maxima of diffraction patterns corresponding to wavelengths 'λ_1' and 'λ_2'. As shown in Fig. 5.13(a), the diffraction angle for the central maximum of the image B is larger than the diffraction angle for the central minimum of the image A. The principal maximum of A is far away from the first minimum of B and vice versa. Hence, their principal maxima are separately visible. The intensity between these two principal maxima is zero. So, the two spectral lines corresponding to the two wavelengths are well resolved.

As shown in Fig. 5.13(b), let the central maximum of *A* coincide with the first minimum of *B* and vice versa. The resultant intensity is shown by a thick curve with a small dip in between the two central maxima. There is a small noticeable decrease in intensity between the two central maxima. This indicates the presence of two wavelengths. Thus, according to Rayleigh's condition, the spectral lines can be just resolved.

Lastly as shown in Fig. 5.13(c), when the diffracted central maxima of the two wavelengths are still closer, then the resultant intensity curve has no dip in the middle. The resultant intensity peak is higher than the individual intensities. Thus, the two images overlap and cannot be distinguished as separate images. So, we cannot resolve them as two spectral lines.

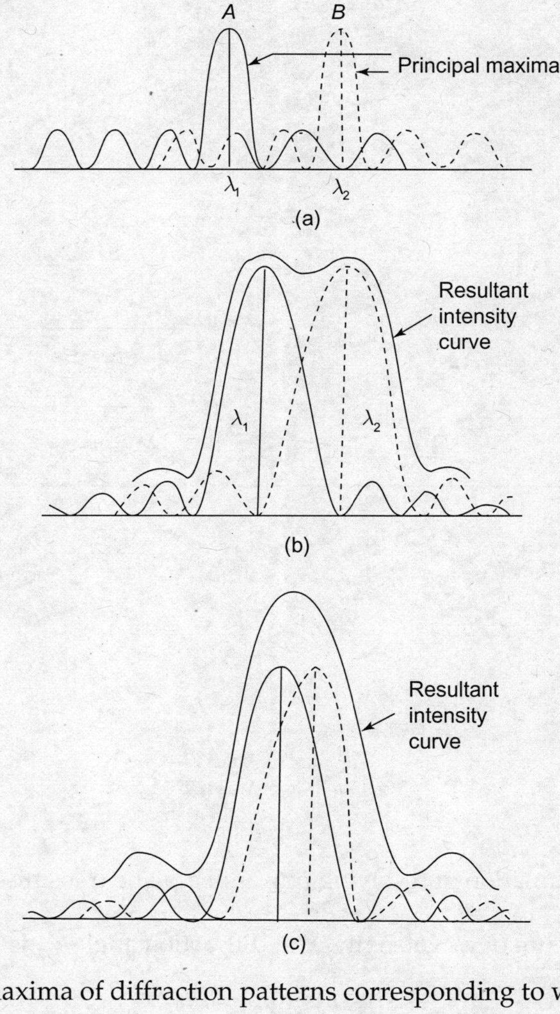

FIGURE 5.13 Central maxima of diffraction patterns corresponding to wavelengths λ_1 and λ_2

5.8.1 Resolving Power of a Plane Transmission Grating

The capacity of an optical instrument to show separate images of very closely placed two objects is called resolving power. The resolving power of a diffraction grating is defined as its ability to form separate diffraction maxima of two closely separated wavelengths.

The expression for resolving the power of a grating can be obtained with the aid of Fig. 5.14.

In Fig. 5.14, XY represents a plane transmission grating having $(l+b)$ as the grating element and N lines per unit width of grating. A light beam having two slightly different wavelengths 'λ' and $\lambda+d\lambda$ is incident normally on the surface of the grating. SS_1 is the screen and P_1 represents the position of nth order primary maximum spectral line of wavelength 'λ' diffracted at an angle 'θ_n'. Similarly, P_2 represents the position of nth order primary maximum spectral line of wavelength $\lambda+d\lambda$, diffracted at an angle $\theta_n+d\theta$. These n^{th} order maxima lines can be resolved if P_2 corresponds to the first minimum of wavelength 'λ'.

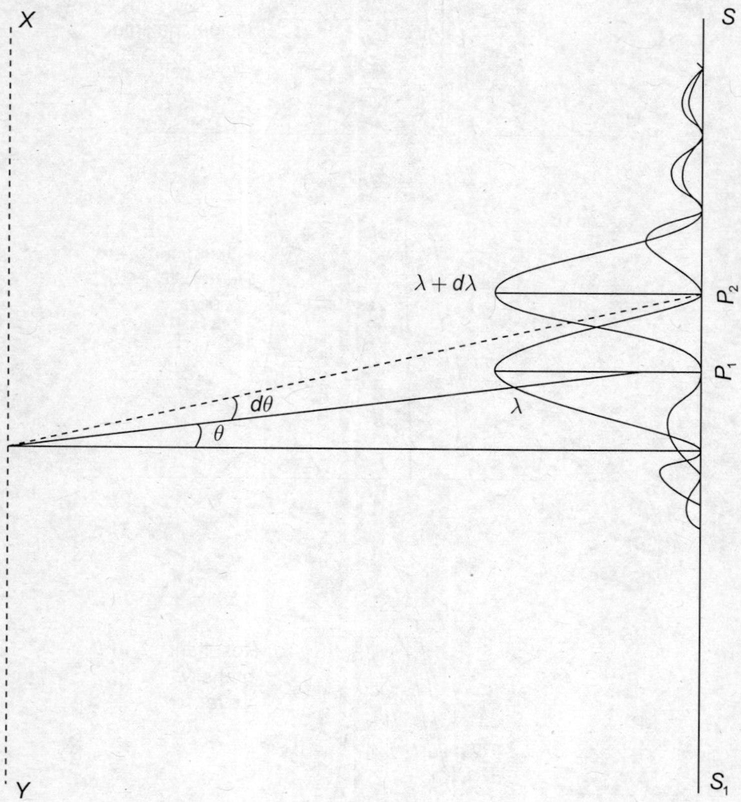

FIGURE 5.14 Calculation of resolving power of a plane transmission grating

The principal maximum of wavelength 'λ' at diffraction angle θ_n is

$$(l+b)\sin\theta_n = n\lambda \tag{5.23}$$

The equation for the first minimum of wavelength 'λ' in the direction $(\theta_n + d\theta)$ is

$$N(l+b)\sin(\theta_n + d\theta) = (nN+1)\lambda \tag{5.24}$$

Next, the nth principal maximum for wavelength $(\lambda + d\lambda)$ in the direction $\theta_n + d\theta$ is

$$(l+b)\sin(\theta_n + d\theta) = n(\lambda + d\lambda) \tag{5.25}$$

Multiplying Eq. (5.25) by N, we have

$$N(l+b)\sin(\theta_n + d\theta) = nN(\lambda + d\lambda) \tag{5.26}$$

The LHS of Eqs. (5.24) and (5.26) are the same.

$$\therefore \qquad (nN+1)\lambda = nN(\lambda + d\lambda)$$

or

$$nN\lambda + \lambda = nN\lambda + nNd\lambda$$

or

$$\lambda = nNd\lambda \quad \text{or} \quad \boxed{\frac{\lambda}{d\lambda} = nN} \tag{5.27}$$

Equation (5.27) represents the expression for the resolving power of the grating. From Eq. (5.27), we know that the resolving power is directly proportional to (i) the diffraction order of the spectrum and (ii) the total number of lines per unit width of the grating.

Note

A prism spectrum due to dispersion of light can also be obtained but unlike grating spectrum it gives only one spectrum. The spectral colours are in the order from red to violet.

EXAMPLE 5.6

Find the resolving power of a grating having 6000 lines/cm in the first-order diffraction. The ruled length of the grating is 15 cm.

Solution: Lines per cm length of grating = 6000 lines/cm

$$\text{Total number of lines on the grating} = N = 15 \times 6000$$

$$= 90{,}000 \text{ lines}$$

Resolving power is

$$\frac{\lambda}{d\lambda} = nN = 1 \times 90{,}000$$

$$= \mathbf{90{,}000}$$

EXAMPLE 5.7

Find the minimum number of lines required in a grating to resolve two spectral lines of wavelength 5890 Å in the second-order diffraction.

Solution: Average wavelength,

$$\lambda = \left(\frac{5890 + 5896}{2}\right)$$

$$\lambda = 5893 \text{ Å}$$

$$= 5893 \times 10^{-10} \text{ m}$$

Difference in wavelength,

$$d\lambda = (5896 - 5890)$$

$$= 6 \text{ Å}$$

$$= 6 \times 10^{-10} \text{ m}$$

$n = 2,$

$$\frac{\lambda}{d\lambda} = nN \Rightarrow N = \frac{1}{n}\frac{\lambda}{d\lambda}$$

$\Rightarrow$

$$N = \frac{1}{2\times}\frac{5893 \times 10^{-10}}{6 \times 10^{-10}}$$

$\Rightarrow$

$$\mathbf{N = 419.18 \text{ lines}}$$

5.9 ▶ Fresnel's Half-period Zones

Let $ABCD$ be a plane wave front of wavelength λ proceeding in the direction of arrow. Q is a point where the resultant intensity to be computed (Fig. 5.15).

The calculation of resultant intensity is done by dividing the wave front into a number of zones called Fresnel's half-period zones. Suppose OQ is normal to wave front ($OQ = d$) with sphere having radii $\left(d + \frac{\lambda}{2}\right)$, $\left(d + \frac{2\lambda}{2}\right)$, $\left(d + \frac{3\lambda}{2}\right)$, ..., $\left(d + \frac{n\lambda}{2}\right)$ is drawn.

The plane $ABCD$ cut these spheres into concentric circles with centre O and radii OP_1, OP_2, OP_3, ..., OP_n.

Here,

(i) the area of the first innermost circle is called first half-period zone.

(ii) the area enclosed between the first and second circle is called second half-period zone and so on. A resultant wave starts from each zone.

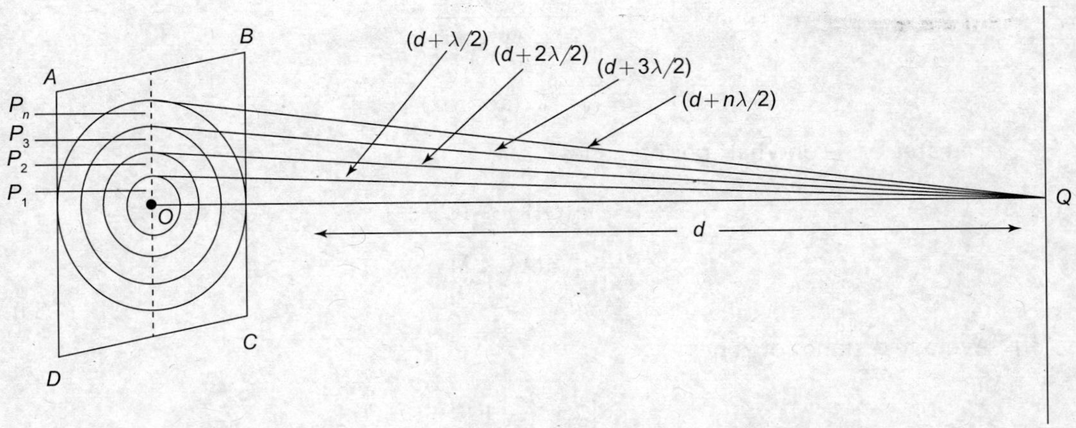

FIGURE 5.15 Fresnel's half-period zones

5.9.1 Relative Phase of the Zones

As $P_1Q = OQ + (\lambda/2)$, hence the ray reaching at Q from O and the circumference of the first circle are in opposite phase, i.e. the path difference between the two is of $\lambda/2$ or phase difference is of the order of π. The mean phase of secondary wavelets of the first zone is $\pi/2$. Similarly, the mean phase of secondary zones is $3\pi/2$.

Similarly, the mean phases due to third, fourth, etc. zones are $\frac{5\pi}{2}, \frac{7\pi}{2}, \ldots$. Thus, the successive zones differ by a half period $\left(\frac{T}{2}\right)$ that's why they are called half-period zones.

5.9.2 Amplitude Due to a Zone

(i)

$$\text{Area of first zone} = \pi(OP_1)^2$$

$$= \pi[P_1Q^2 - OQ^2]$$

$$= \pi\left[\left(d + \frac{\lambda}{2}\right)^2 - d^2\right]$$

$$\boxed{= \pi d\lambda \text{ (approx.)}} \quad (5.28)$$

(ii) The radius of the first half-period zones

i.e., $OP_1 = \sqrt{(d\lambda)}$

The radius of the second zone $= OP_2$

$$= \sqrt{(P_2Q)^2 - (OQ)^2}$$

$$= \sqrt{(d+\lambda)^2 - d^2}$$

$$= \sqrt{2d\lambda} \text{ (approx)} \tag{5.29}$$

The area of second half-period zones

$$= \pi[(OP_2)^2 - (OP_1)^2]$$

$$= \pi[(2d\lambda - d\lambda]$$

$$= \pi d\lambda \tag{5.30}$$

(iii) Average distance of zones

$$= \frac{\left(d+\frac{n\lambda}{2}\right)+\left\{d+(n-1)\frac{\lambda}{2}\right\}}{2}$$

$$\boxed{= \boldsymbol{d+(2n-1)\frac{\lambda}{4}}} \tag{5.31}$$

[Average distance of nth zone from Q]

(iv) **Obliquity factor (θ_n):** The angle between the normal to zone and line joining from zone to Q is called obliquity factor $f(\theta_n)$.

(v) Amplitude to n^{th} zone is proportional to

$$\frac{\pi\left[d+(2n-1)\frac{\lambda}{4}\right]\lambda}{d+(2n-1)\frac{\lambda}{4}} f(\theta_n) = \pi\lambda f(\theta_n) \tag{5.32}$$

As the order of zone increases, $f(\theta_n)$ decreases. Thus, the amplitude of wave from a zone at Q decreases as the order of the zone increases.

(vi) **Resultant amplitude:** The resultant amplitude at Q at any instant is given by (Fig. 5.16)

$$R = R_1 - R_2 + R_3 - R_4 + R_5 + \cdots + R_n \text{ (if } n \text{ is odd)}$$

or

$$R = R_1 - R_2 + R_3 - R_4 + R_5 + \cdots + R_n \text{ (if } n \text{ is even)}$$

If n is odd,

$$R = \frac{R_1}{2} + \frac{R_n}{2} = \frac{R_1 + R_n}{2}$$

But, if n is even, $\quad R = \frac{R_1}{2} + \frac{R_{n-2}}{2} - R_n$

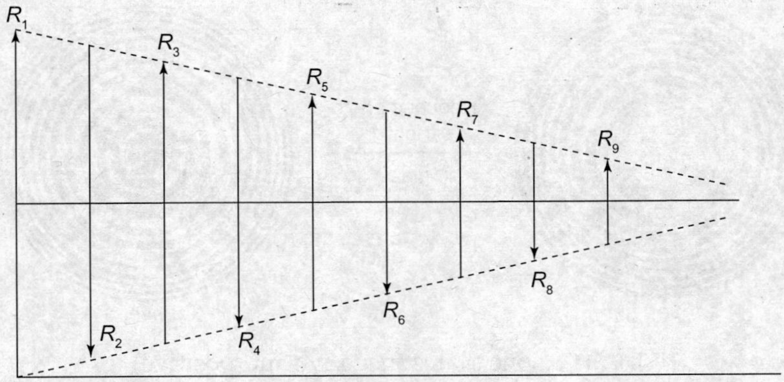

FIGURE 5.16 Resultant amplitude of half-period zones

As $n \to \infty$, R_n and $R_{n-1} \to 0$

Thus,
$$R = \frac{R_1}{2}$$

Hence, the resultant intensity is

$$I \propto \left(\frac{R_1}{2}\right)^2 \propto \frac{R_1^{\,2}}{4}$$

Here, R_1 = amplitude of wave at Q due to first half-period zone
R_2 = amplitude of wave at Q due to second half-period zone and so on.

5.10 ▶ Zone Plate

Zone plate is sort of highly focussed convex lens with multiple radii. So, there are multiple focal lengths in a zone plate. To construct a zone plate, concentric circles are drawn on a white paper so that the radii are proportional to the square roots of n natural number.

$$r_n \propto \sqrt{n}$$

or
$$r_1 \propto \sqrt{1} = 1$$

$$r_2 \propto \sqrt{2} = 1.414$$

$$r_3 \propto \sqrt{3} = 1.732 \text{ and so on.}$$

The odd zones are covered with black ink and then a reduced photograph is taken. When its negative is developed, odd zones become transparent to light while even zones will cut off the light. Such type of plate is known as a zone plate as shown Fig. 5.17.

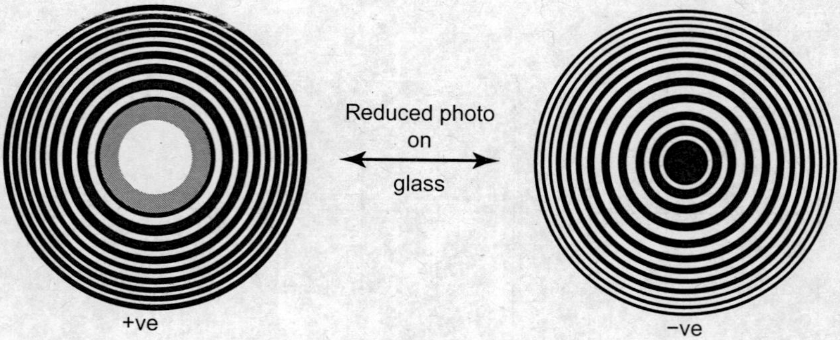

FIGURE 5.17 Zone plates (negative and positive)

Note

The image of a lens is brighter than that of a zone plate. Unlike zone plate a lens produces image at one place only. Also the image by a lens is produced by the phenomenon of refraction and not by diffraction.

5.10.1 Theory of a Zone Plate

(a) When light is coming from a source at infinity

Let us suppose that the different zones coincide with the half-period zones (Fig. 5.18).

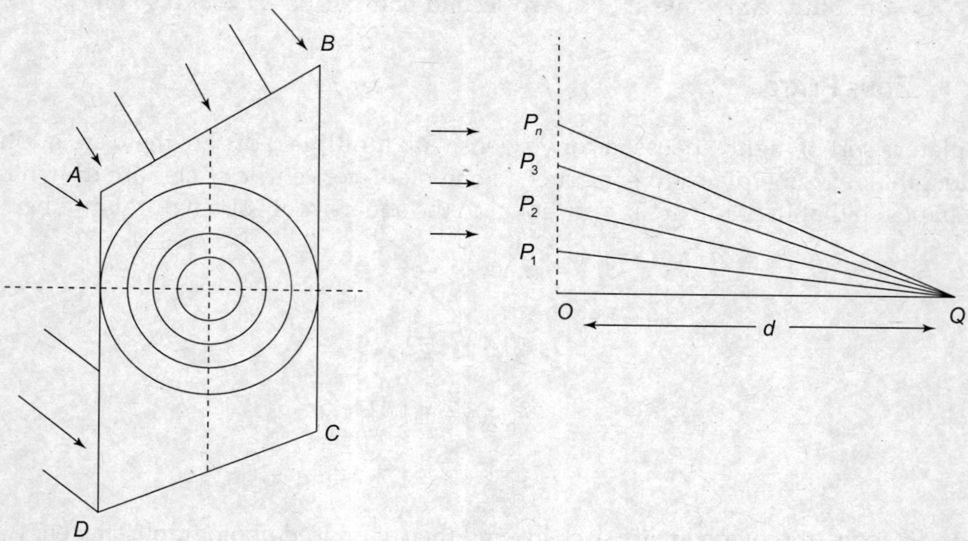

FIGURE 5.18 When light is coming from the source at infinity

Here, $$OQ = d$$

$$P_1Q = \left(d + \frac{\lambda}{2}\right)$$

$$\dots\dots\dots$$

$$P_nQ = \left(d + \frac{n\lambda}{2}\right)$$

Let $OP_1 = r_1,\ OP_2 = r_2,\ OP_3 = r_3,\ \dots,\ OP_n = r_n$, where $r_1, r_2, \dots, r_n$ are radii of different zones on the zone plate.
In ΔOP_nQ,

$$P_nQ^2 = OP_n{}^2 + OQ^2$$

$$\left(d + \frac{n\lambda}{2}\right)^2 = r_n{}^2 + d^2$$

$$d^2 + \frac{n^2\lambda^2}{4} + dn\lambda = r_n{}^2 + d^2$$

$$r_n{}^2 = \frac{n^2\lambda^2}{4} + dn\lambda$$

Neglecting higher power of n,

$$r_n{}^2 = dn\lambda$$
$$r_n = \sqrt{dn\lambda}$$

$$\text{Area of } n^{\text{th}} \text{ zones} = \pi r_n^2 - \pi r_{n-1}^2$$

$$= \pi\{dn\lambda - [d(n-1)\lambda]\}$$

$$= \pi[dn\lambda - dn\lambda + d\lambda] = \pi d\lambda$$

$$\text{Area of } n^{\text{th}} \text{ zone} = \pi d\lambda \tag{5.33}$$

The area of different zones on zone plate is same.

Focal length of zone plate (d)

$$r_n^2 = dn\lambda$$

$$\Rightarrow \qquad d = \frac{r_n^2}{n\lambda} \tag{5.34}$$

(b) When the source and screen are at finite distance

Let us consider a part of zone plate held vertically coincides with half-period zone (Fig. 5.19).

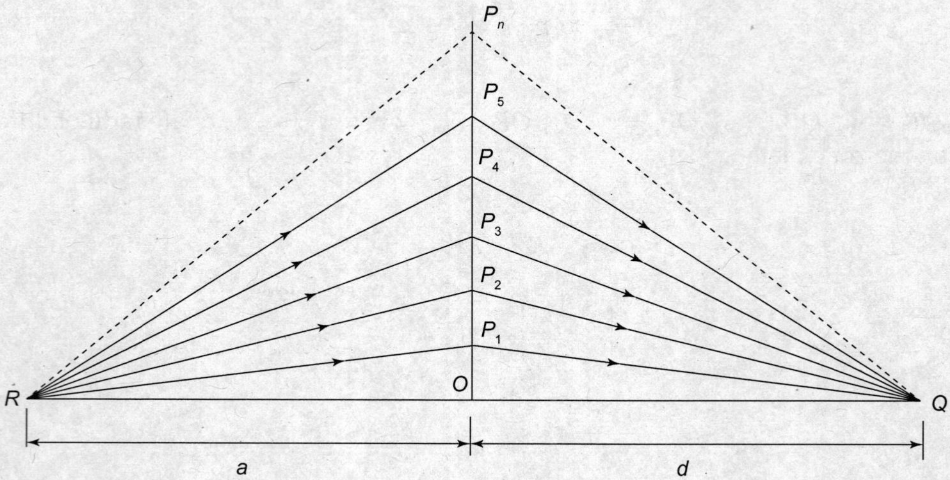

FIGURE 5.19 Zone plate when the source and screen are at finite distances

Here,
$$RO = RP_1 = RP_2 = \cdots = RP_n = a$$

$$P_nQ = \left(d + \frac{n\lambda}{2}\right)$$

Thus,

$$RP_1 + P_1Q = a + d + \frac{\lambda}{2}$$

$$RP_2 + P_2Q = a + d + \frac{2\lambda}{2}$$

$$RP_n + P_nQ = a + b + \frac{n\lambda}{2} \tag{5.35}$$

$$RP_1^2 = a^2 + r_1^2 = a^2\left(1 + \frac{r_1^2}{a^2}\right)$$

$\Rightarrow$
$$RP_1 = a\left(1 + \frac{r_1^2}{a^2}\right)^{\frac{1}{2}} = a + \frac{1}{2}\frac{r_1^2}{a^2}\cdot a$$

Neglecting higher power,

$$RP_1 = a\left(1+\frac{r_1^2}{2a^2}\right) \tag{5.36}$$

In ΔP_1OQ,
$$P_1Q^2 = r_1^2 + d^2$$

$$P_1Q = d\left(1+\frac{r_1^2}{d^2}\right)^{\frac{1}{2}} \quad \text{(neglecting higher power)}$$

$$P_1Q = d\left[1+\frac{r_1^2}{2d^2}\right] \tag{5.37}$$

From Eqs. (5.36) and (5.37),

$$RP_n = a\left(1+\frac{r_n^2}{2a^2}\right)$$

And
$$P_nQ = d\left(1+\frac{r_1^2}{2d^2}\right) \tag{5.38}$$

From Eqs. (5.35) and (5.38)

$$RP_n + P_nQ = a + d + \frac{n\lambda}{2}$$

$$\Rightarrow \quad a+\frac{r_n^2}{2a^2}+d+\frac{r_n^2}{2d^2} = a+d+\frac{n\lambda}{2}$$

$$\Rightarrow \quad \frac{r_n^2}{2}\left(\frac{1}{a^2}+\frac{1}{d^2}\right) = \frac{n\lambda}{2}$$

$$\Rightarrow \quad \frac{1}{a^2}+\frac{1}{d^2} = \frac{n\lambda}{r_n^2} \tag{5.39}$$

If we compare Eq. (5.39) with Eq. (5.40)

$$\frac{1}{v}-\frac{1}{u} = \frac{1}{f} \tag{5.40}$$

we get

$$\frac{1}{f} = \frac{n\lambda}{r_n^2}$$

$$\boxed{f = \frac{r_n^2}{n\lambda}} \tag{5.41}$$

Equation (5.41) shows that zone plate is a sort of lens with multiple focal lengths.

Note

Zone plates are considered as a cheap alternative to expensive targeting lasers.

5.10.2 Comparison of a Zone Plate and Convex Lens

(a) Differences

(i) The zone plate has multiple focal lengths and forms a series of point images of decreasing intensity. Convex lens has one focal length, which can be calculated from lens maker formula.

(ii) For a zone plate, the path difference between the rays from transparent zones is λ, but for a convex lens all the rays for reaching on image point have the same optical path.

(iii) In case of a zone plate, the focal length for violet colour is greater than that for red, i.e. $f_v > f_R$, but in case of a convex lens, it is reverse.

(b) Similarities

(i) The relation between the conjugate distances is similar for both.

(ii) For both, the focal length depends on wavelength λ. Both show chromatic aberration.

5.11 ▶ Fresnel's Diffraction Due to a Straight Edge

Let us consider that S be a narrow slit illuminated with monochromatic light of wavelength λ. WW' is a cylindrical wave front, which is diverging from S. QR is the illuminated region. AB is a straight edge of opaque obstacle, which is placed parallel to the slit. XY is the screen (Fig. 5.20). If there is no diffraction of light, then we would have obtained complete darkness below P but uniform illumination above it. But actually we observe a few unequally spaced bright and dark bonds with poor contrast and running parallel to the length of the slit. In the geometrical shadow below R, the intensity falls rapidly and becomes zero at a finite distance from R.

Let us consider Q, a point of maximum or minimum intensity, i.e. according to path difference between AQ and OQ is equal to an odd or even number of half wavelength (Fig. 5.21). i.e.,

$$AQ - OQ = (2n \pm 1)\frac{\lambda}{2} \quad \text{(for maxima)} \tag{5.42}$$

And

$$AQ - OQ = n\lambda \quad \text{(for minima)} \tag{5.43}$$

$SA = a$, $AR = b$, thus

$$(AQ)^2 = (AR)^2 + (RQ)^2$$

$$(AQ)^2 = (b^2 + x^2) \quad [\because RQ = x]$$

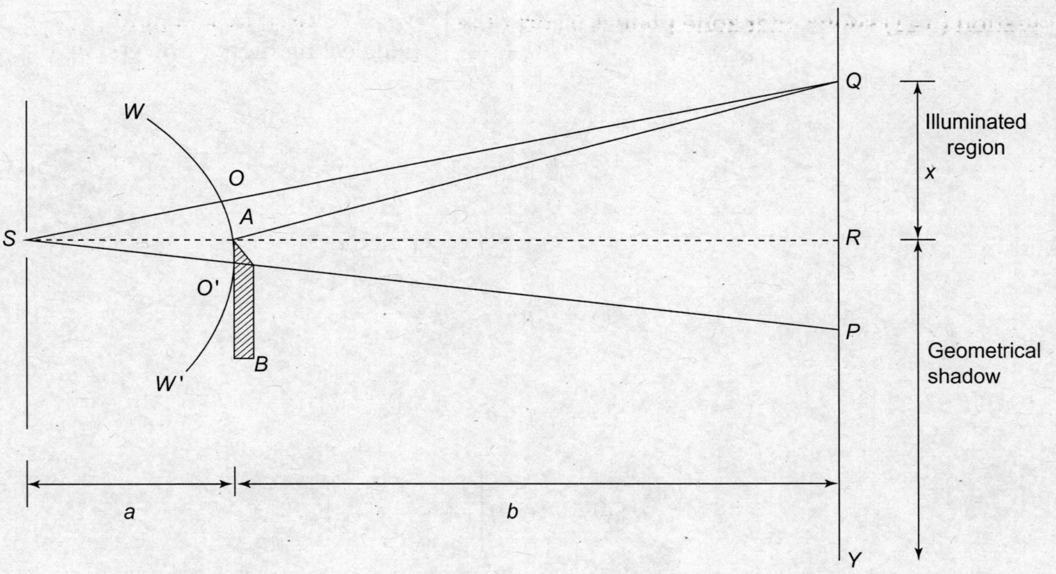

FIGURE 5.20 Fresnel's diffraction due to a straight edge

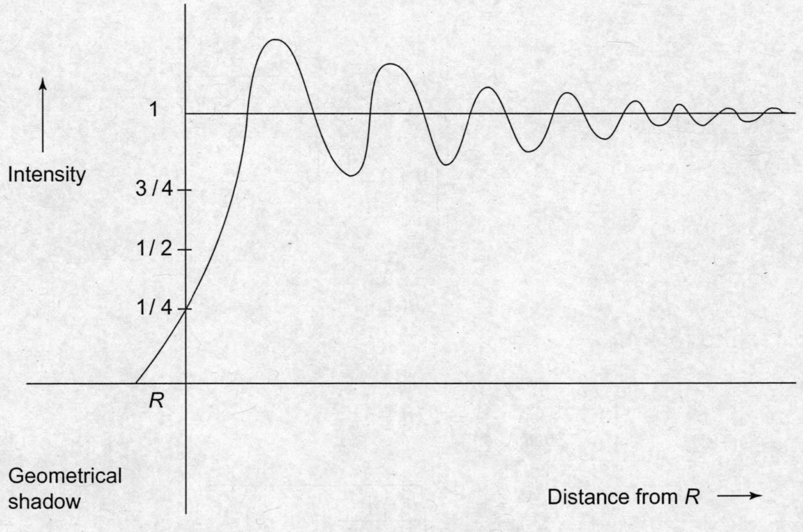

FIGURE 5.21 Variation of intensity with distance

$$AQ = (b^2 + x^2)^{\frac{1}{2}}$$

$$AQ = b\left[1 + \frac{x^2}{b^2}\right]^{\frac{1}{2}}$$

$$AQ = b\left[1 + \frac{x^2}{2b^2}\right] \quad \text{(Neglecting higher powers)}$$

$$AQ = b + \frac{x^2}{2b} \tag{5.44}$$

And

$$OQ = SQ - SO = SQ - SA$$

$$SQ = (SR^2 + RQ^2)$$

$$SQ = \left\{(a+b)^2 + x^2\right\}^{\frac{1}{2}}$$

$$SQ = (a+b)\left\{1 + \frac{x^2}{2(a+b)^2}\right\}$$

$$SQ = (a+b) + \frac{x^2}{2(a+b)}$$

$$OQ = SQ - SO = b + \frac{x^2}{2(a+b)} \tag{5.45}$$

$$AQ - OQ = \left(b + \frac{x^2}{2b}\right) - \left(b + \frac{x^2}{2(a+b)}\right)$$

$$= \frac{ax^2}{2b(a+b)} \tag{5.46}$$

From Eqs. (5.42) and (5.46),

$$\frac{ax^2}{2b(a+b)} = (2n \pm 1)\frac{\lambda}{2}$$

$$x = \sqrt{\frac{b(a+b)}{a}\{(2n \pm 1)\lambda\}}$$

or,

$$x = y\sqrt{(2n \pm 1)} \tag{5.47}$$

where

$$y = \left[\frac{b(a+b)}{a}\lambda\right]^{\frac{1}{2}}$$

The distance of successive maxima from R is given by (on putting $n = 0, 1, 2, 3, \ldots$)

$$x_1 = y$$
$$x_2 = y\sqrt{3}$$
$$x_3 = y\sqrt{5}$$
$$x_4 = y\sqrt{7}$$

The separation between successive maxima is as follows:

$$(x_2 - x_1) = y\sqrt{3-1} = 0.73\, y$$
$$(x_3 - x_2) = y\sqrt{5-3} = 0.50\, y$$
$$(x_4 - x_3) = y\sqrt{7-5} = 0.43\, y$$
$$\dots \dots \dots$$

which are in decreasing order.

Note

The width of central maxima is directly proportional to the wavelength of light used but inversely proportional to the slit width.

RECAP ZONE

POINTS TO REMEMBER

- Bending of light round the corners of an obstacle and spreading of light into geometrical shadow region is called diffraction.
- For observing the phenomenon of diffraction, the obstacle or aperture should be of size of wavelength of light.
- Diffraction is divided into two classes:
  (i) Fresnel's diffraction
  (ii) Fraunhofer's diffraction
- Principal maximum due to a single slit diffraction is given by
$$I = I_0 \frac{\sin^2 \alpha}{\alpha^2}$$
- Diffraction intensity due to double slit arrangement is given by
$$I = \frac{4I_0 \sin^2 \alpha}{\alpha^2} \cdot \cos^2 \beta$$

- Diffraction intensity due to N slits arrangement is

$$I = N^2 I_0 \frac{\sin^2 \alpha}{\alpha^2}$$

- The dispersive power of a grating is directly proportional to the order n and the number of lines on the grating surface.
- The resolving power of a grating depends on the number of lines on the grating and the order of the spectrum.
- Focal length of a particular zone plate is

$$f = \frac{r_n^2}{n\lambda}$$

- The position of nth maxima in Fresnel's diffraction pattern is given by

$$x = y\sqrt{2n \pm 1},\ y = \left[\frac{b(a+b)}{a}\lambda\right]^{\frac{1}{2}}$$

- In Fresnel's diffraction

  (i) The area of the zone is $\pi d\lambda$

  (ii) The obliquity factor is $\pi\lambda f(\theta_n)$

Key Terms with Definitions

- **Diffraction grating** – An array of parallel slits which produces a spectrum by diffraction is called diffraction grating.
- **Grating element** – The distance between corresponding points of adjacent slits is called grating element.
- **Plane transmission grating** – If the ruling is obtained by drawing parallel lines on plane transparent surfaces, then the grating is called plane transmission grating.
- **Reflection grating** – If the ruling is obtained by drawing parallel lines on silvered surfaces, then the grating is called reflection grating.
- **Resolving power of a diffraction grating** – It is the capacity to form separate diffraction maxima of two wavelengths, which are very close to each other.
- **Dispersive power of grating** – The rate of variation of angle of diffraction with wavelength.
- **Resolution** – Two sources are said to be resolved by an optical instrument when the central maximum in the diffraction pattern of one falls over the first minimum of other.

- **Focal length** – The distance between the centre of a lens and its focus.
- **Zone plate** – A device used to focus light, it uses diffraction to study wave character of light.
- **Fresnel zones** – It is a set of radially symmetric rings on the zone plate.
- **Missing order** – It occurs when the diffraction minima overlaps with interference maximum.

Important Formulae and Equations

Equation Number	Equation	Remarks
(5.2)	$I = R^2 = A^2\left[\frac{\sin\alpha}{\alpha}\right]^2$	Resulatant amplitude for Fraunhofer diffraction at a single slit
(5.9)	$I = 4I_0\frac{\sin^2\alpha}{\alpha^2}\cos^2\beta$	Intensity distribution in case of Fraunhofer diffraction at a double slit
(5.19)	$(l+b)\sin\theta = \pm n\lambda$	Central maxima in plane transmission grating
(5.20)	$n_{\max} \le \frac{1}{N\lambda}$	Maximum number of lines per unit distance of grating
(5.22)	$\frac{d\theta}{d\lambda} = \frac{n}{(l+b)\cos\theta}$	Dispersive power of a grating
(5.27)	$\frac{\lambda}{d\lambda} = nN$	Resolving power of grating
(5.28)	$\pi d\lambda$ (approx.)	Area of first zone
(5.31)	$d + (2n-1)\frac{\lambda}{4}$	Average distance between zones, d = distance of screen from zone plate
(5.41)	$f = \frac{r_n^2}{n\lambda}$	Focal length of zone plate

Review Zone

Short Answer Questions

1. What do you mean by diffraction of light?
2. What is grating element?
3. What should be the order of size of obstacle for diffraction of light?
4. What is resolving power of a grating?
5. What is the difference between resolving power and dispersive power of a grating?
6. The two sodium lines appear as separate in the second order, but not in the first order, why?
7. Discuss the features of Fraunhofer diffraction pattern produced by a double slit.
8. What do you mean by limit of resolution?
9. Give the theory of a plane diffraction grating of the transmission type.
10. Does the resolving power depend on the number of ruling per centimetre on the grating?
11. Distinguish between Fresnel's and Fraunhofer's diffraction
12. What are half-period zones?
13. Show that the area of each zone is independent of number of zone.

Long Answer Questions

1. Discuss the phenomenon of Fraunhofer diffraction due to a single slit.
2. Describe Fraunhofer diffraction at double slit and deduce the position of maxima and minima.
3. Obtain an expression for Fresnel's diffraction due to straight edge.
4. Give the theory of plane transmission diffraction grating.
5. Give an expression for resolving power of a grating.
6. Define dispersive power of a grating and obtain an expression for it.
7. Explain the phenomenon of single slit diffraction.
8. Discuss the formation of primary maxima and secondary minima and maxima, with a plane transmission diffraction grating.
9. Give the theory of zone plate. Show that amplitude of wave from a zone decreases as the order of the zone increases.

NUMERICAL PROBLEMS

1. A screen placed 2 m away from a narrow slit, find the slit width if the first minima lie 5 mm on either side of the central's minimum when the plane wave of $\lambda = 5\times10^{-5}$ cm is incident on the slit.

   Ans. 0.02

2. A grating has 8000 ruling per inch. It is illuminated with light of wavelength 4000 Å at normal incidence. Find the angle at which the first and second order maxima occur?

   Ans. $\theta_1 = 7°22'$ $\theta_2 = 14°42'$

3. A diffraction grating used at normal incident gives a line (5400 Å) in a certain order superposed on the violet line (4050 Å) of the next higher order. If the angle of diffraction is 30°, how many lines per cm are there in the grating?

   Ans. 3086

4. How many orders will be visible if the wavelength of the incidence radiation is 4800 Å and the number of line on the grating is 2500 per inch?

   Ans. 21

5. What should be the minimum number of line per inch in a half inch width grating to resolve D_1 (5896 Å) and D_2 (5896 Å) lines of sodium?

   Ans. 982 lines per cm

6. A diffraction grating used at normal incident gives a yellow line (λ = 6000 Å) in a certain spectral order superposed to a blue line (λ = 4800 Å) of the next higher order. If the angle of diffraction is $\sin^{-1}(3/4)$, calculate the grating elements.

   Ans. 3.2×10^{-4} m

CHAPTER 6 Interference

Learning Objectives

By the end of this chapter, the student will be able:

- To introduce the conditions for sustained interference of light beams
- To demonstrate Young's double slit experiment
- To define coherence of light
- To illustrate the Fresnel's bi-prism experiment
- To discuss the interference in thin and wedge-shaped films,
- To discuss Newton's rings and other applications of thin films
- To explain the working of Michelson interferometer

6.1 ▶ Introduction

When a number of waves passes through a point in a medium at the same time, they combine to produce a resultant wave having a different amplitude and hence different intensity than the individual waves at that point. This phenomenon is known as interference. The resultant intensity at a point is due to the combined influence of all the waves pass through that point. Interference produces modification in the distribution of intensity of light, and it has been observed with light waves, sound waves, water waves, etc. Interference can be explained using the principle of superposition of waves.

6.2 ▶ Conditions for Sustained Interference

To produce sustained interference, the two light sources that we use for interference should be coherent, i.e. the two light waves should have the same frequency and they are to be in phase or they maintain constant phase difference. To produce a stationary interference pattern, the resultant intensity at any point should be constant with time. Two separate light sources (say two sodium lamps) will not produce observable interference pattern on the screen. This is because light is coming from a large number of independent atoms of the sources, and each atom emits light in a time of 10^{-9} sec. Even if the atoms are emitting light under similar conditions, the waves coming from different atoms may differ in their initial phase. The light coming from two independent sources will be coherent for a period of about 10^{-9} sec. So, the interference produced by such sources of light changes for every 10^{-9} sec. We cannot notice such changes of intensity of interference light with our naked eye. Hence, we say that the two independent sources of light do not have a fixed phase relationship over a prolonged period of time, and so they cannot produce any stationary interference pattern, because the phase difference between the two sources does not remain constant. Hence, to produce a sustained interference pattern, the two coherent sources of light used for interference should be derived from a single source of light, usually by amplitude division of a single source of light. The two coherent light beams that have been obtained in this method are used to produce interference. Some more conditions necessary for the interference of light are given below:

(i) The two waves must have equal amplitudes.

(ii) The two coherent sources must be closely placed.

(iii) The light must be continuously emitted.

6.2.1 Phase Difference and Path Difference

Phase difference is defined as the difference in phase angles of two vibrating particles, i.e.

$$\phi = \frac{2\pi}{\lambda} \times (\text{path difference})$$

If μx is the optical path difference between two particles, then the phase difference is

$$\phi = \frac{2\pi}{\lambda} \mu x$$

where the optical path difference μx is the distance travelled in vacuum containing the same number of waves as the actual path travelled in the medium.

6.2.2 Methods of Producing Interference

Interference can be produced by two following methods:

(a) **Division of wavefront**

In this method, the wavefront is divided into two parts, either by reflection, refraction or by diffraction by using mirror, lens, prism or grating. These two parts of the same wave front travel unequal distances and reunite at some angles, thus producing interference fringes, e.g. Young's double slit experiment, Fresnel's biprism and Lloyd's mirror.

(b) **Division of amplitude**

The amplitude of the incoming beam is divided into two parts, either by parallel reflection or by refraction. These two parts travel unequal distances and reunite to produce interference, e.g. Newton's rings and Michelson's interferometer.

6.3 ▶ Young's Double Slit Experiment

In 1801, Thomas young first demonstrated the phenomenon of interference of light. His experimental set-up is shown in Fig. 6.1. As shown in the figure, he passed sun light through a narrow pin hole 'S' on an opaque surface and then through two closely spaced pin holes S_1 and S_2 in another opaque surface. A screen 'XY' was arranged in front of the pin holes S_1 and S_2. He observed a few coloured bright and dark interference bands on

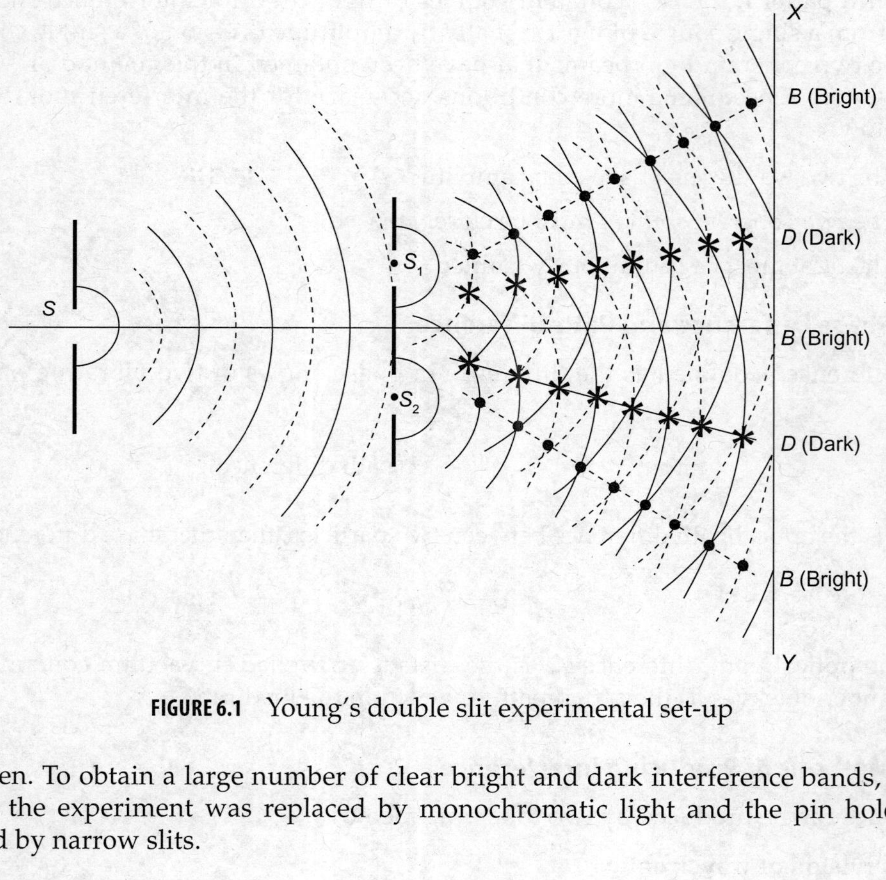

FIGURE 6.1 Young's double slit experimental set-up

the screen. To obtain a large number of clear bright and dark interference bands, the sun light in the experiment was replaced by monochromatic light and the pin holes were replaced by narrow slits.

Explanation of interference

The interference has been explained based on the wave theory of light. As light passes through the pin hole 'S', spherical wave fronts spread out as shown in Fig. 6.1. The solid curves represent crests and dotted curves represent troughs on the waves. According to Huygens wave theory, every point on the surface of the wave front acts as a source of secondary wavelets. As the wave front spreads to pin holes S_1 and S_2, the wave front at S_1 and S_2 acts as secondary sources. The radii of these secondary wave fronts increase as they go away from S_1 and S_2 and superimpose to produce interference. At a point if a crest of one wave falls on the crest of another wave (or a trough of one wave combines with a trough of another wave), then the amplitude of the resultant wave is the sum of amplitudes of the two crests (or troughs). This causes constructive interference at that point and a bright interference fringe is observed at that point. On the other hand, if a crest of one wave falls on the trough of another wave, then destructive interference takes place at that point and the resultant amplitude is a minimum, so a dark interference fringe is observed. Since the intensity (I) of light is proportional to square of amplitude ($I \propto R^2$), the intensity is large at the places where constructive interference takes place and the intensity is minimum at the places where destructive interference takes place. On the screen, a large number of alternatively bright and dark interference bands (or fringes) of equal width are observed.

6.3.1 Analytical Treatment of Interference

As shown in Fig. 6.2, a source of monochromatic light of wave length 'λ' is passed through a pin hole 'S'. Spherical wave fronts were diverging and pass through the pin holes S_1 and S_2. The wave fronts in the pin holes S_1 and S_2 act as a source of secondary wavelets. The secondary wave fronts from S_1 and S_2 combine to produce an interference pattern on the screen 'XY'. Now we will determine the intensity of interference fringes at a point P on the screen. The screen is at a distance D from S_1 and S_2.

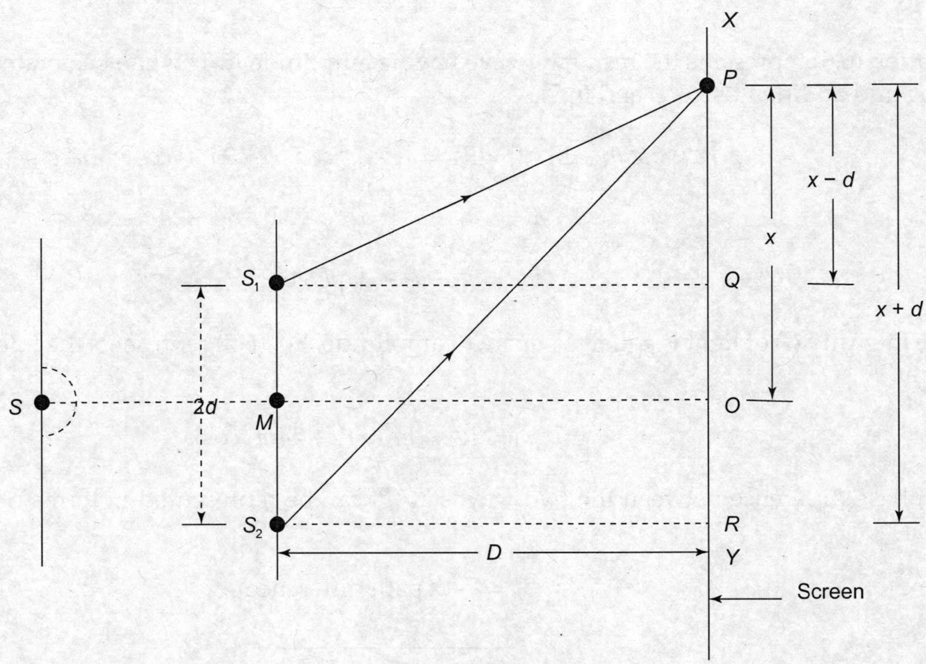

FIGURE 6.2 Schematic representation of interference due to two slits

Let the amplitudes of the waves from S_1 and S_2 be a_1 and a_2. Let the phase difference between the waves at P be ϕ. Suppose that Y_1 and Y_2 are the displacements of these waves at P.

Then $$Y_1 = a_1 \sin \omega t \tag{6.1}$$

And $$Y_2 = a_2 \sin(\omega t + \phi) \tag{6.2}$$

These two waves combine to produce a resultant wave having displacement Y, given as

$$\begin{aligned} Y = Y_1 + Y_2 &= a_1 \sin \omega t + a_2 \sin(\omega t + \phi) \\ &= a_1 \sin \omega t + a_2 \sin \omega t \cos \phi + a_2 \cos \omega t \sin \phi \\ &= \sin \omega t [a_1 + a_2 \cos \phi] + \cos \omega t (a_2 \sin \phi) \end{aligned} \tag{6.3}$$

Let $$R \cos \theta = a_1 + a_2 \cos \phi \tag{6.4}$$

And $$R \sin \theta = a_2 \sin \phi \tag{6.5}$$

Substituting Eqs. (6.4) and (6.5) in Eq. (6.3), we get

$$\begin{aligned} Y &= \sin \omega t \cdot R \cos \theta + \cos \omega t \cdot R \sin \theta \\ &= R \sin(\omega t + \theta) \end{aligned} \tag{6.6}$$

Equation (6.6) represents the resultant wave having amplitude R. This can be obtained by squaring and adding Eqs. (6.4) and (6.5).

$$R^2(\cos^2 \theta + \sin^2 \theta) = R^2 = a_1^2 + a_2^2 \cos^2 \phi + 2a_1 a_2 \cos \phi + a_2^2 \sin^2 \phi$$

$$R^2 = a_1^2 + a_2^2(\cos^2 \phi + \sin^2 \phi) + 2a_1 a_2 \cos \phi$$

$$R^2 = a_1^2 + a_2^2 + 2a_1 a_2 \cos \phi \tag{6.7}$$

Since intensity (I) of light is equal to square of amplitude, Eq. (6.7) represents the intensity of light at 'P'.

Hence, $$I = R^2 = a_1^2 + a_2^2 + 2a_1 a_2 \cos \phi \tag{6.8}$$

The phase difference between the two waves at 'P' can be represented in terms of path difference as

$$\phi = \frac{2\pi}{\lambda} \times \text{path difference}$$

$$\boxed{= \frac{2\pi}{\lambda} \times (S_2P - S_1P)} \tag{6.9}$$

Based on the phase or path difference, we see the following special cases for intensity of interference fringes.

Case I: Condition for maximum intensity: For maximum intensity, $\cos\phi = 1$ in Eq. (6.8), then $\phi = 0, 2\pi, 4\pi, \ldots = 2n\pi$, where $n = 0, 1, 2, \ldots$

or the path difference $S_2P - S_1P = n\lambda$

Then the intensity of light at '*P*' is

$$I_{\text{max}} = a_1^2 + a_2^2 + 2a_1a_2 = (a_1 + a_2)^2$$

This is larger than the sum of the individual intensities of waves, i.e., $a_1^2 + a_2^2$. Suppose $a_1 = a_2 = a$, then $I_{\text{max}} = 4a^2$.

Case II: Condition for minimum intensity: For minimum intensity, $\cos\phi = -1$ in Eq. (6.8), then
$\phi = \pi, 3\pi, 5\pi, \ldots = (2n+1)\pi$, where $n = 1, 2, 3, \ldots$

or the path difference $(S_2P - S_1P) = (2n+1)\dfrac{\lambda}{2}$

Now the intensity of light at '*P*' is

$$I_{\text{min}} = a_1^2 + a_2^2 - 2a_1a_2 = (a_1 - a_2)^2.$$

Suppose $a_1 = a_2 = a$, then the intensity of light is zero.

The above conditions show that the intensity of light at bright points is $4a^2$ whereas at dark points the intensity is zero. According to conservation of energy, the intensity of light is redistributed as shown in Fig. 6.3.

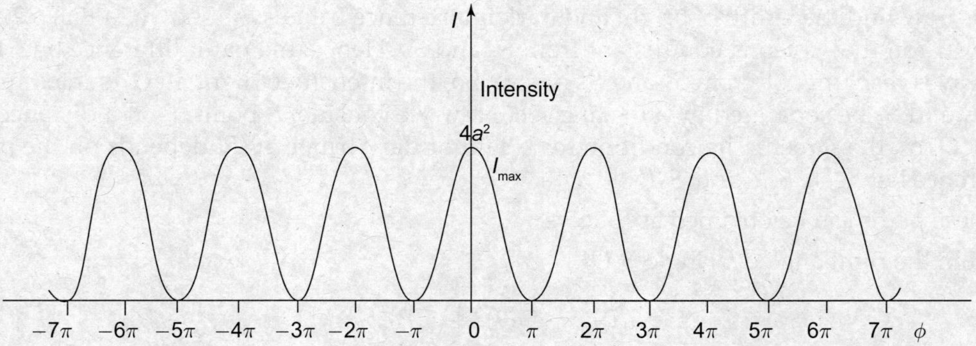

FIGURE 6.3 Distribution of intensity of light at different points due to interference

EXAMPLE 6.1

Two coherent sources of intensities $10\,\omega/\text{m}^2$ and $25\,\omega/\text{m}^2$ interfere to form fringes. Find the ratio of maximum intensity to minimum intensity.

Solution: We know that

Intensity, $I = a^2$ (square of amplitude)

So,
$$\frac{I_1}{I_2} = \frac{a_1^2}{a_2^2} = \frac{10}{25}$$

$$\Rightarrow \quad \frac{a_1}{a_2} = \frac{3.1623}{5}$$

$$\Rightarrow \quad a_1 = 0.6324\, a_2$$

The ratio of I_{max} and I_{min} is

$$\frac{I_{max}}{I_{min}} = \frac{(a_1 + a_2)^2}{(a_1 - a_2)^2}$$

$$= \frac{(0.6324\, a_2 + a_2)^2}{(0.6324\, a_2 - a_2)^2}$$

$$= \frac{\mathbf{2.6647}}{\mathbf{0.1351}} = \mathbf{19.729}$$

6.3.2 Theory of Interference Fringes

Let us now find the width of bright and dark interference fringes. As shown in Fig. 6.2, the point 'O' on the screen is equidistant from S_1 and S_2. Hence, the path difference between the waves reaching 'O' from S_1 and S_2 is zero. So, the intensity of light at 'O' is maximum. Let S_1 and S_2 be separated by a distance $2d$. Now we consider a point 'P' at a distance 'x' from 'O' on the screen. The condition for bright or dark fringe at 'P' depends on the path difference between S_1P and S_2P.

These paths can be obtained as follows:

From the right angled triangle S_1QP,

$$(S_1P)^2 = (S_1Q)^2 + (QP)^2$$

$$= D^2 + (x-d)^2 \quad (\because QP = OP - OQ = x - d)$$

Similarly from the right angle triangle S_2RP

$$(S_2P)^2 = (S_2R)^2 + (RP)^2$$

$$= D^2 + (x+d)^2 \quad (\because RP = RO + OP = d + x)$$

$$\because \quad (S_2P)^2 - (S_1P)^2 = (x+d)^2 - (x-d)^2 = 4xd$$

or $$(S_2P - S_1P)(S_2P + S_1P) = 4xd$$

$$(S_2P - S_1P) \cdot 2D = 4xd \qquad \text{(since } D \approx x \text{ or } d \text{ so } S_2P \approx S_1P \approx D\text{)}$$

or $$(S_2P - S_1P) = \frac{4xd}{2D} = \frac{2xd}{D} \tag{6.10}$$

Now we consider the following cases:

1. **Bright fringes:** Suppose the point P is bright, then the path difference $S_2P - S_1P = n\lambda$, when $n = 0, 1, 2, \ldots$.

   Substituting the above value in Eq. (6.10), we have

   $$n\lambda = \frac{2xd}{D} \quad \text{or} \quad x = \frac{n\lambda D}{2d}$$

   At 'O', a bright fringe is present and the next bright fringes are found when $n = 1, 2, 3, \ldots$.

   i.e., when

   $$n = 1, \quad x_1 = \frac{\lambda D}{2d}$$
   $$n = 2, \quad x_2 = \frac{2\lambda D}{2d}$$
   $$n = 3, \quad x_3 = \frac{3\lambda D}{2d}$$
   $$n = n, \quad x_n = \frac{n\lambda D}{2d}$$

   The distance between any two consecutive bright fringes is

   $$x_{n+1} - x_n = \frac{\lambda D}{2d} = \beta \tag{6.11}$$

2. **Dark fringes:** Suppose the point 'P' is dark, then the path difference $S_2P - S_1P = (2n+1)\frac{\lambda}{2}$, where $n = 1, 2, 3, \ldots$.

   Substituting the above value in Eq. (6.10), we have

   $$\frac{2xd}{D} = (2n+1)\frac{\lambda}{2} \quad \text{or} \quad x = \frac{(2n+1)}{4d}\lambda D$$

   Dark fringes are found at the following distances from 'O', when $n = 0, x_0 = \dfrac{\lambda D}{4d}$

   $$n = 1, \quad x_1 = \frac{3\lambda D}{4d}$$

$$n=2, \quad x_2=\frac{5\lambda D}{4d}$$

$$n=3, \quad x_3=\frac{7\lambda D}{4d}$$

$$n=n, \quad x_n=\frac{(2n+1)\lambda D}{4d}$$

This distance between any two consecutive dark fringes is

$$\boxed{x_n - x_{n-1} = \frac{2\lambda D}{4d} = \frac{\lambda D}{2d} = \beta} \tag{6.12}$$

From Eqs. (6.11) and (6.12), we know that the separation between any two consecutive bright or dark fringes is the same and is known as fringe width 'β'.

$$\therefore \quad \beta=\frac{\lambda D}{2d}$$

Note

If the entire arrangement of double slit experiment is dipped in water, then the fringes become narrower because in water wavelength decreases due to which fringe width decreases.

EXAMPLE 6.2

In a double slit experiment, a light of 5460 Å is exposed to slits which are 0.1 mm apart. The screen is placed 2 m away from the slits. What is the angular position of the 10th maximum and 1st minimum?

Solution: The given data are

Wavelength of light (λ) = 5460 Å = 5460×10^{-10} m

Separation between slits ($2d$) = 0.1 mm = 1×10^{-4} m

Distance of screen (D) = 2 m

Angular position of 10th maximum $\left(\phi_{\max 10}\right) = ?$

Angular position of 1st minimum $\left(\phi_{\min 1}\right) = ?$

Condition for maximum intensity is $x_n = \dfrac{n\lambda D}{2d}$

Distance of 10th maximum,

$$x_{\max 10} = \frac{10 \times 5460 \times 10^{-10} \times 2}{1 \times 10^{-4}} = 0.1092 \text{ m}$$

10th maximum angular position, $\phi_{\max 10} = \dfrac{x_{\max 10}}{D}$

$$\therefore \qquad \tan\theta_{\max 10} = \frac{0.1092}{2} = 0.0546 \text{ radians}$$
$$= 3°7'37''$$

Condition for minimum intensity is $x_n = \dfrac{(2n-1)\lambda D}{4d}$

Distance of 1st minimum,

$$x_{\min 1} = \frac{\lambda D}{4d} = \frac{\lambda D}{2\times 2d}$$
$$= \frac{5460\times 10^{-10}\times 2}{2\times 10^{-4}}$$
$$= 5460\times 10^{-6} \text{ m}$$

Angular position of 1st minimum is

$$\left(\tan\theta_{\min 1}\right) = \frac{x_{\min 1}}{D}$$
$$= \frac{5460\times 10^{-6}}{2}$$
$$= 0.00273 \text{ radians}$$
$$\mathbf{= 0.156° = 0°9'23''}$$

6.4 ▶ Coherence

To produce interference, coherent sources of light are required. When two or more electromagnetic waves are said to be coherent, they have same frequency and are in phase or maintain a constant phase difference between them. In general, the phase can vary from point to point or can change from time to time. So we have two different kinds of coherences, namely (i) temporal coherence and (ii) spatial coherence.

(i) **Temporal coherence:** This refers to the correlation between the field of a wave at a point at some time and the field at the same point at a later time. For example, at a point (x, y, z), let the fields at times t_1 and t_2 be $E(x, y, z, t_1)$ and $E(x, y, z, t_2)$. If the phase difference between the two fields is constant during the period normally covered by observations, then the wave is said to be temporally coherent. On the other hand, if the phase difference changes many times and also there exists irregularity during the short period of time, then it is said to be non-coherent.

(ii) **Spatial coherence:** If a constant phase difference exists at different points in space between the waves over a time t, then they are said to be spatially coherent. Temporal coherence refers to a single beam of light, whereas spatial coherence refers to the relationship between two separate beams of light.

6.5 ▶ Fresnel's Biprism Experiment

Fresnel's biprism is used to obtain coherent sources. The biprism produces two virtual images of a slit by refraction. A biprism consists of two identical prisms of very small vertex angle (nearly 1/2°) with their bases cemented together. In practice, a biprism is made from a single piece of glass by suitably grinding and polishing it. The two sections of the biprism make an obtuse angle of 179°.

A narrow slit S is illuminated by monochromatic light. The biprism is placed in front of the slit with the refracting edge parallel to the slit. Light from S passes through two halves of the biprism and the wave fronts are refracted symmetrically and separately. The refracted rays appear to come from the two virtual images S_1 and S_2 of the slit S. Since they are derived from the same original source, they act as coherent sources and they are formed in the plane of the slit. The interference pattern is obtained on the screen. It is observed in the shaded portion where the two beams overlap. It is observed through an eyepiece as shown in Fig. 6.4. The eyepiece is provided with a micrometer to measure the fringe width and separation between the virtual sources.

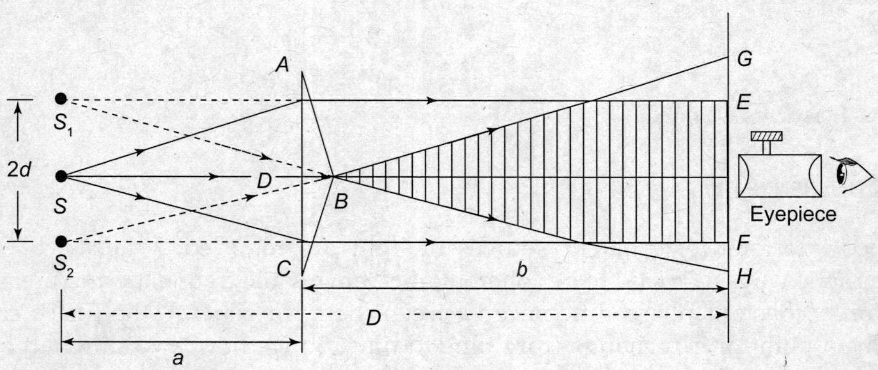

FIGURE 6.4 Fresnel's biprism

6.5.1 Determination of Wavelength of Light

In order to determine the wavelength of monochromatic light with the help of biprism fringes, we use the formula $\beta = \lambda D/2d$.

or
$$\lambda = \frac{\beta 2d}{D} \tag{6.13}$$

The value of fringe width β, the distance $2d$ between the virtual coherent sources S_1 and S_2 and the normal distance D of the plane of observation of the fringes from the slit should be measured after making a few adjustments in the apparatus.

The experiment is performed on a heavy metallic optical bench, about 2 m in length and supported on four leveling of eight crews at the base. The bench is provided with a scale on one side. The bench carries four upright stands for the adjustable slit, the biprism with a high power micrometer, Ramsdson's eyepiece and a convergent lens. These uprights are capable of movement along and perpendicular to the length of the bench and may be adjusted to any desired height.

Determination of distance between the source and the screen

The readings of the stands of the slit and eyepiece on the scale of the optical bench are taken. The difference between the two readings gives D, which is corrected for the bench error.

Determination of distance between the two virtual sources

The distance between the two virtual sources in the biprism experiment can be found by using the fact that for a prism with a very small refracting angle, the deviation δ produced in a ray is given by

$$\delta = (\mu - 1)\alpha \tag{6.14}$$

where μ is the refractive index of the material of the prism and δ is the refracting angle. From Fig. 6.5,

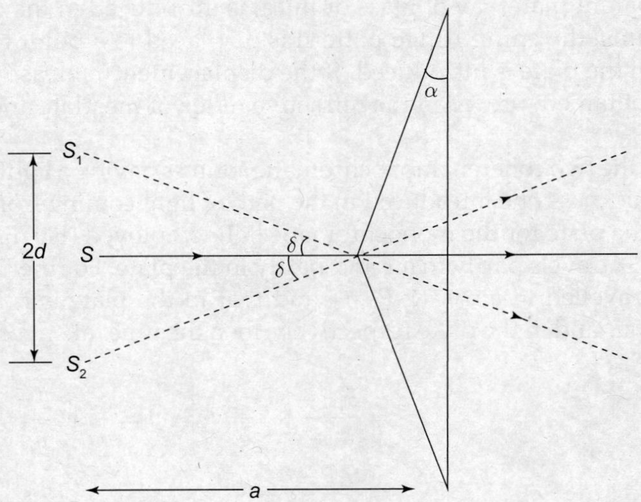

FIGURE 6.5 Angle of deviation from a biprism

$$\delta = \frac{d}{a} \tag{6.15}$$

Equations (6.14) and (6.15), we have

$$2d = 2a(\mu - 1)\alpha \tag{6.16}$$

where α is given in radians.

Measurement of the fringe width (β)

After obtaining the fringe, the vertical cross-wire of the eyepiece is set on a bright fringe on one side of the interference pattern. The readings of the micrometer screw are taken. Then the eyepiece is moved laterally so that the vertical cross-wires coincide with successive bright fringes and the corresponding readings are noted. From these readings, the fringe width (β) is found using the formula

$$\beta = \frac{\text{Distance moved}}{\text{Number of fringes crossed}}$$

Thus, substituting the values of D, d and fringe width, the wavelength can be calculated.

Note

The angular fringe width is defined as the angular separation between consecutive bright or dark fringes.

6.5.2 Displacement of Fringes

When a thin transparent plate, say of glass or mica, is introduced in the path of one of the two interfering beams, the entire fringe pattern is displaced to a point towards the beam in the path of which the plate is introduced. If the displacement is measured, the thickness of the plate can be obtained, provided the refractive index of the plate and the wavelength of light are known.

Let S_1 and S_2 be the two coherent monochromatic sources having a light of wave length λ. Let a thin plate of thickness t be introduced in the path of light coming from S_1. Let μ be the refractive index of the plate for the monochromatic light employed (Fig. 6.6).

Now light from S_1 travels partly in air and partly in the plate. For the light path from S_1 to P, the distance travelled in air is $(S_1P - t)$ and that in the plate is t. If c and v are the velocities of light in air and in the plate respectively, then the time taken for the journey from S_1 to P is given as

$$= \frac{S_1P - t}{c} + \frac{t}{v} \qquad \left[\because \mu = \frac{c}{v} \text{ or } \frac{1}{v} = \frac{\mu}{c}\right]$$

$$= \frac{S_1P - t}{c} + \frac{\mu t}{c}$$

$$= \frac{S_1P + (\mu - 1)t}{c}$$

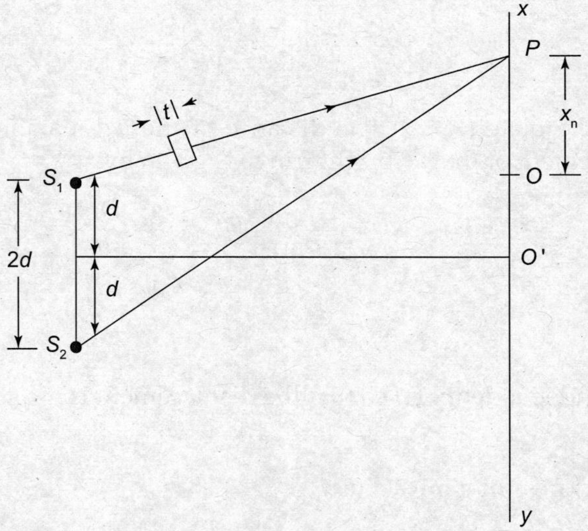

FIGURE 6.6 Displacement of fringes

It follows from this relation that the effective path of air from S_1 to P is $\left[S_1P+(\mu-1)t\right]$, i.e. the air path S_1P is increased by an amount $(\mu-1)t$ as a result of the introduction of the plate. Let O be the position of the central bright fringe in the absence of the plate, with the optical paths S_1O and S_2O being equal. After introducing the plate, the two optical paths become unequal. Therefore, the central fringe is shifted to O such that at O the two optical paths become equal. A similar argument applies for all the fringes. Now, at any point P, the effective path difference

$$= S_2P-\left[S_1P+(\mu-1)t\right]$$
$$= S_2P-S_1P-(\mu-1)t$$

Let $S_1S_2 = 2d$ and the distance of the screen from $S_1S_2 = D$ and $OP = X_n$ so that

$$S_2P-S_1P=\frac{2d}{D}x_n$$

Thus, the effective path difference at P *is*

$$\frac{2d}{D}x_n-(\mu-1)t$$

If the point P is the centre of the nth bright fringe, the effective path difference should be equal to $n\lambda$. That is,

$$\frac{2d}{D}x_n-(\mu-1)t=n\lambda$$

or $$x_n = \frac{D}{2d}[n\lambda + (\mu - 1)t]$$

In the absence of the plate ($t = 0$), the distance of the nth maximum from 0 is $\frac{n\lambda D}{2d}$. Therefore, the displacement of the nth bright fringe is given by

$$x_0 = \frac{D}{2d}[n\lambda + (\mu - 1)t] - \frac{Dn\lambda}{2d}$$

$$x_0 = \frac{D}{2d}(\mu - 1)t \tag{6.17}$$

This expression is independent of n so that the displacement is the same for all the bright fringes.

Determination of thickness of a mica sheet

Equation (6.17) shows that the introduction of the given mica plate in one of the interfering beams produces a shift x_0, i.e.,

$$x_0 = \frac{D}{2d}(\mu - 1)t$$

Thus, $$\boxed{t = \frac{x_0(2d)}{D(\mu - 1)}} \tag{6.18}$$

6.6 ▶ Introduction to Thin Films

When a film of oil spread over the surface of water is illuminated by white light, beautiful colours are seen. This is due to interference between the light waves reflected from the film, and between the light waves transmitted through the film.

A thin film may be a thin sheet of transparent material such as glass, mica, an air film enclosed between two transparent sheets or a soap bubble. When light is incident on such film, it is partially reflected from the upper surface and a major portion is transmitted into the film. Again, a small part of transmitted component is reflected back into the film by lower surface and the rest of it is transmitted out of the film. A small portion of light thus gets partially reflected in succession several times within the film as shown in Fig. 6.7. Therefore, interference in thin films is due to the division of amplitude. Newton and Hook first observed the thin film interference. However, Thomas Young gave the correct explanation of the phenomenon.

Note

If the angle of biprism is increased, the distance between two virtual sources is increased and fringe width is reduced. So the fringes will not be separately visible and ultimately disappear.

6.7 ▶ Interference in Parallel Thin Films

Brilliant colours are often exhibited when a beam of white light from an extended source is reflected from the skin of soap bubble or from a thin layer of or floating on water or from a thin wedge-shaped air film between two glass plates. The explanation of the origin of this colour phenomenon was given by Dr Youngs in 1802 in terms of the interference of light waves reflected from the upper and the lower surfaces of the thin film.

A ray of monochromatic light *SA* be incident at an angle *i* on a parallel-sided transparent thin film of thickness *t* and refractive index $(\mu > 1)$.

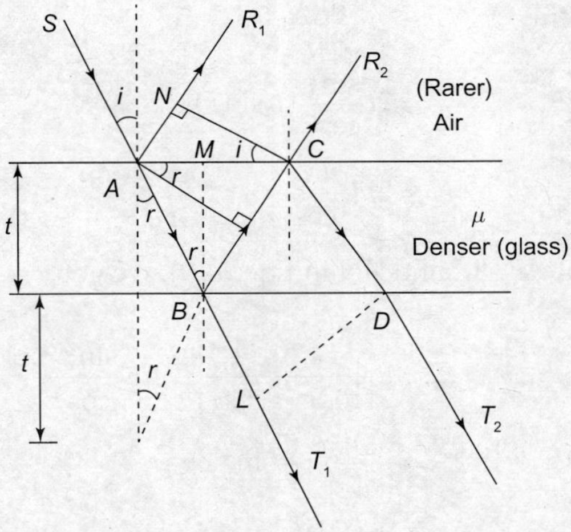

FIGURE 6.7 Interference in thin films

At *A*, it is partly reflected along AR_1 and partly refracted along *AB* at angle *r*. At *B*, it is again partly reflected along *BC* and partly refracted along BT_1. Similar reflections and refractions occur at *C*, *D*, ... as shown in Fig. 6.7. Thus, we get a set of parallel reflected rays AR_1, CR_2, ... and a set of transmitted rays BT_1, DT_2, ... Let us first consider the reflected rays only.

At each of the points *A*, *B*, *C*, *D*, ..., only a small part of light is reflected and the rest is refracted. Therefore, the rays AR_1 and CR_2 each having undergone one reflection, have almost equal intensities, and the rest have rapidly decreasing intensities and can be ignored. The rays AR_1 and CR_2 being derived from the same incident ray, are coherent and in a position to interfere. Let us first calculate the path difference between them.

Let *CN* and *BM* be perpendicular to AR_1 and *AC*. As the path of the rays AR_1 and CR_2 beyond *CN* is equal, the path difference between them is

$$\Delta = \text{path } ABC \text{ in film} - \text{path } AN \text{ in air}$$

$$= \mu(AB + BC) - AN \tag{6.19}$$

From Fig. 6.7,

$$\begin{aligned} AB = BC &= \frac{BM}{\cos r} \\ &= \frac{t}{\cos r} \\ AN &= AC \sin i \\ &= (AM + MC)\sin i \\ &= (BM \tan r + BM \tan r)\sin i \\ &= 2t\frac{\sin r}{\cos r}\sin i \\ &= 2t\frac{\sin r}{\cos r}(\mu \sin r) \\ AN &= 2\mu t\frac{\sin^2 r}{\cos r} \end{aligned}$$

Substituting the values of *AB*, *BC* and *AN* in Eq. (6.19), we get

$$\begin{aligned} \Delta &= \mu\left[\frac{t}{\cos r} + \frac{t}{\cos r}\right] - 2\mu t\frac{\sin^2 r}{\cos r} \\ &= \frac{2\mu t}{\cos r}\left(1 - \sin^2 r\right) \\ &= 2\mu t \cos r \end{aligned}$$

The ray AR_1 having suffered a reflection at the surface of a denser medium undergoes a phase change of π, which is equivalent to a path difference of λ. Hence, the effective path difference between AR_1 and CR_2 is $2\mu t \cos r - (\lambda/2)$.

6.7.1 Interference Due to Reflected Light

The two rays will reinforce each other if the path difference between them is an integral multiple of conditions of maxima and minima in reflected light. The two rays will reinforce each other if the path difference between them is an integral multiple of λ, i.e., for maxima

$$2\mu t \cos r - \frac{\lambda}{2} = n\lambda$$

or

$$\boxed{\mathbf{2\mu t \cos r = (2n+1)\frac{\lambda}{2}}} \tag{6.20}$$

Again the two rays will destroy each other if the path difference between them is an odd multiple of $(\lambda/2)$, i.e. for minima

$$2\mu t\cos r-\frac{\lambda}{2}=(2n-1)\frac{\lambda}{2}$$

$$\boxed{2\mu t\cos r=n\lambda} \tag{6.21}$$

6.7.2 Interference Due to Transmitted Light

Similarly, the path difference between the transmitted rays BT_1 and DT_2 is given by

$$\begin{aligned}\Delta &= \mu(BC+CD)-BL\\ &= 2\mu t\cos r\end{aligned}$$

In this case, there is no phase change due to reflection B or C because in either case the light is travelling from denser to rarer medium. Hence, the effective path difference between BT_1 and DT_2 is also $2\mu t\cos r$.

Conditions for maxima and minima in transmitted light

The two rays BT_1 and DT_2 will reinforce each other if

$$2\mu t\cos r=n\lambda \text{ (condition of maxima)} \tag{6.22}$$

where $n=1,2,3,\ldots$
Again the two rays will destroy each other if

$$2\mu t\cos r=(2n+1)\frac{\lambda}{2} \text{ (condition of minima)} \tag{6.23}$$

where $n=0,1,2,\ldots$
A comparison of equations shows that the conditions for maxima and minima in the reflected light are just the reverse of those in the transmitted light. Hence, the film, which appears bright in reflected light, will appear dark in transmitted light and vice versa. In other words, the appearances in two cases are complementary to each other.

Note

According to Stoke's law, when a light is reflected at the surface of an optically denser medium, it suffers a phase change of π, i.e a path change of $\lambda/2$.

EXAMPLE 6.3

A soap film of refractive index 1.33 and thickness 5000 Å is exposed to white light. What wavelengths in the visible region are reflected?

Solution: The given data are

$$\text{Refractive index of soap } (\mu)=1.33$$

$$\text{Thickness of soap film } (t)=5000 \text{ Å}=5000\times10^{-10} \text{ m}$$

The incident light reflected on both surfaces of film combine to produce interference. So, the condition for constructive interference is used for reflection.

i.e., $$2\mu t\cos r - \frac{\lambda}{2} = n\lambda$$

For maximum reflection, $$r = 1$$

then $$2\mu t - \frac{\lambda}{2} = n\lambda$$

or $$\lambda = \frac{4\mu t}{(2n+1)}$$

Substituting the values,

$$\lambda = \frac{4\times 1.33\times 5000\times 10^{-10}}{(2n+1)}$$

$$= \frac{26,600\times 10^{-10}}{(2n+1)}\text{m}$$

For

$$n=0,\quad \lambda_1 = 26,600\times 10^{-10}\text{ m (IR region)}$$

$$n=1,\quad \lambda_2 = \frac{26,600\times 10^{-10}}{3}\text{m} = 8666.67\times 10^{-10}\text{ m (IR region)}$$

$$n=2,\quad \lambda_3 = \frac{26,600\times 10^{-10}}{5}\text{m} = 5320\times 10^{-10}\text{ m (Visible region)}$$

$$n=3,\quad \lambda_4 = \frac{26,600\times 10^{-10}}{7}\text{m} = 3800\times 10^{-10}\text{ m (UV region)}$$

From the above wavelengths, 5320 Å lies in the visible region. This wavelength of white light is reflected maximum.

6.8 ▶ Interference in Wedge-Shaped Thin Films

6.8.1 Interference Pattern

Consider a thin film in the shape of a wedge whose sides form a small angle θ and illuminated with plane monochromatic light waves. The directly reflected wave BR and the internally reflected wave B_1R_1 originate from the same incident wave propagating along AB and they are capable of producing servable interference effect. The two interfering waves do not reach the eye along parallel paths but as shown in the figure and they appear to diverge from a point O in the rear of the film. Destructive or constructive interference, therefore, occurs at the point O, which is however, virtual. It the two interfering waves BR and B_1R_1 fall on lens, they will cross each other at a real point O. Consequently, real reinforcement or destructive interference would occur at O.

The optical path difference, between the waves BR and B_1R_1, is expressed by

$$\Delta = \mu(BC + CB_1) - BD$$
$$= \mu(BE + EC + CB_1) - BD$$

From Fig. 6.8,

$$BD = \mu BE$$

Thus,

$$\delta = EC + CB_1 \tag{6.24}$$

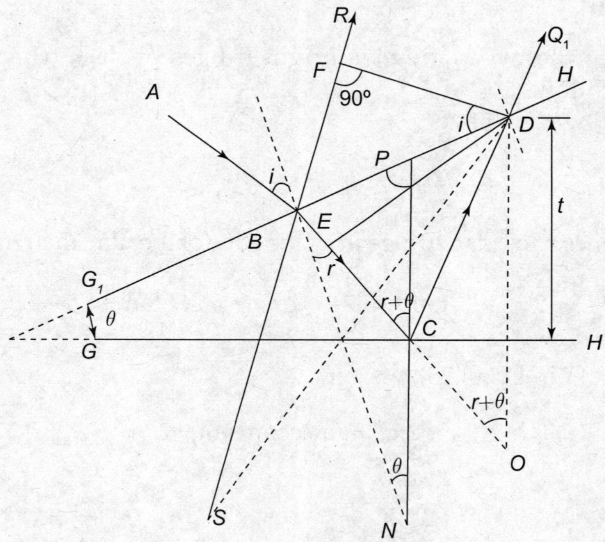

FIGURE 6.8 Interference in wedged-shaped film

In Fig. 6.8, BN and CN are respective normal to the upper and lower surfaces of the film; hence, we get $\angle CNB = \theta$ and $\angle PCB = \theta + r = \angle CBN + \angle CNB$ and $\angle PCB_1 = \angle PCB = \theta + r$.

Also, $\angle CB_1O = \angle PCB_1 = (\theta + r) = \angle B_1OC$. Thus, B_1OC is an isosceles triangle, with $CB_1 = CO$. The path difference reduces to

$$\Delta = \mu(EC + CO) = \mu EO$$
$$= \mu \beta_1 O \cos(\theta + r)$$
$$\Delta = 2\mu t \cos(\theta + r) \tag{6.25}$$

where t stands for the thickness of the film at the point B_1. The path difference D thus varies because of both changing thickness as well as changing angle of incidence provided the broad light source be at finite distance from the film. Now taking into account the change of phase of the wave due to reflection at B, the wave BR and B_1R_1 will interfere constructively when

$$\boxed{2\mu t \cos(\theta + r) = (2n+1)\frac{\lambda}{2}} \tag{6.26}$$

where $n = 0,1,2,3,...$
and they will interfere destructively when

$$\boxed{2\mu t \cos(\theta + r) = 2n\frac{\lambda}{2}} \tag{6.27}$$

As we proceed along the film in the direction in which its thickness increases, alternate dark and bright bands parallel to the edge of the film is observed. Since each band is the locus of constant thickness of the film, these fringes are called fringes of constant thickness.

6.8.2 Fringe Width

The distance between the two successive bright fringes or dark fringes or fringe width may be obtained as follows:

For the nth dark fringe, we have

$$2\mu t \cos(\theta + r) = 2n\frac{\lambda}{2} \tag{6.28}$$

Let X_n be the distance of nth dark fringe from the edge of the film, then from Fig. 6.9, we have

$$\tan\theta = \frac{t}{x_n} \quad \text{or} \quad t = X_n \tan\theta$$

Putting this value of t in Eq. (6.28), we get

$$2\mu x_n \tan\theta \cos(\theta + r) = n\lambda \tag{6.29}$$

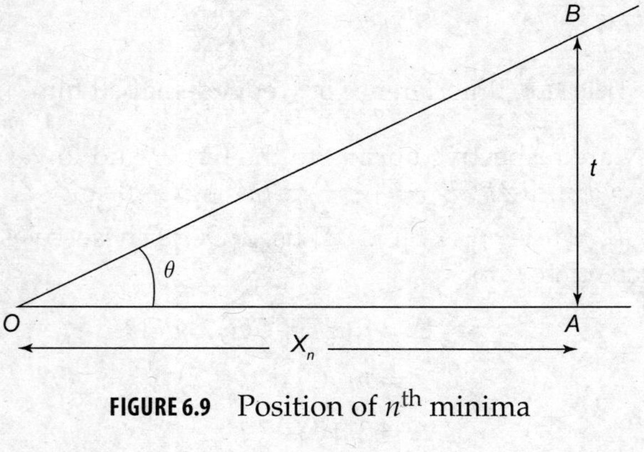

FIGURE 6.9 Position of nth minima

Similarly, if X_{n+1} be the distance of dark fringe, then

$$2\mu x_{n+1} \tan\theta \cos(\theta + r) = (n+1)\lambda \tag{6.30}$$

Subtracting Eq. (6.29) from Eq. (6.30), we have

$$\beta = x_{n+1} - x_n = \frac{\lambda}{2\mu \tan\theta \cos(\theta + r)} \tag{6.31}$$

for normal incidence $i=r=0$, and $\cos(\theta+r)=\cos\theta$

$$\beta=\frac{\lambda}{2\mu\tan\theta\cos\theta}=\frac{\lambda}{2\mu\sin\theta}$$

for very small value of θ

$$\sin\theta\cong\theta$$

$$\beta=\frac{\lambda}{2\mu\theta}$$

For refractive index of unity

$$(\beta)=\frac{\lambda}{2\theta} \tag{6.32}$$

6.8.3 Wedge Angle

The wedge angle may be determined experimentally. The position of two dark fringes located at distances x_1 and x_2 from the apex is noted. Let t_1 and t_2 be the thicknesses of the wedges at x_1 and x_2.

The condition for dark fringe at t_1 and t_2 may be given as

$$2\mu t_1=m\lambda$$

but $$t_1=x_1\tan\theta=x_1\theta$$

$\therefore$ $$2\mu x_1\theta=m\lambda \tag{6.33}$$

Similarly

$$2\mu x_2\theta=(m+N)\lambda \tag{6.34}$$

where N is the number of dark fringes between at x_1 and x_2. Subtracting Eqs. (6.33) and (6.34),

$$2\mu(x_2-x_1)\theta=(N)\lambda$$

$\therefore$ $$\boxed{\theta=\frac{(N)\lambda}{2\mu(x_2-x_1)}} \tag{6.35}$$

6.8.4 Spacer Thickness

The spacer thickness t_3 may be determined experimentally.
From Fig. 6.10,

$$t_3=x_3\tan\theta=x_3\theta$$

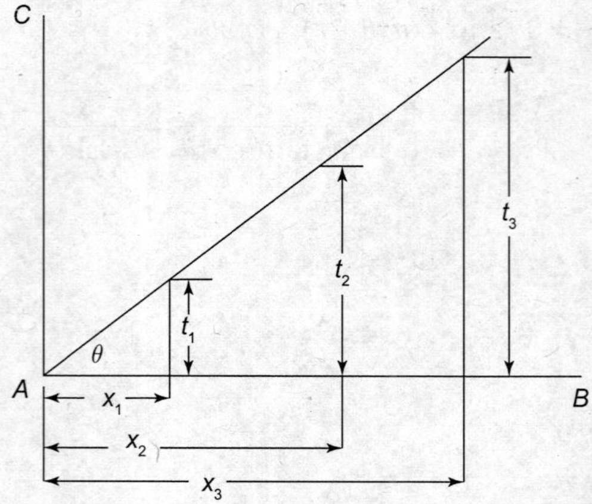

FIGURE 6.10 Determination of wedge angle and thickness of thin object

$$\therefore \qquad t_3 = \frac{x_3 N\lambda}{2(x_2 - x_1)}$$

Note

The interference fringes formed by reflection from well separated plane parallel surfaces which have a nearly normal orientation are called Haidinger fringes. However, when the film is wedge shaped, i.e. of varying thickness, the fringes by reflected light are almost straight and of equal thickness. These are called Fizeau's fringes.

6.8.5 Newton's Rings

As shown in Fig. 6.11, a plano convex lens (L) having long focal length, f ($\approx$100 cm) is placed with its convex surface on a plane glass plate (G). A gradually increasing thickness of air film will be formed between plane glass plate and convex surface of plano convex lens. The thickness of air film will be zero at the point of contact and symmetrically increases as we go radially from the point of contact. A monochromatic light of wavelength 'λ' is allowed to fall normally on the lens with the help of a glass plate 'P' kept at 45° to the incident monochromatic beam. A part of the incident light rays is reflected up at the convex surface of the lens and the remaining light is transmitted through the air film. Again a part of this transmitted light is reflected at on the top surface of the glass plate (G). Both the reflected rays combine to produce an interference pattern in the form of alternate bright and dark concentric circular rings, known as Newton's rings, because Newton first demonstrated and showed these rings. The rings are circular because the air film has circular symmetry. These rings can be seen through the travelling microscope M.

Theory: Newton's rings by reflected light

Let us discuss the interference condition for bright and dark fringes and also the spacing between Newton's rings. To obtain the relation between the radius of Newton's ring and the radius of curvature of the lens, consider Fig. 6.12.

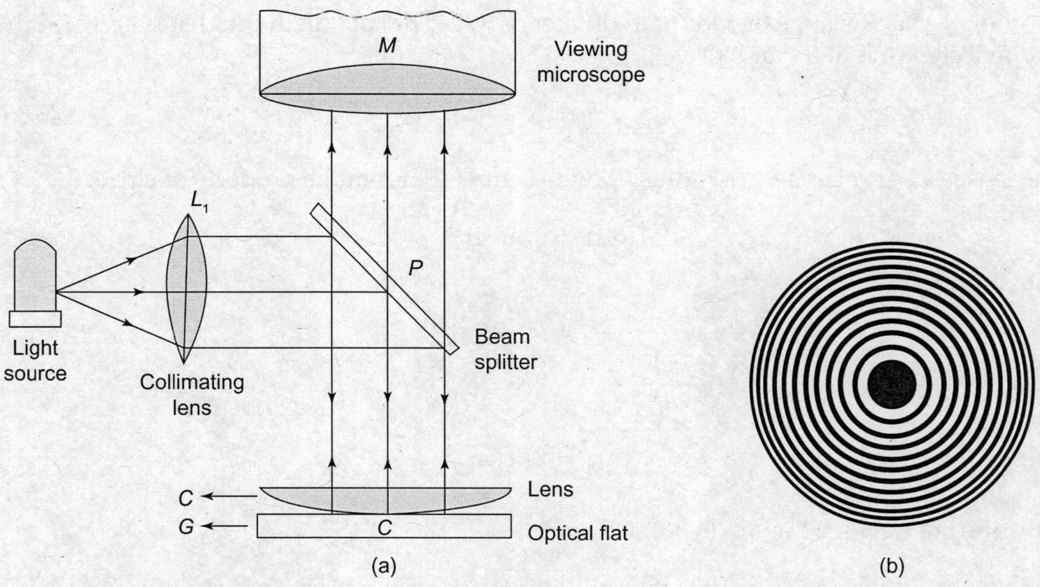

FIGURE 6.11 (a) Experimental arrangement of Newton's rings and (b) Newton's rings as seen through a microscope

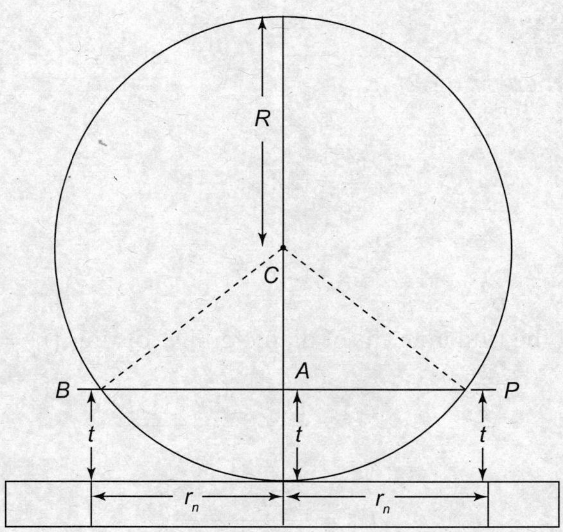

FIGURE 6.12 Calculation of film thickness from radius of curvature of lens

Let the lens be in contact with glass plate at O and let the radius of curvature of the lens be R. Let a vertical light ray be partially reflected and partially transmitted at 'P'. The transmitted light is again reflected at Q on glass plate G. Let the thickness of air film at P be $PQ(=t)$ and the radius of Newton's ring at Q be r_n. The ray reflected at Q suffers an additional phase change of π or path difference $\lambda/2$. The total path difference between the two reflected rays at 'P' and 'Q' is

$$\delta = 2t + \frac{\lambda}{2} \tag{6.36}$$

The lens is a part of sphere of radius 'R' with centre 'C'. From the property of circle,

$$BA \times AP = OA \times AD$$

Substituting the values

$$r_n \times r_n = t \times (2R - t)$$

$$= 2Rt - t^2 \approx 2Rt$$

$\therefore$
$$r_n^2 = 2Rt \text{ or } t = \frac{r_n^2}{2R} \tag{6.37}$$

Substituting Eq. (6.37) in Eq. (6.36), we have

$$\delta = 2\frac{r_n^2}{2R} + \frac{\lambda}{2} = \frac{r_n^2}{R} + \frac{\lambda}{2} \tag{6.38}$$

If the path difference is $n\lambda$, then constructive interference takes place and a bright n^{th} Newton's ring will be formed.

i.e., $\delta = \frac{r_n^2}{R} + \frac{\lambda}{2} = n\lambda$, where $n = 1, 2, 3, \ldots$

or
$$\frac{r_n^2}{R} = (2n-1)\frac{\lambda}{2}$$

or
$$r_n^2 = (2n-1)\frac{\lambda R}{2} \tag{6.39}$$

Expressing radius of the ring in terms of diameter 'D_n' of ring $(r_n = D_n/2)$, we have

$$r_n^2 = \frac{D_n^2}{4} = \frac{(2n-1)\lambda R}{2}$$

or
$$D_n^2 = \frac{4(2n-1)\lambda R}{2}$$

or
$$\boxed{\boldsymbol{D_n = \sqrt{2(2n-1)\lambda R}}} \quad \text{[for bright ring]} \tag{6.40}$$

Similarly for the nth dark ring, the path difference is equal to $(2n-1)\frac{\lambda}{2}$.

∴ From Eq. (2.38),

$$\delta = \frac{r_n^2}{R} + \frac{\lambda}{2} = (2n-1)\frac{\lambda}{2}$$

or

$$\frac{r_n^2}{R} = (2n-1)\frac{\lambda}{2} - \frac{\lambda}{2}$$

∴

$$r_n^2 = n\lambda R \tag{6.41}$$

Expressing radius of ring in terms of diameter of ring $[r_n = D_n / 2]$, we have

$$r_n^2 = \frac{D_n^2}{4} = n\lambda R$$

$$D_n^2 = 4n\lambda R$$

or

$$\boxed{\boldsymbol{D_n = \sqrt{4n\lambda R}}} \quad \text{[for dark ring]} \tag{6.42}$$

At the point of contact of lens and glass plate [at 'O'], the path difference is zero and phase change π takes place due to reflection on glass plate; hence, dark spot will be formed at the centre of ring system.

From Eq. (6.42), we know the diameter of rings is proportional to square root of the order of rings (i.e., $\sqrt{n}$). Hence, the spacing between consecutive rings goes on decreasing with an increase of order of rings.

The theory of Newton's rings can be used to determine the wavelength of monochromatic light and the refractive index of a given liquid.

Application: Determination of the wavelength of monochromatic light

Using the experimental set-up shown in Fig. 6.11(a), the diameters of various dark rings are measured using travelling microscope. To measure the diameter of rings, first the microscope is adjusted so that the centre of cross-wires coincides with the centre of the ring system. The centre of the cross-wires is moved to one direction (say left) so that the vertical cross-wire is tangential to the 21st dark ring (say). Now the reading is noted from the horizontal scale of travelling microscope. The microscope is moved towards right such that the vertical cross-wire is tangential to the 18th ring, then the reading is noted. By moving the microscope in the same direction, readings are noted for every 3 rings. Similar readings are noted for these rings on the right side also. The difference of readings on the left side and right side gives the diameter of the various rings.

Let the diameter of the nth and $(n+m)^{th}$ dark rings be D_n and D_{n+m}. Then

$$D_n^2 = 4n\lambda R \text{ and } D_{n+m}^2 = 4(n+m)\lambda R$$

$\therefore$ $$D_{n+m}^2 - D_n^2 = 4m\lambda R$$

or $$\lambda = \frac{D_{n+m}^2 - D_n^2}{4mR} \quad (6.43)$$

From Eq. (6.43), the wavelength of the light used can be determined. Here, R is the radius of curvature of the lens. Here, $(n + m) = 21$ and $n = 18$, therefore $m = 3$.

Application: Determination of the refractive index of a given liquid

The experimental set-up shown in Fig. 6.11(a) is used to find the refractive index of a given liquid. To find the refractive index of a liquid, the plane glass plate and plano convex lens set-up is placed in a small metal container. The diameter of nth and $(n+m)^{th}$ dark rings is determined, when there is air between planoconvex lens and plane glass plate.

Then we have,

$$D_{n+m}^2 - D_n^2 = 4m\lambda R \quad (6.44)$$

Now, the given liquid whose refractive index (μ) is to be measured is introduced into the space between plano-convex lens and plane glass plate without disturbing the experimental set-up. The optical path between plano convex lens and plane glass, plate is not 't' but 'μt'; hence, the diameters of Newton's rings are changed. Now the diameters of nth and $(n+m)^{th}$ dark rings are measured.

Then, $$D_{n+m}'^2 = D_n'^2 = \frac{4m\lambda R}{\mu} \quad (6.45)$$

where D_n' and D_{n+m}' represent the diameters of nth and $(n+m)^{th}$ dark rings with liquid between piano convex lens and plane glass plate.

From Eq. (6.44) and (6.45), we have

$$\boxed{\mu = \frac{D_{n+m}^2 - D_n^2}{D_{n+m}'^2 - D_n'^2}} \quad (6.46)$$

Using Eq. (6.46), the refractive index of a given liquid can be determined.

EXAMPLE 6.4

In a Newton's rings experiment, the diameter of the 15th ring was found to be 0.59 cm and that of the 5th ring is 0.336 cm. If the radius of curvature of the lens is 100 cm, find the wavelength of the light.

Solution: The given data are

Diameter of Newton's 15th ring $(D_{15}) = 0.59$ cm $= 0.59 \times 10^{-2}$ m

Diameter of Newton's 5th ring $(D_5) = 0.336$ cm $= 0.336 \times 10^{-2}$ m

Radius of curvature of lens $(R) = 100$ cm $= 1$ m

Wave length of light $(\lambda) = ?$

$$\begin{aligned}\lambda &= \frac{D_{n+m}^2 - D_n^2}{4mR} \\ &= \frac{\left(0.59\times10^{-2}\right)^2 - \left(0.336\times10^{-2}\right)^2}{4\times10\times1} \qquad \begin{bmatrix}\because \text{Here}, n+m=15 \\ \text{and} \quad n=5 \\ \Rightarrow \quad m=10\end{bmatrix} \\ &= \frac{0.3481\times10^{-4} - 0.112896\times10^{-4}}{40} \\ &= \frac{0.235204\times10^{-4}}{40} = 0.00588\times10^{-4} \text{ m} \\ &= \mathbf{5880\times10^{-10}\ m = 5880\ Å}\end{aligned}$$

EXAMPLE 6.5

Newton's rings are observed in the reflected light of wavelength 5900 Å. The diameter of 10th dark ring is 0.5 cm. Find the radius of curvature of the lens used.

Solution: The given data are

Wavelength of light $(\lambda) = 5900$ Å $= 5900 \times 10^{-10}$ m

Diameter of 10th Newton's dark ring $(D_{10}) = 0.5$ cm $= 0.5 \times 10^{-2}$ m

Radius of curvature of lens $(R) = ?$

Formula is $D_n^2 = 4n\lambda R$

or

$$R = \frac{D_n^2}{4n\lambda} = \frac{\left(0.5\times10^{-2}\right)^2}{4\times10\times5900\times10^{-10}}$$

$$= \mathbf{\frac{0.25\times10^{-4}}{236\times10^{-7}} = 1.059\ m}$$

6.9 ▶ Applications of Thin Films

6.9.1 Colours in Thin Films

The colours observed in this film in reflected light will be complementary to those observed in transmitted light. This is because the conditions for maxima and minima in the reflected light are just the reverse of those in the transmitted light.

Blackness of an excessively thin film in reflected light

In an extremely thin film when seen in the white light appears to be completely dark in the reflected light. The reason is that if the film is extremely thin, i.e. $t \to 0$, the effective path difference in the reflected part is

$$2\mu t \cos(\theta + r) + \frac{\lambda}{2} \simeq \frac{\lambda}{2}$$

and thus, condition of minima is satisfied for each wavelength of white light. Thus, the film appears dark in the reflected light.

6.9.2 Necessity of Broad Source

When a thin film is illuminated with light from a point source and it is observed with a lens of small aperture such as the rays that can enter the eye are confined to a small range of directions. Thus, the field of view is narrow. But if we employ an extended source, the light from every point of the source after reflection from the film enters the eye in a separate direction. If the accommodation is relaxed, the waves in each pair are brought together at a different point on the retina. Thus, the field of view is wide and the interference effects can be observed over the entire film.

6.9.3 Testing of Flatness of Surface

The flatness of surface may be quickly inspected by putting an optical flat on the surface at an angle and illuminated by monochromatic light. The air wedge will be straight and equidistant band if the surface is smooth. If the fringes are curved towards the contact edge, the surface will be concave and if the fringes are curved away, it will be convex.

6.9.4 Colourful Hoardings

A change in angle of incident of rays causes corresponding change in path difference. The optical path difference decreases with increase in the angle of incidence. Therefore, as the inclination of film is changed, it will appear in different colours. This principle is used in display of hoardings.

6.9.5 Anti-reflecting Coating

Interference-based coatings were invented in November 1935 by Alexander Smakula, who was working for the Carl Zeiss optics company. Anti-reflection coatings were a German military secret until the early stages of World War II. Katharine Burr Blodgett and Irving Langmuir developed organic anti-reflection coatings in the late 1930.

An anti-reflective or anti-reflection (AR) coating is a type of optical coating applied to the surface of lenses and other optical devices to reduce reflection. This improves the efficiency of the system since less light is lost. In complex systems such as a telescope, the reduction in reflections also improves the contrast of the image by elimination of stray light. This is especially important in planetary astronomy. In other applications, the primary benefit is the elimination of the reflection itself, such as a coating on eyeglass lenses that makes the eyes of the wearer more visible to others, or a coating to reduce the glint from a covert viewer's binoculars or telescopic sight.

Many coatings consist of transparent thin film structures with alternating layers of contrasting refractive index. Layer thicknesses are chosen to produce destructive interference in the beams reflected from the interfaces, and constructive interference in the corresponding transmitted beams. This makes the structure's performance change with wavelength and incident angle, so that colour effects often appear at oblique angles. A wavelength range must be specified when designing or ordering such coatings, but good performance can often be achieved for a relatively wide range of frequencies: usually a choice of IR, visible or UV is offered.

Note

The radius R of planoconvex lens can be obtained with the help of spherometer using the following formula:

$$R = \frac{\ell^2}{6h} + \frac{h}{2}$$

EXAMPLE 6.6

Light waves of wavelengths 650 nm and 500 nm produce interference fringes on a screen at a distance of 1 m from a double slit of separation 0.5 mm. Find the least distance of a point from the central maximum where the bright fringe due to both sources coincide.

Solution: The given data are

Wavelength of first source $(\lambda_1) = 650$ nm

Wavelength of second source $(\lambda_2) = 500$ nm

Distance of screen $(D) = 1$ m

Separation between slits $(2d) = 0.5 \text{ mm} = 0.5 \times 10^{-3}$ m

Distance from central maximum where bright fringes due to both sources coincide $(x) = ?$

Let us consider the nth bright fringe of the first source and the mth bright fringes of the second source coincide at a distance of 'x' from central maximum.

Then,

$$x = \frac{n\lambda_1 D}{2d} = \frac{m\lambda_2 D}{2d}$$

or, $$n\lambda_1 = m\lambda_2$$

or, $$\frac{n}{m} = \frac{\lambda_2}{\lambda_1} = \frac{500}{650}$$
$$= \frac{10}{13}$$

Therefore, the 10th bright fringe due to the first source coincides with the 13th bright fringe due to the second source.

Also, $$x = \frac{\lambda_1}{2}$$
$$= \frac{10 \times 650 \times 10^{-9} \times 1}{0.5 \times 10^{-3}}$$
$$= 0.013 \text{ m} = 13 \text{ mm}$$

Therefore, the bright fringes of both sources will coincide at a distance of 13 mm from central maximum.

EXAMPLE 6.7

Calculate the thickness of air film at the 10th dark ring in a Newton's rings system, viewed normally by a reflected light of wavelength 500 nm. The diameter of the 10th dark ring is 2 mm.

Solution: The given data are

Wavelength of light $(\lambda) = 500 \text{ nm} = 500 \times 10^{-9}$ m

Number of the dark ring viewed $(n) = 10$

Diameter of 10th dark ring $(D_{10}) = 2 \text{ mm} = 2 \times 10^{-3}$ m

Radius of 10th dark ring $(r_{10}) = \frac{D_{10}}{2} = 1 \times 10^{-3}$ m

Thickness of air film $(t) = ?$

Condition for dark ring is

$$D_n = \sqrt{4n\lambda R}$$

or $$D_n^2 = 4n\lambda R$$

or $$R = \frac{d_n^2}{4n\lambda} = \frac{\left(2\times10^{-3}\right)^2}{4\times10\times500\times10^{-9}}$$

$$= \frac{4\times10^{-6}}{40\times500\times10^{-9}} = 0.2\text{ m}$$

Also, $$t = \frac{r_n^2}{2R} = \frac{\left(10^{-3}\right)^2}{2\times0.2}$$

$$= 2.5\times10^{-6}\text{ m}$$

$$= \mathbf{2.5\ \mu m}$$

6.10 ▶ Michelson's Interferometer

In Michelson's interferometer, the amplitude of light beam coming from an extended source is divided into two equal parts of equal intensity by partial reflection and transmission. These two beams travel in two different directions at right angles to each other and then brought together after reflecting from two different mirrors, thus producing interference fringes.

Experimental arrangement

As shown in Fig. 6.13, Michelson's interferometer consists of two glass plates G_1 and G_2. Glass plate G_1 is half silvered on its back surface inclined at an angle of 45° to the direction of incident light. Glass plate G_2 is made up of same material as that of G_1 and having same thickness, also mounted parallel to G_1. Two plane mirrors M_1 and M_2 are polished silvered on their front surfaces and mounted at right angles to each other. The mirror M_1 is mounted on a carriage and can be moved forward or backward with the help of a fine screw of uniform pitch, while mirror M_2 is fixed. Levelling screws at the back of mirrors M_1 and M_2 are also provided to change their planes according to need.

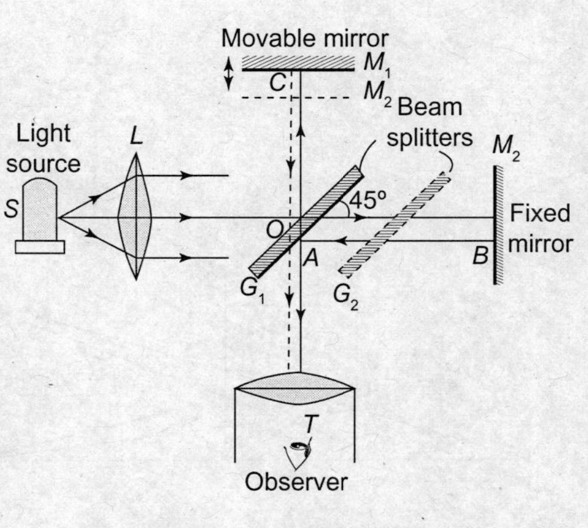

FIGURE 6.13 Michelson's interferometer

Working

A ray of light from extended monochromatic source of light S (as shown in Fig. 6.13) is incident on glass plate G_1 and is partially reflected from its back silvered surface. After being transmitted from G_1, it enters the telescope T. While the transmitted part of the beam moves towards M_2, it gets reflected back towards G_1. At G_1, it is reflected to enter the telescope T. These two beams entering the telescope T are coherent and thus produces interference fringes.

Function of glass plate G_2

The glass plate G_2 is called compensating glass plate. The beam (1) crosses the plate G_1 thrice before entering the telescope T while in the absence of glass plate G_2, beam (2) crosses the glass plate G_1 only once.

Thus, to make the path travelled by two beams equal, G_2 is introduced in the path of beam (2). Now both the beams travel equal paths before entering telescope T, interfere with each other and give rise to interference pattern.

6.10.1 Types of Fringes

Case 1: Circular fringes

The types of fringes in Michelson's interferometer depend on the inclination of M_1 and M_2. Let M_2' be the image of M_2 such that $OM_2 = OM_2'$. Now, the interference pattern will be formed due to the light reflected from the mirror surface M_1 (real) and M_2' (virtual).

The situation will be same as if the source S has two virtual images S_1 and S_2. The rays from S_1 and S_2 will reach telescope T along paths S_1AT and S_2BT. The point T will be dark or bright according to the path difference $S_1T - S_2T (= S_1P)$ is odd or even multiple of $\lambda/2$ (Fig. 6.14).

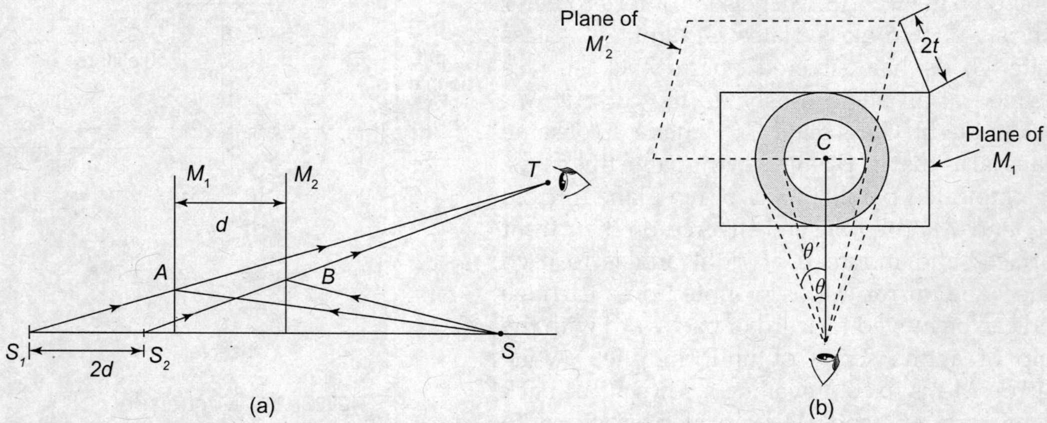

FIGURE 6.14 (a) Formation of circular fringes when M_1 and M_2 are exactly perpendicular to each other; (b) Circular fringes

If d is the distance between M_1 and M_2', then $S_1S_2 = 2d$.

$$\Delta = S_1P = S_1S_2 \cos\theta$$
$$= 2d\cos\theta$$

Thus, for bright fringes, the path difference is

$$2d\cos\theta + \frac{\lambda}{2} = \frac{2n\lambda}{2}$$

$$\Rightarrow \qquad 2d\cos\theta = (2n-1)\frac{\lambda}{2} \tag{6.47}$$

and for dark fringes, the path difference will be

$$2d\cos\theta + \frac{\lambda}{2} = (2n+1)\frac{\lambda}{2}$$

$$\Rightarrow \qquad 2d\cos\theta = n\lambda \tag{6.48}$$

FIGURE 6.15 When M_1 and M_2' coincide

Subcase (i)
If M_1 and M_2 are at right angles to each other and at an equal distance from point O, then M_1 and M_2' will coincide. Thus, the path difference is $\Delta = 0$ as $d = 0$.

The field of view will be completely dark as shown in Fig. 6.15.

Subcase (ii)
When M_1 is parallel to M_2' then widely spaced circular fringes will be formed as shown in Fig. 6.16. These fringes are called fringes of equal inclination and are known as Haidinger fringes.

FIGURE 6.16 Haidinger fringes

Case 2: Localized fringes
If M_1 and M_2' are not parallel to each other, we get localized fringes as shown in Fig. 6.17(a), (b) and (c).

The space between M_1 and M_2' forms wedge-shaped air film.

Case 3: White light fringes
When white light source is used instead of monochromatic light source, the central fringe will be dark. But there will be 8 to 10

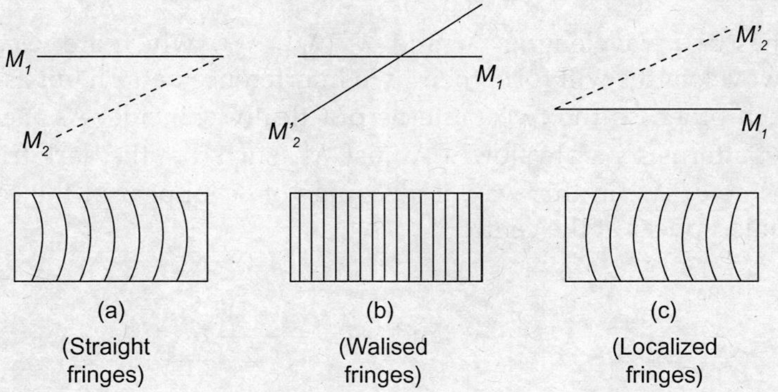

FIGURE 6.17 Localized fringes

coloured fringes around central fringe for a small path difference of the order of a few wavelengths.

Application: Determination of wavelength of light used

Mirrors M_1 and M_2 are adjusted exactly perpendicular to each other. Now, the position of M_1 is adjusted till some bright fringes appear in the field of view. Now, adjust the cross-wire of eyepiece on a particular bright fringe and note down its reading. Move M_1 slowly and measure the distance travelled in displacing some finite number of bright fringes from the field of view (say n fringes). Let this distance travelled be d; then

$$2d = n\lambda$$

$$\Rightarrow \qquad \lambda = \frac{2d}{n}$$

Knowing d and n, the wavelength of light used can be measured very easily.

Application: Determination of refractive index of thin films

Adjust the interferometer to produce white light straight line fringes and then crosswire of telescope is adjusted on the central bright fringe and note down its reading. Now, a film of known thickness is introduced in the path of one of the interfering beams. This will cause a path difference of $2(\mu-1)t$ in two interfering beams, where t is the thickness of the given film. Due to the path difference introduced in beams, the central fringe shifts. Move mirror M_1 till the central fringe is again brought back and note the displacement of the mirror (say x).

Now applying $2x = 2(\mu-1)t$, x can be measured as

$$\Rightarrow \qquad x = (\mu-1)t \tag{6.49}$$

Thus by measuring x and t, the value of μ can be calculated.

Application: Resolution of closely spaced spectral lines

Let the source has two wavelengths λ_1 and λ_2 $(\lambda_1 > \lambda_2)$, which are very close to each other. Both the wavelengths will form their own interference pattern but as λ_1 and λ_2 are very close to each other, so the two patterns practically coincide. As the mirror M_1 is moved, the two patterns separate slowly. Adjust M_1 such that the dark fringe of λ_1 falls on the bright fringe of λ_2. In this way, if n^{th} fringe of λ_1 appear at the centre, $(n+1)^{\text{th}}$ fringe of λ_2 should appear at the centre.
Hence,

$$x = \frac{n\lambda_1}{2} \quad \text{and} \quad x = \frac{(n+1)\lambda_2}{2}$$

$$\Rightarrow \qquad n = \frac{2x}{\lambda_1} \tag{6.50}$$

and

$$(n+1) = \frac{2x}{\lambda_2} \tag{6.51}$$

Subtracting Eq. (6.50) from Eq. (6.51)

$$(n+1) - n = \frac{2x}{\lambda_2} - \frac{2x}{\lambda_1}$$

$$1 = \frac{2x(\lambda_1 - \lambda_2)}{\lambda_1 \lambda_2}$$

$$\Rightarrow \qquad \lambda_1 - \lambda_2 = \frac{\lambda_1 \lambda_2}{2x} \tag{6.52}$$

By measuring the distance x moved by mirror M_1 between two consecutive positions of maximum indistinctness, the two closely spaced spectral wavelengths can be resolved very easily.

EXAMPLE 6.8

Two slits separated by a distance of 0.2 mm are illuminated by a mono-chromatic light of wavelength 550 nm. Calculate the fringe width on a screen at a distance of 1 m from the slits.

Solution: The given data are

Separation between the slits $(2d) = 0.2 \text{ mm} = 0.2 \times 10^{-3}$ m

Wavelength of light $(\lambda) = 550 \text{ nm} = 550 \times 10^{-9}$ m

Distance of the screen $(D) = 1$ m

Fringe width $(\beta) = ?$

$$\beta = \frac{\lambda D}{2d} = \frac{500 \times 10^{-9} \times 1}{0.2 \times 10^{-3}}$$

$$= \frac{500 \times 10^{-6}}{0.2} = 2750 \times 10^{-6} \text{m}$$

$$= \mathbf{2.75 \times 10^{-3} \ m = 2.75 \ mm}$$

EXAMPLE 6.9

Light of wavelength 500 nm forms an interference pattern on a screen at a distance of 2 m from the slit. If 100 fringes are formed within a distance of 5 cm on the screen, find the distance between the slits.

Solution: The given data are

Wave length of light $(\lambda) = 500\,\text{nm} = 500 \times 10^{-9}\,\text{m}$

Distance of screen $(D) = 2\,\text{m}$

$$\text{Fringe width } (\beta) = \frac{5}{100}\,\text{cm}$$

$$= \frac{50}{100}\,\text{mm} = 0.5\,\text{mm}$$

$$= 0.5 \times 10^{-3}\,\text{m}$$

Separation between slits $(2d) = ?$

We know $\beta = \dfrac{\lambda D}{2d}$

$$2d = \frac{\lambda D}{\beta}$$

or

$$= \frac{500 \times 10^{-9} \times 2}{0.5 \times 10^{-3}}$$

$$= 2 \times 10^{-3}\,\text{m}$$

$$= \mathbf{2\,mm}$$

EXAMPLE 6.10

Two coherent sources whose intensity ratio is 36:1 produce interference fringes. Deduce the ratio of maximum intensity to minimum intensity.

Solution: Intensities ratio of coherent sources $= a_1^2 : a_2^2 = 36:1$

$\therefore$

$$a_1 : a_2 = 6:1$$

Minimum intensity of the interference fringe $= (a_1 - a_2)^2$

$$= (6-1)^2 = 25$$

Maximum intensity of the interference fringe $=(a_1-a_2)^2$

$$=(6+1)^2=49$$

The **ratio of maximum intensity to minimum intensity** $= \mathbf{49:25 \approx 2:1}$

RECAP ZONE

POINTS TO REMEMBER

- Interference is the phenomenon of redistribution of light energy due to the superposition of two light waves coming from two coherent sources.
- In Fresnel's biprism, the fringe width is

$$\beta=\frac{D\lambda}{2d}$$

where $2d$ = separation between two sources and D = distance between source and screen.

- When a thin transparent sheet of thickness 't' is introduced in the path of one of the beams in Fresnel's biprism, then interference pattern is displaced by amount

$$\Delta y=\frac{\beta}{\lambda}(\mu-1)t$$

- In Newton's ring experiment:

(i) Diameter of dark fringes:

$$D_n=\sqrt{\frac{4n\lambda R}{\mu}}\text{, here } \mu=1 \text{ for air}$$

(ii) Diameter of bright fringes:

$$D_n=\sqrt{\frac{2(2n-1)\lambda R}{\mu}}\text{, here } \mu=1 \text{ for air}$$

(iii) Wavelength of monochromatic light:

$$\lambda=\frac{D_{n+m}^2-D_n^2}{4mR}$$

- In Michelson's interferometer:

(i) If N fringes shift by displacing one mirror through distance x, then $\lambda=\frac{2x}{N}$

(ii) Measurement of difference of wavelengths:

$$\Delta\lambda=\frac{\lambda_{av}^2}{2x}$$

(iii) Measurement of thickness of thin sheet:

$$t = \frac{x}{(\mu - 1)}$$

- Interferometer is an instrument which employs large path difference between the interfering beams to obtain interference. It is generally used to measure the wavelengths or distance between fine spectral lines.
- As $D_n \alpha \sqrt{n}$ and $D_n \alpha \sqrt{(2n-1)}$ for dark and bright rings in Newton's ring, so the rings get closer as the order of the rings increases.
- The Newton's rings are circular in nature, because these rings are the foci of constant thickness of the air film and these foci are concentric circle and hence fringes are circular.
- The centre of the Newton's ring interference pattern in reflected light is dark.
- In Newton's ring experiment, the interference pattern of transmitted light has a bright centre.

Key Terms with Definitions

- **Coherent sources** – Two interferring sources of light emitting waves, which have always a constant phase difference between them.
- **Constructive interference** – When the crest or trough of one wave falls on the crest or trough of the other wave and two waves are exactly in phase. The resultant amplitude is greater than the amplitude of either wave.
- **Destructive interference** – When the crest or trough of one wave falls on the trough or crest of other wave and two interfering waves are out of phase.
- **Fringe width** – The distance between any two consecutive bright or dark fringes.
- **Nodal lines** – The lines that represent the positions of zero vibrations or minimum intensity.
- **Antinodal lines** – The lines that represent the positions of maximum intensity.
- **Sustained interference** – Well-defined and observable interference pattern.
- **Non-reflecting film** – When the loss of light intensity falling on thin film is quite objectionable.

Important Formulae and Equations

Equation Number	Equation	Remarks
(6.9)	$\phi = \frac{2\pi}{\lambda} \times (\text{path difference})$	**Phase difference between two interfering waves**

(6.12)	$\beta = \dfrac{\lambda D}{2d}$	**Fringe width**
(6.18)	$t = \dfrac{x_0(2d)}{D(\mu - 1)}$	**Thickness of thin mica sheet in fresnel's biprism.**
(6.20)	$2\mu t \cos r = (2n+1)\dfrac{\lambda}{2}$	**Condition of maxima (Newton's rings system)**
(6.21)	$2\mu t \cos r = n\lambda$	**Condition of minima (Newton's rings system)**
(6.26)	$2\mu t \cos(\theta + r) = (2n+1)\dfrac{\lambda}{2}$	**Condition of maxima (Michelson's interferometer)**
(6.27)	$2\mu t \cos(\theta + r) = n\lambda$	**Condition of minima (Michelson's interferometer)**
(6.35)	$\theta = \dfrac{(N)\lambda}{2\mu(x_2 - x_1)}$	**Wedge angle of wedge-shaped film**
(6.40)	$D_n = \sqrt{2(2n-1)\lambda R}$	**Diameter of bright fringe in Newton's ring reflected system**
(6.42)	$D_n = \sqrt{4n\lambda R}$	**Diameter of dark fringe in Newton's ring reflected system**
(6.46)	$\mu = \dfrac{D_{n+m}^2 - D_n^2}{D_{n+m}'^2 - D_n'^2}$	**Refractive index of liquid in Newton's ring system**

REVIEW ZONE

SHORT ANSWER QUESTIONS

1. What do you mean by interference of light?
2. What are coherent sources?
3. What is a Fresnel's biprism?
4. Define fringe width.
5. Give two methods of producing interference.
6. What are Newton's rings?
7. Discuss Haidinger fringes.
8. Give the conditions for sustained interference of light.
9. Explain why a thick film shows no colour in reflected white light.
10. What are achromatic fringes?

LONG ANSWER QUESTIONS

1. What are coherent sources? Discuss how coherent sources can be produced with the help of Fresnel's biprism?
2. What is interference of light? Discuss the necessary conditions for obtaining sustained interference. Give two methods of producing interference.
3. Discuss Young's double slit experiment in detail. Describe its theory and experimental arrangement.
4. Describe Fresnel's biprism with a neat diagram and explain how wavelength of monochromatic light is determined by it.
5. Discuss the formation of Newton's rings by (i) reflected light and (ii) transmitted light. Derive an expression for diameter of nth dark ring in reflected light.
6. Describe and explain the formation of Newton's rings in transmitted light.
7. Discuss the theory, working and experimental arrangement of Michelson's interferometer.
8. Discuss the applications of Newton's ring experiment and Michelson's interferometer in detail.
9. How will you produce circular fringes with Michelson's interferometer?
10. Give a difference between the fringes produced with the help of Fresnel's biprism, Newton's rings and Michelson's interferometer.

NUMERICAL PROBLEMS

1. In Young's double slit experiment, the distance between the slits is 0.5 mm and the distance of the screen from coherent sources is 50 cm. If the fringe width is 0.95 cm. Find the wavelength of light.

   Ans. 5000 Å

2. When a thin sheet of transparent material of thickness 6.3×10^{-4} cm is introduced in the path of one of the beams causing interference, the central bright fringe shifts to the position occupied by the sixth bright fringe. Find the refractive index of the material of the sheet. (Given $\lambda = 5460$ Å).

   Ans. $\mu = 1.52$

3. Magnesium fluoride has a refractive index of 1.38 and is frequently used to coat lenses. How thick this coating should be for maximum transmittance at a wavelength of 5.30 mm?

   Ans. $t = 0.9601 \times 10^{-5}$ cm

4. In a Fresnel's biprism experiment, the fringe width is observed to be 0.087 mm. What will it become if the slit to biprism distance is reduced to (3/4) of the original distance?

   Ans. 0.116 mm

5. In Newton's ring experiment, the diameter of the 5th ring was 0.3 cm and the diameter of 25th ring was 0.8 cm. If the radius of curvature of

the planoconvex lens is 100 cm, find the wavelength of the light used.

Ans. 4.87×10^{-5} cm

6. In Newton's ring experiment, the diameter of the 10th dark ring is 0.433 cm. Find the wavelength of incident light, if the radius of curvature of the lens is 70 cm.

Ans. 6695 Å

7. In a Michelson's interferometer, 200 fringes cross the field of view when the movable mirror is displaced through 0.0589 mm. Calculate the wavelength of monochromatic light used.

Ans. 5890×10^{-8} cm

8. Calculate the distance between two successive positions of a movable mirror of a Michelson's interferometer giving best fringes in the case of sodium having lines of wavelength 5890 Å and 5896 Å.

Ans. 0.02894 cm

CHAPTER 7

Polarization

Learning Objectives

By the end of this chapter, the student will be able:

- To understand the phenomenon of polarization of light
- To distinguish between polarized and unpolarized light waves
- To understand planes of polarization, Brewster's law and Malus law
- To explain the polarization by double reflection and refraction
- To discuss the working of Nicol's prism and Polaroid film
- To describe the working of quarter wave plate and half wave plate
- To explain the theory of plane, circularly and elliptically polarized light
- To explore the applications of polarized light waves
- To discuss optical activity, optical rotation and applications in polarimeters and photo-elasticity

7.1 ▶ Introduction

The phenomenon of interference and diffraction shows the wave nature of light waves but they do not reveal the character of light waves, i.e. whether they are longitudinal or transverse in nature. Light is made up of electric and magnetic field vectors, which are perpendicular to each other and also perpendicular to the direction of propagation of light waves (Fig. 7.1). The phenomena of polarization tells us about the transverse nature of light.

7.2 ▶ Polarization of Light

When ordinary light is normally incident on a pair of tourmaline crystal plates P and P' which have planes at right angles to the direction of propagation of light, it gets polarized. The intensity will be maximum when the axis of P' is parallel to that of P [Fig. 7.2(a)] and the intensity will be minimum when P and P' are at right angles to each other [Fig. 7.2(b)].

This shows that the light is transverse in nature. It is clear from Fig. 7.2(b) that after passing P, the light vibrates only in one direction, i.e. it is said to be polarized. The tourmaline crystal P is used to produce polarized light, and so it is called a polarizer; the tourmaline crystal P' is used to analyze the polarized light, and so it is called an analyzer.

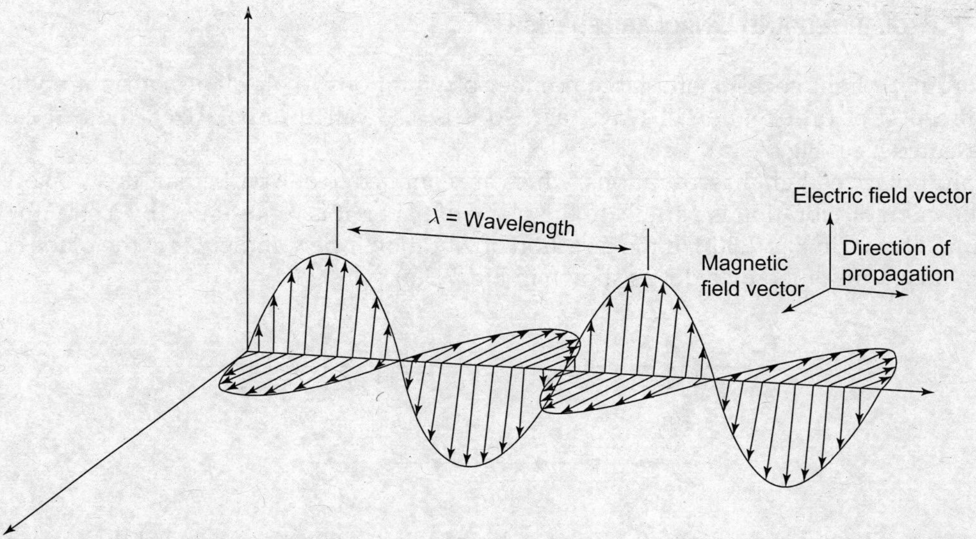

FIGURE 7.1 Light as an electromagnetic wave

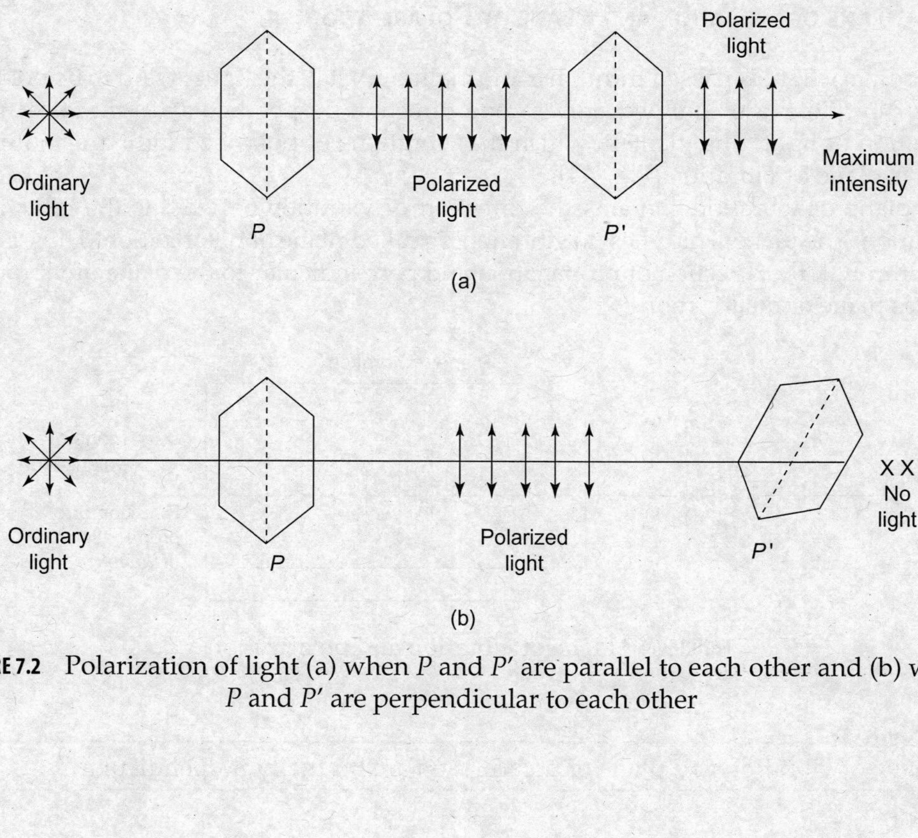

FIGURE 7.2 Polarization of light (a) when P and P' are parallel to each other and (b) when P and P' are perpendicular to each other

7.3 ▶ Polarized and Unpolarized Light

The ordinary light consists of a large number of vibrations in all planes at right angles to the direction of propagation of wave, and so it is also called unpolarized light. It can be represented as in Fig. 7.3(a).

Plane polarized light has vibrations, which are confined to one particular direction only. If the direction of vibration is parallel to the plane of paper, it is represented by a straight line arrow as shown in Fig. 7.3(b). If the direction of vibration is perpendicular to the plane of the paper, it is represented by a dot as shown in Fig. 7.3(c).

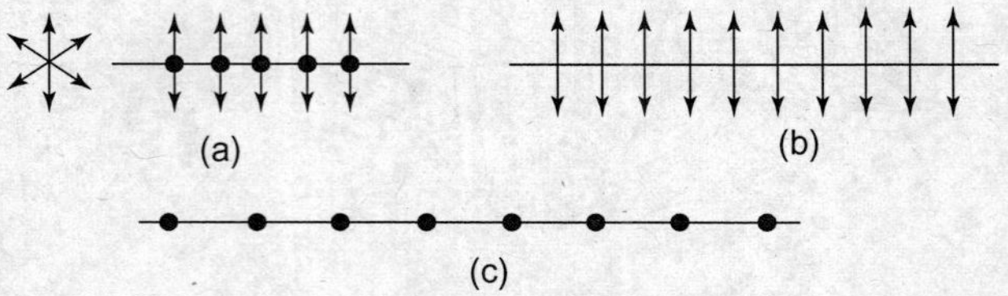

FIGURE 7.3 (a) Unpolarized light; (b) Plane polarized light (vibrations parallel to the plane of paper); (c) Plane polarized light (vibrations perpendicular to the plane of paper)

7.4 ▶ Plane of Vibration and Plane of Polarization

When ordinary light is passed through a tourmaline crystal, the light is said to be polarized and the vibrations are confined only in one direction perpendicular to the direction of propagation of light. The plane in which the vibrations of polarized light are confined is known as plane of vibration [Fig. 7.4].

This plane of vibration contains the direction of vibration as well as the direction of propagation. The plane which has no vibrations is called plane of polarization. Thus, a plane passing through the direction of propagation and perpendicular to the plane of vibration is known as plane of polarization.

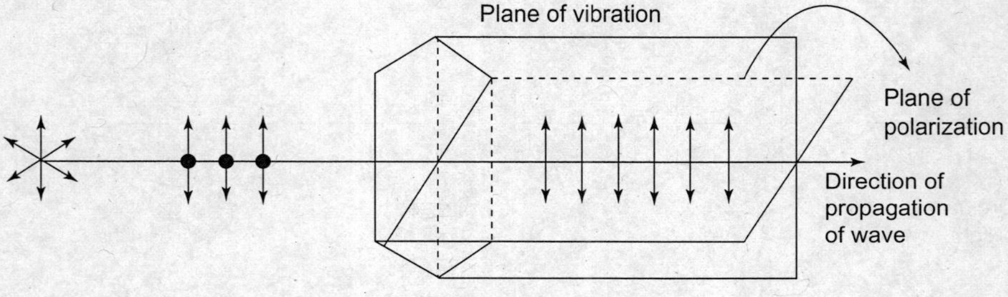

FIGURE 7.4 Plane of vibration and polarization

Note

The plane of vibration and plane of polarization are mutually perpendicular.

7.5 ▶ Brewster's Law

Sir David Brewster in 1811 performed a number of experiments to study the phenomenon of polarization of light by reflection at different surfaces. He observed that at a particular angle i_p, the light is completely polarized in the plane of incident ray. He discovered the relation

$$\mu = \tan i_p \tag{7.1}$$

which is known as Brewster's law (Fig. 7.5). Here, the reflected and refracted rays are perpendicular to each other. Suppose a beam CD of the unpolarized light is incident on the glass surface at the polarizing angle i_p. Let DE be the reflected and DF the refracted ray, and r be the angle of refraction. Then, from Brewster's law,

$$\mu = \tan i_p = \frac{\sin i_p}{\cos i_p} \tag{7.2}$$

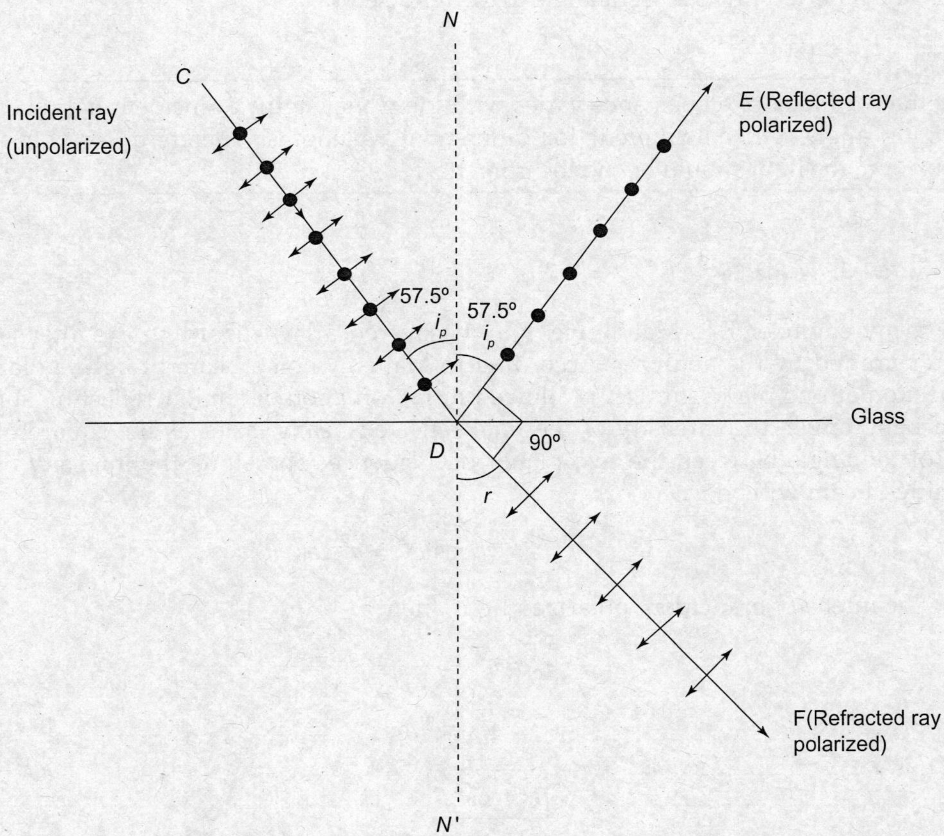

FIGURE 7.5 Brewster's law (polarization by reflection)

From Snell's law,

$$\mu = \frac{\sin i_p}{\sin r} \tag{7.3}$$

From Eqs. (7.2) and (7.3), we get

$$\frac{\sin i_p}{\cos i_p} = \frac{\sin i_p}{\sin r}$$

$$\Rightarrow \quad \frac{\sin i_p}{\sin\left(\frac{\pi}{2} - i_p\right)} = \frac{\sin i_p}{\sin r}$$

$$\Rightarrow \quad \frac{\pi}{2} - i_p = r$$

$$\Rightarrow \quad \boxed{i_p + r = \frac{\pi}{2}} \tag{7.4}$$

Thus, the reflected ray is at right angle to the refracted ray.

Note

The refracted index of a substance varies with the wavelengths of incident light; hence, polarizing angle will be different for different wavelengths. Therefore, for complete polarization, the light should be monochromatic.

7.6 ▶ LAW OF MALUS

French army engineer E.L. Malus (Fig. 7.6) discovered a law regarding the intensity of light transmitted by the analyzer. According to Malus, when a beam of light, polarized by reflection at one plane surface, is allowed to fall on another similar reflecting surface at polarizing angle, the intensity of the final reflected beam varies as the square of the cosine of the angle between the two planes of incidence. Therefore, the intensity of the transmitted beam will be

$$I = (a\cos\theta)^2 = a^2\cos^2\theta \tag{7.5}$$

If I_o be the intensity of incident polarized light, then

$$I_o = a^2$$

$$\Rightarrow \quad I = I_o \cos^2\theta \tag{7.6}$$

$$\Rightarrow \quad \boxed{I \propto \cos^2\theta} \tag{7.7}$$

Case (i): When $\theta = 0$, i.e. when the two planes are parallel to each other, then $\cos\theta = 1$ and from Eq. (7.6) $I = I_o$.

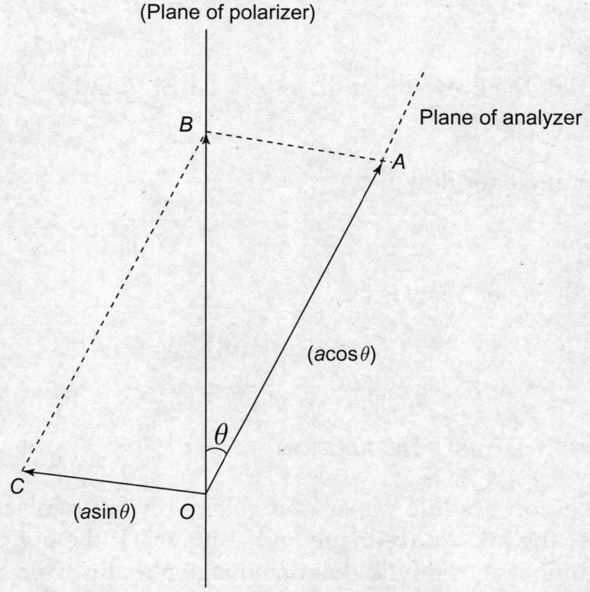

FIGURE 7.6 Malus law (Transmitted light)

Case (ii): When $\theta = \pi/2$, i.e. when the two planes are perpendicular to each other, from Eq. (7.6) $I = 0$.

EXAMPLE 7.1

Calculate the Brewster angle for (i) ethylalcohol for which $\mu = 1.361$ and (ii) carbon tetrachloride for which $\mu = 1.461$.

Solution: Brewster's law is $\tan i_p = \mu$

(i) For ethylalcohol, $\mu = 1.361$

So,
$$i_p = \tan^{-1} 1.361$$
$$= 53°41'36''$$

(ii) For carbon tetrachloride, $\mu = 1.461$

So,
$$i_p = \tan^{-1} 1.461$$
$$= \mathbf{54°36'35''}$$

EXAMPLE 7.2

For flint glass material, the Brewster angle is $i_p = 62°24'$. Find the refractive index of the material.

Solution: Brewster angle for flint glass, $i_p = 62°24'$

Refractive index, $\mu = ?$

Brewster's law is tan $i_p = \mu$

So, $\boldsymbol{\mu = \tan 62°24' = 1.9128}$

7.7 ▶ Polarization by Double Refraction

If an ink dot is marked on a white paper and calcite crystal is placed over it, then we observe two images of the ink dot (o-image and e-image). If the crystal is rotated slowly about a vertical axis, then it is observed that one image remains fixed and the other image rotates with the rotation of the calcite crystal. The fixed image is normal and is called ordinary image while the other one is called extraordinary image.

In 1869, Erasmus Bartholinus discovered that when a ray of unpolarized light is incident on a calcite crystal as shown in Fig. 7.7(b), then inside the crystal, the ray breaks up into o-ray and e-ray. The o-ray travels along QS and makes as angle of refraction r_1 whereas the e-ray travels along QR and makes as angle of refraction r_2. For o-ray, the refractive index

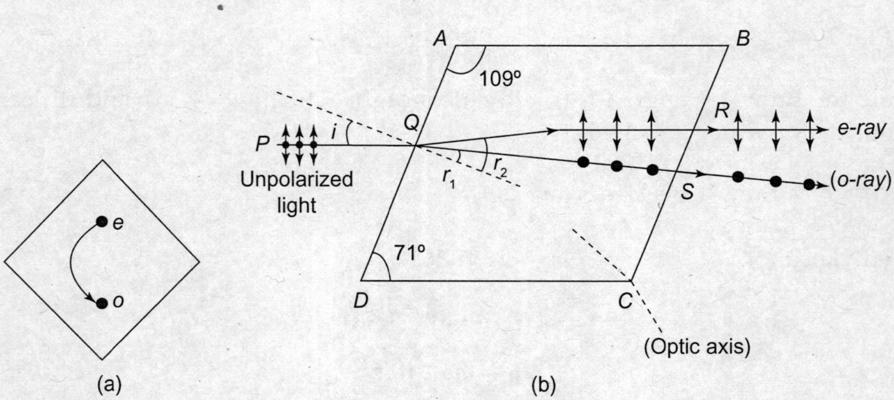

FIGURE 7.7 (a) Double refraction; (b) Double refraction in calcite crystals

$\mu_o = \dfrac{\sin i}{\sin r_1}$ is constant, whereas for e-ray, $\mu_e = \dfrac{\sin i}{\sin r_2}$ is not constant, but varies with the angle of incident i.

In calcite, $\mu_o > \mu_e$ because $r_1 < r_2$. Thus, the velocity of light for ordinary ray inside the crystal will be less than the extraordinary ray.

Note

The crystals that have $\mu_o > \mu_e$, and hence $v_o < v_e$, are called uniaxial negative crystals. The crystals that have $\mu_e > \mu_o$, and hence $v_o > v_e$, are called uniaxial positive crystals.

7.7.1 Geometry of Calcite Crystal

It is also known as Iceland spar ($CaCO_3$). It is a colourless crystal transparent to visible as well as ultraviolet light.

It belongs to the rhombohedral class of hexagonal system. The six faces of rhombohedron are parallelograms each having angles of 78° and 102°. The two opposite corners *A* and *B*, where three obtuse angles (102°) meet are called blunt corners (Fig. 7.8). A line passing through any of the blunt corners (*A* or *B*) and making equal angles with the three faces which meet at this corner, locate the direction of the optic axis of the crystal.

Any plane that contains the optic axis and is perpendicular to two opposite faces is called a principal section.

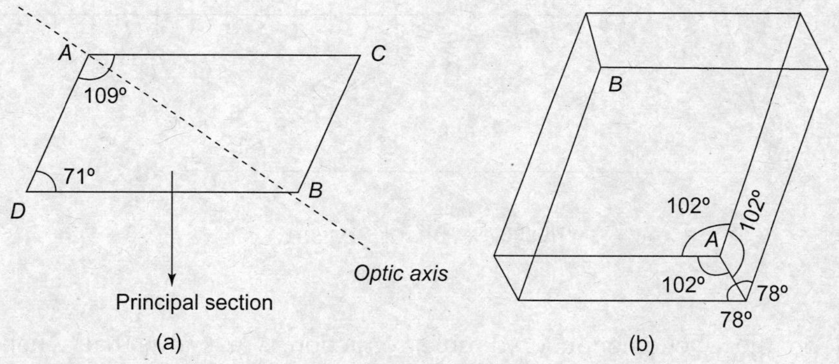

FIGURE 7.8 Calcite crystal (a) and (b)

7.8 ▶ Uniaxial and Bi-axial Crystals

When a ray of light is incident on some certain crystals, it gets split up into two refracted rays. One of the refracted rays follows the ordinary laws of refraction and hence is called the ordinary ray [o-ray], while the other refracted ray does not follow such laws and is called extraordinary ray (e-ray). If an object is viewed from such crystals, two images are observed. One image corresponds to o-ray, while the other corresponds to e-ray. This phenomenon is called double refraction and such crystals that show this phenomenon are called doubly refracting crystals.

These doubly refracting crystals are of two types: (a) uniaxial crystals and (b) biaxial crystals.

In uniaxial crystals, there is one direction, called optic axis, along which the two refracted rays travel with the same velocity. Examples are calcite, quartz and tourmaline. In biaxial crystals, there are two optic axes. Examples are mica, topaz, aragonite, cane sugar, etc.

Note

Crystals having one optic axis are called uniaxial crystals and those having two optic axes are called biaxial crystals.

7.9 ▶ Nicol's Prism

Nicol's prism is an optical device for producing and analyzing a plane polarized light (Fig. 7.9).

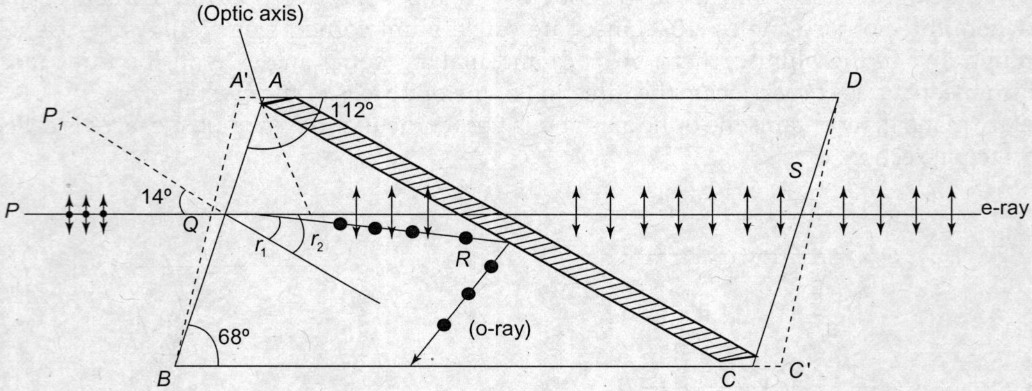

FIGURE 7.9 Nicol's prism

Principle

It is based on the phenomenon of double refraction. We know that when ordinary unpolarized light is transmitted through a calcite crystal, it splits up into o- and e-ray. Nicol prism eliminates o-ray by using the phenomenon of total internal reflection.

Construction

It is constructed from a calcite crystal whose length is three times as its width is taken. The end faces are cut down so as to reduce the angles of principal section to 68° and 112° instead of 71° and 109°. The crystal is now cut along a diagonal, and the two cut surfaces are grounded and polished optically flat and then cemented together again by Canada Balsam, which is a transparent material. It is more denser (optically) than calcite for e-ray but less denser for o-ray.

For sodium light:

Refraction index for o-ray, $\mu_o = 1.658$

Refractive index for e-ray, $\mu_e = 1.486$

Refractive index for Canada Balsam, $\mu = 1.55$

7.9.1 Action of Nicol's Prism

When a ray of light PQ enters the face AB in a direction parallel to the long side, it is doubly refracted into o-ray (QR) and e-ray (QS). The o-ray suffers total internal reflection at the Canada Balsam because Canada Balsam is optically mere denser than calcite for the e-ray and less denser than calcite for o-ray.

The e-ray is transmitted through Canada Balsam. The value of critical angle i_c for the o-ray is

$$i_c = \sin^{-1}\frac{1.55}{1.658} = 69^\circ$$

Limitations

(i) If the incident beam makes an angle smaller than $\angle P_1QP$, the ordinary ray will strike the calcite–Balsom layer at an angle less than 69°. In this case, it will not be reflected but transmitted through the prism. Thus, the angle of incidence on force AB is limited to 14°.

(ii) If the incident beam makes an angle greater than $\angle P_1QP$ with face AB, then the extraordinary ray will refract more and may suffer total internal reflection. Hence, the incident beam should be confined within an angle of 14°.

7.9.2 Nicol Prism as Analyzer and Polarizer

When two Nicols are arranged co-axially, then the first Nicol which produces plane polarized light is called polarizer, whereas the second Nicol that analyzes the polarized light is known as analyzer (Fig. 7.10).

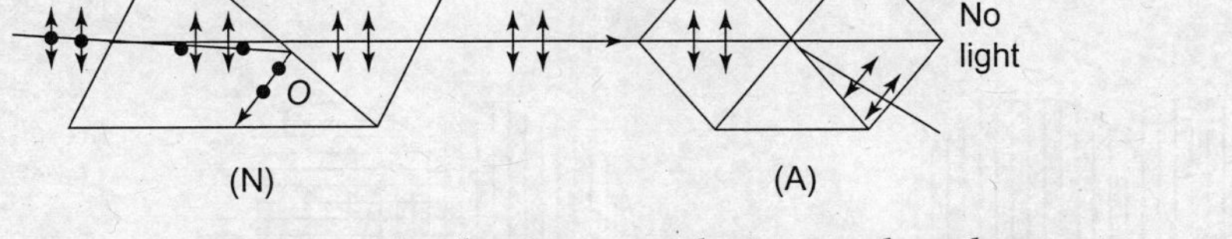

FIGURE 7.10 Nicol prism as polarizer and analyzer

When the principal section of Prism A is parallel to that of N, its vibrations will be parallel to the principal section of A. Thus, the ray behaves as e-ray in Prism A and is therefore completely transmitted. The intensity of emergent light is maximum. When the prism A is rotated, the intensity of the emergent ray decreases. When A is exactly at right angle to the principal section of N, no light emerges from the prism A. Thus, here Prism N produces plane polarized light and prism A detect it.

Note

If a given light ray is observed through Nicol prism and shows a variation in intensity with zero minimum intensity, the given light ray will be plane polarized.

7.10 ▶ Polaroid

Polaroid is an optical device used to produce plane polarized light by the phenomenon of double refraction and selective absorption of light. Polaroid is a thin film containing microcrystals of iodosulphate quinine embedded in nitrocellulose material. The axes of microcrystals are parallel to one another. When unpolarized light is incident perpendicularly on the polaroid sheet, then the electric vectors that are parallel to hydrocarbon chain of the molecules are absorbed, and the electric vectors of light that are perpendicular to the hydrocarbon chain of the molecules are transmitted. Hence, the direction perpendicular to the hydrocarbon chain is called the transmission axis of the polaroid. The transmitted light is linearly polarized as shown in Fig. 7.11(a).

The intensity of light that can be transmitted can be varied by using two polaroid films mounted separately between thin glass plates and kept adjacently as shown in Fig. 7.11(b) and (c).

When the polaroids are placed with their transmission axes parallel as shown in Fig. 7.11(b), then the intensity of polarized light that comes out is maximum, and if one of the polaroids (second) is rotated through 90° as shown in Fig. 7.11(c), then no light comes out through the second polaroid.

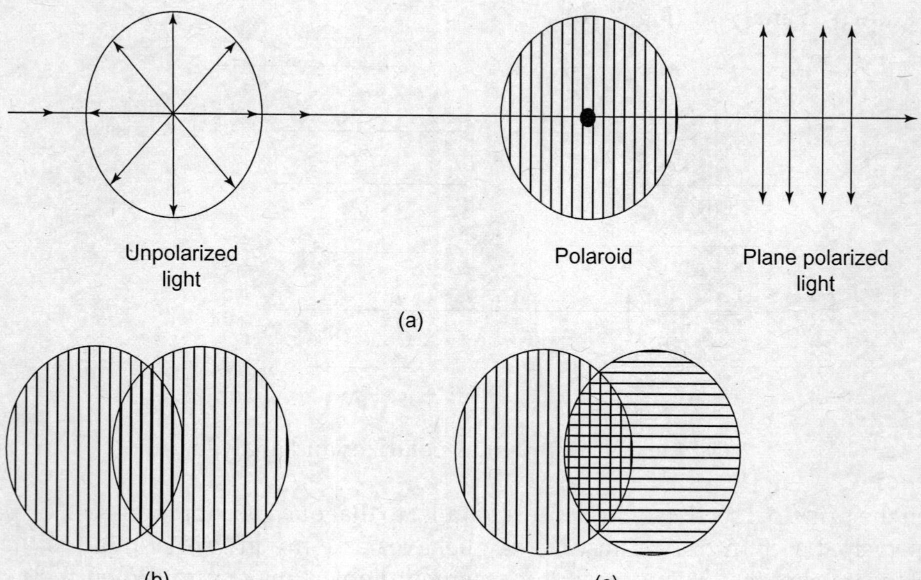

FIGURE 7.11 (a) Transmission of unpolarized light through polaroid, (b) in both the polaroids, the transmission axes are parallel and (c) perpendicular in polaroids

7.10.1 Uses of Polaroids

Polaroids are used

(i) to produce and analyze polarized light,

(ii) to vary the intensity of light,

(iii) to suppress headlight glare in motor vehicles,

(iv) to increase colour contrast in oil paintings,

(v) to produce three-dimensional motion pictures.

7.11 ▶ Quarter-wave Plate

A quarter-wave $\lambda/4$ plate is a thin double refracting crystal of calcite or quartz, cut and polished parallel to its optic axis to a thickness '*d*' such that it produces a path difference of $\lambda/4$ or phase difference of $\pi/2$ between the o-ray and the e-ray when plane polarized light is incident normally on the surface and passes through the plate.

As shown in Fig. 7.12, consider a calcite crystal of thickness '*d*'. The optic axis is parallel to the surface. When a plane polarized light is incident normally on the surface, then the light will split up into o-ray and e-ray. These rays travel with different speeds in the crystal. In calcite crystal, the e-ray travel faster than o-ray. Hence, the refractive index of o-ray (μ_o) is higher than the refractive index of e-ray (μ_e) in the crystal. The optical path covered by the o-ray as it passes through the crystal of thickness '*d*' is $\mu_o d$. Similarly, the optical path covered by the e-ray as it passes through the crystal of thickness '*d*' is $\mu_e d$.

∴ The path difference, $\Delta = \mu_o d - \mu_e d = d(\mu_o - \mu_e)$

As the crystal is a quarter-wave plate, it introduces a path difference of $\lambda/4$ between o-ray and e-ray. Therefore, $\Delta = \lambda/4$ and $\Delta = d(\mu_o - \mu_e)$

So we can write, $$d = \frac{\lambda}{4(\mu_o - \mu_e)}$$

For some other crystals like quartz, where $\mu_e > \mu_o$, the thickness of the quarter-wave plate is

$$\boxed{d = \lambda/4(\mu_e - \mu_o)} \tag{7.8}$$

Using Eq. (7.8), the thickness of the quarter-wave plate can be estimated.

Note

Quarter-wave plates are used to analyze all kinds of polarized light.

7.12 ▶ Half-wave Plate

It produces a path difference of $\lambda/2$ or phase difference of π between o-ray and e-ray. Let '*d*' be the thickness of half-wave plate.

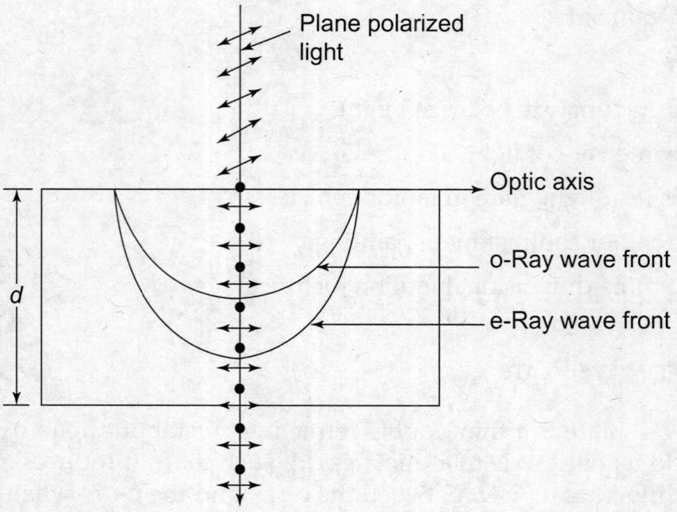

FIGURE 7.12 Propagation of polarized light in a calcite quarter-wave plate

For half-wave plate, $$(\mu_o - \mu_e)d = \frac{\lambda}{2}$$

$\Rightarrow$ $$\boxed{d = \frac{\lambda}{2(\mu_o - \mu_e)}} \quad (7.9)$$

For a half-wave plate, the path difference, $\Delta = \lambda/2 = d(\mu_o - \mu_e)$, $d = \lambda/2(\mu_o - \mu_e)$ in calcite crystal half-wave plate. In the case of quartz, $d = \lambda/2(\mu_e - \mu_o)$. Using this equation, the thickness of half-wave plate can be estimated.

Note

When a plane polarized light is incident on a half-wave plate, the emergent light is plane polarized.

EXAMPLE 7.3

The refractive index of a polarizer is 1.54. Find the polarization angle and angle of refraction.

Solution: The refractive index of a polarizer, $\mu = 1.54$

Brewster's law is $\tan i_p = \mu$

The polarization angle, $$i_p = \tan^{-1}\mu$$

$$= \tan^{-1} 1.54 = 57°$$

We know $i_p + r = \frac{\pi}{2}$

So $r = 90° - 57°$

$= \mathbf{33°}$

EXAMPLE 7.4

The refractive indices of mica for ordinary and extraordinary rays are 1.586 and 1.592 with a wavelength of 5460 Å. Find the thickness of mica sheet to act as a quarter-wave plate.

Solution: Wavelength of light, $\lambda = 5460 \text{ Å} = 5460 \times 10^{-10} \text{m}$

Refractive index of o-ray, $\mu_o = 1.586$

Refractive index of e-ray, $\mu_e = 1.592$

For quarter-wave plate,

$$d = \frac{\lambda}{4(\mu_e - \mu_o)}$$

$$= \frac{5460 \times 10^{-10}}{4(1.592 - 1.586)}$$

$$= \frac{0.546 \times 10^{-6}}{0.024} \text{m}$$

$$= \mathbf{22.75 \times 10^{-6} \text{ m}}$$

7.13 ▶ Theory of Plane, Circularly and Elliptically Polarized Light

A beam of plane polarized light can be obtained from a Nicol prism. This beam of plane polarized light is made incident normally on the surface of a calcite crystal cut parallel to its optic axis.

As shown in Fig. 7.13(a), let the plane of polarization of the incident beam makes an angle θ with the optic axis and let the amplitude of this incident light be A.

As polarized light enters into the calcite crystal, it will split up into two components, e-ray and o-ray. The e-ray amplitude $A \cos\theta$ is parallel to the optic axis and o-ray amplitude $A \sin\theta$ is perpendicular to the optic axis. Inside the crystal, e-ray and o-ray travel in the same direction with different amplitudes. On emerging from the crystal, the rays have a phase difference 'δ' (say), depending on the thickness 'd' of the crystal. Let v be the frequency of light. The e-ray and o-ray can be represented in terms of simple harmonic motions at right angles to each other having a phase difference 'δ'. The e-ray moves faster than o-rays in calcite crystal. Hence, the instantaneous displacements are

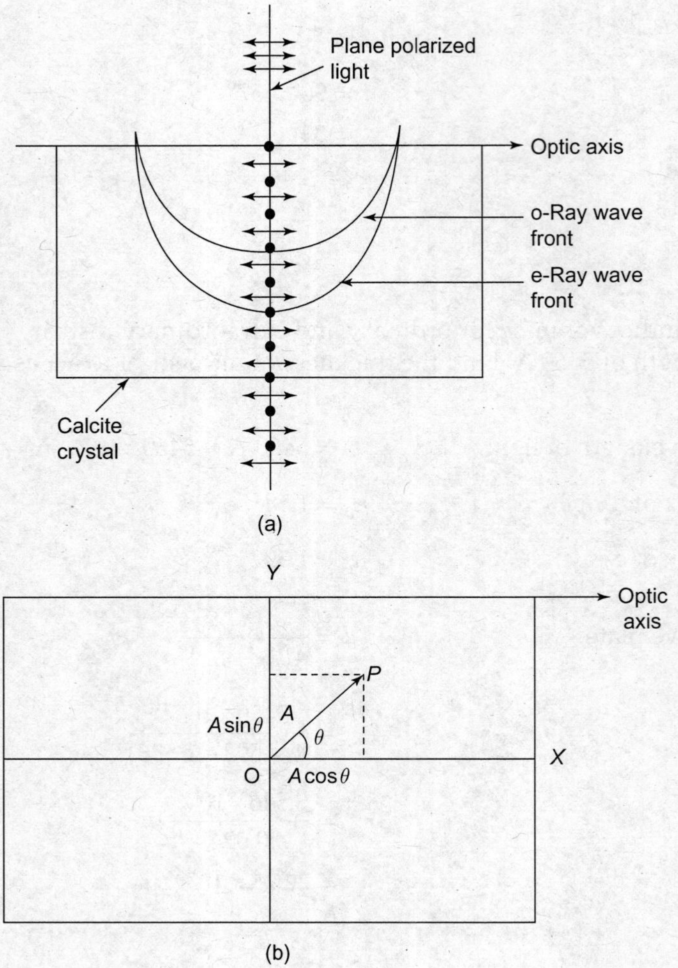

FIGURE 7.13 (a) Plane wave incident on calcite crystal and (b) e-ray and o-ray light amplitudes in calcite crystal

$$x = A\cos\theta\sin(\omega t + \delta) \text{ for e-ray} \tag{7.10}$$

and

$$y = A\sin\theta\sin\omega t \text{ for o-ray} \tag{7.11}$$

where

$$\omega = 2\pi n.$$

Let $A\cos\theta = a$ and $A\sin\theta = b$, then Eqs. (7.10) and (7.11) become

$$x = a\sin(\omega t + \delta) \tag{7.12}$$

$$y = b\sin\omega t \tag{7.13}$$

From Eq. (7.13),

$$\frac{y}{b} = \sin\omega t \text{ and } \cos\omega t = \sqrt{1-\frac{y^2}{b^2}} \tag{7.14}$$

From Eq. (7.12),

$$\frac{x}{a} = \sin\omega t\cos\delta + \cos\omega t\sin\delta \tag{7.15}$$

Substituting Eq. (7.15) in Eq. (7.12), we get

$$\frac{x}{a} = \frac{y}{b}\cos\delta + \sqrt{1-\frac{y^2}{b^2}}\sin\delta$$

$$\frac{x}{a} - \frac{y}{b}\cos\delta = \sqrt{1-\frac{y^2}{b^2}}\sin\delta$$

Squaring both sides,

$$\frac{x^2}{a^2} + \frac{y^2}{b^2}\cos^2\delta - 2\frac{x}{a}\frac{y}{b}\cos\delta = \left(1-\frac{y^2}{b^2}\right)\sin^2\delta$$

or
$$\frac{x^2}{a^2} + \frac{y^2}{b^2}\cos^2\delta + \frac{y^2}{b^2}\sin^2\delta - 2\frac{x}{a}\frac{y}{b}\cos\delta = \sin^2\delta$$

or
$$\frac{x^2}{a^2} + \frac{y^2}{b^2} - \frac{2xy}{ab}\cos\delta = \sin^2\delta \tag{7.16}$$

This is the general equation for an ellipse.

Special cases:

Case I: Suppose the phase difference $\delta = 0$. Then $\delta = 0$ and $\delta = 1$. Equation (7.16) becomes

$$\frac{x^2}{a^2} + \frac{y^2}{b^2} - \frac{2xy}{ab} = 0$$

or
$$\left(\frac{x}{a} - \frac{y}{b}\right)^2 = 0$$

or
$$\frac{x}{a} - \frac{y}{b} = 0$$

or
$$y = \frac{b}{a}x \tag{7.17}$$

This is the equation for a straight line. So the light that comes out of the crystal is plane polarized.

Case II: Suppose $\delta = \frac{\pi}{2}, \frac{3\pi}{2}, \frac{5\pi}{2}, \ldots, \left(\frac{2n+1}{2}\right)\pi,\ n = 0,1,2,3,\ldots$

Then, Eq. (7.16) becomes

$$\frac{x^2}{a^2} + \frac{y^2}{b^2} = 1 \tag{7.18}$$

Equation (7.18) represents an ellipse. So, the emergent light from the crystal will be elliptically polarized.

Case III: Suppose $\delta = \frac{\pi}{2}$ and $a = b$

Then, from Eq. (7.16)

$$x^2 + y^2 = a^2 \tag{7.19}$$

Equation (7.19) represents a circle. So, the emergent light from the crystal will be circularly polarized. Circularly polarized light can also be produced when the incident plane polarized light makes an angle of 45° with the optic axis. The linear, elliptical and circular polarizations for different values of δ are shown in Fig. 7.14.

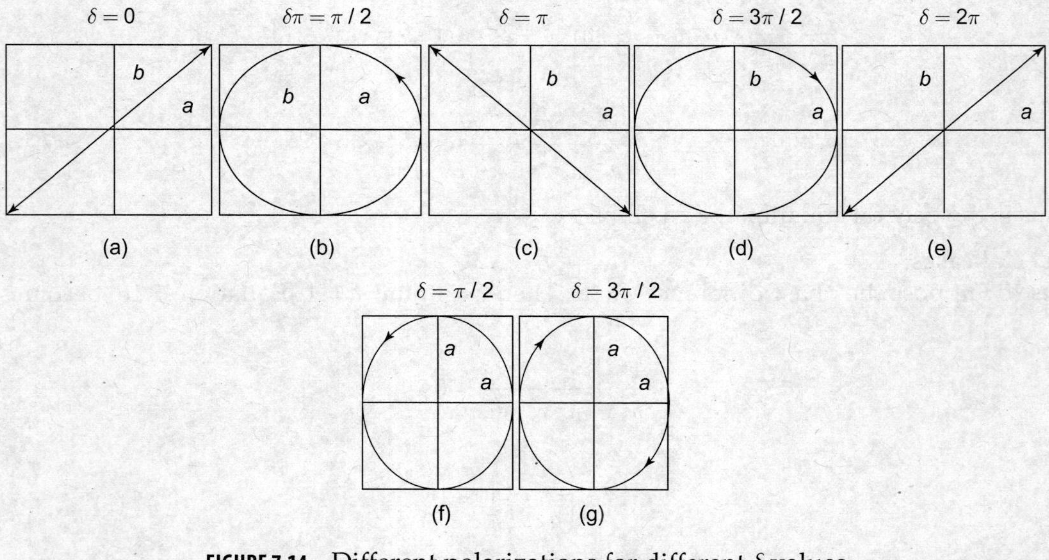

FIGURE 7.14 Different polarizations for different δ values

7.14 ▶ Production of Plane, Circularly and Elliptically Polarized Light

When an ordinary light is passed through a Nicol prism, it splits up into ordinary and extraordinary light beams. The ordinary light follows Snell's law. Therefore, it is totally internally reflected at the Canada balsam layer while the extraordinary ray passes through the Nicol prism. This emergent ray is the plane polarized light.

7.14.1 Circularly Polarized Light

The ordinary polarized light emerging from the Nicol prism is plane polarized; this plane polarized light is passed normally through a quarter-wave plate such that the angle between the optic axis and the vibration of the plane polarized light is 45°. The emergent ray from the quarter-wave plate is circularly polarized.

7.14.2 Elliptically Polarized Light

The ordinary polarized light emerging from the Nicol prism is plane polarized. This plane polarized light is passed normally through a quarter-wave plate such that the angle between the optic axis and the vibration of the plane polarized light is any angle other than 0°, 45° and 90°. The emergent ray from the quarter-wave plate is elliptically polarized.

7.15 ▶ Detection of Plane, Circularly and Elliptically Polarized Light

A given beam of light may be unpolarized, partially plane polarized, plane polarized, elliptically polarized or circularly polarized. The nature of the given light beam may be detected by using another rotating Nicol prism and the intensity variation is observed. The following are the three possibilities:

(a) If there is no change in the intensity, light is either unpolarized or circularly polarized.

To differentiate between unpolarized and circularly polarized light, the light under test is first passed normally through the quarter-wave plate and then viewed through another rotating Nicol prism and the variation of intensity as again observed in the following two ways:

(i) If there is no variation in the intensity, the light under the test is unpolarized.

(ii) If there is variation in the intensity with minimum zero to maximum, twice in each rotation, the light under test is circularly polarized light. These two conditions are shown in Fig. 7.15.

(b) If the intensity is varying but not zero, the light is either partially polarized or elliptically polarized.

To differentiate between partially and elliptically polarized light, the light under test is first passed normally through the quarter-wave plate and then viewed through another rotating Nicol prism and the variation of intensity is again observed in the following two ways:

(i) If there is variation in intensity with minimum intensity equal to zero (twice in each rotation), the light under the test is elliptically polarized.

(ii) If there is a variation in intensity with minimum intensity not equal to zero to maximum (twice in each rotation), the light under the test is partially polarized light. These two conditions are shown in Fig. 7.16.

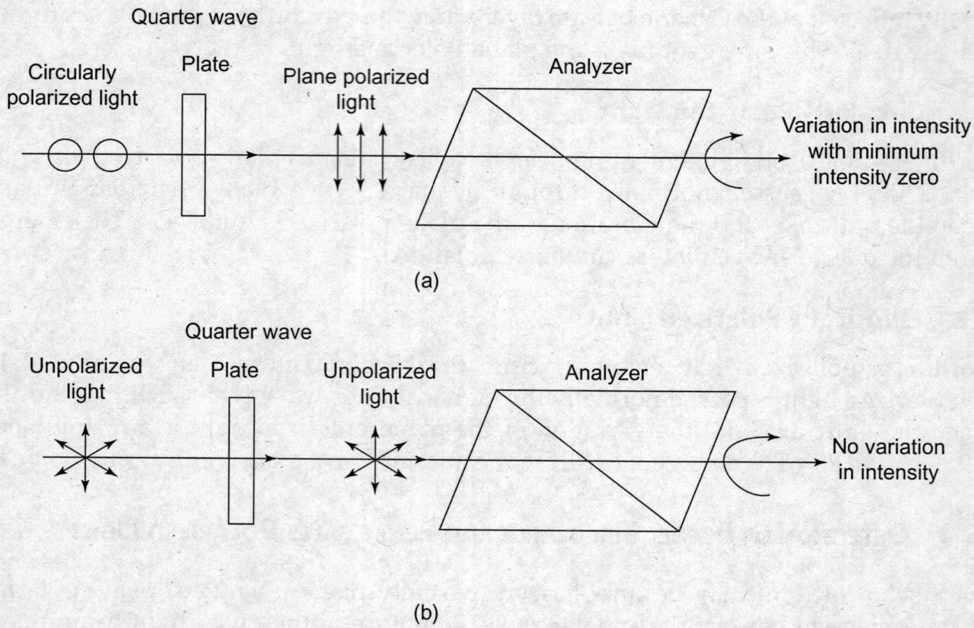

FIGURE 7.15 Detection of the circularly polarized light

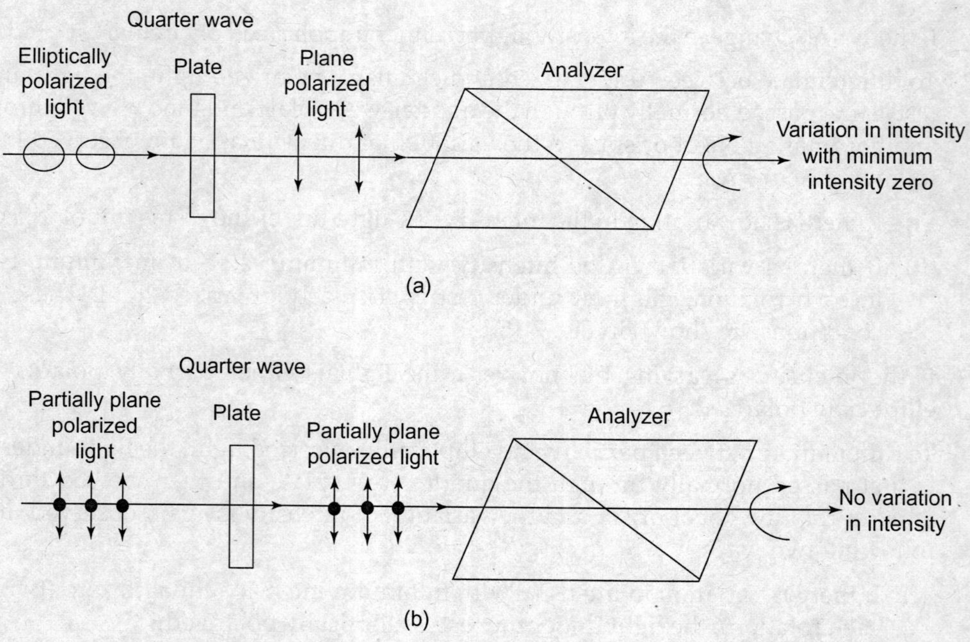

FIGURE 7.16 Detection of elliptically polarized light

(c) If the intensity is varying with minimum zero, the light is plane polarized (see Fig. 7.17).

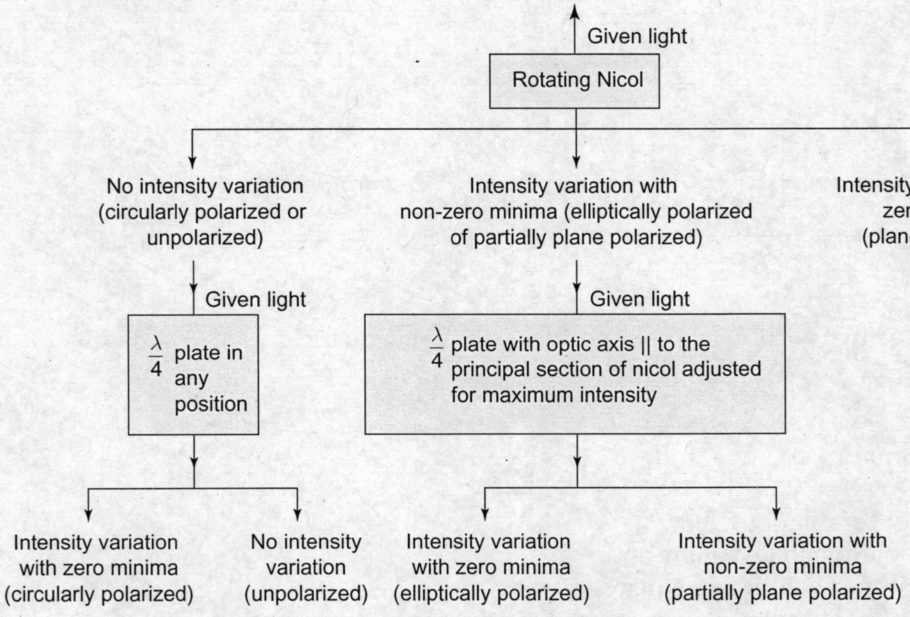

FIGURE 7.17 Detection of the plane polarized light

7.16 ▶ Optical Activity

When a beam of plane polarized light propagates through certain substances like quartz, sugar crystal, etc., the plane of vibration or the plane of polarization of the emergent beam is not the same as that of the incident beam but has rotated through a certain angle. This phenomenon is called rotatory polarization and the property of rotating the plane of vibration or plane of polarization is known as optical activity. The angle, through which the plane of polarization is rotated, is known as angle of rotation. This optical rotation can be demonstrated by the following experiment.

In Fig. 7.18(a), when two Nicol prisms are set in a crossed position, we get no light through analyzer *A*. But when we use a quartz crystal between two crossed Nicols, then the quartz crystal changes the plane of vibration of the light coming from polarizer *N* and we get light through analyzer *A*.

7.16.1 Specific Rotation

The specific rotation of a substance at a particular temperature and for a given wavelength of light is defined as the rotation of plane of vibration of polarized light (in degrees) produced by 1 decimetre length of the solution when its concentration is 1 g per cc, i.e.

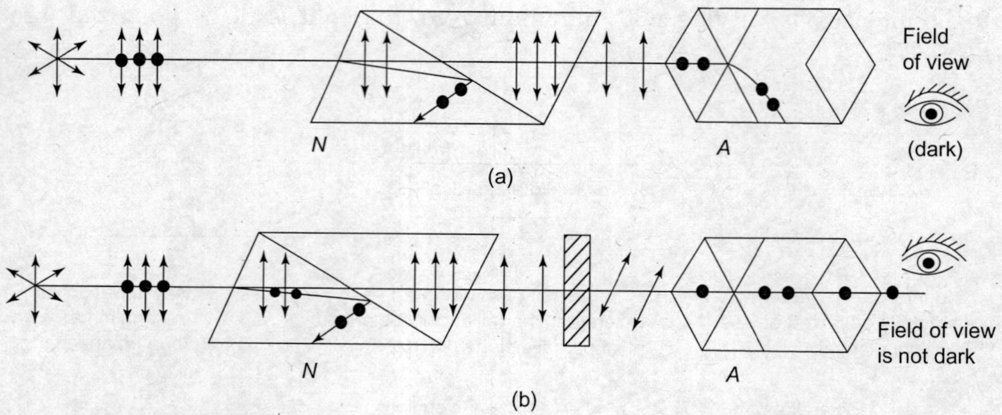

FIGURE 7.18 (a) Two Nicols in crossed positions. (b) Using quartz crystal in between two crossed prisms

Specific rotation,

$$\boxed{S = \frac{\theta}{l \times C}} \tag{7.20}$$

where θ = angle of rotation,
l = length of solution in decimetre and
c = concentration of solution in g per cc.
If the length of the solution is given in cm, then the specific rotation is

$$S = \frac{10 \times \theta}{l \times C}$$

Molecular rotation is the product of specific rotation and the molecular weight.

The amount of optical rotation depends upon the thickness and density of the crystal, concentration in case of solutions, temperature and the wavelength of light used. There are two types of optically active substances. They are as follows:

1. **Right-handed or dextrorotatory:** The substances which rotate the plane of polarization or plane of vibration in the clockwise direction when the observer faces towards the source of light are called right-handed or dextrorotatory or positive substances. Examples are cinnabar, cane sugar, etc.
2. **Left-handed or laevo-rotatory:** The substance which rotates the plane of polarization or plane of vibration in the anti-clockwise direction or left-hand side with respect to an observer facing the light travelling towards him are called left-handed or laevo-rotatory. An example of this kind is fruit sugar.

Calcite does not produce any change in the plane of vibration of the plane polarized light, i.e. it is not an optically active substance. Dextro- and laevorotatries are the mirror images of their orientations. This property is posed by the organic compounds whose molecules are asymmetric.

7.16.2 Fresnel's Theory of Optical Rotation

According to Fresnel, the plane polarized light consists of resultant of two circularly polarized vibrations rotating in the opposite directions with the same angular velocity.

When the plane polarized light enters a crystal along its optic axis, it spits up into two circularly polarized vibrations. In case of an optically inactive substance, these components travel with the same velocities along the optic axis. Hence, at the emergence, they produce a plane polarized light without any rotation of plane of vibration.

In case of optically active substances, these components travel with different velocities. In dextrorotatory, the right-handed or clockwise component travels faster, i.e. $V_R > V_L$.

In laevo-rotatory, the left-handed or anti-clockwise component travels faster, i.e. $V_L > V_R$.

Hence, in an optically active substance, a phase difference is developed in the two beams while traversing through the substances.

Thus, the plane of polarization of the emergent plane polarized light gets rotated through a certain angle relative to that of the incident light. The rotation depends on the phase difference between the two beams.

Consider a plane polarized light beam incident normally on a doubly reflecting crystal, such as quartz plate cut with its faces perpendicular to the optic axis.

Let the vibration in the incident plane of the polarized beam be represented by

$$x = 0$$
$$y = 2a \sin \omega t$$

where $2a$ is the amplitude of the incident vibrations.

These equations may be rewritten as

$$x = a \cos \omega t - a \cos \omega t \tag{7.21}$$

$$y = a \sin \omega t + a \sin \omega t \tag{7.22}$$

But by the principle of superposition, Eqs. (7.21) and (7.22) are the resultant of two vibrations x_1, x_2 and y_1, y_2, where x_1 and x_2 are the components of clockwise circular motion and x_2 and y_2 are the components of anti-clockwise circular motion (left-handed).
Now,

$$x_1 = a \cos \omega t$$
$$y_1 = a \sin \omega t$$
$$x_2 = a \cos \omega t$$
$$y_2 = a \sin \omega t$$

These circularly polarized waves travel along the optic axis of quartz with unequal velocities. Therefore, when they emerge from the plate, there is a phase difference d between them.

The emergent circular vibrations may be represented by

$$x_1 = a \cos \omega t$$

and

$$y_1 = a\sin\omega t$$
$$x_2 = -a\cos(\omega t + \delta)$$

and

$$y_2 = a\sin(\omega t + \delta)$$

The resultant vibrations are given by

$$\begin{aligned} x &= x_1 + x_2 \\ &= a\cos\omega t - a\cos(\omega t + \delta) \\ &= a[\cos\omega t - \cos(\omega t + \delta)] \end{aligned}$$

$$X = a\left[\frac{2\sin\{\omega t + (\omega t + \delta)\}}{\dfrac{2\sin\{(\omega t + \delta) - \omega t\}}{2}}\right]$$

or,
$$X = 2a\sin\left\{\omega t + \frac{\delta}{2}\right\}\cdot\sin\frac{\delta}{2} \tag{7.23}$$

$$\begin{aligned} Y &= y_1 + y_2 \\ &= a\sin\omega t + a\sin(\omega t + \delta) \end{aligned}$$

$$Y = a\left[\frac{2\sin\{\omega t + (\omega t + \delta)\}}{\dfrac{2\cos\{\omega t - (\omega t + \delta)\}}{2}}\right]$$

or,
$$Y = 2a\sin\left\{\omega t + \frac{\delta}{2}\right\}\cdot\cos\frac{\delta}{2} \tag{7.24}$$

$$\frac{X}{Y} = \frac{\sin\dfrac{\delta}{2}}{\cos\dfrac{\delta}{2}} = \tan\frac{\delta}{2} \tag{7.25}$$

This is the equation of straight line inclined at an angle of $\delta/2$ with the y-axis. Let μ_R and μ_L be the refractive indices of the right- and left-handed circularly polarized light.

EXAMPLE 7.5

Calculate the specific rotation if the plane of polarization is turned through 30°, after traversing 25 cm length of 30% sugar solution.

Solution:

$$S = \frac{10\theta}{l \times C}$$

$$S = \frac{\mathbf{10 \times 30}}{\mathbf{25 \times 0.3}} = \mathbf{40°}$$

EXAMPLE 7.6

A sugar solution of specific rotation $52°$ decimetre per g/cc causes a rotation of $12°$ in a column of 10 cm long. What is the concentration of the solution?

Solution: Here, $S = 52°$, $l = 10$ cm, $\theta = 12°$, $C = ?$

But
$$S = \frac{10\theta}{l \times C} \text{ or } C = \frac{10\theta}{sl}$$

$$C = \frac{10 \times 12}{52 \times 10} = \frac{3}{13} = 0.23 \text{ g/cc}$$

$$\mathbf{C = 0.23\ g/cc}$$

7.17 ▶ POLARIMETERS

The device with the help of which the angle of rotation (θ) of plane polarized light can be measured is known as a polarimeter. By measuring the angle θ, the specific rotation S can be calculated, if the concentration and the length of the solution are known.

The polarimeters are of two types:

(a) Laurent's half-shade polarimeter

(b) Bi-quartz polarimeter

When these polarimeters are used to determine the quantity of sugar in a solution, they are also called saccharimeters.

7.17.1 Laurent's Half-shade Polarimeter

It consists of two separate Nicols P (called polarizer) and A (called analyzer) mounted in a brass tube, placed at some distance apart and capable of rotation about a common axis. A glass tube T having a large diameter in the middle contains the active solution under examination. The two ends of the tube are covered by a flat and parallel glass plate. It is mounted between the polarizer P and analyzer A on a rigid iron base [Fig. 7.19(a)].

Monochromatic light of wavelength λ from a source S, rendered parallel by a convex lens L, falls on a Nicol prism P. After passing through P, the light becomes plane polarized.

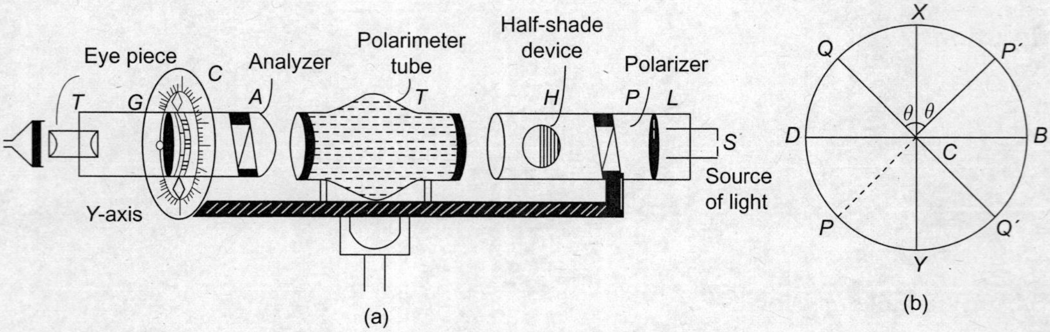

FIGURE 7.19 (a) Laurent's half-shade polarimeters; (b) Action of half-shade

The plane polarized light now passes through a half-shade device H and then through the specific solution which is filled in tube T whose polarity is to be determined. The transmitted light passes through analyzing Nicol A, which can be rotated about the direction of the propagation of light. The emergent light from Nicol A is viewed through a Galilean telescope G.

Action of Laurent's Half-shade plate

The Laurent's system consists of a half-shade plate in two halves: one with the quartz cut parallel to its optic axis and the other a matching plate of glass chosen so as to absorb and reflect the same amount of light as the quartz plate. The quartz plate is a half-wave plate. It introduces a path difference of $\lambda/2$ (or phase difference of π) between the ordinary and extraordinary rays in the transmission normal through it. Let the plane of vibration of the plane polarized light from the polarizer P falls normally on the half-shade plate along CP. The light passing through the glass plate remains unaffected, while that falling on the quartz plate is broken up into two components: e-component (CX parallel to optic axis XY) and O-component (perpendicular to the optic axis, i.e. along CB). As in quartz, the O-component travels faster. Hence on emergence, the O-component has vibrations along CD and the E-component has vibrations still along CX. Therefore, the emergent wave CQ is the resultant of vibrations along CD and CX. Here, $P'CX = QCX = \theta$. Thus, the angle between the vibration planes of light emerging from quartz CQ and that of light emerging from glass CP is 2θ.

Thus, there are two plane polarized lights: one emerging from the glass with vibrations in the plane CP while the other emerging from quartz with vibrations in the plane CQ. If the principal plane of the analyzing Nicol A is parallel to QCQ′, the light from the quartz plate will be passed unobstructed. Thus, the quartz plate half will be brighter than the glass plate half. If the principal plane of the analyzing Nicol A is parallel to PCP′, then the light from the glass plate will be passed unobstructed while the light from the quartz plate will be partially obstructed. Thus, the right half will appear brighter as compared to the left half. But when the principal plane of the analyzing Nicol A is parallel to the optic axis XCY, the two halves appear equally illuminated.

Determination of the specific rotation of sugar solution

To find the specific rotation of cane sugar or of an optically active substance, the glass tube T is first filled with clear water, and the analyzer A is set in the position of equal brightness of the two halves of the field of view. The reading on the verniers is noted. Now, the sugar solutions of known concentration are filled in the tube and the tube is placed again in the same place. Due to optical activity of the solution, the vibrations pass from the quartz half to the glass half and cause rotation in them. Therefore, on the introduction of the tube containing the sugar solution, the field of view is not equally bright. The analyzer is rotated in the clockwise direction and is brought to a position so that the whole field of view again appears equally bright. The new position of the verniers on the circular scale is noted. The difference in the two readings of the analyzer gives the angle of rotation θ produced by the solution. In the actual experiment, the angles of rotation for the solutions of various concentrations are measured. A graph is plotted between the concentration C and the angle of rotation θ and the ratio θ / C is determined. The specific rotation of the cane sugar is then determined from the following relation:

$$S = \frac{10\theta}{lC}$$

where l is the length of the tube in cm, C is the concentration of the solution in gm/cc and θ is the angle of rotation in degrees (Fig. 7.20).

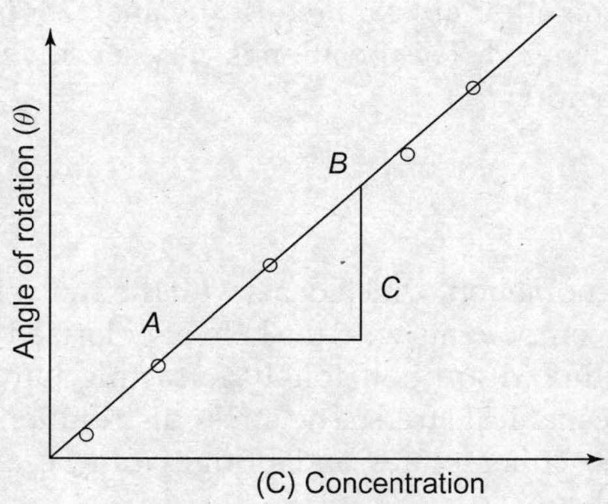

FIGURE 7.20 Example for determination of the specific rotation of sugar solution

7.17.2 Bi-quartz Polarimeter

The arrangement of bi-quartz polarimeter is same as that of Laurent's half-shade polarimeter, but it is more accurate instrument and also more sensitive than half-shade polarimeter. The two main differences between bi-quartz polarimeter and half-shade polarimeter are as follows:

(i) Monochromatic sodium light used in half-shade polarimeter is replaced by white light in bi-quartz polarimeter.

(ii) Bi-quartz polarimeter uses biquartz plate in place of Laurent's half-shade plate.

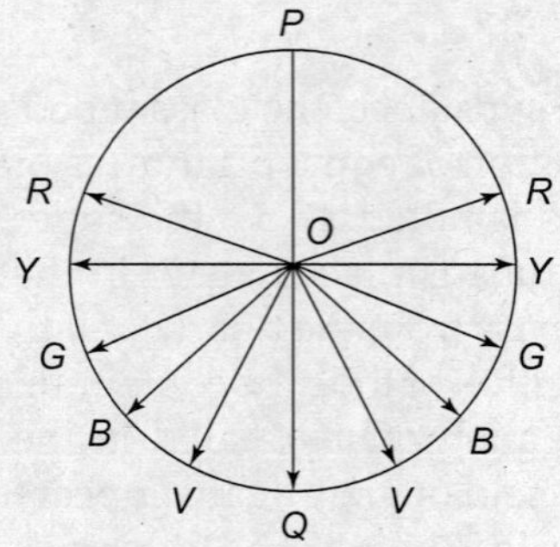

FIGURE 7.21 Action of bi-quartz plate

Action of Bi-quartz plate

It consists of two semi-circular plates of quartz, each of thickness 3.75 mm (one is left-handed quartz and the other is right-handed quartz). Both are cut perpendicular to optic axis and joined together so as to form a complete circular plate (Fig. 7.21). The thickness of both the plates is so adjusted that each rotates the plane of polarization of the yellow light through 90°.

The left-handed and right-handed quartz plates rotate the plane of polarization of the incident light in anti-clockwise and clockwise directions, respectively.

When a beam of white light, rendered plane polarized with a polarizer, travels through a bi-quartz normally, the phenomenon of rotatory dispersion occurs because the polarizing Nicol rotates the different wavelengths of white light to different extents. But the rotation of yellow colour is 90° and hence YOY is a straight line. The red rays are rotated the least and the violet rays are rotated the maximum. If the principal plane of analyzer is parallel to POQ, the yellow beam of light will not be transmitted and the appearance of the two halves is similar. The two halves have a greyish violet tint, called the sensitive tint or tint of the passage. When the analyzer is rotated slightly from this setting, one half of the field of view appears blue while the other appears red. If the analyzer is rotated in the opposite direction, the colours are changed. This position is very sensitive and is used for accurate determination of optical rotation.

7.18 ▶ Photoelasticity

The photoelasticity is a phenomenon, which deals with the analysis of stress distribution in mechanical and structural components with the help of polarized light. Certain substances like Bakelite and glass celluloid are not doubly refracting but show this phenomenon only when subjected to mechanical stress. The stress distribution in complex engineering structures such as girders, boiler plates and bridges can be analyzed by photoelastic studies of transparent models.

Note

(i) Sun glasses and LCDs use polarized light.

(ii) The angle between the direction of propagation and plane of polarization in electromagnetic waves is 0.

RECAP ZONE

POINTS TO REMEMBER

- The light which acquires the property of one-sidedness is called polarized light and the light which consists of vibrations in all the planes is called unpolarized light.
- In calcite crystal, $v_o < v_e$ (velocity of ordinary rays is less than that of extraordinary rays and $\mu_o > \mu_e$. So, calcite is a negative crystal.
- In quartz crystal, $v_o > v_e$ and $\mu_o < \mu_e$. So, it is a positive crystal.
- When a plane polarized light is incident on a half-wave plate, the emergent light is plane polarized. But, when a plane polarized light is incident on a quarter-wave plate such that o- and e- rays make an angle of $45°$ and the emergent light becomes circularly polarized.
- When a plane polarized light is incident on a quartz wave plate such that o- and e-rays make an angle other than $0°$, $45°$ or $90°$, the emergent light becomes elliptically polarized.
- For a half-wave plate, the thickness of plate is

$$d = \frac{\lambda}{2(\mu_o - \mu_e)}$$

and for a quarter-wave plate

$$d = \frac{\lambda}{4(\mu_o - \mu_e)}$$

- The rotation produced by 1 decimeter (or 10 cm) length of optical active solution of concentration 1 g/cc, is given as:

$$S = \underbrace{\frac{\theta}{l \times c}}_{\text{When length is in decimetres}} = \underbrace{\frac{10\theta}{l \times c}}_{\text{When length is in cm}}$$

- Polarimeters are used to measure the angle of rotation of plane of polarization.

KEY TERMS WITH DEFINITIONS

- **Polarized wave** – The wave unsymmetical about a direction is called polarized wave.
- **Nicol prism** – It is made up of calcite, which is a doubly reflecting crystal.
- **Optical activity** – The property of rotating the plane of vibration of a plane polarized light about its direction of propagation by some crystal.
- **Angle of rotation** – The angle through which the plane of polarization is rotated by an optically active substance.

- **Specific rotation** – The rotation produced by one decimetre length of its solution when the concentration is 1 g per cm^3.
- **Saccharimeters** – When the polarimeters are used to determine the quantity of sugar in a solution, they are called saccharimeters.

Important Formulae and Equations

Equation Number	Equation	Remarks
(7.4)	$i_p + r = \frac{\pi}{2}$	Brewster's law
(7.7)	$I \propto \cos^2 \theta$	Law of Malus
(7.8)	$d = \frac{\lambda}{4(\mu_e - \mu_o)}$	Thickness of quarter-wave plate
(7.9)	$d = \frac{\lambda}{2(\mu_o - \mu_e)}$	Thickness of half-wave plate
(7.20)	$S = \frac{\theta}{l \times C}$	Specific rotation

Review Zone

Short Answer Questions

1. Define polarization.
2. Distinguish between polarized and unpolarized light.
3. What is Brewster's law?
4. What do you mean by double refraction?
5. Define law of Malus.
6. What do you mean by ordinary and extraordinary rays?
7. Define a quarter-wave plate and a half-wave plate.
8. What is meant by optical activity?
9. What is a biaxial crystal?
10. Define specific rotation.

Long Answer Questions

1. Explain the terms polarization and Brewster's law. Show from this law, when light is incident on the transparent substance at polarizing angle, the reflected and refracted rays are at right angles to each other.

2. Explain the principle, construction and working of a Nicol prism. Explain the phenomenon of double refraction in uniaxial crystals.
3. What are quarter- and half-wave plates? Mention some of the uses of them.
4. Describe the phenomenon of double refraction in crystals. Explain how a Nicol prism can be used a polarizer and anlayzer?
5. Describe how, with the help of a Nicol prism and quarter-wave plate, plane polarized, circularly polarized and elliptical polarized lights are produced and detected?
6. Discuss the phenomenon of rotation of plane of polarized light by optically active material.
7. Describe the construction and working of a Laurent's half-shade polarimeter. Explain how you would use it to determine the specific rotation of sugar solution?
8. Describe the construction and working of bi-quartz polarimeter. Explain how you would use it to determine the specific rotation of sugar solution?

Numerical Problems

1. A beam of light travelling in water strikes a glass plate which is also immersed in water when the angle of incidence is 51°, the reflected beam is found to be plane polarized. Calculate the refractive index of glass.

   Ans. 1.235
2. The polarizing plates have polarizing directions parallel so as to transmit maximum intensity of light. Through what angle must either plate be turned if the intensity of the transmitted beam is to drop by one-third?

   Ans. 145°18′
3. Two Nicols are oriented with their principal planes making an angle of 30°. What percentage of incident unpolarized light will pass through the system?

   Ans. 37.5%
4. What is the optical thickness of a quarter-wave plate for the light of wavelength 600 nm, the bifringence of the plate ($\mu_e \sim \mu_o$) being 0.172?

   Ans. 0.872 μm
5. A tube of sugar solution 20 cm long is placed between crossed Nicols and illuminated with light of wavelength 6×10^{-5} cm. If the optical rotation produced is 13° and specific rotation is 65°, determine the strength of solution.

   Ans. 10%

CHAPTER 8

Special Theory of Relativity

Learning Objectives

By the end of this chapter, the student will be able:

- To recall the Newtonian relativity, pre-relativistic physics and the Galilean transformations
- To understand consistency of speed of light and the Michelson-Morley experiment
- To describe the postulates of special theory of relativity
- To express Lorentz transformation equations
- To understand implications of special theory of relativity like time dilation and length contraction and mass–energy equivalence
- To explain Doppler effect and red and blue shifts

8.1 ▶ Introduction

It was observed at the begining of the 20th century, that the classical physics, represented by Newton's mechanics, was inadequate to explain all phenomena at microscopic level or those involving bodies moving close to the speed of light. Later, new concepts of quantum mechanics were developed to account for the microscopic scale phenomena. The fact that light travels with the same speed in all directions regardless of the motion of the source or the observer, led Einstein to develop the Special Theory of Relativity in 1905. The Galilean coordinate transformation between the reference frames moving with respect to each other is compatible with the constancy of the speed of light and has been replaced by the Lorentz transformation. In this theory, Einstein postulated that all laws of physics are same in all inertial reference frames, whose coordinates are related to each other by Lorentz transformations.

8.2 ▶ Newtonian Relativity

The Newton's first law of motion states that a body moves with a constant velocity unless acted on by forces. This statement is about a particular reference frame – an inertial frame, in which the second law holds good.

The second law states that a body is accelerated in proportion to and in the direction of the force acted on it. This cannot be justified without the specification of the reference frame provided by the first law. Any frame that is moving uniformly with respect to a given inertial frame also satisfies the first law and is an inertial frame.

In classical mechanics, the coordinates of the inertial frames are related to each other by the Galilean transformations. Newton's second law is observed to be invariant under the Galilean transformation, if forces that depend on the relative positions of the interacting bodies are under consideration. Newton's laws of motion appear the same in all inertial frames, and therefore, as far as Newton's mechanics is concerned, there can be no physical criteria for a preferred inertial frame. The idea of the *rest frame* loses its absolute meaning and the word *velocity* can only retain its relative meaning. These concepts are called Newtonian relativity.

8.2.1 Pre-relativistic Physics

According to Newton's laws of motion:

- Free particles move with constant velocity.
- The vector force **F** is proportional to the rate of change of momentum, i.e.,

$$\mathbf{F} = \frac{d}{dt}(mv).$$

- To every action, there is an equal and opposite reaction.

The first of these laws singles out inertial frames as the non-accelerating ones.

Consider now a frame O [i.e., a set of spatial coordinates (x, y, z) and a time coordinate t] and another frame $\vec{O}$ with coordinates $(\overline{x}, \overline{y}, \overline{z}, \overline{t})$ which moves in the x-direction with uniform speed v relative to the frame O.

It is suggested that the two sets of coordinates are related by

$$\begin{aligned} \overline{x} &= x - vt, \\ \overline{y} &= y, \\ \overline{z} &= z, \\ \overline{t} &= t. \end{aligned} \tag{8.1}$$

These are the Galilean transformations.

If the particle has a velocity **u** with components (u_1, u_2, u_3) in frame O, its velocity in the frame O is given by

$$\begin{aligned} \overline{u}_1 &= u_1 - v, \\ \overline{u}_2 &= u_2, \\ \overline{u}_3 &= u_3, \end{aligned} \tag{8.2}$$

or

$$\overline{\mathbf{u}} = \mathbf{u} - \mathbf{v}, \tag{8.3}$$

where $$u_1 = \frac{dx_1}{dt},\ u_2 = \frac{dx_2}{dt},\ u_3 = \frac{dx_3}{dt} \tag{8.4}$$

In general, if the coordinate axes and the origins of frames O and $\bar{O}$ differ, then

$$\bar{\mathbf{X}} = R\mathbf{X} - \mathbf{v}t - \mathbf{d}, \tag{8.5}$$

where $\mathbf{X}$ has components (x, y, z). Here, R is a rotation matrix aligning O and $\bar{O}$, $\mathbf{v}$ is the relative velocity of O with respect to $\bar{O}$, and $\mathbf{d}$ is the displacement from the origin O.

Since the transformation is linear [constant velocity in $O \Rightarrow$ constant velocity in $\bar{O}$], $\bar{O}$ is inertial if O is inertial.

Thus, there are an infinite set of inertial frames, all moving uniformly with respect to each other.

Newton's laws apply in all inertial frame since

$$\bar{\mathbf{a}} = \frac{d\mathbf{u}}{dt} = \frac{d\bar{\mathbf{u}}}{dt}, \tag{8.6}$$

and $\mathbf{F}$ is invariant. Thus, we have Newtonian (Galilean) relativity.

The laws of mechanics do not allow measurement of absolute velocity; however, measurement of absolute acceleration is possible.

8.3 ▶ Galilean Transformation and the Newtonian Relativity

An event is something that happens at a point in space and at an instant of time. To specify an event, we need four coordinates in a particular reference frame (S), say t, x, y and z. Consider another reference frame (S') which is moving with a constant speed v in the x-direction with respect to S. To specify the same even in S', one needs a different set of coordinates, t', x', y' and z'. The two sets (t, x, y and z) and (t', x', y' and z') are related to each other. According to the Galilean coordinate transformations,

$$\boxed{\left.\begin{aligned} t' &= t \\ x' &= x - vt \\ y' &= y \\ z' &= z \end{aligned}\right\}} \tag{8.7}$$

These transformations leave the time interval between the two events and the length (space interval) unchanged.

An initial reference frame is the one in which the law of inertia (Newton's first law) holds good.

If S is an inertial frame, so is S'.

The special theory of relativity deals only with the description of events by observers in inertial frames of reference. The acceleration of a particle is the same in all reference frames, which move relative to one another with constant velocity.

$$\vec{a}' = \frac{d\vec{u}'}{dt'} = \frac{d(\vec{u} - \vec{v})}{dt}$$
$$= \frac{d\vec{u}}{dt} = \vec{a}$$

If the mass of a particle (m) and the force (F) acting on it are the same in S and S', then the Newtonian laws of motion are the same in both systems (invariant under the Galilean coordinate transformations).

$$m\vec{a}' = m\vec{a} = \vec{F}$$

The laws of mechanics are the same in all inertial frames.

The electromagnetic force not only depends on the relative positions of the charged particles but also on their velocities. The velocity of a particle is not the same in S and S', and therefore the electromagnetic force on a charged particle is not the same in S and S'. Thus, electrodynamics is not invariant under the Galilean coordinate transformations.

$$\Delta t' = t_2' - t_1' = t_2 - t_1$$
$$= \Delta t$$

$$\Delta l' = x_2' - x_1' = (x_2 - vt_2) - (x_1 - vt_1)$$
$$= (x_2 - x_1) - v(t_2 - t_1) = (x_2 - x_1) - v(t_2' - t_1')$$
$$= \Delta l$$

$$\frac{d\vec{u}}{dt} = 0 \text{ if } \vec{F} = 0 \quad \left(u_x = \frac{dx}{dt}, u_y = \frac{dy}{dt}, u_z = \frac{dz}{dt}\right)$$

$$u_x' = \frac{dx'}{dt'} = \frac{d(x - vt)}{dt} = u_x - v \ (v = \text{constant})$$

$$u_y' = \frac{dy'}{dt'} = \frac{dy}{dt} = u_y$$

$$u_z' = \frac{dz'}{dt'} = \frac{dz}{dt} = u_z$$

$$\frac{du_x'}{dt'} = \frac{d(u_x - v)}{dt} - \frac{du_x}{dt}$$

$$\frac{du_y'}{dt'} = \frac{du_y}{dt}$$

$$\frac{du_z'}{dt'} = \frac{du_z}{dt}$$

This proves that Newtonian laws are invariant in inertial frames.

Note

If the mass of a particle (m) and the force (F) acting on it are the same in S and S', then the Newtonian laws of motion are the same in both systems, i.e. invariant under the Galilean coordinate transformations.

$$m\vec{a}' = m\vec{a} = \vec{F}$$

The laws of mechanics are the same in all inertial frames.

8.4 ▶ Speed of Light and the Michelson-Morley Experiment

Electromagnetism is summarized by Maxwell's equations. Maxwell's equations are not invariant under the Galilean coordinate transformations, and therefore, one can choose an absolute rest frame which holds good in Maxwell's equations. The speed of light derived by Maxwell's equations comes to

$$c = \frac{1}{\sqrt{\epsilon_0 \mu_0}} = 2.997925 \times 10^8 \text{ m/sec}$$

regardless of the motion of the source. The medium of light propagation was given the name ether and the absolute rest frame was considered to be the one in which *ether* is at rest.

In a frame 'S' moving at a constant speed v with respect to this ether frame, an observer would measure a different speed for the light ($c \pm v$) according to the Galilean transformations. The Michelson-Morley experiment was designed as an attempt to locate the ether frame (the absolute rest frame) (Fig. 8.1).

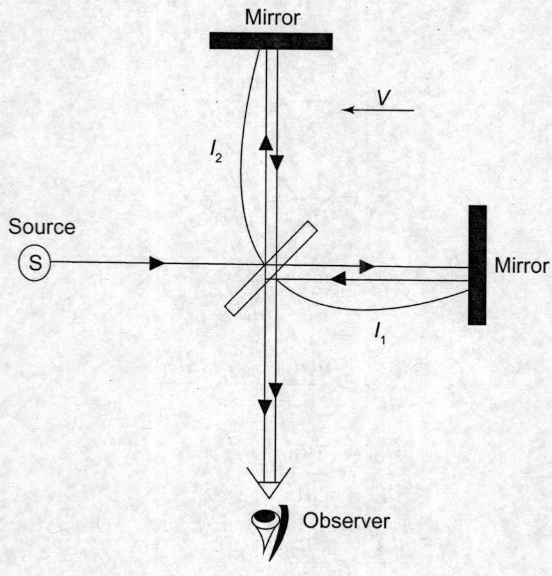

FIGURE 8.1 Michelson-Morley experiment

$$t_1 = \frac{l_1}{c-v} + \frac{l_1}{c+v} = \frac{2l_1}{c} \frac{1}{1-\frac{v^2}{c^2}}$$

$$t_2 = \frac{2l_2}{\sqrt{c^2 - v^2}} = \frac{2l_2}{c} \frac{1}{\sqrt{1-\frac{v^2}{c^2}}}$$

$$\Delta t = t_2 - t_1 = \frac{2}{c}\left(\frac{l_2}{\sqrt{1-\frac{v^2}{c^2}}} - \frac{l_1}{1-\frac{v^2}{c^2}}\right)$$

If we rotate the instrument through 90° (Fig. 8.2):

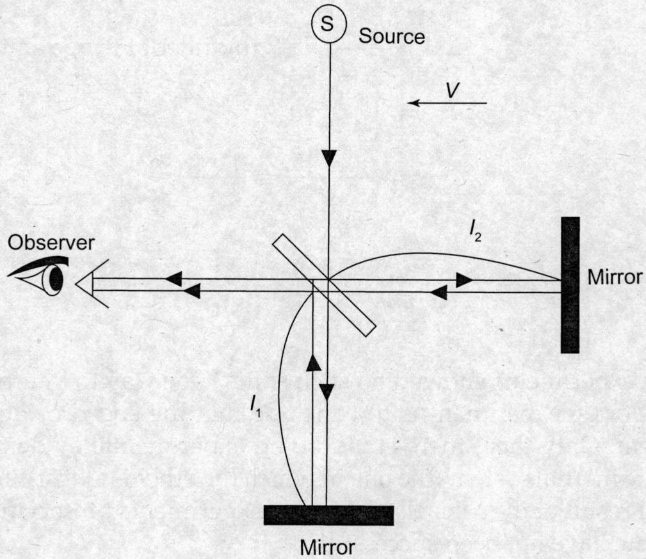

FIGURE 8.2 Rotated frame of observation in Michelson-Morley experiment

$$t_1' = \frac{2l_1}{\sqrt{c^2 - v^2}} = \frac{2l_1}{c} \frac{1}{\sqrt{1-\frac{v^2}{c^2}}}$$

$$t_2' = \frac{l_2}{c-v} + \frac{l_2}{c+v} = \frac{2l_2}{c} \frac{1}{1-\frac{v^2}{c^2}}$$

$$\Delta t' = t_2' - t_1' = \frac{2}{c}\left(\frac{l_2}{1-\frac{v^2}{c^2}} - \frac{l_1}{\sqrt{1-\frac{v^2}{c^2}}}\right)$$

$$\Delta t' - \Delta t = \frac{2}{c}\left(\frac{l_1+l_2}{1-\frac{v^2}{c^2}} - \frac{l_1+l_2}{\sqrt{1-\frac{v^2}{c^2}}}\right)$$

$$= \frac{2(l_1+l_2)}{c}\left(1+\frac{v^2}{c^2}-1-\frac{1}{2}\frac{v^2}{c^2}\right)$$

$$= \frac{l_1+l_2}{c}\frac{v_2}{c_2}$$

The change in the path difference of the beams by one wavelength is equal to shift in one of the fringes.

$$\Delta N = \frac{\Delta t' - \Delta t}{T} \quad \text{(number of fringe shifts)}$$

$$= \frac{\frac{l_1+l_2}{c}\frac{v^2}{c^2}}{\frac{\lambda}{c}} = \frac{l_1+l_2}{\lambda}\frac{v_2}{c_2}$$

For $\lambda = 5.5 \times 10^{-7}$ m, $l_1 = l_2 = 22$ m, $v/c = 10^{-4}$

$$\Rightarrow \qquad \Delta N = 0.4$$

The result of the experiments showed no such shifts. One way to interpret the null result of the Michelson-Morley experiment is to conclude that the speed of light is the same in every inertial system. Only then, there could be no experimental evidence to indicate the existence of a unique inertial system (the one in which the ether is at rest). The other attempts to interpret the null result of the Michelson-Morley experiments preserving the concept of a preferred ether frame have not been successful.

8.5 ▶ Postulates of Special Theory of Relativity

The laws of physics are generally assumed to be invariant under displacement, in time and space, as well as rotation about any axis. Einstein advocated the following additional symmetry for the laws of physics.

Postulate 1:

The laws of physics should be the same for all observers in uniform motion.

Given this symmetrical postulate and Newton's first law (that all objects persist in their state of uniform motion unless acted on by external forces), Einstein came to what we call the postulate of relative motion: All uniform motion are relative.

Consider the following situation. A train stops at a railway station, with another train facing the opposite direction on the other track. Ten minutes before the first train is due to leave, a person looks out of the window towards the second train, and see that it has slowly started to move relative to the first. After sometime, the person realizes that it was the other train that was moving.

To take this example further, suppose that the train is now moving at a constant speed of 100 km/hr. As long as the track is straight, we can pour beverages, juggle, play cards or do anything else without taking into account the fact that with the train everyone else is also moving. According to the postulate of relative motion, if the motion is truly uniform and a person is isolated from the effects of the matter outside the train, there is no experiment that can be performed that will prove whether the train is in motion or standing still. The best one can do is to measure its relative motion compared to some other object.

It is noticeable here that the uniformity of the motion plays an important role. Suppose that the driver of a train spots a cow blocking the tracks, and applies the breaks, causing the train to decelerate rapidly. People immediately detect this change in motion: if people are standing, they get thrown forward, or if they are holding a drink, the liquid suddenly splashes out. If a person is sitting on a seat facing the back of the train, he feels pressed hard against the back of the seat. Even if a person happens to look out of the window and see the change in relative motion between him and the neighbouring train, he would not be tempted to conclude that the other train was accelerating. It is the drink that is spilling, while the people on the other train are completely unperturbed. Thus, we can conclude that sudden change in motion affects the relative motion.

This seems to contradict that 'all motion is relative'. If this were true, why would observers in accelerating motion be different from those in uniform motion? The answer to this can be found in Newton's first law: there can be no acceleration without force. The many effects that you feel as your train decelerates are ultimately due to the forces causing your train to slow down. Although uniform motion is relative, there are differences between observers moving at constant velocity and those that are accelerating.

Einstein needed to make another assumption in order to arrive at the Special Theory of Relativity.

The behaviour of light is correctly described by Maxwell's theory of electromagnetism.

Maxwell's theory describes light as a wave whose velocity in vacuum is determined by fundamental constants of nature. Consequently, the second postulate can be formed as below:

Postulate 2:

The speed of light is the same for all observers in uniform motion. This is called constant speed of light postulate.

Note

The laws of physics are generally assumed to be invariant under displacement, in time and space, as well as rotation about any axis.

8.6 ▶ Derivation of the Lorentz Transformation Equations

Let us consider two inertial frames S and S' with a common x (x')-axis and y', z'-axes parallel to y, z-axes, respectively (Fig. 8.3). And the frame S' moving along $x(x')$ axes with a uniform velocity v. To simplify the algebra, we choose the relative velocity to be along the x (x')-axis (without loss of generality).

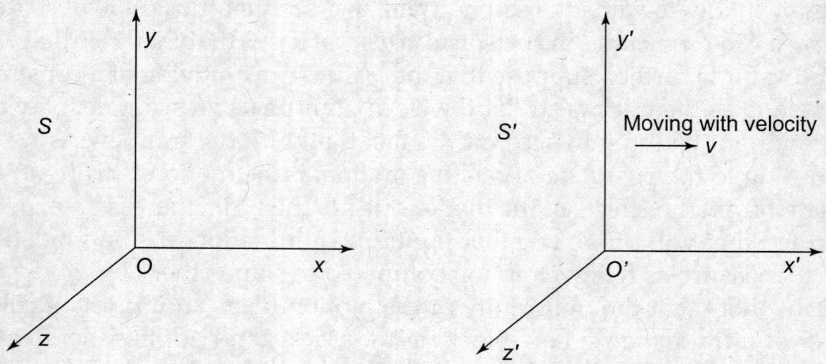

FIGURE 8.3 Derivation of the Lorentz transformation equations using inertial frames S and S'

The space–time coordinates of an event are t, x, y and z in S and t', x', y' and z' in S'.

$$t' = t'(t, x, y, z)$$

$$x' = x'(t, x, y, z)$$

$$y' = y'(t, x, y, z)$$

$$z' = z'(t, x, y, z)$$

The differential form of the above equation is

$$dt' = a_{00}\,dt + a_{01}\,dx + a_{02}\,dy + a_{03}\,dz$$

$$dx' = a_{10}\,dt + a_{11}\,dx + a_{12}\,dy + a_{13}\,dz$$

$$dy' = a_{20}\,dt + a_{21}\,dx + a_{22}\,dy + a_{23}\,dz$$

$$dz' = a_{30}\,dt + a_{31}\,dx + a_{32}\,dy + a_{33}\,dz$$

Now if we accept the assumption of the space–time homogeneity, the coefficients $a_{\alpha\beta}$ ($\alpha, \beta = 0, 1, 2, 3$) cannot depend on coordinates t, x, y and z. Therefore, we can integrate the above equations to obtain:

$$t' = a_{00}\,t + a_{01}\,x + a_{02}\,y + a_{03}\,z + b_0$$

$$x' = a_{10}\,t + a_{11}\,x + a_{12}\,y + a_{13}\,z + b_1$$

$$y' = a_{20}\,t + a_{21}\,x + a_{22}\,y + a_{23}\,z + b_2$$

$$z' = a_{30}\,t + a_{31}\,x + a_{32}\,y + a_{33}\,z + b_3$$

We simplify the situation once more by setting $t = t' = 0$ at the origin at the instant the origins O and O' coincide. Then $b_0 = b_1 = b_2 = b_3 = 0$. Since the x-axis coincides continuously with the x'-axis, it follows that $y' = z' = 0$ for $y = z = 0$, x, t is arbitrary.

$$a_{20} = a_{30} = a_{21} = a_{31} = 0$$

$$y' = a_{22}\, y + a_{23}\, z$$

$$z' = a_{32}\, y + a_{33}\, z$$

Also, the x-y plane should transform over to the x'-y' plane. Then it follows that $z' = 0$, y arbitrary $\Rightarrow a_{32} = 0 \Rightarrow z' = a_{33}\, z$. Likewise, $y' = a_{22} y$.

Now consider a rod lying along the y-axis extending from $y = 0$ to $y = 1$. The length of the rod measured by an S-observer is 1, and to an S'-observer the length is a_{22}. If we fix the same rod along the y'-axis, then an S'-observer should measure the length to be 1 and an S-observer should measure it to be $1/a_{22}$. The symmetry between the two inertial frames requires that

$$a_{22} = \frac{1}{a_{22}} \Rightarrow a_{22} = 1$$

Likewise, $a_{33} = 1$. Also, from the symmetry (or isotropy) argument, we conclude that t' and x' cannot depend on y and z.

$$a_{02} = a_{12} = a_{03} = a_{13} = 0$$

$$t' = a_{00}\, t + a_{01}\, x$$

$$x' = a_{10}\, t + a_{11}\, x$$

For the origin O', $x' = 0$ and $x = vt$

$$0 = a_{10}\, t = a_{11}\, vt = (a_{10} + va_{11})t$$

$$\Rightarrow \qquad a_{10} = -\, va_{11}$$

$$\Rightarrow \qquad x' = a_{11}\, t + (x - vt)$$

Summarizing the results obtained so far,

$$t' = a_{00}\, t + a_{01}\, x$$

$$x' = a_{11}\, (x - vt)$$

$$y' = y$$

$$z' = z$$

Now we assume that the speed of light is same, c, in all directions in both S and S'. Then for a light wave originating at the origin at $t = t' = 0$,

$$x^2 + y^2 + z^2 = c^2 t^2$$

$$x'^2 + y'^2 + z'^2 = c^2 t'^2$$

$$\Rightarrow \qquad a_{11}^2\left(x^2 - 2vxt + v^2t^2\right) + y^2 + z^2 = c^2\left(a_{00}t^2 + 2a_{00}a_{01}tx + a_{01}^2x^2\right)$$

$$\Rightarrow \qquad \left(a_{11}^2 - c^2a_{01}^2\right)x^2 + y^2z^2 = \left(c^2a_{00}^2 - a_{11}^2v^2\right)t^2 + 2\left(c^2a_{00}a_{01} + va_{11}^2\right)tx$$

$$\Rightarrow \qquad a_{11}^2 - c^2a_{01}^2 = 1$$

$$\Rightarrow \qquad c^2a_{00}^2 - a_{11}^2v^2 = 0$$

$$\Rightarrow \qquad a_{00} = \frac{1}{\sqrt{1-\dfrac{v^2}{c^2}}} \qquad a_{01} = \frac{-\dfrac{v}{c^2}}{\sqrt{1-\dfrac{v^2}{c^2}}} \qquad a_{11} = \frac{1}{\sqrt{1-\dfrac{v^2}{c^2}}}$$

$$\Rightarrow \qquad \boxed{t' = \frac{t - \dfrac{v}{c^2}x}{\sqrt{1-\dfrac{v^2}{c^2}}}} \tag{8.8}$$

$$\boxed{\left.\begin{aligned} x' &= \frac{x - vt}{\sqrt{1-\dfrac{v^2}{c^2}}} \\ y' &= y \\ z' &= z \end{aligned}\right\}} \tag{8.9}$$

The inverse transformation equations are

$$\boxed{\left.\begin{aligned} t &= \frac{t' + \dfrac{v}{c^2}x'}{\sqrt{1-\dfrac{v^2}{c^2}}} \\ x &= \frac{x' + vt'}{\sqrt{1-\dfrac{v^2}{c^2}}} \\ y &= y' \\ z &= z' \end{aligned}\right\}} \tag{8.10}$$

Note that the above two sets of transformation equation are identical in form except that v changes to $-v$. For the more general case of an arbitrary direction of v

$$t' = \frac{t - \dfrac{\vec{v}\cdot r'}{c^2}}{\sqrt{1-\dfrac{v^2}{c^2}}} \quad \vec{r} = \vec{r} - \frac{\vec{v}\,t}{\sqrt{1-\dfrac{v^2}{c^2}}} - \left(1 - \frac{1}{\sqrt{1}-\dfrac{v^2}{c^2}}\right)\frac{\vec{r}\,(\vec{v}\cdot\vec{r})}{v^2}$$

8.7 ▶ Principle of Special Theory of Relativity

In 1905, Einstein generalized the Galilean relativity principle (applicable only to Newton's laws) to the whole of physics by postulating that:

(i) All inertial frames are equivalent for all experiments, i.e. no experiment can measure absolute velocity.

(ii) Maxwell's equations and the speed of light must be the same for all observers.

Einstein's motive was to avoid inconsistencies between Maxwell's equations and Galilean relativity. The Lorentz transformations must relate to actual space and time measurements.

Theory of special relativity abolishes the idea of absolute space (e.g., the ether) and absolute time, but it leaves the origin of inertial frames unexplained. Mach's principle states that the inertial frames are determined by the rest of the matter in the universe (i.e., those that are non-accelerating with respect to the rest of the universe). Einstein later tried to incorporate his idea into general theory of relativity.

Note

Theory of special relativity abolishes the idea of absolute space (e.g., the ether) and absolute time, but it leaves the origin of inertial frames unexplained.

8.8 ▶ Consequences of the Special Theory of Relativity

The two major consequences arising out of the theory of relativity are time dilation and length contraction.

8.8.1 Time Dilation

Consider a clock which consists of a laser capable of sending out very short pulses and a mirror at a distance to reflect those pulses back to a detector. The correct time interval will be measured by an observer at rest with respect to this "clock". Consider a second observer watching as this clock goes past him/her at some speed (from his/her point of view). This observer will see the light pulse travel greater distance before it is back to the detector (see Fig. 8.4). According to the postulates of special relativity, the speed of light is the same for both inertial observers, though the other observer will claim a greater time interval.

This is time dilation effect and it can be shown that the two times are related by the following expression:

$$\boxed{\Delta t = \gamma \Delta t_0 = \frac{\Delta t_0}{\sqrt{1 - v^2/c^2}}} \tag{8.11}$$

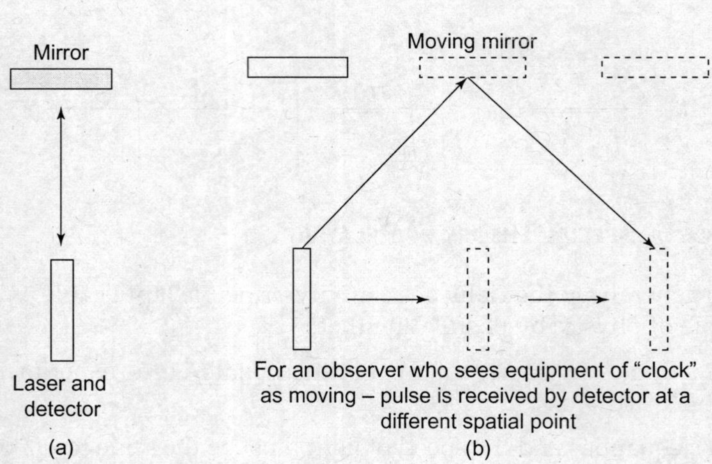

FIGURE 8.4 (a) Pulsed laser detector and mirror at rest; (b) the same equipment in motion towards right

This is the time dilation result. The proper time will be the shortest interval – the time intervals measured by other observers will be longer.

The slowing of clocks in a moving system is a real effect and it applies equally to all kinds of time, for example, biological and chemical reaction rates. The cosmic ray muons, which have short lifetime, travel a very short distance even if travelling at the speed of light. However, they are observed to exist for much longer than this as the time interval is much longer than the correct time which the muon measures. These particles disintegrate spontaneously after an average lifetime of about 2.2×10^{-6} seconds. It is clear that in its short lifetime, a muon cannot, even at the speed of light, travel more than 600 m. Although muons exist in the upper layers of the atmosphere, some 10 km up, they can be detected here on the earth. From the muon's point of view (i.e., from their frame of reference), they only live about 2 μs. However, from our point of view, they live considerably longer, indeed long enough to reach the surface of the earth (by a factor of $1/\sqrt{1-v^2/c^2}$). Particle accelerator experiments have proven the time dilation effect and atomic clocks have been flown at high speed and then compared to a similar clock on the ground to directly confirm the time dilation prediction.

Time in the moving system will be observed by a stationary observer to be running slower by the factor:

$$t' = \frac{t}{\sqrt{1-\frac{v^2}{c^2}}}$$

The amount of time dilation with velocity is shown in Fig. 8.5.

Note

Time in the moving system will be observed by a stationary observer to be running slower by the factor of $1/\sqrt{1-v^2/c^2}$. Since v^2/c^2 will always be non-zero, this factor will always be greater than 1.

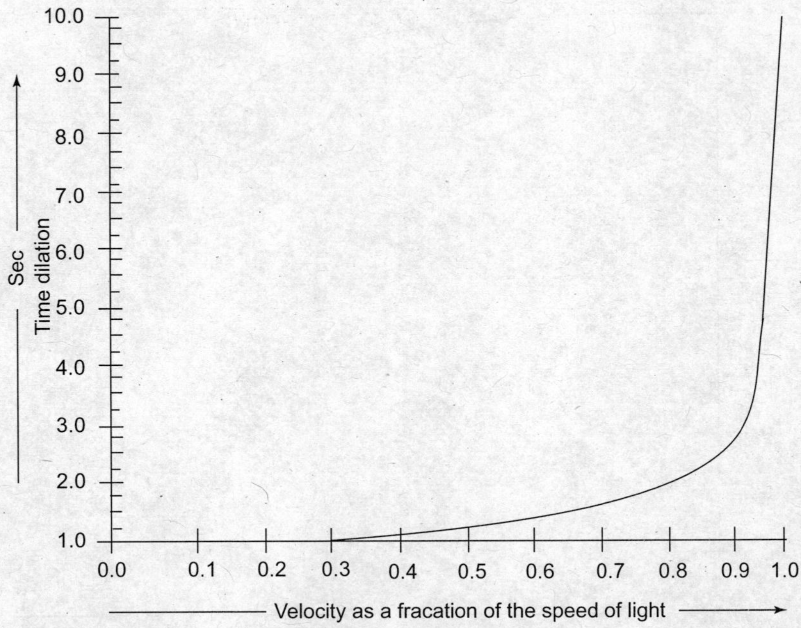

FIGURE 8.5 Variation of time dilation versus velocity

8.8.2 Length Contraction

Similar to an effect on measurement of time interval, the relative motion has an observable effect on measurement of length. An observer moving with respect to an object will observe it to be contracted along the direction of motion by the factor:

$$\boxed{l' = l\sqrt{1-\frac{v^2}{c^2}} = l\gamma, \quad \gamma = \sqrt{1-\frac{v^2}{c^2}}} \tag{8.12}$$

where l' is the length observed by the moving observer, l is the length measured when at rest with respect to the object, v is the velocity of the observer, and c is the speed of light. If we express v as a fraction of the speed of light, the units in which it has to be expressed are not relevant.

But despite this we do not see the people on the sidewalk skinny when we zip past them on our bike. Figure 8.6 shows a graph of this equation and explains the reason.

It is clear from the graph that the length contraction has little effect at small velocities and really starts to kick in only at speed near to that of light. Table 8.1 gives the observed length for velocities of various fractions of the speed of light.

Remember that the speed of light is very large compared to any velocity we encounter in our everyday life. It is about 300,000,000 metres per second, or a thousand million kilometres per hour.

Whizzing along on the bike at 60 km/hr, means travelling at 0.00000005594 of the speed of light. Thus, stationary objects appear to be contracted by one part in 0.0000000000000155, which is not perceptible.

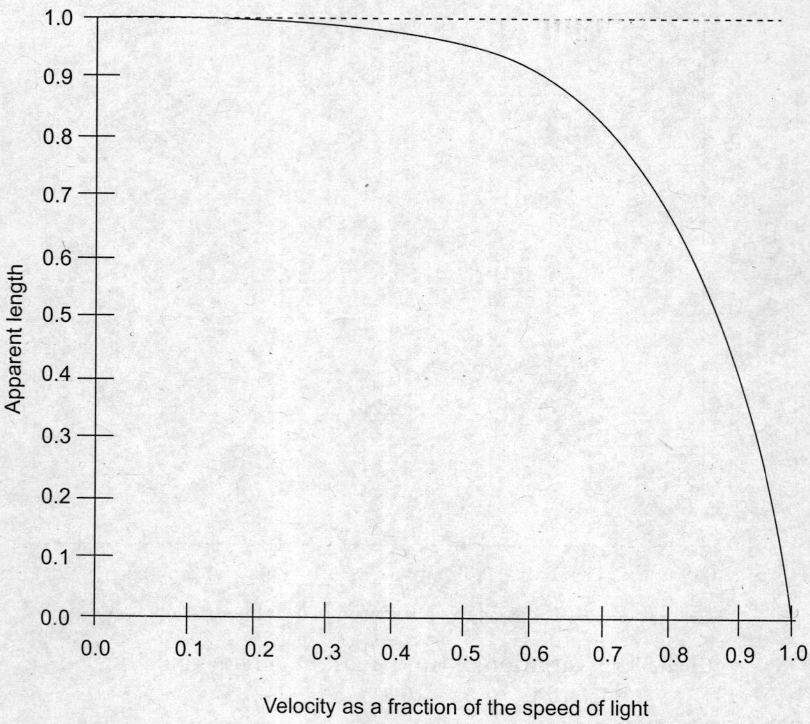

FIGURE 8.6 Variation in apparent length vs velocity

TABLE 8.1 Length contraction at different velocities

Velocity	Observed length
0	1.000
0.1	0.995
0.2	0.980
0.3	0.954
0.4	0.917
0.5	0.866
0.6	0.800
0.7	0.714
0.8	0.600
0.9	0.436
0.95	0.312
0.97	0.243

(Cont'd)

TABLE 8.1 (Continued)

Velocity	Observed length
0.99	0.141
0.995	0.100
0.999	0.045
0.9999	0.014

Now suppose the observer O wants to measure the length of a rod. He/She can only do this by taking an instantaneous measurement of the spatial coordinates of the end of the rod, that is x_1 and x_2.

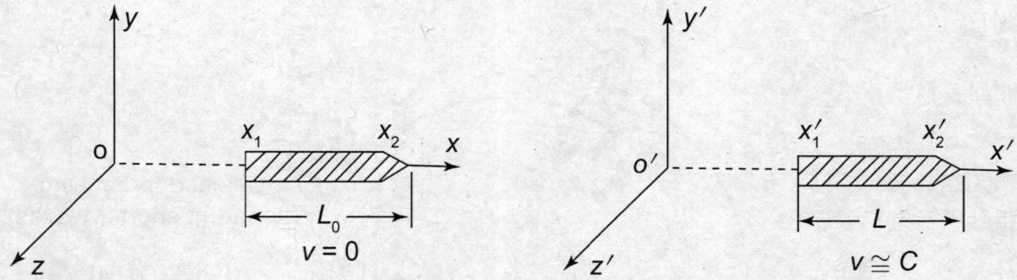

FIGURE 8.7 Length of rod L_0 as observed in frame O at rest appears as L in a moving frame O'

The Lorentz transformations in such a case give

$$x_1' = \gamma(x_1 - vt_1)$$

$$x_2' = \gamma(x_2 - vt_2) \tag{8.13}$$

So the length is given by

$$L_0 = x_2 - x_1$$

$$L_0 = \gamma^{-1}(x_2' - x_1')$$

$$L_0 = \gamma^{-1}L \tag{8.14}$$

since $t_1 = t_2$ (instantaneous measurement by O). Writing this result in terms of v and c, we have

$$L_0 = \sqrt{\frac{1-v^2}{c^2}}L \tag{8.15}$$

Note that this is not a physical effect on the rod, but an effect of space–time itself as observed.

Note

The relative motion has an observable effect on measurement of length. An observer moving with respect to an object will observe it to be contracted along the direction of motion by the factor $1/\sqrt{1-v^2/c^2}$.

8.9 ▶ The Doppler Effect

If you carefully listen to the siren of a fast approaching ambulance, the siren's pitch changes as the vehicle races towards and then away from you. First, the pitch becomes higher and then lower. This change in pitch results from a shift in the frequency of the sound waves, as illustrated in Fig. 8.8. The phenomenon was explained in detail by C. Doppler and is known as Doppler Effect.

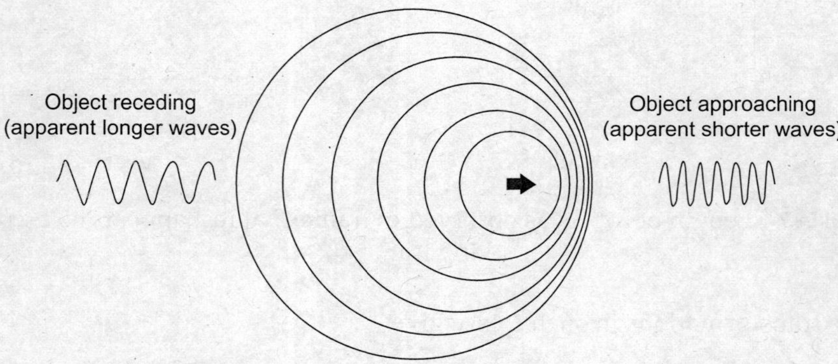

FIGURE 8.8 The Doppler effect

As the ambulance approaches nearer, the sound waves from its siren are compressed towards the observer. The intervals between the waves diminish, which translates into an increase in frequency or pitch. As the ambulance recedes, the sound waves are stretched relative to the observer, causing the siren's pitch to decrease and you can determine if the ambulance is coming nearer or speeding away. If you could measure the rate of change of pitch, an estimate could be made about the ambulance's speed.

By analogy, the electromagnetic radiation emitted by a moving object also exhibits Doppler effect. The radiation emitted by an object moving towards an observer is squeezed; its frequency appears to increase and is therefore said to be blue shifted. In contrast, the radiation emitted by an object moving away is stretched or red shifted. As in the ambulance analogy, blue and red shifts exhibited by stars, galaxies and gas clouds also indicate their motions with respect to the observer.

Note

The pitch of a fast-approaching sound source changes as it races towards and then away from you. First, the pitch becomes higher and then lower. This change in pitch results from a shift in the frequency of the sound waves. The phenomenon was explained in detail by C. Doppler and is known as Doppler Effect.

8.10 ▶ Red and Blue Shifts

Light from a moving object will appear to have different wavelengths depending on the relative motion of the source and the observer (Fig. 8.9).

Observers looking at an object, that is moving away froth them observe that light has a longer wavelength than it had when it was emitted (a red shift), while observers looking at an approaching source view light that has shifted to shorter wavelength (a blue shift).

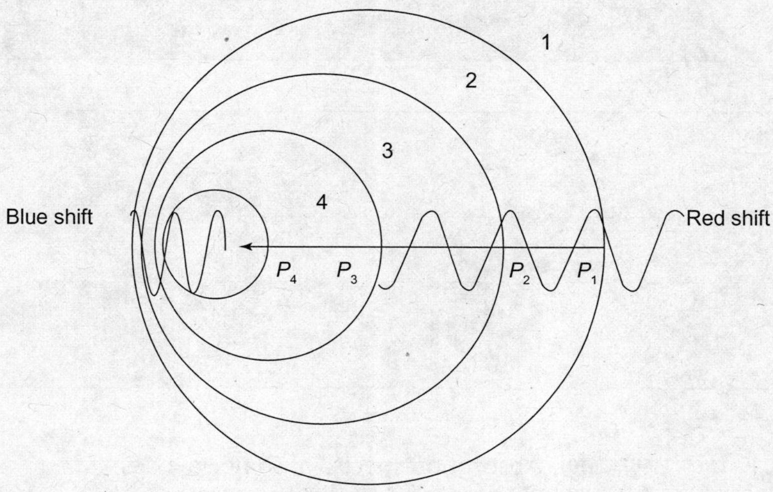

FIGURE 8.9 Red and blue shifts

The spectrum of a galactic star is shown at the bottom in Fig. 8.10. The spectrum shows the dark absorption lines first seen by Fraunhofer. These lines can be used to identify the chemical elements in distant stars. They also tell us the radial velocity of star. The other three spectra from bottom to top show the spectra from a nearby galaxy, a medium distance galaxy and a distant galaxy. The spectra shown are negatives and so the brightest parts of the galaxies are black. Notice how the pattern of absorption lines shift to red as the galaxies get fainter. The numbers above and below the spectra are wavelengths measured in nanometres (nm).

In a star which is at rest with respect to us, or in a laboratory standard, the line wavelengths are 393 and 397 nm from Ca II (ionized calcium); 410, 434, 486 and 656 nm from H I (atomic hydrogen); 518 nm from Mg I (neutral magnesium); and 589 nm from Na I (neutral sodium). By measuring the amount of shift to red, we can determine that the bright galaxy is moving away at 3000 km/sec, which is 1 percent of the speed of light, because its lines are shifted in wavelength by 1 percent to the red. The red shift, z is defined such that:

$$1+z=\frac{\lambda(\text{observed})}{\lambda(\text{emitted})}$$

which is

$$1+z=\frac{397}{393}=\frac{401}{397}=\frac{414}{410}$$

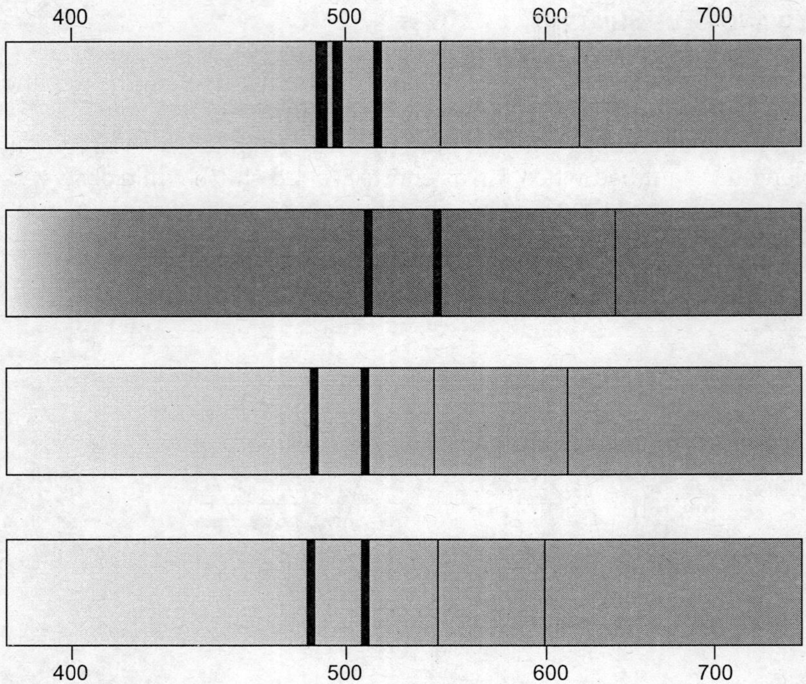

FIGURE 8.10 Absorption spectra of some galaxies

$$= \frac{438}{434} = \frac{491}{486} = \frac{523}{518}$$
$$= \frac{595}{589} = \frac{663}{656} = 1.01$$

In this case, $z = 0.01$ for the bright galaxy. The radial velocity is usually approximated as v (rad) $= cz$, where c is the speed of light. The medium bright galaxy is moving away at 15,000 km/sec with $z = 0.05$, while the faintest and most distant galaxy is moving away at 75,000 km/sec with $z = 0.25$.

Note

Light from a moving object will appear to have different wavelengths depending on the relative motion of the source and the observer. If it shifts towards longer wavelength, it is called red shift and if towards shorter wavelengths it is called blue shift.

8.11 ▶ Mass–energy Equivalence

Another implication of special theory of relativity was in the form of mass–energy equivalence. We know that when the velocity of an object is increased, kinetic energy is imparted. We also know that kinetic energy can be converted into other forms of energy, such as potential energy and heat energy and vice-versa. If increasing the velocity of an object increases its inertial mass, then in some sense, we are converting kinetic energy into mass. Einstein reasoned that mass and energy must somehow be different manifestations of the same thing. In particular, if we can convert kinetic energy into mass, we should also be able to convert mass into energy. The special theory of relativity, in fact, gives an explicit expression for how mass and energy are related:

$$\boxed{E = mc^2 = m_0\gamma c^2} \tag{8.16}$$

Thus, as expected from earlier objects, the energy of an object increases without bound as its velocity approaches the speed of light. The surprising aspect of Eq. (8.7) is that even when an object is at rest, it has energy, for a mass at rest when $v = 0$, $\gamma = 1$ and $m = m_0$,

$$E_0 = m_0c^2 \tag{8.17}$$

The energy of an object at rest is related to its mass by the factor c^2, which is a very large number. The consequence is that very small masses can produce vast amounts of energy, and hence nuclear power plants and bombs can work because they can convert mass into other forms of energy. In fact, if all the mass in a single gram of sand could be converted into energy, it could provide enough electricity to light up a whole city for a year. However, we will see in the section on nuclear physics that it is not easy to convert large amounts of mass into energy.

Recap Zone

Points to Remember

- The study of Newtonian mechanics and laws of relative motion led to the Galilean transformations and the Newtonian relativity.
- The Maxwell's theory of electromagnetism established that light travels with the same speed in all directions irrespective of the motion of source or observer.
- The experimental tests like Michelson-Morley experiment failed to establish the existence of the medium ether.
- Einstein postulated that all laws of physics are same in all inertial frames of reference and the speed of light is same for all observers in uniform motion as the fundamental postulates of relativity.
- Lorentz transformations relate the velocity and co-ordinates in two frames of reference in motion with respect to each other.
- Maxwell's equations of electromagnetism were also found to be invariant under Lorentz transformations.
- The time interval between two events is delayed and the length of an object appears to be contracted when observed from a frame in uniform motion.

- The Doppler effect is observed for the light coming from receding stars in our and other galaxies that establishes the theory of relativity.
- The mass and energy are inter-convertible and are the manifestation of same. These are related by $E = mc^2$.
- The theory of relativity limits the maximum speed of particles up to the speed of light but does not rule out existence of particles already moving with a velocity greater than c, such particles are named as Tachyons.

KEY TERMS WITH DEFINITIONS

- **Doppler effect** – It is the change in frequency or wavelength of a wave (or other periodic event) for an observer moving relative to its source.
- **Time dilation** – In the theory of relativity, time dilation is a difference of elapsed time between two events as measured by observers either moving relative to each other or differently situated from a gravitational mass or masses.
- **Length contraction** – It is the physical phenomenon of decrease in length detected by an observer of objects that travel at any non-zero velocity relative to that observer.

IMPORTANT FORMULAE AND EQUATIONS

Equation Number	Equation	Remarks
(8.7)	$t' = t$ $x' = x - vt$ $y' = y$ $z' = z$	**The two sets (t, x, y, z) and $(t(, x', y', z')$ are related to each other, according to the Galilean transformation as, $t' = t$**
(8.8)	$t' = \dfrac{t - \dfrac{v}{c^2}x}{\sqrt{1 - \dfrac{v^2}{c^2}}}$	**According to Lorentz transformation**
(8.9)	$x' = \dfrac{x - vt}{\sqrt{1 - \dfrac{v^2}{c^2}}};\ y' = y;\ z' = z$	
(8.10)	$t = \dfrac{t' + \dfrac{v}{c^2}x'}{\sqrt{1 - \dfrac{v^2}{c^2}}};\ x = \dfrac{x' - vt'}{\sqrt{1 - \dfrac{v^2}{c^2}}};\ y = y';\ z = z'$	**According to inverse Lorentz transformation**

(8.11)	$\Delta t = \dfrac{\Delta t_0}{\sqrt{1-\dfrac{v^2}{c^2}}}$	**Time dilation**
(8.12)	$l' = l\sqrt{1-\dfrac{v^2}{c^2}}$	**Length contraction**
(8.16)	$E = mc^2$	**Mass energy equivalence when considering (rest mass) $= E_0 = m_0c^2$**

REVIEW ZONE

SHORT ANSWER QUESTIONS

1. Describe Michelson-Morley experiment and physical significance of its results.
2. Explain the basic postulates of special theory of relativity.
3. Discuss Lorentz transformations.
4. Derive the expressions for length contraction and time dilation.
5. What is mass–energy equivalence?
6. If one photon has a speed c in one reference frame, can it be at rest in some other frame of reference?
7. What are Red Shift and Blue Shift? How they provide experimental proofs of theory of relativity?

NUMERICAL PROBLEMS

1. Show that for the values of $v << c$ the Lorentz transformation reduces to Galilean transformation.
2. If (x_1, y_1, z_1, t_1) are event co-ordinates in frame S_1 and the corresponding co-ordinates for the same event in frame S_2 are (x_2, y_2, z_2, t_2) then show that $dS_1^2 = dx_1^2 + dy_1^2 + dz_1^2 - c^2\, dt_1^2$ is invariant under Lorentz transformation.
3. A rod of length 10 m long is moving along its length with a velocity of 0.6 C. Calculate its length as it appears to an observer at rest.
4. What should be the velocity of a rocket to have its length contracted to 98% of its length on ground?
5. Two events take place at an interval of 10 seconds in space. If these are observed by an astronaut in space ship travelling at a speed of 3×10^7 m/s. What would be the apparent interval observed by astronaut?

6. If a mass of 4 kg is fully converted into energy, how much energy is produced?
7. Calculate the kinetic energy of an electron moving with a velocity of 0.98 c in a laboratory.
8. Calculate the percentage contraction of a rod moving with a velocity of 0.8 c in a direction inclined at 60° to its length.
9. A π-meson has a mean life of 2×10^{-8} S when measurement at rest. How far does it go before decaying if moving at a speed of 0.99 c?
10. Find the volume of a cube whose side is I_0 at rest and is moving along one of its side with velocity v.
11. Calculate the kinetic energy of a particle moving with the velocity of light.

CHAPTER 9

Ultrasonics

Learning Objectives

By the end of this chapter, the student will be able:

- To understand ultrasonic waves and their methods of generation
- To learn about piezoelectric and magnetostriction effect
- To explain cavitation effect and its applications
- To demonstrate working of acoustic grating
- To describe applications of ultrasonic waves (e.g., SONAR)

9.1 ▶ Introduction

The human ear can detect the audio frequencies within the range of 20 to 20,000 kHz. The sound waves that are having frequencies beyond 20,000 kHz, i.e. beyond audible range, are called ultrasonics (meaning *ultra*-beyond; *sonic*-sound). These waves are generated and detected by the methods different from those used for sound. Bats can generate ultrasonic waves and detect these by using their body sensors to detect their pray in dark. The bats, some fishes such as dolphin, whales, sharks, some nocturnal insects, crickets, grass hoppers, some species of frogs, rats, snakes, butterflies, etc. have capabilities to generate and detect ultrasonic waves. This, they do to defend themselves from prays or to detect their pray by using sonar mechanism. Human ears cannot detect these waves.

9.2 ▶ Generation of Ultrasonic Waves

We can generate ultrasonic waves by the following two main methods:

1. Piezoelectric generator
2. Magnetostrictive generator

In order to understand the working of these ultrasonic generators, let us first understand about the materials that are specifically used in generation of ultrasonic waves due to their unique properties. There are mainly two types of materials that are used for ultrasonic wave generation. They are piezoelectric and magnetostrictive materials.

9.3 ▶ Piezoelectric Materials

These are the materials that were discovered by Curie brothers in 1880 while studying the properties of asymmetrical crystals. These crystals are found with two different types of axes, one known as electrical axis and the other as mechanical axis. When a mechanical pressure is applied across mechanical axis of the crystal, an electrical signal is generated across electrical axis, or vice-versa. This effect is called piezoelectric effect. Some well-known crystals such as quartz, tourmaline, Rochelle salt, etc. show this type of property.

When a potential difference is applied across the opposite faces along electrical axis of their crystals, then a mechanical expansion or contraction takes place along mechanical axis. This is sometimes called as inverse piezoelectric effect.

The piezoelectric and inverse piezoelectric effects are shown in Fig. 9.1(a) and(b). The induced charge, or mechanical contraction/expansion is used to generate the ultrasonic waves that are mechanical sound waves of a higher frequency than audio range.

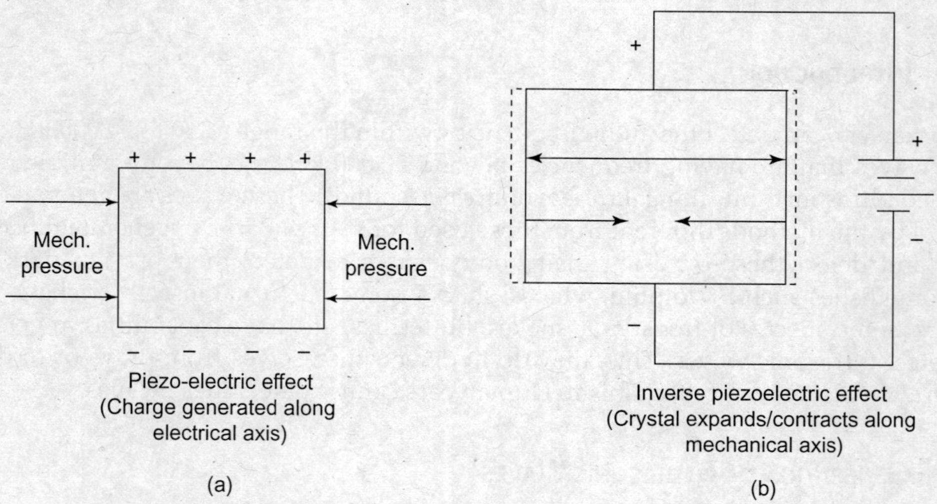

FIGURE 9.1 Piezoelectric (a) and inverse piezoelectric (b) effects

The quartz crystal is a most popular and ideal piezoelectric material as it meets all the required characteristics. This can also be synthesized in lab and can be used up to 400°C. It can easily be cut in different angles and shapes to meet the frequency and power requirements.

9.4 ▶ PIEZOELECTRIC GENERATOR

The inverse piezoelectric effect is the principle of working of these generators. When an oscillating electrostatic field is applied across the two faces of a crystal, then material expansion/contractions takes place across faces perpendicular to the direction of electric field.

The circuit diagram of a piezoelectric oscillator is shown in Fig. 9.2. A crystal of piezoelectric material with electrical surfaces *A* and *B* is subjected to the oscillating electric field generated by a tank circuit comprising coils L_1, L_2, L_3 and capacitor C and transistor *T*. The coil L_2 is the feedback loop and coils L_1, L_2 and L_3 are inductively coupled to each other. The variable capacitor *C* is charged to tune the frequency of oscillator. The frequency of oscillation is given as

$$f = \frac{1}{2\pi\sqrt{L_1 C}}$$

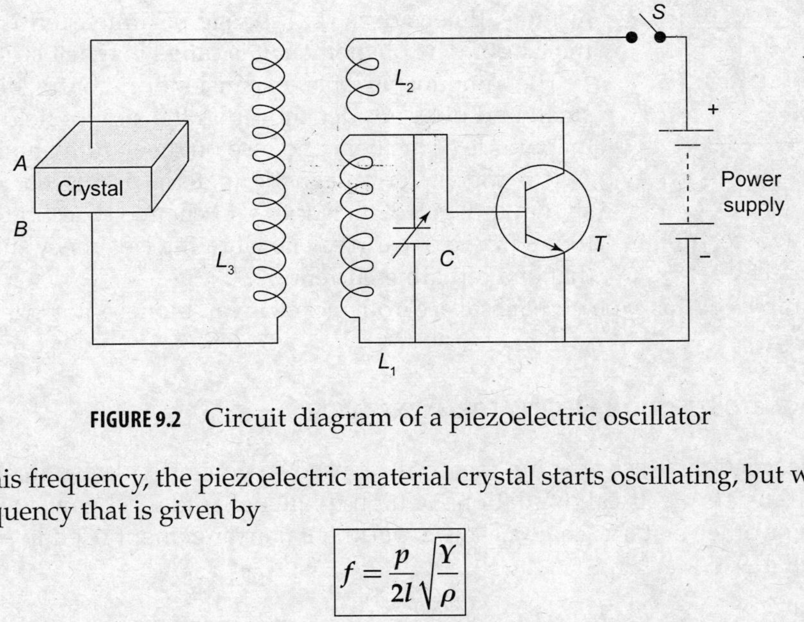

FIGURE 9.2 Circuit diagram of a piezoelectric oscillator

Under this frequency, the piezoelectric material crystal starts oscillating, but with its own natural frequency that is given by

$$f = \frac{p}{2l}\sqrt{\frac{Y}{\rho}} \quad (9.1)$$

where $P = 1,2,3,\ldots$ for fundamental, first overtone, second overtone, etc.
Y = Young's modulus of the material of crystal
ρ = density of the material of crystal.

When natural frequency of oscillation of crystal matches that of the electrical circuit, the resonance takes place and the crystal produces ultrasonic longitudinal waves of large amplitude.

Advantages

(i) The frequencies up to 500 MHz can be produced by this arrangement.

(ii) The output power can also be high

(iii) The frequency of crystal is not affected by temperature and humidity.

Disadvantages

(i) The cost of crystal is high.

(ii) The precise cutting and shaping of crystal is a complex process.

Note

Frequency of oscillation is given by, $f = \dfrac{1}{2\pi\sqrt{LC}}$.

9.5 ▶ Magnetostriction Effect

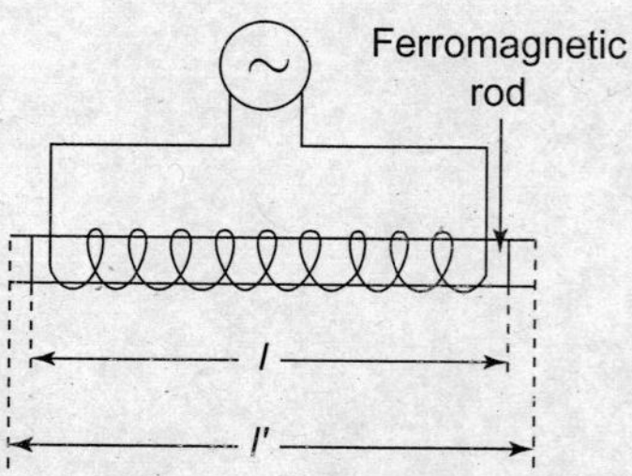

FIGURE 9.3 Magnetostriction effect

Some ferromagnetic materials such as iron or nickel when subjected to alternating magnetic fields show a small change in their dimensions. This change in dimension is parallel to the length. The phenomenon is called magnetostriction effect.

The amount of this expansion or contraction is very small and does not depend upon the sign of field but on the magnitude of the field and also on the nature of material.

If a rod of ferromagnetic material is placed in a coil as shown in Fig. 9.3, its length will change twice in one cycle of frequency or one can say its vibrating frequency shall be twice that of alternating current.

Some examples of this type of material are iron, nickel, invar, monel and perm alloys and cobalt ferrites.

9.6 ▶ Magnetostriction Oscillator/Generator

The magnetostriction effect of some materials is used to generate ultrasonic waves by a device that transfers electrical oscillations of high frequency into mechanical longitudinal waves in air or other material media. This experimental arrangement used in such device is shown in Fig. 9.4.

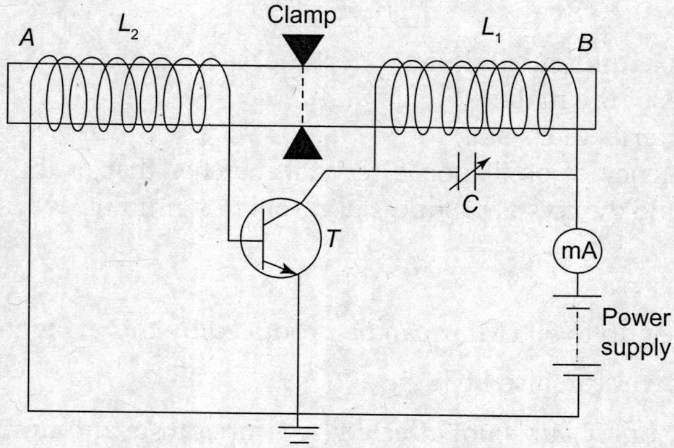

FIGURE 9.4 Experimental arrangement of a magnetostriction oscillator/generator

A ferromagnetic material rod AB is rigidly clamped in the middle. An oscillating current is generated using an LC circuit and a transistor that creates an oscillating magnetic field across coils L_1 and L_2. The frequency of oscillator is controlled by the variable capacitor C. The coil L_1 and the capacitor C make the resonant circuit of the device. The coil L_2 fed to the base of the transistor T acts as a feedback loop.

When the circuit is energized by switching on the power supply, the collector circuit oscillates with the frequency

$$\boxed{f = \frac{1}{2\pi\sqrt{L_1 C}}} \tag{9.2}$$

This oscillating current also produces an oscillating magnetic field of frequency f along the length of the rod. As a result, the ferromagnetic rod also starts oscillating with the frequency given by

$$\boxed{f = \frac{1}{2l}\sqrt{\frac{Y}{\rho}}} \tag{9.3}$$

where l = length of the rod

Y = Young's modulus of material of rod

ρ = density of the rod

The value of capacitor C is adjusted in such a way that the frequencies f in Eqs. 9.2 and 9.3 are same. This condition is called resonance. At this condition, the rod continues to vibrate without damping as per frequency of the L_1C circuit with maximum amplitude, and also generates ultrasonic waves from its both ends, that can be used.

Note

The natural frequency of vibration is given by, $f = \frac{p}{2l}\sqrt{\frac{Y}{\rho}}$

where l = length of the rod, p = 1, 2, 3, ... for fundamental, first overtone, second overtone, Y = Young's modulus and ρ = density of material of the rod.

Advantages

(i) The design and cost are very low.

(ii) At low frequencies, high-power waves can be generated.

(iii) The circuit has long life as no movable/mechanical parts are involved.

Disadvantages

(i) The frequencies of oscillations drift with temperatures.

(ii) There is some energy loss due to hysteresis and eddy cements.

(iii) The frequencies beyond 300 KHz (or MHz) cannot be generated.

9.7 ▶ Cavitation Effect

Cavitation is the formation of gas bubbles of a flowing liquid in a region where the pressure of the liquid falls below its vapour pressure. It is usually divided into two classes of behaviour: inertial (or transient) cavitation and non-inertial cavitation.

Inertial cavitation is the process where a void or bubble in a liquid rapidly collapses, producing a shock wave. It occurs in nature in the strikes of mantis, shrimps and pistol shrimps, as well as in the vascular tissues of plants. In artefacts, it can occur in control valves, pumps, propellers and impellers.

Non-inertial cavitation is the process in which a bubble in a fluid is forced to oscillate in size or shape due to some form of energy input, such as an acoustic field. Such cavitation is often employed in ultrasonic cleaning baths and can also be observed in pumps, propellers, etc.

Since the shock waves formed by cavitation are strong enough to significantly damage moving parts, cavitation is usually an undesirable phenomenon. It is specifically avoided in the design of machines such as turbines or propellers, and eliminating cavitation is a major field in the study of fluid dynamics.

9.7.1 Applications Based on Cavitation

Chemical engineering applications

In industry, cavitation is often used to homogenize, or mix and break down, suspended particles in a colloidal liquid compound such as paint mixtures or milk. Many industrial mixing machines are based upon this principle. It is usually achieved through impeller design or by forcing the mixture through an annular opening that a narrow entrance orifice with a much larger exit orifice. In the latter case, the drastic decrease in pressure as the liquid accelerates into a larger volume induces cavitation. This method can be controlled with hydraulic devices that control inlet orifice size, allowing for dynamic adjustment during the process, or modification for different substances. The surface of this type of mixing valve, against which surface the cavitation bubbles are driven causing their implosion, undergoes tremendous mechanical and thermal localized stress; they are therefore often constructed of super-hard or tough materials such as stainless steel, Stellite alloy or even polycrystalline diamond (PCD).

Cavitating water purification devices have also been designed, in which the extreme conditions of cavitation can break down pollutants and organic molecules. Spectral analysis of light emitted in sonochemical reactions reveal chemical and plasma-based mechanisms of energy transfer. The light emitted from cavitation bubbles is termed as sonoluminesence.

Hydrophobic chemicals are attracted underwater by cavitation as the pressure difference between the bubbles and the liquid water forces them to join together. This effect may assist in protein folding.

Biomedical application

- Cavitation plays an important role for the destruction of kidney stones in shock wave lithotripsy.
- Currently, tests are being conducted as to whether cavitation can be used to transfer large molecules into biological cells (sonoporation).
- Nitrogen cavitation is a method used in research to release proteins from cell membranes while leaving organelles intact.

- Cavitation plays a key role in non-thermal non-invasive fractionation of tissue for treatment of a variety of diseases. It also probably plays a role in HIFU (high intensity focussed ultrasound) a thermal non-invasive treatment methodology for cancer.
- Ultrasound is sometimes used to increase bone formation, for instance post-surgical applications.
- Ultrasound treatments or exposure can create cavitation that can potentially result in a syndrome involving manifestations of nausea, headache, tinnitus, pain, dizziness and fatigue.

Cleaning application

In industrial cleaning applications, cavitation has sufficient power to overcome the particle-to-substrate adhesion forces, loosening contaminants. The threshold pressure required to initiate cavitation is a strong function of the pulse width and the power input. This method works by generating controlled acoustic cavitation in the cleaning fluid, picking up and carrying contaminant particles away so that they do not reattach to the material being cleaned.

Note

Cavitation is useful in HIFU (high intensity focussed ultrasound) a thermal non-invasive treatment methodology for cancer.

9.8 ▶ Applications of Ultrasonic

9.8.1 Echo Sounding

It is the technique of using sound pulses directed from the surface or from a submarine vertically down to measure the distance to the bottom by means of sound waves. Echo sounding can also refer to hydroacoustic "echo sounders" defined as active sound in water (sonar) used to study fish. Hydroacoustic assessments have traditionally employed mobile surveys from boats to evaluate fish biomass and spatial distributions. Conversely, fixed-location techniques use stationary transducers to monitor passing fish. Distance is measured by multiplying half the time from the signal's outgoing pulse to its return by the speed of sound in the water, which is approximately 1.5 km per sec. Echo sounding is effectively a special-purpose application of sonar used to locate the bottom.

9.8.2 Sonar

There are various and wide applications for ultrasonic waves, but none better than SONAR. SONAR is an acronym for "Sound Navigation and Ranging". It employs the principle of Echo Sounding and is mostly used to find the depth of the sea.

Experimental arrangement

An ultrasonic beam is transmitted through the sea and the reflected wave is received (Fig. 9.5) and the time interval is measured. Knowing the velocity of the ultrasonic waves in the sea, the distance travelled inside is calculated and hence the depth of the sea. In most cases, the piezoelectric transducer can act both as a transmitter and as a receiver.

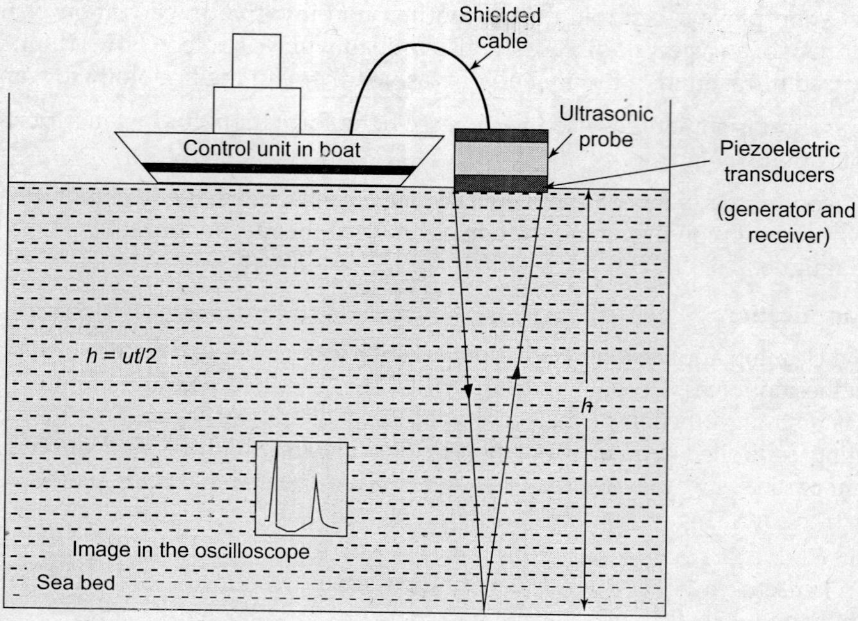

FIGURE 9.5 Illustration of working of SONAR

Working

An ultrasonic beam of narrow pulse produced by the quartz crystal of a piezoelectric oscillator is directed towards the sea bed. The transmitted wave gets reflected from the bottom of the sea bed and gets detected by the crystal itself. The received signal is amplified and fed to an oscilloscope. The image in the oscilloscope (Fig. 9.5) shows two peaks – one transmitted and the other received on a time-based scale.

The time interval between the pulse of the transmitted signal and that of the received signal is noted from the time-based scale. Let the time interval be 't'. If the velocity of the ultrasonic waves in sea water is 'u', then

$$\boxed{u = \frac{d}{t}} \tag{9.4}$$

where d is the total distance travelled (from transmitter to the sea bed and back to the receiver), i.e.,

$$d = 2h \tag{9.5}$$

then,

$$u = \frac{2h}{t} \tag{9.6}$$

or

$$h = \frac{ut}{2} \tag{9.7}$$

where h is the distance or depth at which the reflection of the pulse occurs.

This instrument can be calibrated to directly measure the depth of the sea and is known as fathometer. With necessary ancillary instruments, this instrument can be deployed in order to measure the depth and to map the sea bed continuously.

9.8.3 Non-Destructive Testing

Any engineering product after development needs to be tested for defects. Sometimes, the process of testing a material creates damage on the material. For example, to determine the ultimate strength of a material, it is subjected to high values of stress. The specimen breaks while determining a parameter. This is destructive testing. Many testing procedures reduce the reliability and functioning of the product. They are also destructive testing procedures as they cause some kind of damage on the material.

Non-destructive testing (NDT) is the method of testing or examining a product to identify the defects without causing any damage to the product. The visual inspection is the most common non-destructive test. Human eye is an NDT tool. Other aids used for visual inspection are microscope, telescope, endoscope, fibrescope, gauges, magnifying glasses, flashlights and tape measures, or calipers. For example, fillet gauges are measuring devices used for checking convexity, concavity and flatness of welded areas. There are portable video inspection units available for visual inspection of large tanks, vessels and railway lines. Robotic crawlers fitted with cameras are available for visual inspection in hazardous environments such as nuclear reactors. The major areas of application of ultrasonic NDT methods include the following:

- Detection and location of flaws and leaks
- Thickness measurements
- Estimation/properties of materials
- Characterization of microstructure of materials

9.8.4 Ultrasonic Machining

It is, also known as ultrasonic impact grinding, a machining operation in which an abrasive slurry freely flows between the work piece and a vibrating tool. It differs from most other machining operations because very little heat is produced. The tool never contacts the work piece and as a result the grinding pressure is rarely more than 2 pounds, which makes this operation perfect for machining extremely hard and brittle materials, such as glass, sapphire, ruby, diamond and ceramics.

Surface finish

The surface finish of ultrasonic machining depends upon the hardness of the workpiece/tool and the average diameter of the abrasive grain used. Up close, this process simply utilizes the plastic deformation of metal for the tool and the brittleness of the workpiece. As the tool vibrates, it pushes down on the abrasive slurry (containing many grains) till the grains impact the brittle workpiece. The workpiece is broken down while the tool bends very slightly. Commonly used tool material consists of nickel and soft steels.

Machine time

Machine time depends upon the frequency at which the tool is vibrating, the grain size and hardness (which must be equal or greater than the hardness of the workpiece), and the viscosity of the slurry fluid. Common grain materials used are silicon carbide and boron carbide/because of their hardness. The less viscous is the slurry fluid, the faster it can carry away used abrasive.

9.8.5 Ultrasonic Welding/Soldering

It is an industrial technique whereby high-frequency ultrasonic acoustic vibrations are locally applied to workpieces being held together under pressure to create a solid-state weld. It is commonly used for plastics and especially for joining dissimilar materials. In ultrasonic welding, there are no connective bolts, nails, soldering materials or adhesives necessary to bind the materials together.

- For joining complex injection molded thermoplastic parts, ultrasonic welding equipment can be easily customized to fit the exact specifications of the parts being welded. The parts are sandwiched between a fixed shaped nest (anvil) and a sonotrode (horn) connected to a transducer, and a 20 kHz low-amplitude acoustic vibration is emitted.
- When welding plastics, the interface of the two parts is specially designed to concentrate the melting process. One of the materials usually has a spiked energy director, which contacts the second plastic part. The ultrasonic energy melts the point contact between the parts, creating a joint. This process is a good automated alternative to glue, screws or snap-fit designs. It is typically used with small parts (e.g., cell phones, consumer electronics, disposable medical tools, toys, etc.) but it can be used on parts as large as a small automotive instrument cluster.
- Ultrasonics can also be used to weld metals, but are typically limited to small welds of thin, malleable metals, e.g. aluminium, copper, nickel.
- Ultrasonics would not be used in welding the chassis of an automobile or in welding pieces of a bicycle together, due to the power levels required.
- Ultrasonic welding of thermoplastics causes local melting of the plastic due to absorption of vibration energy. The vibrations are introduced across the joint to be welded.
- In metals, ultrasonic welding occurs due to high-pressure dispersion of surface oxides and local motion of the materials. Although there is heating, it is not enough to melt the base materials. Vibrations are introduced only along the joint being welded.

9.8.6 Ultrasonic Cleaning

In fabrication of electronic devices, it becomes highly essential to clean the surface of the parts and components at different stages of production. Cleaning of the surfaces is commonly carried out in either organic solvents or weakly alkaline aqueous solution containing surface active agent. To make the scrubbing of the surface more effective, the phenomenon of cavitation is utilized. The chief advantage of this method is that it enables cleaning the surfaces of small products of complex configuration. Jewelers make use of ultrasonic bath to clean jewellery.

9.8.7 Ultrasonic Imagining and Medical Diagnosis

Diagnostic sonography (ultrasonography)

It is an ultrasound-based diagnostic imaging technique used for visualizing subcutaneous body structures including tendons, muscles, joints, vessels and internal organs for possible pathology or lesions. Obstetric sonography is commonly used during pregnancy and is widely recognized by the public.

Note

In physics, the term "ultrasound" applies to all acoustic energy (longitudinal, mechanical wave) with a frequency above the audible range of human hearing. The audible range of sound is 20 hertz–20 kilohertz. Ultrasound is frequency greater than 20 kilohertz.

Typical diagnostic sonographic scanners operate in the frequency range of 1–18 megahertz, though frequencies up to 50–100 megahertz have been used experimentally in a technique known as biomicroscopy in special regions, such as the anterior chamber of eye. The above frequencies are hundreds of times greater than the limit of human hearing, which is typically accepted as 20 kilohertz. The choice of frequency is a trade-off between spatial resolution of the image and imaging depth: lower frequencies produce less resolution but image deeper into the body.

Sonography (ultrasonography) is widely used in medicine. It is possible to perform both diagnosis and therapeutic procedures, using ultrasound to guide interventional procedures (for instance biopsies or drainage of fluid collections). Sonographers are medical professionals who perform scans for diagnostic purposes. They typically use a hand-held probe (called a transducer) that is placed directly on and moved over the patient.

Sonography is effective for imaging soft tissues of the body. Superficial structures such as muscles, tendons, testes, breast and the neonatal brain are imaged at a higher frequency (7–18 MHz), which provides better axial and lateral resolution. Deeper structures such as liver and kidney are imaged at a lower frequency 1–6 MHz with lower axial and lateral resolution but greater penetration.

9.8.8 Other Industrial Applications

Ultrasonic waves are used for detecting external (surface) and internal faults in materials.

They are used for welding metals that cannot be welded by conventional methods. A metal hammer vibrated at ultrasonic frequency is directed towards two metals to be welded, wherein the molecules of the metals diffuse into each other without being heated and thus the metals get welded. This method of welding is known as cold welding.

A mechanical drill when oscillated with ultrasonic frequency bores holes in metals and alloys. By a similar technique, when the drill is moved along the metal at a constant speed, it can cut the metal.

Aluminium cannot be soldered by normal methods. Because when heated for soldering, aluminium immediately forms an oxide layer which acts as a resistance to the electric current. The ultrasonic waves when used in conjugation with normal solder remove the oxide layer as soon as it forms and hence facilitates good soldering.

Ultrasonic waves accelerate detergent molecules in a cleaning tank when passed through it. These accelerated detergent molecules strike off dirt particles from the cloth being washed and thus it is cleaned. The same wave when passed through the wet cloth vibrates off the water molecules and thus drying them. The above processes are known as acoustical cleaning and drying, respectively.

EXAMPLE 9.1

A quartz crystal with a thickness of 0.5 mm and a density of 2650 kgm^{-3} vibrates longitudinally producing ultrasonic waves. Find the fundamental frequency of vibration if the Young's modulus of quartz is $7.9\times10^{10}\,\text{Nm}^{-2}$.

Solution: The fundamental frequency of vibration is $f=\frac{p}{2t}\sqrt{\frac{Y}{\rho}}$, where $p=1$

$$=\frac{1}{2\left(0.5\times10^{-3}\right)}\sqrt{\frac{7.9\times10^{10}}{2650}}$$

$$=10^{3}\sqrt{\frac{7.9\times10^{10}}{0.265\times10^{4}}}$$

$$=10^{3}\times\frac{10^{5}}{10^{2}}\sqrt{29.81}$$

$$=5.46\times10^{6}\text{ Hz}$$

$$=5.46\text{ MHz}$$

The fundamental frequency of vibration is 5.46 MHz.

EXAMPLE 9.2

An ultrasonic generator has a quartz crystal vibrating at its fundamental frequency. The thickness of the crystal is 2 mm, density is 2650 kg m^{-3}, and Young's modulus is $7.9\times10^{10}\text{ Nm}^{-2}$. Calculate the fundamental frequency of vibration and the second overtone. If the inductance value of the tank circuit of the generator is 5 Henry, find the value of the capacitance under resonance.

Solution: The fundamental frequency of vibration is $f=\frac{p}{2t}\sqrt{\frac{Y}{\rho}}$, where $p=1$

$$=\frac{1}{2\left(2\times10^{-3}\right)}\sqrt{\frac{7.9\times10^{10}}{2650}}$$

$$=0.25\times10^{3}\sqrt{\frac{7.9\times10^{10}}{0.265\times10^{4}}}$$

$$=0.25\times5.46\times10^{6}$$

$$=1365\times10^{6}\,\text{Hz}$$

$$=1.365\text{ MHz}$$

The fundamental frequency of vibration is 1.365 MHz. For the second overtone, $p=3$.

$$=\frac{3}{2\left(2\times10^{-3}\right)}\sqrt{\frac{7.9\times10^{10}}{2650}}$$

$$=1.5\times10^{3}\sqrt{\frac{7.9\times10^{10}}{0.265\times10^{4}}}$$

$$=1.5\times10^{3}\times\frac{10^{5}}{10^{2}}\sqrt{29.81}$$

$$=1.5\times5.46\times10^{6}$$

$$=8.19\times10^{6}\,\text{Hz}$$

$$=8.19\text{ MHz}$$

The frequency of the second overtone is 8.19 MHz.

The frequency of the tank circuit is $f_1=\frac{1}{2\pi\sqrt{LC}}$.

The resonance condition is $f=f_1$

$$\frac{1}{2\times3.14\sqrt{5}\times C}=1.365\times10^{6}$$

Squaring on both the sides and by rearranging,

$$\frac{1}{C}=\left(1.365\times10^{6}\right)^{2}\times(2\times3.14)^{2}\times5$$

$$=367.4\times10^{12}$$

$$C=\frac{1}{367.4\times10^{12}}$$

$$= 0.0027 \times 10^{-12} = 2.7 \times 10^{-9} F$$
$$= 2.7 \text{ nF}$$

The value of the capacitance is 2.7 nF.

EXAMPLE 9.3

An ultrasonic beam is used to determine the length of a swimming pool. The adjacent harmonics are separated by 50 Hz. What is the length of the pool if the velocity of ultrasonic waves in water is 1330 ms^{-2}?

Solution: Let the length of the pool be 'l'. Since the waves setup a standing wave pattern between the ends of the pool, the wavelength will be '$2l$' (because nodes are formed at either ends, the fundamental frequency is half the wavelength, i.e. 1/2).

The difference in adjacent waves is $f = 50$ Hz, which is equal to the frequency of the fundamental note. We know that

$$f = \frac{v}{\lambda} = \frac{v}{2l}$$

$$\therefore \quad l = \frac{v}{2f}$$
$$= \frac{1330}{2 \times 50}$$
$$= 13.3 \text{ m}$$

The length of the pool is 13.3 m.

EXAMPLE 9.4

A quartz crystal vibrates at its natural frequency of 30 kHz. Find its thickness if the density is 2650 kg m^{-3} and the Young's modulus is 7.9×10^{10} Nm^{-2}.

Solution: The fundamental frequency of vibration is $f = \frac{p}{2t}\sqrt{\frac{Y}{\rho}}$.

Squaring and rearranging for t, we get

$$t = \frac{p^2}{(2f^2)}\left(\frac{Y}{\rho}\right)$$
$$= \frac{1}{(2 \times 30 \times 10^3)^2}\left(\frac{7.9 \times 10^{10}}{2650}\right)$$

$$= 0.000278 \times 10^{-6} \times 0.003 \times 10^{10}$$
$$= 8 \times 10^{-7} \times 10^{4} \text{ m}$$
$$= 8 \times 10^{-3} \text{ m}$$
$$= 8 \text{ mm}$$

The thickness of the quartz crystal is 8 mm.

RECAP ZONE

POINTS TO REMEMBER

- Ultrasonics is the branch of physics that deals with the propagation and detection of ultrasonic waves.
- Ultrasonic waves are sound waves that have frequencies above 20,000 Hz or 20 KHz.
- Ultrasonic waves are reflected, refracted and absorbed like ordinary sound waves.
- Ultrasonic waves are produced by different devices, which include magnetostriction oscillator and piezoelectric oscillator.
- When ultrasonic waves are passed through a liquid, the refractive index changes at the places of variation in pressure inside the liquid. This change in refractive index resembles a grating (acoustical grating).
- The refractive index increases at nodes (compression – maximum pressure) and decreases at antinodes (rarefactions – minimum pressure).
- Frequencies upto 500 MHz can be produced by piezoelectric generator.
- Cavitation, in industries, is often used to homogenize or mix and break down, suspended particles in a colloidal liquid compound such as paint mixtures or milk.
- SONAR is an acronym for "Sound Navigation and Ranging". It employs the principle of Echo Sounding and is mostly used to find the depth of the sea.
- Human eye is an NDT (Non-destructive Testing) tool.
- Ultrasonics can also be used to weld metals, but are typically limited to small welds of thin, malleable metals, e.g. aluminium, copper, nickel.
- Ultrasonic waves are used for detecting external and internal faults in materials.

KEY TERMS WITH DEFINITIONS

- **Cavitation** – It is the formation of gas bubbles of a flowing liquid in a region where the pressure of the liquid falls below its vapour pressure. It is divided into two classes of behaviour: inertial and non-inertial.

- **Echo sounding** – It is the technique of using sound pulses directed from the surface or from a submarine vertically down to measure the distance to the bottom by means of sound waves.
- **Ultrasonic soldering** – It is an industrial technique whereby high-frequency ultrasonic acoustic vibrations are locally applied to work pieces being held together under pressure to create a solid-stage weld.
- **Diagnostic sonography** – It is an ultrasound-based diagnostic imaging technique used for visualizing subcutaneous body structures including tendons, muscles, joints, vessels and internal organs for possible pathology or lesions.
- **NDT** – Non-Destructive Testing is the method of testing or examining a product to identify the defects without affecting the serviceability of the product. Human eye is an NDT tool.
- **Ultrasonic machining** – It is known as ultrasonic impact grinding, and is a machining operation in which an abrasive slurry freely flows between the work piece and a vibrating tool.
- **Magnetostriction effect** – It is defined as the change in the dimensions of a magnetized ferromagnetic rod (e.g., iron, nickel), on application of an alternating magnetic field parallel to its length.
- **Piezoelectric effect** – It states that when pressure or mechanical stress is applied to and along a certain axis (mechanical axis) of a crystal (e.g., quartz, tourmaline), equal and opposite charges are produced along the perpendicular axis (electrical axis) to the optic axis of the crystal.
- **Inverse piezoelectric effect** – It states that 'when potential difference is applied to opposite faces of a crystal along a certain axis (electrical axis), the crystal starts to vibrate along the perpendicular axis (mechanical axis) of the crystal. This is the basis for the production of ultrasonic waves.

Important Formulae and Equations

Equation Number	Equation	Remarks
(9.1)	$f = \frac{p}{2l}\sqrt{\frac{Y}{\rho}}$ $p = 1, 2, ...$ for fundamental, first overtone, ... $Y =$ Young's modulus of material of crystal $\rho =$ density of the material	Natural frequency of a piezoelectric material

(9.2)	$f = \dfrac{1}{2\pi\sqrt{LC}}$ L = Inductance C = Capacitance	**Frequency of oscillation in piezoelectric and magnetostriction oscillator**
(9.3)	$f = \dfrac{1}{2l}\sqrt{\dfrac{Y}{\rho}}$ l = Length of the rod Y = Young's modulus of the material of the rod ρ = Density of the rod	**Frequency of the ferromagnetic rod in magnetostriction generator**
(9.4)	$u = \dfrac{d}{t}$ d = Total distance travelled (from transmitter to the sea bed and back to the receiver) $d = 2h$ u = velocity of the ultrasonic waves in sea water $\dfrac{ut}{2} = h$ h = distance/depth at which the reflection of pulse occurs t = time	**Height/depth of sea bed**

Review Zone

Short Answer Questions

1. What are ultrasonic waves?
2. Why are ultrasonic waves not audible to humans?
3. Mention the properties of ultrasonic waves.
4. Mention any two methods of production of ultrasonic waves.
5. Define magnetostriction effect.
6. Write a short note on the properties of the magnetostriction effect.
7. What is a tank circuit?
8. What are the advantages of the magnetostriction oscillator?
9. What are the disadvantages of the magnetostriction oscillator?
10. What are the ways in which a quartz crystal is cut? Why?

11. Define piezoelectric effect.
12. What is inverse piezoelectric effect? What is its importance?
13. What are the advantages of piezoelectric effect?
14. What is the limitation of the piezoelectric effect?
15. What is SONAR?
16. Name some methods for the detection of ultrasonic waves.

Long Answer Questions

1. What do you mean by ultrasonic waves? Mention the properties of ultrasonic waves.
2. Discuss the methods of generating ultrasonic waves.
3. Explain magnetostriction effect briefly.
4. List out the advantages and disadvantages of piezoelectric generator.
5. Define piezoelectric effect.
6. What is inverse piezoelectric effect? What is its importance?
7. Tabulate the difference between inertial and non-inertial cavitation effects.
8. Enlist few biomedical applications of cavitation effect.
9. What is SONAR? Explain the working and principle of SONAR with the help of a schematic diagram.
10. Explain NDT and mention its applications.
11. What do you mean by ultrasonic soldering?
12. What is ultrasonic machining?
13. Write a short note on diagnostic sonography.
14. Tabulate the difference between ultrasonic and piezoelectric materials.
15. Mention the industrial applications of ultrasonic waves.

Numerical Problems

1. An ultrasonic generator consists of a quartz crystal of thickness 1 mm and density $2790\ \text{kg/m}^3$. If the crystal undergoes longitudinal vibrations, find its fundamental frequency of vibration and its first overtone. $(E = 8.8 \times 10^{10}\ \text{N/m}^2)$

   Ans. 2.8 MHz and 5.6 MHz

2. An ultrasonic beam is used to determine the thickness of a steel plate. As the beam is reflected of the front and back surfaces of the plate, the difference in frequency of the waves is found to be 75 kHz. If the velocity of sound in steel is 5000 m/s, find the thickness of the steel plate. $\Delta f = v/2d$

   Ans. 0.03 m

CHAPTER 10

Shock Waves

Learning Objectives

By the end of this chapter, the student will be able:

- To understand the shock waves
- To distinguish between sound and shock waves
- To discuss applications of shock waves
- To derive Rankine-Hugoniot equations
- To learn methods of generation of shock waves in lab
- To illustrate use of shock waves for medical applications

10.1 ▶ Introduction

When the speed of a source of sound (e.g., an airplane) equals the speed of sound ($u = v$), the generated wave fronts cannot escape the source (Fig. 10.1). The resulting pile up of waves forms a large amplitude sound barrier that makes sustained flight at this speed difficult and risky.

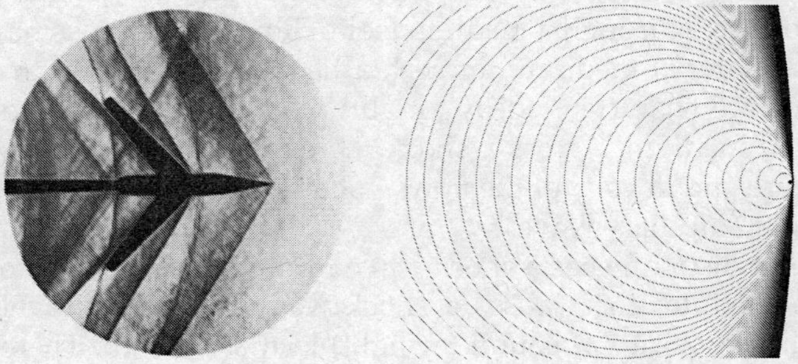

FIGURE 10.1 Sound barrier forms at $u = v$ to produce shock waves

The term "sound barrier" or "sonic barrier" first came into use during second World War. Fighter pilots engaged in high-speed dives noticed that aerodynamic drag increased much more than normal with increased speed. When the speed of a source exceeds the speed of sound ($u > v$), the wave fronts lag behind the source in a cone-shaped region with the source at the vertex as shown in Fig. 10.2. The edge of the cone forms a supersonic wave front with a unusually large amplitude called a "shock wave". When a shock wave reaches an observer, a "sonic boom" is heard.

FIGURE 10.2 Wave front lagging behind the source if $u > v$

The bursting sound of crackers and shattering of window and door pans when a fighter jet passes above your house are the simple examples of the shock waves.

Note

When the speed of a source exceeds the speed of sound ($u > v$), the wave fronts lag behind the source in a cone-shaped region with the source at the vertex as shown in Fig. 10.2. The edge of the cone forms a supersonic wave front with a unusually large amplitude called a "shock wave". When a shock wave reaches an observer, a "sonic boom" is heard.

A shock wave is a type of propagating disturbance. Like an ordinary wave, it carries energy and can propagate through a medium (solid, liquid gas or plasma) or in some cases in the absence of a material medium, through a field such as an electromagnetic field in space. Shock waves are characterized by an abrupt, nearly discontinuous change in the characteristics of the medium. Across a shock there is always an extremely rapid rise in pressure, temperature and density of the flow. A shock wave travels through most media at a higher speed than an ordinary wave.

The first thing that distinguishes a shock wave from an ordinary wave is that the initial disturbance in the medium that causes a shock wave is always traveling at a velocity greater than the phase velocity of sound (or light) in the medium. There is also a kind of electromagnetic analogue to a shock wave known as Cherenkov radiation that is created when a charged particle travels through a medium at a velocity faster than that of the phase velocity of light in the medium.

In case of shock waves, the process of compression and rarefaction caused by a shock wave is an irreversible process as it leads to a change in entropy of the gas or the medium. The zone of discontinuity is quite sharp between the disturbance and the shock waves, and the changes in pressure, temperature and density are large enough. The boundary conditions involved in analyzing shock waves are known as the Rankine-Hugoniot conditions (Section 10.8).

Note

Unlike ordinary sound waves, the speed of a shock wave varies with its amplitude. The speed of a shock wave is always greater than the speed of sound in the medium and decreases as the amplitude of the wave decreases. When the shock wave speed equals the normal speed, the shock wave dies and is reduced to an ordinary sound wave.

10.2 ▶ Distinctions and Types of Sound Waves

Acoustics

These are the waves in human audible frequency range and deal with the study of waves through gases, liquids and solids.

Ultrasonics

These are a type of acoustic waves that have frequencies higher than the audible sound (> 20 kHz).

Infrasonic waves

These are the sound waves that have frequencies lower than the human audible range (< 20 Hz).

Subsonic

Waves that travel with speeds less than the speed of sound in a medium are called subsonic.

Supersonic

Waves that travel with speeds greater than the speed of sound in a medium are called supersonic waves.

10.3 ▶ Why to Study Shock Waves?

The study of shock waves helps physical scientists understand the general principles according to which bursts of energy produce waves in nature and how it can be utilized for other applications. The phenomenon of shock waves has applications in various disciplines, such as astronomy, medicine plasma physics, space science and geology.

Note

A shock wave is a large-amplitude wave in an elastic medium formed by a sudden compression of that medium. A shock wave travels faster than sound, its speed varies directly with amplitude, and as it passes a point, its intensity falls off more quickly than that of a normal wave.

10.4 ▶ Mach Number

The Mach number is defined as the ratio of speed of the object to the speed of the sound in the surrounding medium. It is a dimensionless quantity. Let us consider an object moving in a medium with the velocity v_o and the velocity of sound in the medium is v_s.

Mathematically, the Mach number M is given by

$$\boxed{M = \frac{v_o}{v_s}} \tag{10.1}$$

Mach 0.5 represents half the speed of sound and Mach 2 represents twice the speed of sound. The Mach number is named after the scientist Ernst Mach.

The following is a table representing the regimes of fluid flow of moving object in a fluid with regard to Mach number.

Mach No.	< 0.8	0.8–1.2	1.0	1.0–5.0	5.0–10.0	> 10
Regime	Subsonic	Transonic	Sonic	Supersonic	Hypersonic	High-hypersonic

10.5 ▶ Sonic Boom

Sonic boom is a common name for the loud noise that is created by the "shock wave" produced by the air-plane that is travelling at speeds greater than that of sound (speed of sound is approximately 332 m/s or 1195 km/hr or 717 miles/hr). These speeds are called supersonic speeds; hence, this phenomenon is sometimes called the supersonic boom.

Normally, for a plane that is going at subsonic speeds (lower than that of sound), the sound of the plane is radiated in all directions. However, the individual sound wavelets are compressed at the front of the plane and further spread at the back of the plane because of the forward speed of the plane. The effect is known as the Doppler effect and accounts for the change of the "pitch" of the plane's sound as it passes us (Fig. 10.3(a)). When the plane is approaching us, its sound has a higher pitch than if it is going away from us.

(a) (b)

FIGURE 10.3 (a) $u < v$ Doppler effect. (b) $u > v$ Shock wave

Because the source is now travelling at the same velocity as the waves, the waves never get ahead of the source, and all waves pile up at one point (Fig. 10.3(b)). The distance between the waves goes to zero and so the frequency becomes very high. More importantly, all of the energy gets concentrated into a very small distance – this is called a shock wave. In this case, the observer does not hear the approaching source at all until the shock wave hits with all the energy in the wave. For sound waves, this can cause a very loud noise, called a sonic boom. Any time a source exceeds the speed of the wave, a shock wave will be formed. If the source is travelling faster than the waves, the waves never catch upto the source, and a different sort of pattern is formed. The shape of the shock wave is called a Mach cone and the opening angle of the cone is given by

$$\boxed{\sin\theta = \frac{v_{\text{sound}}}{v_{\text{source}}}} \tag{10.2}$$

Each kind of wave has its own example of shock waves and Mach cones.

10.6 ▶ Applications of Shock Waves

The following are the applications of the shock waves.

1. The reflection of a shock wave is used in labs to remove micron size dust particles from the surface of silicon wafers in semiconductor industry.
2. Shock waves are used in medical therapy in orthopedics and for breaking kidney and gallbladder stones.
3. Shock waves are used in pencil industries to impregnate preservatives into wood slabs.
4. Shock waves are used in sandal oil extraction.
5. Shock waves are used in cell transformation.
6. Shock waves are used in geophysics for rock structure studies.

10.7 ▶ Conservation of Mass, Energy and Momentum in Shock Waves

10.7.1 Conservation of Mass

The basic law of conservation is defined as mass can neither be created nor be destroyed. For a closed system the total mass of the system remains constant. Mathematically

$$M_{\text{sys}} = \text{constant} \tag{10.3}$$

Hence,

$$\boxed{\frac{dM_{\text{sys}}}{dt} = 0} \tag{10.4}$$

For an inertial frame of reference, it is a volume fixed in space or moving with constant velocity through which the fluid flows. The surface enclosing the volume is referred to as the control surface. For a control volume or open system conservation of mass is expressed in the rate form as

$$M_{in} - M_{out} = \frac{dM_{cv}}{dt}$$

Here, M_{in} and M_{out} are the masses flow into and out of the control volume, respectively. dM_{cv}/dt is the rate of change of mass within the control volume boundaries. In fluid mechanics, the conservation of mass relation written for a differential control volume is usually called the continuity equation.

10.7.2 Conservation of Momentum

It is given as the total momentum of the system is always conserved. By applying Newton's second law of motion to the control volume, the conservation of momentum is "any change in momentum of the fluid within a volume be due to the net flow of fluid into the volume and the action of external forces on the fluid within the volume."

10.7.3 Conservation of Energy

It deals with energy can neither be created nor be destroyed. In other words, total energy of a system remains constant. In case of a fluid system under adiabatic conditions, the total energy entering the system must be same as the total energy leaving the system. The total energy of the control volume remains constant.

All above conservation laws are obeyed in case of shock waves of acoustics as well as e.m. waves.

10.8 ▶ Rankine-Hugoniot Equations for Normal Shock Waves

10.8.1 Normal Shock Waves

Consider a flow with speeds greater than sonic speed and hence shock waves are generated. Normal shock waves are the basic type of shock waves that are in a direction perpendicular to the flow.

10.8.2 Rankine-Hugoniot Equations

Consider a rectangular region of fluid (say air) around a normal shock wave as shown in Fig. 10.4. To the left of the shock wave, we have the initial and the final conditions to the right. M_1, V_1, p_1, T_1 and ρ_1 represent the initial conditions of the flow and M_2, V_2, p_2, T_2 and ρ_2 represent the final conditions of the flow.

The following are the assumptions with regard to the flow:

1. It is a steady flow.
2. It is an adiabatic system.
3. There are no viscous forces acting between the system and boundaries.
4. There are no body forces acting.

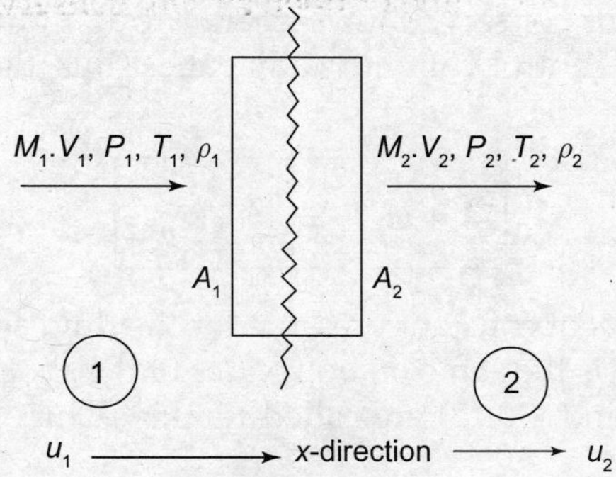

FIGURE 10.4 Normal shock wave

Let us determine the mass flow of the rectangular fluid system of the equation of continuity. The mass flow into the rectangular system is given by $\rho_1 u_1 A_1$. Here, u_1 is the inflow velocity along the x-direction. Similarly, the mass flow out of the rectangular system is given by $\rho_2 u_2 A_2$. Here, u_2 is the outflow velocity along the x-direction.

Applying the law of conservation of energy, we have

$$\boxed{\rho_1 u_1 A_1 = \rho_2 u_2 A_2} \tag{10.5}$$

For a rectangular system, $A_1 = A_2$; hence, Eq. (10.6) becomes

$$\rho_1 u_1 = \rho_2 u_2 \tag{10.6}$$

Let us derive the momentum equation using the law of conservation of momentum. The rate of momentum flow entering the system is given by $(\rho_1 u_1 A_1)u_1$. Similarly, the rate of momentum flow out of the system is given by $(\rho_2 u_2 A_2)u_2$. Hence, the net change in momentum is given by

$$(\rho_1 u_1 A_1)u_1 - (\rho_2 u_2 A_2)u_2 \tag{10.7}$$

The net force acting on the system in terms of pressure and area is given by

$$p_1 A_1 - \rho_2 A_2 \tag{10.8}$$

Since the rate of flow is considered per second and using the law of conservation of momentum, we have

$$(\rho_1 u_1 A_1)u_1 - (\rho_2 u_2 A_2)u_2 = p_1 A_1 - p_2 A_2 \tag{10.9}$$

Simplifying the above equation results in

$$\left(p_1 + \rho_1 u_1^2\right) A_1 = \left(p_2 + \rho_2 u_2^2\right) A_2 \tag{10.10}$$

Since $A_1 = A_2$ for a rectangular system, we get

$$\boxed{p_1 + \rho_1 u_1^2 = p_2 + \rho_2 u_2^2} \tag{10.11}$$

Lastly, consider the law of conservation of energy to the flow across the shock wave. Hence, the flow through the control volume is adiabatic. Thus, the energy equation is given by

$$\boxed{\frac{u_1^2}{2}+C_pT_1=\frac{u_2^2}{2}+C_pT_2} \tag{10.12}$$

Here, C_p is the specific heat at constant pressure for the fluid system. The above equation states that the total enthalpy is the same on both sides of the shock wave.

Equations (10.5), (10.11) and (10.12) are called Rankine-Hugoniot equations for normal shock waves.

10.9 ▶ Generation of Shock Waves

10.9.1 Naturally Produced Shock Waves

Natural events, such as earthquakes, cause shock waves by releasing energy through the ground, air or water at a speed of three to two miles per second. Shock waves also happen in outer space. NASA scientists have captured images of shock waves caused by solar flares on the surface of the sun.

10.9.2 Aeronautically Produced

When an aircraft exceeds the speed of sound, the air particles are completely undisturbed, having had no advanced warning of the airplane's approach, and in the next instant the same air particles are forced to undergo sudden and drastic changes in temperature, pressure, density and velocity. These forces create a shock wave.

10.9.3 Mechanically Produced

Piezoelectric, electromagnetic or electrohydraulic transducers can produce small shock waves in a laboratory by introducing a pressure wave through a conductive material. Veterinarians and physicians use these shock waves for certain types of medical procedures including kidney stone removal, treatment of soft tissue injuries and cancer.

10.10 ▶ Methods of Creating Shock Waves in Laboratory

Shock waves find applications extending into the fields of engineering, manufacturing, medical, biological and scientific research. Shock waves are produced by sudden release of energy like in explosions, supersonic flows, impact of projectiles at high speeds, etc. Even though shock waves have such a wide range of applications and common occurrence, it is not a topic in the school or college studies. This is due to the safety issues involved with regard to the high pressure associated with the production and propagation of shock waves.

Note

Piezoelectric, electromagnetic or electrohydraulic transducers can produce small shock waves in a laboratory by introducing a pressure wave through a conductive material.

10.10.1 Shock Tubes

A device used to produce shock waves of required strength in the laboratory is called shock tube. It consists of a long simple tube which is separated into two regions (sections) by a meta diaphragm. The regions are driver and driven sections. The metal diaphragm is commonly driven by compressed high pressure driver gas using either a compressor or by detonation of small quantity of explosive.

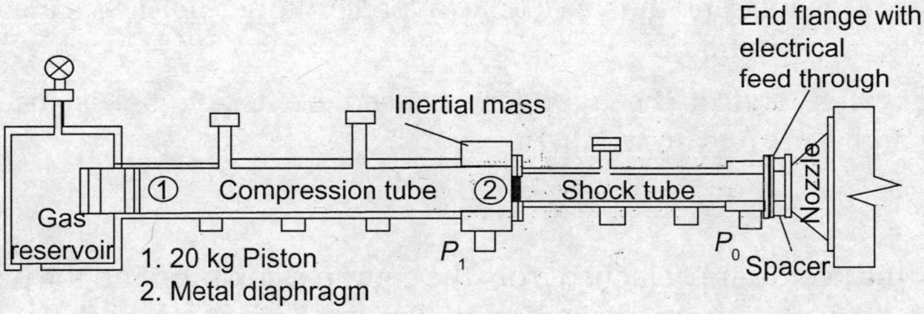

FIGURE 10.5 Compression type shock tube

Thus, driver side ruptures the diaphragm creating a shock wave which travels into the driven section along the length of the shock tube. Such shock tubes can be used for classroom teaching. The compression driven shock tube is as shown in Fig. 10.5.

10.10.2 Reddy's Hand Operated Shockwave Tube

Dr K.P.J. Reddy and associates have developed a miniaturized simple hand-operated pressure-driven shock tube named as "Reddy tube" which is very simple to operate. The main aim of this capability of this convenient is to operate simple, mini shock tube which is more versatile than the traditional compression-driven and blast-driven shock tubes. Reddy shock tube is a special hand-held piston driven shock tube capable of operating in traditional compression driven mode using manual energy. Reddy tube is the first such devise which is capable of producing shock waves with Mach numbers exceeding 1.5 by simple hand operation. The construction of Reddy's shock tube is as shown in Fig. 10.6.

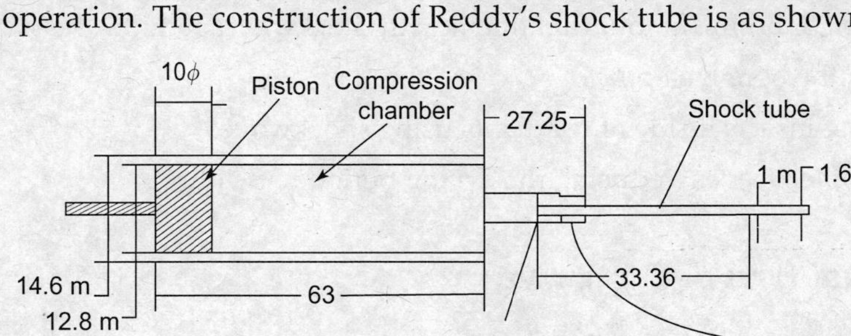

FIGURE 10.6 Reddy's hand operated shockwave tube

Construction

Reddy's tube operates on the principle of conventional piston driven shock tube. Reddy shock tube is made of simple medical syringe with the following dimensions:

- Compression chamber is of length $\sim$ 60 mm and diameter $\sim$ 12.8 mm.
- A plastic piston with plunger with outer diameter 12.8 mm fitted closely inside the compression chamber.
- O-rings on the piston head prevent the leakage of leaking of gas or air.
- The driven shock tube is a 30 mm long SS tube of 1 mm internal diameter and wall thickness of 0.8 mm.
- The compression tube and the shock tube (hypodermic needle) are separated by a 50 mm thick plastic diaphragm.
- The elliptic free end of the hypodermic needle (shock tube) is made circular by grinding and is open to atmospheric air.

Working

Diaphragm and the needle are detached from the compression tube and the piston is moved outwards to fill the compression chamber with ambient air at atmospheric pressure. The diaphragm and the needle are attached to the compression tube. The piston is pushed into the compression chamber so that the air inside is compressed and pressure builds. As the piston moves the pressure increases and finally the diaphragm ruptures resulting in the very high-speed flow of the compressed air. Schlieren images of the flow at the exit of the Reddy tube confirm the generation of shock waves. Thus, shock waves of Mach 1.5–2.0 are generated at the tip of the needle.

10.10.3 Applications of Reddy Shockwaves Tubes

1. Testing aerospace materials tolerance on shockwaves.
2. Efficient artificial insemination to cow.
3. Needleless drug delivery (painless injection).
4. Study on high temperature chemical kinetics studies.
5. Treatment of Brain Tumour (treatment with shockwaves reduces the size of tumour).
6. Clearing the bore well pipes.
7. Removing insects inside of tea plants using shockwaves.
8. Drying of tea leaves (reducing the drying period to 8 hours).

10.11 ▶ Other Types of Shockwave

There are a number of examples of other shock waves, broadly grouped with similar shock phenomena:

10.11.1 Water Waves

The easiest waves to visualize are water waves. In this case, you just need to drive a boat faster than the speed of water waves, which is not hard to do (Fig. 10.7).

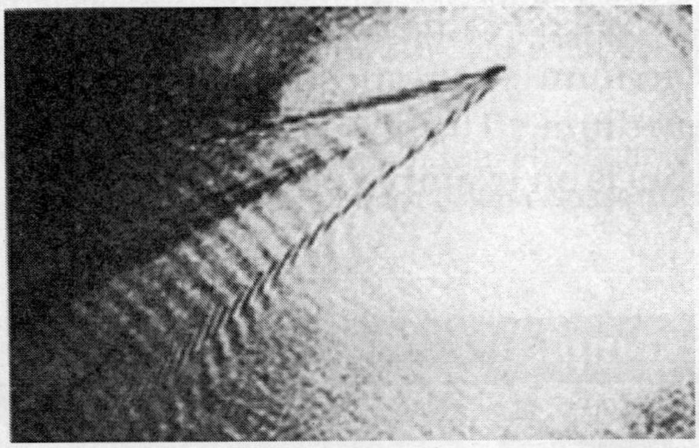

FIGURE 10.7 Water waves by a fast-moving boat produce shock

The Mach cone is better controlled in experiment (Fig. 10.8) with water waves.

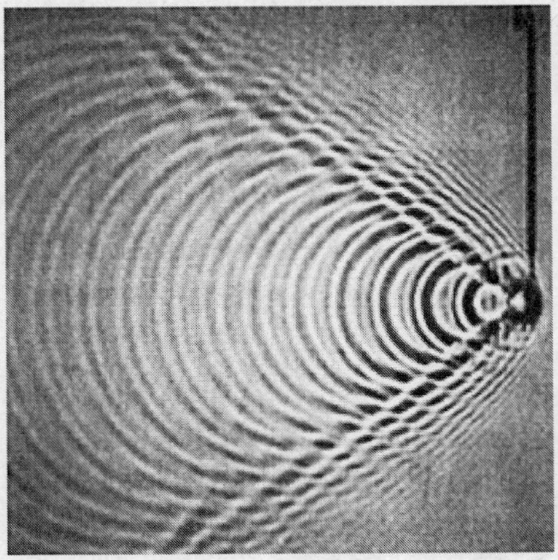

FIGURE 10.8 Controlled experiment with water waves

Sound waves are harder to visualize, but they are easy to hear. If a jet exceeds the speed of sound, it produces a loud sonic boom, which can even rattle or break windows. Jets are no longer allowed to fly faster than the speed of sound, over cities, at least.

10.11.2 Electromagnetic Shock Waves

Can you produce a shock wave of light? Most people will answer no, because to create a shock wave, something must travel faster than the wave. Since nothing can travel faster than the speed of light this could never happen. However, this is not quite true. In a transparent material, like glass, light does slow down by a factor called the "index of refraction". The index of refraction of most types of glass is about 1.5. So, inside of a piece of glass, only travels about 67% of its normal speed. This means that a very fast particle can actually exceed the speed of light in these materials. If you accelerate electrons to a very high velocity and fire them into a piece of glass or plastic, they will produce a shock wave of light that you can see. This light is called "Cherenkov radiation". In the reactor, uranium atoms give off energetic neutrons. These neutrons collide with electrons and give them very high velocities. In the water, the speed of light is reduced, just like in glass. The electrons then produce Cherenkov radiation, which appears as a bluish glow.

10.11.3 Shock Waves in Plasma

Laser pulses can produce an expanding shock wave in a tiny plasma of high-energy ions and electrons. The researchers used a sequence of two laser pulses – one to vapourize a nanoparticle into plasma and a second to generate the shock wave. The first laser pulse transforms a nanoparticle into plasma, and the second drives a shock wave through the expanding plasma. In outer space, most of the matter is in plasma state; therefore, the solar or star bursts keep creating shock waves quite frequently.

10.11.4 Moving Shockwave

It usually consists of a shock wave propagating into a stationary medium. The gas ahead of the shock is stationary (in the laboratory frame) and the gas behind the shock can be supersonic in the laboratory frame. The shock propagates with a wave front which is normal (at right angles) to the direction of flow. Moving shocks are usually generated by the interaction of two bodies of gas at different pressures, with a shock wave propagating into the lower pressure gas and an expansion wave propagating into the higher pressure gas. Example is balloon bursting.

10.11.5 Detonation Shockwave

A detonation wave is essentially a shock supported by a trailing exothermic reaction. It involves a wave travelling through a highly combustible or chemically unstable mixture or a high explosive as shown in Fig. 10.9.

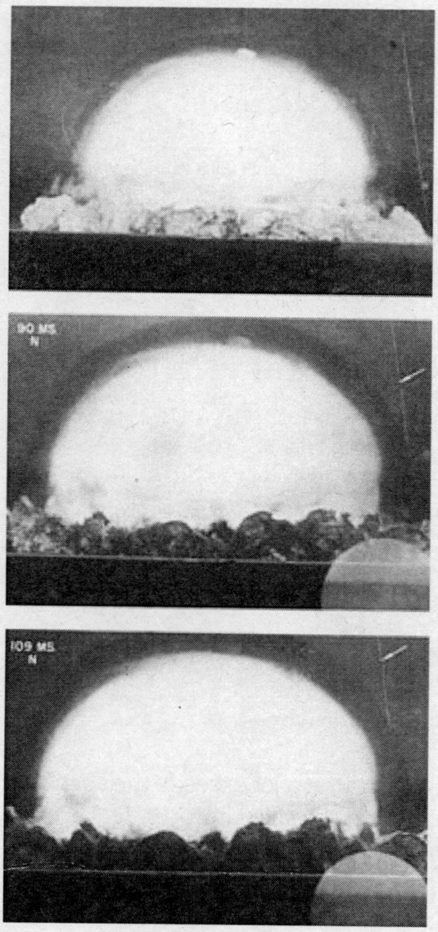

FIGURE 10.9 Detonation shock after bomb blast

Figure 10.10 shows *Schlieren photograph* of the detached shock on a bullet in supersonic flight, published by Ernst Mach and Peter Salcher in 1887.

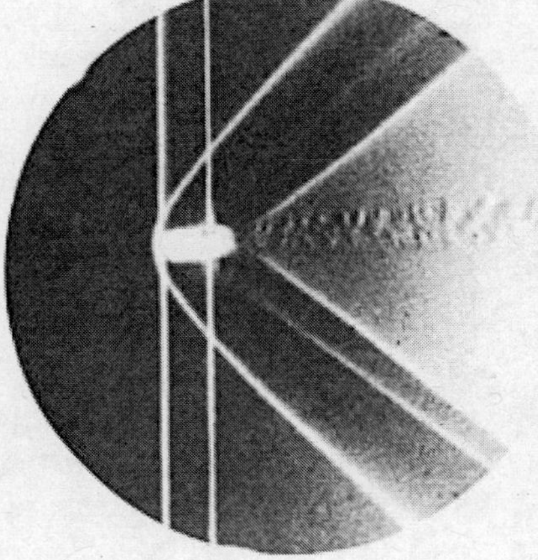

FIGURE10.10 Shadowgram of shock waves from supersonic bullet fired from rifile

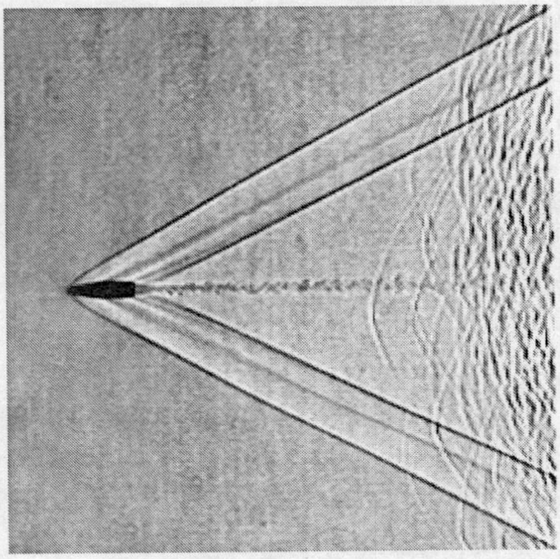

FIGURE10.11 Shadowgram of shock waves from a supersonic bullet fired from a rifle

The shadowgraph optical technique reveals that the bullet is moving at about a Mach number of 1.9. Left- and right-running bow waves and tail waves stream back from the bullet and its turbulent wake is also visible. Patterns at the far right (Fig. 10.11) are from unburned gunpowder particles ejected by the rifle.

10.11.6 Bow Shockwave (Detached Shock)

These shocks are curved and form a small distance in front of the body. Directly in front of the body, they stand at 90 degrees to the oncoming flow and then curve around the body. Examples are space return vehicles (Apollo, Space shuttle), bullets, the boundary (Bow shock) of a magnetosphere.

10.11.7 Attached Shockwave

These shocks appear as attached to the tip of sharp bodies moving at supersonic speeds. Examples are supersonic wedges and cones with small apex angles.

10.11.8 Shockwave in Rapid Granular Flows

Shock waves can also occur in rapid flows of dense granular materials down inclined channels or slopes such as glaciers. Strong shocks in rapid dense granular flows can be studied theoretically and analyzed to compare with experimental data.

10.11.9 Shockwave in Astrophysics

Astrophysical environments feature many different types of shock waves. Some common examples are supernovae shock waves or blast waves travelling through the interstellar medium, the bow shock caused by the Earth's magnetic field colliding with the solar wind and shock waves caused by galaxies colliding with each other.

10.11.10 Shockwave by Meteor Entering Atmosphere

When the 2013 meteor entered into the Earth's atmosphere with an energy release equivalent to 100 or more kilotons of TNT, dozens of times more powerful than the atomic bomb dropped on Hiroshima, the meteor's shock wave produced damages as in a supersonic jet's flyby (directly underneath the meteor's path) and as a detonation wave, with the circular shock wave centred at the meteor explosion, causing multiple instances of broken glass in the city of Chelyabinsk and neighbouring areas.

10.11.11 Combustion Engines

The radial Internal Combustion Wave Rotor also called wave disk engine is a kind of pistonless rotary engine that utilizes shock waves to transfer energy between a high-energy fluid to a low-energy fluid, thereby increasing both temperature and pressure of the low-energy fluid.

10.12 ▶ Shockwaves for Medical Applications

In medicine, shock waves were first used for kidney stone fragmentation in the 1970s and have since become the method of choice for most kidney and ureteral stones. In the 1990s, these shock waves were successfully utilized for the treatment of several musculoskeletal disorders. Shock waves pass through the surface of a body without causing injury and may act therapeutically in determined areas within the body. Focussed shock waves can be generated with electrohydraulic, piezoelectric and electromagnetic generators.

10.12.1 Piezoelectric

Piezoelectric shock waves principles generate small needle like focus areas, which may carry very high intensity, similar to a burning glass effect; thus caution needs to be observed at some treatment regions. Piezoelectric elements are arranged on a spherical surface and are synchronously excited by an electrical pulse to emit a pressure wave in the direction of the centre of the spherical surface. The process is self-focussing and the targets can be precisely set.

10.12.2 Electromagnetic

The method of electromagnetic shock wave generation is based on the physical principle of electromagnetic induction, as used for example in loudspeakers. The arrangement of coils and membranes is optimized to generate powerful and short acoustical pulses. This method is good for optimized intensities and focus zones, with long lasting constant dosage outputs, resulting in a gentle, less painful thus well tolerable and low noise treatment.

RECAP ZONE

POINTS TO REMEMBER

- When the speed of a source exceeds the speed of sound ($u > v$), the wave fronts lag behind the source in a cone-shaped region with the source at the vertex. The edge of the cone forms a supersonic wave front with a unusually large amplitude called a "shock wave"
- Shock waves are characterized by an abrupt, nearly discontinuous change in the characteristics of the medium. Across a shock there is always an extremely rapid rise in pressure, temperature and density of the flow.
- There is also a kind of electromagnetic analogue to a shock wave known as Cherenkov radiation that is created when a charged particle travels through a medium at a velocity faster than that of the phase velocity of light in the medium.
- In case of shock waves, the process of compression and rarefaction caused by a shock wave is an irreversible process as it leads to a change in entropy of the gas or the medium.
- The speed of a shock wave is always greater than the speed of sound in the medium and decreases as the amplitude of the wave decreases. When the shock wave speed equals the normal speed, the shock wave dies and is reduced to an ordinary sound wave.
- The phenomenon of shock waves has applications in various disciplines, such as astronomy, medicine and geology.
- Shock waves also happen in outer space plasma or e.m. waves.
- Piezoelectric, electromagnetic or electrohydraulic transducers can produce small shock waves in a laboratory by introducing a pressure wave through a conductive material.
- The Mach number is defined as the ratio of speed of the object to the speed of the sound in the surrounding medium. It is dimensionless quantity.
- Sonic boom is a common name for the loud noise that is created by the "shock wave" produced by the air-plane that is travelling at speeds greater than that of sound.
- Shock waves are also used as standard method of kidney and uretheral stone fragmentation.
- Shock waves are also utilized for the treatment of several musculoskeletal disorders as these pass through the surface of a body without causing injury and may act therapeutically in specified location within the body.

KEY TERMS WITH DEFINITIONS

- The Mach number is defined as the ratio of speed of the object to the speed of the sound in the surrounding medium. It is a dimensionless quantity. Mathematically, the Mach number M is given by $M = v_0/v_S$.
- The boundary conditions involved in analyzing shock waves are known as the *Rankine-Hugoniot conditions.*
- *Ultrasonics* are type of acoustic waves that have frequencies higher than the audible sound (> 20 kHz).
- *Infrasonic waves* are the sound waves that have frequencies lower than the human audible range.
- The waves that travel with speeds less than the speed of sound in a medium are called *subsonic waves.*
- Waves which travel with speeds greater than the speed of sound in a medium are called *supersonic waves.*
- Dr. Reddy's hand operated shock tube can be used for generation of shock waves in laboratory. The shock waves of Mach 1.5–2.0 can be created in lab with this tube.
- There are many types of shock waves that exist in nature such as water waves in front of a fast moving boat, cerenhov radiation due to accelerating charged particles in crystals and detonation shock waves.
- The shock waves find various applications in diagnostic and treatment of kidney and gallbladder stone fragmetation and some musculoskeletal disorders

IMPORTANT FORMULAE AND EQUATIONS

Equation Number	Equation	Remarks
(10.1)	$M = v_0/v_s$	Mach number M
(10.2)	$\sin\theta = v_{\text{sound}} / v_{\text{source}}$	Opening angle of the Mach cone
(10.4)	$dM_{sys} / dt = 0$	Conservation of mass
(10.5)	$\rho_1 u_1 A_1 = \rho_2 u_2 A_2$	Rankine-Hugoniot Equations
(10.11)	$p_1 + \rho_1 u_1^2 = p_2 + \rho_2 u_2^2$	
(10.12)	$\frac{u_1^2}{2} + C_p T_1 = \frac{u_2^2}{2} + C_p T_2$	

Review Zone

Short Answer Questions

1. What are the types of sound wave?
2. Distinguish ultrasonic, sonic and infrasonic waves.
3. What are shock waves? Define their characteristics.
4. What are the conditions for production of shock waves?
5. What is Mach number?
6. What do you mean by sonic boom?
7. Write a note on application of shock wave.
8. What are Rankine-Hugoniot equations for normal shock waves?
9. Discuss the methods of production of shock waves.
10. What is Reddy's shock wave tube? Discuss its working and applications.
11. What type of shock waves exists in nature? Discuss with example.
12. Describe the medical applications of shock waves.

CHAPTER 11

Acoustics

Learning Objectives

By the end of this chapter, the student will be able:

- To learn about production and propagation of sound in the medium
- To characterize properties of sound
- To understand human perception of sound
- To learn about reflection and absorption of sound
- To define decibel as unit
- To describe the Sabine's formula
- To understand acoustics of buildings, defects and remedies

11.1 ▶ What is Sound?

We all hear several types of sounds around us like music, machines, human voice, etc. For all these sounds to reach us, there must be a medium of propagation, production and detection. Thus, the sound has two stages, first is its production and the second is its perception or detection. Let us first discuss its production.

11.1.1 Production of Sound

Sound is produced by the rapid variation in the average density or pressure of air molecules above and below the atmospheric pressure. There are minute changes in atmospheric pressure in the air that are referred to as sound pressure and the fluctuations in pressure as sound waves. Sound waves are produced by vibrating bodies such as human vocal chords and musical instruments such as drums, guitar string, loudspeakers, jet engine and almost every mechanically vibrating object. The vibrating objects cause a disturbance to the surrounding air causing them to bounce off each other with a force proportional to the disturbance. These disturbances are propagated in the air as alternate compressions and rarefactions that are called longitudinal waves.

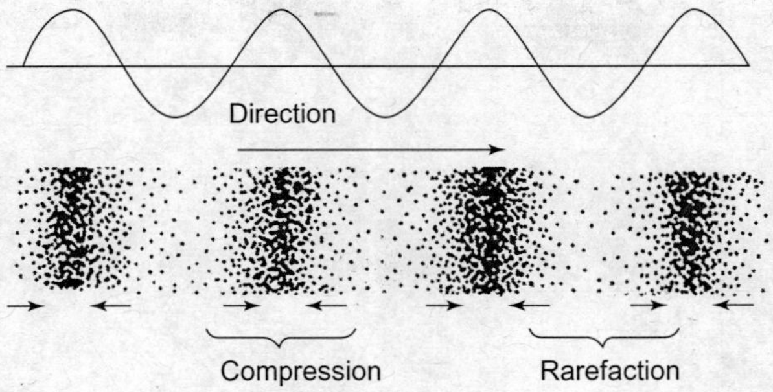

FIGURE 11.1 Propagation of sound as compressions and rarefactions

11.1.2 Detection or Sensing the Sound

We perceive sound as these pressure variations in air that cause our ear drums to vibrate. The energy of their interaction with our eardrum creates ripples above and below the normal atmospheric pressure. When the molecules are pushed closer together, it is called compression; when they are pulled apart, it is called rarefaction (Fig. 11.1). The back and forth oscillation of pressure produces sound waves. These vibrating motions are transmitted via the bone chain to the inner ear, disturbing the liquid there, which in turn awakens the auditory nerves (Fig. 11.6).

11.2 ▶ Speed of Sound

The speed at which sound travels from its source is directly influenced by both the medium through which it travels and the factors affecting the medium, such as altitude, humidity and temperature for the medium like air. At room temperature (25°C) and at atmospheric pressure, the speed of sound in air is 332 m/sec.

It is important to note that the speed of sound in air is determined by the conditions of the air itself (e.g., humidity, temperature, altitude). It is not dependent upon the sound's amplitude, frequency or wavelength. To calculate the approximate speed of sound in dry air at sea level, we use the formula $V = 331.4 + 0.6T_c$ where V = velocity (m/s), T_c = temperature in Celsius.

11.3 ▶ Characteristics of Sound Waves

Though there are many qualities that are related to musical sounds. Sound waves are often characterized by four basic qualities. The main ones are the frequency, amplitude, wave shape and phase. The sound waves are periodic waves in air that change from equilibrium (average atmospheric pressure) to maximum compression to maximum rarefaction back to equilibrium repetitively. The complete round trip back to the starting point is called a cycle (Fig. 11.2).

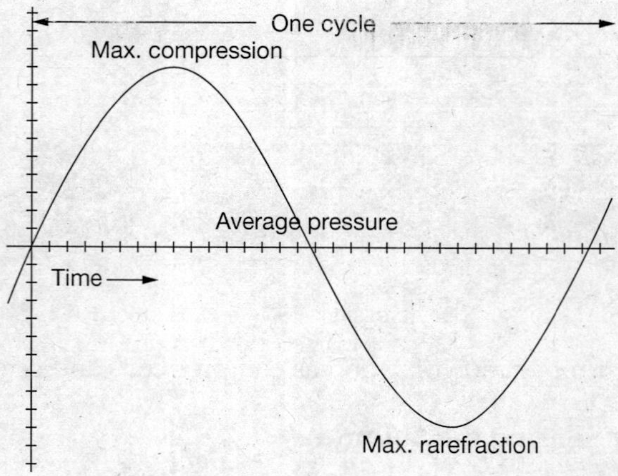

FIGURE 11.2 Wave form of a complete cycle

11.3.1 Frequency

It is possible to measure the number of these cycles (or frequency) per second (Fig. 11.3). The number of cycles per unit of time is called the frequency. For convenience, frequency is most often measured in cycles per second (cps) or the interchangeable Hertz (Hz) (60 cps = 60 Hz), named after the physicist Hertz. 1000 Hz is often referred to as 1 kHz (kilohertz).

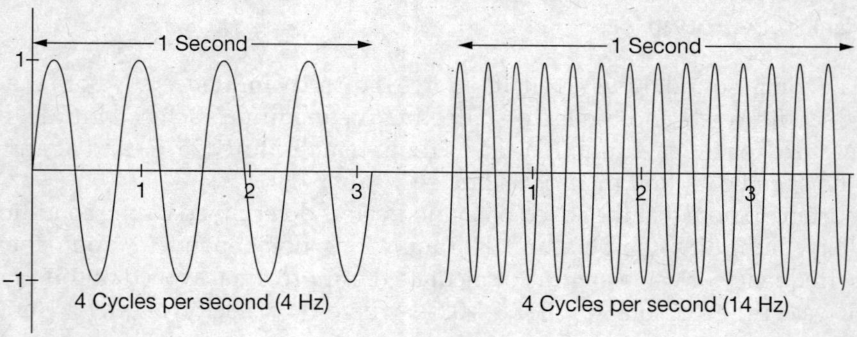

FIGURE 11.3 More wave cycles per second in higher frequency sound

Note

At room temperature and at atmospheric pressure, the speed of sound in air is 332 m/s.

11.3.2 Amplitude

Amplitude is a measure of the degree of change (positive or negative) in atmospheric pressure (the compression and rarefaction of air molecules) in the sound waves. Sounds with greater amplitude will produce greater changes in atmospheric pressure from high pressure to low pressure. Amplitude is almost always a comparative measurement, since at the lowest amplitude end (silence), some air molecules are always in motion and at the highest end, the amount of compression and rarefaction, though finite, is extreme.

In electronic circuits, amplitude may be increased by expanding the degree of change in an oscillating electrical current. A woodwind player may increase the amplitude of their sound by providing greater force in the air column, i.e., blowing harder (Fig. 11.4).

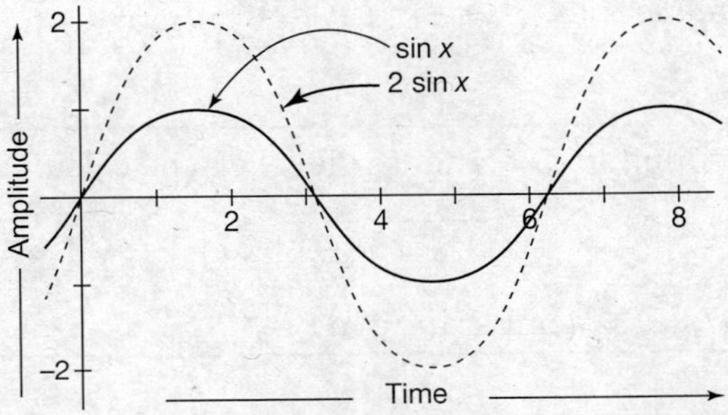

FIGURE 11.4 More displacement from mean position in higher amplitude sound

11.3.3 Phase

Phase denotes a particular point in the cycle of a waveform, measured as an angle in degrees. It is normally not an audible characteristic of a single wave (but can be when we use very low-frequency waves as controls in synthesis). Phase is a very important factor in the interaction of one wave with another, either acoustically or electronically.

When waveforms of either the same or differing phases are combined [Fig. 11.5(a)], they interfere with each other, and their instantaneous amplitudes are summed to create a new composite wave. Two waves with the same frequency and phase will combine to create a single sound of greater amplitude; this is called constructive interference.

Two identical waves 180° out of phase [Fig. 11.5(b)] will completely cancel each other out in a process called phase cancellation or destructive interference. Similarly, those out of phase

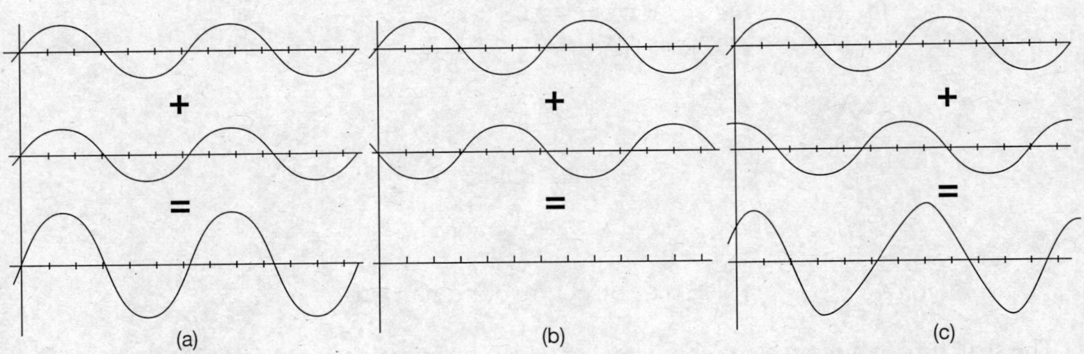

FIGURE 11.5 (a) Two waves in perfect phase add up; (b) the waves completely out of phase cancel out; (c) waves with arbitrary phase difference will add up to result in an arbitrary wave

[Fig. 11.5(c)] will result in arbitrary waves. In real-world acoustic environments, constructive and destructive interferences occur constantly due to room acoustics and other factors. In fact, interference between the sound source and reflected waves is key to producing standing waves. Sometimes taking a small step to the side may completely change the characteristic of a sound because it alters the phase relationship of the source and its reflections. Stereo microphone pairs that are improperly placed can inadvertently lead to unwanted phase cancellations at certain frequencies.

Note

Approximate speed of sound in dry air at sea level can be calculated as

$$\boxed{V \approx 331.4 + 0.6T}$$

where, V = velocity (m/s), T = temperature in °C.

11.4 ▶ Detection of Sound by Human Ear

Sound is detected by ears in humans and animals. Some creatures detect sound through their bodies or some special organs that are sensitive to the vibrations of the surroundings. Different animals and creatures have different frequency and intensity ranges of sensitivity. Microphones are electrical/electronic equivalent of human ear. These can be designed for different frequency ranges and sensitivities.

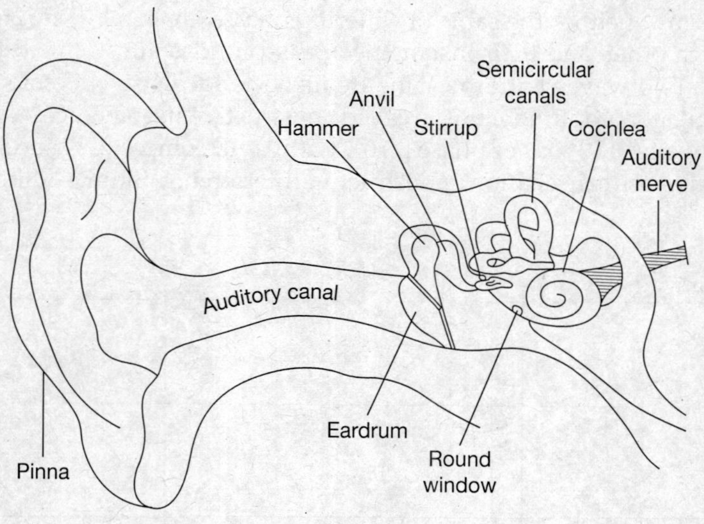

FIGURE 11.6 Structure of human ear

The structure of human ear is shown in Fig. 11.6. The human ear consists of three parts, the outer ear, the middle ear and the inner ear. The inner ear is also called the cochlea. ('Cochlea' means 'snail' in Latin; the cochlea gets its name from its distinctive coiled up shape.)

1. The outer ear consists of the pinna, ear or auditory canal and eardrum.
2. The middle ear consists of the hammer, anvil and stirrup and eardrum.
3. The inner ear consists of the cochlea, the auditory (hearing) nerve and the brain.

There is a hair growth and wax production inside ear canal that has a purpose of self-cleaning and to stop or repel insects and water from the ear. Sound waves enter the ear canal and make the eardrum vibrate. This action moves the tiny chain of bones (hammer, anvil and stirrup) in the middle ear. The last bone in this chain stirrup knocks on the membrane window of the cochlea and makes the fluid in the cochlea vibrate. The fluid vibration then triggers a response in the hearing nerve connected to brain. The semicircular canals connected to cochlea are full of a fluid that changes the levels and guides human brain like a gyroscope and gives us the sense of bending or any tilt in our body.

11.4.1 Sensitivity of Human Ear

Human ear can detect the sound frequencies in the range of 20–20,000 Hz only. The frequencies beyond 20 kHz are called Ultrasonic waves and those below 20 Hz are called Infrasonic waves. Different types of musical instruments generate sound waves of different frequencies and range. Animals have a different type of ear response to sound frequencies, e.g. like the humans, the cats, dogs and elephants can detect very low frequency sounds.

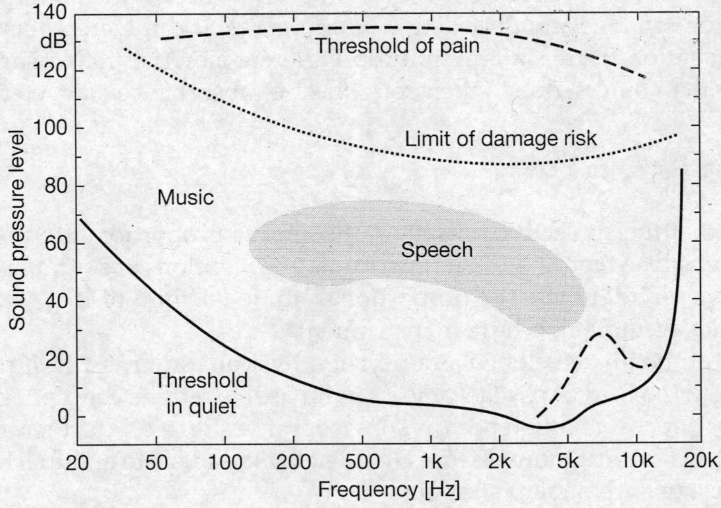

FIGURE 11.7 Human ear response to sound frequencies

Humans can hear only above a minimum intensity level shown (Fig. 11.7) as threshold in quite that is not uniform over all frequency range. The range of frequency response is different to music and voice sounds. If the intensity of sound or pressure increases beyond limit of damage risk, the ears can be damaged and beyond threshold of pain severe pain can be felt by the person.

Table 11.1 shows the ranges of some common animals compared to humans.

TABLE 11.1 Ranges of some common animals compared to humans

	Lower frequency (Hz)	Upper frequency (Hz)
Humans	20	20,000
Dogs	50	45,000
Cats	45	85,000
Bats	20	120,000
Dolphins	0.25	200,000
Elephants	5	10,000

Note

Human ear can detect the sound frequencies in the range of 20–20,000 Hz only.

11.5 ▶ Reflection of Sound

Sound waves reflect from the surfaces and objects in the same way light reflects, i.e. the angle of incidence equals the angle of reflection. A sound wave hitting a flat wall at 45° will reflect off it at 45°. The reflected wave can interfere with the incident (original) wave, producing the constructive and destructive interferences. It can increase its amplitude or, with phase cancellation, decrease its amplitude. Different materials reflect some frequencies more efficiently than others, due to their roughness or absorbance characteristics.

11.6 ▶ Weber-Fechner Law

This law states that the subjective sensation of sound is proportional to the logarithm of the stimulus intensity. According to this law, human perceptions of sound work as follows: Perceived loudness/brightness is proportional to logarithm of the actual intensity measured with an accurate non-human instrument.

This law was originally developed as a psycho-physical law. Ernest Heinrich Weber was an early pioneer in the field of psycho-physics, who developed the concept of the difference threshold or just noticeable difference (JND). Weber investigated it further and found that the size of the JND for most human senses (e.g., sight, sound, taste and touch) is a constant fraction of the size of the standard stimulus.

$$k = \frac{JND}{S} \tag{11.1}$$

Expressed mathematically, this is known as Weber's law, where k is a constant called the Weber fraction and S is the value of the standard stimulus. This equation is usually expressed in the form of Fechner's law. Gustav Fechner derived a relationship between the intensity of a specific stimulus and the perceived (estimated) magnitude. To derive this relationship, Fechner made two important assumptions:

1. The JND is a constant fraction of the stimulus (i.e., Weber's law holds) and
2. The JND is the basic unit of perceived magnitude, so that one JND is perceptually equal to another JND.

By accepting these assumptions, Fechner hypothesized that the magnitude of a stimulus can be determined by starting at the detection threshold (JND) and then adding JNDs. From this, Fechner derived the following mathematical relationship between perceived magnitude (*P*) and stimulus intensity (*I*), where *k* is a constant fraction (Weber's Law).

$$P = k\log(I) \tag{11.2}$$

This relation is called Weber-Fechner law.

11.7 ▶ Absorption Coefficient

The absorption coefficient is defined as the ratio of the sound energy absorbed by the surface to that of the total sound energy incident on the surface.

i.e., $$\text{Absorption coefficient}\ 'a' = \frac{\text{Sound energy absorbed by the surface}}{\text{Total sound energy incident on the surface}}$$

Absorption coefficient is a quantity that is essential in designing an acoustically good building. The value of the absorption coefficient affects reverberation and reverberation time. If the absorption value is higher, then more sound energy is absorbed and the sound dies quickly. If the value is lower, then there is less absorption and the reverberation time increases. For example, if an auditorium's intended use is for musical concert, then the value of the absorbent coefficient for the auditorium should be lower. This allows the sound to persist longer and makes the musical notes continuous.

If the auditorium is designed for public speaking, then the value of absorption coefficient should be higher (but not very high). This absorbs a large portion of the sound energy and thus there is no overlapping of individual words. Typically for a musical hall, the absorption coefficient value is such that the reverberation time is of the order of 1.5 to 2 s. In the case of the auditorium being used for the purpose of speech, the reverberation time should be between 0.5 and 1 s.

Alternatively, the absorption coefficient is defined as the reciprocal of an area which absorbs the same amount of sound energy as absorbed by a unit area of an open window.

The unit for absorption coefficient is W/m^2 or open window units (OWU). For example, if a wall of 4 m^2 area absorbs the same sound energy as that of a 1 m^2 area of an open window, then the absorption coefficient of the wall is $= 1/4 = 0.25$ OWU.

The absorption coefficients (at 512 Hz) of some common materials are given below.

Materials	Absorption coefficient
Marble	0.01
Glass	0.027

(Cont'd)

TABLE (Continued)

Materials	Absorption coefficient
Concrete	0.17
Asbestos	0.26
Heavy curtains	0.50
Fibre glass	0.75

11.8 ▶ Reverberation

A sound produced inside a hall will propagate in all directions. Sound waves incident on the surfaces of walls, floor, ceiling and furniture inside a hall will be multiply reflected. A listener inside the hall will receive the sound waves directly from the source, as well as the reflected waves. As the source of sound is turned off, the listener hears the sound with gradually reducing intensity for some time due to the persistence of sound by multiple reflections at different places in the room. The persistence of audible sound even after the source of sound is turned off is called reverberation.

11.9 ▶ Sabine's Formula

The following are Sabine's conclusions:

(i) The reverberation time is directly proportional to the volume (V) of the hall.

(ii) The reverberation time is inversely proportional to the coefficient of absorption of different materials and surfaces inside the hall.

(iii) Reverberation time depends on the frequency of the sound waves, because absorption coefficient for most of the materials increases with frequency.

Reverberation time,
$$T \propto \frac{\text{Volume of the hall, } V}{\text{Absorption of sound, } A} \tag{11.3}$$
or
$$T = \frac{KV}{A} \tag{11.4}$$

where K = proportionality constant, the value of K is 0.161.

$$\therefore \quad T = 0.161\frac{V}{A} \tag{11.5}$$

where

$$A = \sum_{1}^{n} a_i s_i = a_1 s_1 + a_2 s_2 + \cdots + a_n s_n \tag{11.6}$$

where $s_1, s_2, s_3, ..., s_n$ are surface areas of different materials and $a_1, a_2, a_3, ..., a_n$ are their absorption coefficients, respectively.

11.10 ▶ Derivation of Sabine's Formula Using Growth and Decay Method

Sabine derived a mathematical equation for reverberation time by measuring the rise and decay of sound energy inside a closed hall. The following assumptions are considered in the derivation.

(i) The enclosure is a big one so that sound energy is uniformly distributed in it.
(ii) Sound travels uniformly in all directions from the source.
(iii) Absorption of sound energy by air is negligible.
(iv) Standing wave formation is negligible.

When sound is produced inside a hall, the sound energy is spread uniformly inside the hall (Fig. 11.8).

Decay of sound energy in the hall is given by

$$E_M = E e^{na}$$

where
E_M is the maximum sound energy density in the hall
E is the energy per unit volume of the hall, i.e. energy density
a is the absorption coefficient of the surface of the wall
t is the time
Hence at time $t = 0, E = E_M$
If this energy density remains constant, then it is known as steady state.
or

$$E = E_m e^{-at} \tag{11.7}$$

Equation (11.7) shows the decay of sound energy density with time when the source of sound is turned off. A graph can be plotted between the sound density versus time as shown in Fig. 11.9.

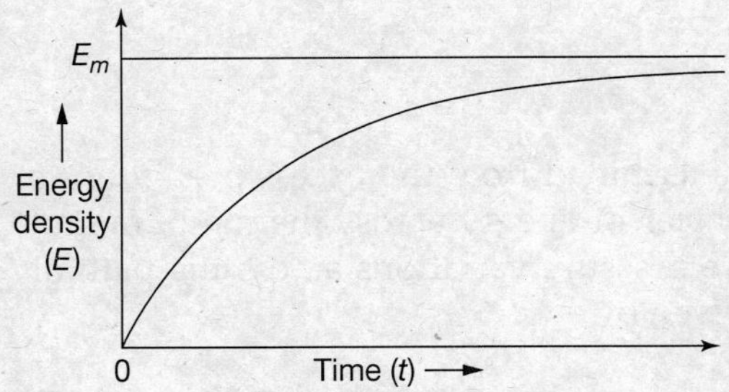

FIGURE 11.8 Growth of sound energy density in a hall

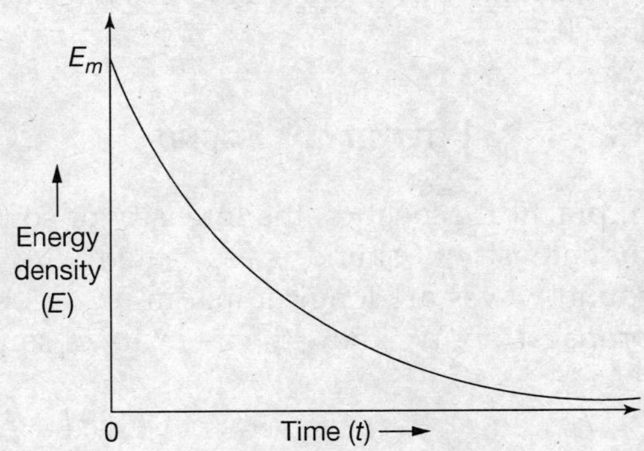

FIGURE 11.9 Decay of sound energy density

Sabine's formula

According to the definition of reverberation time, T, we have

$$\frac{E}{E_m} = 10^{-6} \text{ at } t = T$$

From Eq. (11.7), we can write

$$\frac{E}{E_m} = e^{-at} = 10^{-6} \text{ at } t = T$$

Taking logarithms on both sides of the above equation,

$$aT = 6\log 10 = 6 \times 2.3026$$

It has been proved that

$$a = \frac{vA}{4V} \tag{11.8}$$

where v = velocity of sound in air at room temperature, i.e. 344 M/sec
A = total absorption area of the hall
V= total volume of the hall

$$aT = \frac{vA}{4V} T = 6 \times 2.3026$$

Taking v = velocity of sound in air = 344 m/s

$$T = \frac{4 \times 6 \times 2.3026V}{344 \times A}$$

or

$$T = \frac{0.161V}{A} \tag{11.9}$$

This is Sabine's formula for reverberation time.

Equation (11.9) is applicable for large halls, when the average absorption coefficient is less than 0.2.

11.11 ▶ Intensity of Sound

In practical acoustics, the intensity of sound is determined from the excess pressure. As the intensity of sound is associated with the transfer of energy across the medium and sound waves are longitudinal in nature, there are pressure variations across the path of propagation. We know the intensity of sound is given by

$$I = 2\pi^2 n^2 a^2 v\rho \tag{11.10}$$

Multiplying and dividing Eq. (11.10) by $2\rho v$ and rearranging it, we get

$$I = \frac{(2\pi n a v \rho)^2}{2v}$$

As a simple harmonic pressure wave (sound wave) travels through a medium, the maximum excess pressure is given by

$$P_{\text{max}} = 2\pi n a v \rho \tag{11.11}$$

Comparing the previous equation with Eq. (11.4), we get

$$I = \frac{(P_{\text{max}})^2}{2\rho v} \tag{11.12}$$

This relation shows that the intensity of sound varies directly as the square of the excess pressure. (For ordinary conversation, P_{max} is 0.1 N/m^2 and the sound output is $1.13 \times 10^{-5}\,\text{W/m}^2$.)

Bel and Decibel

To measure sound, a new term known as intensity level, which is the ratio of intensity (to be measured) with respect to a standard intensity, is used practically. The unit for the intensity level of sound is *bel*. One bel is the change in intensity of the sound when the intensity changes by a factor of 10 from the standard intensity. The value of standard intensity is $10^{-12}\,\text{W/m}^2$ or $10^{-16}\,\text{W/cm}^2$ and is denoted as I_0. Since 'bel' (denoted as B) is a large unit, a smaller and convenient unit called *decibel* (dB), which is one-tenth of bel, is used for measuring the intensity level. Thus, we can say that the dynamic range of hearing for humans is 12 bels or 120 dB (for an intensity of $1\,\text{W/m}^2$).

$$\text{Intensity level} = \log_{10}\left(\frac{I}{I_0}\right)\text{bel} \tag{11.13}$$

$$= 10\log_{10}\left(\frac{I}{I_0}\right)\text{dB} \tag{11.14}$$

Let L_1 be the loudness of sound of intensity I and L_0 be the loudness of the standard intensity I_0. Then the intensity level is the difference in loudness of the two intensities given by the Weber-Fechner law as

$$L \propto \log_{10} I$$

$$\therefore \quad L = K \log_{10} I \tag{11.15}$$

If $L_1 = K\log_{10} I$ and $L_0 = K\log_{10} I_0$, then the intensity level (IL) is given as

$$IL = L_1 - L_0$$
$$IL = K\log_{10} l - K\log_{10} I_0$$

$$IL = K \log_{10}\left(\frac{I}{I_0}\right)$$

When $K = 1$,

$$\text{IL} = \log_{10}\left(\frac{I}{I_0}\right)\text{bel}$$

or

$$IL = \log_{10}\left(\frac{I}{I_0}\right)\text{dB}$$

Suppose the intensity level (IL) changes by 1 dB, then the change in intensity

$$1\,\text{dB} = 10\log_{10}\left(\frac{I}{I_0}\right) \tag{11.16}$$

$$\therefore \qquad \log_{10}\frac{I}{I_0} = 0.1$$

Taking antilog, we get

$$\left(\frac{I}{I_0}\right) = 1.26$$

$$\therefore \qquad \frac{I - I_0}{I_0} = \frac{1.26 - 1}{1} = \frac{0.26}{1} \tag{11.17}$$

i.e., the change in IL is 1 dB when the intensity of sound changes by 26%.

Loudness

The loudness is an auditory property for human ear. A sound may appear loud for one person but may be soft for the other. However, it can be related to the intensity or average amplitude of the sound wave (Fig. 11.10) and can be expressed in dB (decibels).

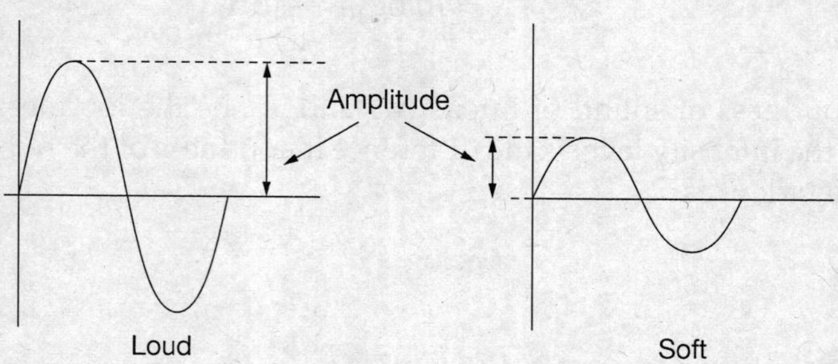

FIGURE 11.10 The amplitude determines loudness

A sound of frequency 2 KHz may have an amplitude of x and appear very loud to human ear, but another frequency, e.g. 20 Hz may be unbearable despite having an amplitude of $2x$. It means that the loudness is property of perception of sound.

Figure 11.11 shows the equal loudness curves for different frequencies. These curves are called Fletcher-Munson curves.

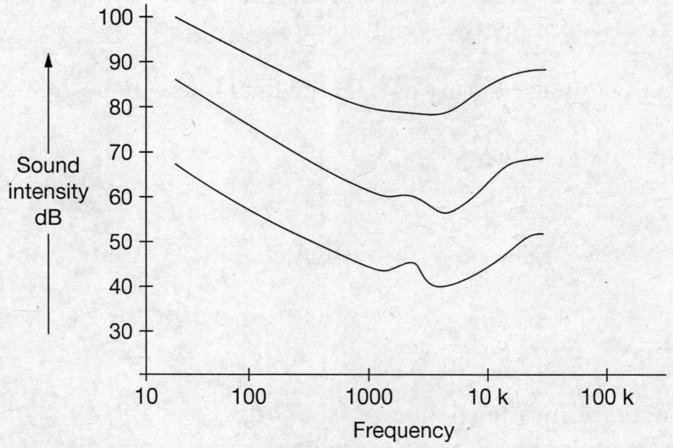

FIGURE 11.11 Equal loudness curves for different frequencies (Fletcher-Munson curves)

EXAMPLE 11.1

The intensity of sound produced by a machine is $0.2\,\text{W/m}^2$. Calculate its intensity level in decibel.

Solution: The intensity of sound produced $= 0.2\ \text{W/m}^2$.
We know that

$$
\begin{aligned}
IL &= 10\log_{10}\left(\frac{I}{I_0}\right)\\
&= 10\log_{10}\left(\frac{0.2}{10^{-12}}\right)\\
&= 10\log_{10}\left(\frac{2\times10^{-1}}{10^{-12}}\right)\\
&= 10\log_{10}\left(2\times10^{-1}\times10^{12}\right)\\
&= 10\log_{10}\left(2\times10^{11}\right)\\
&= 10\log 2 + 10\log 10^{11}\\
&= 10\times0.3010 + 110\\
&= 113\ \text{dB}
\end{aligned}
$$

The intensity level of the machine is 113 dB.

EXAMPLE 11.2

A generator has a sound intensity of 7 dB in a room. If the sound intensity is tripled, what will be the relative intensity level in the room?

Solution: Given the initial intensity as I, let the final intensity be I_1.

Then according to the data given, $\frac{I_1}{I} = 3$.
Therefore,

$$IL = 10\log_{10}\left(\frac{I_1}{I}\right)$$
$$= 10\log_{10}(3) = 10\times(0.4771)$$
$$= 4.77 \text{ dB}$$

Therefore, the increase in intensity level is 4.77 dB.

EXAMPLE 11.3

What is the intensity level of a sounding instrument which radiates energy at the rate of 4.81 W to a person standing at a distance of 15 m?

Solution:

$$\text{Intensity of sound} = \frac{\text{Radiating power}}{\text{Area}}$$
$$= \frac{P}{4\pi r^2} = \frac{4.81}{4\pi(15)^2}$$
$$= 0.0017 \text{ W/m}^2$$
$$IL = 10\log_{10}\left(\frac{I}{I_0}\right)$$
$$IL = 10\log_{10}\left(\frac{0.0017}{10^{-12}}\right)$$
$$= 10\log_{10}\left(\frac{1.7\times10^{-3}}{10^{-12}}\right)$$
$$= 10\log_{10}\left(1.7\times10^{-3}\times10^{12}\right)$$
$$= 10\times0.23 + 10\times\log_{10}\left(10^9\right) = 2.3 + 90$$
$$= \mathbf{92.3 \text{ dB}}$$

EXAMPLE 11.4

Calculate the intensity levels for Example 11.3, if the standard intensities are (i) $0.0001\ \text{W/m}^2$, (ii) $1\ \text{W/m}^2$ and (iii) $10\ \text{W/m}^2$. Find the intensity levels of the same source, if the person stands at distances of (a) 10 m, (b) 20 m and (c) 25 m for a standard intensity of $10^{-12}\ \text{W/m}^2$.

Solution:

(i) Intensity of sound has been calculated as $0.0017\ \text{W/m}^2$ from Example 11.3. Standard intensity is $0.0001\ \text{W/m}^2$.

$$IL = 10\log_{10}\left(\frac{I}{I_0}\right)$$

$$IL = 10\log_{10}\left(\frac{0.0017}{0.0001}\right)$$

$$= 10\log_{10}\left(\frac{1.7\times10^{-3}}{1\times10^{-4}}\right)$$

$$= 10\log_{10}\left(1.7\times10^{-3}\times10^{4}\right)$$

$$= 10\log 1.7 + 10\log 10$$

$$= 2.3 + 10$$

$$= 12.3\ \text{dB}$$

(ii) Standard intensity is $I\ \text{W/m}^2$

$$IL = 10\log_{10}\left(\frac{I}{I_0}\right)$$

$$IL = 10\log_{10}\left(\frac{0.0017}{1}\right)$$

$$= 10\log_{10}\left(1.7\times10^{-3}\right)$$

$$= 10\times(-2.77)$$

$$= -27.7\ \text{dB}$$

(iii) Standard intensity is $10\ \text{W/m}^2$.

$$IL = 10\log_{10}\left(\frac{I}{I_0}\right)$$

$$IL = 10\log_{10}\left(\frac{0.0017}{10}\right)$$

$$= 10\log_{10}\left(1.7\times10^{-4}\right)$$
$$= 10\times(-3.77)$$
$$= -37.7 \text{ dB}$$

(a) At a distance of 10 m

$$\text{Intensity of sound} = \frac{\text{Radiating power}}{\text{Area}}$$
$$= \frac{P}{4\pi r^2}$$
$$= \frac{4.81}{4\pi(10)^2}$$
$$= 0.00383 \text{ W/m}^2$$
$$IL = 10\log_{10}\left(\frac{I}{I_0}\right)$$
$$IL = 10\log_{10}\left(\frac{0.00383}{10^{-12}}\right)$$
$$= 10\log_{10}\left(\frac{3.83\times10^{-3}}{10^{-12}}\right)$$
$$= 10\log_{10}\left(3.83\times10^{-3}\times10^{12}\right)$$
$$= 10\times0.58 + 10\times\log_{10}\left(10^{9}\right)$$
$$= 5.8 + 90 = 95.8 \text{ dB}$$

(b) At a distance of 20 m

$$\text{Intensity of sound} = \frac{\text{Radiating power}}{\text{Area}}$$
$$= \frac{P}{4\pi r^2} = \frac{4.81}{4\pi(20)^2}$$
$$= 0.0009575 \text{ W/m}^2$$
$$IL = 10\log_{10}\left(\frac{I}{I_0}\right)$$
$$IL = 10\log_{10}\left(\frac{0.0009575}{10^{-12}}\right)$$
$$= 10\log_{10}\left(\frac{9.575\times10^{-4}}{10^{-12}}\right)$$

$$= 10\log_{10}\left(9.575\times10^{-4}\times10^{12}\right)$$

$$= 10\times0.9811+10\times\log_{10}\left(10^{8}\right)$$

$$= 9.811+80$$

$$= 89.811 \text{ dB}$$

(c) At a distance of 25 m

$$\text{Intensity of sound} = \frac{\text{Radiating power}}{\text{Area}}$$

$$= \frac{P}{4\pi r^2} = \frac{4.81}{4\pi(25)^2}$$

$$= 0.000613 \text{ W/m}^2$$

$$IL = 10\log_{10}\left(\frac{I}{I_0}\right)$$

$$IL = 10\log_{10}\left(\frac{0.000613}{10^{-12}}\right)$$

$$= 10\log_{10}\left(\frac{6.13\times10^{-4}}{10^{-12}}\right)$$

$$= 10\log_{10}\left(6.13\times10^{-4}\times10^{12}\right)$$

$$= 10\times0.787+10\times\log_{10}\left(10^{8}\right)$$

$$= 7.87+80$$

$$= \mathbf{87.87\ dB}$$

EXAMPLE 11.5

What is the intensity of sound for an intensity level of 65 dB? Standard intensity being 10^{-12} W/m^2.

Solution: The intensity level

$$IL = 10\log_{10}\left(\frac{I}{I_0}\right)$$

$$65dB = 10\log_{10}\left(\frac{I}{10^{-12}}\right)$$

$$\frac{65}{10} = \left[\log_{10}(I) - \log_{10}\left(10^{-12}\right)\right]$$

$$6.5 = \left[\log_{10}(I)\right] - \left[(-12)\log_{10}(10)\right]$$

$$6.5 = \log_{10}(I) + 12(1)$$
$$6.5 - 12 = \log_{10}(I)$$
$$-5.5 = \log_{10}(I)$$

Taking antilog on both sides,

$$0.000003162 = 1$$

i.e.,
$$\mathbf{I = 3.162 \times 10^{-6}\ W/m^2}$$

EXAMPLE 11.6

The change in the value of intensity level of a source is observed to be 20 dB, when its sound power is decreased from 600 mW.

(i) What is the final power of the source?

(ii) Also find the value of intensity level of the source if its power is increased to 800 mW.

Solution:

(i) Let the standard, initial and final values of power be P_0, P_1 and P_2, respectively.

$$\text{Then the initial intensity level} = 10\log_{10}\left(\frac{P_1}{P_0}\right)$$

$$\text{The final intensity level} = 10\log_{10}\left(\frac{P_2}{P_0}\right)$$

$$\text{The change in intensity level} = \text{Final intensity} - \text{Initial intensity}$$
$$= 20\ \text{dB}$$
$$= 10\log_{10}\left(\frac{P_2}{P_0}\right) - 10\log_{10}\left(\frac{P_1}{P_0}\right)$$
$$= 10\log_{10}\left(\frac{P_2}{P_1}\right) = -20\ \text{dB}$$

(−ve sign ∵ the power decreases)

∴
$$\log_{10}\left(\frac{P_2}{P_1}\right) = -2$$

or
$$\left(\frac{P_2}{P_1}\right) = 0.01$$

$$\therefore \quad P_2 = 0.01P_1$$
$$= 0.01 \times 600$$
$$= 6 \text{ mW}$$

The final power of sound = 6 mW.

(ii) The change in the intensity level = Final intensity level – Initial intensity level

$$= 10\log_{10}\left(\frac{P_2}{P_0}\right) - 10\log_{10}\left(\frac{P_1}{P_0}\right)$$
$$= 10\log_{10}\left(\frac{P_2}{P_1}\right)$$
$$= 10\log_{10}\left(\frac{800}{600}\right)$$
$$= 10\log_{10}(1.33)$$
$$= 10(0.124)$$
$$= 1.24 \text{ dB}$$

Therefore, the change in the intensity level = 1.24 dB.

11.12 ▶ Acoustics of Buildings

Sound engineering becomes very important when designing a hall either for speech or music purposes. The bottom line in designing an acoustically good building is to see that the sound is distributed uniformly inside the room with every single person hearing the sound clearly.

A hall or auditorium designed for lecture or concert should have right acoustical balance and it should have the following advantages:

- The initial sound from source should have adequate intensity.
- The sound should be evenly spread over the whole area covered by the audience.
- The successive sound in the speech or music should be clear and distinct

There are several factors which determine the acoustical quality of a hall. Some of them are been discussed here:

(i) Shape of wall and ceiling

The side wall and ceiling are potentially useful reflecting surface; thus, it should be properly designed to maximize their usefulness. The rear wall and floor are source for harmful reflection which should be avoided. Parallel hard wall create echo problem. Thus, splayed side wall should be used which greatly reduces the echo problem and enhances the acoustical quality of the hall. In view of this, a fan-shaped floor plan should be used.

(ii) Surface of balconies

The concave surface within the hall is not desirable because it focusses sound reflections. Such surfaces must be broken up with smaller convex surfaces so that sound is diffused in all directions. However, the surface of balconies should be concave to reflect the sound within the hall.

(iii) Floor plan with diverging side walls

The interior surface of the hall should be given utmost attention to make the hall acoustically satisfactory. If the side wall is parallel, they are to be covered with absorbing materials. As the reflection from the rear wall is of no use, it should be covered with absorbents. In a large hall, a false ceiling should be provided. The false ceiling be positioned near the proscenium and should be constructed of reflective material and inclined in proper way to help reflections of sound from the stage to reach the rear seats in the hall. The floor should be covered with carpet which covers the useless reflecting surface and greatly reduces audience noise.

(iv) Seating arrangement

The seats should be arranged in concentric arcs of the circle. Flat floor seating of more than few rows obstructs the good visibility and good hearing. Sloped seating is essential for large audience to avoid visibility and provides good acoustics. The successive rows of the seats have to be raised over the preceding ones; thus, the floor level rises towards the rear end. The rise in level may be about 8 to 12 cm per row. The angle subtended by the horizontal from the stage with the person in the first row in front of the stage should not exceed 30°. The first row distance from the stage should not be less than 4.5 m.

11.13 ▶ Acoustic Requirements of a Good Hall/Auditorium

An auditorium or hall is said to be acoustically better if it satisfies the following conditions:

- Uniform distribution of sound throughout the hall with correct loudness and quality even when the hall is full or empty.
- No overlapping of syllables.
- The quality of music or speech should be unchanged.
- No external or internal noise disturbances.
- No echoes or resonance.
- Having an optimum reverberation time.
- No maxima or minima of sound inside the hall.
- No echelon effect.

11.13.1 Optimum Reverberation Time

Reverberation time is used to determine how quickly the sound decays in a room. It depends on the physical volume and the surface materials (chairs, curtains, etc.) present in a room. The value of the reverberation time plays a crucial part in describing the acoustical nature of a building, because if the reverberation time is too large, overlapping of successive sound notes takes place, resulting in loss of clarity. Large spaces, such as cathedrals and indoor gymnasiums, generally have longer reverberation times and hence sound lively or boomy depending on the listener.

On the other hand, if the reverberation time is too small in a hall, the loudness is insufficient and a speaker finds no response from the hall. Such a hall is said to be acoustically dead. For example, small rooms, such as bedrooms and recording studios, are less reverberant and are thus dry or dead. Hence, it becomes very important to set the value of reverberation time neither large nor low, but at a satisfactory value known as optimum reverberation time.

A deviation of 10% from the optimum value will be satisfactory if all other important attributes for room acoustics have been achieved. For music, reverberation adds to the fullness of the tone, blended sound and richness of bass frequencies. In general, larger rooms should have higher reverberation time ranges than smaller rooms of the same type.

11.13.2 Control of Reverberation Time

Long reverberation times can degrade speech, perception by hearing-impaired persons, far more than the normal-hearing persons. For such persons, the reverberation times should be well below the accepted values for most of the situations.

A large reverberation time can be reduced by the following ways:

- Providing windows and ventilators.
- Using heavy curtains with folds.
- Decreasing the volume of the hall.
- Covering the walls of the room with sound-absorbing materials like felt, fibre board, glass wool, etc.
- Covering the floor with carpets.
- Having a large audience.

Reverberation times required for some common activities are listed in Appendix B.

11.13.3 Uniform Loudness

It is a degree of sensation produced in the ear. If the intensity (and hence the loudness) is weakened, then it is possible for the sound to go below the level of audibility. It is thus required that sufficient loudness in every part of a hall is present for satisfactory hearing.

Remedies

The increase in loudness can be achieved by the following ways:

- To get good loudness, maximum reflection from the stage is desirable. This can be done by placing a large sounding board behind the speaker facing the audience.
- Large polished reflecting surfaces immediately above the speaker are also helpful.
- Use of good quality loudspeakers.

- Low ceilings can also help to reflect the sound towards the audience.
- The wall surface at the speaker's end can be given a parabolic shape, which enables almost uniform spread of sound intensity in every part of the room (Fig. 11.12).

11.13.4 Focussed Sound at Audience

Concentration of sound waves at any particular region in a hall can happen if the hall contains curved surfaces on the ceiling. This focussing of sound waves will produce a maximum intensity of sound at the focal point of the curved surface and zero intensity at some other place. This non-uniform distribution of sound intensity is unwanted and hence has to be corrected.

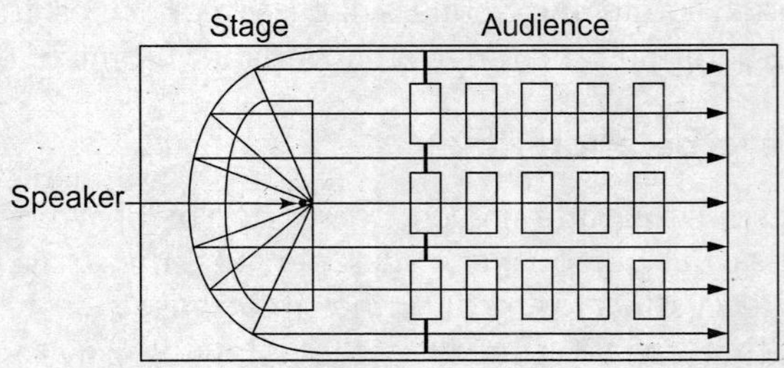

FIGURE 11.12 Uniform spreading of sound

As shown in Fig. 11.13(a), the listener receives sound directly from the speaker. He also receives: reflected and focussed sounds from the ceiling. Thus, the intensity of sound is comparatively higher for him than others at other places in the hall. Sometimes the direct sound and the reflected sound may not be in phase. This may result in the loss of sound and a minimum intensity is heard. Further, the direct and the reflected sounds may form a stationary wave form. All the above effects result in an uneven distribution of sound.

Remedies

To correct the problem, concave surfaces are to be avoided and either flat [Fig. 11.13(b)] or convex [Fig. 11.13(c)] surfaces should be used while designing an auditorium. In case the aesthetic design requires a concave shape, then the surface should not be highly polished or it has to be covered with sound-absorbing materials.

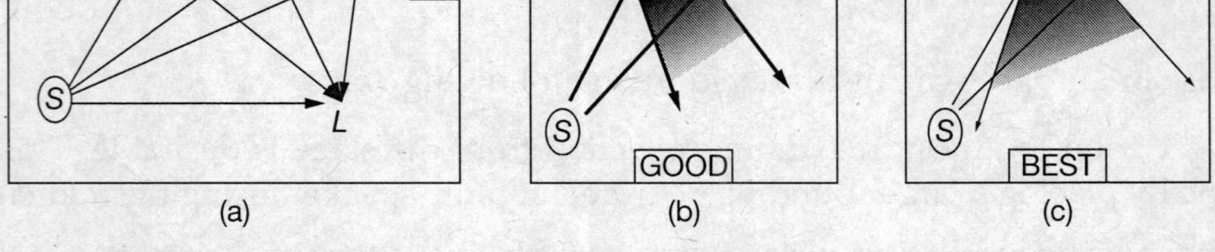

FIGURE 11.13 Focussing of sound waves

11.13.5 Minimum Echo

An echo is the repetition of the original sound by a reflecting surface. An echo is produced when the reflecting surface is more than 17 m away from the sound source or the time interval between the two sounds is about 1/10th of a second. In both the cases, the reflected sound is heard little later as a distinct repetition of the original sound causing confusion to the listener (The reflected sound is heard along with the original sound.) This effect is more common when the reflecting surface is curved.

Another problem with the echoes is the occurrence of discrete echoes. Here, the echoing sound is heard distinctly after the original sound had died down. Echoes heard in canyons, mountains belong to this category. Flutter echoes are another type of echoes that occur when sound is rapidly bounced back and forth between two parallel walls (also between ceiling and floor), creating a ringing effect.

Remedies

Echoes can be avoided by covering the ceilings and floors with sound-absorbing materials. Also, they are prevented by avoiding curved surfaces in the design of the hall.

11.13.6 Minimum Echelon Effect

The sound produced in front of a structure with regular spacing like staircases [Fig. 11.14(a)] or a set of railings may produce a musical note due to the regular repetition of sound (echoes) along with original sound. This makes the original sound unintelligible and confusing. Such an effect is known as echelon effect. If the frequency of this note is within the audible range, then the listener will hear only this note prominently.

Remedies

Echelon effect can be minimized and avoided by avoiding regular repeating structures or covering them with sound-absorbing materials like carpets, etc. [Fig. 11.14(b)].

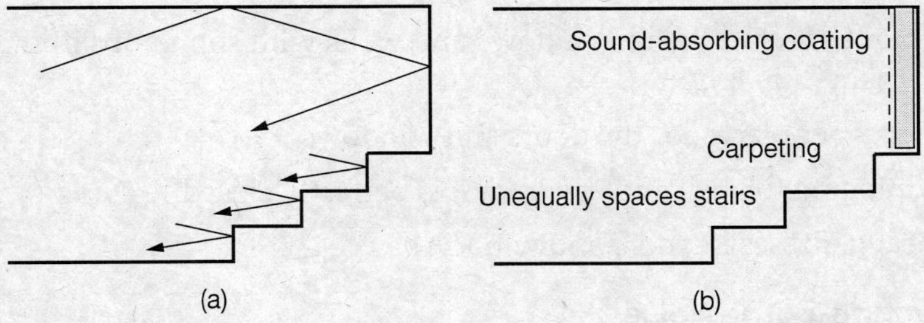

FIGURE 11.14 Focussing of sound waves

11.13.7 No Resonance

Sometimes window panes, loosely fitted wooden portions and wall separators, hollows and crevices, etc. start vibrating by absorbing frequencies from the sound produced in the hall. When the natural frequency of the vibrating surface is matched by a frequency of the sounding note, the surface is set into forced vibrations and an undesirable resonance is created. This resonance amplifies the vibrations and thereby produces a new sounding body that creates a jarring effect in the hall. The original sound is thus distorted due to the resulting interference and hence this leads to unpleasant hearing.

Remedies

This effect can be rectified by hanging a large number of curtains in the hall. Convex cylindrical segments of the walls and ceilings diverge the sound and thus produce a more uniform distribution.

Note

Even an acoustically good hall can produce reverberations and echoes if not occupied by sufficient number of audiences.

11.14 ▶ Noise Pollution

Any unwanted sound that reaches our ear can be termed as a noise. There are three types of noises:

- Air-borne noises,
- Structure-borne noises,
- Inside noises.

11.14.1 Air-borne Noises

Noises that come from outside through open windows, ventilators, doors, etc. are known as air-borne noises or outside noises. This is more common in a densely populated area or a room beside an area with heavy traffic.

Remedies

- Using double-layered doors, windows and walls with sound-absorbing materials in between them [Fig. 11.15(a)].
- Allotting proper places for the doors and windows.
- Air-conditioning the hall and ensuring it is perfectly closed.
- Use of heavy glasses in doors and windows.

11.14.2 Structure-borne Noise

Any noises that can be conveyed through the structure of the building are known as structure-borne noises. The sources of this type of noise are produced either inside the building or outside the building which create structural vibrations.

Remedies

Main reason for the structure-borne noises is the continuous medium available for their travel. The continuity is broken by introducing a new layer of sound-insulating materials

[Fig. 11.15(a)] which reduces the extent of their travel through the structure. Usage of double walls and suitable sound-absorbing materials [Fig. 11.15(b) and (c)] along with anti-vibration mounts (layer of wood, springs, shock absorbers, etc.) for vibrating surfaces (engines, motors, etc.) also reduces structure-borne noises.

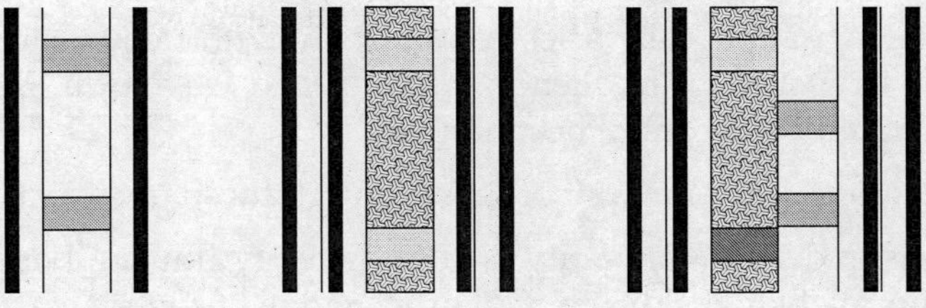

(a) (b) (c)

FIGURE 11.15 (a) One layer gypsum wall board (GWB); studs, one layer GWB – poor insulation; (b) two layers GWB, glass fibre insulation and studs, two layer GWB – better insulation; (c) two layers GWB, glass fibre insulation, two sets of studs, two layers GWB – best insulation

11.14.3 Inside Noises

The noises that are produced inside the same room or in the adjacent rooms are known as inside noises. Moving people, movement of furniture, working machines (typewriters, air-conditioners, fans, etc.) are the sources for this type of noise.

Remedies

Covering the floor with carpets, padding the walls and ceilings with sound-absorbing materials, and placing sound-producing machineries on sound absorbent pads to reduce the inside noises.

Recap Zone

Points to Remember

- Acoustics is a branch of physics that deals with the process of generation, reception, propagation and analysis of sound.
- Audible frequencies are those frequencies that lie between 20 and 20,000 Hz, producing a sensation of hearing in the human nervous system.
- Inaudible frequencies are those that lie out of this range and are not audible.
- The sound frequencies that lie outside the upper limit of the audible frequencies are known as ultrasonic frequencies (> 20,000 Hz) and those that lie outside of the lower limit are known as infrasonic frequencies (< 20 Hz).
- Musical sound produces a pleasing effect to the ears, the waveform is regular and repeatable, and there are no sudden changes in amplitude.

- Noise is unpleasant and even though its frequency spectrum is continuous, the waveform is irregular and lacks periodicity, also it undergoes sudden changes in amplitude.
- The musical sound is characterized by three distinct qualities: pitch, intensity or loudness, and quality or timbre.
- Pitch is a mental sensation that varies parabolically with frequency between 20 and 20,000 Hz. It is a function of frequency, intensity and waveform. But the pitch is a physiological quantity and cannot be measured.
- Intensity of the sound is the amount of sensation produced in the ear.
- Intensity is measured by the quantity of sound energy flowing per second across a unit area held normally to the direction of propagation of sound wave (unit: J/m^2 s or W/m^2J). It is given by $I = 2\pi^2 n^2 a^2 vp$.
- Weber-Fechner law states the intensity dependence of loudness as

$$L \propto \log I$$

i.e.,

$$\boxed{L = K \log I}$$

i.e., the loudness produced is directly proportional to the logarithm of intensity.
- Quality/timbre determines the quality of sound, which is its ability to distinguish between different sounds of the same frequency.
- Bel is the change in intensity of the sound when the intensity changes by a factor of it from the standard intensity.
- As bel is a large unit, a smaller and convenient unit called decibel, which is 1/10th of bel is used for measuring the intensity level.
- Persistence of sound even after the source is cut off is known as reverberation.
- The time taken for the sound to decrease to one millionth ($1/10^6$, i.e., by 60 dB) of its initial value after the source had been cut off is known as reverberation time.
- Absorption coefficient is defined as the ratio of the sound energy absorbed by the surface is that of the total sound energy incident on the surface. Its unit is open window units (OWU).

i.e., $$\text{Absorption coefficient } 'a' = \frac{\text{Sound energy absorbed by the surface}}{\text{Total sound energy incident on the surface}}$$

Alternatively, it is also defined as the reciprocal of its area which absorbs the same sound energy as absorbed by a unit area of an open window.
- Reverberation time is given by

$$T = \frac{0.161V}{\sum aS}$$

According to the equation, the reverberation time is directly proportional to the volume of the hall, inversely proportional to the area of absorbing surfaces, and inversely proportional to the absorption coefficient of the individual surfaces.

KEY TERMS WITH DEFINITIONS

- **Sound** – Sound is a form of energy which is produced due to the vibration of objects and it causes a sensation of hearing in our ears.
- **Compressions** – When a vibrating object moves forward, it pushes and compresses the air in front of it forming a region of high pressure called compression.
- **Rarefaction** – When vibrating objects moves backwards, it forms a region of low pressure called rarefaction.
- **Cycle** – A complete round trip back to the starting point by a vibrating particle is called a cycle.
- **Amplitude** – The amplitude of sound wave is the height of the crest or rough.
- **Wavelength** – Wavelength is the distance between two consecutive compressions or two consecutive rarefactions.
- **Frequency** – The number of vibrations completed by a particle in one second is frequency.
- **Time period** – The time taken by the particle of the medium for completing one oscillation is called the time period.
- **Loudness** – Loudness of sound is the measure of sound energy reaching the ear per second. Loudness or softness of a sound wave is the sensation that depends upon its amplitude.
- **Weber-Fectmer law** – It states that the subjective sensation is proportional intensity.

$$L \propto \log I \text{ i.e, } L = K \log I$$

- **Absorption coefficient** – The reciprocal of an area which absorbs the same amount of sound energy as absorbed by a unit area of an open window.
- **Sabine's formula** – The reverberation time is directly proportional to be coefficient of absorption of different materials and surfaces inside the hall.
- **Reverberation** – Persistence of sound wave for a long time because of repeated reflections of sound are called reverberation.
- **Reverberation time** – It is perceived as the time for the sound to die away after the sound source ceases, but that of course depends upon the intensity of the sound.
- **Echo** – Repetition of sound due to reflection of original sound by a large and hard obstacle.

$$V \approx 331.4 + 0.6T_c$$

IMPORTANT FORMULAE AND EQUATIONS

Equation Number	Equation	Remarks
(1)	$V \approx 331.4 + 0.6T$	**Approximate speed of sound in dry air at sea level. *T* is temperature**

(2)	$R = \frac{\text{Sound energy absorbed by the surface}}{\text{Total sound energy incident on the surface}}$	Absorption coefficient
(3)	$T = \frac{0.161V}{\sum as}$	Sabine's formula for reverberation time
(4)	$I = 2\pi^2 n^2 a^2 vp$	Intensity of sound
(5)	$IL = 10\log_{10}\left(\frac{I}{I_0}\right)$	Intensity level
(6)	$P = k \log (I)$	Weber-Fectmer law

REVIEW ZONE

SHORT ANSWER QUESTIONS

1. What are sound, ultrasonics and infrasonics?
2. What do you mean by the standard intensity of sound?
3. How is audible sound classified?
4. Mention the differences between the musical sound and noise.
5. What are the three qualities by which a musical sound is characterized?
6. What do you mean by intensity level?
7. Define intensity of sound.
8. State reverberation time.
9. What is reverberation?
10. What do you mean by absorption coefficient?
11. What are the assumptions of the Sabine's reverberation time theory?
12. What are the salient features of the Sabine's formula for the reverberation time?
13. On what factors (any four) do the acoustics of a building gets affected?
14. What are the different sources for noises in a building?
15. What do you mean by echelon effect?
16. Define reverberation time. Derive with necessary diagrams the Sabine's reverberation time formula.
17. Explain in detail the factors affecting the acoustics of a building.
18. Write short notes on (i) pitch, (ii) loudness and (iii) quality.
19. Describe in detail the growth and decay of sound in an enclosed space.
20. What are bad acoustics? Describe the ways of remedying the problem.

CHAPTER 12

Wave Particle Duality and Uncertainty

Learning Objectives

By the end of this chapter, the student will be able:

- To understand dual nature of light
- To describe photoelectric effect and Compton effect
- To quantify particle waves and their experimental observation in Davisson-Germer experiment
- To define particle velocity, wave velocity and group velocity
- To analyze problems related to classical and quantum physics
- To introduce the concept of quark and gluons

12.1 ▶ Introduction

As light was thought to behave as particles and sometimes as waves, similarly electrons and other particles were considered to possess a wave particle dual nature. The evidence for the existence of light as waves had already been established when the photoelectric effect gave firm evidence of the particle nature as well (Fig. 12.1). On the other hand, the particle properties of electrons were well proved beyond doubt but the de-Broglie hypothesis and the subsequent experiments by Davisson and Germer also established the wave nature of electron.

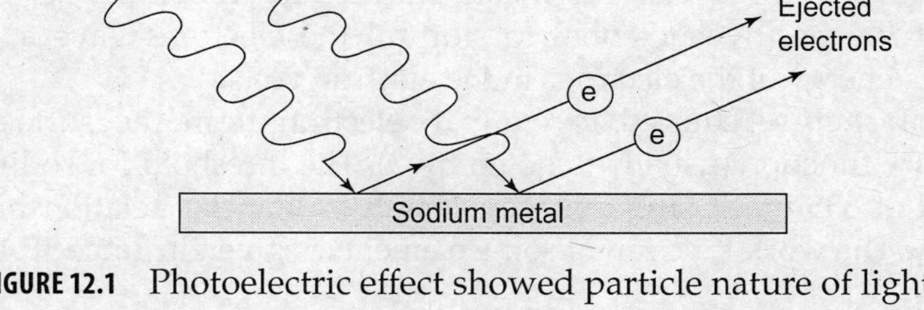

FIGURE 12.1 Photoelectric effect showed particle nature of light

12.2 ▶ Photoelectric Effect

Certain materials like sodium metal emitted electrons from surface when bombarded with light. The phenomenon was called photoelectric effect (Fig. 12.2). The observed remarkable aspects of photoelectric effect are as follows:

1. The electrons were emitted immediately without any time lag.
2. On increasing the intensity of light, there was an increase in the number of photoelectrons, but not to their maximum kinetic energy.
3. Red light did not cause the emission of electrons, no matter what the intensity is.
4. A weak violet light ejected only a few electrons, but their maximum kinetic energies were greater than those for intense light of longer wavelengths.

The details of the photoelectric effect were in direct contradiction to the expectations of classical physics. The explanation marked one of the major steps towards quantum theory.

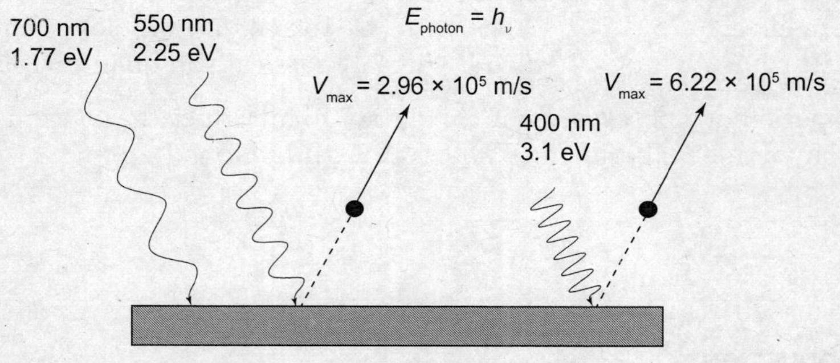

FIGURE 12.2 Photoelectric effect with potassium metal (for potassium, ~ 2.0 eV needed to emit electrons)

Analysis of data from the photoelectric experiment shows that the energy of the emitted electrons is proportional to the frequency of the illuminating light. This shows that whatever was knocking the electrons out had energy proportional to the light frequency. The remarkable fact that the emission energy is independent of the total energy of illumination shows that the interaction must be like that of a particle which gives away its entire energy to an electron. This fits well with Planck's hypothesis that light in the blackbody radiation experiment can exist only in discrete bundles of energy $E = h\nu$ (photons).

Electrons emitted from a sodium metal surface are measured as electric current. Measurement of the positive potential to stop all the electrons gave a measure of the maximum kinetic energy of the electrons in the electron volts.

The minimum energy required to emit an electron from the surface is called the photoelectric work function (usually denoted by ϕ). The threshold for sodium corresponds to a wavelength of 683 nm. Using this wavelength in Planck's relationship gives photon energy of 1.82 eV. The work function of some elements is given in Table 12.1.

TABLE 12.1 Work function for photoelectric effect

Element	Work function (eV)	Element	Work function (eV)
Aluminium	4.08	Beryllium	5.0
Cadmium	4.07	Calcium	2.9
Carbon	4.81	Caesium	2.1
Cobalt	5.0	Copper	4.7
Gold	5.1	Iron	4.5
Lead	4.14	Magnesium	3.68
Mercury	4.5	Nickel	5.01
Niobium	4.3	Potassium	2.3
Platinum	6.35	Selenium	5.11
Silver	4.73	Sodium	5.11
Uranium	3.6	Zinc	4.3

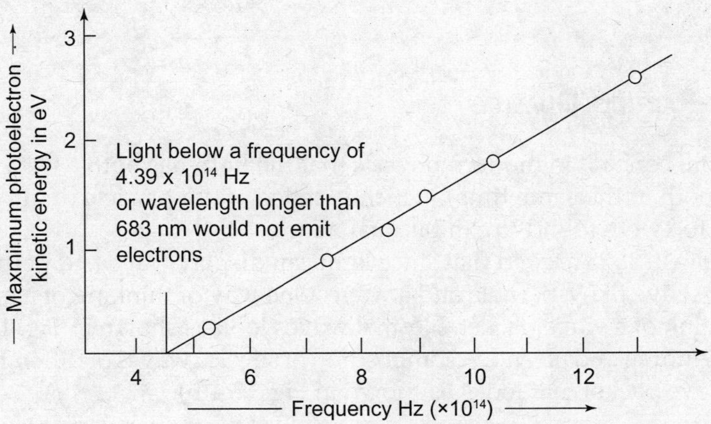

FIGURE 12.3 Plot of kinetic energy of photoelectron (in terms of reverse voltage) vs frequency of incident photons

The fact that the plot was not dependent upon the intensity of the incident light implies that the interaction is like a particle, which gives all its energy to an ejected electron minus the binding energy with the surface (Fig. 12.3).

Einstein explained the photoelectric effect by assuming that light exists in a particle-like form, packets of energy (quanta) called photons. There is no current flow in case of red light because the packets of energy carried by each individual red photon are too weak to knock the electrons off the atoms. It is not dependent on the number of red photons bombarded on to the cathode. But the individual UV photons are strong enough to release the electrons and cause a current flow.

EXAMPLE 12.1

Calculate the work function in electron volts of a metal, given that photoelectric threshold is (i) 6200 Å and (ii) 5000 Å.

Solution: Given (i) $\lambda_0 = 6.2 \times 10^{-7}$ m

(ii) $\lambda_0 = 5.0 \times 10^{-7}$ m

(i)
$$\phi_0 = h\nu_0 = \frac{hc}{\lambda_0}$$

$$\phi_0 = \frac{6.62 \times 10^{-34} \times 3 \times 10^8}{6.2 \times 10^{-7} \times 1.6 \times 10^{-19}} \text{eV}$$

$$= \mathbf{2.0\ eV}$$

(ii)
$$\lambda_0 = 5.0 \times 10^{-7} \text{ m}$$

$$\phi_0 = \frac{6.62 \times 10^{-34} \times 3 \times 10^8}{5 \times 10^{-7} \times 1.6 \times 10^{-19}} \text{ eV} = \mathbf{2.5\ eV}$$

12.3 ▶ Wave-Particle Duality

It is a fundamental concept in modern physics that the light has both a wave and a particle nature (but not both at the same time), which is called wave-particle duality.

The electrons too were found to exhibit dual nature.

Louis de-Broglie in 1923 argued that since light can display wave and particle properties, matter can also behave like a particle and a wave. One way of thinking of a matter wave (or a photon) is to think of a wave packet. Normal waves look as shown in Fig. 12.4(a).

Having no beginning and no end. A composition of several waves of different wavelengths can produce a wave packet that looks as shown in Fig. 12.4(b).

FIGURE 12.4 (a) Normal wave (continuous); (b) Wave packet

A photon, or a free-moving electron, can be thought of as a wave packet, having both wave-like properties and also the single position and size which we associate with a particle. There are some problems, such as the wave packet does not stop at a finite distance from its peak, it also goes on for each and every point, indicating that an electron may exists at all points in its path.

De-Broglie produced a simple formula that the wavelength of a matter particle is related to the momentum of the particle. Hence, energy is also connected to the wave property of matter.

While de-Broglie waves are difficult to accept in particles and solid things with definite size and positions, electron waves were confirmed in the laboratory by running electron beams through slits and demonstrating that interference patterns are formed. The length of the wave diminishes in proportion to the momentum of an object. So greater the mass of the object involved, shorter the waves. The wavelength of a human, for example, is only one millionth of a centimetre that is why people are not able to observe their wave behaviour.

EXAMPLE 12.2

Calculate the wavelength associated with an electron subjected to a potential difference of 1.25 kV.

Solution: The de-Broglie wavelength λ is given by

$$\lambda = \frac{h}{mv} = \frac{h}{\sqrt{2mE}}$$

$$m = 9.1\times10^{-31}\ \text{kg},\ e^{-} = 1.6\times10^{-19}\,\text{C}$$

$$\text{eV} = E = 1.25\text{K electron volts}$$

$$\lambda = \frac{6.62\times10^{-34}}{\sqrt{\left(2\times9.1\times10^{-31}\times1.25\times10^{3}\times1.6\times10^{-19}\right)}}$$

$$\lambda = 1.10\times10^{-9}\,\text{m}$$

$$\boldsymbol{\lambda = 11\,\text{Å}}$$

EXAMPLE 12.3

A spectral line has wavelength 4000 Å. Calculate frequency and the energy in eV of the photon associated with it.

Solution:

$$\lambda = 4000\times10^{-10}\,\text{m}$$

$$c = 3\times10^{8}\,\text{m/s}$$

$$h = 6.62 \times 10^{-34}\,\text{J-s}$$

$$\nu = \frac{c}{\lambda}$$

$$= \frac{3 \times 10^8}{4000 \times 10^{-10}}$$

$$= \mathbf{0.75 \times 10^{15}\,Hz}$$

12.4 ▶ De-Broglie Hypothesis

De-Broglie hypothesis (matter waves)

According to de-Broglie hypothesis, a moving particle is associated with a wave which is known as de-Broglie wave. The wavelength of matter wave is given by

$$\boxed{\lambda = \frac{h}{mv} = \frac{h}{p}} \tag{12.1}$$

where m = mass of the material particle, v = velocity and p = momentum.

According to de-Broglie, the material particle in motion involved two different velocities, one which is the velocity of particle (v) and the other is the velocity of propagation of the wave associated with the particle, v_p, i.e. phase velocity of wave.

If E is the energy of particle, ν is the frequency of wave

$$E = h\nu$$

$$\nu = \frac{E}{h} \tag{12.2}$$

But according to mass–energy relationship of Einstein,

$$E = mc^2 \tag{12.3}$$

From Eqs. (12.2) and (12.3), we get

$$\nu = \frac{mc^2}{h} \text{ and } v_p = \nu\lambda = \frac{mc^2}{h} \times \frac{h}{mv}$$

$$\Rightarrow \quad \boxed{v_p = \frac{c^2}{v}} \tag{12.4}$$

Equation (12.4) shows that the phase velocity of matter waves is greater than the velocity of light. But, this is an unexpected result. This problem was removed by Schrodinger by postulating that a single wave is not attached with the particle but the material particle in motion is equivalent to a wave packet (a group of waves).

EXAMPLE 12.4

The de-Broglie wavelength of the electron is 30 Å. Calculate the phase velocity and group velocity of the matter wave associated with the electron.

Solution: For matter waves,

$$\lambda = \frac{h}{mv} \text{ or } v = \frac{h}{m\lambda}$$

$$v = \frac{6.62 \times 10^{-34}\,\text{J-s}}{\left(30 \times 10^{-10}\,\text{m}\right) \times \left(9.1 \times 10^{-31}\,\text{kg}\right)}$$

$$= 2.42 \times 10^{5}\,\text{m/s}$$

Phase velocity,

$$v_p = \frac{v}{2} = 1.21 \times 10^{5}\,\text{m/s}$$

And,

Group velocity $v_g = v = 2.42 \times 10^5$ m/s.

12.4.1 Phase Velocity (v_p) or Wave Velocity

Consider a wave

$$y = a\sin(wt - kx) \tag{12.5}$$

Here, a = amplitude, w = angular frequency $(2\pi\nu)$

$k = \frac{2\pi}{\lambda}$ propagation constant of wave

For planes of constant phase,

$$\Rightarrow \quad (wt - kx) = \text{constant}$$

$$\Rightarrow \quad w\frac{dt}{dt} - k\frac{dx}{dt} = 0$$

$$\Rightarrow \quad w - k\frac{dx}{dt} = 0$$

$$\Rightarrow \quad \frac{dx}{dt} = \frac{w}{k} = v_p$$

Here, $v_p = w/k \rightarrow$ Phase velocity or wave velocity.

Thus, phase velocity is the velocity with which the planes of constant phase advances through the medium.

12.4.2 Group Velocity

The pulse consists of a number of waves slightly differing in frequency from one another. The superposition of such waves is known as wave group (wave packet) as shown in Fig. 12.5. Thus, a wave packet comprises a group of waves slightly differing in velocity, wavelength with phases and amplitudes such that they interfere constructively over a small region of space, where the particle can be located, outside this region they interfere destructively so that the amplitude reduces to zero.

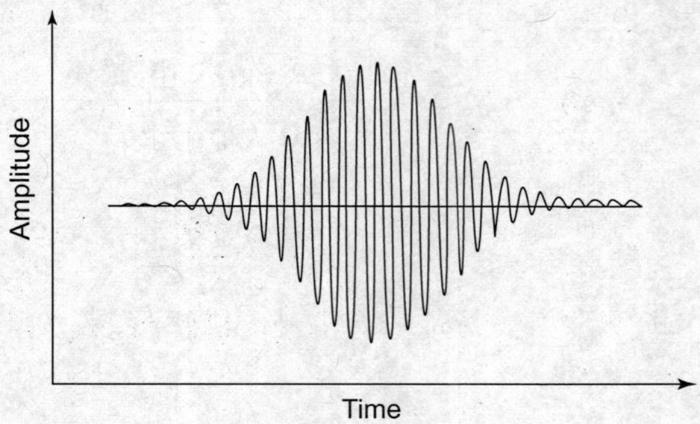

FIGURE 12.5 Wave packet

$$y_1 = a\sin(wt - kx) \tag{12.6}$$

$$y_2 = a\sin(w't - k'x) \tag{12.7}$$

$$\begin{aligned} y &= y_1 + y_2 \\ &= a\sin(wt - kx) + a\sin(w't - k'x) \end{aligned}$$

$$y = 2a\cos\left[\left(\frac{w - w'}{2}\right)t - \left(\frac{k - k'}{2}\right)x\right] \sin\left[\left(\frac{w + w'}{2}\right)t - \left(\frac{k + k'}{2}\right)x\right]$$

Here,
$$\begin{aligned} \text{amplitude} &= 2a\cos\left[\left(\frac{dw}{2}\right)t - \left(\frac{dk}{2}\right)x\right] \\ &= 2a\cos\frac{dw}{2}\left[t - \frac{dk\,x}{dw}\right] \\ &= 2a\cos\frac{dw}{2}\left[t - \left(\frac{dk}{dw}\right)x\right] \end{aligned}$$

Here,
$$\frac{dw}{dk} = v_g = \text{group velocity}$$

and
$$v_g = \frac{dw}{dk} = \frac{w - w'}{k - k'}$$

(a) Relation between v_g and v_p

We know that

$$v_p = \frac{w}{k}$$

$\Rightarrow$
$$w = kv_p \tag{12.8}$$

$$dw = kdv_p + v_p dk \tag{12.9}$$

$$\frac{dw}{dk} = k\frac{dv_p}{dk} + v_p \tag{12.10}$$

Here,

$$v_g = v_p + k\frac{dv_p}{dk} \quad \left[\begin{array}{c} k = \frac{2\pi}{\lambda}, \lambda = \frac{2\pi}{k} \\ \frac{d\lambda}{dk} = \frac{-2\pi}{k^2} \end{array} \right]$$

$\Rightarrow$
$$v_g = v_p + k\frac{dv_p}{d\lambda}\frac{d\lambda}{dk}$$

$$\boxed{v_g = v_p - \lambda\frac{dv_p}{d\lambda}} \tag{12.11}$$

(b) Relation between v_g and v

$$\frac{1}{2}mv^2 = E - V$$

$$v^2 = \frac{2(E-V)}{m}$$

$$v = \left[\frac{2(E-V)}{m}\right]^{\frac{1}{2}} \tag{12.12}$$

But
$$\lambda = \frac{h}{mv}$$

$$\frac{1}{\lambda} = \frac{mv}{h} = \frac{m}{h}\left[\frac{2(E-V)}{m}\right]^{\frac{1}{2}} \quad \text{[from (12.12)]}$$

But

$$v_g = v_p - \lambda\left(\frac{dv_p}{d\lambda}\right)$$

$$\Rightarrow \qquad v_g = \lambda^2\left[\frac{v_p}{\lambda^2} - \frac{1}{\lambda}\left(\frac{dv_p}{d\lambda}\right)\right]$$

$$\Rightarrow \qquad v_g = -\lambda^2 \frac{d}{d\lambda}\left(\frac{v_p}{\lambda}\right) = -\lambda^2 \frac{d\nu}{d\lambda} \tag{12.13}$$

$$\Rightarrow \qquad \frac{1}{v_g} = -\frac{1}{\lambda^2}\frac{d\lambda}{d\nu} = \frac{d}{d\nu}\left(\frac{1}{\lambda}\right)$$

$$\Rightarrow \qquad \frac{1}{v_g} = \frac{d}{d\nu}\left[\frac{m}{h}\left\{\frac{2(E-V)}{m}\right\}^{\frac{1}{2}}\right]$$

$$= \frac{1}{h}\left[\frac{d}{d\nu}\{2m(h\nu - V)\}^{\frac{1}{2}}\right]$$

$$= \frac{1}{h}\frac{1}{2}\{2m(h\nu - V)\}^{-\frac{1}{2}} \cdot 2mh$$

$$= \frac{m}{\left\{2m(E-V)^{\frac{1}{2}}\right\}}$$

$$= \left[\frac{m}{2(E-V)}\right]^{\frac{1}{2}}$$

$$\Rightarrow \qquad \frac{1}{v_g} = \frac{1}{v} \quad \text{or} \quad v_g = v$$

It means that a material particle in motion is equivalent to group of waves or a wave packet.

(c) Show that for a non-relativistic particle, the phase velocity is half of the group velocity

For particles which are equivalent to a wave packet, the phase velocity v_p is given by

$$v_p = \nu\lambda \tag{12.14}$$

According to de-Broglie,

$$\lambda = \frac{h}{mv} = \frac{h}{mv_g} \tag{12.15}$$

The total energy of the particle is given by

$$E = \frac{1}{2} m v_g^2$$

Now,

$$\nu = \frac{E}{h} = \frac{\frac{1}{2} m v_g^2}{h}$$

Substituting the values of v and λ in Eq. (12.14), we get

$$\boxed{v_p = \frac{\frac{1}{2} m v_g^2}{h} \times \frac{h}{m v_g} = \frac{1}{2} v_g} \tag{12.16}$$

$$\therefore \qquad \text{Phase velocity} = \frac{1}{2} \text{ group velocity}$$

Note

Wave nature cannot be observed in daily life. We daily come across macroscopic objects for which the de-Broglie wavelength is much smaller than the size of object. This wavelength is usually too small to show any effect.

12.5 ▶ Experimental Evidences of Matter Waves

12.5.1 Davisson–Germer Experiment

The Davisson–Germer experiment demonstrates the wave nature of the electrons, confirming the earlier hypothesis of de-Broglie. Putting wave-particle duality on a firm experimental footing, the Bragg's law for diffraction had already been applied to X-ray diffraction, but this was the first application to particle waves [Fig. 12.6(a)].

Davisson and Germer designed and built an apparatus with vacuum chamber for the purpose of measuring the energies of electrons scattered from a metal surface. Electrons from a heated filament are accelerated by a voltage and allowed to strike the surface of nickel metal crystal.

The electron beam is directed towards the nickel which acts as a target, which could be rotated to observe angular dependence of the scattered electrons. An electron detector (called a Faraday cage) is mounted along an arc so that it could be rotated to capture electrons at different angles. At certain angles, there is a peak in the intensity of the scattered electron beam. This peak indicates wave behaviour of the electrons, which could be interpreted by Bragg's law to give values for the lattice spacing in the nickel crystal [Fig. 12.6(b)].

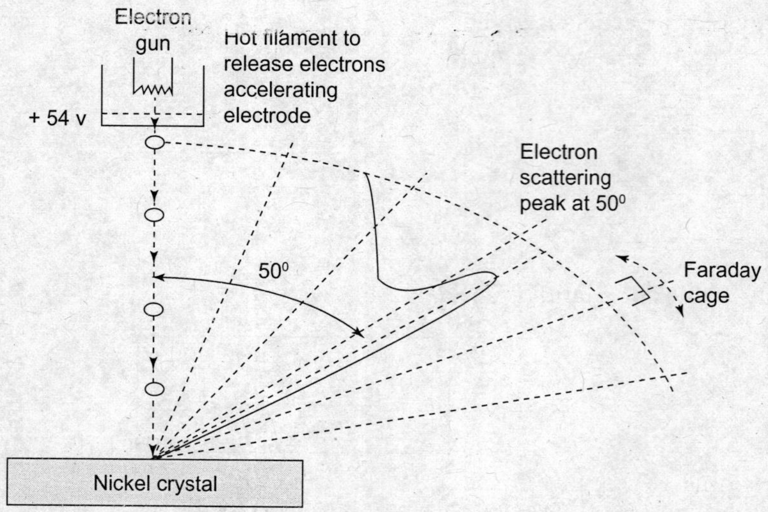

FIGURE 12.6 (a) Experimental arrangement used by Davisson–Germer

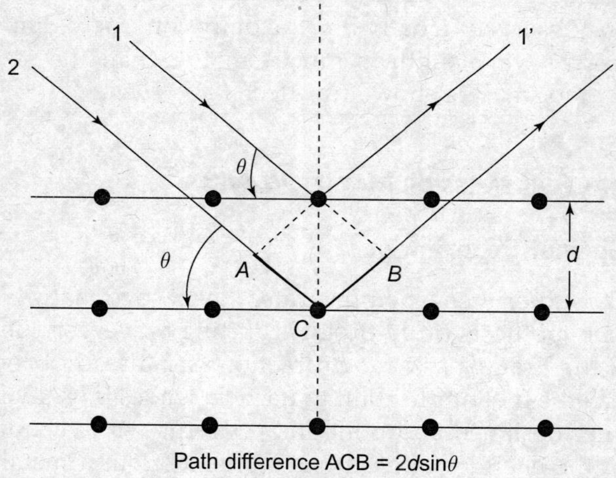

FIGURE 12.6 (b) Bragg's scattering from lattice

The experimental data of Davisson–Germer experiment shows repeated peaks of scattered electron intensity with increasing accelerating voltage. This data is then collected at a fixed scattering angle. Using the Bragg's law, the de-Broglie wavelength expression, and the kinetic energy of the accelerated electrons gives the relationship

$$\underbrace{\frac{1}{\lambda}}_{\text{Electron wavelength}} = \underbrace{\frac{n}{2d\sin\theta}}_{\text{Bragg's law}}$$

or

$$2d\sin\theta = n\lambda$$

or
$$\frac{1}{\lambda}=\frac{p}{h}$$
de Broglie relationship

or
$$\frac{1}{\lambda}=\frac{\sqrt{2mE}}{h}$$

$$\Rightarrow \quad \frac{1}{\lambda}=\frac{\sqrt{2meV}}{h} \quad \text{or} \quad \lambda \propto \frac{1}{\sqrt{V}}$$
Acceleration through voltage V

An accelerating voltage of 54 volts gave a definite peak at a scattering angle of 50°. The angle θ in the Bragg's law corresponding to that scattering angle is 65°, and for that angle, the calculated lattice spacing is 0.092 nm. For that lattice spacing and scattering angle, the relationship for wavelength as a function of voltage is empirically

$$\frac{1}{\lambda(nm)}=\frac{n}{2d\sin\theta}=0.815\sqrt{V}$$

For $n = 1, 2, 3$ gives values for the square root of voltage 7.36, 14.7 and 22, which appear to agree with the first, third and fifth peaks above (Fig. 12.7).

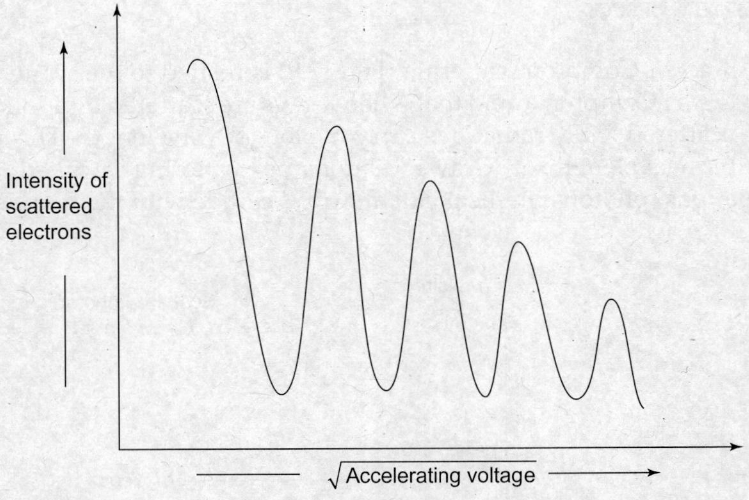

FIGURE 12.7 Plot of intensity of scattered electrons vs square root of accelerating voltage

12.5.2 G. P. Thomson's Experiment

Davisson and Germer used slow electrons but G. P. Thomson extended the research on electron waves to high-speed electrons having energy 60,000 eV (Fig. 12.8). He used a fine highly energetic beam of electrons to incident upon a thin film of gold. Gold consists of micro crystals oriented at random. The electron diffracted by crystals making the same glancing angle will move in a cone and produce a circular ring on the photographic plate. Several rings are formed on the plate for different values of n and θ in Bragg's equation

$$2d\sin\theta = n\lambda$$

and λ can be calculated from the de-Broglie relation $\lambda = h/mv$.

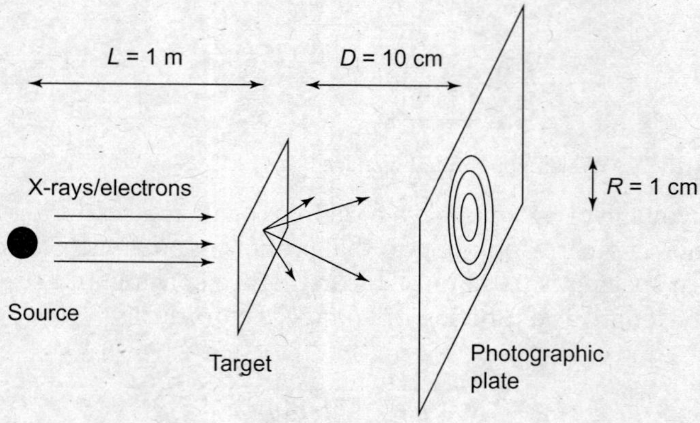

FIGURE 12.8 G. P. Thomson experiment

12.6 ▶ Compton Effect

The Compton effect or Compton scattering (Fig. 12.9) is related to the scattering of X-rays by free electrons. A.H Compton found that when X-rays are scattered by a solid target (such as carbon), the scattered X-ray radiations carry the longer wavelength. This phenomenon of increase in the wavelength of X-ray radiations by scattering is called the Compton effect. The radiations (photons) in the incident X-rays collide with the free electrons of the

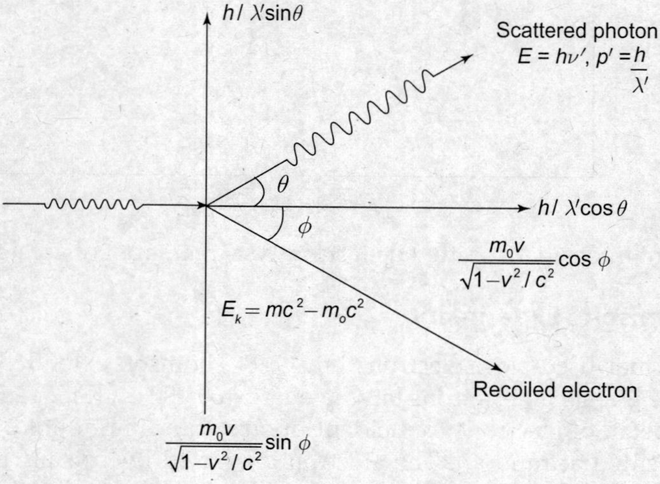

FIGURE 12.9 Compton scattering

target. The incident photon transfers some of the energy to the electron (if the collision is inelastic) and if the energy of wavelength of the scattered proton remains the same then the collision is elastic. Thus, as a result of this collision, the energy of scattered X-ray photon decreases or wavelength increases.

Let us assume

$\lambda =$ wavelength of incident X-rays

$\lambda' =$ wavelength of scattered X-rays

$\Delta\lambda = \lambda' - \lambda =$ Compton shift

$$\text{Energy of incident X-ray photon} = h\nu = \frac{hc}{\lambda}$$

$$\text{Momentum of incident X-ray photon} = \frac{h}{\lambda}$$

$$\text{Energy of scattered X-ray photon} = h\nu' = \frac{hc}{\lambda'}$$

$$\text{Momentum of X-ray (scattered) photon} = \frac{h}{\lambda'}$$

$$\text{Kinetic energy of the recoiled electron} = (m - m_0)c^2$$

$$= \left[\frac{m_0}{\sqrt{1 - \frac{v^2}{c^2}}} - m_0\right]c^2$$

$$\text{Momentum of the recoiled electron} = mv = \frac{m_0 v}{\sqrt{1 - \frac{v^2}{c^2}}}$$

Now, according to law of conservation of energy,

$$E = E' + E_k$$

Energy of incident photon = Energy of scattered photon + Energy of the recoiled electron

$$\frac{hc}{\lambda} = \frac{hc}{\lambda'} + (m - m_0)c^2$$

$$\Rightarrow \quad \frac{h}{\lambda} - \frac{h}{\lambda'} + m_0 c = \frac{m_0 c}{\sqrt{1 - \frac{v^2}{c^2}}} \qquad (12.17)$$

According to law of conservation of momentum,

$$\frac{h}{\lambda'}\sin\theta = \frac{m_0 v}{\sqrt{1-\frac{v^2}{c^2}}}\sin\phi$$

$$\Rightarrow \qquad \frac{h}{\lambda} = \frac{h}{\lambda'}\cos\theta + \frac{m_0 v}{\sqrt{1-\frac{v^2}{c^2}}}\cos\phi$$

$$\Rightarrow \qquad \frac{h}{\lambda} - \frac{h}{\lambda'}\cos\theta = \frac{m_0 v}{\sqrt{1-\frac{v^2}{c^2}}}\cos\phi \tag{12.18}$$

Squaring and adding Eqs. (12.17) and (12.18), we get

$$\frac{h^2}{\lambda'^2} + \frac{h^2}{\lambda'^2} - \frac{2h^2}{\lambda\lambda'}\cos\theta = \frac{m_0{}^2 v^2}{\sqrt{1-\frac{v^2}{c^2}}} \tag{12.19}$$

On squaring and solving Eq. (12.17), we get

$$\frac{h^2}{\lambda^2} + \frac{h^2}{\lambda'^2} - \frac{2h^2}{\lambda\lambda'} + 2hm_0 c\left(\frac{\lambda' - \lambda}{\lambda\lambda'}\right) = \frac{m_0^2 v^2 c^2}{c^2 - v^2} \tag{12.20}$$

On comparing Eqs. (12.19) and (12.20), we get

$$\frac{h^2}{\lambda^2} + \frac{h^2}{\lambda'^2} - \frac{2h^2}{\lambda\lambda'} + 2hm_0 c\frac{(\lambda' - \lambda)}{(\lambda\lambda')} = \frac{h^2}{\lambda^2} + \frac{h^2}{\lambda'^2} - \frac{2h^2}{\lambda\lambda'}\cos\theta$$

$$\Rightarrow \qquad (\lambda' - \lambda) = \frac{h}{m_0 c}(1 - \cos\theta)$$

$$\Rightarrow \qquad \boxed{\boldsymbol{\Delta\lambda = \frac{h}{m_0 c}(1 - \cos\theta)}} \tag{12.21}$$

Equation (12.21) gives the Compton shift.

Note

The Compton effect is observed significantly with the X-rays, which are very short wavelength radiations. If we use visible light instead of X-rays, then we get

$$\Delta\lambda = 0.0484\text{Å} = 0.001\%$$

Thus, the Compton effect is not observable with visible light.

EXAMPLE 12.5

Calculate the Compton shift if X-rays of wavelength 1.0 Å are scattered from a carbon block. The scattered radiation is viewed at 90° to the incident beam.

Solution: Given $\lambda = 1.0\,\text{Å} = 10^{-10}\,\text{m}$ and $\phi = 90^{\circ}$

$$\Delta\lambda = \frac{h}{m_0 c}(1-\cos\phi)$$

$$= \frac{6.62\times10^{-34}}{9.1\times10^{-31}\times3\times10^{8}}(1-\cos 90^{\circ})$$

$$= \mathbf{0.0242\,Å}$$

EXAMPLE 12.6

X-rays with $\lambda = 1\,\text{Å}$ are scattered from a carbon block. The scattered radiation is viewed at 90° to the incident beam. What kinetic energy is imparted to the recoil electron?

Solution: Given $\lambda = 1\times10^{-10}\,\text{m}$

We know that

$$\frac{hc}{\lambda} = \frac{hc}{\lambda'} + E_k$$

$$\Rightarrow \quad \frac{hc}{\lambda} = \frac{hc}{\lambda+\Delta\lambda} + E_k$$

$$= \frac{6.62\times10^{-37}\times3\times10^{8}\times2.425\times10^{-12}}{1\times10^{-10}\times(1+0.02425)\times10^{-10}}$$

$$= 47.02\times10^{-18}\,\text{J}$$

$$= \mathbf{294\ eV}$$

12.7 ▶ Uncertainty Principle

According to this principle, "It is impossible to measure simultaneously the position of a particle along a particular direction, say x, and also its momentum in the same direction, p_x, with unlimited accuracy."

If Δx is the uncertainty in position measurement and Δp_x is the uncertainty in momentum measurement, then

$$\boxed{\Delta x \cdot \Delta p_x \approx \hbar} \tag{12.22}$$

Similarly,
$$\Delta y \cdot \Delta p_y \approx \hbar \text{ and}$$
$$\Delta z \cdot \Delta p_z \approx \hbar$$

It means that the smaller is the value of Δx, more exactly, we can measure the position; the larger is the value of Δp_x less exactly we can determine momentum and vice-versa. Similarly, the uncertainty in measurement of energy and time can be written as

$$\boxed{\Delta E \cdot \Delta t \approx \hbar} \tag{12.23}$$

And the angular momentum-angle uncertainty can be written as

$$\Delta J \Delta \theta \approx \hbar$$

More accurately, Eqs. (12.22)–(12.23) can be written as

$$\Delta x \Delta p_x \geq \frac{\hbar}{2}$$

$$\Delta E \Delta t \geq \frac{\hbar}{2}$$

$$\Delta J \Delta \theta \geq \frac{\hbar}{2}$$

Heisenberg conceived a hypothetical gamma ray microscope to demonstrate his uncertainty principle (Fig. 12.10). The scattering of gamma rays was used to probe the position of an electron as it traverses the microscope. The wavelength " λ " of the gamma rays was very small and they carried a large amount of energy $E = hc/\lambda$. The trajectory of the electron was agitated by collisions with the gamma ray probes. With Bohr's help, Heisenberg showed that the uncertainty in the measurement of the position of the electron is approximately equal to the wavelength of the gamma rays due to the effects of diffraction when the gamma rays were viewed through the aperture of the telescope. In other words, $\Delta x \sim \lambda$. The uncertainty on the momentum of the electron is approximately equal to the momentum of the single photon used to illuminate the electron, or $\Delta p_x \sim h/\lambda$. If these two uncertainties are multiplied together, we obtain the familiar uncertainty principle, we get $\Delta x \; \Delta p_x \geq h$.

Note

According to uncertainty principle, the electrons cannot exist inside the nucleus. Also the spectral lines can never be infinitely sharp but must have a natural finite width.

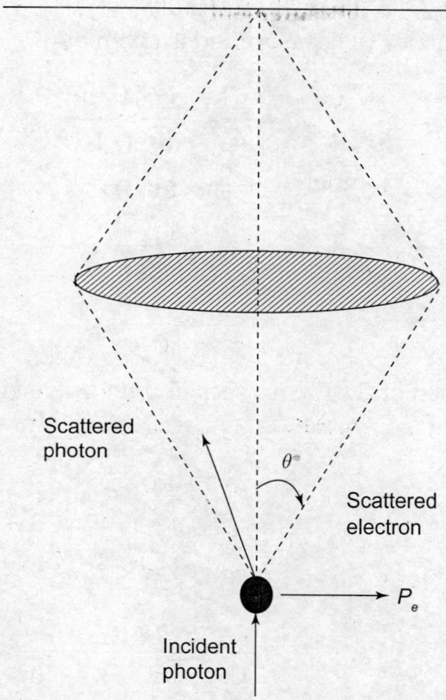

FIGURE 12.10 Gamma ray microscope experiment

EXAMPLE 12.7

The average period that elapses between the excitation of an atom and the time it emits radiation is 10^{-8} sec. Find the uncertainty in energy emitted and the uncertainty in the frequency of light emitted.

Solution:

$$h = 6.63 \times 10^{-34}\text{ J-s}$$

The uncertainty in the energy is given by

$$\Delta E = \frac{\hbar}{\Delta t}$$

$$= \frac{6.63 \times 10^{-34}}{2\pi \times 10^{-8}}$$

$$= 1.054 \times 10^{-26}\text{ J}$$

The uncertainty in the frequency of light emitted is given by

$$\Delta\nu = \frac{\Delta E}{h} = \frac{1.054\times10^{-26}}{6.63\times10^{-34}}$$

$$= \mathbf{1.59\times10^{7}\ Hz}$$

EXAMPLE 12.8

A proton is moving with a speed of 2×10^{8} m/sec. Find the wavelength of matter wave associated with it.

Solution: Given

$$v = 2\times10^{8}\,\text{m/sec}$$

And

$$\lambda = \frac{h}{mv}$$

$$= \frac{6.62\times10^{-34}}{1.67\times10^{-27}\times2\times10^{8}}$$

$$= \mathbf{1.98\times10^{-15}\,m}$$

EXAMPLE 12.9

Prove uncertainty relation for energy and time from position-momentum uncertainty relation.

Solution: We know that

$$\Delta x\Delta p \geq \frac{\hbar}{2}$$

Consider a free particle of mass 'm' moving with velocity v.

Its kinetic energy is,

$$E = \frac{1}{2}mv^{2} = \frac{p^{2}}{2m}$$

Uncertainty in energy is given by

$$\Delta E = \frac{2p\Delta p}{2m}$$

$$\Delta E = \frac{p\Delta p}{m}$$

$$\Delta E = \frac{mv\Delta p}{m} = v\Delta p$$

But $$v = \frac{\Delta x}{\Delta t}$$

$\Rightarrow$ $$\Delta E = \frac{\Delta x}{\Delta t}\Delta p$$

$\Rightarrow$ $$\Delta E\Delta t = \Delta x\Delta p$$

But $$\Delta x\Delta p \geq \frac{\hbar}{2}$$

So, $$\boldsymbol{\Delta E\Delta t \geq \frac{\hbar}{2}}$$

12.8 ▶ Elementary Idea of Quantum Statistics

The position of a particle in a three-dimensional static system is completely defined by three position coordinates x, y, z. This three-dimensional space is called the position space. Similarly, the momentum p of the particle is completely specified by the three momentum coordinates p_x, p_y, p_z in three-dimensional space called momentum space. Combination of position and momentum spaces is called the phase space. Particle in phase space is represented by a point having six coordinates x, y, z, p_x, p_y, p_z. The available volume in phase space is divided in to large number of compartment and each compartment is further divided in to large number of elementary cells of equal size. The distribution of particles depends on statistics.

There are two types of statistics:

(a) Classical statistics or Maxwell-Boltzmann statistics

(b) Quantum statistics

12.8.1 Classical Statistics

In classical statistics, the particles are distinguishable. The volume of the elementary cell can be made as small as we want.

Any number of particles can occupy a single cell in phase space. The most probable distribution of particles among the various possible individual energies (Ei's) is given by

$$\eta_i = \frac{gi}{e^{\alpha+\beta\varepsilon_i}}$$

12.8.2 Quantum Statistics

Quantum statistics is of two types:

(i) Bose-Einstein (B-E) statistics

The particles are treated as indistinguishable and any number of the particles can occupy a single cell in phase space. This statistics is applied to the particles having integral spin angular momentum (in unit of $\hbar$), e.g. photons, α-particles and mesons. Such particles are called Bosons. Most probable distribution of particles is given as

$$\eta_i = \frac{g_i}{e^{\alpha+\beta\varepsilon_i} - 1}$$

(ii) Fermi-Dirac statistics (F-D)

The particles are treated as indistinguishable, but obey Pauli-Exclusion principle, according to which no two particles can occupy the same cell in phase space, i.e., not more than one particle can occupy a single cell. Particles obeying F-D statistics have half integral spin angular momentum and are called fermions. Such particles are protons, electrons, neutrons, etc., all having $1/2$. Most probable distribution of the particles among various energy levels for a system obeying Bose-Einstein statistics is given as

$$\eta_i = \frac{g_i}{e^{\alpha+\beta\varepsilon_i} + 1}$$

12.9 ▶ Quarks and Gluons

Gellman and G. Ziverg in 1964 independently proposed that all strongly interacting particles are built up of three new undiscovered particles called quarks. Quarks and Gluons are the building blocks for particles such as neutrons and protons. But these particles behave differently than protons or neutrons. Quarks are confined within the larger particles and so cannot be separated and found in isolation. The force between two quarks becomes larger as they move farther apart, whereas the force between a nucleus and an electron, or two nucleons in a nucleus grows weaker as their separation increases. Quarks combine to form composite particles called hadrons, the most stable of which are protons and neutrons, the components of atomic nuclei (Fig. 12.11). Quarks have intrinsic properties like electric charge, mass, colour charge and spin. There are six types of quarks: up, down, top, bottom, strange and charm. Up and down have lowest masses of all quarks. These are generally stable in nature and mostly found in universe, whereas the other quarks can be produced only in high energy collisions (such as in cosmic rays and in particle accelerators). The properties of these quarks are given in Table 12.2.

The quarks of a given colour create a field (called gluon field) around them so that they emit or reabsorb a type of hypothetical particles such as emission or reabsorption of virtual protons by electrically charged particles. These hypothetical particles are called gluons.

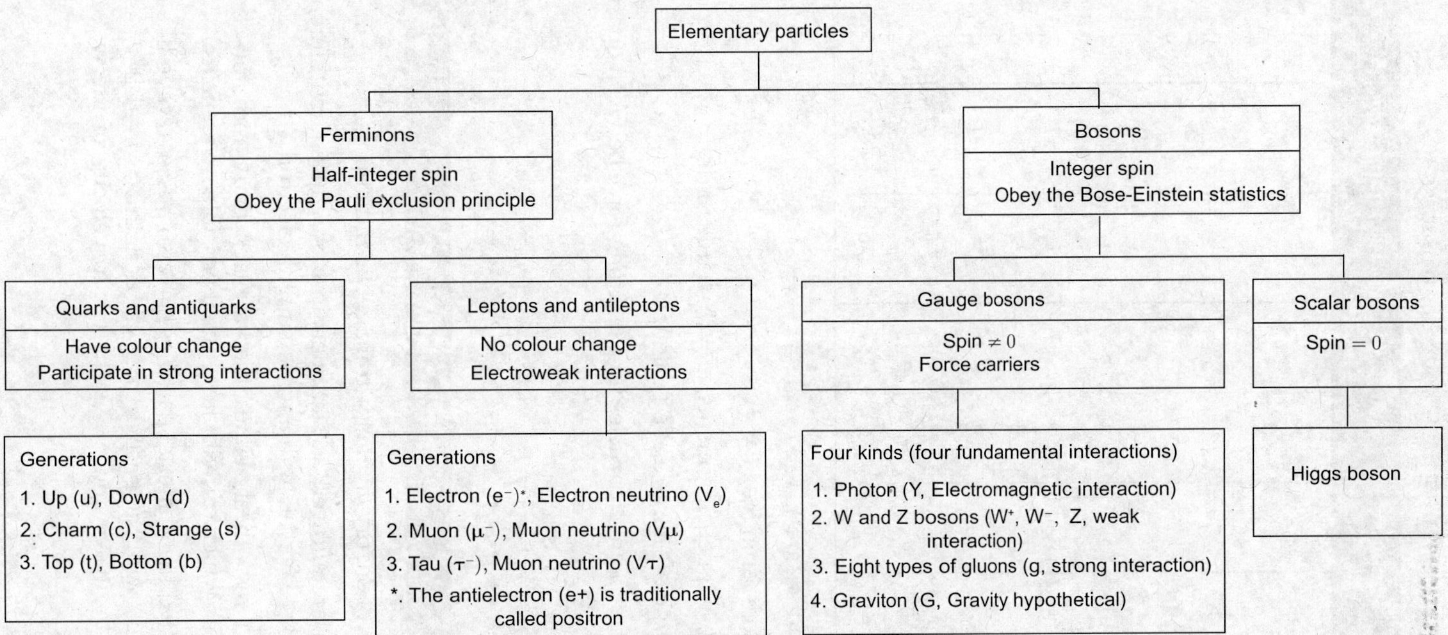

FIGURE 12.11 Classification of elementry particles

TABLE 12.2 Properties of quarks

Quark/ Antiquark	Symbol		Charge, *c*		Baryan number, *B*		Strangeness, *S*	
Up	u	$\bar{u}$	$+\frac{2}{3}$	$-\frac{2}{3}$	$\frac{1}{3}$	$-\frac{1}{3}$	0	0
Down	d	$\bar{d}$	$-\frac{1}{3}$	$+\frac{1}{3}$	$\frac{1}{3}$	$-\frac{1}{3}$	0	0
Charm	e	$\bar{e}$	$+\frac{2}{3}$	$-\frac{2}{3}$	$\frac{1}{3}$	$-\frac{1}{3}$	0	0
Strange	s	$\bar{s}$	$-\frac{1}{3}$	$+\frac{1}{3}$	$\frac{1}{3}$	$-\frac{1}{3}$	1	1
Top	t	$\bar{t}$	$+\frac{2}{3}$	$-\frac{2}{3}$	$\frac{1}{3}$	$-\frac{1}{3}$	0	0
bottom	b	$\bar{b}$	$-\frac{1}{3}$	$+\frac{1}{3}$	$\frac{1}{3}$	$-\frac{1}{3}$	0	0

The properties of gluons are as follows.

(i) They have zero isospin (i.e., $I = 0$).

(ii) Charge, $Q = 0$; magnetic moments, $m = 0$; charge parity, $c = -1$.

(iii) They carry a type of charge, called colour.

(iv) They act as the exchange particles for the strong force between quarks.

RECAP ZONE

POINTS TO REMEMBER

- Photon is a packet of energy $h\nu$ of electromagnetic radiation.
- Only those orbits are allowed as stationery orbits whose circumference is integral multiple of de-Broglie wavelength associated with electron.
- Planck's quantum hypothesis states that a black body radiator contains simple harmonic oscillators which can vibrate with all possible frequencies.
- De-Broglie suggested that material particles can also possess wave-like character and the wave associated with the particle has wavelength $\lambda = h/p$. If v_p is the phase or wave velocity of the wave and v_g is the group velocity of the wave packet, then $v_g = \frac{dw}{dk}$ and also $v_p = \frac{1}{2}v_g$.

- The uncertainty principle states that "It is impossible to measure simultaneously the position of a particle along a particular direction and also its momentum in same direction with unlimited accuracy."
- There are two types of quantum statistics – Bose-Einstein and Fermi-Dirac statistics.
- In B-E statistics, particulars are indistinguishable but any number of particles can occupy a cell.
- In F-D statistics, particulars are indistinguishable but obey Pauli-exclusion principle, i.e. not more than one particle can occupy a cell.
- Quarks and gluons are the building block of larger particles like protons and neutrons, which in turn form atoms.
- The total energy emitted by a black body per unit area per unit time is infinity at all temperatures.
- In some experiments, light behaves as wave while in other it behaves as particle; hence, it has dual nature.
- Davisson and Germer gave the first experimental evidence of wave-like properties of beams of material particles.
- G. P. Thomson exhibited the wave nature of high energetic electrons.
- The concept of Bohr's orbit violates the uncertainty principle.

KEY TERMS WITH DEFINITIONS

- **Electromagnetic radiation** – It refers to the waves of the electromagnetic field, propagating through space carrying electromagnetic radiation energy. It includes radio waves, microwaves, infrared, ultraviolet, X-rays and gamma radiation.
- **Photoelectrons** – When light strikes on the surface of the material, the electrons gain sufficient energy from light to overcome the work function of the material and hence emitted from the surface.
- **Blackbody** – A hypothetical perfect absorber and radiator of energy with no reflecting power.
- **Matter wave** – All matter can exhibit wave-like behaviour. For example, a beam of electrons can be diffracted just like a beam of light or a water wave.
- **Phase velocity** – The phase velocity is equal to the product of the frequency multiplied by the wavelength.
- **Group velocity** – The speed at which the energy of a wave travels.
- **Quantum statistics** – The fundamental feature of quantum mechanics that distinguishes it from classical mechanics is that particles of a particular type are indistinguishable from one another.

- **Boson** – A subatomic particle, such as a photon, which has zero or integral spin and follows the statistical description given by S. N. Bose and Einstein.
- **Hadron** – A subatomic particle of a type including the baryons and mesons, which can take part in the strong interaction.
- **Mesons** – These are hadronic subatomic particles composed of one quarks and one antiquark, bound together by the strong interaction.
- **Baryon** – A subatomic particle, such as a nucleon or hyperon, which has a mass equal to or greater than that of a proton.
- **Proton** – The proton is a baryon and is considered to be composed of two up quarks and one down quark.
- **Neutron** – The neutron is a baryon and is considered to be composed of two down quarks and one up quark.
- **Electrons** – Electrons and positrons are not made up of quarks. They are in a separate "family" of particles known as LEPTONS. Leptons and quarks are "fundamental" (indivisible) particles.

Important Formulae and Equations

Equation Number	Equation	Remarks
(12.1)	$\lambda = \frac{h}{mv} = \frac{h}{p}$	de-Broglie wavelength of matter wave
(12.4)	$v_p = \frac{c^2}{v}$	Phase velocity of matter wave
(12.11)	$v_g = v_p - \lambda \frac{dv_p}{d\lambda}$	Group velocity of matter waves
(12.16)	$v_p = \frac{\frac{1}{2} m v_g^2}{h} \times \frac{h}{m v_g} = \frac{1}{2} v_g$	Relation between phase velocity and group velocity
(12.21)	$\Delta\lambda = \frac{h}{m_0 c}(1 - \cos\theta)$	Compton shift
(12.22)	$\Delta x \cdot \Delta p_x \approx \hbar$	Uncertainity in position and momentum measurement
(12.23)	$\Delta E \cdot \Delta t \approx \hbar$	Uncertainity in measurement of energy and time

Review Zone

Short Answer Questions

1. What is Planck's hypothesis?
2. Explain de-Broglie hypothesis.
3. What is the difference between phase velocity and group velocity?
4. What is de-Broglie wavelength?
5. What is Heisenberg uncertainty principle?
6. What do you understand by uncertainty principle?
7. Write a short note on F-D statistics.
8. Explain in brief B-E statistics.
9. What are Bosons and Fermions?
10. What do you mean by Compton effect?
11. What are quarks, gluons and hadrons?
12. What do you mean by Quantum Statistics?
13. Write a short note on photoelectric effect.
14. Name the experiments which show wave nature of light.
15. Explain the following terms: position space, momentum space and phase space

Long Answer Questions

1. What is Planck's quantum hypothesis to explain the observed spectrum of a black body? Give the shortcomings of old quantum theory.
2. What are the shortcomings of classical theory? Derive formula for the de-Broglie wavelength of a particle.
3. What is the difference between phase and group velocities? Show that the de-Broglie group velocity associated with the wave packet is equal to the velocity of the particle.
4. Compare Bose-Einstein and Fermi-Dirac statistics.
5. Show that for a non-relativistic particle, the phase velocity is half of the group velocity.
6. Discuss and derive de-Broglie hypothesis. How it has been verified experimentally?
7. What are matter waves? Mention any three properties of matter waves. Describe the experiment of Davisson and Germer to demonstrate the wave nature of the electrons.
8. Derive an expression for group velocity with which a wave group travels.
9. What is Planck's constant?
10. Show that the group velocity and phase velocity are same in non-dispersive medium.
11. Describe G. P. Thomson experiment to prove the validity of the uncertainty principle and to show the wave nature of the electrons.
12. Explain Heisenberg's uncertainty principle and its important applications.
13. How do the group velocity and wave velocity differ? Show that the phase

velocity can exceed the velocity of light in vacuum.

14. Why can electron not exist in the nucleus? Why Heisenberg's principle is important for microscopic bodies alone?

15. State time energy uncertainty principle. Obtain its expression from position-momentum uncertainty principle.

16. Show that the group velocity of the particle is equal to the velocity of the particle. Derive the relation between ν_g and ν_P.

Numerical Problems

1. Calculate the smallest possible uncertainty in the position of an electron moving with velocity 3×10^7 m/s.

   Ans. 0.038 Å

2. The average life time of hydrogen in excited state is 2.5×10^{-14} s. Calculate the uncertainty in the measurement of energy in this state.

   Ans. 0.025 eV

3. Calculate the de-Broglie wavelengths of a 2 eV photon and an electron with kinetic energy of 2 eV.

   Ans. 8.6 Å

4. What will be the energy of a gamma ray photon having wavelength 1Å?

   Ans. 1.24×10^4 eV

5. A proton and a deuteron have the same kinetic energy. Which has a longer wavelength?

   Ans. Proton

6. Find the phase and group velocities of an electron whose de-Broglie wavelength is 1.2 Å.

   Ans. 3.03×10^6 m/sec and 6.06×10^6 m/sec

7. A certain spectral line has wavelength 4000 Å. Calculate the energy of the photon.

   Ans. 4.9×10^{-19} J

8. The threshold frequency for photoelectric emission in copper is 1.1×10^{15} Hz. Find the maximum energy in eV when $\gamma = 1.2 \times 10^{15}$ Hz (frequency).

   Ans. 0.414 eV

9. Show that the de-Broglie wave velocity is a function of wavelength in free space.

10. An electron has de-Broglie wavelength 2.0×10^{-12} m. Find its kinetic energy. Also find the phase and group velocity of de-Broglie waves.

CHAPTER 13

Quantum Mechanics

Learning Objectives

By the end of this chapter, the student will be able:

- To discuss phenomena that led to the development of quantum theory
- To identify incompatibilities of classical physics with the quantum theory
- To compute one-dimensional Schrodinger wave equation for simple potential types
- To understand the concept of uncertainty and its role in quantum physics

13.1 ▶ Introduction

Classical Physics deals mainly with the study of motion of fluids and particles. During the turn of 19th century, it was observed that the phenomena on the atomic and sub-atomic shells do not fit well into the framework of classical physics and that the explanation requires entirely new principles. In microscopic worlds, the dynamical variables like energy and momentum did not have the same meaning as in classical dynamics. These variables were found to have a discrete value in a particular state of an atom and did not change in continuous manner from one state to another. Thus, these new concepts led to the formulation of new theory called Quantum Mechanics, which was successful in giving satisfactory explanation to many observed phenomena. Quantum mechanics is the theory of atomic and nuclear systems.

13.2 ▶ Difficulties with Classical Physics

Classical mechanics is mainly based on Newton's laws of motion: (i) the law of inertia, (ii) the law of force and (iii) the law of action and reaction. The classical mechanics explains correctly the motion of celestial bodies like planets, stars, etc. moving with non-relativistic speed (i.e., $v \ll c$). The events are generally continuous or predictable in classical mechanics. Classical mechanics is used to explain most of the phenomena we come across in day-to-day life activities. Also, in very complex applications like launching satellites and rockets too, the classical mechanics plays very important role. The inadequacies of classical mechanics are

(i) It could not explain the stability of atoms.

(ii) It could not explain observed spectrum of black body radiations.

(iii) It could not explain the observed variation of specific heat of metals and gases.

(iv) It could not explain the origin of discrete spectra of atoms since according to classical mechanics the energy changes are always continuous.

13.3 ▶ Need and Origin of Quantum Concept

When we deal with microscopic particles, i.e. particles with dimension smaller or comparable to the de-Broglie wavelength, then the mechanics of the particles will be governed by quantum theory. In quantum mechanics, one can never simultaneously determine the position as well as the momentum of a particle (Heisenberg's uncertainty principle). It is the fundamental theory of nature at small scales and low energies of atoms and subatomic particles. Every measured quantity is quantized in quantum mechanics. Without quantum mechanics, there would be no transistor, and hence no computer and no laser. The events are unpredictable in this theory. The development of quantum mechanics took place in two stages. The first stage is based on Max Planck's hypothesis according to which the radiation is emitted or absorbed in discrete quanta of energy. This theory was a mixture of classical and non-classical concepts. The stage second of quantum mechanics began with two points of views: Matrix mechanics given by Heisenberg and wave mechanics developed by Schrodinger. The complete theory of particles called quantum field theory was accepted in 1947.

13.4 ▶ Assumptions of Quantum Theory

Instead of two sets of physical principles, one for microscopic universe and the other for macroscopic quantities, there is only single set for both in quantum theory. The basic assumptions of the theory are as follows:

(i) The probability of occurrence of an event is given by the square of the absolute value of a complex number ψ (wave function or probability amplitude). If ρ is the probability of the event, then

$$\rho = |\psi|^2 \tag{13.1}$$

(ii) When event occurs in different alternative ways, the probability amplitude for the event is the sum of probability amplitude for each way, i.e.,

$$\psi = \psi_1 + \psi_2 \text{ and } \rho = |\psi_1 + \psi_2|^2 \tag{13.2}$$

(iii) If in an experiment, it can be determined whether one or another alternative is used, the probability of the event is the sum of the probabilities for each alternative, i.e.,

$$\rho = \rho_1 + \rho_2 \tag{13.3}$$

13.5 ▶ Wave Function

Waves are normally associated with quantities that vary periodically. But in case of matter waves the quantity that varies periodically is called wave function. While ψ itself has no physical interpretation but the square of its absolute magnitude $|\psi|^2$ evaluated at a point at a particular time gives the probability of finding the particle there at that time, i.e. $|\psi|^2$ is the probability density. The probability of finding the particle in volume $dv = dx\, dy\, dz$ is $|\psi|^2\, dx\, dy\, dz$ and

$$\iiint_V |\psi|^2\, dx\, dy\, dz = 1 \tag{13.4}$$

Thus, it is normalizable.

13.5.1 Properties of Wave Function

(i) A wave function must be finite everywhere. If it is infinite for a particular point, it means that there is infinitely large probability of finding the particle at that point which violates uncertainty principle.

(ii) It must be single valued. If ψ has more than one value at a particular point, it means that there is a probability of finding the particle at more than one place, which is again not allowed.

(iii) It must be continuous and must have a continuous derivative everywhere. We know that $\frac{d^2\psi}{dx^2}$ must be finite everywhere.

(iv) If $\frac{d\psi}{dx}$ is continuous, then ψ is also continuous. Thus, ψ and $\frac{d\psi}{dx}, \frac{d\psi}{dy}, \frac{d\psi}{dz}$ must be finite, continuous and single valued everywhere.

Note

The wave function should be physically acceptable.

EXAMPLE 13.1

Show that $y = Ae^{-i\omega(t-x/v)}$ is a solution of the wave equation.

Solution: We know

$$\frac{dy}{dx} = \frac{\partial}{\partial x} Ae^{-i\omega(t-x/v)} = \frac{i\omega}{v} y$$

$\Rightarrow$
$$\frac{\partial^2 y}{\partial x^2}=\frac{i\omega}{v}\frac{\partial y}{\partial x}=\frac{-\omega^2}{v^2}y \tag{1}$$

or
$$\frac{\partial^2 y}{\partial t^2}=i^2\omega^2 y=-\omega^2 y \tag{2}$$

From Eqs. (1) and (2)

$$\frac{\partial^2 y}{\partial x^2}=\frac{1}{v^2}\frac{\partial^2 y}{\partial t^2}\frac{-\omega^2}{v^2}y$$
$$=\frac{1}{v^2}(-\omega^2 y)$$

Thus, the displacement equation is the solution of the wave equation.

13.6 ▶ Time Independent Schrodinger Wave Equation

Consider a system of stationary waves associated with a particle where x, y, z be the coordinates of the particle and ψ be the wave function.

The differential equation of the wave motion is

$$\frac{\partial^2 \psi}{\partial t^2}=v^2\left(\frac{\partial^2 \psi}{\partial x^2}+\frac{\partial^2 \psi}{\partial y^2}+\frac{\partial^2 \psi}{\partial z^2}\right) \tag{13.5}$$

$$\frac{\partial^2 \psi}{\partial t^2}=v^2\nabla^2\psi$$

where
$$\nabla^2=\frac{\partial^2}{\partial x^2}+\frac{\partial^2}{\partial y^2}+\frac{\partial^2}{\partial z^2}$$

The solution of Eq. (13.5) is given by

$$\psi=\psi_0\sin wt=\psi_0\sin 2\pi vt$$

$$\frac{\partial \psi}{\partial t}=\psi_0(2\pi v)\cos 2\pi vt$$

$$\frac{\partial^2 \psi}{\partial t^2}=-\psi_0(2\pi v)^2\sin 2\pi vt$$

$$\frac{\partial^2 \psi}{\partial t^2}=-4\pi^2 v^2\psi$$

$$\frac{d^2 \psi}{dt^2}=\frac{-4\pi v^2}{\lambda^2}\psi \tag{13.6}$$

Put Eq. (13.6) in Eq. (13.5)

$$v^2\nabla^2\psi = \frac{-4\pi^2 v^2}{\lambda^2}\psi$$

$$\nabla^2\psi + \frac{4\pi^2}{\lambda^2}\psi = 0 \qquad \left[\because \lambda = \frac{h}{mv} \quad \lambda = \frac{h}{p}\right]$$

$$\nabla^2\psi + \frac{4\pi^2}{h^2}m^2v^2\psi = 0 \tag{13.7}$$

Here, $$\frac{1}{2}mv^2 = E - V$$

$\Rightarrow$ $$m^2v^2 = 2m(E-V)$$

Thus, from Eq. (13.7), we get

$$\nabla^2\psi + \frac{4\pi^2}{h^2}2\text{m}(\text{E}-\text{V})\psi = 0$$

$\Rightarrow$ $$\nabla^2\psi + \frac{8\pi^2 m}{h^2}(E-V)\psi = 0 \quad \left[\because \hbar = \frac{h}{2\pi}\right]$$

$$\boxed{\boldsymbol{\nabla^2\psi + \frac{2m}{\hbar^2}(E-V)\psi = 0}} \tag{13.8}$$

Also, $$\frac{\hbar^2}{2m}\nabla^2\psi + (E-V)\psi = 0$$

$$\left[\frac{-\hbar^2}{2m}\nabla^2 + V\right]\psi = E\psi$$

$$\hat{H}\psi = E\psi$$

where $\hat{\text{H}}$ = Hamiltonian operator. For a free particle, $V = 0$.
From Eq. (13.8), the Schrodinger equation reduces to

$$\nabla^2\psi + \frac{2mE\psi}{\hbar^2} = 0 \quad \text{for free particle.}$$

13.7 ▶ Schrodinger Time Dependent Wave Equation

Schrodinger time dependent wave equation can be obtained by eliminating E from time independent Schrodinger equation.

The differential equation of one dimensional motion is

$$\frac{\partial^2\psi}{\partial t^2} = v^2\nabla^2\psi$$

Its solution is

$$\psi = \psi_0 e^{-iwt}$$

$$\frac{d\psi}{dt} = \psi_0(-iw)e^{-iwt}$$

$$= \psi_0(-\mathrm{i}\,2\pi\nu)e^{-iwt}$$

$$\frac{d\psi}{dt} = -2\pi\nu i\psi$$

$$= -2\pi i\left(\frac{E}{h}\right)\psi \quad [\because E = h\nu]$$

$$= -\frac{iE}{\hbar}\psi$$

$$\frac{d\psi}{dt} = -\frac{iE}{\hbar}\psi$$

$\Rightarrow$
$$E\psi = i\hbar\frac{d\psi}{dt} \tag{13.9}$$

We know that
$$\nabla^2\psi + \frac{2m}{\hbar^2}(E-V)\psi = 0$$

Put Eq. (13.9) in the above equation, we get

$$\nabla^2\psi + \frac{2m}{\hbar^2}\left[i\hbar\frac{d\psi}{dt} - V\psi\right] = 0$$

$$\frac{-\hbar^2}{2m}\nabla^2\psi + V\psi = i\hbar\frac{d\psi}{dt}$$

This is Schrodinger time dependent equation.
The above equation can also be written as

$$\boxed{\left(\frac{-\hbar^2}{2m}\nabla^2 + V\right)\psi = i\hbar\frac{\partial\psi}{\partial t}} \tag{13.10}$$

$$\hat{\mathrm{H}}\psi = \hat{\mathrm{E}}\psi$$

where $\hat{H}=\left(\frac{-h^2}{2m}\nabla^2+V\right)$ is the Hamiltonian operator and $\hat{E}=i\hbar\frac{\partial}{\partial t}$ is the enegy operator.

13.8 ▶ Operators

An operator is defined as a rule, which changes a given function into another function, i.e. if z operates on $f(x)$ and gives a function $g(x)$, i.e.,

$$zf(x)=g(x)$$

then z is said to be an operator.

There is an operator corresponding to every observable quantity. If we consider an operator A corresponding to an observable quantity a, then

$$A\psi = a\psi$$

The wave function ψ that satisfies the above equation is called eigen function and the corresponding observable quantity is called eigen value. The equation is called eigen value equation.

Some of the operators are given in Table 13.1.

TABLE 13.1 Some classical quantities and their corresponding operators

S. No.	Quantity	Operator
1	Position x, y, z	x, y, z
2	Momentum p	$-i\hbar\nabla$
3	Energy E	$i\hbar\frac{\partial}{\partial t}$
4	Hamiltonian H	$\frac{-\hbar^2}{2m}\frac{\partial^2}{\partial x^2}+V(r)$
5	KE	$\frac{-\hbar^2}{2m}\frac{\partial^2}{\partial r^2}$

Note

An operator is generally anything which is capable of doing something to a function.

13.9 ▶ Expectation Values

The average or expectation value $<\alpha>$ of any observable α is given as

$$< \alpha > = \frac{\int_{-\infty}^{+\infty} \psi^*(r,t)\alpha_{op}\psi(r,t)dV}{\int_{-\infty}^{+\infty} \psi^*\psi dV} \tag{13.11}$$

where α_{op} is the operator associated with the dynamical variable α and the volume element $dV = dx\,dy\,dz$. The expectation value of linear momentum is

$$< p_x > = \frac{\int_{-\infty}^{+\infty} \psi^*\left(-i\hbar\frac{\partial}{\partial x}\right)\psi\,dx}{\int_{-\infty}^{+\infty} \psi^*\psi\,dV} \tag{13.12}$$

Normalization condition is

$$\int_{-\infty}^{+\infty} \psi^*(r,t)\,\alpha_{op}\psi(r,t)dV = 1 \tag{13.13}$$

The expectation values of position vector r are

$$< x > = \int_{-\infty}^{+\infty} \psi^* x\,\psi\,dV \tag{13.14}$$

$$< y > = \int_{-\infty}^{+\infty} \psi^* y\,\psi\,dV \tag{13.15}$$

$$< z > = \int_{-\infty}^{+\infty} \psi^* z\,\psi\,dV \tag{13.16}$$

Similarly, the expectation values of potential energy and momentum are

$$< V > = \int_{-\infty}^{+\infty} \psi^*(r,t)\,V(r,t)\,\psi(r,t)dV \tag{13.17}$$

$$< p > = \int_{-\infty}^{+\infty} \psi^*(r,t)\,p\,\psi(r,t)dV \tag{13.18}$$

Also,

$$< p_x > = \int_{-\infty}^{+\infty} \psi^*(x,t)\,p_x\,\psi(x,t)dx \tag{13.19}$$

$$< p_y > = \int_{-\infty}^{+\infty} \psi^*(y, t)\, p_y\, \psi(y, t) dy \tag{13.20}$$

$$< p_z > = \int_{-\infty}^{+\infty} \psi^*(z, t)\, p_z\, \psi(z, t) dz \tag{13.21}$$

A wave function is said to be normalized if its normalization constant is unity, i.e.,

$$\text{if } \boxed{\int_{-\infty}^{+\infty} \int_{-\infty}^{+\infty} \int_{-\infty}^{+\infty} |\psi|^2\, dx\, dy\, dz = 1} \tag{13.22}$$

and if two different wave functions ψ_1 and ψ_2 satisfy the Schrodinger wave equation and if $\int_V \psi_1^* \psi_2\, dV$ or $\int_V \psi_1 \psi_2^*\, dV$ vanishes over the total volume V, then these are called orthogonal wave functions, i.e.,

$$\boxed{\int_V \psi_1^* \psi_2\, dV = \int_V \psi_1 \psi_2^*\, dV = 0} \tag{13.23}$$

for $1 \neq 2$ and the set of wave functions which are normalized and also orthogonal are called orthonormal wave functions. If the wave functions ψ_1 and ψ_2 are simultaneously normalized, then we have

$$\int \psi_1 \psi_1^*\, dV = \int \psi_2 \psi_2^*\, dV = 1 \tag{13.24}$$

Equations (13.23) and (13.24) may collectively be written as

$$\int \psi_1^* \psi_2 dV = 0 \text{ if } 1 \neq 2 \tag{13.25}$$

$$= 1 \text{ if } 1 = 2 \tag{13.26}$$

EXAMPLE 13.2

Find the eigen function and eigen values for wave function

$$\psi(x) = e^{4x}$$

Solution:

$$\frac{\partial^2}{\partial x^2}\left(e^{4x}\right) = 4\frac{\partial}{\partial x}\left(e^{4x}\right) = 16e^{4x}$$

Here, $\frac{\partial^2}{\partial x^2}$ is the operator, e^{4x} is the eigen function and 16 is referred to as the eigen value of the given function.

EXAMPLE 13.3

An eigen function of the operator $\frac{\partial^2}{\partial x^2}$ is cos nx, where n = 1, 2, 3, ... Find the corresponding eigen values.

Solution: Operating the operator on eigen function x, we get

$$\frac{\partial^2}{\partial x^2}\cos nx = n\frac{\partial}{\partial x}\sin nx$$

$$= n^2 \cos nx$$

$$\Rightarrow \qquad \frac{\partial^2}{\partial x^2}\cos nx = n^2 \cos nx$$

Here, the values of $n = 1, 2, 3, \ldots$

Thus, the corresponding eigen values are 1, 4, 9....

EXAMPLE 13.4

Find the values of momentum for an electron in a box of length 1 Å for $x = 1, 2$.

Solution Let $a = 1\ \text{Å} = 10^{-10}\ \text{m}$

Momentum, $$p_n = \frac{n\hbar\pi}{a}$$

For $n = 1$, $$p_1 = \frac{\hbar\pi}{a} = \frac{h}{2a} = \mathbf{3.3 \times 10^{-24}\ kgm/s}$$

For $n = 2$, $$p_2 = \frac{2\hbar\pi}{a} = \frac{h}{a} = \mathbf{6.63 \times 10^{-24}\ kgm/s}$$

13.10 ▶ Ehrenfest Theorem

This theorem states that the Schrodinger equation leads to the Newton's (classical) laws of motion on the average.

As Newton's law may be written as

$$\frac{d\vec{r}}{dt} = \frac{\vec{p}}{m} \tag{13.27}$$

and
$$\frac{d\vec{p}}{d\vec{t}} = -\text{grad } V \tag{13.28}$$

Also, the components of $\vec{p}$ can be written as

$$\frac{dp_x}{dt} = -\frac{\partial V}{\partial x}, \frac{dp_y}{dt} = -\frac{\partial V}{\partial y} \text{ and } \frac{dp_z}{dt} = -\frac{\partial V}{\partial z}$$

But in quantum theory, it is not possible to define x, y, z, p_x, p_y and p_z in the classical sense. The approximate values of x, y, z, p_x, p_y and p_z may be found by considering the time rate of change of average values. If the width of the wave packet is neglected, the values of these derivatives must be equal to the classically calculated value.

We know that the time dependence of the expectation value of the position is

$$\boxed{\frac{d}{dt}\langle x\rangle = \frac{i}{\hbar}\left\langle \left[\hat{H}, \hat{x}\right]\right\rangle + \left\langle \frac{\partial \hat{x}}{\partial t}\right\rangle} \tag{13.29}$$

But $\hat{x}$ operator does not depend explicitly on time, thus $\left\langle \frac{\partial \hat{x}}{\partial t}\right\rangle = 0$, i.e.,

$$\frac{d}{dt}\langle x\rangle = \frac{i}{\hbar}\left\langle \left[\left(\frac{\hat{p}^2}{2\text{m}} + V(x)\right), \hat{x}\right]\right\rangle$$

but $V(x) = f(x)$ and $\left\langle\left[V(x), \hat{x}\right]\right\rangle = 0$

Thus from the last equation

$$\begin{aligned}
\frac{d}{dt}\langle x\rangle &= \frac{i}{\hbar}\left\langle \left[\frac{\hat{p}^2}{2m}, \hat{x}\right]\right\rangle \\
&= \frac{i}{2m\hbar}\left\langle \left[\hat{p}^2, \hat{x}\right]\right\rangle \\
&= \frac{i}{2m\hbar}\langle \hat{p}[\hat{p}, \hat{x}] + [\hat{p}, \hat{x}]\hat{p}\rangle \\
&= \frac{-i}{2m\hbar}\langle \hat{p}i\hbar + i\hbar\hat{p}\rangle \\
&= \frac{2\hbar}{2m\hbar}\langle \hat{p}\rangle
\end{aligned}$$

Similarly,

$$\frac{d}{dt}\langle x\rangle = \frac{\langle \hat{p}_x\rangle}{m},\ \frac{d}{dt}\langle y\rangle = \frac{\langle \hat{p}_y\rangle}{m} \text{ and } \frac{d}{dt}\langle z\rangle = \frac{\langle \hat{p}_z\rangle}{m} \tag{13.30}$$

Also, the Hamiltonian operator is

$$\hat{H} = \frac{\hat{p}^2}{2m} + V(x)$$

$$\Rightarrow \qquad \frac{d}{dt}\langle p\rangle = \frac{i}{\hbar}\left\langle\left[\left(\frac{\hat{p}^2}{2m} + V(x)\right), \hat{p}\right]\right\rangle$$

but,

$$\left\langle\left[\frac{\hat{p}^2}{2m}, \hat{p}\right]\right\rangle = 0$$

$$\Rightarrow \qquad \frac{d}{dt}\langle p\rangle = \frac{i}{\hbar}\langle[V(x), \hat{p}]\rangle \tag{13.31}$$

and

$$\begin{aligned}\langle[V(x), \hat{p}]\rangle &= \langle\psi\,|\,V\hat{p}\psi\rangle - \langle\psi\,|\,\hat{p}V\psi\rangle \\ &= \left\langle\psi\left|V\frac{\hbar}{i}\frac{\partial\psi}{\partial x}\right.\right\rangle - \left\langle\psi\left|\frac{\hbar}{i}\frac{\partial}{\partial x}(V\psi)\right.\right\rangle \\ &= -\left\langle\psi\left|\frac{\hbar}{i}\frac{\partial V}{\partial x}\psi\right.\right\rangle\end{aligned}$$

From Eq. (13.31),

$$\frac{d}{dt}\langle p\rangle = \frac{-i}{\hbar}\left\langle\psi\left|\frac{\hbar}{i}\frac{\partial V}{\partial x}\psi\right.\right\rangle$$

$$\boxed{\frac{d}{dt}\langle p\rangle = -\left\langle\frac{\partial V}{\partial x}\right\rangle} \tag{13.32}$$

These reductions [given by Eqs. (13.30) and (13.32)] of quantum mechanical dynamics to the classical trajectories are known as Ehrenfest's theorem.

Note

Ehrenfest's theorem holds for expectation values and not for the eigen values themselves.

13.11 ▶ Solution of Some Problems in Quantum Mechanics

13.11.1 Free Particle

We know that the time-independent Schrodinger equation is

$$\nabla^2\psi+\frac{2m}{\hbar^2}(E-V)\psi=0 \tag{13.33}$$

For free particle,

$$V(r)=0 \quad (\therefore \text{ potential energy}=0)$$

Thus,

$$\frac{\partial^2\psi}{\partial x^2}+\frac{\partial^2\psi}{\partial y^2}+\frac{\partial^2\psi}{\partial z^2}+\frac{2m}{\hbar^2}E\psi=0 \tag{13.34}$$

This equation can be solved by putting

$$\psi(x,y,z)=u(x)\,v(y)\,w(z) \tag{13.35}$$

Putting Eq. (13.35) into Eq. (13.34), we get

$$vw\frac{\partial^2 u}{dx^2}+uw\frac{\partial^2 v}{\partial y^2}+uv\frac{\partial^2 w}{\partial z^2}+\frac{2\text{m}}{\hbar^2}Euvw=0$$

Dividing this equation by uvw,

$$\frac{1}{u}\frac{\partial^2 u}{\partial x^2}+\frac{1}{v}\frac{\partial^2 v}{\partial y^2}+\frac{1}{w}\frac{\partial^2 w}{\partial z^2}+\frac{2m}{\hbar^2}E=0 \tag{13.36}$$

Also,

let $$\frac{1}{u}\frac{\partial^2 u}{\partial x^2}=k_x$$

$$\frac{1}{v}\frac{\partial^2 u}{\partial y^2}=k_y$$

$$\frac{1}{w}\frac{\partial^2 w}{\partial z^2}=k_z$$

Thus, from Eq. (13.36),

$$k_x+k_y+k_z+\frac{2mE}{\hbar^2}=0$$

$\Rightarrow$ $$\frac{1}{u}\frac{\partial^2 u}{\partial x^2}=-\frac{2m}{\hbar^2}E_x$$

$$\Rightarrow \qquad \frac{\partial^2 u}{\partial x^2} + \frac{2mE_x u}{\hbar^2} = 0$$

The general solution is

$$u(x) = A_x \frac{\sin\sqrt{2mE_x}}{\hbar}(x - x_o)$$

$$v(y) = A_y \frac{\sin\sqrt{2mE_y}}{\hbar}(y - y_o)$$

$$w(z) = A_z \frac{\sin\sqrt{2mE_z}}{\hbar}(z - z_0)$$

Thus, the complete solution is

$$\psi(x, y, z, t) = A \frac{\sin\sqrt{2mE_x}}{\hbar}(x - x_o)e^{-iE_x t/h} \frac{\sin\sqrt{2mE_y}}{\hbar}(y - y_o)\frac{\sin\sqrt{2mE_z}}{\hbar}(z - z_o)e^{-iE_z t/h}$$

where $A = A_x A_y A_z$

$$\therefore \quad \psi(x, y, z, t) = A \frac{\sin\sqrt{2mE_x}}{\hbar}(x - x_o)\frac{\sin\sqrt{2mE_y}}{\hbar}(y - y_o)\frac{\sin\sqrt{2mE_z}}{\hbar}(z - z_o)e^{-(E_x + E_y + E_z)t/h}$$

$$= N \frac{\sin\sqrt{2mE_x}}{\hbar}(x - x_0)\frac{\sin\sqrt{2mE_y}}{\hbar}(y - y_0)\frac{\sin\sqrt{2mE_z}}{\hbar}(z - z_0) \times e^{iEt/h}$$

where $E = E_x + E_y + E_z$.

Let us consider the propagation in x-direction only.

i.e., $$E_y = E_z = 0 \text{ and } E_x = E$$

$$\Rightarrow \qquad \psi(x, t) = A \frac{\sin\sqrt{2mE}}{\hbar}(x - x_o)e^{-iEt/h} \tag{13.37}$$

Equation (13.37) gives a set of standing waves with wavelength normal to x-axis. The general equation of plane wave along the x-direction is

$$\boxed{\boldsymbol{\psi = A \sin\frac{2\pi}{\lambda}xe^{-iEt/h}}} \tag{13.38}$$

where $\lambda = \dfrac{h}{\sqrt{2mE}}$ and $E = \dfrac{1}{2}mv^2$

also $\lambda = \dfrac{h}{mv}$ is the deBroglie expression for the wavelength associated with a particle of mass m and velocity V.

13.11.2 Particle in a Three-Dimensional Box

Let us consider a particle having mass m trapped in a box with edges parallel to x, y and z respectively and having sides a, b and c, respectively (Fig. 13.1).

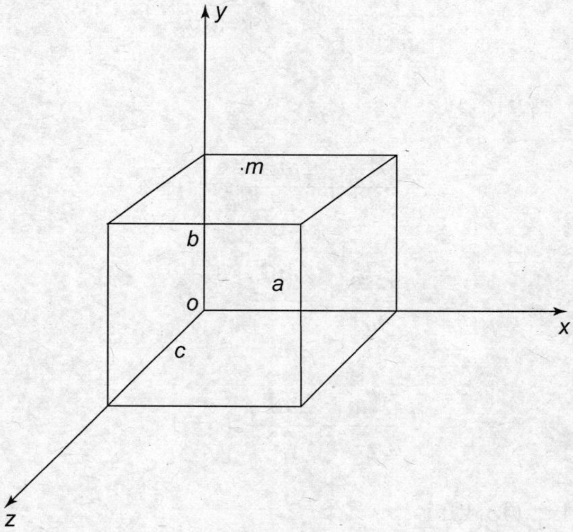

FIGURE 13.1 Particle in a three-dimensional box

The particle moves freely within the region $0 < x < a$, $0 < y < b$ and $0 < z < c$, i.e.,

$$V(x, y, z) = 0, \quad a > x > 0$$

$$V(x, y, z) = 0, \quad b > y > 0$$

$$V(x, y, z) = 0, \quad c > z > 0$$

We know that the Schrodinger wave equation for this case can be written as

$$\psi(x, y, z) = u(x)\, v(y)\, w(z)$$

or

$$\psi = uvw$$

Thus, we get

$$\frac{1}{u}\frac{\partial^2 u}{\partial x^2} + \frac{1}{v}\frac{\partial^2 v}{\partial y^2} + \frac{1}{w}\frac{\partial^2 w}{\partial z^2} + \frac{2mE}{\hbar^2} = 0$$

and

$$\frac{1}{u}\frac{\partial^2 u}{\partial x^2}+\frac{1}{v}\frac{\partial^2 v}{\partial y^2}+\frac{1}{w}\frac{\partial^2 w}{\partial z^2}+\frac{2mE}{\hbar^2} \tag{13.39}$$

$\Rightarrow$ let $\frac{1}{u}\frac{\partial^2 u}{\partial x^2}=k_x,\ \frac{1}{v}\frac{\partial^2 v}{\partial y^2}=k_y$ and $\frac{1}{w}\frac{\partial^2 w}{\partial z^2}=k_z$

Thus,

$$\frac{-2mE}{\hbar^2}=k_x+k_y+k_z$$

Also, Eq. (13.39) can be rewritten as

$$\frac{\partial^2 u}{\partial x^2}+\frac{2mE_x}{\hbar^2}u=0 \tag{13.40}$$

$$\frac{\partial^2 v}{\partial y^2}+\frac{2mE_y}{\hbar^2}v=0 \tag{13.41}$$

$$\frac{\partial^2 w}{\partial z^2}+\frac{2mE_z}{\hbar^2}w=0 \tag{13.42}$$

The general solution of Eq. (13.40) is

$$u(x)=A_1\sin kx+A_2\cos kx \tag{13.43}$$

where A_1 and A_2 are two arbitrary constants.

Here, $|u(x)|^2$ is the probability of finding the particle along the x-axis.
Thus, boundary conditions for the particle can be written as

$$|u(x)|=0 \quad \text{when } x=0 \text{ and } x=a \tag{13.44}$$

and

$$u(x)=0 \quad \text{when } x=0 \text{ and } x=a \tag{13.45}$$

From Eq. (13.43),

$$u(x)=0 \quad \text{when } x=0, \text{ so } B=0$$

and

$$u(x)=0 \quad \text{when x}=a$$

$\therefore$

$$A_1\sin ka=0$$

$\Rightarrow$

$$ka=n_x\pi$$

$$\Rightarrow \qquad k = \frac{n_x \pi}{a}$$

$$\Rightarrow \qquad u(x) = A_1 \sin \frac{n_x \pi}{a} x \tag{13.46}$$

Applying normalization condition,

$$\int_0^a |u(x)|^2 \, dx = 1$$

$$\Rightarrow \qquad \int_0^a \left| A_1 \sin \frac{n_x \pi}{a} x \right|^2 dx = 1$$

$$\Rightarrow \qquad A_1^2 \int_0^a \sin^2 \frac{n_x \pi}{a} x \, dx = 1$$

$$\Rightarrow \qquad A_1^2 \times \frac{a}{2} = 1 \quad \text{or} \quad A_1 = \sqrt{\frac{2}{a}}$$

$$\Rightarrow \qquad u(x) = \sqrt{\frac{2}{a}} \sin \frac{n_x \pi}{a} x \tag{13.47}$$

$$\Rightarrow \qquad v(y) = \sqrt{\frac{2}{b}} \sin \frac{n_y \pi}{b} y \tag{13.48}$$

$$w(z) = \sqrt{\frac{2}{c}} \sin \frac{n_z \pi}{c} z \tag{13.49}$$

The complete solution will be

$$\psi(x, y, z) = \sqrt{\frac{2}{a}} \sin \frac{n_x \pi}{a} x \sqrt{\frac{2}{b}} \sin \frac{n_y \pi}{b} y$$

$$= \sqrt{\frac{2}{a}} \sin \frac{n_z \pi}{c} z$$

$$\Rightarrow \qquad \psi(x, y, z) = \frac{2\sqrt{2}}{\sqrt{abc}} \sin \frac{n_x \pi}{a} x \sin \frac{n_y \pi}{b} y \sin \frac{n_z \pi}{c} z \times e^{-iEt/h} \tag{13.50}$$

where $$\frac{\partial^2 u}{\partial x^2} = \left(\frac{n_x \pi}{a} \right)^2 \sqrt{\frac{2}{a}} \sin \frac{n_x \pi}{a} x = -\left(\frac{n_x \pi}{a} \right)^2 u(x) \tag{13.51}$$

Putting Eq. (13.51) into Eq. (13.47), we get

$$-\left(\frac{n_x\pi}{a}\right)^2 u(x)+E_x u(x)\frac{2m}{\hbar^2}=0$$

$$E_x=\frac{1}{2m}\left(\frac{n_x\pi\hbar}{a}\right)^2$$

$$E_x=\frac{n_x^2h^2}{8ma^2}\quad\left[\because E=E_x+E_y+E_z\right]$$

Similarly,

$$E_y=\frac{n_y^2h^2}{8mb^2}\ \text{and}\ E_z=\frac{n_z^2h^2}{8mc^2}$$

Total energy is $E=E_x+E_y+E_z$

$$\Rightarrow\qquad \boxed{E=\frac{h^2}{8m}\left[\frac{n_x^2}{a^2}+\frac{n_y^2}{b^2}+\frac{n_z^2}{c^2}\right]}\qquad(13.52)$$

where n_x, n_y and n_z are three positive numbers. The energy levels and eigen functions for the particle in a box are shown in Fig. 13.2.

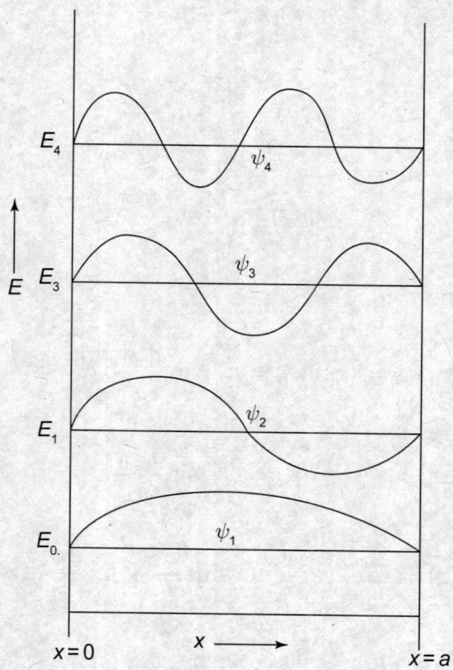

FIGURE 13.2 Energy levels and eigen functions for a particle in a box

13.11.3 Single Step Barrier Potential

A potential barrier corresponds to the force field acting on a particle being zero everywhere except in a limited region as shown in Fig. 13.3.

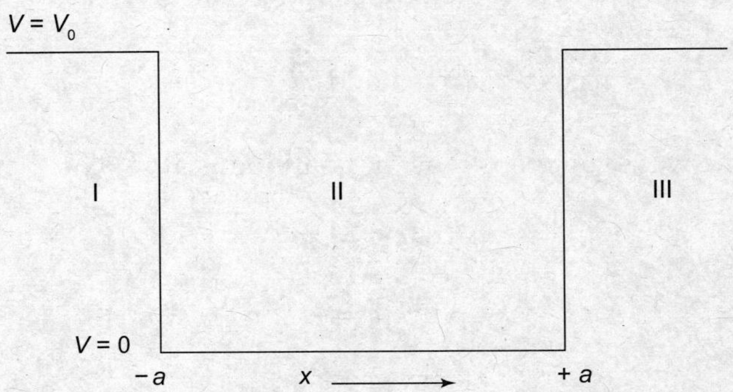

FIGURE 13.3 Single step potential barrier

Here,

$$V = 0, \quad x < |a|$$

$$V = V_0, \quad x > |a|$$

We restrict E as $0 < E < V_0$

Thus, for case (i) when $|x| < a$ (i.e., $V = 0$)

$$\frac{d^2\psi}{dx^2} + k_1^2\psi = 0 \tag{13.53}$$

where
$$k_1^2 = \frac{2mE}{\hbar^2}$$

For case (ii), when $|x| > a$, then

$$\frac{d^2\psi}{dx^2} + \frac{2m}{\hbar^2}(E - V_0)\psi = 0$$

$\Rightarrow$
$$\frac{d^2\psi}{dx^2} - k_2^2\psi = 0$$

where
$$k_2^2 = \frac{2m}{\hbar^2}(V_0 - E)$$

Here,
$$k_1^2 + k_2^2 = \frac{2mV_0}{\hbar^2}$$

The solution of the two equations is

$$\psi = a_1 \cos k_1 x + b \sin k_1 x;\ |x| < a \text{ and } \psi = C_1 e^{-k_2 x} + d_1 e^{+k_2 x};\ |x| > a \tag{13.54}$$

The symmetric parts of the above written functions can be given as

$$\psi = a_1 \cos k_1 x;\ |x| < a \text{ and } \psi = C_1 e^{-k_2 x};\ |x| > a \tag{13.55}$$

But we know that the wave functions must be continuous at $x = \pm a$

$$\Rightarrow \quad a_1 \cos k_1 a = c_1 e^{-k_2 a}$$

$$\Rightarrow \quad -k_1 a_1 \sin k_1 a = -k_2 c\, e^{-k_2 a}$$

$$\Rightarrow \quad k_1 \tan k_1 a = k_2$$

Put $k_1\, a = \alpha$ and $k_2\, a = \beta$ (13.56)

and since

$$k_1^2 + k_2^2 = \frac{2mV_0}{\hbar^2}$$

so put,

$$\left(\frac{2mV_0}{\hbar^2}\right)^{1/2} a = r$$

$$\Rightarrow \quad \alpha^2 \tan^2 \alpha = \beta^2$$

and

$$\alpha^2 + \beta^2 = r^2$$

$$\Rightarrow \quad \beta^2 = r^2 - \alpha^2$$

$$\Rightarrow \quad \alpha^2 \tan^2 \alpha = r^2 - \alpha^2$$

$$\Rightarrow \quad \alpha |\sec \alpha| = r \tag{13.57}$$

Similarly, with the anti-symmetric part, we will get the continuity condition as

$$\alpha |\operatorname{cosec} \alpha| = r \tag{13.58}$$

Here, the complete wave function can be obtained by normalizing it.

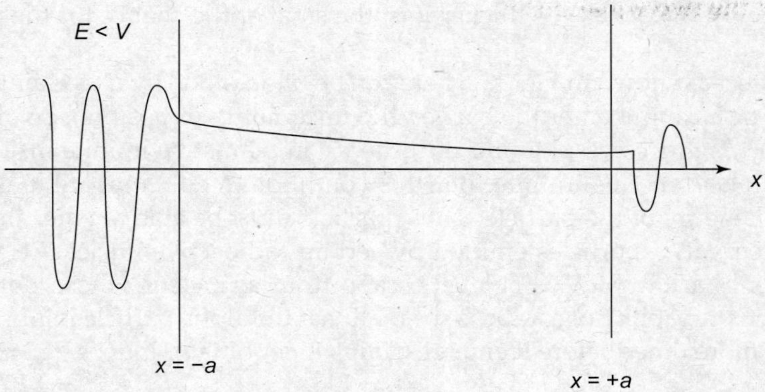

FIGURE 13.4 Single step potential barrier

When the wave functions are plotted as shown in Fig. 13.4, it can be seen that these do not fall to zero within the box but extend a little bit out of it. Thus, the probability of particle existing outside the box is not zero but finite. The phenomenon here exists is completely quantum mechanical effect and is called tunnelling effect or barrier penetration.

13.11.4 Potential Barrier of Finite Width or Rectangular Potential Barrier

If a particle is impinging on a barrier with energy less than the height of the potential barrier, it will not necessarily be totally reflected by it but there is always a probability that it may cross the barrier and continue its forward motion. Let us consider a particle incident on the barrier of type as shown in Fig. 13.5, where $E < V_0$.

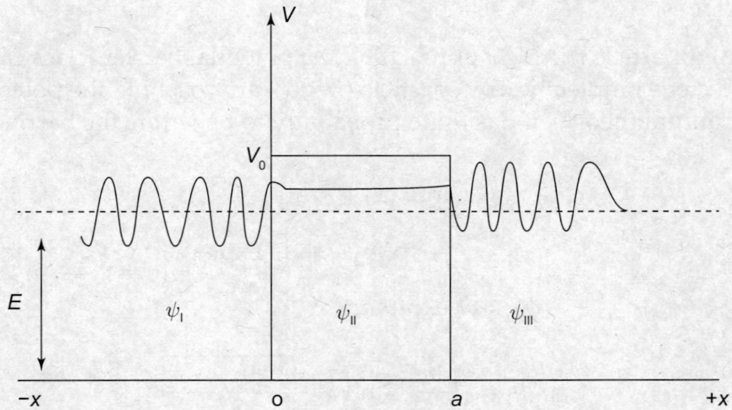

FIGURE 13.5 Rectangular potential barrier

$$
\begin{aligned}
V(x) &= 0, \quad x < 0 \quad &\text{(region I)}\\
V(x) &= V_0, \quad 0 < x < a \quad &\text{(region II)}\\
V(x) &= 0, \quad x < a \quad &\text{(region III)}
\end{aligned}
$$

Here, the higher the barrier, the thicker it is, the smaller the chance for the particle to get through it.

The tunnel effect as shown in Fig. 13.6 here can be understood by uncertainty principle. If the incident particle cannot enter the barrier, the uncertainty in position, Δx must be 0. But since, $\Delta x \ \Delta p \geq \hbar/2$, the corresponding uncertainty in p must be infinite inside the barrier, which means E also must be infinite. But this condition is not compatible with the finite momentum and energy of the particle. Thus, particle must be able to enter the barrier. The tunnel effect occurs in α-particles emitted by certain radioactive nuclei. An alpha particle whose K.E. is only a few MeV is able to escape from a nucleus whose potential wall is 25 MeV high. The probability of escape is so small that the alpha particle might have to strike the wall 10^{38} or more times before it emerges, but sooner or later it does get out.

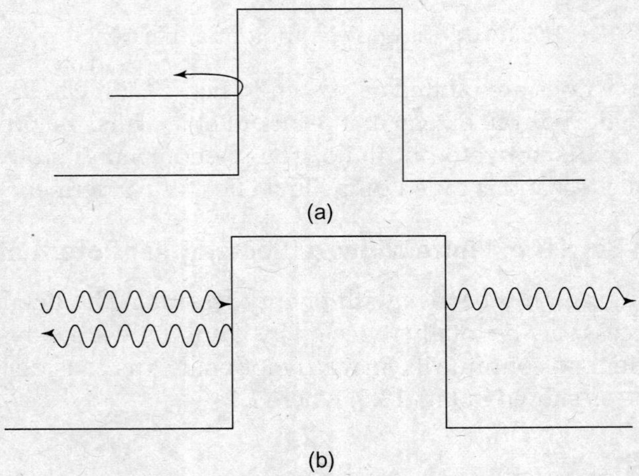

FIGURE 13.6 (a) A particle ($E < V$) approaching the potential barrier (classical picture), it must be reflected completely. (b) A particle ($E < V$) approaching the potential barrier (quantum theory), it has finite probability to penetrate the barrier

$$\psi_{\text{I}} = \exp(ik_1x) + R\exp(-ik_1x); \ \ x \leq 0$$

$$\psi_{\text{II}} = A_1\exp(ik_2x) + A_2\exp(-ik_2x); \ \ 0 < x < a$$

$$\psi_{\text{III}} = T\exp(ik_1x); \ \ x \geq a$$

Here, $$k_1 = \sqrt{\frac{2mE}{\hbar^2}} \text{ and } k_2\sqrt{\frac{2m(E-V_0)}{\hbar^2}}$$

Boundary conditions are

(i) $\psi_{\text{I}}(0) = \psi_{\text{II}}(0)$ and $\psi_{\text{III}}(a) = \psi_{\text{III}}(a)$

(ii) $\dot{\psi}_{\text{I}}(0) = \dot{\psi}_{\text{II}}(0)$ and $\dot{\psi}_{\text{III}}(a) = \dot{\psi}_{\text{III}}(a)$

where $\dot{\psi} = \frac{d\psi}{dx}$ (13.59)

Thus,
$$1 + R = A_1 + A_2$$

$$ik_1(1-R) = ik_2(A_1 - A_2) \tag{13.60}$$

$$\Rightarrow \quad A_1e^{ik_2a} + A_2e^{-ik_2a} = Te^{ik_1a} \text{ and } ik_2(A_1\,e^{ik_2a} - A_2\,e^{-ik_2a}) = ik_1Te^{ik_1a} \tag{13.61}$$

From Eq. (13.60),

$$ik_2 + ik_2R = ik_2A_1 + ik_2\,A_2$$

$$ik_1 - ik_1R = ik_2A - ik_2\,A_2$$

$$\Rightarrow \quad 2A_1 = \left(1+\frac{k_1}{k_2}\right) + R\left(1-\frac{k_1}{k_2}\right) \text{ and } 2A_2 = \left(1-\frac{k_1}{k_2}\right) + R\left(1+\frac{k_1}{k_2}\right) \tag{13.62}$$

From Eq. (13.58),

$$\Rightarrow \quad 2A_1e^{ik_2a} = Te^{ik_1a}\left(1+\frac{k_1}{k_2}\right) \text{ and } 2A_2e^{ik_2a} = Te^{ik_1a}\left(1-\frac{k_1}{k_2}\right) \tag{13.63}$$

Comparing Eqs. (13.62) and (13.63), we get

$$\left\{\left(1+\frac{k_1}{k_2}\right) + R\left(1-\frac{k_1}{k_2}\right)\right\}e^{ik_2a} = Te^{ik_1a}\left(1+\frac{k_1}{k_2}\right) \tag{13.64}$$

and
$$\left\{\left(1-\frac{k_1}{k_2}\right) + R\left(1+\frac{k_1}{k_2}\right)\right\}e^{-ik_2a} = Te^{ik_1a}\left(1-\frac{k_1}{k_2}\right) \tag{13.65}$$

Dividing Eq. (13.64) by Eq. (13.65), we get

$$R\left\{\left(1-\frac{k_1}{k_2}\right)^2 e^{ik_2a} - \left(1+\frac{k_1}{k_2}\right)^2 e^{-ik_2a}\right\} = \left(1+\frac{k_1}{k_2}\right)\left(1-\frac{k_1}{k_2}\right)\left\{e^{ik_2a} - e^{-ik_2a}\right\}$$

Put
$$\frac{k_1}{k_2} = r$$

$$\Rightarrow \quad R\left\{e^{ik_2a} - 2re^{ik_2a} + r^2e^{ik_2a} - e^{-ik_2a} - 2re^{-ik_2a} - r^2e^{-ik_2a}\right\} = (1-r^2)(2i\sin k_2\,a)$$

$$\Rightarrow \quad R\left[2ik_2a - 2r \times 2\cos k_2a + r^2 2i\sin k_2a\right] = \left[(1-r^2)2i\sin k_2a\right]$$

$$\Rightarrow \qquad R = \frac{(1-r^2)\sin k_2 a}{(r^2+1)\sin k_2 a + 2i\, r\cos k_2 a}$$

and

$$T = \frac{2i\, r\, e^{-ik_2 a}}{(r^2+1)\sin k_2 a + 2i\, r\cos k_2 a}$$

The reflection and transmission coefficients are complex and their amplitude can be evaluated as

$$\boxed{|T|^2 = \frac{(2r)^2}{(r^2+1)^2 \sin^2 k_2 a + (2r)^2 \cos k_2 a}} \tag{13.66}$$

and

$$\boxed{|R|^2 = \frac{(1-r^2)^2 \sin^2 k_2 a}{(r^2+1)^2 \sin^2 k_2 a + (2r)^2 \cos k_2 a}} \tag{13.67}$$

Now,

(i) When $E > V_0$, k_2 under the barrier will be real and positive. Hence, r is real and positive.

(ii) When $E > V_0$, k_2 will be imaginary and thus sin and cosine terms become hypothetic.

$$|T|^2 = \frac{-\left(2\frac{k_1}{k_2}\right)^2}{\left(1-\frac{k_1^2}{k_2^2}\right)\sin^2 h\, k_2 a - \left(\frac{2k_1}{k_2}\right)^2 \cos^2 h\, k_2 a}$$

The transmission coefficient will be zero at $E = 0$ as $k_1 = 0$ and it will increase gradually till at $E = V_0$,

$$k_1 = \sqrt{\frac{2\text{m}V_0}{\hbar^2}}, k_2 = \sqrt{\frac{2\text{m}V_0}{\hbar^2}}$$

i.e.,

$$k_1 = k_2$$

$$\Rightarrow \qquad |T|^2 = \frac{-4}{0 - 4\cos^2 h\, k_2 a}$$

$$= \frac{1}{\left\{1+\left(\frac{k_1 a}{2}\right)^2\right\}}$$

This phenomenon of transmission of a particle through a potential barrier of finite width and height is called wave mechanical tunnelling.

The major applications of the tunnelling effect are as given below.

(i) **Alpha decay:** For an alpha particle to escape from the nucleus, it penetrates through the potential barrier created by the combination of the nuclear force and the Coulomb's repulsion between the alpha particle and rest of the nucleus.

(ii) Scanning tunnelling microscopes (STM's)

EXAMPLE 13.5

An electron is bound in one-dimensional box of size 4×10^{-10}m. What will be the minimum energy?

Solution: The quantized energy levels of a particle in one-dimensional box of size a is

$$E_n = \frac{n^2 h^2}{8ma^2}$$

For minimum energy, $n = 1$

$$E_{\min} = \frac{h^2}{8ma^2}$$

Here, $h = 6.6 \times 10^{-34}$ J-s, $m = 9.1 \times 10^{-31}$kg, $a = 4 \times 10^{-10}$ m

$$\mathbf{E_{min} = 3.77 \times 10^{-19}\, J}$$

EXAMPLE 13.6

Find the probability that a particle trapped in a box L cm wide can be found between $0.45L$ and $0.55L$ for the ground and first excited state.

Solution: For the ground state,

$$P_{x_1 x_2} = \int_{x_1}^{x_2} |\psi_n|^2 \, d\lambda$$

$$= \frac{2}{L}\int_{x_1}^{x_2} \sin^2 \frac{n\pi x}{L} dx$$

$$= \left[\frac{x}{L} - \frac{1}{2n\pi}\sin\frac{2n\pi x}{L}\right]_{0.45L}^{0.55L}$$

(i) For ground state, $n = 1$

$$P_{x_1x_2} = 0.198 = 19.8\%$$

(ii) For first excited state, $n = 2$

$$P_{x_1x_2} = 0.0065 = 0.65\%$$

EXAMPLE 13.7

Find the expectation value <x> of the position of a particle trapped in a box L cm.

Solution:

$$\langle x \rangle = \int_{-\infty}^{+\infty} x|\psi|^2\, d\lambda$$

$$= \frac{2}{L}\int_0^L x \sin^2 \frac{n\pi x}{L}\, dx$$

$$= \frac{2}{L}\left[\frac{x^2}{4} - x\frac{\sin(2n\pi x/L)}{4n\pi/L} - \frac{\cos(2n\pi x/L)}{8(n\pi/L)^2}\right]_0^L$$

since $\sin n\pi = 0$, $\cos 2n\pi = 1$, $\cos 0 = 1$, for all values of n.

$$\langle x \rangle = \frac{2}{L}\left(\frac{L^2}{4}\right) = \frac{L}{2}$$

EXAMPLE 13.8

Evaluate the lowest energy that a neutron possesses while it is confined inside the nucleus of the atom.

Solution: We know, for a neutron,

$$a = 10^{-14}\text{m}$$

$$m = 1.67 \times 10^{-27}\text{ kg}$$

and

$$h = 6.6 \times 10^{-34}\text{J-s}$$

Thus,

$$E_n = \frac{n^2h^2}{8ma^2} = 3.2^9 \times 10^{-13}\text{ J}$$

EXAMPLE 13.9

A particle is moving in one-dimensional potential box (of infinite height) of width 25 Å. Calculate the probability of finding the particle within an interval of 5 Å at the centres of the box when it is in its state of least energy.

Solution: We know, the wave function is

$$\psi = \sqrt{\frac{2}{a}} \sin \frac{n\pi x}{a}$$

Here, $n = 1$, so,

$$\psi = \sqrt{\frac{2}{a}} \sin \frac{\pi x}{a}$$

At the centre of the box, $x = a/2$ and the probability is

$$|\psi(x)|^2 = \left[\sqrt{\frac{2}{a}} \sin \frac{\pi(a/2)}{a}\right]^2$$

$$= \frac{2}{a} \sin^2 \frac{\pi}{2} = \frac{2}{a}$$

The probability ρ in the interval Δx will be

$$\rho = |\psi(x)|^2 \Delta x$$

$$= \frac{2}{a} \Delta x$$

Here, $a = 25$ Å

$$= 25 \times 10^{-10} \text{ m}$$

$$\Delta x = 5 \text{ Å}$$

$$= 5 \times 10^{-10} \text{ m}$$

$\Rightarrow$

$$\rho = \frac{2 \times 5 \times 10^{-10}}{25 \times 10^{-10}}$$

$$= 0.4 = 4\%$$

RECAP ZONE

POINTS TO REMEMBER

- The probability density or the probability of finding the particle per unit volume is given by $|\psi|^2$.
- In quantum mechanics, each observable is represented by an operator which acts on a wave function ψ to give a new wave function.
- Schrodinger equation is the guiding equation to locate the position of the particle within the wave packet.
- Expectation value of a dynamical quantity is defined as the average of the result of a large number of measurements on independent systems.
- Ehrenfest's theorem states that the average motion of a wave packet agrees with the motion of the corresponding classical particle.
- Quantum mechanics allows a fluctuation in the measurements of observables while classical mechanics assumes that every observable in principle is absolutely determinate.
- The eigen value corresponding to a eigen function is the characteristic or proper energy value for that equation.
- The quantum tunnelling of particles predicts a definite probability for the transmission of the electrons through thin barrier.
- Basics of quantum theory were developed by Planck, Einstein, Schrodinger and Heisenberg.
- According to Heisenberg, it is proved that
  - (i) The electron cannot reside inside the nucleus
  - (ii) Radius of Bohr's first orbit can be calculated
  - (iii) Energy of a particle in a box can be evaluated
  - (iv) Ground state energy of a liner harmonic oscillator can be calculated.
- The wave function ψ is a complex quantity and has no physical meaning, but $\psi\psi^*$ is a complex conjugate and represents probability.

KEY TERMS WITH DEFINITIONS

- **Quantum mechanics** – Description of motion and interaction of particles at the small scales.

- **Matric mechanics** – Where the observed quantities like frequencies and intensities of spectral lines are taken into account.
- **Wave mechanics** – Where the concepts of classical wave theory and de-Broglie's wave particle relationship are combined with each other.

Important Formulae and Equations

Equation Number	Equation	Remarks
(13.8)	$\nabla_{\psi}^{2}+\frac{2m}{\hbar^{2}}(E-V)\psi=0$	**Time-independent Schrodinger wave equation**
(13.10)	$\left(\frac{-\hbar^{2}}{2m}\nabla^{2}+v\right)\psi=i\hbar\frac{\partial\psi}{\partial t}$	**Time-dependent Schrodinger wave equation**
(13.11)	$<\alpha>=\frac{\int_{-\infty}^{+\infty}\psi^{*}(r,t)\alpha_{op}\psi(r,t)dV}{\int_{-\infty}^{+\infty}\psi^{*}\psi dV}$	**Expectation value of any observable α**
(13.17)	$<V>=\int_{-\infty}^{+\infty}\psi^{*}(r,t)\,V(r,t)\,\psi(r,t)dV$	**Expectation value of potential energy**
(13.18)	$<p>=\int_{-\infty}^{+\infty}\psi^{*}(r,t)\,p\,\psi(r,t)dV$	**Expectation value of momentum**
(13.22)	$\int_{-\infty}^{+\infty}\int_{-\infty}^{+\infty}\int_{-\infty}^{+\infty}\lvert\psi\rvert^{2}\,dx\,dy\,dz=1$	**Normalization condition for a wave function ψ**
(13.23)	$\int_{V}\psi_{1}^{*}\psi_{2}dV=\int_{V}\psi_{1}\psi_{2}^{*}dV=0$	**Orthogonal wave functions**
(13.29)	$\frac{d}{dt}\langle x\rangle=\frac{i}{\hbar}\left\langle\left\lvert\hat{H},\hat{x}\right\rvert\right\rangle+\left\langle\frac{\partial\hat{x}}{\partial t}\right\rangle$	**Time dependence of the expectation value of position**
(13.32)	$\frac{d}{dt}\langle p\rangle=-\left\langle\frac{\partial V}{\partial x}\right\rangle$	**Ehrenfest's equation**

(13.38)	$\psi = A \sin \frac{2\pi}{\lambda} x e^{-iEt/\hbar}$	The general solution of plane wave along x-direction
(13.52)	$E = \frac{h^2}{8m}\left[\frac{n_x^2}{a^2} + \frac{n_y^2}{b^2} + \frac{n_z^2}{c^2}\right]$	Total energy of a particle confined in three-dimensional box
(13.66) and (13.67)	$\lvert T\rvert^2 = \frac{(2r)^2}{(r^2+1)^2 \sin^2 k_2 a + (2r)^2 \cos k_2 a}$ and $\lvert R\rvert^2 = \frac{(1-r^2)^2 \sin^2 k_2 a}{(r^2+1)^2 \sin^2 k_2 a + (2r)^2 \cos k_2 a}$	Reflection and transmission coefficients, when a particle is assumed in rectangular potential barrier

REVIEW ZONE

SHORT ANSWER QUESTIONS

1. Write the physical significance of a wave function.
2. Differentiate between ψ and $|\psi|^2$.
3. What are the conditions for an acceptable wave function?
4. What is an operator?
5. Write the operators associated with energy and momentum.
6. What do you mean by tunnelling through a barrier?
7. What do you mean by expectation values of dynamical quantities?
8. What is Bohr's probability interpretation of wave function?
9. Write time-independent Schrodinger wave equation.
10. Write time-dependent Schrodinger wave equation.
11. State Etherfest's theorem.
12. How a free particle wave function signifies a particle in space and momentum?
13. What is the difference between the normalized and orthogonal wave functions?
14. Write two major differences between classical mechanics and quantum mechanics.
15. What are the conditions and limitations, a wave function must obey?

LONG ANSWER QUESTIONS

1. How do you predict the energy of a particle in closed box from
   (a) Classical theory
   (b) Quantum theory
2. Derive time-independent Schrodinger wave equation.
3. Derive time-dependent Schrodinger wave equation for a free particle.
4. State and prove Ehrenfest's theorem.
5. Write down Schrodinger's wave equation for a particle in a box. Solve it to obtain eigen functions and show that the eigen values are discrete.
6. Obtain the Schrodinger wave equation for a step potential barrier. Hence, show that there is a distinct possibility of transmission of a particle having energy less than the potential barrier height.
7. Explain the phenomenon of quantum tunnelling across a thin barrier of width L using the Schrodinger wave equation.
8. Establish the three-dimensional Schrodinger's wave equation for a particle confined in a box. Derive an equation for energy.
9. Discuss the motion of a particle across a rectangular potential barrier and deduce an expression for the probability that the particle will penetrate the barrier. Discuss the application of this result.
10. Calculate the transmission coefficient of a rectangular potential barrier and discuss the physical phenomenon which is explained on the basis of leakage through a barrier.
11. Obtain Schrodinger's wave equation for a particle in square well potential and discuss the energy levels when the well is infinitely deep.
12. Derive time-independent and time-dependent equations for a non-relativistic particle.
13. Starting from the wave equation and introducing energy and momentum of particle, obtain an expression for three-dimensional Schrodinger equation in time-dependent form.

NUMERICAL PROBLEMS

1. Find the height of the potential barrier in eV for α-particles emitted from an element (At. No. 86). Assume the effective nuclear radius $= 10^{-14}$m.

   Ans. 24.16 MeV

2. Calculate the probability of transmission for a proton of energy 1 MeV through a 4 MeV high rectangular potential energy barrier of width 10^{-2} cm.

   Ans. 0.0015

3. Calculate the transmission probability for an electron of total energy 2 eV incident upon a rectangular potential barrier of height 4 eV and width 10^{-9} m.

   Ans. 2×10^{-6}

4. Calculate the first two energy levels of an electron confined to a box 1 Å wide ($h = 6.254 \times 10^{-34}$J-s).

   Ans. – 37.7 eV and 156.8 eV

5. An electron is confined to a one-dimensional box of side 2 Å. Obtain the first four eigenvalues of the electron eV.

Ans. 3.77×10^{-19} J

6. Explain the concept of operator and eigen function, and eigen value operating $\frac{\partial^2}{\partial x^2}$ on the wave function $\psi(x) = e^{2x}$.

Ans. $\left[\frac{\partial^2}{\partial x^2} e^{2x}, 4\right]$

7. Find the probability that a particle trapped in a box of width L can be found between 0.1 L and 0.2 L for the ground and the first excited states.

Ans. 10%

8. Determine the energy of the lowest three level for an electron in a square well of width 2 Å.

$\therefore$ 0.150×10^{-17} J, 0.60×10^{-17} J, 1.35×10^{-17} J

9. A 10 eV electron is incident on a barrier of height 20 eV. What is the probability that the electron will tunnel through the barrier if its width is (a) 0.5 nm and (b) 0.10 nm.

Ans. 7.028, 10^{-15}

10. An eigen function of the operator d^2/dx^2 is $\psi = e^{2x}$. Find the corresponding eigen value.

Ans. 4

11. A particle is in motion along a line between $x = 0$ and $x = a$, the PE is infinite.The wave function for the particle in the nth state is given by $\psi_n = A \sin\left(\frac{n\pi x}{a}\right)$. Find the expression for normalized wave function.

Ans. $\left[\sqrt{\frac{2}{a}} \sin\frac{n\pi x}{a}\right]$

12. A particle limited to the x-axis has the wave function $\psi = ax$ between $x = 0$ and $x = 1$, $\psi = 0$ elsewhere.

(a) Find the probability that the particle can be found between $x = 0.45$ and $x = 0.55$.

(b) Find the expectation value $\langle x \rangle$ of the particle's position.

Ans. (a) $0.0251a^2$, (b) $a^2/4$

CHAPTER 14

Lasers

Learning Objectives

By the end of this chapter, the student will be able:

- To understand working principle of laser
- To differentiate stimulated and spontaneous emissions
- To describe the working of different types of lasers
- To discuss the application of lasers
- To recognize important industrial applications of lasers

14.1 ▶ Introduction

LASER is an acronym of "Light Amplification by Stimulated Emission of Radiation". Einstein in 1917 proved the existence of stimulated emission and then the idea of LASER became feasible. In 1953, Charles H. Townes produced the first ammonia MASER (Microwave Amplification by Stimulated Emission of Radiation), a device similar to LASER but produces microwaves rather than optical radiations. The LASER is the outgrowth of MASER. In 1960, T.H. Maiman built the first laser by using ruby crystals as the active medium. It was the pulsed laser. After that in 1961, He–Ne laser was developed by Ali Javan and his associates, which gives continuous output rather than pulses.

In 1964, Townes, Basov and Prokhorov were awarded the noble prize for the invention of MASER. LASERS are used in communication, holography, recording the data and medicine, etc. Laser-guided missiles are the dangerous modern weapons.

14.2 ▶ Spontaneous and Stimulated Emission

When an atom lies initially in lower state (1) having energy E_1, it can be raised to an excited state (2), having energy E_2 by absorbing photon of frequency ν, where $\nu = (E_2 - E_1)/h$. This process is known as absorption (Fig. 14.1).

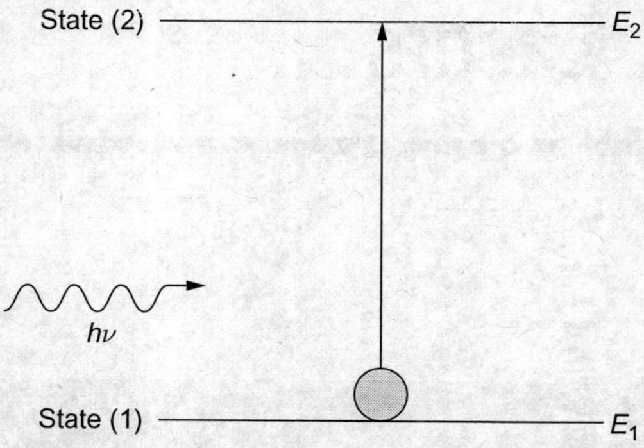

FIGURE 14.1 Absorption

The transition probability of atom from state (1) to state (2) is proportional to the energy density $u(\nu)$ of the radiation incident on the atom and is given as follows:

$$P_{12} = B_{12}\, u(\nu) \tag{14.1}$$

where B_{12} is known as Einstein's coefficient of radiation.

The atom in excited state (2) does not stay there for longer time because the excited state with higher energy is inherently unstable. It returns to lower energy level, by emitting a photon of frequency ν. This is called spontaneous emission. It is a random and incoherent process. The probability of spontaneous emission from state (2) to state (1) is given as follows:

$$\boxed{(P_{21})_{\text{spon.}} = A_{21}} \tag{14.2}$$

where A_{21} is known as Einstein's coefficient of spontaneous emission of radiation (Fig. 14.2).

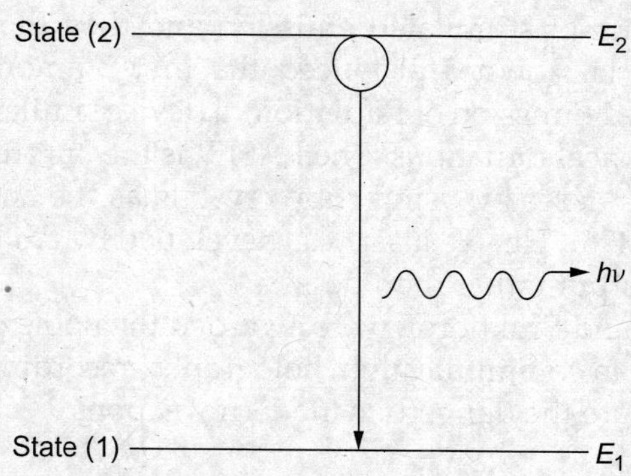

FIGURE 14.2 Spontaneous emission

If there are a number of atoms in excited state (2), then the photons emitted by each atom have random directions and random phases and hence incoherent.

But if there exists a population inversion, i.e. the atoms in excited state are more than that in lower state, then before returning to lower state, the atoms in excited state are induced to interact with photons having energy $h\nu = E_2 - E_1$. These atoms now return to the lower states by emitting photons which are in the same phase, travelling in the same direction and are coherent. This type of emission is known as stimulated emission (Fig. 14.3).

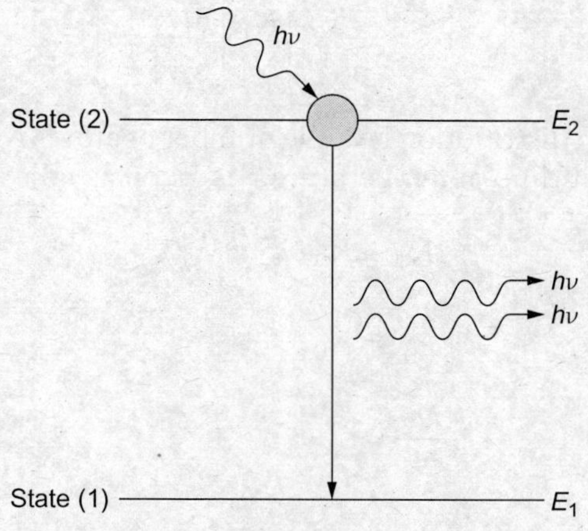

FIGURE 14.3 Stimulated emission

The probability of stimulated emission is proportional to energy density $u(\nu)$ of the stimulating radiation and is given by

$$\boxed{(P_{21})_{\text{sti}} = B_{21}\, u(\nu)} \tag{14.3}$$

where B_{21} is the Einstein's coefficient of stimulated emission.

14.3 ▶ Relation Between Einstein's *A* and *B* Coefficients

Let N_1 and N_2 be the number of atoms in energy states (1) and (2), respectively. The atoms in energy state (1) can absorb photons and give rise to absorption per unit time given as follows:

$$N_1\, P_{12} = N_1\, B_{12}\, u(\nu) \tag{14.4}$$

The atoms in energy state (2) can return to energy state (1) through emission (spontaneous + stimulated) process. This emission per unit time can be given as follows:

$$N_2\, P_{21} = N_2\, [A_{21} + B_{21}\, u(\nu)] \tag{14.5}$$

In thermal equilibrium, the absorption and emission rates must be equal,

i.e., $$N_1\, P_{12} = N_2\, P_{21}$$

or $$N_1\, B_{12}\, u(\nu) = N_2\, [A_{21} + B_{21}\, u(\nu)]$$

or $$u(\nu)\, [N_1\, B_{12} - N_2\, B_{21}] = N_2\, A_{21}$$

or $$u(\nu) = \frac{N_2 A_{21}}{N_1 B_{12} - N_2 B_{21}}$$

$$= \frac{A_{21}}{B_{21}} \frac{1}{\left[\frac{N_1}{N_2}\left\{\frac{B_{12}}{B_{21}}\right\} - 1\right]} \quad (14.6)$$

According to Boltzmann distribution law, the number of atoms N_1 and N_2 in energy states E_1 and E_2 in thermal equilibrium at temperature T is given as follows:

$$N_1 = N_0 e^{-E_1/K_B T}$$

And $$N_2 = N_0 e^{-E_2/K_B T}$$

$\Rightarrow$ $$\frac{N_2}{N_1} = e^{-(E_2 - E_1)/K_B T} \quad (14.7)$$

$\Rightarrow$ $$\frac{N_2}{N_1} = e^{-h\nu/K_B T}$$

$\Rightarrow$ $$\frac{N_1}{N_2} = e^{h\nu/K_B T} \quad (14.8)$$

From Eqs. (14.6) and (14.8), we get

$$u(\nu) = \frac{A_{21}}{B_{21}} \frac{1}{\left(e^{h\nu/k_B T} \frac{B_{12}}{B_{21}} - 1\right)} \quad (14.9)$$

Here, $$B_{12} = B_{21}$$

$$u(\nu) = \frac{A_{21}}{B_{21}} \frac{1}{(e^{h\nu/k_B T} - 1)} \quad (14.10)$$

But according to Planck's relation,

$$u(\nu) = \frac{8\pi h\nu^3}{c^3} \frac{1}{(e^{h\nu/k_B T} - 1)} \quad (14.11)$$

Comparing Eqs. (14.10) and (14.11), we get

$$\frac{A_{21}}{B_{21}} = \frac{8\pi h\nu^3}{c^3}$$

$$\boxed{\frac{A_{21}}{B_{21}} \propto \nu^3} \quad (14.12)$$

Equation (14.12) shows that the probability of spontaneous emission dominates over induced or stimulated emission. Also, the ratio of spontaneous to stimulated emission is proportional to the frequency of radiations.

Note

Due to exchange of energy between an atom of the medium and the radiation field, the transfer of momentum also occurs there.

EXAMPLE 14.1

If the energy of the first principle line of sodium atom is 2.1 eV, find the wavelength of the light emitted when a transition occurs between the first excited state and the ground state.

Solution: The energy of the first excited level is 2.1 eV = $2.1 \times 1.6 \times 10^{-19}$ J

$$E = h\nu = \frac{hc}{\lambda}$$

$$\therefore \quad \lambda = \frac{hc}{E}$$

$$\lambda = \frac{6.626 \times 10^{-34} \times 3 \times 10^{8}}{2.1 \times 1.6 \times 10^{-19}} \text{ m}$$

$$= \frac{19.878 \times 10^{-26}}{3.36 \times 10^{-19}} \text{ m}$$

$$\therefore \quad \lambda = 5.91 \times 10^{-7} \text{ m}$$

or $\quad \boldsymbol{\lambda = 5910}$ **Å**

EXAMPLE 14.2

Show that the stimulated emission is not possible for (i) sodium D line at 300°C and (ii) at optical frequencies under thermal equilibrium.

Solution:

(i) The ratio of stimulated emission to spontaneous emission is

$$\frac{\text{Stimulated emission}}{\text{Spontaneous emission}} = \frac{N_{ST}}{N_{SP}} = \frac{1}{e^{h\nu/KT} - 1}$$

Substituting the values, we get

$$= \frac{1}{\exp\left[\dfrac{6.62\times10^{-34}\times3\times10^{8}}{5890\times10^{-10}\times1.38\times10^{-23}\times573}\right]-1} \quad (\therefore \nu = c/\lambda)$$

$$= \frac{1}{\exp\left[\dfrac{19.86\times10^{-26}}{4657458.6\times10^{-33}}\right]-1}$$

$$= \frac{1}{\exp(42.64)-1}$$

$$= \frac{1}{3.3\times10^{18}-1} \approx 0.303\times10^{-18}$$

As the ratio is negative, it means that spontaneous emission has a greater probability than stimulated emission.

(ii) Similarly at optical frequencies $\sim \lambda = 5000$ Å and at a temperature of $T = 300$ K, the ratio is given by

$$\frac{\text{Stimulated emission}}{\text{Spontataneous emition}} = \frac{N_{ST}}{N_{SP}}$$

$$= \frac{1}{e^{h\nu/KT}-1}$$

Once again substituting the values, we get

$$= \frac{1}{\exp\left[\dfrac{6.62\times10^{-34}\times3\times10^{8}}{5000\times10^{-10}\times1.38\times10^{-23}\times300}\right]-1}$$

$$= \frac{1}{\exp\left[\dfrac{19.86\times10^{-26}}{2,070,000\times10^{-33}}\right]-1}$$

$$= \frac{1}{\exp(95.94)-1}$$

$$= \frac{1}{4.64\times10^{41}-1} \approx 0.215\times10^{-41}$$

As in the previous case, the ratio is negative, and hence stimulated emission is not possible at optical frequencies and at thermal equilibrium.

14.4 ▶ Population Inversion

The energy state of an atom is the result of energy levels of individual electrons of that atom. The ground state of an atom has minimum energy and the state that has more energy than the ground state is called an excited state. At thermal equilibrium, the number of atoms N_1 in lower energy state having energy E_1 are related, to the number of atoms N_2 in next higher energy state and is given by Boltzmann relation, i.e.,

$$N_2 = N_1 \exp(-\{E_2 - E_1\} / k_B T)$$

$$\Rightarrow \quad N_2 = N_1 e^{-(E_2 - E_1)/k_B T} \tag{14.13}$$

Here, $E_2 > E_1$

So, $N_2 < N_1$

But, if the process of stimulated emission dominates over the process of spontaneous emission, then

$$N_2 > N_1$$

i.e., The population of higher state is more than that of lower state. This condition is known as population inversion. In this state of population inversion, the higher levels are more populated than lower levels. Population inversion is the necessary condition for the operation of laser (Fig. 14.4).

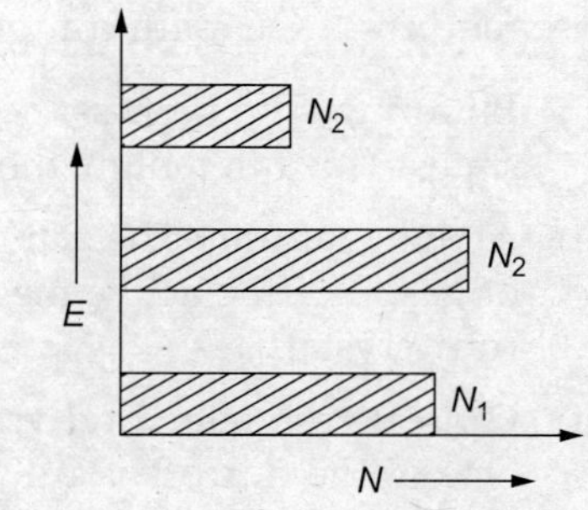

FIGURE 14.4 Inverted population of a system ($N_2 > N_1$)

14.5 ▶ Pumping

The process by which population inversion is achieved, is called pumping. It can be done by some external source of energy, e.g. gas discharge, flash lamp, etc. Some of the commonly used methods for pumping are as follows:

(i) **Optical pumping:** If photons are supplied to the active medium for causing population inversion, the pumping is called optical pumping. This type of pumping was used in ruby laser.

(ii) **Electrical discharge:** The pumping by electrical discharge is commonly preferred in the materials whose higher energy states have a narrow bandwidth. A potential difference is applied between cathode and anode in a discharge tube; the electrons emitted from cathode are accelerated towards anode. Some of the electrons ionize the active medium, which produces population inversion. This process is also called direct-electron excitation, e.g. argon-ion laser.

(iii) **Direct conversion:** In this type of pumping, a direct conversion of electrical energy into radiant energy takes place, e.g. semiconductor laser.

(iv) **Chemical reaction:** In CO_2 laser, this type of pumping is used. Here, energy for population inversion comes from a chemical reaction without any need for other energy sources. E.g., Hydrogen can combine with fluorine to form hydrogen fluoride.

$$H_2 + F_2 \rightarrow 2HF$$

(v) **Inelastic atom–atom collision:** In this type of pumping, one type of atoms is excited to higher state with the help of electrical discharge. Now, these excited atoms collide inelastically with another type of atoms and transfer their energy to the latter atoms. These latter atoms provide the population inversion needed for lasing action, e.g. He–Ne laser.

14.6 ▶ Main Components of a Laser

A laser requires three main components. These are as follows:

(i) **Pumping:** The process by which the atoms from ground state are excited to higher states. The different methods of pumping are explained well in Section 14.5.

(ii) **Active medium:** The active medium means laser material. It may be a solid, liquid or gas. E.g., the active medium in He–Ne laser is He–Ne gas and in ruby laser, it is ruby crystals.

(iii) **Optical resonator cavity:** It consists of two mirrors M_1 and M_2 facing each other. One of the mirrors is fully reflecting and the other mirror is partially transparent. The active medium is enclosed by the cavity. It increases the intensity of a laser beam (Fig. 14.5).

Once the population inversion is achieved by pumping, stimulated emission is triggered by a photon of appropriate frequency. Now, more photons are emitted than entered, thus the input is amplified by active medium. When gains are higher than losses in the cavity, the laser beam is produced.

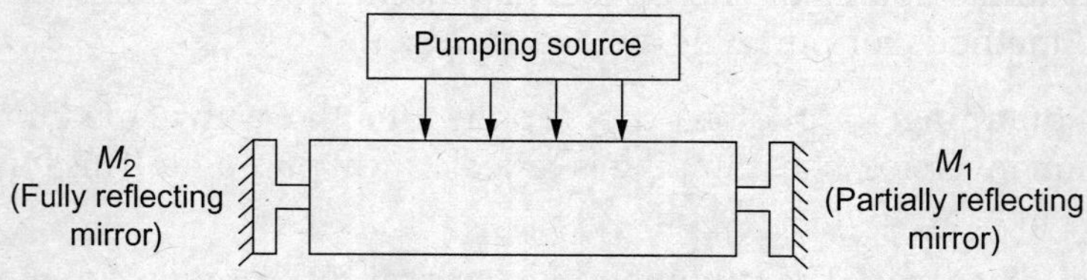

FIGURE 14.5 Optical resonator cavity

14.7 ▶ Characteristics of a Laser Beam

Laser light differs from conventional light sources in a number of ways. The main characteristics of a laser beam are as follows:

(i) Monochromaticity
(ii) Coherence
(iii) Directionality
(iv) Intensity

14.7.1 Monochromaticity

It is the ability of emitted radiation or photon to retain its frequency or wavelength. Monochromaticity means that the laser light contains only one specific wavelength which will be equal to the energy difference between the two levels. The emission from laser is sharp and narrow. It means that the energy levels are also discrete and sharp. But in actual practice, the frequency of the emitted radiation lies between ν and $\nu + d\nu$. This results in broadening of the spectral line, which is known as spectral broadening. This is due to the following major processes:

- *Doppler broadening*: The atoms at the time of emission are constantly moving with different velocities and directions, which results in the broadening of the spectral line.
- *Collision broadening*: A change in frequency results if an atom undergoes collision at the time of emission, which broadens the spectral line.
- *Natural broadening*: A photon emitted by an electron wave results in the damping of the amplitude of the wave, which shortens the wave and broadens the spectral line.

The frequency spread $\Delta\nu$ of the emitted line is related to its wavelength spread $\Delta\lambda$ as

$$\Delta\lambda = -\left[\frac{c}{\nu^2}\right]\Delta\nu \tag{14.14}$$

The laser has a low frequency spread of $\Delta\nu = 0.001$ nm, compared to the ordinary light of $\Delta\nu$ as high as -300 nm.

No light source, ordinary or laser emits radiation at single frequency, i.e. no light source is absolutely, monochromatic. So the term monochromatic is used in relative sense only. There may be a light source emitting a line of bandwidth $\Delta\bar{\nu}$ at an average frequency ν that gives the monochromaticity of the radiation.

To determine the relative degree of monochromaticity of different sources of light, we can study fringe visibility defined by V as

$$\boxed{V = \frac{I_{\text{max}} - I_{\text{min}}}{I_{\text{max}} + I_{\text{min}}}} \tag{14.15}$$

Also,

$$\boxed{\Delta\bar{\nu} = \frac{1}{\tau_c}} \tag{14.16}$$

This relation shows that longer the coherence time, smaller the value of monochromaticity.

Note

Longer the coherence time, the smaller is the frequency width of the wave.

14.7.2 Coherence

Coherence is a property of a wave being in phase with itself and also with another wave over a period of time and space or distance. In other words, coherence is the predictability of the amplitude and phase at any point on the wave knowing the amplitude and phase at any other point either on the same wave or on a neighbouring wave. Lasers have the following two types of coherences.

- *Temporal coherence*: The condition in which the light waves are monochromatic, i.e. each cycle of the wave takes exactly the same time to pass a given point in space.
- *Spatial coherence*: The condition in which the light waves travelling through space is not only of the same frequency but also in phase with each other.

In reality, both temporal and spatial coherences are due to the optical cavity. The cavity produces temporal coherence by excluding any light on wavelength which is not an even multiple of the distance between mirrors. And it produces spatial coherence (by acting as a spatial filter with gain) by reflecting the light with a spherical wavefront of a particular radius.

The relation between coherence lengths of time is given by

$$L = C\tau_C \tag{14.17}$$

14.7.3 Directionality

Laser radiations are highly directional unlike ordinary light that spreads radiations in all directions. The output beam is circularly symmetric, in that it is more intense at the centre and falls off as

$$\exp\left[-\left(\frac{r}{\omega}\right)^2\right] \tag{14.18}$$

At $r = 0$, the intensity is maximum and decreases as the radius increases. This is called the Gaussian beam. The outer edge of the beam is the radial distance from the axis at which the intensity of the beam has dropped to $1/e$ of its value at the axis (Fig. 14.6).

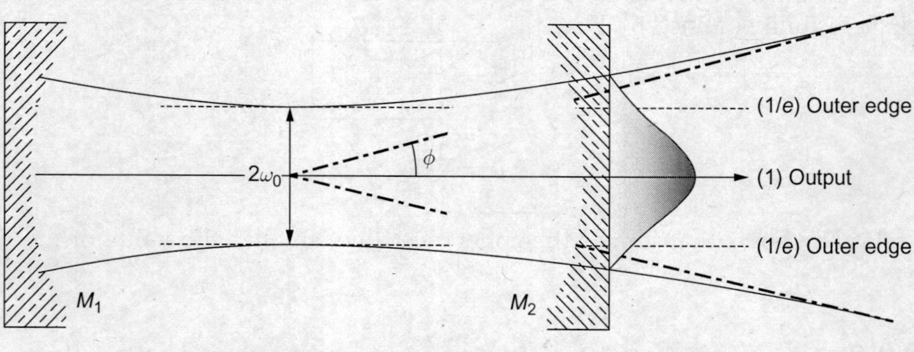

FIGURE 14.6 The Gaussian beam and the output

Laser radiation can travel as a light parallel beam up to a distance of d^2/λ (d is the diameter of the aperture through which the laser radiation of wavelength λ comes out) and spread angularly after travelling that distance. The full-angle beam divergence (i.e., twice the angle of the outer edge makes with the axis of the beam) expresses the directionality of the laser radiation as

$$\boxed{\phi = \frac{1.27\lambda}{2\omega_0}} \tag{14.19}$$

The angular spread of the laser beam ($\Delta\theta$ or $\phi = \lambda/d$) is < 0.01 mrad, which means that the beam spread is only 1 mm/m. An ordinary light, on the contrary (sodium vapour, search light, etc.), spreads at 1 m/m. Because of this exceptional directionality, it is even possible to focus a laser beam on the surface of Moon (with a beam spread of only a few kms). The angle of divergence (deg) (Fig. 14.7) of a laser beam is expressed as

$$\phi = \frac{\text{arc length}}{\text{radius}} \tag{14.20}$$

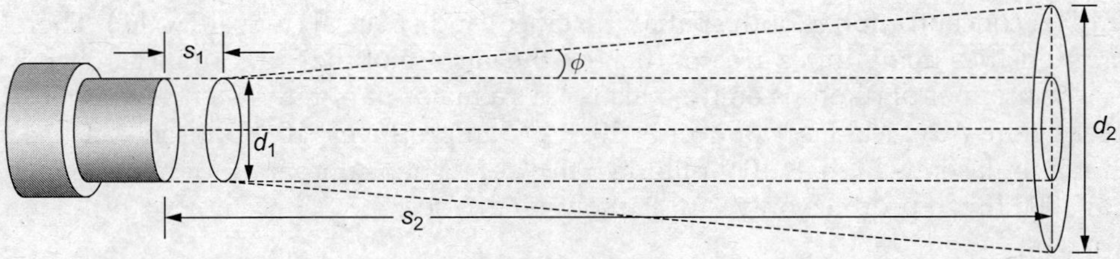

FIGURE 14.7 Calculation of the divergence of a laser beam

i.e.,

$$\phi = \frac{(d_2 - d_1)}{2(s_2 - s_1)} \tag{14.21}$$

where d_1 and d_2 are the diameters of the spots measured at distances s_1 and s_2 from the laser aperture.

The beam often has a small divergence (highly collimated), but a perfectly collimated beam cannot be created due to the effect of diffraction. Nonetheless, a laser beam will spread much less than a beam of incoherent light. Thus, a beam generated by a small laboratory laser like a helium–neon (HeNe) laser spreads to approximately 1 mile (1.6 km) in diameter if shone from the Earth's surface to the Moon. By comparison, the output of a typical semiconductor laser, due to its small diameter, diverges almost immediately on exiting the aperture, at an angle that may be as high as 50°. However, such a divergent beam can be transformed into a collimated beam by means of a lens.

EXAMPLE 14.3

Calculate the divergence of a laser beam at distances of 1 and 10 m from the laser spot and whose diameters are 2 and 4 mm, respectively.

Solution:

$$\text{Divergence } \phi = \frac{d_2 - d_1}{2(s_2 - s_1)}$$

where

d_1 and d_2 are diameters of the laser beam

s_1 and s_2 are the distances of spots from the laser output

$$\therefore \qquad \phi = \frac{(4-2)\times 10^{-3}}{2(10-1)} = \frac{2\times 10^{-3}}{18}$$

0.11×10^{-3} rad = 0.11 mrad

14.7.4 Intensity

As the laser beam is narrow, almost all of its energy is highly concentrated into a small region. The concentration is both spatial (distance) and spectral (wavelengths). Therefore, power intensities many times greater than on the surface of the Sun can be achieved with lasers. The number of photons emitted also plays a major part in the increased intensity of the laser beam. A typical laser output of 10^{-3} W can produce $\sim 10^{24}$ photons/m^2/sec and a high-power laser as high as 10^{34} photons/m^2/sec. When compared to this, even a black body at 1000 K can produce only $\sim 10^{16}$ photons/m^2/sec.

Note

The intensity of 100 W laser is 10,000 times larger than that of 100 W bulb.

14.8 ▶ Types of Lasers

Depending upon the operational principles, lasers can be classified as solid state or gas lasers. Some commonly used lasers are discussed here.

14.8.1 Ruby Laser

The first type of laser constructed, was a ruby laser (Fig. 14.8), in 1960 by T.H. Maiman. The ruby mineral (Corundum) is aluminium oxide with a small amount (about 0.05%) of chromium, which gives it characteristic pink or red colour by absorbing green and blue light. It is a three-level laser.

It consists of three parts:

(i) The working substance in the form of a rod of ruby crystal.

(ii) The resonator cavity made up of fully reflecting mirror on the left and a partially reflecting mirror on the right.

(iii) The pumping system consists of a helical xenon discharge tube.

Construction

Ruby laser consists of a single crystal in the form of a cylindrical rod, which is 8 cm long and 5 mm in diameter. Ruby crystal is enclosed in a gas tube, which is coated with an electronic flash lamp L filled with xenon. The lamp L produces flashes of 550 mm light. The system is cooled by using a coolant around the ruby rod.

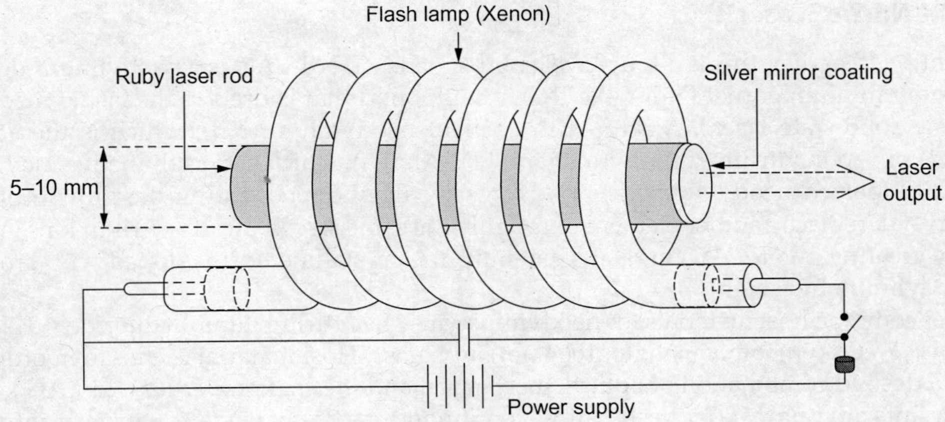

FIGURE 14.8 Ruby laser

Working

When the flash lamp *L* is switched on, intense light of wavelength 550 mm falls on the ruby, which excites the ruby's atoms.

A large number of chromium ions in the ground state are excited to the excited state $4F_1$.

In case of ruby laser, the energy bands are 0.1 μm wide. The excited chromium ions lose some of the energy and make radiation less transitions of the pairs of levels associated with 2E metastable state within 10^{-8} seconds. Population inversion is created in between ground state and metastable state. We get lasing transition in between these two states, which produces laser beam (Fig. 14.9).

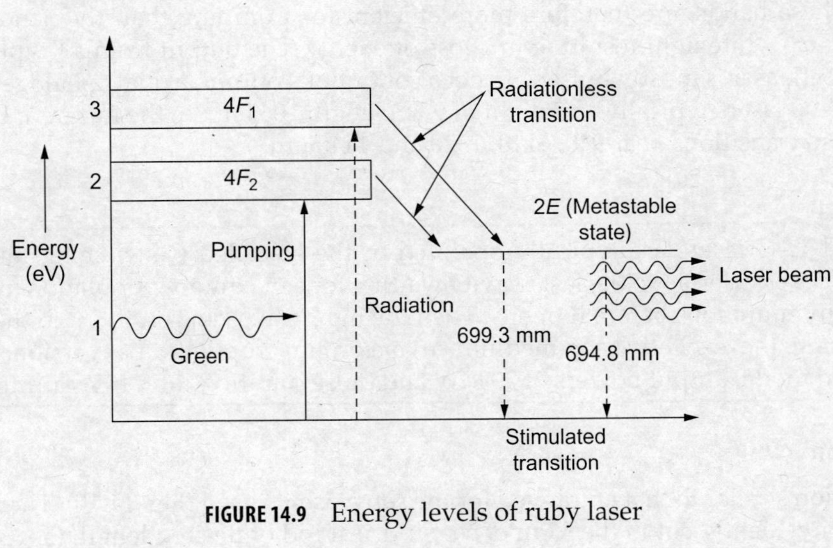

FIGURE 14.9 Energy levels of ruby laser

The light travels along the axis of the ruby laser, which gets reflected back and forth by mirrors and amplifies laser light. The laser beam comes out in pulses.

14.8.2 Nd-YAG Laser

As mentioned earlier, this is a four-level solid-state laser. Nd-YAG is a neodymium-doped yttrium-aluminium garnet (Nd-$Y_3Al_5O_{12}$), a compound that is used as the lasing medium for some solid-state lasers. A rare earth element neodymium (Nd), which is the active centre, is mixed in yttrium-aluminium garnet ($Y_3Al_5O_{12}$), which acts as the active medium. The triply ionized neodymium ions (Nd^{3+}), selectively doped during the growth of the YAG crystal, replace another element of roughly the same size, typically yttrium ions (Y^{3+}), thereby creating an Nd-YAG crystal. Generally, the crystalline host is doped with around 1% neodymium by weight.

Other common host materials for neodymium are YLF (yttrium lithium fluoride, 1047 and 1053 nm), YVO_4 (yttrium vanadate, 1064 nm) and glass. Host material is chosen in order to obtain a desired combination of optical, mechanical and thermal properties. Nd-YAG lasers and variants are pumped by flash lamps, continuous gas discharge lamps or near infrared laser diodes (DPSS lasers).

Note

The amount of the neodymium dopant in the material varies between 0.5 and 1.4 molar%. Higher dopant concentration is used for pulsed lasers, and lower concentration is suitable for continuous wave lasers. Nd-YAG is pinkish purple in colour; lightly white-doped rods, are less intensely coloured, almost white than heavier doped ones.

Pumping

As this is a solid-state laser, pumping could be done by optical means. Nd-YAG absorbs mostly in the bands between 730–760 nm and 790–820 nm. Krypton flash lamps, with high output at these bands, are therefore more efficient for pumping than the xenon lamps, which produce white light resulting in a loss of energy. The output from a krypton-flash-lamp-pumped laser is mostly pulsed. To get a continuous output, a quartz-halogen lamp is used. Nd-YAG lasers typically emit light of wavelength 1064 nm in the infrared. However, there are also transitions near 940, 1120, 1320 and 1440 nm.

Note

Pulsed Nd-YAG lasers are typically operated in the so-called Q-switching mode. An optical switch is inserted in the laser cavity waiting for a maximum population inversion in the neodymium ions before it opens. Then the light wave can run through the cavity, depopulating the excited laser medium at maximum population inversion. In this Q-switched mode, output powers of 20 MW and pulse durations of < 10 ns are achieved.

Construction

A construction similar to that employed for the ruby laser is used (Fig. 14.10). The Nd-YAG crystal is conveniently cut in the form of a cylindrical rod of desired length (5–10 cm) and diameter (6–9 mm).

Unlike a ruby laser where the xenon flash lamp was coiled around the ruby rod, the Nd-YAG rod is placed inside at one of the foci of an elliptical reflector cavity. At the other

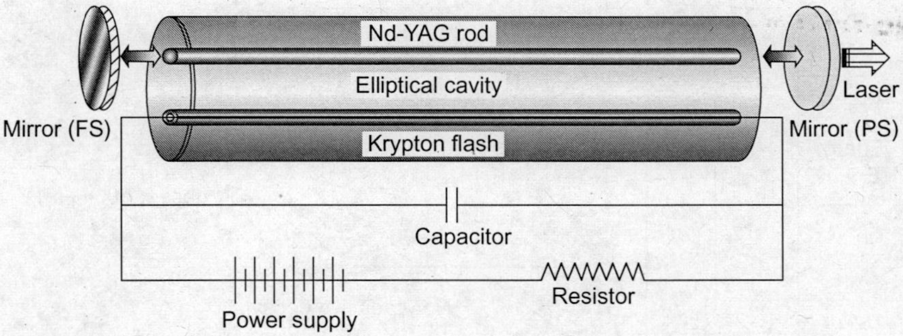

FIGURE 14.10 Experimental set-up of an Nd-YAG laser system

focus, a krypton flash tube, which is connected to the power supply via a capacitor connected in parallel and a resistor connected in series, is placed. The reason for this arrangement is to maximize the pumping power (Fig. 14.11). Almost all the power emitted by the flash lamp kept at one focus reaches the rod kept at the other focus. In this way, maximum pumping takes place, and hence a large population inversion is achieved. This results in an intense laser beam.

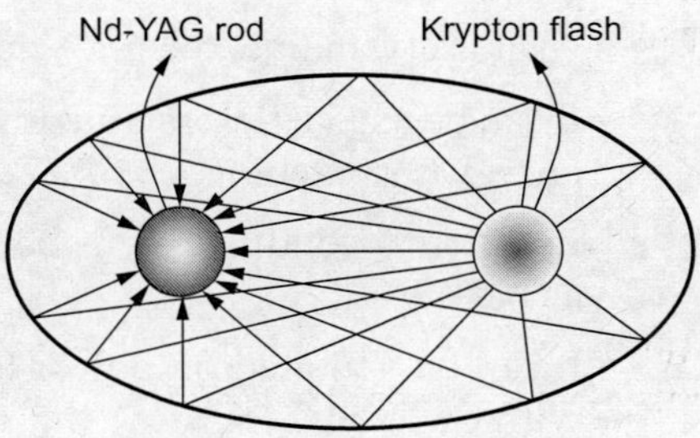

FIGURE 14.11 Cross-sectional view of the experimental arrangement of the Nd-YAG laser system

Working

The laser action is achieved by means of optical pumping utilizing a flash lamp (either a xenon flash lamp at moderate pressure or a krypton flash lamp at high pressure). Only a small fraction of the energy given out by the flash lamp is used for excitation and the rest merely heats up the apparatus. This requires a separate cooling mechanism to cool the laser apparatus.

The neodymium ions absorb radiations at around 730 nm (E_1) and 800 nm (E_2) and go to the respective excited states (Fig. 14.12). From these excited absorption levels, the atoms decay by means of a rapid non-radiative transition to the metastable level (E_4), where they populate and achieve population inversion. Once this condition is achieved, photons are emitted following stimulated emission. The emitted photons are allowed to pass back and forth upwards for a million times through a set of optical resonators where their number builds up and the amplitude increases. Finally, a laser output at 1064 nm is obtained corresponding to the transition between the levels E_4 and E_1.

Advantages

The advantages of Nd-YAG laser are as follows:

- It has a high output and repetition rate.
- It is much easier to achieve population inversion.

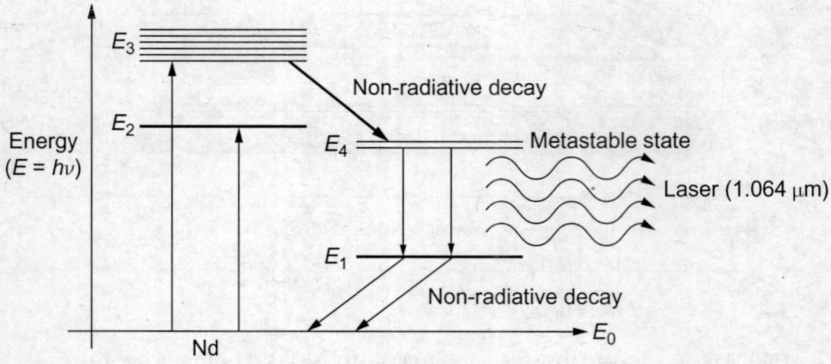

FIGURE 14.12 Energy level diagram of Nd-YAG laser

- As it is crystalline, the line width is small and therefore it has lower thresholds.
- It can be used in lasers utilizing frequency doubling and frequency tripling, and high-energy Q-switching.
- Its thermal conductivity is better and its fluorescence lifetime is about twice as long as Nd-YVO_4.
- It can be operated on power levels of up to kilowatts. It can be directly Q-switched with Cr^{4+}-YAG.
- Nd-YAG lasers at 1064 nm and its best absorption band for pumping is 1 nm wide and located at 807.5 nm.

Disadvantages

- Its electronic energy level structure is complicated.

Applications

Nd-YAG laser is the most widely used active laser medium in solid-state lasers.

- It is used for transmitting signals over longer distances as long-haul communication systems and in remote sensing.
- In the medical field, it is used in endoscopic applications and as a means of correcting posterior capsular opacification.
- It is used in manufacturing as a means of engraving, etching or marking a variety of metals and plastics.
- Cerium-doped YAG (YAG-Ce) is used as a phosphor in applications ranging from cathode ray tubes to white LEDs.
- Nd-YAG laser is widely used for cutting, welding and marking of metals and other materials, and also in spectroscopy and for pumping dye lasers.

EXAMPLE 14.4

An Nd-YAG laser emits light at a wavelength of 1.063×10^{-6} m. If the output power is 20 W, then how many photons are emitted in 10 min when the laser is in operation?

Solution: Given

$$\text{Wavelength } \lambda = 1.063 \times 10^{-6} \text{ m}$$

$$\text{Output power } P = 20 \text{ W}$$

Find the ratio of the energy of the laser output to the energy of the emitted photon.

$$\text{Energy of the emitted light (photon)} = E = h\nu = \frac{hc}{\lambda}$$

$$E = \frac{6.626 \times 10^{-34} \times 3 \times 10^{8}}{1.063 \times 10^{-6}}$$

$$= \frac{19.878}{1.063} \times 10^{-24}$$

$$= 18.699 \times 10^{-24} \text{ J}$$

$$\text{Energy of the laser output} = 20 \text{ W} = 20 \text{ J/s}$$

$$= 20 \times 10 \times 60 \text{ J/s (as the output is for 10 min)}$$

$$\text{Hence, the number of photons emitted} = \frac{20 \times 10 \times 60}{18.699 \times 10^{-24}}$$

$$= \frac{12{,}000}{18.699 \times 10^{-24}}$$

$$= \mathbf{641.7 \times 10^{24}} \textbf{ photons in 10 min}$$

14.8.3 He-Ne Gas Laser

Helium-Neon gas laser is a continuous four-level gas laser. It consists of a long, narrow cylindrical tube made up of fused quartz. The diameter of the tube will vary from 2 to 8 mm and length will vary from 10 to 100 cm. Flat quartz plates are sealed at the ends of the tube; the plates are sealed at Brewster angle with the axis of the tube to obtain polarized laser light as shown in Fig. 14.13. So, the plates are called Brewster windows. The tube is filled with helium and neon gases in the ratio of 10:1. The partial pressure of helium gas is 1 mm of Hg and neon gas is 0.1 mm of Hg, so that the pressure of the mixture of gases inside the tube is nearly 1 mm of Hg.

Laser action is due to the neon atoms. Helium is used for selective pumping of neon atoms to upper energy levels. Two electrodes are fixed near the ends of the tube to pass electric discharge through the gas. Two optically plane mirrors are fixed at the two ends of the tube normal to its axis. One of the mirrors is fully silvered so that nearly 100% reflection takes place and the other is partially silvered so that 1% of the light incident on it will be transmitted. Optical resonance column is formed between these mirrors.

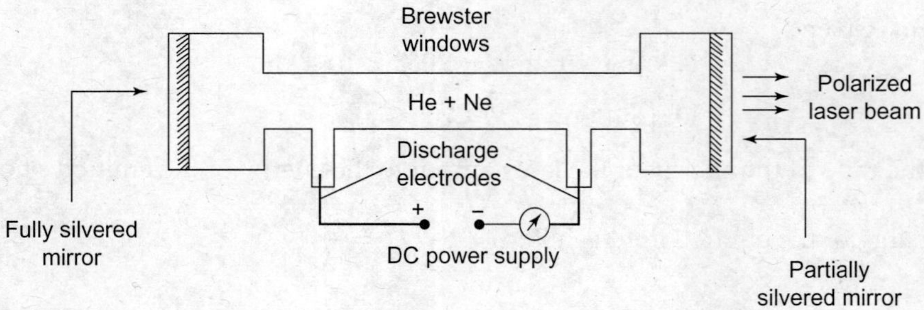

FIGURE 14.13 Helium-Neon gas laser

Working

When a voltage of about 1000 V is applied between the electrodes, then electrical discharge takes place through the gas in the tube. The free electrons accelerate towards the positive electrode. In their journey, some of these electrons collide with the majority helium gaseous atoms in the tube. When a fast-moving electron collides with a ground-state He atoms, then the helium atoms are pumped to, two metastable energy levels 2^1s and 2^3s of helium as shown in Fig. 14.14. In the metastable state, the atoms remain relatively long time. So, more number of helium atoms will be present in metastable state than in ground state, which leads to an increase of population in each of these metastable states.

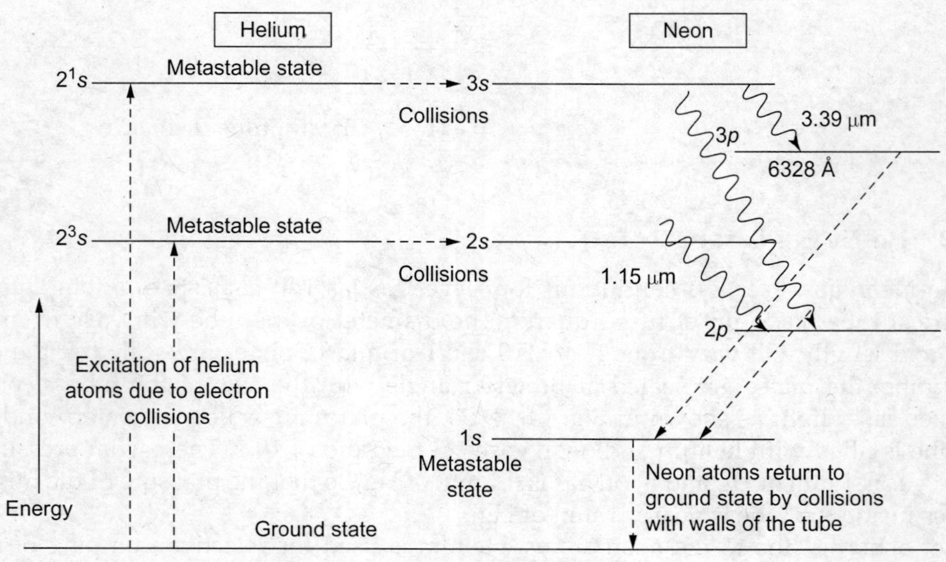

FIGURE 14.14 He-Ne energy level diagram

Inside the tube, the helium atoms present in metastable states may make collisions with the neon atoms present in the ground state and excite them to 2s and 3s levels. During collisions, the metastable helium atoms transfer their energy to ground state neon atoms and the helium atoms come back to the ground state. During collision, resonance transfer of energy from He to Ne atoms takes place because the 2^1s and 2^3s energy levels of helium atoms are very close with 3s and 2s energy levels of neon. Thus, the population inversion in neon atoms takes place. The excited neon atoms transit to ground state in three different ways leading to three lasers of different wavelengths. They are: (a) transition from 3s level to 3p level gives rise to radiation of wavelength 3.39 μm, this lies in the infrared region, (b) transition from 3s level to 2p level gives rise to visible radiation of wavelength 6328 Å, this lies in red region and (c) the transition from 2s level to 2p level gives rise to a wavelength of 1.15 μm, this again lies in the infrared region. The atoms in 3p and 2p level undergo spontaneous transitions to 1s level, this is a metastable state (3s and 2s levels are not metastable states). The photons emitted by the atoms coming down from 3p or 2p level to 1s level are likely to excite the 1s atoms back to 3p or 2p levels. This affects the population inversion in 3s and 2s levels. The atoms in 1s level return back to the ground level mainly by collisions with the walls of the discharge tube. This effect makes the gain of He-Ne laser to be inversely proportional to the diameter of the discharge tube, so the discharge tubes are made only to a few millimetres of diameter. The mirrors placed outside the tube produces optical pumping in the resonance column inside the tube, which enhances stimulated emissions. Red laser light comes out of the partially silvered mirror.

EXAMPLE 14.5

If the half-width of the He–Ne laser operating at a wavelength of 6328 Å is 1500 MHz, what must be the length of the laser cavity to ensure that only one longitudinal mode oscillates?

Solution: The length of cavity is given by

$$L = \frac{mc}{2\Delta\nu} = 0.1 \text{ m}$$

14.8.4 Semiconductor Laser

Semiconductor or diode laser was invented in 1961. The transitions in semiconductor lasers are associated with the energy states in valence and conduction bands.

The advantages of semiconductor lasers over other gaseous lasers are as follows:

(i) These are very small in size.

(ii) In gaseous lasers, only the 1% working substance is active while in diode lasers whole of the active medium is active.

(iii) The efficiency is very high.

(iv) They require a very low operating power.

Construction

Figure 14.15 shows the basic structure of a GaAs P-N junction diode. P-region is heavily doped which is produced by diffusing Zn into N-type gaAs (10^{24} atoms per cubic metre). One pair of faces perpendicular to the junction plane is cleared and polished so that they act as reflecting mirrors. The other two faces are roughened to eliminate lasing in these directions.

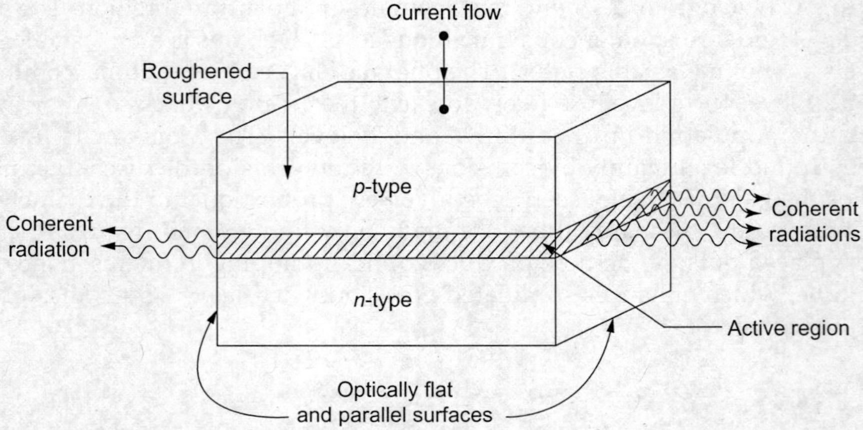

FIGURE 14.15 Semiconductor laser

Working

In n-type semiconductor, electrons are donated by impurities to fill the conduction band upto fermi level E_f, which falls below conduction band. In p-type semiconductor, holes are added by the acceptors down to energy level E_f, which falls above valence band. In the absence of external biased voltage, a potential barrier is built up. But, when the external voltage is applied (forward biasing), the electrons flow to p-side and holes to n-side. As the applied voltage is increased, diode current is gradually increased; a point is reached where significant population inversion exists.

The stimulated emission occurs at frequencies corresponding to normal modes of the cavity whose wavelengths are as follows:

$$\lambda = \frac{2L}{n};$$

where L = length of the cavity and n = 1, 2, 3.

Above threshold current, the diode emits stimulated radiation with a narrow spectral emission (~ few Å) (Fig. 14.16).

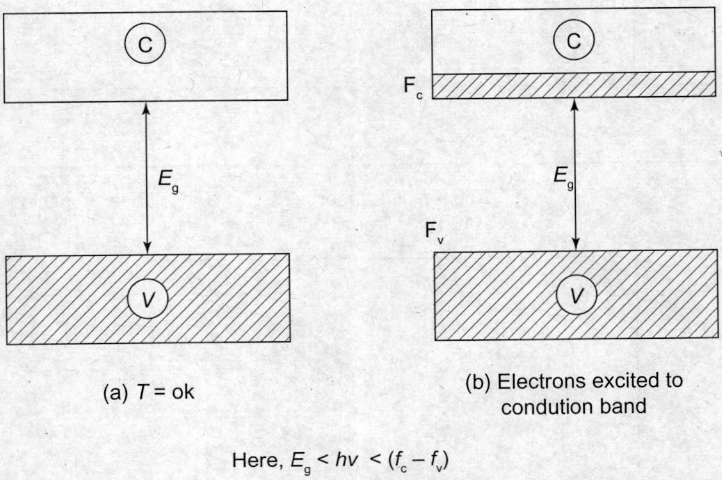

FIGURE 14.16 Energy level diagram of semiconductor laser

14.8.5 CO_2 Laser: Molecular Laser

The molecules have energy levels corresponding to rotational and vibrational motions. The molecular laser has three vibrational modes as shown in Fig. 14.17.

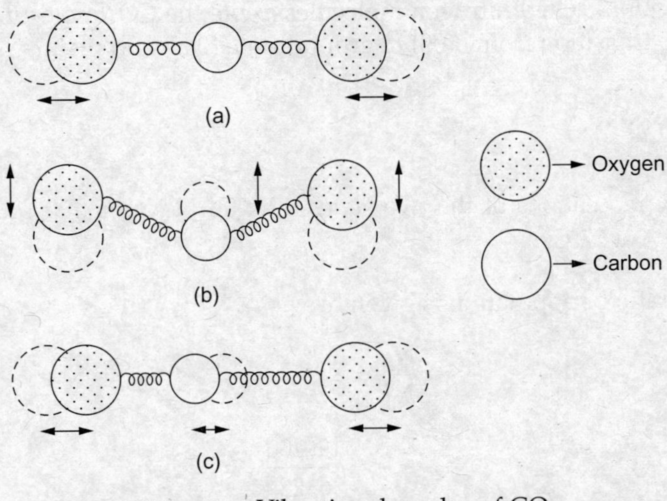

FIGURE 14.17 Vibrational modes of CO_2

(a) Symmetric stretch
(b) Bending and
(c) Asymmetric stretch

The energy state of molecule is represented by some number (001) say, the molecule is in pure asymmetric stretch mode, no energy is associated with symmetric stretch represented

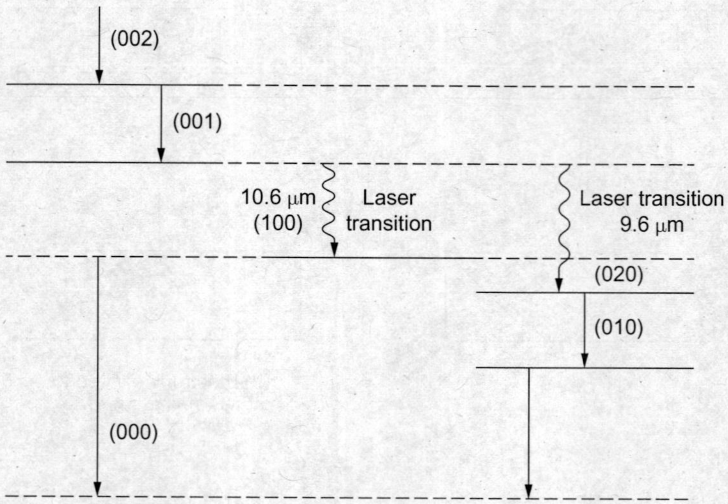

FIGURE 14.18 Energy level diagram of CO_2 laser

by (000) (Fig. 14.18). Due to rotational states, vibrational states split into fine sublevels, but these splittings are smaller as compared to the separation of electronic states of the molecule. The energy state (001) is a metastable state and population inversion is created between (001) and (100). To increase the efficiency of the laser, nitrogen and helium are added with CO_2 (CO_2: N_2: He as 1:4:5). Excited nitrogen molecules excite the CO_2 molecules. Helium helps in speeding up the transition from (100) to ground state (000) and maintains the population inversion.

EXAMPLE 14.6

Find the ratio of populations of the two states in a He–Ne laser that produces light of wavelength 6328 Å at 27°C.

Solution: The ratio of population is given by

$$\frac{N_2}{N_1} = e^{-(E_2-E_1)/KT}$$

$$\Rightarrow \quad E_2 - E_1 = \frac{12,400}{6328}\text{ eV}$$
$$= 1.96 \text{ eV}$$

$$\Rightarrow \quad \frac{N_2}{N_1} = \exp\left[\frac{-19.6 \text{ eV}}{(8.61\times10^{-5}\text{ eV})(300 \text{ K})}\right]$$

$$\Rightarrow \qquad \frac{N_2}{N_1} = e^{-75.88}$$

$$\Rightarrow \qquad \frac{N_2}{N_1} = \mathbf{1.1 \times 10^{-33}}$$

EXAMPLE 14.7

A 10 mW laser has a beam diameter of 1.6 mm. What is the intensity of the light assuming that it is uniform across the beam?

Solution:

$$I = \frac{P}{A} = \frac{10 \times 10^{-3}\ \mathrm{W}}{3.14 \times \left(0.8 \times 10^{-3}\ \mathrm{m}\right)^2}$$

$$= \mathbf{4.97\ kW / m^2}$$

14.9 ▶ Applications of Lasers

Lasers find applications in various fields of science and technology. They are described as follows:

1. **In communications**:
   (a) Lasers are used in optical fibre communications. In optical fibre communications, lasers are used as light source to transmit audio, video signals and data to long distances without attenuation and distortion.
   (b) The narrow angular spread of laser beam recommends that laser beam can be used for the communication between earth and moon or to other satellites.
   (c) As laser radiation is not absorbed by water, so laser beam can be used in under water (inside sea) communication networks.
2. **Industrial applications:** Lasers are used in metal cutting, welding, surface treatment and hole drilling.
   (a) When a laser beam is focussed on a very small area, then laser light energy is converted into heat energy, so the material may be heated, melted and evaporated. Using these techniques, holes can be drilled in steel, ceramics, diamond and alloys. Using lasers, controlled orifices and aerosol nozzles are drilled with controlled precision. Holes of micron order can be easily drilled using lasers. These techniques are used in cutting materials such as metal sheets and clothes. In mass production of stitched clothes, the cutting in the desired

dimension is done by stock filing a large number of cloth materials and cutting them all at once by exposing a laser beam.

Using lasers, cutting can be obtained to any desired shape and the cut surface is very smooth.

(b) Welding has been carried by using laser beam. A laser beam is exposed to the place where welding has to be done, at that place the material melts due to the heat produced by the beam and on cooling the material makes a stronger joint.

(c) Dissimilar metals can be welded and microwelding is done with great ease.

(d) Laser beam is used in selective heat treatment for tempering the desired parts in automobile industry.

(e) Lasers are widely used in electronic industry in trimming the components of ICs.

3. **Medical applications:**

(a) Lasers are used in eye surgery, especially in detached retina. Under certain abnormal conditions, the retina may get detached from the choroid; this results in blindness at the detached part of the retina. The retina can be attached to the choroid by heating it over a number of spots almost of the size of tissues. The heating can be achieved by focussing a laser beam of predetermined intensity on to the retina. The laser beam reaches the right spot where the welding of the retina to the choroid occurs. The flash of laser beam lost in short time (1 millisecond). The patient does not feel any pain, so anaesthesia is not required.

(b) Lasers are used for treatments such as plastic surgery, skin injuries and to remove moles and tumours developed in skin tissue.

(c) Lasers are used in stomatology—the study of mouth and its disease. Where a laser beam is used for selective destroying, the part of the tooth affected by caries. Mouth ulcers can be cured by exposing it to a laser beam.

(d) Laser radiation is sent through optical fibre to open the blocked artery region; here the laser rays burn the excess growth in the blocked region and regulates blood flow without any requirement for bypass surgery.

(e) Lasers are used to destroy kidney stones and gall stones. The laser pulses are sent through optical fibres to the stoned region. The laser pulses break the stones into small pieces.

(f) Lasers are used in cancer diagnosis and therapy.

(g) Lasers are used in blood loss less surgery. During operation, the ctit blood veins are fused at their tips by exposing to infrared laser light, so that there is no blood loss.

(h) Lasers are used to control haemorrhage.

(i) Using organ and CO_2 lasers, liver and lung treatment can be carried out.

(j) Lasers are used in endoscopes to detect hidden parts.

(k) Laser Doppler velocimetry is used to measure blood velocity in the blood vessels.

4. **Military applications:** The various military applications are:

(a) Death rays: By focussing high-energetic laser beam for few seconds to aircraft, missile, etc. can be destroyed. So, these rays are called death rays or war weapons.

(b) Laser gun: The vital part of enemy body can be evaporated at short range by focussing a highly convergent laser beam from a laser gun.

(c) LIDAR (Light Detecting And Ranging): In place of RADAR, we can use LIDAR to estimate the size and shape of distant objects or war weapons. The difference between RADAR and LIDAR is that, in case of RADAR, radio waves are used whereas in case of LIDAR light is used.

5. **In computers:** By using lasers, a large amount of information or data can be stored in CD-ROM or their storage capacity can be increased. Lasers are also used in computer printers.

6. **In thermonuclear fusion:** To initiate nuclear fusion reaction, very high temperature and pressure are required. This can be created by concentrating large amount of laser energy in a small volume. In the fusion of deuterium and tritium, irradiation with a high-energy laser beam pulse of 1 nanosecond duration develops a temperature of 10^{17} °C; this temperature is sufficient to initiate nuclear fusion reaction.

7. **In scientific research:**

(a) Laser beam can initiate or fasten chemical reactions. Laser beam helps us to study the nature of chemical bonds. An intense laser beam can break molecules.

(b) Lasers are used in counting of atoms in isotope separation and to separate isotopes of uranium.

(c) Lasers are used to estimate the size and shape of biological cells such as erythrocytes.

(d) Lasers can be used in air pollution, to find the size of dust particles.

(e) Lasers are used in holography for recording and reconstruction of a hologram. Using holograms, the three-dimensional images of objects can be recorded.

(f) To measure the constantly changing distance between moon and earth by astronomers. This gives the day-to-day changes in the rotation of earth about its axis and slight wobbles.

(g) In plastic industries, polymers are obtained by irradiating monomers. During laser irradiation, the monomers are united to form polymers.

(h) By using lasers, the stimulated Raman spectrum is obtained for small biological samples.

(i) Lasers are used to develop hidden finger prints and to clean delicate pieces of art.

EXAMPLE 14.8

For ordinary source, the coherence time is $\tau_c = 10^{-10}$ sec. Obtain the degree of non-monochromaticity for $\lambda_0 = 5400$ Å.

Solution: Given

$$\tau_c = 10^{-10} \text{ sec}$$

$$\Delta\nu = \frac{1}{\tau c} = \frac{1}{10^{-10}} = 10^{10} \text{ Hz}$$

for $\lambda_0 = 5400$ Å,

$$\nu_0 = \frac{C}{\lambda_0} = \frac{1}{18} \times 10^{16}$$

Degree of non-monochromaticity is

$$\frac{\Delta\nu}{\nu_0} = \frac{18 \times 10^{10}}{10^{16}}$$

$$= 18 \times 10^{-6}$$

$$= \mathbf{0.000018}$$

EXAMPLE 14.9

Calculate the power per unit area delivered by a laser pulse of energy 4.0×10^{-3} J and the pulse length in time as 10^{-9} sec. When the pulse is focussed on target to a very small spot of radius 1.5×10^{-5} m?

Solution: Given

$$P = 4.0 \times 10^{-3} \text{ J}$$

$$r = 1.5 \times 10^{-5} \text{ m}$$

Power delivered per unit area is

$$I = \frac{P}{A}$$

$$= \frac{4.0 \times 10^{6} \text{ W}}{7.065 \times 10^{-10} \text{m}^2}$$

$$= \mathbf{5.7 \times 10^{15} \text{ W/m}^2}$$

EXAMPLE 14.10

A 0.1 W laser beam with an aperture of 5.0 mm emits a light of wavelength 6943 Å. Calculate the areal spread and intensity of the image when the beam is focussed with a lens having focal length 100 mm.

Solution: Given radius of aperture = diameter/2

or $$r = 2.5 \times 10^{-3}\text{ m}$$

$$\lambda = 6.943 \times 10^{-7}\text{ m}$$

$$f = 0.1\text{ m} = D$$

$$P = 0.1\text{ W}$$

$$\text{Angular spread }(\theta) = \frac{0.637\lambda}{r}$$

$$\Rightarrow \quad \theta = 1.769 \times 10^{-4}\text{ radians}$$

$$\text{Areal spread} = (\theta \cdot D)^2 = (\theta \cdot f)^2$$

$$= 3.129 \times 10^{-10}\text{ m}^2$$

and the intensity is given by

$$I = \frac{P}{A} = \frac{0.1\text{ W}}{3.129 \times 10^{-10}\text{ m}^2}$$

$$\Rightarrow \quad \mathbf{I = 3.196 \times 10^{8}\ W/m^2}$$

Recap Zone

Points to Remember

- MASER is defined as Microwave Amplification by Stimulated Emission of Radiation.
- LASER is defined as Light Amplification by Stimulated Emission of Radiation.
- Laser radiation is monochromatic (contains only one specific wavelength), coherent ("organized" in that each photon moves in step with the others), directional (emitted beam of light is intense and concentrated) and intense.
- Coherence is a property of a wave being in phase with itself and also with another wave over a period of time and space or distance. Coherency is of two types: temporal coherence and spatial coherence.
- Pumping is the process of exciting an atom to an excited state.
- Optical resonator is the entire assembly of a pair of end-reflecting mirrors, commonly used to reflect the photons and undergo multiple reflections and travel through the active medium.
- The length (L) between the mirrors is known as the resonating length and the volume in between the mirrors is known as the resonating cavity.

- Saturation is the condition when the resonator balances the loss of the radiation with the gain within the cavity.
- Rate of absorption = rate of emission (spontaneous + stimulated).
- The transitions between the energy states take place via three distinct processes, namely, stimulated absorption, spontaneous emission and stimulated emission.
- Nd-YAG is a neodymium-doped yttrium-aluminium garnet (Nd-$Y_3Al_5O_{12}$).
- The CO_2 molecule undergoes three types or modes of vibration, namely, symmetric mode or stretching, asymmetric mode or stretching, and bending mode.
- On the nature of the band gap, the semiconductors are classified as indirect band gap semiconductor and direct band gap semiconductor.
- Depending on nature of construction and working, the diode lasers are divided into two types, namely, homojunction diode laser and heterojunction diode laser.

Key Terms with Definitions

- **LASER** – An acronym of light amplification by stimulated emission of radiation.
- **Spontaneous emission** – Emission of radiation without any external causation.
- **Stimulated emission** – Emission of radiation due to interaction of external electromagnetic field with the excited atom.
- **Active species** – The atoms (ions or molecules) involved in the process of stimulated emission.
- **Pumping** – The process by which the upper level of the active species is populated more rapidly than its lower level.
- **Directionality** – The laser emit light along a particular direction coinciding with the axis of the resonator.
- **Fringe visibility** – Contrast of interference fringes.
- **Coherence time** – The wave has a definite phase relationship only for a limited time τ_c known as coherence time.
- **Doping** – The process of adding impurities to intrinsic semiconductors to alter their properties.
- **Line broadening** – The radiations are not strictly monochromatic but occupies a finite frequency bandwidth, which is called line broadening.
- **Life time** – The time during which the number of atoms in an excited state decreases by a factor of $1/e$ of its initial value.
- **Resonator cavity** – The laser medium is placed between a pair of mirrors forming the resonator cavity.

IMPORTANT FORMULAE AND EQUATIONS

Equation Number	Equation	Remarks
(14.2)	$(P_{21})_{\text{spon.}} = A_{21}$	The probability of spontaneous emission from state (2) to state (1)
(14.3)	$(P_{21})_{\text{sti}} = B_{21}\, u(\nu)$	The probability of stimulated emission
(14.12)	$\dfrac{A_{21}}{B_{21}} \propto \nu^3$	The ratio of spontaneous to stimulated emission
(14.15)	$V = \dfrac{I_{\max} - I_{\min}}{I_{\max} + I_{\min}}$	Fringe visibility
(14.16)	$\Delta\bar{\nu} = \dfrac{1}{\tau_c}$	Relation between coherence time (τ_c) and frequency bandwidth ($\Delta\bar{\nu}$)
(14.19)	$\phi = \dfrac{1.27\lambda}{2w_0}$	Directionality of laser beam where ϕ gives the full angle beam divergence of a laser beam

REVIEW ZONE

SHORT ANSWER QUESTIONS

1. Define the terms MASER and LASER.
2. What do you mean by a photon?
3. What are the properties of laser radiation?
4. What are the different types of broadening processes?
5. What do you mean by coherence?
6. What are the different types of coherence?
7. What do you mean by an optical resonator?
8. What are the three distinct processes by which a transition can take place?
9. What is the equilibrium condition for transitions between energy levels?
10. What are Einstein's coefficients?
11. What do you mean by population inversion?

LONG ANSWER QUESTIONS

1. What do you mean by lasers? Describe main components of a laser and explain the characteristics of a laser beam.

2. Discuss Einstein's coefficients. Derive a relation between them.
3. Give the description of He-Ne laser and explain the construction and working of it.
4. Discuss semiconductor laser. What are the necessary conditions for the lasing action to take place?
5. Describe the construction, principle and working of a ruby laser.
6. Write a note on gaseous lasers.
7. Discuss principle, construction and working of an Nd-YAG laser. Also give its advantages.
8. Discuss some of the applications of lasers.

NUMERICAL PROBLEMS

1. A certain ruby laser emits 1.00 J pulses of light whose wavelength is 694 nm. What is the minimum number of cr^{+3} ions in the ruby?

   Ans. 3.49×10^{18} ions
2. A laser beam having a wavelength of 400 nm and an aperture of 0.25 cm is sent to identify an object. Calculate the angular spread and area of the beam. Given the distance between the laser source and the object as 10 km.

   Ans. 1.6×10^{-4} rad, 2.56 m^2
3. In He-Ne laser, the two plane mirrors forming the resonant cavity are placed at 0.8 m. What is the mode of separation of longitudinal cavity in terms of frequency?

   Ans. 187.5 MHz
4. The sodium line at λ = 589 nm produced a low-pressure discharge, has a Doppler width of 0.0194 Å. Calculate

   (a) the frequency of the light,

   (b) the line width in Hz and

   (c) the coherence length in cms

   Ans. 5.093×10^{14}Hz, 1.67×10^{9}Hz, 5.72 cm
5. A laser beam can be focussed on an area equal to the square of its wavelength. For a He-Ne laser, the wavelength of emitted light is 6328 Å. If the laser radiates energy at the rate of 1 mω, find out the intensity of the focussed beam.
6. The relative population of atoms in two energy levels E_1 and E_2 is 10^{-20}. Calculate the energy level difference in eV. Assume $T = 300$K.

   Ans. 1.2 eV
7. Calculate the relative population of Na atoms in a sodium lamp in the first excited state and in the ground state at a temperature of 300°C. Assume $\lambda = 590$ nm.

   Ans. 3.15×10^{-19}
8. Compute the coherence length of yellow light with 5893 Å in 10^{-12} seconds pulse duration. Find the bandwidth also.

   Ans. 59 mm, 5.1×10^{11} Hz
9. If the pulse width of a laser (λ = 10640 Å) is 25 ms and average output power per pulse is 0.8 W, how many photons does each pulse contain?

   Ans. 107×10^{15}
10. A laser beam $\lambda = 6000$ Å on earth is focussed by a lens of diameter 2 m on to a crater on the moon. The distance of the moon is 4×10^{8} m. How big is the sport on the moon?

    Ans. 3×10^{-7} rad, 1.4×10^{4} m^2

CHAPTER 15

Holography

Learning Objectives

By the end of this chapter, the student will be able:

- To distinguish optical photograph and holograph
- To understand the process of construction and reconstruction of holograms
- To explore the applications of holography
- To visualize the three-dimensional feature of hologram
- To understand the types of holograms

15.1 ▶ Introduction

The conventional photography records only the intensities of light coming from an object. It fails to record the phases of the waves that come from the objects. Hence, they show two-dimensional images of three-dimensional objects. In 1948, Dennis Gabor, a British scientist, developed a method of recording and producing three-dimensional images of objects through the phenomena of interference, which is known as holography. In Greek, 'holo' means 'whole' or 'complete', and 'holography' means 'complete recording', i.e. the intensities and phase of the waves that come from the objects are recorded. In 1971, he received noble prize for his efforts.

15.2 ▶ Basic Principle of Holography

An object is illuminated with a beam of coherent light (object beam). Then every point on the surface of the object acts as a source of secondary waves. These secondary waves spread in all directions. Some of these waves are allowed to fall on a recording plate (holographic plate). Simultaneously, another beam of same coherent light (reference beam) is allowed to fall on this holographic plate. In the holographic plate, both the beams combine and interference pattern is formed. This interference pattern is recorded on the holographic plate. The three-dimensional image of the object can be seen by exposing the recorded holographic plate (hologram) to coherent light. This is the basic principle of holography.

Note

In holography, the hologram is recorded when the light source is a laser and the image is an interferogram.

15.3 ▶ Recording of Image on a Holographic Plate

Figure 15.1 shows the method of recording an image on a holographic plate. The monochromatic light from a laser has been passed through a 50% beam splitter so that the amplitude division of the incident beam into two beams takes place. One beam falls on mirror M_1 and the light reflected from M_1 falls on the object. This beam is known as an object beam.

The object scatters this beam in all directions, so that a part of the scattered beam falls on the holographic plate. The other beam is reflected by mirror M_2 and falls on the holographic plate. This beam is known as the reference beam. Superposition of the scattered rays from the object and the reference beam takes place on the plane of the holographic plate, so that interference pattern is formed on the plate H and it is recorded. The recorded interference pattern contains all the information of the scattered rays, i.e. the phases and intensities of the scattered rays. For proper recording, the holographic plate has to be exposed to the interference pattern for a few seconds. After exposing, the holographic plate is to be developed and fixed as like in the case of ordinary photograph. The recorded holographic plate is known as hologram or Gabor zone plate. The hologram does not contain a distinct image of the object. It contains information in the form of interference pattern or interferogram.

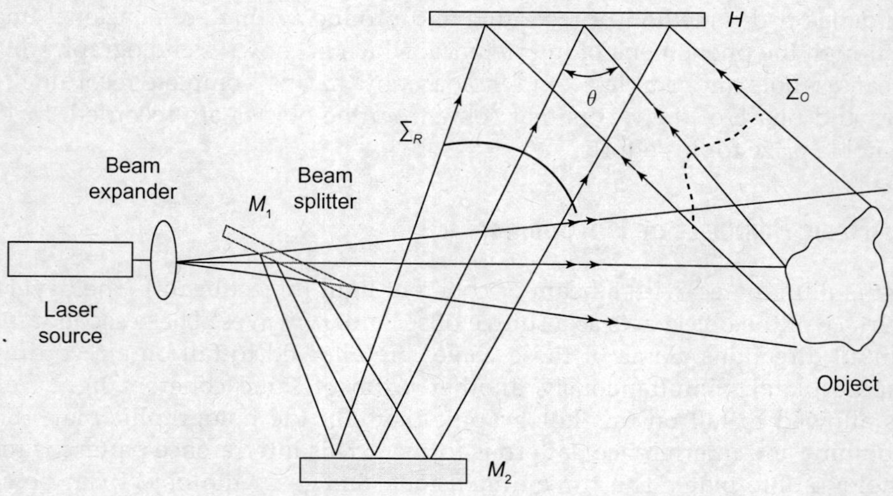

FIGURE 15.1 Recording of hologram

15.4 ▶ Reconstruction of Image From a Hologram

As shown in Fig. 15.2, the hologram is exposed to the laser beam (that used during construction or identical to the reference beam used for construction) from one side and it can be viewed from the other side. This beam is known as reconstruction beam. The reconstruction beam illuminates the hologram at the same angle as the reference beam. The hologram acts as a diffraction grating, so constructive interference takes place in some directions and destructive interference takes place in other directions. A real image is formed in front of the hologram and a virtual image is formed behind the hologram. It is identical to the object and hence it appears as if the object is present. The three-dimensional effect in the image can be seen by moving the head of the observer. During recording, the secondary waves from every point of the object expose complete plate. So, each bit of the plate contains complete information of the object. Hence, image can be constructed using a small piece of hologram.

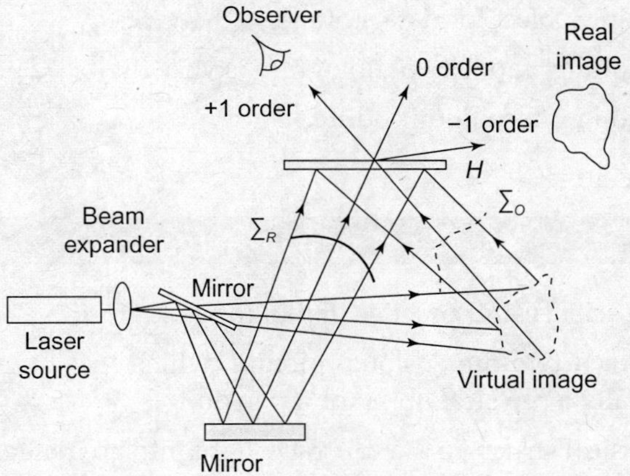

FIGURE 15.2 Image reconstruction in a hologram

15.5 ▶ Applications of Holography

1. The three-dimensional images produced by holograms have been used in various fields, such as technical, educational also in advertising, artistic display, etc.
2. Holographic diffraction gratings: The interference of two plane wavefronts of laser beams on the surface of holographic plate produces holographic diffraction grating. The lines in this grating are more uniform than in case of conventional grating.
3. Hologram is a reliable source for data storage, because even a small broken piece of hologram contains complete data or information about the object with reduced clarity.
4. The information-holding capacity of a hologram is very high because many objects can be recorded in a single hologram, by slightly changing the angle between reference beam and holographic plate. For each different angle, different images can be stored.
5. In hospitals, holography can be used to view the working of inner organs three dimensionally, i.e. the beating of the heart, the foetus of the pregnant lady and flowing blood based on motion holography.

6. Holographic interferometry is used in non-destructive testing of materials to find flaws in structural parts and minute distortions due to stress or vibrations etc. in the objects.
7. Holography is used in information coding.

RECAP ZONE

POINTS TO REMEMBER

- The process of recording of three-dimensional images through interference is called holography.
- The recorded three-dimensional images can be reproduced by diffraction process.
- Hologram is a reliable object for data storage.
- The information-holding capacity of hologram is very high.
- Holography is used in information coding.

KEY TERMS WITH DEFINITIONS

- **Holography** – The science and practice of making holograms.
- **Image** – A photo or a two-dimensional picture, which has a similar appearance to some subject - usually a physical object or a person.
- **Hologram** – A physical structure that diffracts light into an image.
- **Holography interferometry** – A technique that enables static and dynamic displacements of objects with optically rough surfaces to be measured to optical interferometric precision.

REVIEW ZONE

SHORT ANSWER QUESTIONS

1. What is a hologram?
2. What is the major difference between photography and holography?
3. Why can visible light not be used to produce hologram of a large size?
4. Explain the principle of holography.
5. Describe the uses of holography.

Long Answer Questions

1. What is the basic principle of holography? Explain.
2. How does one construct and reconstruct a hologram?
3. Mention some of the applications of holography.
4. Describe the phenomenon of recording of image on a holographic plate.

CHAPTER 16

Photonics and Fibre Optics

Learning Objectives

By the end of this chapter, the student will be able:

- To understand the working principle of optical fibres
- To recognize the light sources used for transmission of optical signals
- To differentiate between graded index and multimode optical fibres
- To understand structural and performance difference of different types of optical fibres
- To understand the complete fibre optic communication system

16.1 ▸ INTRODUCTION

Optical fibres are thin, transparent cylindrical glass or plastic objects whose diameter is very smaller than its length and which allows light to be guided from one end to the other end with minimal loss.

Even though the invention of laser in 1960 intensified the interest in light as a means of communication, the reliability of long-distance links was limited due to some natural difficulties in transmitting light. These were overcome with the advent of dielectric waveguides or optical fibres fabricated from glass, which reduced the degradation of optical signal. Since the invention of fibre optics, long-distance communication links using light have grown tremendously. Although the uses of optical fibre are quite numerous, they are widely used in the field of telecommunication, medicine and automotive industries. In the telecommunication front, approximately 10 billion digital bits can be transmitted per second through an optical fibre link in a commercial network, enough to carry millions of telephone calls. Refinements in optical fibres, along with the development of new lasers and diodes, may one day allow commercial fibre-optic networks to carry trillions of bits of data per second.

16.2 ▶ Optical Fibres

16.2.1 History

Fibre-optic cable is a cylindrical pipe made of glass, plastic or a combination of both glass and plastic. It is fabricated in such a way that this pipe can guide light from one end to the other. Total internal reflection (TIR) of light in a beam of water – essentially guided light – was demonstrated by the physicist John Tyndall in 1854. Tyndall showed that light could be bent around a corner while it travelled through a jet of pouring water. In 1880, Alexander Graham Bell invented the photophone. Bell demonstrated that a membrane in response to sound could modulate an optical signal, light. But, this was a free space transmission system. The light was not guided. Guided optical communications had to wait for the advent of 20th century.

Even after the first patent on guided optical communications over glass, obtained by AT&T in 1934, there were really no materials to fabricate glass or other type of transparent materials. Hence, the technique of guided optical communications using fibre-optic cable with sufficiently low attenuation had to wait for another 30 years. During 1960s, researchers working in different academic, industrial and government laboratories obtained a much better understanding of the loss mechanisms in glass fibre-optic cable. This resulted in a continued decrease in attenuation of the optical fibres through 1970s, from 1000 dB/km to less than 20 dB/km. Such a decrease in loss in fibres allowed practical guided light communications using glass fibre-optic cable to take off in a big way. Moreover, in the late 1980s and 1990s, this momentum increased with the even lower cost plastic fibre-optic cable and plastic clad silica (PCS).

16.2.2 Materials Used for Optical Fibres

Usually, optical fibres are flexible, thin and cylindrical and are made of transparent materials such as glass and plastic. The most abundant and widespread material used to make optical fibre is glass and most often this is an oxide glass based on silica (SiO_2) along with some additives. The desired properties of a fibre are optical quality, mechanical strength and flexibility. Owing to these reasons, plastic optical fibres are made with polymethylmethacrylate (PMMA) polymer and its copolymers.

Plastic fibres have many advantages over glass fibres.

- After processing of plastic fibres is much simpler and less demanding than that of all glass fibres.

Note

The idea of having light guided through bent glass is not new or high-tech. An unverified assertion says that Leonardo Da Vinci actually mentioned such a means for guiding light in one of his notebooks.

Using the technique of molecular alignment, a 0.27 mm diameter PMMA-based image guide has exhibited unlimited 180° flexing cycles with a bending radius of 1.5 mm. This is to be compared with an 8 mm bending radius of the same diameter glass fibre.

- They are lighter and are of lower cost than glass fibres.
- They are more flexible than glass fibres.

On the other hand, plastic fibres cannot withstand high temperature. However, PMMA plastic fibres generally can operate within a continuous maximum temperature of 167°F (75°C). For short duration, it may be subjected to a maximum of 212°F (100°C).

16.3 ▶ PRINCIPLE OF OPTICS IN FIBRES

Light transmission through any transparent material depends mainly on an important fundamental quantity, i.e. its refractive index. The laws of refraction and reflection, which are the guiding factors for a ray of light inside a material, are dependent on the refractive index; hence, it becomes the starting point of our discussion.

Refractive index (n or μ) is defined as the ratio of the velocity of light in vacuum (c) to that of the velocity of light in a medium (v), i.e.,

$$n \text{ or } \mu = \frac{c}{v}$$

This is expressed as a positive real number greater than 1. As the velocity of light in air is the same as in vacuum, the refractive index of air is 1. Similarly, the refractive index of glass is 1.5 and that of water is 1.33.

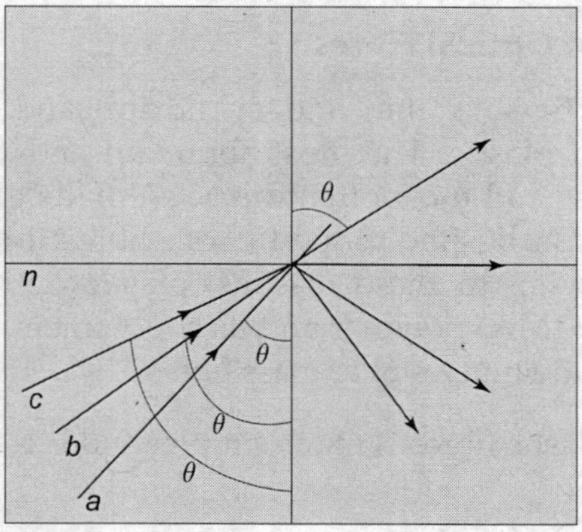

FIGURE 16.1 Refraction and reflection at the interface between two media ($n_1 > n_2$) (a) incident angle $\theta_1 < \theta_c$ (critical angle), (b) incident angle $\theta_2 = \theta_c$, (c) incident angle $\theta_3 > \theta_c$ (condition for total internal reflection)

Light undergoes refraction and/or reflection whenever there is a change of refractive index in its path. If a ray of light propagates at a certain angle θ_1 (where $\theta_1 < \theta_c$) from a high refractive index medium (n_1) to a low refractive medium (n_2), then a portion of the light will be reflected back to medium 'n_1' and another part of the light will be refracted into medium 'n_2'. As shown in Fig. 16.1, the ray 'a' incident at an angle θ_1 at a boundary between two layers of refractive indices n_1 and n_2 undergoes both refraction and reflection. This behaviour of light can be expressed by Snell's law as

$$n_1 \sin\theta_1 = n_2 \sin\theta_2 \tag{16.1}$$

But strangely, as the angle of incidence increases, the ray refracts away from the normal. If the angle θ_1 is increased to θ_c (ray 'b'), then θ_2 reaches 90°. The critical angle, θ_c, is then defined as the angle of incidence of light for which the refracted light just graces the boundary between the two media and is expressed as

$$\sin\theta_c = \frac{n_2}{n_1}(\sin 90) = \frac{n_2}{n_1} \tag{16.2}$$

or

$$\boxed{\theta_c = \sin^{-1}\left(\frac{n_1}{n_2}\right)} \tag{16.3}$$

For any angle above this critical angle, where $\theta_3 > \theta_c$ (ray 'c'), we can observe that the ray is totally reflected back into the same higher refractive index medium n_1 itself, i.e. the ray is totally internally reflected. It is this property of TIR that is utilized to guide the light to travel through an optical fibre.

16.3.1 Total Internal Reflection

The phenomenon in which the incident light is completely reflected within the incident medium itself is called *total internal reflection* (Fig. 16.2). Total internal reflection is the most important phenomenon for the guiding of light in optical fibres. Under the condition of TIR, light can be completely reflected at a dielectric interface without any reflective coating. For TIR to occur, the ray of light has to be incident on a dielectric interface from a high refractive index region to a low refractive index region.

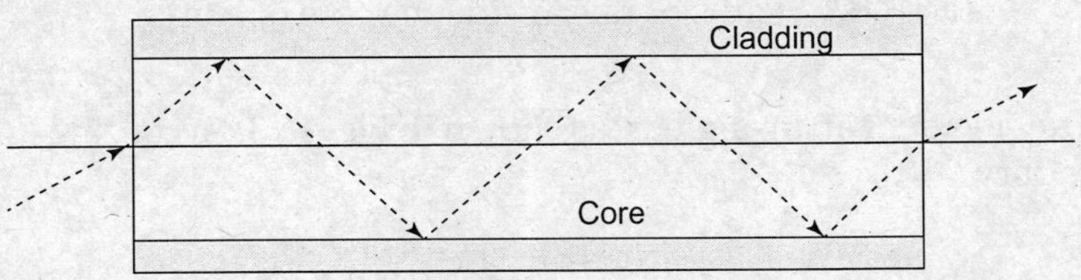

FIGURE 16.2 Occurrence of total internal reflection in an optical fibre

However, the TIR in the transmitting medium (core) causes some electromagnetic field disturbance in the second medium (cladding) resulting in leakage of light, which is known as the phenomenon of frustrated total internal reflection. This normally occurs between any two closely placed media (e.g., two fibre cores placed together).

Note

In a fibre-optic image bundle, this phenomenon can be significant between adjacent fibres even when the diameter of a fibre is many times greater than the wavelength of light. This results in coupling of light between fibres causing light leakage between adjacent fibres over a region where the separation between the fibres is up to a few wavelengths of light.

16.3.2 Acceptance Angle

Consider Fig. 16.3 in which a ray is incident on the face of a step index (SI) fibre at an angle ϕ_1. The ray will be reflected inside the core and refracted into the cladding. The refraction inside the cladding is considered to be loss of energy in the fibre. However, when the angle is decreased to ϕ_c known as the critical angle of incidence, the ray just graces the boundary inside the fibre. For any angle of incidence within the angle $\phi_{c'}$ the ray of light will be totally internally reflected. Hence, the angle $\phi_{c'}$ known as the maximum acceptance angle (ϕ_{max}) is the maximum limiting angle of a fibre that allows light to travel through the fibre. This angle is defined as the maximum angle of incidence at the face of a fibre, which allows the light to travel through it by total internal reflection.

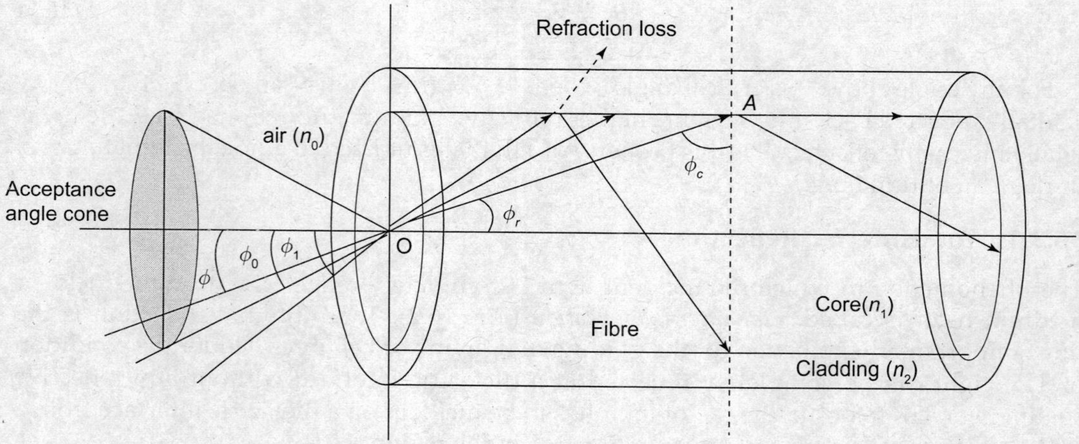

FIGURE 16.3 Paths of rays inside the core of a fibre

Consider the ray incident at an angle ϕ as shown in Fig. 16.3. According to Snell's law, at the air-core boundary

$$n_0 \sin\phi = n_1 \sin\phi_r$$

where n_0, n_1 and n_2 are the refractive indices of air, core and cladding, respectively, i.e.,

$$\sin\phi = \frac{n_1}{n_0}\sin\phi_r$$

From ΔOAB, $\phi_r = (90 - \phi_c)$

$$\therefore \quad \sin\phi = \frac{n_1}{n_0}\sin(90-\phi_c)$$

or

$$\sin\phi = \frac{n_1}{n_0}\cos\phi_c$$

Since $\cos^2\phi + \sin^2\phi = 1$, $\cos^2\phi = 1 - \sin^2\phi$, or $\cos\phi = \sqrt{(1-\sin^2\phi)}$

$$\sin\phi = \frac{n_1}{n_0}\sqrt{(1-\sin^2\phi_c)}$$

From Eq. (16.2),

$$\sin\phi = \frac{n_1}{n_0}\sqrt{\left[1-\left(\frac{n_2}{n_1}\right)^2\right]}$$

$$= \frac{n_1}{n_0}\left(\frac{\sqrt{n_1^2-n_2^2}}{n_1}\right)$$

$$= \frac{\left(n_1^2-n_2^2\right)^{1/2}}{n_0}$$

$$\sin\phi_{\max} = \left(n_1^2-n_2^2\right)^{1/2} \tag{16.4}$$

This maximum angle of incidence is known as acceptance angle, and is given by

$$\boxed{\boldsymbol{\phi_{\max} = \sin^{-1}\left(n_1^2-n_2^2\right)^{1/2} (\text{for } n_0 = 1)}} \tag{16.5}$$

16.3.3 Numerical Aperture

It is well known that an optical fibre allows light to propagate through it. For efficient transmission, the optical fibre has to maximize the amount of light travelling through it. To measure the amount of light gathered by the fibre, a quantity known as numerical aperture is used. *Numerical aperture* (NA) is defined as a measure of the light-gathering ability of an optical fibre; also it is defined as a quantity that is numerically equal to sine of the acceptance angle ($\sin\phi_{\max}$), i.e.,

$$\boxed{\boldsymbol{\text{NA} = \sin\phi_{\max} = \sqrt{n_1^2-n_2^2}}} \tag{16.6}$$

Numerical aperture is a term that is commonly used in optical fibre communication, relating the media involved (air, core and cladding) and the acceptance angle. Numerical aperture can also be defined in terms of the difference in relative refractive index (Δ) between the core and the cladding (for a step index fibre).

$$\text{NA} = n_1\sqrt{2\Delta} \tag{16.7}$$

or

$$\Delta = \frac{n_1-n_2}{n_1} \quad (\text{for } \Delta = 1) \tag{16.8}$$

The NA for a graded index fibre is given by

$$\text{NA}_{\text{graded}} = \sin \phi_a \cos \gamma \tag{16.9}$$

where ϕ_a is the acceptance angle for the graded index fibre and γ is half the angular change of every reflection (Fig. 16.4).

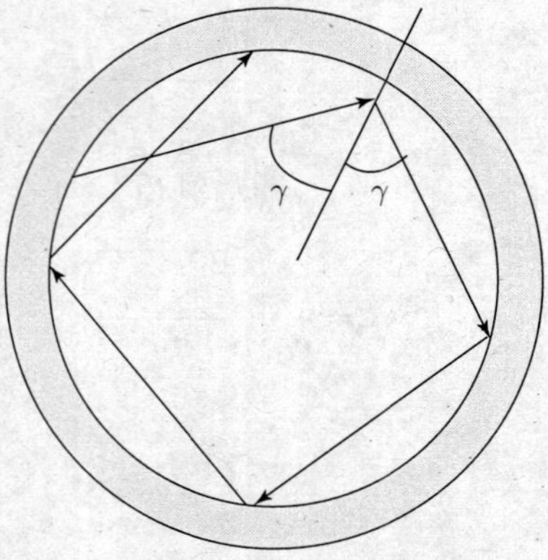

FIGURE 16.4 Cross-sectional view of a graded index fibre

Note

Δ is always positive because $n_1 > n_2$ for total internal reflection conditions, in order to guide light effectively through a fibre, $\Delta << 1$. Typically, Δ is of the order of 0.01 fibres with large Δ will not be useful for optical communication due to the occurring of multipath dispersion.

EXAMPLE 16.1

Determine the numerical aperture of a step index fibre when the core refractive index $n_1 = 1.5$ and the cladding refractive index $n_2 = 1.48$. Find the maximum angle for entrance of light if the fibre is placed in air.

Solution:

(i) We know that

$$\begin{aligned} \text{NA} &= \sqrt{n_1^2 - n_2^2} \\ &= \sqrt{(1.5)^2 - (1.48)^2} \\ &= 0.244 \end{aligned}$$

(ii) The maximum entrance angle is

$$\phi_{\max} = \sin^{-1} \sqrt{n_1^2 - n_2^2}$$

$$= \sin^{-1}\sqrt{NA/n} \quad [n = \text{refractive index of air} = 1]$$

$$= \sin^{-1}\sqrt{0.244/n}$$

$$= \mathbf{14.13^\circ}$$

EXAMPLE 16.2

An optical fibre has fibre index 1.36 and relative difference in index $\Delta = 0.025$. Find the numerical aperture and the acceptance angle.

Solution: The numerical aperture is

$$NA = n_1\sqrt{2\Delta}$$

$$= 1.36\sqrt{2\times 0.025}$$

$$= 0.304$$

Acceptance angle is,

$$\phi_{max} = \sin^{-1}(NA)$$

$$= \mathbf{\sin^{-1}(0.304) = 17.7^\circ}$$

16.4 ▶ Structure of an Optical Fibre

A fibre-optic cable is a collection of optical fibres bundled together, in which the light entering at one end leaves through the other end without any loss in energy. A fibre is cylindrical in nature with each fibre constructed identically having three major layers (Fig. 16.5). They are

- Core (inner cylinder) made of glass or plastic with a diameter of $\sim 50\ \mu m$.
- Cladding (cylinder surrounding the core) made of glass or plastic having a diameter of $125-200\ \mu m$.
- Jacket (outer cylindrical lavers).

Core: Core is made up of silica or silicon dioxide (SiO_2) with other materials such as arsenic, quartz, etc. The core has a higher refractive index, n_1, than that of its outer layer cladding, n_2 (i.e., $n_1 > n_2$) and has a diameter of $10-50\ \mu m$.

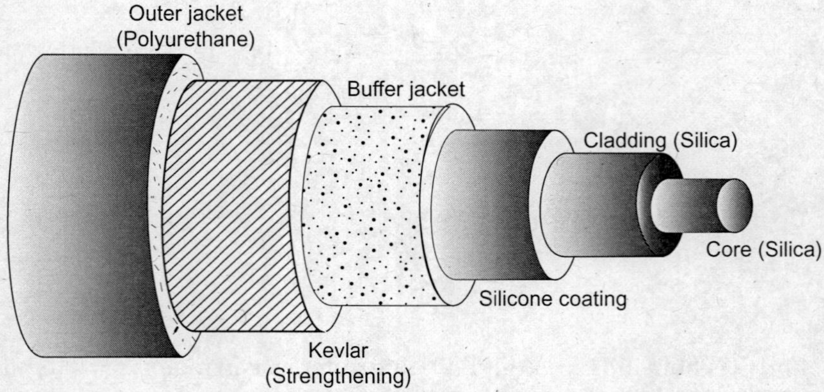

FIGURE 16.5 Typical structure of an optical fibre

Cladding: Cladding is the outer dielectric medium of the fibre. Its main property is to retain the light wave within the core and to some extent provide mechanical strength and safety to the core. This outer layer is coated with small amounts of boron, germanium or phosphorus to reduce its refractive index than that of the core. This facilitates the TIR of light within the core and thus helps in the propagation of light. The cladding has a diameter of 125–200 μm.

Jacket: Since both the core and the cladding are made of glass, it becomes necessary to have a protective layer that prevents the fibre from breaking. The jacket usually consists of one or more layers of polymer. Its role is to protect the core and the cladding from shocks that might affect their optical or physical properties by acting as a shock absorber. It also provides protection from abrasions, solvents and other contaminants. The jacket does not have any optical properties that might affect the propagation of light within the fibre-optic cable. Just inside the jacket, a strength member (like a layer of Kevlar) that provides toughness and tensile strength to the fibre may be added to the fibre-optic cable, so that it can be pulled during installation. A buffer is also added between the strength member and the cladding. A polyurethane outer jacket covers all the inner layers and provides flexibility and ruggedness to the fibre.

Note

Optical fibres are preferred over metal wires because they are extremely light, small and can be accommodated in small place.

16.5 ▸ Types of Optical Fibres

Optical fibres are generally classified according to the following:

(i) Material of construction

(ii) Number of modes of travel of light rays within the fibre

(iii) Refractive index variation within the fibre

16.5.1 Types of Fibres According to the Material of Construction

Optical fibres are normally made with two types of materials, namely glass (silica) and plastic. Based on this, there are three types of fibres, they are

- Glass fibres
- Plastic fibres
- Plastic clad silica fibres

Glass fibres

Silica is the commonly used material for the manufacture of a glass fibre. The glass employed in a fibre-optic cable is ultrapure, ultra transparent, silicon dioxide or fused quartz. Glass fibres are made mostly by fusing silica and metal oxides. A pure glass fibre-optic cable has a glass core and a glass cladding and is by far the most widely used fibre. During the glass fibre-optic cable fabrication process, impurities are purposely added to the pure glass so as to obtain the desired indices of refraction needed to guide light.

For example, diffusing various oxides to the silica rod, such as P_2O_5, B_2O_3 and GeO_2, forms the core of the fibre while the undiffused pure form of silica remains as the cladding. Germanium or phosphorous is added to increase the index of refraction. Boron or fluorine is added to decrease the index of refraction. Other impurities that may somehow remain in the glass cable after fabrication (residual impurities) may increase the attenuation by either scattering or absorbing light.

Glass fibre-optic cable has the lowest attenuation and comes at the highest cost. Fibre-optic cable sizes are usually expressed by giving the core size followed by the cladding size. Consequently, 50/125 indicates a core diameter of 50 μm and a cladding diameter of 125 μm, and 100/140 indicates a core diameter of 100 μm and a cladding diameter of 140 μm.

Plastic fibres

These are low-cost optical fibres. They have a large NA and a large acceptance angle, because of the ease of having a large refractive index variation between the core and the cladding. Some examples of the plastic fibres are as follows.

- Polystyrene (core) – methylmethacrylate (cladding)
- Polymethylmethacrylate (core) – copolymer (cladding)

Plastic fibre-optic cable has the highest attenuation, because of the nature of the material they offer larger signal attenuation than glass fibres. However, this drawback is overcome by their ruggedness and durability. Plastic fibre-optic cable has a plastic core and plastic cladding. This fibre-optic cable is quite thick. Typical dimensions are 480/500, 735/750 and 980/1000. The core generally consists of PMMA coated with a fluoropolymer.

Plastic fibre-optic cable does have a problem with flammability. Because of this, it may not be appropriate for certain environments, and hence care has to be taken when it is run through such environments. Otherwise, plastic fibre is considered extremely rugged with a tight-bend radius and the ability to withstand abuse.

Plastic clad silica fibres

Plastic clad silica fibre is another type of optical cable. It has an attenuation that lies between glass and plastic and a cost that lies between their costs as well. PCS fibre-optic cable has a glass core that is made of silica, while the cladding is plastic (a low refractive index silicone elastomer). PCS fabricated with a silicone elastomer cladding suffers from three major defects. It has considerable plasticity (this makes connector application difficult), adhesive bonding is not possible, and it is practically insoluble in organic solvents. Although all these drawbacks make this type of fibre-optic cable not particularly popular, there have been some improvements in it in recent years.

16.5.2 Types of Fibres According to Number of Modes Guided Through Fibres

A mode is a path taken by a ray of light through an optical fibre. It is a mathematical concept of explaining the propagation of light waves in an optical fibre. The number of modes of propagation within a fibre depends on the diameter of the fibre core. Normally, narrower the fibre core lesser is the number of modes of propagation. The condition for propagation of a mode is given by

$$V = \frac{\pi d n_1}{\lambda_0}\left[1 - \left(\frac{n_2}{n_1}\right)^2\right]^{1/2} \tag{16.10}$$

$$= \frac{\pi d}{\lambda_0}(\text{NA}) \tag{16.11}$$

Note

In 1984, the IEC standardized PCS fibre-optic cable to have the following dimensions: core, 200 mm; silicone elastomer cladding, 380 mm; and jacket, 600 mm.

where V is known as the V-number, and it depends on the characteristics of the fibre and the wavelength of the light propagating through it. For single-mode propagation, the value of V-number should be ≤ 2.405. At this condition, the wavelength of the light is known as the cut-off wavelength (λ_c) for propagation and is given by

$$\lambda_c = \frac{\pi d}{2.405}(\text{NA}) \tag{16.12}$$

The number of modes (N) that can propagate in an SI fibre is

$$\boxed{N \cong \frac{V^2}{2}} \tag{16.13}$$

The number of modes that can propagate in a graded index fibre is half of that of an SI fibre, i.e.,

$$\boxed{N \cong \frac{V^2}{4}} \tag{16.14}$$

Hence, on the basis of propagation of number of modes, fibres are classified as

- Single-mode fibres
- Multimode fibres

Single-mode fibres

A single-mode fibre [Fig. 16.6(a)] has a narrow core (8–10 μm) and the refractive index between the core and the cladding does not change very much throughout the fibre. Because of the narrower core, light travels almost parallel to the axis, thus creating little or almost no pulse dispersion. Hence, they can carry a higher bandwidth (up to 50 times more distance than a multimode fibre) but require a light source with a narrow spectral width. The small core and single mode virtually eliminate any distortion that could result from overlapping light pulses, providing the least signal attenuation and the highest transmission speed of any fibre cable type. However, the cost of a single-mode fibre is high. Typically, in a single-mode fibre, lowest-order-bound mode can propagate at the wavelength of approximately 1300–1320 nm.

Multimode fibres

As the name suggests, this type of fibre allows more than one mode to travel through it. As it has to allow more number of modes to travel, the fibre has to have a larger core [Fig. 16.6(b)]. A typical multimode fibre has a core diameter in the order of 150–200 μm. Such a large core not only allows more number of modes to travel but also makes it easier to launch the light into the fibre and to connect two fibres together. On the other hand, larger the number of modes, larger will be the dispersion of light inside the fibre, i.e. some of the light rays that make up the digital pulse may travel a direct route (i.e., parallel and close to the axis), whereas other rays zig-zag as they bounce off the cladding.

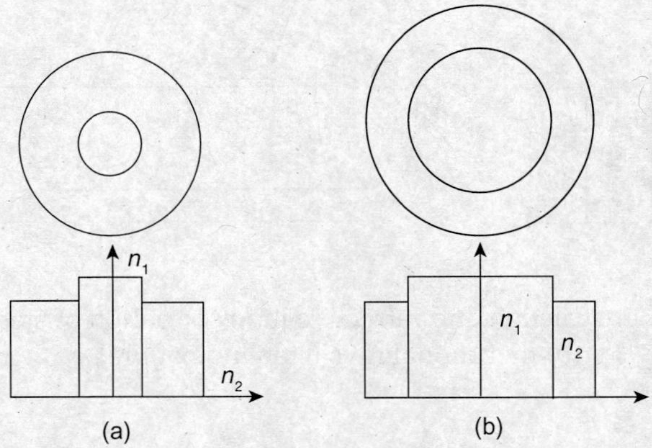

FIGURE 16.6 (a) Single-mode fibre; (b) multimode fibre

These alternative pathways cause the different groupings of light rays, referred to as modes, to arrive separately at a receiving point, representing a loss known as intermodal dispersion, in longer cable (> 3000 ft), multiple paths of light cause signal distortion at the receiving end, resulting in an unclear and incomplete data transmission On the contrary, a multimode fibre can provide a higher bandwidth at high speeds over medium distances.

Note

Single-mode fibres are also known as monomode fibres or unimode fibres or single-mode optical waveguides.

16.5.3 Types of Fibres According to Refractive Index Variation

Optical fibres have two layers namely core and cladding with different refractive indices. The variation of the refractive index within the core determines the nature of the fibre. An optical fibre is characterized according to the variation of the core refractive index as

- Step index fibres
- Graded index fibres

Step index fibres

In an SI fibre, the refractive index remains constant throughout the core and decreases at the core-cladding boundary and then remains constant throughout the cladding [Fig. 16.7(a)]. As the refractive index varies as a 'step' at the core-cladding boundary, the fibre is known as an SI fibre. The light rays travel through the fibre as meridional rays [Fig. 16.7(b)], i.e. the rays zig-zag through the length of the fibre, crossing the fibre axis for every reflection.

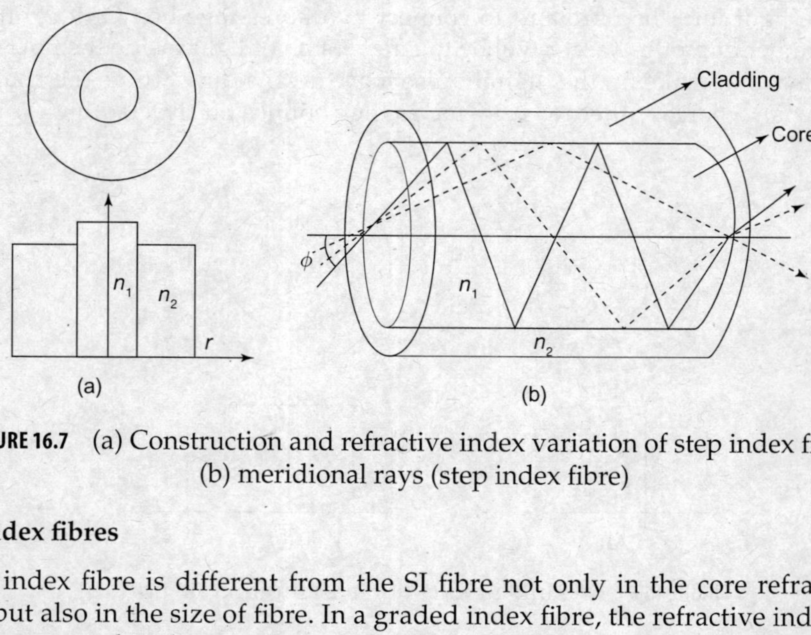

FIGURE 16.7 (a) Construction and refractive index variation of step index fibre; (b) meridional rays (step index fibre)

Graded index fibres

A graded index fibre is different from the SI fibre not only in the core refractive index variation but also in the size of fibre. In a graded index fibre, the refractive index does not remain constant within the core, but instead, it gradually decreases on either side from the axis of the core towards the core-cladding boundary [Fig. 16.8(a)]. At the boundary, the refractive index of the core is the minimum and equals the refractive index of the cladding.

Because of the gradual variation of the refractive index along the radius of the core, the fibre is known as graded index fibre.

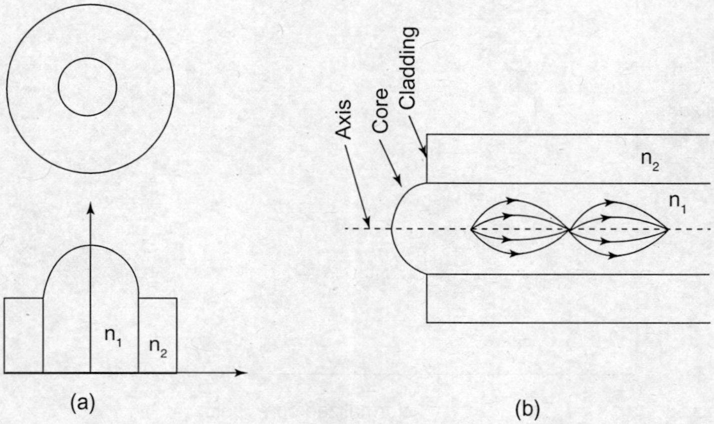

FIGURE 16.8 (a) Construction and refractive index variation of a graded index fibre; (b) helical or skew rays

As the refractive index variation resembles concentric cylinders of equal refractive indices around the axis, any ray of light entering any one of the cylindrical layers get trapped by it and remains within that layer for the rest of the travel. This means that the ray of light does not and will not cross the axis of the fibre and merely goes in a helical path around the axis. Therefore, the rays of light in a graded index fibre are known as helical rays [Fig. 16.8(b)]. As the paths of the light rays are skewed, they are also known as skew rays.

Moreover, a higher refractive index at the centre makes the light rays moving down the axis travel more slowly than those near the cladding. Because of the graded refractive index variation, light inside the core travels helically, thus reducing its travel distance. The shortened path and the higher speed ray near the core-cladding boundary allow light to arrive at a receiver at about the same time as the slow but straight-travelling ray close to the core axis. This allows almost all the rays of light to travel the same amount of time inside the core of the fibre, which result in less dispersion for a digital pulse.

The refractive index variation at the core of a graded index fibre is given by

$$n(r) = n_1\left[1-2\Delta\left(\frac{r}{a}\right)^{\alpha}\right]^{1/2} \tag{16.15}$$

The refractive index variation at the cladding is given by

$$n_2 = n_1\,(1-2\Delta) \tag{16.16}$$

where n_1 and n_2 are the refractive indices of the core and cladding, r is the radial distance from the axis of the fibre core, a is the radius of the fibre core, Δ is the relative refractive index variation, and α is the refractive index of the profile.

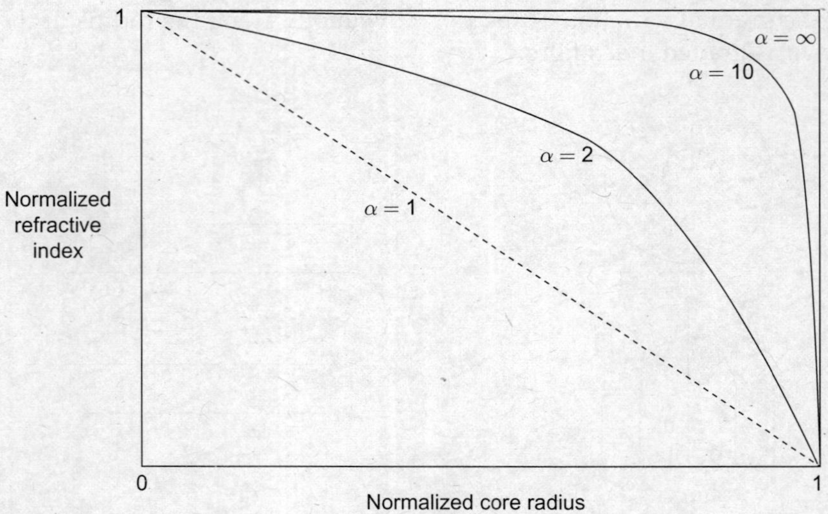

FIGURE 16.9 Refractive index profiles for a graded index fibre with different α-values

The refractive index profile for a linear type graded index fibre is $\alpha = 1$ and parabolic type graded index fibre is $\alpha = 2$. When $\alpha = \infty$, the rays of light travel as they would through an SI fibre.

Figure 16.9 shows the profile of refractive index from the core centre to the outside edge of a grade index fibre with different α-values. In communication fibres, the α-value is very important because the bandwidth of the optical fibre depends on this value. The bandwidth is maximized when the refractive index profile is parabolic (i.e., $\alpha = 2$).

In general, based on all these factors, optical fibres are broadly classified into three major types:

- Single-mode step index (SMSI) fibre
- Multimode step index (MMSI) fibre
- Multimode graded index (MMGI) fibre

16.5.4 Differences Between Step Index Fibres and Graded Index Fibres

TABLE 16.1 Differences between step and graded index fibres

	Step index fibre	Graded index fibre
1.	The refractive index of the core is uniform and step or abrupt change in refractive index takes place at the interface of core and cladding in step index fibres.	The refractive index of core is non-uniform, the refractive index of core decreases parabolically from the axis of the fibre to its surface.

(Cont'd)

TABLE 16.1 (Continued)

	Step index fibre	Graded index fibre
2.	The light rays propagate in zig-zag manner inside the core. The rays travel in the fibre as meridional rays and they cross the fibre axis for every reflection.	The light rays propagate in the form of skew rays or helical rays. They will not cross the fibre axis.
3.	Signal distortion is more in case of high-angle rays in multimode step index fibre. In single-mode step index fibre, there is no distortion.	Signal distortion is very low even though the rays travel with different speeds inside the fibre.
4.	The bandwidth is about 50 MHz km for multimode step index fibre whereas it is more than 1000 MHz km in case of single-mode step index fibre.	The bandwidth of the fibre lies in between 200 MHz km to 600 MHz km even though theoretically it has an infinite bandwidth.
5.	Attenuation of light rays is more in multimode step index fibres but for single-mode step index fibres, it is very less.	Attenuation of light rays is less in graded index fibres.
6.	NA of multimode step index fibre is more whereas in single-mode step index fibres, it is very less.	NA of graded index fibres is less.

16.5.5 Differences Between Single-Mode Fibres and Multimode Fibres

TABLE 16.2 Difference between single-mode and multimode fibres

	Single-mode fibres	Multimode fibres
1.	In single-mode fibres, there is only one path for ray propagation.	In multimode fibres, a large number of paths are available for light ray propagation.
2.	Single-mode step index fibres have less core diameter (<10 μm) and the difference between the refractive indices of core and cladding is very small.	Multimode step index fibres have larger core diameter (50–200 μm) and the difference between the refractive indices of core and cladding is large.
3.	In single-mode fibres, there is no dispersion.	There is signal distortion and dispersion takes place in multimode fibres.
4.	Signal transmission capacity is less but the single-mode fibres are suitable for long distance communication.	Signal transmission capacity is more in multimode fibres. Because of large dispersion and attenuation, they are less suitable for long distance transmission.
5.	Launching of light into single-mode fibres is difficult.	Launching of light into multimode fibres is easy.
6.	Fabrication cost is very high.	Fabrication cost is less.

(Cont'd)

TABLE 16.2 (Continued)

	Single-mode fibres	Multimode fibres
7.	The V-number of a fibre $\left[V = \frac{2\pi}{\lambda} n_1 r \sqrt{2\Delta}\right]$ is less than 2.405 for single-mode fibre. n_1, r are the refractive index and radius of core, respectively, λ = wavelength of light that propagates through the fibre.	The V-number of a multimode fibre is greater than 2.405.

16.6 ▶ Fibre-Optic Communication System

Communication is an exchange of information between source and user that is located away from the source. A typical exchange is where the information generated at one end (source) is transported to the other end (user). The information is transported or transmitted by means of a transmission medium. The medium is a physical entity and has to be located between the source and the user and is also accessible to both (Fig. 16.10). The transmission medium has to have a set of properties, which exists in a quiescent state.

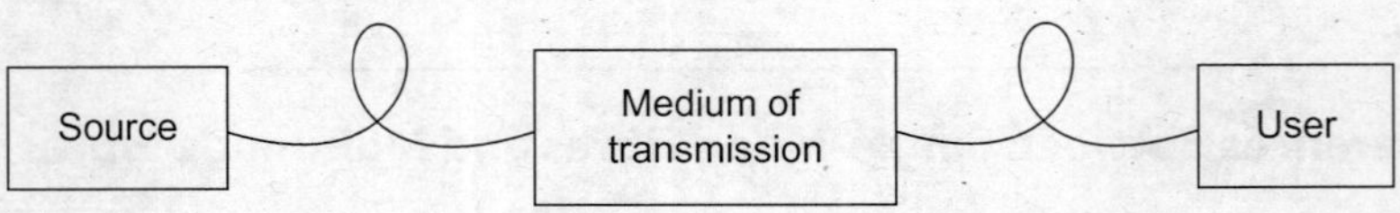

FIGURE 16.10 Source, medium and user

The combination of transmitter, transmission medium and receiver is termed as the communication link (Fig. 16.11) and because of the limitation placed on the information to be a sequence of bits (digital transmission), this combination is generally referred to as a data link. Attenuation or loss increases with the distance through the transmission medium. In fact, the amplitude attenuation is measured in dB/km. As propagation continues, attenuation increases. Ultimately, the propagating signal is attenuated until it is at some minimal detectable level. The distance at which the signal reaches this minimal level could be quite significant. The transmission medium must be able to deliver at least the minimal detectable level of output signal to the receiver by the user. If it cannot, communications between the source and the user will not take place.

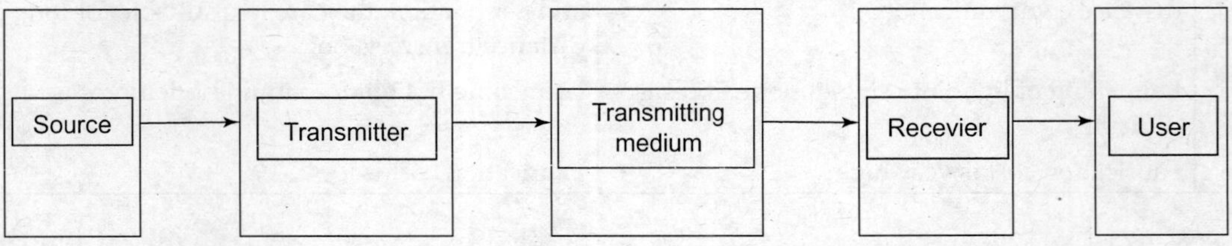

FIGURE 16.11 Fundamental block of a communication link

There are some tricks to getting around the attenuation problem. For example, if the disturbance has been attenuated to a minimal detectable level yet it has still not arrived at the receiver/user, then the output signal can then be regenerated at this location, i.e. the signal can be boosted back to its original energy level. It can be repeated and then allowed to propagate on its way to the receiver/user. This is shown in Fig. 16.12.

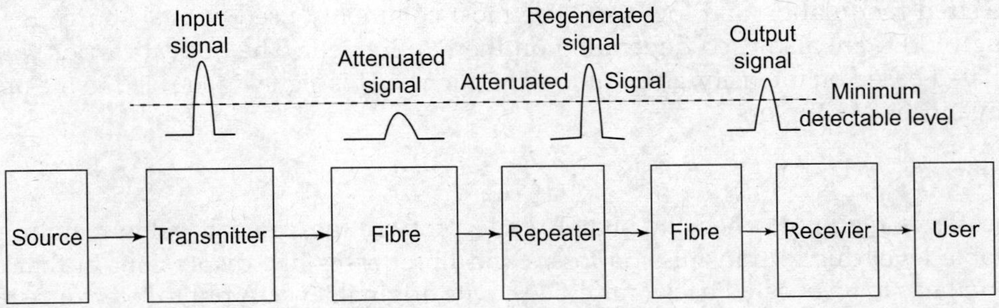

FIGURE 16.12 Regenerating and repeating an attenuated signal in order to reach the user

A fundamental digital fibre-optical communication system is shown in Fig. 16.13. As such, it consists of three major system blocks. They are

- Transmitter
- Repeater
- Receiver

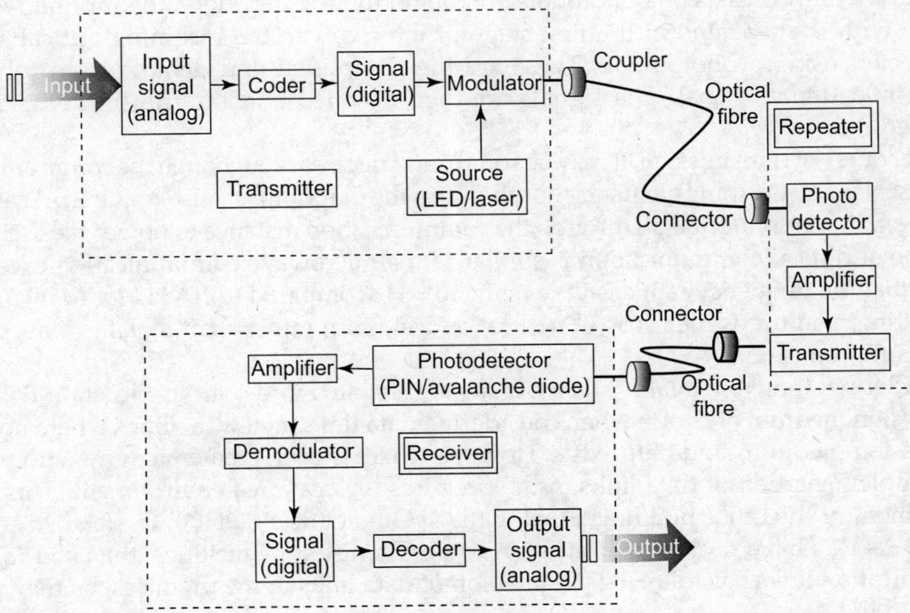

FIGURE 16.13 Functional components of a basic digital fibre-optical communication system

Transmitter

The transmitter consists of the source, coder and modulator. The input analog signal (data, voice, music, etc.) is converted to a digital signal by means of a coder or an encoder. The converted digital signal is then modulated into an optical carrier and transmitted. The sources are chosen such that they have high efficiency, excellent power output, long life, desired modulation and low cost. The most commonly used optical source is laser, although LEDs are also used depending on the requirement. The optical carrier wave is modulated based on intensity, amplitude or frequency. This signal is coupled to the optical fibre by means of couplers.

Repeater

A repeater is required when the signal while propagating can fall below a minimum detectable level due to transmission losses and fibre losses like dispersion. In that case, the signal has to be boosted back (amplified) to its original level. A photodetector receives the optical signal and converts it into an electrical signal. This is done using the PIN (P-Intrinsic-N) photodiode or avalanche photodiode. This works on the principle of creation of electron–hole pairs at the junction of a PN diode by successive collisions of the impinging photons. The released excessive electrons output a current, which is in direct relationship with the impinging photons. This current is then amplified by a suitable electronic circuit. The amplified current is then converted back to an optical signal and then transmitted.

Receiver

The receiver unit consists of a photodetector, demodulator, decoder and amplifier. As was the case with the repeater, in the first stage of the receiver, the incoming optical signals are detected using a photodiode. These are then amplified and demodulated to obtain the required digital signal. The signals are then decoded and the transmitted signal is outputted.

In the course of transmission, it may be some times necessary to change the communication channels. Hence, switching elements (not shown in the diagram) are also required. A separate power feed line is connected all through the communication distance to power these circuits. This type of optical communication is similar to the microwave communication except for the fact that the frequency range is very high (10^{15} Hz compared to 10^9 Hz for MW) and the transmitting medium is dielectric in nature (optical fibre) rather than metallic waveguides in MW communication.

Considering these in totality, there is no argument. Fibre-optic cable should be the transmission medium of choice when considering data links in new facilities where no other transmission medium candidate exists. However, there is even greater activity with respect to the implementation of data links in the premises or local area environment. This is the environment of the office building, Small Office Home Office (SOHO), the factory and the campus. As PCs have proliferated throughout all premise type facilities, the need for data communication links has followed. Installation of data links in the premises, be they point-to-point, multipoint, part of a local area network (LAN), or whatever, is a major agenda item for many business concerns.

16.7 ▶ Attenuation in Optical Fibres

A very important parameter of an optical fibre is the attenuation of light signal in the fibre. Attenuation decreases light transmittance. Usually, the power of light at the output end of optical fibre is less than the power launched at the input end, then the signal is said to be attenuated. The signal attenuation is defined as the ratio of the input optical power (P_i) into the fibre to the power of light coming out at the output end (P_0). The attenuation coefficient is given as:

$$\boxed{\alpha = \frac{10}{L}\log_{10}\frac{P_0}{P_i} \text{ dB/km},} \tag{16.17}$$

where L is the length of the fibre.

The causes of attenuation are numerous; some of them are waveguide structure, material compositions, material dispersion, material scattering, microbending losses, mode coupling radiation losses, etc. The attenuation is the function of wavelength and material. Optical communication wavelengths are 0.8, 1.3 and 1.55 μm. The attenuation is mainly due to (i) absorption and (ii) scattering.

Absorption losses

In glass fibres, three different absorptions take place. They are ultraviolet absorption, infrared absorption and ion resonance absorption. Ion resonance absorption losses in pure fused silica are shown in Fig. 16.14.

Absorption of UV radiation around 0.14 μm results in the ionization of valence electrons. Absorption of IR photons by atoms within the glass molecules causes heating. This gives absorption peak at 8 μm, also minor peaks at 3.2, 3.8 and 4.4 μm. The OH^- ions of water trapped during manufacturing causes absorption at 0.95, 1.25 and 1.39 μm as shown in Fig. 16.14. The presence of other impurities such as iron, copper and chromium also causes absorption. All these absorptions result in absorption loss in the fibre.

Scattering losses

The molten glass, when drawn into a very thin fibre under proper tension causes sub-microscopic variation in the density of glass in the fibre takes place. The dopants added to glass to vary the refractive index also leads to inhomogenities in the fibre. The microscopic variation of density and inhomogenities act as reflecting and refracting facets; these scatter a small portion of light passing through the glass. Thus, the scattering losses. If the size of density-fluctuating regions is of the order of $\lambda/10$ or less, then they act as point source scattering centre. This kind of scattering is known as Rayleigh scattering. The scattering loss is proportional to $1/\lambda^4$. On this basis, the scattering loss at a wavelength of 1.3 μm is

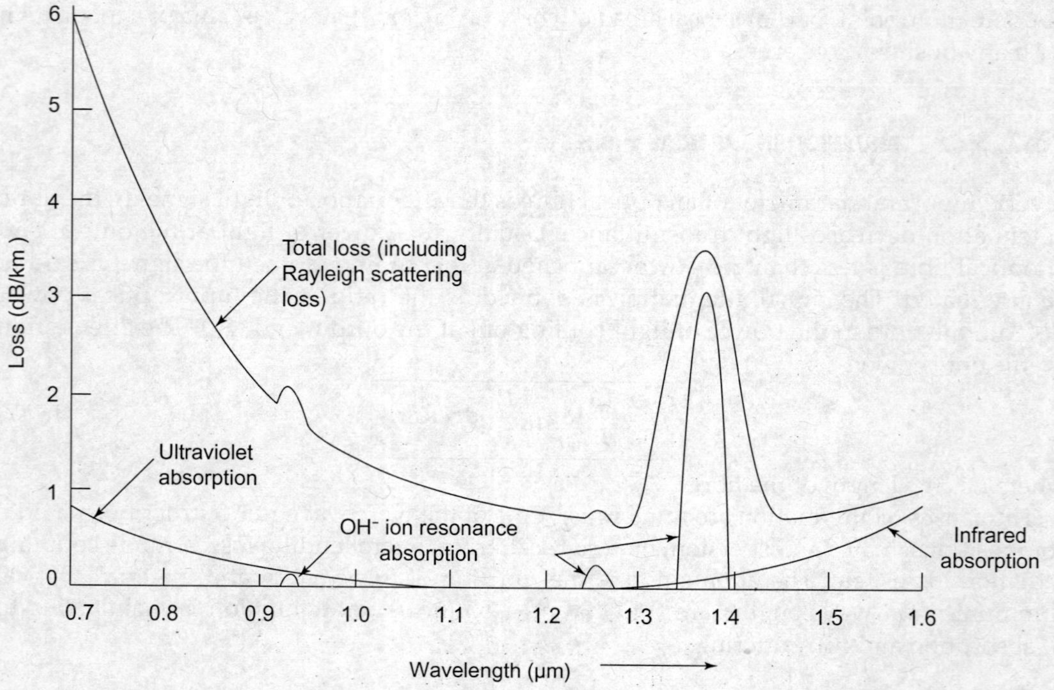

FIGURE 16.14 Ion resonance absorption loss effects in fused silica glass fibres

about 0.3 dB/km whereas at a wavelength of 0.7 μm it is about 5 dB/km. The Rayleigh scattering losses for silica is shown in Fig. 16.15.

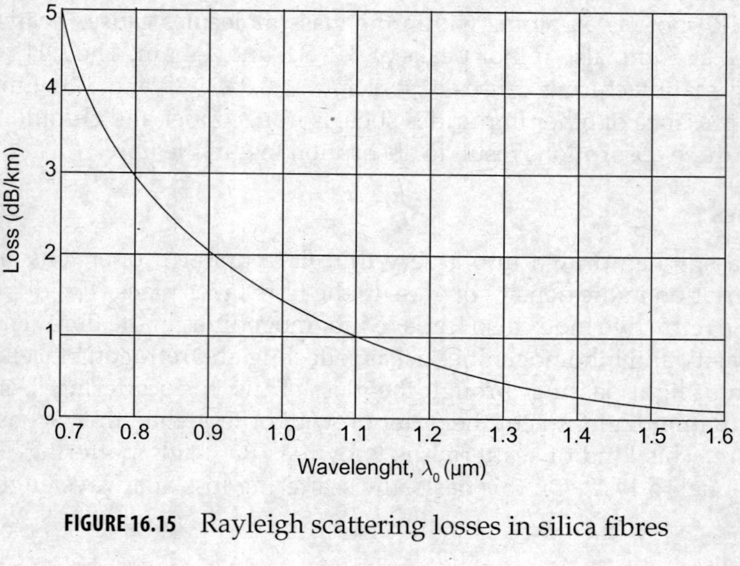

FIGURE 16.15 Rayleigh scattering losses in silica fibres

Bending losses

In a bent fibre, there is loss in power of the transmitted signal called bending losses. Einstein explained the bending losses. According to Einstein's theory of relativity, the part of the ray that enters into cladding will travel faster. The energy associated with this part of the ray is lost. This loss can be represented by absorption coefficient (α)

$$\boxed{\alpha = C \exp \frac{-R}{R_c}}, \qquad (16.18)$$

where C = constant

R = radius of curvature of fibre bend and $R_c = \dfrac{r}{(\mathrm{NA})^2}$

r = radius of the fibre. The bends with radius of curvature is of magnitude of the fibre radius gives rise to heavy losses.

Microbending and wave guide losses

A large number of small bends present in the fibre cause large attenuation in the signal transmission. This is known as microbending loss. Usually, microbends are formed when an unsheathed fibre is wound in tension on a drum during manufacture. These bends will be more if the surface of drum is non-uniform.

During manufacturing, if proper care is not taken, then a continuous small variation in the fibre diameter or circularity is formed. This gives scattering loss, known as waveguide losses.

EXAMPLE 16.3

Find the diameter of the core for single-mode transmission at 8500 Å whose refractive indices for core and cladding are 1.48 and 1.47, respectively.

Solution: For a single mode propagation, V-number is 2.405 and also

$$V = \frac{\pi d n_1}{\lambda_0}\left[1 - \left(\frac{n_2}{n_1}\right)^2\right]^{1/2}$$

or

$$V = \frac{\pi d}{\lambda_0}\sqrt{n_1^2 - n_2^2}$$

$\Rightarrow$

$$2.405 = \frac{3.14d}{8500 \times 10^{-10}\sqrt{(1.48)^2 - (1.47)^2}}$$

or

$$\mathbf{d = 3.79 \times 10^{-6}m}$$

EXAMPLE 16.4

The optical power, after propagating through a fibre that is 500 m long is reduced to 25% of its original value. Calculate the fibre loss in dB/km.

Solution: The loss per km is given as

$$\text{dB} = -\frac{10}{L}\log_{10}\left(\frac{p_o}{p_i}\right)\text{dB / km}$$

$$\text{Loss} = -\frac{10}{(1/2)}\log_{10}\left(\frac{25}{100}\right) \quad [\because L = 500\text{ m} = 1/2\text{ km}]$$

$$= -20\,(\log 25 - \log 100)$$

$$= \mathbf{12.042\ dB/km}$$

EXAMPLE 16.5

A communication system uses 10 km fibre having a loss of 2.3 dB/km. Compute the output power if the input power is 400 μW.

Solution: We know the loss per km is

$$\frac{-10}{L}\log_{10}\left(\frac{p_0}{p_i}\right)$$

$$\Rightarrow \quad 2.3 = \frac{-10}{10}\log_{10}\left(\frac{p_0}{p_i}\right) \text{ or } \log_{10}\left(\frac{p_0}{p_i}\right) = -2.3$$

$$\Rightarrow \quad \frac{p_0}{p_i} = \text{anti}\log\text{of}(-2.3) = 0.00199$$

$$\Rightarrow \quad p_0 = \mathbf{0.00199 \times p_i = 0.795\ \mu W}$$

16.8 ▶ Losses in Fibres

The biggest problem with any optical fibre has to do with losses in one form or other. Some losses, such as the absorption of some light energy by the glass itself, cannot be avoided. But there are other kinds of problems that can be reduced by careful design of the fibre.

If a narrow pulse of light (light of short duration, such as that might be produced by a flashbulb) is applied to one end of an optical fibre, the pulse of light at the far end will have a lower amplitude and a longer duration (i.e., a wider and weaker pulse). The resulting phenomenon called pulse dispersion or pulse spreading has three basic causes. These are

- Modal dispersion
- Material dispersion
- Waveguide delay distortion

Bandwidth of a fibre is its maximum allowable rate of transmission of information. The product of bandwidth and dispersion is known as bandwidth-dispersion product (BDP) and is used as a quality factor of the fibre.

16.8.1 Modal Dispersion

Higher and more complex propagation modes take a longer time to travel through a given distance in any medium. Because of this they require a wider path than a simple basic mode. Since an optical fibre with a larger core diameter can allow more different propagation modes, the more pronounced the modal dispersion effect will be. On the other hand, if the core diameter is small enough, it blocks all but the basic mode and the effect gets minimized (e.g., single-mode fibre).

16.8.2 Material Dispersion

The velocity of propagation through the core is not the same for all colours (or wavelengths) of light. Hence, a narrow white pulse entering the fibre will produce a series of overlapping pulses of different colours at the far end. This can be avoided by using a monochromatic light source such as an LED or a laser diode, so that all of the light passing through the fibre is at very nearly the same wavelength.

16.8.3 Waveguide Delay Distortion

This was first observed with microwaves travelling through waveguides. An optical fibre essentially being a waveguide for light waves exhibits the same behaviour, wherein each propagating mode within the fibre experiences a slight dispersion effect simply because of the confining waveguide, which cannot behave as if it were open space. This dispersion effect is quite small, which cannot be eliminated but can be ignored.

The basic problem caused by any dispersion effect is that the rate at which data are transmitted through the fibre gets limited. For example, if the modulation of the light amplitude is at a higher rate, then the dispersion levels out the changes, so that the output light at the far end of the fibre has nearly constant amplitude. The final result is that the modulations become indecipherable, and all data are lost.

16.9 ▶ Dispersion in Optical Fibres

Dispersion is the phenomenon in which phase velocity of a wave depends on its frequency. Media having such a property are termed as dispersive media. Dispersion is a problem in fibre communication. It limits the bandwidth of the fibre. When a pulse of light travels along a fibre, the signal is not only attenuated but is also distorted. This is due to dispersion. Dispersion is defined as the signal broadening or spreading while it propagates inside the fibre. Information which is to be transmitted through the fibre is first coded in the form of light pulses, and then these light pulses are fed into the fibre. A light pulse sent into an optical fibre broadens in time as it passes through the fibre. This phenomenon of broadening or spreading of pulses is known as pulse dispersion. This occurs because the different rays excited at the input end at one time after traversing through the fibre, exit the fibre at different times. Dispersion is sometimes called chromatic dispersion to emphasize its wavelength-dependent nature. There are many types of dispersion in optical fibre.

Modal dispersion or intermodal dispersion

Modal dispersion is a distortion mechanism occurring in multimode fibres and other waveguides, in which the signal is spread in time because the propagation velocity of the optical signal is not the same for all modes.

Modal dispersion is the dominant source of dispersion in multimode fibres. It does not exist in single-mode fibres. It occurs because each mode travels a different distance over the same time span. As light travels through the fibre, it spreads to fill all the possible modes or paths through the core.

16.9.1 Multimode Step Index Fibres

In order to obtain an expression for the amount of dispersion or pulse broadening in a multimode step index fibre, let us suppose that a ray OA be incident on the entrance aperture of the fibre making an angle α with the axis. This ray is refracted into the core along AB and strikes the upper core-cladding interface at B as shown in Fig. 16.16. If the ray AB is totally internally reflected back into the core and after suffering total internal reflection at the point C on the lower interface and further strikes the upper interface at D, then the distance BD covered by the ray in time t is given by

$$t = \frac{BC + CD}{v} \quad \text{but } v = \frac{c}{n}$$

where v is the velocity of the ray in the core of refractive index n_1 and c is the velocity of light,

$$t = \frac{BC + CD}{c/n_1}$$

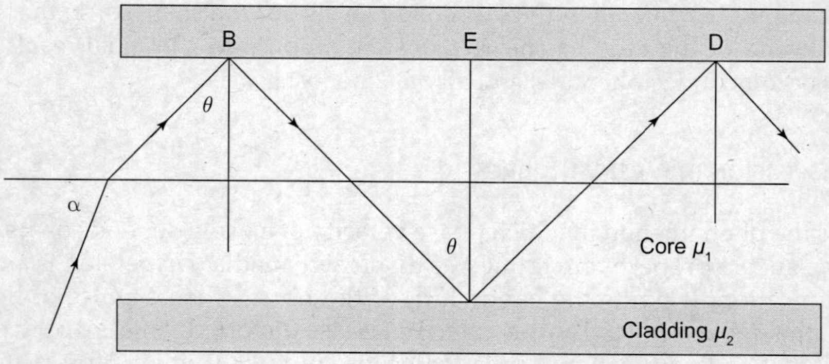

FIGURE 16.16 Propagation of ray through multimode step index fibre

From $\angle EBC$, $BE/BC = \cos\theta$ and in $\angle DEC$, $DE/DC = \cos\theta$

$$BC = \frac{BE}{\cos\theta} \text{ but } DC = \frac{DE}{\cos\theta}$$

$$\therefore \qquad BC + DC = \frac{BE + DE}{\cos\theta} = \frac{BD}{\cos\theta}$$

Therefore, the above equation becomes

$$t = \frac{BDn_1}{c\cos\theta}$$

As the ray in the fibre propagates by a series of total internal reflections at the interface, the time taken by the ray in traversing an axial length l of the fibre will be

$$\tau = \frac{ln_1}{c\cos\theta} \tag{16.19}$$

Equation (16.19) shows that the time taken by a ray is the function of the angle θ made by the ray with the core axis, which leads to the pulse dispersion. If we consider that all the light rays lying between angles θ and θ_c (critical angle) are present, then the time taken by the rays making zero angle with the fibre axis would be minimum and is given by putting $\theta = 0$ in Eq. (16.19) as

$$\tau_{\min} = \frac{l\,n_1}{c} \tag{16.20}$$

Similarly, the time taken by rays making an angle with the fibre axis is equal to the critical angle, that is $\theta = \theta_c$, would be maximum and is obtained by putting $\theta = \theta_c$ in Eq. (16.19) as

$$\tau_{\min} = \frac{l\,n_1}{c\cos\theta} \text{ but } \cos\theta_c = \frac{n_2}{n_1}$$

where μ_2 is the refractive index of cladding

$$\tau_{\min} = \frac{l\,n_1^2}{cn_2} \tag{16.21}$$

Hence, if all the rays enter the fibre at once, the rays will occupy a time interval $\Delta\tau$ at the output end of duration.

$$\Delta\tau = \tau_{\max} - \tau_{\min}$$

$$= \frac{l}{c}\frac{n_1^2}{n_2} - \frac{l\,n_1}{c}$$

$$\boxed{\Delta\tau = \frac{n_1 l}{c}\left[\frac{n_1}{n_2}\right] - 1 = \frac{l n_1}{c}\Delta} \tag{16.22}$$

This is the required expression for the amount of pulse broadening.

EXAMPLE 16.6

A step index fibre has a core refractive index of 1.48. If the core diameter and the numerical aperture of the fibre are, respectively, 50 and 0.5 μm, find the refractive index of the cladding, the acceptance angle and maximum number of modes of light of wavelength 1 μm the fibre can carry.

Solution:

(i) Numerical aperture, $\mathrm{NA} = \sqrt{\left(n_1^2 - n_2^2\right)}$

i.e.,

$$0.5 = \sqrt{\left(1.48^2 - n_2^2\right)}$$

Upon squaring and rearranging, we get $n_2^2 = 1.48^2 - 0.5^2 = 1.94$

$$\therefore \quad n_2 = \sqrt{1.94} = 1.39$$

(ii) From the definition for NA, we get, $\mathrm{NA} = \sin F_{\max}$

$$\therefore \quad \phi_{\max} = \sin^{-1}(\mathrm{NA}) = \sin^{-1}(0.5)$$

i.e.,

$$\phi_{\max} = 30$$

(iii) For a step-index fibre, the number of modes N is given by

$$N_{\text{step}} = 4.9\left[\frac{\mathrm{NA}\times d}{\lambda}\right]^2$$

$$= 4.9\left[\frac{0.5\times 50\times 10^{-6}}{1\times 10^{-6}}\right]^2$$

$$= 4.9[25]^2 = 4.9 \times 625$$

$$= 3062.5 \cong 3062 \text{ modes}$$

A graded index fibre of the same diameter allows half the number of modes of a step index fibre, i.e.,

$$N_{grad} = \left[\frac{N_{step}}{2}\right]$$

$$\therefore \quad \mathbf{N_{grad} = \frac{3062}{2} = 1531 \ \ modes}$$

EXAMPLE 16.7

What is the NA of a fibre, which has a relative refractive index difference of 0.05 and a core refractive index of 1.5?

Solution:

$$\text{NA} = n_1\,(2\Delta)^{1/2}$$

$$= 1.5(2 \times 0.05)^{1/2}$$

$$= 1.5(0.1)^{1/2} = 1.5 \times 0.316$$

$$\therefore \quad \mathbf{NA = 0.47}$$

EXAMPLE 16.8

Find the refractive index of the core and cladding if the NA is 0.3 and the relative refractive index difference is 3%.

Solution:

$$\text{NA} = n_1(2\Delta)^{1/2}$$

$$\therefore \quad n_1 = \frac{\text{NA}}{\sqrt{2\Delta}} = \frac{0.3}{\sqrt{3/100}}$$

$$= \frac{0.3}{\sqrt{0.03}} = \frac{0.3}{0.173}$$

$$n_1 = 1.73$$

Also

$$n_2 = n_1(1-\Delta)$$

$$= 1.73\,(1 - 0.03)$$

$$= 1.73(0.97)$$

$$\mathbf{n_2 = 1.68}$$

16.10 ▶ Applications of Optical Fibres

Due to the various advantages, optical fibres widely used in many systems. Some of these are as follows:

1. Digital transmission system
2. Close circuit TV (CCTV) links for traffic control and security, etc.
3. Optical fibre sensors
4. The public telecommunication networks, such as telephone, cable TV, videophone, multimedia, desktop teleconferencing, etc.
5. Undersea cable system for international telecommunication network
6. In military mobiles such as air-craft, tanks, ships, etc., fibre-guided missiles, short and long distance communication links.
7. The single multiplexing and transmission in automotive electronics, centralized locking, door, lamps, power, windows, seat mechanism, etc., in automobiles optical fibres are widely used.

Advantages

The optical fibre has many advantages over conventional cable or wireless system. Some of the advantages are discussed below.

(a) **Information capacity and low attenuation:** Information-carrying capacity of optical fibre cable is much more than copper, coaxial cable and microwave systems. It is normally expressed in terms of the amount of data that can be transferred through the cable in a given period of time or information bandwidth. The higher the bandwidth, the greater the information-carrying capacity.

(b) **Small size:** The fibre optic cable is a lot smaller in size than conventional wire cables. It is light weight and occupies much less space. A fibre optic cable of 0.25-inch diameter will carry the same information as a 3-inch bundle of 900 pains of copper cable.

(c) **Security and privacy:** Problems with placing so many copper cable together is called cross-talk, defined as the interference of signals between adjacent wire pairs. There is no cross-talk in optical fibres and the transmission is secure and private.

(d) **Electromagnetic isolation:** The electromagnetic field present around high voltage lines do not affect the data being transferred in fibre optic cable.

(e) **Electrical isolation:** Fibre optic cable is not an electrically conductive medium. Thus, fibre optic transmitter and receiver do not have to have a common ground between them. Fibre cable is immune to lighting, radiations and nuclear explosions.

(f) **Reliability and safety:** Optical fibre can withstand environmental hazard such as pollution, corrosion due to salts. They have longer life span of 20–30 year than copper cables.

(g) **Low cost:** The fibre optic cable has higher bandwidth than that required for the signal being transmitted. When communication link change is required, i.e. higher bandwidth is needed, the fibre cable will not be replaced, only transmitter or receiver will be changed.

Disadvantages

The disadvantages of fibre optic cable are few but must be mentioned.

(a) Some types of cable are harder to join, especially if using splicing. Splicing can be more time consuming and thus increasing the cost of installation of the cable.

(b) All fibre optic systems are limited to fix point-to-point ground installation. They cannot leave the ground nor be associated with a mobile communication station.

(c) Since fibre optic is non-conductive, if an electrical communication is needed at remote site, an additional conducting member has to be added the total cable configuration.

(d) Fragility is another major disadvantage of optical fibre. These are easily broken or damaged due to age and vibrations.

(e) Lack of standards and knowledge are a problem. Standards for implementation of networks and signal level are still under progress. Lack of knowledge of the industry and installation techniques have been a major issue.

All these disadvantages can be solved or overcome but must not be over looked.

Applications of optical fibres

There are lot of applications of optical fibres. Some of them are discussed below:

(a) **Optical fibre communication system through optical fibre cable:** Fibre optics deal with communication of audio as well as video signals. It essentially consists of a transmitter, an optical fibre cable and a receiver. The transmitter unit consists of a suitable electronics circuit that convert audio or video input into electrical signals. By using optical fibre, the transmission loss is low. There is complete isolation between the sender and the receiver.

The fibre itself can withstand environment conditions such as salt, pollution and radiation with no resulting corrosion and minimal nuclear radiation effects, so it is more reliable. The transmission is more secure and private.

(b) **Optical fibres in information technology:** A modern large complete system is composed of a large number of interconnections ranging in length over 16 orders of magnitude from the micrometre dimension (VLSI) connections to thousands of kilometre for terrestrial link in computer networks. The transmission line features of fibre optics are potentially attractive for many of these computer connections.

(c) **Optical fibre as sensors:** Sensor is a transducer which is used to convert one physical variable into another. Sensors are of two types, i.e.,

(i) active optical fibre sensors

(ii) passive optical fibre sensors.

(i) **Active optical fibre sensors:** The device consists of an optical fibre placed between a pair of ridged plates, impart a periodic perturbation to the fibre. The quantity to be measured acts directly on the fibre to modify the radiation passing through it. Because of displacement, the plates induce microbend losses. The intensity of the light output varies. By measuring the change in intensity, displacement is measured. Sensors can be designed to record the change in phase or polarization modulation as shown in Fig. 16.17.

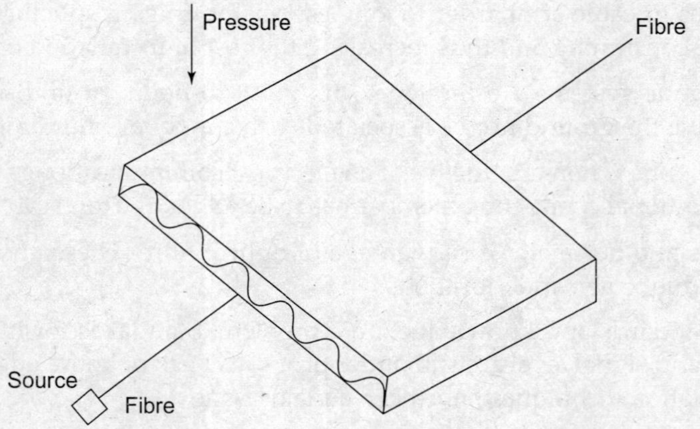

FIGURE 16.17 Displacement or pressure sensor

(ii) **Passive optical fibre sensors:** The device consists of two fibres, one to transmit light from source to object and the other to collect light from the object. The quantity to be measured acts indirectly on the fibre to modify the radiation. Change in the position of the object will result in changes in the amount of light collected by the detector. By recording the change in intensity, displacements (Δd) are estimated as shown in Fig. 16.18.

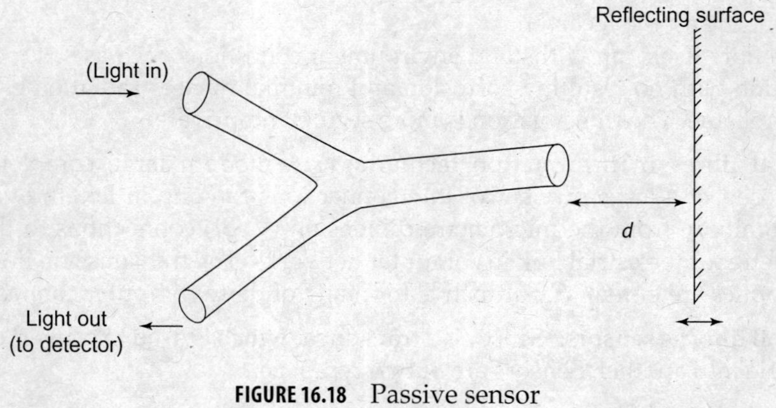

FIGURE 16.18 Passive sensor

(d) **Military applications:** Optical fibres are finding a lot of uses in many military operations. If conventional copper cables are to be transported by an aircraft, a ship or a tank for control mechanisms or communications, then a lot of weight has to be carried. However, with the optical fibre, weight is enormously reduced and also communication network is greatly improved, as with fibre optics, secrecy is also maintained. During recent wars, fibre-guided missiles to collect the video information would be used. This information is passed on to the ground control van, which sends further commands to the missiles through sensors. Control van constantly monitors the course of missiles and corrects the directions of the missiles to hit the target precisely.

RECAP ZONE

POINTS TO REMEMBER

- Optical fibres are flexible, thin and cylindrical structures and are made of transparent materials such as glass and plastic, which allows light to be guided from one end to the other end with minimal loss.
- Refractive index (μ or n) is defined as the ratio of the velocity of light in vacuum (c) to that of the velocity of light in a medium (v), $'\mu'$ or $'n' = c/v$.
- The critical angle, θ_c, is then defined as the angle of incidence of light for which the refracted light just graces the boundary between the two media and is expressed as

$$\sin\theta_c = \frac{n_2}{n_1}$$

- The phenomenon during which the incident light is completely reflected within the incident medium itself is called total internal reflection.
- Total internal reflection occurs when the ray of light is incident on a dielectric interface from a high refractive index region to a low refractive index region.
- Acceptance angle is defined as the maximum angle of incidence at the face of a fibre, which allows the light to travel through it by total internal reflection.
- There are three types of fibres, namely glass fibre, plastic fibre and plastic clad silica (PCS) fibre.
- A mode is a path taken by a ray of light through an optical fibre and is a mathematical concept of explaining the propagation of light waves in an optical fibre.
- The condition for propagation of a mode is given by

$$V = \frac{\pi d n_1}{\lambda_0}\left[1 - \left(\frac{n_2}{n_1}\right)^2\right]^{1/2} = \frac{\pi d}{\lambda_0}(\text{NA})$$

- The cut-off wavelength (λ_c) for propagation is given by

$$\lambda_c = \frac{\pi d}{2.405}(\text{NA})$$

- The number of modes (N) that can propagate in a step index fibre is

$$N \cong \frac{V^2}{2}$$

- The number of modes that can propagate in a graded index fibre is half of that of a step index fibre, which is given by

$$N \cong \frac{V^2}{4}$$

- Hence, on the basis of propagation of number of modes, fibres are classified as single-mode fibre and multimode fibre.
- A single-mode fibre has a narrow core ($8-10\ \mu$m) and the refractive index between the core and the cladding does not change very much throughout the fibre.
- A multimode fibre allows more than one mode to travel through it and therefore has a larger core. Typically, the core diameter of a multimode fibre is in the order of 150–200 μm.
- In general, optical fibres are broadly classified into three major types as single-mode step index (SMSI) fibre, multimode step index (MMSI) fibre and multimode graded index (MMGI) fibre.
- The loss in fibre is due to a phenomenon called pulse dispersion or pulse spreading and has three basic causes. They are modal dispersion, material dispersion and waveguide delay distortion.
- Communication is an exchange of information between source and user that is located away from the source.
- A sensor is a device that is either physical or biological (human organs such as eye and skin) in nature, which can detect and/sense a signal, physical condition, etc.
- There are two major types of optical sensors: they are active or intrinsic sensors and passive or extrinsic sensors.

KEY TERMS WITH DEFINITIONS

- **Optics** – The scientific study of light or the properties of transmission and deflection of other forms of radiations.
- **Fibre optics** – Technology related to transportation of optical energy in guiding media specifically glass fibres.

- **Transmission** – Process of transmitting something or the state of being transmitted.
- **Transmitter** – Equipment used to generate and transmit electromagnetic waves carrying messages or signals, especially those of radio or television.
- **Receiver** – Equipment used to convert electrical signals into sounds.
- **Total internal reflection** – Reflection (complete) of a light ray reaching an interface with a less dense medium when the angle of incident exceeds the critical angle.
- **Numerical aperture** – Measure of amount of light that can be accepted by the fibre.
- **Acceptance angle** – Maximum angle that a light ray can have relative to the axis of fibre so that it may propagate inside the fibre.
- **Bandwidth** – A range of frequencies within a given band, in particular that is used for transmitting a signal.
- **Modulator** – It converts the electrical message into proper format and impresses this signal onto the wave generated by carrier source.
- **Carrier wave** – Carrier source generates the wave on which the information is transmitted.
- **Detector** – The function of detector is to separate the information from the carrier wave.
- **Demodulation** – The process of separating the information from the carrier wave.
- **Attention** – Reduction in amplitude and intensity of a signal.
- **Intrinsic absorption loss** – Due to interaction of light with one or more major components of glass.
- **Extrinsic absorption loss** – Due to impurities within the glass.
- **Dispersion** – Distortion of light pulse as it travels from one end of the fibre to the other end of fibre.

Important Formulae and Equations

Equation Number	Equation	Remarks
(16.3)	$\theta_c = \sin^{-1}\left(\frac{n_1}{n_2}\right)$	Critical angle
(16.5)	$\phi_{max} = \sin^{-1}(n_1^2 - n_2^2)^{1/2}$ where n_1 = refractive index of core n_2 = refractive index of cladding	Acceptance angle
(16.6)	$NA = \sqrt{n_1^2 - n_2^2}$	Numerical aperture

(16.13)	$N \cong \frac{V^2}{2}$	**Number of modes in step index fibre**
(16.14)	$N \cong \frac{V^2}{4}$	**Number of modes in graded index fibre**
(16.17)	$\frac{10}{L} \log_{10} \frac{p_0}{p_i} \frac{\text{dB}}{\text{km}}$ where p_0 = output power p_i = input optical power	**Attention coefficient**
(16.18)	$\alpha = C \exp . \frac{-R}{R_C}$ where R = radius of curvature in a bent fibre and $R_C = \frac{r}{(\text{NA})^2}$, r = radius of the fibre	**Scattering loss in silica optical fibre**
(16.22)	$\Delta\tau = \frac{l\, n_1}{C} \Delta$	**Amount of pulse broadening**

Review Zone

Short Answer Questions

1. What are optical fibres?
2. What is the basic principle of fibre optics?
3. What do you mean by refractive index of a medium?
4. Define total internal reflection.
5. State critical angle.
6. Define acceptance angle.
7. Define numerical aperture.
8. How do you construct a fibre?
9. What are the ways in which an optical fibre is categorized?
10. What do you mean by mode in a fibre?
11. What are the types of fibres based on the modes of travel?
12. Give the merits and demerits of optical fibres in communication system.

Long Answer Questions

1. Explain basic principle of optical fibres. Discuss fibre classification.
2. Explain acceptance angle and acceptance cone of a filter. What do

you mean by numerical aperture of a filter? Derive expressions for them.

3. Explain single-mode and multimode fibre. What are the advantages of optical fibres over copper wire?
4. Discuss the phenomena of dispersion in optical fibre.
5. Write a short note on the structure of an optical fibre.
6. Derive and explain the losses in fibres.
7. Enumerate the advantages and disadvantages of optical fibres.
8. With the help of a block diagram, explain how communication is possible using optical fibres.
9. What are the inherent qualities of a sensor that makes them useful for many applications?
10. Describe with neat diagrams, the working of displacement sensor and pressure sensor.

NUMERICAL PROBLEMS

1. The numerical aperture in a fibre is 0.48. What is the acceptance angle of it?

   Ans. $\phi_{max} = 28°41'$

2. The core diameter of multimode step index fibre is 50 μm. The numerical aperture is 0.25. Calculate the number of guided modes at operating wavelength of 0.75 μm.

   Ans. $N = 1370$

3. Find the core radius necessary for single-mode operation at 850 μm in singe index fibre with $n_1 = 1.48$ and $n_2 = 1.47$. What are the numerical aperture and maximum acceptance angle of this fibre?

   Ans. $r = 1.89$ μm, NA $= 0.1717$, $\theta_{max} = 9°53'12''$

4. An optical fibre has a coating of glass ($n_3 = 1.512$) to protect the optical surface of the fibre if the fibre itself has an index of reflection $n_2 = 1.70$. What is the critical angle for total reflection of ray inside the fibre?

   Ans. $\phi_c = 62.8°$

5. A multimode graded index fibre has a core with a parabolic refractive index profile which has a diameter of 60 μm and NA of 0.25. If the fibre is operating at a wavelength of 1.1 μm, then determine the total number of guided modes that the fibre will support.

   Ans. $N = 459$

6. An optical fibre made of silica glass has a relative refractive index difference of 0.45% and acceptance angle for the fibre in the air is 0.2 radian. Find the speed of light in core.

   Ans. V core $= 2.5 \times 10^8$ m/s

7. A silica optical fibre has numerical fibre of 0.40 and core refractive index (n_1) of 1.50. If the relative refractive index difference is 1%. Determine the acceptance angle for the fibre if the launching takes place from air.

   Ans. $\phi_{max} = 23.6°$

8. Calculate the refractive indices of the core (n_1) and cladding (n_2) material of a fibre if NA $= 0.22$ and

   $$\Delta = \frac{n_1 - n_2}{n_1} = 0.011.$$

   Ans. $n_1 = 1.42, n_2 = 1.40$

CHAPTER 17 Electrostatics

Learning Objectives

By the end of this chapter, the student will be able:

- To explain the basics of electrostatics and vector calculations
- To understand the concepts of scalars, vectors, divergence and curl
- To define electric field and the relation between potential and electric field
- To understand the applications and physical ideas contained in Gauss divergence theorem, Green's theorem and Stoke's theorem
- To explain the significance of Laplace's equation

17.1 ▶ Introduction

Electrostatics i.e. the static electric field is associated with charges that are stationary. An atom is electrically neutral but it becomes positively charged when loses an electron and negatively charged when gains an electron. Two important phenomena are associated with charged particles. Like charges repel each other and unlike charges attract each other. The materials, in electrostatics, fall under two categories:

(i) Conductors that allow the charges to pass through them and

(ii) Insulators/dielectrics that do not allow the charges to pass through them.

In this chapter, we will discuss the fundamentals of electrostatics, scalars, vectors and their applications, through examples. The mathematics is the language of physics. The physical formulae are the applications of laws of physics. Vector analysis provides the efficient way to present these laws. The physical quantities that can be defined by magnitude completely are called scalar quantities, e.g., mass, charge, potential and time etc. whereas, the quantities that are completely defined by a magnitude and direction are called vector quantities. The vector analysis brings to the forefront the physical ideas involved in equations.

17.2 ▶ Gradient of a Scalar Field

The gradient of a scalar V (the electric potential) point function is given by

$$\boxed{\text{grad } V = \vec{\nabla} V = \frac{\partial V}{\partial x}\hat{i} + \frac{\partial V}{\partial y}\hat{j} + \frac{\partial V}{\partial z}\hat{k}} \tag{17.1}$$

The gradient is a differential operator. It is a vector and respected as $\vec{\nabla}$ (del bar) and is given by

$$\vec{\nabla} = \hat{i}\frac{\partial}{\partial x} + \hat{j}\frac{\partial}{\partial y} + \hat{k}\frac{\partial}{\partial z}$$

where $\hat{i}, \hat{j}$ and $\hat{k}$ are the unit vectors along X, Y and Z directions, respectively.

In Eq. (17.1), the component $\partial V / \partial x$ represents the rate of change of $\vec{\nabla} V$ along the X-direction. Similarly, the components $\partial V / \partial y$ and $\partial V / \partial z$ represent the rate of change of $\vec{\nabla} V$ along Y and Z directions, respectively.

Note

Gradient tells how much something changes as we move from one point to another.

17.3 ▶ Relationship Between Electric Field and Potential

The intensity and the electric field $(\vec{E})$ is a vector quantity and the electric potential (V) is a scalar. There is no electric field in the region, if the electric potential in that region is same at different points. But if it is different at different points in that region then the electric field exists there. The direction of electric field is in the direction of decrease of potential. Thus, the relation between the electric potential and the electric field can be given by

$$\boxed{\vec{E} = -\frac{\partial V}{\partial r}\hat{r}} \text{ or } \boxed{\vec{E} = -\vec{\nabla} V} \tag{17.2}$$

where $\hat{r}$ is the unit vector along the field direction and the negative sign shows that the field is in the direction of decreasing potential.

Note

Potential is a property of the field that describes the action of the field upon an object.

17.4 ▶ Curl of a Vector Field

The curl of any vector $\vec{A}$ is defined as the circulation of the vector along the contour (l) enclosed by unit surface area (Fig. 17.1). It is represented as $\vec{\nabla}\times$ (del bar cross).

The curl of a vector actually gives the rotation of that vector in vector field. Curl of a vector $\vec{A}$ can be written as

$$\vec{\nabla}\times\vec{A} = \textbf{curl } \vec{\mathbf{A}} = \begin{vmatrix} \hat{i} & \hat{j} & \hat{k} \\ \frac{\partial}{\partial x} & \frac{\partial}{\partial y} & \frac{\partial}{\partial x} \\ A_x & A_y & A_z \end{vmatrix} \tag{17.3}$$

$$= \hat{i}\left(\frac{\partial Az}{\partial y} - \frac{\partial Ay}{\partial Z}\right) + \hat{j}\left(\frac{\partial Ax}{\partial Z} - \frac{\partial Az}{\partial x}\right) + \hat{k}\left(\frac{\partial Ay}{\partial x} - \frac{\partial Ax}{\partial y}\right) \tag{17.4}$$

where $\vec{A} = A_x\hat{i} + A_y\hat{j} + A_z\hat{k}$ and $\vec{\nabla} = \hat{i}\frac{\partial}{\partial x} + \hat{j}\frac{\partial}{\partial y} + \hat{k}\frac{\partial}{\partial z}$.

The circulation of $\vec{A}$ over a surface area S is represented as

$$\int_S \left(\text{curl } \vec{A}\right) d\vec{S} = \int_S \left(\vec{\nabla}\times\vec{A}\right)\cdot d\vec{S}$$

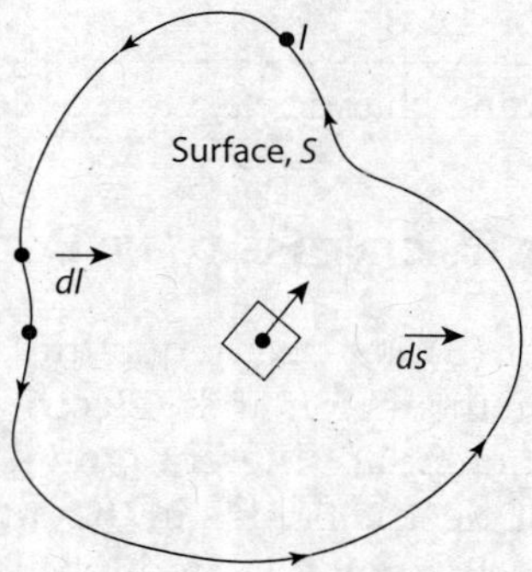

FIGURE 17.1 Surface enclosed by a closed path 'l'

Note

The curl of a vector field measures the tendancy for the vector field to swirl around.

17.5 ▶ Divergence of a Vector Point Function

The divergence of a vector point function $\vec{A}$ is denoted by div $\vec{A}$ and is defined as

$$\text{div } \vec{A} = \nabla\cdot\vec{A} = \left(\hat{i}\frac{\partial}{\partial x} + \hat{j}\frac{\partial}{\partial y} + \hat{k}\frac{\partial}{\partial z}\right)\vec{A}$$

if $\vec{A} = A_1\hat{i} + A_2\hat{j} + A_3\hat{k}$, then

$$\nabla\cdot\vec{A} = \left(\hat{i}\frac{\partial}{\partial x} + \hat{j}\frac{\partial}{\partial y} + \hat{k}\frac{\partial}{\partial z}\right)\cdot\left(A_1\hat{i} + A_2\hat{j} + A_3\hat{k}\right)$$

$$\boxed{\nabla \cdot \vec{A} = \frac{\partial A_1}{\partial x} + \frac{\partial A_2}{\partial y} + \frac{\partial A_3}{\partial z}} \tag{17.5}$$

Here, $\hat{i} \cdot \hat{i} = \hat{j} \cdot \hat{j} = \hat{k} \cdot \hat{k} = 1$ and $\hat{i} \cdot \hat{j} = \hat{j} \cdot \hat{k} = \hat{k} \cdot \hat{i} = 0$.

Note

Divergence talks about the rate of change along the three axes.

17.6 ▶ Gauss Divergence Theorem

If $\vec{A}$ is a vector point function, enclosed in a volume V bounded by a closed surface S, having continuous first-order partial derivatives, then

$$\boxed{\iiint_V \nabla \cdot \vec{A}\, dV = \iint_S \vec{A} \cdot \hat{n}\, dS} \tag{17.6}$$

where $\hat{n}$ is the unit vector drawn normal to surface S.

Here,

$$\vec{A} = A_1\hat{i} + A_2\hat{j} + A_3\hat{k}$$

$$\nabla \cdot \vec{A} = \text{div}\, \vec{A} = \frac{\partial A_1}{\partial x} + \frac{\partial A_2}{\partial y} + \frac{\partial A_3}{\partial z}$$

$$\hat{n} = \cos\alpha\hat{i} + \cos\beta\hat{j} + \cos\gamma\hat{k}$$

where α, β and γ are the angles drawn by unit vector $\hat{n}$ with the positive directions of x, y and z, respectively.

Thus, from Eq. (17.6),

$$\iiint_V \left(\frac{\partial A_1}{dx} + \frac{\partial A_2}{dy} + \frac{\partial A_3}{dz}\right) dxdydz = \iint_S (A_1\cos\alpha + A_2\cos\beta + A_3\cos\gamma)\, dS$$

or

$$\boxed{\iiint_V \left(\frac{dA_1}{dx} + \frac{dA_2}{dy} + \frac{dA_3}{dz}\right) dxdydz = \iint_S (A_1 dydz + A_2 dzdx + A_3 dxdy)} \tag{17.7}$$

where $\cos\alpha\, dS = dydz$

$\cos\beta\, dS = dzdx$

$\cos\gamma\, dS = dxdy$

17.7 ▶ Green's Theorem in the Plane

If $P(x, y)$ and $Q(x, y)$ be the continuous functions of x and y having continuous partial derivatives $\frac{\partial P}{\partial y}$ and $\frac{\partial Q}{\partial x}$ in a region R bounded by a closed curve, then

$$\oint_C (Pdx + Qdy) = \int\int_R \left(\frac{dQ}{dx} - \frac{dP}{dy}\right) dxdy \tag{17.8}$$

where C is traversed in the counterclockwise direction.

Note

Green's theorem states that the sum of fluid outflows from a volume is equal to the total outflow summed about an enclosing area.

17.8 ▶ Stoke's Theorem

If $\vec{A} = A_1\hat{i} + A_2\hat{j} + A_3\hat{k}$ be a vector point function having continuous first-order partial derivatives and S be an open surface bounded by a closed curve C, then

$$\oint_C \vec{A}\cdot d\vec{r} = \iint_S \text{curl}\,\vec{A}\cdot\hat{n}\,ds \tag{17.9}$$

Here,

$$\vec{A} = A_1\hat{i} + A_2\hat{j} + A_3\hat{k}$$

$$d\vec{r} = dx\hat{i} + dy\hat{j} + dz\hat{k}$$

or

$$\vec{A}\cdot d\vec{r} = A_1dx + A_2dy + A_3dz$$

So,

$$\oint_C \vec{A}\cdot d\vec{r} = \oint_C (A_1dx + A_2dy + A_3dz) \tag{17.10}$$

and

$$\text{curl} = \vec{A}\begin{vmatrix} \hat{i} & \hat{j} & \hat{k} \\ \frac{\partial}{\partial x} & \frac{\partial}{\partial y} & \frac{\partial}{\partial z} \\ A_x & A_y & A_z \end{vmatrix}$$

$$= \left(\frac{\partial A_3}{\partial y} - \frac{\partial A_2}{\partial z}\right)\hat{i} + \left(\frac{\partial A_1}{\partial z} - \frac{\partial A_3}{\partial x}\right)\hat{j} + \left(\frac{\partial A_2}{\partial x} - \frac{\partial A_1}{\partial y}\right)\hat{k}$$

$$\therefore \quad \text{curl}\,\vec{A}\cdot\hat{n} = \left(\frac{\partial A_3}{\partial y} - \frac{\partial A_2}{\partial z}\right)\cos\alpha + \left(\frac{\partial A_1}{\partial z} - \frac{\partial A_3}{\partial x}\right)\cos\beta + \left(\frac{\partial A_2}{\partial x} - \frac{\partial A_1}{\partial y}\right)\cos\gamma$$

Also, $$\iint_S \text{curl}\,\vec{A}\cdot\hat{n}\,dS = \iint_S \left[\left(\frac{\partial A_3}{\partial y}-\frac{\partial A_2}{\partial z}\right)\cos\alpha + \left(\frac{\partial A_1}{\partial z}-\frac{\partial A_3}{\partial x}\right)\cos\beta + \left(\frac{\partial A_2}{\partial x}-\frac{\partial A_1}{\partial y}\right)\cos\gamma\right] dS$$

$$= \iint_S \left(\frac{\partial A_3}{\partial y}-\frac{\partial A_2}{\partial z}\right)\cos\alpha\, dS + \iint_S \left(\frac{\partial A_1}{\partial z}-\frac{\partial A_3}{\partial x}\right)\cos\beta\, dS + \iint_S \left(\frac{\partial A_2}{\partial x}-\frac{\partial A_1}{\partial y}\right)\cos\gamma\, dS \quad (17.11)$$

Note

Stoke's theorem says that the surface integral of curl *A* over a surface *S* is the circulation of *A* around the boundary of the surface.

EXAMPLE 17.1

Find grad ϕ at the points (1,–2,–1) when ϕ is given as $\phi = 3x^2 y - y^2 z^2$.

Solution:

$$\text{Grad}\,\phi = \nabla\phi = \left(\hat{i}\frac{\partial}{\partial x}+\hat{j}\frac{\partial}{\partial y}+\hat{k}\frac{\partial}{\partial z}\right)\phi$$

$$= \left(\hat{i}\frac{\partial}{\partial x}+\hat{j}\frac{\partial}{\partial y}+\hat{k}\frac{\partial}{\partial z}\right)(3x^2y - y^2z^2)$$

$$= \hat{i}(6xy)+\hat{j}(3x^2-2yz^2)+\hat{k}(-2y^2z)$$

$$= \mathbf{-12\hat{i}+7\hat{j}-8\hat{k}}$$

EXAMPLE 17.2

Show that (i) grad $\left(\frac{1}{r}\right) = \frac{-\vec{r}}{r^3}$

(ii) grad $r = \frac{\vec{r}}{r}$ if $\vec{r} = x\hat{i}+y\hat{j}+z\hat{k}$

Solution: (i)

$$\text{grad}\left(\frac{1}{r}\right) = \nabla\left(\frac{1}{r}\right) = \left(\hat{i}\frac{\partial}{\partial x}+\hat{j}\frac{\partial}{\partial y}+\hat{k}\frac{\partial}{\partial z}\right)$$

$$= \hat{i}\left(\frac{\partial}{\partial x}\frac{1}{r}\right)+\hat{j}\left(\frac{\partial}{\partial y}\frac{1}{r}\right)+\hat{k}\left(\frac{\partial}{\partial z}\frac{1}{r}\right)$$

$$= \hat{i}\left(\frac{-1}{r^2}\frac{\partial r}{\partial x}\right)+\hat{j}\left(\frac{-1}{r^2}\frac{\partial r}{\partial y}\right)+\hat{k}\left(\frac{-1}{r^2}\frac{\partial r}{\partial z}\right)$$

$$\text{grad}\,(1/r) = \hat{i}\left(\frac{-1}{r^2}\frac{x}{r}\right) + \hat{j}\left(\frac{-1}{r^2}\frac{y}{r}\right) + \hat{k}\left(\frac{-1}{r^2}\frac{z}{r}\right)$$

Here, $\vec{r} = x\hat{i} + y\hat{j} + z\hat{k}$

$$r^2 = x^2 + y^2 + z^2$$

$$2r\frac{dr}{dx} = 2x$$

$$\frac{dr}{dx} = \frac{x}{r},\ \frac{dr}{dy} = \frac{y}{r},\ \frac{dr}{dz} = \frac{z}{r}$$

$$= \frac{-1}{r^3}\left(x\hat{i} + y\hat{j} + z\hat{k}\right) = \frac{-1}{r^3}(\vec{r}) = -\frac{\vec{r}}{r^3}$$

(ii) $$\text{grad}\, r = \nabla r = \left(\hat{i}\frac{\partial}{\partial x} + \hat{j}\frac{\partial}{\partial y} + \hat{k}\frac{\partial}{\partial z}\right)r$$

$$\nabla r = \hat{i}\frac{\partial r}{\partial x} + \hat{j}\frac{\partial r}{\partial y} + \hat{k}\frac{\partial r}{\partial z} \qquad (1)$$

We know $\vec{r} = x\hat{i} + y\hat{j} + z\hat{k}$

and $r^2 = x^2 + y^2 + z^2$

$$2r\frac{\partial r}{\partial x} = 2x \quad \Bigg| \quad 2r\frac{\partial r}{dy} = 2y \quad \Bigg| \quad 2r\frac{\partial r}{\partial z} = 2z$$

$$\frac{dr}{dx} = \frac{x}{r} \quad \Bigg| \quad \frac{\partial r}{dy} = \frac{y}{r} \quad \Bigg| \quad \frac{\partial r}{\partial z} = \frac{z}{r}$$

Putting the values of $\frac{\partial r}{\partial x}, \frac{\partial r}{\partial y}$ and $\frac{\partial r}{\partial z}$ in Eq. (1), we get

$$\nabla r = \hat{i}\left(\frac{x}{r}\right) + \hat{j}\left(\frac{y}{r}\right) + \hat{k}\left(\frac{z}{r}\right)$$

$$\nabla r = \frac{1}{r}\left(x\hat{i} + y\hat{j} + z\hat{k}\right)$$

$$= \frac{\vec{\boldsymbol{r}}}{\boldsymbol{r}}$$

EXAMPLE 17.3

Find the divergence and curl of a vector function $\vec{V} = (xyz)\hat{i} + (3x^2y)\hat{j} + (xz^2 - y^2z)\hat{k}$ at the point (2, −1, 1).

Solution:

$$\text{div}(\vec{V}) = \frac{\partial}{\partial x}(xyz) + \frac{\partial}{\partial y}(3x^2y) + \frac{\partial}{\partial z}(xz^2 - y^2z)$$

$$\text{div}(\vec{V}) = yz + 3x^2 + 2xz - y^2 = -1 + 12 + 4 - 1$$

$$\text{div}(\vec{V}) = 14 \text{ at } (2, -1, 1)$$

$$\text{curl } \vec{V} = \begin{vmatrix} \hat{i} & \hat{j} & \hat{k} \\ \dfrac{\partial}{\partial x} & \dfrac{\partial}{\partial y} & \dfrac{\partial}{\partial z} \\ xyz & 3x^2y & (xz^2 - y^2z) \end{vmatrix}$$

$$= \hat{i}(-2yz - 0) + \hat{j}(xy - z^2) + \hat{k}(6xy - xz)$$

$$= 2\hat{i} - 3\hat{j} + (-14)\hat{k}$$

$$\textbf{curl } \vec{V} = 2\hat{i} - 3\hat{j} - 14\hat{k} \textbf{ at } (2, -1, 1)$$

EXAMPLE 17.4

A vector field is given by $\vec{A} = (x^2 + xy^2)\hat{i} + (y^2 + x^2y)\hat{j}$. Show that the field is irrotational.

Solution:

$$\text{curl } \vec{A} = \nabla \times \vec{A} = \begin{vmatrix} \hat{i} & \hat{j} & \hat{k} \\ \dfrac{\partial}{\partial x} & \dfrac{\partial}{\partial y} & \dfrac{\partial}{\partial z} \\ (x^2 + xy^2) & (y^2 + x^2y) & 0 \end{vmatrix}$$

$$\text{curl } \vec{A} = \hat{i}(0 - 0) - \hat{j}(0 - 0) + \hat{k}(2xy - 2xy) = 0$$

curl $\vec{A}$ = 0, this shows that the field is irrotational.

EXAMPLE 17.5

If the vector $\vec{F}=\left(ax^2y+yz\right)\hat{i}+\left(xy^2-xz^2\right)\hat{j}+\left(2xyz-2x^2y^2\right)\hat{k}$ is solenoidal, find the value of a.

Solution:

Here, $$\vec{F}=\left(ax^2y+yz\right)\hat{i}+\left(xy^2-xz^2\right)\hat{j}+\left(2xyz-2x^2y^2\right)\hat{k} \quad (1)$$

If $\vec{F}$ is solenoidal, then div $\vec{F}=0$
So, from Eq. (1)

$$\text{div}\,\vec{F}=\nabla\cdot\vec{F}=\left(\hat{i}\frac{\partial}{\partial x}+\hat{j}\frac{\partial}{\partial y}+\hat{k}\frac{\partial}{\partial z}\right)$$

$$\left[\left(ax^2y+yz\right)\hat{i}+\left(xy^2-xz^2\right)\hat{j}+\left(2xyz-2x^2y^2\right)\hat{k}\right]$$

$$=\frac{\partial}{\partial x}(ax^2y+yz)+\frac{\partial}{\partial y}(xy^2-xz^2)+\frac{\partial}{\partial z}(2xyz-2x^2y^2)$$

$$\nabla\cdot\vec{F}=2axy+2xy+2xy=2axy+4xy$$

$$\nabla\cdot\vec{F}=2(a+2)xy \quad (2)$$

Since $\vec{F}$ is solenoidal, so $\nabla\cdot\vec{F}=0$
From Eq. (2),

$$2(a+2)\,xy=0$$

$$\boldsymbol{a=-2}$$

EXAMPLE 17.6

Evaluate $\iint_S \vec{A}\cdot\hat{n}\,dS$, where $\vec{A}=\left(x+y^2\right)\hat{i}-2x\hat{j}+2yz\hat{k}$ and S is the surface of plane $2x+y+2z=6$ in the first octant.

Solution: Given that

$$\nabla\left(2x+y+2z\right)=2\hat{i}+\hat{j}+2\hat{k}$$

$\hat{n}=$ unit vector normal to S

$$=\frac{2\hat{i}+\hat{j}+2\hat{k}}{\sqrt{(2)^2+(1)^2+(2)^2}}=\frac{2}{3}\hat{i}+\frac{1}{3}\hat{j}+\frac{2}{3}\hat{k}$$

$$\hat{n}\cdot\hat{k} = \hat{k}\left(\frac{2}{3}\hat{i} + \frac{1}{3}\hat{j} + \frac{2}{3}\hat{k}\right) = \frac{2}{3}$$

$$\therefore \qquad \iint_S \vec{A}\cdot\hat{n}\,dS = \iint_R \vec{A}\cdot\vec{n}\frac{dxdy}{|\hat{n}\cdot\hat{k}|} \qquad (1)$$

Now,

$$\vec{A}\cdot\hat{n} = \left[(x+y^2)\hat{i} - 2x\hat{j} + 2yz\hat{k}\right]\cdot\left(\frac{2}{3}\hat{i} + \frac{1}{3}\hat{j} + \frac{2}{3}\hat{k}\right)$$

$$= \frac{2}{3}(x+y^2) - \frac{2}{3}x + \frac{4}{3}yz$$

$$= \frac{2}{3}y^2 + \frac{4}{3}yz \quad \left[\begin{matrix}2x+y+2z=6\\ z=\dfrac{6-2x-y}{2}\end{matrix}\right]$$

$$= \frac{2}{3}y^2 + \frac{4}{3}y\left(\frac{6-2x-y}{2}\right)$$

$$= \frac{2}{3}y(y+6-2x-y)$$

$$\vec{A}\cdot\hat{n} = \frac{4}{3}y(3-x)$$

Hence,

$$\int\int_S \vec{A}\cdot\hat{n}\,dS = \int\int_R \vec{A}\cdot\hat{n}\frac{dxdy}{(\hat{n}\cdot\hat{k})}$$

$$= \int_0^3\int_0^{6-2x}\left\{\frac{4}{3}y(3-x)\frac{3}{2}\right\}dxdy$$

$$= \int_0^3\int_0^{6-2x} 2y(3-x)dydx$$

$$= \int_0^3 2(3-x).\left[\frac{y^2}{2}\right]_2^{6-2x} dx$$

$$= \int_0^3 (3-x)(6-2x)^2\,dx$$

$$= 4\int_0^3 (3-x)^3\,dx$$

$$= 4\left[\frac{(3-x)^4}{4(-1)}\right]_0^3$$
$$= -(0-81) = 81$$

EXAMPLE 17.7

Find $\iint \vec{F}\cdot\hat{n}\,ds$ where $\vec{F} = (2x+3z)\hat{i} - (xz+y)\hat{j} + (y^2+2z)\hat{k}$ and S is the surface of the sphere having centre at (3, −1, 2) and radius 3.

Solution: Using Gauss divergence theorem,

$$\iint_S \vec{F}\cdot d\vec{S} = \iiint_V \text{div}\,\vec{F}\,dV$$
$$= \iiint_V \left[\frac{\partial}{\partial x}(2x+3z) + \frac{\partial}{\partial y}(-xz-y) + \frac{\partial}{\partial z}(y^2+2z)\right]dV$$
$$= \iiint_V (2-1+2)dV = 3\iiint_V dV = 3V$$

But V = volume of sphere $= \frac{4}{3}\pi r^3 = \frac{4}{3}\pi(3)^3 = 36\pi$

Hence, $$\iint_S \vec{F}\cdot d\vec{S} = 3V = 3\times 36\pi = 108\pi$$

EXAMPLE 17.8

Evaluate $\iint_S \vec{F}\cdot\hat{n}\,dS$ over the entire surface of the region above the xy-plane bounded by the cone $Z^2 = x^2 + y^2$ and the plane $Z = 4$, if $\vec{F} = 4xz\hat{i} + xyz^2\hat{j} + 3z\hat{k}$.

Solution: If V is the volume enclosed by S, then V is bounded by $Z = 0, Z = 4, Z^2 = x^2 + y^2$
Using Gauss divergence theorem,

$$\iint_S \vec{F}\cdot\hat{n}\,dS = \iiint_V \text{div}\,\vec{F}\,dV$$
$$= \iiint_V \left[\frac{\partial}{\partial x}(4xz) + \frac{\partial}{\partial y}(xyz^2) + \frac{\partial}{\partial z}(3z)\right]dV$$

$$= \iiint_V (4z + xz^2 + 3)dV$$

$$= \int_0^4 \int_{-z}^{z} \int_{-\sqrt{z^2-y^2}}^{\sqrt{z^2-y^2}} (4z + xz^2 + 3)dxdydz$$

$$= 2\int_0^4 \int_{-z}^{z} \int_0^{\sqrt{z^2-y^2}} (4z + 3)dxdydz \qquad \left[\text{since, } \int_{-\sqrt{z^2-y^2}}^{\sqrt{z^2-y^2}} xdx = 0\right]$$

$$= 2\int_0^4 \int_{-z}^{z} (4z+3)\sqrt{z^2 - y^2}\,dydz$$

$$= 4\int_0^4 \int_0^{z} (4z+3)\sqrt{z^2 - y^2}\,dydz$$

$$= 4\int_0^4 (4z+3)\left[\frac{y\sqrt{z^2 - y^2}}{2} + \frac{z^2}{2}\sin^{-1}\frac{y}{z}\right]_0^z dz$$

$$= 4\int_0^4 (4z+3)\left[\frac{z^2}{2}\sin^{-1} 1\right]dz$$

$$= 4 \times \frac{\pi}{4}\int_0^4 (4z^3 + 3z^2)dz$$

$$= \pi\left[z^4 + z^3\right]_0^4 = \pi(256 + 64) = 320\pi$$

EXAMPLE 17.9

If $\vec{r} = x\hat{i} + y\hat{j} + z\hat{k}$, show that curl $\vec{r} = 0$.

Solution: Here, $\vec{r} = x\hat{i} + y\hat{j} + z\hat{k}$

$$\text{curl}\,\vec{r} = \nabla \times \vec{r}$$

$$= \begin{vmatrix} \hat{i} & \hat{j} & \hat{k} \\ \dfrac{\partial}{\partial x} & \dfrac{\partial}{\partial y} & \dfrac{\partial}{\partial z} \\ x & y & z \end{vmatrix}$$

$$= \hat{i}[0-0]+\hat{j}[0-0]+\hat{k}[0-0]$$

$$= \mathbf{\nabla \times \vec{r} = 0}$$

EXAMPLE 17.10

Verify Green's theorem in the plane $\oint_C (3x^2-8y^2)\,dx+(4\text{y}-6\text{xy})\text{dy}$ where C is the boundary of the region defined by

(a) $y=\sqrt{x}, y=x^2$

(b) $x=0, y=0, x+y=1$

Solution: (a) $y=\sqrt{x}$

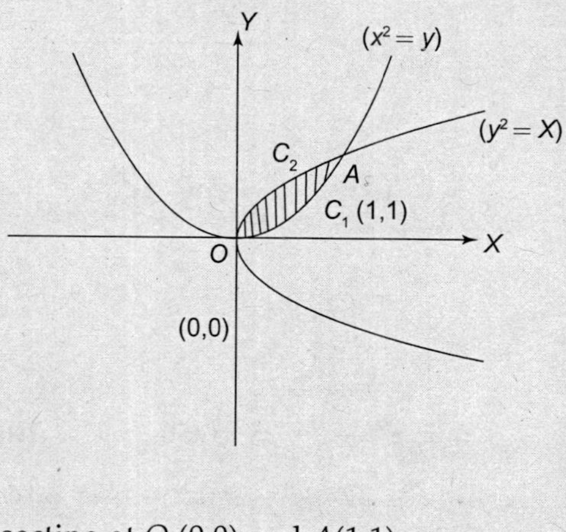

i.e., $y^2 = x$ and $y = x^2$

are two parabolas intersecting at O (0,0) and A(1,1).

Here,

$$P = 3x^2-8y^2, Q = 4y-6xy$$

$$\frac{\partial P}{\partial y} = -16y, \frac{\partial Q}{\partial x} = -6y$$

$$\frac{\partial Q}{\partial x} - \frac{\partial P}{dy} = 10y$$

If R is the region bounded by C, then

$$\iint_R \left(\frac{\partial Q}{\partial x} - \frac{\partial P}{\partial y}\right) dx\,dy$$

$$= \int_0^1 \int_{x^2}^{\sqrt{x}} 10y\,dy\,dx$$

$$= 5\int_0^1 \left[y^2\right]_{x^2}^{\sqrt{x}} dx$$

$$= 5\int_0^1 (x - x^4)\,dx$$

$$= 5\left[\frac{x^2}{2} - \frac{x^5}{5}\right]_0^1$$

$$= 5\left(\frac{1}{2} - \frac{1}{5}\right) = \frac{3}{2} \quad (1)$$

Also, $$\oint_C (Pdx + Qdy) = \int_{C_1} (Pdx + Qdy) + \int_{C_2} (Pdx + Qdy)$$

Along C_1, $x^2 = y$, $2xdx = dy$

Line integral along C_1,

$$= \int_{C_1} (Pdx + Qdy)$$

$$= \int_0^1 \left(3x^2 - 8x^4\right) dx + \left(4x^2 - 6x.x^2\right) 2xdx$$

$$= \int_0^1 \left(3x^2 + 8x^3 - 20x^4\right) dx$$

$$= \left[x^3 + 2x^4 - 4x^5\right]_0^1$$

$$= -1$$

Along C_2, $y^2 = x$, $2ydy = dx$

Line integral along C_2,

$$= \int_{C_2} (Pdx + Qdy)$$

$$= \int_1^0 \left(3y^4 - 8y^2\right)2y\,dy + \left(4y - 6y^2 \cdot y\right)dy$$

$$= \int_1^0 \left(4y - 22y^3 + 6y^5\right)dy$$

$$= \left[2y^2 - \frac{11}{2}y^4 + y^6\right]_1^0 = \frac{5}{2}$$

Line integral along $C = -1 + \frac{5}{2} = \frac{3}{2}$

i.e. $$\oint_C (P\,dx + Q\,dy) = \frac{3}{2} \qquad (2)$$

The equality of Eqs. (1) and (2) verifies Green's theorem.

(c) Here, $\iint_R \left(\frac{\partial Q}{\partial x} - \frac{\partial P}{\partial y}\right) dx\,dy = \int_0^1 \int_0^{1-x} 10y\,dy\,dx = \int_0^1 5\left[y^2\right]_0^{1-x} dx$

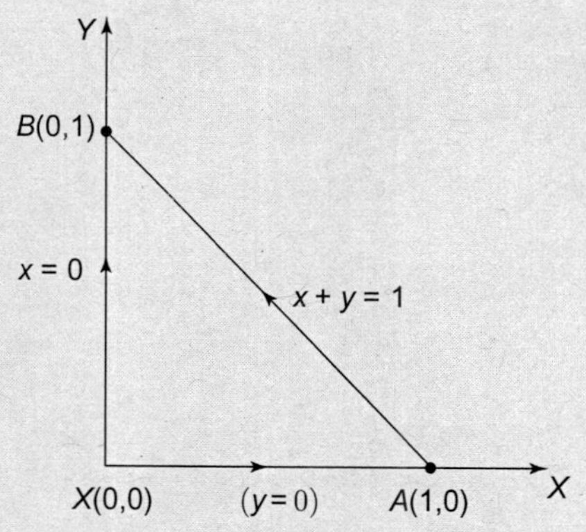

$$= 5\int_0^1 (1-x)^2\,dx$$

$$= 5\left[\frac{(1-x)^3}{-3}\right]_0^1 = \frac{-5}{3}(0-1) = \frac{5}{3} \qquad (1)$$

Along OA, $y = 0$, $dy = 0$

Line integral along $OA = \int_0^1 3x^2\,dx = \left[x^3\right]_0^1$

Along AB, $y = 1 - x$, $dy = -dx$

Line integral along AB

$$= \int_1^0 \left[3x^2 - 8(1-x)^2\right]dx + [4(1-x) - 6x(1-x)](-dx)$$

$$= \int_1^0 \left(3x^2 - 8 + 16x - 8x^2 - 4 + 4x + 6x - 6x^2\right)dx$$

$$= \int_1^0 \left(-12 + 26x - 11x^2\right)dx$$

$$= \left[-12x + 13x^2 - \frac{11}{3}3x^3\right]_1^0$$

$$= -1\left[-12 + 13 - \frac{11}{3}\right]$$

$$= \frac{8}{3}$$

Along BO, $x = 0$, $dx = 0$

Line integral along $BO = \int_1^0 4y\,dy = \left[2y^2\right]_1^0$

$$= -2$$

Line integral along $C = 1 + 8/3 - 2 = 5/3$

i.e., $$\oint_C (Pdx + Qdy) = 5/3 \qquad (2)$$

The equations (1) and (2) verifies Green's theorem in the plane.

EXAMPLE 17.11

Verify Stoke's theorem for $\bar{F} = \left(x^2 + y^2\right)\hat{i} - 2xy\hat{j}$ taken round the rectangle bounded by the lines $x = \pm a, y = 0, y = b$

Soultion: Let C denotes the boundary of the rectangle ABED, then

$$\oint_C \vec{F}\cdot d\vec{r} = \oint_C \left[\left(x^2 + y^2\right)\hat{i} - 2xy\hat{j}\right]\left(\hat{i}\,dx + \hat{j}\,dy\right)$$

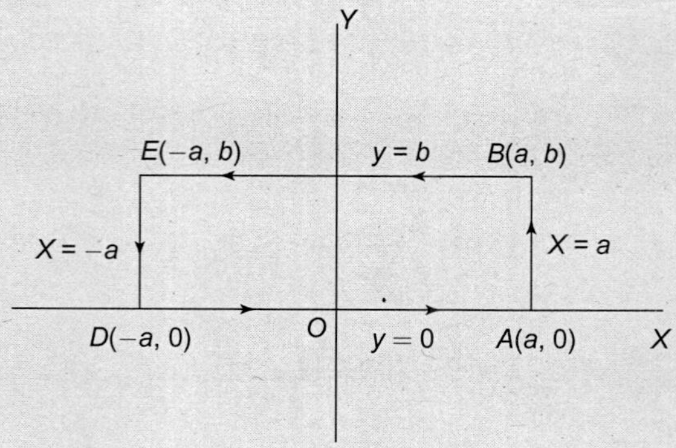

$$= \oint_C \left[\left(x^2 + y^2\right)dx - 2xy\,dy\right]$$

The curve C consists of four lines AB, BE, ED and DA.
Along AB, $x = a$, $dx = 0$ and y varies from 0 to b.

$$\int_{AB} \left[\left(x^2 + y^2\right)dx - 2xy\,dy\right]$$

$$= \int_0^b -2ay\,dy = -a\left[y^2\right]_0^b$$

$$= -ab^2 \qquad (1)$$

Along BE, $y = b$, $dy = 0$, x varies a to $-a$

$$\int_{BE} \left[\left(x^2 + y^2\right)dx - 2xy\,dy\right]$$

$$= \int_a^{-a} \left(x^2 + y^2\right)dx$$

$$= \left[\frac{x^3}{3} b^2 x\right]_a^{-a}$$

$$= \frac{-2a^2}{3} - 2ab^2 \qquad (2)$$

Along ED, $x=-a$, $dx=0$, y varies from b to 0,

$$\int_{ED}\left[\left(x^2+y^2\right)dx-2xy\,dy\right]$$

$$\int_b^0 2ay\,dy=a\left[y^2\right]_b^0$$

$$=-ab^2 \tag{3}$$

Along DA, $y=0$, $dy=0$, x varies from $-a$ to a.

$$\int_{DA}\left[\left(x^2+y^2\right)dx-2xy\,dy\right]$$

$$=\int_{-a}^{a} x^2dx=\frac{2a^3}{3} \tag{4}$$

Adding Eqs. (1) and (2), (3) and (4) we get

$$\oint_C \vec{F}\cdot d\vec{r}=-ab^2-\frac{2a^3}{3}-2ab^2-ab^2+\frac{2a^3}{3}$$

$$=-4ab^2 \tag{5}$$

and

$$\operatorname{curl}\vec{f}=\begin{vmatrix}\hat{i} & \hat{j} & \hat{k}\\ \dfrac{\partial}{\partial x} & \dfrac{\partial}{\partial y} & \dfrac{\partial}{\partial z}\\ \left(x^2+y^2\right) & (-2xy) & (0)\end{vmatrix}$$

$$=(-2y-2y)\hat{k}=-4y\hat{k}$$

For the surface S, $\hat{n}=\hat{k}$

$$\operatorname{curl}\overline{F}\cdot\hat{n}=-4y\hat{k}\cdot\hat{k}$$

$$=-4y$$

$$\iint_S \operatorname{curl}\overline{F}\cdot\overline{n}\,dS=\int_a^b\int_{-a}^{a}-4y\,dx\,dy$$

$$=\int_a^b -4y\,[x]_{-a}^{a}\,dy$$

$$= -8a \int_0^b y dy$$

$$= -8a \left[\frac{y^2}{2} \right]_a^b = -4ab^2 \quad (6)$$

The equality of Eqs. (5) and (6) Verifies the stroke's theorem.

EXAMPLE 17.12

Evaluate the surface integral $\iint_S \text{curl}\,\bar{F}\hat{n}\,dS$ by transforming it into a line integral, S being that part of the surface of the paraboloid $z = 1 - x^2 - y^2$, for which $z \geq 0$ and $\bar{F} = y\hat{i} + Z\hat{j} + x\hat{k}$.

Solution: The boundary C of the surface S is the circle $x^2 + y^2 = 2$, $z = 0$ whose parametric equations are $x = \cos t$, $y - \sin t$, $z = 0, 0 \leq 1 < 2\pi$

By Stroke's theorem,

We have

$$\iint_S \text{curl}\,\bar{F} \cdot \hat{n}\,dS = \oint_C \bar{F} \cdot d\vec{r}$$

$$= \oint_C (y\hat{i} + z\hat{j} + x\hat{k}).(\hat{i}\,dx + \hat{j}dy + \hat{k}dz)$$

$$= \oint_C (ydx + Zdy + xdz)$$

$$= \oint_C ydx, \text{ since on } C, z = 0, dz = 0$$

$$= \oint_C y \frac{dx}{dt} dt$$

$$= \int_0^{2\pi} \sin t(-\sin t)\,dt$$

$$= -4 \int_0^{\pi/2} \sin^2 t\,dt$$

$$= -\frac{4}{2} \times \frac{1}{2} \times \frac{\pi}{2}$$

$$= -\pi$$

RECAP ZONE

POINTS TO REMEMBER

- If ϕ is a constant scalar point function, then $\nabla\phi = 0$.
- If ϕ_1 and ϕ_2 are two scalar point functions, then

  (i) $\nabla(\phi_1 \pm \phi_2) = \nabla\phi_1 \pm \nabla\phi_2$

  (ii) $\nabla(\phi_1\phi_2) = \phi_1\nabla\phi_2 + \phi_2\nabla\phi_1$

  (iii) $\nabla\left(\dfrac{\phi_1}{\phi_2}\right) = \dfrac{\phi_2\nabla\phi_1 - \phi_1\nabla\phi_2}{\phi_2^2}$; $\phi_2 \neq 0$

- A unit vector $\hat{n}$ normal to the given surface ϕ can be written as $\hat{n} = \dfrac{\nabla\phi}{|\nabla\phi|}$.
- Divergence of a differentiable vector point function $\vec{V}$ is given by

$$\nabla\cdot\vec{V} = \frac{\partial V_1}{\partial x} + \frac{\partial V_2}{\partial y} + \frac{\partial V_3}{\partial z}$$

- The curl (or rotation) of a differentiable vector point function $\vec{V}$ is denoted by curl $\vec{V}$ or $\nabla\times\vec{V}$.
- If curl $\vec{V} = 0$, then $\vec{V}$ is said to be an irrotational vector, otherwise rotational.

  (i) $\text{div}(\vec{A}+\vec{B}) = \text{div}\,\vec{A} + \text{div}\,\vec{B}$

  (ii) $\nabla\times(\vec{A}+\vec{B}) = \nabla\times\vec{A} + \nabla\times\vec{B}$

  (iii) If $\vec{A}$ is a vector function and ϕ is a scalar function, then

  $\text{div}(\phi\vec{A}) = \phi\,\text{div}\,\vec{A} + (\text{grad}\,\phi)\cdot\vec{A}$

  (iv) $\text{div}(\vec{A}\times\vec{B}) = \vec{B}\cdot\text{curl}\,\vec{A} - \vec{A}\cdot\text{curl}\,\vec{B}$

  (v) $\nabla(\vec{A}\cdot\vec{B}) = (\vec{A}\cdot\nabla)\vec{B} + (\vec{B}\cdot\nabla)\vec{A} + \vec{A}\times(\nabla\times\vec{B}) + \vec{B}\times(\nabla\times\vec{A})$

  (vi) $\nabla\times(\vec{A}\times\vec{B}) = (\nabla\cdot\vec{B})\vec{A} - (\nabla\cdot\vec{A})\vec{B} + (\vec{B}\cdot\nabla)\vec{A} - (\vec{A}\cdot\nabla)\vec{B}$

- $\text{div}(\text{grad}\,\phi) = \nabla^2\phi$ where ∇^2 is called Laplacian operator.
- $\nabla^2\phi = 0$ is called Laplace's equation.
- $\text{curl}(\text{grad}\,\phi) = \nabla\times\nabla\phi = 0$

- $\text{div}(\text{curl}\,\vec{V}) = \nabla\cdot(\nabla\times\vec{V}) = 0$
- $\nabla\times(\nabla\times\vec{V}) = \text{curl}(\text{curl}\,\vec{V})$

  $= \nabla(\nabla\cdot\vec{V}) - \nabla^2\vec{V}$
- $\text{grad}(\text{div}\,\vec{V}) = \text{curl}(\text{curl}\,\vec{V}) + \nabla^2\vec{V}$
- The examples of scalar fields are temperature, sound intensity in a hall and refractive index of a medium, etc.
- Velocity field and force field are the examples of vector field.

Key Terms with Definitions

- **Point or position function** – A physical quantity that can be expressed as a function of the position of the point in a region of space.
- **Field** – It is a function that specifies a particular quantity everywhere in a region.
- **Vector (or scalar) field** – If the quantity is a vector (or scalar), its region is known as vector (or scalar) field.
- **Potential** – The potential or electric potential at a given point is the magnitude of potential point at that point per unit charge.
- **Potential difference** – The difference of electric potential between two points.
- **Electric potential** – Amount of work done at a point in moving a unit positive charge from infinity to that point.
- **Gradient** – It is a vector operation which operates on a scalar function to produce a vector whose magnitude is the maximum rate of change of the function at that point of gradient.
- **Curl** – It is a vector operator that describes the infinitesimal rotation.
- **Physical quantities** – It is a physical property that can be measured, e.g. mass of length, etc.
- **Electric field** – A region around a charged particle within which a force would be exerted on other charged particles.

Important Formulae and Equations

Equation Number	Equation	Remarks
(17.1)	$\text{grad}\,V = \frac{\partial V}{\partial x}\hat{i} + \frac{\partial V}{\partial y}\hat{j} + \frac{\partial V}{\partial z}\hat{k}$	Gradient of a scalar V

(17.2)	$\vec{E} = -\vec{\nabla}V$	**Relation between electric field $\vec{E}$ and potential V**
(17.3)	$\vec{\nabla}\times\vec{A} = \begin{vmatrix} \hat{i} & \hat{j} & \hat{k} \\ \partial/\partial x & \partial/\partial y & \partial/\partial z \\ A_x & A_y & A_z \end{vmatrix}$	**Curl of a vector field $\vec{A}$**
(17.5)	$\nabla\cdot\vec{A} = \dfrac{\partial A_1}{\partial x} + \dfrac{\partial A_2}{\partial y} + \dfrac{\partial A_3}{\partial z}$	**Divergence of a vector point function**
(17.6)	$\iiint\limits_V \nabla\cdot\vec{A}\,dV = \iiint\limits_S \vec{A}\cdot\hat{n}\,dS$	**Gauss divergence theorem**
(17.7)	$\iiint\limits_V \left(\dfrac{dA_1}{dx} + \dfrac{dA_2}{dy} + \dfrac{dA_3}{dz}\right) dx\,dy\,dz = \iint\limits_S (A_1 dy\,dz + A_2 dz\,dx + A_3 dx\,dy)$	**Gauss theorem**
(17.8)	$\oint\limits_C (P dx + Q dy) = \iint\limits_R \left(\dfrac{dQ}{dx} - \dfrac{dP}{dy}\right) dx\,dy$	**Green's theorem in a plane**
(17.9)	$\oint\limits_C \vec{A}\cdot\vec{dr} = \iint\limits_S \text{curl}\vec{A}\cdot\hat{n}\,dS$	**Stoke's theorem**

REVIEW ZONE

SHORT ANSWER QUESTIONS

1. What do you mean by gradient of a scalar field?
2. Derive a relationship between electric field and potential.
3. Define curl of a vector field.
4. What do you understand by divergence of a vector point function?
5. State Gauss divergence theorem.
6. Explain Stoke's theorem.
7. State and prove Green's theorem.
8. What do you mean by a scalar quantity?
9. What is a vector quantity?
10. Define electric field.
11. Define electrical potential.
12. What is the difference between a potential and potential difference?

Long Answer Questions

1. State and explain Gauss divergence theorem.
2. State and prove Stoke's theorem.
3. Explain gradient of a scalar field, curl of a vector field and divergence of a vector point function. Also, describe their properties.
4. What do you understand by electric field and electric potential. Establish a relation between them.
5. Derive a relation between electric potential and electric field. Also show that the direction of electric field is the direction of decrease of potential.

Numerical Problems

1. Prove that (i) $\nabla^2(r^n) = n(n+1)r^{n-2}$

   (ii) $\nabla^2\left(\frac{x}{r^3}\right) = 0$

   (iii) $\text{div}\left(\frac{\vec{r}}{r^3}\right) = 0$

   where $\vec{r} = x\hat{i} + y\hat{j} + z\hat{k}$

2. Show that (i) $\text{grad}\left(\frac{1}{r}\right) = \frac{-\vec{r}}{r^3}$

   (ii) $\text{grad}\ (\vec{a}\cdot\vec{r}) = \vec{a}$,

   where $\vec{a}$ is a constant vector

   and $\vec{r} = x\vec{i} + y\vec{j} + z\vec{k}$

3. Find curl (curl $\vec{V}$) where

   $\vec{V} = (2xz^2)\vec{i} - yz\vec{j} + 3xz^3\hat{k}$ at (1, 1, 1).

   Ans. $(5\hat{i} + 3\hat{k})$

4. Show that div(grad ϕ)= $\nabla^2\phi$

   where $\nabla^2 = \frac{d^2}{dx^2} + \frac{d^2}{dy^2} + \frac{d^2}{dz^2}$

5. Prove that $\vec{a}\cdot\nabla\left(\vec{b}\cdot\nabla\frac{1}{r}\right)$

   $= \frac{3(\vec{a}\cdot\vec{r})(\vec{b}\cdot\vec{r})}{r^5} - \frac{\vec{a}\cdot\vec{b}}{r^3}$

   where $\vec{a}$ and $\vec{b}$ are constant vectors.

6. Evaluate $\iint_S \vec{A}\cdot\hat{n}\,dS$,

   where $\vec{A} = z\hat{i} + x\hat{j} - 3y^2z\hat{k}$ and S is the curved surface of the cylinder $x^2 + y^2 = 16$ included in the first octant between Z = 0 and Z = 5.

   Ans. 90

7. Verify Gauss divergence theorem for $\vec{F} = 4x\hat{i} - 2y^2\hat{j} + Z^2\hat{k}$ taken over the region bounded by the cylinder $x^2 + y^2 = 4$, $z = 0$, $z = 3$.

   Ans. 84π

8. Prove that $\nabla^2 f(r) = f''(r) + \frac{2}{r}f'(r)$.

   Hence evaluate $\nabla^2(\log r)$.

9. Use Green's theorem in a plane to evaluate the integral

   $\oint_C \left[(2x^2 - y^2)dx + (x^2 + y^2)dy\right]$,

   where C is the boundary in the xy-plane of the area enclosed by the

x-axis and the semi-circle $x^2 + y^2 = 1$ in the upper half xy-plane.

Ans. 4/3

10. Using Stoke's theorem for the vector function

$$\vec{A} = (2x + y - 2z)\hat{i} + (2x - 4y + z^2)\hat{j} + (x - 2y + z^2)\hat{k},$$

evaluate the integral $\int_C \vec{A} \cdot d\vec{r}$, where C is the circle with centre at (0, 0, 3) and radius 5 in the plane $Z = 3$.

Ans. 25π

11. Find the directional derivative of div($\vec{u}$) at the point (1, 2, 2) in the direction of the outer normal to the sphere $x^2 + y^2 + z^2 = 9$ for

$$\vec{u} = x^4\hat{i} + y^4\hat{j} + z^4\hat{k}.$$

Ans. 68

12. Find the angles between the surfaces $x^2 + y^2 + z^2 = 9$ and $z = x^2+y^2 - 3$ at the point (2, −1, 2).

Ans. $\cos^{-1} \dfrac{8}{3\sqrt{21}}$

CHAPTER 18

Electromagnetic Theory

Learning Objectives

By the end of this chapter, the student will be able:

- To discuss physical significance of Maxwell equations
- To derive expressions for magnetostatic and electrostatic fields
- To derive poynting theorem from Maxwell equations
- To formulate the potential problems within electrostatics and magnetostatics
- To define and make calculations of plane electromagnetic waves in different media
- To interpret and solve real electrodynamic problems

18.1 ▶ Introduction

In early days, electricity and magnetism were treated as entirely separate subjects. But in 1820, Oersted noticed that an electric current could deflect a magnetic compass needle. Later, Ampere postulated that all magnetic phenomena are due to electric charges in motion. Then in 1831, Faraday discovered that a moving magnet generates an electric current. Thus, time-varying electric field gives rise to magnetic field and vice-versa. It was Maxwell and Lorentz who argued that electricity and magnetism are interconnected and presented a theory combining these two subjects, called electromagnetism. The variations of electric and magnetic intensity are transverse in nature. The variations of $\vec{E}$ and $\vec{H}$ are perpendicular to each other and also to the directions of wave propagation. The wave patterns of $\vec{E}$ and $\vec{H}$ for a travelling electromagnetic wave obey the fundamental equations, called Maxwell equations. These equations are mathematical abstractions of experimental results.

18.2 ▶ Basic Laws of Electricity and Magnetism

(i) Gauss's Law in Electrostatics

The flux of electric displacement vector $\vec{D}$ is given by

$$\boxed{\int_S \vec{D}\cdot d\vec{s} = \int_V \rho\, dV} \tag{18.1}$$

where S is the surface integral over the surface enclosing volume V.
Applying Gauss divergence theorem to Eq. (18.1),

$$\int_V \text{div}\,\vec{D}\, dV = \int_V \rho\, dV \tag{18.2}$$

$$\int_V \left(\text{div}\,\vec{D} - \rho\right) dV = 0 \tag{18.3}$$

As volume is arbitrary

$$\text{div}\,\vec{D} - \rho = 0$$

$$\text{div}\,\vec{D} = \rho$$

$$\Rightarrow \qquad \nabla\cdot\vec{D} = \rho \tag{18.4}$$

(ii) Gauss's Law in Magnetism

The flux of magnetic flux density $\vec{B}$ is given by

$$\boxed{\int \vec{B}\cdot d\vec{S} = 0} \tag{18.5}$$

Using Gauss divergence theorem to Eq. (18.5),

$$\int_V \left(\text{div}\,\vec{B}\right) dV = 0 \tag{18.6}$$

As volume is arbitrary,

$$\text{div}\,\vec{B} = 0$$

or

$$\nabla\cdot\vec{B} = 0 \tag{18.7}$$

(iii) Universal Form of Faraday's Law of Electromagnetic Induction

$$\boxed{\textbf{Curl}\ \vec{B} = -\frac{\partial \vec{B}}{\partial t}} \tag{18.8}$$

(iv) Ampere's Law in Magnetostatics

$$\oint_C \vec{H}\cdot d\vec{l} = \text{Current enclosed by path}$$

$$\boxed{= \int_V \vec{J} \cdot d\vec{S}} \tag{18.9}$$

where $\vec{H}$ is the magnetic field intensity and $\vec{J}$ is the current density.
Using Stoke's theorem on Eq. (18.9)

$$\int_S \text{Curl}\,\vec{H} \cdot d\vec{S} = \int_S \vec{J} \cdot d\vec{S} \tag{18.10}$$

or
$$\int \left(\text{Curl}\,\vec{H} - \vec{J}\right) d\vec{S} = 0 \tag{18.11}$$

As surface S is arbitrary,

$$\text{Curl}\,\vec{H} - \vec{J} = 0$$

or
$$\text{Curl}\,\vec{H} = \vec{J} \tag{18.12}$$

In addition to these four laws, we have

(i) Ohm's law

$$\vec{J} = \sigma \vec{E}$$

where σ is the electrical conductivity and $\vec{E}$ is the electric field strength.

(ii) $\vec{D} = \in \vec{E}$ and $\vec{B} = \mu \vec{H}$

where $\in$ = permittivity of medium and μ = permeability of medium

EXAMPLE 18.1

Find out the current density and electric field intensity for aluminium conductor with conductivity $\sigma = 3.82 \times 10^7$ siemen/m and the mobility of the electron, $\mu_e = 0.0014$ m^2/volt-sec and drift velocity 5.3×10^{-4} m/s.

Solution: The drift velocity of an electron is

$$\vec{v}_d = \mu_e \vec{E}$$

or
$$\vec{E} = \frac{\vec{v}_d}{\mu_e}$$

or
$$\vec{E} = \frac{5.3 \times 10^{-4}}{0.0014}$$

$$= 0.3786 \text{ V/m}$$

$$\text{Current density } \vec{J} = \sigma \vec{E}$$

$$= 3.82 \times 10^7 \times 0.3786$$

$$= \mathbf{1.446 \times 10^7\ Amp/m^2}$$

18.3 ▶ DERIVATION OF MAXWELL EQUATIONS

1. Derivation of first equation $\rightarrow$ $\operatorname{div} \vec{D} = \nabla \cdot \vec{D} = \rho$

Let us consider a surface S bounding total charge consisting of free charge plus polarization. If ρ and ρ_p are the charge densities of free charge and polarization charge, then Gauss's law can be expressed as

$$\int_S \vec{E} \cdot d\vec{S} = \frac{1}{\varepsilon_0} \int_V \left(\rho + \rho_p\right) dV \tag{18.13}$$

But

$$\rho_p = -\operatorname{div} \vec{P} \tag{18.14}$$

From Eqs. (18.13) and (18.14),

$$\int_S \varepsilon_0 \vec{E} \cdot d\vec{S} = \int_V \rho\, dV - \int_V \operatorname{div} \vec{P}\, dV$$

i.e.,

$$\int_V \operatorname{div}\left(\varepsilon_0 \vec{E} + \vec{P}\right) dV = \int_V \rho\, dV \tag{18.15}$$

$$\left[\because \int_S \varepsilon_0 \vec{E}\, d\vec{S} = \int_V \operatorname{div}\left(\varepsilon_0 \vec{E}\right) dV\right]$$

or

$$\int_V \left\{\operatorname{div}\left(\vec{D}\right) - \rho\right\} dV = 0 \quad \left[\because \varepsilon_0 \vec{E} + \vec{P} = \vec{D}\right] \tag{18.16}$$

or

$$\left(\operatorname{div} \vec{D} - \rho\right) = 0$$

$\Rightarrow$

$$\boxed{\nabla \cdot \vec{D} = \rho} \tag{18.17}$$

2. Derivation of second equation

$$\operatorname{div} \vec{B} = \nabla \cdot \vec{B} = 0$$

We know that the flux of magnetic induction $\vec{B}$ across the closed surface is always zero.

i.e.,

$$\int_S \vec{B} \cdot d\vec{S} = 0 \tag{18.18}$$

Using Gauss's divergence theorem,

$$\int_V \left(\operatorname{div}\vec{B}\right) dV = 0 \tag{18.19}$$

or $$\operatorname{div}\vec{B} = 0$$

or $$\boxed{\nabla \cdot \vec{B} = 0} \tag{18.20}$$

3. Derivation of third equation

$$\text{Curl } \vec{E} = -\frac{\partial \vec{B}}{\partial t}$$

or $$\nabla \times \vec{E} = -\frac{\partial \vec{B}}{\partial t}$$

According to Faraday's law of electromagnetic induction, it is known that e.m.f. induced in a closed loop is defined as negative rate of change of magnetic flux

$$e = -\frac{d\phi}{dt} \tag{18.21}$$

But magnetic flux $$\phi = \int_S \vec{B} \cdot d\vec{S}$$

where S is any surface having loop as boundary.

$$e = -\frac{d}{dt}\int_S \vec{B} \cdot d\vec{S}$$

$$e = -\int_S \frac{\partial \vec{B}}{\partial t} d\vec{S} \tag{18.22}$$

But e.m.f. 'e' can also be computed by calculating the work done in carrying a unit charge round the closed loop.
Thus,

$$e = \int_C \vec{E} \cdot d\vec{l} \tag{18.23}$$

From Eqs. (18.22) and (18.23),

$$\int \vec{E} \cdot d\vec{l} = -\int_S \frac{\partial \vec{B}}{\partial t} d\vec{S} \tag{18.24}$$

Using Stoke's theorem,

$$\int_S \left(\nabla \times \vec{E}\right) d\vec{S} = -\int_S \frac{\partial \vec{B}}{\partial t} d\vec{S} \tag{18.25}$$

or $$\int_S \left[\left(\nabla\times\vec{E}+\frac{\partial\vec{B}}{\partial t}\right)\right] d\vec{S}=0$$

or $$\boxed{\nabla\times\vec{E}=-\frac{\partial\vec{B}}{\partial t}} \tag{18.26}$$

4. Derivation of fourth equation

$$\text{Curl }\vec{H}=\vec{J}+\frac{\partial\vec{D}}{\partial t}$$

We know that the Ampere's circuital law is

$$\text{Curl }\vec{B}=\nabla\times\vec{B}=\mu_0\vec{J} \tag{18.27}$$

or $$\text{Curl }\vec{H}=\vec{J} \qquad \left[\because \vec{B}=\mu_0\vec{H}\right] \tag{18.28}$$

Taking divergence of Eq. (18.28),

$$\text{div}\left(\text{Curl}\,\vec{H}\right)=\text{div}\,\vec{J}$$

$$0=-\frac{\partial\rho}{\partial t} \qquad \left[\because \text{eqn of continuty, div}\,\vec{J}=-\frac{\partial\rho}{\partial t}\right]$$

or $$\frac{\partial\rho}{\partial t}=0 \tag{18.29}$$

Equation (18.29) is valid only for steady currents but for time-varying fields

$$\text{Curl }\vec{H}=\vec{J}+\vec{J}' \tag{18.30}$$

or $$\text{div}\left(\text{Curl }\vec{H}\right)=\text{div}\,\vec{J}+\text{div}\,\vec{J}'$$

or $$0=\text{div}\,\vec{J}+\text{div}\,\vec{J}'$$

or $$\text{div}\,\vec{J}'=-\text{div}\,\vec{J} \tag{18.31}$$

or $$\text{div}\,\vec{J}'=-\left(\frac{-\partial\rho}{\partial t}\right)=\frac{\partial\rho}{\partial t}=\frac{\partial}{\partial t}\left(\nabla\cdot\vec{D}\right) \tag{18.32}$$

$$\left[\because \nabla\cdot\vec{D}=\rho\right]$$

or $$\text{div}\,\vec{J}'=\frac{\partial}{\partial t}\left(\nabla\cdot\vec{D}\right)$$

$$=\nabla\cdot\left(\frac{\partial\vec{D}}{\partial t}\right)$$

$$\text{div}\,\vec{J}' = \text{div}\left(\frac{\partial \vec{D}}{\partial t}\right) \tag{18.33}$$

From Eq. (18.30),

$$\boxed{\textbf{Curl } \vec{H} = \vec{J} + \frac{\partial \vec{D}}{\partial t}} \quad \left[\because \text{div}\,\vec{J}' = \text{div}\frac{\partial \vec{D}}{\partial t} \text{ or } \vec{J}' = \frac{\partial \vec{D}}{\partial t}\right] \tag{18.34}$$

where $\vec{J}' = \frac{\partial \vec{D}}{\partial t}$ is called displacement current density which is equally effective in producing magnetic field as conduction current density. This displacement current is responsible for the production of magnetic field in empty space where the conduction current is zero.

Thus, four Maxwell equations are as follows:

1. $\nabla \cdot \vec{D} = \rho$
2. $\nabla \cdot \vec{B} = 0$
3. $\nabla \times \vec{E} = -\frac{\partial \vec{B}}{\partial t}$
4. $\nabla \times \vec{H} = \vec{J} + \frac{\partial \vec{D}}{\partial t}$

Note

The electromagnetic waves travel with the speed of light in free space and are transverse in nature.

18.4 ▶ Physical Significance of Maxwell Equations

(i) Maxwell's first equation is $\rightarrow \nabla \cdot \vec{D} = \rho$

Integrating this over an arbitrary volume V, we get

$$\int_V \left(\nabla \cdot \vec{D}\right) dV = \int_V \rho \, dV \tag{18.35}$$

Using Gauss divergence theorem,

$$\int_S \vec{D} \cdot d\vec{S} = \int_V \rho \, dV \tag{18.36}$$

Equation (18.36) represents Maxwell's first equation in integral form.

But, $\int_V \rho \, dV = q,$ the net charge contained in volume V.

Thus, Eq. (18.36) becomes

$$\boxed{\int_S \vec{D}\cdot d\vec{S} = q} \tag{18.37}$$

Maxwell's first equation signifies that the net out word flux of electric displacement vector through the surface enclosing a volume is equal to the net charge contained within the volume.

(ii) Maxwell's second equation is

$$\nabla\cdot\vec{B} = 0 \tag{18.38}$$

Integrating it over an arbitrary volume V,

$$\int_V \left(\nabla\cdot\vec{B}\right) dV = 0 \tag{18.39}$$

Using Gauss divergence theorem,

$$\int_V \left(\nabla\cdot\vec{B}\right) dV = \int_S \vec{B}\cdot d\vec{S} = 0 \tag{18.40}$$

Equation (18.40) represents Maxwell's second equation in integral form and signifies that the net outward flux of magnetic induction B through any closed surface is zero.

(iii) Maxwell's third equation is

$$\nabla\times\vec{E} = -\frac{\partial\vec{B}}{\partial t} \tag{18.41}$$

Integrating Eq. (18.41) over a surface S bounded by a curve C,

$$\int_S \left(\nabla\times\vec{E}\right) d\vec{S} = \frac{-\partial}{\partial t}\int_S \vec{B}\cdot d\vec{S} \tag{18.42}$$

Using Stoke's theorem,

$$\int_C \vec{E}\cdot d\vec{l} = -\frac{\partial}{\partial t}\int_S \vec{B}\cdot d\vec{S} \quad \left[\because \int_S \left(\nabla\times\vec{E}\right) d\vec{s} = \int_C \vec{E}\cdot d\vec{l}\right] \tag{18.43}$$

Equation (18.43) represents Maxwell's third equation in integral form and signifies that the electromotive force $\left(e = \int_C \vec{E}\cdot d\vec{l}\right)$ around a closed path is equal to the negative rate of change of magnetic flux linked with the path $\left(\because \phi = \int_S \vec{B}\cdot d\vec{s} = \text{magnetic flux}\right)$.

(iv) Maxwell's fourth equation is

$$\nabla \times \vec{H} = \vec{J} + \frac{\partial \vec{D}}{\partial t} \tag{18.44}$$

Taking surface integral over surface S bounded by curve C, we obtain

$$\int_S (\nabla \times \vec{H}) d\vec{S} = \int_S \left(\vec{J} + \frac{\partial \vec{D}}{\partial t} \right) d\vec{S} \tag{18.45}$$

Using Stoke's theorem,

$$\int_S (\nabla \times \vec{H}) d\vec{S} = \oint \vec{H} \cdot d\vec{l} \tag{18.46}$$

From Eqs. (18.45) and (18.46),

$$\oint \vec{H} \cdot d\vec{l} = \int_S \left(\vec{J} + \frac{\partial \vec{D}}{\partial t} \right) d\vec{S} \tag{18.47}$$

Equation (18.47) represents the Maxwell's fourth equation in integral form and signifies that the magnetomotive force $\left(m \cdot m \cdot f = \oint \vec{H} \cdot d\vec{l} \right)$ around a closed path is equal to the conduction current plus displacement current through any surface bounded by the path.

EXAMPLE 18.2

A copper wire 0.254 cm in diameter carries a current of 50 Amp. Find the magnetic field induction $\vec{B}$ at the surface of the wire. The permeability constant $\mu_o = 4\pi \times 10^{-7}$ wb/Amp-m.

Solution: The magnitude of magnetic field induction $\vec{B}$ at a point distant d from a straight wire carrying a current i Amp is given by

$$\vec{B} = \frac{\mu_o i}{2\pi d}$$

At the surface of the wire,

$$d = r, \text{ the radius of wire}$$

$$\vec{B} = \frac{\mu_o i}{2\pi r}$$

Here,

$$r = \frac{0.254}{2} = 0.127 \text{ cm}$$

$$= 0.127 \times 10^{-2} \text{ m}$$

$$i = 50 \text{ Amp}$$

and $$\mu_o = 4\pi \times 10^{-7} \text{ wb/Amp-m}$$

$\therefore$ $$\vec{B} = \frac{4\pi \times 10^{-7} \times 50}{2\pi \times 0.127 \times 10^{-2}}$$
$$= \mathbf{7.87 \times 10^{-3} \text{ wb/m}^2}$$

18.5 ▶ Equation of Continuity

We know that the net amount of charge in an isolated system remains constant. The movement of charge constitutes the current, i.e.,

$$i = \frac{dq}{dt} = \frac{d}{dt}\int \rho \, dV \tag{18.48}$$

where ρ is the charge density. The net amount of charge which crosses a unit area perpendicular is defined as current density.

Let a volume V is enclosed by a surface S, $\vec{J} \cdot d\vec{S}$ will represent the charge leaving volume V through a surface element dS in unit time (Fig. 18.1).

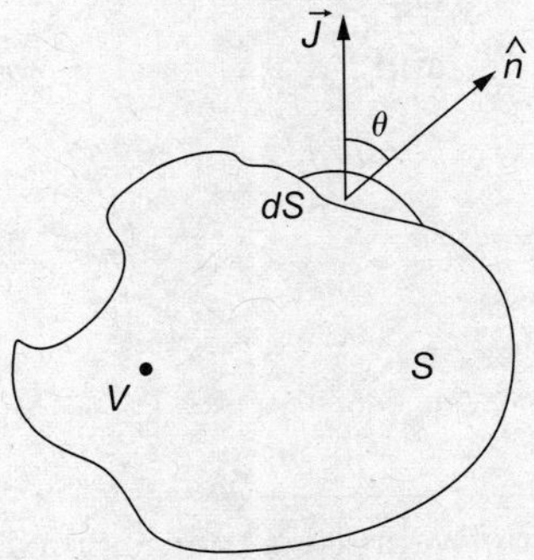

FIGURE 18.1

Now,

$$i = -\oint \vec{J} \cdot d\vec{S} \tag{18.49}$$

Thus,

$$\oint \vec{J} \cdot d\vec{S} = -\frac{dq}{dt} = -\frac{d}{dt}\int_V \rho \, dV = -\frac{\partial \rho}{\partial t} dV \tag{18.50}$$

Using Gauss divergence theorem,

$$\int_V \left(\operatorname{div} \vec{J}\right) dV = -\int_V \frac{\partial \rho}{\partial t} dV \tag{18.51}$$

$$\Rightarrow \qquad \int_V \left(\operatorname{div} \vec{J} + \frac{\partial \rho}{\partial t} \right) dV = 0$$

$$\Rightarrow \qquad \boxed{\operatorname{div} \vec{J} + \frac{\partial \rho}{\partial t} = 0} \tag{18.52}$$

Equation (18.52) is called equation of continuity. It states that current flowing out of a given volume must be equal to the rate of decrease of charge within the volume.

For stationary current, ρ is constant.

So,
$$\frac{\partial \rho}{\partial t} = 0$$

$$\Rightarrow \qquad \boxed{\operatorname{div} \vec{J} = 0} \tag{18.53}$$

18.5.1 Maxwell Equations in Free Space

In free space, the volume charge density ρ and the current density $\vec{J}$ are zero. Hence, the Maxwell equations take the form:

$$\nabla \cdot \vec{D} = 0 \qquad \text{Also, here}$$

$$\nabla \cdot \vec{B} = 0 \qquad \vec{D} = \varepsilon_0 \vec{E} \tag{18.54}$$

$$\text{and}$$

$$\nabla \times \vec{E} = -\frac{\partial \vec{B}}{\partial t} \qquad \vec{B} = \mu_0 \vec{H}$$

$$\nabla \times \vec{H} = \frac{\partial \vec{D}}{\partial t}$$

Note

An electromagnetic interaction between two charges means an exchange of energy and momentum between them.

18.6 ▶ MAXWELL EQUATIONS IN LINEAR ISOTROPIC MEDIUM

In a Linear Isotropic Medium

$$\vec{D} = \varepsilon \vec{E}$$

and

$$\vec{B} = \mu \vec{H} \tag{18.55}$$

Maxwell equations for linear isotropic medium take the form

$$\nabla \cdot \vec{D} = 0 \tag{18.56}$$

$$\nabla \cdot \vec{B} = 0 \tag{18.57}$$

$$\nabla \times \vec{E} = -\mu \frac{\partial \vec{H}}{\partial t} \tag{18.58}$$

$$\nabla \times \vec{H} = \varepsilon \frac{\partial \vec{E}}{\partial t} \tag{18.59}$$

If electromagnetic fields vary harmonically with time, then

$$\vec{D} = D_0 e^{iwt}, \quad \vec{B} = B_0 e^{iwt} \tag{18.60}$$

Also,

$$\frac{\partial \vec{D}}{\partial t} = D_0 (iw) e^{iwt} = iw\vec{D} \tag{18.61}$$

and

$$\frac{\partial \vec{B}}{\partial t} = B_0 (iw) e^{iwt} = iw\vec{B} \tag{18.62}$$

Equations (18.56)–(18.59) for harmonically varying field take the form:

$$\boxed{\nabla \cdot \vec{D} = \rho} \tag{18.63}$$

and

$$\boxed{\nabla \times \vec{E} + iw\vec{B} = 0} \tag{18.64}$$

$$\boxed{\nabla \cdot \vec{B} = 0} \tag{18.65}$$

$$\boxed{\nabla \times \vec{H} - iw\vec{D} = \vec{J}} \tag{18.66}$$

Thus, Eqs. (18.63)–(18.66) represents the Maxwell equations for harmonically varying fields.

EXAMPLE 18.3

Show that equation of continuity $\operatorname{div} \vec{J} + \frac{\partial \rho}{\partial t} = 0$ is contained in Maxwell's equations.

Solution: Maxwell's fourth equation has the differential form

$$\vec{\nabla} \times \vec{H} = \vec{J} + \frac{\partial \vec{D}}{\partial t}$$

Taking divergence of both sides, we have

$$\vec{\nabla} \cdot \left(\vec{\nabla} \times \vec{H}\right) = \vec{\nabla} \cdot \left(\vec{J} + \frac{\partial \vec{D}}{\partial t}\right)$$

But $$\left(\vec{\nabla}\cdot\vec{\nabla}\times\vec{H}\right)=0$$

$\therefore$ $$\vec{\nabla}\cdot\left(\vec{J}+\frac{\partial\vec{D}}{\partial t}\right)=0$$

or $$\vec{\nabla}\cdot\vec{J}+\vec{\nabla}\cdot\left(\frac{\partial\vec{D}}{\partial t}\right)=0$$

or $$\vec{\nabla}\cdot\vec{J}+\frac{\partial}{\partial t}\left(\vec{\nabla}\cdot\vec{D}\right)=0$$

Now from Maxwell's first equation $\vec{\nabla}\cdot\vec{D}=\rho$

So $$\boldsymbol{\vec{\nabla}\cdot\vec{J}+\frac{\partial\rho}{\partial t}=0}$$

18.7 ▶ Poynting Theorem and its Physical Significance

Since, we know that electrostatic potential energy is

$$U_e=\frac{1}{2}\int_V \vec{E}\cdot\vec{D}\,dV \tag{18.67}$$

and energy stored in magnetic field is

$$U_m=\frac{1}{2}\int_V \vec{H}\cdot\vec{B}\,dV \tag{18.68}$$

Maxwell equations are

$$\left.\begin{aligned}\nabla\cdot\vec{D}&=\rho & (a)\\ \nabla\cdot\vec{B}&=0 & (b)\\ \nabla\times\vec{E}&=-\frac{\partial\vec{B}}{\partial t} & (c)\\ \nabla\times\vec{H}&=\vec{J}+\frac{\partial\vec{D}}{\partial t} & (d)\end{aligned}\right] \tag{18.69}$$

Taking scalar product of Eq. (18.69)(c) with $\vec{H}$, we get

$$\vec{H}\cdot\left(\nabla\times\vec{E}\right)=-\vec{H}\cdot\frac{\partial\vec{B}}{\partial t} \tag{18.70}$$

and taking scalar product of Eq. (18.69)(d) with $\vec{E}$, we get

$$\vec{E}\cdot\left(\nabla\times\vec{H}\right)=\vec{E}\cdot\vec{J}+\vec{E}\cdot\frac{\partial\vec{D}}{\partial t} \tag{18.71}$$

Using vector identity

$$\text{div}\left(\vec{E}\times\vec{H}\right)=\vec{H}\cdot\text{Curl }\vec{E}-\vec{E}\cdot\text{Curl }\vec{H} \tag{18.72}$$

Using Eqs. (18.70) and (18.71) in Eq. (18.72),

$$H\cdot\left(\nabla\times\vec{E}\right)-\vec{E}\cdot\left(\text{Curl }\vec{H}\right)=-\vec{H}\frac{\partial\vec{B}}{\partial t}-\vec{E}\cdot\frac{\partial\vec{D}}{\partial t}-\vec{E}\cdot\vec{J} \tag{18.73}$$

$$\Rightarrow \qquad \text{div}\left(\vec{E}\times\vec{H}\right)=-\left(\vec{H}\frac{\partial\vec{B}}{\partial t}+\vec{E}\frac{\partial\vec{D}}{\partial t}\right)-\vec{E}\cdot\vec{J} \tag{18.74}$$

If medium is linear, then

$$\vec{B}=\mu\vec{H},$$

$$\vec{D}=\varepsilon\vec{E} \tag{18.75}$$

and

$$\vec{E}\cdot\frac{\partial\vec{D}}{\partial t}=\vec{E}\cdot\frac{\partial}{\partial t}\left(\varepsilon\vec{E}\right)=\frac{1}{2}\varepsilon\frac{\partial}{\partial t}\left(\vec{E}\right)^2$$

$$=\frac{1}{2}\varepsilon\frac{\partial\left(\vec{E}\right)^2}{\partial t}=\frac{\partial}{\partial t}\left(\frac{1}{2}\vec{E}\cdot\varepsilon\vec{E}\right)$$

$$=\frac{\partial}{\partial t}\left(\frac{1}{2}\vec{E}\cdot\vec{D}\right) \tag{18.76}$$

and

$$\vec{H}\cdot\frac{\partial\vec{B}}{\partial t}=\vec{H}\cdot\frac{\partial}{\partial t}\left(\mu\vec{H}\right)$$

$$=\frac{1}{2}\mu\frac{\partial}{\partial t}\left(\vec{H}\right)^2=\frac{\partial}{\partial t}\left(\frac{1}{2}\mu\vec{H}\cdot\vec{H}\right)$$

$$=\frac{\partial}{\partial t}\left(\frac{1}{2}\vec{H}\cdot\vec{B}\right) \tag{18.77}$$

Putting Eqs. (18.76) and (18.77) in Eq. (18.74), we get

$$\text{div}\left(\vec{E}\times\vec{H}\right)=-\frac{\partial}{\partial t}\left(\frac{1}{2}\vec{E}\cdot\vec{D}\right)-\frac{\partial}{\partial t}\left(\frac{1}{2}\vec{H}\cdot\vec{B}\right)-\vec{E}\cdot\vec{J}$$

$$\operatorname{div}\left(\vec{E}\times\vec{H}\right)=-\frac{\partial}{\partial t}\left[\frac{1}{2}\left(\vec{E}\cdot\vec{D}+\vec{H}\cdot\vec{B}\right)\right]-\vec{E}\cdot\vec{J} \tag{18.78}$$

Taking volume integral of Eq. (18.78),

$$\int_V \operatorname{div}\left(\vec{E}\times\vec{H}\right)dV=-\int_V\left\{\frac{\partial}{\partial t}\frac{1}{2}\left(\vec{E}\cdot\vec{D}+\vec{\mu}\cdot\vec{B}\right)\right\}-\vec{E}\cdot\vec{J} \tag{18.79}$$

Using Gauss divergence theorem,

$$\oint_S\left(\vec{E}\times\vec{H}\right)d\vec{S}=-\frac{d}{dt}\int_V\frac{1}{2}\left(\vec{E}\cdot\vec{D}+\vec{H}\cdot\vec{B}\right)-\int_V\vec{J}\cdot\vec{E}\,dv \tag{18.80}$$

Rearranging Eq. (18.80),

$$\underset{\text{I}}{-\int_V\vec{J}\cdot\vec{E}\,dv}=\underset{\text{II}}{\frac{d}{dt}\int_V\frac{1}{2}\left(\vec{E}\cdot\vec{D}+\vec{H}\cdot\vec{B}\right)dV}+\underset{\text{III}}{\oint_S\left(\vec{E}\times\vec{H}\right)d\vec{S}}$$

Equation (18.80) represents Poynting theorem.
Significance of terms in above equation:

$\int_v \vec{J}\cdot\vec{E}\,dV$ represents the rate of energy transferred into electromagnetic field through the motion of free charge in volume V.

The term

$$\frac{d}{dt}\int_V\frac{1}{2}\left(\vec{E}\cdot\vec{D}+\vec{H}\cdot\vec{B}\right)dV$$

represents the rate of electromagnetic energy stored in volume (18.81).

Here, $\int_V\frac{1}{2}\left(\vec{E}\cdot\vec{D}\right)dV$ represents electrostatic potential energy in volume V

and $\int_V\frac{1}{2}\left(\vec{H}\cdot\vec{B}\right)dV$ = magnetic energy in volume V.

The term $\oint_S\left(\vec{E}\times\vec{H}\right)d\vec{S}$ represents the amount of electromagnetic energy crossing the closed surface per second.

The vector $\left(\vec{E}\times\vec{H}\right)$ is known as Poynting vector and usually represented by the symbol $\vec{S}$, i.e.,

$$\boxed{\textbf{Poynting vector } \vec{S}=\vec{E}\times\vec{H}} \tag{18.81}$$

EXAMPLE 18.4

If the earth receives 2 cal/min-cm^2 solar energy, what are the amplitudes of electric and magnetic field of radiation?

Solution: As Poynting vector,

$$\vec{S} = \vec{E} \times \vec{H}$$

$$= \vec{E}\vec{H} \sin 90^\circ$$

$$= \vec{E}\vec{H}$$

$$\text{Solar energy} = 2 \text{ cal/min-cm}^2$$

$$= \frac{2 \times 4.18 \times 10^4}{60} \text{Jm}^{-2}\text{s}^{-1}$$

Both are energy flux per unit area per second.

Hence,
$$\vec{E}\vec{H} = \frac{2 \times 4.18 \times 10^4}{60} \approx 1400$$

But
$$\frac{\vec{E}}{\vec{H}} = \sqrt{\frac{\mu_0}{E_0}} = 120\pi = 377$$

$\therefore$
$$\vec{E}\vec{H} \times \frac{\vec{E}}{\vec{H}} = 1400 \times 377$$

$$\vec{E} = \sqrt{1400 \times 377} = 726.5 \text{ V/m}$$

Now,
$$\vec{H} = \frac{E}{377} = 1.927 \text{ Amp/m}$$

Amplitudes of electric and magnetic field of radiation are

$$\boldsymbol{\vec{E}_0 = \vec{E}\sqrt{2} = 1024.3 \text{ V/m}}$$

and
$$\boldsymbol{\vec{H}_0 = \vec{H}\sqrt{2} = 2.717 \text{ Amp/m}}$$

EXAMPLE 18.5

Calculate the value of Poynting vector and the amplitude of the electric field if a laser beam of 500 w is concentrated using a bust on a cross-section area of 10^{-10} M^2. Given $E_0 = 9 \times 10^{-12}$ F/M.

Solution: Poynting vector is given by

$$\vec{S} = \frac{\text{Power}}{\text{Area}} = \frac{500}{10^{-10}}$$
$$= 5\times10^{12}\ \text{W/M}^2$$

Average value of Poynting vector is given by relation

$$S_{AV} = \frac{1}{2}C\cdot\varepsilon\cdot E_0^2$$

$\therefore$ $$\text{Electric field } E_0 = \left(\frac{2S_{AV}}{CE}\right)^{1/2}$$

Now $$t = 9\times10^{-12}\ \text{F/M}$$
$$C = 3\times10^{8}\ \text{M/sec}$$
$$\vec{E}_0 = \left[\frac{2\times5\times10^{12}}{3\times10^{8}\times9\times10^{-12}}\right]^{1/2}$$
$$= \mathbf{60.9\times10^{6}\ V/M}$$

EXAMPLE 18.6

A monochromatic plane polarized em-wave is travelling eastwards. The wave is polarized with $\vec{E}$ directed vertically up and down alternately. Calculate $\vec{E}$, $\vec{B}$ and $\vec{S}$ provided that the amplitude of the electric field strength is 0.05 V/m and the frequency is 6 MHz.

Solution: Let the em-wave be propagated along x-axis with its direction of polarization along y-axis. Thus, $E_x = E_z = 0$. The magnetic field vector B is $\perp$ to the direction of propagation of wave and that of E. B will be confined along the z direction. Thus, $B_x = B_y = 0$; the frequency of wave $f = 6\times10^6$ Hz.

We have, $$\lambda = c/f = 3\times10^{8}/6\times10^{6}$$
$$= 50\ \text{m}$$

Thus, the electric field is given by

$$E_y = E_0\sin\frac{2\pi}{\lambda}(ct - x)$$
$$= 0.05\sin(2\pi\times6\times10^{6}t - 2\pi x/50)$$
$$= 0.05\sin(3.77\times10^{6}t - 0.126x)j\ \text{V/m}$$

$$\vec{E} = 0.05 \sin (3.77 \times 10^6 t - 0.126x) j \text{ Volt/m}$$

Since B is $\perp$ to E and the direction of propagation of wave,

$$B_z = E_y/c = 1.67 \times 10^{-10} \sin (3.77 \times 10^6 t - 0.126x)$$

$$\vec{B} = 1.67 \times 10^{-10} \sin (3.77 \times 10^6 t - 0.126x) \text{ K Wb/m}^2$$

Poynting vector S is given by

$$\vec{S} = \frac{\vec{E} \times \vec{B}}{\mu_0} = \frac{E_y \hat{j} \times B_z K}{\mu_0}$$

$$= \frac{E_y B_z \hat{i}}{\mu_0}$$

Thus, S will have only x component. Its magnitude is given by

$$\vec{S} = \frac{0.05 \times 1.67 \times 10^{-10}}{4\pi \times 10^{-7}} \sin^2 \left(3.77 \times 10^6 t - 0.126x\right)$$

$$= 6.65 \times 10^{-6} \sin^2 (3.77 \times 10^6 t - 0.126x) \text{ W/m}^2$$

$$\mathbf{S_{av} = \langle S \rangle = 3.325 \times 10^{-6} \text{ W/m}^2}$$

18.8 ▶ Plane Electromagnetic Waves in Free Space

We know that the Maxwell equations in free space are

$$\nabla \cdot \vec{D} = 0 \tag{18.82a}$$

$$\nabla \cdot \vec{B} = 0 \tag{18.82b}$$

$$\nabla \times \vec{E} = -\mu_0 \frac{\partial \vec{H}}{\partial t} \tag{18.82c}$$

and

$$\nabla \times \vec{H} = \varepsilon_0 \frac{\partial \vec{E}}{\partial t} \tag{18.82d}$$

Taking curl of Eq. (18.82c)

$\Rightarrow$

$$\nabla \times \left(\nabla \times \vec{E}\right) = -\mu_0 \frac{\partial}{\partial t}\left(\text{Curl } \vec{H}\right) \tag{18.83}$$

Put Eq. (18.82d) in Eq. (18.83)

$$\nabla^2 \vec{E} = \mu_0 \frac{\partial}{\partial t}\left[\varepsilon_0 \frac{\partial \vec{E}}{\partial t}\right]$$

$$\Rightarrow \qquad \boxed{\nabla^2 \vec{E} = \mu_0 \varepsilon_0 \frac{\partial^2 \vec{E}}{\partial t}} \qquad (18.84)$$

Similarly, taking curl of Eq. (18.82d),

$$\Rightarrow \qquad \nabla\left(\nabla \times \vec{H}\right) = \varepsilon_0 \frac{\partial}{\partial t}\left(\nabla \times \vec{E}\right) \qquad (18.85)$$

Using Eq. (18.82c) in Eq. (18.85)

$$-\nabla^2 \vec{H} = \varepsilon_0 \frac{\partial}{\partial t}\left(-\mu_0 \frac{\partial \vec{H}}{\partial t}\right)$$

$$\nabla^2 \vec{H} = \mu_0 \varepsilon_0 \frac{\partial^2 \vec{H}}{\partial t^2}$$

or

$$\boxed{\nabla^2 \vec{H} = \mu_0 \varepsilon_0 \frac{\partial^2 \vec{H}}{\partial t^2}} \qquad (18.86)$$

Equations (18.84) and (18.86) represent wave equation governing electric and magnetic fields ($\vec{E}$ and $\vec{H}$) in free space. Equations (18.6) and (18.8) are vector equations of identical form, which means that each of 6 components of $\vec{E}$ and $\vec{H}$ separately satisfies the same scalar wave equation of the form

$$\nabla^2 \phi - \mu_0 \varepsilon_0 \frac{\partial^2 \phi}{\partial t^2} = 0 \qquad (18.87)$$

where,

$$v = \frac{1}{\sqrt{\mu_0 \varepsilon_0}} = c$$

Thus, Eqs. (18.84), (18.86) and (18.87) take the form

$$\nabla^2 \vec{E} = \frac{1}{c^2} \frac{\partial^2 \vec{E}}{\partial t^2} \qquad (18.88)$$

$$\nabla^2 \vec{H} = \frac{1}{c^2} \frac{\partial^2 \vec{H}}{\partial t^2} \qquad (18.89)$$

and

$$\nabla^2 \phi = \frac{1}{c^2} \frac{\partial^2 \phi}{\partial t^2} \qquad (18.90)$$

Note

A plane wave is defined as a wave whose amplitude is the same at any point in a plane perpendicular to a specified direction.

18.9 ▶ Plane Electromagnetic Waves in an Isotropic Dielectric (Plane Electromagnetic Waves Non-Conducting Medium)

A non-conducting medium which has same properties all directions is called an isotropic dielectric.

Here, Maxwell equations are

$$\nabla \cdot \vec{D} = 0 \tag{18.91a}$$

$$\nabla \cdot \vec{B} = 0 \tag{18.91b}$$

$$\nabla \times \vec{E} = -\mu \frac{\partial \vec{H}}{\partial t} \tag{18.91c}$$

$$\nabla \times \vec{H} = \varepsilon \frac{\partial \vec{E}}{\partial t} \tag{18.91d}$$

Taking curl of Eq. (18.91c),

$$\nabla \times \left(\nabla \times \vec{E}\right) = -\mu \cdot \frac{\partial}{\partial t}\left(\nabla \times \vec{H}\right) \tag{18.92}$$

Using Eq. (18.91d)(d) in Eq. (18.92),

$$-\nabla^2 \vec{E} = -\mu \frac{\partial}{\partial t}\left(\varepsilon \frac{\partial \vec{E}}{\partial t}\right)$$

$$\nabla^2 \vec{E} = \mu\varepsilon \frac{\partial^2 \vec{E}}{\partial t^2}$$

$$\boxed{\nabla^2 \vec{E} = \mu\varepsilon \frac{\partial^2 \vec{E}}{\partial t^2}} \tag{18.93}$$

Similarly, taking curl of Eq. (18.91d), we get

$$\nabla \times \left(\nabla \times \vec{H}\right) = \varepsilon \frac{\partial}{\partial t}\left(\nabla \times \vec{E}\right) \tag{18.94}$$

Using Eq. (18.91c) in Eq. (18.94),

$$-\nabla^2 \vec{H} = \varepsilon \frac{\partial}{\partial t}\left(-\mu \frac{\partial \vec{H}}{\partial t}\right)$$

$$\nabla^2 \vec{H} = \mu\varepsilon \frac{\partial^2 \vec{H}}{\partial t^2}$$

$$\Rightarrow \quad \boxed{\nabla^2 \vec{H} = \mu \in \frac{\partial^2 \vec{H}}{\partial t^2}} \tag{18.95}$$

Equations (18.93) and (18.95) are vector equations of identical form which means that each of Eq. (18.96) components of $\vec{E}$ and $\vec{H}$ separately satisfies the same scalar wave equation of the form

$$\nabla^2\phi - \mu\varepsilon\frac{\partial^2\phi}{\partial t^2} = 0 \tag{18.96}$$

Here,
$$v = \frac{1}{\sqrt{\mu\varepsilon}} = \frac{c}{\sqrt{\mu_r\varepsilon_r}}$$

μ_r is the relative permeability of medium and ε_r is the relative permittivity of the medium.

Note

Since $\mu_r > 1$ and $\varepsilon_r > 1$ thereby indication that the speed of electromagnetic waves an isotropic dielectric is less than the speed of electromagnetic waves in free space.

18.10 ▶ Waveguides

At high frequencies ($\lambda < 10$ cm), the loses in a transmission line become quite large. It is convenient to use hollow conducting wave for the propagation of electromagnetic energy at such frequencies. The wave is confined to move in the direction of the tube. Such a wave is called guided wave and the hollow conducting tube as waveguide. In the waveguide, the wave is altered from a plane wave, as is required in order that the electric and magnetic boundary conditions at the surface of the conductor be satisfied. There are many different guides of electromagnetic waves that can satisfy boundary conditions and propagate energy down a guide. There are two most commonly used types, regular and circular. The geometry and size of guide allow only one mode of transmission for a limited range of frequencies. As the magnetic field is always perpendicular to the electric field in electromagnetic waves, hence must be parallel to the board face of the rectangular guide. The electric field configuration is similar to that in a polarized plane wave (Fig. 18.2).

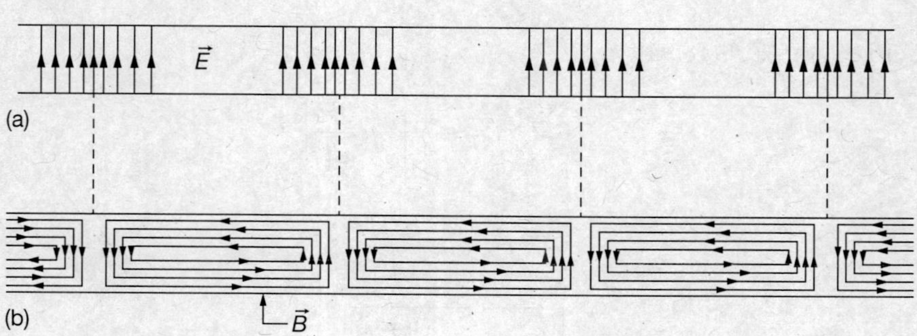

FIGURE 18.2 (a) Showing a side view of the line of $\vec{E}$ and (b) showing the top view of the lines of $\vec{B}$

Note

The waves in a waveguide have larger wavelength in free space.

18.11 ▶ Coaxial Cables

Coaxial cable is a cable consisting of an inner conductor, surrounded by a tubular insulating layer typically made from a flexible material with a high dielectric constant, all of which is then surrounded by another conductive layer and then finally covered again with a thin insulating layer on the outside (Fig. 18.3). The term coaxial comes from the inner conductor and the outer shield sharing the same geometrical axis. Coaxial cables are often used as a transmission line for radio frequency signals. In a hypothetical ideal coaxial cable, the electromagnetic fields carrying the signal exist only in the space between the inner and outer conductors. Practical cables achieve this objective to a high degree.

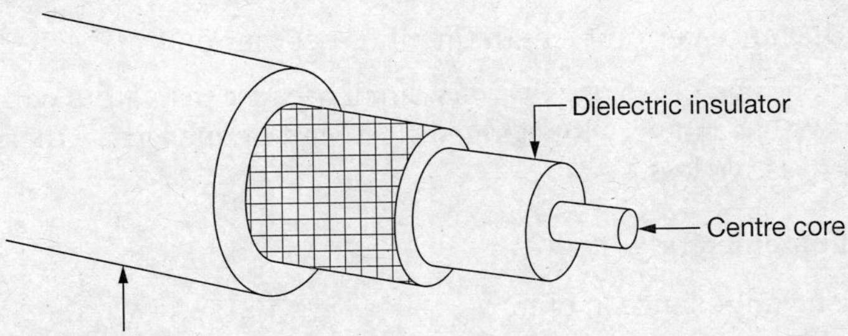

FIGURE 18.3

A coaxial cable provides protection of signals from external electromagnetic interference and effectively guides signals with low emission along the length of the cable.

18.11.1 Parameters of a Coaxial Cable

Coaxial cable is a particular kind of transmission line, so that circuit models developed for general transmission lines are appropriate.

Physical parameters

(a) Outside diameter of inner conductor is d.

(b) Inside diameter of shield is D.

(c) Dielectric constant of the insulator is ε. Dielectric constant of free space is denoted by ε_0, but when the insulator is a mixture of different dielectric materials, then the term effective dielectric constant ε_{eff} is often used.

(d) Magnetic permeability of the insulator is μ, permeability of free space μ_0 and $\mu = \mu_r \mu_0$, where μ_r = relative permeability.

Fundamental electrical parameters

(a) Shunt capacitance per unit length,

$$C = \frac{2\pi\varepsilon}{\ln(D/d)} = \frac{2\pi\varepsilon_0\varepsilon_r}{\ln(D/d)} \quad \text{(in Farad per metre)}$$

(b) Series inductance per unit length,

$$L = \frac{\mu}{2\pi}\ln(D/d)$$

$$= \frac{\mu_0\mu_r}{2\pi}\ln(D/d) \quad \text{(in Henrys per metre)}$$

(c) Series resistance per unit length (in ohms per metre). The resistance per unit length is just the resistance of inner conductor and the shield at low frequencies. At higher frequency, skin effect increases the effective resistance by confining the conduction to a thin layer of each conductor.

(d) Shunt conductance per unit length (in mhos per metre).

The shunt conductance is usually very small because insulators with good dielectric properties are used (a very low tangent). At high frequencies, a dielectric can have a significant resistive loss.

Derived electrical parameters

(a) Characteristic impedance in ohm

$$z_0 = \sqrt{\frac{L}{C}}$$

and $$z_0 = \frac{1}{2\pi}\sqrt{\frac{\mu}{E}}\ln\frac{D}{d} \approx \log_{10}\frac{D}{d}$$

(b) Attenuation (loss) per unit length (in decibels a ohm metre or Nepers per metre). This is dependent on the loss in the dielectric material filling the cable and resistive loss in the centre conductor and outer shield. These losses are frequency dependent, the losses becoming higher as the frequency increases. Skin effect losses in the conductors can be reduced by increasing the diameter of the cable.

(c) Velocity of propagation (in metres per second). The velocity of propagation depends on the dielectric constant and permeability (which is usually 1).

$$v = \frac{1}{\sqrt{\varepsilon\mu}} = \frac{C}{\sqrt{\varepsilon_r\mu_r}}$$

(d) Cut-off frequency is determined by the possibility of exciting other propagation modes of the coaxial cable. The average circumference of the insulator is $\pi(D+d)/2$.

(e) Peak voltage.

(f) Outside diameter, which dictates, where connectors must be used to terminate the cable.

EXAMPLE 18.7

A rectangular waveguide has a broad wall dimension of 2 cm and is fed by 10 MHz carrier from a coaxial cable. Determine whether a TE_{10} mode of wave will be propagated. If so calculate its guide wavelength and phase and group velocities.

Solution: We know that for TE_{10} mode the critical wavelength is

$$\lambda_c = 2a = 2 \times 2 = 4 \text{ cm.}$$

The free space wavelength $\lambda = \dfrac{v}{f} = \dfrac{c}{f} = \dfrac{3\times10^8}{10\times10^9} = 3 \text{ cm}$

Since the critical wavelength λ_c for TE_{10} mode is greater than free space wavelength λ.
∴ TE_{10} wave will propagate

The wavelength of the guide wave g for the TE_{10} mode

$$= \frac{\lambda}{\sqrt{1-(\lambda/\lambda_c)^2}}$$

$$= \frac{3}{\sqrt{1-(3/4)^2}} = 4.536 \text{ cm}$$

∴ Phase velocity,

$$V_{ph} = \lambda_g f = 4.536 \times 10^{-2} \times 10 \times 10^9$$

$$= 4.53 \times 10^8 \text{ ms}^{-1}$$

Group velocity,

$$V_{gr} = \frac{c\lambda}{\lambda_g} = 3\times10^8 \times \frac{3}{4.536}$$

$$= \mathbf{1.984\times10^8 \, ms^{-1}}$$

EXAMPLE 18.8

Show that the average energy density in an electromagnetic plane wave is $\langle \rho_e \rangle = \dfrac{\langle S \rangle}{C}$ where S is the Poynting vector. What is the momentum density in the wave?

Solution: The average energy density is a harmonic electric field and a magnetic field is given as follows:

$$\langle E_{av}\rangle = \frac{1}{4}\in_0 E_0^2;\ \langle B_{av}\rangle = \frac{1}{4}\frac{B_0^2}{\mu_0}$$

We know that
$$B_0 = \frac{E_0}{C}$$

Thus,
$$\langle \rho_e\rangle = \frac{1}{4}\in_0 E_0^2 + \frac{1}{4}\frac{E_0^2}{C^2\mu_0}$$
$$= \frac{1}{2}\in_0 E_0^2,$$

$$\langle S\rangle = \frac{C\in_0 E_0^2}{2} = \frac{\langle \rho_e\rangle}{C}$$

The radiation carries momentum and hence exerts pressure

$$\frac{\langle S\rangle}{C^2} - \frac{\langle \rho_e\rangle}{C} = \langle \rho_m\rangle$$

The analogy can be seen from quantum theory also;

$$E = h\theta = \frac{hc}{\lambda} = pc \Rightarrow \langle \rho_e\rangle = c\langle \rho_m\rangle$$

EXAMPLE 18.9

In free space, $E(z, t) = 10^3\sin(wt - \alpha z)\hat{j}(V/M)$

(a) obtain the magnetic field vector $H(z, t)$

(b) determine the propagation constant κ of the wave with a frequency $v = 95.5$ Hz.

Solution: The phase factor in the electric field vector shows that the direction of propagation is along the positive z direction. This implies that the Poynting vector ($S = E \times H$) is along the positive z direction. Thus, the magnetic field vector is along the negative x axis. The impedance for such field vectors is represented as follows:

$$\frac{E_y}{-H_x} = Z_0 = 120\pi\Omega$$

(a) The magnetic field vector is given as follows:

$$H_x = \frac{-10^3\sin(wt - \alpha z)}{120\pi}\frac{\text{V}}{\Omega\text{m}}$$

$$\Rightarrow \quad H_x = \frac{-10^3\sin(wt - \alpha z)}{120\pi}\frac{\text{A}}{\text{m}}$$

(b) In general, the propagation wave vector is defined as follows:

$$\hat{k} = \sqrt{\hat{i}w\mu\left(5 + \hat{i}w \in\right)}$$

In free space, $\sigma = 0$ so that

$$\hat{k} = \hat{i}w\sqrt{\mu_0 \in_0} = \hat{i}\left(\frac{2\pi\upsilon}{c}\right)$$

$$= i\,2\pi\frac{\left(95.5\times10^6\right)}{3\times10^8}$$

$$= \hat{i}\,2.0\,/\,\text{m}$$

Thus, the space alternation factor is zero as it is free space. In addition, the phase shift constant of the propagation wave vector is given as follows:

$$\hat{\boldsymbol{k}} = \boldsymbol{\alpha} + \hat{\boldsymbol{i}}\boldsymbol{\beta} = \boldsymbol{\alpha} = \mathbf{2\ rad\,/\,M.}$$

EXAMPLE 18.10

Assuming that all the energy from a 1000 watt lamp is radiated uniformly. Calculate the average intensities of electric and magnetic fields of radiation at a distance of 2 m from the lamp.

Solution: Let ρ_0 be the power radial uniformly in all directions, then the power or energy flux per unit area, per second at a distance r from the point source will be

$$S_{av} = \frac{p_0}{4\pi r^2} = \frac{1000}{4\pi(2)^2}\,\text{Watt}\,/\,\text{m}^2$$

Now from Poynting theorem

$$\left|\vec{S}\right| = \left|\vec{E}\times\vec{H}\right| = \vec{E}\vec{H}\sin 90 \quad \left(\because \text{as}\,\vec{E}\perp\vec{H}\right)$$

$$\vec{E}\vec{H} = \frac{1000}{\sqrt{6\pi}} \qquad (1)$$

Now,

$$\frac{\vec{E}}{\vec{H}} = \sqrt{\frac{\mu_0}{\varepsilon_0}} = 376.72\Omega \qquad (2)$$

Product of Eqs. (1) and (2) will be

$$\vec{E}\vec{H}\times\frac{\vec{E}}{\vec{H}} = \frac{376.72\times1000}{16\pi}$$

$$\vec{E} = \left(\frac{376.72 \times 1000}{16 \times 3.14}\right)^{1/2}$$

or

$$\vec{E} = 86.59 \text{ V / m}$$

Putting in Eq. (1), we get

$$\vec{H} = \frac{\mathbf{1000}}{\mathbf{16 \times 3.14}} = \mathbf{0.25 \text{ amp-turn/m}}$$

RECAP ZONE

POINTS TO REMEMBER

- Electric intensity at a point is given by $\vec{E} = -\frac{dV}{dr}$, if V = constant then potential gradient in any direction is zero. Hence, potential can exist where there is no electric field.
- The changing electric field term is known as displacement current D.
- Absolute index of refraction is $\mu = \frac{c}{v} = \sqrt{\varepsilon_r \mu_r}$.
- Faraday's first law of electromagnetism states that whenever the magnetic flux linked with a closed circuit changes, an induced e.m.f is set up in the circuit whose magnitude, at any instant, is proportional to the rate of change of magnetic flux ϕ_B linked with the circuit, i.e., $e \propto \frac{d\phi_B}{dt}$.
- Faraday's second law (Lenz's law) of electromagnetic induction states that the direction of induced e.m.f or current in the circuit is such that it opposes the change in flux that produced it.
- Ampere's law in magnetostatics states that the line integral $\int B \cdot dl$ of magnetic induction B for a closed path is numerically equal to μ_0 times the current through the area bounded by the path, i.e., $\oint B \cdot dl = \mu_0 i$.
- Electromagnetic waves are coupled electric field and magnetic field vectors that move with the speed of light and exhibit typical wave behaviour.
- There are four fundamental equations of electromagnetism known as Maxwell equations, which may be written as

  (a) Differential form of Maxwell equations

  (1) $\vec{\nabla} \cdot \vec{D} = \vec{S}$ (2) $\vec{\nabla} \cdot \vec{B} = 0$ (3) $\vec{\nabla} \times \vec{E} = -\frac{d\vec{B}}{dt}$ (4) $\vec{\nabla} \times \vec{H} = J + \frac{\partial \vec{D}}{\partial t}$

(b) Integral form of Maxwell equations

(1) $\int_S \vec{D}\cdot d\vec{S} = \int_V \rho\, dV$ (2) $\int_S \vec{B}\cdot d\vec{S} = 0$

(3) $\int_C \vec{E}\cdot d\vec{l} = -\frac{d}{dt}\int_S \vec{B}\cdot d\vec{S}$ (4) $\int_C \vec{H}\cdot d\vec{l} = \int_S \left(\vec{J} + \frac{\partial \vec{D}}{\partial t}\right) d\vec{S}$

- Poynting vector is defined as the vector product of electric field and magnetic field vectors, which gives the time rate of flow of wave energy per unit area of the medium.
- The retina of the eye is sensitive only to colours in the visible region, i.e. wavelengths lying between 3900 Å and 7800 Å. This region corresponds to visible part of the spectrum. That's why we cannot see the portion other than visible one of electromagnetic spectrum.
- Maxwell's electromagnetic equations are

(a) $\nabla\cdot\vec{D} = \rho$ $(\vec{D} = \in\vec{E})$ (for free space, $\vec{D} = \in\vec{E}$)

This gives Gauss's law in electrostatics.

(b) $\nabla\cdot\vec{B} = 0$ $[(\vec{B} = \mu\vec{H})$ (for free space $\vec{B} = \mu_0\vec{H})]$

This represents the Gauss's law in magnetostatics absence of magnetic monopoles.

(c) $\nabla\times\vec{E} = \frac{-\partial\vec{B}}{\partial t}$

This represents the Faraday's law of electromagnetic induction.

(d) $\nabla\times\vec{H} = \vec{J} + \frac{\partial\vec{D}}{\partial t}$: Modified Amphere's law

$\frac{\partial\vec{D}}{\partial t} = J_0 =$ displacement current

- Poisson's equation $\nabla^2 V = \frac{-\rho}{E_0}$ (for free space $\rho = 0 \Rightarrow \nabla^2 V = 0$)
- Poynting vector $\vec{S} = \vec{E}\times\vec{H}$ gives the direction and amount of energy flowing per unit area per second.
- Plane electromagnetic waves in free space:

$$\nabla^2\vec{E} = \mu_0 \in_0 \frac{\partial^2\vec{E}}{\partial t}; \ \nabla^2\vec{H} = \mu_0 \in_0 \frac{\partial^2\vec{H}}{\partial t^2}$$

for isotropic medium

$$\nabla^2\vec{E} = \mu\in\frac{\partial^2\vec{E}}{\partial t^2}; \ \nabla^2\vec{H} = \mu\in\frac{\partial^2\vec{H}}{\partial t^2}$$

IMPORTANT FORMULAE AND EQUATIONS

Equation Number	Equation	Remarks
(18.1)	$\int_S \vec{D}\cdot ds = \int_V \rho dV$	Gauss's law electrostatics
(18.5)	$\int \vec{B}\cdot d\vec{s} = 0$	Gauss's law in magnetism
(18.8)	$\text{Curl}\,\vec{B} = \frac{-\partial\vec{B}}{\partial t}$	Faraday's law of electromagnetic induction
(18.9)	$\oint_C \vec{H}\cdot d\vec{l} = \int_V \vec{J}\cdot d\vec{S}$	Amphere's law in magnetostatics
(18.17)	$\nabla\cdot\vec{D} = \rho$	Maxwell equations
(18.20)	$\nabla\cdot\vec{B} = 0$	
(18.26)	$\nabla\times\vec{E} = \frac{-\partial\vec{B}}{\partial t}$	
(18.34)	$\nabla\times\vec{H} = \vec{J} + \frac{\partial\vec{D}}{\partial t}$	
(18.37)	$\int_S \vec{D}\cdot d\vec{s} = q$	Net outward flux of electric displacement vector through the surface enclosing volume is equal to the net change contained within the volume
(18.52)	$\text{div}\vec{J} + \frac{\partial\rho}{\partial t} = 0$	Equation of continuity
(18.53)	$\text{div}\,\vec{J} = 0$	Equation of continuity for stationary current
(18.63)	$\nabla\cdot\vec{D} = \rho$	Maxwell equations in linear isotropic medium
(18.64)	$\nabla\times\vec{E} + i\omega\vec{B} = 0$	
(18.65)	$\nabla\cdot\vec{B} = 0$	
(18.66)	$\nabla\times\vec{H} - i\omega\vec{D} = \vec{J}$	
(18.81)	$\vec{S} = \vec{E}\times\vec{H}$	Poynting vector

(18.84)	$\nabla^2 E = \mu_0 \in_0 \dfrac{\partial^2 \vec{E}}{\partial t^2}$	**Wave equations in free space**
(18.86)	$\nabla^2 \vec{H} = \mu_0 \in_0 \dfrac{\partial^2 \vec{H}}{\partial t^2}$	
(18.93)	$\nabla^2 \vec{E} = \mu\varepsilon \dfrac{d^2 \vec{E}}{\partial t^2}$	**Wave equations in air isotropic medium**
(18.95)	$\nabla^2 \vec{H} = \mu \in \dfrac{\partial^2 \vec{H}}{\partial t^2}$	

REVIEW ZONE

SHORT ANSWER QUESTIONS

1. What are em-waves? How they propagate?
2. Give the importance of displacement current.
3. Give the physical interpretation of differential form of Gauss's law in magnetostatics.
4. Derive an equation for a simple plane wave.
5. Write Maxwell equations for free space.
6. Write Maxwell equations for non-conductive medium.
7. Give the physical significance of equation of continuity.
8. Write short notes on the following:
   (a) Co-axial cables
   (b) Poynting vector
   (c) Waveguide
9. Write the differential form of Maxwell equations.
10. Write Maxwell equations in integral form.

LONG ANSWER QUESTIONS

1. State Maxwell equations for the em-field and obtain the wave equation in free space.
2. Derive Maxwell equations. Explain the physical significance of each equation.
3. Discuss the concept of Poynting vector. How is it related with the sum of energies electric and magnetic field?
4. Derive an equation of simple plane electromagnetic wave in free space.

5. Derive an equation of simple plane em-wave in medium.
6. Write a short note on co-axial cable and waveguide.
7. Give the basic ideas of em-wave theory and state Maxwell equations. Derive simple plane wave equation in an isotropic non-conducting medium.
8. Write Maxwell equations for electromagnetic fields. Solve these equations to get velocity of propagation of electromagnetic waves.
9. What is equation of continuity? Derive Maxwell equations and give their physical interpretation.
10. State and explain Maxwell equations. Give analysis of propagation of electromagnetic waves in dielectric media.
11. Write the Maxwell equations in differential form and integral form.
12. What is Poynting vector? How is the Poynting theorem derived from Maxwell equations? Explain Poynting theorem.

NUMERICAL PROBLEMS

1. The relative permittivity of distilled water is 81. Calculate refractive index and velocity of light in it.

   Ans. $n = (9)$, $v = 3.33 \times 10^7$ m/s

2. When the amplitude of magnetic field in a plane wave is 2 A/m,
   (a) determine the magnitude of the electric field for the plane wave in free space.
   (b) determine the magnitude of electron field when the wave propagates in a medium which is characterized by $\sigma = 0$ and $\mu = \mu_0$ and $\varepsilon_0 = 4\varepsilon_0$.

   Ans. (a) $E = z_0 H = 240\pi$ V/m,

   (b) $z_0 = \sqrt{\frac{\mu}{E}} = 60\pi\ \Omega$ and

   $E = 120\pi$ V/m

3. If the magnitude of H in a plane wave is 1.0 Amp/m, find the magnitude of E for a plane wave in free space.

   Ans. $|E| = 377$ volts/m

4. If the amount of incident solar energy on the earth is 5 cal/min-cm^2, then calculate the amplitudes of electric and magnetic fields of radiations.

   Ans. $E_0 = 16.1 \times 10^2$ V/m,

   $H_0 = 0.04 \times 10^2$ Amp/m^2

5. If the electric amplitude of wave is 5 V/m, what is the magnetic amplitude of the wave.

   Ans. 1.675×10^{-8} Wb/m^2

6. A plane monochromatic linearly polarized lightwave is travelling eastwards, the wave is polarized with E directed vertically up and down. Write the expression for E, H and B if $E_0 = 0.1$ V/m and frequency $= 20$ MHz.

   Ans. $E_z = 0.1 \sin 4\pi \times 10^7 t - 0.419x$;

   $B_y = -E_z/C$

   $H_y = B_y/\mu_0$

7. The electric field in air is given by $E = \rho t e^{-\rho - t} a_\phi$ V/m. Find B and J.

Ans. $(2 - \rho)(1 + t)\, e^{-\rho - t} a_z \text{Wb/m}^2$; $(1 + t)(3 - \rho)(4\pi)^{-1} 10^7\, c^{-\rho - t} a_\phi$ A/m^2

8. The intensity of the em-wave from a distant 1.2 MHz radio station is 3×10^{-11} W/m^2. Find the expressions for the electric and magnetic fields in this region.

Ans. $1.5 \times 10^{-4} \sin(7.5 \times 10^6 t - 0.025x)$ V/m; $5 \times 10^{-13} \sin(7.5 \times 10^6 t - 0.025x) T$

9. A coaxial cable with the dimensions $a = 0.5$ mm, 0.3 mm and $t = 0.4$ mm is filled with a dielectric material having $\epsilon_r = 2.0$, $\sigma_d = 10\ \mu$s/m. The conductors have $\sigma_c = 50$ ms/m. Calculate the parameter values of L, C, G, Rd at 50 MHz.

Ans. 0.358 μH/m, 62 pF/m, 35.1 μs/m, 0.03 Ω/m, 0.743 Ω/m

10. Show that in a source free region ($J = 0$; $\rho_v = 0$), Maxwell's equations can be reduced to zero.

CHAPTER 19

Dielectric Materials

Learning Objectives

By the end of this chapter, the student will be able:

- To understand and distinguish insulators and dielectric materials
- To define polarization of dielectrics
- To discuss the effects of polarization of a dielectric on electric field
- To define the relation between electric field, dielectric constant and polarization and their physical significances
- To discuss the modified form of Gauss's law in dielectrics
- To know difference between the two types of dielectrics at the microscopic level
- To derive the field and energy stored in capacitors
- To explain applications of the dielectric materials

19.1 ▶ Introduction

The dielectrics are non-conductors of electricity because they do not possess free charges. The electrical conductivity of the dielectric materials is either very small or zero. A dielectric material stores electrical energy with a minimum dissipation of power, since the electrons are bound to their parent molecules, and hence, there are no free charges. An important property of dielectric materials is that on application of electric field, they get polarized. Therefore, dielectrics are non-conducting materials which can be polarized by an electric field. Inspite of being poor conductor of electricity, the dielectrics are most interesting and useful electrical engineering materials. Some of the examples of dielectric materials are ceramics, resins, mica, wood and glass.

19.2 ▶ Non-Polar and Polar Molecules

A non-polar molecule is one in which the centre of gravity of the positive charges coincides with the centre of gravity of the negative charges. A non-polar molecule has zero electric dipole moment (Fig. 19.1). For example, H_2, N_2, Cl_2, CH_4, CCl_4, C_6H_6, C_2H_2, CO_2 and C_2H_4.

A polar molecule is one in which the centre of gravity of positive charges is separated from the centre of gravity of the negative charges by a finite distance. The polar molecule has an electric dipole moment (permanent) (Fig. 19.2). For example, HCl, CO, H_2O and NH_3.

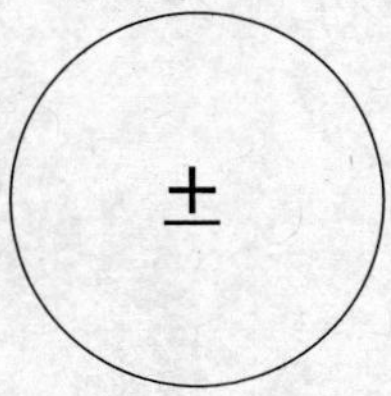

FIGURE 19.1 Non-polar molecule

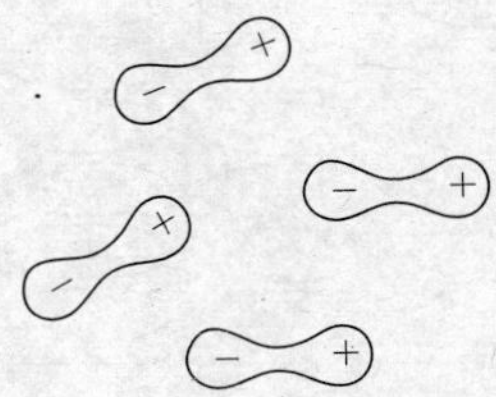

FIGURE 19.2 Polar molecule

Note

Dielectric materials have high resistivity. They have energy gap more than 3 eV. In these materials, all the electrons are localized, i.e. engaged in bonding.

19.3 ▶ Polarization of Dielectrics

When a dielectric material (polar or non-polar) is placed in an electric field, the charge centres of polar molecules get displaced and the molecules become induced dipoles.

For polar molecules which already have randomly oriented permanent dipoles, the effect of the electric field is to orient the dipoles along the direction of the electric field and increase their dipole moment.

The displacement of electric charges results in the formation of electric dipole moment in atoms, ions or molecules of the material (Fig. 19.3).

The types of polarization are as follows.

(i) Electronic polarization

When the electrons in various shells are displaced relative to the nucleus by the external field, an induced electric dipole moment is produced. This is called electronic polarization (Fig. 19.4).

(ii) Ionic polarization

If a molecule contains ions, then the applied field tends to displace the positive and negative ions in opposite direction, causing the change in bond length. This is called ionic or atomic polarization (Fig. 19.5).

(iii) Orientational or dipolar polarization

If the molecule has permanent dipoles, these are oriented randomly in the absence of an external field and results in a complete cancellation of the polarization. But when an external field is applied, these dipoles tend to orient and align with the field, giving an effective polarization, called orientational or dipolar polarization (Fig. 19.6).

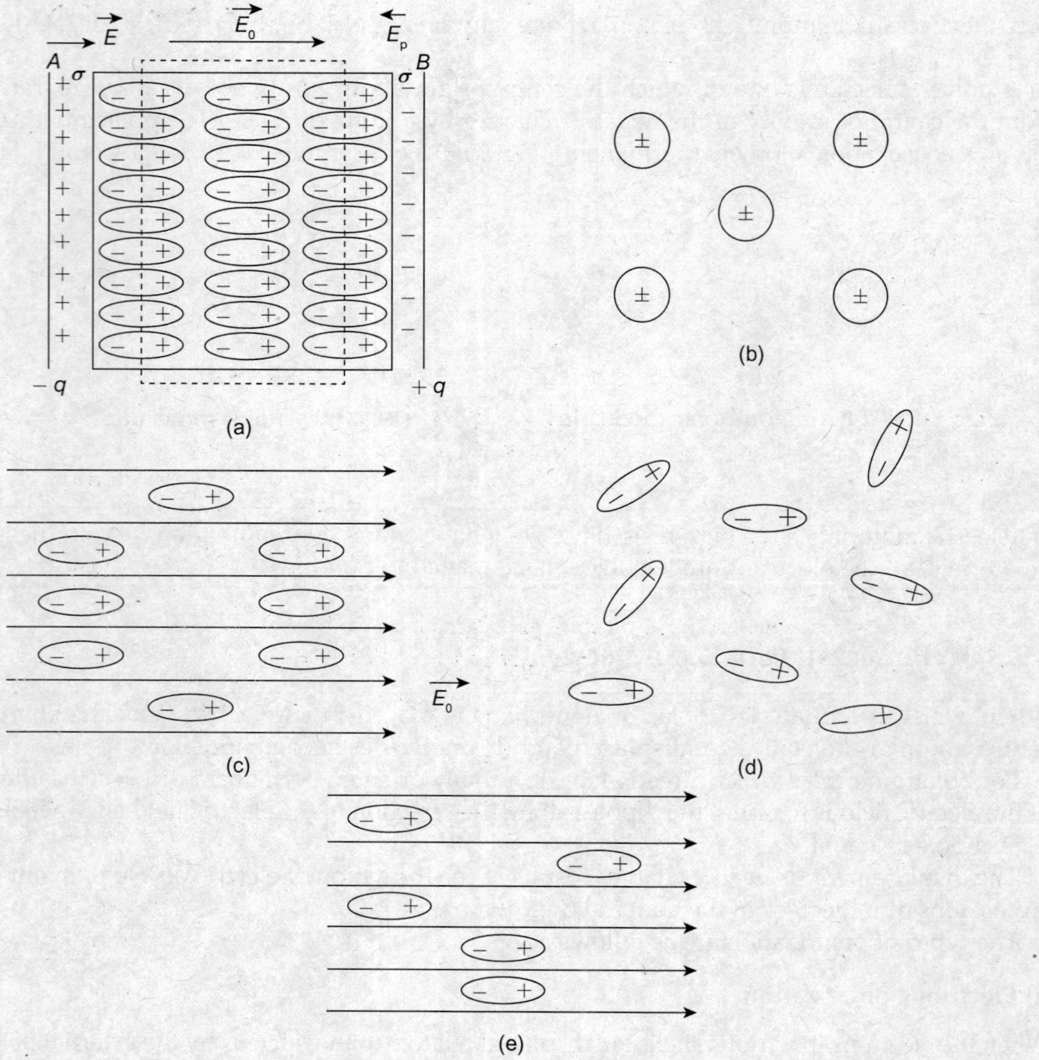

FIGURE 19.3 (a) Dielectric slab in uniform electric field. (b) Before applying electric field on non-polar molecules. (c) After applying electric field on non-polar molecules. (d) Polar molecules without electric field. (e) Polar molecules with electric field

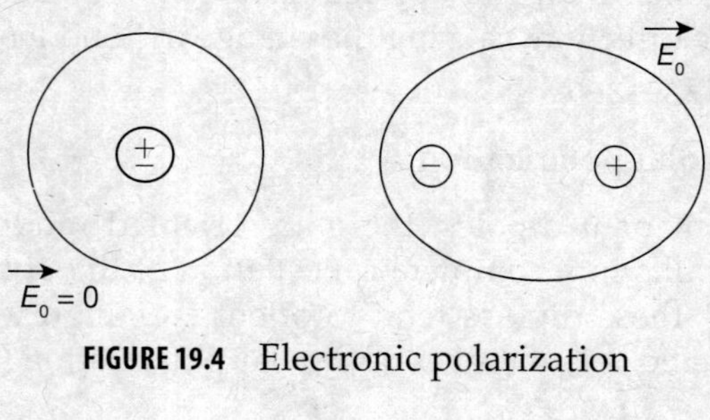

FIGURE 19.4 Electronic polarization

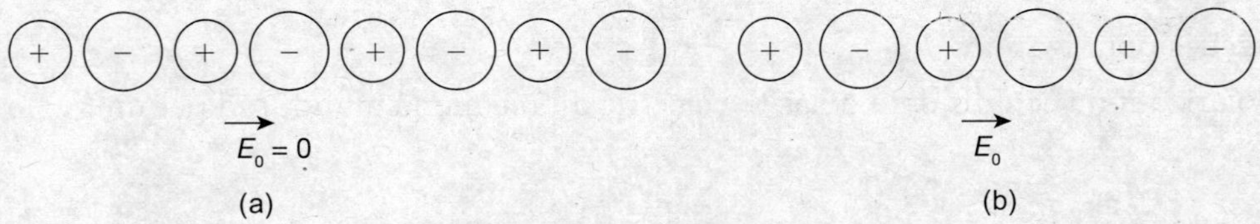

FIGURE 19.5 Ionic polarization

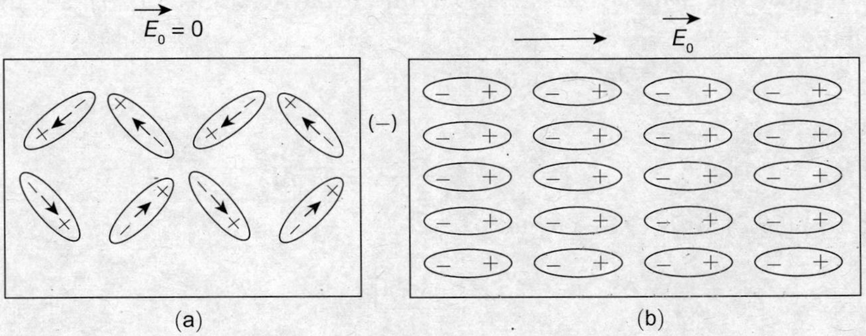

FIGURE 19.6 Orientational polarization

The total polarizability α is the sum of various contributions and can be written as

$$\alpha = \alpha_e + \alpha_i + \alpha_d$$

where α_e = electronic polarizability

α_i = ionic polarizability

α_d = dipolar polarizability

and
$$\alpha = \frac{\vec{p}}{E_{loc}}$$

The total polarizability α is defined as the dipole moment $\vec{p}$, that an atom or molecules acquire in a local electric field E_{loc}. The local E_{loc} field is that field which is actually effective in polarizing the atoms in a solid and is also called molecular field in the dielectric.

19.4 ▶ Relation Between Three Electric Vectors

(i) Electric field intensity $(\vec{E})$

The electric field strength at any point in an electric field is defined as the force experienced per unit infinitesimal charge (q_0). If $\vec{F}$ is the force on small charge q_0, then

$$\vec{E} = \lim_{q_0 \to 0} \frac{\vec{F}}{q_0}$$

The direction of $\vec{E}$ is along the direction of $\vec{F}$. The units of $\vec{E}$ are Newton/Coulomb or volt/metre.

(ii) Electric polarization $(\vec{P})$

The polarization vector is defined as the electric dipole moment induced per unit volume, i.e.,

$$\vec{P} = \frac{\vec{p}}{V} \tag{19.1}$$

Consider a dielectric slab of face area A with surface charges $(\sigma_P A)$ separated by a distance l then

$$\text{Dipole moment } (\vec{p}) = \sigma_P A l \tag{19.2}$$

From Eq. (19.1),

$$\vec{p} = \vec{P}\cdot V = \vec{P}\cdot Al \tag{19.3}$$

Comparing Eqs. (19.2) and (19.3), we get

$$\boxed{\vec{\mathbf{P}} = \sigma_{\mathbf{P}}} \tag{19.4}$$

Thus, the polarization vector $\vec{P}$ depends on the induced surface density.

(iii) Electric displacement vector $(\vec{D})$

When a dielectric placed in the parallel plates of a capacitor is subjected to an external electric field $\vec{E}_0$, then dielectric is said to be polarized. The charges are induced on the plates of capacitor, and q' is the induced charge on the boundary of dielectric, the resultant field on the dielectric is given as

$$\vec{E} = \vec{E}_0 - \vec{E}_P$$

where $\vec{E}_0 = \dfrac{q_0}{\varepsilon_0 A}$ and $\vec{E}_P = \dfrac{q'}{\varepsilon_0 A}$ where A is the area of the dielectric slab.

Thus, the above equation becomes

$$\vec{E} = \frac{q_0}{\varepsilon_0 A} - \frac{q'}{\varepsilon_0 A}$$

$$\Rightarrow \qquad \varepsilon_0 \vec{E} = \frac{q_0}{A} - \frac{q'}{A} \tag{19.5}$$

$$\Rightarrow \qquad \frac{q_0}{A} = \varepsilon_0 \vec{E} + \vec{P} \qquad \left[\vec{P} = \frac{q'}{A}\right]$$

$$\Rightarrow \qquad \boxed{\vec{\mathbf{D}} = \varepsilon_0 \vec{\mathbf{E}} + \vec{\mathbf{P}}} \tag{19.6}$$

where $\vec{D}$ = electric displacement vector and is defined as the number of lines of force received by a unit area.

Equation (19.6) represents the relation between $\vec{D}, \vec{E}$ and $\vec{P}$ (Fig. 19.7).

In free space, when there is no dielectric,

$\vec{P}=0,$ we get $$\vec{D}=\varepsilon_0\,\vec{E} \tag{19.7}$$

Also, $$\vec{D}=\frac{q_0}{A}=K\varepsilon_0\left(\frac{q_0}{K\varepsilon_0 A}\right)=K\varepsilon_0\left(\frac{q_0}{\varepsilon A}\right)$$

$\Rightarrow$ $$\vec{D}=K\varepsilon_0\vec{E} \tag{19.8}$$

$\Rightarrow$ $$\vec{D}=\varepsilon\vec{E} \tag{19.9}$$

where $K\varepsilon_0=\varepsilon$ permittivity of the medium and $K=$ dielectric constant.

If C_0 and V_0 are the capacitance and potential difference between the plates without dielectric and C_d and V_d the corresponding values with dielectrics within the plates of a capacitor, then

$$\boxed{\frac{C_d}{C_0}=\frac{V_0}{V_d}=K=\frac{\varepsilon}{\varepsilon_0}=\varepsilon_r} \tag{19.10}$$

Here, dielectric constant K is defined as the ratio of capacitance of a capacitor with dielectric to the capacitance of the same capacitor without dielectric. K for air or vacuum is unity.

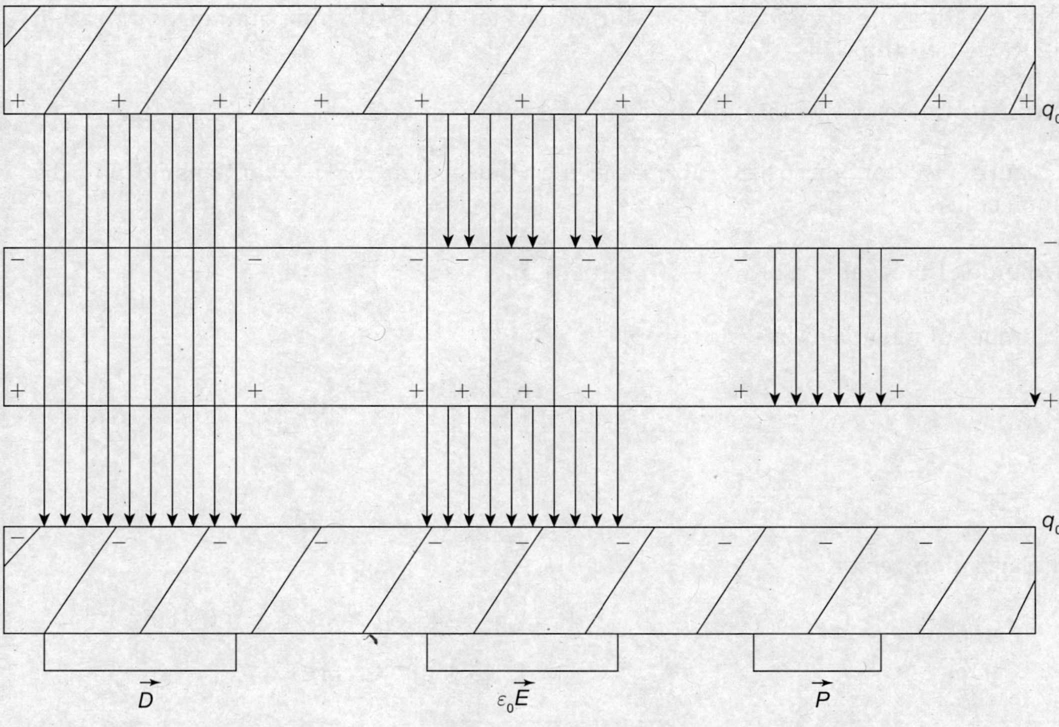

FIGURE 19.7 Relation between three $(\vec{E},\vec{D},\vec{P})$ vectors

Now, we have

$$\vec{D} = \vec{P} + \varepsilon_0 \vec{E} \tag{19.11}$$

But,

$$\vec{D} = K\varepsilon_0 \vec{E} \tag{19.12}$$

From Eqs. (19.11) and (19.12),

$$K\varepsilon_0 \vec{E} = \vec{P} + \varepsilon_0 \vec{E}$$

$$\Rightarrow \quad \vec{P} = (K-1)\varepsilon_0 \vec{E}$$

$$\Rightarrow \quad \vec{P} = \chi_e \, \varepsilon_0 \, \vec{E}$$

where $\chi_e = (k-1)$ is called electrical susceptibility of dielectric and is characteristics of the material. It is a dimensionless quantity.

EXAMPLE 19.1

A water molecule has a dipole moment 5×10^{-25} C-cm, where all the molecular dipoles point in the same direction. If the radius of a water drop is 0.5 mm, find the polarization in the water drop.

Solution: We know that the molecular mass of water is $(2 \times 1) + (8 \times 2) = 18$ g.

18 gm of H_2O contains 6.023×10^{23} molecules, thus $\frac{18}{10^3}\,\text{m}^3$ of H_2O contains 6.023×10^{26} molecules

Volume of the water drop $= \frac{4\pi}{3}\left(0.5 \times 10^{-3}\right)^3 \text{m}^3$

Number of molecules in water drop

$$= \frac{6.023 \times 10^{26}}{18} \times \frac{4\pi}{3}\left(0.5 \times 10^{-3}\right)^3$$

$$= 0.245 \times 10^{17}$$

$$= 2.45 \times 10^{16}$$

Polarization vector

$$\vec{P} = V\vec{p} = 2.45 \times 10^{16} \times 5 \times 10^{-25}\,\text{C/cm}^2$$

$$= \mathbf{12.25 \times 10^{-9}\,C/cm^2}$$

EXAMPLE 19.2

The polarization vector of a gas at NTP is 4.0×10^{-10} C/cm^2. Find the induced dipole moment due to each atom.

Solution: The polarization vector is given as

$$\vec{P} = V\vec{p}$$

$$\Rightarrow \quad 4.0\times10^{-10}\,\frac{\text{C}}{\text{m}^2} = \frac{6.023\times10^{23}}{22.4\times10^{-3}\,\text{m}^3}\times\vec{p}$$

$$\Rightarrow \quad \vec{p} = \frac{22.4\times10^{-3}}{6.023\times23}\,\text{m}^3\times4.0\times10^{-10}\,\frac{\text{C}}{\text{m}^2}$$

$$\Rightarrow \quad \vec{p} = \mathbf{14.87\times10^{-36}\ C\text{-}m}$$

EXAMPLE 19.3

The parallel plates of a capacitor have an area of 1.00×10^{-1}m^2 each and are 1.00×10^{-2} m apart. The capacitor is connected to a power supply of 2.00 KV. A thin insulating plastic sheet is inserted between the plate drops to 1.00 kV, whereas the charge on the capacitor plates remains constant. Calculate the original capacitance, the dielectric constant and electric field $\vec{E}$ after the dielectric is inserted.

Solution: We know that the capacitance is given as

$$C_0 = \varepsilon_0\frac{A}{d}$$

$$= \frac{8.85\times10^{-12}F}{\text{m}}\,\frac{\left(1.00\times10^{-1}\right)\text{m}^2}{\left(1.00\times10^{-2}\right)\text{m}}$$

$$= 8.85\ \text{pF}$$

and

$$K = \frac{C_d}{C_0} = \frac{V_0}{V_d}$$

Here,

$$C_d = \frac{q}{V} = \frac{C_0V_0}{V}$$

$$= \frac{8.85\times10^{-12}F\times2\times10^3 V}{1.0\times10^3 V}$$

$$= \frac{17.7\times10^{-9} C}{1.0\times10^{3} V}$$

$$= 17.7 \text{ pF}$$

and

$$K = \frac{C_d}{C_0} = \frac{17.7 \text{ pF}}{8.88 \text{ pF}} = 2$$

or

$$K = \frac{V_0}{V_d}$$

$$= \frac{2.0\times10^{3} V}{1.0\times10^{3} V} = 2$$

Also,

$$\vec{E} = \frac{q}{\varepsilon A} = \frac{\vec{E}_0}{K} = \frac{V}{d}$$

$$= \frac{1.0\times10^{3} V}{1.00\times10^{-2} \text{m}}$$

$$= \mathbf{10^5\ V/m}$$

19.5 ▶ Gauss's Law and Dielectrics

Gauss's law in electrostatics states that the surface integral of the normal component of electric field over any closed surface is equal to $1/\varepsilon_0$ times the total charge enclosed by the surface, i.e.,

$$\boxed{\oint \vec{E}_0 \cdot dA = \frac{q_0}{\varepsilon_0}} \tag{19.13}$$

$$\vec{E}_0 A = \frac{q_0}{\varepsilon_0} \text{ and } \vec{E}\cdot KA = \frac{q_0}{\varepsilon_0}$$

$$\Rightarrow \quad \vec{E} = \frac{q_0}{\varepsilon_0 AK} \tag{19.14}$$

where $\vec{E}_0$ is the external field applied on dielectrics. If $\vec{E}_P$ is the field produced due to induced surface charges, then $\vec{E} = \vec{E}_0 - \vec{E}_P$ is the net effective field within the dielectric (Fig. 19.8) and the Gauss's law can be written as

$$\oint \vec{E}\cdot dA = \frac{(q_0 - q')}{\varepsilon_0} \tag{19.15}$$

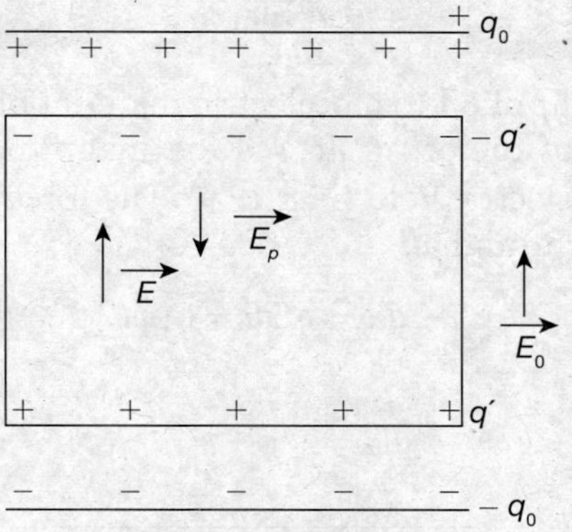

FIGURE 19.8 Gauss law in dielectrics

$$\vec{E}\cdot A = \frac{q_0}{\varepsilon_0} - \frac{q'}{\varepsilon_0}$$

$$\vec{E} = \frac{q_0}{\varepsilon_0 A} - \frac{q'}{\varepsilon_0 A} \tag{19.16}$$

From Eqs. (19.14) and (19.16),

$$\frac{q_0}{\varepsilon_0 AK} = \frac{q_0}{\varepsilon_0 A} - \frac{q'}{\varepsilon_0 A}$$

$$\Rightarrow \qquad (q_0 - q') = \frac{q_0}{K} \tag{19.17}$$

From Eqs. (19.15) and (19.17),

$$\oint \vec{E}\cdot dA = \frac{q_0}{\varepsilon_0 K} \tag{19.18}$$

or

$$\oint \vec{E}\cdot(\varepsilon_0 K)\,dA = q_0$$

$$\Rightarrow \qquad \oint \vec{E}\cdot\varepsilon\, dA = q_0$$

$$\Rightarrow \qquad \boxed{\oint \vec{D}\cdot dA = q_0} \tag{19.19}$$

Equation (19.19) represents the Gauss law in dielectrics, which states that the surface integral of electric displacement vector over a closed surface is equal to free charge enclosed within the surface.

Note

Electric displacement vector depends on the substance being considered.

19.6 ▶ ENERGY STORED IN ELECTROSTATIC FIELD

Consider an electric field $\vec{E}$ applied to a dielectric material block. Let l be the length and A the area of cross-section of dielectric. If q is the magnitude of charges and dl is the displacement caused by the electric field (Fig. 19.9). The force acting on the charges is $q\vec{E}$. The work done by the force through dl is

$$dw = q\vec{E}dl = \vec{E}(qdl)$$

$$dw = \vec{E}\,d\vec{p} \tag{19.20}$$

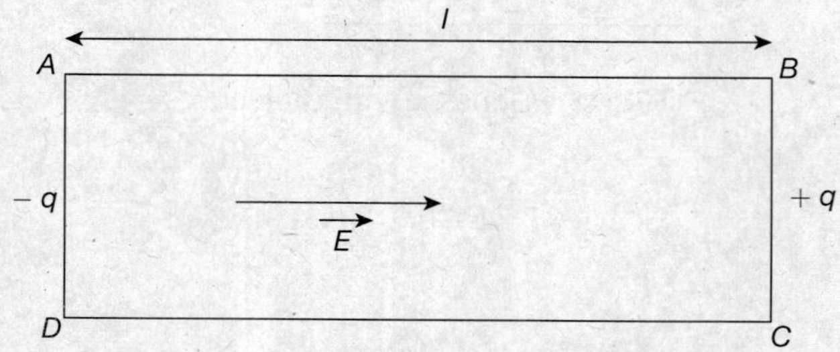

FIGURE 19.9 Energy stored in electrostatic field

This work is stored as electrostatic energy of the dielectric and this energy changes of the internal energy of the molecules. Polarisation vector is

$$\vec{P} = \frac{\vec{p}}{V} = \frac{\vec{p}}{lA}$$

or
$$d\vec{P} = \frac{d\vec{p}}{lA}$$

or
$$d\vec{p} = Vd\vec{P} \tag{19.21}$$

But we know that $\vec{P} = \vec{D} - \varepsilon_0\vec{E}$

or
$$d\vec{P} = d\vec{D} - \varepsilon_0 d\vec{E} = K\varepsilon_0 d\vec{E} - \varepsilon_0 d\vec{E} \quad \left[\because \vec{P} = K\varepsilon_0\vec{E} - \varepsilon_0\vec{E}\right]$$

or
$$d\vec{P} = (K-1)\varepsilon_0 d\vec{E} \tag{19.22}$$

From Eqs. (19.21) and (19.22),

$$d\vec{P} = \frac{d\vec{p}}{V} = (K-1)\varepsilon_0 d\vec{E}$$

or
$$d\vec{p} = (K-1)V\varepsilon_0 d\vec{E} \tag{19.23}$$

From Eqs. (19.20) and (19.23),

$$dw = \vec{E}\cdot(K-1)V\varepsilon_0 d\vec{E} \tag{19.24}$$

Total work done for polarization of dielectric is

$$W = \int dw = \int \vec{E}(K-1)V\varepsilon_0 d\vec{E}$$

$$W = (K-1)V\varepsilon_0 \frac{E^2}{2}$$

Energy density due to polarization is

$$U = \frac{W}{V} = \frac{1}{2}\varepsilon_0 (K-1)E^2 \tag{19.25}$$

Equation (19.25) gives the energy density due to polarization.
Total energy density of the dielectric is

U = energy density in free space + energy density due to polarization

$$U = \frac{1}{2}\varepsilon_0 E^2 + \frac{1}{2}\varepsilon_0 (K-1)E^2$$

or

$$\boxed{U = \frac{1}{2}\varepsilon_0 K E^2} \tag{19.26}$$

Equation (19.26) gives the energy stored per unit volume in dielectric in electrostatic field.

Note

Since the capacitance of a capacitor always increases after filling the dielectric medium, thus K is always greater than unity.

19.7 ▶ Behaviour of Dielectrics in Alternating Field

When a dielectric material is placed in an alternating field, the orientation of the dipoles and hence the polarization will tend to reverse when the polarity of the field changes. As long as the frequency remains low (< 10^6 C/s), there is no significant lag in polarization with alteration of field. The permittivity is independent of frequency and has same magnitude as in a static field. When the frequency is increased, the dipoles will not be able to rotate rapidly and their oscillators will lag behind those of the field. With further increase in frequency, the permanent dipoles will be unable to follow the field, and the orientational polarization stops. This happens in radio frequency range 10^6–10^{11} Hz. At still higher frequencies, i.e. in the infrared range 10^{11}–10^{14} Hz, the ions cannot follow the field, and hence ionic polarization stops. Thus, only electronic polarization contributes at high frequencies.

The permittivity of the dielectric material decreases with increase in frequency and this phenomenon is called anomalous dielectric dispersion.

Dielectric absorption – Dispersion arising during the transition from full atomic polarization at radio frequencies to negligible atomic polarization at optical frequency is called dielectric absorption.

Dielectric relaxation – Dispersion arising during the transition from full orientational polarization at zero or low frequencies to negligible orientational polarization at high frequency is called dielectric relaxation.

19.8 ▶ Claussius-Mossotti Relation

The dipole moment of an atom per unit polarizing field is called atomic polarizability. Thus, the dipole moment is

$$\vec{P} = \alpha \vec{E}_l \tag{19.27}$$

If there are N atoms per unit volume, then

$$\vec{P} = N\vec{p} = N\alpha \vec{E}_l \tag{19.28}$$

But as we know,

$$\vec{E}_l = \left(\vec{E} + \frac{\vec{P}}{3\varepsilon_0}\right) \tag{19.29}$$

Putting value of $\vec{E}_l$ from Eq. (19.29) to Eq. (19.28),

$$\vec{P} = N\alpha\left(\vec{E} + \frac{\vec{P}}{3\varepsilon_0}\right) \tag{19.30}$$

Also,

$$\vec{P} = \varepsilon_0 (K-1)\vec{E} \tag{19.31}$$

From Eqs. (19.30) and (19.31),

$$\varepsilon_0 (K-1)\vec{E} = N\alpha\left[\vec{E} + \frac{\varepsilon_0 (K-1)\vec{E}}{3\varepsilon_0}\right] \tag{19.32}$$

or

$$\frac{N\alpha}{3\varepsilon_0} = \frac{K-1}{K+2}$$

or

$$\boxed{\boldsymbol{\alpha = \frac{3\varepsilon_0}{N}\frac{(K-1)}{(K-2)}}} \tag{19.33}$$

Equation (19.33) is the required Claussius-Mossotti relation.
Multiplying both sides of Eq. (19.33) by molar volume, we get

$$\left(\frac{N\alpha}{3\varepsilon_0}\right)\frac{M_A}{\rho} = \left(\frac{K-1}{K+2}\right)\frac{M_A}{\rho} \tag{19.34}$$

but $\frac{NM_A}{\rho} = N_A$, the avogadro's number

$$\therefore \qquad \frac{N_A\alpha}{3\varepsilon_0} = \left(\frac{K-1}{K+2}\right)\frac{M_A}{\rho} = P_m, \text{ called molar polarization.}$$

But the refractive index of medium is $n = \sqrt{K}$, thus from Eq. (19.33,) we get

$$\boxed{\alpha = \frac{3\varepsilon_0}{N}\frac{\left(n^2-1\right)}{\left(n^2+2\right)}} \tag{19.35}$$

This is called Lorentz-Lorentz relation.

EXAMPLE 19.4

If the electric field strength is 10^4 V/m due to a dielectric constant of 3, find the polarization vector.

Solution:

$$\vec{P} = (K-1)\varepsilon_0\vec{E}$$

$$= (3-1)\times 8.85\times 10^{-12}\times 10^4\,\frac{\text{C}}{\text{m}}$$

$$\mathbf{= 17.7 \times 10^{-8}\ C/m}$$

EXAMPLE 19.5

A thick sheet of polythene of dimension 0.5 mm possessing a relative dielectric constant of 2.0, is subjected to 110 V. Calculate the polarization.

Solution: We know that $\vec{P} = \varepsilon_0(\text{K}-1)\vec{E}$

Here, $K = 2$

and $\vec{E} = \frac{V}{d} = \frac{110\text{ V}}{5\times 10^{-3}\text{m}}$

$$= 220 \times 10^3 \text{ V/m}$$

Thus, $\vec{P} = 8.85\times 10^{-12}\text{C}\frac{\text{m}^2}{\text{V}}(2-1)\times\frac{220\times 10^3\text{ V}}{\text{m}}$

$$\vec{P} = \mathbf{1.94\times 10^{-6}\ C\text{-}m}$$

EXAMPLE 19.6

If the electronic polarization of a material is ($Z \times 10^{-23}$) $C^2/N^{-1}/m$ at NTP, what will be dielectric constant of the material?

Solution: Using Claussius-Massotti relation,

$$K - 1 = N\alpha_e$$

$\Rightarrow$

$$K = (6.023 \times 10^{23})(Z \times 10^{-23})\, C^2 / N / m$$

$$\mathbf{K = Z \times 6.023}$$

19.9 ▶ Internal Field or Local Field in Liquids and Solids

The electric field which an atom sees or the total electric field at the atom site is called the internal field or the local field. As long as the density of molecules is high, field acting on a given molecule consists partly of the applied field and partly of the field resulting from the mutual interaction of the other molecules, polarized under the influence of the external filed [Fig. 19.10(a)].

Figure 19.10(b) shows an array of equispaced atomic dipoles separated by a distance a. Here, the aim is to find the local field $\vec{E}_l$ which a representative atom A sees. If $\vec{p}_i$ is the induced dipole moment in each atom due to applied field $\vec{E}$, the internal field seen by atom A is the sum of the applied and the fields by other atoms P, Q, R and S.

Here,

$$\vec{E}_{AP} = \frac{2\vec{p}_i}{4\pi\varepsilon_0 a^3} \tag{19.36}$$

and

$$\vec{E}_{AR} = \frac{1}{4\pi\varepsilon_0}\frac{2(-\vec{p}_i)}{(-a)^3} \tag{19.37}$$

and

$$\vec{E}_{AP} + \vec{E}_{AR} = \frac{\vec{p}_i}{\pi\varepsilon_0 a^3} \tag{19.38}$$

Similarly,

$$\vec{E}_{AQ} + \vec{E}_{AS} = \frac{\vec{p}_i}{\pi\varepsilon_0 \left(2a^3\right)} \tag{19.39}$$

The internal field at atom A is given by

$$\vec{E}_l(A) = \vec{E} + \frac{\vec{p}_i}{\pi\varepsilon_0 a^3} + \frac{\vec{p}_i}{\pi\varepsilon_0 (2a)^3} + \cdots. \tag{19.40}$$

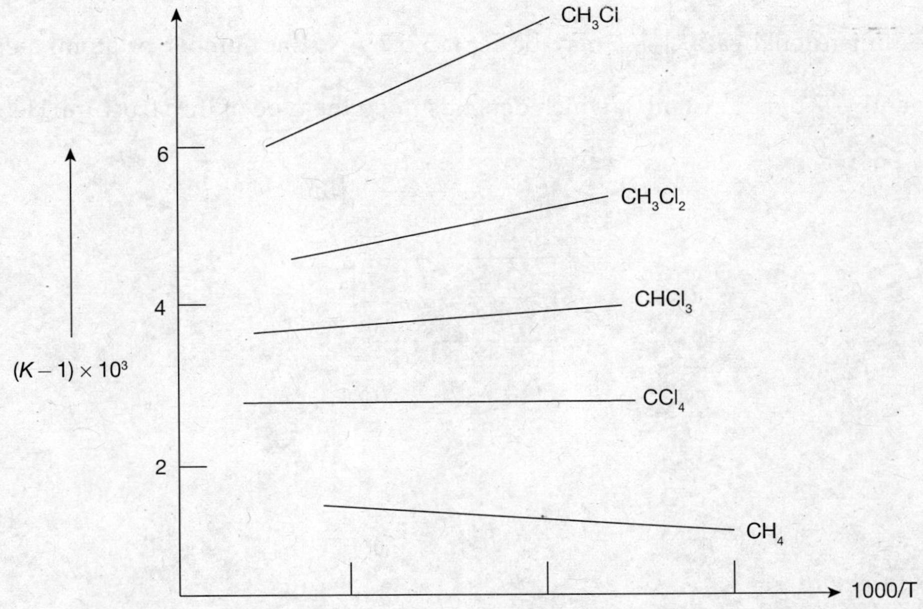

(a)

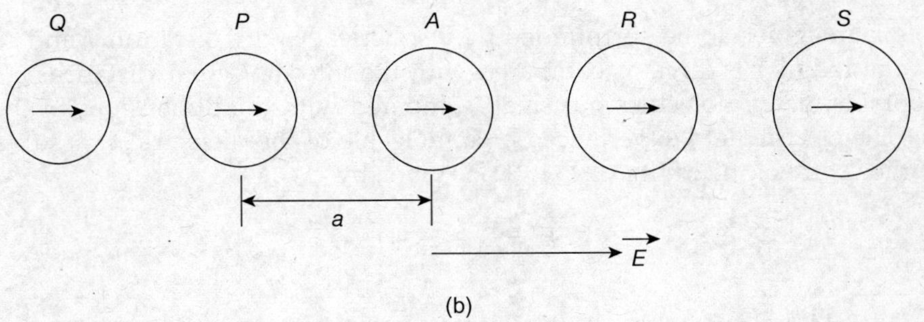

(b)

FIGURE 19.10 (a) Temperature variation of the static dielectric constant of some vapours. (b) String of atoms (one-dimensional atomic array)

$$\vec{E}_l = \vec{E} + \frac{\vec{p}_i}{\pi\varepsilon_0 a^3}\left[\sum_{n=1}^{\infty}\frac{1}{n^3}\right] \tag{19.41}$$

But $\sum_{n=1}^{\infty}\frac{1}{n^3} \simeq 1.2$ with n as an integer.

Equation (19.41) can be written as

$$\boxed{\vec{E}_l = \vec{E} + \frac{1.2\vec{p}_i}{\pi\varepsilon_0 a^3}} \tag{19.42}$$

In three-dimensional case, $\left(\frac{1}{a^3}\right)$ may be replaced by N, the number of atoms per unit volume and $\frac{1.2}{\pi}$ by a constant γ, which depends upon the type of the structure. Hence,

$$\vec{E}_l = \vec{E} + \frac{\gamma N \vec{p}_i}{\varepsilon_0} \qquad \left[\text{But } N\vec{p}_i = \vec{P}\right]$$

$$\Rightarrow \qquad \boxed{\vec{E}_l = \vec{E}_i = \vec{E} + \frac{\gamma \vec{P}}{\varepsilon_0}} \tag{19.43}$$

Now,

$$\vec{P} = N\alpha\vec{E}_i = N\alpha\vec{E}_l$$
$$= \vec{E}\varepsilon_0 (K-1)$$

$$\Rightarrow \qquad (K-1) = \frac{\vec{P}}{\vec{E}\varepsilon_0} = \frac{N\alpha\vec{E}_l}{\vec{E}\varepsilon_0} \tag{19.44}$$

19.10 ▶ Evaluation of the Local Field for Cubic Structure

The dipole is imagined to be surrounded by a spherical cavity of radius r and centre A, such that the radius r is large as compared with the intermolecular distance so that the sphere contains many molecules but small compared with the dimensions of the whole dielectric. The electric field experienced by a molecule of the dielectric, if assumed to be placed at the centre of the cavity (Fig. 19.11) is given by

$$\vec{E}_l = \vec{E}_1 + \vec{E}_2 + \vec{E}_3 + \vec{E}_4 \tag{19.45}$$

where,

(i) $\vec{E}_1 =$ Field intensity due to charge density on the plates of the capacitor (with no dielectric)

(ii) $\vec{E}_2 =$ Field at the atom due to the polarized charges on the plane surfaces of the dielectric (dipolarizing field)

(iii) $\vec{E}_3 =$ Field due to polarized charge on the surface of the spherical cavity

(iv) $\vec{E}_4 =$ Field due to all the dipoles inside the spherical cavity

Now,

$$\vec{D} = \vec{E}_1\varepsilon_0 = \vec{E}\varepsilon_0 + \vec{P} \tag{19.46}$$

$$\vec{E}_1 = \frac{\vec{D}}{\varepsilon_0} = \vec{E} + \frac{\vec{P}}{\varepsilon_0} \tag{19.47}$$

$$\vec{E}_2 = -\frac{\vec{D}}{\varepsilon_0} = -\frac{\vec{P}}{\varepsilon_0} \tag{19.48}$$

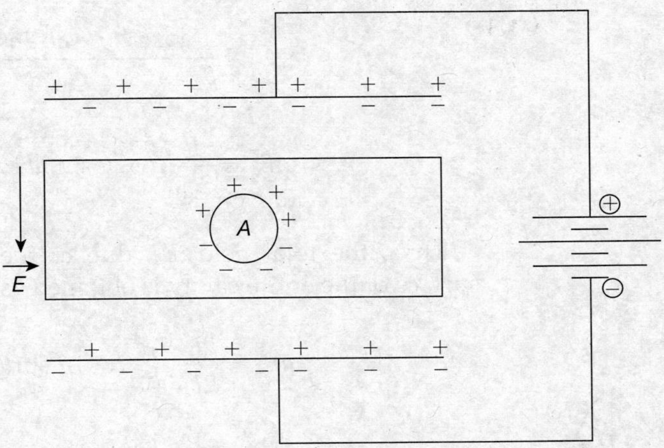

FIGURE 19.11 (a) Calculation of local field

$\vec{E}_3$ is the field intensity at A due to other atoms contained in the cavity. We are assuming a cubic structure, so $\vec{E}_3 = 0$ because of symmetry. $\vec{E}_4$ is the field intensity due to polarization charges on the surface of the cavity and can be calculated.

If dA is the surface area of sphere of radius r lying between θ and $\theta + d\theta$, where θ is the direction with the reference to the direction of the applied field, then

$$dA = 2\pi\,(PQ)\,(QR)$$

But, $$\sin\theta = \frac{PQ}{r},\ PQ = r\sin\theta$$

and $$d\theta = \frac{QR}{r},\quad QR = rd\theta$$

Hence,

$$dA = 2\pi r \sin\theta r\, d\theta$$

$$dA = 2\pi r^2 \sin\theta\, d\theta$$

The charge dq on the surface dA is equal to the normal component of the polarization multiplied by the surface area.

Then, $$dq = \vec{P}\cos\theta dA$$

$$= \vec{P}\left(2\pi r^2 \sin\theta\cos\theta d\theta\right)$$

The field due to this charge at A denoted by dE_4 in the direction $\theta = 0$ is

$$dE_4 = \frac{dq \times 1 \times \cos\theta}{4\pi\varepsilon_0 r^2}$$

$$= \frac{\vec{P} \cos\theta \, 2\pi r^2 \sin\theta \, d\theta \, \cos\theta}{4\pi\varepsilon_0 r^2}$$

$$dE_4 = \frac{\vec{P}}{2\varepsilon_0} \cos^2\theta \, \sin\theta d\theta$$

Thus, the total field $\vec{E}_4$ due to the charges on the surface of the entire cavity is obtained as

$$\int dE_4 = \frac{\vec{P}}{2E_0} \int_0^{\pi} \cos^2\theta \, \sin\theta d\theta$$

$$= \frac{\vec{P}}{2\varepsilon_0} \int_0^{\pi} \cos^2\theta d(-\cos\theta)$$

$$= -\frac{\vec{P}}{2\varepsilon_0} \left[\frac{\cos^3\theta}{3} \right]_0^{\pi}$$

$$\vec{E}_4 = -\frac{\vec{P}}{6\varepsilon_0}[-1-1] = \frac{\vec{P}}{3\varepsilon_0} \qquad (19.49)$$

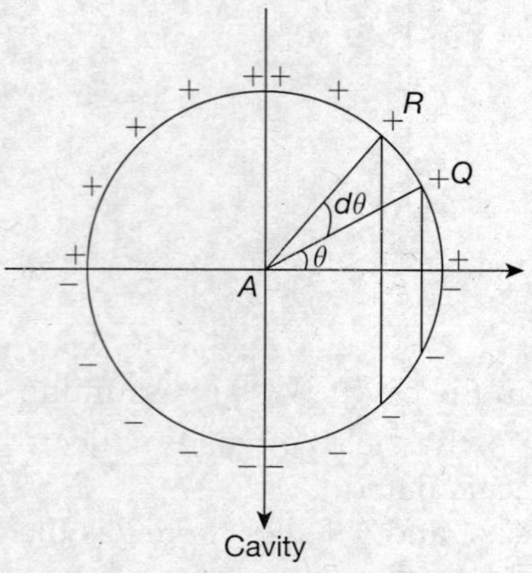

FIGURE 19.11(b)

Thus,

$$\vec{E}_l = \vec{E} + \frac{\vec{P}}{\varepsilon_0} - \frac{\vec{P}}{\varepsilon_0} + 0 + \frac{\vec{P}}{3\varepsilon_0}$$

$$\boxed{\vec{E}_l = \vec{E} + \frac{\vec{P}}{3\varepsilon_0}} \qquad (19.50)$$

Equation (19.50) gives the internal field at an atom in a cubic structure. $\vec{E}_l$ is also called Lorentz field. The Lorentz field is larger than the applied field by an amount that is directly proportional to the polarization density.

19.11 ▶ FERRO-ELECTRICITY

Below a certain temperature, it is found that some materials spontaneously acquire an electric dipole moment. Such materials are called ferroelectrics. These crystals exhibit hysteresis curve. There are three main types of crystal structure, which exhibits ferroelectricity: (1) Rochelle salt structure, $Nak(C_4H_4O_6).4H_2O$, (2) the perovskite group, consisting mainly of titanates and niobates $BaTiO_3$ and (3) the dihydrogen phosphates and arsenates, e.g. KH_2PO_4.

When an electric field is applied to a specimen of a ferroelectric crystal, the polarization first rises rapidly with applied field to a value above which the dependence is linear. Linear extrapolation to zero field gives P_s, the saturation or spontaneous polarization. On subsequently reducing the field to zero, remanent (residual) polarization P_r remains. The negative field to reduce the polarization to zero is called coercive field and denoted by E_c. The hysteresis loop of a ferroelectric material (Fig. 19.12) changes it as the temperature is increased. The height and width decrease with the increase of temperature. At a certain temperature known as ferroelectric Curie temperature, the loop merges in to a straight line and the ferroelectric behaviour of the material disappears.

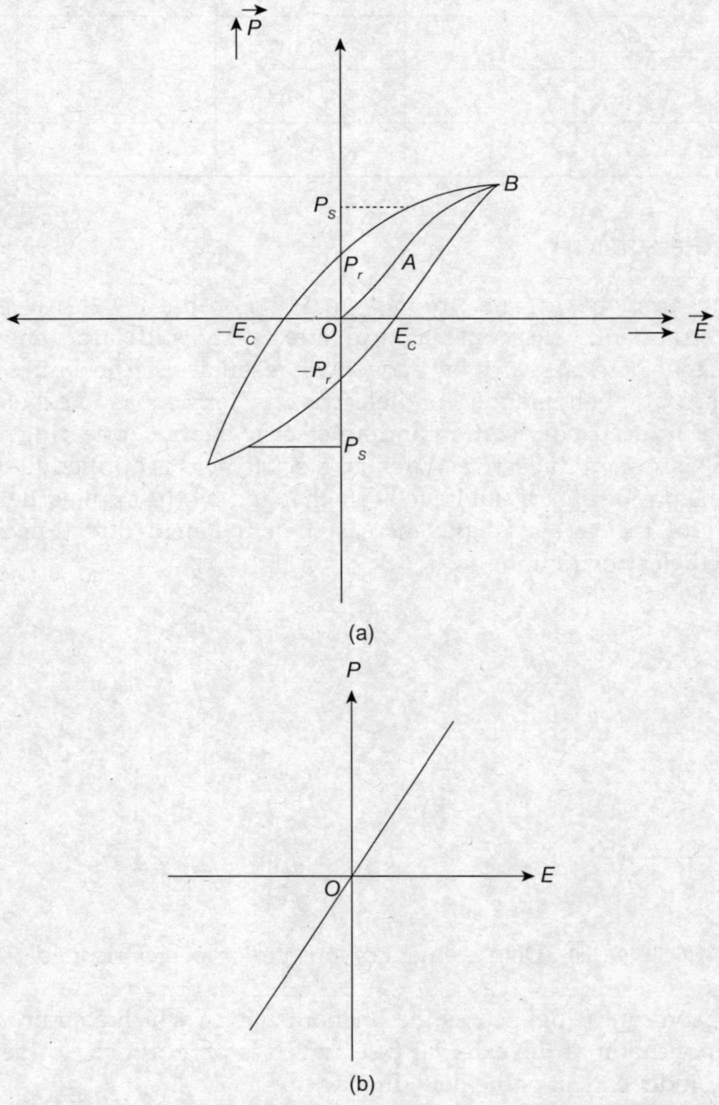

FIGURE 19.12 (a) Hysteresis loop of ferroelectric materials. (b) $P-E$ relation above Curie temperature

TABLE 19.1 Some crystals with their properties

Crystal	Curie temperature (K)	Spontaneous polarization, P_s (Coulomb/m^2)	E_r in weak fields
Rochelle	297 to 255	0.24×10^{-2}	200
$BaTiO_3$	391	26×10^{-2}	1000 to 1700
KH_2PO_4	123	4.95×10^{-2}	–
$KNbO_3$	708	26.6×10^{-2}	–
KD_2PO_4	213	9×10^{-2}	–

19.12 ▶ PIEZO-ELECTRICITY

Most dielectrics are weakly anisotropic, weakly non-linear and weakly sensitive to mechanical strain, but electromechanical effects are still large enough to be of practical value in electrically driven acoustical resonators. The entire phenomenon of electromechanical phenomenon in dielectrics is termed as piezoelectricity. It is evidenced by an induced polarization and an electric charge appearing on the surface of a mechanically strained dielectric. When an electric field is applied to dielectrics, the structural dimensions of the crystal have been changed by the applied field (Fig. 19.13). The dimensions of the crystals have increased in the field direction. This physical property is called electrostriction.

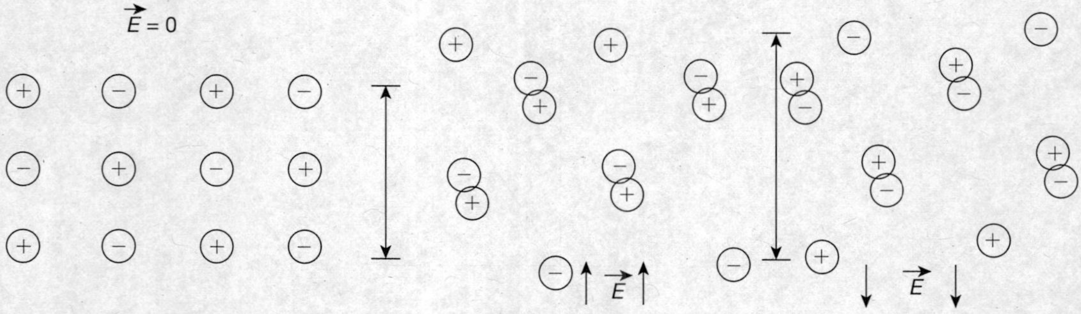

FIGURE 19.13(a) Displacement of ions under an electric field

So electrostriction can be defined as the phenomenon in which a strain is produced in a solid on the application of the external field which is proportional to the square of the applied field and independent of the field direction.

Now, let us consider the effect of mechanical strain on the dimensions of the crystal. We consider two kinds of crystals namely, those with a centre of symmetry and those without.

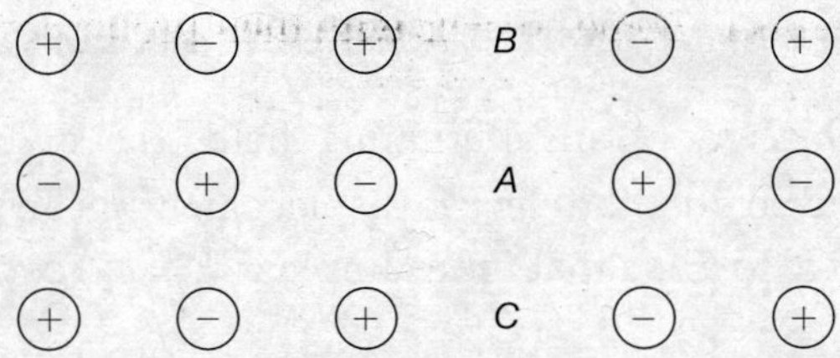

FIGURE 19.13 (b) Ionic structure with a centre of symmetry

In Fig. 19.13(b), the position of the ion *A* is a centre of symmetry.

When the array has been compressed by a mechanical force, the ion *B* has been moved closer to ion *A* thus decreasing the *BA* dipole moment. However, the ion C has also been moved closer to ion *A* which decreases the *CA* dipole moment by same amount in the opposite direction to the decrease in *BA* moment. Thus, no net change in polarization results from mechanical deformation of a crystal with centre of symmetry.

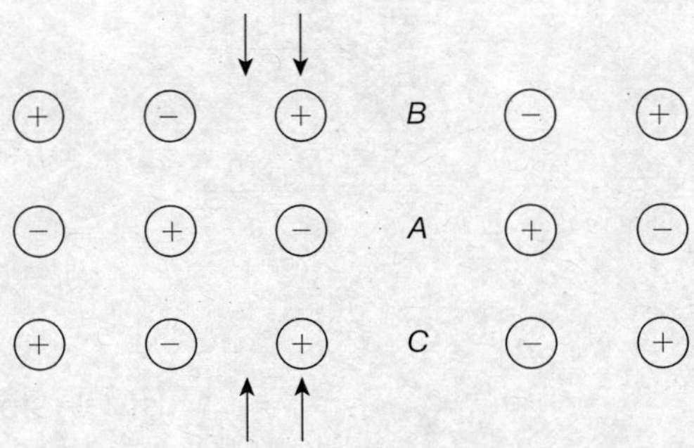

FIGURE 19.13 (c) Structure compressed by a force, no net change of polarization

But, if we have a structure without centre of symmetry, when strained, this structure develops net dipole moments due to each ion and its neighbours. Since both the strain and polarization are characteristically directional in a crystal, complex relationship exists between these quantities. For example, quartz crystals cut in wafers whose faces are parallel to the basal plane (x-cut) will contract or example in a direction parallel to an electric field and perpendicular to the faces.

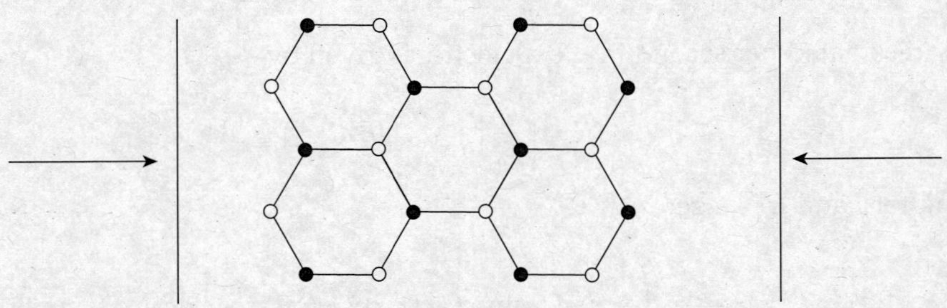

FIGURE 19.13 (d) Ionic structure without a centre of symmetry

19.13 ▶ Complex Dielectric Constant and Dielectric Loss

When a dielectric is subjected to an alternating field, the polarization $\vec{P}$ also varies periodically with time, and so does the electric displacement $\vec{D}$. In general, $\vec{P}$ and $\vec{D}$ may lag behind in phase relative to $\vec{E}$ so that for example if $E = E_0 \cos \omega t$, we have

$$\vec{D} = D_0 \cos(\omega t - \delta) \tag{19.51}$$

$$\vec{D} = D_0 \cos\delta \cos wt + D_0 \sin\delta \sin wt$$

$$\vec{D} = D_1 \cos\omega t + D_2 \sin\omega t$$

where δ = phase angle, $D_1 = D_0 \cos\delta$ and $D_2 = D_0 \sin\delta$

For most dielectrics, D_0 is proportional to E_0.

Here,
$$\varepsilon_r' = \frac{D_1}{E_0} = \frac{D_0 \cos\delta}{E_0}$$

$$\varepsilon_r'' = \frac{D_1}{E_0} = \frac{D_0 \sin\delta}{E_0}$$

and
$$\overset{*}{\varepsilon}_r = \varepsilon_r' - i\varepsilon_r'' \tag{19.52}$$

Thus,
$$D = \overset{*}{\varepsilon}_r E_0 \, e^{iwt} = \overset{*}{\varepsilon}_r (\cos\omega t + i \sin\omega t) E_0 \tag{19.53}$$

Also,
$$\tan\delta = \frac{\varepsilon_r{''}}{\varepsilon_r{'}} \tag{19.54}$$

Both $\varepsilon_r{'}$ and $E_r{''}$ are frequency dependent, the phase angle δ is also frequency dependent. We shall now calculate the energy loss in the dielectric in the form of heat which is proportional to $\varepsilon_r{''}$. The current density in the capacitor is equal to $\frac{d}{dt}\left(\vec{D}\right)$.

Thus,
$$I = \omega(-D_1 \sin\omega t + D_2 \cos\omega t)$$

The energy dissipated per second in the dielectric is given by

$$W = \left(\frac{\omega}{2\pi}\right) \int_0^{2\pi/\omega} IE \, dt$$

Substituting for I and E, we get

$$W = \frac{\omega}{2\pi} \left[\int_0^{2\pi/\omega} \omega(-D_1 \sin\omega t + D_2 \cos\omega t) E_0 \cos\omega t \, dt \right]$$

$$= \frac{\omega}{2\pi}\left[-\omega \int_{0}^{2\pi/\omega} E_0\, D_1 \,\sin\omega t \cos\omega t\, dt + \omega \int_{0}^{2\pi/\omega} E_0 D_2 \cos^2 \omega t\, dt\right]$$

The value of integral containing D_1 is equal to 0,

$$W = \left(\frac{\omega}{2}\right) D_2 E_0$$

$$\boxed{W = \left(\frac{\omega}{2}\right) E_0^2\, \varepsilon''} \qquad (19.55)$$

The energy losses are thus proportional to sin δ, since $\varepsilon'' = \left(\dfrac{D_0}{E_0}\right)\sin\delta$. For this reason, sin δ is called the loss factor and δ is loss angle. For small values of δ, tan $\delta \approx$ sin $\delta \approx \delta$. The dielectric loss at low frequencies is mainly due to DC resistivity. But at high frequencies, the dielectric loss is mostly due to dipole rotations. The losses which fall in the infrared region are called as infrared absorption and in the optical region are referred to as optical absorption.

19.14 ▶ APPLICATIONS OF DIELECTRIC MATERIALS

The following are some of the applications of dielectric materials:

(i) Quartz crystals are used for the preparation of ultrasonic transducers, crystal oscillators, decay lines and filters, etc.

(ii) Barium titanate is used for the preparation of accelerometers.

(iii) Lead zirconate titanate $(PbZr_xTi_{1-x}O_3)$ is used for the preparation of earphones, microphones, spark generators (gas lighters, car ignition, etc.), displacement transducers and accelerometers.

(iv) The insulating dielectric liquids are used in transformers, switchgears and generators.

(v) Dielectric materials are used as insulating material in power cables, signal cables, electric motors and electric irons.

(vi) Dielectric materials are used in radiation detectors, thermionic valves, strain gouges, capacitors, resistors and many other electric devices.

(vii) The electro-optic devices are prepared using dielectric materials.

EXAMPLE 19.7

A dielectric slab of thickness b = 0.50 cm and dielectric constant $K = 7$, is placed between the plates of a parallel plate capacitor plate area $A = 100\ \text{cm}^2$ and separation d = 1 cm. A potential difference $V_0 = 100\,V$ is applied with no dielectric present. The battery is then disconnected and the dielectric slab is inserted. Calculate the three electric vectors $\vec{E}, \vec{D}$ and $\vec{P}$ in the dielectric.

Solution: The electric intensity in air between the two plates is

$$E_0 = \frac{V_0}{d} = \frac{100}{10^{-2}} = 10^4 \text{ V/m}$$

When the dielectric $(K=7)$ is inserted, the electric intensity in the dielectric becomes

$$E = \frac{E_0}{K} = \frac{1.0\times10^4}{7} = 1.43\times10^3 \text{ V/m}$$

The electric displacement in the dielectric is

$$\vec{D} = K\varepsilon_0\vec{E} = 7\times\left(8.85\times10^{-12}\right)\times\left(1.43\times10^3\right) = 8.85\times10^{-8} \text{ C/m}^2$$

The electric polarization in the dielectric is

$$\vec{P} = (K-1)\varepsilon_0\vec{E} = (7-1)\times8.85\times10^{-12}\times1.43\times10^3 = 7.6\times10^{-8} \text{ C/m}^2$$

In air, the three electric vectors are

$$\vec{E} = 1.43\times10^3 \text{ V/m}$$

$$\vec{D} = 8.85\times10^{-8} \text{ C/m}^2$$

and $$\vec{P} = 7.6\times10^{-8} \text{ C/m}^2$$

EXAMPLE 19.8

Calculate the electronic polarizability of Ar atom. Given $K = 1.0024$ at NTP and $N = 2.7 \times 10^{25}$ atoms/m^3.

Solution:

$$\vec{P} = \varepsilon_0 (K-1)\vec{E} \quad (1)$$

as

$$\vec{P} = N\alpha\vec{E} \quad (2)$$

From Eqs. (1) and (2),

$$N\alpha\vec{E} = \varepsilon_0 (K-1)\vec{E}$$

$$\Rightarrow \quad \alpha = \frac{\varepsilon_0 (K-1)}{N}$$

$$= \frac{\left(8.85 \times 10^{-12}\,\text{Fm}^{-1}\right)(1.0024-1)}{2.7 \times 10^{25}\ \text{m}^{-3}}$$

$$\Rightarrow \quad \boldsymbol{\alpha = 7.9 \times 10^{-40}\ \text{Fm}^2}$$

EXAMPLE 19.9

For a gas, the value of dielectric constant at 0°C is 1.000038. Calculate the electric susceptibility at this temperature.

Solution: We know that

$$K = 1 + \chi_e$$

$$\Rightarrow \quad \chi_e = K - 1$$

$$= 1.000038 - 1$$

$$= \mathbf{0.000038}$$

EXAMPLE 19.10

A parallel plate capacitor is filled with insulating material of dielectric constant K. What effect does this have on its capacitance?

Solution: As the field is confined to the space between the plates, the dielectric will reduce $\vec{E}$ and hence potential difference V will be reduced by a factor of $1/K$. Thus, the capacitance $C = q/V$ is increased by a factor of dielectric constant, i.e. $C = KC_0$, where C_0 is the capacitance with dielectric.

RECAP ZONE

POINTS TO REMEMBER

- The dielectrics are non-conductors of electricity as they do not possess free charges. Their electrical conductivity is almost zero.
- A molecule in which the centres of gravity of positive and negative charges coincide is called non-polar molecule. They do not possess any dipole moment.
- A molecule in which the centres of gravity of positive and negative charges do not coincide but separated by a small distance, is called polar molecule. They possess permanent electric dipole moment.
- When a dielectric is placed in an electric field, the positive and negative charges are separated and this process is called polarization of dielectric.
- The relation between three electric vectors $\left(\vec{P}, \vec{E} \text{ and } \vec{D}\right)$ is given as

$$\vec{D} = \varepsilon_0 \vec{E} + \vec{P}$$

- Gauss law in dielectrics states that the surface integral of electric displacement vector over a closed surface is equal to the free charge enclosed within the surface.
- The energy stored per unit volume in dielectric in electrostatic field is given by

$$U = \frac{1}{2}\varepsilon_0 K E^2$$

- When a dielectric is placed in an AC field, there is an energy loss called dielectric loss which arises mostly due to rotation of dipoles.
- Dielectric loss is the loss of electrical energy in the form of heat by a dielectric material during polarization process when subjected to an AC voltage.
- The product of magnitude of a charge and the distance between two charges is called the dipole moment.
- Permittivity represents the dielectric property of a material. It indicates the polarizability nature of the material.
- Polarizability is defined as the ratio of average dipole moment to the applied electric field.
- Ionic polarization takes place due to the displacement of cations and anions in opposite directions.
- Materials exhibit electric polarization even in the absence of an applied electric field are known as ferroelectric materials.

KEY TERMS WITH DEFINITIONS

- **Dielectric** – Having the property of transmitting electric force without conduction.
- **Capacitance** – The ratio of the change in an electric charge in a system to the corresponding change in its electric potential.
- **Electric dipole** – It is a separation of positive and negative charges.
- **Permittivity** – The permittivity is a characteristic of space, and the relative permittivity or "dielectric constant" is a way to characterize the reduction in effective field because of the polarization of the dielectric.
- **Dielectric strength** – The maximum voltage that the dielectric can withstand without rupture.
- **Break down voltage** – The voltage required to puncture a dielectric or insulating material.
- **Potential gradient** – It is defined as the rate of change of potential with distance measured in the direction of the electric force at that point.
- **Dielectric loss** – The loss appearing in the form of heat due to reversal of electric stresses compiling molecular rearrangement.
- **Pyro electricity** – It is interpreted as the ability of certain materials to generate a temporary voltage when they are heated or cooled.
- **Piezoelectricity** – Electric polarization in a substance (especially certain crystals) resulting from the application of mechanical stress.
- **Ferromagnetism** – Ferromagnetism is the basic mechanism by which certain materials form permanent magnets or are attracted to magnets.

IMPORTANT FORMULAE AND EQUATIONS

Equation Number	Equation	Remarks
(19.4)	$\vec{P} = \sigma_p$	Polarization vector
(19.6)	$\vec{D} = \varepsilon_0 \vec{E} + \vec{P}$	Relation between three electric vectors
(19.10)	$K = \frac{C_d}{C_0} = \frac{V_0}{V_d} = \frac{\varepsilon}{\varepsilon_0}$	Dielectric constant of a material
(19.13)	$\oint \vec{E_0} \cdot dA = \frac{q_0}{\varepsilon_0}$	Gauss law in electrostatistic
(19.19)	$\oint \vec{D} \cdot dA = q_0$	Gauss law in dielectrics

(19.26)	$U = \frac{1}{2}\varepsilon_0 KE^2$	**Energy stored per unit volume in dielectric in electrostatic field**
(19.33)	$\alpha = \frac{3\varepsilon_0(\text{K}-1)}{N(\text{K}-2)}$	**Claussius-Mossotti relation**
(19.35)	$\alpha = \frac{3\varepsilon_0(n^2-1)}{N(n^2+2)}$	**Lorentz-Lorentz relation**
(19.50)	$\vec{E}_l = \vec{E} + \frac{\vec{P}}{3\varepsilon_0}$	**Internal field at an atom**
(19.55)	$W = \left(\frac{\omega}{2}\right)E_0^2\varepsilon''$	**Energy dissipated per second in the dielectrics**

REVIEW ZONE

SHORT ANSWER QUESTIONS

1. What is a dielectric material?
2. What are polar and non-polar molecules?
3. State Gauss law in dielectrics.
4. Define electric susceptibility.
5. Derive the relation between electric susceptibility and dielectric constant.
6. Write the Claussius-Mossotti relation.
7. Write Lorentz relation.
8. What do you mean by piezo-electricity?
9. Define dielectric constant.
10. What do you understand by electric field strength?
11. What is the difference between local electric field and applied electric field?
12. Why is a capacitor known to be an energy-string device?

LONG ANSWER QUESTIONS

1. What do you understand by dielectric materials? Define $\vec{D}, \vec{E}$ and $\vec{P}$ and establish a relationship between them.
2. Explain polarization in dielectrics.
3. State and prove Gauss's law in dielectrics.
4. Deduce an expression for energy stored in dielectric in electrostatic field.
5. Explain the behaviour of a dielectric in AC field and discuss the dielectric losses.

6. Derive an expression for Claussius-Mossotti relation.
7. Derive an expression to calculate the energy stored in dielectrics.
8. Write the essential difference between a conductor, an insulator and a dielectric material.
9. Explain the difference between a ferroelectric pyroelectric and a piezoelectric material.
10. Discuss, with proper diagrams, the formation of dipoles in a dielectric material.
11. Explain the different types of solid dielectric materials.
12. Explain the molecular theory of polarization in dielectrics. Also show that $\vec{D} = \varepsilon_0 \vec{E} + \vec{P}$ where the symbols have their usual meanings.

NUMERICAL PROBLEMS

1. If a NaCl crystal is subjected to an electric field of 100 V/m and the resulting polarization is 4.3 × 10^{-8} Cm^2, calculate the relative permittivity of NaCl.

   Ans. 8×10^{-12} F/m

2. The polarizability of ammonia molecule is found approximately by the measurement of dielectric constant as 2.42 × 10^{39} C^2m/N and 1.74 × 10^{-39} C^2m/N at 309 K and 448 K, respectively. Calculate for each temperature the orientation polarizability.

   Ans. 6×10^{-39} Fm^2, 4×10^{-39} Fm^2

3. An electric field of 9×10^4 V/m is applied on a sample of *Ne* gas at NTP. Calculate the dipole moment induced in each atom, if the dielectric constant of neon at NTP is 1.000134. Find the atomic polarizability of *Ne*.

   Ans. 0.000134, 7×10^{-39} Fm^2

4. If the relative permittivity of *Kr* atom is 1.0006612 $C^2/N/m^2$, calculate the electronic polarizability of *Kr* at NTP.

   Ans. 2.176 × 10^{-40} C^2/Nm^2

5. If the electronic polarizability of argon atom is 1.432 × 10^{-40} $C^2/N/m^2$, calculate the relative permittivity of argon at NTP.

   Ans. 1.000435

6. A molecule has a dipole moment of 3 × 10^{-25} C-cm, where all the molecular dipoles point in the same direction. If the radius of the molecule as a drop is 0.25 mm, find the polarization in the molecular drop.

   Ans. 6.78 × 10^{-9} C/cm^2

7. Calculate the maximum potential gradient to which a 0.25 mm thick mica sheet can be subjected. The dielectric strength for mica is 100 × 10^6 V/m.

   Ans. 2.5 × 10^4 V

CHAPTER 20

Magnetic Materials

Learning Objectives

By the end of this chapter, the student will be able:

- To classify various types of magnetic materials
- To discuss properties of dia, para and ferro magnetic materials
- To select materials on hysteresis basis for various applications
- To solve the theoretical and experimental problems of soft and hard materials
- To discuss the applications of magnetic materials

20.1 ▶ Introduction

Magnetic materials get magnetized when placed in a magnetic field. The magnetic properties of solids originate due to the motion of electrons and depending upon the response of materials to the applied magnetic field, all materials are classified into three main categories:

(i) Diamagnetic materials

(ii) Paramagnetic materials

(iii) Ferromagnetic materials

There are also two more classes of magnetic materials, which are given as follows.

(iv) Anti-ferromagnetic materials

(v) Ferrimagnetic materials

In this chapter, we will discuss the properties of the materials on the basis of the magnetic properties exhibited by them.

20.2 ▶ Some Important Terms

There are certain important terms that are frequently used in the study of magnetic properties of the materials. The following are the important terms:

(i) Magnetic induction

When a material is placed in an external magnetic field, the material is said to be magnetized. The number of magnetic lines of force passing per unit area normally in free space is a measure of magnetizing field. The magnetizing field is denoted by H. It is also called magnetic field intensity while the net number of magnetic lines of force within the material medium when it is placed in an external magnetic field is called magnetic induction. It is denoted by B. Its unit is tesla or weber/m^2.

(ii) Intensity of magnetization

When a magnetic material is placed in an external magnetic field, the induced magnetic moment per unit volume is called the intensity of magnetization. It is denoted by I and $I = M/V$.. Its units are ampere/metre, where M = total magnetic moment and V = total volume of magnetized substance.

(iii) Magnetic susceptibility

It is defined as the ratio of intensity of magnetization I to magnetic intensity H, i.e.,

$$I \propto H$$

or

$$I = \chi_m H$$

where χ_m is called magnetic susceptibility and it is a dimensionless constant. It is defined as the characteristic of the medium.

(iv) Relative permeability

It is defined as the ratio of magnetic induction, B, in the medium to the magnetic induction, B_0 in the free space, i.e. $\mu_r = B/B_0$.

Generally, the magnetizing field, i.e. applied external magnetic induction in free space, is expressed by H, i.e. $B_0 = H$ and the flux density is B. So, the ratio B/H is called absolute permeability (μ) of the material medium, i.e. $\mu = B/H$ but the absolute permeability of vacuum is denoted by μ_0 and also

$$\mu_0 = \frac{B_0}{H} = 4\pi \times 10^{-7} \text{ wb/A-m or henry/m}$$

The relative permeability is also defined as the ratio of permeability of medium to the permeability of free space, i.e. $\mu_r = \dfrac{\mu}{\mu_0}$, it has no units.

(v) Relation between μ_r and χ_m

We know that when a sample material is placed in a magnetizing field *H*, due to the alignment of current loop, it gets magnetized. Now, the magnetic flux density *B* of the sample can be written as

$$B = \mu_0 (H + I) \tag{20.1}$$

The units of *H* and *I* are same (ampere-turn/metre) and the unit of *B* is tesla or wb/m^2.

Also,
$$B = \mu H \tag{20.2}$$

From Eqs. (20.1) and (20.2),

$$\mu H = \mu_0 (H + I)$$

$$\Rightarrow \quad \mu = \mu_0 \left(1 + \frac{I}{H}\right)$$

$$\Rightarrow \quad \mu = \mu_0 (1 + \chi_m)$$

$$\Rightarrow \quad \frac{\mu}{\mu_0} = 1 + \chi_m$$

$$\Rightarrow \quad \boxed{\mu_r = 1 + \chi_m} \tag{20.3}$$

Equation (20.3) gives the relation between relative permeability and magnetic susceptibility.

Note

Intensity of magnetization may also be defined as the pole-strength per unit area of cross-section.

EXAMPLE 20.1

The magnetic susceptibility of medium is 9.48×10^{-9}. Calculate the relative permeability.

Solution: Given $\chi_m = 9.48 \times 10^{-9}$

$$\mu_r = 1 + X_m$$

$$= 1 + 9.48 \times 10^{-9}$$

i.e., The relative permeability is slightly greater than 1.

EXAMPLE 20.2

The horizontal component of flux density of the earth's magnetic field is 1.7×10^{-5} weber/m^2. What is the horizontal component of the magnetic intensity?

Solution:

$$B_0 = \mu_0 H$$

$$\Rightarrow \qquad H = \frac{B_0}{\mu_0}$$

$$= \frac{1.7\times10^{-5}\ \text{weber/m}^2}{4\pi\times10^{-7}\ \text{weber/amp-metre}}$$

$$\Rightarrow \qquad \mathbf{H = 13.5\ A/m}$$

EXAMPLE 20.3

The saturation magnetic induction of nickel is 0.65 T. If the density of nickel is 8906 kg/m^3 and atomic weight is 58.7, find out the magnetic moment of the nickel atom in Bohr magneton.

Solution: Number of atoms/m^3

$$N = \frac{\rho N_A}{M}$$

$$= \frac{8906\ \text{kg/m}^3 \times 6.023\times10^{26}\ \text{atoms/k}\cdot\text{mol}}{58.7}$$

$$= 9.14\times10^{28}\,/\text{m}^3$$

Magnetic moment

$$\mu_m = \frac{B}{N\mu_0}$$

$$= \frac{0.65T}{9.14\times10^{28}\,/\,\text{m}^3\times4\pi\times10^{-7}}$$

$$= 5.66\times10^{-24}\ \text{A}\cdot\text{m}^2$$

$$= \frac{\mathbf{5.66\times10^{-24}}}{\mathbf{9.27\times10^{-24}}} = \mathbf{0.61\mu_B}.$$

20.3 ▶ Magnetic Moment of an Atom

In an atom, negatively charged electrons move around the positively charged nucleus in circular orbits. These circulating electrons produce electric current. Actually, each circulating electron is equivalent to a current loop, which produces magnetic field. Hence, an atom can be considered as a magnetic dipole. Therefore, the atom has magnetic moment due to (i) the orbital motion of the electron and (ii) spinning motion of the electron.

We now calculate the magnetic moment of an atom due to both of the motions of electron.

20.3.1 Magnetic Moment of an Atom Due to Orbital Motion of Electron

Let us consider an electron having mass m and charge $-e$ revolving with speed v in a circular orbit of radius r.

The current produced by the electron,

$$i = \frac{e}{T} \tag{20.4}$$

$$\Rightarrow \quad i = \frac{e}{2\pi r / v} \tag{20.5}$$

$$\left[\text{Time period, } T = \frac{2\pi r}{v}\right]$$

Magnetic moment, $$M_l = iA$$

$$\Rightarrow \quad M_l = \frac{ev \times (\pi r^2)}{2\pi r} = \frac{evr}{2} \tag{20.6}$$

Here, area $A = \pi r^2$

Angular momentum of the electron is

$$L = mvr \tag{20.7}$$

Dividing Eq. (20.6) by Eq. (20.7), we get

$$\frac{M_l}{L} = \frac{evr/2}{mvr} = \frac{e}{2m}$$

$$\Rightarrow \quad \frac{M_l}{L} = \frac{e}{2m} \tag{20.8}$$

This ratio of magnetic moment to angular momentum is called orbital gyromagnetic ratio, and it is denoted by g,

i.e., $$\boxed{g = \frac{M_l}{L} = \frac{e}{2m}} \tag{20.9}$$

in vector form equation (20.8) can be written as:

$$M_l = -\frac{e}{2m}L \tag{20.10}$$

Here, the negative sign indicates that the magnetic moment M_l and the angular momentum L are directed opposite to each other.

Also,

$$L = \sqrt{l(l+1)}\frac{h}{2\pi}$$

Thus,

$$M_l = \frac{e}{2m}\sqrt{l(l+1)}\frac{h}{2\pi}$$

$$M_l = \frac{eh}{4\pi m}\sqrt{l(l+1)} \tag{20.11}$$

Here, $l = 0, 1, 2, 3, \ldots$ for s, p, d and f electrons, respectively.

In Eq. (20.11), the quantity $\frac{eh}{4\pi m}$ is called Bohr magneton.

Note

The magnetic moment of an atom can be expressed in terms of an atomic unit of magnetic moment called Bohr magneton and the value of one Bohr magneton is

$$9.27 \times 10^{-24}\, A - m^2$$

EXAMPLE 20.4

Calculate the magnetic moment of an electron moving in a circular orbit of radius R about a proton.

Solution: We know that the magnetic moment of an electron is

$$M = \text{area of loop} \times \text{current}$$

$$\Rightarrow \quad M = \pi R^2\left(\frac{e}{T}\right)$$

$$\Rightarrow \quad M = \pi R^2\left(\frac{ev}{2\pi R}\right) \quad \left[\text{since, } T = \frac{2\pi R}{v}\right] \tag{1}$$

But

$$F = ma$$

$$\Rightarrow \quad \frac{Ke^2}{R^2} = \frac{mv^2}{R}$$

$$\Rightarrow \qquad v = \sqrt{\frac{Ke^2}{mR}} \qquad (2)$$

From (1) and (2),

$$M = \pi R^2 \frac{ev}{2\pi R}$$

$$= \frac{Re}{2}\sqrt{\frac{Ke^2}{mR}}$$

$$\Rightarrow \qquad \mathbf{M = \frac{e^2}{2}\sqrt{\frac{kR}{m}}}$$

20.3.2 Magnetic Moment of an Atom Due to Spinning Motion of Electron

The electron also spins about its own axis, and therefore possesses spin angular momentum, which is given by

$$S = \sqrt{s(s+1)}\frac{h}{2\pi} \qquad (20.12)$$

where s = spin quantum number and its value is always $\frac{1}{2}$.
Thus,

$$S = \sqrt{\frac{1}{2}\left(\frac{1}{2}+1\right)}\frac{h}{2\pi} = \frac{\sqrt{3}}{2}\frac{h}{2\pi}$$

The spin magnetic moment is given by

$$\frac{M_s}{S} = \frac{e}{m_e}$$

or

$$\boxed{M_s = \frac{e}{m_e}S = \sqrt{s(s+1)}\frac{eh}{4\pi m}} \qquad (20.13)$$

Now, the total magnetic moment of the atom can be obtained by the sum of orbital magnetic moment and spin magnetic moment, i.e.,

$$M_{\text{total}} = M_l + M_s \qquad (20.14)$$

But M_s is much greater than M_l.

If the electron spin moments are set free to orient themselves, they will be oriented in the direction of applied field.

Note

Nuclear spin also contributes to magnetic moment of atoms, which arises due to spin of nucleus and given by $\mu_n = \frac{eh}{4\pi M_p}$.

EXAMPLE 20.5

The saturation magnetic induction of nickel is 0.65 T. If the density of nickel is 8906 kg/m^3 and the atomic weight is 58.7, find out the magnetic moment of the nickel atom in Bohr magneton.

Solution: Number of atoms/m^3

$$N = \frac{\rho N_A}{M}$$

$$= \frac{8906\ \text{kg/m}^3 \times 6.023 \times 10^{26}\ \text{atoms/k}\cdot\text{mol}}{58.7}$$

$$= 9.14 \times 10^{28}/\text{m}^3.$$

Magnetic moment

$$\mu_m = \frac{B}{N\mu_0}$$

$$= \frac{0.65T}{9.14 \times 10^{28}\ /\text{m}^3 \times 4\pi \times 10^{-7}}$$

$$= 5.66 \times 10^{-24}\ \text{A}\cdot\text{m}^2$$

$$= \frac{\mathbf{5.66 \times 10^{-24}}}{\mathbf{9.27 \times 10^{-24}}} = \mathbf{0.61}\boldsymbol{\mu_B}$$

20.4 ▶ Classification of Magnetic Materials

Any material when placed in a magnetic field behaves in a different manner than the others. Thus, all materials can be divided into three broad categories namely diamagnetic, paramagnetic and ferromagnetic materials.

Diamagnetic materials

The diamagnetic materials are those which are weakly magnetized by strong magnetic fields in a direction opposite to that of the applied magnetic field. For example, gold, bismuth, sodium, diamond, copper, silver, hydrogen and nitrogen show diamagnetism.

The main characteristics of diamagnetic substances are as follows:

(i) When a bar of diamagnetic material is placed in a strong magnetic field, the total number of magnetic lines of induction in the material is less than that in free space as shown in Fig. 20.1. Thus, the magnetic moment, intensity of magnetization and magnetic susceptibility are negative and relative permeability of the material is less than 1 for these substances.

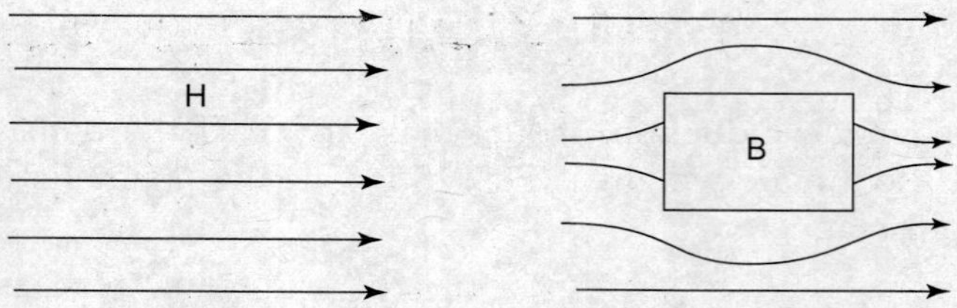

FIGURE 20.1 Diamagnetism

(ii) In a non-uniform field, a diamagnetic substance tends to move from the stronger parts of the field to the weaker parts.

(ii) When a bar of diamagnetic material is suspended between the poles of a magnet, it stays at right angles to the direction of field.

(iv) A diamagnetic liquid when placed in a U-tube and between the pole pieces of strong magnets, always shows a depression in the limb of the U-tube.

(v) The susceptibility of diamagnetic substance is independent of the temperature.

Paramagnetic materials

The paramagnetic materials are those which are weakly magnetized by strong magnetic fields in the same direction as that of applied magnetic field. For example, platinum, aluminium, manganese, oxygen shows paramagnetism.

The main characteristics of paramagnetic substances are as follows:

(i) When a bar of paramagnetic substance is placed in a strong magnetic field, the total number of magnetic lines of induction in the material are slightly greater than that in free space as shown in Fig. 20.2.

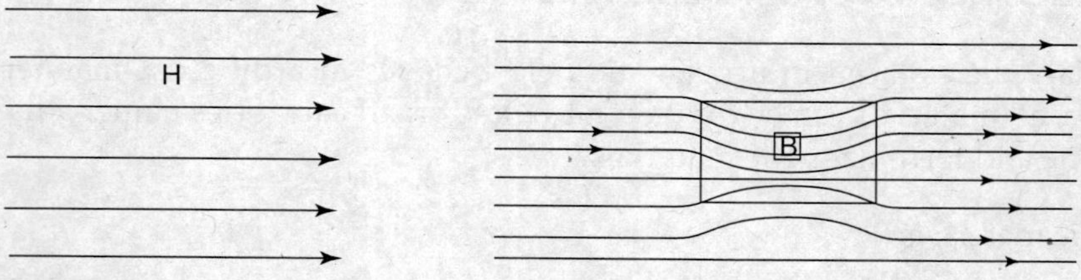

FIGURE 20.2 Paramagnetism

Thus, the magnetic moment, intensity of magnetization and magnetic susceptibility are positive but small and the relative permeability is slightly greater than 1.

(ii) In a non-uniform field, a paramagnetic substance tends to move from the weaker parts of the field to the stronger parts.

(iii) When a bar of paramagnetic material is suspended between the poles of a magnet, it stays along the field.

(iv) A paramagnetic liquid, when placed in a U-tube and between the pole pieces of a strong magnet, always show a rise in the limb of the U-tube.

(v) The susceptibility of a paramagnetic substance decreases with the rise in temperature.

Ferromagnetic materials

The ferromagnetic materials are those which are strongly magnetized by relatively weak magnetic fields in the same direction as that of applied field. These are attracted very fast when brought closer to the ends of a magnet. The main characteristics of ferromagnetic substances are as follows:

(i) When a bar of ferromagnetic substance is placed in a magnetic field, there is a crowding of flux lines within the substance as shown in Fig. 20.3.

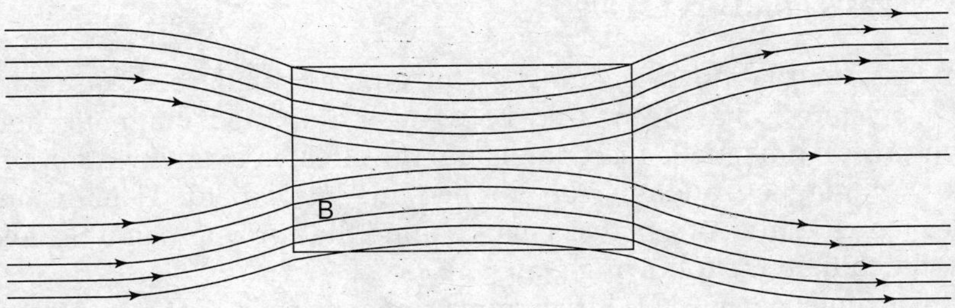

FIGURE 20.3 Ferromagnetism

Thus, these substances possess all the properties of a paramagnetic substance but with greater intensity, i.e. the magnetic moment, intensity of magnetization and magnetic susceptibilities are positive and quite large.

Also, the relative permeability is of the order of hundreds and thousands.

(ii) The susceptibility of ferromagnetic substance decreases with rise in temperature. At a certain temperature, called curie temperature, the permeability μ of a ferromagnetic material becomes practically equal to μ_0. Above Curie temperature, the ferromagnetic material becomes paramagnetic. According to Curie law,

$$\chi_m \propto \frac{1}{T}$$

Note

The Curie temperature of iron is about 770°C.

EXAMPLE 20.6

Diamagnetic Al_2O_3 is subjected to an external magnetic field of 10^5 A/m. Evaluate magnetization and magnetic flux density in Al_2O_3 (Susceptibility of $Al_2O_3 = -5 \times 10^{-5}$).

Solution:

$$\text{Magnetization } M = \chi H = \left(-5\times10^{-5}\right)\left(10^{5}\,\text{A/m}\right)$$
$$= -5\,\text{A/m}$$

Magnetic flux density,

$$B = \mu_0 (H + M)$$
$$= \left(12.57\times10^{-7}\ \text{H/m}\right)\left(10^{5} - 5\,\text{A/m}\right)$$
$$= \mathbf{0.126\ wb\,/\,m^2}$$

20.5 ▸ Hysteresis and B-H Curve

Hysteresis means retardation or lagging of an effect behind the cause of the effect. In magnetism, hysteresis has been used between the applied magnetic field (H) and magnetization (M) of a magnetic material. Here, the effect is magnetization of a material and the cause of magnetization is the applied magnetic field. Usually, in magnetic materials, the magnetization of a material lags behind the applied magnetic field. This can be explained in detail, in the following way.

We start with an unmagnetized ($M = 0$) ferromagnetic specimen. With an increasing applied magnetic field on it, the magnetization of the specimen increases from zero to high values. The increase is non-linear. With small applied fields, the domains pointing approximately in the field direction increase at the expense of those that are not. In other words, their boundaries move so as to expand the favourable domains. This gives rise to a small magnetization corresponding to the initial position of the hysteresis curve shown in Fig. 20.4.

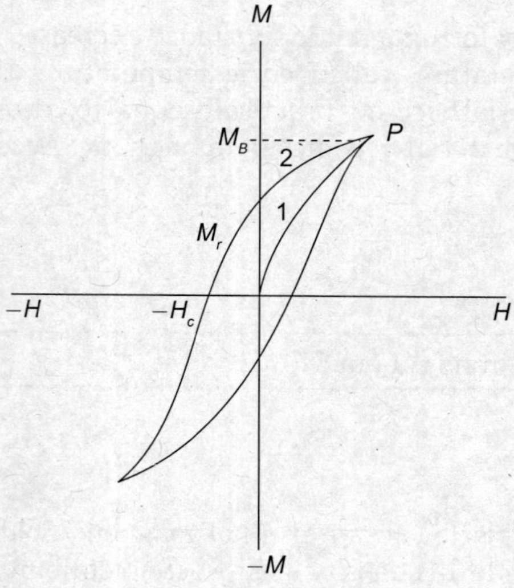

FIGURE 20.4 Hysteresis curve

With higher fields, the magnetization increases rapidly with H. At these field strengths, the boundary moments are often large and irreversible, i.e. the boundaries do not go back into their original position on reducing H. Application of still higher fields rotates the magnetization vectors into the field direction, i.e. all the domains point in the direction of H, then the specimen is said to be saturatly magnetized ($M = M_S$).

The saturated magnetization is represented as point "P" in Fig. 20.4. If the field is decreased, then the magnetization decreases below the value M_S, but this decrease of M does not occur along the same path (curve 1) because the domains do not easily return to the original random arrangement. As H is reduced to zero, M does not decrease in phase but lags behind H. The value of M that remains in the material when H is reduced to zero is called residual magnetization or remanence magnetization (M_r).

To remove the residual magnetism in the material, magnetic field is applied in the opposite direction and gradually increased from zero. The magnetization in the material becomes zero for an applied magnetic field of $-H_C$, called coercive field for the material. Further increase of the applied magnetic field in the opposite direction results in the material magnetization in the opposite direction. Again decreasing the magnetic field in the opposite direction to zero results in the residual magnetism in opposite direction. Again increasing the magnetic field in the forward direction, we will get a curve that completes a closed loop called hysteresis loop. This loop includes some area. This area indicates the amount of energy wasted in one cycle of operation.

20.5.1 Soft and Hard Magnetic Materials

The criterion of selection of a material is based on the hysteresis curve discussed in Section 20.5. Both Ferromagnetic and Ferrimagnetic materials are classified as either soft or hard on the basis of hysteresis characteristics (Fig. 20.5). The properties of hard and soft magnetic materials are given below.

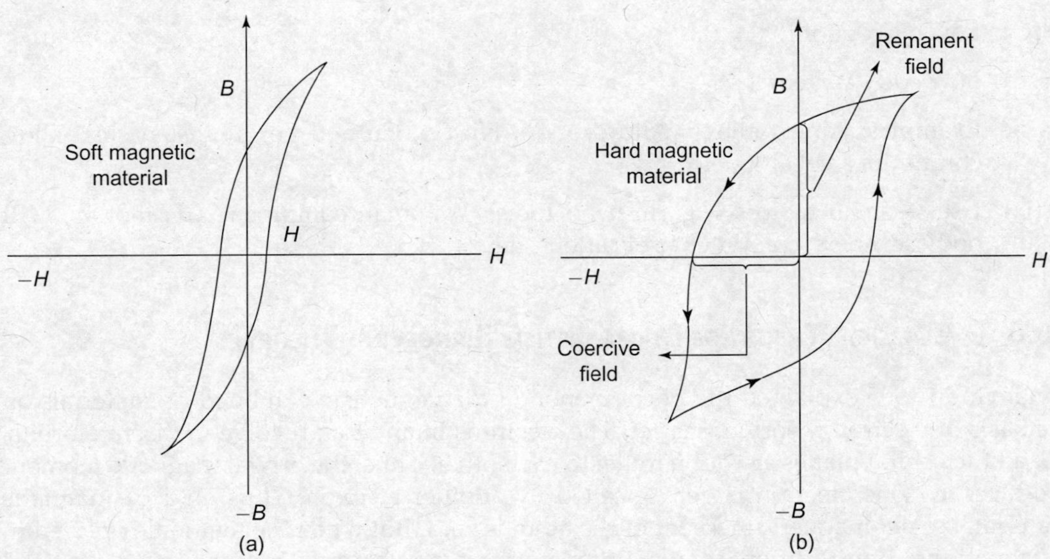

FIGURE 20.5 Soft and hard magnetic materials

Properties of soft magnetic materials

Soft materials can be easily magnetized or easily demagnetized. Some of the properties of soft magnetic materials are

(i) Low coercivity

(ii) Low hysteresis loss

(iii) High permeability

(iv) High susceptibility

(v) Low remanent magnetization

(vi) Examples: Alloys of Fe and Ni, silicon, commercial iron, alloys of Ni, Cu, Cr and Fe, Alloys of Fe, Si and B, etc.

(vii) Uses – core of transformers, power generators, audio frequency applications, etc.

Properties of hard magnetic materials

The hard magnetic materials can neither be easily magnetized nor be easily demagnetized. Some of the properties of hard magnetic materials are

(i) High coercivity

(ii) Low initial permeability

(iii) High hysteresis energy loss

(iv) High remanent magnetization

(v) High saturation flux density

(vi) Low susceptibility

(vii) Low eddy current loss

(viii) Examples: Alnico alloys (Alloys of Al, Ni, Co, Cu and Fe), tungsten steal alloy, chromium steel alloy, etc.

(ix) Uses – small motors (e.g., in hand tools), Walkman equipment, CD motors, MRI body scanners, computer applications, etc.

20.6 ▶ Classical Theory of Diamagnetism (Langevin's Theory)

Langevin in 1905 explained the phenomenon of diamagnetism exhibited by materials on the basis of electron theory of matter. The electrons in an atom revolve in discrete orbits, but all the orbital planes are not parallel to each other. Thus, the orbital magnetic moment produced by one electron is get cancelled by another electron. Thus, in a diamagnetic material, the net magnetic moment of the atom is zero. But when the material is placed in a magnetic field, an additional magnetic force acts on the electron. Thus, the total centripetal force on the electron becomes $F_e \pm F_m$. Here, F_m may be radially inwards or outwards according to the rotation of the electron with respect to the field (as shown in Fig. 20.6).

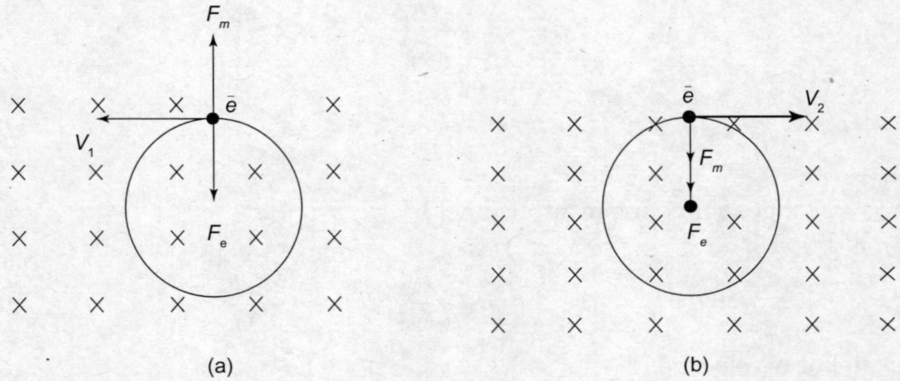

FIGURE 20.6 (a) F_e and F_m pointed in opposite directions and (b) F_e and F_m pointed in same directions

Due to applied magnetic field, F_m changes the angular velocity of the electron.

As shown in Fig. 20.6(a), the velocity is decreased while in Fig. 20.6(b), the velocity increases.

As a result, the magnetic moment in the first case decreases but the magnetic moment in the second case [Fig. 20.6(b)] increases. Therefore, the electronic orbits shift slightly relative to each other to produce a negative magnetic moment of the atom. Let us calculate this change in magnetic moment.

Let us consider a single electron revolving at angular frequency ω_0 in an orbit of radius r, in the absence of magnetic field, centripetal force will be

$$F_e = \frac{mv_0^2}{r} = m\omega_0^2 r \tag{20.15}$$

where v_0 = speed of $\bar{e}$.

When a uniform magnetic field is applied on the material,

$$F_m = evB = e(\omega r)B \tag{20.16}$$

Here, ω and ω_0 are different, because due to the action of F_m, angular velocity changes. When an external field B is applied, the e.m.f induced in the current loop is given by

$$\varepsilon = -\frac{d\phi}{dt} \tag{20.17}$$

This e.m.f induced will result in an electric field E and by Newton's law,

$$m\frac{dv}{dt} = eE \tag{20.18}$$

$$\Rightarrow \qquad m\frac{dv}{dt} = e\frac{\varepsilon}{2\pi r} = -\frac{e}{2\pi r}\frac{d\phi}{dt}$$

$$\Rightarrow \qquad m\frac{\Delta v}{\Delta t} = \frac{-e}{2\pi r}\frac{\Delta\phi}{\Delta t}$$

$$\Delta v = -\frac{e}{2\pi rm}\Delta\phi \tag{20.19}$$

where $\Delta\phi$ = change in flux due to applied field $= \pi r^2 B$
From Eq. (20.19),

$$\Delta v = -\frac{e}{2\pi rm}\left(\pi r^2 B\right) = -\frac{erB}{2m} \tag{20.20}$$

Change in linear velocity is

$$\Delta v = -\frac{erB}{2m}$$

So, change in angular velocity will be

$$\Delta\omega = \frac{\Delta v}{r} = -\frac{eB}{2m} \tag{20.21}$$

The angular velocity given by Eq. (20.21) is known as Larmor's velocity. Now, the magnetic moment of orbiting electron is

$$M = iA = \frac{e\omega}{2\pi}\left(\pi r^2\right) = \frac{er^2\omega}{2}$$

When the angular velocity changes by $\Delta\omega$, the change in magnetic moment is given by

$$\begin{aligned}\Delta m &= \frac{er^2}{2}(\Delta\omega)\\ &= \frac{er^2}{2}\left(-\frac{eB}{2m}\right)\\ &= -\frac{e^2r^2B}{4m}\end{aligned} \tag{20.22}$$

In an atom of atomic number Z, there will be Z electrons and Z electronic orbits. The change in magnetic moment for z electrons will be z times the magnetic moment given by Eq. (20.22)

$$-\frac{e^2r^2BZ}{4m}$$

But all the electrons do not rotate in same orbit, so replacing r^2 by mean value $\overline{r}^2$ and supposing that all the electronic orbits have all possible orientation with respect to the field and taking the spherical symmetry of the atom into account, we can write

$$\overline{x}^2 = \overline{y}^2 = \overline{z}^2 = \frac{R^2}{3}$$

[If x, y, z are the coordinates of any point on an orbit of radius R]

If the direction of field B is along the Z-axis, thus

$$\overline{r}^2 = \overline{x}^2 + \overline{y}^2 + 0$$

$$= \frac{R^2}{3} + \frac{R^2}{3} = \frac{2R^2}{3}$$

Hence, the total magnetic moment of the atom is

$$M = -\frac{e^2 BZ}{4m} \sum \overline{r}^2 = -\frac{Ze^2 R^2 B}{6m} \tag{20.23}$$

If the material has n electrons per unit volume, then the intensity of magnetization can be written as

$$I = \frac{M}{V}$$

$$= -\frac{nze^2 R^2 B}{6m} = -\frac{nze^2 R^2 \mu_0 H}{6m} \quad [\because B = \mu_0 H\ ] \tag{20.24}$$

The susceptibility of the substance is

$$\boxed{\chi_m = \frac{I}{H} = -\frac{\mu_0 n Z e^2 R^2}{6m}} \tag{20.25}$$

From Eq. (20.25), it is clear that the susceptibility of diamagnetic substances is negative and independent of the temperature.

20.7 ▶ Langevin's Theory of Paramagnetism

Langevin explained paramagnetism on the basis of the fact that the atoms, ions or molecules (of a paramagnetic material) have a net permanent magnetic moment. When the paramagnetic substance is placed under the effect of external applied magnetic field, the molecules tend to align themselves in the direction of the field, due to the torque produced in it. But the alignment is not complete due to the thermal random motion of the molecules.

Therefore, average alignment will give rise to a net magnetic moment per unit volume in the same direction as that of applied field.

Let us suppose that the atom consists of a permanent magnetic moment M and θ be the angle between the molecular dipole and external field B, as shown in Fig. 20.7.

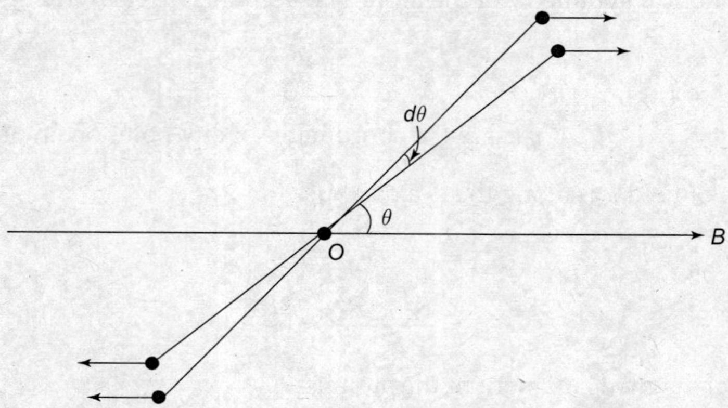

FIGURE 20.7 Torque acting on a dipole

The magnetic potential energy in the field is

$$\omega = -\vec{M} \cdot \vec{B} = -MB\cos\theta \tag{20.26}$$

According to Boltzmann's statistics, the number of molecules per unit time per unit volume having energy same ω and $\omega + d\omega$ is

$$dn = C\, e^{-\omega/kT} d\omega \tag{20.27}$$

From Eq. (20.26),

$$d\omega = MB\sin\theta d\theta \tag{20.28}$$

Putting the values of ω and $\omega + d\omega$ from Eqs. (20.26) and (20.27) into Eq. (20.28), we get

$$dn = C e^{MB\cos\theta/kT} \cdot MB\sin\theta d\theta \tag{20.29}$$

Now, integrate Eq. (20.29) for all possible orientations of the dipole, i.e. for $\theta = 0$ to π.

Thus, the total number of atoms of molecules n per unit volume of the substance is given by

$$n = \int dn = \int_0^{\pi} C e^{MB\cos\theta/kT} \cdot MB\sin\theta d\theta \tag{20.30}$$

$$\Rightarrow \qquad C = \frac{n}{\int_0^{\pi} C e^{MB\cos\theta/kT} \cdot MB\sin\theta\, d\theta} \tag{20.31}$$

The intensity of magnetization I is given by

$$I = \int_0^\infty M \cos\theta dn \tag{20.32}$$

$$= C\int_0^\pi M^2B\, e^{MB\cos\theta/kT} \cdot \sin\theta\cos\theta\, d\theta$$

$$= nM^2B \frac{\int_0^\infty e^{MB\cos\theta/kT} \cdot \sin\theta\cos\theta\, d\theta}{\int_0^\infty e^{Ms\cos\theta/kT} MB\sin\theta\, d\theta}$$

or

$$I = nM \frac{\int_0^\infty e^{MB\cos\theta/kT} \sin\theta\cos\theta d\theta}{\int_0^\infty e^{MB\cos\theta/kT} \sin\theta d\theta} \tag{20.33}$$

Let $\dfrac{MB}{kT} = x,\ \cos\theta = y$

Now Eq. (20.33) can be rewritten as

$$I = nM \frac{\int_{-1}^{+1} e^{xy} y dy}{\int_{-1}^{+1} e^{xy} dy}$$

$$= nM \frac{\left[\dfrac{ye^{xy}}{x} - \dfrac{e^{xy}}{x^2}\right]_{-1}^{+1}}{\left[\dfrac{e^{xy}}{x}\right]_{-1}^{+1}}$$

$$= nM\left[\frac{\left(e^x + e^{-x}\right)}{\left(e^x - e^{-x}\right)} - \frac{1}{x}\right]$$

or

$$I = I_s\left[\coth x - \frac{1}{x}\right] \tag{20.34}$$

or $$I = I_s L(x)$$

where $I_s = nM$ gives the saturation value of I, i.e. when the molecules align themselves parallel to the field.

If we plot a graph between I/I_s and x, we find that for large value of x, the Langevin's function, i.e. $I/I_s = \coth x - (1/x)$ tends to unity. For small values of x, the curve is linear (as shown in Fig. 20.8).

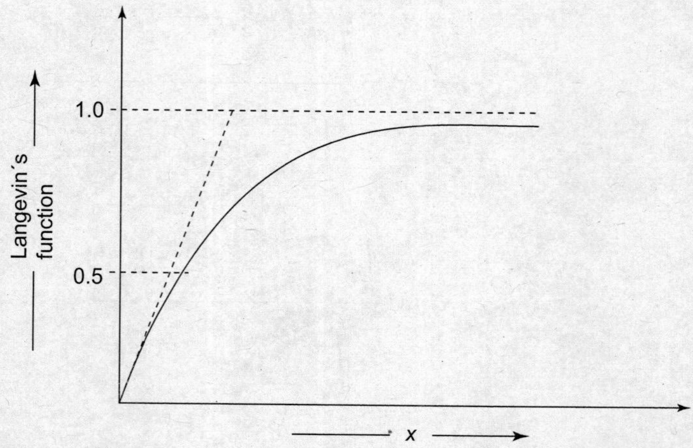

FIGURE 20.8 Langevin's function vs x

We can expand coth x in the power series of x; then Eq. (20.34) becomes

$$I = I_s\left[\left(\frac{1}{x}+\frac{x}{3}\right)-\frac{1}{x}\right]$$

or $$I = I_s\left(\frac{x}{3}\right) = \frac{nM^2B}{3kT} \quad \left[\because\ x = \frac{MB}{KT} \text{ and } I = Mn\right]$$

$\Rightarrow$ $$I = \frac{nM^2}{3kT}(\mu_0 H) \quad [\because\ B = \mu_0 H] \tag{20.35}$$

The susceptibility can be written as

$$\chi_m = \frac{I}{H} = \frac{nM^2(\mu_0 H)}{3kT(H)}$$

$$\boxed{\chi_m = \frac{nM^2\mu_0}{3kT}} \tag{20.36}$$

If we put $\lambda = \dfrac{I_s^2 \mu_0}{3k}$, then Eq. (20.36) can be written as

$$\chi_m = \frac{\lambda}{T}$$

This is same as Curie law (Fig. 20.9).

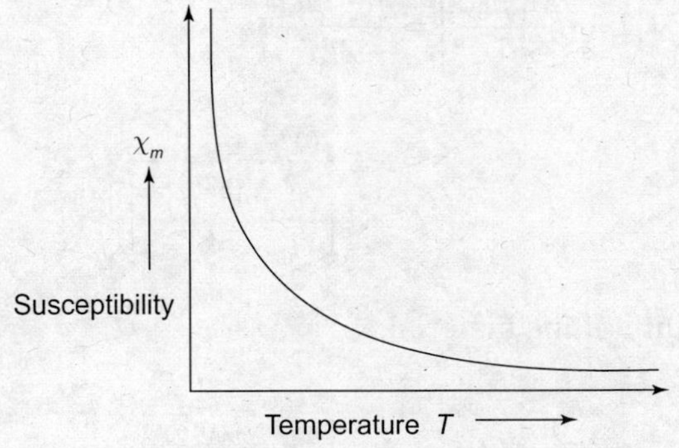

FIGURE 20.9 Curie law

20.8 ▶ WEISS–MOLECULAR FIELD THEORY OF PARAMAGNETISM

The Langevin's theory could not explain the dependence of χ_m on temperature for several paramagnetic materials like highly compressed and cooled gases, very concentrated solutions of salt, crystals and solid salts, for which

$$\chi_m = \frac{\lambda}{T-\theta}, \ \theta \text{ is being a constant.}$$

It also could not explain the relation between ferro and paramagnetism.

Therefore, Weiss in 1907 modified the Langevin's theory by introducing the new concept of internal molecular field. He assumed that, in real gases, there are mutual interactions in the molecules; therefore, there must exist a molecular field H_i which is directly proportional to the intensity of magnetization, i.e.,

$$H_i \propto I \Rightarrow H_i = \alpha I \tag{20.37}$$

Thus, the net effective field acting on the dipole is

$$H_e = H + H_i \tag{20.38}$$

where H = external applied field.

$$\Rightarrow \qquad H_e = H + \alpha I \tag{20.39}$$

Equation (20.35) can be rewritten as

$$I = \frac{nM^2\mu_0}{3kT}(H + \alpha I)$$

$$\Rightarrow \quad I = \frac{nM^2\mu_0 H}{3kT} + \frac{nM^2\mu_0\alpha I}{3kT}$$

$$\Rightarrow \quad I\left(1 - \frac{nM^2\mu_0\alpha}{3kT}\right) = \frac{nM^2\mu_0 H}{3kT}$$

$$\Rightarrow \quad I = \frac{nM^2\mu_0 H}{3kT\left(1 - \frac{nM^2\mu_0\alpha}{3kT}\right)} \tag{20.40}$$

and $\frac{I}{H} = \chi_m$, susceptibility of the material.

Thus,

$$\chi_m = \frac{I}{H} = \frac{nM^2\mu_0}{3kT\left(1 - \frac{nM^2\mu_0\alpha}{3kT}\right)}$$

$$\Rightarrow \quad \chi_m = \frac{nM^2\mu_0 / 3kT}{1 - \frac{nM^2\mu_0\alpha}{3kT}} = \frac{\mu_0 nM^2 / 3k}{T - \frac{nM^2\mu_0\alpha}{3k}}$$

$$\Rightarrow \quad \boxed{\boldsymbol{\chi_m = \frac{\lambda}{T - \theta}}} \tag{20.41}$$

where $\lambda = \mu_0 nM^2 / 3k$

and $\theta = \frac{nM^2\mu_0\alpha}{3k}$

Equation (20.41) represents the Curie-Weiss law. The constant θ is known as Curie temperature. It holds good only for temperatures $T > |\theta|$.

Note

Curie temperature is low for $T < \theta$ and χ_m is negative, i.e. paramagnetism would become diamagnetism. But Curie temperature is quite low and so it is a rare situation.

EXAMPLE 20.7

A sample of paramagnetic material contains 2.0×10^{24} atomic dipoles, each having dipole moments 1.5×10^{-23} J/K. The sample is subjected to a uniform magnetic field of 0.84 T and cooled to a temperature of 4.2 K. The degree of magnetic saturation achieved is equal to 15%. Calculate the total dipole moment of the sample for a magnetic field of 0.98 T and a temperature of 2.8 K.

Solution: Initial dipole moment,

$$M_i = \frac{15}{100}\times1.5\times10^{-23}\times2.0\times10^{24}$$
$$= 4.5\ \text{J/T}$$

According to Curie law,

$$M = \text{constant}\times\frac{H}{T}$$

$$\Rightarrow \qquad \frac{M_f}{M_i} = \frac{H_f}{H_i}\times\frac{T_i}{T_f}$$

$$\Rightarrow \qquad M_f = M_i\frac{H_f}{H_i}\frac{T_i}{T_f}$$

$$= 4.5\times\frac{0.98}{0.84}\times\frac{4.2}{2.8}\ \text{J/T}$$

$$= \mathbf{7.875\ J/T}$$

EXAMPLE 20.8

A paramagnetic gas consists of atoms each with a dipole moment of 1.5×10^{-23} J/T. The temperature of the gas is 27°C and its number density is $2.0\times10^{26}\ /\text{m}^3$. What is the maximum possible magnetization of the sample which it is placed in an external magnetic field? Do you think the sample can achieve close to maximum magnetization for a magnetic field of 3.0 T?

Solution: We know that,

$$I_{\max} = \frac{Mn}{V}$$

$$\Rightarrow \qquad \frac{n}{V} = 2.0\times10^{26}\ /\text{m}^3$$

$$\Rightarrow \qquad M = 1.5\times10^{-23}\,\text{J/T}$$

$$\Rightarrow \qquad I_{max} = 3.0\times10^{3}\,\text{A/m} \qquad \left(1\,\text{J/T} = 1\,\text{A}^2\text{m}\right)$$

The actual magnetization of sample is less than this due to the thermal disorder, which tends to disrupt the alignment of the atomic dipoles with the external field comparing the thermal energy $(=(3/2)K_BT)$ with the potential energy of an atomic dipole when aligned with the field $(= MB)$

$$\Rightarrow \qquad \frac{3}{2}K_BT = \frac{3}{2}\times1.38\times10^{-28}\times300$$

$$= 6.21\times10^{-21}\,\text{J}$$

and

$$MB = 1.5\times10^{-23}\times3.0$$

$$= 4.5\times10^{-23}\,\text{J}$$

$$\Rightarrow \qquad \frac{\frac{3}{2}K_BT}{MB} = 138$$

This shows that the randomizing effect is much greater than the alignment effect.

20.9 ▶ Weiss Molecular Field Theory (Domain Theory) of Ferromagnetism

There are certain transition metals like Fe, Co, Ni exhibit magnetization even when the applied field is removed. This phenomenon is called Ferromagnetism.

Weiss proposed the new concept of magnetic domains in ferromagnetic substances. The ferromagnetic substances like paramagnetic substances have permanent magnetic moments but these substances due to certain mutual interactions, form innumerable small effective regions called "domains".

Each domain has 10^{17} to 10^{21} atoms whose magnetic dipoles are aligned in same directions even in the absence of any external field. However, the direction of magnetization in different domains is different in unmagnetized material; thus, the resultant magnetization is zero.

The cause of spontaneous magnetization within each domain is the existence of the molecular field which tries to produce parallel alignment of dipoles. This internal molecular field (H_i) is proportional to the magnetizational intensity of the material, i.e.,

$$H_i \propto I$$

or $$H_i = \alpha I \tag{20.42}$$

Thus, the effective magnetic field on the dipole becomes

$$H_e = H + H_i$$

or $$H_e = H + \alpha I \tag{20.43}$$

Using Langevin's theory of paramagnetism, we have

$$I = nM\left[\cot h\, x - \frac{1}{x}\right] \tag{20.44}$$

or $$I = nM\,L(x)$$

$$\left[\because\ L(x) = \cot h\, x - \frac{1}{x} = \text{Langevin's function}\right]$$

Also,

$$I = I_s\,L(x) \quad [\because\ I_s = nM\,]$$

$$\Rightarrow \quad \frac{I}{I_s} = L(x) \tag{20.45}$$

$$x = \frac{MB}{kT} = \frac{M\mu_0 H_e}{kT}$$

$$\Rightarrow \quad x = \frac{M\mu_0}{kT}(H + \alpha I)$$

$$\Rightarrow \quad x = \frac{M\,\mu_0}{kT}(\alpha I)$$

[since for spontaneous ferromagnetism $H = 0$]

$$\Rightarrow \quad I = \frac{xkT}{M\,\mu_0\alpha}$$

$$\Rightarrow \quad \frac{I}{I_s} = \frac{xkT}{M\mu_0\alpha(nM)} = \frac{xkT}{n\mu_0\alpha M^2} \tag{20.46}$$

[since for spontaneous ferromagnetism $H = 0$)

Equations (20.45) and (20.46) give the condition of spontaneous magnetization.

We can solve graphically these two equations for determining the two variables I/I_s and x as shown in Fig. 20.10.

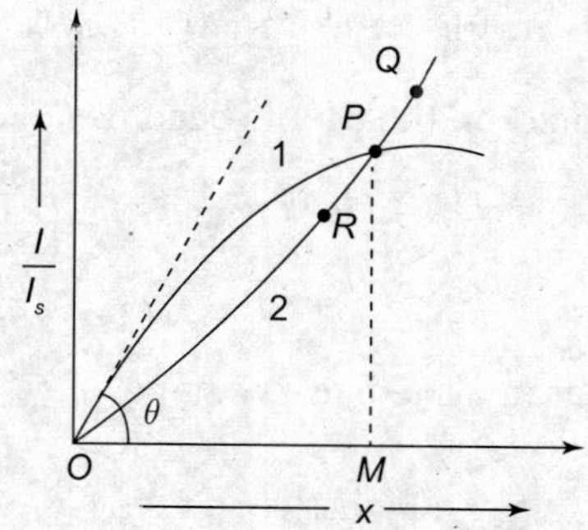

FIGURE 20.10 I/I_s versus x

Here, in Fig. 20.10, Curve (1) is corresponding to Eq. (20.45) is known as Langevin's curve while Curve (2) is corresponding to Eq. (20.46). These two curves meet at points O and P; it means that these have two solutions $I/I_s = 0$ and $I/I_s = PM$ but $I/I_s = 0$ is not true so $I/I_s = PM$ is a correct solution. Point A represents the stable state of spontaneous magnetization. If any molecule in a domain assumes state R, the local magnetization is less than the corresponding equilibrium state A. Similarly, if any molecule in a domain assumes state B, then local magnetization is more than the corresponding equilibrium state A.

$$\text{Slope of the Curve (2)} = \frac{kT}{\mu_0 \alpha n M^2}$$

For $x \to 0$, $L(x) = x/3$, so the slope of this Curve (2) which is a straight line is $1/3$. Thus,

$$\frac{1}{3} = \frac{kT}{\mu_0 \alpha n M^2}$$

$$\Rightarrow \qquad T = \frac{\mu_0\, \alpha\, n M^2}{3k}$$

$$\Rightarrow \qquad \theta = \frac{\mu_0\, \alpha\, n M^2}{3k} \qquad (20.47)$$

which is known as Curie temperature.

Thus, for any temperature below Curie temperature, the state of magnetization is possible even without an applied field. At temperatures above Curie temperature, the ferromagnets become paramagnets.

20.10 ▶ ANTI-FERROMAGNETISM

In ferromagnetic substances, the atomic dipoles are aligned parallel to one another in the absence of external magnetic field. This is because of positive exchange interaction energy between the neighbouring dipoles while this energy is negative in some substances; these substances are known as anti-ferromagnetic substances. The atomic dipoles in these substances are aligned in anti-parallel manner and there will be no spontaneous magnetization. The phenomenon of anti-ferromagnetism was studied by Neel and Bitter and the theory was extended by Van-Veleck. A simple example of anti-ferromagnetic substances is a simple cubic lattice where the dipoles at adjacent corners of each cubic cell point in opposite directions. The system may be supposed as consisting of two interlocking sub-lattices, one is spontaneously magnetized in one direction and the other is magnetized in opposite direction. The spontaneous magnetization occurs below a certain transition temperature, called "Neel point". Above this temperature, the material behaves like paramagnetic material. The material as a whole shows no spontaneous magnetization in zero field, because the two sub-lattices are equally and oppositely polarized. When external field is applied, a small magnetization develops which gives positive susceptibility. Examples are Cr, MnF_2 and MnO.

20.10.1 Molecular Field Theory of Anti-Ferromagnetism

Let us consider an anti-ferromagnetic system having two interlocked sub-lattices P and Q. Let a dipole in lattice P be subjected to external magnetic field H and internal magnetic field proportional to intensity of magnetization of lattice Q, i.e. I_Q. The effective magnetic field on the dipole P is

$$H_P = H - \alpha I_Q \tag{20.48}$$

Similarly, the effective field on dipole Q is

$$H_Q = H - \alpha I_P \tag{20.49}$$

Case (a): At high temperature, i.e. $T > T_N$

The dipole will be randomly oriented above Neel temperature and magnetization of each sub-lattice obeys Curie law (Fig. 20.11) if the effective field instead of external field is taken, i.e.,

$$\left.\begin{aligned} I_P &= \frac{1}{2}\lambda\frac{H_P}{T} \\ I_Q &= \frac{1}{2}\lambda\frac{H_P}{T} \end{aligned}\right| \tag{20.50}$$

The net magnetization is

$$I = I_P + I_Q$$

or

$$I = \frac{1}{2}\frac{\lambda}{T}\left(H_P + H_Q\right)$$

or

$$I = \frac{1}{2}\frac{\lambda}{T}\left[2H - \alpha\left(I_P + I_Q\right)\right] \tag{20.51}$$

$$I = \frac{\lambda}{T}\left[H - \frac{\alpha I}{2}\right]$$

$$\Rightarrow \qquad I + \frac{\lambda}{T}\frac{\alpha I}{2} = \frac{\lambda}{T}H$$

$$\Rightarrow \qquad I\left[1 + \frac{\alpha\lambda}{2T}\right] = \frac{\lambda}{T}H$$

$$\Rightarrow \qquad I = \frac{\lambda H}{T\left[1 + \frac{\alpha\lambda}{2T}\right]} = \frac{\lambda}{T + \frac{\alpha\lambda}{2}}H$$

Magnetic susceptibility is

$$\chi_m = \frac{I}{H} = \frac{\lambda}{T + \frac{\alpha\lambda}{2}}$$

$$\boxed{\boldsymbol{\chi_m = \frac{\lambda}{T + \theta}}} \tag{20.52}$$

where $$\theta = \frac{\alpha\lambda}{2}$$

This equation for susceptibility is similar to the expression for ferromagnetism except the negative sign of θ.

Case (b): At low temperature, $T < T_N$ when an external magnetic field is applied at a temperature below Neel temperature, a positive magnetization takes place (Fig. 20.11). The external magnetic field changes the magnetization of each sub-lattice slightly, so

$$I_P = I_0 + \delta I$$

and $$I_Q = -I_0 + \delta I$$

Also,

$$H_P = H - \alpha I_0 + \delta I = \alpha I_0 + H - \alpha\delta I$$

$$H_Q = H - \alpha I_0 + \delta I = -\alpha I_0 + H - \alpha\delta I$$

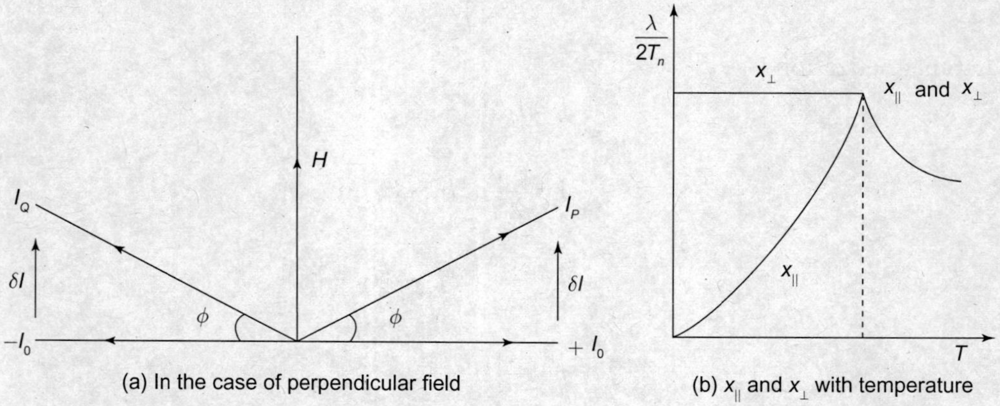

FIGURE 20.11 Anti-ferromagnetism

If the field is applied parallel to I_0, susceptibility decrease to zero as the temperature falls to zero.

But if the external field is applied perpendicular to I_0, the magnetization turns through an angle Q towards H and the susceptibility will be

$$\chi_{\perp} = \frac{\lambda}{2T_n} \tag{20.53}$$

where $\chi_{\perp}$ is constant below Neel temperature.

20.11 ▶ Ferrimagnetism or Ferrites

As discussed in earlier section, the magnetic moments of the sub-lattices of anti-ferromagnetic substances are equal and opposite that is why the net spontaneous magnetization is zero. But in certain substances, the magnetic moments of two sub-lattices are not equal in magnitude which results in small finite spontaneous magnetization. Such substances are called Ferrites and the phenomenon is called Ferrimagnetism (Fig. 20.12).

The behaviour of ferrites is similar to ferromagnetic substances except the temperature dependency of their spontaneous magnetization.

The magnetization of ferrites does not necessarily decrease monotonically with the rise in temperature; it can even pass through zero before Neel temperature. For example, Fe_3O_4. The chemical formula for ferrites in $MO \cdot Fe_2O_3$, where M is a divalent cation and can be Zn, cd, Fe, Ni, Cu, Co or Mg. These substances have high magnetic permeability, small coercive force, high saturation magnetization and high resistivity (of the order of $10^3\,\Omega$-m). The ferrites are very useful in the field of high and ultra-high frequency electronics.

20.11.1 χ_m Versus T for Ferromagnetic, Anti-Ferromagnetic and Ferromagnetic Substances

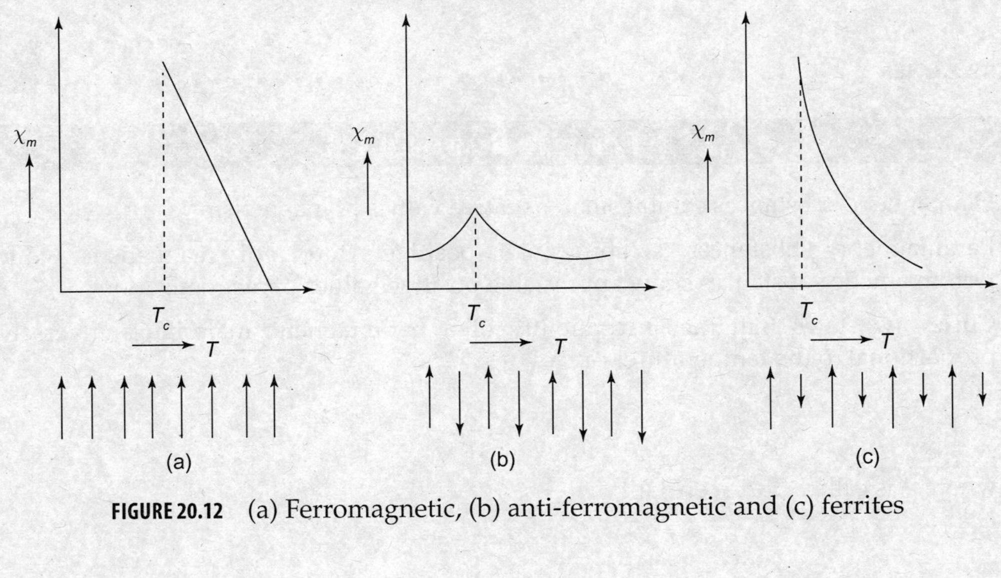

FIGURE 20.12 (a) Ferromagnetic, (b) anti-ferromagnetic and (c) ferrites

20.12 ▶ Ferroelectricity

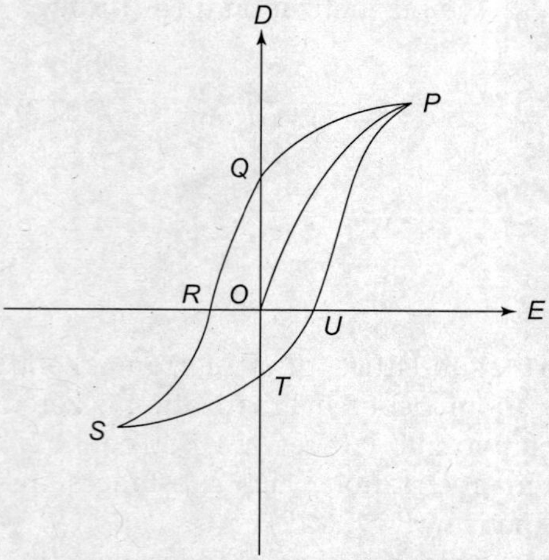

FIGURE 20.13 Ferroelectricity

In certain substances like Rochelle salt and barium titanate, the centre of positive charge does not coincide with the centre of negative charge . When an external electric field is applied, a displacement of centre of positive and negative charges takes place. This polarization associated with the displacement introduces an internal field. This internal field created due to displacement of charge centres increases and stabilizes the polarization. Therefore, a part of polarization exists even in the absence of electric field, i.e. when the electric field is removed. This phenomenon is known as ferroelectricity and the substances that show this phenomenon are called Ferroelectric crystals. Figure 20.13 shows the variation of resultant electric displacement D with the applied electric field E.

If the initial displacement is zero, the relationship between D and E is given by OP, D/E depends upon the value of applied field. As E increases, D acquires its maximum value and saturates. If now field is reduced, the displacement does not fall to zero even if E becomes zero.

This shows residual displacement OQ. In order to remove this displacement OR field is applied which makes $D = 0$ and called coercive field. This complete loop PQRSTUP is the characteristic of ferroelectric state of the crystal.

Ferroelectricity disappears above transition temperature or Curie temperature. Ferroelectric crystals always exhibit the piezoelectric property.

Recap Zone

Points to Remember

- Diamagnetic substances exhibit small negative values of magnetism.
- Paramagnetic substances exhibit small positive values of magnetism while ferromagnetic substances show appreciable positive values of magnetization.
- Curie law states that the susceptibility of a paramagnetic material is inversely proportional to the temperature, i.e.,

$$= \frac{\lambda}{T}$$

where λ is called Curie constant.

- Anti-ferromagnetic materials are those materials in which magnetic dipoles tend to align in a direction opposite to the external magnetic field direction and uncompensated anti-ferromagnetism is called ferrimagnetism.
- Ferrites are generally iron oxides in which the magnetic moments are of unequal magnitudes. Ferrites have the general formula $\chi^{2+}Fe_2^{3+}O_4^{2-}$. These can have two types of structures: regular structure and inverse spinal structure.
- Relation between B, H and I is given by

$$B = \mu_0 (H + I)$$

- Hysteresis loss per cycle = area of $B - H$ loop = μ_0 times the area of $M - H$ loop.
- The susceptibility of ferromagnetic substances decreases with rise of temperature but at a temperature, called Curie temperature, the ferromagnetic substance becomes paramagnetic.
- Diamagnetism is universal property of all substances due to tendency of forming pairs of opposite spins in all substances.
- For magnetic insulation of a device, it should be placed inside a ferromagnetic substance.
- The magnetic materials which are easily magnetized and demagnetized are known as soft magnetic materials.
- The materials which have high resistance to demagnetization are known as hard materials.

Key Terms with Definitions

- **Magnetic shield** – It does not allow the magnetic field to penetrate into space enclosed.
- **Direction of easy magnetization** – The direction in which the magnetization is strongest.
- **Hard magnetization** – The direction corresponding to lowest magnetization.
- **Curie temperature** – The definite temperature above which ferromagnetic behaviour disappears.
- **Hysteresis** – It lags in changes of magnetization behind variations of the magnetic field.
- **Retentivity** – The measure of the magnetic flux density remaining in the material when the magnetizing field is removed.
- **Coercivity** – Measure of the magnetic field strength required to destroy the residual magnetism in the material.
- **Domains** – A macroscopic ferromagnetic specimen is divided into small regions.
- **Magentostriction** – The magnetically induced elastic strain is called magentostriction.
- **Hysteresis loss** – The energy dissipated per cycle of magnetization.

IMPORTANT FORMULAE AND EQUATIONS

Equation Number	Equation	Remarks
(20.3)	$\mu_l = 1 + \chi_m$	**Relation between relative permeability and magnetic susceptibility**
(20.9)	$g = \dfrac{M_l}{L} = \dfrac{e}{2m}$	**Orbital gyromagnetic ratio**
(20.13)	$M_s = \dfrac{e}{m_e} S = \sqrt{s(s+1)}\dfrac{eh}{4\pi m}$	**Spin magnetic moment of atom**
(20.25)	$\chi_m = \dfrac{I}{H} = -\dfrac{\mu_0 n Z e^2 R^2}{6m}$	**Diamagnetic susceptibility of the substance**
(20.36)	$\chi_m = \dfrac{\mu_0 n M^2}{3kT}$	**Paramagnetic susceptibility (as given by Langevin's theory)**
(20.41)	$\chi_m = \dfrac{\lambda}{T-\theta}$	**Weiss theory of paramagnetism**
(20.52)	$\chi_m = \dfrac{\lambda}{T+\theta}$	**Anti-ferromagnetism**

REVIEW ZONE

SHORT ANSWER QUESTIONS

1. Define magnetic induction.
2. What do you mean by magnetic permeability and intensity of magnetization?
3. Define magnetic susceptibility.
4. Is div *B* essentially zero?
5. Is div *H* essentially zero?
6. What is hysteresis? What does the area of hysteresis curve present?
7. Define coercivity and retentivity.
8. What is Curie temperature? Is it unique for all substances?
9. Define dia, para and ferromagnetism.
10. Define anti-ferromagnetism and ferrimagnetism.
11. State Curie law.
12. What is Bohr magneton?
13. Why is diamagnetism almost independent of temperature?
14. How do you account for the magnetic properties of materials?
15. Give a difference between hard and soft magnetic materials.

LONG ANSWER QUESTIONS

1. What are atomic magnetic moments? Define the terms magnetization, permeability susceptibility and flux density.
2. Discuss Langevin's theory of dia-magnetism and prove that the dia-magnetic susceptibility is negative.
3. Distinguish between dia-, para- and ferromagnetic substances. Give an account of Langevin's theory of paramagnetism.
4. Discuss Curie-Weiss law of paramagnetic substances. Explain the effect of temperature on paramagnetic properties of a material.
5. Explain Weiss theory of ferromagnetism. How the susceptibility of a ferromagnetic material varies with temperature, explain it with the help of diagram?
6. Discuss the phenomenon of anti-ferromagnetism. How does an anti-ferromagnetic material differs from a diamagnetic material?
7. Explain the difference between ferro- and anti-ferromagnetism.
8. Write short notes on:
   (i) Hysteresis
   (ii) Curie-Weiss law
   (iii) Ferrites
   (iv) Permanent magnets

NUMERICAL PROBLEMS

1. A bar magnet has a coercivity of 5×10^3 amp/metre. It is desired to demagnetize it by inserting it inside a solenoid 10 cm long and having 50 turns. What current should be sent through the solenoid?

   Ans. 10A

2. An electron in a hydrogen atom moves in a circular orbit of radius $0.5\,\text{Å}$. The electron performs 10^{16} revolutions per second. Determine the magnetic moment associated with the orbital motion of the electron.

   Ans. $1.26 \times 10^{-23}\,\text{Am}^2$

3. A magnetic field of 1800 A/m produces magnetic flux of 3×10^{-5} wb in an iron bar of cross-sectional area $0.2\,\text{cm}^2$. Calculate the permeability.

   Ans. $8.3 \times 10^{-4}\,\text{H/m}$

4. The saturation magnetic induction of nickel is 0.65 wb/m^2. If the density of nickel is 8906 kg/m^3 and the atomic weight is 58.7. Calculate the magnetic moment of nickel in Bohr magneton.

   Ans. $0.61\,\mu_B$

5. Iron has a relative permeability of 5000. Calculate its magnetic susceptibility.

   Ans. 4999

6. Assume that iron atoms have magnetic moment of two Bohr magnetons. Calculate the Curie constant if its density is 7150 kg/m^3 and atomic weight is 55.84.

   Ans. 0.209

7. Assuming a bcc lattice for iron with lattice constant $a = 3A$, calculate the

Curie constant and Weiss constant. Given that the Curie temperature of iron is 1050 K.

Ans. 0.193, 5440

8. A paramagnetic system of electron spin magnetic dipole moment is placed in an external field of 10^5 A/m. Calculate the average magnetic moment per dipole at 300 K.

Ans. 2.6×10^{-27} A·m^2

9. A paramagnetic salt contains 10^{28} ions/m^3 with magnetic moment of one Bohr magneton. Calculate the paramagnetic susceptibility and the magnetization produced in a uniform magnetic field of 10^6 A/m when the temperature is 27°C.

10. An electron is an atom of hydrogen revolves in an orbit of radius 0.51 Å. Calculate the change in magnetic moment for this electron if a magnetic field of induction 2 T acts at right angles to the plane of the orbit.

Ans. 3.66×10^{-29} A·m^2

CHAPTER 21

X-ray Crystallography

Learning Objectives

By the end of this chapter, the student will be able:

- To understand X-rays, their types, properties and production
- To learn about diffraction of X-rays from crystals and Bragg's law
- To explore different methods of crystal structure determination
- To learn about X-ray radiography
- To discuss the advantages and disadvantages of X-ray imaging

21.1 ▶ Introduction

The crystallography is an experimental technique of determining the arrangement of atoms in the crystalline solids. X-ray crystallography methods are used to determine the structure of inorganic crystals and large bio-molecules such as proteins. Before the development of X-ray diffraction based crystallography, the study of crystals was based on physical measurements of their geometry. This involved measuring the angles of crystal faces relative to each other and to theoretical reference axes and establishing the symmetry of the crystal in question.

The modern methods depend on the analysis of the diffraction patterns of a sample targeted by a beam of X-rays. However, some other particle beams such as electrons or neutrons are also used. This is facilitated by the wave properties of these particles. The X-rays and particle beams interact differently with the specimen. Because of these different forms of interaction, radiation type is selected for different crystallographic studies. In this chapter, we shall confine only to the X-ray crystallography that is based upon X-rays as radiation.

21.2 ▶ X-rays

X-rays are a form of electromagnetic radiation having wavelengths ranging from 0.01 to 10 nanometres and energies ranging from 100 eV to 100 keV. They are a part of the electromagnetic spectrum as shown in Fig. 21.1. The X-rays were discovered in 1895 by a German scientist Wilhelm Conrad Röentgen.

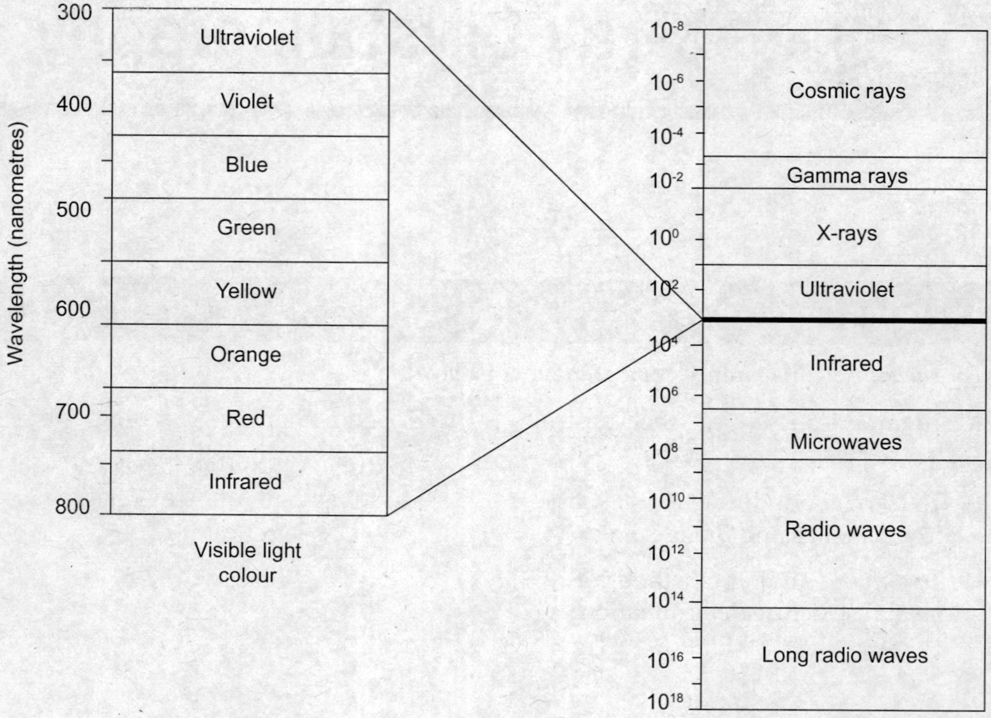

FIGURE 21.1 X-rays in the electromagnetic spectrum

21.2.1 Types of X-rays

Based upon their energies, the X-rays are broadly classified into two categories: soft X-rays and hard X-rays. Another classification of X-rays is based upon their origin.

Soft X-rays can be produced at comparatively low potential difference in an X-ray tube, while the hard X-rays are produced at comparatively high potential difference. Due to their penetrating ability, hard X-rays are widely used in medical radiography and airport security scanners. The attenuation length of hard X-rays in water is about four orders of magnitude longer as compared to soft X-rays. Since the wavelengths of hard X-rays are of the order of the size of atoms, they are scattered by the crystal planes that have the inter-planar distances of the same order. These are therefore useful for determining the crystal structures by X-ray scattering by several methods of X-ray crystallography.

21.3 ▶ PROPERTIES OF X-RAYS

The X-ray photons carry high energy and can easily ionize atoms or break the molecular bonds of living tissues. Therefore, these are very harmful to living tissues. However, the ionizing property of X-rays can be utilized for various purposes including cancer treatment to kill the cells using radiation therapy. X-ray spectroscopy and XRF techniques are used for material characterization. X-rays have much shorter wavelength than visible light that makes it possible to probe structures much smaller than what can be seen using an optical microscope.

The energy of X-ray photon is inversely proportional to its wavelength as given by the Einstein's equation

$$E = h\nu$$

or

$$E = \frac{hc}{\lambda}$$

where E = energy of X-ray photon, h = Planck's constant, ν = frequency of X-ray waves, c = velocity of light and λ = wavelength of X-ray.

Note

X-rays are a form of electromagnetic radiation having wavelengths approximately in the range of 0.01 to 10 nm.

21.4 ▶ Production of X-rays

X-rays can be generated by accelerating electrons with high voltage in a device called an X-ray tube as shown in Fig. 21.2. The tube consists of an evacuated chamber having a tungsten filament at one end of the tube, called a cathode, and a metal target at the other end, called an anode. A large voltage difference (in kV) is applied between the cathode and the anode, causing the electrons to move from the filament to the anode target with a large velocity. Upon colliding with the atoms in the anode target, they knock electrons out of the inner shells of the target atoms. Following these vacancy creations, a series of electronic transitions results in the generation of X-rays. The electrons from higher energy states fill the vacancies, emitting X-ray photons with precise energies. The maximum energy of the X-ray photons is limited by the energy of the incident electron, which is equal to the voltage on the tube times the electron charge.

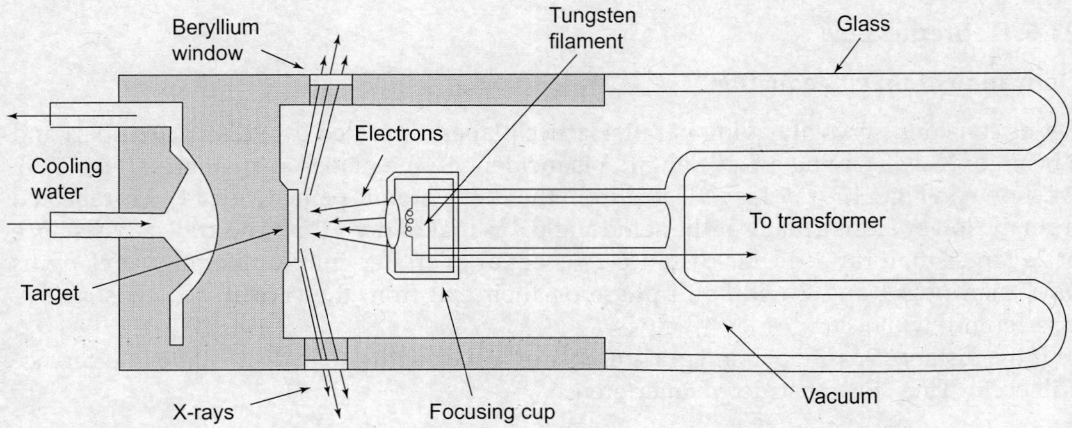

FIGURE 21.2 Production of X-rays in an X-ray tube

Note

The maximum energy an X-ray photon can have in an X-ray tube is equal to the maximum kinetic energy of electron in the tube:

$$E = h\nu = hc/\lambda = \text{eV}$$

EXAMPLE 21.1

Calculate the maximum energy and the minimum wavelength of X-ray photons emitted from an X-ray tube working at 40 kilovolts accelerating voltage.

Solution:

$$E = h\nu = \frac{hc}{\lambda} = \text{eV}$$

$$\therefore \quad \lambda = \frac{hc}{\text{eV}}$$

$$= \frac{6.62\times10^{-34}\times3\times10^{8}}{1.6\times10^{-19}\times40000}$$

$$= 3.10 \times 10^{-9}$$

$$\lambda = \mathbf{3.1\ nm}$$

21.5 ▶ Diffraction of X-rays

X-ray diffraction (XRD) is based on the dual nature of X-rays to obtain information about the structure of crystalline materials. This technique is used for the identification and characterization of compounds based on their diffraction pattern.

21.5.1 Bragg's Law

Derivation of Bragg's equation

Let us consider a crystal having parallel lattice planes separated by a distance d apart and a beam of X-rays having a wavelength λ is incident on the planes at an angle θ (Fig. 21.3). As shown in Fig. 21.3, ray 1 is reflected from the upper atomic plane while ray 2 is reflected from the lower atomic plane at the same angle θ as that of ray 1 but ray 2 travels a distance of $2a$ larger than ray 1. If this distance $2a$ is equal to an integral number of wavelengths ($n\lambda$), then rays 1 and 2 will be in phase on their exit from the crystal and constructive interference will occur.

If the distance $2a$ is not an integral number of wavelengths, then destructive interference will occur. Thus, for constructive interference

$$n\lambda = 2a$$

Also,

$$a = d \sin\theta$$

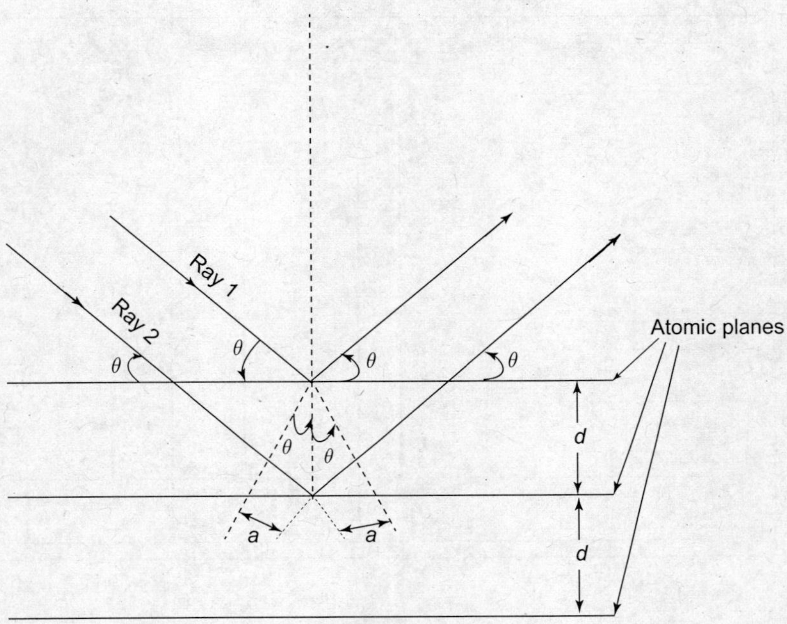

FIGURE 21.3 Bragg's law

So, $$2a = 2d \sin\theta$$

Thus, $$n\lambda = 2d \sin\theta$$

where $n = 1, 2, 3, \ldots$ for the first order, second order, third order, etc., for maxima. This relation is known as Bragg's law.

If we know the wavelength, λ, of the X-rays incident on the crystal, and we can measure the angle θ of the X rays diffracted from the crystal, then we can measure the spacing (d-spacing) between the atomic planes as given below:

$$d = \frac{n\lambda}{2 \sin\theta}$$

We then re-orient the crystal to expose another atomic plane and measure the d-spacing between all atomic planes in the crystal, which helps us to determine the crystal structure of the given material and the size of the unit cell of the crystal.

Theory of Bragg's diffraction

When a beam of monochromatic X-ray is incident on the crystal, each atom of the crystal becomes a source of scattering radiations and scattering of X-ray from these planes can be regarded as reflections known as Bragg's reflections or Bragg's planes. At some particular angles, reflections from parallel planes reinforce each other and produce maximum intensity. For some other angles, these reflections from the parallel planes are out of phase with each other and produce zero intensity. Some typical XRD patterns of KCL, NaCl and ZnO are shown in Figs. 21.4 and 21.5.

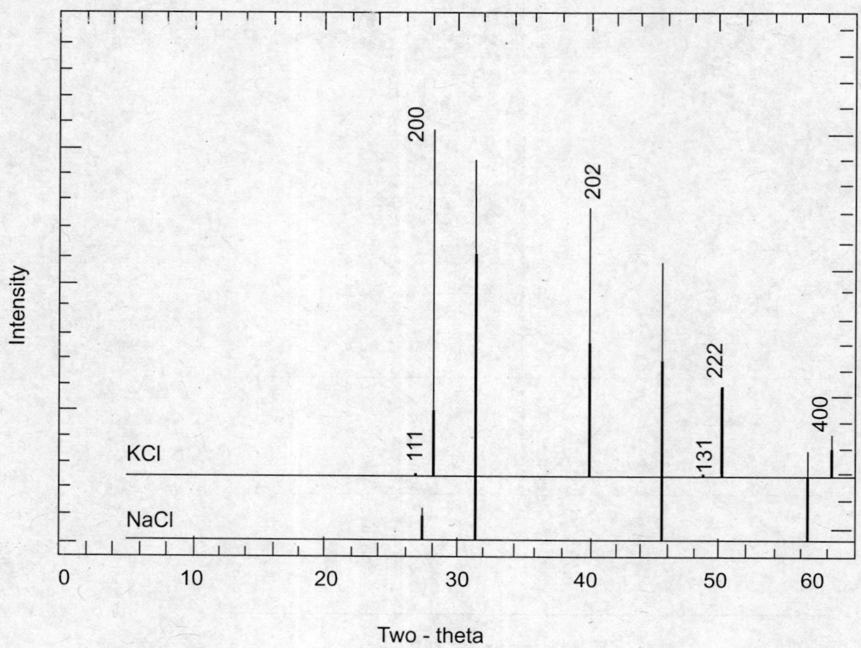

FIGURE 21.4 Typical ideal XRD spectrum of KCl and NaCl

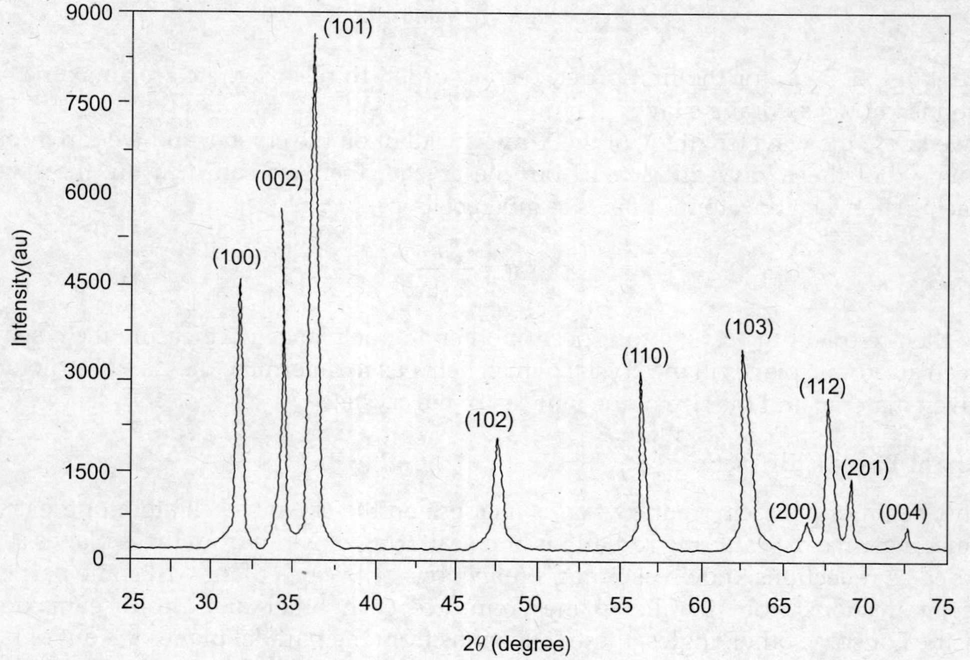

FIGURE 21.5 Experimental X-ray diffraction pattern of ZnO (Intensity vs 2θ)

Bragg's X-ray spectrometer

Bragg's spectrometer is an instrument which is used to determine the wavelength of X-ray as shown in Fig. 21.6.

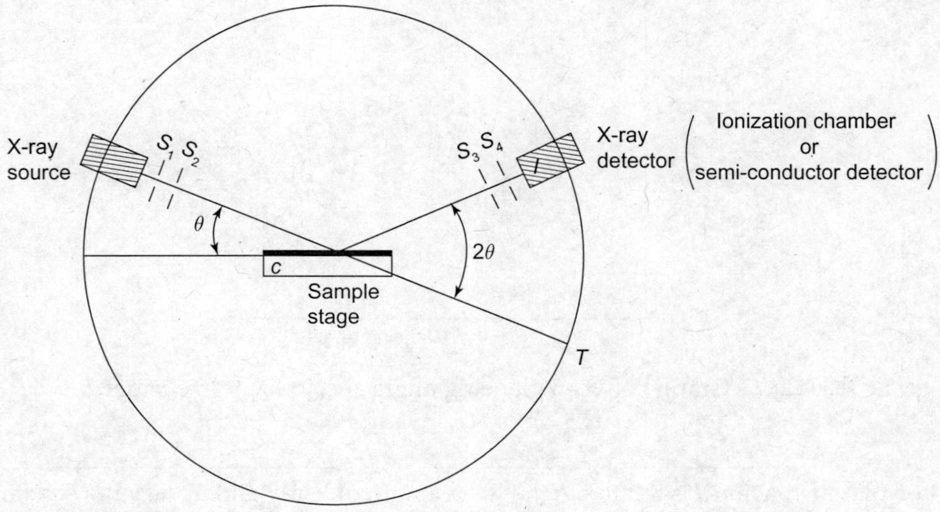

FIGURE 21.6 Bragg's spectrometer

A beam of monochromatic X-ray from an X-ray tube is allowed to pass through two fine narrow slits $S1$ and $S2$ to collimate it into a fine pencil. The pencil is then made to fall upon the crystal '*c*' (usually NaCl with 100 face) mounted on the spectrometer crystal table T which is capable of rotating about a vertical axis and its position can be read on a circular graduated scale S with vernier. The reflected pencil beam after passing through the slits $S3$ and $S4$ enters the ionization chamber I. The X-rays entering the ionization chamber ionize the gas, which causes a current to flow between the electrodes; a galvanometer is used to measure this current. The ionization current gives the intensity of X-rays reflected by the crystal.

The ionization current is measured for different values of glancing angle θ for the incident beam. The glancing angle is increased in small steps from $\theta = 0$. For each rotation of the crystal, the ionization chamber is rotated twice to get the reflected beam. A graph is drawn between the glancing angle θ and the ionization current (Fig. 21.7).

In Fig 21.7, the first, second and third peak correspond to the first, second and third order, respectively. Now, knowing the glancing angle θ and the spacing d for the crystal, wavelength of X-rays can be determined by the formula

$$2d \sin \theta = n\lambda$$

For example, for NaCl crystal, the spacing for 100 face is 2.82 Å. We get first-order reflection at a glancing angle 6°. Thus, for the first-order reflection from Bragg's equation, we have

$$\lambda = \frac{2\sin\theta}{n} = 2 \times 2.82\ \text{Å} \times \frac{\sin 6°}{1}$$
$$= 0.59\ \text{Å}$$

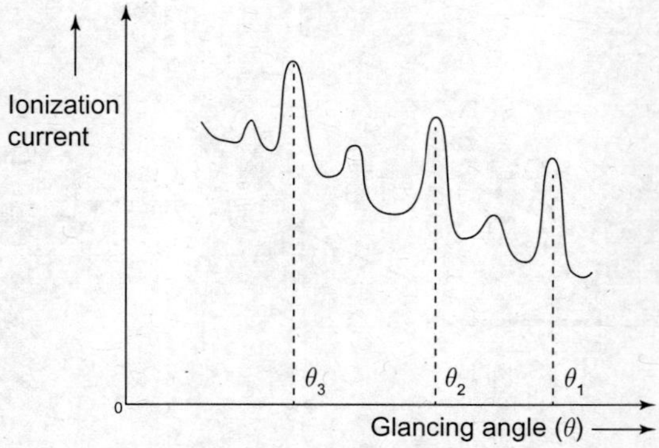

FIGURE 21.7 Graph between glancing angle and ionization current

Note

The inter-planar spacing d is related to the wavelength of X-ray and its order of scattering as $2d \sin \theta = n\lambda$. This relation forms the basis of X-ray crystallography and is known as Bragg's law.

EXAMPLE 21.2

The Bragg's angle in the first order for (220) reflection from nickel (FCC) is 38.2°, when X-rays of wavelength 1.54 Å are employed in a diffraction experiment. Determine the lattice parameter of nickel.

Solution: Order of diffraction, $n = 1$

Diffraction angle, $\theta = 38.2°$

Wavelength of light, $\lambda = 1.54\ \text{Å}$

Plane of reflection = (220)

Lattice parameter, $a = ?$

Bragg's law is $2d \sin \theta = n\lambda$

$$d = \frac{n\lambda}{2\sin\theta} = \frac{1 \times 1.54}{2 \times \sin 38.2}\ \text{Å}$$

Also $$d = \frac{a}{\sqrt{h^2 + k^2 + l^2}}$$

$$a = d \times \sqrt{h^2 + k^2 + l^2}$$
$$= \frac{1 \times 1.54}{2 \times \sin 38.2} \times \sqrt{2^2 + 2^2 + 0^2}$$
$$= \frac{\mathbf{4.35578}}{\mathbf{1.23682}} = \mathbf{3.522\ \text{Å}}$$

Methods of crystal structure determination

There are several methods of crystal structure determination in X-ray crystallography. These are based upon scientific techniques of determining the precise positions/ arrangements of atoms in a crystal where beams of X-ray strike a crystal and cause the beam of light to diffract into many specific directions.

These methods are selected based upon the types of crystals and available X-ray sources. All these methods presume that the small pieces of crystals too have the same planer symmetries as the big size crystals. For most of inorganic salts (that have a few atoms and small size), powder method is employed. In this method, a small bead of powder is placed in the path of fine beam of X-ray and its diffraction pattern is recorded on an X-ray film or a sensitive detector plate.

In case of large molecules such as proteins or biological molecules, the method of scattering is similar but the X-ray intensities are high and detectors are sensitive plates. The analysis of such crystal structures is more complex and is normally done these days by software.

Different types of machines are available in market for specific purposes. These are mainly of two types:

- Powder Diffraction Spectrometer
- Protein XRD Spectrometer

The following steps are involved in the process:

(i) X-ray beams are shot through a crystal of the atom. (The crystal is mounted to a Goniometer to keep it in place during the process of rotation.)

(ii) The crystal causes the beam of X-ray to diffract in a predictable pattern based on their crystal lattice structure.

(iii) The result is a diffraction pattern generally seen as shown in Fig. 21.8.

After we observe the reflections of the photons on the X-ray film or detector, we change the angle at which the X-ray beam hits the crystal. The crystal should be as perfect as possible.

21.6 ▶ Laue Method

It is the oldest method of X-ray crystallography methods, which is used to determine the crystal structure of the material and the electron charge distribution within the solid. The experimental arrangement of this method which consists of a single crystal is held stationary at proper orientation in the beam of well-collimated X-ray radiation of continuous wavelength as shown in Fig. 21.8.

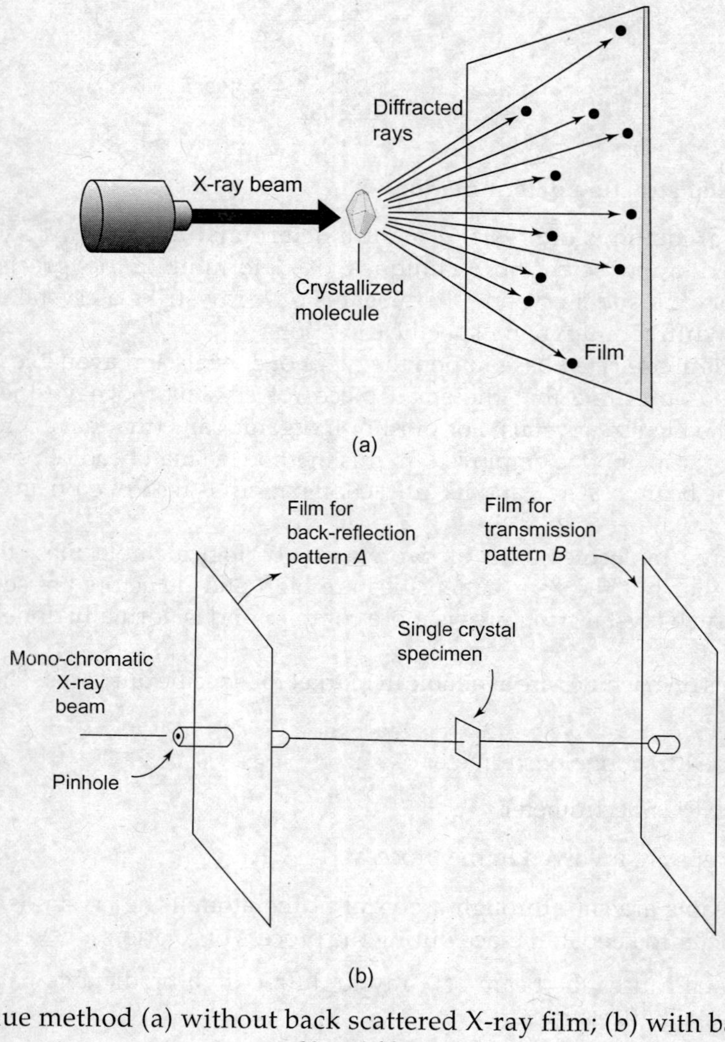

FIGURE 21.8 Laue method (a) without back scattered X-ray film; (b) with back scattered X-ray film

When an X ray beam through pin hole is made to fall on the crystal, the beam is diffracted by the crystal. Now the transmitted-diffracted and reflected-diffracted beams are collected by films B and A, respectively. The transmitted beam forms a series of spots, called Laue spots on the film, which are the characteristics of the crystal structure. The crystal selects out and diffracts the discrete values of θ for which planes exist of spacing d and incidence angle θ satisfying the Bragg relation $2d \sin \theta = n\lambda$. Each spot in Laue pattern corresponds to the inference maxima (Fig. 21.9). By analyzing the position and intensities of these spots, we can

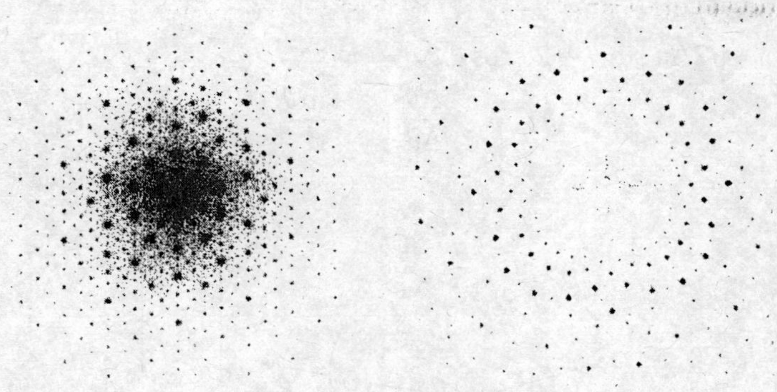

FIGURE 21.9 Laue X-ray diffraction patterns as recorded on X-ray film

determine the crystal structure. This method is very convenient for the rapid determination of crystal orientation and symmetry. It is also useful in studying the extent of crystalline imperfection under mechanical and thermal treatment.

21.7 ▶ Powder or Debye-Scherrer Method

Powder method is a standard and straight forward technique to determine the various d-values for the crystal of a given material. The experimental arrangement for powder method consists of a narrow monochromatic X ray beam, which is allowed to enter a cylindrical chamber through a pin hole and made to fall on the powder sample (here the sample is in powder form instead of a single crystal, which is set along the axis of the cylindrical chamber). The tiny crystals in the powder sample are randomly oriented and make all possible angles with the incident beam. When the X ray beam falls on the sample, for each set of d and n, there is a particular value of θ, which satisfies the Bragg's relation. Thus, for a particular value of θ, all the diffracted rays will lie on a conical surface having its apex at P and semi-vertical angle 2θ. The diffracted X rays are recorded by powder camera. The powder camera (Fig. 21.10) consists of a metallic cylinder and the sample is placed in its centre. The sample is often glued onto a glass rod with clear fingernail varnish. A strip of X-ray film is placed inside the cylinder. Punched into one side of the film is a hole for the beam collimator and punched into the other side 180° away, is another hole through which a beam catcher can be placed. The camera is closed by a light-tight lid and placed in front of the X-ray beam. The recorded pattern on the photographic film is shown on the right-hand side of Fig. 21.10. If L is the distance between the arcs of the same circle on the photographic film and R is the radius of the camera, we have $2\theta_{\text{rad}} = L/R$ or $\theta = L/2R$, thus by measuring the diameters of the arcs ($2L$) and using the above written relation, we can calculate θ.

The value of θ will give the spacing between the planes with the help of Bragg's relation as given below:

$$2d \sin \theta = n\lambda$$

Differentiating it, we get

$$2\Delta d \sin\theta + 2d\cos\theta\,\Delta\theta = n\lambda$$

or
$$\frac{\Delta\theta}{\Delta d} = -\frac{\tan\theta}{d}$$

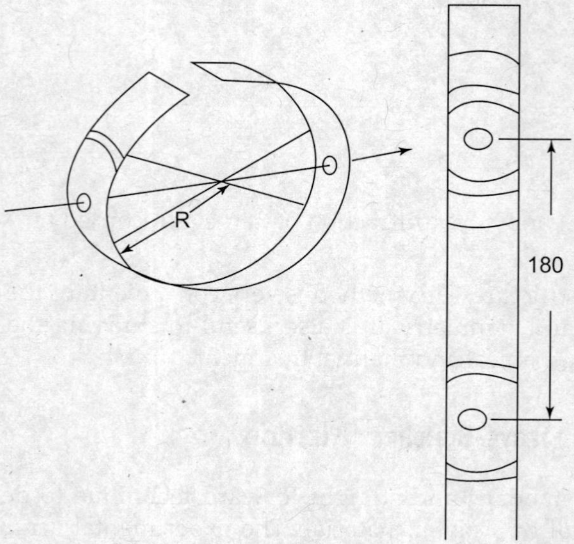

FIGURE 21.10 Powder method

Here, if θ is 90°, then 2θ will be 180°. In this condition, the X rays reflected back along their initial paths, which is called back reflection (Fig. 21.11). Such reflected beams cannot be recorded. However, we can make $\Delta\theta/\Delta d$ very large with a large possible value of θ.

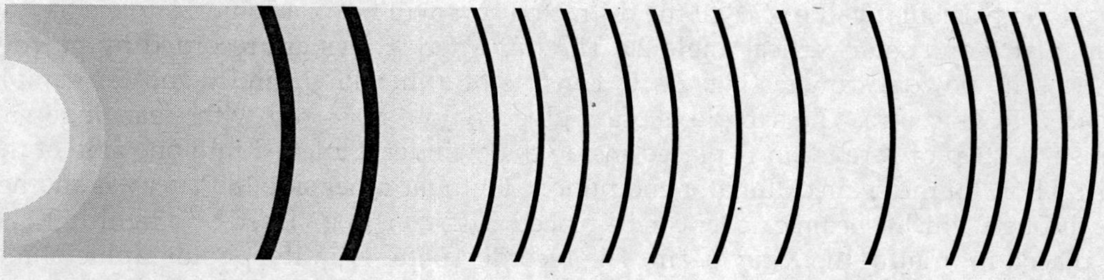

FIGURE 21.11 Diffraction pattern in Powder or Debye-Scherrer method

21.8 ▶ Rotating-Crystal Method

This method is generally used to determine the lattice constant of the crystal and also to determine the unknown crystal structures. In the experimental arrangement, a film is mounted in a cylindrical holder concentric with a rotating spindle on which the single crystal having dimensions not more than 1 mm is mounted about an axis normal to the mono-energetic X-ray beam (Fig. 21.12). This method is used for structure determination when a single-crystal specimen is available. The beam is diffracted from a given crystal plane for some angle θ which satisfies the Bragg equation.

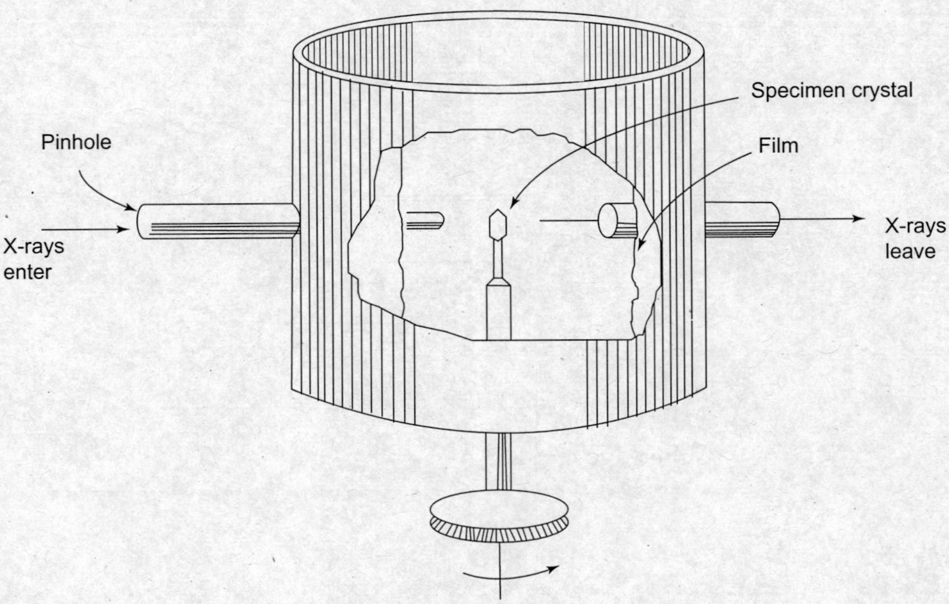

FIGURE 21.12 Rotating crystal method

The reflected beams from all planes parallel to the vertical rotation axis will lie in the horizontal plane, and the planes with other orientations will reflect in layers above and below the horizontal plane. The reflected beams from the crystal planes are located on the surface of imaginary cones. By recording these diffraction patterns in terms of both angles and intensities for various crystal orientations, we can determine the shape and size of a unit cell of the crystal as well as the arrangement of atoms inside the cell.

EXAMPLE 21.3

The Bragg's angle for reflection from the (111) plane in a FCC crystal is 19.2° for an X-ray wavelength of 1.54 A.U. Compute the cube edge of the unit cell.

Solution: The given data are

Bragg's angle, $\theta = 19.2°$

Wavelength of X-rays, $\lambda = 1.54$ Å

Order of diffraction, $n = 1$

Cube edge, $a = ?$

Bragg's law

$$2d \sin \theta = n\lambda$$

$$2d \sin 19.2° = 1 \times 1.54$$

$$d = \frac{1.54}{2 \times \sin 19.2°} = \frac{1.54}{2 \times 0.328}$$

$$= 2.3411 \text{ Å}$$

$$d = \frac{a}{\sqrt{h^2 + k^2 + l^2}}$$

or

$$a = d\sqrt{h^2 + k^2 + l^2}$$

$$= 2.3411 \times \sqrt{1^2 + 1^2 + 1^2}$$

$$= 2.3411 \times \sqrt{3} = 4.05 \text{ Å}$$

Cube edge $a = 4.05$ Å

EXAMPLE 21.4

The first-order diffraction occurs when an X-ray beam of wavelength 0.675 Å incident at a glancing angle of 5°25′ on a crystal. What is the glancing angle for third-order diffraction to occur?

Solution: Wavelength of X-rays (λ) = 0.675 Å
Glancing angle for first-order ($n = 1$) diffraction (θ_1) = 5°25′
The glancing angle for third-order ($n = 3$) diffraction (θ_3) = ?
Bragg's equation is $2d \sin \theta = n\lambda$

For the first-order diffraction, $2d \sin \theta_1 = 1\lambda$

$$2d \sin 5°25' = 0.675 \times 10^{-10} \text{m}$$

$$d = \frac{0.675 \times 10^{-10}}{2 \sin 5°25'} \text{m}$$

$$= \frac{0.675 \times 10^{-10}}{0.1888}$$

$$= 3.575 \times 10^{-10} \text{m}$$

$$= 3.575 \text{ Å}$$

For the third-order diffraction,

$$2d \sin \theta_3 = 3\lambda$$

$$\sin\theta_3 = \frac{3\lambda}{2d}$$

$$= \frac{3\times0.675\times10^{-10}}{2\times3.575\times10^{-10}}$$

$$= 0.283$$

$$\theta_3 = \sin^{-1}(0.283)$$

$$= \mathbf{16.45° = 16°28'}.$$

EXAMPLE 21.5

What is the angle at which the third-order reflection of X-rays of 0.79 Å wavelength can occur in a calcite crystal of 3.04×10^{-8} cm spacing?

Solution: Wavelength of X-rays, $\lambda = 0.79$ Å $= 0.79 \times 10^{-8}$ cm

Inter-planar spacing, $d = 3.04 \times 10^{-8}$ cm

Order of diffraction, $n = 3$

Angle of diffraction, $\theta = ?$

$$\mathbf{2d\ sin\ \theta = n\lambda}$$

EXAMPLE 21.6

A beam of X-rays is incident on a NaCl crystal with lattice plane spacing 0.282 nm. Calculate the wavelength of X-rays if the first-order Bragg reflection takes place at a glancing angle of 8°35′. Also calculate the maximum order of diffraction possible.

Solution: The given data are
NaCl crystal is FCC

Lattice plane spacing (d) = 0.282 nm

Wavelength of rays (λ) = ?

Order of diffraction **(n)** = 1

Glancing angle $\theta = 8°35'$

Bragg's equation is $n\lambda = 2d\sin\theta$

$$1\lambda = 2 \times 0.282 \times 10^{-9} \sin(8°35')$$

$$= 0.0842 \text{ nm}$$

Maximum order of diffraction (n_{max}) = **?**

$$2d\ \sin\theta = n\lambda$$

if $\theta = 90°$ then $n = n_{max}$

$$\therefore \qquad 2d = n_{max}\ \lambda$$

$$n_{max} = \frac{2d}{\lambda}$$

$$= \frac{2 \times 0.282\ \text{nm}}{0.0842\ \text{nm}}$$

$$= \mathbf{6.7 \approx 7}$$

EXAMPLE 21.7

Calculate the glancing angle at (110) plane of a cubic crystal having axial length of 0.26 nm corresponding to the second-order diffraction maximum for the X-rays of wavelength 0.65 nm.

Solution: The given data are
Edge length of cubic system, $a = 0.26$ nm
Wavelength of X-rays, $\lambda = 0.065$ nm
Glancing angle, for plane (110), $\theta = ?$
Order of diffraction, $n = 2$
Separation between (110) planes of a cube,

$$d = \frac{a}{\sqrt{h^2 + k^2 + l^2}} = \frac{0.26}{\sqrt{1^2 + 1^2 + 0^2}}\text{nm}$$

$$= \frac{0.26}{\sqrt{2}} = 0.184\ \text{nm}$$

Bragg's law

$$2d\ \sin\theta = n\lambda$$

$$2 \times 0.184\ \text{nm} \times \sin\theta = 2 \times 0.065\ \text{nm}$$

$$\sin\theta = \frac{0.065}{0.184} = 0.353$$

$$\therefore \qquad \theta = \sin^{-1}(0.353)$$

$$= \mathbf{20°41'13''}$$

RECAP ZONE

POINTS TO REMEMBER

- Crystallography is an experimental technique to determine the crystal structure of materials. It uses X-rays as a tool of investigation of crystals.
- Due to the capital inter-planar gaps being of the order of X-ray wavelengths, these are able to scatter diffract X-rays.
- The X-rays are diffracted by crystals under Bragg's law where

$$n\lambda = 2d \sin \theta$$

- The energy of X-ray is given as

$$E = h\nu = \frac{hc}{\lambda} = eV$$

  e - electronic charge, V - accelerating voltage of X-ray tube.
- X-rays from X-ray tubes are used after filtration and collimation.
- The inter-planar distance d of an unknown crystal d is determined as

$$d = \frac{n\lambda}{2\sin \theta}$$

- XRD spectra are recorded for small angle for powdered crystals to take an average of all intensities scattered from different tiny crystals that have the same lattice planes as a large one.
- Laue method is the oldest method of X-ray crystallography methods, which is used to determine the crystal structure of the material and the electron charge distribution within the solid.
- Rotating crystal method is used to determine the lattice constant of the crystal and also to determine the unknown crystal structures.

KEY TERMS WITH DEFINITIONS

- **Crystallography** – It is an experimental science of determining the arrangement of atoms in the crystalline solids. X-ray crystallography is used to determine the structure of inorganic and large biomolecules such as proteins.
- **XRD** – X-ray diffraction is the technique which is used for the identification and characterization of compounds based on their diffraction pattern.
- **X-rays** – These are very short wavelength electromagnetic waves that have high penetrating power.

- **Bragg's X-ray diffraction law** $2d \sin\theta = n\lambda$
- Energy of X-rays photon is given by

$$E = h\nu = \frac{hc}{\lambda}$$

where h = Plank's constant

c = velocity of light

λ = wavelength of X-ray

REVIEW ZONE

SHORT ANSWER QUESTIONS

1. What are X-rays? Mention their properties.
2. Differentiate between soft and hard X-rays.
3. Derive Bragg's law for X-ray diffraction.
4. Describe Laue method of X-ray crystallography with the help of a schematic diagram.
5. What are the various methods of crystal structure determination?
6. How can X-rays be produced?
7. What is an XRD pattern? What does a peak indicate?

LONG ANSWER QUESTIONS

1. What are main properties of X-rays that make these different from visible rays?
2. Why only crystals can diffract X-rays?
3. Derive Bragg's law for X-ray diffraction.
4. Describe the working of a Bragg's spectrometer.
5. Explain how x-rays can be used to determine crystal structure.
6. What is an XRD pattern? What do the peaks in the spectrum indicate?
7. What are various types of X-ray diffraction spectrometers? Highlight the major differences of these.

CHAPTER 22

Crystal Structure

Learning Objectives

By the end of this chapter, the student will be able:

- To differentiate crystalline and amorphous solids
- To understand the classification system of crystals
- To determine inter-planar spacing of crystals
- To recognize phases of liquid crystals
- To categorize imperfections in crystals

22.1 ▶ Introduction

Matter exists in three different states; they are gaseous, liquid and solid states. In gaseous and liquid states, the atoms or molecules of the substance move from one place to another and there is no fixed position of atoms in the substance. In solids, the positions of the atoms or molecules are fixed and may or may not be present periodically at regular intervals of distance. If the atoms or molecules in a solid are periodical at regular intervals of distances in three-dimensional space, then that solid is known as *crystalline solid.* If the atoms or molecules do not have such a periodicity in a solid, then that solid is known as amorphous solid. When the periodicity of atoms or molecules is extended throughout the solid, then the solid is known as *single crystalline solid.* If the periodicity of atoms or molecules is extended up to small regions called grains and if these grains are very large in number and are of different sizes in the solid, such a material is known as *polycrystalline solid.* Various phases of matter that are found in nature are shown in Fig. 22.1. Table 22.1 shows the comparison of crystaline and amorphous solids based upon some physical properties. The study of geometric form and other physical properties of crystalline solids by using X-rays, electron beams and neutron beams constitutes the science of crystallography.

TABLE 22.1 Distinction between crystalline and amorphous solids

Crystalline solids	Amorphous solids
1. The atoms or molecules of the crystalline solids are periodic in space.	1. The atoms or molecules of the amorphous solids are not periodic in space.

(Cont'd)

TABLE 22.1 (Continued)

Crystalline solids	Amorphous solids
2. Some crystalline solids are anisotropic, i.e. the magnitude of physical properties (such as refractive index, electrical conductivity, thermal conductivity, etc.) are different along different directions of the crystal.	2. Amorphous solids are isotropic, i.e. the magnitude of the physical properties is same along all directions of the solid.
3. Crystalline solids have sharp melting points.	3. Amorphous solids do not possess sharp melting points.
4. Breaks are observed in the cooling curve of a crystalline solid.	4. Breaks are not observed in the cooling curve.
5. A crystal breaks along certain crystallographic planes.	5. When an amorphous solid breaks, the broken surface is irregular because it has no crystal planes.

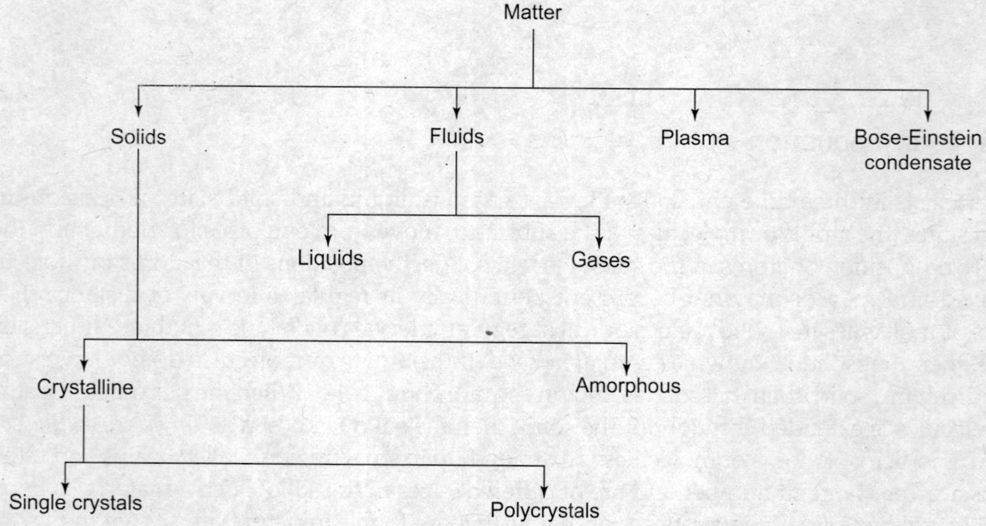

FIGURE 22.1 Matter and its various phases

22.2 ▶ Solid Materials

Solids are broadly classified into two types:

1. Crystalline solids or true solids
2. Amorphous solids or pseudo solids

22.2.1 Crystalline Solids

Crystalline solids are usually built up of a number of crystals, in which the crystals may be of similar or widely varying sizes. They may be made up of *metallic crystal or non-metallic crystal*. A crystalline solid has definite melting point and a definite heat of fusion.

Metallic crystal

Metallic crystals are widely used in various fields due to their favourable properties of strength, ductility, conductivity and reflection. In the formation of metallic crystal, the ions are connected only indirectly, through the free electron surrounding them. As in the formation of metallic bond, no directional properties are involved. So, each atom attracts as many neighbouring atoms as it can, resulting in a closely packed structure with short, strong bond and high density of solid is formed, e.g. Cu, Ag, Al, Ni (FCC), W, Mo, Cr (BCC), Mg, Zn, Ti, Cd (HCP). In the above example, all the crystals belong to crystalline solids and their structure is given in brackets.

Non-metallic crystal

Non-metallic crystals may have covalent or ionic bonding or any great variety of mixed bonds. Atoms of two or more kinds are frequently involved, e.g. crystalline carbon, crystallized polymer or plastics.

Non-metallic crystal changes its form at different temperatures or pressures (polymorphism); e.g., tin crystallizes in a non-metallic diamond structure (grey tin) at low temperature, while at non-temperature, it forms metallic structure (white tin).

In short, crystalline solids have the following characteristics:

1. They have regular and definite geometrical form.
2. They have sharp melting points.
3. They have clean cleavage.
4. They have definite heat of fusion.
5. They are anisotropic, i.e. their mechanical and electrical properties depend on the direction along which they are measured.

22.2.2 Non-Crystalline Solids

Materials in which the molecule is the basic structured solid and which have no regular structure are classified as amorphous.

Amorphous solids

In amorphous materials, the elementary particles are mixed together in a disorderly manner, their positions bearing no fixed relationship to those of their neighbours. An amorphous structure does not generally possess elasticity, but only plasticity (except rubber), e.g. glass, plastic, starch, rubber, etc.

The amorphous solids show the following characteristics:

1. They are regarded as super-cooled liquids.
2. They do not have definite geometrical shape.
3. They melt over a wide range of temperature.
4. They do not have definite heat of fusion.
5. They are isotropic, i.e. they have similar physical properties in all directions, because the constituent particles are arranged in a random manner.

22.3 ▶ Crystal System

Before the discussion about crystal system, we deal with the terms associated with the crystal system.

Crystallography

It is a branch of science which deals with geometrical properties and structure of crystal and crystalline substance.

Interfacial angle

The angle between any two faces of a crystal is called interfacial angle (Fig. 22.2).

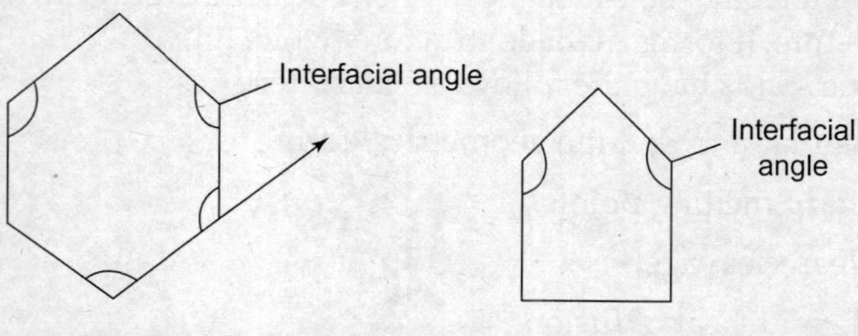

FIGURE 22.2 Crystal (interfacial) angles in a crystal

It may be noted that the size of the faces or even shapes of crystals of a substance may vary widely with conditions of crystallization, but the interfacial angle between any two corresponding faces of a crystal is always constant.

Crystal symmetry

Definite ordered arrangement of atoms in a crystal is known as crystal symmetry. Every crystal must possess three types of symmetry:

(a) Plane of symmetry
(b) Axis of symmetry
(c) Centre of symmetry

Plane of symmetry: It is an imaginary plane which can divide the crystal into half such that one is mirror image of the other (Fig. 22.3).

Axis or line of symmetry: A symmetry axis is a line passing through the crystal so that the definite angular rotation of the crystal produces exactly same original appearance.

For example, Fig. 22.4 shows a crystal and the symmetry axis. If the cube crystal is rotated through 90°, every time it produces exactly the same appearance.

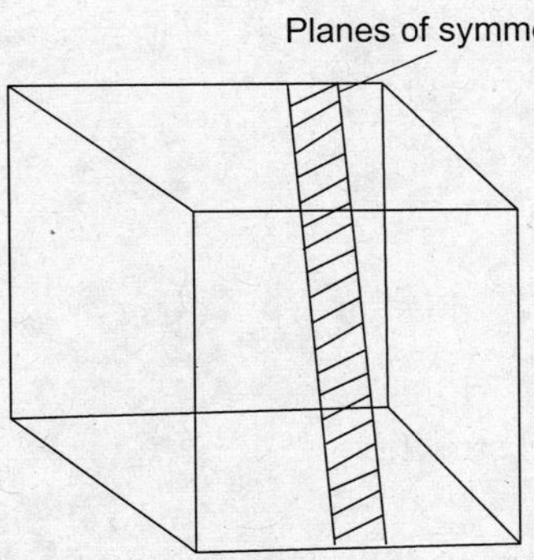

FIGURE 22.3 Planes of symmetry

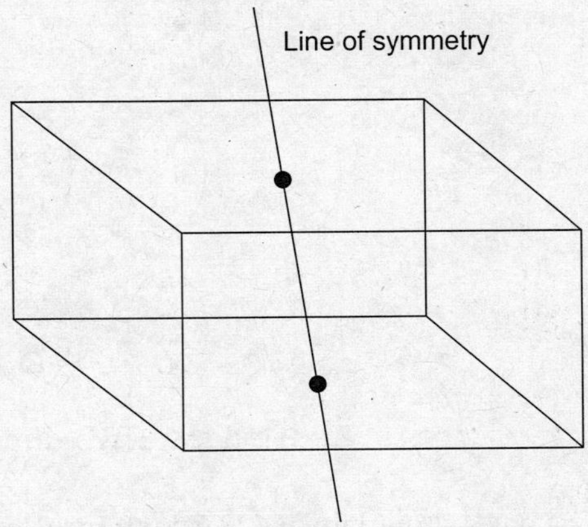

FIGURE 22.4 Axis or line of symmetry

Centre of symmetry: A centre of symmetry is a point in the crystal such that any straight line through it passes through a pair of similar points situated at a same distance but on the opposite side of centre of symmetry (Fig. 22.5).

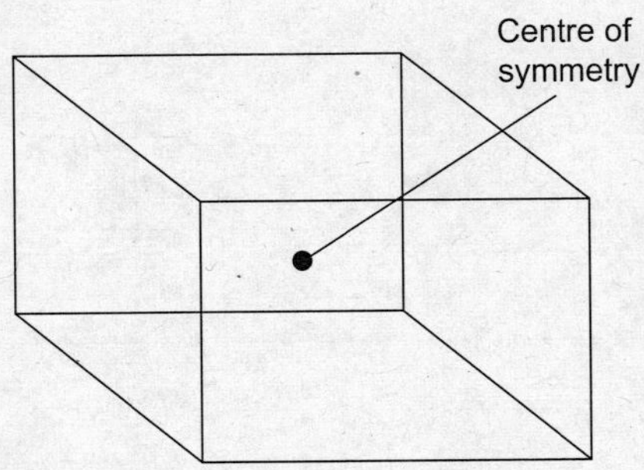

FIGURE 22.5 Centre of symmetry

Lattice is the regular geometrical arrangement of points in the crystal space. Atoms arrange themselves in distinct pattern in space called a space lattice. Space lattice is nothing but an imaginary line (three-dimensional network) connecting the atoms. Every point in space lattice has identical surrounding.

In other words, the regular arrangement of an infinite set of points (atom, ions or molecules) in space is called space lattice.

22.3.1 Lattice Crystal Structure

The crystal structure is formed by associating every lattice point with an assembly of atoms or molecules or ions, which are identical in composition, arrangement and orientation, is called as the *basis*. The atomic arrangement in a crystal is called *crystal structure.* If the basis is substituted for the lattice points, then the resulting structure is called crystal structure which is shown in Fig. 22.6.

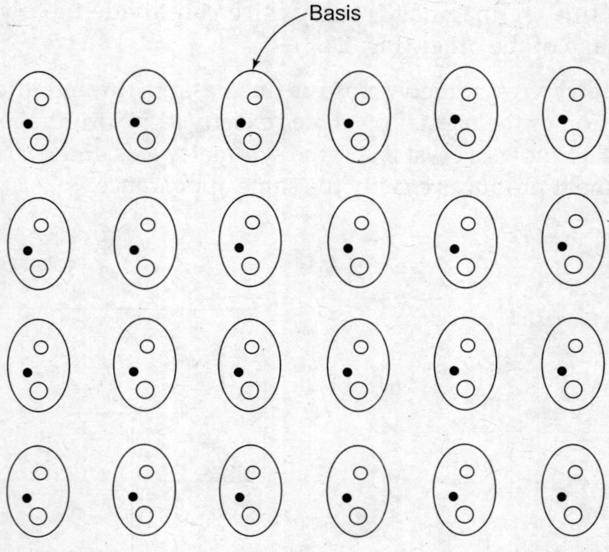

FIGURE 22.6 Two-dimensional crystal structure

The basis shown in Fig. 22.6 contains three different atoms. In copper and sodium crystals, the basis is single atoms; in NaCl, the basis is diatomic and in CaF_2 the basis is triatomic.

Thus, lattice + basis = crystal structure (Fig. 22.7).

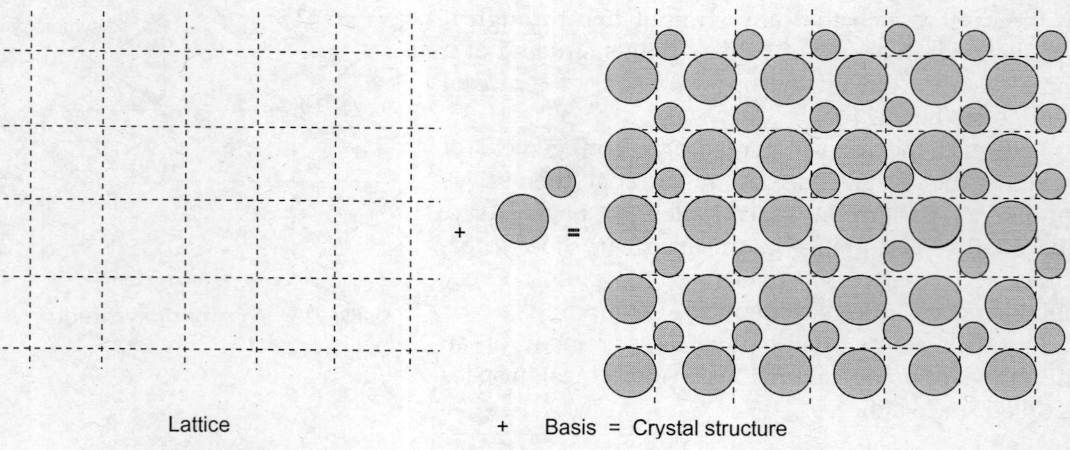

FIGURE 22.7 Lattice and basis of crystal structure

22.3.2 Unit Cell

The smallest repeating pattern (unit) from which the lattice is known is called unit cell. The unit cells are repeated over and over again in three dimensions, and as a result into the whole of space lattice of the crystal.

In other words, the smallest volume that contains the full pattern of repetition is called a unit cell. The unit cell represents the picture of the entire crystal.

For describing unit cell, we must know

(a) The distance a, b and c, i.e. length of edge of unit cell (called lattice parameter)

(b) The angles α, β and γ between three imaginary lines

Types of unit cell

The unit cells are of four types (Fig. 22.8):

1. Primitive or simple
2. Face-centred
3. Body-centred
4. End-centred

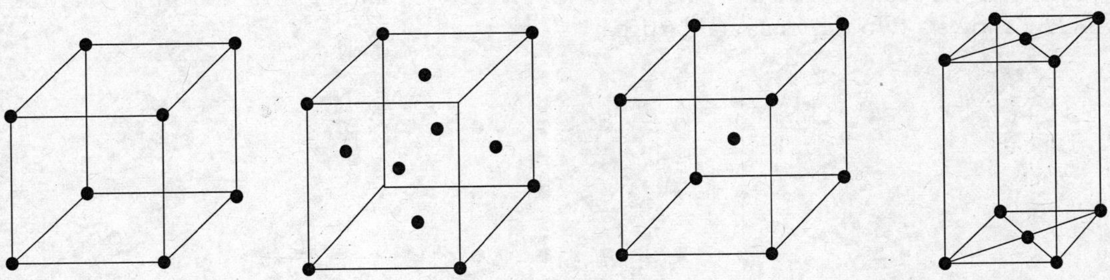

1. Bodv-centred 2. Face-centred 3. Simple 4. End-centred

FIGURE 22.8 Four types of unit cell

Primitive or simple: In this type of unit cell, atoms are present only at the corners of the unit cell.

Face-centred (FCC): In this type, atoms are present at the corners as well as at the centres of each six faces.

Body-centred (BCC): In this type, atoms are present at the corners and an additional point is present in the centre of unit cell.

End-centred: In this type of unit cell, atoms are present at the corners and at the centres of two ends.

22.3.3 Space Lattice and Basis

In a solid crystalline material, the atoms or molecules are arranged regularly and periodically in three dimensions. To explain crystal symmetries easily, it is convenient to represent an atom or a group of atoms that repeats in three dimensions in the crystal as a unit, of atoms or atom in a crystal is replaced by a point in space, then the resultant points in space are called *space lattice*. Each point in space is called a lattice point and each unit of atoms or atom is called basis or pattern. A space lattice represents the geometrical pattern of crystal in which the surroundings of each lattice point is the same.

When atoms are arranged orderly in a definite array, then it is said to have a specific structure, with one atom occupying one lattice point, e.g. iron with BCC structure. However, in some cases, as shown in Fig. 22.7 there might be one or more atoms at a specific orientation from one of the atoms that does not belong to any lattice point (i.e., no lattice point exists for that atom). It can be positioned only by means of coordinates from the closest atom to it. Therefore, lattice atom has a second atom, and hence there are two atoms associated with each lattice point, e.g. magnesium with HCP structure. This unit assembly of atoms at each lattice point is known as the *basis*. Hence, it could be said that the combination of the lattice (point) plus the basis forms the crystal structure.

22.3.4 Crystal Axes

If the surrounding of each lattice point is same or if the atom or all the atoms (of basis) at lattice points are identical, then such a lattice is called *Bravais lattice*. On the other hand, if the atom of the basis at lattice points is not same, then it is said to be a non-Bravais lattice. Fig.22.9 shows a two-dimensional lattice.

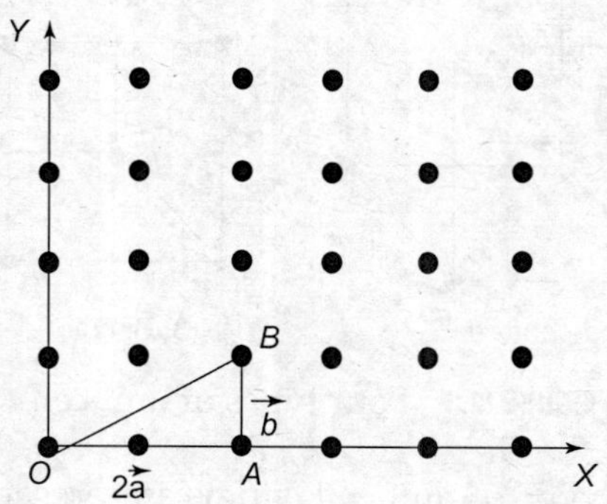

FIGURE 22.9 Two-dimensional lattice

To present translational vectors or basis vectors, consider a co-ordinate system with its origin at the lattice point 'O'. Let $\overline{OA} = 2\vec{a}$ and $\overline{AB} = \vec{b}$, such that $\overline{OB} = 2\vec{a} + \vec{b}$, where $\vec{a}$ and $\vec{b}$ are called translational or basic vectors along X and Y directions. The position vector $\vec{R}$ of any lattice point can be represented as $\vec{R} = n_1\vec{a} + n_2\vec{b}$, where n_1 and n_2 are integers, their values depend on the position of the lattice point under consideration with respect to the origin. In three dimensions, the position vector of a point can be expressed as $\vec{R} = n_1\vec{a} + n_2\vec{b} + n_3\vec{c}$, where $\vec{a}, \vec{b}$ and $\vec{c}$ are the translational or basis vectors along X, Y and Z directions, respectively. They are also called translational primitives.

22.3.5 Primitive Cell

A unit cell which has only one atom or one lattice point is known as a *primitive cell* (Fig. 22.11). Therefore, a primitive cell can be defined as a unit cell formed by the primitives

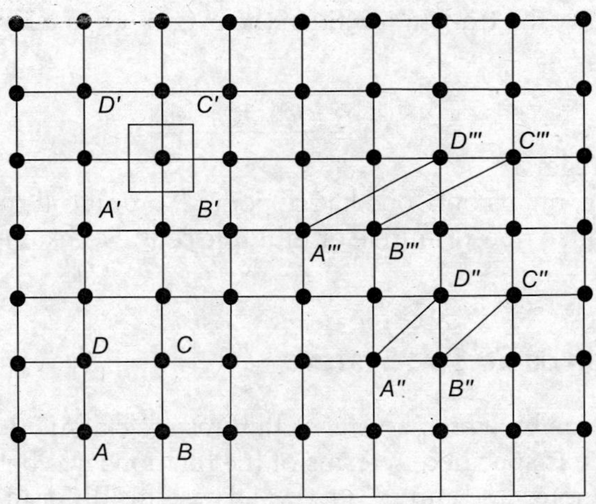

FIGURE 22.10 Unit cells in crystal lattice

a, b and c (distances from the origin) having only one lattice point. On the contrary, a unit cell may contain one or more lattice points or atoms as shown (Fig. 22.10).

Note

Corollary: Thus, a primitive cell can be a unit cell, but a unit cell need not always be a primitive cell.

In three dimensions, a primitive cell is visualized as a cube of unit length formed by the axes having eight atoms at its eight corners (Fig. 22.11). All these eight atoms are shared equally with eight other adjacent unit cells, thus leaving only one atom for each cell (8 atoms $\times$ 1/8th of each atom $=$ 1 atom). Unit cell geometry is completely defined in terms of six parameters: the three non-coplanar or translational vectors a, b and c (edge lengths of a unit cell formed on the crystallographic axes X, Y and Z, respectively) and the three inter-axial or crystallographic angles α, β and γ (Fig. 22.12).

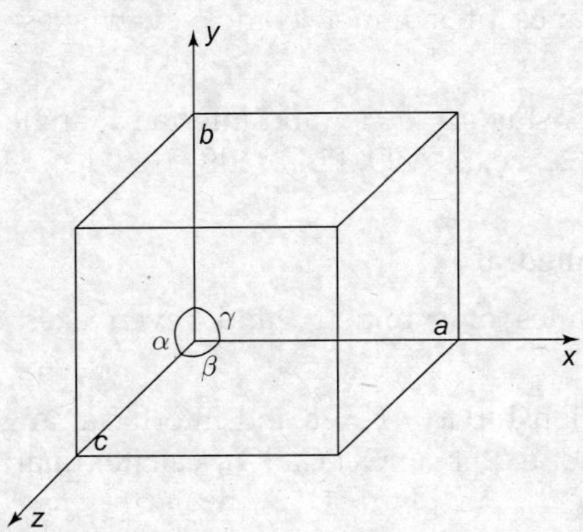

FIGURE 22.11 A primitive cell and unit cells

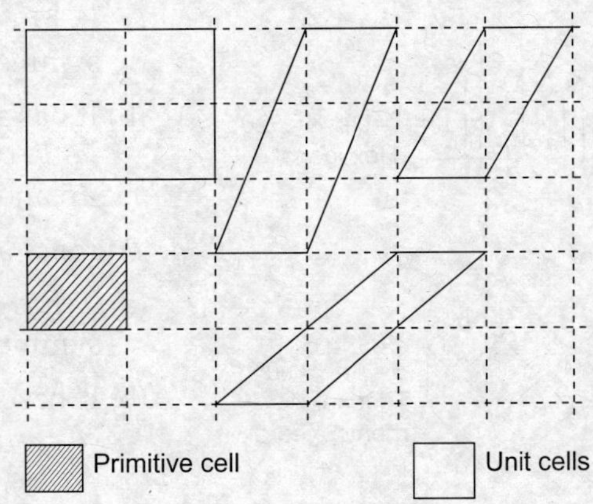

FIGURE 22.12 Unit cell parameters

Within a crystal structure, the translation from the origin 'O' to a lattice point 'P' is given by the translation vector as

$$\boxed{\vec{r} = la + mb + nc}$$

where l, m and n are integers.

A primitive unit cell contains only one lattice point. If a unit cell contains more than one lattice point, then it is called non-primitive or multiple cells. For example, BCC and FCC are non-primitive unit cells.

22.4 ▶ Classification of Crystal Systems

Crystal system is a scheme by which crystal structures are classified according to unit cell geometry. This geometry is specified in terms of the relationships between edge lengths a, b and c and interfacial angles α, β and γ. The crystal systems differ in the length of the unit cell edges (a, b, c) and the angle between the unit cell edges.

Basically, there are seven different crystal systems:

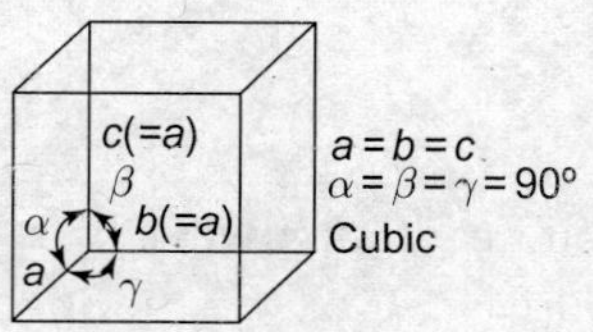

1. **Cubic**
   (a) It has a plane of symmetry and 13 axes of symmetry.
   (b) It has axial relationship $a = b = c$, and angles $\alpha = \beta = \gamma = 90°$. e.g., NaCl, KCl, CaF_2, Pb, Ag, Au, etc.

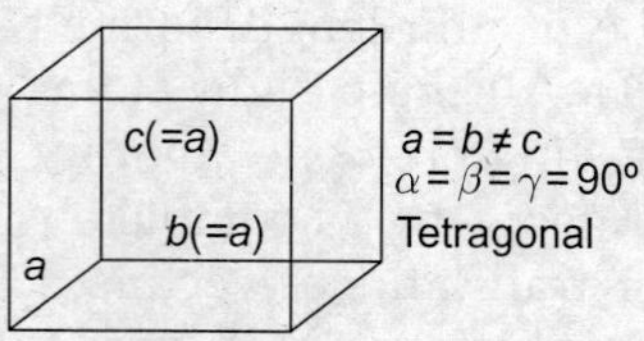

2. **Tetragonal**
   (a) It has five planes of symmetry and five axes of symmetry.
   (b) It has axial relationship $a = b \neq c$ and interfacial angle $\alpha = \beta = \gamma = 90°$. e.g., SnO_2, ZnO_z, TiO_2, white Sn, $PbWO_4$, etc.

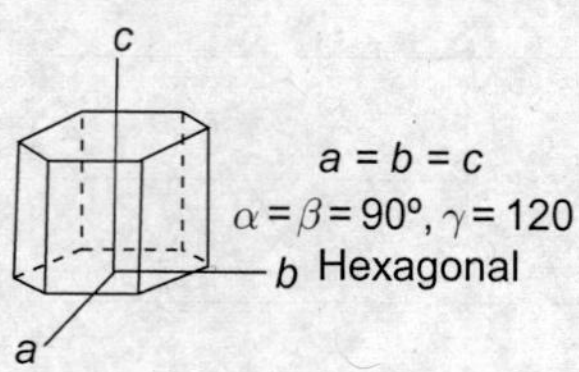

3. **Hexagonal**
   (a) It has seven planes of symmetry and seven axes of symmetry.
   (b) It has axis relationship $a = b \neq c$, and interfacial angle $\alpha = \beta = 90°$, $\gamma = 120°$. e.g., ZnO, PbI_2, CdS, HgS, ice, Mg, Zn and Cd.

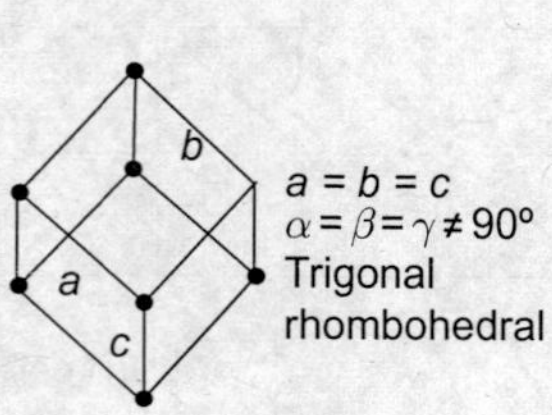

4. **Trigonal or rhombohedral**
   (a) It has seven planes of symmetry and seven axes of symmetry.
   (b) It has axial relationship $a = b = c$ and interfacial angle $\alpha = \beta = \gamma \neq 90°$. e.g., $NaNO_3$, $CaSO_4$, calcite, quartz, As, Sb, Bi.

5. **Orthorhombic or (rhombic)**

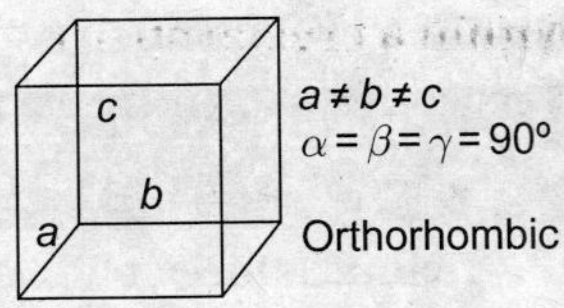

(a) It has three planes of symmetry and three axes of symmetry.

(b) It has axial relationship $a \neq b \neq c$ and interfacial angle $\alpha = \beta = \gamma = 90°$. e.g., KNO_3, K_2SO_4 $PbCO_3$, $BaSO_4$ Rhombic sulphur, $MgSO_4 \cdot 7H_2O$.

6. **Monoclinic**

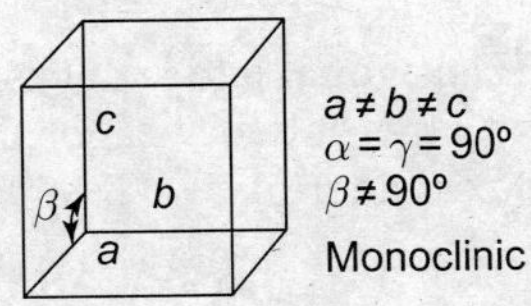

(a) It has a plane of symmetry and an axis of symmetry.

(b) It has axial relationship $a \neq b \neq c$ and interfacial angle $\alpha = \gamma = 90°$, $\beta \neq 90°$. e.g., $Na_2SO_4 \cdot 10H_2O$, $Na_2B_4O_7 \cdot 10H_2O$, Monoclinic sulphur.

7. **Triclinic**

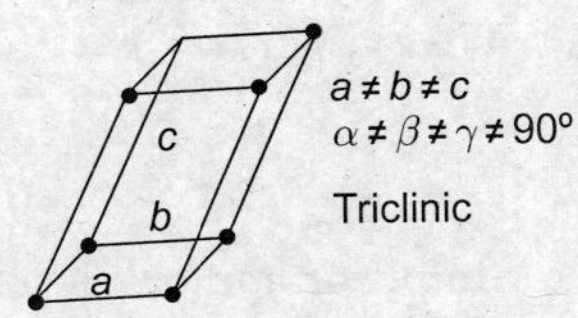

(a) It has no plane of symmetry and no axes of symmetry but have a centre of symmetry.

(b) It has an axial relationship $a \neq b \neq c$ and interfacial angle $\alpha \neq \beta \neq \gamma \neq 90°$. e.g., $CuSO_4.5H_2O$, $K_2Cr_2O_7$, H_2BO_3.

Note

All the seven crystal systems have a centre of symmetry.

22.4.1 Bravais Lattices

Bravais lattices are named after their originator Auguste Bravais. These are nothing but 14 distinguishable three-dimensional space lattices that can be generated by repeated translation of three non-coplanar vectors, *a*, *b* and *c* of a unit cell in three-dimensional space. Each crystal system is based on any one of Bravais lattices.

The 14 Bravais lattices are described in Table 22.2.

TABLE 22.2 The list of 14 Bravais lattices

Types of lattices	Corresponding structures
1. **Simple monoclinic lattice:** It has lattice points at eight corners of unit cell. It has vectors $a \neq b \neq c$ and inter-axial angles $\alpha = \gamma = 90° \neq \beta$.	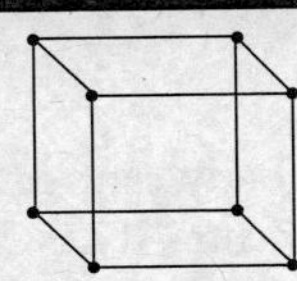
2. **End-centred monoclinic lattice:** It has lattice points at the eight corners and at two face centres opposite to each other. It has vectors $a \neq b \neq c$ and interfacial angles $\alpha = \beta = 90° \neq \gamma$.	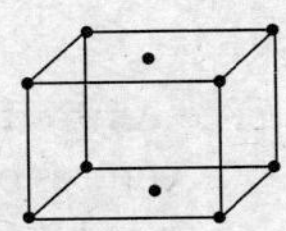

(Cont'd)

TABLE 22.2 (Continued)

Types of lattices	Corresponding structures
3. **Triclinic lattice:** It has lattice point at the eight corners of unit cell. It has vectors $a \neq b \neq c$ and inter-axial angles $\alpha \neq \beta \neq \gamma = 90°$.	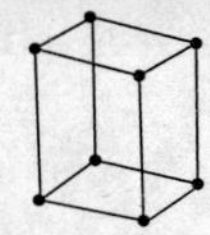
4. **Hexagonal lattice:** It has points at the 12 corners of hexagonal prism and at the centres of two hexagonal faces of unit cell. It has vectors $a = b \neq c$ and inter-axial angles $\alpha = \beta = 90°$ and $\gamma = 120°$.	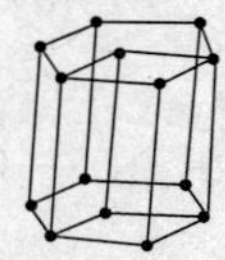
5. **Rhombohedral lattice:** It has lattice points at the eight corners of unit cell and it has vectors $a = b = c$ and inter-axial angles $\alpha = \beta = \gamma \neq 90°$.	
6. **Simple orthorhombic lattice:** It has lattice points at the eight corners of unit cell. It has vectors $a \neq b \neq c$ and inter-axial angles $\alpha = \beta = \gamma = 90°$.	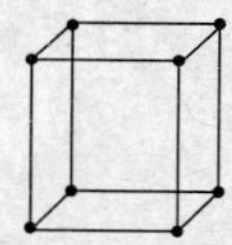
7. **Body-centred orthorhombic lattice:** It has lattice points at the eight corners and at the body centre. It has vectors $a \neq b \neq c$ and inter-axial angles $\alpha = \beta = \gamma = 90°$.	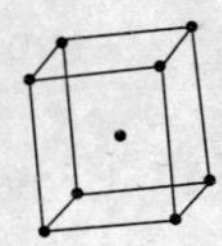
8. **End-centred orthorhombic lattice:** It has lattice points at eight corners and at two face centres opposite to each other. It has vectors $a \neq b \neq c$ and inter-axial angles $\alpha = \beta = \gamma = 90°$.	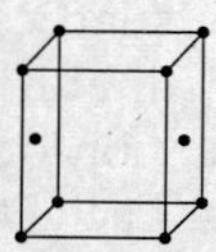
9. **Face-centred orthorhombic lattice:** It has lattice points at the eight corners and at the six faces of unit cell. It has vectors $a \neq b \neq c$ and interfacial angles $\alpha = \beta = \gamma = 90°$.	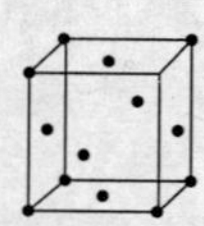
10. **Simple cubic lattice:** It has lattice points at eight comers of unit cell. It has vectors $a = b = c$ and inter-axial angles $\alpha = \beta = \gamma = 90°$.	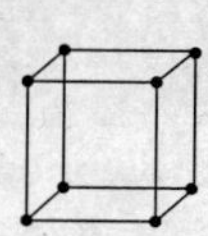
11. **Body-centred cubic lattice:** It has lattice points at eight corners and at the body centre. It has vectors $a = b = c$ and inter-axial angle $\alpha = \beta = \gamma = 90°$.	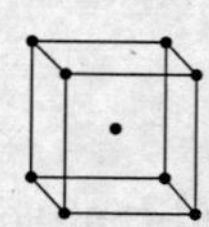
12. **Face-centred cubic lattice:** It has lattice points at the eight corners and at the face centres of the unit cell. It has vectors $a = b = c$ and inter-axial angles $\alpha = \beta = \gamma = 90°$.	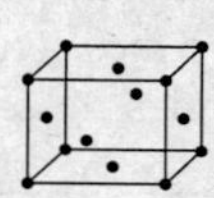

(Cont'd)

TABLE 22.2 (Continued)

Types of lattices	Corresponding structures
13. **Simple tetragonal lattice:** It has lattice points at eight corners of unit cell. It has vectors $a = b \neq c$ and inter-axial angles $\alpha = \beta = \gamma = 90°$.	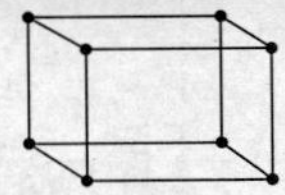
14. **Body-centred tetragonal lattice:** It has lattice points at the eight corners of the unit cell and body centre. It has vectors $a = b \neq c$ and inter-axial angles $\alpha = \beta = \gamma = 90°$.	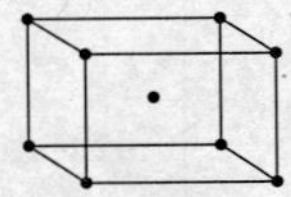

TABLE 22.3 Seven crystal systems with Bravais lattices based upon geometry

Crystal system	Bravias lattices	Parameter of unit cell	
		Cell dimensions	Crystal angles
1. Cubic	Primitive, FCC, BCC = 3	$a = b = c$	$\alpha = \beta = \gamma = 90°$
2. Orthorthombic	Primitive, FCC, BCC, end-centred = 4	$a \neq b \neq c$	$\alpha = \beta = \gamma = 90°$
3. Tetragonal	Primitive, body-centred = 2	$a = b \neq c$	$\alpha = \beta = \gamma = 90°$
4. Monoclinic	Primitive, end-centred = 2	$a \neq b \neq c$	$\alpha = \beta = 90°, \gamma \neq 90°$
5. Triclinic	Primitive = 1	$a \neq b \neq c$	$\alpha \neq \beta \neq \gamma \neq 90°$
6. Hexagonal	Primitive = 1	$a = b \neq c$	$\alpha = \beta = 90°, \gamma = 120°$
7. Rhombohedral or trigonal	Primitive = 1	$a = b = c$	$\alpha = \beta = \gamma \neq 90°$

22.5 ▶ Lattice Parameters and Crystal Systems

For representing the type of distribution of lattice points in space, seven different co-ordinate systems are required. These co-ordinate systems are called *crystal systems*. The crystal systems are named on the basis of geometrical shape and symmetry (Table 22.3). The seven crystal systems are: (1) Cubic, (2) Tetragonal, (3) Orthorhombic, (4) Monoclinic, (5) Triclinic, (6) Rhombohedral (or Trigonal) and (7) Hexagonal. Space lattices are classified according to their symmetry. In 1948, Bravais showed that 14 lattices are sufficient to describe all crystals. These 14 lattices are known as *Bravais lattices* and are classified into seven crystal systems based on cell parameters (Table 22.4). The Bravais lattices are categorized as primitive lattice (*P*), body-centred lattice (*I*), face-centred lattice (*F*) and base-centred lattice (*C*). These seven crystal systems and Bravais lattices are described below.

Cubic crystal system

In this crystal system, all the unit cell edge lengths are equal and are at right angles to one another, i.e. $a = b = c$ and $\alpha = \beta = \gamma = 90°$. In cubic system, there are three Bravais lattices; they are simple (primitive), body-centred and face-centred (Fig. 22.13). Examples for cubic system are Au, Cu, Ag, NaCl, diamond, etc.

TABLE 22.4 The examples of crystal systems and Bravais lattice

Sr. No.	Crystal system	Lattice parameter	Bravais lattice	Example of crystal
1.	Cubic	$a = b = c$ $\alpha = \beta = \gamma = 90°$	1. Simple 2. Body-centred	Ag, Cu, NaCl, Na
2.	Tetragonal	$a = b \neq c$ $\alpha = \beta = \gamma = 90°$	3. Face-centered 4. Simple 5. Body-centred	TiO_2, SnO_2, $NaSO_4$
3.	Orthorhombic	$a \neq b \neq c$ $\alpha = \beta = \gamma = 90°$	6. Simple 7. Base-centred 8. Body-centred 9. Face-centred	Ga, KNO_3, $BaSO_4$, $PbCO_3$
4.	Monoclinic	$a \neq b \neq c$ $\alpha = \beta = 90° \neq \gamma$	10. Simple 11. Base-centred	$FeSO_4$, gypsum
5.	Triclinic	$a \neq b \neq c$ $\alpha = \beta \neq \gamma \neq 90°$	12. Simple	$K_2Cr_2O_7$, $CuSO_4$, $5H_2O$
6.	Trigonal	$a = b = c$ $\alpha = \beta = \gamma \neq 90°$	13. Simple	As, Sb, Bi
7.	Hexagonal	$a = b = c$ $\alpha = \beta = 90°$ $\gamma = 120$	14. Simple	Mg, Zn, Cd

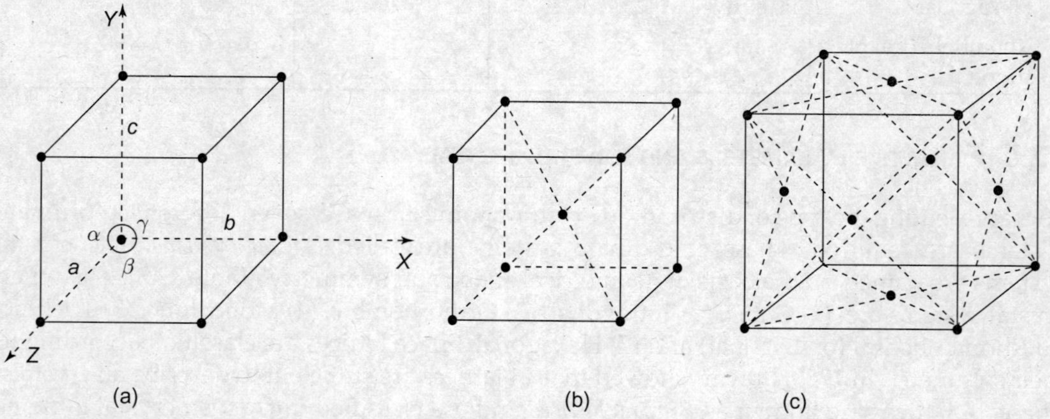

FIGURE 22.13 Cubic crystal system: (a) simple cubic (*P*), (b) body-centred cube (*I*) and (c) face-centred cube (*F*)

In simple cubic lattice, lattice points or atoms are present at the corners of the cube. In body-centred cube, atoms are present at the corners and one atom is completely present at the centre of the cube. In the case of face-centred cube, atoms are present at corners and at the centres of all faces of cube.

Tetragonal crystal system

In this crystal system, two lengths of the unit cell edges are equal whereas the third length is different. The three edges are perpendicular to one another, i.e. $a = b = c$ and $\alpha = \beta = \gamma = 90°$. In tetragonal system, there are two Bravais lattices; they are simple and body-centred. These are shown in Fig. 22.14. Examples for tetragonal crystal systems are TiO_2, SnO_2, etc.

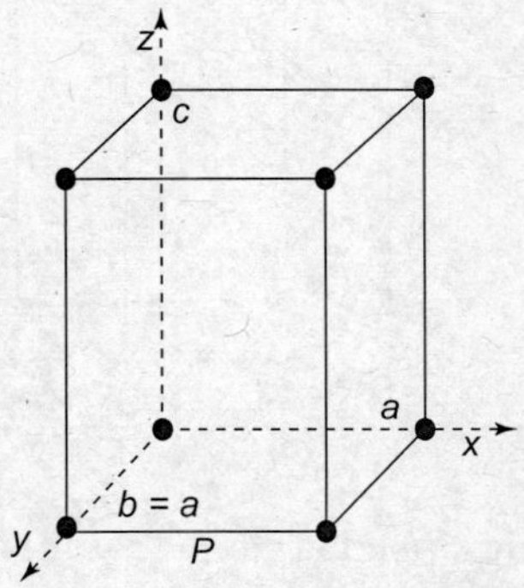

FIGURE 22.14 Tetragonal crystal system

Orthorhombic crystal system

In this crystal system, unit cell edge lengths are different and they are perpendicular to one another, i.e. $a \neq b \neq c$ and $\alpha = \beta = \gamma = 90°$. There are four Bravais lattices in this system. They are simple, face-centred, body-centred and base-centred. These are shown in Fig. 22.15. Examples for orthorhombic crystal system are $BaSO_4$, K_2SO_4, $SnSO_4$, etc.

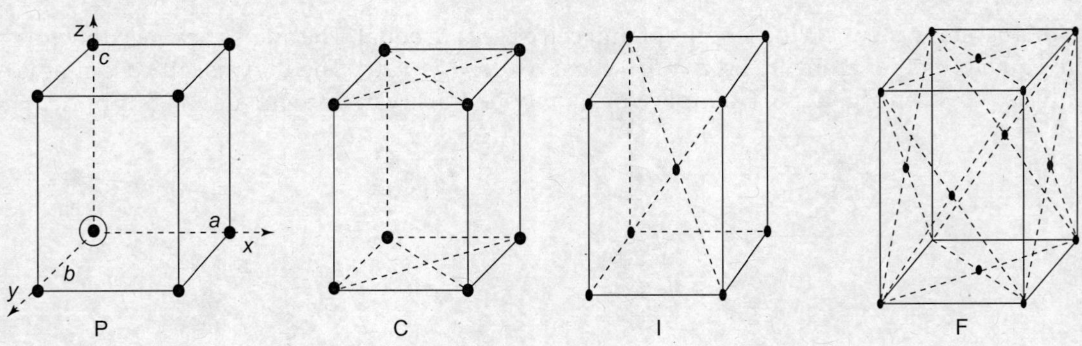

FIGURE 22.15 Orthorhombic crystal system

Monoclinic crystal system

In this crystal system, the unit cell edge lengths are different. Two unit cell edges are not perpendicular, but they are perpendicular to the third edge, i.e. $a \neq b \neq c$; $\alpha = \gamma = 90° \neq \beta$. This crystal system has two Bravais lattices; they are simple and base-centred. These are shown in Fig. 22.16. Examples for monoclinic crystal system are $CaSO_4 \cdot 2H_2O$ (gypsum), Na_3AlF_6 (cryolite), etc.

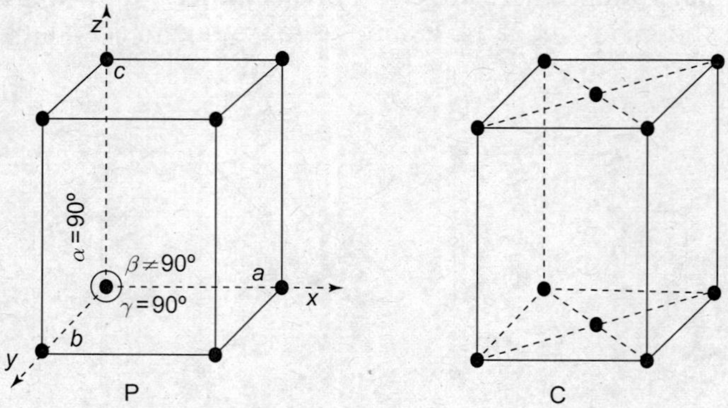

FIGURE 22.16 Monoclinic crystal system

Triclinic crystal system

In this crystal system, the unit cell edge lengths are different and are not perpendicular, i.e. $a \neq b \neq c$ and $\alpha \neq \beta \neq \gamma \neq 90°$ and all the angles are different. This crystal exists in primitive cell only. This is shown in Fig. 22.17. Examples for triclinic crystal system are $K_2Cr_2O_7$, $CuSO_r$ $5H_2O$, etc.

Rhombohedral (Trigonal) crystal system

In this crystal system, all the lengths of unit cell edges are equal. The angles between the axes are equal but other than 90°, i.e. $a = b = c$ and $\alpha = \beta = \gamma \neq 90°$. The Bravais lattice is simple only as shown in Fig. 22.18. Examples for rhombohedral crystal system are As, Bi, Sb, etc.

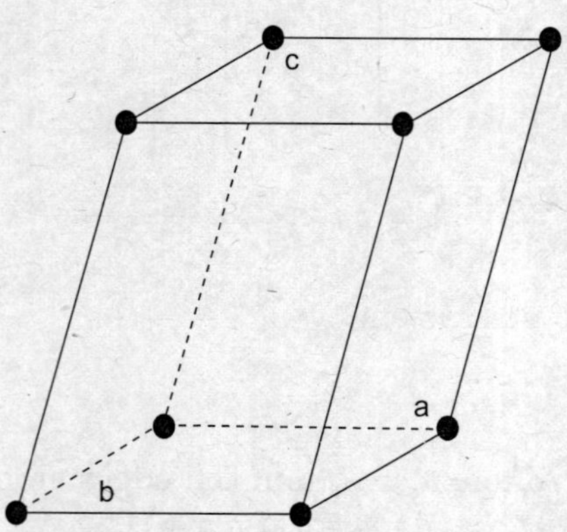

FIGURE 22.17 Triclinic crystal system

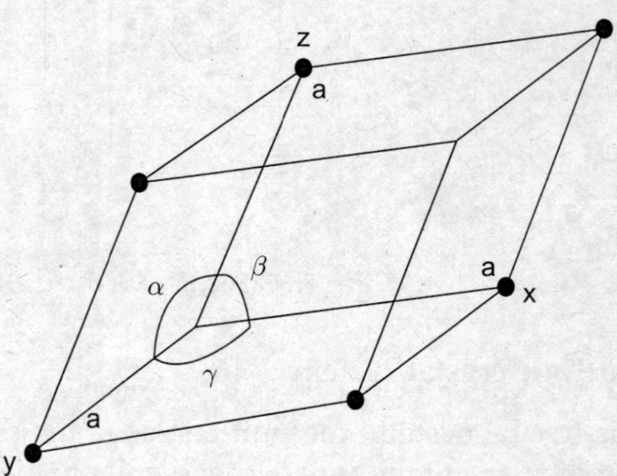

FIGURE 22.18 Rhombohedral crystal system

Hexagonal crystal system

In this crystal system, two sides of the unit cell edge lengths are equal and the angle between these edges is 120°. These two edges are perpendicular to the third edge and are not equal in length, i.e. $a = b \neq c$ and $\alpha = \beta = 90°, \gamma = 120°$. The Bravais lattice is primitive only. This is shown in Fig. 22.19. The atoms in this crystal system are arranged in the form of a hexagonal close pack.

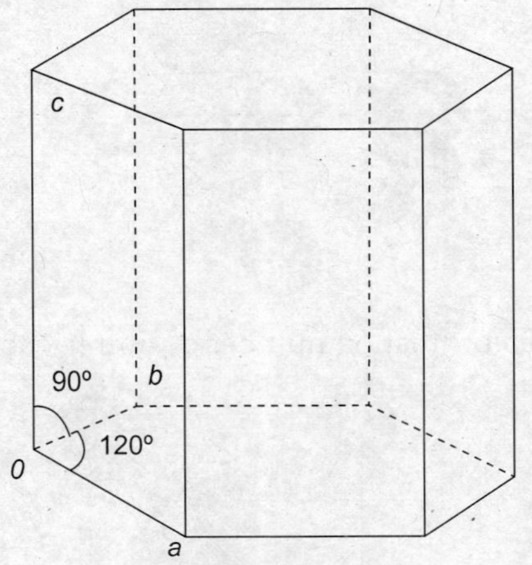

FIGURE 22.19 Hexagonal crystal system

22.5.1 Crystal Parameters

Nearest neighbour distance (2r)

The distance between the centres of two nearest neighbour atoms (Fig. 22.20) is known as the nearest neighbour distance. For example, when two atoms are touching each other, then the nearest neighbour distance will be the distance between the centres of the two atoms, i.e. $a = 2r$, where r is the radius of the atom.

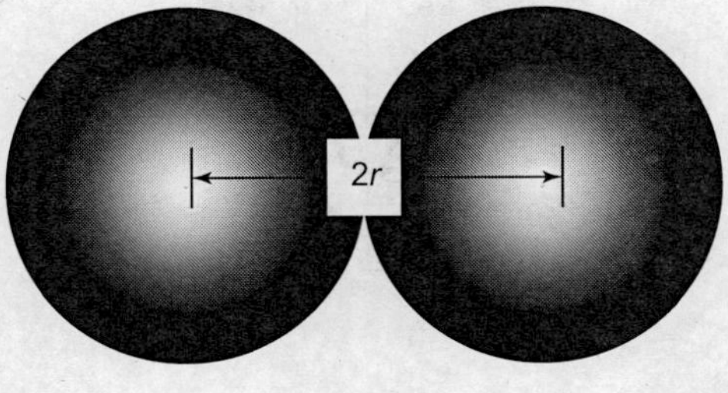

FIGURE 22.20 Distance between the centres of two atoms

Atomic radius (r)

From the definition of nearest neighbour distance, we can observe that the atomic radius is half the distance between the nearest neighbouring atoms that are touching each other, i.e. $r = a/2$.

Co-ordination number (CN)

In a crystal, the number of equidistant nearest neighbouring atoms that a reference atom has in the lattice is known as the co-ordination number, *N*. The co-ordination numbers for the three types of cubic systems are shown in Figs. 22.21–22.23.

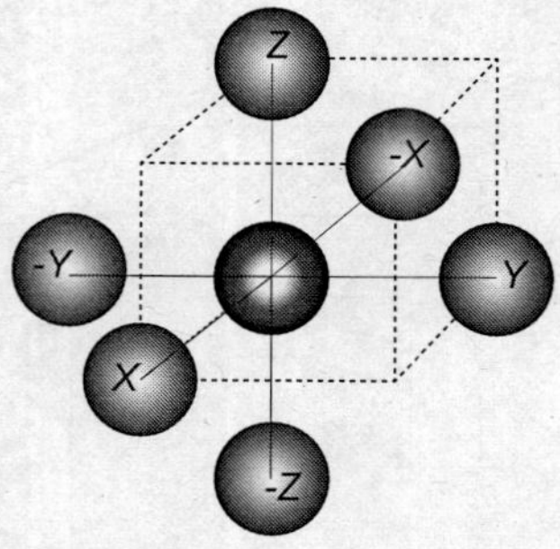

FIGURE 22.21 Co-ordination number of simple cubic structure

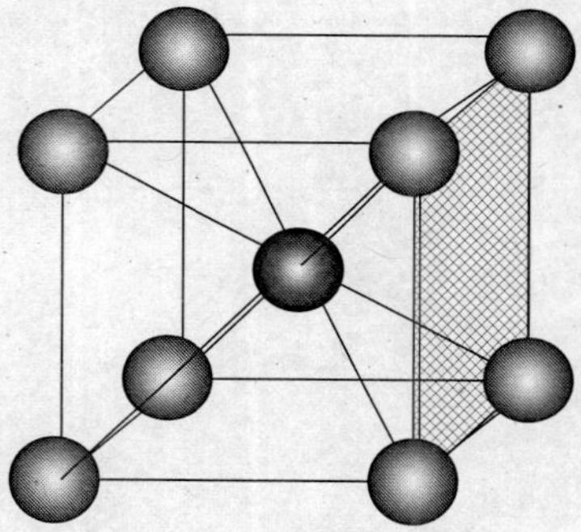

FIGURE 22.22 Co-ordination number for body-centred cubic structure

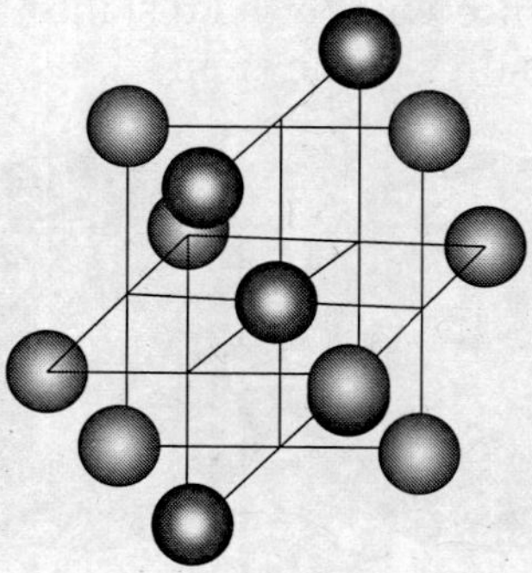

FIGURE 22.23 Co-ordination number for face-centred cubic structure

Atomic packing factor (packing fraction)

Atomic packing factor is defined as the ratio of the volume (v) occupied by the effective number of atoms in a unit cell to the total volume (V) of the unit cell, i.e.,

$$\text{Atomic packing fraction (APF)} = \frac{\text{Volume of all (effective number) the atoms in the unit cell}}{\text{Volume of the cell}} \quad (22.1)$$

22.6 ▶ Calculation of Atomic Packing Fraction for Various Systems

To find the atomic packing fraction for the crystal structures, the following steps are to be followed:

- Find the relationship between the radius (r) and the lattice constant (a) of the structure.
- Determine the effective number of atoms within the structure.
- Find the total volume of the effective number of atoms.
- Divide the effective atom volume with the total volume of the structure.
- Convert the ratio in terms of percentage by multiplying with 100.

22.6.1 Simple Cubic

A simple cubic structure is formed by arranging eight atoms touching each other at the eight corners of a cube (Fig. 22.24a). Since the atoms touch each other in simple cubic, the nearest neighbour distance (Fig. 22.24b) is

$$2r = a$$

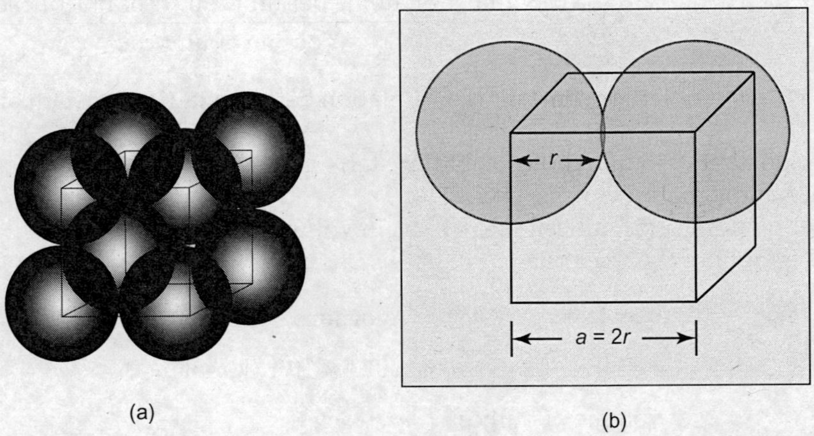

FIGURE 22.24 (a) Simple cubic structure; (b) arrangement of atoms on the face of the SC structure

Therefore, the lattice constant is

$$\boxed{a = 2r} \quad (22.2)$$

1. **Co-ordination Number (CN):** Number of equidistance nearest neighbour points that an atom has in the given structure. Greater the co-ordination number more closely the structure will be packed.

2. **Nearest Neighbour Distance (2s):** The distance between the centres of the two nearest neighbouring atoms is known as nearest neighbour distance or lattice constant (a)

$$r = \text{Atomic radius}$$

$$a = \text{Nearest neighbour distance} = \text{Lattice constant}$$

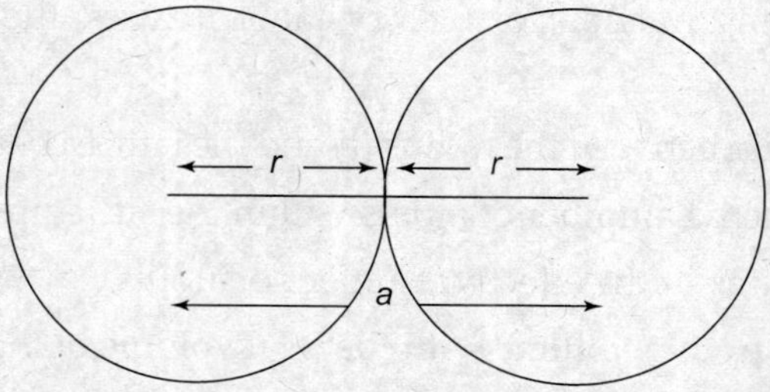

3. **Atomic Packing Factor or Atomic Packing Density (APF):** Ratio of volume of atoms per unit cell to the total volume of the unit cell.

OR

The fraction of space occupied by atoms in a unit cell, i.e.,

i.e., $$\text{APF} = \frac{\text{Volume of all atoms in unit cell}}{\text{Volume of unit cell}}$$

$\therefore$ $$\text{APF} = \frac{\text{Number of atoms per unit cell} \times \text{Volume of each atom}}{\text{Volume of unit cell}}$$

*Calculation of lattice constant (OR) Relation between lattice constant and density

Let

a = Lattice constant of material (i.e., cubic crystal)
δ = *Density of the material of the crystal
n = Number of atoms per unit cell (i.e., atoms in volume a^3 of the unit cell)
M = Atomic weight of the material

$$N_a = \text{Avogadro's number}$$
$$= 6.023 \times 10^{26}/\text{k mole}$$

$$\text{Volume of unit cell} = a^3$$

$\therefore$ $$\boxed{\textbf{Mass in each unit cell} = \delta a^3} \qquad \left[\because \; \delta = \frac{M}{V}\right] \tag{22.3}$$

$$\text{Mass of each atom} = \frac{M}{N_A}$$

$\therefore$ $$\boxed{\textbf{Mass in each unit cell} = n \times \frac{M}{N_A}} \tag{22.4}$$

Equating Eqs. (22.3) and (22.4), we get

$$\delta a^3 = n \times \frac{M}{N_A}$$

$$\therefore \quad a^3 = \frac{n}{\delta}\frac{M}{N_A}$$

$$\therefore \quad \boxed{a = \left[\frac{n}{N_A}\frac{M}{\delta}\right]^{1/3}} \tag{22.5}$$

This is the required relation between the lattice constant and the density of the material of crystal.

Note

$\frac{M}{\delta} = VA =$ molar volume contained N_A atoms

22.6.2 Simple Cubic (SC) Structure

The unit cell edge lengths of this structure along the crystallographic axes and inter-axial angles are equal (i.e., $a = b = c$ and $\alpha = \beta = \gamma = 90°$). Atoms are present only at the corners of this unit cell. A corner atom is shared by eight unit cells, so that the contribution of a corner atom to a unit cell is 1/8. The cube has eight corners; hence, the contribution of eight corner atoms to a unit cell or the number of atoms per unit cell $= \frac{1}{8} \times 8 = 1$.

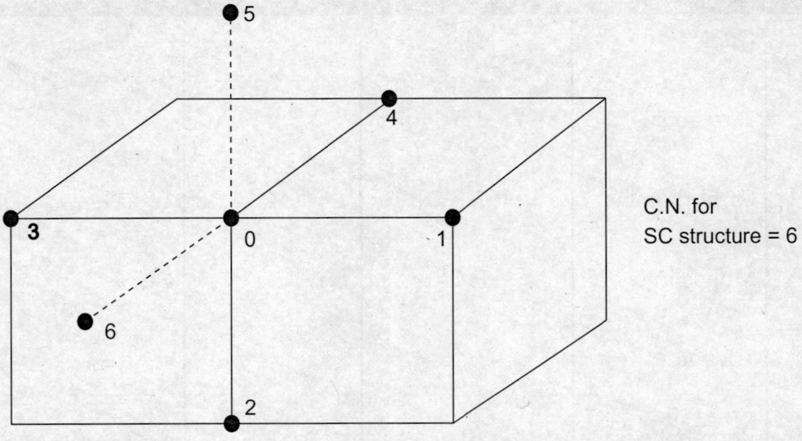

Let 'r' be the radius of an atom. The surfaces of the atoms touch along the cube edges. So, the distance between the centres of two neighbouring atoms or the nearest neighbour distance ($2r$) is equal to the lattice constant 'a'. In simple cubic cell, the number of nearest neighbour atoms to an atom or co-ordination number is 6. Since atoms are present at a distance of 'a' along $\pm X$, $\pm Y$ and $\pm Z$ directions. The number of nearest equidistant

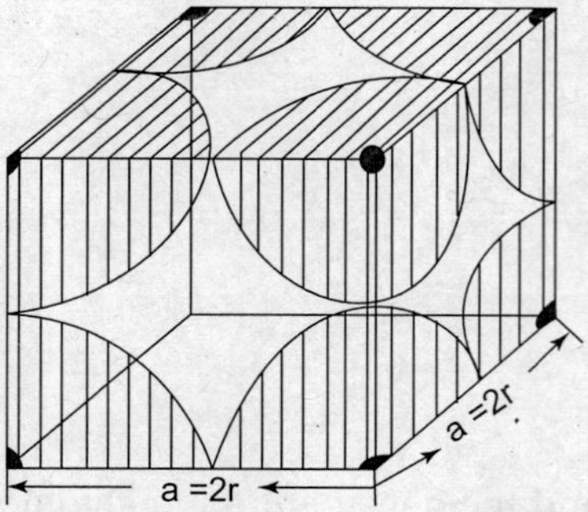

FIGURE 22.25 Simple cubic structure

neighbouring atoms to an atom in the structure is called co-ordination number. Figure 22.25 shows the simple cubic structure. Next, we find the fraction of the unit cell volume occupied by the atoms. The simple cubic structure contains only one atom per unit cell.

The volume occupied by atoms in the unit cell $(v) = 1\times\frac{4}{3}\pi r^3$ and

The volume of unit cell $(V) = a^3$. Hence, the packing factor or density of packing in the unit cell

$$\boxed{(\mathbf{PF}) = \frac{v}{V} = \frac{\frac{4}{3}\pi r^3}{a^3}} \tag{22.6}$$

$$= \frac{4}{3}\frac{\pi r^3}{(2r)^3} = \frac{\pi}{6}$$

$$= 0.52 \text{ or } 52\%$$

Atomic packing factor is defined as the fraction of the space occupied by atoms in the unit cell or it is the ratio of the volume occupied by atoms in the unit cell to the unit cell volume. An example for simple cubic structure is polonium crystal.

TABLE 22.5 The 14 Bravais lattices of 7 crystal systems

Sl. No.	Crystal system	Types of Bravais lattices	No. of Bravais lattices	Relation between lengths and angles
1.	Cubic	P, I, F	3	$a = b = c$ $\alpha = \beta = \gamma = 90°$
2.	Tetragonal	P, I	2	$a = b \neq c$ $\alpha = \beta = \gamma = 90°°$
3.	Orthorhombic	P, I, F, C	4	$a \neq b \neq c$ $\alpha = \beta = \gamma = 90°$
4.	Monoclinic	P, C	2	$a \neq b \neq c$ $\alpha = \gamma = 90° \neq \beta$
5.	Triclinic	P	1	$a \neq b \neq c$ $\alpha \neq \beta \neq \gamma$
6.	Rhombohedral (trigonal)	P	1	$a = b = c$ $\alpha = \beta = \gamma \neq 90°$
7.	Hexagonal	P	1	$a = b \neq c$ $\alpha = \beta = 90°$ $\gamma = 120°$

22.6.3 Body-Centered Cubic (BCC) Structure

For this unit cell, atoms are present at the corners of the cube and one atom is completely present at the centre of the unit cell. The centre of the unit cell is defined as the intersecting point of two body diagonals (AD and BE as shown in Fig. 22.26). A corner atom is shared by eight unit cells so that the contribution of a corner atom to a unit cell is 1/8. Therefore, the number of atoms per unit cell $= (1/8) \times 8 + 1 = 2$. The centre atom is surrounded by eight corner atoms, so the co-ordination number is 8. The surfaces of unit cell corner atoms may not touch, but they are in contact with the centre atom, i.e. the surfaces of atoms are in contact along a body diagonal of the unit cell. Half the distance between the centres of a corner atom and central atom is equal to the radius (r) of an atom. The relation between unit cell edge length (a) and radius (r) of an atom can be obtained with reference to Fig. 22.26 (b).

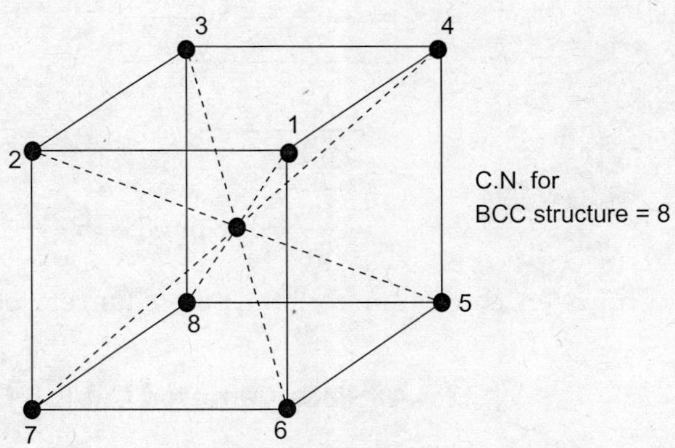

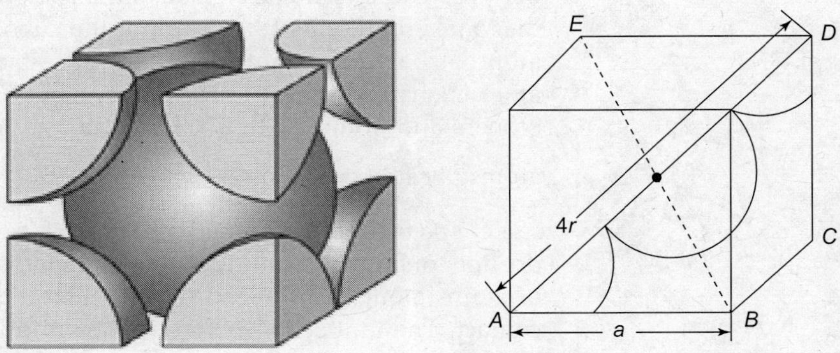

(a) (b)

FIGURE 22.26 Body-centred cubic structure

The length of the body diagonal AD $= 4r$

$$\therefore \quad AD^2 = AC^2 + CD^2 = AB^2 + BC^2 + CD^2$$

$$= a^2 + a^2 + a^2 = 3a^2$$

$$(4r)^2 = 3a^2$$

$$4r = \sqrt{3}a$$

or $$a = \frac{4}{\sqrt{3}}r$$

$$\text{Lastly, packing factor (PF)} = \frac{\text{Volume of all atoms in unit cell}}{\text{volume of unit cell}}$$

$$= \frac{v}{V}$$

$$\boxed{= \frac{2 \times \frac{4}{3}\pi r^3}{a^3} = \frac{8\pi r^3}{3a^3}} \tag{22.7}$$

$$= \frac{8\pi r^3 3\sqrt{3}}{3(4r)^3}$$

$$= \frac{\sqrt{3}\pi}{8} = 0.68 \text{ or } 68\%$$

The elements such as tungsten, chromium, sodium, potassium, etc. possess bcc structure.

22.6.4 Face-Centered Cubic (FCC) Structure

Atoms are present at the corners and at the face centres of this cubic structure. The intersection of face diagonals represents face centre of the cube. A corner atom is shared by eight unit cells and a face-centred atom is shared by two unit cells. The cube has eight corners and bounded by six faces; so, the number of atoms per unit cell $= \frac{1}{8} \times 8 + \frac{1}{2} \times 6 = 4$.

Let r be the radius of an atom. The surfaces of atoms do not touch along unit cell edges but the surfaces of atoms along face diagonals of this structure are in contact. The unit cell structure is shown in Fig. 22.27. Half of the nearest neighbour distance along the face diagonal is equal to radius of an atom.

The relation between the radius of an atom and unit cell edge length of a unit cell can be obtained with reference to Fig. 22.27.

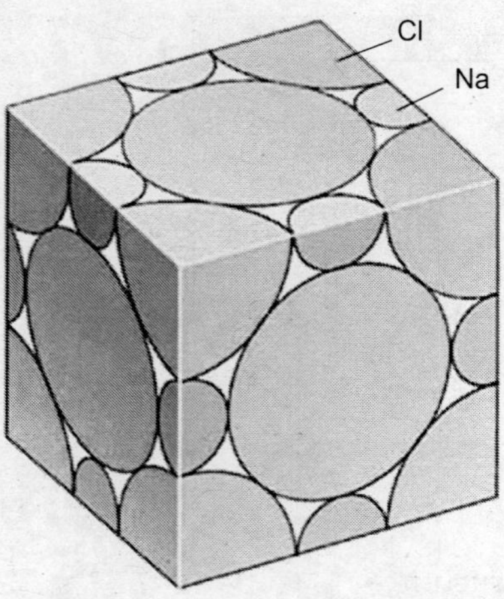

FIGURE 22.27 FCC structure

$$AC^2 = AB^2 + BC^2.$$

$$(4r)^2 = a^2 + a^2 = 2a^2$$

$$4r = \sqrt{2}a \quad \text{or} \quad a = 2\sqrt{2}r$$

The co-ordination number is 12, and this can be explained in Fig. 22.28.

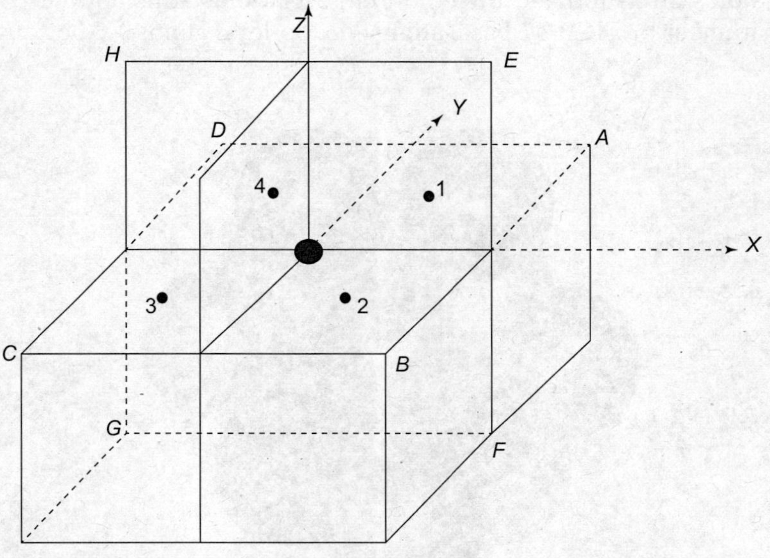

FIGURE 22.28 C.N. for FCC structure = 12

Note

There are three planes, *XY* plane named as *ABCD*, *XZ* plane named as *EFGH* and the third is *YZ* plane (not named). Each plane has 4 FCC atoms (numbered as 1234 for *XY* plane, not numbered for others).

A face-centred atom of the cubic structure is surrounded by four corner atoms of the face of a unit cell, four surrounding face-centred atoms of the unit cell and four surrounding face-centred atoms of the adjacent unit cell. Therefore, the co-ordination number is 12.

The packing factor PF of the unit cell

$$= \frac{\text{volume occupied by all the atoms in a unit cell}}{\text{unit cell volume}}$$

$$\boxed{= \frac{v}{V} = \frac{4 \times \frac{4}{3}\pi r^3}{a^3} = \frac{16\pi r^3}{3(2\sqrt{2}r)^3}} \qquad (22.8)$$

$$= \frac{16\pi r^3}{3 \times 8 \times 2\sqrt{2}r^3} = \frac{\pi}{3\sqrt{2}} = 0.74 \text{ or } 74\%$$

The packing factor of FCC structure is 74%. Examples for this structure are Cu, Ag, Al, etc.

22.6.5 HCP Crystal Structure

The co-ordination number for HCP structure (Fig. 22.29) is as follows: the base-centred atom has six atoms on the basal plane touching it. At the same distance, there are three middle-level atoms on its top and three middle-level atoms at its bottom. Therefore, the co-ordination number for HCP is 6 basal atoms + 3 top-level atoms + 3 bottom-level atoms = 12 atoms.

FIGURE 22.29 Co-ordination number for HCP structure

No.	Characteristics (Parameters)	SC	BCC	FCC	HCP
1.	Co-ordination number (CN)	6	8	12	12
2.	Atoms per unit cell (Z)	1	2	4	6
3.	Atomic radius (r)	$\frac{a}{2}$	$\frac{a\sqrt{3}}{4}$	$\frac{a}{2\sqrt{2}}$	$\frac{a}{2}$
4.	Unit cell volume (V)	a^3	a^3	a^3	$3\sqrt{2}a^3$
5.	Atomic packing factor (APF)	$\frac{\pi}{6} = 0.52$	$\frac{\pi\sqrt{3}}{8} = 0.68$	$\frac{\pi}{3\sqrt{2}} = 0.74$	$\frac{\pi}{3\sqrt{2}} = 0.74$
6.	Atomic packing fraction	52%	68%	74%	74%
7.	Void space [(1 – APF) ↔ 100]%	48%	32%	26%	26%
8.	Density (δ)	$\frac{M}{N_A a^3}$	$\frac{2M}{N_A a^3}$	$\frac{4M}{N_A a^3}$	$\frac{\sqrt{2}M}{N_A a^3}$

EXAMPLE 22.1

Chromium has BCC structure. Its atomic radius is 0.1249 nm. Calculate the free volume/unit cell.

Solution: Given data are
Atomic radius of chromium, $r = 0.1249$ nm.
Free volume/unit cell = ?
If 'a' is the BCC unit cell edge length, then the relation between 'a' and 'r' is

$$a = \frac{4}{\sqrt{3}} r = \frac{4}{\sqrt{3}} \times 0.1249 \text{ nm}$$

$$= 0.28845 \text{ nm} = 0.28845 \text{ Å}.$$

Volume of unit cell,

$$V = a^3 = (0.28845)^3 \text{ nm}^3$$

$$= 0.024 \text{ nm}^3 = 24 \text{ (Å)}^3$$

Number of atoms in BCC unit cell = 2

Hence volume of atoms in unit cell,

$$v = \frac{4}{3}\pi r^3 \times 2 = 0.01633 \text{ nm}^3$$

$$= 16.33 \text{ Å}$$

$$\text{Free volume/unit cell} = V - v = 0.00767 \text{ nm}^3$$

$$= \mathbf{7.67 \ (Å)^3}$$

22.7 ▶ Some Important Crystal Structure

22.7.1 Diamond Crystal Structure

The structure of diamond crystal unit cell is considered as the interpenetration of two FCC carbon unit cells along their body diagonals by a distance of $1/4^{\text{th}}$ the diagonal length. The origin of one unit cell is at (000) and the origin of another unit cell is at $\left(\frac{a}{4}, \frac{a}{4}, \frac{a}{4}\right)$. The diamond lattice structure and the projection of lattice points on bottom face of the cube are shown in Fig. 22.30. The fractions denote height of lattice points above the bottom face, in terms of cube edge.

In diamond, each carbon atom is bonded covalently with other surrounding four carbon atoms so that the number of nearest neighbours or the co-ordination number is 4. The surfaces of diamond unit cell corner atom placed at point X and the nearest diagonal atom placed at point Y touch each other, and hence $XY = 2r$, where r is the radius of each atom.

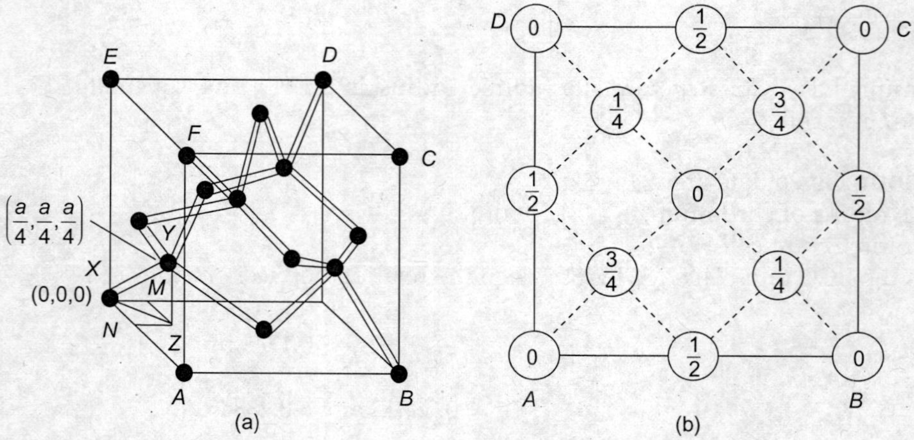

FIGURE 22.30 (a) Diamond lattice structure; (b) Projection of diamond lattice points on bottom face

The relation between radius of an atom and unit cell edge length 'a' can be given with reference to Fig. 22.30(a).

$$(XY)^2 = (XZ)^2 + (ZY)^2 = (XN)^2 + (NZ)^2 + (ZY)^2$$

$$(2r)^2 = \left(\frac{a}{4}\right)^2 + \left(\frac{a}{4}\right)^2 + \left(\frac{a}{4}\right)^2 = 3\left(\frac{a}{4}\right)^2$$

$$2r = \sqrt{3}\,\frac{a}{4} \quad \text{(or)} \quad a = \frac{8r}{\sqrt{3}}$$

The number of atoms in the unit cell can be estimated in the following way. Atoms are present at eight corners of the cube, six face centres of the cube and four atoms are completely present in the unit cell along the four body diagonal at $1/4^{\text{th}}$ distance. Out of the four, two start from one diagonally opposite atoms (X and B) of the bottom face while the other two from diagonally J opposite atoms (D and F) from the top face to the cube. Therefore, the total number of atoms per unit cell $= \frac{1}{8} \times + \frac{1}{2} \times 6 + 4 = 8.$

Finally,
$$\text{PF} = \frac{\text{volume occupied by all atoms of the unit cell } (v)}{\text{volume of unit cell } (V)}$$

$$\text{PF} = \frac{\text{Number of atoms per unit cell} \times \text{volume of each atom}}{\text{volume of unit cell}}$$

$$= \frac{8 \times \frac{4}{3}\pi r^3}{a^3}$$

$$= \frac{32}{3} \frac{\pi r^3}{\left(\frac{8\pi}{\sqrt{3}}\right)}$$

$$= \frac{32\pi r^3 3\sqrt{3}}{3 \times 8 \times 8 \times 8 r^3}$$

$$= \frac{\pi\sqrt{3}}{16} = 0.34 \text{ or } 34\%.$$

This is a loosely packed structure; C, Ge and Si crystallize in this form.

1. The strength of the structure lies in its covalent bonds.
2. While most minerals are made from ionic bonds, diamond is made of pure carbon and has only covalent bonds.
3. Diamond has an FCC structure, where each carbon atom is surrounded by four other carbon atoms, arranged in a tetrahedral matrix that offers the most stable natural shape.

Diamonds have a range of unique properties:

- Diamond is the hardest known natural material on Earth.
- Diamond conducts heat five times faster than copper.
- Pure diamond is the most transparent material known. It reflects visible light, ultraviolet light and infrared light and also light of all the frequencies in between.
- Diamond can be a conductor or an insulator; so, it can either allow electricity to pass through it or block it.

Applications of diamond

- **Telecommunications:** Diamond cubes can allow several signals at once to pass along a cable, all at different wavelengths of light. This makes them useful in telecommunications. They can also withstand high-energy conditions and changes in temperature.
- **Heat spreaders:** Heat passes through diamonds far more quickly than it passes through copper. This makes it useful as heat spread in places where lots of heat is generated in a small space. Microelectronic devices are an example.
- **Surgery:** Diamond blades on scalpels have an ultrafine edge that can cut very precisely, which is important for surgery. Diamond also remains sharp for much longer than steel. As it cuts, it can transmit laser light, which allows it to cauterize the wound to control bleeding.
- **Health care:** Diamond is made of carbon and so it is the ideal material to use in our carbon-based bodies. Scientists are developing diamond implants that could help monitor health or take over the role of failing tissues.

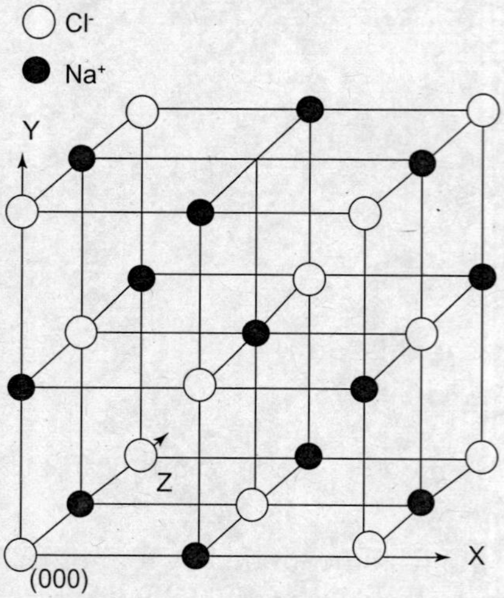

FIGURE 22.31 NaCl lattice structure

22.7.2 Sodium Chloride Crystal Structure

The Bravais lattice of NaCl crystal is a face-centred cube as shown in Fig. 22.31. NaCl is an ionic crystal. It consists of two FCC sub-lattices, one of Cl^- ions having origin at (000) and the other of Na^+ ion having origin at one half of the body diagonal of the unit cell. The Na^+ and Cl^- ions are present at equilibrium distances in the crystal. The attractive force between Na^+ and Cl^- ions in the crystal is balanced by the repulsive forces between the same type of charges. In this crystal, either Na^+ or Cl^- ions could occupy the corner and face-centred positions with the ions of the opposite type occupying the alternate positions in between them. Each ion is surrounded by six other type ions so that the co-ordination number is 6. If Cl^- ions occupy corners of the unit cell, one corner is taken as the origin and the co-ordinates of ions are expressed in fractions of the edge length of the cube. The co-ordinates of Cl^- ions are

$$(0,0,0), \left(\frac{1}{2},0,\frac{1}{2}\right), \left(0,\frac{1}{2},\frac{1}{2}\right), \left(\frac{1}{2},\frac{1}{2},0\right).$$

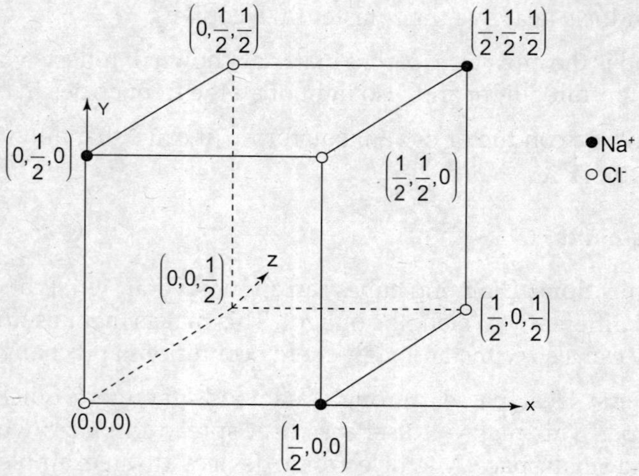

FIGURE 22.32

In the same co-ordinate system, the co-ordinates of the four Na^+ ions are

$$\left(\frac{1}{2},\frac{1}{2},\frac{1}{2}\right), \left(\frac{1}{2},0,0\right), \left(0,\frac{1}{2},0\right), \left(0,0,\frac{1}{2}\right).$$

The unit cell of NaCl crystal contains four sodium and four chlorine ions. Therefore, the unit cell contains four NaCl molecules. The other crystals, which show this type of structure, are KC1, KBr, MgO, AgBr, etc.

22.7.3 Zinc Blende Crystal Structure

This is an ionic compound. This structure consists of two different types of atoms. The Bravais lattice of ZnS is a face-centred cube. The crystal structure of ZnS is considered as the interpenetration of two FCC sub-lattices along the body diagonal such that the origin of one sub-lattice is at a distance of $1/4^{th}$ the diagonal length. One sub-lattice contains Zn atoms and the other sub-lattice contains S atoms. The ZnS crystal structure is similar to diamond crystal structure. The unit cell contains four ZnS molecules. Each atom is surrounded by four other kinds of atoms at the corners of a regular tetrahedron. The co-ordinates of S atom are $\left(\frac{1}{4},\frac{1}{4},\frac{1}{4}\right)$ and that of Zn atom co-ordinates are

$$(0,0,0),\left(\frac{1}{2},0,\frac{1}{2}\right),\left(\frac{1}{2},\frac{1}{2},0\right) \text{ and } \left(0,\frac{1}{2},\frac{1}{2}\right).$$

Other examples of this structure are CdS, InSb, CaCl, GaAs, GaP, etc.

22.8 ▶ Miller Indices

22.8.1 Miller Indices for Direction in Crystal

Crystal planes are defined as some imaginary planes inside a crystal in which large concentration of atoms is present. Inside the crystal, there exists certain directions along which large concentration of atoms exists. These directions are called *crystal directions.* Figure 22.33 shows a two-dimensional lattice with different orientations of crystal planes.

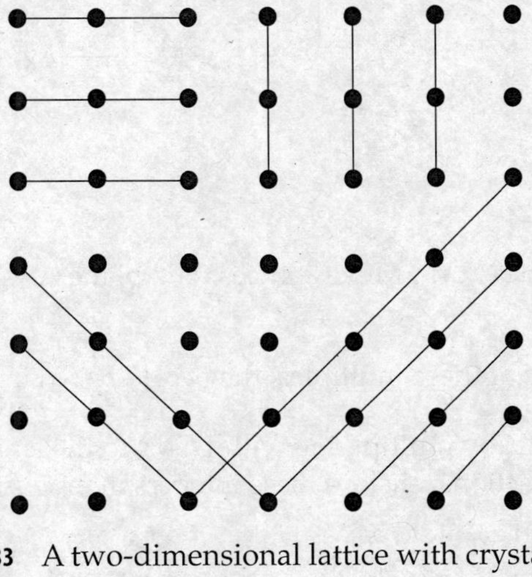

FIGURE 22.33 A two-dimensional lattice with crystal planes

Crystal planes and directions can be represented by a set of three small integers called Miller indices (because Miller derived a method of representing crystal planes). These integers are represented in general as *h*, *k* and *l*. If these integers are enclosed in round brackets as (*hkl*), then it represents a plane. On the other hand, if they are enclosed in square brackets as [*hkl*], then it represents crystal direction perpendicular to the above-said plane. Next, we will see the way of obtaining Miller indices for a plane.

(i) As shown in Fig. 22.34, take a lattice point as origin '*O*' of crystallographic axes *X*, *Y* and *Z* in a space lattice. The unit cell translational distances or lattice constants along *X*, *Y* and *Z* directions are *a*, *b* and c, respectively. Let a crystal plane ABC intersect these three axes at 2*a*, 3*b* and *c*. In general, the intercepts can be represented as *pa*, *qb* and *rc*.

(ii) Divide these intercepts with lattice points translational distances along the axes to obtain intercepts of the plane in terms of multiples of unit cell translational lengths.

i.e., $\frac{2a}{a}, \frac{3b}{b}, \frac{c}{c}$, in general $\frac{pa}{a}, \frac{qb}{b}, \frac{rc}{c}$

2, 3, 1 $\qquad$ p, q, r

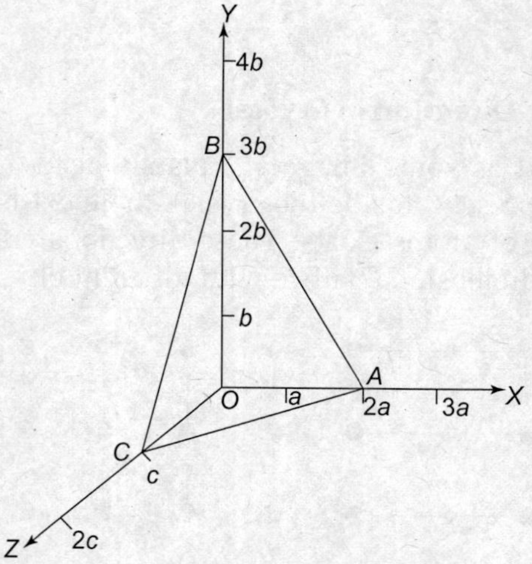

FIGURE 22.34 Miller indices for a plane ABC

(iii) Take the reciprocals of these multiples, they are $\frac{1}{2}, \frac{1}{3}, \frac{1}{1}$, in general $\frac{1}{p}, \frac{1}{q}, \frac{1}{r}$

(iv) Clear these fractions (by multiplying with LCM) to the smallest integers having the same ratio as the fractions, enclose these integers in brackets.

$\frac{1}{2}\times 6, \frac{1}{3}\times 6, \frac{1}{1}\times 6$ in general $\frac{pqr}{p}, \frac{pqr}{q}, \frac{pqr}{r}$

3 2 6 $\qquad$ $qr \quad pr \quad pq$

∴ Miller indices of the plane ABC is $(hkl) = (326)$.
In general, indices of a plane are represented as $(hkl) = (qr\ pr\ pq)$

or
$$\boxed{\frac{1}{p} : \frac{1}{q} : \frac{1}{r} = h : k : l} \tag{22.9}$$

Miller indices may be defined as a set of three small integers obtained by clearing the reciprocals of the three intercepts (in terms of multiples of unit cell edges) made by a plane on crystallographic axes.

22.8.2 Important Features of Miller Indices

(i) Miller indices represent a set of equidistant parallel planes.

(ii) If the Miller indices of a plane represent some multiples of Miller indices of another plane, then these planes are parallel. For example (844) and (422) or (211) are parallel planes.

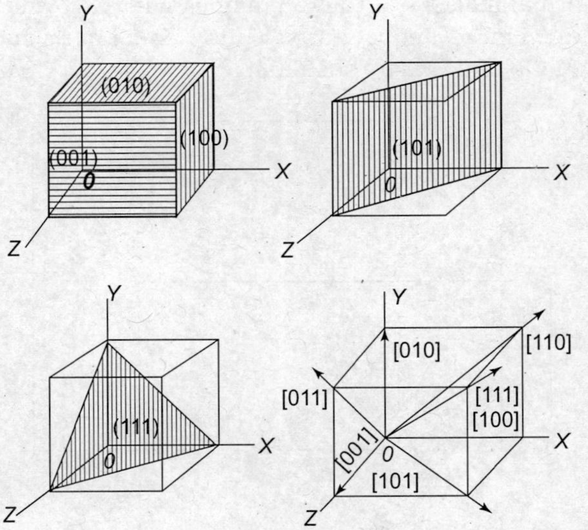

FIGURE 22.35 Some important crystal planes and directions in a cubic crystal

(iii) If (hkl) are the Miller indices of a plane, then the plane divides the lattice constant 'a' along the X-axis into h equal parts, 'b' along the Y-axis into k equal parts and 'c' along the Z-axis into l equal parts.

(iv) If a plane is parallel to one of the crystallographic axes, then the plane intersects that axis, at infinity and the Miller indices along that direction is zero.

(v) If a plane cuts an axis on the negative side of the origin, then the corresponding index is negative and is indicated by placing a minus sign above the index. For example, if the plane cuts on the negative Y-axis, then the Miller indices of the plane is $(h\bar{k}l)$.

(vi) When Miller indices are enclosed in curly brackets, {*hkl*}, they refer to planes which in the crystal are equivalent even though their Miller indices may differ (Fig. 22.35). For example in a cubic lattice, all cube faces are equivalent, they are (100), (010), (001), $(\bar{1}00)$, $(0\bar{1}0)$, $(00\bar{1})$; these planes are represented as {100}. Similarly, a full set of equivalent directions in a crystal is represented by a symbol <*hkl*>. For example, the eight body diagonals of a cube [111], $[\bar{1}\bar{1}\bar{1}]$, $[\bar{1}11]$, $[1\bar{1}1]$,$[11\bar{1}]$, $[\bar{1}\bar{1}1]$, $[\bar{1}1\bar{1}]$, $[1\bar{1}\bar{1}]$ are designated as <111>.

22.9 ▶ INTER-PLANAR SPACING

The separation between successive parallel planes in rectangular axes crystal system can be extracted easily. Let us consider a rectangular [cartesian] co-ordinate system with origin '*O*' at one of the lattice points. Let (*hkl*) be the Miller indices of a plane ABC, which makes intercepts OA, OB and OC on *X*, *Y* and *Z* axes, respectively, as shown in Fig. 22.36. A normal to this plane from the origin passes through a point *N* in the plane ABC, such that ON $= d_1$. This normal makes α', β' and γ' angles with *X*, *Y* and *Z* axes, respectively. Since the plane divides '*a*' into '*h*' equal parts, *b* into *k* equal parts and *c* into *l* equal parts, then the intercepts OA, OB and OC are such that

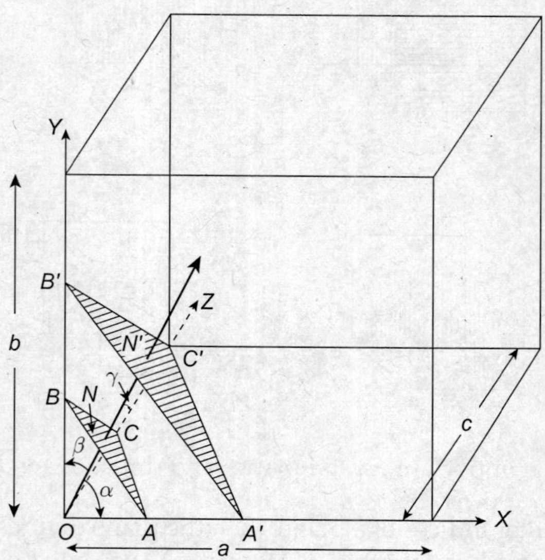

FIGURE 22.36 Orthorhombic crystal

$$\text{OA} = \frac{a}{h},\ \text{OB} = \frac{b}{k} \text{ and } \text{OC} = \frac{c}{l}$$

where *a*, *b*, *c* are the unit cell edge lengths along *X*, *Y* and *Z* axes, respectively. From Fig. 22.36,

$$\cos\alpha' = \frac{d_1}{OA}, \ \cos\beta' = \frac{d_1}{OB} \ \text{and} \ \cos\gamma' = \frac{d_1}{OC} \tag{22.10}$$

Let the co-ordinates of *N* be *x*, *y* and *z* along X, Y and Z axes, then

$$(ON)^2 = d_1^2 = x^2 + y^2 + z^2$$

Also from Fig. 22.36,

$$\cos\alpha' = \frac{x}{d_1}, \ \cos\beta' = \frac{y}{d_1}, \ \cos\gamma' = \frac{z}{d_1}$$

Substitute Eq. (22.10) in above equation gives

$$d_1^2 = d_1^2\cos^2\alpha' + d_1^2\cos^2\beta' + d_1^2\cos^2\gamma'$$

$$= d_1^2[\cos^2\alpha' + \cos^2\beta' + \cos^2\gamma']$$

Therefore,

$$\cos^2\alpha' + \cos^2\beta' + \cos^2\gamma' = 1$$

Substitute value of $(ON)^2$ in Eq. (22.10) gives

$$\frac{d_1^2}{(OA)^2} + \frac{d_1^2}{(OB)^2} + \frac{d_1^2}{(OC)^2} = 1 \tag{22.11}$$

Again substitute Eq. (22.10) in Eq. (22.11)

$$\frac{d_1^2 h^2}{a^2} + \frac{d_1^2 k^2}{b^2} + \frac{d_1^2 l^2}{c^2} = 1$$

(or)

$$d_1^2\left[\frac{h^2}{a^2} + \frac{k^2}{b^2} + \frac{l^2}{c^2}\right] = 1$$

$$d_1 = \frac{1}{\sqrt{\frac{h^2}{a^2} + \frac{k^2}{b^2} + \frac{l^2}{c^2}}} \tag{22.12}$$

Let $\left(\frac{h}{2}\frac{k}{2}\frac{l}{2}\right)$ be the Miller indices of the next plane A′ B′ C′, this plane makes intercepts OA′, OB′ arid OC′ on *X*, *Y* and *Z* axes, respectively. A normal from the origin to this plane passes through a point N′, so that $ON' = d_2$. As the extension of d_1 is d_2, it makes same angles α', β' and γ' with *X*, *Y* and *Z* axes, respectively. Since the plane segments '*a*' into $h/2$ equal parts, *b* into $k/2$ equal parts and *c* into $l/2$ equal parts, then the intercepts OA′, OB′ and OC′ are such that

$$\text{OA}' = \frac{a}{\left(h/2\right)} = \frac{2a}{h}, \ \text{OB}' = \frac{b}{\left(k/2\right)} = \frac{2b}{k} \ \text{and} \ \text{OC}' = \frac{c}{\left(l/2\right)} = \frac{2c}{l} \tag{22.13}$$

From Fig. 22.36,

$$\cos\alpha' = \frac{d_2}{(\text{OA}')}, \ \cos\beta' = \frac{d_2}{(\text{OB}')} \ \text{and} \ \cos\gamma' = \frac{d_2}{(\text{OC}')} \tag{22.14}$$

Let the co-ordinates of N' are x', y' and z' along X, Y and Z axes, respectively.

$$(\text{ON}')^2 = d_2^2 = x'^2 + y'^2 + z'^2 \tag{22.15}$$

Also from Fig. 22.36,

$$\boxed{\cos\alpha' = \frac{x'}{d_2}, \cos\beta' = \frac{y'}{d_2} \ \text{and} \ \cos\gamma' = \frac{z}{d_2}} \tag{22.16}$$

Substitute Eq. (22.15) in Eq. (22.14) gives

$$d_2^2 = d_2^2\cos^2\alpha' + d_2^2\cos^2\beta' + d_2^2\cos^2\gamma'$$

$$= d_2^2\left[\cos^2\alpha' + \cos^2\beta' + \cos^2\gamma'\right]$$

or

$$\cos^2\alpha' + \cos^2\beta' + \cos^2\gamma' = 1 \tag{22.17}$$

Substitute Eq. (22.14) in Eq. (22.17) gives

$$\frac{d_2^2}{(\text{OA}')^2} + \frac{d_2^2}{(\text{OB}')^2} + \frac{d_2^2}{(\text{OC}')^2} = 1$$

Again substitute Eq. (22.13) in Eq. (22.14) gives

$$\frac{d_2^2 h^2}{(2a)^2} + \frac{d_2^2 k^2}{(2b)^2} + \frac{d_2^2 l^2}{(2c)^2} = 1 \ \text{(or)} \ d_2^2\left[\frac{h^2}{(2a)^2} + \frac{k^2}{(2b)^2} + \frac{l^2}{(2c)^2}\right] = 1$$

or

$$d_2 = \frac{1}{\sqrt{\frac{h^2}{(2a)^2} + \frac{k^2}{(2b)^2} + \frac{l^2}{(2c)^2}}} \ \text{(or)} \ d_2 = \frac{2}{\sqrt{\frac{h^2}{a^2} + \frac{k^2}{b^2} + \frac{l^2}{c^2}}} \tag{22.18}$$

Let the separation between the planes ABC and A′B′C′ is ***'d'***

$$\therefore \quad d = d_2 - d_1 = \frac{1}{\sqrt{\frac{h^2}{a^2}+\frac{k^2}{b^2}+\frac{l^2}{c^2}}} \tag{22.19}$$

Using Eq. (22.19), we can determine the inter-planar separation in orthorhombic crystals. For tetragonal crystal $a = b \neq c$, substitute these values in Eq. (22.19), we have

$$d = \frac{1}{\sqrt{\frac{h^2}{a^2}+\frac{k^2}{a^2}+\frac{l^2}{c^2}}} = \frac{1}{\sqrt{\frac{h^2+k^2}{a^2}+\frac{l^2}{c^2}}} \tag{22.20}$$

For cubic crystals, $a = b = c$, substitute these values in Eq. (22.19), we have

$$\textbf{Inter-planer spacing } d = \frac{1}{\sqrt{\frac{h^2}{a^2}+\frac{k^2}{a^2}+\frac{l^2}{a^2}}} \quad \textbf{(or)} \quad d = \frac{a}{\sqrt{h^2+k^2+l^2}} \tag{22.21}$$

22.10 ▶ Liquid Crystals and Phases

Liquid crystals (LCs) are a state of matter that has properties between those of conventional liquid and solid crystals. LC may flow like a liquid, but its molecules may be oriented in a crystal-like way. The liquid crystal state is a distinct phase of matter observed between the crystalline (solid) and isotropic (liquid) states. There are many phase of liquid crystal states, depending upon the amount of order in the material. Some of them have been described below.

22.10.1 Phases of Liquid Crystal

Nematic phases

The nematic liquid crystal phase is characterized by molecules that have no positional order but tend to point in the same direction (along the director). In Fig. 22.37, notice that the molecules point vertically but are arranged with no particular order.

The nematic liquid crystals are composed of rod-like molecules with the long axes of neighbouring molecules aligned approximately parallel to one another. A dimension unit vector n called the director is introduced to represent the direction of preferred orientation of molecules in the neighbourhood of any point.

Liquid crystals are anisotropic materials, and the physical properties of the system vary with the average alignment with the director. If the alignment is large, the material is very anisotropic. Similarly, if the alignment is small, the material is almost isotropic.

FIGURE 22.37 Nematic phases

A special class of nematic liquid crystals is called chiral nematic. It refers to the unique ability to selectively reflect one component of circularly polarized light. The term chiral nematic is used interchangeably with cholesteric.

Smectic phases

The word "smectic" is derived from the Greek word for soap. This seemingly ambiguous origin is explained by the fact that the thick, slippery substance often found at the bottom of a soap dish is actually a type of smectic liquid crystal.

The smectic state is another distinct mesophase of liquid crystal substances. Molecules in this phase show a degree translational order not present in the nematic. In the smectic state, the molecules maintain the general orientational order of nematics, but also tend to align themselves in layers or planes. Motion is restricted to within these planes, and separate planes are observed to flow past each other. The increased order means that the smectic state is more "solid-like" than the nematic.

In the smectic-A mesophase (Fig. 22.38), the director is perpendicular to the smectic plane, and there is no particular positional order in the layer. Similarly, in the smectic-B mesophase, molecules are arranged as in the smectic-A mesophase, but the director is at a constant tilt angle measured normally to the smectic plane.

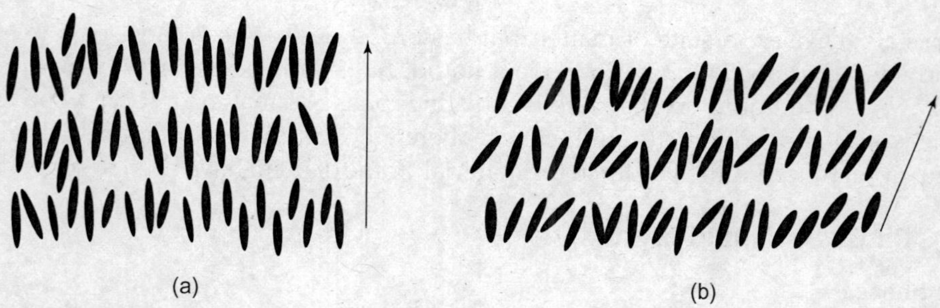

FIGURE 22.38 (a) Picture of the smectic-A phase. (b) Picture of the smectic-B phase

As in the nematic, the smectic-B (Fig. 22.38), mesophase has a chiral state designated C*. Consistent with the smectic-B, the director makes a tilt angle with respect to the smectic layer. The difference is that this angle rotates from layer to layer forming a helix. In other words, the director of the smectic-C* mesophase is not parallel or perpendicular to the layers, and it rotates from one layer to the next. Notice the twist of the director, represented by the black arrows, in each layer in Fig. 22.39.

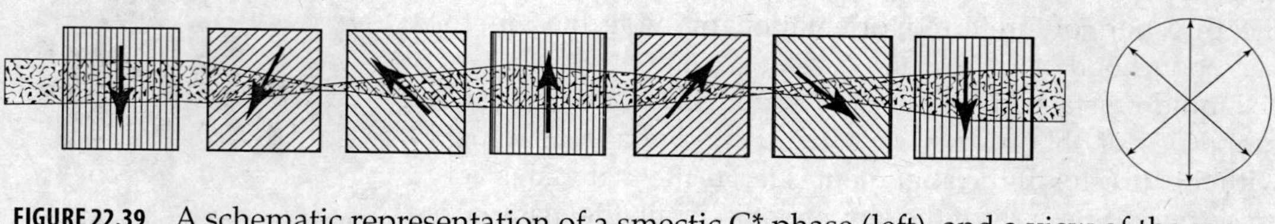

FIGURE 22.39 A schematic representation of a smectic C* phase (left), and a view of the same phase, but along the axis (right)

In some smectic mesophases, the molecules are affected by the various layers above and below them. Therefore, a small amount of three-dimensional orders is observed. Smectic-G is an example demonstrating this type of arrangement.

Cholesteric phases

The cholesteric (or chiral nematic) liquid crystal phase is typically composed of nematic mesogenic molecules containing a chiral centre, which produces intermolecular forces that favour alignment between molecules at a slight angle to one another. This leads to the formation of a structure which can be visualized as a stack of very thin 2-D nematic-like layers with the director in each layer twisted with respect to those above and below. In this structure, the directors actually form in a continuous helical pattern about the layer normal as illustrated by the black arrow in Fig. 22.39 and animation. The black arrow represents director orientation in the succession of layers along the stack.

The molecules shown are merely representations of the many chiral nematic mesogens lying in the slabs of infinitesimal thickness, with a distribution of orientation around the director. This is not to be confused with the planar arrangement found in smectic mesophases.

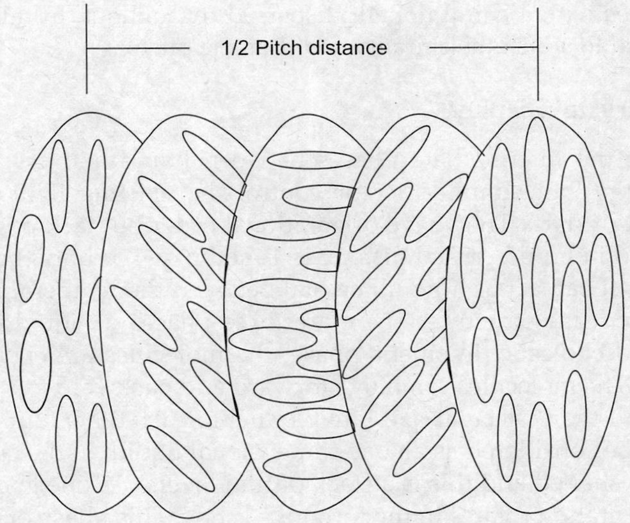

FIGURE 22.40 Cholesteric mesophase

An important characteristic of the cholesteric mesophase is the pitch. The pitch, *p*, is defined as the distance it takes for the director to rotate one full turn in the helix as illustrated in Fig. 22.40. A byproduct of the helical structure of the chiral nematic phase is its ability to selectively reflect light of wavelengths equal to the pitch length, so that a colour will be reflected when the pitch is equal to the corresponding wavelength of light in the visible spectrum, the effect is based on the temperature dependence of the gradual change in director orientation between successive layers (illustrated above), which modifies the pitch length resulting in an alteration of the wavelength of reflected light according to the temperature. The angle at which the director changes can be made larger, and thus tighten the pitch, by increasing the temperature of the molecules, hence giving them more thermal energy. Similarly, decreasing the temperature of the molecules increases the pitch length of the chiral nematic liquid crystal. This makes it possible to build a *liquid crystal thermometer* that displays the temperature of its environment by the reflected colour. Mixtures of various types of these liquid crystals are often used to create sensors with a wide variety of responses to temperature change. Such sensors are used for thermometers often in the form of heat-sensitive films to detect flaws in circuit board connections, fluid flow patterns, condition of batteries, the presence of radiation or in novelties such as "mood" rings.

In the fabrication of films, since putting chiral nematic liquid crystals directly on a black background would lead, to degradation and perhaps contamination, the crystals are micro-encapsulated into particles of very small dimensions. The particles are then treated with a binding material that will contract upon curing so as to flatten the microcapsules and produce the best alignment for brighter colours. An application of a class of chiral nematic liquid crystals which are less temperature sensitive is to create materials such as clothing, dolls, inks and paints.

The wavelength of the reflected light can also be controlled by adjusting the chemical composition, since cholesterics can either consist of exclusively chiral molecules or consist of nematic molecules with a chiral dopant dispersed throughout. In this case, the dopant concentration is used to adjust the chirality and thus the pitch.

22.10.2 Liquid Crystal Display

Liquid crystals are the intermediate phases between liquid and crystal. LCDs do not generate light energy, but simply alter or control the existing light to make selected areas appear bright or dark. They have orientational order but lack positional order. The material that is used in liquid crystal display is 4-methoxy-4′ n-butyl benzylidene aniline (MBBA) molecules. It can act as a liquid crystal between the temperatures of 21–48°C. It has an elongated rod-like structure. There are three phases in liquid crystals. They are smetic, nematic and cholestic. In smetic phase, the molecules are cigar shaped and are arranged in layers. The molecules can move forward and backward but not up and down. In nematic phase, all the molecular axes are parallel to each other, but it is not a layered structure. This can be considered as an one-dimensional liquid. In the cholestic phase, the molecules are plate shaped and this is a stack of thin layers. As one goes down the stack, the direction of orientation rotates in the form of a screw. This phase of crystals possesses double refraction. The other liquid crystal is pentyl cyano biphenyl. This has nematic form in the temperature range of 18–35°C.

22.10.3 Types of LC Display

There are two types of liquid crystal displays: (i) Dynamic scattering display and (ii) Twisted nematic field effect display. The dynamic scattering display is not presently used because of its short life time and larger power consumption during operation. In case of twisted nematic field effect display, a thin layer of liquid crystal material of 10–20 pm thick is kept in between two glass plates coated with transparent tin oxide on the inner side surfaces, which acts as electrodes as shown in Fig. 22.41(a). In the absence of applied electric field, the top glass plate is rotated through 90°; this causes the liquid crystal molecules also to be twisted through 90°. Above the top glass plate, a polarizer and below the bottom glass plate, an analyzer are kept in crossed positions. When light is allowed to pass through the liquid crystal cell and through the crossed polarizer and analyzer, the cell appears bright due to additional phase difference introduced by twisting. When an electric field is applied, the liquid crystal molecules orient themselves parallel to the field direction as shown in Fig. 22.41(b). Hence, the cell appears dark due to the crossed polarizer and analyzer.

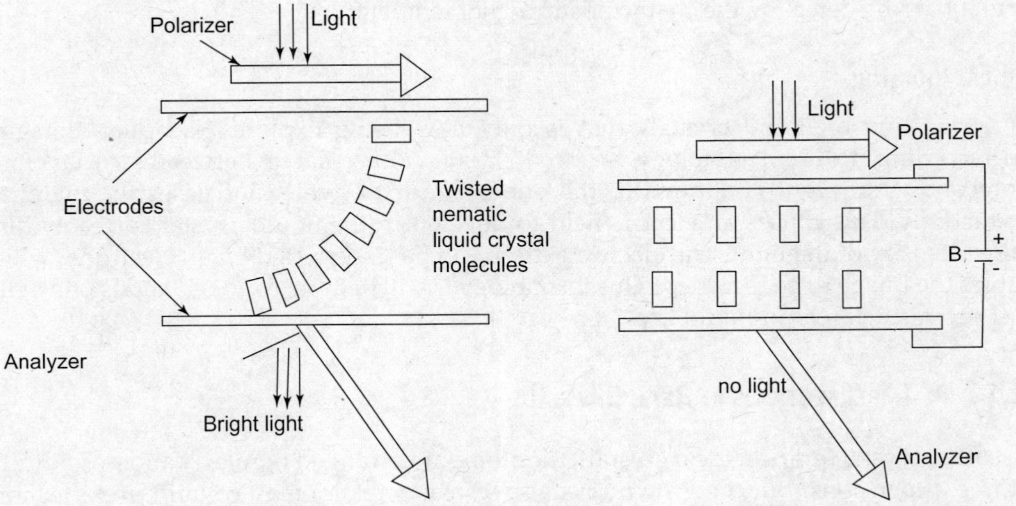

FIGURE 22.41 (a) Twisted nematic field effect display in the absence of applied electric field. (b) Twisted nematic field effect display under applied electric field

22.10.4 Application of Liquid Crystals

Liquid crystal technology has had a major effect in many areas of science and engineering, as well as device technology. Applications for this special kind of material are still being discovered and continue to provide effective solutions to many different problems.

Liquid crystal displays

The most common application of liquid crystal technology is liquid crystal displays (LCD). This field has grown into a multi-billion dollar industry, and many significant scientific and engineering discoveries have been made. Some of them have been discussed in the following sections.

Liquid crystal thermometers

As demonstrated earlier, chiral nematic (cholesteric) liquid crystals reflect light with a wavelength equal to the pitch. Because the pitch is dependent upon temperature, the colour reflected also is dependent upon temperature. Liquid crystals make it possible to accurately gauge temperature just by looking at the colour of the thermometer. By mixing different compounds, a device for practically any temperature range can be built.

The "mood ring", a popular novelty a few years ago, took advantage of the unique ability of the chiral nematic liquid crystal. More important and practical applications have been developed in such diverse areas such as medicine and electronics. Special liquid crystal devices can be attached to the skin to show a "map" of temperatures. This is useful because often physical problems, such as tumours, have a different temperature than the surrounding tissue. Liquid crystal temperature sensors can also be used to find bad connections on a circuit board by detecting the characteristic higher temperature.

Optical imaging

An application of liquid crystals that is only now being explored is optical imaging and recording. In this technology, a liquid crystal cell is placed between two layers of photoconductor. Light is applied to the photoconductor, which increases the materials conductivity. This causes an electric field to develop in the liquid crystal corresponding to the intensity of the light. The electric pattern can be transmitted by an electrode, which enables the image to be recorded. This technology is still being developed and is one of the most promising areas of liquid crystal research.

22.11 ▶ Imperfections in Real Crystals

In a sound crystal (or in an ideal crystal), the atoms are arranged regularly and periodically in three dimensions. But the grown crystals (or real crystals) may contain imperfections or defects. These defects are mainly divided into point, line, surface and volume defects. They are described below:

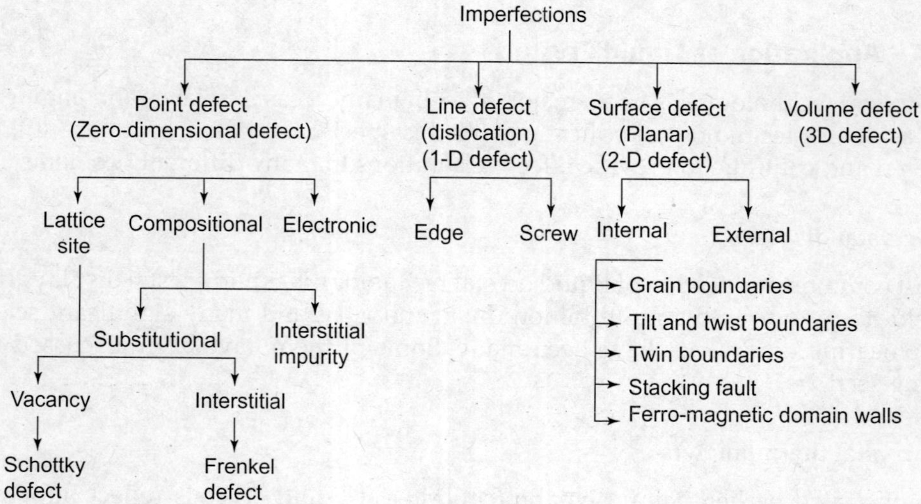

FIGURE 22.42 Imperfections in crystals

22.11.1 Point Defects

As the name indicates, these defects are at some points in the crystal. So, these are also called zero-dimensional defects. The point defects are divided into three categories: (a) lattice site defects; this includes vacancies (Schottky defect) and interstitialcies (Frenkel defect), (b) compositional defects; this includes substitutional impurity and interstitial impurity and (c) electronic defects.

(a) **Lattice site defects:** In this type of defects, some atoms may not be present in their regular atomic sites. They are as follows:

(i) *Vacancies:* As shown in Fig. 22.43, at a lattice point, one or two or three atoms are missed, and this is referred to as single or double or triple vacancies, respectively. The vacancies are formed due to the imperfect packing during crystallization or due to thermal vibrations at high temperatures.

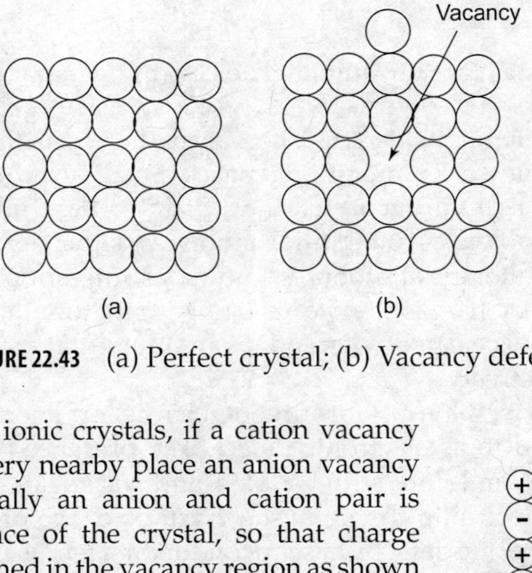

FIGURE 22.43 (a) Perfect crystal; (b) Vacancy defect

Schottky defect: In ionic crystals, if a cation vacancy exists, then in the very nearby place an anion vacancy also exists, i.e. usually an anion and cation pair is moved to the surface of the crystal, so that charge neutrality is maintained in the vacancy region as shown in Fig. 22.44. This is known as Schottky defect. Crystals such as NaCl, KCl, KBr, etc. show Schottky defect.

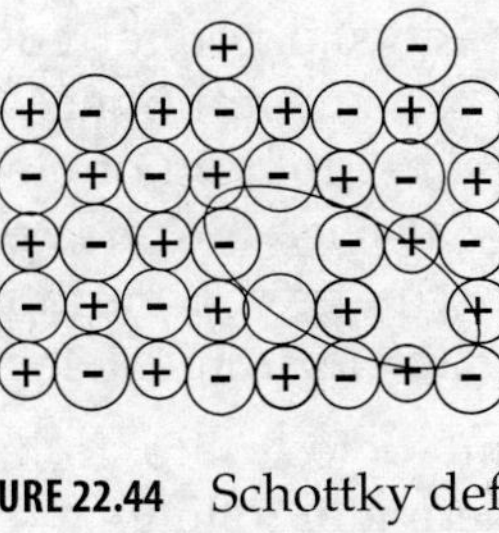

FIGURE 22.44 Schottky defect

(ii) *Interstitial defect:* If an atom is moved to an interstitial space in the crystal, then the defect is known as interstitial defect.

Frenkel defect: In ionic crystals, if a cation (positive ion) moves to an interstitial space, then a vacancy is formed in its atomic position. Here, charge neutrality is maintained in the defective region as shown in Fig. 22.45. This type of defect is known as Frenkel defect. Crystals such as CaF_2, AgBr, AgI, etc. show Frenkel defect.

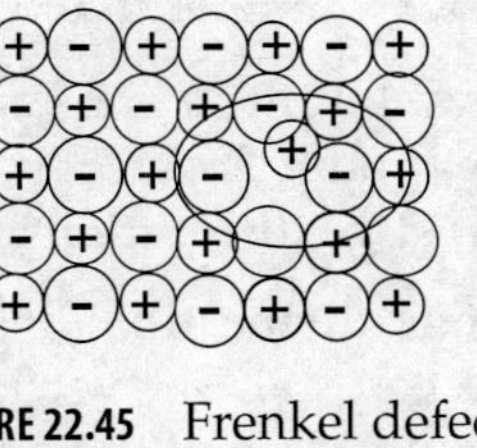

FIGURE 22.45 Frenkel defect

(b) **Compositional defect:** The presence of impurity atoms in the crystal leads to compositional defects. Impurity atoms are present at the sites of regular parent atoms or in the interstitial space. These defects are described below.

(i) ***Substitutional defect:*** As shown in Fig. 22.46(a), during crystallization few foreign atoms occupy the regular parental atoms site. For example, in extrinsic semiconductors either third or fifth group atoms occupy the sites of silicon or germanium atoms.

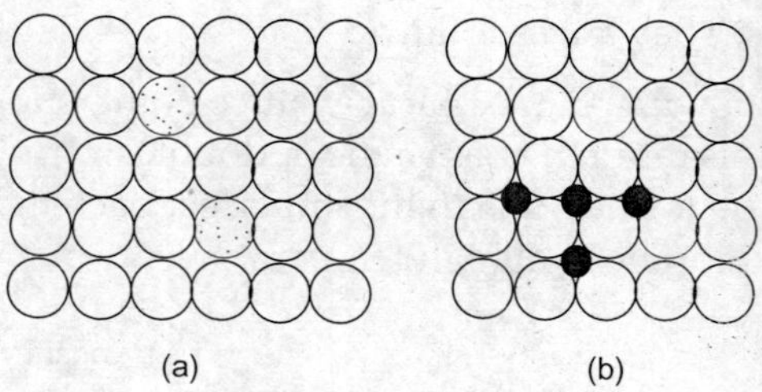

FIGURE 22.46 (a) Substitutional defect; (b) Interstitial defect

(ii) ***Interstitial impurity defect:*** The spaces between the parental atoms in a crystal are known as interstitial spaces. Small-sized (lower atomic number) atoms, such as hydrogen etc., may fit into these interstitial spaces. These atoms are known as interstitial atoms and the defect formed due to the presence of interstitial atoms is known as interstitial defect. This is shown in Fig. 22.46(b). If 'r' is the radius of a parent atom, then an octahedral and a tetrahedral space can accommodate an interstitial atom of radius $0.414r$ and $0.225r$, respectively.

(c) **Electronic defects:** Non-uniformity of charge or energy distribution in the crystal is referred to as electronic defect. The presence of impurity atoms such as substitutional and interstitial atoms and vacancies can vary the uniform distribution of electronic charge in the crystal. So, the presence of these defects also leads to electronic defects. In semiconductors, temperature variation changes charge concentration, so the variation of temperature (i.e., thermal energy) leads to electronic defects.

Point defects are formed by thermal fluctuations, by severe deformation (i.e., by hammering **or** rolling) and by bombarding with high energetic particles.

22.11.2 Line and Screw Dislocations

If a crystal plane ends somewhere in the crystal, then along the edge of that incomplete plane produces defect in the crystal called line defect. The line defect is of two types: they are (i) edge dislocation and (ii) screw dislocation.

(i) Edge dislocation: Figure 22.47(a) shows three-dimensional view and front face of a perfect crystal. The vertical crystal planes are parallel to side faces of a crystal as shown in the figure. One of the crystal planes does not pass from top to bottom face of the crystal, but ends somewhere in the crystal as shown in Fig. 22.47(b). In this crystal, just above the edge of incomplete plane, the atoms are in a state of compression so that the bond distances are less than normal values and below the edge of incomplete plane, the atoms are far apart, so the bond distances are larger than normal values. This situation extends all along the edge of this incomplete plane producing edge dislocation.

The extra plane indicated in Fig. 22.47(b) can be either above or below the slip plane shown as dotted line in Fig. 22.47(c). If the incomplete extra plane is above the slip plane, then the edge dislocation is positive and is represented by the symbol ⊥; on the other hand, if it is below the slip plane, then the edge dislocation is negative and is represented by the symbol *T*. If one plane of atoms glides over another separated by an integral multiple of interatomic distance is called *slip,* and the slip plane is the plane in which slip has taken place. Thus, the crystal consists of slipped and normal regions.

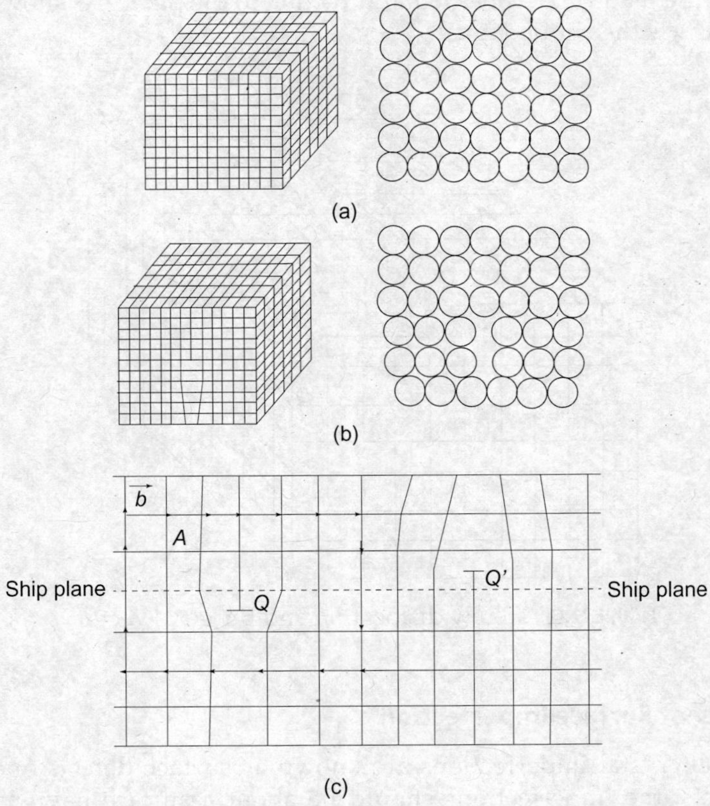

FIGURE 22.47 (a) Three-dimensional view of perfect crystal; front view of perfect crystal. (b) Three-dimensional view of edge dislocation crystal; front view of edge dislocation crystal. (c) Positive and negative edge dislocations

The magnitude and direction of the displacement of crystal planes due to edge dislocation can be represented by a vector called *Burger's vector,* which is perpendicular to the dislocation line. This indicates how much and in what direction the lattice above the slip plane is shifted with respect to the lattice below the slip plane. Figure 22.47(c) shows a method of determining Burger's vector for edge dislocation. To find the magnitude and direction of Burger's vector, one starts arbitrarily from a lattice point A, drawing atom-to-atom vectors round the dislocation in clockwise direction to form a closed circuit. Here, the number of vectors in horizontal direction at the top and bottom and vertical vectors at the left and right are equal, but the circuit is not closed unless we put the vector $\vec{b}$, as shown in the circuit. This is the Burger's vector for the above-said edge dislocation.

Screw dislocation: The crystal planes spiral about a line in the crystal, called dislocation line. The screw dislocation is shown in Fig. 22.48. Due to the spiralling of crystal planes, the atoms at one end of the plane are displaced by one atomic distance with respect to the other end of the plane in perpendicular direction to the plane. As shown in Fig. 22.48, the planner ABCD is the slipped area.

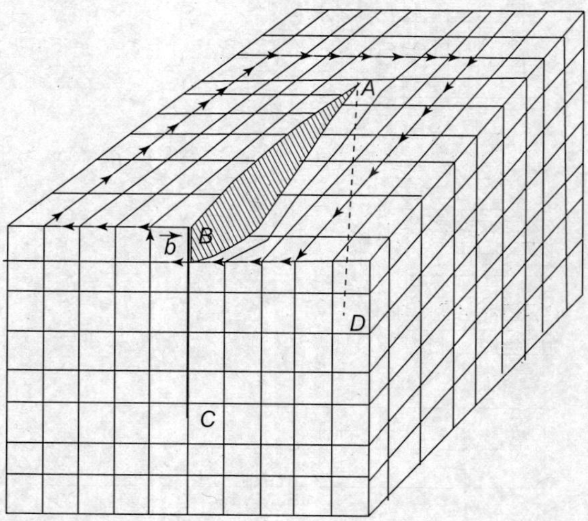

FIGURE 22.48 Screw dislocation and Burger's vector

22.11.3 Planer or Surface Imperfection

Two-dimensional crystal imperfections are known as surface defects or plane defects. In surface defects, the imperfections should lie about a surface having a few atomic dimensions thick. Surface defects are of two types: (a) external surface defects and (b) internal surface defects.

(a) **External surface imperfections:** Every atom present inside the crystal has a large number of surrounding atoms, whereas the atoms present on the surface of the crystal has neighbouring atoms only on one side of it. Hence, atomic bonds do not extend beyond the surface of the crystal. Because of this, the surface atoms possess larger energy than the interior atoms. This larger energy at the surface causes imperfection at the external surface itself.

(b) **Internal surface imperfections:** The change in stacking of atomic planes across a boundary in the crystal is known as internal surface imperfection. Some of the internal surface imperfections are explained below:

(i) **Grain boundaries:** The non-periodicity of atoms between the crystallets (grains) of a polycrystalline material causes grain boundary surface defect.

The thickness of this non-periodic region is of the order of 2 to 10 atomic distances or more. This boundary region is called a *crystal boundary or a grain boundary* and is shown in Fig. 22.49. The orientation of the crystallets changes sharply at the grain boundary. If the misorientation angle between the crystallets is greater than 10° to 15°, then it is called a *high-angle grain boundary.* On the other hand, if the misorientation angle between the adjacent crystals is of the order of a few degrees or less than 10°, then it is called a *low-angle grain boundary.*

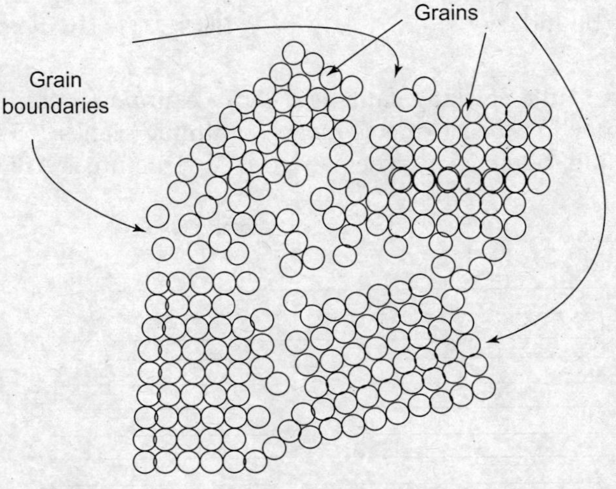

FIGURE 22.49 High-angle grain boundaries

(ii) **Tilt and twist boundaries:** Tilt boundary has an array of edge dislocations as indicated by (⊥) in Fig. 22.50. In the figure, 'h' is the vertical spacing between two consecutive edge dislocations and 'b' is the length of Burger's vector. Here, $\tan\theta \approx \theta = \dfrac{b}{h}$ is the angle of tilt or misorientation.

Twin boundaries: This is a surface imperfection that separates two mirror orientations of a crystal.

As shown in Fig. 22.51, the atomic arrangement on one side of a twin boundary is a mirror reflection of the atomic arrangement on the other side of the twin boundary. The region between the two boundaries is called the twinned region. The twin boundaries can be seen under an optical microscope.

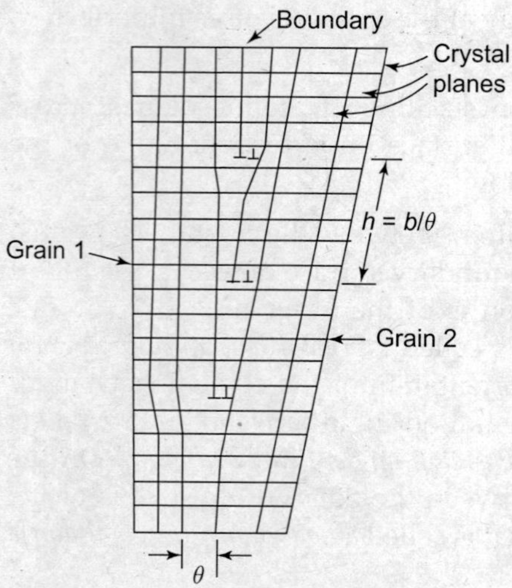

FIGURE 22.50 Tilt boundary

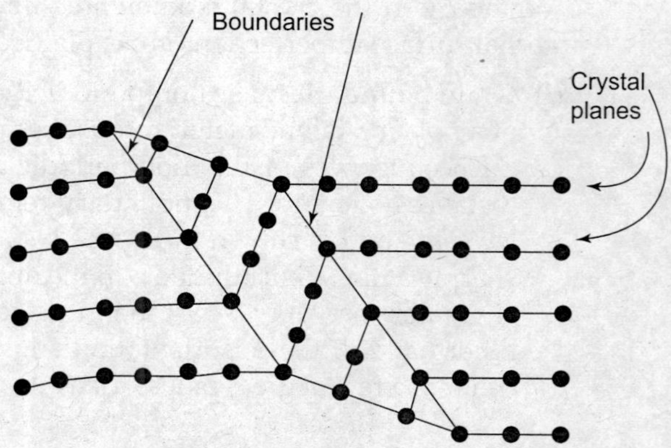

FIGURE 22.51 Twin boundaries

(iii) **Stacking fault:** Stacking fault is a surface imperfection in which there is a discrepancy in the stacking sequence of atomic planes. As shown in Fig. 22.52, the stacking sequence in close packed *ECC* structure is ABCABCABC….

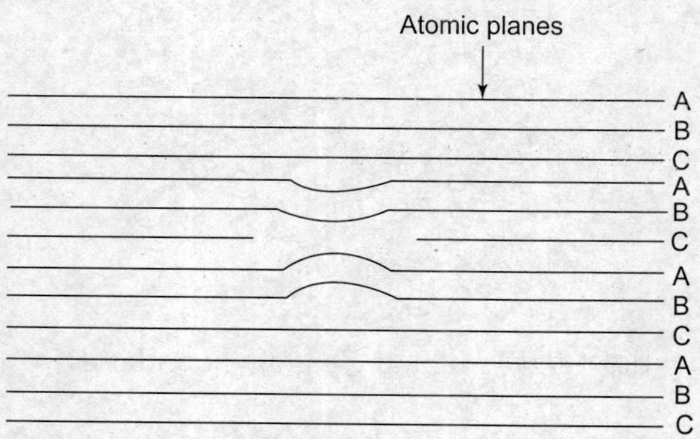

FIGURE 22.52 Stacking fault

Suppose that in a small region in atomic layer '*C*', the atoms are not positioned properly. Then at this region, the stacking sequence is different and here the stacking sequence becomes …*ABAB*.... This is the stacking sequence of *HCP* structure. Thus, the missing atoms in a small area of atomic layer '*C*' give rise to a stacking fault in close packed *FCC* crystal. The crystal will be sound on both sides of the fault.

(iv) **Ferro-magnetic domain walls:** A ferro-magnetic material contains a large number of ferro-magnetic domains. Each domain is magnetized to saturation in a particular direction inside the material. The intensity of magnetic field and hence the magnetic field energy is almost uniform inside the domains. However, the intensity of magnetic field and the magnetic field energy is more at the surface of the domains. This large magnetic field energy on the surface of the domains gives rise to a surface imperfection known as magnetic domain-wall imperfection.

22.11.4 Volume Imperfections

The cracks that are formed due to small electrostatic dissimilarity between the stacking layers, or due to sudden thermal waves or by using the material for some applications cause volume defects.

Recap Zone

Points to Remember

- Solids are broadly classified into two types:
  - Crystalline solids or true solids
  - Amorphous solids or pseudo solids
- Crystalline solids are usually built up of a number of crystals, in which the crystals may be of similar or widely varying size.
- Materials in which the molecule is the basic structured solid and which have no regular structure are classified as amorphous.
- Lattice is the regular geometrical arrangement of points in the crystal space. Atoms arrange themselves in distinct pattern in space called a space lattice.
- The smallest volume that contains the full pattern of repetition is called a unit cell.
- The unit cells are of four types:
  - Primitive or simple
  - Face-centred
  - Body-centred
  - End-centred
- Bravais lattices are named after their originator Auguste Bravais. These are nothing but 14 distinguishable three-dimensional space lattices that can be generated by repeated translation of three non-coplanar vectors, a, b and c of a unit cell in three-dimensional space. Each crystal system is based on the any one of Bravais lattice.
- Crystalline solids are characterized by the presence of a long-range ordering of the atoms with a sharp and precise melting point.
- Amorphous solids, on the other hand, do not have the long-range ordering, and have no sharp and precise melting point.
- Single crystals are those in which the orientation of the atoms/molecules is all uniform and continuing throughout the entire crystal.

- Polycrystals are made up of smaller crystallites known as grains.
- The arrangement of points having identical geometrical orientations is known as space lattice.
- Bravais lattices are lattices which are uniquely different from each other in crystallographic orientation.
- On the basis of the different combinations of cube lengths a, b and c and the angles α, β and γ, the crystals can be grouped into seven crystal systems.
- The 14 Bravais space lattices are grouped under these seven crystal systems, and those space lattices which can be described by the same co-ordinate system are said to be in the same crystal system.
- A unit cell is the smallest volume that contains the full pattern of a crystal. They are also known as the smallest, self-repeating geometrical patterns of a crystal, conveniently describing the symmetry of the crystal and can be considered the building blocks of the crystal itself.
- A primitive cell is a unit cell which has only one atom or one lattice point.
- Lattice planes are a large number of parallel equidistant planes that pass through the lattice points and make up the crystal. Lattice planes exist in different directions and in different interplanar distances.
- Interplanar distance (d) is the perpendicular distance between any two adjacent planes.
- Miller indices (written within parentheses '()' without commas in between) are the representation of lattice planes which are identified by their intercepts on the coordinate axes.
- Zero can never be an intercept for any plane.
- Miller indices are defined as the reciprocal of the intercepts made by the plane on the crystallographic axes, which are then reduced to their smallest numbers.
- Alternatively, Miller indices are the three smallest possible integers, which have the same ratio as the reciprocals of the intercepts of the plane concerned along the three axes.
- A family of planes is represented by placing the Miller indices in braces '{ }'.
- The Miller indices of direction are enclosed in square brackets '[ ]'.
- A family of the Miller indices directions is enclosed in carets '⟨ ⟩'.
- For a cubic system $a = b = c$, the inter-planar distance is given by

$$d = \frac{a}{\sqrt{h^2 + k^2 + l^2}}$$

- In a crystal, the co-ordination number, N, is the number of equidistant nearest neighbour atoms that a reference atom has in a lattice.
- The co-ordination number for SC structure is 6, BCC structure is 8, and FCC structure is 12.
- Atomic, packing factor is defined as the ratio of the volume (v) occupied by the effective number of atoms in a unit cell, to the total volume (V) of the unit cell

Key Terms with Definitions

- Crystals are solids with regular structure of atoms or molecules.
- Crystals can be metallic as well non-metallic.
- Solids also exist as amorphous structures that have no regular structure.
- Smallest repeating pattern in a solid structure is called unit cell.
- Crystal system is classified into seven major types according to unit cell geometries.
- There are 14 distinguishable 3D space lattices that can be generated using three co-ordinates of a unit cell.
- A unit cell with only one atom is called a primitive cell.

Important Formulae and Equations

Equation Number	Equation	Remarks
(22.1)	Atomic packing fraction $= \dfrac{\text{Volume of all atoms in cell}}{\text{Volume of cell}}$	
(22.2)	Atomic radius $r = \dfrac{a}{2}$	$a =$ lattice constant
(22.5)	$a = \left[\dfrac{n}{N_A}\dfrac{M}{\delta}\right]^{1/3}$	Relation between lattice constant a and density of crystal material
(22.6)	Packing fraction for cubic structure $= \dfrac{\frac{3}{4}\pi r^3}{a^3}$	
(22.7)	Packing fraction for BCC structure $= \dfrac{8\pi r^3}{3a^3}$	
(22.8)	Packing fraction for FCC structure $= \dfrac{16\pi r^3}{3(2\sqrt{2}r)^3}$	
(22.9)	Miller indices $h:k:l = \dfrac{1}{p}:\dfrac{1}{q}:\dfrac{1}{r}$	
(22.21)	Inter-planer spacing $d = \dfrac{1}{\sqrt{\dfrac{h^2}{a^2}+\dfrac{k^2}{a^2}+\dfrac{l^2}{a^2}}} = \dfrac{a}{\sqrt{h^2+k^2+l^2}}$	

REVIEW ZONE

SHORT ANSWER QUESTIONS

1. Calculate the number (n) of atoms contained within
   (a) Cubic cell
   (b) Body-centred cubic cell
   (c) Face-centred cubic cell
2. Identify the following crystal system:

$a = 4$ nm	$b = 5$ nm	$c = 6$ nm	$\alpha = \beta = \gamma = 90°$
$a = 7$ nm	$b = 7$ nm	$c = 7$ nm	$\alpha = \beta = \gamma = 90°$
$a = 6$ nm	$b = 7$ nm	$c = 8$ nm	$\alpha \neq \beta \neq \gamma \neq 90°$
$a = 7$ nm	$b = 7$ nm	$c = 10$ nm	$\alpha = \beta = \gamma = 90°$
$a = 6.5$	$b = 6{:}5$	$c = 6$	$\alpha = \beta = 90°$ $\gamma = 120°$

(1st Sem., BPUT, 2007)

3. Calculate Miller indices of crystal plane, which is cut through the crystal axes at $3a, -3b, -3c$?
4. A crystal plane intercepts the crystallographic axes at $\frac{1}{2}a$, $\frac{3}{4}b$ and $\frac{3}{2}c$, where a, b and c are unit lengths along x, y and z, respectively. What are the Miller indices of plane?
5. A substance crystallizes with a body-centred cubic lattice. Calculate the distance between the 200, 110 and 222 planes. The length of side of unit cell is 4.34 Å.
6. A compound formed by elements X and Y crystallizes in cubic structure where X atoms are at the corner of cube and Y atoms are at face centre. What is the formula of compound?
7. A closed-packed structure of uniform sphere has the cell edge = 0.8 nm. Calculate the radius of the molecule if it has
   (a) BCC lattice, (b) FCC lattice
8. Sodium metal crystallizes in body-centred cubic lattice with cell edge = 4.29 Å. What is the radius of sodium atom?
9. Metallic gold crystallizes in a face-centred cubic lattice. The length of cubic unit cell is a = 4.070 Å (atomic mass gold = 197 amu). Calculate density and packing fraction.
10. Titanium metal has a density of 4.54 g cm^3 and edge length of 412.6 pm. Determine unit cell where titanium crystallizes.
11. A metal X crystallizes in two cubic phases face-centred cubic (FCC) and body-centred cubic (BCC) whose unit cell lengths are 3.5 and 3.0 Å. Calculate the ratio of densities of FCC and BCC.
12. Calcium crystallizes in a face-centred cubic unit cell with a = 0.556 nm. Calculate the density. If
    (a) it contained 0.1% Frenkel defects
    (b) it contained 0.1% Schottky defects
13. Predict the co-ordination number (CN) of cation in crystals of each of the following compound.

Compound	Radius cation	Radius anion
MgO	0.65 Å	140 Å
MgS	0.65 Å	1.84 Å
CsCI	1.69 Å	1:81 Å

14. Define interfacial angle. What is its significance?
15. Distinguish between crystalline solid and amorphous solid.
16. NaCl has FCC structure. How many Na^+ and Cl^- ions are there in unit cell?
17. What are the miller indices if the plane intersect the crystal axis at 2a, b, 2c?
18. Silver has an atomic radius of 0.144 nm. Calculate the densities of silver if it crystallizes (a) simple cube and (b) FCC structure. (at. wt. = 108 gm)
19. A substance $A_x B_y$ crystallizes in FCC lattice in which atom A occupies each corner of cube and atom B occupies the centres of each face of cube. Identify the correct composition.
20. In the compound AX, the radius of A^+ ion is 95 pm and that of X^- is 181 pm. Predict the crystal structure of AX and write the co-ordination number of each of the ions.
21. Name the Bravais lattices encountered in cubic system. Sketch their unit cells and find for each type unit cell
    (a) the number of atoms per unit cell.
    (b) co-ordination number for each atom in the lattice.

Long Answer Questions

1. Show that FCC is the most closely packed of the three cubic structures by working out of packing factors.
2. Describe the structure of NaCl.
3. Explain the terms: (i) basis, (ii) space lattice and (iii) unit cell.
4. Describe seven crystal systems with diagrams.
5. Obtain the relations between the edge of the unit cell and atomic radius for the BCC and FCC lattices.
6. What are Bravais lattices?
7. What is packing fraction? Deduce packing factors for simple cubic and BCC structures.
8. Define co-ordination number and packing factor of a crystal.
9. Describe FCC crystal structure.
10. Obtain an expression for the packing factor of FCC structure.
11. Explain the crystal structure of diamond with a two-dimensional diagram.
12. Define crystal lattice, unit cell, lattice parameter and co-ordination number.
13. Explain the unit cell and lattice parameters. What is a primitive cell and how does it differ from unit cell?
14. Explain the terms: (i) basis, (ii) space lattice, (iii) lattice parameters and (iv) unit cell.
15. Describe BCC structure, with suitable example.

16. Describe in detail, the seven crystal systems with diagrams.
17. Prove that which type of the cubic crystal structure has closest packing of atoms. Describe the relation between the atomic radius and the unit cell dimension of the crystal, mentioned above.
18. Tabulate the characteristics of the unit cells of different crystal systems.
19. Illustrate Bravais lattices.
20. Describe the crystal structures of diamond and sodium chloride.
21. Illustrate simple cubic, FCC and BCC crystal structures.
22. What is space lattice? Find the packing fraction for BCC and FCC crystals.
23. Describe in detail the structure of diamond.
24. Explain various types of bondings in solids with suitable examples.
25. Show that FCC crystals are closely packed than BCC crystals
26. Classify various lattice types in the crystal system.
27. Describe in detail the structure of ZnS.
28. What is a Bravais lattice? What are the different space lattices in the cubic system?
29. Describe Bragg's X-ray spectrometer and explain how Bragg's law can be verified.
30. Explain the influence of point defects in crystals and how do they affect the properties of materials.
31. Obtain an expression for the energy required to create a vacancy in the crystal.
32. Derive an expression for the inter-planar spacing in the case of a cubic structure.
33. Derive an expression for the energy change due to creation of vacancies inside a solid.
34. Derive an expression for the concentration of Frenkel defects present in a crystal at any temperature.
35. Sketch the planes (120), $(2\ (00\bar{1})\ 3)$ and directions [100] and [211].
36. Explain how the X-ray diffraction can be employed to determine the crystal structure. Give the ratio of inter-planar distances of (100), (110) and (111) planes for a simple cubic structure.
37. Distinguish between Frenkel defects and Schottky defects.
38. Explain edge dislocation, screw dislocation and significance of Burger's vector.
39. Write short notes on Burger's vector in dislocations.
40. What are Miller indices? Derive an expression for the inter-planar spacing between two adjacent planes of Miller indices (*hkl*) in a cubic lattice of edge length '*a*'.
41. Explain and illustrate, with neat sketches, the edge and screw dislocations, and show the Burger's vector in them.

CHAPTER 23 Solid State Materials

Learning Objectives

By the end of this chapter, the student will be able:

- To understand the nature of bonding in solids
- To describe the free electron theory of solids
- To identify its limitations in explaining the physical properties of metals
- To explain thermal conductivity and other bulk properties of solids
- To differentiate metals, semiconductors and insulators
- To explore the elementary theory for Brillion zones in solids in one, two or three dimensions

23.1 ▶ Introduction

Solids are congregation of atoms. These atoms are connected to each other by bonds. The particular type of bonding plays a major role in determining the physical, chemical and electrical properties of a material. The metals contain a large number of free electrons which are free to move about the whole volume of the element. These electrons are very good conductors. We have to study the classical model to explain the properties (electrical and thermal conductivities) of metals.

Thus, to understand the behaviour and characteristics of materials, we shall study bonding in solids, electron theory and band theory of solids in this chapter.

23.2 ▶ Bonding in Solids

A solid is composed of billions of atoms packed together and the forces of attraction that binds the atoms together are very strong. If this is not so, then the atoms can easily crumble or disintegrate. The four main types of forces are gravitational, electrical, nuclear and weak forces. Gravitational forces are responsible to bind the stars and planets; electrical forces are responsible for atoms and molecules. But nuclear and weak forces bind the nucleons.

When two atoms are brought together, electrons in both the atoms are rearranged to form a stable configuration. This type of rearrangement of electrons gives rise to different types of bonds, which hold the atoms together in a solid.

The bonding in solids can be divided into four categories on the basis of the nature of forces. These are

(a) Ionic bonding
(b) Covalent bonding
(c) Molecular bonding
(d) Metallic bonding

23.2.1 Ionic bonding

Ionic or heteropolar bonding occurs between electropositive elements and electronegative elements. Examples are NaCl and MgO. The reason of ionic bonding is the difference in electronegativity of two elements. The solids in which ionic bonding takes place are known as Ionic Solids.

The ionic bond is an electrostatic force of attraction between two oppositely charged ions formed due to the transfer of electrons between the related atoms.

Consider the formation of ionic bond in NaCl as shown in Fig. 23.1. Na has electronic configuration 2, 8, 1 and Cl has electronic configuration 2, 8, 7. Thus, sodium atom gives out an electron and becomes Na^+ ion. Cl atom gains this electron and becomes Cl^- ion. The two ions attract each other and results in the formation of an ionic or electrovalent bond.

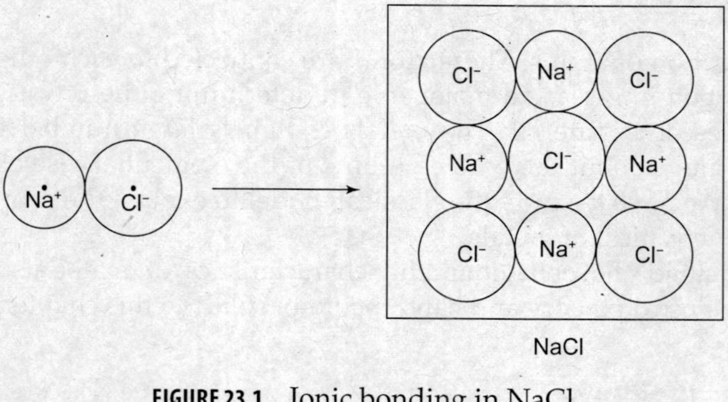

FIGURE 23.1 Ionic bonding in NaCl

Characteristics of ionic solids

The ionic solids have the following characteristics:

(i) The ionic solids are generally crystals in nature.
(ii) These are rigid and have high boiling and melting points.
(iii) These are generally non-conductors but their solutions conduct electricity.

(iv) They crystallize in relatively closed packed structures and ionic bonds are non-directional.

(v) They are transparent for all frequencies upto fundamental absorption frequencies but for higher frequencies, they are opaque.

23.2.2 Covalent Bonding

Covalent bond is formed by sharing valence electrons between two similar or dissimilar atoms to achieve stability. Examples are N_2, O_2, HCl, etc. The simplest case of single covalent bond is Hydrogen molecule. Consider the hydrogen, H_2 molecule. When two H atoms approach each other, each H atom shares one electron with its neighbour and covalent bond is formed between them as shown in Fig. 23.2.

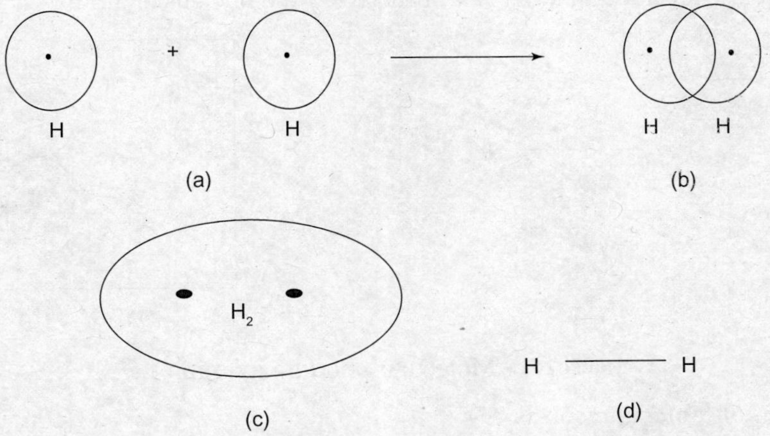

FIGURE 23.2 Covalent bonding in H_2

In covalent bonding, the electrons are not transferred from one atom to another atom like ionic bonding but mutually shared between two atoms.

Characteristics of covalent solids

Some of the characteristics of covalent solids are as follows:

(i) Covalent solids are hard and brittle having high binding energies.

(ii) They have high boiling and melting points and covalent bonds are generally directional.

(iii) Covalent crystals are generally insulators. However, some are semi-conductors like germanium.

(iv) They generally dissolve in non-polar solvents such as carbon tetra chloride and benzene.

(v) They are transparent to long wavelengths but opaque to short wavelengths.

23.2.3 Molecular Bonding

The weak short range forces that hold the molecules together to form a solid are known as Vanderwaal forces. These forces are caused due to electrostatic force of attraction existing between nucleus (positvely charged) of one atom and electrons (negatively charged) of other atom. A molecular bond is formed due to Vanderwaal's forces of attraction. However, these also exists a weak repulsive force between positively charged nuclei of two atoms and negatively charged electrons of two atoms.

The crystal formed due to Vanderwaal's force of attraction is called molecular crystal. Examples are Argon, CO_2, ice and calomel, etc. Let us consider the case of a noble gas. The outermost shells are full and hence rearrangement of electrons is not possible. Thus, noble gases like helium and neon remain monoatomic under ordinary conditions. Now, consider the case of noble gas argon. Let the electron clouds of two atoms is distributed as shown in Fig. 23.3. The electron cloud of atom A is attracted by the nucleus of atom B.

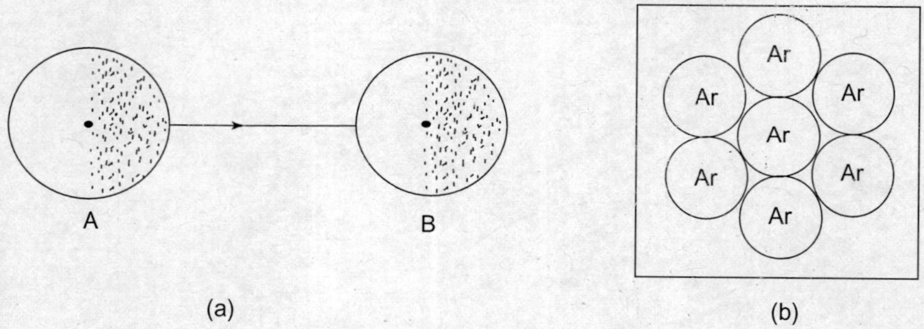

FIGURE 23.3 Molecular bonding in argon

Characteristics of molecular solids

(i) The molecular solids have small binding energy and low conductivity.

(ii) The melting point and density both are low.

(iii) The molecular bonds are formed in molecules which form dipoles.

(iv) They are usually transparent or coloured.

(v) Molecular solids can be found in both crystalline and non-crystalline structures. The molecular bonds are directional for dipole and hydrogen bonds while non-directional for dispersion bonds.

Note

Molecular bonds are also known as secondary bonds because they do not involve the valence electrons at all.

23.2.4 Metallic Bonding

In metallic bonding, the valence electrons which hold the atoms together are not bound to any particular atom and free to move throughout the whole metal. These mobile and free electrons form a kind of electron cloud or electron gas.

The valence electrons of atoms in a metal are loosely held by their atoms. The metallic bond is formed as a result of electrostatic interaction between the positive ions and the electron gas holds the metal together. Thus, in metallic bonding, the force of attraction between the positive metal ions and the electron gas exceeds the mutual repulsion of the electrons in that electron gas. Examples are Na, Al, Cu, etc.

Let us consider the case of metallic bonding in Na (sodium) metal as shown in Fig. 23.4.

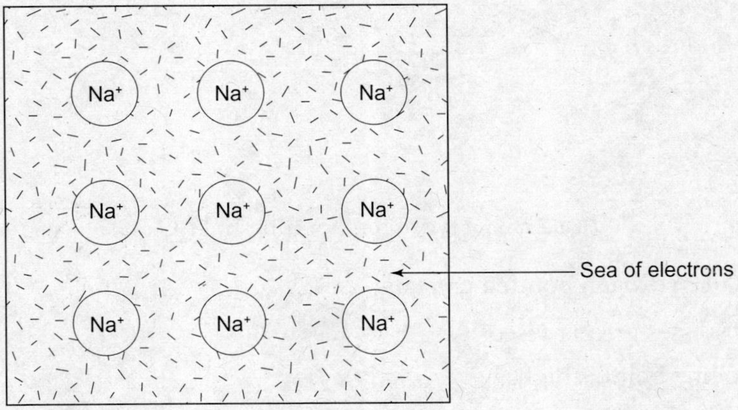

FIGURE 23.4 Metallic bonding in Na metal

Characteristics of metallic solids

(i) Metallic solids have low ionization energies and high electrical conductivity due to availability of a large number of free electrons.

(ii) Due to the symmetrical arrangements of positive ions, metals have a crystalline structure.

(iii) They are opaque to light due to absorption of light energy by free electrons.

(iv) The melting point of metals is intermediate (–39°C to 3410°C), but density is high.

(v) The metallic bonds are comparatively weaker than the ionic and covalent bonds. But they have high thermal conductivity.

23.2.5 Hydrogen Bonding

The hydrogen bond is considered as a special type of dipole bond but is considerably stronger. This bond exists in molecules which has hydrogen atom at one end. When hydrogen atom is covalently bonded to a relatively large atom such as nitrogen, oxygen or fluorine, a permanent dipole is formed.

In this type of molecules, the electron cloud tends to concentrate around nitrogen or oxygen or fluorine nucleus, thus leaving the hydrogen nucleus unprotected. Thus, a strong permanent dipole is formed. The best example of hydrogen bonding is H_2O molecule as shown in Fig. 23.5. Other examples of hydrogen bonding are NH_3 and HF_2.

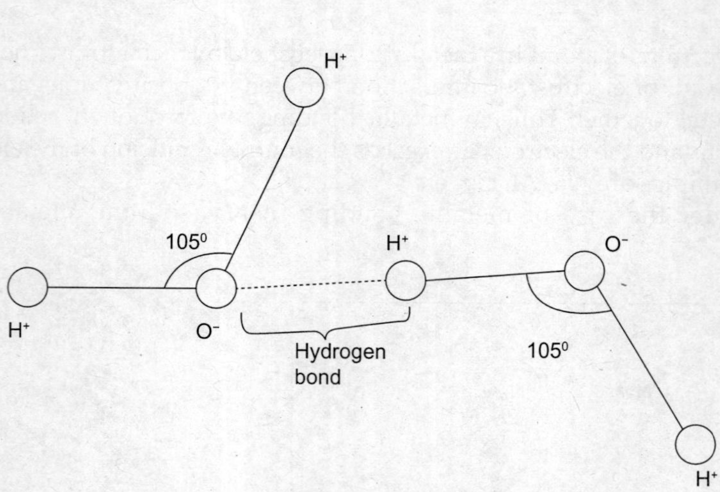

FIGURE 23.5 Hydrogen bonding in H_2O

Characteristics of hydrogen bonded crystals

(i) They possesses greater viscosity.

(ii) Their boiling point is high.

(iii) They are ionic in character but stronger than Vanderwaal's bonds and weaker than ionic or covalent bonds.

(iv) They show abnormal thermal expansion and generally transparent to light.

(v) They have low electrical and thermal conductivities and show peculiar dielectric properties.

Note

Hydrogen bonded crystals always contain H and an electronegative element.

23.3 ▶ LATTICE ENERGY OF A CRYSTAL

The lattice energy is the amount of energy required to separate the ionic crystal into its constituent ions.

Let us consider two ions (one is positively charged and the other is negatively charged) separated by an infinite distance. When ions approach each other, because of considerable overlapping between the electronic shells of ionic crystals, the interatomic binding forces are of two types:

(i) There will be electrostatic attraction between the ions which is described by Coulomb's law. The potential energy is negative in this case because the work is done by the ions. The variation of Coulomb's pot energy is shown by the curve.

(a) In Fig. 23.6, the Coulomb's potential energy is given by

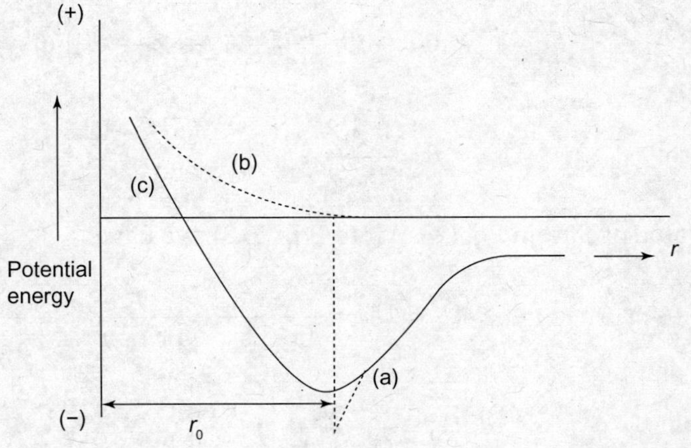

FIGURE 23.6 Lattice energy of a crystal

$$U_{CP} = -A\left(\frac{e^2}{4\pi \in_0 \cdot r}\right) \tag{23.1}$$

(ii) There will be repulsive force due to the repulsion between electron clouds of two ions. This is shown by curve (b) in Fig. 23.6. The pot energy will be positive because the work is done on the ions. The repulsive pot energy is given by

$$U_{Re} = +\frac{B}{r^n} \tag{23.2}$$

Here in Eqs. (23.1) and (23.2), constant A is the moduling constant and B is the repulsive coefficient and n is the repulsive exponent.

The total energy U_T is given by

$$U_T = U_{Re} + U_{CP}$$

$$\Rightarrow \qquad U_T = \frac{B}{r^n} - A\left(\frac{e^2}{4\pi \in_0 \cdot r}\right) \tag{23.3}$$

At equilibrium separation, $r = r_0$, the potential energy U_T must be minimum.

i.e., $$\frac{dU_T}{dr} = 0 \text{ at } r = r_0$$

Thus, differentiating Eq. (23.3), we get

$$\frac{dU_T}{dr} = -n \cdot Br^{-n-1} + A\frac{e^2}{4\pi \in_0 r^2}$$

or, $$0 = -nBr_0^{-n-1} + A\frac{e^2}{4\pi \in_0 r^2} \quad [\text{Put } r = r_0]$$

$\Rightarrow$ $$B = \frac{Ae^2}{4\pi \in_0 n} r_0^{n-1} \tag{23.4}$$

Substituting the value of B from Eq. (23.4) into Eq. (23.3), we have

$$U_0 = U_T = \frac{-Ae^2}{4\pi \in_0 r_0} + \frac{Ae^2}{4\pi \in_0 n} r_0^{n-1} \frac{1}{r_0^n}$$

or, $$\boxed{U_0 = \frac{-Ae^2}{(4\pi \in_0) r_0}\left[1 - \frac{1}{n}\right]} \tag{23.5}$$

Equation (23.5) gives the lattice energy or equilibrium energy of a crystal.

Note

The value of moduling constant A depends on the crystal structure.

EXAMPLE 23.1

The equilibrium distance between ions of NaCl crystal is 2.81 Å. Take $n = 9$, calculate the potential energy per ion pair ($A = 1.748$).

Solution: We know that

$$U_0 = \frac{-Ae^2}{(4\pi \in_0) r_0}\left[1 - \frac{1}{n}\right]$$

Substituting the values, we get

$$U_0 = \frac{-(1.748)(9 \times 10^9)(1.6 \times 10^{-19})^2}{\left(2.81 \times 10^{-10}\right)}\left(1 - \frac{1}{9}\right)$$

or, $$U_0 = -1.27 \times 10^{-18} \text{ Joule}$$

or, $$U_0 = -7.97 \text{ eV}$$

The contribution per ion is half of the value of U_0,

i.e., $$\frac{U_0}{2} = -3.99 \text{ eV}$$

This is called lattice energy per ion in the NaCl crystal.

EXAMPLE 23.2

The potential energy function for the force between two atoms in a diatomic molecule is given as

$$U(x) = \frac{a}{x^{12}} - \frac{b}{x^6}$$

where a and b are positive constants and x is the distance between atoms.

(i) At what value of x is $U(x)$ equal to zero?

(ii) At what value of x is $U(x)$ a maximum?

(iii) Derive an expression for the force between two atoms.

Solution: Given $U(x) = \frac{a}{x^{12}} - \frac{b}{x^6}$

If $U(x) = 0$, then $\frac{a}{x^{12}} - \frac{b}{x^6} = 0$

$$\Rightarrow \quad \frac{1}{x^6}\left[\frac{a}{x^6} - b\right] = 0$$

$$\Rightarrow \quad x = (a/b)^{\frac{1}{6}} \quad \text{or} \quad x = \infty$$

(ii) For minimum value of $U(x)$, $\frac{dU}{dx} = 0$

$$\Rightarrow \quad \frac{1}{dx}\left[\frac{a}{x^{12}} - \frac{b}{x^6}\right] = 0$$

$$\Rightarrow \quad \frac{1}{x^7}\left\{6b - \frac{12a}{x^6}\right\} = 0$$

$$\Rightarrow \quad x = \infty \quad \text{or} \quad x = \left(\frac{2a}{b}\right)^{\frac{1}{6}}$$

(iii) Force F between two atoms is given by

$$F = -\frac{dU}{dx} = -\frac{d}{dx}\left\{\frac{a}{x^{12}} - \frac{a}{x^6}\right\}$$

$$= \frac{12a}{x^{13}} - \frac{6b}{x^7}$$

The force F is repulsive if F is positive

i.e., $$\frac{12a}{x^{13}} - \frac{6b}{x^7} > 0 \quad \text{or} \quad x < \left(\frac{2a}{b}\right)^{\frac{1}{6}} \quad \text{i.e., } x < x_0$$

where $x_0 = \left(\frac{2a}{b}\right)^{\frac{1}{6}}$

The force F is attractive if F is negative, i.e.,

$$\frac{12a}{x^{13}} - \frac{6b}{x^7} > 0$$

This gives $x > x_0$

Thus, the atoms repel each other for x less than x_0 and attract each other for x greater than x_0 where

$$x_0 = \left(\frac{2a}{b}\right)^{\frac{1}{2}}$$

23.4 ▶ Electrical Properties of Materials

In solids, the carriers of electric current are the electrons. Each metal contains a large number of free electrons which can freely move about the whole volume of the element. In order to explain the electrical and thermal properties of metals, Drude and Lorentz in 1900 proposed free electron theory of metals. This theory is applicable for both metals and non-metals.

23.4.1 Classical Free Electron Theory of Metals (Drude–Lorentz Theory)

Drude in 1900 proposed a simple theoretical explanation of electrical conductivity of metals based on electron gas model. Later on, Lorentz modified the theory of Drude. The basic assumptions of Lorentz–Drude theory are as follows.

(i) There are a large number of valence electrons in a metal which move about the whole volume of the metal like the molecules of a perfect gas enclosed in a container.

(ii) The motion of these free electrons is random. They move in all possible directions with different velocities similar to thermal agitation of a perfect gas. The free electrons in a metal constitute electron gas.

(iii) The distribution of velocities of electron gas is in accordance with Maxwellian distribution for a gas. The random speed is a function of temperature. The average kinetic energy of an electron is $\frac{3}{2}kT$, where k is the Boltzmann's constant and T is the absolute temperature.

(iv) The free electrons during their random motion collide with fixed positive ions in the lattice and also among themselves. But the collisions among electrons themselves do not contribute to the electrical conductivity.

(v) In the absence of electric field, the current density vector is zero due to random motion of electrons, but when an electric field is applied to the metal, the free electrons drift slowly with some average velocity known as average drift velocity opposite to the direction of applied electric field. On the basis of this theory, many properties of solids can be explained as given below:

Drift velocity (V_d): When an electric field is applied on the metal, due to drift velocity (V_d), the electrons move opposite to the field direction. The force experienced by a free electron is

$$\vec{F} = -e\vec{E} \tag{23.6}$$

If V_d is the drift velocity of $\overset{-s}{e}$, m is the effective mass of the electron and τ be the time between two consecutive collisions, the acceleration of electrons decreases due to collisions and frictional force can be given as mV_d / τ. Thus, Eq. (23.6) can be modified as

$$m\left(\frac{d\overrightarrow{V_d}}{dt}\right) = -e\vec{E} - \frac{m\overrightarrow{V_d}}{\tau} \tag{23.7}$$

As the electron will acquire a steady motion and have a net drift velocity, so $\dfrac{d\overrightarrow{V_d}}{dt} = 0$

Equation (23.7) becomes

$$0 = -e\vec{E} - m\left(\frac{\overrightarrow{V_d}}{\tau}\right)$$

$$\Rightarrow \qquad \overrightarrow{V_d} = -\frac{e\vec{E}\tau}{m} \tag{23.8}$$

Equation (23.8) gives the steady state velocity of the electron. The variation of drift velocity with time is shown in Fig. 23.7.

Relaxation time (τ): If the applied field is cut off when drift velocity reached its steady value. Equation (23.7) can be written as

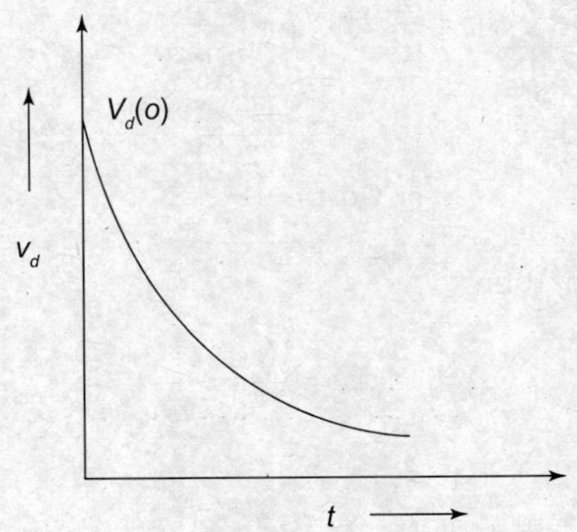

FIGURE 23.7 Variation of V_d with time

$$m\frac{\overrightarrow{V_d}}{dt} = -m\frac{\overrightarrow{V_d}}{\tau}$$

$$\Rightarrow \quad \frac{d\overrightarrow{V_d}}{V_d} = -\frac{dt}{\tau}$$

$$\Rightarrow \quad V_d(t) = V_d(o)e^{\frac{-t}{\tau}}. \tag{23.9}$$

At $t = \tau$,

$$V_d(t) = \frac{V_d(o)}{e} \tag{23.10}$$

Thus, relaxation time is defined as the time taken for the drift velocity to decay $(1/e)$ times of its initial value.

Conductivity (σ): Let us consider that n be the number of free electrons per unit volume. The current density of electrons is given by

$$\vec{J} = -ne\overrightarrow{V_d} = \frac{ne^2\tau}{m}\vec{E} \tag{23.11}$$

But according to Ohm's law

$$\vec{J} = \sigma\vec{E} \tag{23.12}$$

From Eqs. (23.11) and (23.12),

$$\sigma = \frac{ne^2\tau}{m} \tag{23.13}$$

If λ is the mean free path and $\overline{v}$ is the mean speed of free electrons, then

$$\tau = \frac{\lambda}{\overline{v}} \text{ and } \sigma = \frac{-ne\lambda}{m\overline{v}}$$

Also, electrical conductivity is

$$\boxed{\boldsymbol{\sigma = \frac{ne^2\lambda}{m\overline{v}}}} \tag{23.14}$$

If μ is the mobility of electron, then

$$\mu = \frac{\overrightarrow{V_d}}{\vec{E}} = \frac{e\tau}{m} \tag{23.15}$$

From Eqs. (23.14) and (23.15),

$$\Rightarrow \quad \sigma = ne\mu \tag{23.16}$$

The resistivity of the solid is given by

$$\boxed{\rho = \frac{1}{\sigma} = \frac{1}{ne\mu}} \tag{23.17}$$

Ohm's law: We know that the resistivity of a metal is defined as

$$\rho = \frac{RA}{l}$$

but

$$\rho = \frac{1}{\sigma} \text{ and } \vec{E} = \frac{V}{l}$$

Thus,

$$\vec{J} = \sigma \vec{E} = \frac{V}{\rho l} = \frac{V}{RA}$$

$\Rightarrow$

$$JA = \frac{V}{R} = I$$

$\Rightarrow$

$$V = IR \tag{23.18}$$

Note

The great difference in the conductivities of different metals is due to the difference in number of free electrons per unit volume.

23.5 ▶ Thermal Conductivity of Metals

The free electrons also contribute in the conduction of heat along the metal. The electrons have greater KE of thermal agitation at the hot end of a metal bar and low KE at the cold end. If there are n electrons moving randomly in all directions, their motion can be resolved parallel to OX, OY and OZ directions. Let a plane B of unit area be considered in a metal as shown in Fig. 23.8. Consider two planes A and C at temperatures T_1 and T_2 respectively. Both are at a distance λ (mean free path) from the plane T (maintained at temperature T). Let $T_1 > T_2$. So there will be a transfer of energy from A to C. The number of electrons from $A \rightarrow C$ per unit area per second is $n\bar{v}/6$.. Each electron has energy

$$\frac{1}{2} m\bar{v}^2 = \frac{3}{2} k_B T_1$$

Hence, total energy transferred from A to C per unit area per second is

$$= \frac{n\bar{v}}{6} \cdot \frac{3}{2} k_B T_1 \tag{23.19}$$

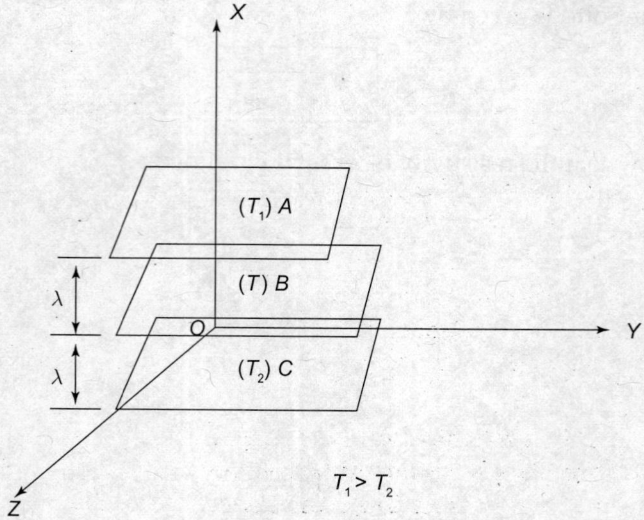

FIGURE 23.8 Flow of heat in a metal

Similarly, the energy transferred from C to A will be

$$= \frac{n\bar{v}}{6} \cdot \frac{3}{2} k_B T_2 \tag{23.20}$$

Net transfer of energy from C to A (through B) per unit area per second is

$$Q = \frac{n\bar{v}}{6}\left(\frac{3k_BT_1}{2} - \frac{3k_BT_2}{2}\right) = \frac{n\bar{v}k_B}{4}(T_1 - T_2) \tag{23.21}$$

But, also

$$Q = \frac{K(T_1 - T_2)}{2\lambda} \tag{23.22}$$

where K = thermal conductivity.
From Eqs. (23.21) and (23.22), we get

$$\frac{n\bar{v}k_B}{4}(T_1 - T_2) = \frac{K(T_1 - T_2)}{2\lambda}$$

$$\Rightarrow \qquad \boxed{K = \frac{n\bar{v}k_B\lambda}{2}} \tag{23.23}$$

Equation (23.23) gives the expression for thermal conductivity of a metal.

23.6 ▶ Wiedemann and Franz Law

Wiedemann and Franz in 1854 discovered that the ratio of thermal to electrical conductivity of metals is proportional to absolute temperature.

From Eqs. (23.13) and (23.23), we get

$$\frac{K}{\sigma}=\frac{n\overline{v}k_B\lambda(m\overline{v})}{2\left(ne^2\lambda\right)}$$

or
$$\frac{K}{\sigma}=\frac{n\overline{v}k_B\lambda m\overline{v}}{2ne^2\lambda}\quad\left[\because \tau=\frac{\lambda}{\overline{v}}\right]$$

$\Rightarrow$
$$\frac{K}{\sigma}=\frac{\overline{v}^2k_B\lambda m}{2e^2\lambda}.$$

$\Rightarrow$
$$\frac{K}{\sigma}=\frac{k_B\lambda m\overline{v}.\overline{v}}{2e^2\lambda}$$

$\Rightarrow$
$$\frac{K}{\sigma}=\frac{k_B\lambda\overline{v}\left(\dfrac{3k_BT}{\overline{v}}\right)}{2e^2\lambda}\quad\left[\overline{v}^2=\frac{3k_BT}{m}\right]$$

or
$$\frac{K}{\sigma}=\frac{3}{2}\frac{k_B{}^2}{e^2}T \tag{23.24}$$

or
$$\boxed{\frac{k}{\sigma}\propto T} \tag{23.25}$$

Equation (23.25) is known as Wiedemann–Franz law, where $\frac{K}{\sigma}=LT$ and $L=$ Lorentz number.

And also
$$L=\frac{3}{2}\frac{k_B^2}{e^2}=1.12\times10^{-8}\ \text{w}\Omega\text{K}^{-2}$$

23.7 ▶ Failures of Classical Free Electron Theory

The free electron theory has the following limitations:

(i) **Specific heat:** The free electron theory has been successful in explaining electrical and thermal conductivities of metals but could not explain why the low temperature specific heat should be dominated by the electronic contribution.

The free electron theory assumes that all valence electrons in a metal can absorb thermal energy. According to law of equipartition of energy, the energy of one kilomole of the metal is

$$U=\frac{3}{2}N_ak_BT$$

The molar heat capacity is

$$(C_v)_{\text{electronic}} = \frac{dU}{dT} = \frac{3}{2}N_a k_B$$

$$= \frac{3}{2}R = 1.5R$$

$$= 12.5 \times 10^3 \text{ J/mol/R}$$

The electronic heat capacity calculated on the basis of free $\overline{e}$ theory is about hundred times greater than the experimentally observed value.

(ii) **Electrical conductivity:** According to free electron theory, the electrical conductivity σ is given by

$$\sigma \propto \frac{1}{\sqrt{T}} \tag{23.26}$$

But the experimentally observed value of σ is proportional to temperature.

$$\sigma_{\text{exp}} \propto \frac{1}{T} \tag{23.27}$$

Equations (23.26) and (23.27) show that the temperature dependence of electrical conductivity could not be explained well by free electron theory.

(iii) **Dependence of electrical conductivity on electron concentration:** According to free electron theory,

$$\sigma = \frac{ne^2\tau}{m} \text{ or } \sigma \propto n$$

where n is electron concentration. But experimentally, this result $(\sigma \propto n)$ does not hold good.

(iv) **Paramagnetism of metals:** Experimental fact that the paramagnetism of metals is nearly independent of temperature which could not be explained by free electron theory.

(v) **Wiedemann–Franz law:** According to Wiedemann–Franz law,

$$\frac{K}{\sigma T} = L = \text{Constant}$$

where $$L = \frac{3}{2}\frac{k_B^2}{e^2}$$

$$= 1.12 \times 10^{-8} \text{ w}\Omega\text{K}^{-2}$$

For copper at 20°C, the electrical resistivity and thermal conductivity are 1.72×10^{-8} Ωm and 386 $\text{Wm}^{-1}\text{K}^{-1}\text{m}^{-1}\text{K}^{-1}$ respectively. Substituting these values, we get

$$\frac{K}{\sigma T} = \frac{386}{5.81 \times 10^7 \times 293}$$

$$= 2.26 \times 10^{-8} \text{ w}\Omega\text{K}^{-2}$$

Thus, the value of Lorenz number does not agree with the value calculated from classical formula. The results obtained by free electron theory agree well only at high as well as at very low temperature.

(vi) **Hall coefficient:** According to free electron theory, the density of electrons has a constant value which is independent of temperature, τ and magnetic field strength but experimentally observed hall coefficient depends upon high magnetic field strength and the temperature.

(vii) **Directional dependence of electrical conductivity:** In metals, sometimes DC electrical conductivity depends on the orientation of the specimen w.r.t. the electric field. The current density $\vec{J}$ need not to be parallel to the electric field $\vec{E}$.

EXAMPLE 23.3

If a metal has 5.82 × 10^{28} conduction electrons/m^3, find the relaxation time in a metal of resistivity 1.543 × 10^{-8} Ω-m.

Solution: We know that

$$\rho = \frac{m}{ne^2\tau}$$

or

$$\tau = \frac{m}{ne^2\rho}$$

Given

$$\rho = 1.543 \times 10^{-8} \Omega\text{-m}$$

$$n = 5.82 \times 10^{28} \text{ electrons/m}^3$$

or

$$\tau = \frac{9.1 \times 10^{-31}}{\left(5.82 \times 10^{28}\right)\left(1.602 \times 10^{-19}\right)^2 \left(1.543 \times 10^{-8}\right)}$$

$$\boldsymbol{\tau = 3.97 \times 10^{-14} \text{ s}}$$

EXAMPLE 23.4

A uniform silver wire has a resistivity of 1.54×10^{-8} Ω-m at room temperature. For an electric field of 1 V/cm, calculate (i) Relaxation time τ, (ii) Drift velocity V_d and (iii) mobility μ.

Solution: The relaxation time is

$$\tau = \frac{m}{ne^2\rho}$$

$$\tau = \frac{9.1\times10^{31}}{\left(5.8\times10^{26}\right)\left(1.6\times10^{-19}\right)^2\left(1.54\times10^{-8}\right)}$$

or

$$\tau = 3.97\times10^{-14}\,\text{s}$$

(ii) The drift velocity is

$$\vec{V}_d = \frac{e\vec{E}\tau}{m}$$

$$= \frac{\left(1.6\times10^{-19}\right)\left(10^2\right)\left(3.97\times10^{-14}\right)}{9.1\times10^{31}}$$

$$= 0.7\ \text{m/s}$$

(iii)

$$\mu = \frac{\vec{V}d}{\vec{E}} = \frac{0.7}{10^2}$$

$$= \mathbf{0.7\times10^{-12}\,m^2/V\text{-}S}$$

23.8 ▶ Sommerfeld or Quantum Theory of Free Electrons

Sommerfeld in 1928 modified the free electron theory on the basis of quantum statistics. The main assumptions of the model are as follows:

(i) The metal contains a large number of free electrons, which are not bound to any particular atom but to the metal as a whole.

(ii) The forces between free electrons and ion cores are neglected. The total energy of electron is wholly kinetic since the potential energy is negligible.

(iii) The interior of the metal has uniform potential whereas the potential outside the metal is very high. The free electrons remain confined inside the metal at room temperature.

(iv) The electrons are assumed to obey Pauli's exclusion principle; hence, they obey Fermi-Dirac statistics. According to this statistics, an energy level can accommodate at the most two electrons, one with spin up and the other with spin down.

Particle in one-dimensional box

Consider an electron of mass m confined in a box of length L as shown in Fig. 23.9. Suppose the particle is confined between $x = 0$ and $x = L$. The potential energy V of the particle is infinitely high on both sides of box while the potential energy within the box is uniform.

Here,

$$V = 0, \text{ for } 0 < x < L$$
$$= \infty, \text{ for } x \le 0, x \ge L$$

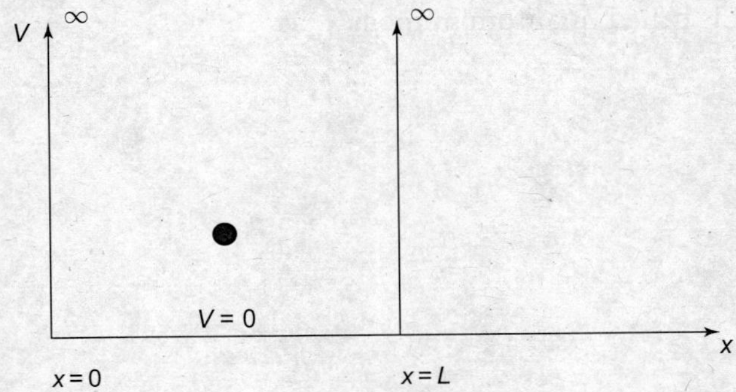

FIGURE 23.9 Particle in one-dimensional box

Schrodinger equation for the region when $V = 0$ is

$$\frac{d^2\psi_n}{dx^2} + \frac{2mE_n}{\hbar^2}\psi_n = 0 \tag{23.28}$$

or

$$\frac{d^2\psi_n}{dx^2} + \alpha^2\psi_n = 0$$

where

$$\alpha^2 = \frac{2mE_n}{\hbar^2} \tag{23.29}$$

and E_n is the energy of electrons in the nth state and ψ_n is the corresponding wave function. The general solution of Eq. (23.28) is

$$\psi_n = A_1 \sin\alpha x + A_2 \cos\alpha x \tag{23.30}$$

Applying the condition, $\psi_n = 0$ at $x = 0$
We get

$$0 = A_1 \sin 0 + A_2 \cos 0 = A_2 \tag{23.31}$$

From Eq. (23.30),

$$\psi_n = A_1 \sin\alpha x \tag{23.32}$$

Now, applying boundary condition,

$$\psi_n = 0 \text{ at } x = L$$

From Eq. (23.30),

$$0 = A_1 \sin\alpha L \tag{23.33}$$

Here, $A_1 \neq 0$. Thus, $\sin \alpha L = 0$ or $\alpha L = n\pi$.

$$\Rightarrow \qquad \alpha = \frac{n\pi}{L}$$

where $n = 1, 2, 3, \ldots$ is called quantum number

$$\text{or} \qquad \alpha^2 = \frac{n^2\pi^2}{L^2} \tag{23.34}$$

From Eqs. (23.29) and (23.34), we get

$$\frac{2mE_n}{\hbar^2} = \frac{n^2\pi^2}{L^2}$$

$$\Rightarrow \qquad E_n = \frac{n^2\pi^2\hbar^2}{2mL^2}$$

The lowest energy is for $n = 1$,

$$\text{i.e.,} \qquad \boxed{\boldsymbol{E_1 = \frac{\pi^2\hbar^2}{2mL^2}}} \tag{23.35}$$

This is called zero point energy.
In general,

$$E_n = n^2 E_1 \tag{23.36}$$

This shows that energy is square function of quantum number n (Fig. 23.10). The energy is quantized and cannot vary continuously (Fig. 23.11).

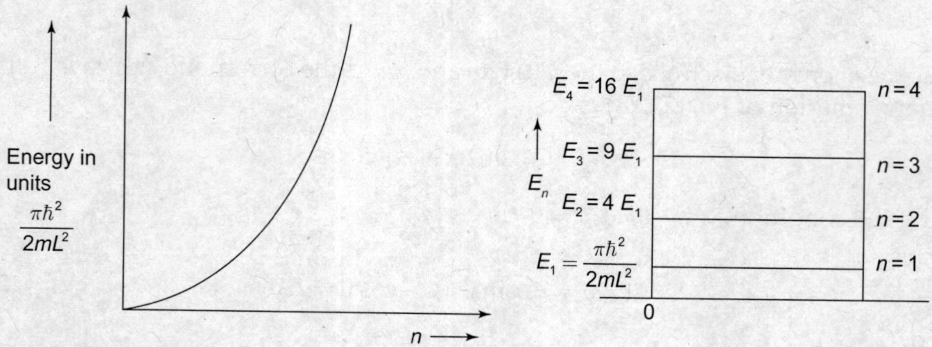

FIGURE 23.10 Energy as a function of quantum number n

Here,

$$E_2 = 4E_1$$
$$E_3 = 9E_1$$
$$E_4 = 16E_1$$

and so on.

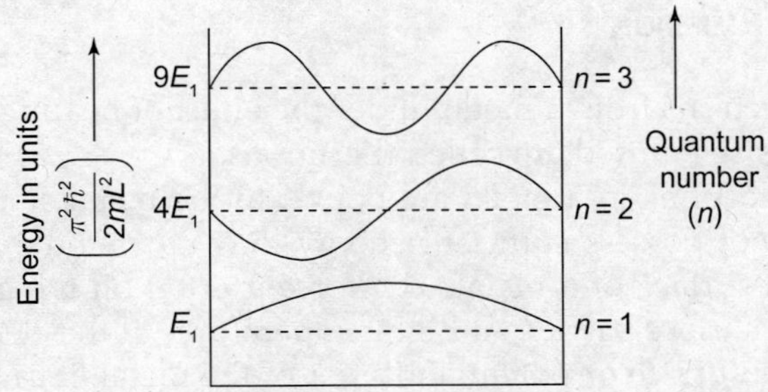

FIGURE 23.11 Lower energy wave functions

For three-dimensional box, the energy is given by

$$E_n = \frac{\pi^2\hbar^2}{2mL^2}\left(n_x^2 + n_y^2 + n_z^2\right) \tag{23.37}$$

Let us consider the wave function associated with different states as

$$\psi_n = A_1 \sin \alpha x = A_1 \sin\left(\frac{n\pi x}{L}\right) \tag{23.38}$$

To calculate the value of A_1, apply normalization condition,

$$\int_0^L \psi_n^* \psi_n \, dx = 1 \tag{23.39}$$

$$\Rightarrow \quad \int_0^L A_1 \sin\left(\frac{n\pi x}{L}\right) \cdot A_1 \sin\left(\frac{n\pi x}{L}\right) dx = 1$$

$$\Rightarrow \quad A_1^2 \int_0^L \sin^2\left(\frac{n\pi x}{L}\right) dx = 1$$

$$\Rightarrow \quad A_1^2 \times \frac{L}{2} = 1$$

$$\Rightarrow \quad A_1 = \sqrt{\frac{2}{L}} \tag{23.40}$$

or

$$\psi_n = \sqrt{\frac{2}{L}} \sin\left(\frac{n\pi x}{L}\right) \tag{23.41}$$

Equation (23.41) gives the energy state functions for a particle in one-dimensional box.

23.9 ▶ Thermionic Emission

The emission of the electrons from a metal under the effect of heating is called thermionic emission. The electrons so emitted are called thermions.

We know that the electrons in a metal are bound to the metal surface and at ordinary temperatures they do not possess sufficient energy to overcome the attractive forces and to escape out from the surface of a metal. The force exerted on the electron to prevent it from escaping is called surface barrier. At high temperatures, if the electron gains sufficient energy to overcome the surface or potential barrier of the metal, it can escape from it. The electrons at ordinary temperature possess fermi energy E_f which is not sufficient to escape out the electron from the surface of the metal. Therefore, the minimum energy (ϕ) must be supplied to the electron for its emission.

At absolute zero, all the levels upto fermi level E_f are filled (as shown in Fig. 23.12). With the increase of temperature, many electrons move to higher energy levels. The energy difference between fermi energy (E_f) and topmost energy level (W) acts as potential barrier and works as work function (ϕ). When thermal energy or heat energy supplied to a metal exceeds ϕ, it results in the emission of electrons by the metal surface which is known as thermionic emission.

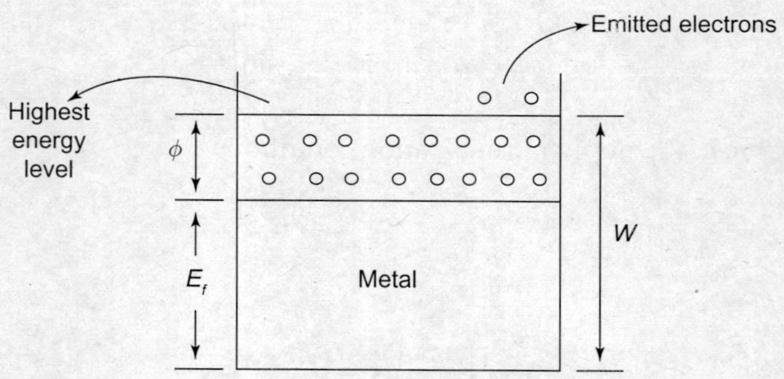

FIGURE 23.12 Thermionic emission

Note

When electrons are emitted from the metallic surface by supplying heat energy, the emission is called thermionic emission. But when energy supplied to the metal gives emission of photons, it is called photoelectric emission.

23.10 ▶ Richardson's Equation

We know that the electrons are emitted from the metallic surface due to thermionic emission. The electron's flux coming out from unit area of metallic surface per second gives the current density. The expression which gives the relation between current density and temperature for a particular metal (derived by Richardson) is known as Richardson's equation.

According to Fermi–Dirac statistics, the number of electrons per unit volume having the momentum in the range p and $(p + dp)$ is

$$dn = g_s \frac{4\pi p^2 dp}{h^3} \frac{1}{e^{\frac{(E-E_f)}{k_B T}} + 1} \tag{23.42}$$

For electrons, $g_s = (2s+1) = 2$, It is called spin degeneracy factor.

Therefore,

$$dn = \frac{2}{h^3} \frac{4\pi p^2 dp}{e^{\frac{(E-E_f)}{k_B T}} + 1} \tag{23.43}$$

If p_x, p_y, p_z are the components of momentum along x, y and z axes and

$$p^2 = p_x^2 + p_y^2 + p_z^2$$

then the volume in the momentum space lying between p_x and $p_x + dp_x$, p_y and $p_y + dp_y$, p_z and $p_z + dp_z$ will be

$$dp_x \, dp_y \, dp_z = 4\pi p^2 dp$$

Thus, the number of electrons per unit volume whose momentum components lie between p_x and $p_x + dp_x$, p_y and $p_y + dp_y$, p_z and $p_z + dp_z$ is

$$dn = \frac{2}{h^3} \frac{dp_x \, dp_y \, dp_z}{e^{\frac{(E-E_f)}{k_B T}} + 1}$$

If v_x, v_y and v_z are the velocity components along x, y and z–axis,

$$d_n = \frac{2}{h^3} \frac{(mdv_x)(mdv_y)(mdv_z)}{e^{\frac{(E-E_f)}{k_B T}} + 1}$$

or

$$dn = \frac{2m^3}{h^3} \frac{dv_x \, dv_y \, dv_z}{e^{(E-E_f)/k_B T} + 1} \tag{23.44}$$

Equation (23.44) gives the number of electrons per unit volume having velocity components in the range v_x and $(v_x + dv_x)$, v_y and $(v_y + dv_y)$ and v_z and $(v_z + dv_z)$

As $k_B T = 0.3\,\text{eV}$ and $(E - E_f)$ is atleast 20 times greater than $k_B T$, thus

$$dn = 2\frac{m^3}{h^3} e^{(E_f - E)/k_B T} dv_x dv_y dv_z$$

$$= 2\frac{m^3}{h^3}e^{\left(E_f/k_BT\right)}\cdot e^{-E/k_BT}dv_x dv_y dv_z \quad \left[\text{neglect 1 in comparison to } e^{\left(E-E_f\right)/k_BT}\right]$$

$$= 2\frac{m^3}{h^3}e^{\left(E_f/k_BT\right)}\cdot e^{-\frac{1}{2}m\left(v_x^2+v_y^2+v_z^2\right)/k_BT} dv_x dv_y dv_z \quad \left[\begin{aligned} &\text{here } E=\frac{1}{2}mv^2 \\ &=\frac{1}{2}m\left(v_x^2+v_y^2+v_z^2\right)\end{aligned}\right]$$

or
$$dn = 2\frac{m^3}{h^3}e^{E_f/k_BT}e^{-mv_x^2/2k_BT}\cdot dv_x\, e^{-mv_y^2/2k_BT}\cdot dv_y e^{-mv_z^2/2k_BT}\cdot dv_z \tag{23.45}$$

Now, suppose the surface of the metal lies in the X–Z plane and X-axis is normal to this surface, the number of electrons per unit volume in velocity range is

$$dn_x = n(v_x)dv_x$$

$$= 2\frac{m^3}{h^3}e^{E_f/k_BT}e^{-mv_x^2/2k_BT}\cdot dv_x \int_{-\infty}^{\infty} e^{-mv_y^2/2k_BT}\cdot dv_y \int_{-\infty}^{+\infty} e^{-mv_z^2/2k_BT}\cdot dv_z \tag{23.46}$$

Here, $\int\limits_{-\infty}^{+\infty} e^{-mv_y^2/2k_BT}\cdot dv_y = \sqrt{\frac{2\pi k_BT}{m}}$ and $\int\limits_{-\infty}^{+\infty} e^{-mv_z^2/2k_BT}\cdot dv_z \sqrt{\frac{2\pi k_BT}{m}}$

From Eq. (23.46),

$$dn_x = 2\frac{m^3}{h^3}e^{E_f/k_BT}e^{-mv_x^2/2k_BT}\cdot dv_x\left(\frac{2\pi k_BT}{m}\right)$$

$$dn_x = \frac{4\pi m^2}{h^3}k_BT\cdot e^{E_F/k_BT}e^{-mv_x^2/2k_BT}\cdot dv_x$$

When the metal is heated, only the electrons escape from the surface along the x-axis which fulfill the condition

$$E = \frac{1}{2}mv_x^2 \geq W$$

i.e.,
$$v_x \geq \sqrt{\frac{2W}{m}}$$

Thus, the number of electrons leaving unit area of the surface in unit time is

$$n_x = \int\limits_{\sqrt{\frac{2W}{m}}}^{\infty} n(v_x)v_x dv_x \tag{23.47}$$

$$= \frac{4\pi m^2 k_B T}{h^3} \cdot e^{E_f/k_B T} \int_{\sqrt{\frac{2W}{m}}}^{\infty} e^{-mv_x^2/2k_B T} v_x dv_x$$

$$= \frac{4\pi m k_B^2}{h^3} \cdot T^2 e^{-(W-E_f)/k_B T}$$

$$n_x = \frac{4\pi m k_B^2}{h^3} \cdot T^2 e^{-\phi/k_B T}$$

where work function $\phi = W - E_f$

The current density is given as

$$J = n_x \cdot e$$

or

$$J = \frac{4\pi m e k_B^2}{h^3} T^2 e^{-\phi/k_B T}$$

or

$$J = AT^2 e^{-\phi/k_B T} \tag{23.48}$$

where $A = \dfrac{4\pi m e k_B^2}{h^3}$ is a constant which is independent of the nature of the metal.

The expression (23.48) is known as Richardson's equation. This equation shows that the emission current density depends upon the work function of the metal and square of the absolute temperature.

Experimental verification

Equation (23.48) may be expressed as

$$\frac{J}{T^2} = Ae^{-\phi/k_B T} \tag{23.49}$$

$$\log_e \frac{J}{T^2} = \log_e A - \frac{\phi}{k_B T}$$

$$\Rightarrow \quad \boxed{\log_e \frac{J}{T^2} = \log_e A - \frac{B}{T}} \tag{23.50}$$

where $B = \dfrac{\phi}{k_B}$ is a constant for a metal.

If we plot a graph between $\dfrac{J}{T^2}$ and $\dfrac{1}{T}$, it is a straight line having slope $-B = -\dfrac{\phi}{k_B}$ and intercept $\log_e A$. This graph (Fig. 23.13) verifies the theoretical results.

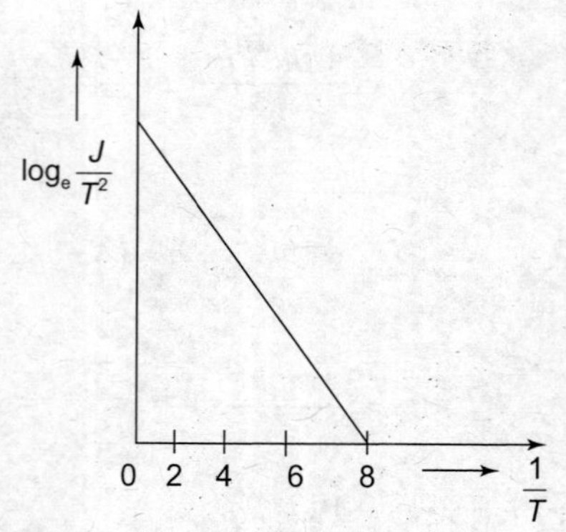

FIGURE 23.13 Experimental verification of Richardson's equation

EXAMPLE 23.5

Find the lowest energy of an electron confined in a box of side 0.1 nm each.

Solution: The possible energies of a particle in cubical box of each side L are

$$E_{n1,n2,n3} = \frac{h^2}{8\pi L^2}\left(n_1^2 + n_2^2 + n_3^2\right)$$

Here, $n_1 = n_2 = n_3 = 1$

$$\Rightarrow \qquad E_1 = \frac{3h^2}{8mL^2} = \frac{3\times\left(6.62\times10^{-34}\right)^2}{8\times9.1\times10^{-31}\times\left(0.1\times10^{-9}\right)}$$

$$= 18.06\times10^{-18}\,\text{Joule}$$

$$= \frac{18.06\times10^{-18}}{1.6\times10^{-19}}\,\text{eV}$$

$$= \mathbf{112.9\ eV}$$

EXAMPLE 23.6

Calculate the energy difference between the ground state and first excited state for an electron in one-dimensional rigid box of length 1 Å. Mass of the electron $= 9.1\times10^{-31}$ kg and $h = 6.626\times10^{-34}$ J-S.

Solution: We know that

$$E_n = \frac{n^2h^2}{8mL^2}; \quad n = 1,2,3,\ldots$$

$$= n^2 \times \frac{\left(6.6\times10^{-34}\,\text{J-S}\right)^2}{8\times9.1\times10^{-31}\,\text{kg}\times\left(10^{-10}\,\text{m}\right)^2}$$

$$= 0.60\times10^{-17}\,n^2 \text{ joule}$$

$$= \frac{0.60\times10^{-17}\,\text{J}}{1.60\times10^{-19}\,\text{J/eV}}\,n^2 = 37.5n^2 \text{ eV}$$

For ground state, $n = 1$

$$E_1 = 37.5\times(1)^2 = 37.5 \text{ eV}$$

For first excited state, $n = 2$

$$E_2 = 37.5\times(2)^2 = 150 \text{ eV}$$

The energy difference between the ground state and the first excited state is $= (E_2 - E_1)$

$$= 150 - 37.5$$

$$= \mathbf{112.5\ eV}$$

EXAMPLE 23.7

The thermionic emission density from an emitter is $0.29\times10^4\ \text{amp/m}^2$ at a temperature of 2500 K. The work function for the emitter is 4.52 eV. Calculate the emission constant A for the emitter. Given $\frac{e}{k_B} = 11600$.

Solution: We know that

$$J = AT^2 \exp\left(\frac{-\phi}{k_BT}\right)$$

Given

$$J = 0.29\times10^4\ \text{amp/m}^2$$

$$T = 2500\,\text{K}$$

$$\phi = 4.52\,\text{eV} = 4.52\times1.6\times10^{-19}\,\text{Joule}$$

$$k_B = \frac{e}{11600} = \frac{\left(1.6\times10^{-19}\right)}{11600}$$

$\Rightarrow$ $$0.29\times10^4 = A(2500)^2 \exp\left[\frac{-4.52\times11600}{(2500)}\right]$$

Solving this, we get

$$A = \frac{0.29}{625}\exp(-20.97)$$

$$= \mathbf{59.40\times10^4\ amp/m^2\text{-}K^2}$$

EXAMPLE 23.8

Fermi velocity of electron in C_s metal is 0.73×10^6 m/s. Calculate its Fermi energy. Also calculate its Fermi temperature.

Solution: We know that

$$\text{KE} = \frac{1}{2}mv_f^2 = E_f$$

$\Rightarrow$ $$E_f = \frac{1}{2}\times9.1\times10^{-31}\times\left(0.73\times10^6\right)^2$$

$$E_f = 1.51\ \text{eV}$$

Also $$k_B T_f = E_f$$

$\Rightarrow$ $$T_f = \frac{E_f}{k_B} = \frac{2.42\times10^{-19}}{1.38\times10^{-23}}$$

$$= \mathbf{1.75\times10^4\ K}$$

23.11 ▶ Band Theory of Solids

Classical physics deals with the macroscopic properties of solid materials like thermal, optical and mechanical properties whereas quantum or modern physics deals with microscopic properties of solids at the atomic and subatomic levels. The band theory of solids deals with the motion of electrons in solids. It explains well the electrical conductivity of solids. On the basis of electrical conductivity, the solids can be classified as conductors, semiconductors and insulators.

23.11.1 Origin of Band Theory

The free or classical electron theory assumed that the conduction electron experience a constant (zero) potential inside a solid. The electrons are free to move inside a solid but restrained at the surface due to infinite potential. This results in quantization of energy in the form of discrete and sharp energy levels as shown in Fig. 23.14(a).

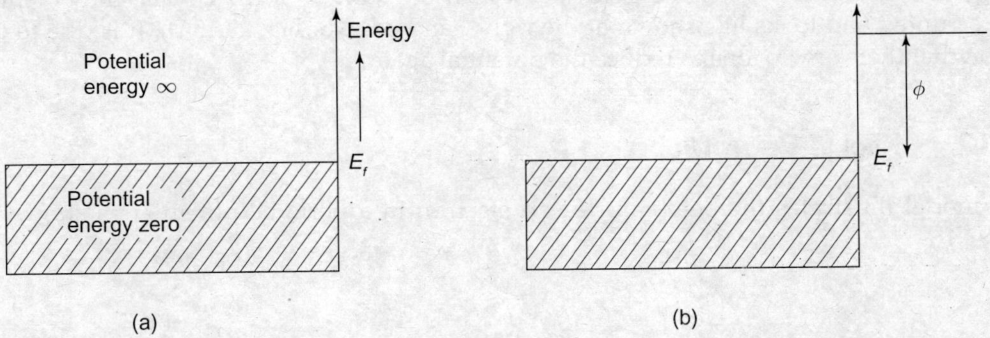

FIGURE 23.14 Potential energy variation in free electron gas model

As the energy inside solid is zero and rises to infinity at the surface, so electrons cannot escape from the surface of the metal. While the phenomenon like thermionic emission and photoelectric effect, there is a probability of emission of electrons from the surface of the metal by applying some energy. This difficulty can be removed by changing the infinite potential at the surface into finite value as Fig. 23.14(b).

According to band theory of solids, instead of single energy levels, in a solid there is a band of energy levels. A set of such closely spaced energy levels is called energy band. The bands of energy levels are referred to the entire solid as a whole and not to single individual atom. Hence, the individual energies within the band are so close together that the energy band may be considered to be continuous.

Inside a real crystal, the motion of a representative electron is shown in Fig. 23.15(a). The potential of electron at the positive ion site is zero and is maximum in between. Thus, electron is considered to be moving in a periodic potential. Figure 23.15(b) shows potential energy curve in the periodic potential of the crystal. The potential experienced by an electron in passing through the crystal in one which is perfectly periodic with the period of lattice as shown in Fig. 23.15(b).

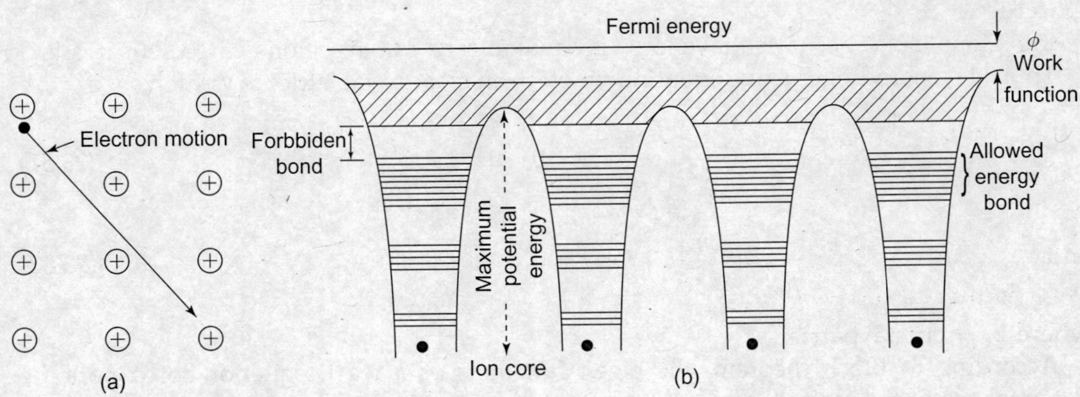

FIGURE 23.15 (a) Motion of electron in a crystal; (b) potential energy curve

At small distances from the ions, the potential energy of valence electron is low and the kinetic energy is high. As the distance of electron from ion increases, its potential energy increases and kinetic energy decreases. At the surface of the crystal, the potential is interrupted and looks like shown at the right-hand edge of Fig. 23.15(b). It is due to the reason that there are no ions to reduce the potential again.

23.12 ▶ Kronig-Penny Model

This model illustrates the behaviour of an electron in a periodic potential as shown in Fig. 23.16.

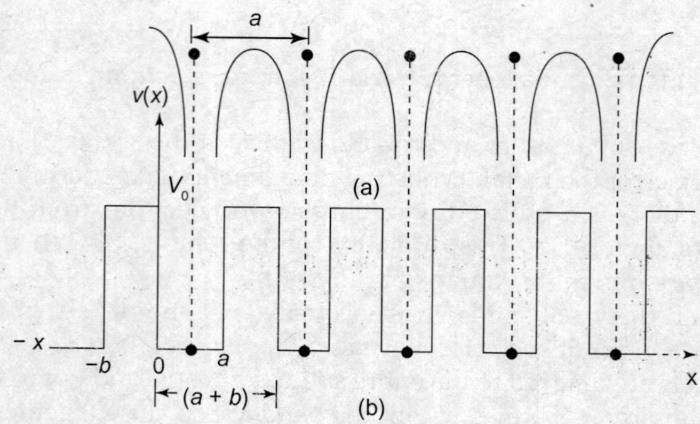

FIGURE 23.16 Periodic square well potential (Kronig–Penny model)

In this model, it is assumed that the potential near the ion core is zero and in between the cores it is constant, i.e. the potential of an electron in a linear array of positive nuclei has a form of periodic square potentials, where an infinite row of rectangular potential wells is separated by barriers of width b, with space periodicity a. Here, a is called the periodicity of the lattice.

In regions $0 < x < a$, potential energy is equal to zero, but in regions $-b < x < 0$, it is V_0. Thus, the Schrodinger wave equation for electron in a crystal lattice is written as

$$\frac{d^2\psi}{dx^2} + \frac{2m}{\hbar^2} E\psi = 0, \quad 0 < x < a \tag{23.51}$$

and

$$\frac{d^2\psi}{dx^2} + \frac{2m}{\hbar^2}[E - V_0]\psi = 0, \quad -b < x < 0 \tag{23.52}$$

where V_0 = crystal barrier.

According to Bloch theorem, the eigen functions of a wave equation for a periodic potential are of the form of a plane wave exp (ikr) times a function $u_k(r)$ having the same crystal lattice periodicity. So the solution of Eq. (23.50) can be written as

$$\psi(x) = u_k(x)e^{ikx} \tag{23.53}$$

Assuming that the total energy E of the electron is less than the potential energy V_0, thus we define two real quantities α and β as

$$\alpha^2 = \frac{8\pi^2 mE}{h^2} = \frac{2mE}{\hbar^2}$$

and
$$\beta^2 = \frac{8\pi^2 m(V_0 - E)}{h^2} = \frac{2m(V_0 - E)}{\hbar^2} \tag{23.54}$$

On differentiating Eq. (23.53), we get

$$\frac{d\psi}{dx} = e^{ikx}\frac{du_k}{dx} + u_k ike^{ikx}$$

and
$$\frac{d^2\psi}{dx^2} = e^{ikx}\frac{d^2u_k}{dx^2} + ike^{ikx}\frac{du_k}{dx} + ike^{ikx}\frac{du_k}{dx} - k^2e^{ikx}u_k$$

i.e.,
$$\frac{d^2\psi}{dx^2} = -k^2e^{ikx}u_k + 2ike^{ikx}\frac{du_k}{dx} + e^{ikx}\frac{d^2u_k}{dx^2}$$

Substituting the values of $\frac{d\psi}{dx}$ and $\frac{d^2\psi}{dx^2}$ in Eqs. (23.51) and (23.52), we get

$$\frac{d^2u_1}{dx^2} + 2ik\frac{du_1}{dx} + \left(\alpha^2 - k^2\right)u_1 = 0 \quad \text{for } 0 < x < a \tag{23.55}$$

and
$$\frac{d^2u_2}{dx^2} + 2ik\frac{du_2}{dx} - \left(\beta^2 + k^2\right)u_2 = 0 \quad \text{for } -b < x < 0 \tag{23.56}$$

where u_1 gives values of $u_k(x)$ in the interval $0 < x < a$, and u_2 in $-b < x < 0$.

The solution of the differential equation (23.55) is of the form

$$u_1 = e^{mx}$$

Thus,
$$\frac{du_1}{dx} = me^{mx}, \quad \frac{d^2u_1}{dx^2} = m^2e^{mx}$$

From Eq. (23.55),

$$m^2e^{mx} + 2ikme^{mx} + \left(\alpha^2 - k^2\right)e^{mx} = 0$$

On solving this equation, we get two roots as

$$m_1 = i(\alpha - k) \quad m_2 = -i(\alpha + k)$$

and
$$u_1 = Ae^{m_1 x} + Be^{m_2 x}$$

$$\Rightarrow \quad u_1 = Ae^{i(\alpha-k)x} + Be^{-i(\alpha+k)x} \tag{23.57}$$

Similarly,

$$u_2 = Ce^{(\beta-ik)x} + De^{-(\beta+ik)x} \tag{23.58}$$

Applying boundary conditions,

$$(u_1)_{x=0} = (u_2)_{x=0}\,; \left(\frac{du_1}{dx}\right)_{x=0} = \left(\frac{du_2}{dx}\right)_{x=0} \tag{23.59}$$

And
$$(u_1)_{x=a} = (u_2)_{x=-b}\,; \left(\frac{du_1}{dx}\right)_{x=a} = \left(\frac{du_2}{dx}\right)_{x=-b} \tag{23.60}$$

Applying these conditions, we get

$$A + B = C + D \tag{23.61}$$

$$Ai(\alpha - k) - Bi(\alpha + k) = C(\beta - ik) - D(\beta + ik) \tag{23.62}$$

$$Ae^{i(\alpha-k)a} + Be^{-i(\alpha+k)a} = Ce^{-(\beta-ik)b} + De^{(\beta+ik)b} \tag{23.63}$$

And $$Ai(\alpha-k)e^{i(\alpha-k)a} - Bi(\alpha+k)e^{-i(\alpha+k)a} = C(\beta-ik)e^{-(\beta-ik)b} - D(\beta+ik)e^{(\beta+ik)b} \tag{23.64}$$

Equations (23.61) to (23.64) have a solution if the determinants of A, B, C and D vanish, i.e.,

$$\begin{vmatrix} 1 & 1 & -1 & -1 \\ i(\alpha-k) & -i(\alpha+k) & -(\beta-ik) & (\beta+ik) \\ e^{i(\alpha-k)a} & e^{-i(\alpha+k)a} & -e^{-(\beta-ik)b} & -e^{(\beta+ik)b} \\ i(\alpha-k)e^{i(\alpha-k)a} & -i(\alpha+k)e^{-i(\alpha+k)a} & -(\beta-ik)e^{(\beta-ik)b} & (\beta+ik)e^{(\beta+ik)b} \end{vmatrix} = 0$$

After solving the above equation, we get

$$\frac{\beta^2 - \alpha^2}{2\alpha\beta} \sinh \beta b \sin \alpha a + \cosh \beta b \cos \alpha a = \cos k(a + b) \tag{23.65}$$

Kroning and Penny considered the possibility that V_0 tends to infinity and b approaches zero in such a way that the product $V_0 b$ remains finite.

Here, $\sinh \beta b \to \beta b$ and $\cos \beta b \to 1$ as $b \to 0$

Thus, Eq. (23.65) can be written as

$$\frac{\beta^2 b}{2\alpha} \sin \alpha a + \cos \alpha a = \cos ka \tag{23.66}$$

Equation (23.66) can also be written as

$$\frac{\beta^2 ab}{2} \frac{\sin \alpha a}{\alpha a} + \cos \alpha a = \cos ka$$

or

$$\boxed{\frac{\boldsymbol{P \sin \alpha a}}{\boldsymbol{\alpha a}} + \boldsymbol{\cos \alpha a = \cos ka}} \tag{23.67}$$

where

$$P = \frac{\beta^2 ab}{2} = \frac{2mV_0 ab}{2\hbar^2} = \frac{mV_0 ab}{\hbar^2} \tag{23.68}$$

Since V_0 tends to infinity, the quantity P is the measure of $V_0 b$ which is the area of potential barrier. The physical significance of P is that as it increases, the area of potential barrier increases and electron is bound more strongly to a particular potential well. On the other side, when P tends to zero, the potential barrier becomes very weak, i.e. the electron becomes free.

If $\dfrac{P \sin \alpha a}{\alpha a} + \cos \alpha a$ is plotted against αa for $P = \dfrac{3}{2}$, we get a curve as shown in Fig. 23.17.

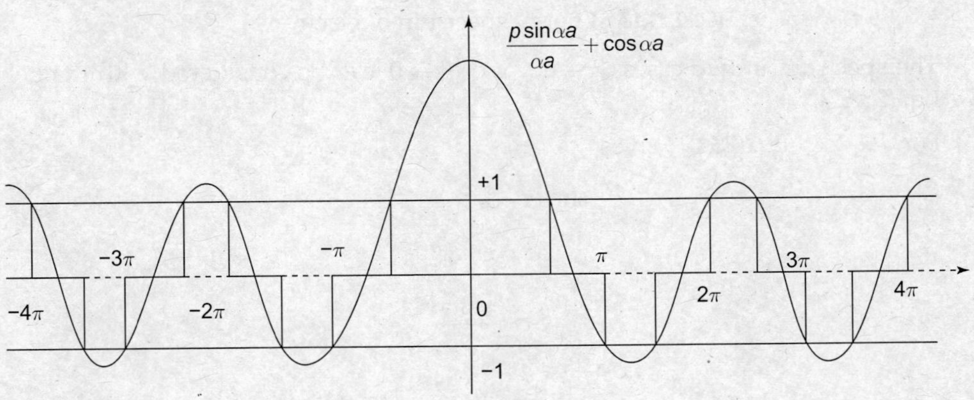

FIGURE 23.17

$$\frac{P \sin \alpha a}{\alpha a} + \cos \alpha a \text{ against } \alpha a \text{ for } P = \frac{3}{2}$$

The following interesting conclusions can be drawn from Fig. 23.17.

(a) The energy spectrum (Fig. 23.18) of the electron consists of alternate regions of allowed energy which are drawn heavily and unallowed energy which are drawn lightly. These regions are known as allowed and forbidden energy bands. The boundaries of allowed ranges of αa corresponds to the value of $\cos ka = +1$ or $ka = n\pi$, i.e. $k = \dfrac{n\pi}{a}$.

(b) As the value of αa increases, the width of energy bands increases and the width of forbidden bands decreases.

(c) When $P \to \infty$, the allowed energy bands are compressed into energy levels and we get a line spectrum. Whereas, when $P \to 0$, we have energy spectrum according to free electron model. In between two limits, we have allowed and forbidden bands.

(d) At any value of P shown by dotted (vertical) line, the shaded areas correspond to allowed bands while unshaded portions correspond to forbidden bands.

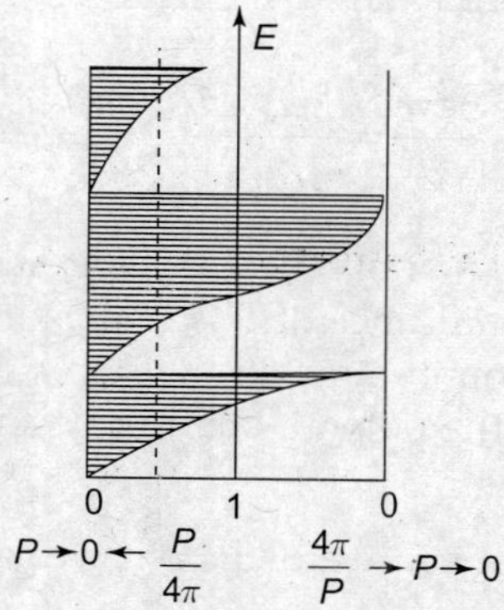

FIGURE 23.18 Energy spectrum of electrons

(e) The spectrum in two cases $P \to \infty$ and $P \to 0$ can be considered with the help of Eq. (23.67).

For $P \to \infty$, Eq. (23.67) gives

$$\sin \alpha a = 0$$

$$\Rightarrow \qquad \alpha a = \pm n\pi$$

$$\Rightarrow \qquad \alpha = +\frac{n\pi}{a},$$

$$\alpha^2 = \frac{n^2\pi^2}{a^2} = \frac{2mE}{\hbar^2}$$

$$\Rightarrow \qquad E = \frac{n^2\pi^2\hbar^2}{2ma^2} \tag{23.69}$$

The energy levels given by Eq. (23.69) are discrete and the electron is completely bound. It will be within potential and moves only in one cell of width a.

On other side, when $P \to 0$, we have

$$\cos \alpha a = \cos ka$$

Now, $\alpha = k$,

$$\alpha^2 = k^2 = \frac{2mE}{\hbar^2}$$

$$\Rightarrow \qquad E = \frac{\hbar^2 k^2}{2m} \tag{23.70}$$

No energy level in this case exists. All energies are allowed to the electrons. The energy corresponds to a completely free particle.

(f) From Eq. (23.70), it is possible to obtain E as a function of wave number k (Fig. 23.19).

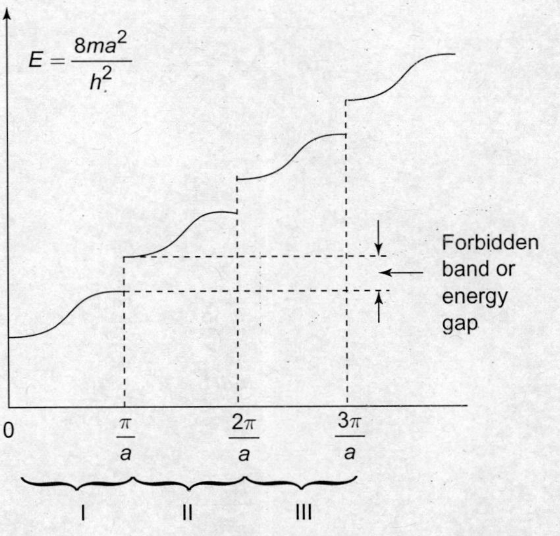

FIGURE 23.19 E as a function of k

Here, discontinuity occurs at $k = \frac{n\pi}{a}$, where $n = 1, 2, 3, \ldots$. The k values define boundaries of I, II and III etc. Brillouin zones. The second Brillouin zone consists of two parts, one from $\frac{+\pi}{a}$ to $\frac{+2\pi}{a}$ and the second from $\frac{-\pi}{a}$ to $\frac{-2\pi}{a}$.

EXAMPLE 23.9

For Kronig–Penney potential with $P << 1$, prove that the lowest energy band at $k = 0$ is $E = \frac{\hbar^2 P}{ma^2}$.

Solution: We know that

$$\frac{P\sin\alpha a}{\alpha a}+\cos\alpha a=\cos ka$$

$$\Rightarrow \quad \frac{P\sin\alpha a}{\alpha a}+\cos\alpha a=1 \qquad (\because k=0)$$

$$\Rightarrow \quad \frac{P}{\alpha a}=\frac{1-\cos\alpha a}{\sin\alpha a}$$

But
$$\cos\alpha a=1-\frac{(\alpha a)^2}{2!}+\cdots=1-\frac{\alpha^2a^2}{2}$$

and
$$\sin\alpha a=\alpha a-\frac{(\alpha a)^2}{3!}+\cdots=\alpha a$$

$$\therefore \quad \frac{P}{\alpha a}=\frac{\alpha^2\,\frac{a^2}{2}}{\alpha a}$$

$$\Rightarrow \quad P=\frac{\alpha^2a^2}{2}$$

$$=\frac{2mE}{\hbar^2}\times\frac{a^2}{2}=\frac{ma^2E}{\hbar^2}$$

$$\Rightarrow \quad \boldsymbol{E=\frac{\hbar^2P}{ma^2}}$$

23.13 ▶ E–K DIAGRAM

The electrons moving in a periodic potential lattice have energy values between allowed energy bands. The allowed energy bands are separated by forbidden energy bands. With the help of Eq. (23.67), it is possible to plot the total energy of the electron E versus the wave number or the propagation vector K and the plot is shown in Fig. 23.20. The right side of Eq. (23.67) becomes $\pm$ 1 for $K=n\pi/a$ and hence discontinuities in the E versus K graph occur at $K=n\pi/a$, where n takes the values ±1, ±2, ±3, ...

From the graph, it is clear that the electron has allowed energy values in the region or zero extending from $K=\frac{-\pi}{a}$ to $K=\frac{+\pi}{a}$. The zone is called the first Brillouin zone.

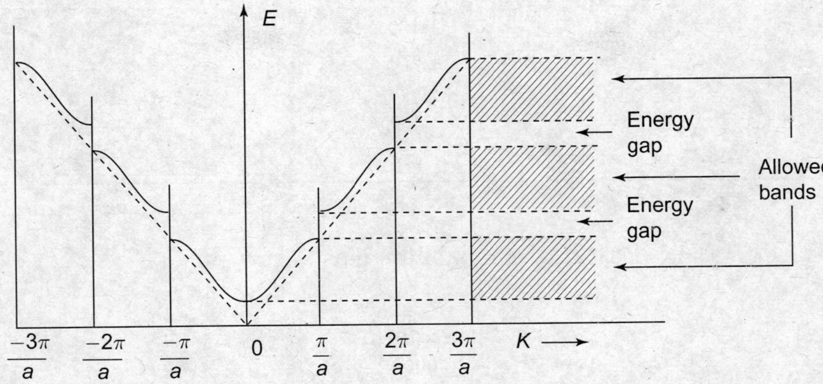

FIGURE 23.20 *E* versus *K* graph

After a break in energy values called forbidden region or band or zone, we get another allowed zone of energy values from $K = \frac{-\pi}{a}$ to $-\frac{2\pi}{a}$ and $K = \frac{+\pi}{a}$ to $+\frac{2\pi}{a}$.

This zone is called second Brillouin zone. Similarly, the other higher order Brillouin zones can also be defined.

Wave functions

At these k values, the electron waves are Bragg reflected according to the formula $2a\sin\theta = n\lambda$ back and forth. Therefore, a standing wave solution exists. The eigen functions of incident and corresponding reflected waves for $k = \pm\left(\frac{n\pi}{a}\right)$ are $e^{in\left(\frac{\pi}{a}\right)x}$ and $e^{-in\left(\frac{\pi}{a}\right)x}$ respectively.

When $n = 1$, then

$$\psi_1 = e^{i(\pi/a)x} + e^{-i(\pi/a)x}$$

$$= \cos\frac{\pi}{a}x$$

and

$$\psi_2 = e^{i(\pi/a)x} - e^{-i(\pi/a)x}$$

$$= \sin\frac{\pi}{a}x$$

Hence, two standing waves are obtained. The probability density curves are shown in Fig. 23.21.

23.13.1 Brillouin Zones in One Dimension

The first Brillouin zone is obtained for $k = -\frac{\pi}{a}$ to $+\frac{\pi}{a}$. The second Brillouin zone extends from $k = -\frac{\pi}{a}$ to $-2\frac{\pi}{a}$ and $k = \frac{\pi}{a}$ to $2\frac{\pi}{a}$ as shown in Fig. 23.22.

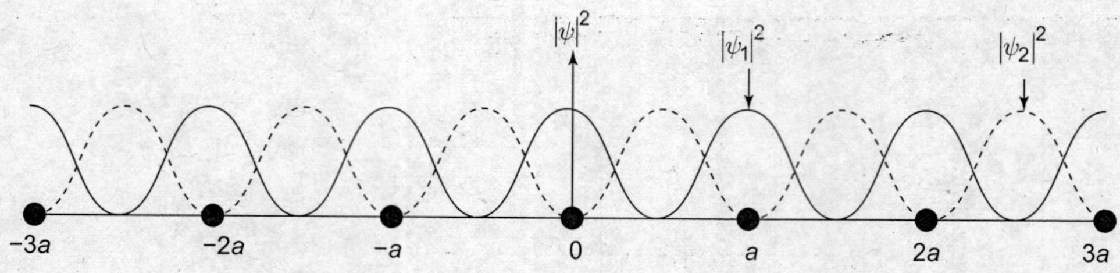

FIGURE 23.21 Probability density curves

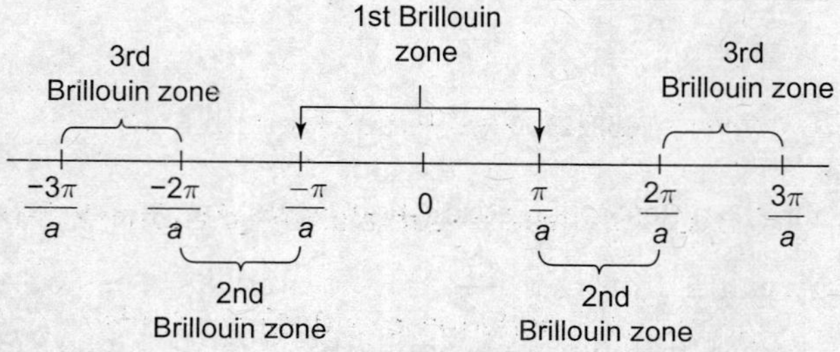

FIGURE 23.22

23.13.2 Brillouin Zone in Two Dimensions

Let us consider the electron moving in the field of two-dimensional square lattice. The wave number k can be analyzed into component along x and y axes, i.e. k_x and k_y.

So, for the first B. zone, $k_x = \pm\frac{\pi}{a}$ and $k_y = \pm\frac{\pi}{a}$. The condition for any energy discontinuity is

$$k = \pm n\frac{\pi}{a}$$

Thus, for two dimensions, the condition leads

$$\boxed{k_x n_1 + k_y n_2 = \frac{\pi}{a}\left(n_1^2 + n_2^2\right)} \tag{23.71}$$

$\Rightarrow n_1 = \pm 1, \quad n_2 = 0$ giving $k_x = \pm\frac{\pi}{a}$

$n_1 = 0, \quad n_2 = \pm 1$ giving $k_y = \pm\frac{\pi}{a}$

Thus, a square passing through A, B, C and D in Fig. 23.23 is called the first B. zone.

The second B. zone should pass through E, F, G and H with $n_1 = \pm 1, n_2 = \pm 1$ the equations give the boundary of second B. zone as

$$n_1 = +1, n_2 = +1; k_x + k_y = \frac{2\pi}{a}$$

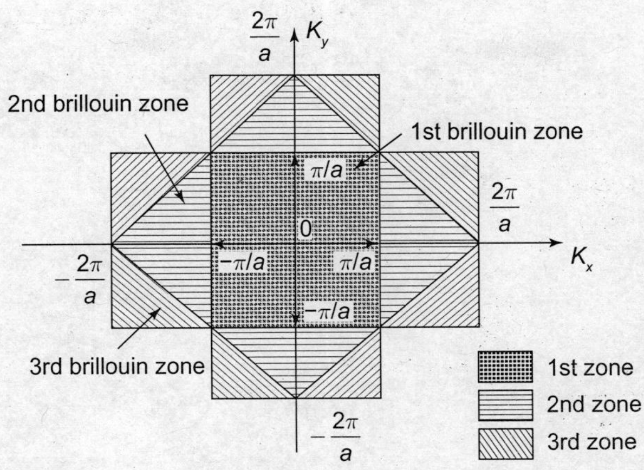

FIGURE 23.23 Brillouin zone in two dimensions

$$n_1 = -1, n_2 = +1; -k_x + k_y = \frac{2\pi}{a}$$

$$n_1 = +1, n_2 = -1; k_x - k_y = \frac{2\pi}{a}$$

$$n_1 = -1, n_2 = -1; -k_x - k_y = \frac{2\pi}{a}$$

Also, the equation used to describe the zones in three dimensions is given by

$$\boxed{k_x n_1 + k_y n_2 + k_z n_3 = \frac{\pi}{a}\left(n_1^2 + n_2^2 + n_3^2\right)} \tag{23.72}$$

EXAMPLE 23.10

Consider two-dimensional square lattice of side 3.0 Å. At what electron momentum values do the sides of first Brillouin zone appear? What is the energy of free electron with this momentum?

Solution: We know $p = \hbar k$

For 1st B. zone, $$k = \pm\frac{\pi}{a}$$

$$p = \frac{\hbar\pi}{a} = \frac{h}{2a}$$

$\Rightarrow$ $$p = \frac{6.6\times10^{-34}}{2\times3\times10^{-10}} = 1.1\times10^{-24}\text{kgm/sec}$$

and $$E = \frac{p^2}{2m} = \frac{\left(1.1\times10^{-24}\right)^2}{2\times9.1\times10^{-31}}$$

$$= 6.6\times10^{-19}\text{J}$$

$$= \frac{\mathbf{6.6\times10^{-19}}}{\mathbf{1.6\times10^{-19}}}\mathbf{eV = 4.2\ eV}$$

Recap Zone

Points to Remember

- Ionic bond is formed due to electrostatic attraction of oppositely charged ions.
- Covalent bond is formed due to mutual sharing of valence electrons between atoms.
- Metallic bond is formed due to attraction between the ion cores and conduction electron gas.
- Vanderwaal bond is formed due to Vanderwaal force between permanent dipoles between similar molecules.
- Vanderwaal bonds are weaker bonds.
- Drude considered the free electrons in a metal to be similar to the molecules of gas.
- Wiedemann–Franz law states that the ratio of thermal conductivity K to electrical conductivity σ is a constant quantity at a given temperature.
- According to quantum theory, the energy of free electron is $E_n = \dfrac{n^2h^2}{8mL^2}$.
- The thermionic current density is given by $J = AT^2\exp(-\phi/K_BT)$. This is called Richardson–Dushman equation.
- Kronig–Penny model explains the energy bands in solids to discrete energy levels in isolated atoms by variation of single parameter P called the potential barrier strength. The value of P is $P = \dfrac{mV_0ab}{\hbar^2}$.
- The E–K graph shows the variation of energy of an electron with the wave vector k. The discontinuities in E–K curve lead the representation of band gap.
- The forbidden gap represents the energies which the electrons cannot occupy.

KEY TERMS WITH DEFINITIONS

- **Bonding** – The process of holding adjacent atoms together.
- **Electrical forces** – Responsible for binding the atoms and molecules giving different solid structures.
- **Cohesive energy** – The energy of a crystal is lower than that of free atoms by amount equal to energy required to pull the crystal into a set of free atoms. This is called cohesive or binding energy.
- **Electro negativity** – Tendency to acquire electrons.
- **Lattice energy** – The equilibrium U_0 energy.
- **Fermi energy** – The energy of the highest occupied level at 0 K and the energy level is called fermi level.
- **Thermionic emission** – The emission of electrons by metals due to heating is called thermionic emission
- **Fermi level** – It is the state at which the probability of electron occupation is 1/2 at any temperature above 0 K.
- **Specific heat** – Heat capacity per unit mass of a substance.

IMPORTANT FORMULAE AND EQUATIONS

Equation Number	Equation	Remarks
(23.5)	$U_0 = \frac{-Ae^2}{4\pi \in_0 (r_0)}\left[1-\frac{1}{n}\right]$	Lattice energy
(23.14)	$\sigma = \frac{ne^2\lambda}{m\bar{v}}$	Electrical conductivity of metals
(23.17)	$\rho = \frac{1}{\sigma} = \frac{1}{ne\mu}$	Resistivity of a solid
(23.23)	$K = \frac{n\bar{v}k_B\lambda}{2}$	Thermal conductivity of a metal
(23.25)	$\frac{K}{\sigma} = LT$	Wiedemann–Franz law
(23.35)	$E_1 = \frac{\pi^2\hbar^2}{2mL^2}$	Zero point energy of a solid
(23.50)	$\log_e \frac{J}{T^2} = \log_e A - \frac{\phi}{k_B T}$	Richardson–Dushmann equation

(23.67)	$\dfrac{P\sin\alpha a}{\alpha a}+\cos\alpha a=\cos ka$	**Kronig–Penny model**
(23.71)	$k_x n_1 + k_y n_2 = \dfrac{\pi}{a}\left(n_1^2 + n_2^2\right)$	**Brillouin zones in two dimensions**
(23.72)	$k_x n_1 + k_y n_2 + k_z n_3 = \dfrac{\pi}{a}\left(n_1^2 + n_2^2 + n_3^2\right)$	**Brillouin zones in three dimensions**

REVIEW ZONE

SHORT ANSWER QUESTIONS

1. What are directional bonds?
2. What is moduling constant? Give its value for NaCl structure.
3. Give two examples of ionic solid and covalent solid.
4. Classify the various types of inter atomic bonds.
5. What are ionic bonds?
6. Explain the terms (a) drift velocity, (b) mobility and (c) relaxation time.
7. Discuss the salient features of free electron theory.
8. What are Brillouin zones?
9. What is E–K curve?
10. Write a short note on band theory of solids.
11. Define fermi energy.
12. What do you mean by lattice energy?

LONG ANSWER QUESTIONS

1. Explain the term binding energy. How is it calculated for NaCl structure?
2. What are ionic and covalent bonds? Explain the characteristics of ionic and covalent crystals.
3. What is the difference between metallic and hydrogen bonds? Also explain their characteristics.
4. What is free electron theory? Derive an expression for conductivity of metals on the basis of Drude–Lorentz theory.
5. Define Wiedemann and Franz law. Derive the ratio of thermal conductivity to electrical conductivity.
6. What do you mean by thermionic emission? Also derive Richardson–Dushmann equation.
7. Discuss the origin of energy band structure in solids.
8. Discuss in detail the band theory of solids and explain the formation of bands.

9. Explain Kronig–Penny model. Prove that $\frac{P \sin \alpha a}{\alpha a} + \cos \alpha a = \cos ka$.

10. Discuss the formation of discrete energy levels in solids using Kronig–Penny model.

11. Illustrate the three Brillouin zones for a two-dimensional square lattice.

12. Explain the significance of Brillouin zones with particular reference to any cubic lattice.

NUMERICAL PROBLEMS

1. Obtain the volume density of Si atoms with its lattice constant of 5.3 Å.

   Ans. 5×10^{22} atoms/cm^3

2. Calculate the drift velocity and mean free path of copper when it carries a steady current of 10 ampere and whose radius is 0.08 cm. Assume that the mean thermal velocity = 1.6×10^6 m/s and the resistivity of copper = 2×10^{-8} ohm-m.

   Ans. 36.6×10^{-5} m/s, 3.94×10^{-8} m

3. A copper wire of length 0.5 m and diameter 0.3 mm has a resistance 0.12 Ω at 20°C. If the thermal conductivity of copper at 20°C is 39° W/mK. Calculate the Lorentz number.

   Ans. 1.12×10^{-8} wΩK^{-2}

4. Find the fermi energy in copper on the assumption that each copper atom contributes one free electron to the electron gas. The density of copper is 8.94×10^3 kg/m^3 and its atomic mass is 63.5 u.

   Ans. 7.04 eV

5. Find the least energy of an electron moving in one dimension in an infinitely high potential box of width 0.05 nm.

   Ans. 497.6×10^4 amp/m^2, 3.6×10^{-4} m/s

6. The relaxation time of conduction electron is 10^{-14}s. Compare the thermal conductivity of the material at 0°C and 100°C. Assume density of electrons to be 6×10^{28}/m^3.

   Ans. 290 K

7. A copper wire has a resistivity of 1.8×10^{-8} Ω at room temperature (300 K). Assuming copper is very pure, estimate the resistivity at 700°C and the percentage change in the resistivity from room temperature to 700°C.

   Ans. 5.84×10^{-8} Ω-m, 22.4%

8. Find the ratio of kinetic energies of electron in two-dimensional square lattice which has $k_x = k_y = \frac{\pi}{a}$ and an electron has $k_x = \frac{\pi}{a}, k_y = 0$.

   Ans. 2

9. The electronic heat capacity of zinc is $\cong 1.5 \times 10^{-4}$ T cal/mol K. Find the fermi energy of zinc.

   Ans. 11.4 eV

CHAPTER 24

Semiconducting Material

Learning Objectives

By the end of this chapter, the student will be able:

- To understand the band theory of solids
- To differentiate between intrinsic and extrinsic semiconductors
- To examine the effect of temperature on Fermi function and Fermi level
- To study semiconductor carrier concentrations
- To describe the hall effect and its applications

24.1 ▶ Introduction

When Edison, in 1883, observed that electric current would pass between the heated filament and positive metal plate, then electronic science began. Sir J.J. Thomson in 1897 identified the current as a flow of negative charges, which were named as electrons by Johnstone Stoney a little later.

The current in a solid semiconductor flows due to the result of both electrons and positive charges (holes). By addition of impurity elements to pure Silicon or Germanium semiconductors, we can control the densities of free charges as well as the type of semiconductors. In this chapter, we will study the conduction in solids and semiconductors. Also, we will learn the basic concepts of semiconductors and their applications.

24.2 ▶ Crystalline and Amorphous Solids

The substances in which particles are very closely packed and which have a definite shape and volume are called solids.

The solids may be divided into two broad categories (Fig. 24.1):

1. Amorphous solids
2. Crystalline solids

Amorphous solids

The amorphous solids are those which lack the regular arrangement of atoms or molecules in their structure. These are also called super-cooled liquids with a short range order. These are rigid and have stiffness unlike liquids. The bond between atoms is not equally strong. That's why these solids do not have sharp melting points because on heating, the weak bonds are ruptured first at lower temperatures but the strong bonds rapture at high temperature. Examples: glass, pitch and a few plastic substances.

Crystalline solids

The crystalline solids are those which contain the regular and repeated pattern of atoms or ions or molecules in their structures. These are single crystals having long range orders. All the bonds between their atoms are equally strong due to long range order symmetry. That's why these have sharp melting points. Examples: rock salt, calcite, sugar and quartz.

Note

There are some solids like B_2O_2 which can exist in both crystalline and amorphous forms.

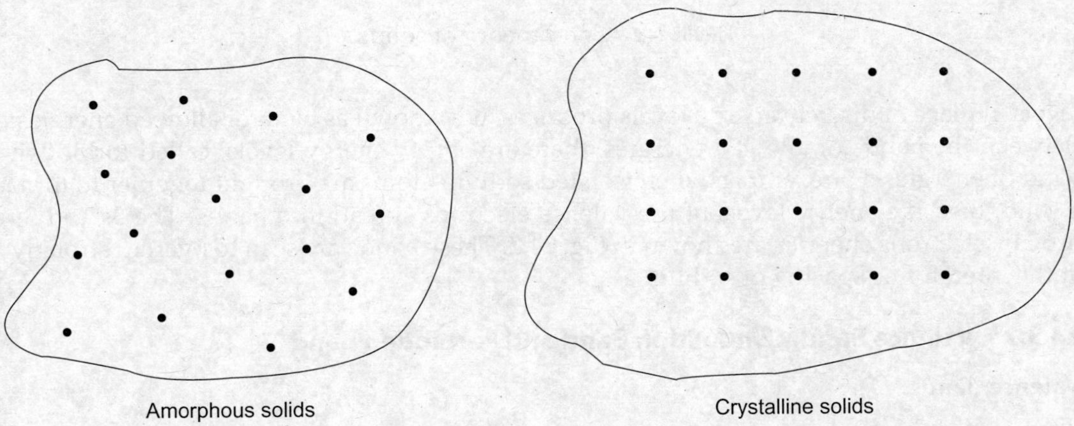

FIGURE 24.1 Types of solids

24.3 ▶ Band Theory of Solids

24.3.1 Energy Bands

An isolated atom possesses discrete energies of different electrons. Suppose two isolated atoms are brought to very close proximity, then the electrons in the orbits of two atoms interact with each other. So, in the combined system, the energies of electrons will not be in the same level but changes and energies will be slightly lower and larger than the original value. Thus, instead of each energy level, a closely spaced two energy levels exist. If 'N' number of atoms is brought together to form a solid, then 'N' number of closed spaced

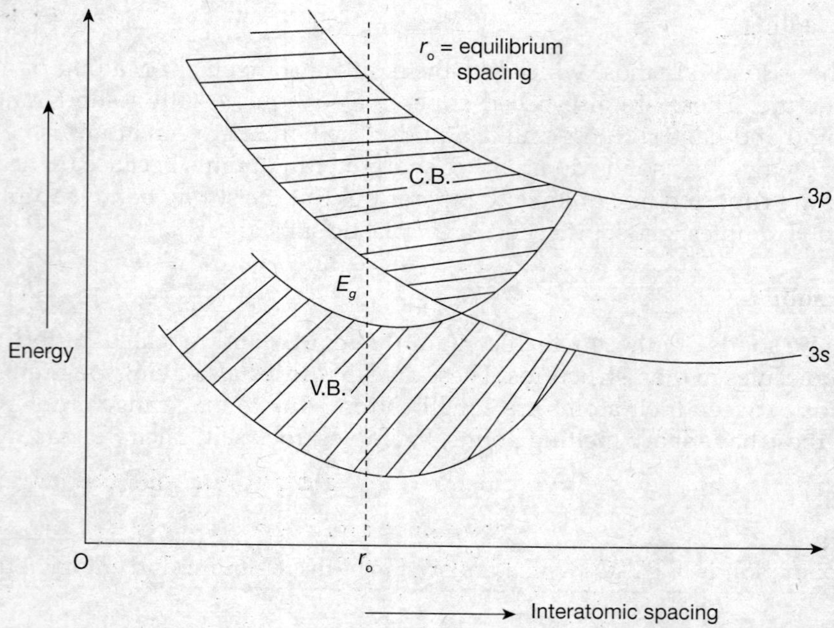

FIGURE 24.2 Band theory of solids

levels in place of discrete energy levels produces, it is known as band of allowed energies. Between the bands of allowed energies, there are empty energy levels, called forbidden band of energies. For example, when isolated sodium atoms are brought together to form a solid, then the energy levels of the valence electrons spread into bands. The 3*s* and 3*p* orbitals electrons energies are shown in Fig. 24.2. These bands are seen to overlap strongly at the interatomic spacing of sodium.

24.3.2 Valence Band, Conduction Band and Forbidden Band

Valence band

The valence band corresponds to the valence electrons present in different atoms of the material.

Conduction band

The electrons in the conduction band are free electrons. The energy band occupying the conduction electrons is known as conduction band. Conduction band electrons are actually responsible for the flow of current. More the number of electrons in the conduction band, more will be the flow of current.

Forbidden band

Forbidden band is the energy gap that separates the conduction and valence bands. No electrons can normally exist in the band gap. The band gap is measured in electron volts (eV). The substances are classified as conductors, semiconductor and insulators based on the width of forbidden gap.

24.4 ▶ Classification of Solids on the Basis of Band Theory

24.4.1 Conductors, Semiconductors and Insulators

If an electron has to experience acceleration in the presence of an electric field, then it must be able to move to slightly higher energy states. This demands the presence of empty and allowed, but slightly higher energy states. If any band is relatively empty and there are a few electrons in that band, then a large number of unoccupied states are available for them into which they can move if subjected to an electric field. If they move, they contribute to current and the material is a good conductor.

As the electrons have to obey Pauli's exclusion principle, the number of electrons that can be placed in a given band is limited. As '*s*' band can accommodate a maximum of $2N$ electrons in $2N$ different energy states, within the band. A '*p*' band can accommodate a maximum of $6N$ electrons in $6N$ different energy states, a '*d*' band can accommodate a maximum of $10N$ electrons within the band. The highest energy band containing the electrons is called the *valence band*. If a material has partially filled valence band, then it can conduct electricity. Also, if a vacant higher energy band (conduction band) overlaps with valence band, then the electrons can easily move to the vacant states and conduct electricity on application of field. As conduction band overlaps with valence band for metals, they are good conductors.

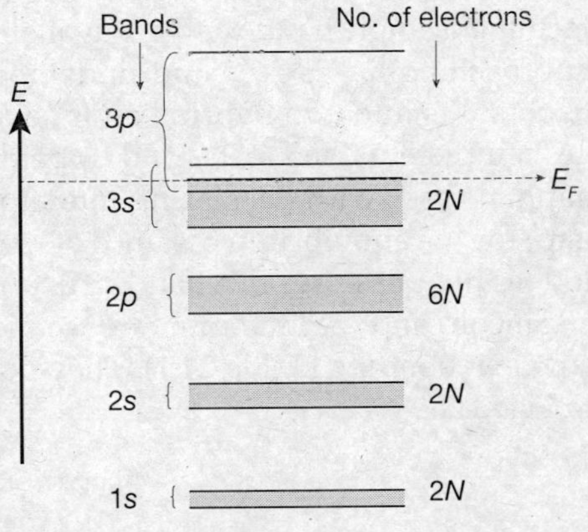

FIGURE 24.3 Energy bands of magnesium solid. Magnesium is a metal due to the overlap of 3*s* and 3*p* bands

Energy bands overlap for metals. Consider another example, magnesium (Mg). The atomic number is $Z = 12$. The electronic configuration is $1s^2 2s^2 2p^6 3s^2$ (refer Fig. 24.3). The 3*s* band is completely filled with $2N$ valence electrons. The next higher energy band (conduction band) 3*p* overlaps with the 3*s* band. 3*p* band has $6N$ unoccupied states. Some states from the top of 3*s* (not completely filled) move to the bottom of 3*p* (partially filled) due to the overlap.

Energy bands do not overlap for semiconductors and insulators. There exist a band gap between the valence band and the next higher energy vacant band (conduction band). This energy gap is called forbidden energy gap, E_g. If the band gap is smaller, i.e. $E_g \leq 1$ eV, then the material is a semiconductor. Good insulators have large band gaps, i.e. $E_g > 5$ eV.

Now consider carbon in its diamond structure, which is a perfect insulator. The electronic configuration of carbon is $1s^2 2s^2 2p^2$. In a solid of N atoms, there are $6N$ electrons. The 2*s* band is completely filled with all $2N$ electrons ($2N$ states), 2*p* band has $6N$ states but only $2N$ are filled states; the remaining $4N$ are vacant states. Then, it looks like carbon is a conductor, but it is not. The reason is explained as follows.

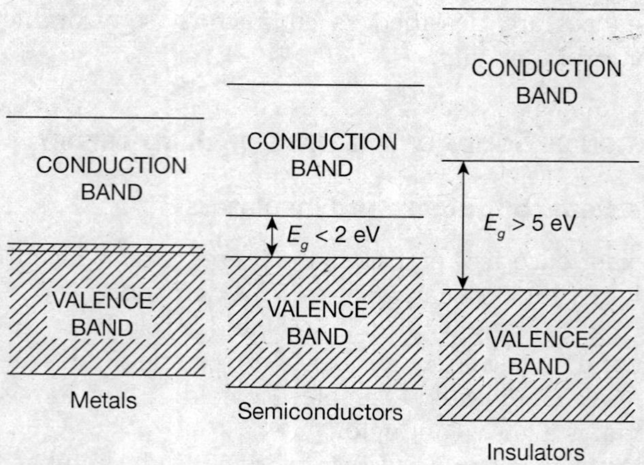

$E_g < 2$ eV $\quad E_g > 5$ eV

FIGURE 24.4 Band gaps of solids

As interatomic distance decreases (refer Fig. 24.4), the 2*s* and 2*p* energy bands overlap, forming a single band with 8*N* available states. As interatomic separation decreases further approaching r_0, '2*s* 2*p*' band undergoes a complex process called *hybridization*. The single band splits into two hybrid bands separated by an energy gap, E_g. Lower hybrid band has 4*N* completely filled states and upper hybrid band has 4*N* completely vacant states. Energy gap increases with decreasing interatomic separation. For diamond, E_g is 6 eV (large band gap) for the equilibrium distance of $r_0 \approx 1.5$ Å. Graphite has a different crystal structure and it does not possesses this band structure.

Silicon and germanium are semiconductors having diamond cubic structure with covalent bonding (Table 24.1). They possess similar hybrid energy band structure as shown in Fig. 24.5.

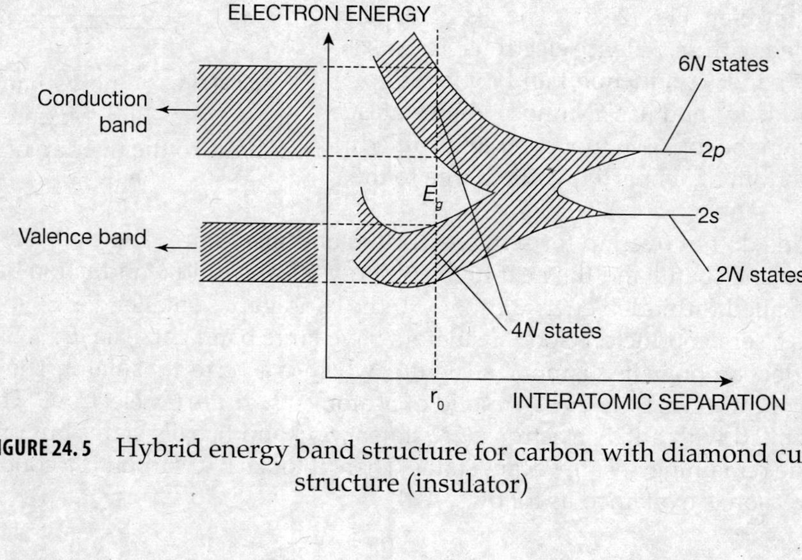

FIGURE 24.5 Hybrid energy band structure for carbon with diamond cubic structure (insulator)

TABLE 24.1 Elements having hybrid energy band structure

Element	Electronic configuration	Energy band hybridization	Band gap	Classification
Carbon (diamond)	$1s^2 2s^2 2p^2$	2s-2p	6 eV	Insulator
Silicon	$1s^2 2s^2 2p^6 3s^2 3p^2$	3s-3p	1.1 eV	Semiconductor
Germanium	$1s^2 2s^2 2p^6 3s^2 3p^6 4s^2 4p^2$	4s-4p	0.7 eV	Semiconductor

Note

The energy band gap E_g of a semiconductor is of the order of 1 eV, whereas insulators have band gaps in the range 5–15 eV.

24.4.2 Single Crystal and Compound Semiconductors

In general, semiconductors are of two types: single crystal and compound.

Single crystal semiconductors, such as gallium and silicon, have a repetitive crystal structure.

Compound semiconductors such as gallium arsenide (GaAs), gallium nitride (GaN), cadmium sulphide (Cds), etc. are made up of two or more semiconducting materials of different atomic structures. The three semiconductors that are frequently used in electronic devices are Ge, Si and GaAs.

24.5 ▶ Types of Semiconductors

24.5.1 Intrinsic Semiconductors

Based on electrical conductivity, materials are divided into conductors, insulators and semiconductors. Usually, metals are good conductors of electricity and all dielectrics are insulators. The electrical conductivity of semiconductors lies in between metals and dielectrics. Good examples for semiconductor are germanium and silicon. These elements belong to IV group in the periodic table. At 0 K, these elements are insulators, whereas at room temperatures they possess certain amount of conductivity. Pure germanium and silicon are called intrinsic semiconductors. By adding a small quantity of either III group or V group element atoms as impurity into pure Ge or Si, the electrical conductivity of the material increases. This impure semiconductor is called an extrinsic semiconductor.

Pure germanium or silicon crystal is called an intrinsic semiconductor. Each semiconductor atom possesses four valence electrons in the outermost orbit. To get stability, each of these atoms has to get eight electrons in the outermost orbit, so that each atom makes four covalent bonds with the surrounding four other atoms in the crystal. A two-dimensional representation of the crystal structure of silicon (or germanium) at 0 K is shown in Fig. 24.6(a). The band diagram of this material is shown in Fig. 24.6(b).

At 0 K, all the valence electrons of Si atoms are participating in covalent bonds and their energies constitute a band of energies called valence band. So, at 0 K, valence band is completely filled; conduction band is empty of electrons. The allowed band of energies above valence band is called conduction band. Suppose, if we increase the temperature of

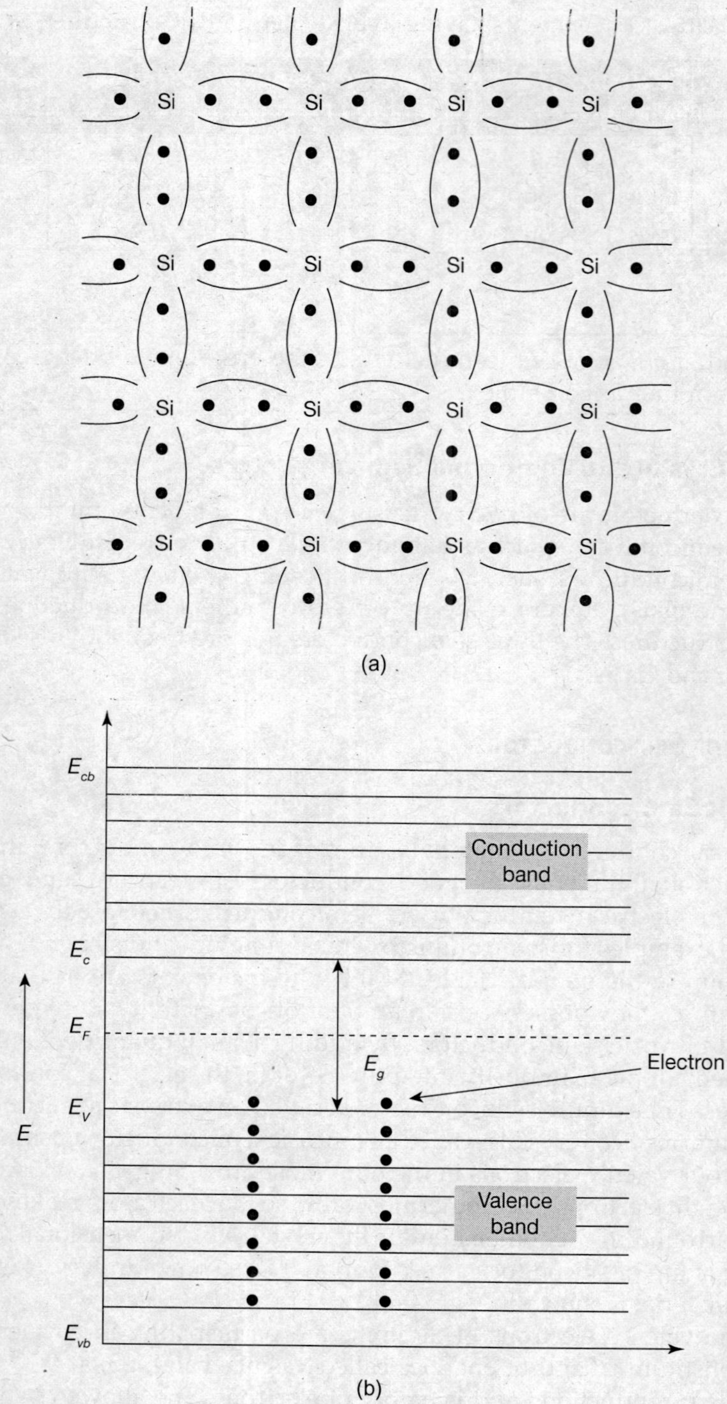

FIGURE 24.6 (a) Crystal structure of Si at 0 K; (b) Band diagram of Si at 0 K

the semiconductor to some room temperature *TK*, at this temperature some of the electrons which are participating in covalent bonds and present in the top energy levels of valence band will take thermal energies. If the increase in the thermal energy of electrons present in top energy levels of valence bond, is equal to or greater than energy gap of the semiconductor, then electrons come away from bonding and move freely inside the crystal as shown in Fig. 24.7(a). Now these electrons possess energies equal to the lower energy levels of conduction band. These free electrons participate in electrical conduction; hence, the band in which these electrons present is named as conduction band. If an electron comes away from bonding, then that atom acquires one unit positive charge, then it participates in electrical conduction. This electron vacancy or electron deficiency of an intrinsic semiconductor is called hole. The electron vacancies in valence band will exist as holes in the valence band as shown in Fig. 24.7(b).

24.5.2 Extrinsic Semiconductors

The conductivity of an intrinsic semiconductor can be increased enormously by adding small amounts of impurity atoms (such as III or V group atoms). For example, the conductivity of silica is increased by 1000 times on adding 10 parts of boron per million part of Si. The process of adding impurities is called doping and the impurity added is called dopant. The doping is done during crystallization process (i.e., when the substance is in molten state). On crystallization, the impurity atoms replace some of the intrinsic atoms.

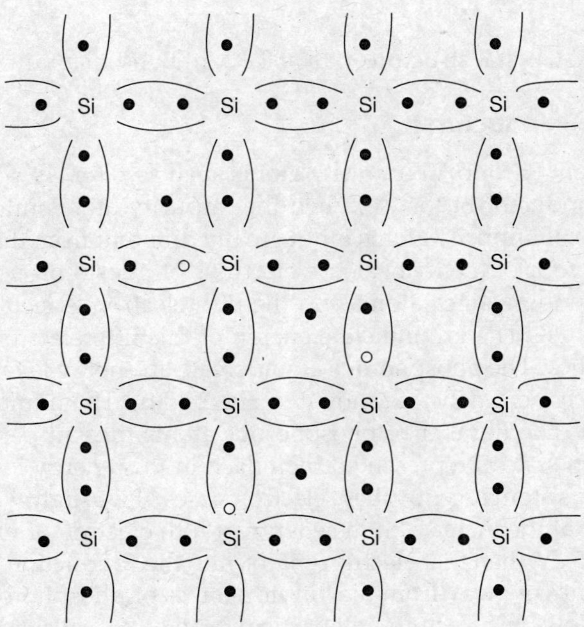

(a)

(Cont'd)

FIGURE 24.7 (Continued)

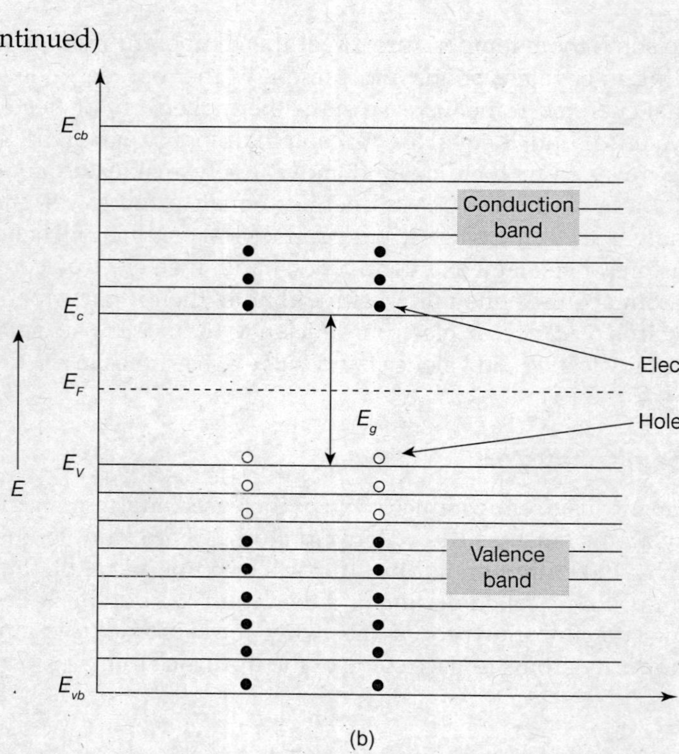

FIGURE.24.7 (a) Crystal structure of Si at *T* K. (b) Band diagram of Si at *T* K

24.5.3 *n*-Type Semiconductors

When some pentavalent (Group V) impurity atoms such as *P* (or *As*, *Sb*, *Bi*, etc.) are added to an intrinsic semiconductor Si (or Ge), then the impurity atoms interlock in the crystal lattice because the size of impurity atoms is not greatly different from that of silica atoms. As shown in Fig. 24.8(a), four of the five valence electrons of phosphorous will make covalent bonds with the surrounding silica atoms and the fifth electron is feebly attached with the phosphorous atom at 0 K. The ground state energy of this fifth electron lies just below the conduction band of silica. The phosphorous atoms create an energy level that lies at 0.045 eV below the conduction band of silica as shown in Fig. 24.8(b). The number of electron states in this level is equal to the number of phosphorous atoms present per unit volume of the material. Each electron state is represented by a dash in this energy level. At 0 K, the fifth electron of phosphorous atoms occupy these electron states. If we slightly rise the temperature of the material such that the increase in the energy of fifth electron of phosphorous is equal to or greater than 0.045 eV, then that electron enters, into the conduction band of silica. In the crystal, that electron moves away from the binding forces of phosphorous atom and moves freely in the crystal. Now, this electron will participate in electrical conduction.

The increase in temperature to free the fifth electron of phosphorous is very much less compared to the increase in temperature to break a covalent bond in the crystal. This means that the phosphorous atoms readily give electrons for conduction, and also the energy level

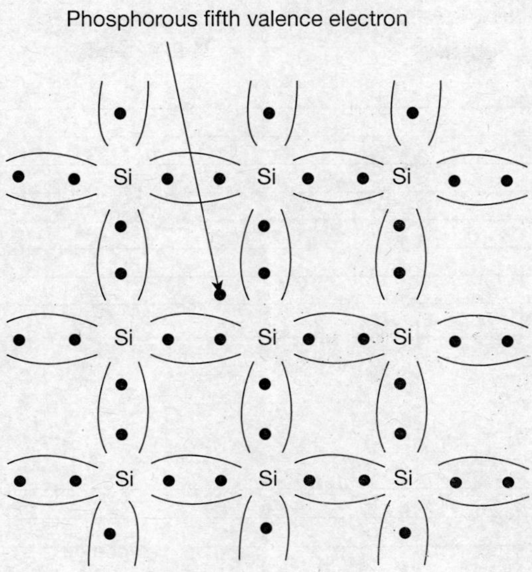
Phosphorous fifth valence electron
Si

(a)

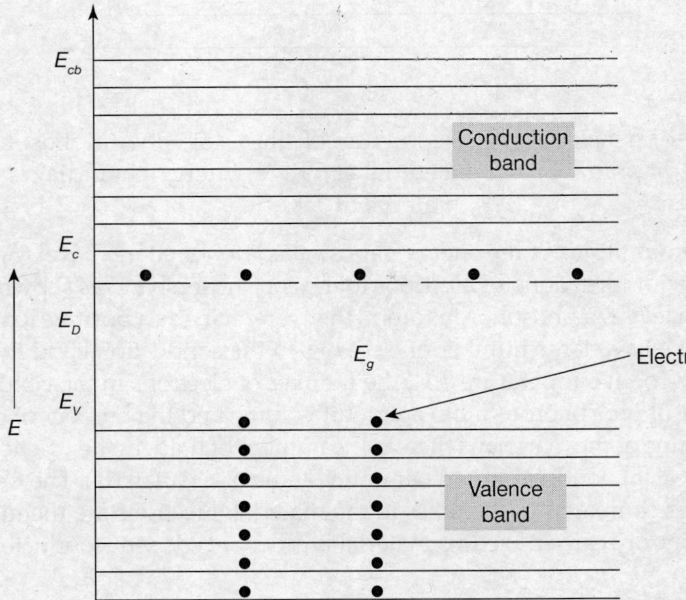
E_{cb}
Conduction band
E_c
E_D
E_g
Electron
E_V
E
Valence band
E_{vb}

(b)

(Cont'd)

FIGURE 24.8 (Continued)

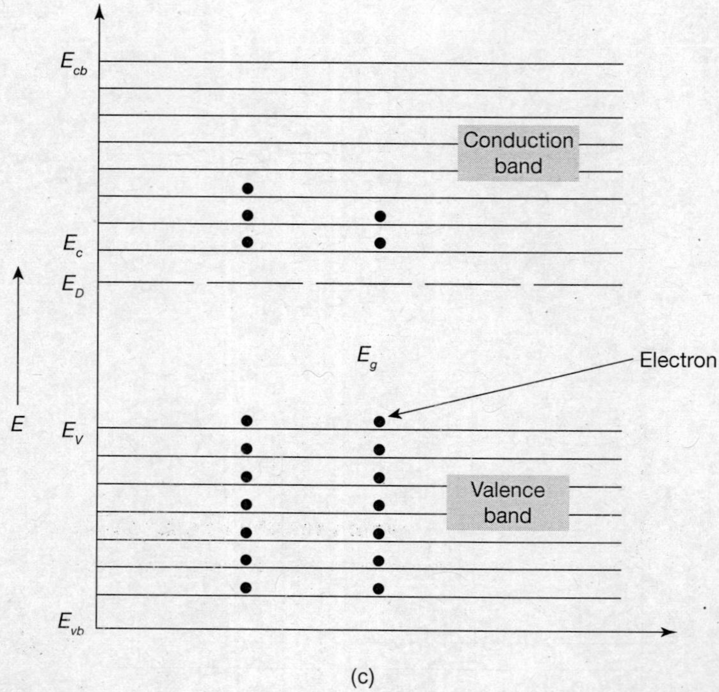

FIGURE 24.8 (a) Two-dimensional crystal structure of silica doped with phosphorous atoms. (b) Energy band diagram of n-type material at 0 K. (c) Energy band diagram of n-type material at T K

created by phosphorous atoms in the energy gap is called donor energy level (E_D). Even at low temperatures, the donor level donates electrons to the conduction band as shown in Fig. 24.8(c).

When an electron leaves a phosphorous atom, then it remains as a positive ion, but holes will not be created. So, we have large number of electrons in the conduction band than the holes in the valence band. At room temperature T K, the number of electrons in the conduction band is equal to the number of phosphorous atoms per unit volume and the number of covalent bonds broken per unit volume of the material, whereas the number of holes in the valence band is equal to the number of covalent bonds ruptured per unit volume of the material. The free electrons are very large in number compared to the holes in the material. Electrons are the majority carriers and holes are the minority carriers, so this material is called n-type semiconductor.

24.5.4 *p*-Type Semiconductors

Instead of adding pentavalent impurity atoms, if trivalent (Group III) impurity atoms such as *B* (or *Al*, *Ga*, *In*, etc.) atoms are doped in an intrinsic semiconductor Si (or Ge), then the impurity atoms occupy some of the silica atom sites because the size of impurity atoms and Si atoms is almost the same. As shown in Fig. 24.9(a), with three valence electrons each boron atom makes three covalent bonds with the surrounding three silica atoms. To attain stability, it completes fourth covalent bond by accepting a stray electron in its vicinity. Since this stray electron belongs to a silica atom, silica atom acquires a hole. The boron atoms introduce an energy level just

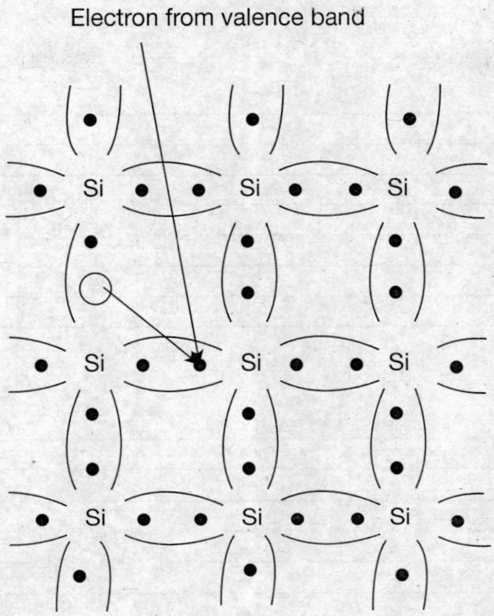
Electron from valence band
Si
Si
Si
Si
Si
Si
Si
Si
Si

(a)

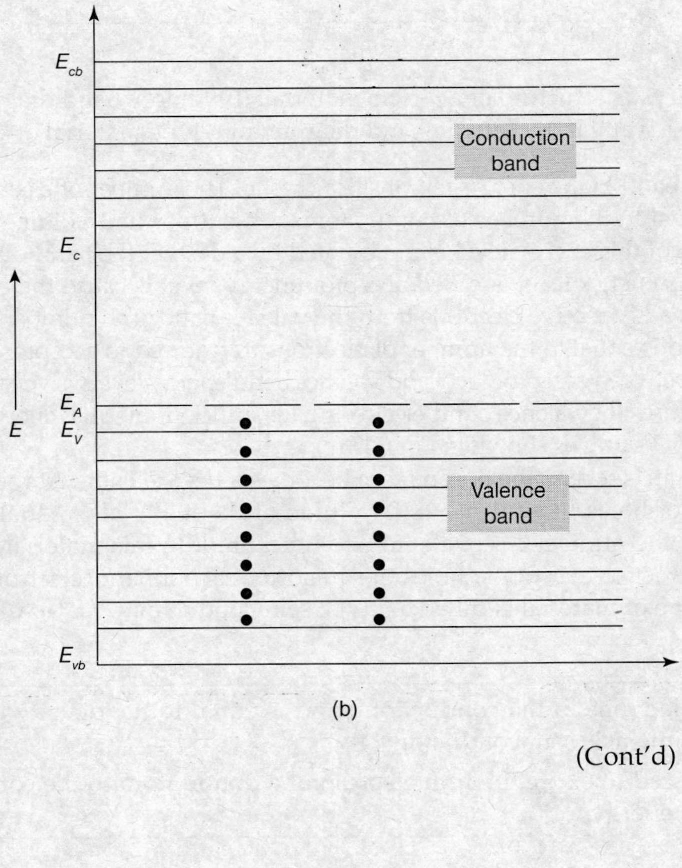
E_{cb}
Conduction band
E_c
E_A
E
E_V
Valence band
E_{vb}

(b)

(Cont'd)

FIGURE 24.9 (Continued)

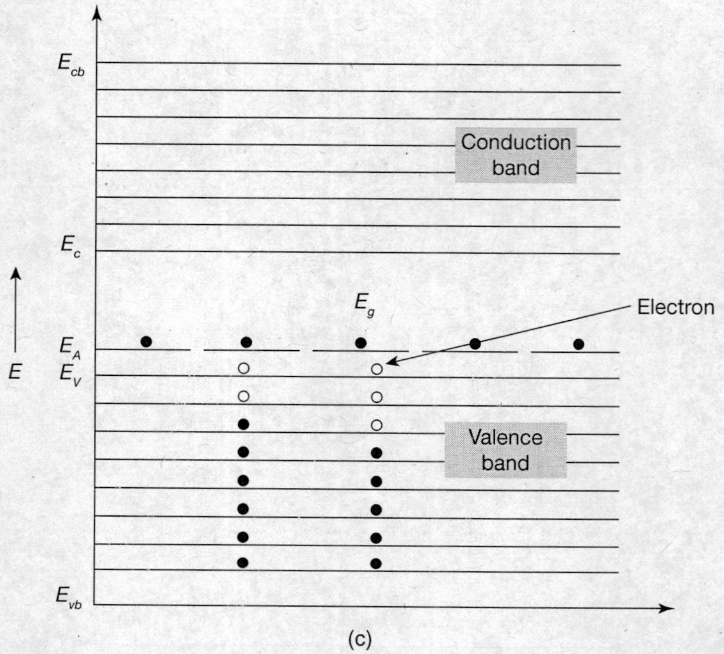

FIGURE 24.9 (a) Crystal structure of a *p*-type material. (b) Energy band diagram of *p*-type material at 0 K. (c) Energy band diagram of *p*-type material at *T* K

above the valence band in the energy gap of silica crystal. The number of electron states in this energy level is equal to the number of boron atoms present per unit volume of silica crystal. Each electron state of this energy level is represented by a dash in the band diagram as shown in Fig. 24.9(b). This energy level is called acceptor energy level because the electron states in this energy level readily receive electrons from the valence band. The number of holes formed in the valence band is equal to the number of electrons transferred to acceptor states. Since the energy difference between acceptor level and valence band energy levels is very small so even at very low temperatures the valence band electrons gain sufficient thermal energy to occupy the acceptor energy level states as shown in Fig. 24.9(c).

After receiving an electron, the boron atom becomes a negative ion. At room temperature *T* K, the number of holes present per unit volume of the material is equal to the number of boron ions and the number of covalent bonds ruptured. In this material, holes are large compared to conduction electrons; holes are majority carriers and electrons are minority carriers, so this type of material is called a *p*-type semiconductor.

Note

The density of electrons in the conduction band is equal to the density of holes in the valence band in intrinsic semiconductor.

The energy required to excite the fifth electron of a donor atom to the conduction band is called binding energy.

EXAMPLE 24.1

Calculate the approximate donor binding energy for GaAs. Assume that $\varepsilon_r = 13.2$ and $m_n^* = 0.0615\ m_e$.

Solution: The binding energy is given as

$$E = \frac{m_n^* e^4}{8(\varepsilon_0 \varepsilon_r)^2 h^2}$$

$$= \frac{0.06245 \times 9.1 \times 10^{-31} \times (1.6 \times 10^{-19})^4}{8 \times (8.85 \times 10^{-12} \times 13.2)^2 \times (6.63 \times 10^{-34})^2}$$

or

$$E = 8.34 \times 10^{-22}\ \text{J}$$

$$= 0.0052\ \text{eV}$$

It means that 5.2 meV energy is required to excite the donor atom.

24.6 ▶ Fermi Distribution Function

Maxwell-Boltzmann distribution formula gives relative probability of occupancy of two states i and j with energies E_i and E_j as

$$N_i = N_j e^{-\frac{E_i - E_j}{kT}} \quad \text{Maxwell-Boltzmann distribution formula}$$

According to Pauli's exclusion principle, no two electrons can occupy the same quantum state in an atom. This is the fundamental principle that governs the electronic configuration of atoms having more than one electron. Electrons obey Pauli's exclusion principle in solids also. Each electron in the solid is described by a unique wave function called a *state.* The presence of an electron in a certain state prevents any other electron to be in that state. Fermi and Dirac formulated a distribution function for the occupancy of electrons in a solid as

$$\boxed{F(E_i) = \frac{1}{e^{\frac{E_i - E_F}{kT}} + 1}} \quad \text{Fermi distribution function} \tag{24.1}$$

E_F is a constant known as *Fermi energy.* This function gives the probability of finding the electrons in various energy levels at any given temperature. This Fermi distribution function can be applied to all particles which are identical, indistinguishable and obey Pauli's exclusion principle. Such particles are called *fermions*.

24.6.1 Effect of Temperature on Fermi Function

At $T = 0$ K,

For $$E < E_F, \ f(E) = 1$$

For $$E > E_F, \ f(E) = 0$$

This means that all the energy levels before the Fermi energy, E_F, are completely filled with electrons, whereas all the levels above E_F are vacant.
At $T > 0$ K, for $E = E_F$,

$$f(E) = \frac{1}{2} = 0.5$$

At higher temperatures, the probability of electrons occupying Fermi energy state is 50%. From this, *Fermi Energy* can also be defined as the energy at which the probability of occupation of states is 0.5 or 50%. As the temperature is increased, the electrons below E_F get thermally excited and move above E_F to slightly higher energy states in metals.

The thermal energy at a temperature, T, is kT, where k is the Boltzmann's constant.
When

$$E = E_F - kT$$

$$F(E) = \frac{1}{e^{(-1)} + 1} = 0.73 \rightarrow 73\% \tag{24.2}$$

of energy states are occupied by electrons. The remaining 27% of energy states below E_F are vacant because the electrons moved to states above E_F. This is verified below.
When

$$E = E_f + kT$$

$$F(E) = \frac{1}{e^{(1)} + 1} = 0.27 \rightarrow 27\%, \tag{24.3}$$

of energy states are occupied by electrons above E_F. Thus, in an energy span of $2kT$, the probability of occupancy of electrons falls from 73% to 27% around E_F. Fermi energy for metals are about a few electron volts. The thermal energy at room temperature 300 K is about 0.025 eV which is much smaller than the Fermi energy for most of the metals (Fig. 24.10).

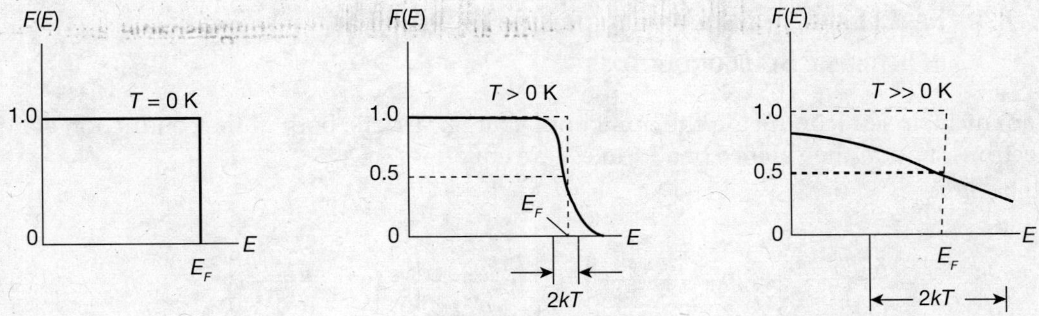

FIGURE 24.10 Fermi distribution function (a) at absolute zero $T = 0$ K, all the energy states upto E_F are fully occupied (b) at a higher temperature $T > 0$ K (c) at *very* high temperatures

EXAMPLE 24.2

An intrinsic semiconductor has an energy gap of 0.4 eV. Calculate the probability of occupation of the lowest level in conduction band at (i) 0°C and (ii) 50°C.

Solution: We know that the probability of occupation is

$$F(E) = \frac{1}{1 + \exp\left[\frac{E - E_F}{KT}\right]}$$

but

$$E_F = \frac{1}{2} E_g$$

Thus,

$$F(E) = \frac{1}{1 + \exp\left[\frac{E_g}{2KT}\right]}$$

(i) At $T = 0°\text{C} = 273$ K, $E_g = 0.4 \text{ eV} = 0.4 \times 1.6 \times 10^{-19}$ J

$$= 6.4 \times 10^{-20} \text{ J}$$

$$F(E) = \frac{1}{1 + \exp\left[\frac{6.4 \times 10^{-20}}{2 \times 1.38 \times 10^{-23} \times 273}\right]} = \mathbf{2.025 \times 10^{-4}\ eV}$$

(ii) At T = 50°C = 323 K

$$F(E) = \frac{1}{1 + \exp\left[\frac{6.4 \times 10^{-20}}{2 \times 1.38 \times 10^{-23} \times 323}\right]} = \mathbf{7.55 \times 10^{-4}\ eV}$$

24.7 ▶ Fermi Level and its Variation with Temperature in Intrinsic Semiconductors

In an intrinsic semiconductor, the number of charge carriers, both in the conduction band (electrons) and in the valence band (holes) are equal,
i.e., $n = p$

$$2\left(\frac{2\pi m_e{}^* kT}{h^2}\right)^{3/2} \exp\left(\frac{E_F - E_C}{kT}\right) = 2\left(\frac{2\pi m_h{}^* kT}{h^2}\right)^{3/2} \exp\left(\frac{E_v - E_F}{kT}\right) \tag{24.4}$$

$$m_e{}^{\frac{3}{2}} \exp\left(\frac{E_F - E_C}{kT}\right) = m_h{}^{*\frac{3}{2}} \exp\left(\frac{E_v - E_F}{kT}\right)$$

$$\left(\frac{m_h{}^*}{m_e{}^*}\right) = \frac{\exp\left(\frac{E_F - E_C}{kT}\right)}{\exp\left(\frac{E_v - E_F}{kT}\right)}$$

$$\left(\frac{m_h{}^*}{m_e{}^*}\right)^{\frac{3}{2}} = \exp\left[\frac{(E_F - E_C) - (E_v - E_F)}{kT}\right]$$

$$\left(\frac{m_h{}^*}{m_e{}^*}\right)^{\frac{3}{2}} = \exp\left(\frac{2E_F - E_C - E_V}{kT}\right)$$

Taking log on both sides,

$$\frac{3}{2}\log\left(\frac{m_h{}^*}{m_e{}^*}\right) = \left(\frac{2E_F - E_C - E_V}{kT}\right)$$

$$\frac{3}{2}kT\log\left(\frac{m_h{}^*}{m_e{}^*}\right) = [2E_F - (E_C + E_V)]$$

$$E_F = \frac{3}{4}kT\log\left(\frac{m_h{}^*}{m_e{}^*}\right) + \left(\frac{E_C + E_V}{2}\right)$$

Theoretically, the effective masses of the electrons and the holes are equal, $m_h{}^* = m_e{}^*$

$$\therefore \quad E_F = \left(\frac{E_C + E_V}{2}\right) \tag{24.5}$$

where E_C is the bottom of the conduction band and E_V the top of the valence band.

Hence in an intrinsic semiconductor in which the number of electrons and holes is equal, the Fermi level lies exactly at the midway between the conduction and valence bands as given by Eq. (24.5). Also, when the charge carriers have equal effective masses, the Fermi level is independent of temperature lines rather than solid lines.

24.7.1 *n*-Type Extrinsic Semiconductors

We know that the density of electrons in the conduction band is given by

$$n = 2\left(\frac{2\pi m_e{}^* kT}{h^2}\right)^{\frac{3}{2}} \exp\left[\frac{E_F - E_C}{kT}\right] \tag{24.6}$$

An *n*-type semiconductor is characterized by the presence of a donor energy level (E_d), which lies just below the conduction band. The donor levels consist of donor ions, namely, the electrons. Let N_d be the number of donor ions per unit volume of the material at the energy level E_d. Figure 24.11 shows the energy level diagram of an N-type semiconductor along with the donor energy level.

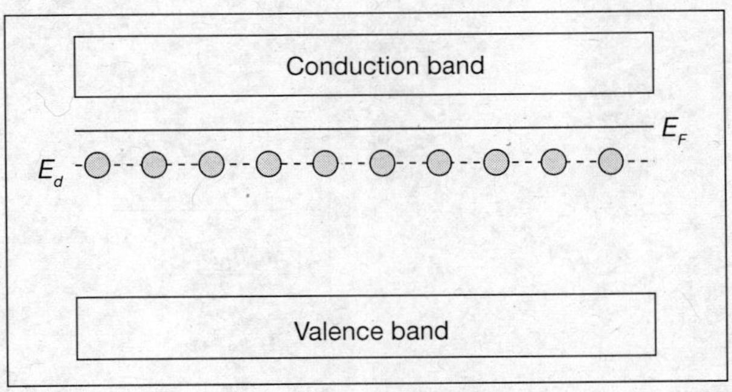

FIGURE 24.11 Energy level diagram of an *n*-type semiconductor

At lower temperatures, when the available thermal energy is not enough to excite the valence electrons, then it is only the donor level that donates electrons to the conduction band. This means that the number of vacant donor ion states (holes) in the donor level and the number of electrons in the conduction band will be equal. Therefore, it will be safe to say that the electrons in the conduction band are only due to the electrons donated from the donor level at low temperatures. But at very low excitations, the number of donor electrons that get excited into the conduction band is very less, and practically all the donor level sites are filled with electrons without any empty sites. The density of holes in the donor level is given by

$$\approx N_d \exp\left[\frac{E_d - E_F}{kT}\right] \tag{24.7}$$

Since the density of electrons in the conduction band is equal to the density of holes, i.e., $$2\left(\frac{2\pi m_e^* kT}{h^2}\right)^{3/2} \exp\left[\frac{E_F - E_C}{kT}\right] = N_d \exp\left[\frac{E_d - E_F}{kT}\right] \tag{24.8}$$

Taking log for this equation and rearranging it, we get

$$\left[\frac{E_F - E_C}{kT}\right] - \left[\frac{E_d - E_F}{kT}\right] = \log\left[\frac{N_d}{2\left[\frac{2\pi m_e^* kT}{h^2}\right]^{3/2}}\right]$$

$$\frac{E_F - E_C - E_d + E_F}{kT} = \log\left[\frac{N_d}{2\left[\frac{2\pi m_e^* kT}{h^2}\right]^{3/2}}\right]$$

$$2E_F - (E_C + E_d) = kT \log\left[\frac{N_d}{2\left[\frac{2\pi m_e^* kT}{h^2}\right]^{3/2}}\right]$$

$\therefore$ $$E_F = \frac{(E_C + E_d)}{2} + \frac{kT}{2} \log\left[\frac{N_d}{2\left[\frac{2\pi m_e^* kT}{h^2}\right]^{3/2}}\right] \tag{24.9}$$

At $T = 0$ K, the second term of Eq. (24.9) becomes zero, and the equation becomes

$$\boxed{\mathbf{E_F = \frac{(E_C + E_d)}{2}}} \tag{24.10}$$

i.e., the Fermi level lies exactly in between the donor level and the conduction band as shown in Fig. 24.11.

24.7.2 Fermi Level for *p*-Type Extrinsic Semiconductors

We know that the density of holes in the valence band is given by

$$p = 2\left(\frac{2\pi m_h^* kT}{h^2}\right)\exp\left[\frac{E_V - E_F}{kT}\right] \tag{24.11}$$

A *p*-type semiconductor is characterized by the presence of acceptor energy level (E_a), which lies just above the valence band. The acceptor levels consist of the presence of acceptor ions, namely, the holes. Let N_a be the number of acceptor ions per unit volume of the material at the energy level of E_a. Figure 24.12 shows the energy level diagram of a *p*-type semiconductor with the presence of the acceptor energy level.

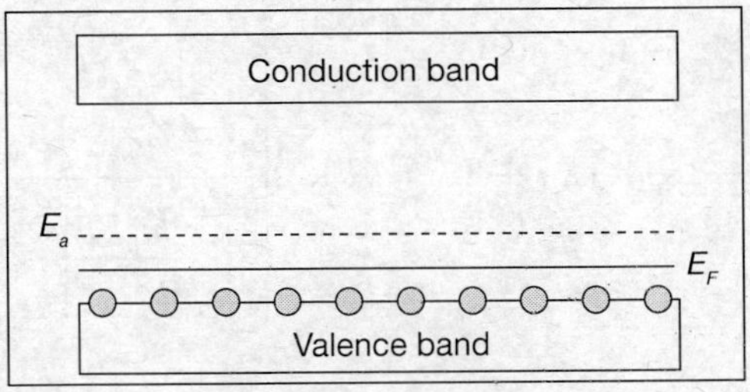

FIGURE 24.12 Energy level diagram of a *p*-type semiconductor

As in the earlier case at lower temperatures, the available thermal energy does not excite the valence electrons to the conduction band. Since the acceptor level is only slightly above the valence band, it is only the acceptor level that accepts the electrons from the valence band that get excited from the valence band. The upward movement of the electrons from the valence band leaves behind holes in the valence band, which can then take part in conduction. Since the accepted electron occupies a vacant site in the acceptor level, it is not free. Therefore, the charge carriers are mostly the holes in the valence band. Then the number of accepted electrons in the acceptor level and the number of holes in the conduction band should be equal, since the electrons in the acceptor level are only due to the electrons donated from the valence band at low temperatures. But at very low excitations the number of electrons that get excited from the valence band to the acceptor level are very less and practically all the acceptor levels are empty.

The density of electrons in the acceptor level is given by

$$\approx N_a \exp\left[\frac{E_F - E_a}{kT}\right] \tag{24.12}$$

Since the density of electrons in the acceptor level is equal to the density of holes in the valence band, i.e.,

$$2\left(\frac{2\pi m_h^* kT}{h^2}\right)^{3/2}\exp\left[\frac{E_V - E_F}{kT}\right] = N_a \exp\left[\frac{E_F - E_a}{kT}\right] \tag{24.13}$$

Taking log for Eq. (24.13) and rearranging it, we get

$$\left[\frac{E_V - E_F}{kT}\right] - \left[\frac{E_F - E_a}{kT}\right] = \log N_a - \log 2\left(\frac{2\pi m_h^* kT}{h^2}\right)^{3/2}$$

$$\frac{E_V - E_F - E_F + E_a}{kT} = \log\left[\frac{N_a}{2\left[\frac{2\pi m_h^* kT}{h^2}\right]^{3/2}}\right]$$

$$-2E_F + (E_V + E_a) = kT\log\left[\frac{N_a}{2\left[\frac{2\pi m_h^* kT}{h^2}\right]^{3/2}}\right]$$

$$\therefore \quad E_F = \frac{(E_V + E_a)}{2} - \frac{kT}{2}\log\left[\frac{N_a}{2\left[\frac{2\pi m_h^* kT}{h^2}\right]^{3/2}}\right]$$

At $T = 0$ K, the second term of this equation becomes zero and we get

$$\boxed{E_F = \frac{(E_V + E_a)}{2}} \tag{24.14}$$

i.e., the Fermi level lies exactly in between the acceptor level and the valence band, as shown in Fig. 24.12.

Note

The density of electrons in the conduction band n_e, density of holes in the valence band, n_h increases exponentially with increase in temperature.

EXAMPLE 24.3

Calculate the number of donor atoms per m^3 of n-type material having resistivity of 0.25 Ω-m, the mobility of electrons is 0.3 $m^2/v\text{-}s$.

Solution: We know

$$\rho = \frac{1}{\sigma} = \frac{1}{ne\mu_n}$$

So,

$$n = \frac{1}{\rho e\mu_n}$$

$$= \frac{1}{0.25 \times 1.6 \times 10^{-19}}$$

$$= \mathbf{8.3 \times 10^{19} \text{ per m}^3}$$

24.8 ▶ Electrons and Holes Concentration in Intrinsic Semiconductors at Thermal Equilibrium

24.8.1 Concentration of Electrons in Conduction Band

The electrons density of states in the conduction band is equal to that for free electrons. i.e.,

$$N_n(E) = \frac{1}{2\pi^2}\left(\frac{2m_n}{h^2}\right)^{\frac{3}{2}}\left(E - E_g\right) \text{per unit volume} \tag{24.15}$$

but the fermi function is

$$f(E) = \frac{1}{1 + e^{(E-E_f)/KT}} \tag{24.16}$$

Total number of electrons in conduction band will be

$$n = \int_{E_g}^{\infty} \frac{1}{2\pi^2}\left(\frac{2m_n}{\hbar^2}\right)^{\frac{3}{2}}\left(E - E_g\right)^{\frac{1}{2}} f(E)dE \tag{24.17}$$

But $E_g - E_f \gg KT$
So,

$$f(E) = \left[1 + e^{(E-E_f)/KT}\right]^{-1} = e^{-(E-E_f)/KT}$$

and

$$n = \frac{1}{2\pi^2}\left(\frac{2m_n}{\hbar^2}\right)^{\frac{3}{2}} \int_{E_g}^{\infty}\left(E - E_g\right)^{\frac{1}{2}} e^{-(E-E_f)/KT}$$

$$\Rightarrow \qquad \boxed{n = 2\left(\frac{m_n KT}{2\pi\hbar^2}\right)^{\frac{3}{2}} e^{(E-E_f)/KT}} \qquad (24.18)$$

Equation (24.18) gives the density of electrons in the conduction band of an intrinsic semiconductor.

24.8.2 Concentration of Holes in Valence Band

The density of holes in valence band is given as

$$N_p(E) = \frac{1}{2\pi^2}\left(\frac{2m_p}{\hbar^2}\right)^{\frac{3}{2}} E^{\frac{1}{2}} \text{ per unit volume}$$

The probability of finding a hole is $\{1 - f(E)\}$.
Thus,

$$f_p(E) = 1 - f(E)$$

$$= 1 - \frac{1}{1 + e^{(E-E_f)/KT}}$$

$$= 1 - \left[1 + e^{(E-E_f)/KT}\right]^{-1}$$

$$= 1 - \left[1 + e^{(E-E_f)/KT}\right]$$

$$f_p(E) = e^{(E-E_f)/KT}$$

Hence, density of holes in the valence band will be

$$p = \int_{-\infty}^{0} \frac{1}{2\pi^2}\left(\frac{2m_p}{\hbar^2}\right)^{\frac{3}{2}} E^{\frac{1}{2}} e^{(E-E_f)/KT} dE \qquad (24.19)$$

$$\therefore \qquad \boxed{p = 2\left(\frac{m_p KT}{2\pi\hbar^2}\right)^{\frac{3}{2}} e^{-E_f/KT}} \qquad (24.20)$$

Equation (24.20) gives the hole concentration in valence band.

EXAMPLE 24.4

For an Si semiconductor with a band gap 1.12 eV, determine the position of the Fermi level at 300 K if $m_n^* = 0.12\, m_e$ and $m_p^* = 0.28\, m_e$.

Solution: We know that

$$E_f = \frac{E_g}{2} + \frac{3KT}{4}\left[\log \frac{m_p^*}{m_n^*}\right]$$

$$E_f = \frac{1.12}{2} + \frac{3 \times 1.38 \times 10^{-23} \times 300}{4 \times 1.6 \times 10^{-19}} \times \log\left[\frac{0.28 m_e}{0.12 m_e}\right]$$

$$= \mathbf{0.56 + 15.14 \times 10^{-3}\ eV}$$

24.9 ▶ Conductivity in Semiconductors

24.9.1 Conductivity of Intrinsic Semiconductors

In a semiconductor, the conduction band electrons and valence band holes participate in electrical conduction. Consider a rectangular bar of intrinsic semiconductor connected to a battery as shown in Fig. 24.13. Then, electric field exists along the X-direction. This field accelerate the free electrons along negative X-direction and holes along positive X-direction.

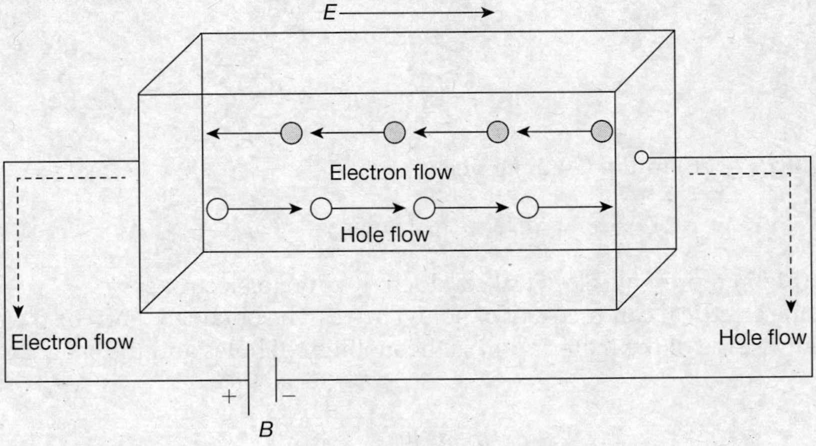

FIGURE 24.13 Electric current in intrinsic semiconductors

So, the velocity of electrons along negative X direction increases and attains some constant resultant velocity. This constant velocity is called drift velocity, represented as v_{de}. This drift velocity is superimposed with their random thermal motion. The total electrical current through the semiconductor is the sum of electron current I_e, and hole current I_h.

$$\text{The total current } (I) = I_e + I_h.$$

To find an expression for electrical conductivity, first we shall consider electron flow in the semiconductor. Let A be the area of cross-section of the bar, v_{de} be the drift velocity of electrons and 'n' be the number of conduction band electrons per unit volume of the semiconductor. Then, the free electrons present in a volume Av_{de} will cross an imaginary plane assumed to be perpendicular to the bar in 1 second.

$$\text{The number of electrons crossing the imaginary plane in 1 s} = nAv_{de} \tag{24.21}$$

$$\text{The electron charge which crosses the plane in 1 s} = I_e = neAv_{de} \tag{24.22}$$

$$\text{The electron current density, } J_e = \frac{I_e}{A} = nev_{de} \tag{24.23}$$

From Ohm's law, the current density (J_e) due to electrons is given as

$$J_e = \sigma_e E \tag{24.24}$$

where σ_e is the electrical conductivity of electrons. From Eqs. (24.23) and (24.24), we have

$$\sigma_e E = nev_{de} \tag{24.25}$$

The drift velocity produced per unit applied electric field is called the mobility of electrons represented as

$$\mu_e = \frac{v_{de}}{E} \text{ or } v_{de} = \mu_e E \tag{24.26}$$

Substituting Eq. (24.26) in Eq. (24.25) gives

$$\sigma_e E = ne\mu_e E \text{ or } \sigma_e = ne\mu_e \tag{24.27}$$

Equation (24.27) represents electrical conductivity due to electrons.

Similarly, the electrical conductivity of holes (σ_h) can be obtained. Let p be the number of holes per unit volume of the material, μ_h is the mobility of holes and the charge on a hole is e, then

$$\sigma_h = pe\mu_h \tag{24.28}$$

The total conductivity of a semiconductor is given by the sum of Eqs. (24.27) and (24.28).

i.e.,

$$\sigma = \sigma_e + \sigma_h = ne\mu_e + pe\mu_h$$

$$= e[n\mu_e + p\mu_h] \tag{24.29}$$

For an intrinsic semiconductor, $n = p = n_i$, where n_i is called intrinsic density. So, Eq. (24.29) can be represented as

$$\boxed{\sigma_i = n_i e[\mu_e + \mu_h]} \tag{24.30}$$

24.9.2 Conductivity of Extrinsic Semiconductors

(a) Conductivity of *n*-type semiconductors

For n-type semiconductor, electron concentration is much greater than hole concentration.

i.e., $n \gg p$ $$\sigma = en\mu_e \tag{24.31}$$

If n_d is the concentration of donor atoms, then $n = n_d$

$$\boxed{\sigma = en_d\mu_e} \tag{24.32}$$

(b) Conductivity of *p*-type semiconductors

For *p*-type semiconductor, electron concentration is negligible as compared to hole concentration.

i.e., $p \gg n$ $$\sigma = ep\mu_h \tag{24.33}$$

If n_a is the concentration of acceptor atoms, then $p = n_a$.

$$\boxed{\sigma = en_a\mu_h} \tag{24.34}$$

EXAMPLE 24.5

Find the resistivity of an intrinsic semiconductor with intrinsic concentration of 2.5×10^{19} per m^3. The mobilities of electrons and holes are 0.40 m^2/V-s and 0.20 m^2/V-s.

Solution: Given data are:

Intrinsic concentration $(n_i) = 2.5 \times 10^{19}/\text{m}^3$

Mobility of electrons $(\mu_n) = 0.40\ \text{m}^2/\text{V-s}$

The mobility of holes $(\mu_p) = 0.20\ \text{m}^2/\text{V-s}$

The conductivity of an intrinsic semiconductor $\sigma_i = n_i e[\mu_n + \mu_p]$

$$\text{The resistivity } (p_i) = \frac{1}{\sigma_i} = \frac{1}{n_i e[\mu_n + \mu_p]}$$

$$= \frac{1}{2.5\times10^{19}\times1.3\times10^{-19}[0.40+0.20]}$$

$$= \frac{\mathbf{1}}{\mathbf{2.5\times1.6\times0.6}} = \mathbf{0.4166\ \Omega\text{-m}}$$

24.10 ▶ Mobility and Conductivity

We know that the electrical conductivity (σ) of a semiconductor is the product of the number of charge carriers, their mobility (μ) and the total charge present in them,

i.e.,
$$\sigma = ne\mu_e + ph\mu_h$$

where n, p are the number of electrons and holes, respectively, and μ_e, μ_h are the mobilities of the electrons and holes, respectively.

Since the charge on an electron is equal to the charge on a hole, the above equation can be rewritten as

$$\sigma = ne\mu_e + pe\mu_h$$

We know that intrinsic semiconductor, the number of electrons is equal to the number of holes, i.e., $n = p = n_i$; therefore, the equation becomes

$$\sigma = n_i e(\mu_e + \mu_h) \tag{24.35}$$

Substituting for n_i, Eq. (24.35) becomes

$$\sigma = 2\left(\frac{2\pi kT}{h^2}\right)^{3/2} \left(m_e^{*} m_h^{*}\right)^{3/2} \exp\left(\frac{-E_g}{2kT}\right)\left[e((u_e + u_h)\right] \tag{24.36}$$

From Eq. (24.35), we can observe the mobilities of the carriers that depend inversely on the temperature as $T^{-3/2}$.

Since except for the temperature T, all other values are constant in Eq. (24.36), we can replace the equation as

$$\sigma_i = A\exp\left[\frac{-E_g}{2kT}\right] \tag{24.37}$$

where σ_i is the intrinsic carrier conductivity, and

$$A\text{ (a constant)} = 2\left(\frac{2\pi kT}{h^2}\right)^{3/2} \left(m_e^{*} m_h^{*}\right)^{3/4} \left[e(u_e + u_h)\right]$$

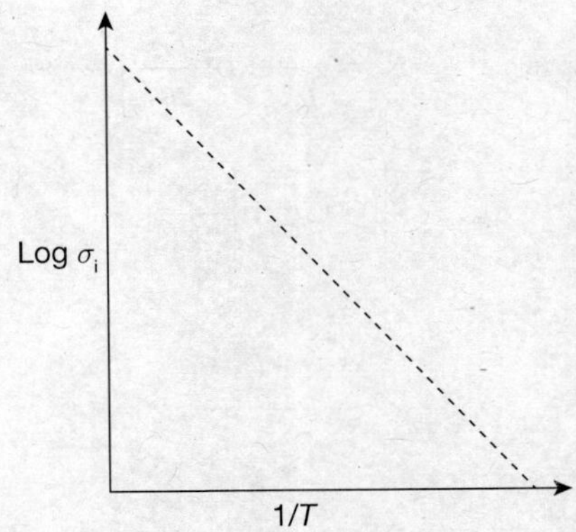

FIGURE 24.14 Variation of electrical conductivity with temperature

Taking log on both sides of Eq. (24.37), we obtain

$$\log \sigma_i = \log A - \frac{E_g}{2kT} \tag{24.38}$$

From the equation, we can observe that the conductivity increases with temperature $(1/T)$. The experimental verification of the logarithm of σ_i and $1/T$ is a straight line as shown in Fig. 24.14.

24.11 ▶ Determination of Band Gap

Band gap is defined as the gap in energy level between the valence and the conduction band. More specifically, a band gap defines the energy difference between the top of the valence band and the bottom of the conduction band, i.e. the energy required for a charge carrier in the valence band to go to the conduction band. The band gap in a semiconductor can be easily derived from the resistivity of the material and the conductivity equation.

From Eq. (24.37), the conductivity of a semiconductor is

$$\sigma_i = A \exp\left(\frac{-E_g}{2kT}\right) \tag{24.39}$$

Since the resistivity (p_i) is the inverse of conductivity (σ_i), Eq. (24.39) becomes

$$\rho_i = \frac{1}{A} \exp\left(\frac{E_g}{2kT}\right)$$

i.e.,

$$\frac{R_i a}{l} = \frac{1}{A} \exp\left(\frac{E_g}{2kT}\right) \qquad \left[\because \rho_i = \frac{R_i a}{l}\right]$$

or

$$R_i = \frac{l}{aA}\exp\left(\frac{E_g}{2kT}\right)$$

$$R_i = C\exp\left(\frac{E_g}{2kT}\right), \tag{24.40}$$

where $C = \dfrac{l}{aA}$.

Taking log on both sides of Eq. (24.40), we get

$$\log R_i = \log C + \left(\frac{E_g}{2kT}\right) \tag{24.41}$$

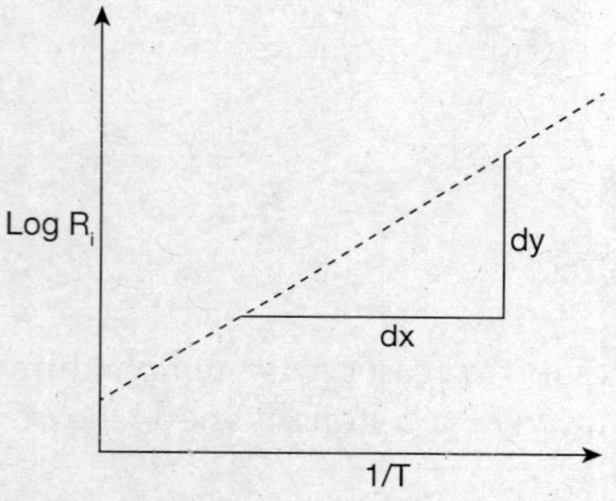

FIGURE 24.15 Variation in resistance with temperature in an intrinsic semiconductor

Careful observation of Eq. (24.41) reveals that the resistance R helps in determining the band gap of a semiconductor. The resistance R of the material is measured at various temperatures and a plot between the log of the resistance and the temperature $(1/T)$ is drawn (Fig. 24.15). The slope of the resultant curve provides the band gap of the material.

From Fig. 24.15, the slope is $\dfrac{dy}{dx} = \dfrac{E_g}{2k}$.

Therefore,

$$E_g = 2k\left(\frac{dy}{dx}\right) \tag{24.42}$$

The slope detemined from the graph helps in determining the band gap of a semiconductor .

Recap Zone

Points to Remember

- **Conductors:** These are the materials that allow the current to flow very easily.
- **Semiconductors:** They have conductivity in between conductors and insulators.
- **Insulators:** The forbidden gap between the conduction band and valence band is extremely wide.
- Intrinsic semiconductors are those substances in which the number of holes is equal to the number of electrons.
- Due to doping, the conductivity of intrinsic semiconductor increases and it becomes extrinsic semiconductor.
- When pentavalent impurities are added to pure semiconductors, n-type extrinsic semiconductors are formed.

- When trivalent impurities are added to intrinsic semiconductors, p-type extrinsic semiconductors are formed.
- Fermi-dirac distribution function is given by

$$f(E) = \frac{1}{1 + e^{(E - E_f)/KT}}$$

- Fermi energy for an intrinsic semiconductor is given by

$$E_F = \left(\frac{E_C + E_V}{2}\right)$$

- Fermi energy for n-type semiconductor is given by

$$E_F = \frac{(E_C + E_d)}{2} + \frac{KT}{2}\log\left[\frac{N_d}{2\left[\frac{2\pi meKT}{h^2}\right]^{3/2}}\right]$$

- Fermi energy for p-type semiconductor is given by

$$E_F = \left(\frac{E_v + E_a}{2}\right) - \frac{KT}{2}\log\left[\frac{N_a}{2\left[\frac{2\pi m_h^* KT}{h^2}\right]^{3/2}}\right]$$

- Concentration of electrons in conduction band is given by

$$n = 2\left(\frac{m_n KT}{2\pi\hbar^2}\right)^{3/2} e^{(E_f - E_g)/KT}$$

- Concentration of holes in valence band is given by

$$p = 2\left(\frac{m_p KT}{2\pi\hbar^2}\right)^{3/2} e^{-E_f/KT}$$

- Conductivity of intrinsic semiconductor is given by

$$\sigma_i = n_i e[\mu_e + \mu_h]$$

- Conductivity of n-type extrinsic semiconductor is given by

$$\sigma = e n_d \mu_e$$

- Conductivity of p-type extrinsic semiconductor is given by

$$\sigma = e n_a \mu_h$$

Key Terms with Definitions

- **Energy bands** – The collection of the individual energy levels of electrons surrounding each atom.
- **Discrete energy levels** – Energy levels of electrons in atoms, molecules or ions which are bound by the electric field of the nucleus.
- **Majority carriers** – The more abundant charge carriers are called majority carriers, which are primarily responsible for current transport in semiconductors.
- **Donor atom** – A donor is a dopant atom that when added to a semiconductor can form *n*-type region.
- **Acceptor atom** – An acceptor is an atom when added to a semiconductor can form a *p*-type region.
- **Fermi level** – It is the term used to describe the top of the collection of electron energy levels at absolute zero temperature .
- **Conductivity** – The degree to which a specified material conductors electricity.
- **Mobility** – Electrical mobility of charged particles in a fluid under an applied electric field.

Important Formulae and Equations

Equation Number	Equation	Remarks
(24.1)	$F(E_i)=\dfrac{1}{e^{(E_i-E_F)/KT}+1}$	Fermi-distribution function
(24.10)	$E_F=\dfrac{E_C+E_d}{2}$	Fermi level for *n*-type extrinsic semiconductor
(24.14)	$E_F=\dfrac{E_V+E_a}{2}$	Fermi level for *p*-type extrinsic semiconductor
(24.18)	$n=2\left(\dfrac{m_n KT}{2\pi\hbar^2}\right)^{3/2} e^{(E-E_f)/KT}$	Density of electrons in conduction band of an intrinsic semiconductor
(24.20)	$p=2\left(\dfrac{m_p KT}{2\pi\hbar^2}\right)^{3/2} e^{-E_f/KT}$	Hole concentration in valence band
(24.30)	$\sigma_i=n_i e[\mu_e+\mu_h]$	Conductivity of intrinsic semiconductors
(24.32)	$\sigma=e\,n_d\mu_e$	Conductivity of *n*-type semiconductor
(24.34)	$\sigma=e\,n_a\,\mu_h$	Conductivity of *p*-type semiconductor

REVIEW ZONE

SHORT ANSWER QUESTIONS

1. Differentiate between amorphous and crystalline solids.
2. What is the difference between valence band, conduction band and forbidden energy gap?
3. Give a difference between single crystal semiconductor and compound semiconductor.
4. What are two types of semiconductors?
5. What is the difference between intrinsic and extrinsic semiconductors?
6. What are p-type semiconductors?
7. What are n-type semiconductors?
8. Write Fermi distribution function.
9. Write the equation of conductivity for intrinsic semiconductors.
10. Write the equation of conductivity for extrinsic semiconductor.
11. What is conductivity?
12. Define mobility.
13. What is the charge carried by a hole?
14. What is the binding energy of an electron?
15. What are majority carriers?
16. Write the units of mobility.
17. State some Column II impurities.
18. Explain the process of formation of electron-hole pairs.
19. Write an expression for the ground-state energy of an electron.
20. How does a hole move in a semiconductor?

LONG ANSWER QUESTIONS

1. Give diagram giving energy band structure of (a) a conductor, (b) a semiconductor, (c) an insulator and hence explain the difference in their electrical conductivities.
2. What do you mean by semiconductors? What are the types of semiconductors? When the semiconductors work as an insulator?
3. Derive an expression for the concentration of electrons in conduction band and holes in valence band.
4. Derive an expression for the Fermi energy for an intrinsic semiconductor.
5. Derive an expression for Fermi energy for n-type and p-type semiconductors.
6. Discuss the effects of temperature on Fermi distribution function.
7. Discuss the concept of holes and effective mass.
8. Derive an expression for the conductivity of intrinsic semiconductors and hence for n-type and p-type extrinsic semiconductors.
9. Derive an expression for band gap energy (E_g) for a semiconductor.
10. Discuss mobility and conductivity in semiconductors.

11. How does the Fermi level plays a significant role in semiconductors? Explain with a minimum of three points of importance.
12. Explain the significance of Fermi energy level. Mention its position in intrinsic and extrinsic semiconductors at 0 K.
13. Derive an expression for the density of electrons in the condition band of an intrinsic semiconductor.
14. Derive an expression for the density of holes in the valence band of an intrinsic semiconductor.

NUMERICAL PROBLEMS

1. Find the resistivity of intrinsic Si, if $n_i = 1.5 \times 10^{16}/\text{m}^3$, $\sigma_n = 0.13$, $\sigma_p = 0.05\ \text{m}^2/\text{V-s}$.

   Ans. 2.314 Ω-m
2. A particular sample of Ge has a donor density of $N_d = 10^{13}$ atoms/cm^3. Assuming all donor atoms to be ionized, calculate the resistivity of the sample.

   Ans. 160.3 Ω-m
3. Mobilities of electrons and holes in a sample of intrinsic Ge at 300 K are 0.45 and 0.2 m^2/V-s, respectively. If the resistivity of the specimen is 3.15 Ω-m, calculate the forbidden energy gap for Ge.

   Ans. 0.73 eV
4. For a Si semiconductor with a band gap 0.9 eV, determine the position of the Fermi level at 400 K if $m_n^* = 0.13\, m_e$, $m_p^* = 0.30\, m_e$.

   Ans. 0.64×10^{-4}/Ω-m
5. The Fermi energy of copper at 0 K is 7.04 eV. Calculate the Fermi energy at 300 K.

   Ans. 1.64×10^{-10} mV-s
6. Mobilities of electrons and holes in a sample of intrinsic germanium at room temperature are 3600 cm^2/V-s and 1700 cm^2/V-s, respectively. If the electron and hole densities are each equal to 2.5×10^{13} per cm^3, calculate the conductivity.

   Ans. 2.12 mho/m
7. Compare the density of charge carriers in a pure silicon crystal at the two temperatures 27°C and 57°C. Eg for Si is 1.1 eV.

   Ans. $2.54 \times 10^{15}/\text{m}^3$, $2.8 \times 10^{15}/\text{m}^3$
8. Calculate the resistivity of an n-type Ge sample at 300 K. The sample has a donor density $N_d = 10^{20}$ atoms/m^3. Assume all donors to be ionized and $\mu_n = 0.38\text{m}^2/\text{V-s}$.

   Ans. 1.64×10^{-10} mV-s
9. Find the resistance of an intrinsic Ge rod (1.5 cm long, 0.5 mm wide and 0.5 mm thick) at 400 K. For Ge, $n_i = 3.0 \times 10^{19}/\text{m}^3$, $\mu_n = 0.40\ \text{m}^2/\text{V-s}$, $\mu_n = 0.20\ \text{m}^2/\text{V-s}$.

   Ans. 0.459 eV
10. In an n-type semiconductor, the Fermi level lies 0.3 eV below the conduction band at 300 K. If the temperature is increased to 330 K. Find the new position of the Fermi level.

    Ans. 0.33 eV below the conduction band

CHAPTER 25 Semiconductor Devices

Learning Objectives

By the end of this chapter, the student will be able:

- To describe the working of a *p-n* junction as diode
- To explain the use of *p-n* diode as rectifier
- To discuss the construction and working of LED, solar cell, photodiode and Zener diode

25.1 ▶ Introduction

The word electronics means the study of the electron behaviour under different conditions of externally applied fields. Electronics plays a major role in almost every sphere of life. It deals in the micro and milli range of voltage, current and power, but it is also capable of controlling mega volts, amperes and watts. All electronic circuits use active and passive components. Passive components are resisters, capacitors and inductors. Active components may be classified again in two categories: (i) tube devices like vacuum diode and gas diode and (ii) semiconductor devices like *p-n* junction diodes, semiconductors, Zener diodes and transistors. In modern electronic systems, the whole electronic circuit is fabricated on a single chip, called integrated circuit. The diode is one of many components used in these circuits. Another important devices are transistors, light-emitting diodes, Zener diodes and photodiodes. In this chapter, we will study about these devices one by one for the better understanding of integrated circuits.

25.2 ▶ P-N Junction Diode

By using special fabrication techniques, if a piece of *p*-type semiconductor is joined with a piece of *n*-type semiconductor, *p-n*-junction diode or crystal diode is formed (Fig. 25.1).

It has the ability to conduct current in one direction. In this, the *p*-side is called anode and the *n*-side is called cathode.

The majority charge carriers in *n*-region are electrons and the minority charge carriers are holes, whereas in *p*-region, holes are majority charge carriers and electrons are minority charge carriers.

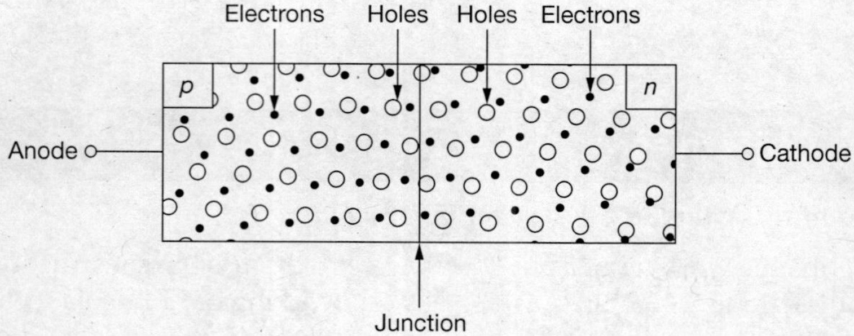

FIGURE 25.1 *p-n* junction diode

Due to the difference in concentration of electrons and holes in *p*- and *n*-regions, diffusion of charge carriers across the junction takes place. The free electrons from *n*-region diffuse into *p*-region leaving behind a positive immobile ion on *n*-side. However, the holes from *p*-region diffuse into the *n*-region leaving behind a negative immobile ion on *p*-side. This diffusion occurs for a very short time. In the immediate neighbourhood of the junction, after a few recombinations of holes and electrons, a restoring force is set up automatically which is called barrier. A depletion region is created (Fig. 25.2) in between *p*- and *n*-regions due to the presence of immobile ions on both sides. Now, the barrier or depletion layer discourages the further diffusion of majority charge carriers across the junction.

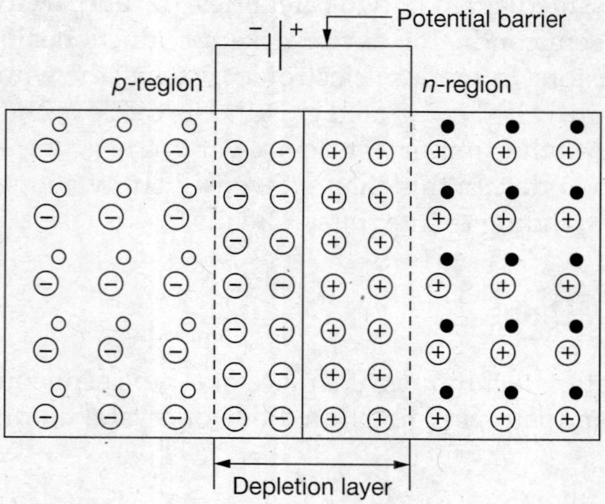

FIGURE 25.2 Formation of depletion layer in *p-n* diode

The electric field between the donor and the accepter ions is called barrier. When no external voltage is connected, then barrier height is of the order of tenths of a volt. For a silicon *p-n* junction, the barrier potential is about 0.7 volts, whereas for a germanium *p-n* junction it is about 0.3 volts.

25.2.1 Width of the Depletion Region

The width of the depletion layer is very thin as compared to the widths of *p-* and *n*-regions. Practically, the width of depletion layer is very small, of the order of 0.5 to 1 micron (1×10^{-6}m). The symbolic representation of *p-n* diode is shown in Fig. 25.3.

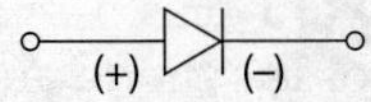

FIGURE 25.3 Symbolic representation of *p-n* diode

25.2.2 Potential Barrier

Potential barrier opposes the flow of electrons and holes across the junction. It is measured in volts. To cross the junction, an external voltage has to be applied in order to overcome the potential barrier (Fig. 25.4).

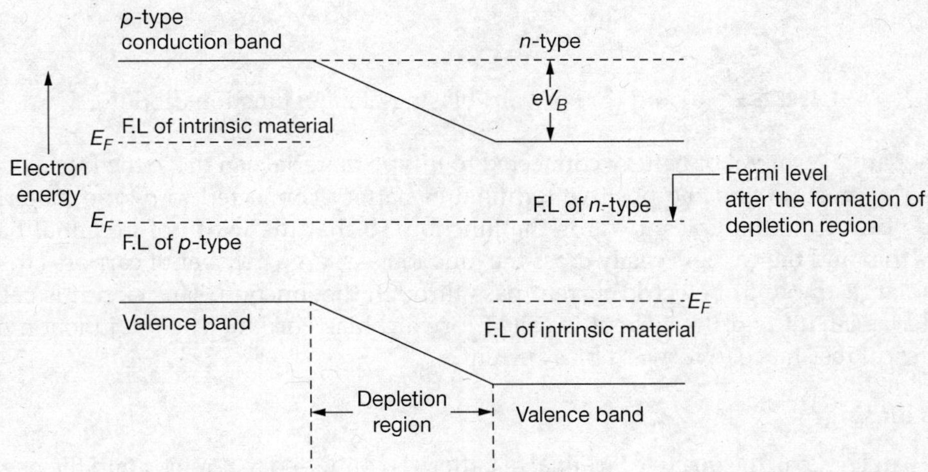

FIGURE 25.4 Energy diagram of *p-n* diode

25.3 ▶ P-N Junction Biasing

The free carriers can be made to cross the junction by increasing their kinetic energy by some external means. The kinetic energy (KE) of carriers can be increased by applying electric potential externally across the junction. As the electric potential across the junction increases, the KE of free electrons also increases when the KE of carriers attain the energy of the potential barrier (eV_B) between *p*-region and *n*-region, the carriers start crossing the junction.

Suppose the positive terminal of a battery is connected to the *p*-type and the negative terminal to the *n*-type of a *p-n* junction, large current flows through the junction, and the *p-n* junction is said to be forward biased. If terminals of battery are interchanged, then very small current flows through the junction and it is said to be reverse biased. The forward and reverse biasing of a *p-n* diode are described below.

Forward biasing

To forward bias a *p-n* junction, the positive terminal of the battery is connected to the *p*-side of the diode and the negative terminal of the battery is connected to the *n*-side of the diode as shown in Fig. 25.5(a). The symbolic representation of *p-n* diode in the forward bias circuit is shown in Fig. 25.5(b).

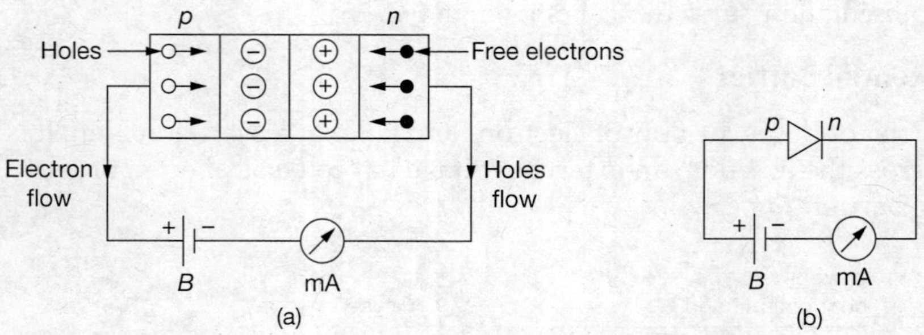

FIGURE 25.5 (a) and (b) Forward biasing of a *p-n* junction diode

The negative terminal of battery connected to *n*-type material can drive the free electrons towards the junction and the positive terminal of battery connected to *p*-type material of the diode can drive the holes towards the junction, so that the width of potential barrier becomes thin and the carriers easily cross the junction. As large number of carriers cross the junction, large amount of electric current pass through the junction. This circuit is called a forward bias circuit and the current is called forward bias current. The *p-n* junction offers low electrical resistance to forward bias current.

Reverse biasing

As shown in Fig. 25.6, the positive terminal of battery is connected to *n*-region and the negative terminal of battery is connected to *p*-region of the diode, in a reverse biased *p-n* junction. The negative terminal of the battery attracts the holes in the *p*-region and the positive terminal

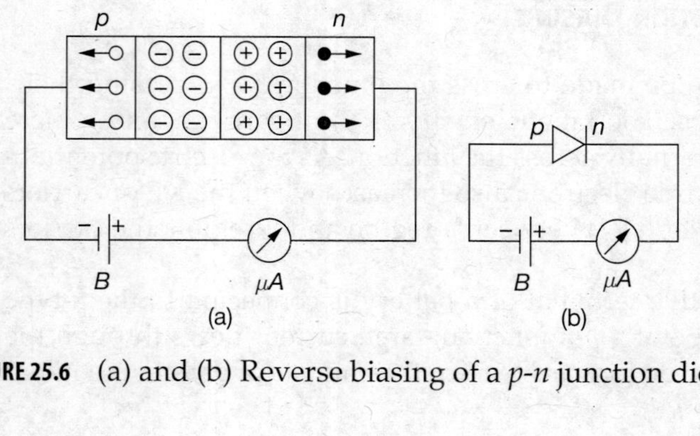

FIGURE 25.6 (a) and (b) Reverse biasing of a *p-n* junction diode

of the battery attracts the electrons in the n-region. Hence, the width of the potential barrier increases and hence the barrier potential increases. This prevents the flow of charge carriers across the junction. Thus, the diode offers high resistance to the current. However, very small leakage current passes through the junction due to minority charge carrier flow.

25.3.1 V-I Characteristics of a *p-n* Junction Diode

It is a graph between the voltage applied across its terminals and the current flows through it as shown in Fig. 25.7.

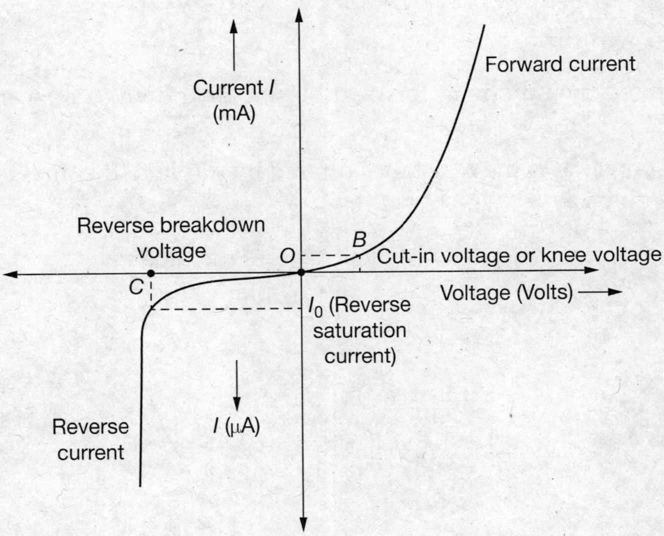

FIGURE 25.7 V-I characteristics of a p-n junction diode

Forward characteristics

Below point B, the forward voltage is small; thus, the forward current through diode is also small. As soon as the voltage increases, the forward current increases. The forward current is a conventional current and the nature of this current is exponential. The forward current is limited by connecting 'R' in series with the p-n diode. Forward current is of the order of a few mA.

The voltage at which the forward current starts increasing rapidly, is called knee or cut-in voltage. For silicon, its value is 0.7 V and for germanium it is 0.3 V. Forward resistance is of the order of few ohms.

Reverse characteristics

The current in reverse biasing flows due to minority charge carriers so its value is very small.

The reverse current is of the order of few μA. This current is treated as negative current. As the reverse voltage reaches the breakdown voltage value, excessive current flows through the p-n diode. The reverse breakdown voltage for a diode is in the range of 50–100 volts. The resistance in reverse biasing is very high and is of the order of few hundred kΩ. The operation of diode in breakdown region should be avoided because the diode may be damaged. Reverse saturation current (I_0) is temperature dependent; it does not depend on reverse voltage.

25.3.2 Static and Dynamic Resistance of a Diode

As we have already studied in the last topic, the forward resistance of a diode is quite small as compared to the reverse resistance.

(a) Forward resistance

The resistance offered by a diode in forward biasing is known as forward resistance. It is of two types:

DC forward resistance: It is the resistance offered by diode to the direct current as shown in Fig. 25.8 and is given as

Static or DC resistance, $$R_f = \frac{\text{DC voltage across the diode}}{\text{DC current}} \tag{25.1}$$

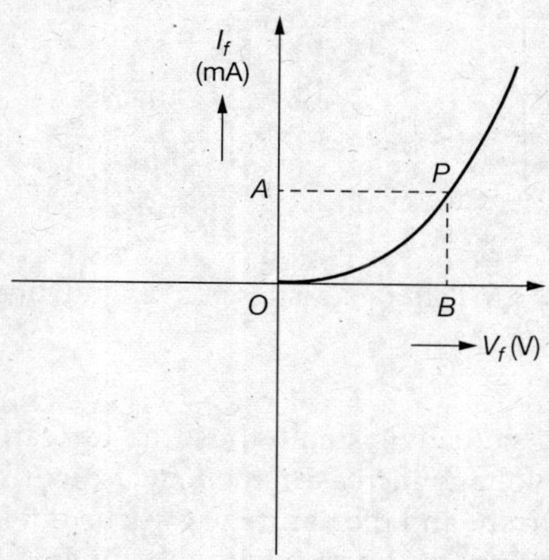

FIGURE 25.8 DC forward resistance of a diode

From Fig. 25.8, it is clear that $R_f = \dfrac{OB}{OA}$

AC forward resistance: It is the resistance offered by diode to the changing forward current. This resistance is more significant because the diodes are generally used with alternating voltages.

In Fig. 25.9, it is clear that, for forward voltage *od*, forward current is *oa*. Similarly, for forward voltage *of*, forward current is *oc*. So, by definition, AC forward resistance is given by

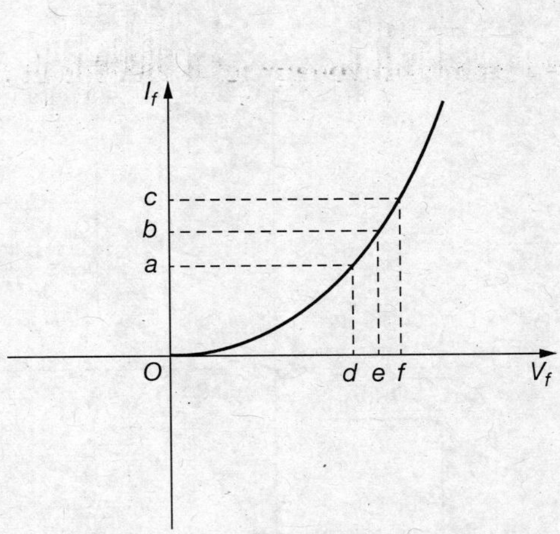

FIGURE 25.9 AC forward resistance of a diode

$$\text{Dynamic or AC forward resistance} = \frac{of - od}{oc - oa} = \frac{df}{ac}$$

$$r_f = \frac{\text{Change in forward voltage}}{\text{Change in forward current}}$$

or
$$r_f = \frac{\Delta V}{\Delta I} \tag{25.2}$$

Forward resistance of a *p-n* diode is very small, generally 1–25 Ω.

(b) Reverse resistance

In reverse biasing, even for a large reverse voltage, the reverse current is very small. It may be 1 μA at a voltage of 5 V (Fig. 25.10).

The static resistance of diode in reverse biasing is

$$R_R = \frac{5\text{ V}}{1\,\mu\text{A}} = 5\text{ M}\Omega \tag{25.3}$$

It is sufficiently high. The diode curve is almost horizontal up to point *A*. So, its dynamic resistance r_r will be extremely high in reverse biasing.

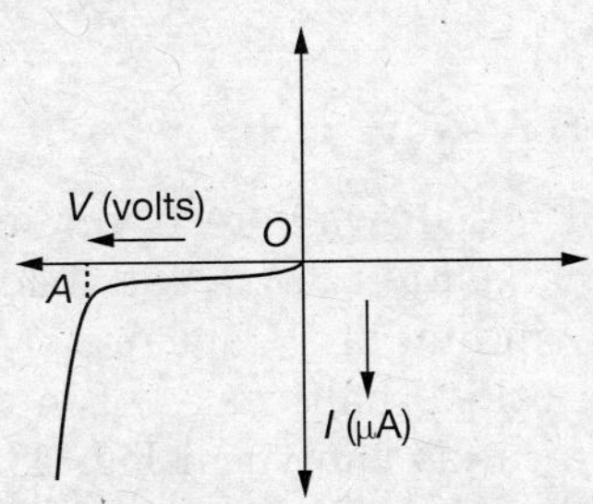

FIGURE 25.10 Reverse resistance of a diode

25.3.3 Comparison of V-I Characteristics of Silicon and Germanium Diodes

The V-I characteristics of silicon and germanium are shown in Fig. 25.11.

- Cut-in voltages for Si and Ge are 0.7 V and 0.3 V, respectively.
- Breakdown voltage for Si diode is higher than that of Ge diode.
- The reverse saturation current for Ge diode is of the order of μA whereas for Si diode, it is of the order of nA.

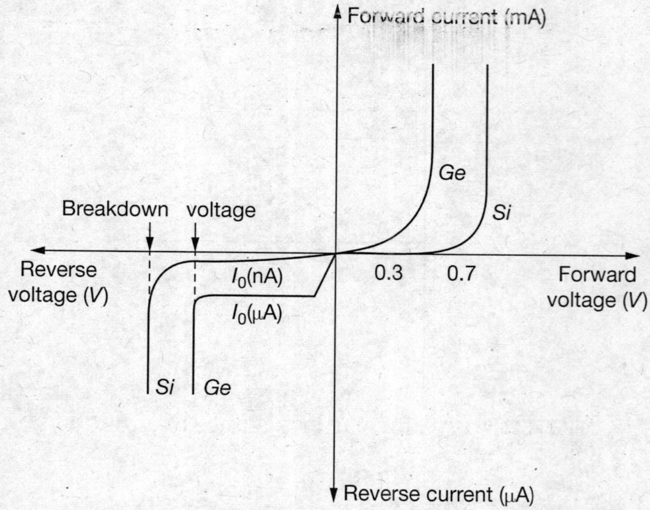

FIGURE 25.11 V-I characteristics of silicon and germanium

25.4 ▶ P-N Junction Diode as Rectifier

A *p-n* diode is used in a circuit to convert AC supply into DC supply. The process of converting AC supply into DC supply is called rectification. A rectifier is a device that permits current to flow through it in one direction only. We have mainly two different types of rectifiers: (i) half-wave rectifier and (ii) full-wave rectifier.

Half-wave rectifier

The half-wave rectifying circuit is shown in Fig. 25.12(a). The circuit consists of a single crystal *p-n* diode and load resister, R_L. These are connected in series with the secondary winding of a transformer (or AC source). The rectifier conducts current only during the positive half-cycles of input AC supply. Let the input AC supply have sinusoidal wave form as shown in Fig. 25.12(b). During each positive half-cycle of AC supply voltage, the diode is in forward biasing and conducts current through the circuit in the direction shown in Fig. 25.12(a).

The diode conducts only for the positive half-cycles of input current, for which it offers practically no resistance. Below the cut-in voltage, there is no current in the circuit. So, here onwards we consider an ideal diode. During each negative half-cycle of input voltage, the diode is in reverse bias and it offers infinite electrical resistance. Thus, no current passes through the diode and hence through the load resistance, R_L. The output voltage across the load, V_{out}, is shown in Fig. 25.12(c). It consists of a series of positive half-cycle voltage pulses of input. The current due to V_{out} always flows through the load, R_L in the same direction. Hence, pulses of DC output are obtained across R_L. These pulses of output are smoothened with the help of a filter circuit.

Here, we see some more points relating to half-wave rectifier.

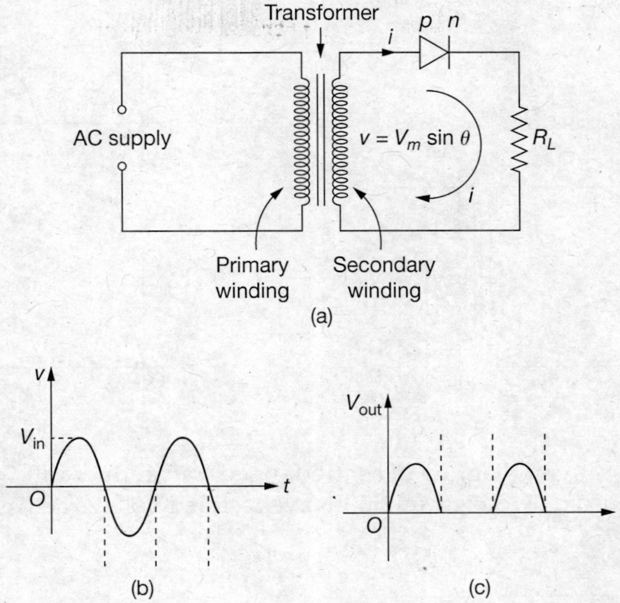

FIGURE 25.12 (a) Half-wave rectifier. (b) Input voltage of half-wave rectifier. (c) Output voltage of half-wave rectifier

In full-wave rectifier, current flows through load resister, R_L in the same direction for both half-cycles of AC input. Usually, there are two configurations for full-wave rectification: (a) centre-tapped full-wave rectifier and (b) bridge rectifier.

Centre-tapped transformer full-wave rectifier

The circuit diagram of centre-tapped full-wave rectifier is shown in Fig. 25.13. It consists of two diodes D_1 and D_2 connected to the secondary of centre-tapped transformer (tapped at "O") so that each diode uses one half-cycle of input AC supply. The sinusoidal input voltage is shown in Fig. 25.13(b) and the output voltage across the load resistor, R_L is shown in Fig. 25.13(c).

During the positive half-cycles of input voltage, the end A of the secondary winding is positive and the end B of secondary winding is negative. This makes the diode D_1 forward biased and the diode D_2 reverse biased. So the diode D_1 conducts current through load resistor, R_L and in the upper half of secondary winding, whereas D_2 does not conduct current during positive half-cycles. During the negative half-cycle of input voltage, the end B of the secondary winding is positive and the end A of secondary winding is negative. This makes the diode D_2 forward biased and the diode D_1 reverse biased. So, the diode D_2 conducts current through load resistor, R_L and in the lower half of the secondary winding, whereas D_1 does not conduct current during negative half-cycles. It is seen that for both half-cycles of input AC voltage, the current that is shown in Fig. 25.13(c) will flow in the same direction through the load resistor R_L. So we say that DC is obtained across the load resistor.

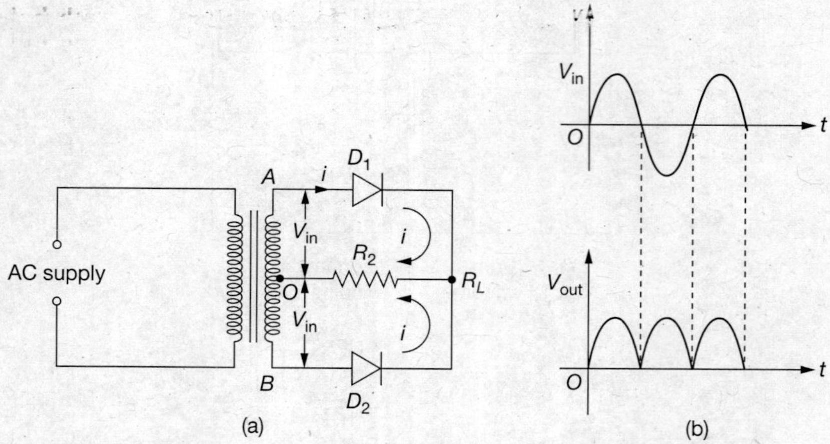

FIGURE 25.13 (a) Circuit diagram of centre-tapped transformer full-wave rectifier. (b) Sinusoidal voltage. (c) Full-wave rectified voltage across load

EXAMPLE 25.1

The applied input AC power to a half-wave rectifier is 100 watts. The DC output power obtained is 40 watts. What is its rectification efficiency?

Solution:

$$\text{Rectification efficiency} = \frac{\text{DC output power}}{\text{AC input power}} = \frac{40}{100}$$

$$= \mathbf{0.4 = 40\%}$$

25.5 ▶ Light Emitting Diode

On forward biasing a *p-n* diode, energy is given off in the form of heat in some diodes, while photons are emitted in addition to heat in some other diodes. A *p-n* diode specially prepared to produce visible and IR light on forward biasing is called light emitting diode. On forward biasing a LED, the majority carriers present in the respective regions of diode across the *p-n* junction. The free electrons at the *n*-side move towards the *p*-side and holes at the *p*-side move towards the *n*-side of the diode. The free electrons that enter the *p*-side from the *n*-side are called minority carriers in the *p*-region and vice versa.

This increases the local minority carrier population than the normal value. This is known as minority carrier injection. The excess minority carriers diffuse away from the junction and produce recombination with majority carriers. For example, the excess minority electrons in the conduction band of the *p*-region recombine with the majority holes in the valence band of the *p*-region and emit photons. Here, the electrons make downward transition from conduction band to valence band for recombination with holes and the difference of energy will be emitted in the form of photons of energy E_g. Similar action takes place in the n-region also. Under reverse bias, no photons are emitted. The above process has been described with the energy band diagram shown in Fig. 25.14. The wave length of emitted photon is given by

$$\boxed{\lambda = \frac{hc}{E_g}} \tag{25.4}$$

where h = Planck's constant = 6.626×10^{-34} Js.

The basic structure of LED is shown in Fig. 25.15(a), and its standard symbol is shown in Fig. 25.15(b). Some of the LED materials are given in Table 25.1.

The contact area of electrode on p-type material is smaller to permit the emergence of a large number of photons when the device is forward biased. The Si and Ge semiconductor diodes will not emit photons on forward biasing but heat is dissipated at the junction.

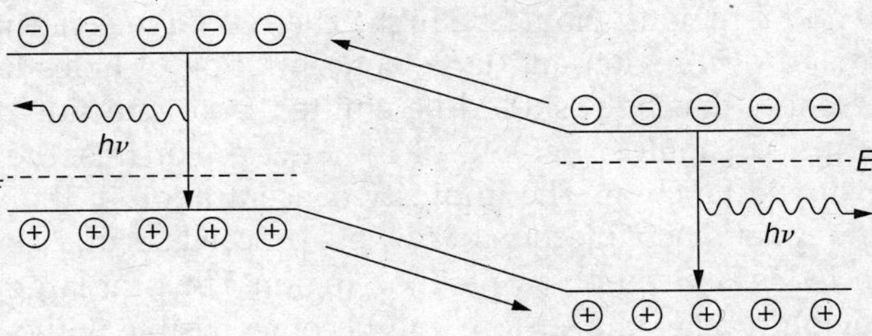

FIGURE 25.14 Band diagram showing injection electro luminescence in forward-biased LED

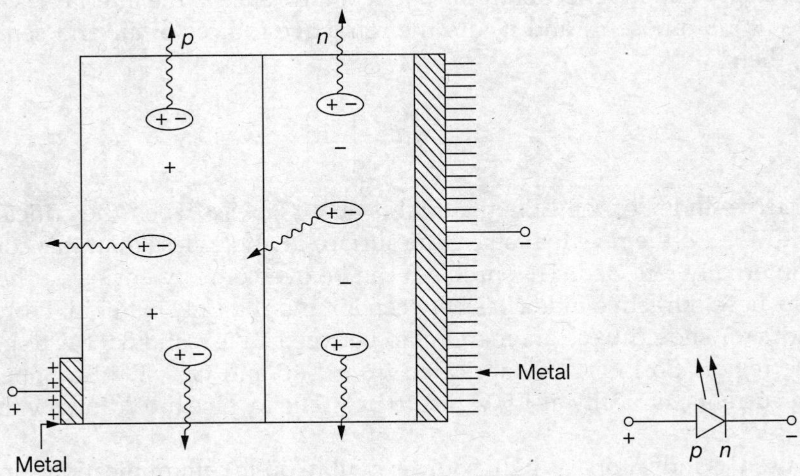

FIGURE 25.15 (a) Basic structure of LED; (b) Symbol of LED

Table 25.1 gives the various LED materials and the colour of emitted waves.

TABLE 25.1 Different LED materials in visible region

Material	Colour
GaP	Green
SiC	Blue
GaN	Blue
Al_nGaP	Yellow
GaAsP (Direct band gap)	Red
GaAsP (Indirect band gap)	Yellow

The emitted photons move in random directions. Some of them will be absorbed in the material. To get large photon emissions per second, a large number of electron-hole recombination should take place per second. To have a large recombination rate, the doping concentration in n- and p-regions should be high. There is large concentration of electrons in the conduction band of n-region and large concentration of holes in the valence band of p-region. It is essential that light should be emitted from one side of the junction and most of the light emitted should come out of the device. For this, the device is made of an asymmetrically doped junction. The impurity concentration in the n-region should be higher than in the p-region. Then injection of carriers proceeds in one direction.

Electrons are injected in large numbers into the p-region. Thus, the large number of photons released in the p-region reaches the surface and becomes visible without loss. As photons pass through the p-region, some of them get absorbed. This can be reduced by making the p-region very thin. Also, the photons that fall on the interface between the p-type material and air cause total internal reflection if the angle of incidence at the interface is greater than critical angle (θ_c). Suppose n_1 and n_2 are the refractive indices of air and semiconductor, respectively, then

$$\theta_c = \sin^{-1}\left(\frac{n_1}{n_2}\right) \tag{25.5}$$

If the refractive index of the LED material is about 3.3–3.8. Then the critical angle will come to 15° to 18°. So the rays that strike the surface at angles less than this come out and the rest get internally reflected. This problem can be overcome by enclosing the LED in an epoxy resin, whose refractive index lies between air and the LED material. From Eq. (25.1), the semiconductor should have an energy gap between 1.7 eV and 3.0 eV, to get emission in the visible region. So LEDs are fabricated from GaP and GaAsP. LEDs operate at low voltages and currents, typically at 1.5 V and 10 mA. The reverse breakdown voltage is very low, usually 3 V.

LEDs are used as indicators, as light sources in fibre optic communication, etc. A number of LEDs are grouped to form a seven-segment display as shown in Fig. 25.16(a). The decimal numbers 0–9 can be displayed using the seven segments. The LEDs can also be arranged in the form of a 5×7 matrix as shown in Fig. 25.16(b) to get a decimal number or an alphabetical character. Such a display is known as an alpha numeric display. The number 3 is shown in Fig. 25.16(a) and the alphabet "C" is shown in Fig. 25.16(b).

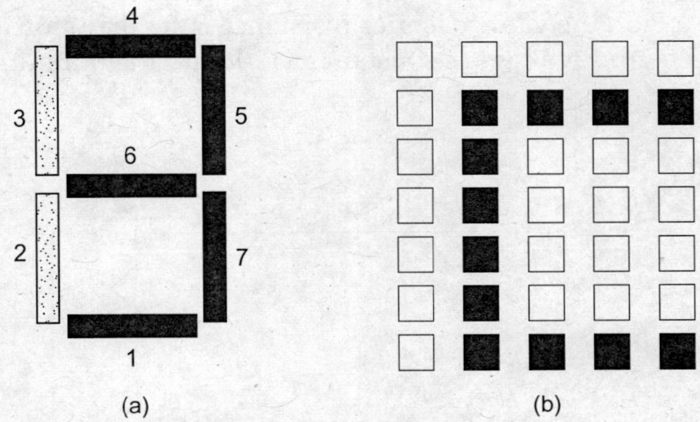

FIGURE 25.16 (a) Seven-segment display; (b) LED array displaying alphabet C

25.6 ▶ Zener Diode

Zener diodes are fabricated to operate in the breakdown region without damage (Fig. 25.17). Zener diodes can be produced with breakdown voltage from 2 V to 200 V by varying the doping level. A Zener diode is specified by its breakdown voltage and the maximum power dissipation.

The most common application of a Zener diode is in the voltage stabilization.

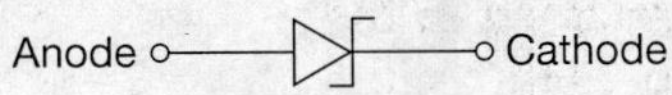

FIGURE 25.17 Zener diode

25.6.1 Zener Diode as Voltage Regulator

The voltage regulator circuit consists of a register R_s connected in series with the input voltage, and a Zener diode is connected in parallel with the load R_L. As long as the voltage across R_L is less than the Zener breakdown voltage V_Z, it will not conduct. If it does not conduct, then R_s and R_L make a potential divider across V_i. The resistance R_s limits the Zener current (Fig. 25.18). The current from the unregulated supply, at the junction of Zener diode splits as

$$I_s = I_z + I_L$$

Now, if load current I_L increases, due to decrease in R_L, the current I_z falls by same percentage through the Zener diode in order to maintain the current I_s constant.

The voltage drop across R_s remains constant. Hence, the output voltage V_o across R_L also remains constant.

On the other hand, if load current decreases, the Zener diode passes an extra current I_z such that I_s current is kept constant. Thus, the output voltage across R_L again remains constant.

EXAMPLE 25.2

A 7.2 V Zener is used in the circuit shown in Fig. 25.18, and the load current is to vary from 12 to 100 mA. Find the value of series resistance R_S to maintain a voltage of 7.2 V across the load. The input voltage is constant at 12 V and the minimum Zener current is 10 mA.

Solution:

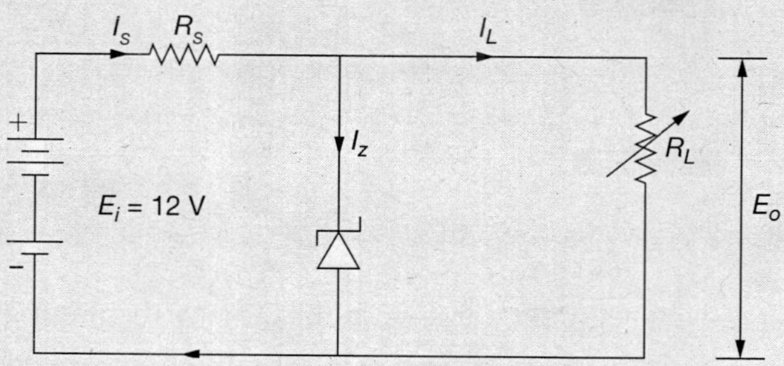

FIGURE 25.18

Here, $E_i = 12$ V, $V_z = 7.2$ V

$$R_S = \frac{E_i - E_o}{(I_z)_{\min} + (I_L)_{\max}}$$

$$= \frac{12 - 7.2}{(10 + 100)\ \text{mA}}$$

$$= \frac{\mathbf{4.8\ V}}{\mathbf{110\ mA}} = \mathbf{43.5\Omega}$$

25.7 ▶ Photovoltaic Process in Semiconductors

In an open-circuited *p*-*n* junction, the height of the potential barrier adjusts itself so that the resultant current becomes zero. The electric field at the junction in such a condition repels the majority carriers. When light is incident on diode, minority current increases.

But the resultant current must remain zero since the diode is open circuited. Thus, the majority current should be increased by the same amount as the minority carrier current.

If the retarding electric field at the junction is reduced, the increase in majority current is possible. This also results in decrease in barrier height. Therefore, photovoltaic emf is developed across the junction. It is of the order of 0.1 volt for Ge cell and 0.5 volt for Si cell.

25.7.1 Solar Cells or Photovoltaic Cells

In 1839, Becquerel discovered the phenomenon of photovoltaic effect. According to this effect, when a pair of electrodes is immersed in an electrolyte and light is allowed to incident on one of the electrodes, a potential difference is created between these electrodes. The devices based on this effect are known as photovoltaic cells.

The potential difference developed is directly proportional to the intensity and frequency of the incident light.

Construction and working

The cell is a *p-n* junction diode with doped semiconductors. Doping of *p* junction is very high. *p* junction is made very thin so that light radiations can penetrate the junction. The exposed area to sun is maximum. *p*-type junction is surrounded by nickel-plated ring which serves as positive terminal of the cell. A metallic contact with *n*-type material is provided so that it can serve as negative terminal (Fig. 25.19).

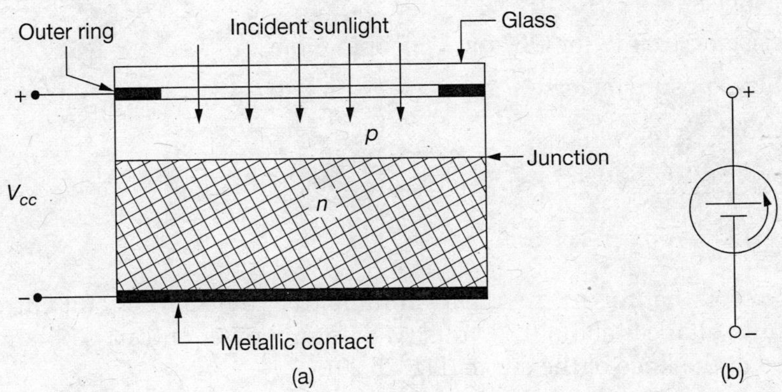

FIGURE 25.19 (a) Solar cell; (b) Symbol

When a photon falls on cell and collides with electrons, it imparts sufficient energy to the electron. As a result, electrons and holes are generated in each junction. With no applied voltage, electrons move freely from *p*-region across the junction. Similarly, holes move across the junction from *n*-region.

This results in the increase of minority charge carrier flow. Depletion region potential causes the photocurrent to flow through the external load.

Advantages

(i) The solar cell does not require any external voltage.

(ii) It is a pollution-free energy conversion device.

(iii) The current generated by a solar cell and its internal emf can be measured by galvanometer.

(vi) These cells are used as a power source in many space satellites for a long period of time.

(v) It can be operated over a wide range of temperature.

Limitations of solar cell

(i) The efficiency of solar cell is very low. It cannot convert all the solar radiations into electrical energy.

(ii) The efficiency of solar cells depends on temperature. For example, the efficiency of silicon cells is low at high temperatures.

Applications

The solar cells are extensively used in

(i) space satellites,

(ii) low resistance relays for ON and OFF operations,

(iii) portable exposure meters,

(iv) infra-red detectors and

(v) data processing industry.

Characteristics

Here, it can be seen that for 100 mw/cm^2 illumination, the short circuit current is 50 mA and open circuit voltage is about 0.57 volts. Maximum power output is obtained when the cell is operated at the knee of the curve (Fig. 25.20).

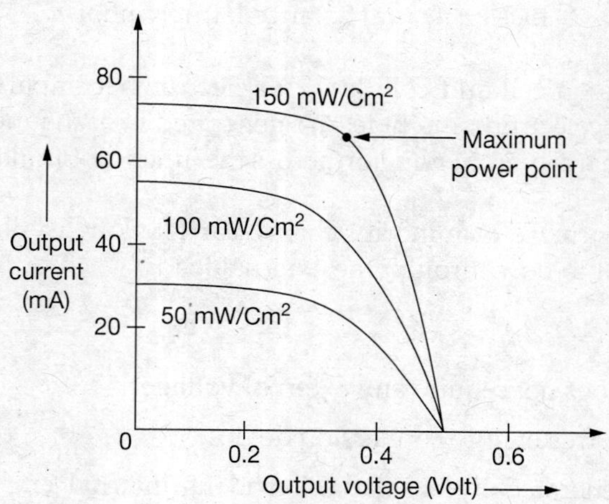

FIGURE 25.20 Solar cell

EXAMPLE 25.3

For a solar cell, $V_m = 0.54\text{V}$, $I_m = 0.024$ amp. The open-circuited voltage is 0.62 V and short-circuit photocurrent is 25 mA. Calculate the fill factor of this cell.

Solution: The fill factor is given by

$$FF = \frac{I_m V_m}{I_L V_{oc}}$$

$$= \frac{0.024 \times 0.54}{0.025 \times 0.62} = 83\%$$

25.7.2 Photodiodes

A photodiode is a semiconductor *p-n* junction, which operates in reverse biasing and respond to photon absorption. The output current of a reverse biased *p-n* junction changes when it is exposed to illumination. The change in output current is liners with respect to luminous flux.

Construction and working

A *p-n* junction diode with light permitted to fall on one of the surface of the device across the junction makes the photodiode. A lens is placed on the junction to focus the incident on the surface. The whole device is kept in a sealed plastic or glass casing (Fig. 25.21).

This case is so designed that the rays are allowed to fall only on one surface but the remaining sides are painted black to restrict the penetration of light rays.

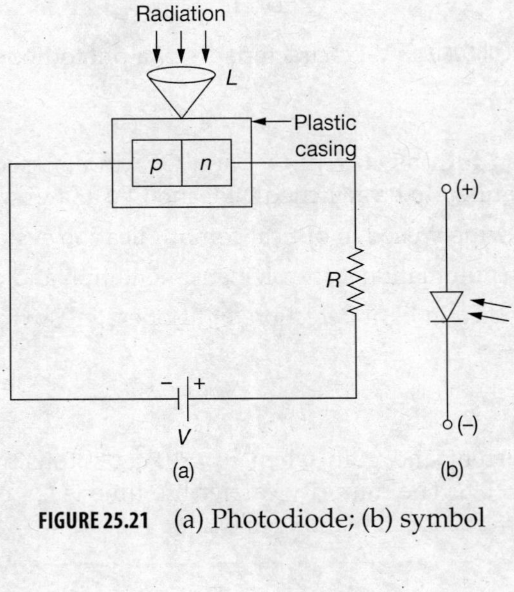

FIGURE 25.21 (a) Photodiode; (b) symbol

When no light is incident on the photodiode, then constant current flows, which is due to the thermally generated minority carriers and this current is called dark current. When light is allowed to be incident on the diode, then additional electron-hole pairs are formed. The photons are absorbed in depletion region and the $\overline{e}$ -hole pairs are generated. The electrons are swept into n-region and holes are swept into p-region and this gives a photocurrent.

With any bias voltage V, the dark current is given by

$$I_d = I_o\,(1 - e^{Ve/\eta KT}) \tag{25.6}$$

and the V-I characteristics of a photodiode (Fig. 25.22) are given by

$$\boxed{I = I_s + I_o\,(1 - e^{Ve/\eta KT})} \tag{25.7}$$

where $\eta = 1$ for Ge and $\eta = 2$ for Si.

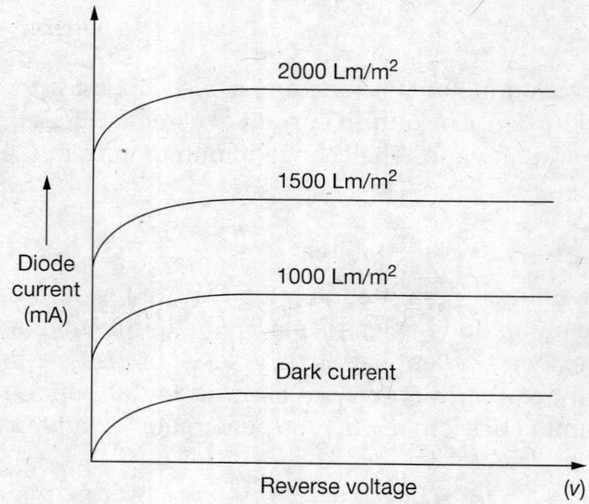

FIGURE 25.22 V-I characteristics of a photodiode

Uses

1. A photodiode can turn its current ON and OFF in nanoseconds; hence, it is used where light is required to be switched OFF and ON at a very fast rate.
2. Photodiodes are widely used in optical communication systems.
3. These are used in automation, control, instrumentation and communication.
4. Photodiodes are used in light detection, light-operated switches, etc.

EXAMPLE 25.4

Calculate the photocurrent and gain when 5×10^{12} photons arriving at surface of photoconductor of $\eta = 0.8$. The minority carrier life time is 0.5 ns and the device has $\mu_n = 2500\ \text{cm}^2/\text{V-s}$, $E = 5000\ \text{V/cm}$ and $l = 10\ \mu\text{m}$.

Solution: The photocurrent is given by

$$I_p = q\left(\eta \frac{\rho_{op}}{hv}\right)\left(\frac{\mu n\ \tau E}{l}\right)$$

$$=(1.6\times10^{-19})(0.8\times5\times10^{12})\left[\frac{25000\times0.5\times10^{-9}\times5000}{10\times10^{-4}\ \text{cm}}\right]$$

$$= 4\times10^{-6}\ \text{A}$$

$$= 4\ \mu\text{A}$$

$$\text{Gain} = \frac{\mu_n\ \tau E}{l}$$

$$= \frac{2500\times(0.5\times10^{-9})\times5000}{10\times10^{-4}}$$

$$= \mathbf{6.25}$$

25.8 ▶ Transistors

A transistor is a multi-electrode semiconductor device, which is formed by sandwitching either *p*-type or *n*-type semiconductor between a pair of opposite types. These can be of two types (Fig. 25.23):

(i) *n-p-n* transistor

(ii) *p-n-p* transistor

The working of both the transistors is identical excepting the type of charge carriers. In *n-p-n* transistors, the mobility of electrons is more than that of *p-n-p*, that's why *n-p-n* are preferred. Hence, the conductivity of *n-p-n* is more than that of *p-n-p* transistors.

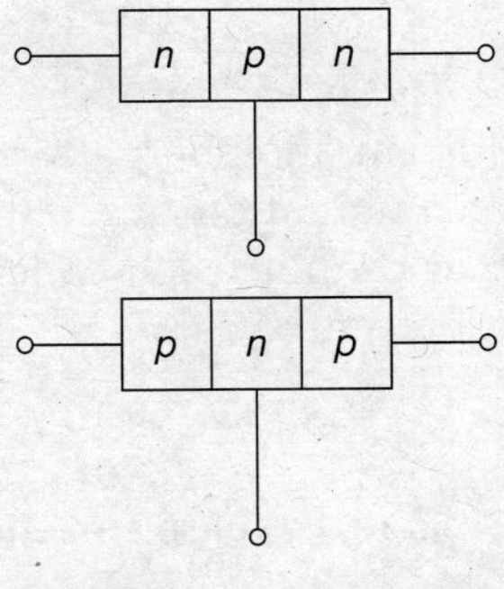

FIGURE 25.23 *n-p-n* and *p-n-p* transistor

25.8.1 Transistor Terminals

There are three sections of transistors, and hence three electrodes. The section on one side (left) is the emitter. Its function is to emit or supply charges.

It is always forward biased w.r.t base. In case of *n-p-n* transistors, it supplies electron but in case of *p-n-p* transistors, it supplies holes to its junction (Fig. 25.24). The section of the other side is called collector. The function of a collector is to collect charges. It is always reverse biased. Its function is to remove charges from its junction.

Its doping is intermediate, i.e. it is heavily doped than base but lightly doped than emitter. The middle region is called base. It provides low resistance for the emitter circuit. The base-emitter junction is always forward biased, while base–collector junction is reverse biased, thus it provides high resistance to the collector circuit.

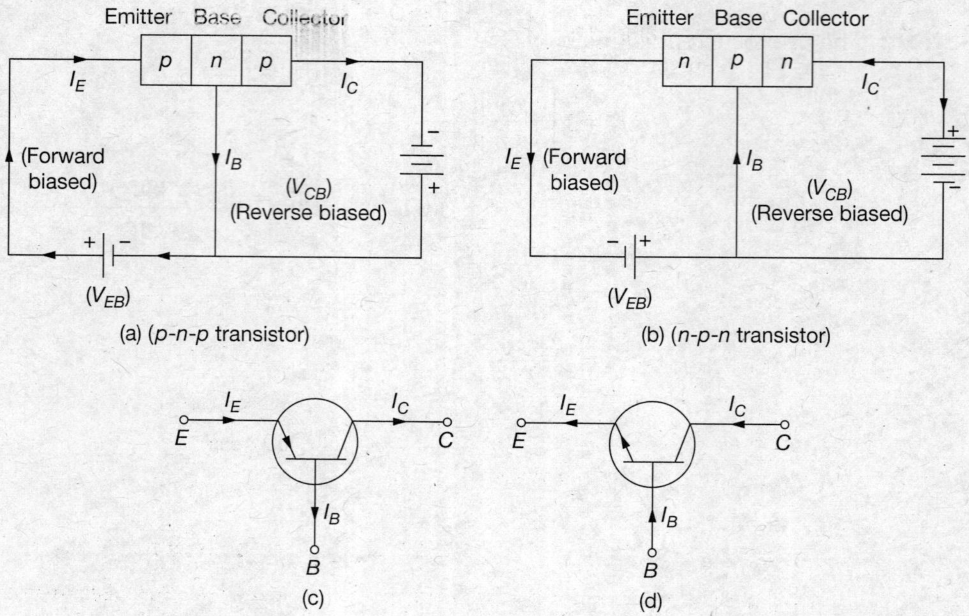

FIGURE 25.24 (a) and (c) *p-n-p* transistor and (b) and (d) *n-p-n* transistor

The base region is made very thin so that it offers a low resistance to the emitter current and charge carriers emitted by emitter are collected by collector. It prevents the recombination of charge carriers so reduces the transit time of carriers. Some charges that cannot reach collector return to emitter via base. Thus, the emitter current can be represented as

$$\boxed{I_E = I_C + I_B} \tag{25.8}$$

Here, the base current is usually very small so Eq. (25.8) can be rewritten as

$$I_E \simeq I_C \tag{25.9}$$

25.8.2 Transistor Configurations

There are three types of transistor circuit configurations: (i) common base (CB) configuration, (ii) common emitter (CE) configuration and (iii) common collector (CC) configuration. There are some advantages of each configuration. The CE configuration is mostly preferred because of the high voltage and current gains of the configuration. Thus, depending upon the requirements of a circuit, a specific configuration out of three is selected. We shall discuss each configuration one by one.

25.8.3 Common Base Configuration (CB)

In this configuration, input is fed between emitter and base while output is obtained between collector and base. A common base *n-p-n* transistor is shown in Fig. 25.25.

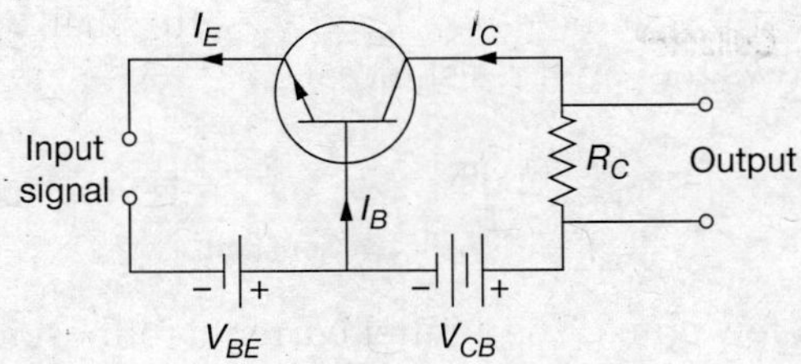

FIGURE 25.25 CB configuration

In this circuit connection, the ratio of change in collector current to the change in emitter current at constant collector base voltage (V_{CB}) is known as current amplification factor, α, i.e.,

$$\boxed{\alpha = \left.\frac{\Delta I_C}{\Delta I_E}\right|_{V_{CB}}} \tag{25.10}$$

The value of α for CB configuration is less than unity. This value can be increased (but not more than unity) by decreasing the base current. Practical values of α range from 0.9 to 0.99.

As a result of electron-hole recombinations in base, whole of the emitter current does not reach the collector thus total collector current

$$I_C = \alpha\, I_E + I_{\text{leakage}} \tag{25.11}$$

where I_{leakage} = current due to minority charge carriers across base–collector junction.

Characteristics of CB configuration

The most important characteristics of common base connection are input and output characteristics.

(a) **Input characteristics:** It is the curve between I_E (emitter current) and V_{EB} (emitter–base voltage) keeping V_{CB} (collector–base voltage) constant (Fig. 25.26).

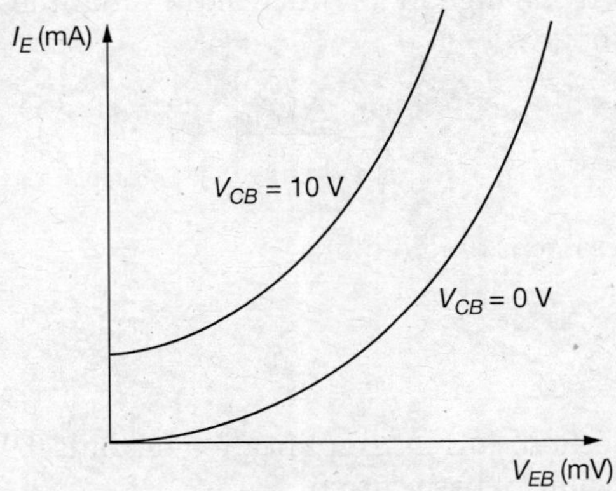

FIGURE 25.26 Input characteristics of CB configuration

Here, the emitter current increases rapidly with small increase in emitter–base voltage, which shows the input resistance, i.e.,

$$r_i = \left.\frac{\Delta V_{BE}}{\Delta I_E}\right|_{V_{CB} = \text{Constant}} \quad \text{is very small} \tag{25.12}$$

It is of the order of few ohms. The emitter current is almost independent of collector–base voltage V_{CB}.

(b) **Output characteristics:** It is the curve between I_C (collector current) and V_{CB} (collector–base voltage) keeping I_E (emitter current) constant (Fig. 25.27).

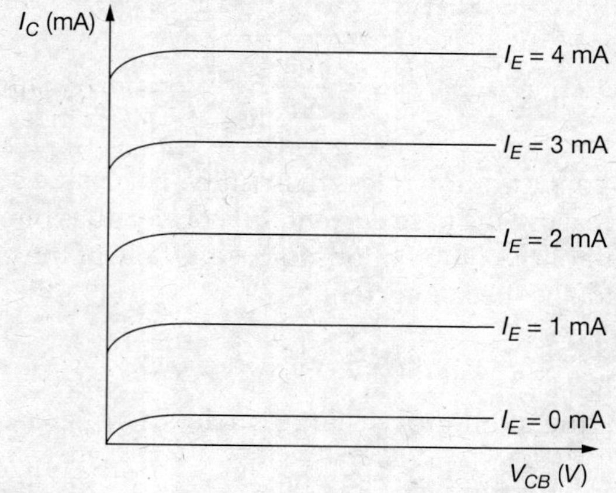

FIGURE 25.27 Output characteristics of CB configuration

In output characteristics, the collector current I_C varies with V_{CB} only at very low voltages (< 1 V). The transistor is never operated in this region. When the value of V_{CB} is increased above 1–2 V, I_C becomes constant and it is independent of V_{CB} and depends only on I_E. The transistor is always operated in this region. A very large change in V_{CB} produces only a tiny change in I_C which shows that the output resistance of the configuration is very high, i.e.,

$$r_o = \left.\frac{\Delta V_{CB}}{\Delta I_C}\right|_{I_E = \text{Constant}}$$

It is of the order of several tens of kilo-ohms.

EXAMPLE 25.5

In a common base connection, current amplification factor is 0.9. If the emitter current is 1 mA, determine the value of base current.

Solution: Here, $\alpha = 0.9$

$$I_E = 1 \text{ mA}$$

Now, $$\alpha = \frac{I_C}{I_E}$$

$\Rightarrow$ $$I_C = \alpha I_E = 0.9 \times 1 = 0.9 \text{ mA}$$

or $$I_E = I_B + I_C$$

$\Rightarrow$ $$I_B = I_E - I_C$$

$$= 1 - 0.9$$

$$= \mathbf{0.1\ mA}$$

25.8.4 Common Emitter Configuration (CE)

In this configuration, input is fed between base and emitter while output is obtained between collector and emitter. A common emitter *n-p-n* transistor is shown in Fig. 25.28.

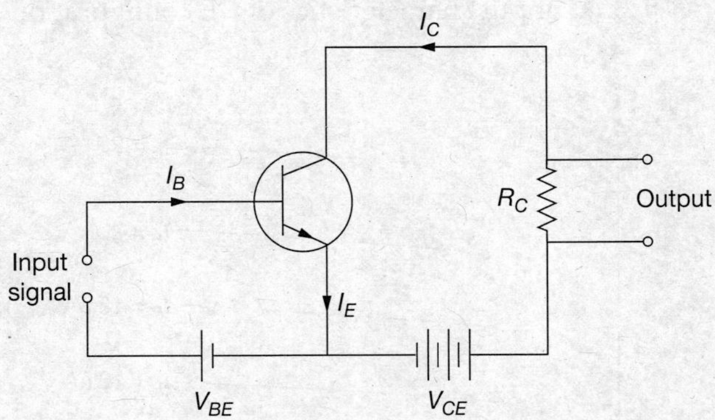

FIGURE 25.28 CE configuration

In this circuit connection, the ratio of change in collector current to the change in base current is known as base current amplification factor,

i.e., $$\boxed{\beta = \frac{\Delta I_C}{\Delta I_B}} \tag{25.13}$$

In every transistor, less than 5% of emitter current flows as the base current. Therefore, the value of β is generally greater than 20. Its value generally ranges from 20 to 500. This type of configuration is frequently used as it gives appreciable current gain as well as voltage gain.

In common emitter circuit, collector current is given as

$$I_E + I_B \simeq I_E \tag{25.14}$$

$$I_C = \alpha I_E + I_{CBO} \tag{25.15}$$

where I_{CBO} = collector–emitter current with base open.

Characteristics of CE configuration

The input and output characteristics of CE configuration are given in Figs. 25.29 and 25.30.

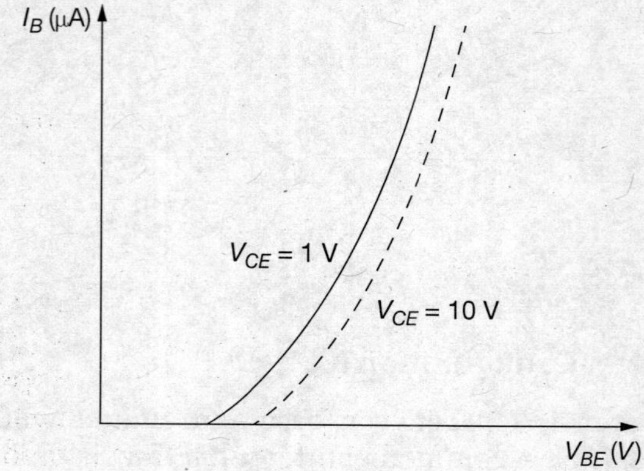

FIGURE 25.29 Input characteristics of CE configuration

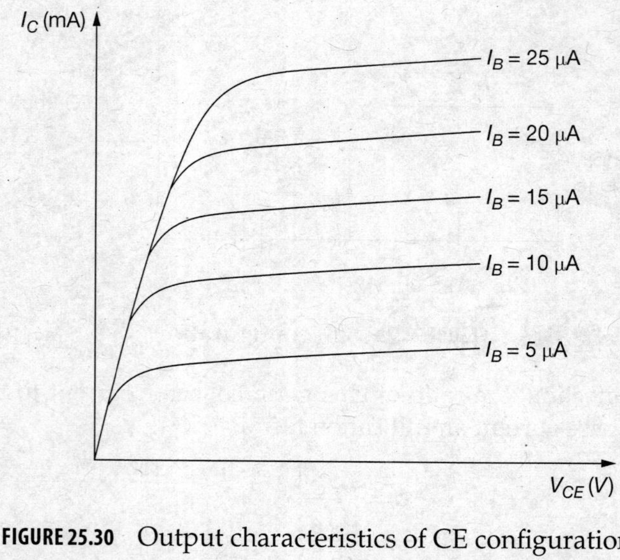

FIGURE 25.30 Output characteristics of CE configuration

(a) **Input characteristics:** It is the curve between base current I_B and base–emitter voltage V_{BE} at constant collector–emitter voltage V_{CE}.

Here, as compared to CB configuration, I_B increases less rapidly with V_{BE}. Therefore, the input resistance

$$\boxed{r_i = \left.\frac{\Delta V_{BE}}{\Delta I_B}\right|_{V_{CE}\,=\,\text{Constant}}} \tag{25.16}$$

of CE circuit is higher than that of CB circuit.

The value of input resistance for a CE circuit is of the order of a few hundred ohms.

(b) **Output characteristics:** It is the curve between I_C (collector current) and V_{CE} (collector–emitter voltage) keeping I_B (base current) constant.

In output characteristics, the collector current I_C varies with V_{CE} for V_{CE} between 0 V and 1 V only. After this value, collector current becomes almost constant and independent of V_{CE}. The value of V_{CE} at which collector current I_C changes with V_{CE} is called knee voltage (V_{knee}). The transistor is always operated in the region above knee voltage. Above knee voltage, I_C is almost constant. Also, for any value of V_{CE} above knee voltage, the value of I_C is approximately equal to βI_B.

The output resistance for CE configuration is given as

$$\boxed{r_o = \left.\frac{\Delta V_{CE}}{\Delta I_C}\right|_{I_B\,=\,\text{Constant}}} \tag{25.17}$$

The output resistance of CE circuit is less than that of a CB circuit. It is of the order of 50 KΩ.

25.8.5 Common Collector Configuration (CC)

In this configuration, input is fed between base and collector while output is obtained between emitter and collector. A common collector *n-p-n* transistor is shown in Fig. 25.31.

In this circuit connection, the ratio of change in emitter current to the change in base current is known as current amplification factor in common collector configuration, i.e.,

$$\gamma = \frac{\Delta I_E}{\Delta I_B}$$

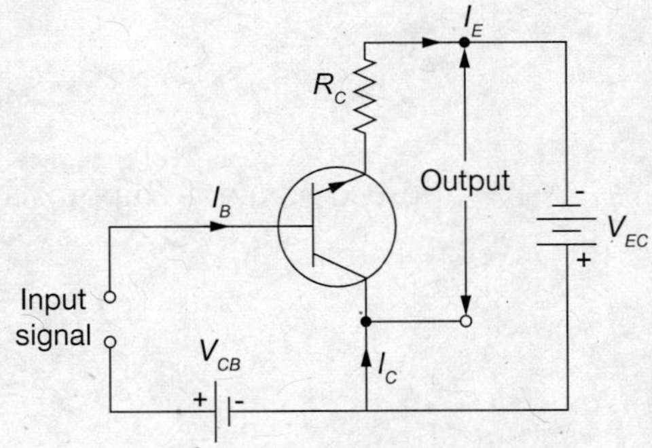

FIGURE 25.31 CC configuration

This configuration gives the same current gain as the common emitter circuit as $\Delta I_E \simeq \Delta I_C$.

Its voltage gain is always less than 1. In common collector circuit, collector current is given as

$$I_C = \alpha\, I_E + I_{CBO} \tag{25.18}$$

where I_{CBO} = collector–emitter current with base open.

The common collector configuration has very high input resistance (about 750 KΩ) and very low output resistance (about 25Ω). This circuit is primarily used for impedance matching.

25.8.6 Relationship Between α, β and γ

(i) We know that

$$\alpha = \frac{\Delta I_C}{\Delta I_E} \tag{25.19}$$

$$\beta = \frac{\Delta I_C}{\Delta I_B} \tag{25.20}$$

But
$$I_E = I_B + I_C$$
$$\Delta I_E = \Delta I_B + \Delta I_C$$
$$\Delta I_B = \Delta I_E - \Delta I_C \tag{25.21}$$

By putting the value of ΔI_B from Eq. (25.21) to Eq. (25.20),

$$\beta = \frac{\Delta I_C}{\Delta I_E - \Delta I_C} \tag{25.22}$$

Also,
$$\beta = \frac{\Delta I_C / \Delta I_E}{1 - \frac{\Delta I_C}{\Delta I_E}} = \frac{\alpha}{1-\alpha}$$

$$\beta = \frac{\alpha}{1-\alpha} \tag{25.23}$$

If α approaches unity, β approaches infinity.

(iii) We know that

$$\gamma = \frac{\Delta I_E}{\Delta I_B} \tag{25.24}$$

But
$$I_E = I_B + I_C$$
$\Rightarrow$
$$\Delta I_E = \Delta I_B + \Delta I_C$$
$\Rightarrow$
$$\Delta I_B = \Delta I_E - \Delta I_C \tag{25.25}$$

By putting the value of ΔI_B from Eq. (25.25) to Eq. (25.24),

$$\gamma = \frac{\Delta I_E}{\Delta I_E - \Delta I_C}$$

$\Rightarrow$
$$\gamma = \frac{\Delta I_E / \Delta I_E}{1 - \frac{\Delta I_C}{\Delta I_E}} = \frac{1}{1-\alpha}$$

$$\Rightarrow \qquad \gamma = \frac{1}{1-\alpha} \tag{25.26}$$

25.9 ▶ Hall Effect

When a current-carrying (conducting) metal or semiconductor is placed inside a magnetic field applied perpendicular to the faces of the material (transverse field), an electric field that is normal to both the magnetic field and the current is produced. This is defined as Hall effect.

The potential difference that is measured across the faces of the material constitutes the Hall voltage (V_H).

25.9.1 Theory

Figure 25.32 illustrates the occurrence of Hall effect in a slab of an n-type material that is subjected to an external magnetic field (B_y) along the y-direction and an electric field (E_x) along the x-direction, both acting in perpendicular directions. If the electric field is due to the electrons only, then the electrons move in an opposite direction with a velocity v_x, as shown in Fig. 25.32. Because of this, the majority charge carriers, the electrons, are subjected to a deflecting force (magnetic Lorentz force) in the upward direction. The drifting is mutually perpendicular to both the fields, and the electrons get accumulated on top of the slab as shown.

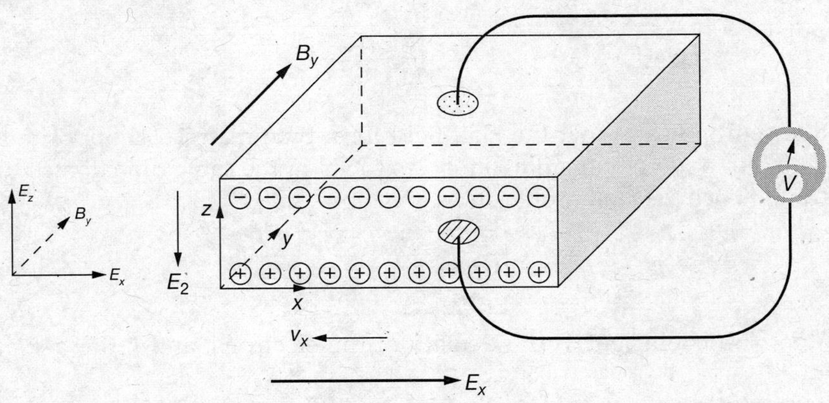

FIGURE 25.32 Diagram illustrating the occurrence of Hall effect

The minority holes, on the other hand, get accumulated at the bottom of the slab. This charge separation across the ends of the slab faces is measured as a potential difference externally. Also, this separation of charges creates a transverse electric field (E_H), in the z-direction. This field, known as the Hall field (E_z) opposes the drifting of electrons.

When the upward deflecting force ($-ev_xB_y$) and the downward opposing force ($-eE_H$) become equal, the situation is said to be under equilibrium. At this junction, the drifting of the electrons stops, because the net force acting on the electron becomes zero. Therefore,

$$-eE_H = -ev_xB_y$$

i.e., $$-eE_H + ev_xB_y = 0 = F_z$$

or, $$eE_H = ev_xB_y$$

$$E_H = v_{xAy} \tag{25.27}$$

The drift velocity v_x of the electrons is related to the current density J_x as

$$J_x = -nev_x$$

∴ $$v_x = -\frac{J_x}{ne} \tag{25.28}$$

where n is the concentration of the electrons.

Substituting Eq. (25.28) in Eq. (25.27), we obtain

$$E_H = -\frac{J_xB_y}{ne}$$

or

$$\frac{E_H}{J_xB_y} = -\frac{1}{ne} \tag{25.29}$$

The term on the right-hand side of the equation describes the Hall coefficient R_H, i.e.,

$$R_H = \frac{E_H}{J_xB_y} = -\frac{1}{ne} \tag{25.30}$$

The negative sign indicates that the Hall field for n-type material is opposite to that of the applied field. For a p-type material, the holes travel in the same direction as that of the applied field, and hence the Hall coefficient is given by

$$\boxed{\mathbf{R_H = \frac{E_H}{J_xB_y} = \frac{1}{pe}}} \tag{25.31}$$

where R_H is the Hall coefficient, n the density of the electrons, and p the density of the holes.

Since from these equations, the quantities E_H, J_x and B_y are measurable, the Hall coefficient R_H can be determined.

25.10 ▶ Determination of Hall Coefficient

Consider a rectangular slab of a material of thickness t and width b, as shown in Fig. 25.33. If the Hall voltage (V_H) is measured across the width of the sample, then the electric field developed is given by (because electric field is the ratio of the potential developed across a distance)

$$E_H = \frac{V_H}{b} \tag{25.32}$$

Substituting Eq. (25.32) in Eq. (25.31), we get

$$R_H = \frac{V_H}{bJB}$$

$$V_H = R_H JBb \tag{25.33}$$

The current density (J) across the material is given by

$$J = \frac{I}{bt} \tag{25.34}$$

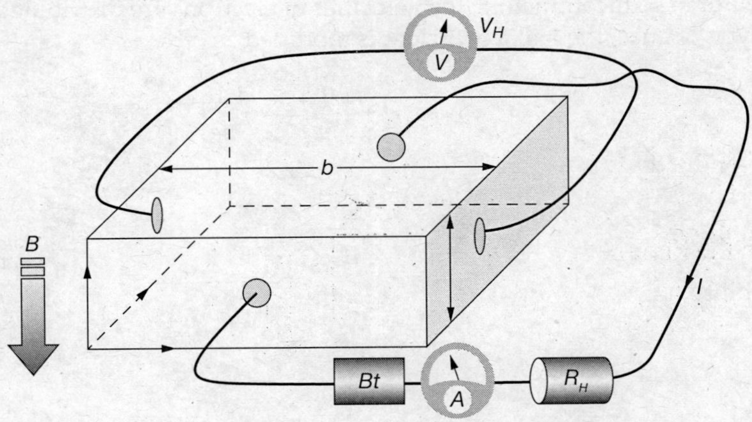

FIGURE 25.33 Experimental set-up for the determination of Hall coefficient

Substituting Eq. (25.34) in Eq. (25.33) and rearranging for R_H, we get

$$V_H = R_H Bb\left(\frac{I}{bt}\right) \quad \left(\because R_H = \frac{tV_H}{IB}\right) \tag{25.35}$$

Knowing the values of the thickness (t), the magnetic field (B), the current and the Hall voltage developed, the Hall coefficient can be determined.

25.10.1 Experiment and Results

As shown in Fig. 25.33, a sample of width b and thickness t is taken. Through the sample, a current I (in amp) is passed through a battery Bt and a rheostat Rh. A magnetic flux density B (in Wb m^{-2}) is produced using the pole pieces of a magnet, and the resultant Hall voltage V_H (in volts) is measured. Using these values, the Hall coefficient R_H (in m^3 C^{-1}) is obtained as given by Eq. (25.35).

Results

We know that the mobility and conductivity are related by (for n-type materials)

$$\sigma_n = \mu_e ne \tag{25.36}$$

or

$$\mu_e = \frac{\sigma_n}{ne} \tag{25.37}$$

i.e.,

$$\mu_e = -\sigma_n R_H \quad \left(\because R_H = -\frac{1}{ne}\right) \tag{25.38}$$

These equations hold good only when the electrons in the conduction band move with a steady drift velocity (i.e., they are not subjected to Coulomb attractions). This is mostly the case in metals. But in semiconductors, the velocities of electrons are distributed over a range (Maxwellian type); hence, the Hall coefficient is modified as

$$R_H = -\frac{3\pi}{8}\left(\frac{1}{ne}\right) = -\frac{1.18}{ne} \tag{25.39}$$

Hence, the conductivity

$$\sigma_n = -\frac{1.18\mu_e}{R_H} \tag{25.40}$$

and the mobility

$$\mu_e = -\frac{\sigma_n R_H}{1.18} \tag{25.41}$$

For a p-type material, by similar argument, we can get

the conductivity

$$\sigma_p = -\frac{1.18\mu_h}{R_H} \tag{25.42}$$

and the mobility

$$\mu_h = -\frac{\sigma_p R_H}{1.18} \tag{25.43}$$

25.11 ▶ Applications of Hall Effect

1. ***Determination of mobility (μ_e)***

   The mobility of a conductor is related to its conductivity as

$$\mu_e = \frac{\sigma_n}{ne} \tag{25.44}$$

   But the concentration of the charges n is related to the Hall coefficient as

$$n = \frac{1.18}{eR_H} \tag{25.45}$$

Substituting for n in Eq. (25.43), we get

$$\mu_e = \frac{\sigma_n(eR_H)}{e(1.18)} = -\frac{\sigma_n R_H}{1.18} \tag{25.46}$$

2. ***Determination of type of semiconductor***

   The Hall coefficient can take either a positive or a negative value. It is positive for a p-type semiconductor and negative for an n-type semiconductor. Hence, from the direction of the Hall voltage developed, we can find out the type of semiconductor. Also, we can determine whether the given material is a conductor or an insulator.

3. ***Carrier concentration calculation***

   The Hall coefficient R_H is related to the carrier concentration n as

$$R_H = \frac{1.18}{ne} \tag{25.47}$$

   Using this equation, knowing the Hall coefficient, the carrier concentration can be determined.

4. ***To measure the magnetic flux density***

   If the Hall coefficient (R_H) is known, then the magnetic flux density (B) of a semiconductor is calculated from the formula

$$B = \frac{V_H t}{IR_H} \tag{25.48}$$

   where I is the Hall current, V_H the Hall voltage, and t the thickness of the sample.

EXAMPLE 25.6

The single carrier holes in a shaped silicon sample is $2.05 \times 10^{22}\ \text{m}^{-3}$. Calculate its Hall coefficient. Electronic charge $= 1.602 \times 10^{-19}$ C.

Solution:

$$R_H = \frac{1}{ne}$$

$$= \frac{1}{\text{charge carrier density} \times \text{electronic charge}}$$

$$= \frac{1}{(2.05 \times 10^{22}) \times (1.602 \times 10^{-19})}$$

$$= \mathbf{3.405 \times 10^{-4}\ m^3/C}$$

EXAMPLE 25.7

Using a free electron model, find the hall coefficient of sodium BCC structure for the Na of cell side 4.28 Å.

Solution: Unit cell of sodium (Na) atom of volume a^3 has 2 atoms.

Therefore,

$$n = 2 \times \left(\frac{1}{a^3}\right) = 2 \times \frac{1}{(4.28 \times 10^{-10})^3}$$

$$= 2.55 \times 10^{28}/\text{m}^3$$

Now, Hall coefficient,

$$R_H = -\frac{1}{ne}$$

$$= \frac{1}{(2.55 \times 10^{28}) \times (1.6 \times 10^{-19})}$$

$$= \mathbf{-\ 0.245 \times 10^{-9}\ m^3/C}$$

EXAMPLE 25.8

A magnetic field of 1 wb m^{-2} is applied along the thickness of a semiconducting crystal of length 10 mm, breadth 5 mm and thickness 1 mm. A current of 10 mA flows along the longest part of the specimen and the Hall voltage across the width is found to be 40 μV. What will be the Hall coefficient of the semiconducting crystal?

Solution: The Hall coefficient is

$$R_H = \frac{1}{ne} = \frac{E_y}{J_x B}$$

$$R_H = \frac{V_y}{bJ_x B}$$

$$\therefore \quad E_y = \frac{V_y}{b}$$

The current density $J_x = \frac{I_x}{bt} = \frac{\text{current}}{\text{area}}$

$$\therefore \quad R_H = \frac{V_y}{bB} \times \frac{bt}{I_x} = \frac{V_y t}{BI_x}$$

Substituting the values

$$R_H = \frac{40 \times 10^{-6} \times 1 \times 10^{-3}}{1 \times 10 \times 10^{-3}}$$

$$= \mathbf{4 \times 10^{-6} \times m^3C^{-1}}$$

Recap Zone

Points to Remember

- A *p-n* junction conducts easily when forward biased and practically no current flows when reverse biased.
- When external voltage is applied to the junction in such a direction that it cancels the potential barrier and permits the current flow, it is called forward biasing.
- When the external voltage is applied to the junction in such a direction that potential barrier is increased, it is called reverse biasing.
- The resistance offered by the diode to forward bias is called forward resistance and the resistance offered by the diode to the reverse bias is known as reverse resistance.
- *p-n* junction diode can be used as a rectifier which can convert AC into DC. It can be of two types – half-wave rectifier and full-wave rectifier.
- The ratio of DC power output to the applied input AC power is known as rectifier efficiency.
- The ratio of rms value of AC component to the DC component in the rectifier output is called ripple factor.
- A Zener diode is an ordinary diode except it is properly doped so as to have a sharp breakdown voltage.
- A transistor consists of two *p-n* junctions formed by sandwitching either *p*-type or *n*-type semiconductor between a pair of opposite types.
- In the light emitting diodes, the radiated energy is in the form of light (or photons).
- When a pair of electrodes is immersed in an electrolyte and light is incident on one of them, then a potential difference is created between the electrodes and the phenomenon is called photovoltaic effect.

- A solar cell is a *p-n* junction illuminated by sunlight with no voltage applied directly. This converts photon power into electric power and delivers this power to load.
- When a specimen carrying a current I is placed in a transverse magnetic field B, then a potential is developed in the specimen perpendicular to both I and B, which is called Hall effect.
- Two terminal devices designed to respond to photon absorption are called photodiodes.

KEY TERMS WITH DEFINITIONS

- **Potential barrier** – It opposes the flow of electrons and holes across the junction.
- **V-I characteristics** – Graph between the voltage applied across the terminals and the current flowing through the same terminals.
- **LED** – A *p-n* diode specially prepared to produce visible and IR light on forward biasing.
- **Photovoltaic effect** – When light is incident on one of the two electrodes immersed in electrolyte and a potential difference is created there.
- **Solar cell** – A device used to convert solar energy into electrical energy.
- **Zener diode** – A *p-n* diode operates in reverse biasing.
- **Photodiode** – A *p-n* diode operates in reverse biasing and respond to photon absorption.
- **Transistor** – Multi-electrode semiconductor device.
- **Common base transistor** – Input is fed between emitter and base and output is obtained between collector and base.
- **Common emitter transistor** – Input is fed between base and emitter and output is obtained between collector and emitter.
- **Common Collector Configuration** – Input is fed between base and collector and output is obtained between emitter and collector.
- **Hall voltage** – The potential difference that is measured across the faces of the material constitutes the hall voltage (V_H).

IMPORTANT FORMULAE AND EQUATIONS

Equation Number	Equation	Remarks
(25.4)	$\lambda = \dfrac{hc}{E_g}$	**Wavelength of emitted photon from LED**
(25.7)	$I = I_s + I_o\left(1 - e\dfrac{V_e}{\eta^{kT}}\right)$	**Diode equation of a photodiode**
(25.8)	$I_E = I_C + I_B$	**Emitter current of a transistor**

(25.10)	$\alpha = \left.\dfrac{\Delta I_C}{\Delta I_E}\right\|_{V_{CB}}$	**Current amplification factor of CB configuration of a transistor**
(25.13)	$\beta = \dfrac{\Delta I_C}{\Delta I_B}$	**Current amplification factor of CE configuration of a transistor**
(25.16)	$r_i = \left.\dfrac{\Delta V_{BE}}{\Delta I_B}\right\|_{V_{CE} = \text{Constant}}$	**Input resistance of CE transistor**
(25.17)	$r_o = \left.\dfrac{\Delta V_{CE}}{\Delta I_C}\right\|_{I_B = \text{Constant}}$	**Output resistance of CE transistor**
(25.31)	$R_H = \dfrac{1}{pe}$	**Hall coefficient**

REVIEW ZONE

SHORT ANSWER QUESTIONS

1. What is a crystal diode?
2. What do you understand by DC and AC resistance of a crystal diode?
3. Draw the V-I characteristics of a diode.
4. Why is Zener diode used as voltage regulator?
5. Why is PIV important in rectifier service?
6. What is a ripple factor?
7. What is Hall effect? Obtain an expression for Hall coefficient for an intrinsic semiconductor.
8. Explain the effect of temperature on resistivity of a semiconductor.
9. What do you mean by fill factor and conversion efficiency of a solar cell?
10. Write a short note on
    (a) Photodiode; (b) Photovoltaic cell
11. Draw V-I characteristics of a photodiode.
12. What is a transistor? Why is it so called?
13. Draw the symbol of *n-p-n* and *p-n-p* transistors and specify the leads.
14. Define α and β and give a relation between them.
15. Draw input and output characteristics of CE transistor.
16. Discuss input and output characteristics of CB transistor.

LONG ANSWER QUESTIONS

1. Draw the graphic symbol of crystal diode and explain its significance. How will you determine the V-I characteristics of a *p-n* diode.
2. Describe a half-wave rectifier using a *p-n* diode.
3. Describe a full-wave rectifier using a crystal diode.
4. Write a short note on Zener diode. Explain how Zener diode maintains constant voltage across the load.
5. Draw the input and output characteristics of CB connection. What do you infer from these characteristics? Define α.
6. How will you determine the input and characteristics of CE connection experimentally?
7. Explain the construction and working of a photovoltaic cell.
8. Explain the construction and working of a solar cell.
9. Explain Hall effect. Show that for an *n*-type semiconductor the Hall coefficient $R_H = -\frac{1}{ne}$.
10. Discuss the applications of Hall effect.
11. In what respect is an LED different from an ordinary *p-n* junction diode? State applications of LEDs. Why should you prefer LEDs over conventional incandescent lamps.
12. Draw a sketch showing the structure of an *n-p-n* junction transistor, label the emitter, base and collector–base junctions.

NUMERICAL PROBLEMS

1. A power supply *A* delivers 10 V DC with a ripple of 0.5 V_{rms}, which the supply *B* delivers 25 V DC with a ripple of 1 mV_{rms}. Which is better power supply?

   Ans. *B*

2. In a common base connection, $I_C = 0.95$ mA, $I_B = 0.05$ mA. Find the value of α.

   Ans. 0.95

3. The Zener diode shown in Fig. 25.34 has $V_Z = 18$ V. The voltage across the load stays at 18 V as long as I_Z is maintained between 200 mA and 2 A. Find the value of series resistance *R* so that E_o remains 18 V while input voltage E_i is free to vary between 22 V and 28 V.

   Ans. 3.33Ω

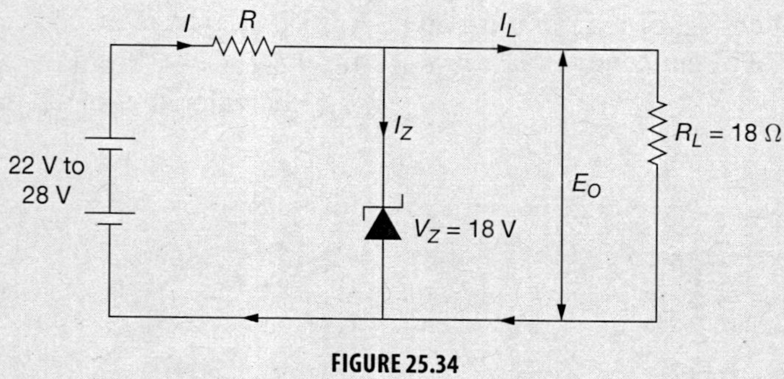

FIGURE 25.34

4. For a transistor, $\beta = 45$ and voltage drop across 1 kΩ which is connected in the collected circuit is 1 volt. Find the base current for common emitter connection.

Ans. 0.022 mA

5. Calculate the open circuit voltage and output power at a voltage 0.35 V for a solar cell whose characteristics are shown in Fig. 25.35.

Ans. 347×10^{-4}w

6. Assuming these are 5×10^{28} atoms/m^3 in copper, and the hall coefficient.

Ans. -0.125×10^{-9} m^3/C

7. The resistivity of a sample semi-conductor is 9 milli-ohm-metre. The holes have mobility of 0.03 m^2/V and calculate Hall coefficient.

Ans. 2.7×10^{-4} m^3/C

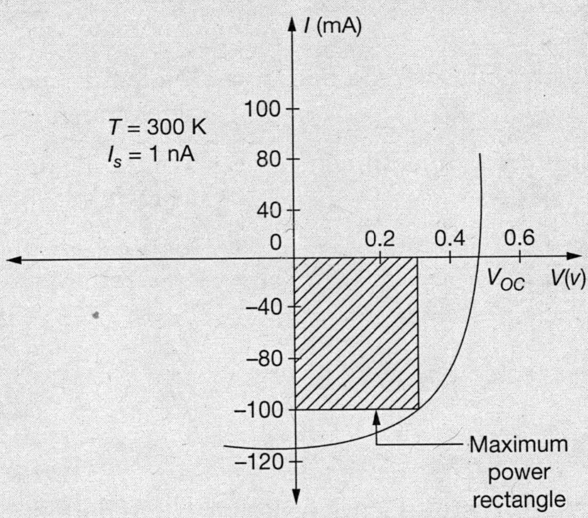

FIGURE 25.35

CHAPTER

26 Nanotechnology

Learning Objectives

By the end of this chapter, the student will be able:

- To understand basic concepts of nanotechnology
- To classify nanostructures based upon dimensions
- To define quantum well, quantum wire and quantum dots
- To illustrate properties of nanoparticles and nanomaterials
- To discuss methods of synthesis of nanomaterials
- To identify carbon allotropes, nanotubes and graphene
- To describe the methods of characterization of nanostructures
- To discuss applications of nanotechnology

26.1 ▶ Introduction

It all started with a lecture "There's Plenty of Room at the Bottom" delivered by Richard P. Feynman at Caltech on December 29, 1959 in which he predicted the possibilities of manipulating atoms and making atomic level machines. But the actual work in the area of nanotechnology started only after the year 2000. The new technological refinements have taken up in past few decades that have simplified and created new methods of manipulations at atomic and molecular level. With the help of development of electron, tunnelling and atomic force microscopes, we can now manipulate or arrange the atoms as per desired spatial positions within a few nanometres. These methods have opened up a full area of research and technology that finds applications in almost every field. Nanomaterials or nanoparticles are the materials which are made of grains that are about 100 nm or less in diameter and contain less than a few ten thousands of atoms. The grain sizes of the conventional materials vary from a few microns to a few millimetres and contain several billions of atoms.

26.2 ▶ Nanoscience and Nanotechnology

Nanoscience is a branch of science that deals with phenomena, processes and materials at atomic, molecular and nanoscales. The fabrication of materials at nanoscale is becoming the key to new developments for new materials and devices for almost every application.

Nanotechnology is basically the design, characterization, production and control of properties of materials at nanoscale and largely finds the new and emerging application of nanoscience. It deals with functional systems based on the use of either sub-units with specific size-dependent properties or individual or combined functionalized sub-units. It can also be defined as the technology to create materials on an atomic scale. Several branches of Science and Technology are now using nanotechnology.

Physics: The construction of specific molecules is governed by the physical forces between the individual atoms composing them. Nanotechnology will involve the continued design of novel molecules for specific purposes. Researchers need to understand how quantum physics affects the behaviour of matter below a certain scale.

Chemistry: The interaction of different molecules is governed by chemical forces. Nanotechnology will involve the controlled interaction of different molecules, often in solution. Understanding how different materials interact with each other is a crucial part of designing new nanomaterials to achieve a given purpose.

Biology: A major focus of nanotechnology is the creation of small devices capable of processing information and performing tasks on the nanoscale. The process by which information encoded in DNA is used to build proteins, which then go on to perform complex tasks including the building of more complex structures, offers one possible template.

Computer Science: As per Moore's law the price performance, speed and capacity of almost every component of the computer. Communications industry has grown exponentially over the last several decades and has been accompanied by steady miniaturization.

Electrical Engineering: To operate independently, nanodevices will need a steady supply of power. Moving power into and out of devices at that scale represents a unique challenge. Within the field of information technology, control of electric signals is also vital to transistor switches and memory storage. A great deal of research is also going into developing nanotechnologies that can generate and manage power more efficiently.

Mechanical Engineering: Even at the nanoscale, issues such as load bearing, wear, material fatigue and lubrication still apply. Detailed knowledge of how to actually build devices that do what we want them to do with an acceptable level of confidence will be a critical component of future research.

26.3 ▶ Dimensional Classification of Nanomaterials

Nanoparticles are generally classified based on their dimensionality, morphology, composition, uniformity and agglomeration.

1D nanomaterials: These are one dimensional in the nanometre scale and are typically thin wires, tubes, rods and include the circuitry of computer chips.

2D nanomaterials: Two-dimensional nanomaterials have two dimensions in the nanometre scale. These include 2D nanostructured films, with nanostructures firmly attached to a substrate. Nanofilms, graphene, nanolayers and nano coatings on surface are examples of 2D nanostructures.

3D nanomaterials: Materials that are nanoscaled in all three dimensions are considered as 3D nanomaterials. These include bulk powders, dispersions, colloids, bundles of wires, fibres and other shape structures.

26.4 ▶ Quantum Well, Quantum Wire and Quantum Dots

- When the size or dimension of a material is continuously reduced from a large or macroscopic size, such a metre or centimetre, to a very small size, the properties remain the same at first, and then small changes begin to occur, until finally when the size drops below 100 nm, dramatic changes in properties can occur.
- If one dimension is reduced to the nano-range while the other dimensions remain large, then we obtain a structure known as *quantum well*.
- If two dimensions are so reduced and one remains large, the resulting structure is referred to as a *quantum wire*.
- The extreme case of this process of size reduction in which all three dimensions reach the low nanometre range is called a *quantum dot*.

The word quantum is associated with the above three types of nanostructures because the changes in properties arise from the quantum mechanical nature of physics in the domain of the ultrasmall.

Figure 26.1 represents the processes of diminishing the size for the case of rectilinear geometry and the corresponding reductions in curvilinear geometry. The conduction electrons are confined in a narrow dimension and such a configuration is referred to as *quantum well*.

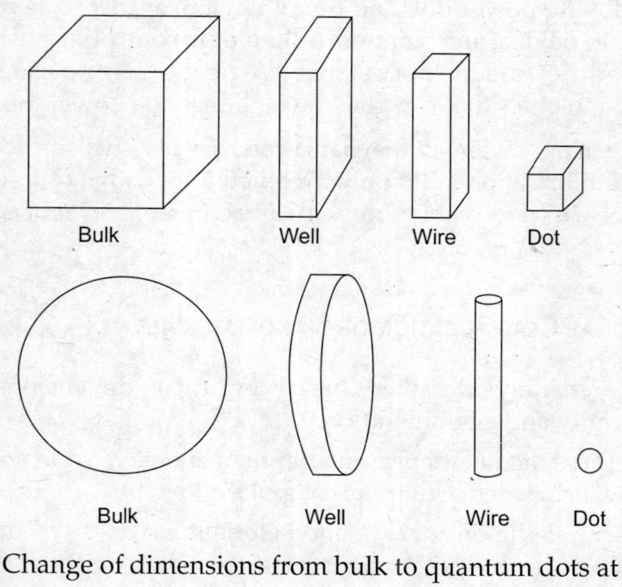

FIGURE 26.1 Change of dimensions from bulk to quantum dots at nano level

A *quantum wire* is a structure such as a copper wire that is long in one dimension, but has a nanometre size as its diameter. In this case, the electrons move freely along the wire but are confined in the transverse directions.

The *quantum dot* may have the shape of a tiny cube, a short cylinder or a sphere with low nanometre dimensions.

The quantum structures are useful in the fabrication of high-efficiency solar cells, infrared detectors, quantum dot lasers, etc. The semiconductors such as PbS, GaAs, CdS, etc. can be synthesized in the nanometre level and they are called semiconductor quantum dots. Their properties like band gap, luminescence, etc. always differ from their bulk counterpart.

26.5 ▶ Physical Properties of Nanoparticles

Nanoparticles are unique because of their large surface area and this dominates the contributions made by the small bulk of the material. Zinc oxide particles have been found to have superior UV blocking properties compared to its bulk substitute. This is one of the reasons why it is often used in the preparation of sunscreen lotions.

Other examples of the physical properties of nanoparticles:

- Colour – Nanoparticles of yellow gold and grey silicon are red in colour.
- Gold nanoparticles melt at much lower temperatures (~300°C for 2.5 nm size) than the gold slabs (1064°C).
- Absorption of solar radiation in photovoltaic cells is much higher in nanoparticles than it is in thin films of continuous sheets of bulk material – since the particles are smaller, they absorb greater amount of solar radiation.

26.5.1 Optical Properties of Nanoparticles

Nanoparticles also often possess unexpected optical properties as they are small enough to confine their electrons and produce quantum effects. One example of this is that gold nanoparticles appear deep red to black in solution.

26.5.2 Formation of Suspensions

An important physical property of nanoparticles is their ability to form suspensions. This is possible since the interaction of the particle surface with the solvent is strong enough to overcome density differences. In bulk materials, this interaction usually results in a material either sinking or floating in a liquid.

26.5.3 Magnetization and Other Properties of Nanoparticles

Other properties that are unique among nanoparticles are quantum confinement in semiconductor particles, surface plasmon resonance in some metal particles and superparamagnetism in magnetic materials.

For example, ferroelectric materials smaller than 10 nm can switch their magnetization direction using room temperature, thermal energy, thus making them unsuitable for memory storage. Thus, this property is not always desired in nanoparticles.

26.5.4 Diffusion Properties of Nanoparticles

At elevated temperatures especially, nanoparticles possess the property of diffusion. Sintering can take place at lower temperatures, over shorter time scales than for larger particles. Although this does not affect the density of the final product, there is a chance of agglomeration.

26.5.5 Hardness of Nanoparticles

Clay nanoparticles, when incorporated into polymer matrices, increase reinforcement, leading to stronger plastics. These nanoparticles are hard, and impart their properties to the polymer (plastic). Nanoparticles have also been attached to textile fibres in order to create smart and functional clothing, such as self-cleaning fabrics or water-repellent cloths.

26.5.6 Semi-solid or Soft Nanoparticles

Semi-solid and soft nanoparticles have been manufactured. Of these notable is the liposome. Various types of liposome nanoparticles are currently used clinically as delivery systems for anticancer drugs, antibiotics and antifungal drugs and vaccines.

26.6 ▶ Unique Properties of Nanomaterials

Most of the nanoparticles possess the following unique properties that are mostly different from their bulk material.

- They have very high magneto resistance.
- They have lower melting point.
- They have high solid state phase transition pressure.
- They have lower Debye temperature.
- They have high self-diffusion coefficient.
- They have high catalytic activity.
- They have lower ferroelectric phase transition temperature.

Apart from the above properties, the following properties may also be different from the bulk but exceptions are always there.

26.6.1 Electron Affinities and Chemical Properties

Variation in electronic properties with size occurs only when there is a variation in inter particle spacing and geometry. As the size is reduced from the bulk, the electronic bands in metals become narrower and the delocalized electronic states are transformed to more localized molecular bonds.

26.6.2 Variation of Physical Properties with Size

It is well established that mechanical, electrical, optical, chemical, semiconducting and magnetic properties of a material depend strongly upon the size and the arrangement of constituent clusters or grains. Variation in electronic properties with size occurs only when there is a variation in inter particle spacing and geometry. As the size is reduced from the bulk, the electronic bands in metals become narrower and the delocalized electronic states are transformed to more localized molecular bonds.

- Figure 26.2 shows the ionization potential and reactivity of the clusters as a function of size.
- The ionization potential is higher at smaller sizes than at the bulk structure.
- The large surface-to-volume ratio and the variation in geometry and electronic structure have a strong effect on catalysis properties.

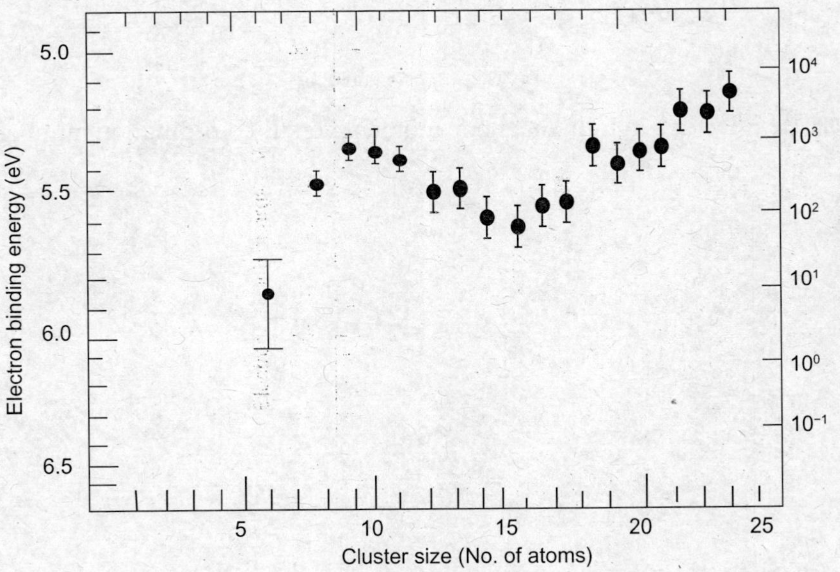

FIGURE 26.2 Size dependence of binding energy

26.6.3 Magnetic Properties

- Nanoparticles of magnetic and even non-magnetic solids exhibit a totally new class of magnetic properties.
- Ferromagnetic and anti-ferromagnetic multi-layers have been found to exhibit *Giant Magneto Resistance* (GMR).
- Small particles differ from the bulk; in that these atoms will have lower co-ordination number.
- From Fig. 26.3, it is inferred that the particles are more magnetic than the bulk material.

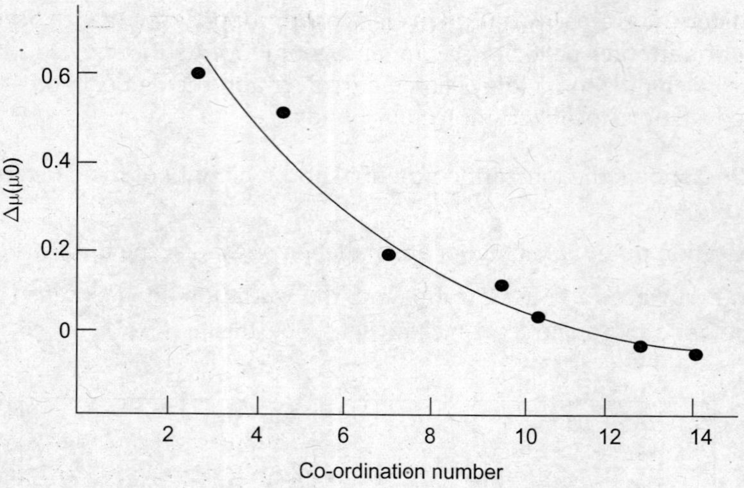

FIGURE 26.3 Change in bulk magnetic moment versus co-ordination number

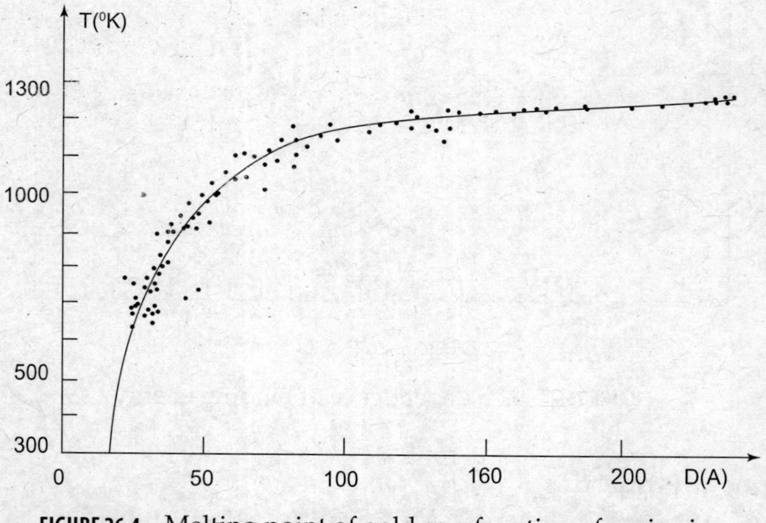

FIGURE 26.4 Melting point of gold as a function of grain size

26.6.4 Thermal Properties

The melting point of materials changes from bulk to nano size (Fig. 26.4) for gold.

26.6.5 Mechanical Behaviour

From Fig. 26.5, it is evident that the melting point reduction is not really significant until the particle size is less than about 10 nm.

- Nanophase metals with their exceptionally small grain size are found to be exceptionally strong.
- It is because clusters and grains in nanophase material are mostly free from dislocations.
- The variation of hardness with diameter of copper nanocrystals is shown in Fig. 26.5.
- From Fig. 26.5, it is revealed that when the grains size was 50 nm in diameter, the copper was twice hard than normal.

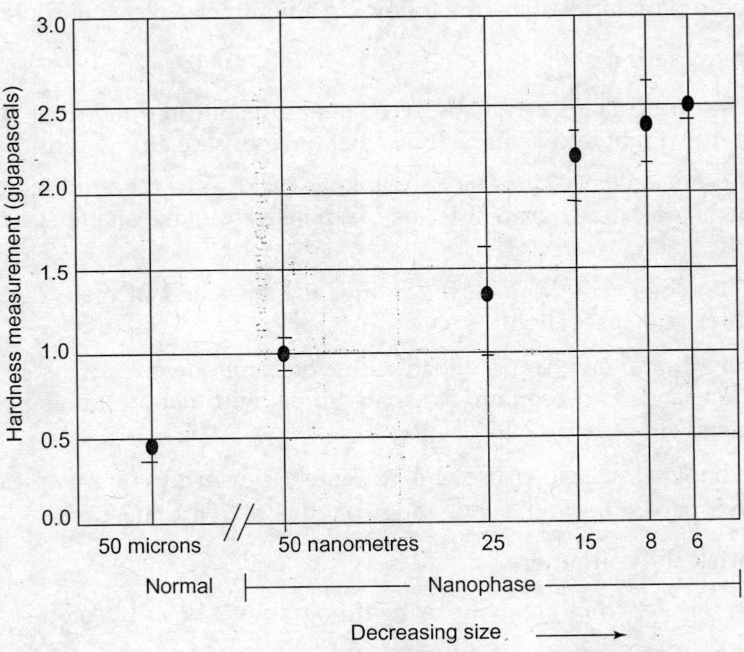

FIGURE 26.5 Strength of nanophase copper as a function of grain size

- The basic mechanical application of nanotechnology is positional control at atomic level.
- At the macroscopic scale, it is easy to hold parts in our hands and assemble them by properly positioning them with respect to each other.
- At the molecular scale, the idea of holding and positioning molecules is new and unimaginable.
- It is possible to continue the revolution in computer hardware right down to molecular gates and wires, something that today's lithographic methods (used to make computer chips) could never hope to do.
- One can inexpensively make very strong and very light materials: shatterproof diamond in precisely the shapes we want, by the ton, and over 50 times lighter than steel of the same strength.

- It is possible to make surgical instruments with high precision and deftness that one could operate on the cells and even molecules from which we are made something well beyond today's medical technology.
- Nanotechnology makes almost every manufactured product faster, lighter, stronger, smarter, safer and cleaner.

26.7 ▶ Technological Advantages of Nanotechnology and Nanomaterials

The nanostructured material offers the following advantages over their bulk counterparts.

Improved transportation

- Today, most airplanes are made from metal despite the fact that diamond has a strength-to-weight ratio over 50 times that of aerospace aluminium.
- Diamond is expensive, it is not possible to make it in the required shapes, and it shatters. Nanotechnology will let us inexpensively make shatter-proof diamond in exactly the shapes we want.
- Nanotechnology will dramatically reduce the costs and increase the capabilities of space ships and space flight.
- The strength-to-weight ratio and the cost of components are absolutely critical to the performance and economy of space ships: with nanotechnology, both of these parameters will be improved.
- Nanotechnology will also provide extremely powerful computers to control and guide the space ships and a wide range of other activities in space.

Atomic memories and computers

- Today, computer chips are made using lithography – literally, "stone writing".
- If the computer hardware revolution is to continue at its current pace, in a decade or so we will have to move beyond lithography to some new post lithographic manufacturing technology. Ultimately, each logic element will be made from just a few atoms.
- Designs for computer gates with less than 1000 atoms have already been proposed and each atom in such a small device has to be in exactly the right place.
- To economically build and interconnect trillions upon trillions of such small and precise devices in a complex three-dimensional pattern, we will need a manufacturing technology well beyond today's lithography, we will need nanotechnology.
- With it, we should be able to build mass storage devices that can store more than a hundred billion billion bytes in a volume the size of a sugar cube.
- RAM can store a mere billion billion bytes in such a volume, and massively parallel computers of the same size that can deliver a billion billion instructions per second.

Military applications

- Today, "smart" weapons are fairly big; we have the "smart bomb" but not the "smart bullet". It will be a possibility in future.
- In future, even weapons as small as a single bullet could pack more computer power than the largest supercomputer in existence today, allowing them to perform real time image analysis of their surroundings and communicate with weapons tracking systems to acquire and navigate to targets with greater precision and control.
- We will also be able to build weapons both inexpensively and much more rapidly, at the same time taking full advantage of the remarkable material properties of diamond.
- Rapid and inexpensive manufacture of great quantities of stronger more precise weapons guided by massively increased computational power will alter the way we fight wars.

Solar energy

- Nanotechnology will cut costs both of the solar cells and of the equipment needed to deploy them, making solar power economical.
- In this application, we need not make new or technically superior solar cells or making inexpensively what we already know.

Medical applications

- If we had surgical tools that were molecular both in their size and precision, we could develop a medical technology that would let us directly heal the injuries at the molecular and cellular level.
- We can have the materials or the particles designed in such a way that these are acceptable to the human body. These can be used to transport the drug to the desired location with precision with minimum side effect.
- Several nutrients can be converted to nanoparticles or conjugated with normal proteins so that these are effectively absorbed and used by our body.
- With the precision of drugs combined with the intelligent guidance of the surgeon's scalpel, we can expect a quantum leap in our medical capabilities.

26.8 ▶ Synthesis of Nanomaterials

The most common method of making tiny micro-structures that can be imagined quickly is by machining the large blocks of material using fine lathes, etching or lithography. These methods work in different size ranges that are of the order of centimetres, microns or few hundreds of nanometres.

Top-down approach

This method of conversion of big size object into a small size object by carving or etching is called *Top-Down approach* of fabrication/synthesis. It is similar to carving a cave in a large mountain or making small items or statues from large blocks of rock or wood. The method starts with a large scale object or pattern and gradually reduces its dimension or dimensions. This can be accomplished by a technique called lithography which shines radiation through a template on to a surface coated with a radiation-sensitive resist; the resist is then removed and the surface is chemically treated to produce the nanostructure.

Bottom-up approach

Similarly, a reverse method can be followed when smaller blocks are put together to construct a larger structure. It is similar to constructing a house by arranging or laying the bricks in a particular order using some adhesive. Such a method of fabrication is called Bottom-up approach. In this approach, the nanostructure is created by collecting, consolidating and arranging individual atoms and molecules into the structure. This is carried out by a sequence of chemical reactions controlled by series of catalysts. It is a process that is widespread in biology where, for example, catalysts called enzymes assemble amino acids to construct living tissue that forms and supports the organs of the body.

There exist a number of methods to synthesize the nanomaterials, which are categorized in to two techniques; "top-down and bottom-up". Solid state route, ball milling comes in the category of top-down approach, while wet chemical routes like sol-gel, co-precipitation, etc. come in the category of bottom-up approach. Secondly, characterization of nanomaterials is necessary to analyze their various properties. Therefore, this chapter describes the various methods of synthesis and characterization of nanomaterials. Characterization techniques include XRD, SEM, TEM, EDAX, UV-Visible spectroscopy, FTIR spectroscopy, etc.

Synthesis routes

Fabrication of nanomaterials with strict control over size, shape and crystalline structure has become very important for the applications of nanotechnology in numerous fields including catalysis, medicine and electronics. Synthesis methods for nanoparticles are typically grouped into two categories: "top-down" and "bottom-up" approaches. The first involves the division of a massive solid into smaller and smaller portions, successively reaching to nanometre size. This approach may involve milling or attrition. The second, "bottom-up", method of nanoparticle fabrication involves the condensation of atoms or molecular entities in a gas phase or in solution to form the material in the nanometre range. The latter approach is far more popular in the synthesis of nanoparticles owing to several advantages associated with it. Figure 26.6 shows the general overview of the two approaches. There are many bottom-up methods of synthesizing metal oxide nanomaterials, such as hydrothermal, combustion synthesis, gas-phase methods, microwave synthesis and sol-gel processing. However, an overview of other techniques usually employed for the synthesis of nanomaterials is also discussed as follows.

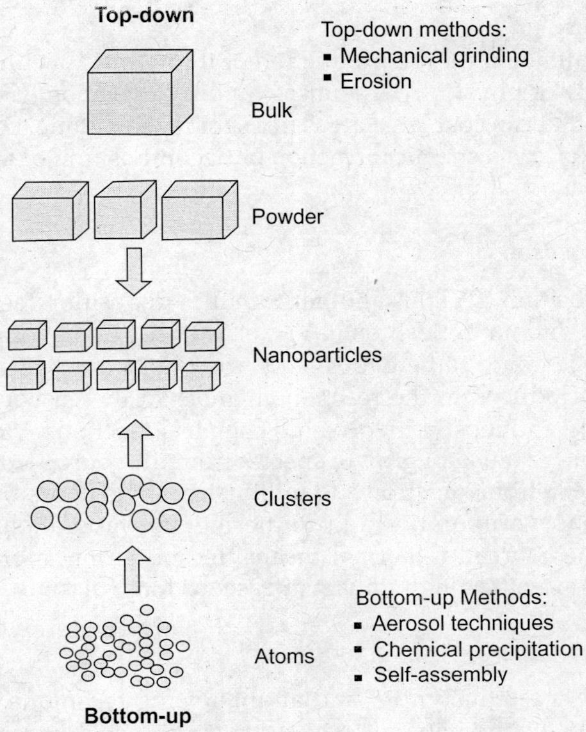

FIGURE 26.6 Schematic of bottom-up and top-down approaches

Combustion route

Combustion synthesis leads to highly crystalline particles with large surface areas. The process involves a rapid heating of a solution containing redox groups. During combustion, the temperature reaches approximately 650°C for one or two minutes making the material crystalline.

Hydrothermal method

Hydrothermal synthesis is typically carried out in a pressurized vessel called an autoclave with the reaction in aqueous solution. The temperature in the autoclave can be raised above the boiling point of water, reaching the pressure of vapour saturation. Hydrothermal synthesis is widely used for the preparation of metal oxide nanoparticles which can easily be obtained through hydrothermal treatment of synthesized precipitates of a metal precursor with water. The hydrothermal method can be useful to control grain size, particle morphology, crystalline phase and surface chemistry through regulation of the solution composition, reaction temperature, pressure, solvent properties, additives and aging time.

Gas phase methods

Gas phase methods are ideal for the production of thin films. Gas phase synthesis can be carried out chemically or physically. Chemical vapour deposition (CVD) is a widely used industrial technique that can coat large areas in a short span of time. During the procedure, metal oxide is formed from a chemical reaction or decomposition of a precursor in the gas phase.

Physical vapour deposition

Physical vapour deposition (PVD) is another thin film deposition technique. The process is similar to chemical vapour deposition (CVD) except that the raw materials/precursors, i.e. the material that is going to be deposited starts out in solid form, whereas in CVD, the precursors are introduced to the reaction chamber in the gaseous state. The process proceeds atomistically and mostly involves no chemical reactions. Various methods have been developed for the removal of growth species from the source or target. The thickness of the deposits can vary from angstroms to millimetres. In general, these methods can be divided into two groups: evaporation and sputtering. In evaporation, the growth species are removed from the source by thermal means. In sputtering, atoms or molecules are dislodged from solid target through impact of gaseous ions (plasma).

Microwave synthesis

Microwave synthesis is relatively new and an interesting technique for the synthesis of oxide materials. Various nanomaterials have been synthesized in remarkably short time under microwave irradiation. Microwave techniques eliminate the use of high temperature calcination for extended periods of time and allow fast, reproducible synthesis of crystalline metal oxide nanomaterials. Utilizing microwave energy for the thermal treatment generally leads to a very fine particle in the nanocrystalline regime because of the shorter synthesis time and a highly focussed local heating.

Sol-gel method

The sol-gel process is a capable wet chemical process to make ceramic and glass materials. This synthesis technique involves the conversion of a system from a colloidal liquid, named sol, into a semi-solid gel phase. The sol-gel technology can be used to prepare ceramic or glass materials in a wide variety of forms: ultra-fine or spherical-shaped powders, thin film coatings, ceramic fibres, microporous inorganic membranes, monolithic or extremely porous arogels. An overview of the sol-gel process is illustrated in Fig. 26.7.

This technique offers many advantages including the low processing temperature, the ability to control the composition on molecular scale and the porosity to obtain high surface area materials, the homogeneity of the final product up to atomic scale. Moreover, it is possible to synthesize complex composition materials, to form higher purity products through the use of high purity reagents. The sol-gel process allows obtaining high quality films up to micron thickness, difficult to obtain using the physical deposition techniques. Moreover, it is possible to synthesize complex composition materials and to provide coatings over complex geometries.

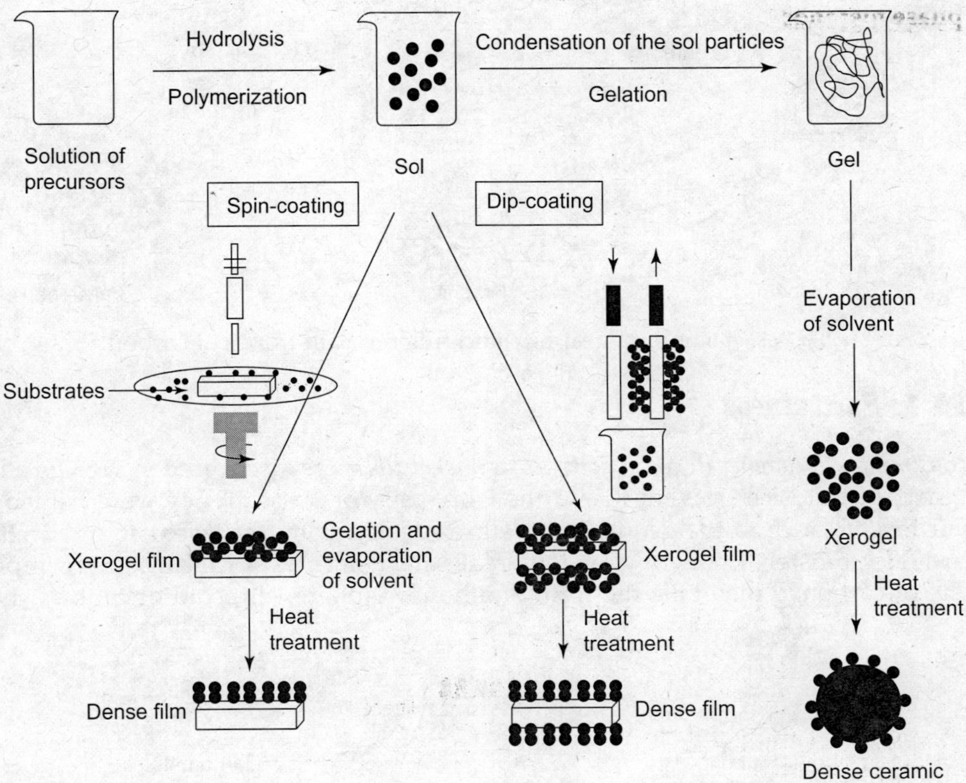

FIGURE 26.7 Mechanism of sol-gel process

The starting materials used in the preparation of the sol are usually inorganic metal salts or metal organic compounds, which by hydrolysis and poly-condensation reactions form the sol. Further processing of the sol enables one to make ceramic materials in different forms. Thin films can be produced by spin-coating or dip-coating. When the sol is cast into a mould, a wet gel will form. By drying and heat-treatment, the gel is converted into dense ceramic or glass materials. If the liquid in a wet gel is removed under a supercritical condition, a highly porous and extremely low density aerogel material is obtained. As the viscosity of a sol is adjusted into a suitable viscosity range, ceramic fibres can be drawn from the sol. Ultra-fine and uniform ceramic powders are formed by precipitation, spray pyrolysis or emulsion techniques.

26.9 ▶ Carbon Allotropes

Carbon has many allotropes due to its valency. These include diamond and graphite. In past few decades, many more allotropes and forms of carbon have been discovered including ball shapes such as Buckminsterfullerene and sheets such as Graphene. Other structures of carbon include nanotubes (single-walled and multi-walled), nanorods and nanoribbons. Some other unusual forms of carbon can also exist at very high temperatures or extreme pressures. As on today about 350 allotropes of carbon have been theoretically predicted.

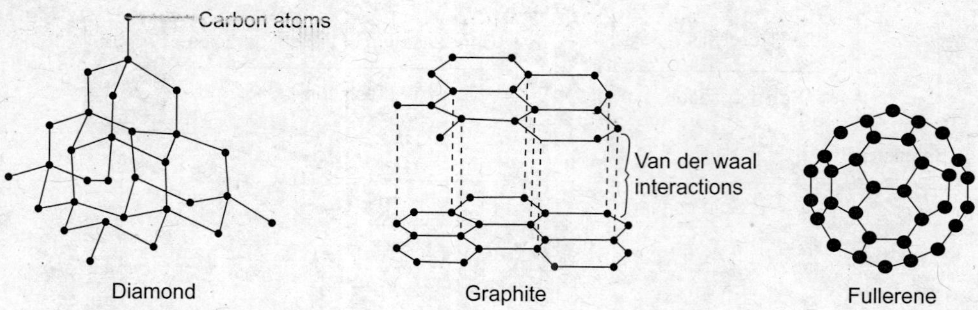

FIGURE 26.8 Diamond, graphite and fullerene allotropes of carbon

26.10 ▶ FULLERENES

The *Buckminsterfullerenes* or just *fullerenes* or *buckyballs* were discovered in 1985 by a team of scientists from Rice University and the University of Sussex. They were named due to their resemblance to the geodesic structures designed by the scientist and architect Richard Buckminster "Bucky" Fuller. Fullerenes are molecules of varying sizes composed entirely of carbon, which take the form of a hollow sphere, ellipsoid or tube as shown in Fig. 26.9.

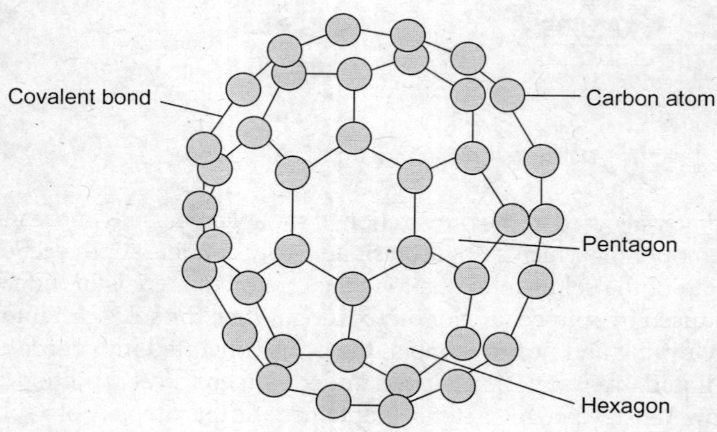

FIGURE 26.9 The Buckyball or fullerene structure of carbon

The chemical and physical properties of fullerenes are still being studied. These have found several applications including the potential for medicinal by binding drugs to the structure to target certain area or cells.

26.11 ▶ CARBON NANOTUBES

Carbon nanotubes are cylindrical structures that are made of only carbon atoms and possess novel properties. These tubes have a variety of applications such as nanoelectronics, optics, engineering materials, etc. These tubes exhibit extraordinary strength, unique electrical properties and are efficient conductors of heat. A nanotube is a member of the fullerene structural family, which also includes buckyballs. Whereas buckyballs are spherical in shape, a nanotube is cylindrical, with at least one end typically capped with a hemisphere of the buckyball structure. Their name is derived from their size, since the diameter of a nanotube is on the order of a few nanometres, while they can be up to several centimetres in length. There are three main types of nanotubes (Fig. 26.10):

1. Single-walled nanotube (SWNTs),
2. Double-walled nanotube (DWNTs) and
3. Multi-walled nanotube (MWNTs).

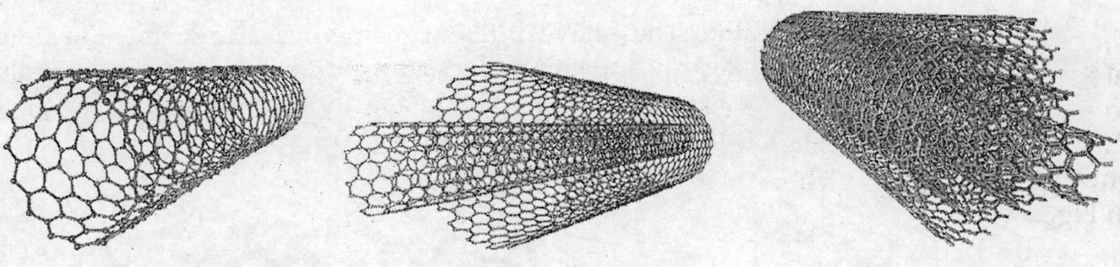

(a) (b) (c)

FIGURE 26.10 (a) Single-walled, (b) double-walled and (c) multi-walled nanotubes

26.12 ▶ Properties of Carbon Nanotubes

The carbon nanotubes (CNTs) have some typical properties as follows.

- CNTs have High Electrical Conductivity.
- CNTs have Very High Tensile Strength.
- CNTs are Highly Flexible – can be bent considerably without damage.
- CNTs are Very Elastic – 18% elongation to failure.
- CNTs have High Thermal Conductivity.
- CNTs have a Low Thermal Expansion Coefficient.

One property of carbon nanotubes is that they are really very strong. Tensile strength is a measure of the amount of force an object can withstand without tearing apart. The tensile strength of carbon nanotubes is approximately 100 times greater than that of steel of the same diameter. There are two things that account for this strength. The first is the strength provided by the interlocking carbon-to-carbon covalent bonds. The second is the fact that each carbon nanotube is one large molecule. This means that it does not have the weak spots found in other materials, such as the boundaries between the crystalline grains that form steel. Nanotubes are not only strong but are also elastic. This means that it takes a lot of force to bend a nanotube if bent these will spring right back to original shape. Young's modulus for carbon nanotubes is about 5 times higher than for steel. There is no other element with a lattice structure in the whole periodic table that bonds to itself with as much strength as carbon atoms. And, since carbon nanotubes have such a perfect structure, they avoid the degradation of strength that you get with other materials. Carbon nanotubes are also lightweight, with a density about one quarter that of steel. The carbon nanotubes also conduct heat well. They have a high thermal conductivity that is more than 10 times that of silver. While metals depend upon the movement of electrons to conduct heat, carbon nanotubes conduct heat by the vibration of the covalent bonds holding the carbon atoms together. The stiffness of the carbon bond helps transmit this vibration throughout the nanotube.

26.13 ▶ Nanotube Geometry

All nanotubes are not exactly alike. These have different geometrical arrangement of atoms and are classified as armchair, zig-zag and chiral. These three different geometries are also referred to as flavors. The three flavors are armchair, zig-zag and chiral [e.g., zig-zag (n, 0); armchair (n, n) and chiral (n, m)].

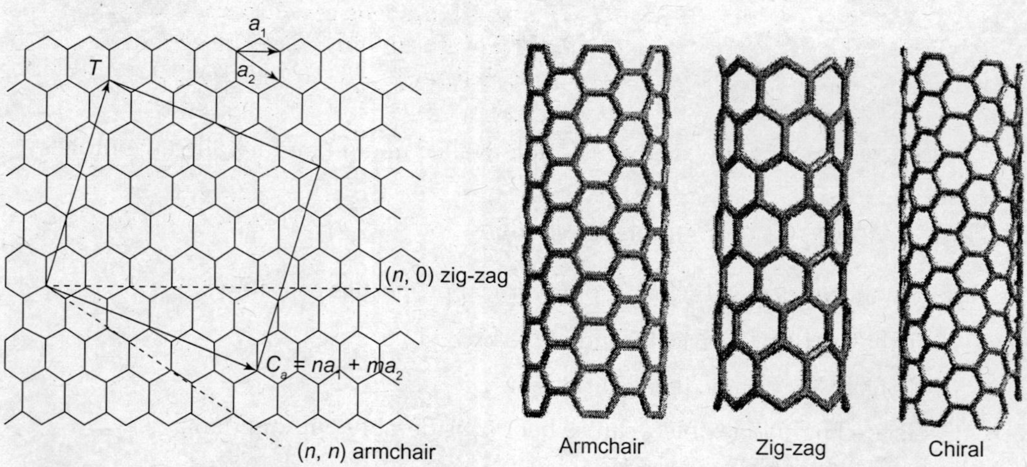

FIGURE 26.11 Three different geometries of nanotubes.

When the nanotube is rolled on a plane surface, then the arrangement of carbon atom that follows a given pattern determines a structure of these flavours. All these flavours differ in some of the properties. Armchair nanotubes have electrical properties like metals but only about a third of all zig-zag and chiral nanotubes have electrical properties like metal. The other two third of nanotube structures have electrical properties like semiconductors. A carbon nanotube conducts electricity just like a metallic wire.

Carbon nanotubes conduct electricity better than metals because the electrons travel through the nanotube like a wave with no atoms to bump into. This type of quantum movement of an electron within nanotubes is called ballistic transport.

26.13.1 Metal or Semiconductor Behaviour of Carbon Nanotube

The diameter of a carbon nanotube and the amount of twist in its lattice determine whether it will be metallic or semiconducting. Similar to electrons in atoms, the electrons in carbon nanotubes can exist only at certain energy levels. A nanotube is metallic if the conduction band is right above the energy level used by electrons attached to atoms (the valance band). In such a case, the electrons can easily move to the conduction band. A nanotube is semiconducting if the energy level of the conduction band is high enough so that there is an energy gap between it and the valance band. In this case, additional energy, such as heat or light is needed for an electron to jump that gap to move to the conduction band. While there is no gap between the valance and conduction bands for armchair nanotubes (which makes them metallic), an energy gap does exist between the valance and conduction bands in about two thirds of zig-zag and chiral nanotubes that makes them semiconducting.

26.14 ▶ GRAPHENE

A single layer (two dimensional) of graphite is called graphene and has extraordinary properties. It is superstrong and stiff, amazingly thin, almost completely transparent, extremely light and an amazing conductor of electricity and heat. It also has some extremely unusual electronic properties. Graphene was first discovered in 2004. It can be produced by epitaxy on an insulating or conducting substrate or by mechanical exfoliation (repeated peeling) from graphite. Its applications may include replacing silicon in high-performance electronic devices and several engineering applications.

FIGURE 26.12 (a) Graphene has a flat crystal lattice made from interlinked hexagons of carbon atoms (balls) tightly bonded together (black lines). (b) The graphene powder synthesized in lab. (c) Multi-layer graphene film as seen under SEM

26.14.1 Properties of Graphene

Graphene is an amazingly pure substance, thanks largely to its simple, orderly structure based on tight, regular, atomic bonding. Carbon is a non-metal, so you might expect graphene to be one too. In fact, it behaves much more like a metal (though the way it conducts electricity is very different), and that has led some scientists to describe it as a semi-metal or a semiconductor (a material like silicon and germanium).

26.14.2 Strength and Stiffness

When we write on a paper or wall with a soft pencil (4B), we find that graphite is quite soft. But unlike graphite, graphene is super-strong, even stronger than diamond. Graphene is believed to be the strongest material yet discovered, it is about 200 times stronger than steel. It is both stiff and elastic so you can stretch it by an amazing amount (20–25 per cent of its original length) without breaking. Because the flat planes of carbon atoms in graphene can flex relatively easily without the atoms breaking apart.

It looks similar to carbon powder in appearance (Fig. 26.12(b)), and like a piece of crumbbled cloth under SEM (Fig. 26.12(c)).

This super-strong property of graphene can find several applications in future. There is a possibility of making its composites that are stronger, tougher, thinner and lighter, than many other presently available materials. One can imagine an energy-saving car with super-strong, super-thin, super-light plastic body panels reinforced with graphene.

26.14.3 Thinness and Lightness

Graphene can be synthesized in very thin layers, sometimes only one atom thick. One can cover a football field with a sheet of graphene weighing less than a gram. This property can be easily exploited for space and aviation related applications.

26.14.4 Heat Conductivity

Graphene has a very high thermal conductivity even far better than silver and copper and much better than either graphite or diamond. We are yet to discover the methods of utilizing this property in adding extra heat-resistance or conductivity to plastics or other materials.

26.15 ▶ Quantum Dots

Quantum dots (QD) are semiconductor nanocrystals, which exhibit quantum mechanical behaviour. The interesting electronic properties of quantum dots arise from the specific size of their energy band gaps.

Properties of quantum dots

These tiny nanoparticles have a diameter, which ranges from 2 to 10 nanometres, with their electronic characteristics depending on their size and shape. Manufacturers are able to accurately control the size of a quantum dot and as a result they are able to "tune" the wavelength of the emitted light to a specific colour.

Quantum dots find applications in a number of areas such as solar cells, transistors, LEDs, medical imaging and quantum computing, thanks to their unique electronic properties.

Optical applications

The high extinction coefficient of a quantum dot makes it perfect for optical uses. Quantum dots of very high quality can be ideal for applications in optical encoding and multiplexing due to their narrow emission spectra and wide excitation profiles.

Light emitting diodes

Quantum dot light emitting diodes (QD-LED) and "QD-White LED" are very useful when producing the displays for electronic devices due to the fact that they emit light in highly specific Gaussian distributions. QD-LED displays can render colours very accurately and use much less power than traditional displays.

Photodetectors

Quantum dot photodetectors (QDPs) can be produced from traditional single-crystalline semiconductors or solution-processed. Solution-processed QDPs are ideal for the integration of several substrates and for use in integrated circuits. These colloidal QDPs find use in machine vision, surveillance, spectroscopy and industrial inspection.

Solar cells and photovoltaics

Traditional solar cells are made of semiconductors and expensive to produce. Theoretical upper limit is 33% efficiency for conversion of sunlight to electricity for these cells. Utilizing quantum dots allows realization of third-generation solar cells at ~60% efficiency in electricity production. Quantum dot solar cells are efficient due to quantum dots' ability to preferentially absorb and emit radiation that results in optimal generation of electric current and voltage. Quantum dot solar cells can be produced using simple chemical reactions and can help to save manufacturing costs.

Anti-counterfeiting capabilities

Inject dots into liquid mixtures, fabrics, polymer matrices, etc. Ability to specifically control absorption and emission spectra to produce unique validation signatures. Almost impossible to mimic with traditional semiconductors.

Biological applications

The latest generation of quantum dots has great potential for use in biological analysis applications. They are widely used to study intracellular processes, tumour targeting, in vivo observation of cell trafficking, diagnostics and cellular imaging at high resolutions. Quantum dots have been proved to be far superior to conventional organic dyes as a result of their high quantum yield, photostability and tunable emission spectrum. They are 100 times more stable and 20 times brighter than traditional fluorescent dyes. The extraordinary photostability exhibited by quantum dots makes them ideal for use in ultrasensitive cellular imaging. This allows several consecutive focal-plane images to be reassembled into three-dimensional images at very high resolution. Quantum dots can target specific cells or proteins using peptides, antibodies or ligands and then observed in order to study the target protein or the behaviour of the cells. Researchers have found out that quantum dots are far better at delivering the gene-silencing tool to target cells than currently used methods. This remarkable phenomenon enables information processing speeds and memory capacity to both be greatly improved when compared to conventional computers.

Future of quantum dots

Quantum dots are zero dimensional and exhibit sharper density of states than structures of higher dimensions. This explains their excellent optical and transport properties, which are currently being studied for potential uses in amplifiers, biological sensors and diode lasers. The broad range of real-time applications of quantum dots in the field of biology is expected to be very useful in many research disciplines such as cancer metastasis, embryogenesis, lymphocyte immunology and stem cell therapeutics. In the future, researchers also believe that quantum dots can be used as the inorganic fluorophore in intra-operative tumour detection when performed using fluorescence spectroscopy.

26.16 ▶ Characterization of Nanostructures/Particles

Characterization refers to study of materials features such as its composition, structure and various properties such as physical, electrical, magnetic, etc. Nanoparticle properties vary significantly with size and shape; accurate measurements of nanoparticles size and shape are therefore critical to their applications. The characterization of nanomaterials is performed at different levels. Some characterization methods are used to study the size, shape and morphology of nanostructure, others are used to obtain detailed structural and physical properties. The structures of materials can be studied at various levels of sophistication, including crystal structure, microstructure, atom-level structure and electronic structure. Surface properties like surface energy, charge on a particle (i.e., its ability to have static electricity), catalytic ability (ability to cause reactions) and ability to nucleate are also investigated. Several bulk properties like mechanical strength, stiffness and flexibility, electrical conductivity, density, etc. are also measured.

The characterization of nanoparticles or structures depends on the type of particle and the nature of the following desired information.

- Semiconducting
- Electronic
- Optoelectronic
- Magnetic
- Dielectric
- Metallic
- Organic
- Biological
- Functional

Besides, visual characterstics like colour, size, shape etc.

Some of most common tools and instrumentation used to characterize nanoparticles include the following techniques:

1. FTIR and FTIR-Raman Spectroscopy
2. Powder X-Ray Diffraction

3. Electron Microscopies (SEM, TEM and HRTEM)
4. Scanning Probe Microscopies (AFM, SPM)
5. Small-Angle X-Ray Scattering (SAXS)
6. X-Ray Fluorescence (EDXRF and WDXRF)
7. X-Ray Photoelectron Spectroscopy (XPS)
8. Photoluminescence (PL)
9. Raman Spectroscopy
10. Magnetic Characterization (NMR, PPMS, Hall effect, B-H)

UV-Vis spectrophotometer

It allows identification, characterization and analysis of metallic nanoparticles (e.g., silver, gold). It can be used to determine size and evaluate the dispersion and local structure of NPs synthesized with metal oxides, selenides and sulphides. It observes the absorption/transmission pattern in the suspension/solution of the particles.

Raman spectrum

It gives information about the molecular structure of the sample (Fig. 26.13).

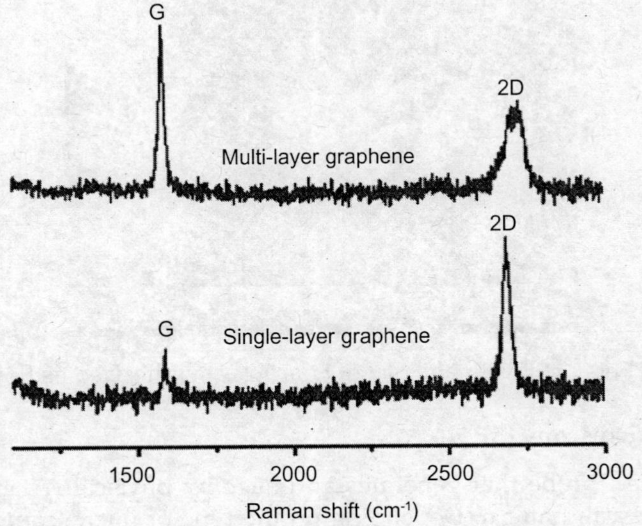

FIGURE 26.13 The Raman spectrum can be used to characterize the graphene

Transmission electron microscope (TEM)

It allows analysis in three dimensions including determination of shape and size. Histograms from the TEM images shown in Fig. 26.14 can be used to count single particles.

FIGURE 26.14 TEM images of ZnO particles of two different samples

Scanning electron microscope (SEM)

It is used to check the surface morphology, shape and size of the particles and structure as shown in Fig. 26.15.

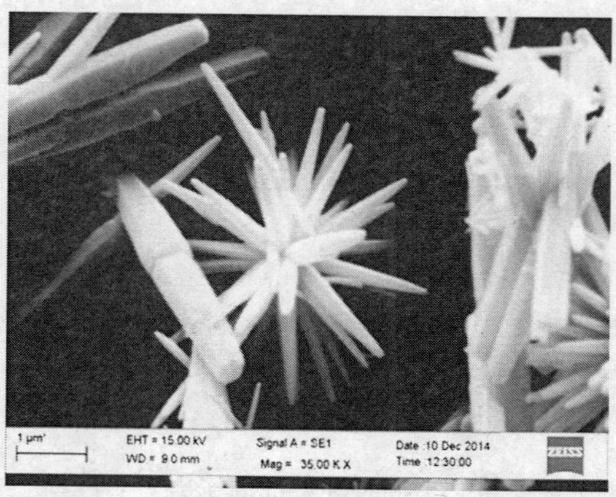

FIGURE 26.15 SEM image of ZnO particles synthesized as flower

Scanning probe microscopy (SPM)

It uses some type of probe that generates an image by physically scanning the sample surface in a raster scan pattern. Depending on the type of microscope, several different surface characteristics can be analyzed by the probe microscopes:

Atomic force microscope (AFM)

It allows visualization and analysis in three dimensions including determination of shape, size and distribution. AFM also allows surface characterization.

X-Ray Diffraction (XRD)

XRD spectrum as shown in Fig. 26.16 gives the information about the crystal structure of the material/particles present in the sample. X-ray and neutron diffraction are routinely used for determination of crystal structure. Synchrotron X-ray radiation and neutrons along with Rietveld analysis have made these techniques more powerful. X-ray diffraction (XRD) is used to obtain the average size of the particles via the Scherrer formula: $a = 0.9A/B \cos \theta$.

Dynamic light scattering (DLS)

It measures hydrodynamic size, size distribution and polydispersity.

Zeta potential

It is the physical property exhibited by any particle in suspension. It is defined as the difference in potential between the bulk solution (dispersing medium) and the surface of the hydrodynamic shear (slipping plane). It can be used to optimize the nano-particle formulations and to predict their long-term stability.

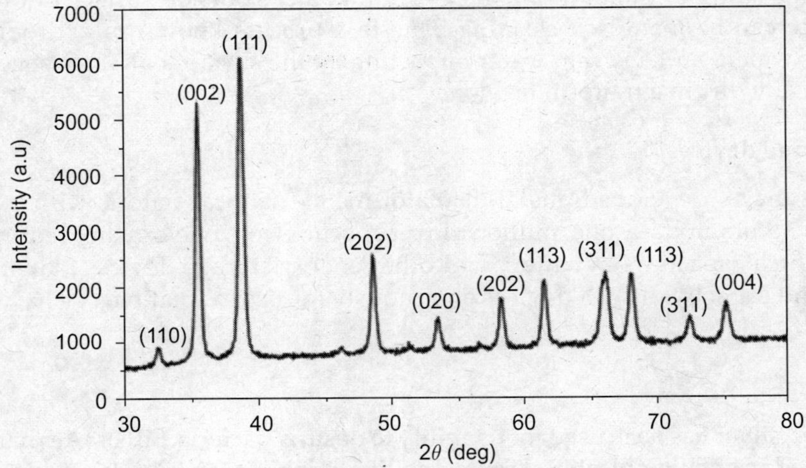

FIGURE 26.16 XRD spectrum of Cu_2O nanoparticles

26.17 ▶ Applications of Nanotechnology

As mentioned earlier, the applications of the nanomaterials are there in almost every field. Some of the important ones are listed below.

Clean fabrics

Nanotechnology can be used to improve the quality of fabrics with superior performance without compromising the look, feel or comfort of the fabric. The nanomaterials can be added or coated on to the fibres of the fabric and make it stain resistant or water repellent.

Solar cells

Using nanoparticles in the manufacture of solar cells can reduce manufacturing costs. They can reduce installation costs by producing flexible rolls instead of rigid crystalline panels. Currently available nanotechnology-based solar cells are not as efficient as traditional ones. In the long-term use of quantum dots, it should reach higher efficiency levels than the silicon-based solar cells.

Skin protection

The bulk titanium dioxide is used as a high-protection sunscreen material against UV rays but have undesirable whitening look at skin. However, its nanoparticles have a comparable UV protection property to the bulk material without whitening effect since the particle size is decreased.

Self-cleaning glass

Using a thin coating of titanium dioxide $\sim$ 25 nm on the outside surface of the glass, the glass surface can be made self-cleaning. Due to a process known as hydrophilicity, the glass "loves water" and any rain water impacting on the surface will form sheets that will wash down any dirt in a uniform fashion.

Energy storage devices

Researchers have developed metal-insulator-metal nanocapacitors with which it is possible to accommodate one million tiny capacitors on one square centimetre area. The use of such capacitors in battery and other energy storage devices may increase the efficiency and capacity of such devices enormously for use in solar energy harvesting and electric cars.

Anti-bacterial

For centuries, silver has been used for its ability to destroy bacteria. Silver (Ag) nanoparticles are embedded in sticking plasters for their ability to inhibit the transmission of viruses.

Diagnostic tools

Pancreatic cancer has a devastatingly low survival rate because it is usually diagnosed at an advanced stage. By attaching a molecule that binds specifically to pancreatic cancer cells to iron oxide nanoparticles, an early diagnosis of pancreatic cancer is made possible with magnetic resonance imaging (MRI).

Drug delivery systems

The drug is attached or encapsulated in a suitable nanoparticle that helps it pass through the stomach to deliver the drug into the bloodstream or take it to the desired part of the body. The administration of different types of drugs uses a variety of nanoparticles.

RECAP ZONE

POINTS TO REMEMBER

- Nanoscience is a branch of science that deals with phenomena, processes and materials at atomic, molecular and nanoscales.
- Nanotechnology is the design, characterization, production and control of properties of materials at nanoscale. It deals with functional systems with specific individual or combined functionalized sub-units.
- Nanoparticles are generally classified based on their dimensionality, morphology, composition, uniformity and agglomeration. Based upon their dimensions, these are classified as quantum well, quantum wire and quantum dots.
- Most of nanoparticles possess unique properties that are mostly different from their bulk material because of their large surface area and this dominates the contributions made by the small bulk of the material.
- The mechanical, electrical, thermal, optical, magnetic and chemical properties are different from their bulk state.
- Due to their unique and controllable properties, the nanoparticles find wide applications in almost every area of science and engineering.
- There are two approaches of synthesis of nanoparticles: one is by reducing the size of bulk material by mechanical or chemical processes, it is called top-down approach; the other is by assembling the desired structure from individual atoms or molecules, it is called bottom-up approach.
- There are many chemical methods of synthesis, these are called routes. Some of them are chemical etching, combustion, hydrothermal, gas phase, physical and chemical vapour deposition, microwave and sol gel.
- Fullerenes are molecules of varying sizes composed entirely of carbon, which take the form of a hollow sphere, ellipsoid or tube.
- Carbon nanotubes are cylindrical structures that are made of only carbon atoms and possess novel properties. These tubes have variety of applications such as nanoelectronics, optics, engineering materials etc.
- The strength of carbon nanotubes is approximately 100 times than that of steel of the same diameter. The first thing that account for this is the strength provided by the interlocking carbon-to-carbon covalent bonds and the second is the fact that each carbon nanotube is one large molecule.
- Carbon nanotubes conduct electricity better than metals because the electrons travel through the nanotube like a wave with no atoms to bump into. This type of quantum movement of an electron within nanotubes is called ballistic transport.

- A single layer of graphite is called graphene and has extraordinary properties. It is super-strong and stiff, amazingly thin, almost completely transparent, extremely light, and an amazing conductor of electricity and heat.
- Quantum dots (QD) are semiconductor nanocrystals, which exhibit quantum mechanical behaviour. The interesting electronic properties of quantum dots arise from the specific size of their energy band gaps.
- The characterization of nanomaterials is performed at different levels. Some characterization methods are used to study the size, shape and morphology of nanostructure; others are used to obtain detailed structural and physical properties.

Key Terms with Definitions

- **Nanoscience** – It is a branch of science that deals with phenomena, processes and materials at atomic, molecular and nanoscales.
- **Nanotechnology** – It is the design, characterization, production and control of properties of materials at nanoscale. It deals with functional systems with specific individual or combined functionalized sub-units.
- **1D nanomaterials** – These are one dimensional in the nanometre scale and are typically thin films or surface coatings, and include the circuitry of computer chips and the anti-reflection and hard coatings on eyeglasses.
- **2D nanomaterials** – These have two dimensions in the nanometre scale. These include 2D nanostructured films, with nanostructures firmly attached to a substrate, or nanopore filters used for small particle separation and filtration.
- **3D nanomaterials** – There are nanoscaled in all three dimensions and are considered 3D nanomaterials. These include thin films deposited under conditions that generate atomic-scale porosity, colloids and free nanoparticles.
- **Quantum well** – If one dimension of the structure is reduced to the nanorange while the other dimensions remain large, then we obtain a structure known as quantum well.
- **Quantum wire** – If two dimensions of the structure are reduced to nanorange and only one dimension remains large, the structure is referred to as a quantum wire.
- **Quantum dot** – In a case where all three dimensions reach the nanometre range, the structure is called a quantum dot.

Review Zone

Short Answer Questions

1. What is nanotechnology?
2. What are quantum dots?
3. What are the unique properties of nanoparticles?
4. Write a note on the dimensional classification of nanostructures.
5. What are physical properties of nanomaterials?
6. Discuss the technological advantages of nanomaterials.
7. What are the methods of synthesis of nanoparticles?
8. What are allotropes of carbon?
9. Write a note on carbon nanotubes?
10. What are the applications of carbon nanotubes?
11. What is graphene? Discuss its applications.
12. What are the common methods of characterization of nanoparticles?
13. What are the applications of nanotechnology in our daily life?

CHAPTER 27

Superconductors

Learning Objectives

By the end of this chapter, the student will be able:

- To differentiate superconductors from conductors
- To determine the upper limit of resistivity
- To understand behavior of magnetic flux in superconductor
- To understand London equations and Meissner's effect
- To explain BCS theory and Josephson's effect
- To describe the applications of superconductivity

27.1 ▶ Introduction

The phenomenon of superconductivity was observed by Kamerlingh Onnes in 1911, while he was measuring the resistivity of mercury at very low temperature. The electrical resistivity of pure mercury drops to zero at a temperature of about 4.2 K. He observed that mercury has passed into a new state and he called it as superconducting state.

Thus,"Superconductivity is a phenomenon in which the electrical resistivity of a substance drops suddenly to zero, when it is cooled below a certain temperature. The substances which show this property are called superconductors."

27.1.1 Characteristics of Superconductors in Superconducting States

The characteristics of superconductors in superconducting state have two aspects:

(a) that do not change in superconducting transition

(b) that change in superconducting state

(a) Characteristics that do not change

(i) The photo electric properties are unchanged in new state. There is no change in the absorption of slow or fast electrons.

(ii) The thermal expansion and elastic properties are also unchanged.

(iii) There is no change in the crystal structure as shown by X-ray diffraction studies.

(iv) In the absence of magnetic field, there is no change of volume and latent heat.

(b) Characteristics that change

(i) The superconducting state can be destroyed by applying a magnetic field equal to critical field H_c on the specimen.

(ii) At transition, the magnetic flux is ejected out by a superconductor and it becomes perfectly diamagnetic.

(iii) The specific heat does not vary linearly, but it varies exponentially (Fig. 27.1) in superconducting state (as shown in big $C_s = Ae^{-\Delta/K_BT}, \Delta = bK_BT_c, b = \text{constant}$ $\Rightarrow C_s = Ae^{-(bT_c/T)}$)

(iv) All thermoelectric effects disappear in superconducting state.

(v) The magnetic properties change by addition of impurities to the specimen.

(vi) The thermal conductivity of superconductors changes. It is lower than the normal state (Fig. 27.2) in superconducting state.

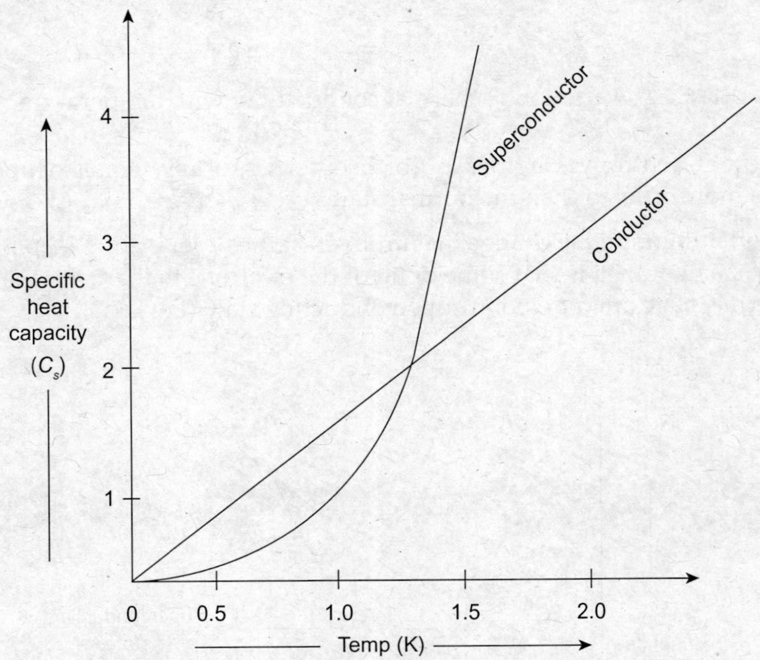

FIGURE 27.1 Variation of specific heat capacity with temperature

(vii) The entropy of a superconductors decrease rapidly on cooling below the transition temperature (Fig. 27.3).

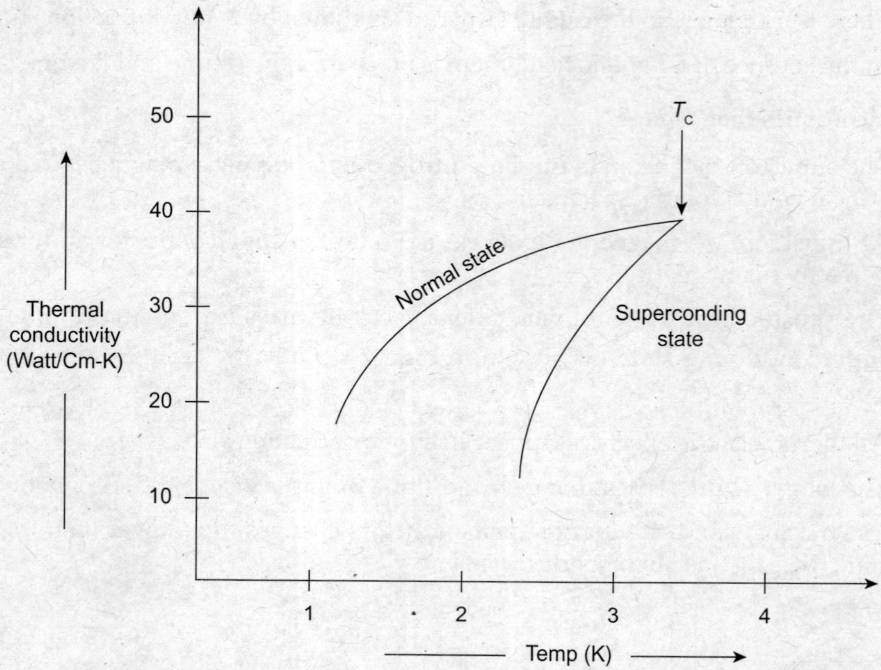

FIGURE 27.2 Variation of thermal conductivity with temperature

We know that entropy is the measure of disorderness of a system. The superconducting state is more ordered than the normal state.

e.g., For aluminum, the change in entropy is of the order of 10^{14} K_B per atom, which is very small. It means that some or all of the electrons that are thermally excited in the normal state are ordered in superconducting state (Fig. 27.3).

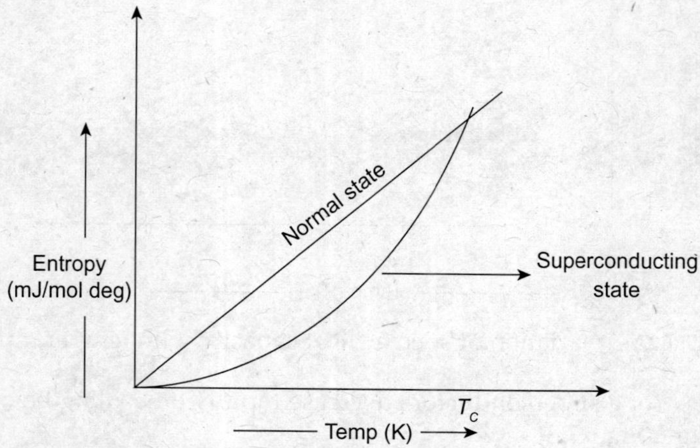

FIGURE 27.3 Variation of entropy with temperature

In type I superconductors, there is a spatial order, which extends over a distance of order of 10^{-6} metre. This range is called coherence length.

(vii) There is an appreciable change in the reflectivity of metals in the infrared region or visible region.

(ix) The critical temperature of superconductors varies with the isotopic mass as $T_c \propto M^{-1/2}$. For example, mercury has T_c from 4.185 K to 4.16 K when M changes from 199.5 to 203.4 amu. Lattice vibrations are actually involved in causing superconductivity (Fig. 27.4).

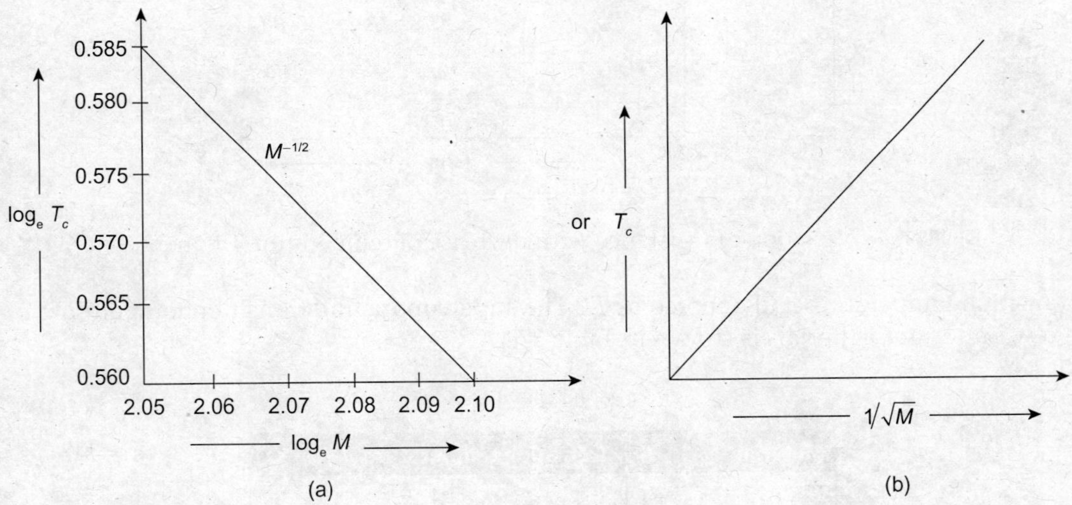

FIGURE 27.4 (a) Isotopic effect for tin metal. (b) Variation of critical temperature with isotopic mass

(x) At very high frequencies, the zero resistance of the superconductor is modified. But the transition temperature remains unaffected.

(xi) **Persistent current**: When a superconductor, at its critical temperature, is placed in a magnetic field, then an induced current starts flowing in the material which opposes the applied magnetic field. This induced current (which is permanent and continuous due to zero resistance) produced in the superconductor is called persistent current.

But when this current increased beyond a critical value I_c (T), the superconductor again becomes a normal conductor. Here, $I_c = 2\pi r\, H_c$. This is known as Silsbee's rule, r being radius of superconducting wire.

27.1.2 Critical Temperature

In 1911, Kamerlingh Onnes found that the electrical resistance of pure mercury vanishes suddenly at 4.2 K as shown in Fig. 27.5. This temperature is called its superconducting

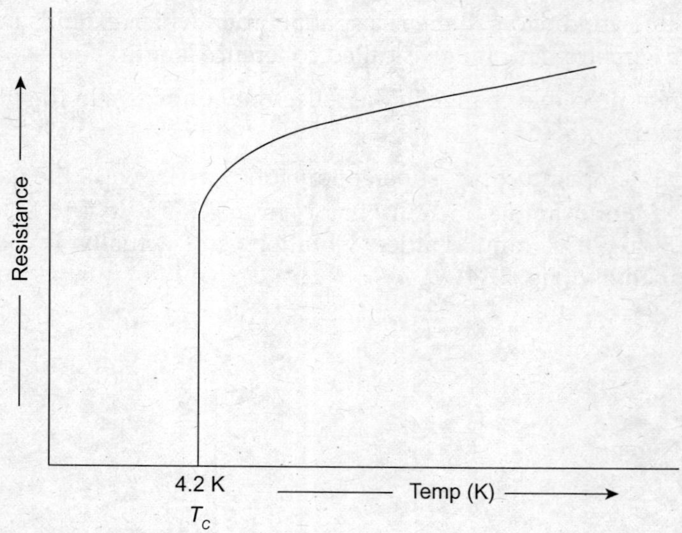

FIGURE 27.5 Variation of resistance with decrease of temperature for mercury

transition temperature and denoted as T_C. The superconducting transition temperature of a few important materials is shown in Table 27.1.

TABLE 27.1

Material	Transition temperature (T_C) (in K)
Hg	4.15
CuS	1.6
Nb	9.3
Nb_3Au	11.5
Nb_3Sn	18.0
Nb_3Ge	23.2
$YBa_2Cu_30_7$	90
$TI_2Ga_2Ba_2Cu_4O_{10}$	125

Above the superconducting transition temperature, the material possesses normal resistance and is said to be in the normal state. Below this temperature, the resistance of the material becomes zero and its conductivity reaches infinity. This state of the material is called superconducting state. Now, the *superconducting transition temperature* can be defined as the temperature at which the material changes from normal state to superconducting state as it is cooled. The total disappearance of electrical resistance of these few substances is called *superconductivity* and the materials that exhibit this property are called *superconductors*. Figure 27.6 shows the variation of electrical resistivity of a superconductor and a normal conductor with temperature.

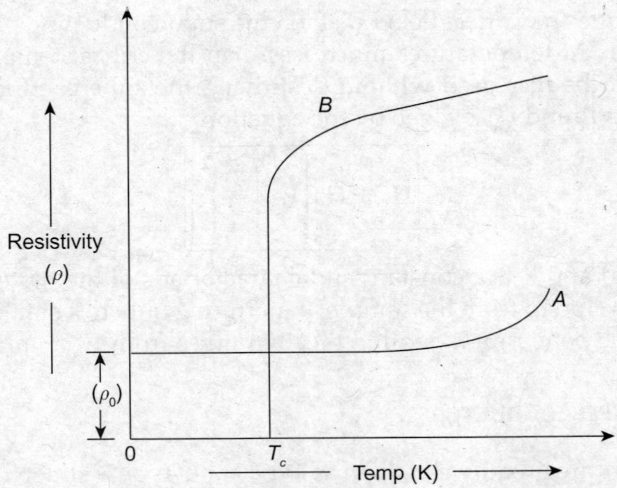

FIGURE 27.6 Variation of electrical resistivity with decrease of temperature. Curve A for a conductor and curve B for a superconductor

27.1.3 Critical Magnetic Field

The magnetic susceptibility of a solid is defined as to what extent the solid can be magnetized. A superconducting material is affected by the presence of a magnetic field and disappears if a sufficiently strong magnetic field is applied. The critical field at which the superconductivity disappears is denoted by $H_c(T)$. This field acts like a boundary and the material is in superconducting state below the field and is normal above the field. The variation of the magnetic field with temperature is shown in Fig. 27.7.

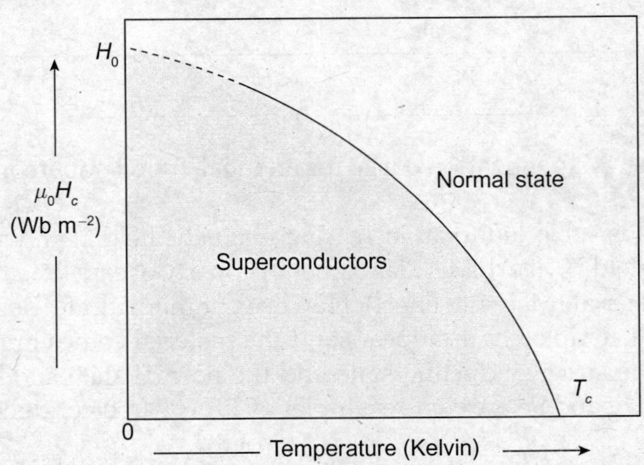

FIGURE 27.7 Critical magnetic field H_c as a function of temperature

When the temperature is within the transition temperature T_c and no magnetic field is applied, the material is in the superconducting state. If the magnetic field is applied and the strength is increased to the critical field H_c, then the superconducting property disappears, i.e. at $T = T_c, H_c = 0$. At temperatures much less than the critical temperature, the critical magnetic field could be increased without destroying the superconducting property. The relation between the T_c and H_c is given by the equation

$$H_c = H_0\left[1-\left(\frac{T}{T_c}\right)^2\right] \tag{27.1}$$

where H_0 is the field at 0 K is a constant and characteristic of the material. For fields less than H_c the flux is excluded from the sample, penetrating only to a smaller depth 'λ' called *penetration depth*. The penetration depth typically ranges from 10^{-5} cm to 10^{-6} cm.

27.2 ▶ Type I Superconductors

The superconductors are broadly classified as Type I and Type II superconductors based on their ability to expel the magnetic field when taken in and out of the transition temperature and magnetic field ranges. The materials that expel the magnetic field completely and sharply at a precise critical magnetic field are called Type I superconductors (Fig. 27.8). They are also known as *soft superconductors*.

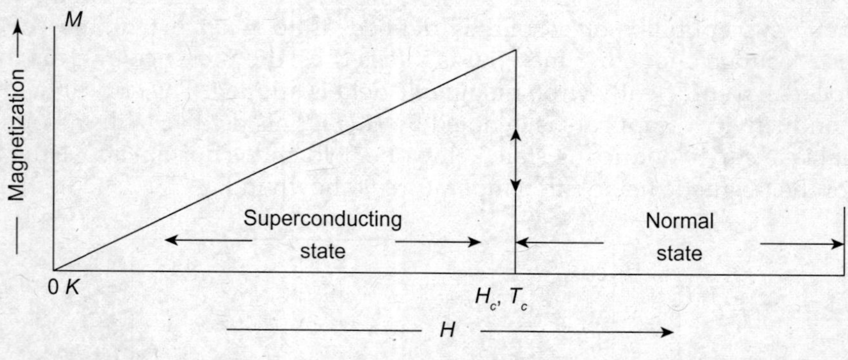

FIGURE 27.8 Magnetization curve for an ideal Type I superconductor

When the material is subjected to an increasing magnetic field from well below the critical field H, to the critical field H_c, it exhibits diamagnetic property of expelling the magnetic flux out of the body of the material (Meissner effect). However, at the critical field H_c, the diamagnetic nature of the material completely disappears and the material comes in the normal state, i.e. the transition from the superconducting state and the normal state sudden and sharp. Pure specimens of Al, Zn, Hg and Sn are some examples of Type I superconductors.

27.2.1 Type II Superconductors

Type II superconductors are characterized by the presence of a mixed superconducting state, when taken through an increasing magnetic field. As shown in Fig. 27.9, up to a magnetic field of H_{c1} known as the lower critical field, the material behaves like a perfect

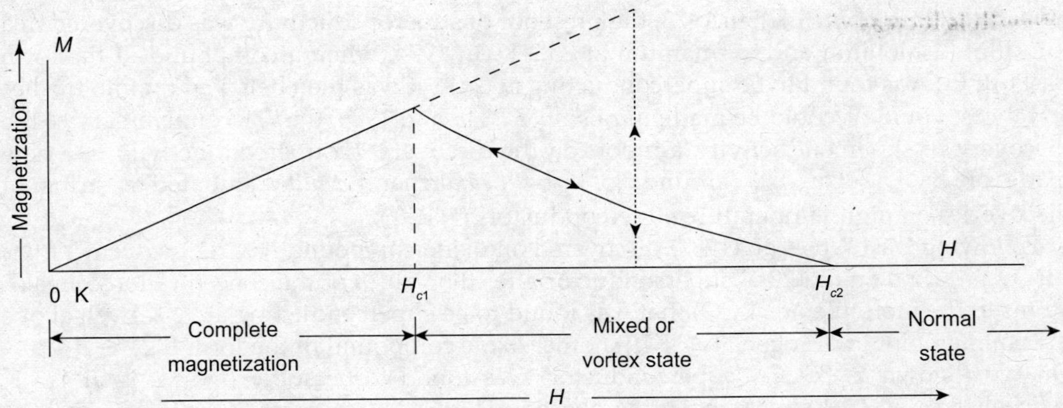

FIGURE 27.9 Magnetization curve for an ideal type of superconductor

superconductor (i.e., it completely expels the flux lines). Above H_{c1}, however, the flux lines start to penetrate the material and the penetration is completed only at H_{c2}, known as the upper critical field. In between the lower and upper critical fields, the material is in a mixed state or vortex state. The difference between type I and type II superconductors is given as follow.

Type I

1. They have only one critical magnetic field.
2. The material loses magnetization abruptly at H_c.
3. They exhibit complete Meissner effect.
4. They show perfect diamagnetic behaviour.
5. They are known as soft superconductors.
6. Example: Al, Hg, Zn and Sn.

Type II

1. They have two critical magnetic fields (H_{c1} and H_{c2}).
2. The material loses magnetization gradually from H_{c1} and H_{c2}.
3. They do not exhibit complete Meissner effect.
4. They do not show perfect diamagnetic behaviour.
5. They are known as hard superconductors.
6. Example: Nb_3Sn, Nb_3Al

27.3 ▶ High-Temperature Superconductors

Ever since the discovery of superconductivity in 1911, scientists have been trying to get superconductivity at a much higher temperature so that they could avoid using liquid helium (boiling point 4 K). In 1973, the presence of superconductivity was discovered in a thin film of niobium and germanium at 23.2 K. In 1974, when a compound of the form Ba-PbBi-O_3 was found to be superconducting at 38 K, it was then believed that no further advances would or could be made in raising T_c. Then in 1986, the 75th anniversary of the discovery of superconductivity, a report on the discovery of superconductivity in a new oxide of $Ba_xLa_{5-x}Cu_5O_{3-y}$ around 35 K by *Bednorz* and *Muller* initiated a string of discoveries on high-temperature superconductors (HTSC).

Following this, Wu et al. (1987) discovered an oxide compound having the form Y_1 Ba_2 Cu_3 $O_{7-\delta}$ (known as 123 – superconductor after the values of the concentration of major atoms in the molecule or YBCO) that was found to be superconducting at 92 K (well above the boiling point of nitrogen at 77 K). Another oxide compound of the form Bi – Sr – Ca – Qi – ft (known as BSCCO superconductor) was found to be superconducting at 115 K (Michel et al. 1987). A superconducting phase in Tl – Ba – Ca – Cu – O with a T_c onset at an all-time high of 125 K followed quickly (Parkin et al. 1988), which remains the highest transition temperature reported till now. All the above-mentioned copper oxide based superconductors are known as *Cuprite oxides*.

Even when the most optimistic suggestion is that the room temperature superconductive and its large-scale application will take years to develop, that superconductivity can exist above liquid N_2 temperature was itself a remarkable development. Liquid nitrogen (77 K) is far safer and easier to use than liquid helium (4 K) or liquid hydrogen (23 K). Scientists and engineers since then are working on the bits and pieces of ceramics to put them to whatever use possible. Of this, superconducting wires, thin sheets and thin fibres have come in for considerable attention. Even though it has been difficult to establish the actual mechanism of occurrence of high-temperature superconductivity, it has been understood that the oxygen atoms play a major role in HTSCs. Most of them have layered structures of copper and oxygen atoms. Some of the important properties of HTSCs are as follows:

- They are highly anisotropic.
- They have the presence of CuO_2 layers.
- They have inherent metallic properties.

Note

Superconductivity at temperatures considerably above the earlier limit of about 30 K, up to about 130 K, is in general known as high-temperature superconductivity.

27.4 ▶ MEISSNER EFFECT

In 1933, Meissner and Ochsenfeld found the exclusion of magnetic flux lines by a superconductor below T_c. They reduced the temperature of a long superconductor in magnetic field. They observed that the superconductor pushes the magnetic lines of force out of the body at some low temperature, T_c as shown in Fig. 27.10. When the material is in normal state, the magnetic force of lines passes through it. The magnetic induction (B) inside the material is given as

$$\boxed{B = \mu_0(H + M) = \mu_0 H(1 + \chi)} \tag{27.2}$$

where μ_0 is the magnetic permeability of free space, H is the intensity of applied magnetic field, M is the magnetization of the material and χ is the magnetic susceptibility given as $\chi = M/H$. Figure 27.10 shows the superconductor in applied magnetic field (H).

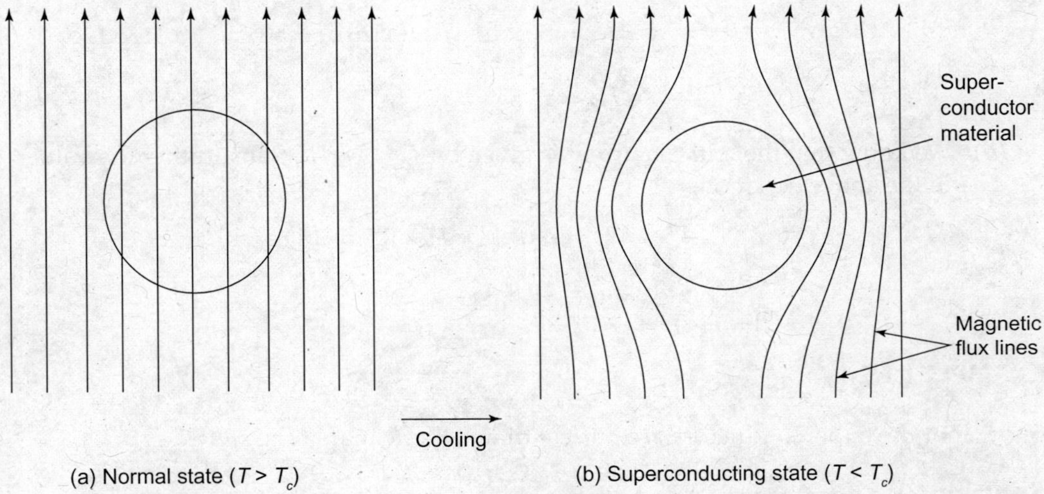

FIGURE 27.10 Superconductor in applied magnetic field (H)

When the temperature reaches below T_c, the material enters into superconducting state and it expels the magnetic force lines. So, $B = 0$ inside the material. From the above equation, we write $0 = \mu_0 (H + M)$.

$$\Rightarrow \qquad H = -M \quad \text{or} \quad \chi = -1 \qquad (27.3)$$

In superconducting state, inside the material magnetization takes place which is equal in magnitude and opposite in direction to the applied field. The superconductor is a perfect diamagnetic material (since $\chi = -1$). The exclusion of magnetic lines from a superconducting when it is cooled in magnetic field to below its transition temperature is called Meissen effect.

EXAMPLE 27.1

Lead has a superconducting transition temperature of 7.26 K. If the initial field at 0 K is 50×10^3 Am^{-1}, calculate the critical field at 6 K. If the lead is in the form of a wire with diameter 1 mm, then calculate the critical current it can carry.

Solution:

(i)
$$H_c = H_0 \left[1 - \left(\frac{T}{T_c} \right)^2 \right]$$

$$= 50\times10^3\left[1-\left(\frac{6}{7.26}\right)^2\right]$$

$$= 50\times10^3\left[1-(0.826)^2\right] = 50\times10^3(0.317)$$

$$H_c = 15.85\times10^3\ \text{A/m}$$

(ii) We know that the critical current is given by, $I_c = 2\pi r H_c$. Substituting the values in the above equation,

$$I_c = 2\times3.14\times0.5\times10^{-3}\times15.85\times10^3$$

$$\mathbf{I_c = 49.8\ A}$$

EXAMPLE 27.2

Show that the superconductors are perfect diamagnetic materials.

Solution: We know that $B = \mu_0(M + H)$
Since $B = 0$, inside a superconductor, we get $M + H = 0$ or $H = -M$.
Therefore, the susceptibility

$$\chi = \frac{M}{H} = -1 \tag{1}$$

Also,

$$B = \mu_r\mu_0 H = \mu_0(M+H)$$

or

$$H_c = H_0\left[1-\left(\frac{T}{T_c}\right)^2\right]$$

which gives

$$\mu_r = 0 \tag{2}$$

Equations (1) and (2) prove that the superconductor is perfectly diamagnetic.

EXAMPLE 27.3

The critical field for niobium is 1×10^5 A/m at 8 K and 2×10^5 A/m at 0 K. Calculate the transition temperature of the element.

Solution:

$$T_c = \frac{T}{\left[1-\frac{H_c(T)}{H_c(0)}\right]^{1/2}} = \frac{8}{\left[1-\frac{1\times10^4}{2\times10^5}\right]^{1/2}}$$

$$= \frac{8}{\left[1-\frac{1}{20}\right]^{1/2}}$$

$$= \frac{8}{(19/20)^{1/2}} = \mathbf{7.08\ K}$$

27.5 ▶ BCS Theory

The existence of energy gap and long-range electronic order in superconducting state pointed that electrons in superconductor are somehow bound together. The positively charged ions screen the Coulomb repulsive forces between electrons. In 1950, Frohlich and Bardeen concluded that a moving electron inside a crystal distorts the crystal lattice and this distortion is quantized in terms of virtual phonons. That means the reaction between an electron and lattice phonons represents the vibrations of crystal lattice in a solid. The electron–phonon interaction can cause resistance or superconductivity. The interaction of electrons and virtual phonons causes superconductivity. We know generally that superconductors are always poor conductors at room temperature and the best conductor does not become superconductors. For example, gold, silver and copper do not become superconductors at low temperatures.

In 1957, Bardeen, Cooper and Schriffer put forward a theory (called BCS theory) (John Bardeen received noble prize twice in physics; in 1947, he invented transistor and later he developed the key concepts of photocopy machine), which explains very well for all the properties shown by superconductors, such as zero resistance, Meissner effect, etc. This theory evolves electron interactions through phonons. The basis for BCS theory is: (i) isotopic effect and (ii) specific heat of superconductors. Isotopic effect, $T_c\, M^{1/2} =$ constant, infers that the transition to superconducting state must involve the dynamics of motions, lattice vibrations and phonons. Also as $T_c \rightarrow 0$, then M approaches infinity. This suggests non-zero transition temperature and finite mass of ions.

Description

Suppose an electron approaches a positive ion core in the crystal, then the electron makes an attractive interaction with a positive ion. This attractive interaction sets in motion the positive ion and this ion motion distorts the lattice. This distortion of lattice is quantized in terms of phonons. At that instant, if another electron approaches the distorted lattice, then the interaction between this second electron and distorted lattice takes place; this interaction lowers the energy of second electron. Now, the two electrons interact through the lattice distortion or the phonon field and results in the lowering of energy of the electrons. The lowering of energy indicates that an attractive force exists between the electrons. This attractive interaction is larger if the two electrons have opposite spin and momenta. This interaction is called electron–lattice–electron interaction or electron–electron interaction through phonons as a mediator. Cooper stated that the presence of an attractive interaction even weak in between two electrons in a superconductor makes them to form a bound pair. He showed that the lowering of energy leads to the formation of a bound state. Such bound pairs of electrons formed by the interaction between the electrons with opposite spin and momenta are known as *Cooper pairs.* This interaction can be represented in terms of the wave vector of electrons as shown in Fig. 27.11. Let an electron having wave vector K_1 emits a virtual phonon q and this phonon is absorbed by another electron having wave vector K_2, then K_1 is scattered as $K_1 - q$ and K_2 as $K_2 + q$. Conservation of energy is not satisfied in this reaction. This process is called *virtual* because virtual phonons are involved in this process.

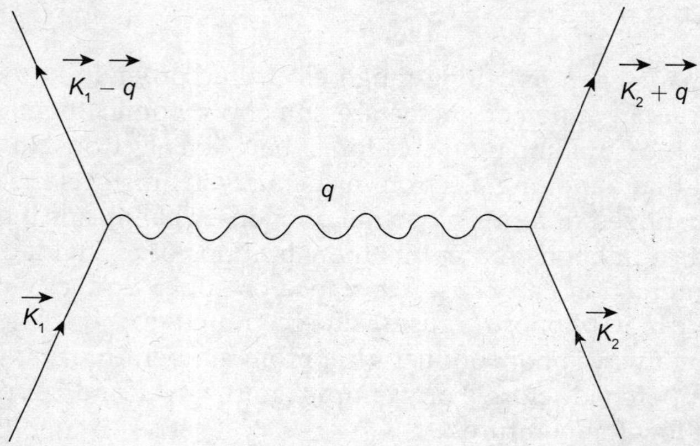

FIGURE 27.11 Electron–electron interaction through phonons

In this interaction, phonon exchange takes place between electrons. If the phonon energy exceeds electronic energy, then the interaction is attractive and the attractive force between these two electrons exceeds the usual repulsive force. These two electrons which interact attractively in the phonon field are called Cooper pairs. The Cooper pair of electrons are said to be in the *bound state or in the condensed state* so that their energy is less than in the free state. The difference of energy of these electrons between these two states is equal to the binding energy of Cooper pair. Below critical temperature, the electron–lattice–electron interaction is stronger than electron–electron Coulomb interaction, so electrons tend to pair up. The pairing is complete at 0 K and is completely broken at critical temperature. According to quantum theory, a wave function could be associated with a Cooper pair by treating it as a single entity. The Cooper pairs do not get scattered in the material and the conductivity becomes infinite which is named as superconductivity. The best conductors such as gold, silver and copper do not exhibit superconductivity because the electrons in these metals move freely in the lattice that, the electron–lattice interaction is virtually, absent and the Cooper pairs will not form. Hence, these metals will not show superconductivity.

BCS theory explains the energy gap in superconductors in the following way: The energy gap at the Fermi surface is the energy difference between the free state of the electron and its paired state. The energy gap is a function of temperature. The energy gap is maximum at 0 K because pairing is complete at this temperature. At transition temperature, the energy gap reduces to zero because pairing is dissolved. The existence of energy gap in superconductor can be proved by the absorption of electromagnetic waves in microwave region. At temperature close to 0 K, a superconductor does not absorb energy of incident radiation until the energy of the incident radiation exceeds the width of the gap (2A). After absorption of energy, the electrons become free or normal.

27.6 ▶ Josephson Effect

A Josephson (*Brian David Josephson*) junction is a type of electronic circuit capable of switching (oscillating) at very high speeds when operated at temperatures approaching absolute zero. The junction is made up of two superconductors, separated by a non-superconducting (*insulating*) layer so thin, that electrons can pass through (*tunnelling*) the insulating barrier.

Electrons flow between two superconducting junctions which are separated by an insulating layer by means of tunnelling even in the absence of an applied voltage. But, when a voltage is applied, then the current flow stops and the current begins to oscillate–Josephson effect.

This flow of current between the superconductors in the absence of an applied voltage is called a *Josephson current*, and the movement of electrons across the barrier is known as *Josephson tunnelling*. Two or more junctions joined by superconducting paths form a *Josephson interferometer.*

A series of Josephson devices are used to define the ST unit of volt. Depending on the particular mode of operation, a Josephson junction can be used as a photon detector or as mixer. Josephson junctions are used in certain specialized instruments such as highly sensitive MV detectors, magnetometers and QUIDs (quantum interference devices). The Josephson effect is influenced by magnetic fields in the vicinity. This enables the Josephson junction to be used as building blocks in devices that measure extremely weak magnetic fields, such as superconducting quantum interference devices (SQUIDs), the most sensitive of magnetometers known till date.

27.6.1 DC Josephson Effect

Consider a Josephson junction consisting of two superconducting metal films separated by a thin oxide barrier. Let it be connected in a circuit as shown in Fig. 27.12. If a constant current is made to flow through the junction, no voltage drop is detected across it, as long as the current stays below some critical value. This is known as DC Josephson effect.

Josephson showed that the super current through the junction is given by

$$I_s = I_c \sin \psi_0 \tag{27.4}$$

where

ψ_0 = Phase difference between the wave functions describing Cooper pairs on both sides of the barrier

I_c = Critical current at zero voltage condition.

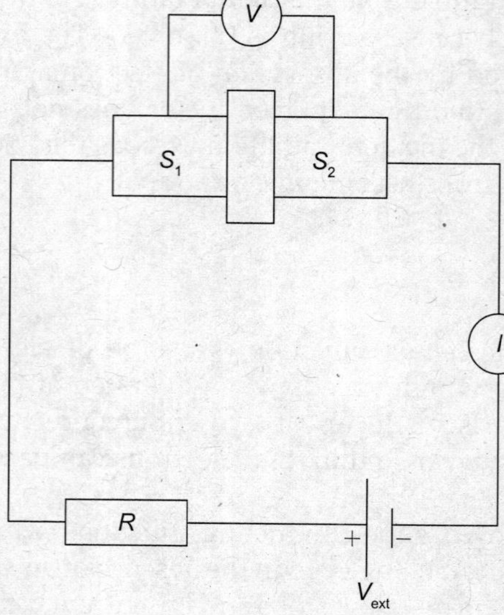

FIGURE 27.12 DC Josephson effect

27.6.2 AC Josephson Effect

If a voltage is applied across the junction, a small oscillating current starts flowing back and forth through the junction, without equilibrating the two sides. This is known as the AC Josephson effect.

The DC voltage generates an alternating current I given by

$$\boxed{I = I_0 \sin(\psi_0 + \Delta\psi)} \tag{27.5}$$

27.7 ▶ Applications of Superconductors

27.7.1 SQUID

Superconducting quantum interference device (SQUID) is the most sensitive type of detector available at present (Fig. 27.13). It consists of a superconducting loop with two Josephson junctions. Using a device called a Josephson junction, a SQUID can detect a change of energy as much, 100 billion times weaker than the electromagnetic energy that moves a compass needle.

SQUIDs consist of tiny loops of superconductors employing Josephson junctions to achieve superposition: each electron moves simultaneously in both directions.

SQUIDs are usually made of either a lead alloy (with 10% gold or indium) and/or niobium often consisting of the tunnel barrier sandwiched between a base electrode of niobium and the top electrode of lead alloy. A radio frequency (RF) SQUID is made up of one Josephson junction, which is mounted on a superconducting ring. An oscillating current is applied to an external circuit, whose voltage changes as an effect of the interaction between it and the ring. The magnetic flux is then measured.

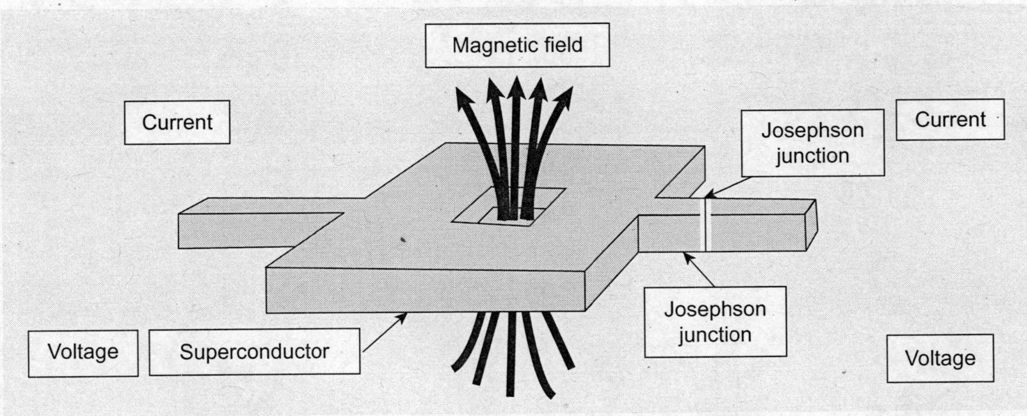

FIGURE 27.13 Construction of a SQUID

A direct current (DC) SQUID, which is much more sensitive, consists of two Josephson junctions employed in parallel so that electrons tunnelling through the junctions demonstrate quantum interference, dependent upon the strength of the magnetic field within a loop. DC SQUIDs demonstrate resistance in response to even tiny variations in a magnetic field, which is the capacity that enables detection of such minute changes.

There are two main types of SQUID – direct current (DC) and radio frequency (RF).

27.7.2 DC SQUID

It was invented in 1964. It has two Josephson junctions in parallel in a superconducting loop. It is based on the DC Josephson effect. In the absence of any external magnetic field, the input current splits into two branches equally. Now consider a small amount of external field applied to the superconducting loop. This results in the screening currents that generate the magnetic field to cancel this applied external field. The current in one of the branches of the superconducting loop is in the direction of I, and is equal to $\left(\frac{I}{2}+\frac{I_S}{2}\right)$ and in the second branch it is in the opposite direction of I and is equal to $\left(\frac{I}{2}+\frac{I_S}{2}\right)$. As soon as the current in any one of the branches exceeds the critical current for the Josephson junction, the superconducting ring becomes resistive and a voltage appears across the junction (Fig. 27.14).

Note

Because the current is moving in two opposite directions, the electrons have the ability to perform as QUIDS (which theoretically could be used to enable quantum computing.

27.7.3 RF SQUID

It was invented in 1965. It is based on the AC Josephson effect and uses only one Josephson junction. It is less sensitive compared to DC SQUID, but is cheaper and easier to manufacture.

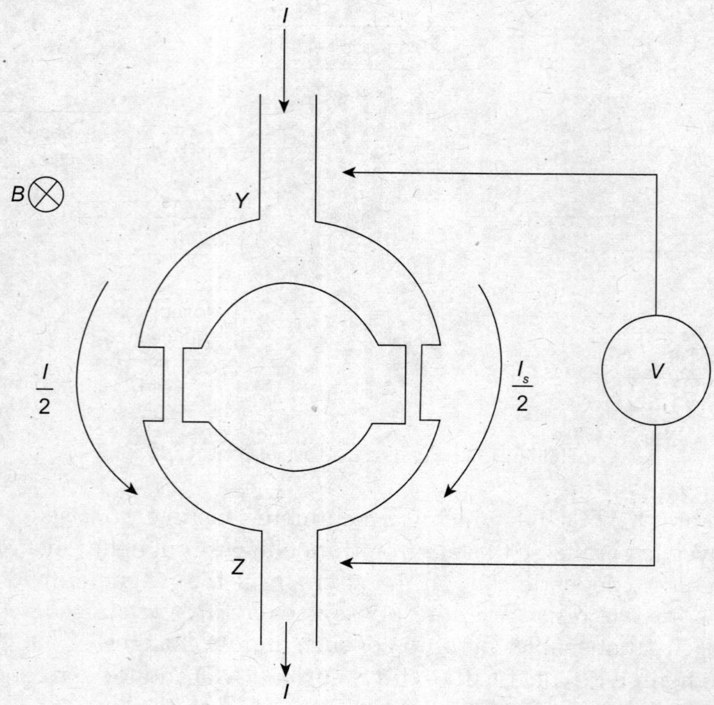

FIGURE 27.14 DC SQUID

The RF SQUID is inductively coupled to a resonant tank circuit. Depending on the external magnetic field, as the SQUID operates in the resistive mode, the effective inductance of the tank circuit changes, thus changing the resonant frequency of the tank circuit.

Applications of SQUIDs

- SQUIDs have been used for a variety of testing purposes that demand extreme sensitivity, including engineering, medical and geological equipment. (Because they measure changes in a magnetic field with such sensitivity, they do not have to come in contact with a *system* that they are testing.)
- One of the devices most promising uses is in *magnetoencephalography* (MEG), the process of measuring magnetic fields to enable brain imaging.
- DC SQUIDs, contained in a helmet-like device, measure the currents created by neural activity.
- The possible SQUID neuroscience applications are myriad.
- SQUIDs are capable of sensing a change in a magnetic field over a billion times weaker than the force that moves the needle on a compass. With this technology, the body can be probed to certain depths without the need for the strong magnetic fields associated with MRIs.

Medical applications

- Superconductors are used to make the most powerful electromagnets known to man including those used in MRI machines and the beam-steering magnets used in particle accelerators.
- An area where superconductors can perform a life-saving function is in the field of bio-magnetism. By impinging a strong superconductor-derived magnetic field into the body, hydrogen atoms that exist in the body's water and fat molecules are forced to accept energy from the magnetic field. They then release this energy at a frequency that can be detected and displayed graphically by a computer.
- Magnetic resonance imaging (MRI) (Fig. 27.15) was actually discovered in the mid-1940s. But the first MRI exam on a human being was not performed until July 3, 1977.

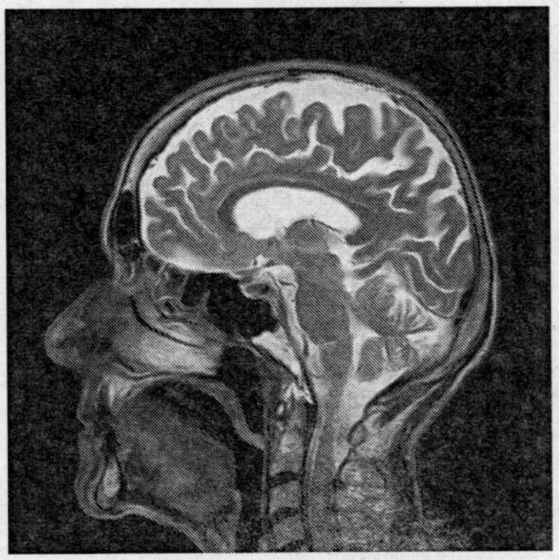

FIGURE 27.15 MRI of a human skull

27.7.4 Plasma Confinement

Plasmas consist of charged particles – positive nuclei and negative electrons – that can be shaped and confined by magnetic forces. Like iron filings in the presence of a magnet, particles in the plasma will follow magnetic field lines. The magnetic field acts as a recipient that is not affected by heat like an ordinary solid container.

Physicists have been exploring the properties of plasmas with Tokamak devices since the 1960s. The doughnut-shaped torus of the Tokamak represented a major break-through in plasma science at the time: here, temperature levels and plasma confinement times reached levels that had never been attained before.

The ITER Tokamak chamber will be twice as large as any previous Tokamak, with a plasma volume of 830 cubic metres. Left to itself, the plasma would occupy all of the space in the chamber; however, no material could withstand contact with the extreme-temperature plasma. Scientists are able to contain or "confine" the plasma away from the walls by exploiting certain of its physical properties.

In ITER, different types of magnetic fields will work in subtle combination to shape the plasma into the form of a ring – or torus – and isolate the very hot plasma from the relatively cold vessel walls in order to retain the energy for as long as possible. The vacuum vessel is the first safety confinement barrier and will not be in contact with the plasma.

Note

A recent study used SQUID-enabled magnetoencephalography (MEP) to measure the surprisingly large level of activity in consumer's brains that is evoked in choosing between any two brands of products.

27.7.5 Maglev (Magnetic Levitation)

Magnetic levitation is an application where superconductors perform extremely well. A small magnet when brought near a superconductor will be repelled because of the induced supercurrents (Fig. 27.16). Since the electrical resistance is zero, supercurrents are generated in the material which will produce mirror images of each pole. Thus, when the magnet is placed above the superconductor such as Yttrium barium copper oxide, it will be levitated by this repulsive force. The levitation process is quite remarkable. Since the levitating currents in the superconductor meet no resistance, they can adjust almost instantly to maintain the levitation. By tapping with a sharp instrument, the suspended magnet can be caused to oscillate or rotate. This motion is found to be damped and will come to rest in a few seconds.

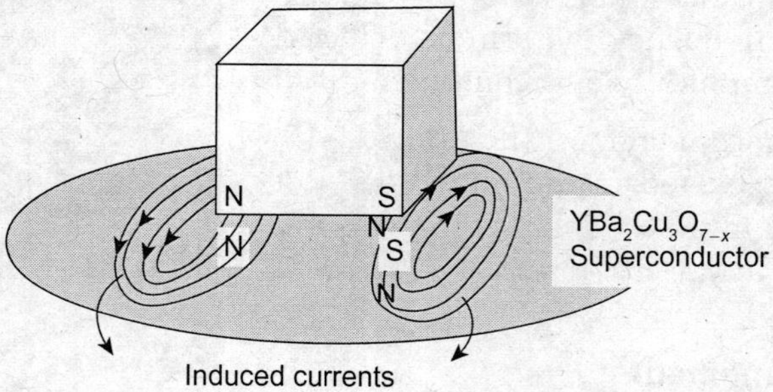

FIGURE 27.16 Magnetic levitation in a superconductor

Maglev trains

Magnetic levitation can be utilized to propel trains and can thus be used for mass transportation. The magnetic field created by the electrified coils in the guide way walls and the track combine to propel the train. This magnetized coil running along the track repels the large magnets on the train's undercarriage, allowing the train to levitate between 0.39 inch and 3.93 inch (1 to 10 cm) above the guide way (500 kmph).

Once the train is levitated, power is supplied to the coils to create a unique system of magnetic fields that pull and push the train along. The electric current supplied to the coils is constantly alternating to change the polarity of the magnetized coils. This change in polarity causes the magnetic field in front of the train to pull the vehicle forward, while the magnetic field behind the train adds more forward thrust. Maglev trains float on a cushion of air, eliminating friction. This lack of friction and the trains' aerodynamic designs allow these trains to reach unprecedented ground transportation speeds of more than 310 mph (500 kmph).

Note

Although the technology has now been proven, the wider use of Maglev vehicles has been constrained by political and environmental concerns (strong magnetic fields can create a bio-hazard). The world's first Maglev train to be adopted into commercial service is a shuttle in Birmingham, England/shut down in 1997 after operating for

11 years. A Sino-German Maglev is currently operating over a 30 km course at Pudong International Airport in Shanghai, China. The Yamanashi Maglev Test Line was opened on April 3, 1997 and in December 2003, a test vehicle (MLX01) attained an incredible speed of 361 mph (581 kph).

Advantages of Maglev

1. It is cheaper, faster, not congested and has a much longer service life.
2. Maglev is very energy efficient. Unlike autos, trucks and airplanes, Maglev does not burn oil, but instead consumes electricity, which can be produced by coal-fired, nuclear, hydro or solar power plants.
3. Maglev vehicles emit no pollution when they consume electricity; hence, no carbon dioxide is emitted.
4. Maglev vehicles have major safety advantages over highway vehicles, trains and airplanes.

Other applications of superconductors are as follows:

1. Type II superconductors can be used as very fast electronic switches due to the way in which a magnetic field can penetrate into the superconductor.
2. Superconducting solenoids produce very strong magnetic fields. They are small in size and are less cumbersome. They do not need either large power supplies or the means of removing heat. The low power requirement and simple cooling technique lead to large savings in cost. Hence, the superconductors can be used in areas of research where high magnetic fields are required.
3. In the electronic industry, ultra-high performance filters are how being built. Since superconducting wire has near zero resistance, even at high frequencies, many more filter stages can be employed to achieve the desired frequency response. This translates into an ability to pass desired frequencies and block undesirable frequencies in high-congestion ratio frequency applications such as cellular telephone systems.
4. Superconducting coils in transformers and electrical machine generate much stronger magnetic fields than the magnetic circuits employing ferromagnetic materials produce. The normal eddy current losses and hysteresis losses will not be present in superconducting devices and hence the size of motors and generators will be drastically reduced.

27.8 ▶ LONDON EQUATIONS

F. London and H. London in 1934 have derived two field equations, which explain zero resistivity and Meissner effect. These equations are known as London equations. The theory given by London brothers was based on the idea that a superconductor has two types of electrons – normal electrons and superconducting electrons.

At transition temperature, the conduction is only due to superconducting electrons. Above critical temperature, the number of normal electrons increases and the number of superconducting electrons decreases. i.e., Above critical temperature, all electrons are normal.

Below critical temperature, the current density is

$$\boxed{\vec{J} = -en_s\,\vec{v}_s} \tag{27.6}$$

where n_s = number of superconducting electrons per unit volume and v_s = velocity of superconducting electrons.

If E is the electric field applied on the materials, then

$$\vec{F} = -e\vec{E} = m\vec{a} = m\frac{d\vec{v}_s}{dt} \tag{27.7}$$

But from Eq. (27.6),

$$\frac{d\vec{J}}{dt} = -en_s\frac{d\vec{v}_s}{dt} = -en_s\left(\frac{-e\vec{E}}{m}\right)$$

or

$$\boxed{\frac{d\vec{J}}{dt} = \frac{e^2 n_s \vec{E}}{m}} \tag{27.8}$$

Equation (27.8) is called the first London equation. It shows that even in the absence of electric field, the current density is constant in the material. The first London equation shows the absence of resistance.

Taking curl of Eq. (27.8),

$$\nabla \times \frac{d\vec{J}}{dt} = \frac{e^2 n_s (\nabla \times \vec{E})}{m}$$

$$= \frac{e^2 n_s}{m}\left(-\mu_0\frac{d\vec{H}}{dt}\right)$$

$$\nabla \times \frac{d\vec{J}}{dt} = \frac{-\mu_0 n_s e^2}{m}\left(\frac{d\vec{H}}{dt}\right) \tag{27.9}$$

On integrating Eq. (27.9),

$$\nabla \times \vec{J} = -\frac{\mu_0 n_s e^2}{m}(H - H_0) \tag{27.10}$$

where H_0 is the constant of integration that represents the field inside the superconductor at $t = 0$. Thus, $H_0 = 0$ and Eq. (27.10) becomes

$$\boxed{\nabla \times \vec{J} = -\frac{\mu_0 e^2 n_s}{m}\vec{H}} \tag{27.11}$$

Equation (27.11) is known as the second London equation.

Meissner effect prevents the presence of magnetic field inside the superconductor. According to Maxwell equation,

$$\text{curl } \vec{H} = (\vec{J} + \vec{J}_D) \tag{27.12}$$

where $\vec{J}_D$ = displacement current and here $\vec{J}_D$ is negligibly small than $\vec{J}$ because the field is varying rapidly with time. So, Eq. (27.12) becomes

$$\text{curl } \vec{H} = \vec{J} \tag{27.13}$$

Taking curl of Eq. (27.13),

$$\text{curl curl } \vec{H} = \text{curl } \vec{J}$$

$$\Rightarrow \quad \text{grad div } \vec{H} - \nabla^2 \vec{H} = \text{curl } \vec{J}$$

$$\Rightarrow \quad -\nabla^2 \vec{H} = \text{curl } \vec{J} \quad (\text{because } \nabla \cdot \vec{H} = 0)$$

$$\Rightarrow \quad -\nabla^2 \vec{H} = -\frac{\mu_0 e^2 n_s}{m} \vec{H} \quad [\text{using Eq. (27.11)}]$$

$$\Rightarrow \quad -\nabla^2 \vec{H} = \frac{\mu_0 e^2 n_s \vec{H}}{m} \tag{27.14}$$

or

$$\nabla^2 \vec{H} = \frac{\vec{H}}{\lambda^2} \tag{27.15}$$

where,

$$\lambda^2 = \frac{m}{\mu_0 n_s e^2} \tag{27.16}$$

or

$$\boxed{\lambda = \sqrt{\frac{m}{\mu_0 n_s e^2}}} \text{ is the penetration depth.} \tag{27.17}$$

Equation (27.15) is a differential equation having solution

$$H = A_1 e^{-x/\lambda} + A_2 e^{x/\lambda} \tag{27.18}$$

where $A_2 e^{x/\lambda}$ show that the field increases onside the superconductor; hence, the term is discarded.

H_0 is the field at the surface of superconductors. The field penetrates in the superconductor. The thickness of this penetration is measured by J. For $x = \lambda$,

$$H = \frac{H_0}{e} \tag{27.19}$$

Thus, the London perpetration on depth is defined as the measure of the distance from the surface of the superconductor at which the magnetic field decays to $1/e$ of its value at the surface.

EXAMPLE 27.4

For a specimen of a superconductor, the critical fields are 1.4×10^5 and 4.2×10^5 A/m, for temperatures 14 K and 13 K, respectively. Calculate the transition temperature and critical fields at 0 K and 4.2 K.

Solution:

$$(H_c)_1 = H_0\left[1-\left(\frac{14}{T_c}\right)^2\right]$$

$$= 1.40\times10^5 \tag{1}$$

$$(H_c)_2 = H_0\left[1-\left(\frac{13}{T_c}\right)^2\right] = 4.2\times10^5 \tag{2}$$

Dividing Eq. (2) by Eq. (1)

$$\frac{(H_c)_2}{(H_c)_1} = \frac{4.2}{1.4} \tag{3}$$

and

$$\frac{(H_c)_2}{(H_c)_1} = \frac{T_c^2 - 13^2}{T_c^2 - 14^2} \tag{4}$$

and from Eqs. (3) and (4), $T_c = 14.5$ K

Substituting the value of T_c in Eq. (1), we get

$$H_0\left[1-\left(\frac{14}{14.5}\right)^2\right] = 1.40\times10^5$$

or

$$H_0 = \frac{1.4\times10^5}{\left[1-\left(\frac{14}{14.5}\right)^2\right]} = 20.67\times10^5 \text{ A/m}$$

Now, $$(H_c)\,4.2 = \boldsymbol{H_0\left[\frac{1-(4.2)^2}{(14.5)^2}\right] = 18.9\times 10^5\ \text{A/m}}$$

EXAMPLE 27.5

The critical temperature for a metal with isotopic mass 199.5 is 4.185 K. Calculate the isotopic mass if the critical temperature falls to 4.133 K.

Solution: We know that

$$T_c M^{1/2} = \text{Constant}$$

Then, $$T_{c_1} M_1^{1/2} = T_{c_2} M_2^{1/2}$$

or $$M_2^{1/2} = (199.5)^{1/2} \times 1.013$$

$\Rightarrow$ $$M_2 = 199.5 \times (1.013)^2$$

$\Rightarrow$ $$\boldsymbol{M_2 = 204.55}$$

EXAMPLE 27.6

The London penetration depth for a superconductor at 3 K and 7.1 K are 39.6 nm and 173 nm, respectively. Determine superconducting transition temperature.

Solution: $\lambda(T_1) = 39.6$ nm, $\lambda(T_2) = 173$ nm

$T_1 = 3$ K, $T_2 = 7.1$ K,

and we know that

$$\frac{\lambda(T)}{\lambda(0)} = \left[1 - \left(\frac{T}{T_c}\right)^4\right]^{-1/2}$$

$$\left[\frac{\lambda(T_1)}{\lambda(T_2)}\right]^2 = \frac{T_c^4 - T_2^4}{T_c^4 - T_1^4}$$

or $$\left(\frac{39.5}{173}\right)^2 = \frac{T_c^4 - (7.1)^4}{T_c^4 - (3)^4}$$

$\Rightarrow$ $$\boldsymbol{T_c = 7.194\ \text{K}}$$

RECAP ZONE

POINTS TO REMEMBER

- "Superconductivity" is a phenomenon in which a conductor has zero resistance below a certain temperature.
- The temperature at which this phenomenon occurs is known as "transition temperature" or "critical temperature".
- "Phonons" are the quantized lattice vibrations.
- Meissner effect is the expulsion of magnetic flux lines from the core of the sample when the sample is in superconducting state.
- The materials that expel the magnetic field completely and sharply at a precise critical magnetic field are called Type I superconductors.
- Type II superconductors are characterized by the presence of a mixed superconducting state, when taken through an increasing magnetic field.
- Because of the presence of a mixed state, Type II superconductors can carry high current densities and therefore are of great practical interest as generators of high magnetic fields for use in plasma fusion.
- Superconductivity at temperatures considerably above the limit of about 30 K and up to about 130 K is generally known as high-temperature superconductivity.
- Some properties of the HTSCs are: they are highly anisotropic, have the presence of CuO_2 layers, and have inherent metallic properties.
- The electrical nature of the YBCO material is strongly dependent on the value of the oxygen deficiency δ, i.e.,

  For $0 < \delta < 0.5$, YBCO is a superconductor

  and for $0.5 < \delta < 1.0$, YBCO becomes an insulator.
- A pair of electrons, interacting attractively through the lattice by means of phonon exchange, overcoming the usual Coulomb repulsion is known as "Cooper pair".
- A Josephson junction is a type of electronic circuit capable of switching (oscillating) at very high speeds when operated at temperatures approaching absolute zero.
- Josephson effect is defined as "electron flow between two superconducting junctions which are separated by an insulating layer by means of tunnelling even in the absence of an applied voltage. But, when a voltage is applied, the current flow stops and the current begins to oscillate".
- The flow of current between the superconductors in the absence of an applied voltage is called a Josephson current, and the movement of electrons across the barrier is known as Josephson tunnelling.

- A Josephson interferometer is formed by two or more junctions joined by superconducting paths.
- Superconducting quantum interference device (SQUID) is the most sensitive type of detector available, consisting of a superconducting loop with two Josephson junctions.
- A radio frequency (RF) SQUID is made up of one Josephson junction while a direct current (DC) SQUID, which is much more sensitive, consists of two Josephson junctions employed in parallel.
- The transition temperature varies according to equation $T_c M^{1/2} = \text{Constant}$.
- The specific heat of a superconductor varies exponentially with temperature.
- All thermoelectric effects disappear in a superconducting state.
- All high-temperature superconductors have a particular type of crystal structure called Perovskite structure.

KEY TERMS WITH DEFINITIONS

- **Critical temperature** – The temperature at which a phase transition occurs.
- **Persistent current** – The induced current produced in superconductor at critical temperature in the presence of field.
- **Critical magnetic field** – The minimum magnetic field required to regain the normal resistivity of a superconductor.
- **Isotopic effect** – The variation of critical temperature of an element with its isotopic mass is known as isotopic effect.
- **Magnetic levitation** – When negative magnetic moment develops in superconducting specimen after placing it in magnetic field.
- **Cooper pair** – Two electrons which interact alternatively in the phonon field.
- **Coherence length** – The maximum distance upto which the electron pairs are corelated to produce superconductivity.
- **Maglev vehicles** – Vehicles which are magnetically levitated are called Maglev vehicles.
- **SQUID** – A thin film of superconducting material deposited on substrate.

Important Formulae and Equations

Equation Number	Equation	Remarks
(27.1)	$H_c = H_0\left[1-\left(\frac{T}{T_c}\right)^2\right]$	Critical magnetic field
(27.2)	$B=\mu_0=(H+M)$ $=\mu_0 H(1+\chi)$	Meissner's effect
(27.5)	$I = I_0 \sin(\psi_0 + \Delta\psi)$	AC Josephson's effect
(27.6)	$\vec{J} = -en_s\vec{V}_s$	Current density
(27.8)	$\frac{d\vec{J}}{dt} = \frac{e^2 n_s \vec{E}}{m}$	First London equation
(27.11)	$\nabla\times\vec{J} = \frac{-\mu_0 e^2 n_s \vec{H}}{m}$	Second London equation
(27.17)	$\lambda = \sqrt{\frac{m}{\mu_0 n_s e^2}}$	London penetration depth

Review Zone

Short Answer Questions

1. Write a short note on:
   (a) Meissner's effect
   (b) Type I and Type II superconductors
2. Explain Maglev vehicles.
3. Explain BCS theory.
4. Define superconductivity.
5. Explain AC and DC Josephson effect in superconductors.
6. What are soft supercondutors?
7. What is the effect of high magnetic field on superconductors?
8. Derive London equations.
9. What is penetration depth?
10. Describe the phenomenon on specific heat and isotopic effect in superconductors.
11. Give a few applications of superconductors.
12. Explain the important characteristics of superconductors.

Long Answer Questions

1. What is superconductivity?
2. Describe the phenomena of Meissner's effect.
3. What are the different types of superconductors?
4. What are high-temperature superconductors?
5. What are the properties of HTSCs?
6. What are SQUIDs?
7. What are the properties of superconductors? Write short notes on each of them.
8. Distinguish between Type I and Type II superconductors with necessary diagrams.
9. Write short notes on HTSCs, Josephson junctions and SQUIDs.
10. Mention some of the applications of superconductors.
11. Derive London equations.
12. What is the significance of critical temperature and critical magnetic field on superconductors?
13. Give the salient features of BCS theory. Describe briefly the formation of Cooper pairs.
14. Discuss the characteristics of superconductors in superconducting state.

Numerical Problems

1. Compute the superconducting electron density of mercury at 3.5 K. Given transition temperature of mercury is 4.22 K.

   Ans. $2.138 \times 10^{28}/\text{m}^3$

2. Calculate the critical current for lead having a diameter of 1 mm at 4.2 K. Critical temperature for lead is 7.18 K and $H_c(0) = 6{\cdot}5 \times 10^4$.

   Ans. 134.26 A

3. The penetration depth of mercury at 3.5 K is 750 Å. Estimate the penetration depth at 0 K.

   Ans. 1033 Å

4. Find London's penetration depth for lead having superconducting electron density of $3 \times 10^{28}\,\text{m}^{-3}$ at 4 K. The transition temperature for lead is 7.22 K.

   Ans. 34.18 nm

5. Critical temperature of a superconductor when no magnetic field is present is T_c. Compute the temperature at which the field becomes half of its value at 0 K.

   Ans. $0.24T_c$

6. A superconducting material has a critical temperature of 3.7 K at zero magnetic field and 0.03067 at 0 K. Find the critical field at 2 K.

   Ans. $H_c = 0.22$ T

7. Compute the critical temperature of a superconductor having isotopic mass of 204.87.

   Ans. 4.3 K

8. London's penetration depth for a sample at 5 K and 7 K are 40.2 nm and 180 nm, respectively. Find its transition temperature and penetration depth at 0 K.

   Ans. $\lambda(0) = 29.5$ nm

CHAPTER 28

Non-Destructive Testing

Learning Objectives

By the end of this chapter, the student will be able:

- To appraise the non-destructive testing
- To explain the working of a non-destructive testing and their classification
- To compare various NDT methods
- To discuss industrial uses of NDT
- To describe the NDT methods using X-ray and gamma rays
- To find applications of NDT in art and archaeology
- To discuss the working of ground penetrating radar and its applications

28.1 ▶ INTRODUCTION

Under normal testing methods, the samples are destroyed partially or completely by physical or chemical procedures followed during the testing. But there are methods available for some tests that can be performed without causing any harm to the original samples. Such methods are of non-invasive in nature. These techniques are now developed enough to determine the integrity of a material, component or structure or quantitatively measure many characteristic of an object. All such methods are called non-destructive testing or NDT techniques. Non-destructive testing (NDT) is the way of obtaining the information about the state, properties and physical defects of examined structure or material without interfering with the attributes of the object. It is worth emphasizing that the non-destructive testing has wider range of application than conventional destructive testing. The most important advantage of NDT over DT is that it is possible to extract the physical properties and to obtain quite detailed physical description of the examined object. By applying appropriate algorithms, one can determine the thermal, strength or electromagnetic features of material without burning, overheating, breaking, bending, loading or just sample destroying procedures.

Thus, the NDT methods are applied when the aspects during reliability and quality investigations of materials are of interest. NDT may be used either during the manufacturing process or on already made items. This technique is used in almost every field of manufacturing and process industry, laboratories, research and development and scientific units. The main fields where the non-destructive methods are employed are aircraft, energy, vehicle and petrochemical industry.

28.2 ▶ Non-Destructive Testing

The non-destructive testing is a way of testing anything without destroying it physically or chemically. This means that the material shape and chemical composition remain unchanged or it can continue to be used and that the testing method has done no harm to the sample. For example, when any material is X-rayed it is not harmed, and so X-ray is a non-destructive method. Similarly, ultrasonic scanning is also a type of non-destructive technique of testing. NDT can be used to ensure the quality right from raw material stage through fabrication and processing to pre-service and in-service inspection. Apart from ensuring the structural integrity, quality and reliability of components and plants, NDT finds extensive applications for production condition monitoring, product life assessment and defects during manufacturing process.

28.3 ▶ Working of a Non-Destructive Testing System

Figure 28.1 presents the general scheme of a non-destructive testing system.

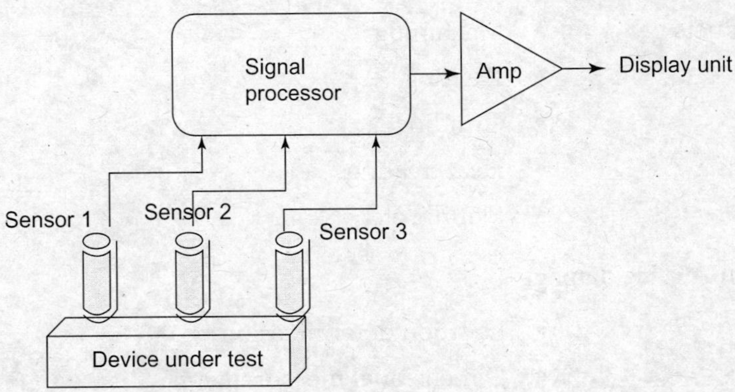

FIGURE 28.1 Working of a non-destructive system

The probes may play the dual role that sometimes can be executed by just one element (in case of two-way physical phenomenon, e.g. piezoelectric effect). The first task for the probe is to excite the material, what could be understood as putting the energy to the item in the form it responses. This energy may be vibration, heat, light or other form of radiation and may be provided in continuous or impulse mode. In case the investigations are more of static character (no energy is provided externally), the exciting probes are not used and the internal energy of the system is taken into consideration. This energy comes from the process that is connected directly with the activities performed by the instrumentation. In such situation, only sensors (measuring probes) are applied.

The next element of NDT system is the measurement and control module. It manages both input (exciting) and output (measuring) probes and the signals they work with. Depending on the applied NDT method, static or dynamic, such a module may consist of signal generator, power stage, signal conditioner and AC/DC converter. The whole testing process is controlled by the operator working with control panel embedded in the device or implemented as a software on a computer. Very often this software also provides sophisticated algorithms for processing the measured signals. These algorithms may operate as comparators (measured vs reference response) or use complicated mathematical operations that extract from input signals the answer for stated problem. The output message may inform just about the state of the examined object; thus if it is sound or not, sometimes about the qualitative and quantitative nature of the fault or about searched physical property.

28.4 ▶ Classification of NDT Methods

There are many types of NDT techniques/methods used, depending on four main criteria:

- Material type
- Defect size
- Defect type
- Defect location

The NDT methods can be used for inspection of

- Raw products
- Machining
- Forgings
- Welding
- Castings
- Grinding
- Extrusions
- Heat treating
- Plating, etc.

Inspection for in-service damage

- Cracking
- Location determination
- Corrosion
- Dimensional measurements
- Erosion/Wear
- Structure and microstructure characterization
- Heat damage
- Estimation of mechanical and physical properties
- Flaw detection and evaluation
- Stress (strain) and dynamic response measurements
- Leak detection

These methods are used for

- To assist in product development
- To screen or sort incoming materials

- To monitor, improve or control manufacturing processes
- To verify proper processing such as heat treating
- To verify proper assembly
- To inspect for in-service damage

28.5 ▶ Non-destructive Testing Methods

There are many non-destructive testing methods, the new ones are still being found, and many already existing and developed. Quite often it happens that two or more methods are applied simultaneously. The most common methods are visual testing (VT), ultrasonic testing (UT), thermography (IR), electromagnetic testing (with the use of eddy current – ET or magnetic particles – MP), radiography, acoustic emission (AE) or testing based on laser technology. Some of them are explained in more details in the following sections.

The most common NDT methods are as follows:

- Visual inspection
- Ultrasonic testing
- Liquid penetrant
- Eddy current
- Magnetic tests
- X-ray or XRF

28.5.1 Visual Inspection

The oldest of all the methods. Components are scanned visually, sometimes with the aid of low- or high-power lenses, fibrescopes, cameras and video equipment, to determine surface condition.

Most basic and common inspection method and tools include fibrescopes, borescopes, magnifying glasses and mirrors. Robotic crawlers permit observation in hazardous or tight areas, such as air ducts, reactors and pipelines. Portable video inspection unit with zoom allows inspection of large tanks and vessels, railroad tank cars, sewer lines, etc.

28.5.2 Liquid Penetration Inspection

In liquid penetration, the test object or material is coated with a visible or fluorescent dye solution. The excess dye is removed from the surface and a developer which acts like a blotter is applied drawing penetrant out of imperfections open to the surface. With visible dyes, the vivid colour contrast between the penetrant and the developer is used. With fluorescent dyes, an ultraviolet lamp is used to make the 'bleed out' fluoresce brightly allowing the imperfection to be seen readily.

A liquid with high surface wetting characteristics is applied to the surface of the part and allowed time to seep into surface breaking defects.

- The excess liquid is removed from the surface of the part.
- A developer (powder) is applied to pull the trapped penetrant out of the defect and spread it on the surface where it can be seen.

- Visual inspection is the final step in the process. The penetrant used is often loaded with a fluorescent dye and the inspection is done under UV light to increase test sensitivity.

28.5.3 Magnetic Particle Inspection

The part under inspection is magnetized. Finely milled iron or other magnetic material particles coated with a dye pigment are then applied to the specimen. These particles are attracted to magnetic flux leakage fields and will cluster to form an indication directly over the discontinuity as shown in Fig. 28.2. This indication can be visually detected under proper lighting conditions.

Magnetic particle inspection is used to identify surface and near surface discontinuities in ferromagnetic materials such as steel and iron. The technique uses the principle that magnetic lines of force (flux) will be distorted by the presence of a discontinuity. Discontinuities (e.g., cracks) are located from the flux distortion following the application of fine magnetic particles to the area under test.

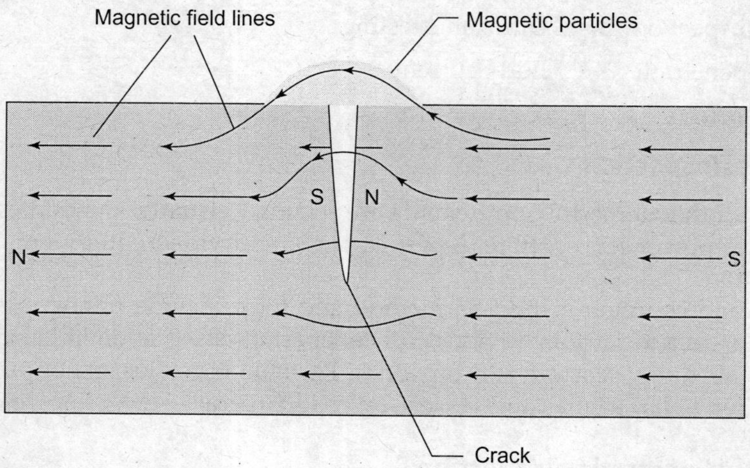

FIGURE 28.2 Disruption in magnetic field indicates a defect in magnetic material

28.5.4 Process Radiography

The process radiography uses an X-ray device or radioactive isotope as a source of radiation, which passes through the material and its image is captured on film or digital device. After processing the film, an image of varying density is obtained. Possible imperfections are identified through density changes. The radiation used in radiography testing is a higher energy (shorter wavelength) electromagnetic waves. The radiation can come from an X-ray generator or a radioactive source. The subassembly components fabricated by fillet welding are radiographed, where the load acting areas as prescribed by the relevant drawing and suspected weld area during visual inspection are indicated.

28.5.5 Film Radiography

The part under inspection is placed between the radiation source and a piece of film. The part will stop some of the radiation. Thicker and more dense area will stop more of the radiation causing lesser exposure. The film darkness (density) will vary with the amount of radiation reaching the film through the test object. Figure 28.3 shows such radiograph of a crack in structure of a casting.

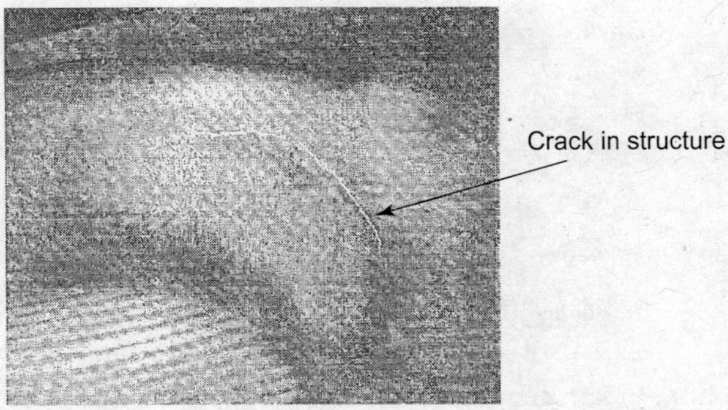

FIGURE 28.3 X-ray/gamma ray radiograph of a crack

28.5.6 Eddy Current Testing

Eddy current testing is particularly well suited for detecting surface cracks but can also be used to make electrical conductivity and coating thickness measurements. Here, a small surface probe is scanned over the part surface in an attempt to detect a crack as shown in Fig. 28.4. In eddy current testing, electrical currents are generated in a conductive material

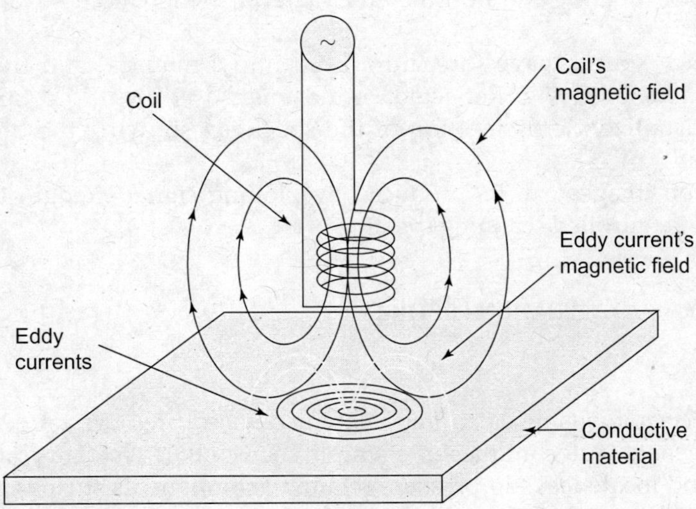

FIGURE 28.4 Use of eddy current for flow detection

by an induced magnetic field. Distortions in the flow of the electric current (eddy currents) caused by imperfections or changes in material's conductive properties will cause changes in the induced magnetic field. These changes, when detected, indicate the presence of the imperfection or change in the test material.

Probe signals are produced by various amounts of corrosion and thinning of pipes. Periodically, the power plants are shut down for inspection. Inspectors feed eddy current probes into heat exchanger tubes to check for corrosion damage as shown in Fig. 28.5.

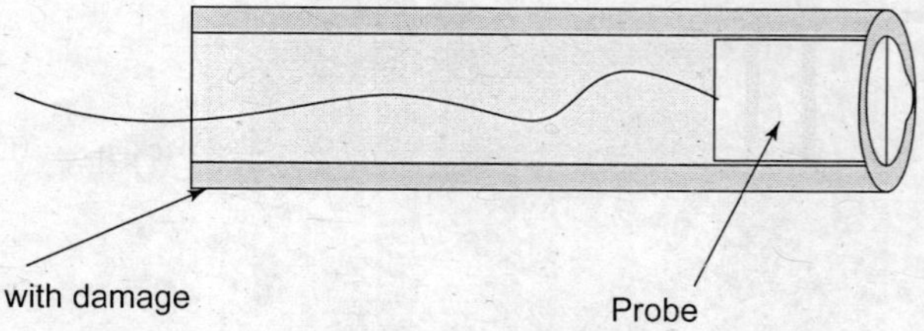

FIGURE 28.5 Eddy current probe in a pipe

Wire Rope Inspection Electromagnetic devices and visual inspections are used to find broken wires and other damage to the wire rope that is used in chairlifts, cranes and other lifting devices. Storage Tank Inspection Robotic crawlers use ultrasound to inspect the walls of large above ground tanks for signs of thinning due to corrosion. Cameras on long articulating arms are used to inspect underground storage tanks for damage.

28.5.7 Ultrasonic Inspection (Pulse-Echo)

Ultrasonic inspection uses high frequency sound waves to detect imperfections or changes in properties within the materials. It can also be used to measure the thickness of a wide range of metallic and non-metallic materials where access from one side only is available.

High frequency sound waves are introduced into a material and they are reflected back from surfaces or flaws. Reflected sound energy is displayed versus time, and inspector can visualize a cross-section of the specimen showing the depth of features that reflect sound.

High resolution images can be produced by plotting signal strength or time-of-flight using a computer-controlled scanning system.

28.6 ▶ Integrity Examination Method

28.6.1 Leak Test

Leaks can be detected by using electronic listening devices, pressure gauge measurements, liquid and gas penetrant techniques or a simple soap-bubble test. Several techniques are used to detect and locate leaks in pressure retaining components such as pressure vessels and pipelines.

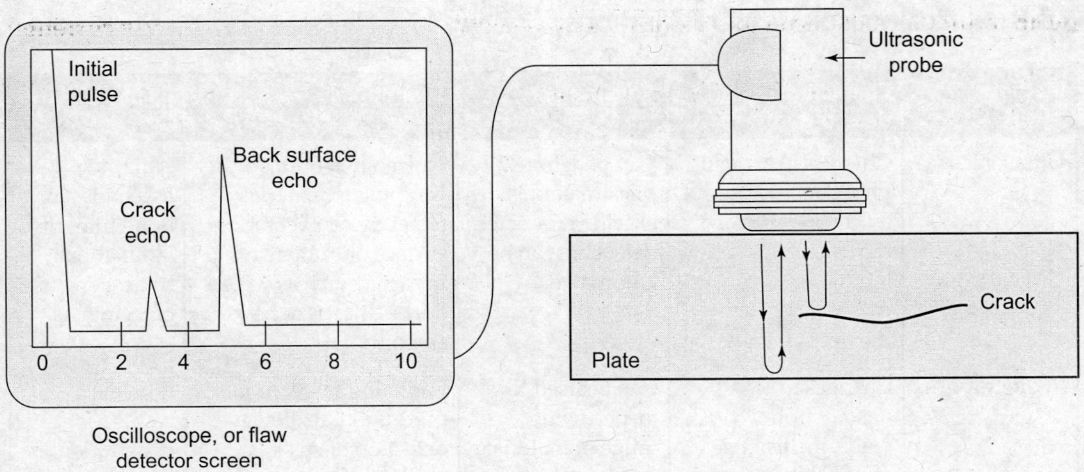

FIGURE 28.6 Ultrasonic testing of structure

28.6.2 Acoustic Emission Test

When a solid material is stressed, growing imperfections (if any) within the material emit short bursts of acoustic energy called 'emissions'. As in ultrasonic testing, acoustic emissions can be detected by special receivers. Emission sources can be evaluated through the study of their intensity, rate and other characteristics. The growing defects can be located by triangulation technique (similar to earthquake epicentre location) as shown in Fig. 28.6.

28.7 ▶ Condition Monitoring Methods

28.7.1 Thermography – Infrared (IR) Testing

Thermography gives the thermal profile of an item, machine or building to be presented in a graphic form which allows a working temperature assessment. From this, variations in the material or component temperature are identified, enabling working limits or corrective actions to be taken.

28.7.2 Vibration Analysis

The rotary machines produce vibration noise. By monitoring the frequency, amplitude, etc. of the vibration, the condition of the machine can be estimated.

28.7.3 Special NDT Methods

NDT engineers and technicians also use magnetic resonance imaging, vibration monitoring, laser ultrasonics, holography, computed tomography as well as many other specialized methods for specialized applications. Discussion of all these methods is beyond the scope of this book.

28.8 ▶ Comparison of NDT Methods

Methods	Characteristics detected	Advantages	Limitations	Example of use
Ultrasonics	Changes in acoustic impedance caused by cracks, voids, inclusions or interfaces	Can penetrate thick materials excellent for crack detection, can be automated	Normally requires coupling to material either by contact to surface or immersion in a fluid such as water. Surface needs to be smooth	Adhesive assembles for bond integrity: laminations; hydrogen cracking
Radiography	Changes in density from voids, inclusions, material variations; placement of internal parts	Can be used to inspect wide range of materials and thicknesses; versatile; film provides record of inspection	Radiation safety requires precautions: expensive; detection of cracks can be difficult unless perpendicular to X-ray film	Pipeline welds for penetration, inclusions and voids; internal defects in casting
Visual (optical)	Surface characteristics such as finish, scratches, cracks or colour; strain in transparent materials; corrosion	Often convenient; can be automated	Can be applied only to surfaces, through surface openings, or to transparent material	Paper, wood or metal for surface finish and uniformity
Eddy current	Changes in electrical conductivity caused by material variations, cracks, voids or inclusions	Readily automated: moderate cost	Limited to electrically conducting materials; limited penetration depth	Heat exchanger tubes for wall thinning and cracks
Liquid current	Surface openings due to cracks, porosity, seams, or folds	Inexpensive, easy to use, readily portable, sensitive to small surface flaws	Flow must be open to surface. Not useful on porous materials or rough surfaces	Turbine blades for surface cracks or porosity; grinding cracks
Magnetic particles	Leakage magnetic flux caused by surface or near-surface cracks voids inclusions or material or geometry changes	Inexpensive or moderate cost, sensitive both to surface and near-surface flaws	Limited to ferromagnetic material; surface preparation and post-inspection demagnetization may be required	Railroad wheels for cracks; large castings

28.9 ▶ Industrial Uses of NDT

28.9.1 Aircraft Inspection

Non-destructive testing is used extensively during the manufacturing of aircraft. NDT is also used to find cracks and corrosion damage during operation of the aircraft.

28.9.2 Jet Engine Inspection

Aircraft engines are overhauled after being in service for a period of time. They are completely disassembled, cleaned, inspected and then reassembled. Fluorescent penetrant inspection is used to check many of the parts for cracking.

28.9.3 Pressure Vessel Inspection

The failure of a pressure vessel can result in the rapid release of a large amount of energy. To protect against this dangerous event, the tanks are inspected using radiography and ultrasonic testing.

28.9.4 Rail Inspection

Special cars are used to inspect thousands of miles of rail to find cracks that could lead to a derailment of railway carriages.

28.9.5 Bridge Inspection

Corrosion, cracking and other damages can all affect a bridge's performance. Some bridges are permanently fitted with acoustic emission sensors that 'listen' for sounds of cracks growing for continuous monitoring.

28.9.6 Pipeline Inspection

NDT is used to inspect oil pipelines to prevent leaks that could damage the environment. Visual inspection, radiography and electromagnetic testing are some of the NDT methods used. A device, known as pig, is placed in the pipeline that collects data on the condition of the pipe as it is pushed along the fluid that is being transported.

28.10 ▶ NDT by X-Ray and Gamma Ray Spectroscopy

Non-destructive testing with hand-held XRF analyzers can be achieved, i.e. in seconds or minutes for a variety of applications. The samples can be in raw form without any processing, and results can be obtained and analyzed with onsite laptops or built-in software. There are also applications in industries where non-destructive testing is not required yet it makes life a lot easier if the sample is not destroyed and can be confirmed with back up analysis in a laboratory.

28.10.1 Soil Remediation

The XRF and gamma rays can be used to test soil for minerals and moisture too.

28.10.2 Mining Exploration

The onsite testing for minerals in rocks can be quickly done using hand-held XRF and gamma ray detectors and spectrometers without any mining.

28.10.3 Water and Air Pollution Monitoring

The XRF and gamma rays can be used for real-time monitoring of toxic metals and minerals in water bodies, rivers and canals.

28.11 ▶ XRF Spectrometers for Non-Destructive Investigations in Art and Archaeology

X-ray fluorescence (XRF) has been used for the investigation of archaeological and historical materials for some 50 years. Though a great variety of spectrometers has been successfully employed, it is only by portable systems that non-destructivity, one of the most attractive features of this technique can be fully exploited; thanks to the possibility of working at site, portable spectrometers have extended the range of use of XRF to any type of object. The fact that archaeological and historical objects are often unique, artistically relevant and not easily movable encourage the use of non-destructive techniques and portable instruments. For this reason, X-ray fluorescence (XRF), which is an elemental analytical technique, is one of the most widely used.

28.11.1 Elemental Composition in Art and Archaeology

Elemental compositional data are used to study archaeological and historical materials with regard to provenance, fabrication technology and manufacturing technique; they can help to distinguish non-original material and, in lucky cases, to spot fakes. Less frequent applications can be found in conservation, where the bare elemental information is sufficient and much has to be known about the aggregation of the components. The materials of archaeological and historical interest are investigated in detail by the information provided by elemental analysis. The layer-by-layer analysis of famous Monalisa painting is one such example.

28.11.2 Ceramics

The close chemical connection between the object and the raw material, as well as the homogeneity of the trace element composition within a single deposit, allow in many cases a precise identification of clay sources that can be used to study provenance and trade routes.

28.11.3 Copper Alloys

Opposite to ceramics, metals do not keep a close chemical connection between the raw materials and the alloy; moreover they can be re-melted; this precludes in general the possibility of compositionally based provenancing, nevertheless differences in elemental composition can be significant due to different fabrication processes for instance, non-original parts can be identified in statues, doors and other complex objects used in combination with microstructural techniques, elemental analysis can provide information on the fabrication technology, with special regard to casting, soldering and repairing.

28.11.4 Glass

Similar to metals, the composition of glass cannot be univocally related to provenance because it can be and actually was recycled; however, elemental analysis can provide important information on fabrication technology, with special regard to the use of different fluxes which can be related to different geographical areas and historical periods decolourants, colourants and opacifiers.

28.11.5 Paintings

A relevant part of painting materials, such as canvas, wood, glue, binding media, some pigments, varnishes, etc., is organic and therefore out of the elemental analysis concern. except for trace elements; another relevant part, mainly base and pigments, is inorganic and can be characterized by elemental analysis; as simple qualitative analyses are able to identify pigments. Table 28.1 shows examples in which the colour pigment compound is identified by the presence of one or two characteristic elements.

TABLE 28.1 Qualitative identification of pigments by the presence of one or two characteristic elements

Colour	Key element	Class of pigments	Compound
Yellow Brown Red Green	Fe	Earths, ochres	Minerals composed of iron oxides and silicates
Blue Green	Cu	Copper-based pigments	Natural compounds (azurite, malachite, atacamite) and artificial compounds (acetates, resinates, chlorides, silicates, sulfates)
White	Pb	Lead white	Basic lead carbonate $2PbCO_3Pb(OH)_2$
Blue	Co	Smalt	Glass SiO_2 (65–70%), K_2O (10–20%), Al_2O_3 (0–8%), CoO (1–18%)
Red	Hg	Cinnabar (vermilion)	Mercuric sulfide (cinnabar) HgS
Red	Pb	Red lead	Lead tetroxide Pb_3O_4

(Cont'd)

TABLE 28.1 (Continued)

Colour	Key element	Class of pigments	Compound
Yellow	As	Orpiment	Arsenic trisulfide As_2O_3
Dark brown	Mn	Manganese brown	Manganese dioxide MnO_2 (pyrolusite)
Yellow	Cd	Cadmium yellow	Cadmium sulfide CdS
White	Ti	Titanium dioxide	Titanium dioxide TiO_2
Yellow	Pb + Sn	Lead-tin yellow, giallolino	Lead-tin oxide Pb_2SnO_4 and $PbSnO_3$
Yellow	Pb + Sb	Lead antimonate yellow, antimony y., Naples y.	Lead antimonate $Pb_3(SbO_4)_2$
Brown	Fe + Mn	Umber	Manganese dioxide (8–10%) + iron hydroxide (45–55%)
Yellow	Pb + Cr	Chrome yellow	Lead chromate $PbCrO_4$
Red	Pb + Cr	Chrome red	Basic lead chromate $PbCrO_4Pb(OH)_2$
Green	As + Cu	Emerald green, Scweinfurt g.	Copper acetoarsenite $Cu(CH_3COO)_2$–$3Cu(AsO_2)_2$
White	Zn + Ba	Lithopone	Zinc sulfide + barium sulfate ZnS+ $BaSO_4$

28.12 ▶ GROUND PENETRATING RADAR

Radar techniques, developed originally for the detection of targets in the sky or on the surface of land or sea, are now being adapted as a means of investigating the composition and integrity of non-conducting materials and structures. The radar technique principally detect back-scattered energy from a target; anomalies within a material give rise to reflections and if the radar antenna is scanned over the material, an image of the anomalies can be generated.

Principle

The principles of radar are well understood and radar for detecting buried objects or Ground Penetrating Radar (GPR) uses the same fundamental physical principles as conventional radar. GPR is the general term applied to techniques which employ radio waves, typically in the 1 to 5000 MHz frequency range, to map structure and features buries in the ground or in man-made structures. Last time GPR has been used in non-destructive testing of non-metallic structures.

A major advantage of the radar technique over the non-destructive testing methods such as ultrasonic is that it is possible to use an antenna (transducer) which is non-conducting. Thus, it is technically feasible to scan areas of interest extremely quickly (up to 30 km/h). The majority of targets sought using subsurface radar methods are non-metallic so that their scattering cross-section is dependent upon the properties of the surrounding dielectric medium. Most targets and voids in particular have a lower relative permittivity and there is not the phase change at an interface that is observed when the scattering is from a metallic boundary. This feature may be used as way of distinguishing between conducting and non-conducting targets. The most common mode of operation is single-fold common-offset reflection profiling as illustrated in Fig. 28.7. In such a reflection survey, a system with a fixed antenna geometry is transported along survey line to map reflections versus position. This mode of operation gives rise to data. The GPR method measures the travel time of electromagnetic impulses in subsurface materials. An impulse radar system radiates repetitive electromagnetic impulses into the earth. A broad bandwidth antenna is usually placed in close proximity and electromagnetic coupled to the ground surface. It is able to detect and measure the depth of reflecting discontinuities in subsurface soils and other earth materials to within a few centimetres depending upon the electromagnetic parameters of the earth medium, the depth to and size of the target and the frequency of operation. As the antenna is pulled across the surface, a continuous profile record is generated and displayed as a two-way travel time versus distance plot on a strip chart recorder and/or colour monitor, LCD for immediate evaluation. Depth calculation can be derived from this data once the signal velocity has been determined. The signal from subsurface radar is transmitted through the ground or a material which attenuates the signal and this causes the range in depth to be limited by the type of material. There is an optimum choice of frequency of operation to achieve best performance in terms of depth and ability to see details in the target structure. This choice is between 1 and 1000 MHz. Generally, low frequencies are used for deep probing (> 50 m) and high frequencies are used for shallow probing (< 50 m).

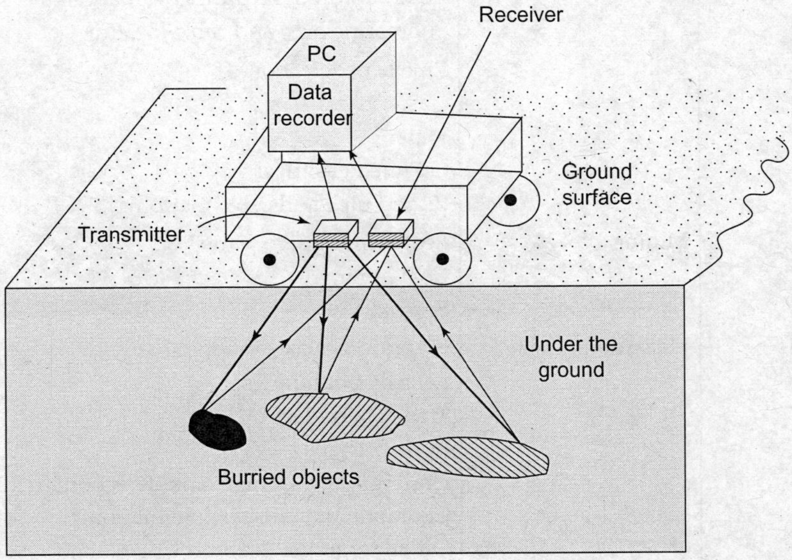

FIGURE 28.7 Working of ground penetrating rador

A GPR system is made up of three main components:

- Control unit,
- Antenna and
- Power supply.

GPR equipment can be run with a variety of power supplies ranging from small rechargeable batteries to vehicle batteries and normal 110/220-volt. Connectors and adapters are available for each power source type.

The control unit contains the electronics which triggers the pulse of radar energy that the antenna sends into the ground. It also has a built-in computer and hard disk/solid state memory to store data for examination after fieldwork. Some systems are controlled by an attached Windows laptop computer with pre-loaded control software. This system allows data processing and interpretation without having to download radar files into another computer.

The antenna receives the electrical pulse produced by the control unit, amplifies it and transmits it into the ground or other medium at a particular frequency. Antenna frequency is one major factor in depth penetration. The higher the frequency of the antenna, the shallower into the ground it will penetrate. A higher frequency antenna will also 'see' smaller targets. Antenna choice is one of the most important factors in survey design. Table 28.2 shows antenna frequency, approximate depth penetration and appropriate application.

TABLE 28.2 Typical applications of ground penetrating radars

Engineering	Miscellaneous	Void detection
		Pavement thickness
		Reinforcing bar locating/evaluation
		Submarine pipe and cable locating
Civil engineering	Utilities	Pipe leak detection
		Gas
		Water
		Oil filled electrical
		Buried pipe and cable mapping
Geological	Mining	Peat profiling
		Coal mining
		Bauxite
Oil and fluids	Hazardous waste	Hazardous waste mapping
		Landfill boundaries
		Trench boundaries
		Buried drums
		Underground storage tank detection
		Contamination instruction mapping
		Oil under ice detection

(Cont'd)

TABLE 28.2 (Continued)

Engineering	Miscellaneous	Void detection
Geological	Miscellaneous	Geological strata profiling
		Earthquake fault mapping
		Highway and other excavation projects
		Radar ground truth
		Bedrock mapping: Subway tunnel excavations
		Rock fracture mapping: Safety and other purposes
		Crevasse detection: Safety and other purposes
		Borehole profiling
		River and lake bottom profiling
		Ice thickness profiling: Sea ice, river ice
		Permafrost mapping
		Sinkhole prediction
		Water table detection/mapping
Military	Ordnance detection	Mines: Metallic and non-metallic
		Unexploded bombs
	Runway integrity movement	Runway integrity movement
	Troop/Equipment movement	Troop/Equipment movement
	Save ice roads	Save ice roads
Transportation	Roads	Pavement base/subgrade
		Thickness
		Voids under payment
	Railroads	Railroad bed profiling
		Tie evaluation
	Airports	Runway integrity testing
	Law	Buried body detection
	Enforcement	Buried weapons
Archaeology	Archaeology	Archaeological prospecting
		Cavity or chamber detection
		Treasure detection

28.12.1 Working of GPR

GPR works by sending a tiny pulse of energy into a material and recording the strength and the time required for the return of any reflected signal. A series of pulses over a single area make up what is called a scan. Reflections are produced whenever the energy pulse enters into a material with different electrical conduction properties or dielectric permittivity from the material it left. The strength, or amplitude, of the reflection is determined by the contrast in the dielectric constants and conductivities of the two materials. This means that a pulse which moves from dry sand (dielectric of 5) to wet sand (dielectric of 30) will produce a very strong reflection, while moving from dry sand (5) to limestone (7) will produce a relatively weak reflection.

While some of the GPR energy pulse is reflected back to the antenna, energy also keeps travelling through the material until it either dissipates (attenuates) or the GPR control unit has closed its time window (Fig. 28.7). The rate of signal attenuation varies widely and is dependent on the properties of the material through which the pulse is passing.

Materials with high dielectric will slow the radar wave and it will not be able to penetrate as far. Materials with high conductivity will attenuate the signal rapidly. Water saturation dramatically raises the dielectric of a material, so a survey area should be carefully inspected for signs of water penetration.

Metals are considered to be a complete reflector and do not allow any amount of signal to pass through. Materials beneath a metal sheet, fine metal mesh or pan decking will not be visible. Radar energy is not emitted from the antenna in a straight line. It is emitted in a cone shape (Fig. 28.7). The two-way travel time for energy at the leading edge of the cone is longer than for energy directly beneath the antenna. This is because that leading edge of the cone represents the hypotenuse of a right triangle.

Because it takes longer for that energy to be received, it is recorded farther down in the profile. As the antenna is moved over a target, the distance between them decreases until the antenna is over the target and increases as the antenna is moved away. It is for this reason that a single target will appear in a data as a hyperbola, or inverted 'U'. The target is actually at the peak amplitude of the positive wavelet as shown in Fig. 28.8.

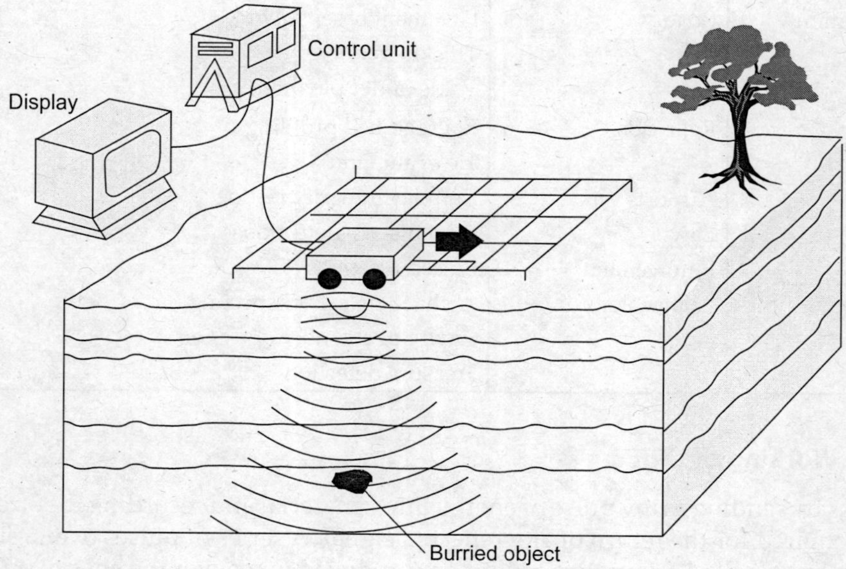

FIGURE 28.8 Detection of underground object by GPR

Data are collected and put together in their appropriate locations for computer processing in a specialized software program. The computer then produces a horizontal surface at a particular depth in the record. This is referred to as a depth slice, which allows operators to interpret a plan view of the survey area.

28.12.2 Data Processing

In many situations, a GPR operator will simply note the location of a target so that it can be avoided. For these clients, it may only be necessary to use a simple linescan format in order to mark the approximate area of the target on the survey surface. Other clients may require detailed subsurface maps and depth to features. These situations will require the operator to use GSSI GPR processing software, which applies mathematical functions to the data in order to remove background interference, migrate hyperbolas, calculate accurate depth and much more.

Table 28.3 lists various ranges of applications of ground penetrating radars.

TABLE 28.3 Distance range applications of ground penetrating radars

Very short range	Short range	Medium range	Long range
Road reinforcement	Tunnel structures	Tunnel structures	Construct surveying
Road structures	Road reinstatement	Construct surveying	Chemical waste
Material integrity	Road structures	Chemical waste	Borehole investigation
Material production	Material integrity	Borehole investigation	Geological surveying
Bridge structures	Material production	Geological surveying	Resource assessment
Building structures	Bridge structure	Resource assessment	Archaeological sites
Fibre optic cables	Building structure	Archaeological sites	
Buried objects	Pipes-cables		
Borehole investigation	Buried objects		
Security investigation	Concrete integrity		
	Borehole investigation		
	Security investigation		

Recap Zone

Points to Remember

- The methods of testing material and object without chemically or physically destroying the sample are called non-destructive testing methods.
- The NDT methods are based upon sensors or properties of materials that can be used indirectly to test.

- Different objects, equipment and materials need different methods of testing. These are selected based upon the desired tests and conditions of sample.
- There are several methods of NDT such as visual, magnetic, electrical, ultrasonic, thermal, eddy current X- ray and gamma ray based, etc.
- A separate set of NDT methods is available for in-process liquid/fluid flow, tanks, vessels, car tanks, sewer lines, air ducts, hazardous chemicals, etc.
- The mechanical or structural NDT methods are there for testing of rail lines, jet engines internal combustion engines, forgings, castings, extrusions, weldings, heat treatment and defence productions.
- The ultrasound scanning, MRI, NMR and PPMS are also some types of NDT methods.
- The ground penetrating radar is one of very important and useful NDT methods that are used in civil engineering, archaeology and history to locate the buried structures, materials and objects.
- The X-ray and gamma ray induced X-ray emission and detection is used to do elemental analysis of archaeology samples without damaging these.

Key Terms with Definitions

- Non-destructive testing methods are those that give desired information about sample without physically or chemically damaging the sample.
- Several types of indirect sensors are the key to such testing methods.
- The NDF methods are widely employed in almost every sector of engineering, business and production industry.
- The condition monitoring NDT is employed in all chemical process industry. E.g., pH monitoring during reaction process is also a type of NDT.
- The X-ray fluorescence method is used to detect elements in samples of minerals, rocks and rare artefacts of archaeological interests.
- The ground penetrating radar or GPR is a method based upon reflection and detection of radio waves from the underground objects.

Review Zone

Short Answer Questions

1. What is NDT?
2. Discuss the principles of NDT.
3. What are various methods of NDT?

4. Give classification of NDT methods based upon types of sensors.
5. What are the radiographic techniques of NDT?
6. Write a note on NDT in fluid/oil industry.
7. What are NDT methods that can be used to detect the defects in a mechanical industry?
8. What is GPR? Write a note on it.
9. Discuss the working principles of GPR.
10. Give details of NDT method to analyze the elemental characteristics of an old painting using X-rays.
11. Is carbon dating also an NDT method?
12. How gamma rays can be used for NDT?

CHAPTER 29

Electron Optics

Learning Objectives

By the end of this chapter, the student will be able:

- To understand the basics of electron optics
- To compare the electrostatic and electromagnetic focusing of electron beams
- To describe the principle and working of an electron microscope
- To analyze the behavior of the electron under electric and magnetic fields
- To discuss the working and use of CRO and CRT
- To have an overview of TEM, SEM and Bainbridge mass spectrograph

29.1 ▶ Introduction

Electron optics is the branch of physics which is concerned with electron and its deflection by electric and magnetic fields. Electron optics is also concerned with the wave properties of electron beams like interference of electron beams when they cross each other and it is also concerned with the deflection of beam of electrons when they pass through the spacing in its submicroscopic structure.

29.2 ▶ Specific Charge on Electron

Specific charge on electron is the ratio between the charge and mass of the electron. In order to find the specific charge of electron, the deflection of an electron in a constant magnetic field is analyzed, when a beam of electrons enters in a direction perpendicular to the magnetic field. Now, electrons move in a circle which is called trajectory of electrons, which is due to the force exerted by magnetic field called Lorentz force. If m, q and v are the mass, charge and velocity of the electron respectively and if it moves in a circle of radius r, then

Lorentz force is given as

$$\boxed{\vec{F} = q\vec{v}\vec{B}} \tag{29.1}$$

Here, Lorentz force will compensate the centripetal force, so

$$\frac{mv^2}{r} = q\vec{v}\vec{B}$$

$$\Rightarrow \quad \boxed{\frac{q}{m} = \frac{\vec{v}}{\vec{B} \cdot r}} \tag{29.2}$$

Thus, we can measure q/m, if we know $\vec{B}$, q and r.

29.3 ▶ Determination of Specific Charge by Thomson's Method

The method to determine the specific charge of an electron was given by J.J. Thomson in 1897. When a beam of electrons moves perpendicular to electric and magnetic fields, now, if the forces due to electric and magnetic fields are equal and opposite, then the beam of electrons will remain undeflected.

As shown in Fig. 29.1, the experimental arrangement consists of a highly evacuated gas container A. The electrons are accelerated from the cathode C. Due to potential difference V created between two anodes P_1 and P_2, a beam of electrons is formed. An electric field is applied on beam of electrons by using two metal plates Q_1 and Q_2. A magnetic field B is also applied on electrons beam. Now, after passing through electric and magnetic fields, the beam of electrons strikes at point O on the screen S. The screen S is coated with a fluorescent material.

In case, when electric and magnetic fields are not present, the electron beam remains undeflected and strikes the screen at point O. Now, adjust the electric and magnetic field strength such that electric and magnetic fields balance each other so the electron beam again strikes at point O on the screen S.

Let us consider an electron having mass m and charge q, it moves with velocity v.

Now, the force on electron due to electric field is

$$\boxed{\vec{F} = q\vec{E}} \tag{29.3}$$

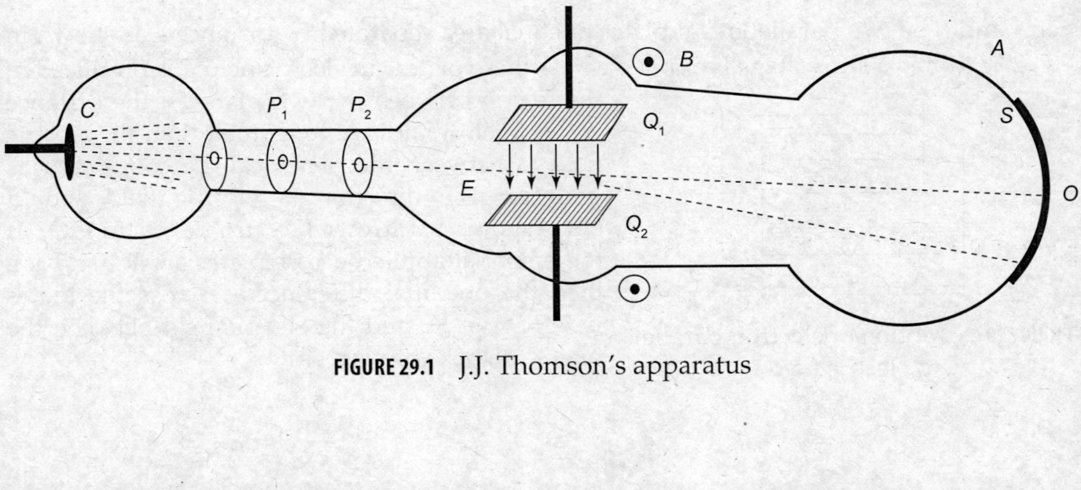

FIGURE 29.1 J.J. Thomson's apparatus

and the force on electron due to the magnetic field B is

$$\vec{F}_M = q\vec{v}\vec{B} \tag{29.4}$$

When these two forces balance each other, then

$$q\vec{E} = q\vec{v}\vec{B}$$

$$\Rightarrow \quad \vec{v} = \frac{\vec{E}}{\vec{B}} \tag{29.5}$$

When the beam is accelerated from cathode to anode, then its potential energy at cathode appears as a gain in the kinetic energy at the anode.
Hence,

$$\frac{1}{2}mv^2 = qV$$

$$\Rightarrow \quad \frac{q}{m} = \frac{1}{2}\frac{v^2}{V}$$

$$\Rightarrow \quad \frac{q}{m} = \frac{1}{2V}\left(\frac{\vec{E}}{\vec{B}}\right)^2 \quad [\because \vec{v} = \frac{\vec{E}}{\vec{B}}]$$

$$\Rightarrow \quad \frac{q}{m} = \frac{E^2}{2VB^2} \tag{29.6}$$

By knowing the quantities $\vec{E}$, V and $\vec{B}$, we can determine q/m ratio.

29.4 ▶ Motion of Electron in Electric Field

When an electron moves in an electric field, there can be three cases: (i) it moves parallel to the electric field, (ii) in perpendicular direction to the electric field or (iii) at an arbitrary angle to the electric field. Let us discuss all these cases.

Case-I Motion of electron parallel to electric field

Let A and B be two parallel metal plates of equal area separated by distance d as shown in Fig. 29.2. When a DC voltage is applied across these plates, an electric field $\vec{E}$ is produced in the region between the two plates. If the distance 'd' between plates is very small than the electric field in between the plates can be considered as uniform. The direction of electric field should be as shown by arrows (i.e., from $+Q$ to $-Q$). If the potential applied across plates are V_A and V_B, then the potential difference V across the plates is $V = V_A - V_B$ and the electric field between the plates can be expressed as

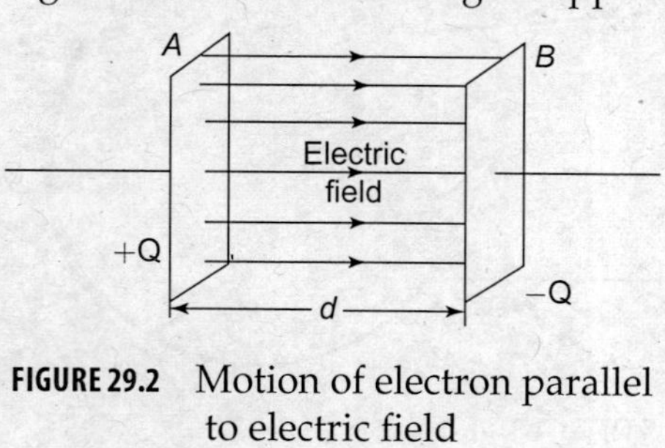

FIGURE 29.2 Motion of electron parallel to electric field

$$\vec{E} = \frac{V}{d}$$

When an electron is released at rest in between these plates, it experiences a force $\vec{F}$ given by

$$\vec{F} = -q\vec{E}$$

where q is the electronic charge. Negative sign indicates that the direction of force on electron is opposite to the electric field. The acceleration produced by this force in electron can be expressed from Newton's second Law as

$$\vec{a} = \frac{\vec{F}}{m} = -\frac{q\vec{E}}{m} \tag{29.7}$$

Since for a uniform electric field all parameters on the right side, i.e. q, $\vec{E}$ and m are constant, the acceleration a is constant. The motion of electron is parallel to uniform electric therefore resembles a free fall of bodies under gravity 'g'. If the initial velocity $v_0 = 0$ at starting point $x = 0$, the final velocity v at x can be written as

$$v^2 = 2ax = 2q\frac{\vec{E}_x}{m} \tag{29.8}$$

The kinetic energy attained by the electron after moving by a distance x can be explained as

$$\text{K.E.} = \frac{1}{2}mv^2 = \frac{1}{2}\cdot\frac{2q\vec{E}_x}{m}$$

$$\text{K.E.} = q\vec{E}_x \tag{29.9}$$

Case-II Motion of electron perpendicular to electric field

Let us consider two parallel plates A and B of length 'x' and separated by a distance 'd'. A uniform electric field $\vec{E}$ is produced between these plates under a potential difference V applied across the two plates and written as

$$\vec{E} = \frac{V}{d}$$

Let an electron enters the electric field from the left side perpendicular to the direction of electric field with initial velocity v_0.

As the electron enters in the electric field, it experiences an electrostatic force $F = qE$ along the y axis. The electron exits the electric field after time 't' acquiring a velocity v that has x and y components. The distance travelled along x axis can be given as

$$x = \vec{v}_0 t \tag{29.10}$$

and the displacement along y axis can be written as

$$y = \frac{1}{2}\vec{a}_g t^2$$

$$y = \frac{q\vec{E}}{2m}t^2 \tag{29.11}$$

Substituting t from Eq. (29.11), we get

$$y = \frac{q\vec{E}}{2mv_0}x^2 \tag{29.12}$$

or

$$y = kx^2 \tag{29.13}$$

where $k = \dfrac{q\vec{E}}{2mv_0{}^2}$ is a constant. Equation (29.13) represents an equation of parabola. The electron thus moves in a parabolic path under an electric field perpendicular to its direction when within the plate region. After it exits the electric field it travels with a velocity 'v' that comprises horizontal and vertical components as

$$v = \sqrt{v_0^2 + v_y^2} \tag{29.14}$$

Deflection of electron

If a fluorescent screen is placed in front of electron path as shown in Fig. 29.3, the electron will strike the screen at point 'P' at a distance 'D' from its original position 'O' without an electric field. From the simple geometry of Fig. 29.3,

$$\tan\theta = \frac{y}{DC} = \frac{OP}{OD}$$

$$\therefore \qquad OP \text{ (Deflection along } y \text{ axis)} = OD \tan\theta \tag{29.15}$$

(If point D lies in the middle of plate, i.e. $DC = \dfrac{x}{2}$ and $x = 1$ unit.)

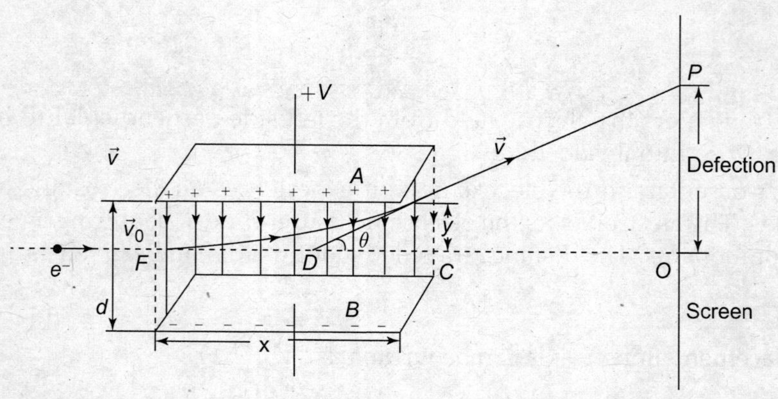

FIGURE 29.3 Motion of electron in electric field perpendicular to its direction of motion

EXAMPLE 29.1

In a Thomson's set up for determining e/m, the same high tension DC supply provides potential to the anode of accelerating column, as also to the positive deflecting plate in the region of crossed fields. If the supply voltage is double, by what factor should the magnetic field is increased to keep the $\bar{e}$ beam undeflected?

Solution: We know that in J.J. Thomson's experiment

$$\frac{e}{m} = \frac{E^2}{2VB^2} \text{ and } \vec{E} = \frac{V}{d}$$

$$\Rightarrow \qquad \frac{e}{m} = \frac{V}{2B^2d^2}$$

This is the condition for undeflected beam.
Since d is fixed,

$$\therefore \qquad B^2 \alpha V$$

$$\Rightarrow \qquad B \alpha \sqrt{v}$$

$$\Rightarrow \qquad B'' = \sqrt{2v} = \sqrt{2}B \quad [\because V^1 = 2V]$$

If V is doubled, B should be increased to $\sqrt{2}B$ to keep the electron beam undeflected.

EXAMPLE 29.2

What is the momentum of acceleration of an electron of speed 2.5×10^6 m/sec in a magnetic field of 2.0 G? Given that $e/m = 1.76\times10^{11}$ C/kg?

Solution: Here,

$$\vec{v} = 2.5\times10^6 \text{ m/sec}$$

$$\vec{B} = 2.0\text{G} = 2.0\times10^{-4}\text{T}$$

$$e/m = 1.76\times10^{11} \text{ C/kg}$$

But,

$$\frac{mv^2}{r} = \vec{B}e\vec{v}$$

$$\Rightarrow \qquad \frac{v^2}{r} = \frac{\vec{B}e\vec{v}}{m}$$

$$\Rightarrow \qquad = 2.0\times10^{-4} \times 1.76\times10^{11} \times 2.5\times10$$

$$= \mathbf{8.80\times10^{13}\ m/sec^2}$$

29.5 ▶ Electrostatic Focussing

We know that the electron's motion can be controlled by using electrostatic and magnetostatic field. A very good example of electrostatic focussing is electron gun, as shown in Fig. 29.4.

Here, a cathode K is used to generate electrons and control grid controls electron flow. The focussing anode focusses the electrons into a narrow beam on the screen. Accelerating anode accelerate the electrons when they pass through it.

The negative voltage of the grid can be varied such that it can control the electron flow or stop it completely. The accelerating anode consists of two cylinders that contain plates with

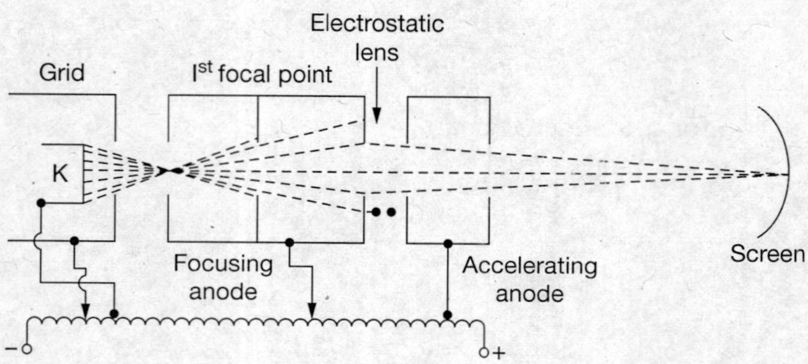

FIGURE 29.4 Electrostatic focussing

small holes in their centres. A high positive potential on the anodes pulls the electrons in the grid. Here, two electrostatic fields exist which focus the electron beam, first electrostatic field exits between control grid and the first anode while the second electrostatic field exits between first and second anodes. The cathode K is given positive voltage with respect to the ground. The grid is at variable negative voltage with respect to cathode. Accelerating anode is provided with several thousand volts (+ve). The potential on focussing anode is positive but less than that on accelerating anode.

The electrostatic fields bend the electron beams as the optical lenses bend the light rays; thus the fields are called lenses. The first electrostatic lens make the electrons to cross at the first focal point in the field but the second lens bends the spreading steams of electrons and returns them to a new focal point on the screen.

Note

In electrostatic focussing, the electron beam is deflected towards the normal as refraction of light.

29.6 ▶ Motion of Electron in Magnetic Field

When electron moves in a uniform magnetic field, a force $\vec{F}$ is experienced by electron that is given by

$$\vec{F} = q\vec{B}v\sin\theta$$

where q is the electronic charge, $\vec{B}$ is the magnetic field, $\vec{v}$ is the velocity of electron and θ is the angle between the direction of motion of electron and the magnetic field. If any of parameters $\vec{B}$, $\vec{v}$ or θ is zero, then no force will be experienced by electron. Since $\vec{B}$ is non-zero for any magnetic field and electron is also in motion, i.e. $v \neq 0$, therefore with θ two extreme cases arise.

Case-I Direction of electron motion is parallel to the magnetic field

In this, two conditions are possible, i.e. either $\theta = 0°$ or $\theta = 180°$.
If electron moves along the direction of magnetic field (north pole to south pole), then $\theta = 0°$. But for reverse direction $\theta = 180°$. For both conditions, sin $\theta = 0$; therefore, no force shall be experienced by electron as shown in Fig. 29.5.

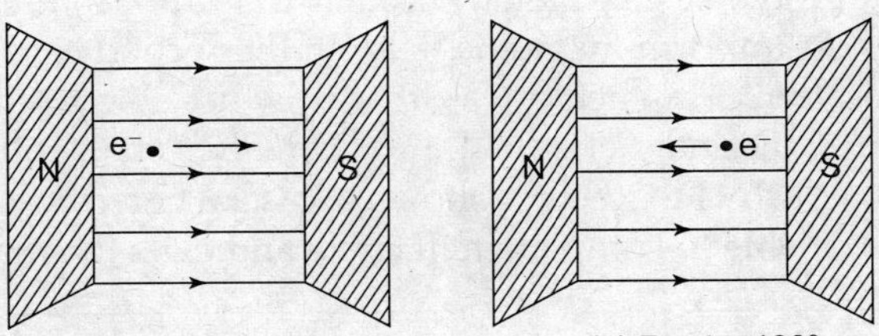

(a) For θ = 0°   (b) For θ = 180°

FIGURE 29.5 Force of electron is zero for both cases

Case-II Direction of electron motion is perpendicular to the magnetic field

In such a condition, the electron experiences maximum force given by

$$\vec{F} = q\vec{B}v\sin(90°) = q\vec{B}v$$

Since this amount of a constant force shall be exerted by the magnetic field, the electron shall be forced to move in a circle. The magnetic force provides the centripetal force required for circular motion given by

$$\frac{mv^2}{R} = q\vec{v}\vec{B}$$

The radius of circle R can be given as

$$R = \frac{m\vec{v}}{q\vec{B}}$$

Using simple equations of circular motion, we can write the time period 'T' of revolution as

$$T = \frac{2\pi R}{\vec{v}} = \frac{2\pi m}{q\vec{B}}$$

Frequency of revolution ν (nu) is given as

$$\nu = \frac{1}{T} = \frac{q\vec{B}}{2\pi m}$$

Angular speed is given as

$$w = 2\pi\nu = \frac{2\pi q\vec{B}}{2\pi m} = \frac{q\vec{B}}{m}$$

It may be noted that the time period 'T', frequency 'ν' and angular speed 'w' are all independent of the speed of electron (or particle). This aspect is used in cyclotron to accelerate charged particles.

The trajectory of particle

The trajectory or path traversed by an electron or charged particle depends on the location of particle in the magnetic field. In the first case shown in Fig. 29.6(a), the particle starts from point S and has a sufficiently large area to cover the first circle traced by the particle, then the electron or particle shall traverse completely circular trajectory in a uniform magnetic field or an helical path in a non-uniform magnetic field. In the second case shown in Fig. 29.6(b), the particle enters a location S in such a way that it is just able to complete its half circle, in such a case it will exit the magnetic field and thereafter shall follow the motion in the direction of its exit. Similarly, Fig. 29.6(c) shows a case where the particle enters at such location that it is even not able to complete half circle and exits the magnetic field.

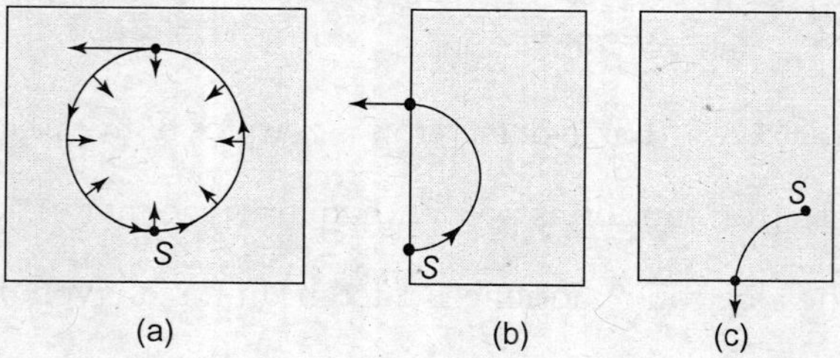

FIGURE 29.6

The helical or spiral motion of electron in non-uniform magnetic field Fig. 29.7(b) is used as a principle of electron beam focussing in electron microscopes, televisions and several other devices and instruments.

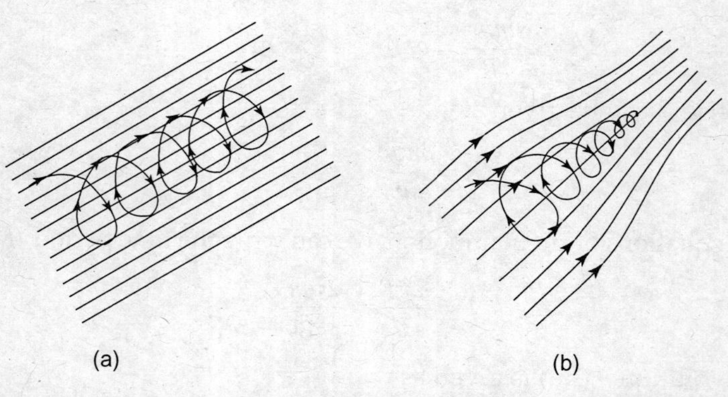

FIGURE 29.7 Motion of charged particle in (a) uniform magnetic field; (b) Non-uniform magnetic field

29.7 ▶ MAGNATOSTATIC FOCUSSING

Electron microscope is a good example to explain the magnetic focussing. As shown in Fig. 29.8, a coil of copper wire produces a magnetic field, which is shaped by the surrounding iron fixture to produce lensing action.

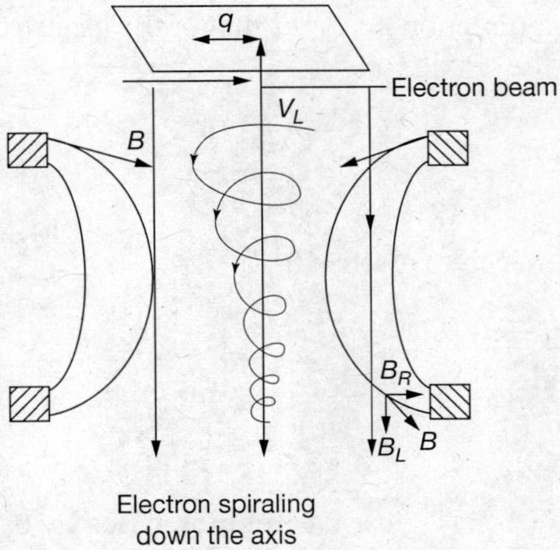

FIGURE 29.8 Magnetostatic focussing

When electron is moving under the effect of magnetic field, it experiences a radial inward force, which is proportional to the Lorentz force ($\vec{v} \times \vec{B}$), where $\vec{v}$ is the electron velocity and $\vec{B}$ is the magnetic flux density.

In electromagnetic lens, the lasing action is similar to that of an optical lens but the focal length depends on two factors – the amount of current through the coil, which determines the magnetic field $\vec{B}$, and the gun voltage which determines the electron velocity $\vec{v}$. We can control the focal length of the lenses by adjusting the currents supplied. A current increases, the radial force experienced by the beam increases which reduces the force length. The focal length *f*, of a lens as shown in Fig. 29.8, is given by

$$f = \frac{CV}{n^2 I^2}$$

where V is the accelerating voltage, C is a constant, n is the number of turns in the coil and I is the current in the coil.

Note

For achieving good lens characteristics, it is required to have constant current through lens and constant accelerating voltage. The source of electrons also needs to be coherent.

EXAMPLE 29.3

A proton, with initial velocity of 5×10^6 m/sec, passes through an electric field (transverse) of 200 V/cm. Calculate the transverse deflection in travelling a distance of 1 m.

Solution: Let the proton direction be x and that of the electric field be y. Time taken to travel a distance L along the x-axis,

$$t = \frac{L}{v} = \frac{1}{5} \times 10^6$$
$$= 2 \times 10^{-7} \text{ sec}$$

Force acting on the proton due to electric field $= qE_y = mF_y$

Proton acceleration $f_y = qE_y/m$

Hence, the transverse deflection

$$y = \frac{1}{2} f_y t^2 = \frac{1}{2} qE_y t^2 / m$$

$= \mathbf{0.04}$ **m molecules by 0.2 mm on the film.**

EXAMPLE 29.4

An electron moves in the earth's magnetic field of $5 \times 10^{-5} T$ with the energy of 10 keV. Find the Larmour's radius of the electron neglecting its velocity component parallel to the magnetic field.

Solution:

$$E = \frac{1}{2} m v_{\perp 0}^2$$
$$= \sqrt{2E / m}$$

and

$$r_L = \frac{V_{\perp 0}}{w_c} = \frac{m V_{\perp 0}}{e\, B}$$

or

$$V_{\perp 0} = \sqrt{\frac{2 \times 10 \times 10^3 \times 1.6 \times 10^{-19}}{9.1 \times 10^{-31}}}$$
$$= 5.93 \times 10^7 \text{ m/s}$$

$\therefore$

$$r_L = \frac{\mathbf{9.1 \times 10^{-31} \times 5.93 \times 10^7}}{\mathbf{1.6 \times 10^{-19} \times 5 \times 10^{-5}}} = \mathbf{6.75 \text{ m}}$$

29.8 ▶ Cathode Ray Oscilloscope (CRO)

CRO is an electronic instrument used for measuring voltage, time, frequency, phase shifts and for observing the shape of input waveform. It is also known a fast x-y plotter which displays an input signal versus another signal or versus time. The heart of CRO is CRT, i.e. cathode ray tube as shown in Fig. 29.9.

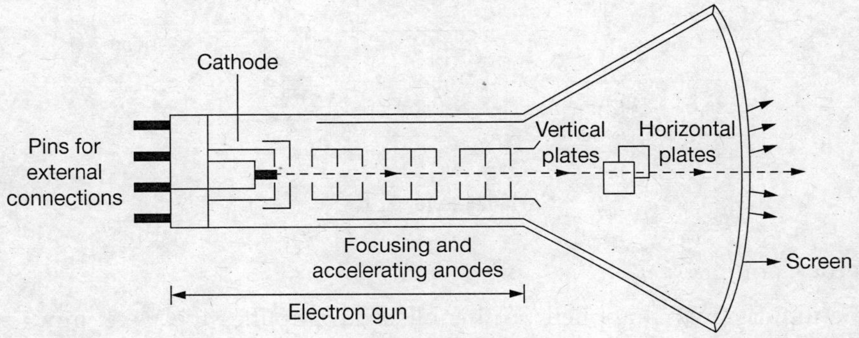

FIGURE 29.9 Cathode ray tube

Construction and operation of CRT

The cathode ray tube (CRT) can be divided into three parts, electron gun assembly, deflection plates and screen. As shown in Fig. 29.9, the electron gun assembly consists of a heater, cathode, control grid, focussing anode and accelerating anode. The cathode emits electrons. A negative voltage is applied to the control grid so that it controls the number of electrons emitted from the cathode. In this way, the intensity of a beam can be controlled by controlling the negative voltage applied to the grid. The sharpness of beam is controlled by potential applied to the focussing anode and the beam is accelerated by the high positive voltage applied to the accelerating anode.

Now the focussed and accelerated electron beam is passed through the vertical and horizontal deflection plates. A voltage is applied to the horizontal plates to deflect the beam in the 'x' direction while the voltage is applied to vertical plates in order to deflect the beam in the 'y' direction. But the simultaneous force exerted in the x and y directions deflects the beam in such a way that a waveform is displayed on the screen. Now, the focussed, accelerated and deflected beam strikes the screen. The screen is coated with phosphorous material which converts the electrical energy into light. All the working parts of CRT are enclosed in an evacuated glass envelope. The electrons striking the screen release secondary electrons. These secondary electrons are collected by Aquadag which is connected to the second anode, collection of secondary electrons is required to keep CRT screen in electrical equilibrium state.

CRO is also called a single trace CRO as it displays only one waveform on its screen. Let us discuss each block of CRO one by one (Fig. 29.10).

Block diagram of a CRO

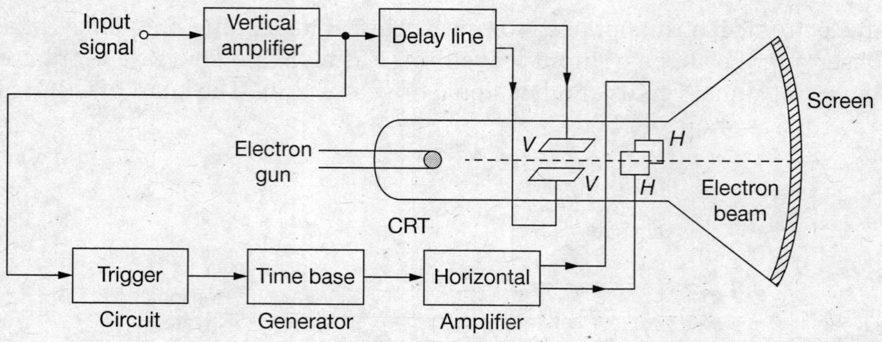

FIGURE 29.10 CRO

(i) Vertical amplifier:

The input signal is applied on this block. It amplifies the weak input signals to produce measurable deflection on the CRO screen. Output of it, is applied to the trigger circuit for triggering the time base generator. Vertical amplifier of a CRO often decides its sensitivity and bandwidth.

(ii) Delay line:

If delay line is not used, then the signal in the horizontal section will reach the horizontal plates slightly late as compared to the signal reaching the vertical plates. But the signal at both the plates (horizontal and vertical) should arrive exactly at the same instant of time.

(iii) Trigger circuit:

The role of trigger circuit in CRO is to establish and maintain a synchronization between the input signal and the horizontal deflection circuitry.

(iv) Time-base generator:

The time-base generator generates a sawtooth voltage waveform and sends it between the horizontal deflection plates. A sawtooth voltage has a linear relation with time so the spot moves with a uniform velocity across the screen in horizontal direction. The x-axis of CRO can be calibrated in terms of time and the input waveform can be displayed with respect to time.

(v) Horizontal amplifier:

It is used to amplify the signal (sawtooth signal) available at the output of the time-base generator. It applies then this signal to the horizontal deflection plates.

(vi) CRT:

The construction and working of CRT are already explained. The phosphor screen of CRT converts the electron beam striking the screen into light and the input signal becomes visible on the CRO screen.

(viii) Power supply:

It generates two levels of DC voltages: first is a low-voltage power supply, which is generated for the operation of electronic circuits and second is high voltage which is generated for the operation of anodes in CRT. The high voltage is of the order of 1000 to 15000 V.

Applications of CRO

Some of the important applications of CRO are given below.

1. To measure the voltage, time, frequency, duty cycle and phase shifts.
2. To measure modulation index in A.M. waves.
3. For testing diodes, transistors, etc.
4. For the analysis of circuits, by observing waveforms at various points of the circuit.
5. It is used in curve tracers, which is used to display directly the V-I characteristics of a device like transistor.
6. In non-destructive testing of materials.
7. In radars, it is used to display transmitted and received pulses (echoes).
8. It is also used in many medical equipment like I.C.U. of a hospital and for displaying ECG.

EXAMPLE 29.5

A solar wind proton is streaming with velocity $v_{\perp 0} = 3\times10^5\,\text{m/sec}$ in the magnetic field of $5\times10^{-9}\,\text{T}$. Compute the Larmour radius, if $m_p = 1.67\times10^{-27}\,\text{kg}$ is given.

Solution:

$$B = 5\times10^{-9}\,\text{T}$$

$$m_p = 1.67\times10^{-27}\,\text{kg}$$

$$v_{\perp 0} = 3\times10^5\,\text{m/sec}$$

Larmour's radius,

$$r_L = \frac{m_p v_{\perp 0}}{eB}$$

$$= \frac{1.67\times10^{-27}\times3\times10^5}{1.6\times10^{-19}\times5\times10^{-9}}$$

or

$$\mathbf{r_L = 6.26\times10^5\,m}$$

EXAMPLE 29.6

An ion engine has a 1T magnetic field, and a collection of H^+ is to be shot out at an $\vec{E}\times\vec{B}$ velocity of 10^6 m/s. How much internal electric field must be present in the collection of ions (H^+)?

Solution: $\vec{E}\times\vec{B}$ drift is the drift of guiding centre, given by

$$\vec{v}_E = \frac{\vec{E}\times\vec{B}}{B^2}$$

$$\Rightarrow \qquad V_E = \frac{\vec{E}}{\vec{B}} \text{ or } \vec{E} = \vec{B}v_E$$

$$\Rightarrow \qquad \vec{E} = 1\times10^6\times1 = 10^6 \text{ V/m}$$

29.9 ▶ Bainbridge Mass Spectrograph

Aston in 1919 developed the first mass spectrograph. It is a device used for measuring the masses of isotopes. The apparatus gives accuracies of one part in 1000 (Fig 29.11).

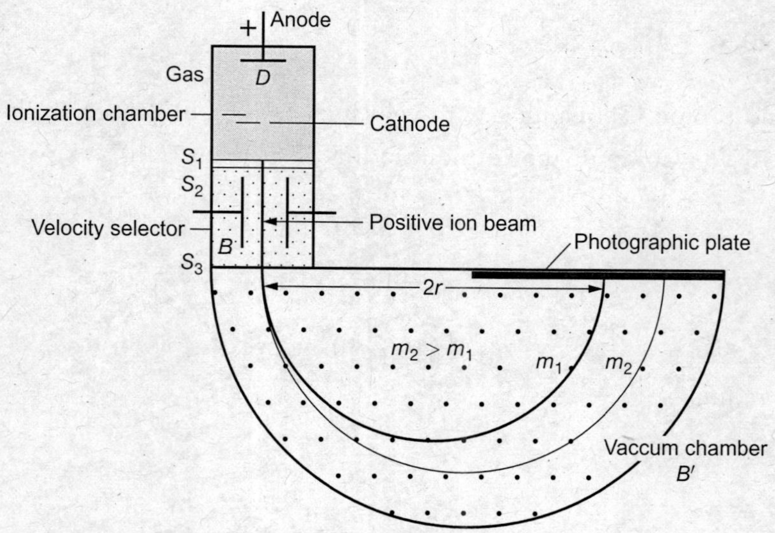

FIGURE 29.11 Bainbridge mass spectrograph

Construction and working

The ions are formed at point D and pass through slit S_1 through cathode C. A potential V is applied between two plates A and B, where the ions are reached. A magnetic field B

is applied at right angles to the electrostatic field and thus the electrostatic and electromagnetic forces act in opposite directions to each other.

A charged particle having charge q and having velocity v will pass through slit S_2 if the resultant force acting on this particle is O, i.e. it travels along a straight line. Here,

$$q\vec{v}\vec{B} = q\vec{E}$$

and the velocity v of the particle is

$$\vec{v} = \frac{\vec{E}}{\vec{B}}$$

Only the particles having a certain velocity can enter the deflection chamber F. Thus, the combination of slits and reflecting plates in the instrument is called velocity selector. In the chamber which is also known as deflection chamber, the ions are under the effect of magnetic field, only thus these ions move in circular orbits. The lighter ions have larger path radius while the heavy ones have smaller path radius. If mass of an ion is m, then

$$\vec{B}q\vec{v} = \frac{mv^2}{r}$$

but $$r = \frac{m\vec{v}}{\vec{B}q}$$

and

mass of an ion $$(m) = \frac{rB^2q}{\vec{E}}$$

Here, it is clear that the radius of path of an ion is directly proportional to the mass of the ion.

The deflection is produced by either a photographic plate or a collector. This produces a small current when ions fall on it. We can vary the magnetic field, so the radii of particles path changes and they can fall on a fixed collector (according to their masses). This method of anlaysis is very accurate and can detect mass differences as one part in 10^9.

Applications

Mass spectrometers are sensitive detectors of isotopes based on their masses. They are used in carbon dating and other radioactive dating process. A number of space crafts and satellites used mass spectrometers to identify the small particles intercepted in space. E.g. SOHO satellite uses a mass spectrometer to analyze the solar wind. These are also used to identify and analyze the residual gases in high vacuum systems.

EXAMPLE 29.7

A mass spectrometer is being used to monitor air pollutants. It is difficult to separate molecules with nearly equal mass such as CO (28.0106u) and N_2 (28.0314u). How large a radius of curvature must a spectrometer have to separate these molecules by 0.20 mm on the film?

Solution: In spectrometer,

$$\frac{mv^2}{r} = \vec{q}v\vec{B}$$

and
$$\frac{\Delta r}{r} = \frac{\Delta m}{m}$$

where the two lines on the film are separated by a distance Δx is

$$\Delta x = 2(r_1 - r_2)$$
$$= 2\Delta r$$

$$\Rightarrow \quad r = \frac{\Delta r}{m} m = \frac{\Delta x \cdot m}{2\Delta m}$$
$$= \frac{0.20 \times 10^{-3} \times 28.0120}{2 \times 0.0028}$$
$$= 1 \text{ m}$$

Thus, the spectrometer must have a radius of curvature of 1 m to separate the given molecules, substituting these values in the relation for q/m of an ion in Bainbridge mass spectrograph, we get

$$\frac{q}{m} = \frac{E}{BB'r}$$

$$\Rightarrow \quad m = \frac{qBB'r}{E}$$

$$\Rightarrow \quad m = 21.06\text{u}$$

Since the mass number is an integer closest to the atomic mass in the atomic mass scale. Therefore, the mass number of the neon isotope = 21.

EXAMPLE 29.8

In a Bainbridge mass spectrograph, the electric field between the plates of a velocity selector is 1000 V/cm and the magnetic induction in both magnetic fields is 0.5 Wb/m^2. A stream of singly charged neon ions moves in a circular path of 8.74 cm radius in the magnetic field. Find the mass number of the neon isotope.

Solution: The neon ions are singly charged, thus charge $q = 1.6 \times 10^{-19}$ Coulomb magnetic fields are equal. Therefore,

$$B = B' = 0.5 \text{ Wb/m}^2.$$

$$\text{Electric field } E = 1000 \text{ V/cm} = 10^5 \text{ V/m}$$

$$\text{Radius of the circular path} = 8.74 \text{ cm}$$

$$= \mathbf{8.74 \times 10^{-2} \text{ m}}$$

RECAP ZONE

POINTS TO REMEMBER

- Electron optics is the branch of physics which deals wih the beam of electrons and their deflection under the effect of electric and magnetic fields.
- The concept of gyratory motion, guiding centre and cycloid motion were introduced and explained in this chapter.
- Electrostatic focussing is done using electric fields, e.g., electron gun.
- Magnetic focussing is achieved by using magnetic field, e.g., electron microcope.
- CRO is a very useful electronic instrument used for measurement of voltage, time, frequency, phase shifts and for observing the shapes of waveforms.
- Cathode ray tube is known as the heart of a CRO. It generates electron beam and displays input waveform on the CRO screen.
- SEM is a microscope that uses electrons rather than light to form an image. It is used due to higher magnification, larger depth of field, greater resolution and to collect compositional and crystallographic information.
- The first TEM was built by max knoll and Ernst Ruska in 1931, Ruska was awarded with the Nobel prize in physics for the development of TEM. It is used to study the growth layer, their composition and defects in semiconductors. It is a powerful tool for analysis in material sciences.
- Bainbridge mass spectrograph is a sensitive detector of isotopes based on their masses. It has a large number of applications in industry and academic labs.
- The resolving power of the electron microscope is 10^5 times greater than that of optical microscope.
- The basic principle of a mass spectrograph is to get different spectral lines on a photographic plate corresponding to the different masses of the isotopes of an element.
- To magnify very minute objects, electron beam is used instead of the light rays because (i) the electrons have a wave nature as that of light rays but have much shorter wavelength, (ii) electrons can be focussed by electric and magnetic fields as light rays are focussed by glass lenses.
- The combination of a mass spectrometer and a gas chromatograph makes a powerful tool for the detection of trace quantities of contaminants or toxins.

Key Terms with Definitions

- **Lens** – Used to converge or diverge the optical rays.
- **Optical microscope** – Device to magnify small objects.
- **CCD** – A charged coupled device is a device for the movement of electrical charge, usually from within the device to an area where the charge can be manipulated.
- **Unipotential lens** – Electrostatic lens where electrons enter and leave it at same potential.
- **Ionization chamber** – Used to ionize the gas whose mass or isotope is to be determined and positive ions are produced.
- **Linear separation** – Method of analysis to detect differences in the masses of two ions as small as one part in 10^9.

Important Formulae and Equations

Equation Number	Equation	Remarks
(29.1)	$\vec{F} = q\vec{v}\vec{B}$	**Lorentz force**
(29.2)	$\frac{q}{m} = \frac{\vec{v}}{\vec{B}r}$	**Change to mass ratio**
(29.3)	$\vec{F} = q\vec{E}$	**Force on e^- due to electric field**

Review Zone

Short Answer Questions

1. What do you understand by electron optics?
2. Define specific charge of an electron.
3. Discuss the motion of an electron under uniform magnetic field.
4. Discuss electrostatic focussing of electron beam.
5. Discuss the principle and image formation of SEM.
6. Write a short note on TEM.
7. What is CRT?
8. Describe the working of a CRO.
9. Write a short note on Bainbridge mass spectrometer.
10. What is the function of Aquadag coating in CRO?
11. What is the purpose of focussing anode in CRO?
12. What do you mean by magnetostatic focussing?

13. What would be the effect of a magnet, if it is brought close to a television?
14. On which principle does the electron microscope work?
15. How do you know whether a moving electron in a certain region is being deflected by an electric field or by a magnetic field or by both?

LONG ANSWER QUESTIONS

1. Explain J.J. Thomson's method to determine the specific charge of an electron.
2. Prove that the motion of an electron in uniform electric and magnetic fields is the sum of simple gyratory motion and the motion of guiding centre when both $\vec{E}$ and $\vec{B}$ fields are perpendicular to each other.
3. Describe electrostatic and magnetostatic focussing.
4. Describe the construction and working of a CRO.
5. What is a CRT? What are the main parts of CRT? Explain each block of CRT.
6. Discuss the principle and working of scanning electron microscope.
7. Discuss the working and experimental arrangement of TEM.
8. Explain the working of Bainbridge mass spectrometer.
9. What type of motion do you expect, when the proton moves (a) along $\vec{B}$, (b) $\perp$ to $\vec{B}$ or (c) along a line at an angle of 0° to the direction of $\vec{B}$?
10. What do you mean by electron optics? How can you determine e/m of an electron by Thomson's method?
11. What type of motion do you expect, when the electron moves (a) along $\vec{B}$ and $\perp$ to $\vec{B}$?
12. What type of motion do you expect, when the proton is placed in uniform electric field?

NUMERICAL PROBLEMS

1. The voltage across the electrodes of a cathode ray gun is 500 eV. Calculate
   (i) the energy gained by electron
   (ii) the speed of the electron
   (iii) the momentum of electron, if mass of electron is given as 9×10^{-31} kg

   Ans. 8×10^{-17} J, 1.33×10^{7} m/s, 12×10^{-24} kg/sec
2. A mono energetic electron beam with electron speed of 5.20×10^{6} m/sec is subject to a magnetic field of 1.30×10^{-4} T, normal to the beam velocity. What is the radius of the circle traced by the beam? (Given $e/m = 1.76\times10^{11}$ c/kg)

   Ans. 22.7 cm
3. In a Thomson set up for determination of e/m, electrons accelerated by 2.5 kV enter the region of crossed electric and magnetic fields of strengths 3.6×10^{4} V/m and 1.2×10^{-3} T, respectively

and go through undeflected. Determine the e/m of an electron.

Ans. 1.8×10^{11} c/kg

4. Calculate the area traced by the trajectory of a 1 KeV He^+ ion in the solar atmosphere near a sunspot, where $B = 5 \times 10^{-2} T$.

Ans. 0.105 m^2

5. In a Bainbridge mass spectrograph, the electric field between the plates is 2.18 × 1.4 V/m and $B = B' =$ 0.57 T. The source contains boron isotopes ^{10}B and ^{11}B. How far apart are the lines formed by the singly charged ions of each isotope on the photographic plate?

Ans. 1.04 mm

6. A proton, a deuteron and an α-particle have equal kinetic energies. Compare the radii of their paths when a normal magnetic field is applied.

Ans. $r_d = \sqrt{2} r_p = \sqrt{2} r_\alpha$

7. An electron is moving in uniform electric and magnetic fields which are perpendicular to each other. Find the drift of the guiding centre if the magnitudes of the electric and magnetic fields be 100 v/m and $10^{-3} T$, respectively.

Ans. 10^5 m/sec

CHAPTER 30 Atomic Structure

Learning Objectives

By the end of this chapter, the student will be able:

- To recall historical aspects of models of atoms
- To understand Rutherford's scattering experiment and its theory
- To interpret emission and absorption spectra of hydrogen on the basis of Bohr's model
- To apply Bohr's hydrogen model to other atoms and discuss its limitations
- To introduce the concept of quantum numbers
- To illustrate Pauli's exclusion principle and Afbau law

30.1 ▸ Introduction

The nature of matter at microscopic level has always been on human mind since the beginning. Ancient philosophers pondered upon the existence of the smallest component of matter. When we look at the objects around us, we try to find their constitution and the way they are made up. The curiosity and exploration go further till we are able to physically divide or separate the constituent building block. But the level of exploration is limited by the resolving power of eyes and the tools that we can use. Before the invention of microscope only large grains of e.g. sand could be seen by naked eyes but with optical microscopes one could see up to order of microns. But with the invention of electron microscopes, we can see the shape of the particles up to a few nanometres. The objects or particles below this size are observed indirectly through some effects that are observable by using some tricks and techniques. The inter-atomic distances of crystals are determined by X-ray diffraction patterns, which are an example of indirect method. The materials around us are now well known to be constituted of molecules which in turn are made up of atoms. The atoms are made up of electrons, protons and neutrons. The further studies indicate that the proton and neutron are made up of particles with fractional charges called quarks that are held together with gluons. Here we will review some models and theories on the basic concepts of atomic structure.

30.1.1 Historical Aspects

In ancient Greek philosophy, the word atom was used to describe the smallest bit of matter. Atom was thought to be indestructible because the Greek word for atom means 'not divisible'. In India, the vedic period had a concept of Anu (molecule) and Parmanu (atom). The rishi Kanad had mentioned the concept of Anu and parmanu. The matter as a whole was divided into five *tatwas* (elements) known as earth, fire, water, space and air. It was presumed that all material things constituted of these five elements mixed in a particular proportion.

Looking at the experimental proofs through modern scientific tools, the thinkers and philosophers of that time argued their assertions as they were conceived. It was after the discovery of current electricity that several new techniques like discharge of electricity through gases, heated filaments, X-rays and so on could become possible. The foundations of modern concept of atomic structure were laid with the help of the law of conservation of mass and law of definite proportions.

Note

In India, the vedic period had a concept of Anu (molecule) and Parmanu (atom). The rishi Kanad had mentioned the concept of Anu and parmanu.

30.1.2 Dalton's Theory

In 1808, an English chemist, John Dalton, proposed that matter consisted of tiny indivisible particles known as atoms. In fact, Dalton was not the first person to propose the existence of atoms. But what distinguished Dalton from others was that he presented evidence for atom's existence.

The founding principles of Dalton's theory of atom are as follows:

1. Matter consists of indivisible entities called atoms.
2. All atoms of a given element are identical in respect of mass and all other physical properties.
3. Different elements have different kinds of atoms having different masses.
4. Atoms are indestructible and retain their identities even in chemical reactions.
5. The formation of a compound from elements occurs through the combination of atoms of unlike elements in whole number ratios.

Throughout the nineteenth century, atomic theory of Dalton dominated. Then in 1879, Sir J.J. Thomson discovered electron. He found that electrons had a negative charge and negligible mass. Thomson proposed a model of atom known as Plum Pudding or Current-Bun model (Fig. 30.1). According to this model, the atom consists of a sphere of radius $\sim 10^{-10}$ m or ~ 1 Å (Augstrom). The positive and negative particles fill the whole sphere uniformly in such a way that atom is electrically neutral.

This model could explain the stability and electrical neutrality of an atom. It could also explain the process of ionization of atoms and residual charge on atoms. This model, however, could not explain the emission of spectral lines from atoms and the large angle scattering of α-particles by heavy atoms as demonstrated later by Rutherford's scattering experiment.

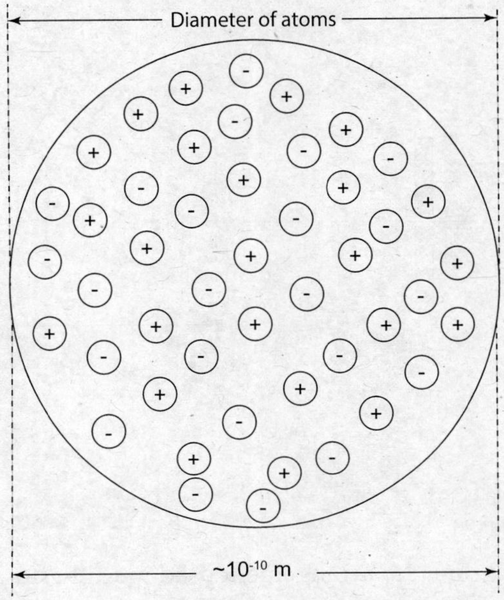

FIGURE 30.1 Thomson's plum pudding model of atom

Note

According to J. J. Thomson's model of atom known as Plum Pudding or Current-Bun model, the atom consists of a sphere of radius $\sim 10^{-10}$ m or ~ 1 Å (Augstrom). The positive and negative particles fill the whole sphere uniformly in such a way that atom is electrically neutral.

30.2 ▶ Rutherford Model of Atom

The current concept of atomic structure is based upon the model proposed by Ernest Rutherford in 1911. During his several experiments with α-particle emission from radioactive materials, he found that most of the α-particles were able to penetrate through the atoms without any deflection. This indicated that most of the volume inside the atom is hollow. To conclude it, he repeated the experiments systematically with gold atoms as target with the help of his students Geiger and Marsden. The experimental arrangement used by Rutherford was shown in Fig. 30.2.

The radioactive material placed inside lead shielding emits α-particles of high energy. These are made to pass through a narrow pin hole screen to make a fine particle beam which is bombarded on a very thin gold foil. On striking the gold foil screen, the α-particles are scattered through different angles. The scattered particles are detected with the help of the ZnS (Zinc Sulphide) screen which gives fluorescent spikes when α-particles fall on it. The number of such spikes is counted in a dark room with the help of ZnS and microscope detector. This whole experimental arrangement was placed inside a vacuum chamber to increase the range of α-particles because α-particles are absorbed in air and to travel long to make such experiment feasible.

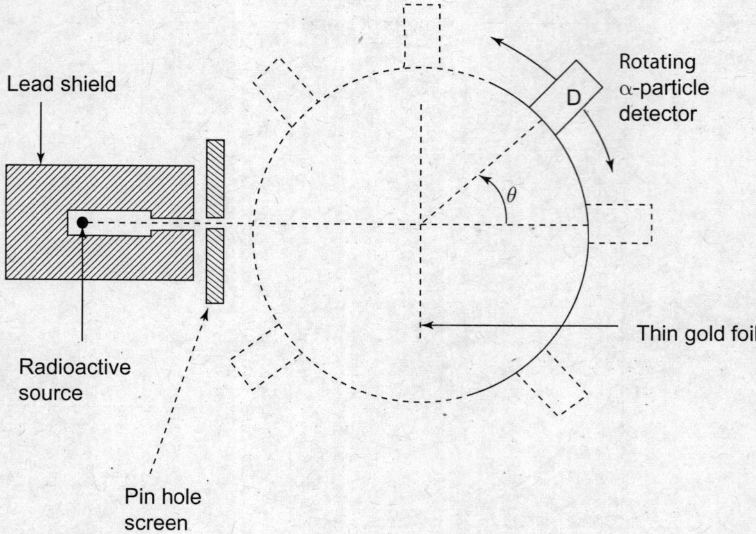

FIGURE 30.2 Experimental arrangement used in Rutherford experiment

Geiger and Marsden, sat for hours and days together to watch and record the tiny spikes of light through microscopes. If Thomson's model of atomic structure was true, the results of scattering should have shown less than four degree scattering as shown in Fig. 30.3(a).

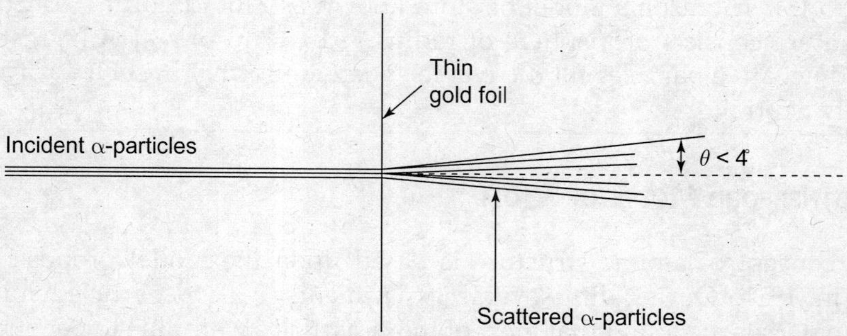

FIGURE 30.3 (a) Expected result of α-scattering experiment according to Thomson's model of atom

Most of the α-particles passed through gold foil with little or no deflection. But some were found to scatter at large angles and few even at 180° that is just turned to move in opposite direction in a straight line [Fig. 30.3(b)]. This was an observation contrary to their expectations. This anomalous scattering led to an explanation, which later became the basis of nuclear model of atom or Rutherford's model of atom.

According to this model, the positive charge and almost the entire mass of an atom are located at the centre of the atom. This dense and central part of an atom is called the nucleus. To support this theory, the size of nucleus was also determined and it was suggested that the

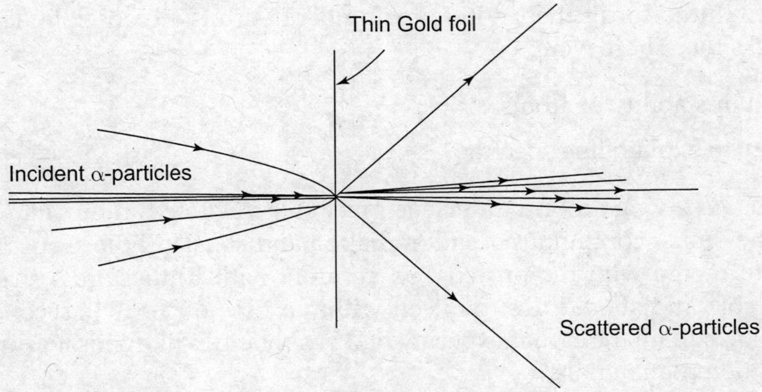

FIGURE 30.3 (b) Observed result of α-scattering experiment by Rutherford (1911)

size should be of the order of 15×10^{-15} m. Compared to the size of an atom ($\sim 10^{-10}$ m), this is very small. Therefore, most of the space inside an atom must be empty. In order to neutralize the electrical charge, the nucleus must be surrounded by negatively charged electrons in such a way that the total negative charge on electrons is equal to the total positive charge on nucleus. For the stability of an atom, Rutherford assumed that electrons revolve around the nucleus in fixed orbits similar to the revolution of planets around the sun (Fig. 30.4). These orbits can be at any distance from the nucleus.

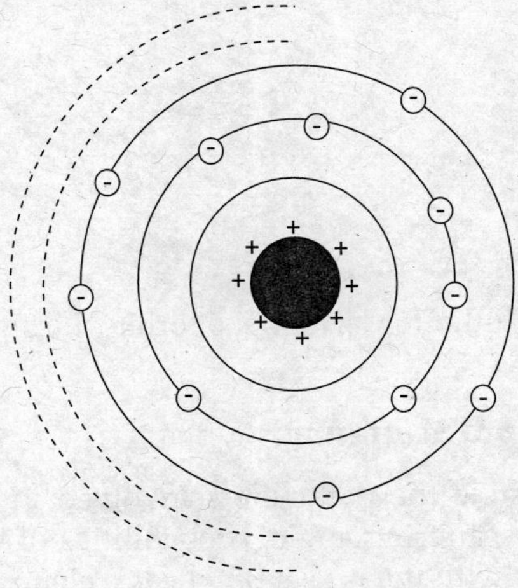

FIGURE 30.4 Rutherford's model of atom

Note

According to Rutherford's model, the positive charge and almost the entire mass of an atom are located at the centre of the atom; the dense and central part of an atom is called the nucleus. To support this theory, the size of nucleus was also determined and it was suggested that it should be of the order of 15×10^{-15} m, which is very small compared to the size of an atom ($\sim 10^{-10}$ m).

According to Rutherford's theory, infinite number of orbits are possible. This model had some drawbacks too. These were:

1. Inherent instability of atoms
2. Inability to explain line spectra

The electrons revolving around a nucleus must be accelerated to move in orbits, which indicate that they must continuously emit radiation and constantly lose energy and follow a spiral orbit to merge with the nucleus. So an atom with Rutherford's structure should be highly unstable and should destroy itself within a time of $\sim 10^{-9}$ seconds (Fig. 30.5). Further, it cannot explain the earlier experimental observations like emission and absorption spectra from different materials.

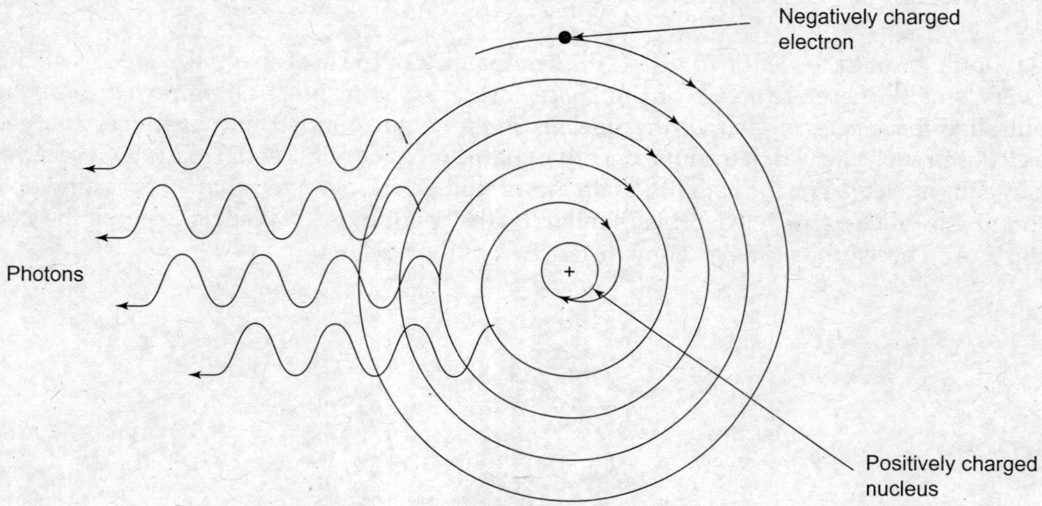

FIGURE 30.5 Collapse of Rutherford's model of orbits during emission of radiation

30.2.1 Theory of Rutherford Scattering

In order to precisely determine the parameters involved in the scattering process, the mathematical estimation of scattering angle, probabilities of scattering should be made. The encounter of α-particles with the nuclei of a heavy element (like gold) is assumed to be only elastic and electrostatic (other forces of interaction are ignored). The geometry of scattering can be represented as shown in Fig. 30.6.

Let us presume that an α-particle of mass m and velocity v_0 approaches a stationary nucleus Z of positive charge ($+Ze$; Z-atomic number of element, e-electronic charge) along a trajectory. The path of the α-particle is a hyperbola asymptotic to lines BO and OC.

It is presumed that mass of the nucleus is so large that it does not get displaced under Coulomb repulsion of α-particle and the total angular momentum of system remains constant.

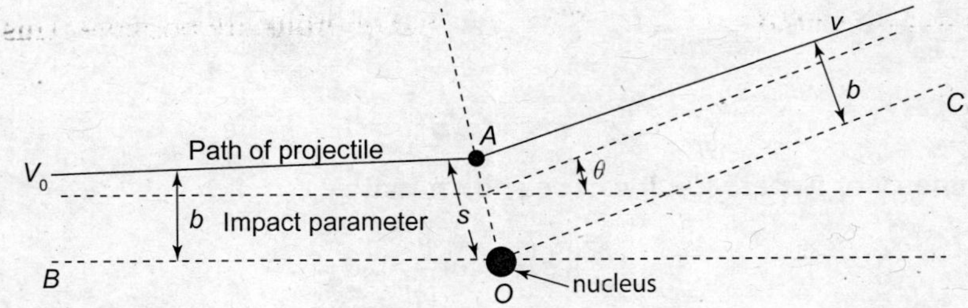

FIGURE 30.6 Scattering of α-particle-trajectory and impact parameter

Let $OA = s$ be the minimum distance of α-particle from the nucleus during its trajectory. This distance is called the distance of closest approach. Since angular momentum is conserved along the trajectory of α-particle

$$mv_o b = mvs$$

where v is the velocity of α-particle after scattering under the repulsive force

$$\frac{1}{4\pi \in_0}\frac{Ze2e}{r^2} \qquad \because \; v = \frac{v_o b}{s}$$

At point A, the kinetic energy of α-particle is $\frac{1}{2}mv^2$ and the potential energy is $\frac{1}{4\pi \in_0}\frac{Ze\cdot 2e}{s}$ $(\because r = s)$.

From conservation of energy,

$$\frac{1}{2}mv_o^2 = \frac{1}{2}mv^2 + \frac{1}{2\pi \in_0}\frac{Ze^2}{s}$$

$$\frac{1}{2}mv_o^2 = \frac{1}{2}m\left(\frac{v_o b}{s}\right)^2 + \frac{1}{2\pi \in_0}\frac{Ze^2}{s}$$

$$\frac{1}{2}mv^2 = \frac{1}{2}mv_o^2\frac{b^2}{s^2} + \frac{1}{2\pi \in_0}\frac{Ze^2}{s}$$

Multiplying both sides by $\frac{2}{mv^2}$, we get

$$1 = \frac{b^2}{s^2} + \frac{Ze^2}{\pi \in_0 mv^2 \cdot s}$$

$$1 = \frac{b^2}{s^2} + \frac{ps}{s^2} \quad \text{where} \left(p = \frac{Ze^2}{\pi \in_0 mv^2}\right)$$

$$s^2 = b^2 + ps$$

$$b^2 = s^2 - ps = s(s-p)$$

From the geometry of hyperbola, it can be shown that

$$s = b\cot\frac{\theta}{2} \tag{30.1}$$

Substituting the value of s in Eq. (30.1), we can obtain

$$p = 2b\cot\theta \tag{30.2}$$

or

$$\boxed{\theta = \tan^{-1}\left(\frac{2b}{p}\right)} \tag{30.3}$$

This equation gives the angle of scattering of α-particle θ in terms of impact parameter b.

30.3 ▶ Emission and Absorption Spectra

When gases or any other elements are heated to very high temperatures, their atoms emit a characteristic visible light, which is used as a tool in chemical analysis. But such an emission is not always in the visible region of spectrum. The emission of this light can be in *IR* (Infra-Red), visible or *UV* (Ultra-violet) region. The finer details of spectrum can be seen if it is made to pass through a fine slit and then a prism or grating so that all components (wavelengths) are resolved. Such an instrument to observe the spectra is called spectrophotometer, the schematics of which are shown in Fig. 30.7.

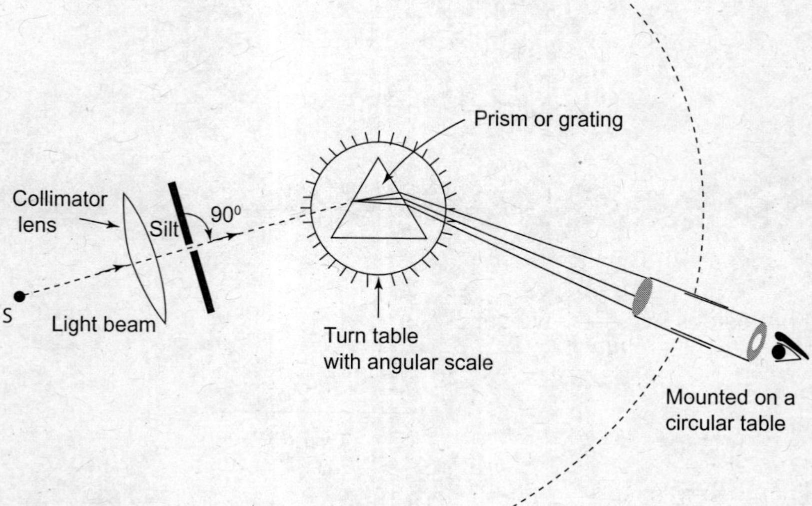

FIGURE 30.7 Schematics of a spectrophotometer

The spectra of light observed through a spectrophotometer appear as band of lines (few or more depending upon source) which is called line spectrum or band spectrum. The colour of these lines is different from each other and corresponds to different wavelengths of light. For example, the spectrum of light from the sodium vapour lamp has only two lines of yellow colour called D_1 and D_2 corresponding to wavelengths of 5890 Å and 5898 Å (Fig. 30.8).

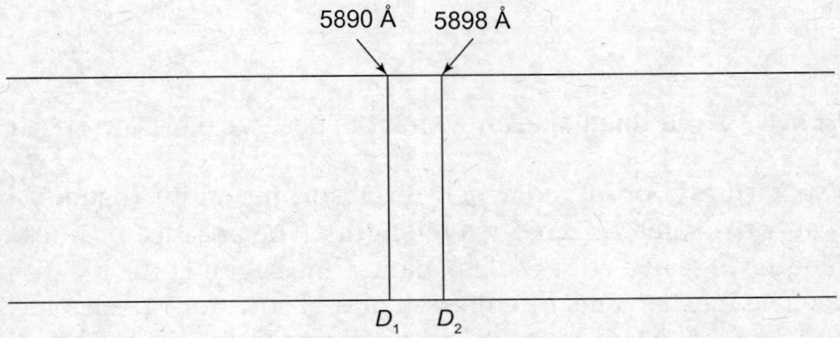

FIGURE 30.8 D_1 D_2 lines of sodium spectrum

Since there is a very small difference of wavelength (only 8Å) between D_1 and D_2 lines, therefore both appear as yellow lines. Some other spectra of elements are shown in Fig. 30.9.

FIGURE 30.9 Sample spectra of some elements

The spectra from atoms of elements can broadly be divided into the following two types:

1. Emission spectrum
2. Absorption spectrum

The emission spectrum is the one when the light originating from a source is seen by a spectrometer. For example, if sodium metal is put into flame, D_1 and D_2 lines are observed in the spectrum.

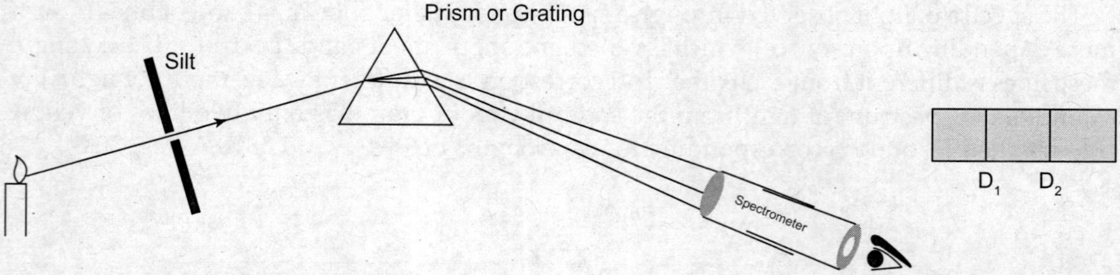

FIGURE 30.10 Experimental arrangement to observe emission spectrum

Absorption spectrum is observed when an absorbing atom (usually in the form of vapour or gas) absorbs some selected wavelengths. The absence of lines corresponding to these wavelengths is seen as dark/black band/lines against the bright part of the rest of the spectrum. The mechanisms of emission and absorption spectra have been shown in Figs. 30.10 and 30.11. The emission and absorption spectra are further divided into the following types:

1. Continuous spectra
2. Line spectra
3. Band spectra

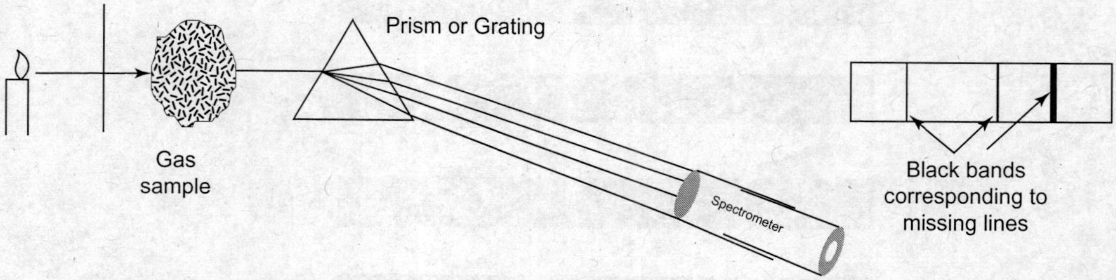

FIGURE 30.11 Experimental arrangement to observe absorption spectrum of a gas

30.3.1 Hydrogen Spectra

The spectrum of hydrogen is observed by using a gas discharge tube with hydrogen gas filled in at low pressure, similar to observation of spectra of chemical salts in flame tests (Fig. 30.12).

The rest of arrangement is similar to other experiments in which the emitted radiation is passed through a narrow slit and prism to observe the spectrum on screen in a dark room or with a spectrometer. The spectral lines in hydrogen spectra show a systematic arrangement. Five bands of lines are observed which extend from *IR* to *UV* region. These bands are shown below and the frequency ranges of each band are shown in Fig. 30.13.

In an attempt to explain the hydrogen spectrum, J.J. Balmer in 1885 found that the wavelengths of four visible lines can be expressed as

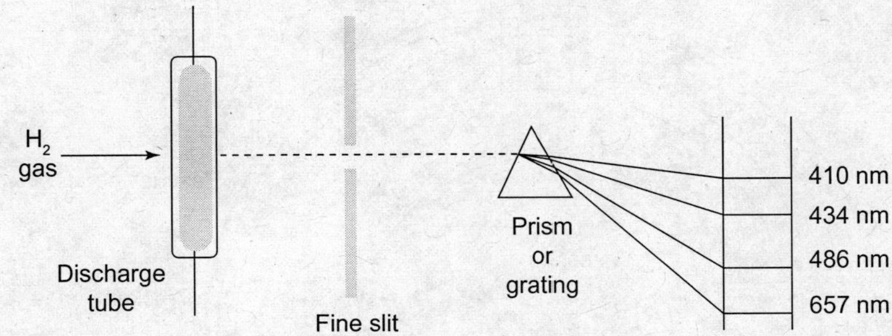

FIGURE 30.12 Typical arrangement to observe emission spectra of H_2 gas

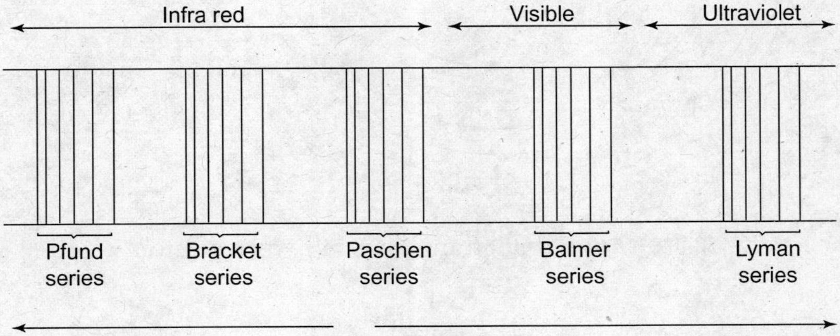

FIGURE 30.13 Observed spectral series of hydrogen (named after scientists)

$$\boxed{\frac{1}{\lambda} = R\left(\frac{1}{2^2} - \frac{1}{n^2}\right)} \tag{30.4}$$

where $n = 3, 4, 5, 6$ and R is a constant. The value of constant R from experiments has been found to be $1.097 \times 10^7 \text{ m}^{-1}$. This constant is known as Rydberg constant. Further experiments showed that Eq. (30.4) can be written for Lyman series in UV region as

$$\frac{1}{\lambda} = R\left(\frac{1}{1^2} - \frac{1}{n^2}\right) \quad n = 2, 3, 4, \ldots$$

For Paschen series, it can be expressed as

$$\frac{1}{\lambda} = R\left(\frac{1}{3^2} - \frac{1}{n^2}\right) \quad n = 4, 5, 6, \ldots$$

Similar expressions can be written for Brackett and Pfund series and hence it can be generalized as

$$\frac{1}{\lambda} = R\left(\frac{1}{n_1^2} - \frac{1}{n_2^2}\right) \quad n_1 = 1, 2, 3, \ldots$$

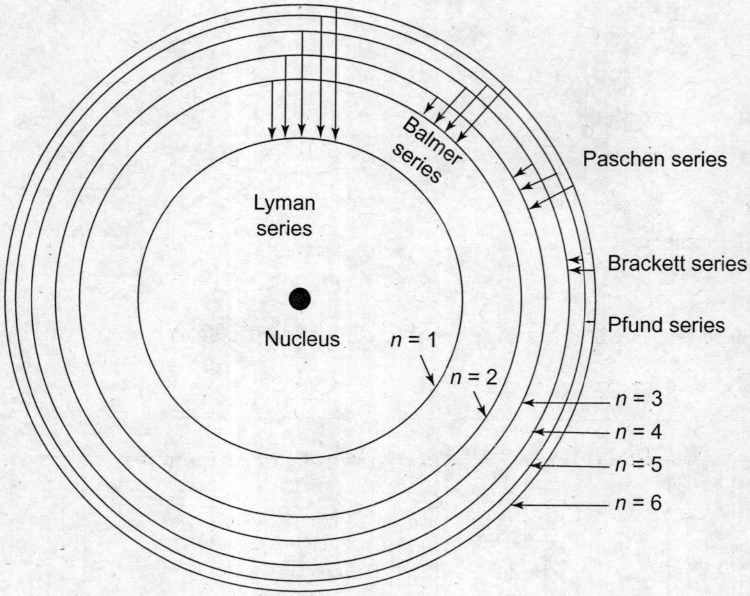

FIGURE 30.14 Orbital transitions in hydrogen atom

$$n_2 = (n_1 + 1)$$

$\frac{1}{\lambda}$ is also expressed as $\bar{\nu}$ called wave number and the expression becomes

$$\boxed{\bar{\nu} = R\left(\frac{1}{n_1^2} - \frac{1}{n_2^2}\right)} \tag{30.5}$$

This relation explains successfully the origin of line spectrum of hydrogen atom. The series of lines are produced by the transition of electrons from higher energy orbits to lower energy orbits identified by n_2 and n_1 respectively. The orbit and energy level diagram for hydrogen atom are given in Fig. 30.14.

30.4 ▶ Bohr's Model of Hydrogen Atom

Even after the failure of Rutherford's model, many scientists were busy in formulating new models or tried to explain the failures of Rutherford model. Niels Bohr, a Danish physicist, in 1913, proposed a model which had some features of Rutherford model and also incorporated some features of Planck's theory of radiation from black bodies. The important postulates proposed by Bohr in his theory are as follows:

1. Electrons revolve around nucleus only in specific or fixed orbits. (Electrons cannot revolve in any arbitrary orbits around the nucleus to balance the centripetal force). These fixed orbits are called stationary orbits or states.

2. While electrons can stay in any of these stationary orbit, they do not radiate any energy and hence do not jump spirally into the nucleus.
3. The energy possessed by an electron in the nth orbit can be given by

$$\boxed{E_n = -R\left(\frac{1}{n^2}\right)} \tag{30.6}$$

where E_n is the energy of electron in nth orbit.

R is the Rydberg constant in energy form is equal to 2.18×10^{-18} J.

The number n can take the integer values only and is called principle quantum number. The energies E_n for all values of n are negative. The energy state $E_1 = R$ is called ground state or the state of lowest energy and hence most stable.

4. The energy is emitted or absorbed by an atom only when an electron jumps from one energy level to the other. The difference of energy of two orbits is either emitted or absorbed. The energy level diagram of hydrogen atom with Bohr's model is shown in Fig. 30.15.

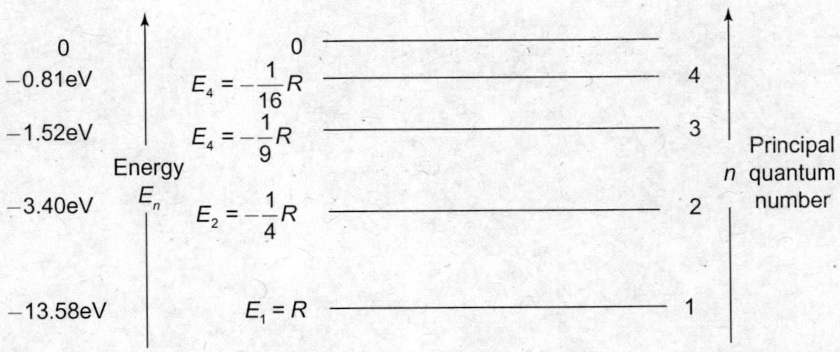

FIGURE 30.15 Energy level diagram of Bohr's model for hydrogen atom (not to the scale)

5. The emission of energy is quantized and not continuous. The energy quanta have magnitude given by

$$\Delta E = E_{n'} - E_n = h\nu \tag{30.7}$$

where h is the energy of an orbit in higher energy than $E_{n'}$, which is less than n'

$$\boxed{\Delta E = R\left(\frac{1}{n^2} - \frac{1}{n'^2}\right) = h\nu} \tag{30.8}$$

If ΔE is positive, then a photon – a packet of energy – will be absorbed. The energy of photon is used to raise the energy level of electron from lower to higher one.

If ΔE is negative, then the difference in energy is emitted as a single photon because of quantization (Fig. 30.16).

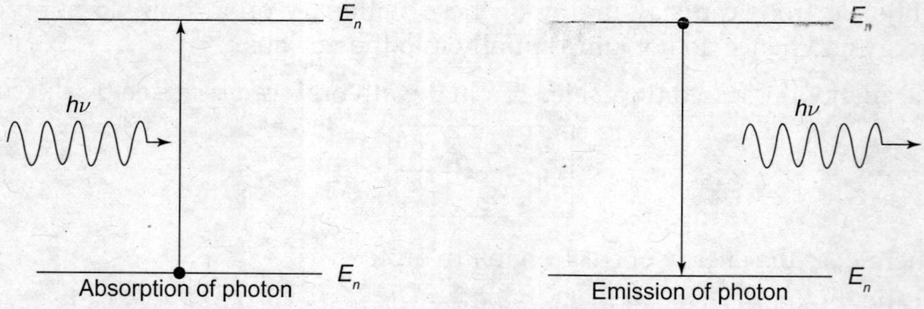

FIGURE 30.16 Absorption and emission of photons in atomic levels

The first postulate of Bohr's theory can be expressed in mathematical form. Classical mechanics shows that the linear momentum of an electron can be expressed in angular momentum as

$$p = mv2\pi r$$

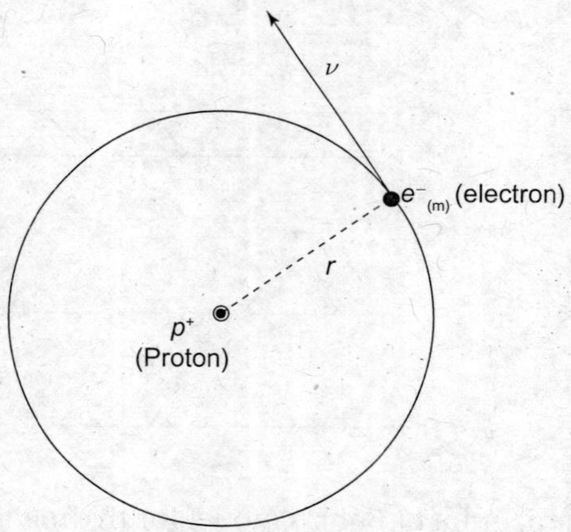

FIGURE 30.17 Electron in circular orbit as assumed by Bohr

For a circular orbit, Bohr assumed it to be quantized, that is, it can have only integer multiple values of h, that is, Planck's constant (Fig. 30.17).

$$\therefore \qquad 2\pi mvr = nh \qquad (30.9)$$

where $n = 1, 2, 3, 4, \ldots\infty$.

The energy of an electron E can be written as

$$E = E_{PE} = E_{KE} \qquad (30.10)$$

where E_{PE} is the potential energy of electron which is mainly due to the electrostatic force between the nucleus (proton in case of hydrogen atom) and the electron in the orbit. This component tends to be zero if electron is removed away from the electrostatic range of nucleus, that is, if atom is ionized. The potential energy component can be expressed as

$$E_{PE} = -\int_r^{\infty} \frac{Ze^2}{r^2} dr = -\frac{Ze^2}{r} \tag{30.11}$$

The kinetic energy component is due to the motion of electron with velocity v and therefore can be expressed as

$$E_{KE} = \frac{1}{2} mv^2$$

The total energy E can therefore be expressed as

$$\boxed{\boldsymbol{E = -\frac{Ze^2}{r} = \frac{1}{2} mv^2}} \tag{30.12}$$

For a stationary orbit, the centripetal force must be balanced by the electrostatic force of attraction, that is,

$$\frac{mv^2}{r} = \frac{Ze \cdot e}{r^2}$$

$$\therefore \quad E = -mv^2 = \frac{1}{2} mv^2 = -\frac{1}{2} mv^2$$

From the above equations, we obtain

$$v = \frac{2\pi Ze^2}{nh} \tag{30.13}$$

$$\therefore \quad E = -\frac{1}{2} m \left(\frac{2\pi Ze^2}{nh} \right)^2 = -\frac{2\pi^2 Z^2 e^4 m}{n^2 h^2} \tag{30.14}$$

$$\therefore \quad hv = E_{n'} - E_n = \frac{2\pi^2 Z^2 e^4 m}{h^2} \left(\frac{1}{n^2} - \frac{1}{n'^2} \right)$$

$$\because \quad v = \frac{c}{\lambda} \tag{30.15}$$

$$\therefore \quad \frac{1}{\lambda} = \frac{2\pi^2 Z^2 e^4 m}{ch^3} \left(\frac{1}{n'^2} - \frac{1}{n^2} \right)$$

$$\frac{1}{\lambda} = R \left(\frac{1}{n'^2} - \frac{1}{n^2} \right) \tag{30.16}$$

$$\therefore \quad \boxed{\boldsymbol{R = \frac{2\pi^2 Z^2 e^4 m}{ch^3}} \text{ **is the Rydberg constant**}} \tag{30.17}$$

For hydrogen atom, $Z = 1$ and all other values are known, then the value of R comes out to be $1.099 \times 10^7\ \text{m}^{-1}$ which collates with the experimentally determined value of $1.097 \times 10^7\ \text{m}^{-1}$ from the Balmer's formula, when applied to the hydrogen spectra.

Note

After the failure of Rutherford model, Niels Bohr, a Danish physicist in 1913, proposed a model which had some features of Rutherford model and also incorporated some features of Planck's theory of radiation from black bodies and quantized orbits of electrons.

30.4.1 Size of Atom

From Eq. (30.17), the radius of the nth orbit can be calculated as

$$r = \frac{Ze^2}{mv^2} = \frac{Ze^2}{m}\left(\frac{2\pi Ze^2}{nh}\right)^2$$
$$= \frac{n^2h^2}{4\pi^2 mZe^2}$$

It indicates that $r \propto n^2$ as all other terms are constant for an atom. For $n = 1$ (i.e., first orbit) of hydrogen atom, $Z = 1$. Then the radius of the orbit is

$$r = 0.528 \times 10^{-8}\ \text{cm} = 0.528\ \text{Å}$$

EXAMPLE 30.1

Calculate the radius of an orbit corresponding to $n = 3$ (r_3) for a hydrogen atom.

Solution:

$$r_3 = \frac{h^2}{4\pi^2 me^2}(3)^2$$

For

$$e = 9.11 \times 10^{-19}\,\text{C}$$
$$m = 9.1 \times 10^{-31}\,\text{kg}$$
$$h = 6.62 \times 10^{-34}\,\text{J.sec}$$
$$r_3 = 0.5299 \times 9 \times 10^{-8}$$
$$r_3 = 4.77 \times 10^{-1}\,\text{m} = 4.77\ \text{Å}$$

Time period of electron in an orbit

The time period of an electron in Bohr's orbits can be calculated from the relation

$$\boxed{T = 2\pi \frac{r_n}{v_n}} \tag{30.18}$$

Fine structure constant

From Eq. (30.13), the velocity of electron in the *n*th orbit is given by

$$v = \frac{2\pi Ze^2}{nh} = \frac{2\pi Ze^2}{ch} \cdot \frac{c}{n}$$

$$= \alpha \frac{c}{hn}$$

where $\boxed{\alpha = \frac{2\pi Ze^2}{ch} \text{ or } \frac{2\pi e^2}{ch} = \frac{1}{137}}$ is the fine structure constant. (30.19)

Bohr's radius can also be written in terms of α as

$$a_0 = \frac{\hbar}{\alpha mc} = 5.28 \times 10^{-11} em \tag{30.20}$$

where $$\hbar = \frac{h}{2\pi}$$

The binding energy can be written as

$$E = -\frac{U}{2} = -\alpha \frac{hc}{2a_0}$$

$$\boxed{= -\alpha^2 \frac{mc^2}{2} = -13.58 \text{ eV}} \tag{30.21}$$

The binding energy of an electron with proton is therefore directly proportional to the electron rest-energy $\frac{mc^2}{2}$.

30.4.2 Extension of Bohr's Model to Other Atoms

Though Bohr's model is specifically applicable to hydrogen atoms, it can be successfully used to explain the spectra of hydrogen like atoms, i.e. the atoms with single electron in orbit but with Z number of protons in the nucleus. Such a condition is satisfied by ionized helium ($Z = 2$), doubly ionized lithium ($Z = 3$) and some other elements ionized in a similar way such that only one electron is left. The observed spectra of such ionized atoms approximately match with the theoretically calculated wavelength using the relation

$$\frac{1}{\lambda} = \frac{2\pi^2 Z^2 e^4 m}{ch^3}\left(\frac{1}{n'^2} - \frac{1}{n^2}\right) \tag{30.22}$$

The failure to explain the spectra of heavier atoms is due to complexity of Coulomb interactions between electrons and protons.

30.4.3 Limitations of Bohr's Model

Bohr's model of atom, in fact, combined Bohr's postulates with Rutherford's nuclear model. Despite explaining spectral of hydrogen atom, this model had some drawbacks such as

1. It cannot explain the spectra of heavier elements.
2. Highly resolved spectra of hydrogen and hydrogen like atoms show a fine structure (i.e., subgroup of spectral lines) that cannot be explained by Bohr's model.
3. In the presence of strong magnetic field, the spectral lines of atoms show their fine structure. This phenomenon is known as the Zeeman Effect, which cannot be explained by Bohr's model.
4. Bohr's model presumes electrons to be particles revolving in stationary circular orbits. This is contrary to de Broglie's dual nature of electrons.
5. Further, the Heisenberg's uncertainty principle raised questions on simultaneous measurement of position and momentum (or velocity). Therefore, the well-defined stationary circular orbits cannot exist.

Due to the above reasons, Bohr's model combines both the particle and the wave nature of electrons. This concept was further elaborated by de Broglie, Heisenberg, Pauli, Schrödinger and Sommerfeld. The result of combining detailed mathematical formulation from their theories led to the concept of quantum numbers and quantum mechanics.

30.5 ▶ Quantum Numbers

Quantum numbers are the numbers which describe the state of an electron in an atom. There are four quantum numbers as given below:

1. **Principal Quantum Number (n):** The principle quantum number 'n' corresponds to the Bohr's orbit and can assume integer values like 1, 2, 3, ..., n. The shells are denoted as K ($n = 1$), L ($n = 2$), M ($n = 3$) and so on.
2. **Orbital Quantum Number (l):** The angular momentum of an electron, L, is given as

$$L = \sqrt{l(l+1)}\frac{h}{2\pi} = \sqrt{l(l+1)}\hbar \tag{30.23}$$

where $l = 0, 1, 2, ..., n - 1$ is called orbital quantum number. It can have total n values from 0 to $(n - 1)$.

3. **Magnetic Quantum Number (m_l):** The z-component of angular momentum can be expressed as

$$L_z = m_l \frac{h}{2\pi} = m_l \hbar$$

where $m_l = -l, -(l-1), \ldots 0, -1 \ldots (l-1)$ or 0 to $\pm\, l$.

4. **Spin Quantum Number (m_s):** The intrinsic angular momentum of an electron is called electron's spin and is expressed as

$$S = \sqrt{\frac{1}{2}\left(\frac{1}{2}+1\right)}\frac{h}{2\pi} \tag{30.24}$$

The z-component of spin of an electron is

$$s_z = m_s \frac{h}{2\pi},$$

where $m_s = +\frac{1}{2}$ and $-\frac{1}{2}$.

The only two possible values of m_s are $+\frac{1}{2}$ and $-\frac{1}{2}$ which are two spin states of an electron.

The four quantum numbers (n, l, m_l, m_s) must be defined to properly identify an electron in an atom. For three principal quantum numbers $n = 1, 2$ and 3, the other quantum numbers are given in Fig. 30.18.

	Principal quantum number n	Orbital quantum number l	Orbital symbol	Magnetic quantum number m_l	Spin quantum number m_s	Number of electrons
I orbit (*K*-shell)	1	0	*s*	0	+½, −½	2 electrons
II orbit (*L*-shell)	2	0	*s*	0	+½	8 electrons
	2	0		0	−½	
	2	1	*p*	−1	+½	
	2	1		−1	−½	
	2	1		0	+½	
	2	1		0	−½	
	2	1		1	+½	
	2	1		1	−½	
III orbit (*M*-shell)	3	2	*d*	−2	+½	18 electrons
	3	2		−2	−½	
	3	2		−1	−½	
	3	2		−1	+½	
	3	2		0	−½	
	3	2		0	+½	
	3	2		+1	−½	
	3	2		+1	+½	
	3	2		+2	−½	
	3	2		+2	+½	

FIGURE 30.18

30.6 ▶ Zeeman Effect

A magneto-optical phenomenon was discovered by Zeeman in 1896. In his experiment, he observed the splitting of spectral lines into components when the source of light was placed in a strong magnetic field. When splitting is up to three lines, the phenomenon is called normal Zeeman effect. The normal Zeeman effect can be quantitatively explained by classical theory of Lorentz. In some cases, the splitting of lines into more than three lines occur which is called anomalous Zeeman effect. This cannot be explained by the classical theory of magnetism.

30.6.1 Experimental Evidence for Quantum Numbers

The experimental observations like the Zeeman Effect in 1920 and the Stern-Gerlach experiment in 1922 provided indirect support to the concept of angular momentum and spin of electrons. The theoretical concept was evolved later in order to explain the experimentally observed fine structure of the spectra. When a strong magnetic field is applied across the hydrogen atoms emitting radiations, then it is observed that few transitions of hydrogen spectra split into fine structural lines.

The splitting of lines (or fine structure) can be explained on the basis of interaction of external magnetic field and the magnetic dipole moment of electrons in various orbital states. The electron transitions shown in Fig. 30.19 are governed by a set of rules for allowed and forbidden transitions. Such rules are called Selection rules.

30.6.2 Selection Rules

The energy level diagrams showing electron transitions from one level to the other are used to describe the structure and configuration of electrons in atoms. Such diagrams are also known as Grotrian diagram or Term diagram. But the electron transitions among various energy levels are not always allowed. These are governed by a set of rules which are known as selection rules. There are rules that govern the change in the values of four quantum numbers n, l, j and s.

For radiative transitions to take place, the selection rules are shown in Fig. 30.19.

Change in quantum number	Change allowed by
Δn	1, 2, 3, ...
Δl	± 1
Δm_L	± 1
Δm_s	0

FIGURE 30.19

In order to supplement these rules, some additional rules are formulated which predict the intensity of emitted lines. These rules are as follows:

1. The lines for which the values of l and m_l change by same amount are most intense. But these are faint if the changes in l and m_s are different.
2. The transitions with opposite changes in the values of l and m_l are not allowed

   i.e., if $\Delta l = +1$ and $\Delta m_l = -1$

   or $\Delta l = -1$ and $\Delta m_l = +1$

   then no lines are emitted.

30.6.3 Spectroscopic Notation of Orbitals

The orbits or shells corresponding to principal quantum numbers (n = 1, 2, 3, …) are designated as K, L, M, N, … shells. The orbitals corresponding to orbital quantum number $l = 0, 1, 2, \ldots (n - 1)$ are represented as s (for sharp), p (for principal), d (for diffuse) and f (for fundamental). For values of l greater than 3, the names are given as g, h, i, … in alphabetical order.

30.7 ▶ Electron Spin

Spin is an intrinsic property of an electron which has its experimental evidence in the fine structure of spectra, similar to quantized angular momentum. The spin S is given as

$$S = \sqrt{\frac{1}{2}\left(\frac{1}{2}+1\right)}\hbar = \frac{\sqrt{3}}{2}\hbar \tag{30.25}$$

The fine structure corresponds to the two different values of

$$S_z = \pm\frac{1}{2}\hbar \tag{30.26}$$

The fine structure splitting of lines (Fig. 30.20) is due to the magnetic moment of electron μ_S given as

$$\mu_S = -\frac{e}{2m}gS \tag{30.27}$$

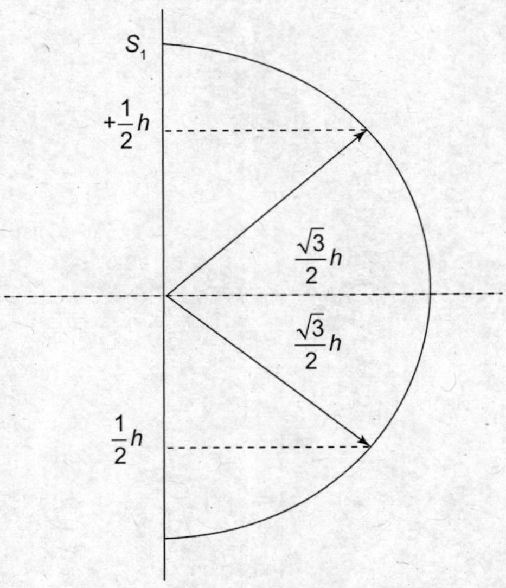

FIGURE 30.20 Splitting of electron spin into quantized states

Here, g is called gyromagnetic ratio and its value is ≈ 2.0023 for electron spin and $g = 1$ for orbital angular momentum. It can be explained on the basis that the electron charge does not flow only around the perimeter of an orbit.

A unit of magnetic moment, μ_B (Bohr's magneton) is defined as

$$\mu_B = \frac{h}{2\pi} \cdot \frac{e}{2m}$$

$$= 9.2732 \times 10^{-24}\ \mathrm{JT}^{-1}$$

$$\mu_B = 5.788 \times 10^{-5}\ \mathrm{eV/T}.$$

30.8 ▶ PAULI'S EXCLUSION PRINCIPLE

Pauli's exclusion principle states that for any two electrons in an atom, all quantum numbers cannot be the same. Out of four quantum numbers n, l, m_l and m_s at least one must be different. In other words, all electrons in an atom lie in different quantum (energy) states. If first three quantum numbers n, l and m_1 are same, then the spin quantum number m_s must be different, that is $+\frac{1}{2}$ for one electron and $-\frac{1}{2}$ for other electron. This principle is similar to the classical concept that two particles cannot occupy the same co-ordinates in space-time. The electrons in an energy state are represented with small arrows, upward (↑) or downward (↓) to indicate spin $+\frac{1}{2}$ or $-\frac{1}{2}$, respectively.

The typical electronic configurations for hydrogen, helium, lithium, beryllium and boron are shown in Fig. 30.21.

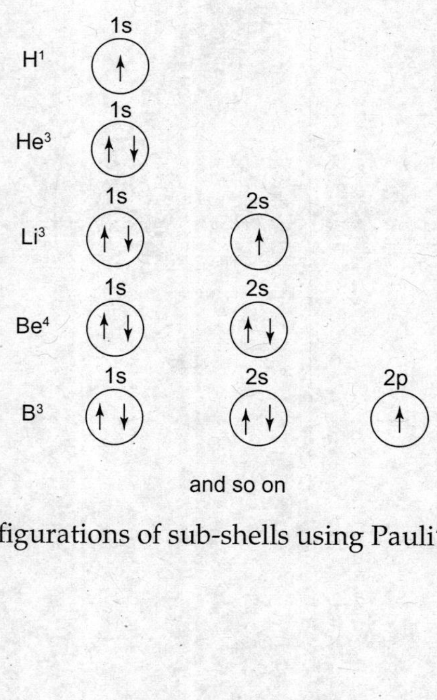

FIGURE 30.21 Spin configurations of sub-shells using Pauli's exclusion principle

30.9 ▶ Aufbau Principle

The available electron energy levels of an atom are filled in the increasing order of sub-shell energy. An atom has its minimum energy in normal or ground state. The Aufbau principle is a rule which determines the order in which electrons are added to different orbitals or sub-shells. According to Aufbau principle, the orbital with lower energy is filled first irrespective of shell to which it belongs.

Hund's rule

According to this rule, electron pairing shall not take place until all vacant sub-shells have at least one electron.

Thus, by taking into account Pauli's exclusion principle, Aufbau principle and Hund's rule, the order of filling vacant sub-shells can be shown in Fig. 30.22.

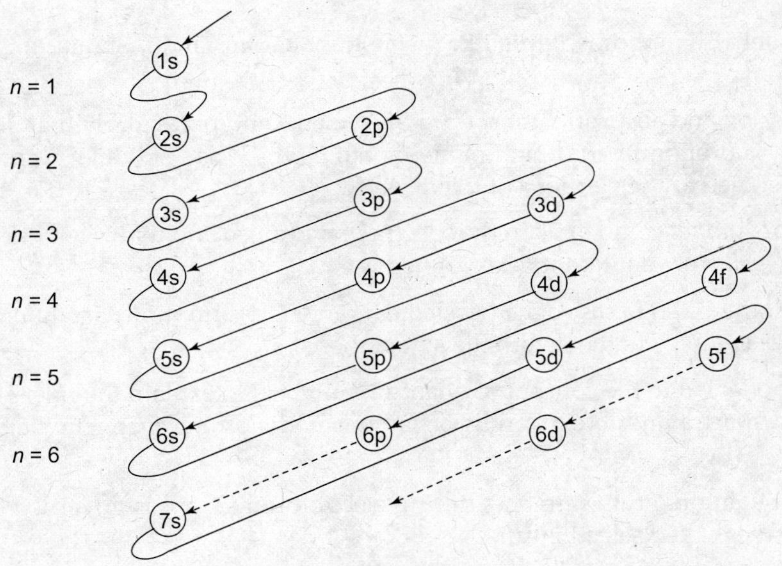

FIGURE 30.22 Use of Hund's rule and Aufbau principle for filling sub-shells

The typical application of the above principles can be seen in cases of C^6, N^7 and O^8. The electronic configurations of these elements are as follows:

C^6	$1s^2$	$2s^2$	$2p_x^1$	$2p_y^1$	$2p_z^0$
N^7	$1s^2$	$2s^2$	$2p_x^1$	$2p_y^1$	$2p_z^1$
O^8	$1s^2$	$2s^2$	$2p_x^2$	$2p_y^1$	$2p_z^1$

In case of C^6, each electron acquires $2p_x$ and $2p_y$ and not that two electrons pair in $2p_x$. This is in conformity with Aufbau principle and Hund's rule. In case of N^7, one electron each acquires p-shell. In case of O^8, the last electron is added to $2p_x$ sub-shell because all sub-shells of p-shell are having one electron each.

Recap Zone

Points to Remember

- Dalton's plum pudding model of atom was not sustainable because it was unable to explain the spectral properties of atoms.
- Rutherford in 1911 suggested nuclear model of atom through his α-scattering experiment.
- This model also had drawbacks like instability of atoms and inability to explain line spectra.
- The emission and absorption of spectra of elements and particularly that of hydrogen led Bohr to propound an orbital model of atom which had stationary, non-radiating orbits assigned to the elections of fixed energy.
- The Bohr's model could explain the spectra of hydrogen and hydrogen like atoms but their fine structure could not be explained.
- Bohr's model failed in case of heavy elements and its assumption of stationary circular orbits was contrary to the de Broglie's hypothesis.
- Bohr's model led to the concept of quantum numbers to explain the finer structure in observed spectra and to explain its behaviour under strong magnetic field (Zeeman effect).
- The change in quantum numbers during electron transitions is governed by a set of defined rules called selection rules.
- The concept of electron spin was experimentally proved by Stern-Gerlach.
- The Pauli's exclusion principle, Aufbau principle and Hund's rule provide guidelines for the order in which the sub-shells of shells (orbits) are filled.

Key Terms with Definitions

- Atom is the smallest invisible component of the matter that cannot be further divided mechanically.
- According to Thomson's model of the atom, the atom consists of a sphere of radius $\sim 10^{-10}$ m full of positive and negative particles in such a way that atom is electrically neutral.

- Rutherford experiment established that the size of atom should be of the order of N 10^{-10} m and the size of nucleus N 10^{-15} m.
- According to Rutherford theory, infinite numbers of orbits and energy levels are possible.
- This model could not explain discrete line spectra and electrons are expected to lose their energy and merge into nucleus.

IMPORTANT FORMULAE AND EQUATIONS

Equation Number	Equation	Remarks
(30.3)	$\theta = \tan^{-1}\left(\frac{2b}{p}\right)$	Rutherford scattering impact factor
(30.4)	$\frac{1}{R}\lambda\left(\frac{1}{n_1^2} - \frac{1}{n_2^2}\right)$	Rydberg constant $n_1 = 1, 2, 3, \ldots, n_2 = 2, 3, 4, \ldots$
(30.5)	$\bar{\nu} = R\left(\frac{1}{n_1^2} - \frac{1}{n_2^2}\right)$	Wave number
(30.6)	$E_n = -R\left(\frac{1}{n^2}\right)$	Bohr's energy E_n – Energy of electron in nth orbit R – Rydberg constant equal to 2.18×10^{-18} J
(30.8)	$\Delta E = E_{n'} - E_n = hv$ $= R\left(\frac{1}{n^2} - \frac{1}{n'^2}\right)$	Energy of quanta
(30.12)	$E = \frac{Ze^2}{r} = \frac{1}{2}mv^2$	Energy is Bohr's orbit
(30.17)	$R = \frac{2\pi^2 Z^2 e^4 m}{ch^3}$	Rydberg constant
(30.18)	$T = 2\pi \frac{r_n}{v_n}$	Period of electron in Bohr's orbit
(30.19)	$\alpha = \frac{2\pi Ze^2}{ch} = \frac{1}{137}$	Fine structure constant

(30.21) $E = \frac{u}{2} = -\alpha \frac{hc}{2a_o}$ **Binding energy**

$= -13.58$ eV

REVIEW ZONE

SHORT ANSWER QUESTIONS

1. What are the features of Rutherford model of an atom?
2. What are major drawbacks of Rutherford model of an atom?
3. Discuss the postulates of Bohr's model of an atom.
4. Bohr's model of an atom is also not a complete model. Elaborate with discussion.
5. How spectral lines of hydrogen can be explained by Bohr's theory? Draw energy level diagrams for spectral lines.
6. How is the spin of electron coupled with the orbital motion of electron? Discuss with the vector model of an atom.
7. What is Zeeman effect? How can it be used to determine the $\frac{e}{m}$ ratio experimentally?
8. What is Pauli's exclusion principle? How is it used to explain the electron configuration of atoms?

NUMERICAL PROBLEMS

1. Calculate the energy needed to excite an electron in hydrogen atom from ground state to the second excited state.
2. An α-particle of 6 MeV energy approaches a nucleus of gold ($Z = 79$) atom with an impact parameter of 2.6×10^{-3} m. Calculate the angle of scattering ($e = 1.6 \times 10^{-19}$ C, 1 eV $= 1.6 \times 10^{-19}$ J).
3. Calculate the distance of closest approach of α-particles to the nucleus when α-particles of 5 MeV are scattered back by a thin sheet of copper ($Z = 29$).
4. Calculate the speed of electron in nth orbit of hydrogen atom. ($e = 1.6 \times 10^{-19}$ C, $h = 6.63 \times 10^{-34}$ J.s)

   $$c = 3 \times 10^8 \text{ m/s}$$

   and $$\frac{1}{4\pi \epsilon_0} = 9 \times 10^9 \text{ Nm}^2/\text{C}^2$$
5. Calculate the wavelength of first line of Balmer series if the Rydberg constant is 1.097×10^7 m^{-1}.

6. The ground state energy of an electron in hydrogen atom is -13.6 eV. Calculate the energy of $n = 3$ state in He^+ and $n = 2$ in Li^{++}.
7. Calculate the wavelength of light emitted by an atom excited to next energy level by 2 eV.
8. Calculate the radius of the electron orbit for (a) hydrogen atom is ground state, (b) He^+ in ground state and (c) Li^{++} in ground state.

CHAPTER 31 X-ray Physics

Learning Objectives

By the end of this chapter, the student will be able:

- To recall the discovery and the properties of X-rays
- To recognize X-rays as a part of electromagnetic spectra
- To identify the types of X-ray spectra
- To discuss the origin of continuous and characteristic spectra
- To understand nomenclature of X-ray lines
- To describe the Moseley's law

31.1 ▶ Historical Background

Wilhelm von Konrad Röentgen in 1895, while studying the green fluorescence in rarefied Crooke's tube, accidently discovered some invisible rays that were spoiling the photographic plates placed nearby. He found these rays to be of higher penetrating power than the visible radiations and emanating from anode of Crooke's tube. These rays were named by him as X-rays (due to unknown nature). The Crooke's tube was later modified as Coolidge X-ray tube (Fig. 31.1) with controlled intensity and hardness that are now commonly used for diagnostic purposes in all X-ray clinics.

The significant properties of X-rays are listed below:

1. They can penetrate deep through solid matter, but the depth of penetration varies from material to material.
2. They ionize the atoms of medium with which they interact.
3. They cause certain chemical compounds to fluorescence.
4. They affect photographic film.
5. They travel in straight lines.
6. These rays are invisible to human eye.
7. These rays are electromagnetic waves of wavelength shorter than ultraviolet light.

8. These rays can be refracted, diffracted and polarized with suitable materials only.
9. They heat the substance on which they fall.

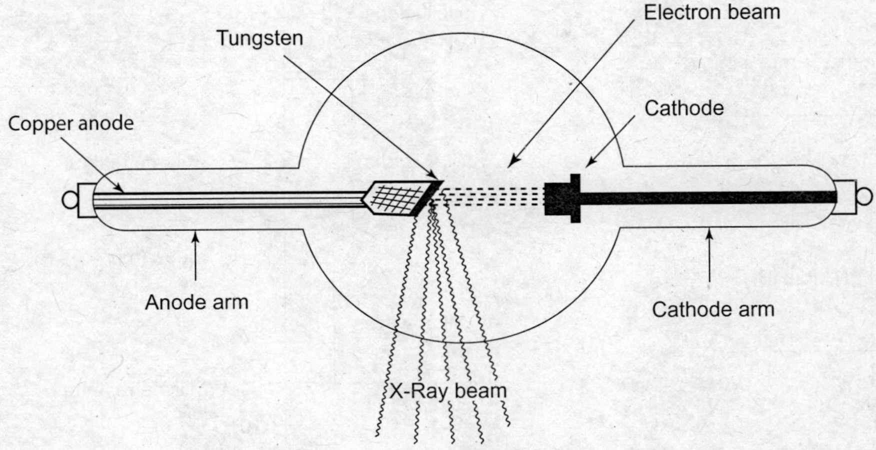

FIGURE 31.1 Schematics of the Coolidge X-ray tube

31.2 ▶ X-RAYS AS A PART OF ELECTROMAGNETIC WAVE SPECTRUM

Electromagnetic wave spectrum extends from radio waves to gamma rays. X-rays too are a part of this spectrum. The complete electromagnetic waves spectrum is shown in Fig. 31.2. X-ray wavelengths (λ) in electromagnetic waves spectrum are conventionally taken from 10^{-3} to 100 Å, but these limits are not sharply defined. There is always an overlap with UV region towards longer wavelength side and with gamma ray on smaller wavelength side. They are called hard X-rays if λ varies from 0.01 to 1 Å, soft X-rays in the range 1 to 10 Å and ultrasoft if λ is greater than 10 Å. X-rays of wavelength greater than 10 Å are called ultraviolet X-rays. Basically, X-rays are distinguished from other e.m. waves by the fact that these waves are produced as a result of bremsstrahlung phenomenon in the electrostatic field of nucleus of an atom or through atom inner shells transitions.

31.3 ▶ X-RAY SPECTRA

An X-ray spectrum is a plot of intensity of X-rays versus wavelength, energy or frequency. The observed intensity variation with respect to energy of X-rays from an X-ray tube is found to be of mainly two types of spectra:

1. Continuous X-ray spectrum
2. Characteristic X-ray spectrum

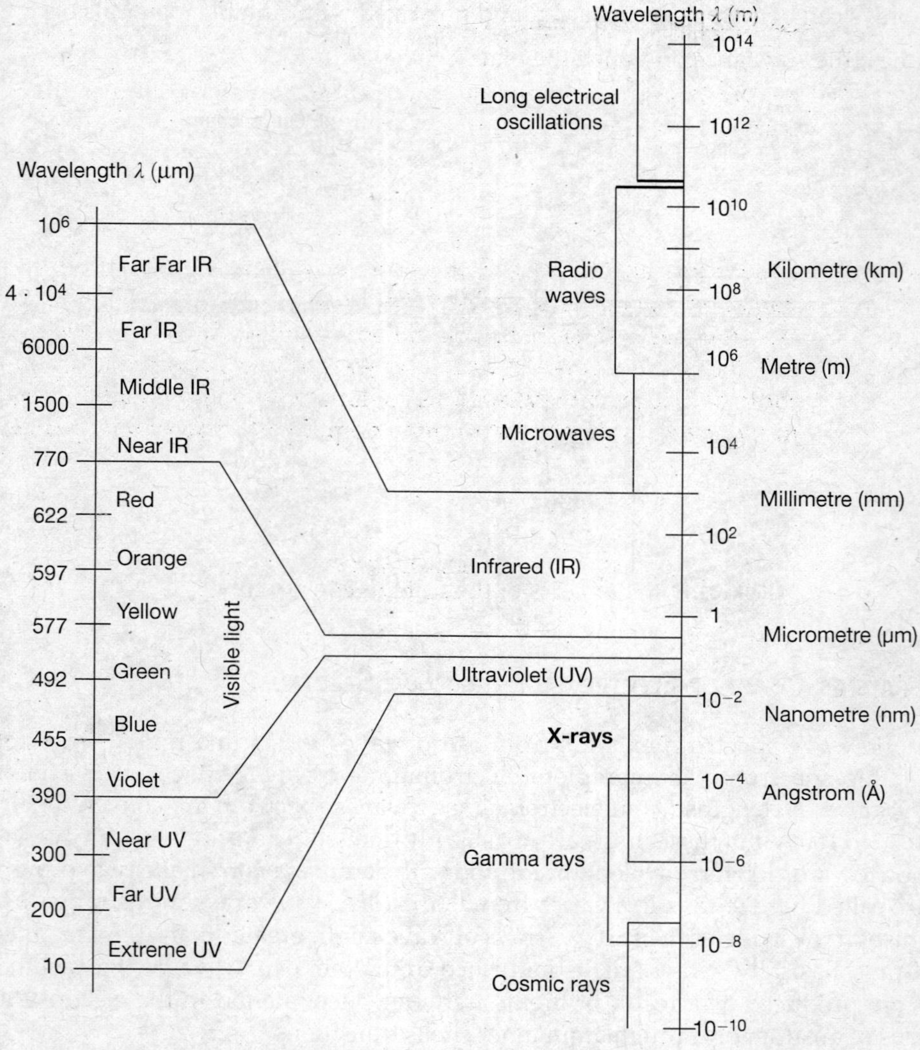

FIGURE 31.2 X-rays in electromagnetic wave spectrum

31.3.1 Continuous Spectrum

Continuous spectrum of X-rays consists of radiations of all possible X-ray wavelengths within a certain range starting from λ_{min} (i.e., the wavelength corresponding to minimum energy) onwards. The continuous spectrum is also known as general spectrum or white spectrum. A typical continuous spectrum is shown in Fig. 31.3. It is evident that by varying the X-ray tube accelerating potential, λ_{min} shifts towards left, i.e. the X-rays attain more energy; the intensity of X-rays also increases with the voltage.

The continuous X-ray spectrum has intensity maxima at λ_{max} where $\lambda_{max} \cong 1.5\lambda_{min}$.

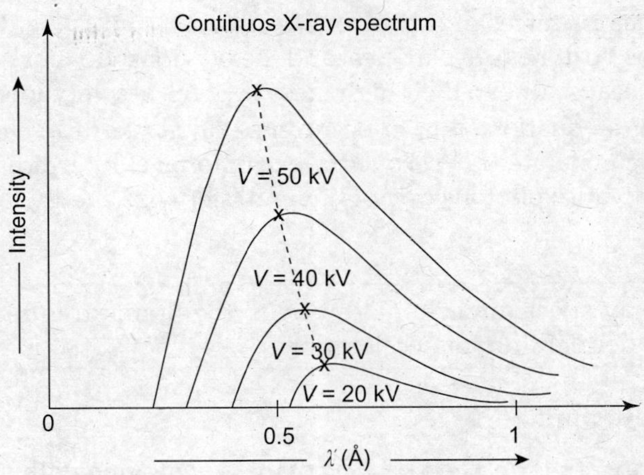

FIGURE 31.3 Continuous spectrum of X-rays

Origin of continuous X-ray spectrum

It was established by Bethe and Heitler in 1934 that when charged particles accelerate or decelerate in a strong electrostatic field they release their energy in the form of radiations. In an X-ray tube, high-speed electrons are suddenly stopped on a target (anode) of high Z atomic number. During this process, electrons pass through the combined field of nucleus and electrons.

Consequently, electrons are decelerated in the field and release their energy in the form of radiation, which lies in the X-ray region. The rate of energy loss according to Bethe-Heitler theory is given as

$$\boxed{-\left.\frac{dE}{dX}\right|_{\text{rad}} = \frac{nz_1^4 z_2^2}{m_0^2} Ef\left(\frac{E}{m_0 c^2}\right)} \tag{31.1}$$

where $-\left.\frac{dE}{dX}\right|_{\text{rad}}$ denotes the rate of loss of energy by particle in the form of radiation,

n – Number of nuclei of the stopping medium per unit volume
z_1 – Charge on the incident particle
z_2 – Atomic number of atoms of the medium
E – Initial total energy of particle

and $f\left(\frac{E}{m_0 c^2}\right)$ is a complex function of E, m_0 and c, the velocity of light, m_0 is the rest mass of the incident particle.

The process of loss of energy by charged particles radiation in a medium is called *Bremsstrahlung* which in German means braking radiation. The nature of radiation depends upon the amount of energy lost in the process, which generally varies from hard X-rays to ultraviolet region.

X-ray tubes are major sources of continuous X-ray spectrum. Other heavy particles such as protons, deutrons, tritones, α-particles and heavy ions do not generate observable continuous X-rays, because due to their high mass they release very low energy radiation as evident from Heitler's equation, due to occurrence of m_0^2 term in the denominator. The continuum cannot be generated by secondary excitation, i.e. by irradiation of matter with high energy X-rays because photons do not undergo stepwise energy loss.

Note

The continuous X-ray spectrum originates due to the Bremsstrahlung, i.e. the electron energy loss during deceleration or acceleration.

Minimum wavelength limit

The total loss of energy of the charged particle of mass m moving with velocity v is given by

$$\Delta E = \frac{1}{2}mv^2 - \frac{1}{2}mv'^2$$

where v' is the final velocity of particle after deceleration. If energy ΔE is converted into electromagnetic wave photon at once, then it will be equivalent to the energy of photon given by $h\nu$.

or $$\Delta E = h\nu$$

or $$\Delta E = \frac{hc}{\lambda}$$

The particle (which is electron in case of X-ray tube) starts from zero velocity and acquires the energy $= eV$ under the applied potential V. If the electron is stopped instantaneously, then whole energy acquired by the electron under potential V is converted to X-ray photon.

i.e., $$\frac{hc}{\lambda_{min}} = \text{eV}$$

or $$\boxed{\lambda_{\mathbf{min}} = \frac{hc}{\mathbf{eV}} = \frac{\mathbf{12396}}{\mathbf{V}}\,\text{Å}} \tag{31.2}$$

where h is the Planck's constant, c is the velocity of light, e is the electronic charge and V is the accelerating voltage in volts.

From Eq. (31.2), it is evident that

$$\lambda_{min} \propto \frac{1}{V} \tag{31.3}$$

Ideally, the energy of an electron is not always lost in continuous fashion but in stepwise series of encounters, giving rise to a number of assorted quanta whose wavelengths are longer than if the energy were lost instantaneously. Consequently, the spectrum from a stream of electrons exhibits a continuous range of wavelengths from λ_{min} upwards. Here, λ_{min} corresponds to the maximum electron energy transferred to an *x*-radiation in a single encounter. Equation (31.3) shows that the minimum wavelength of emitted X-rays is inversely proportional to the voltage applied across the electrodes in order to accelerate electrons. This equation is also called Duane-Hunt law.

31.3.2 Characteristic X-ray Spectrum

Some peaks of grouped wavelengths appear to be superimposed upon the continuous spectrum of X-rays. These spectral lines are characteristic of the material used as target. The group of such lines towards short wavelength region is termed as *K*-series, the next group in the direction of increasing wavelength as *L*-series and *M*-series and so on. The characteristic line spectrum also has some definite relative intensity with respect to other lines. A characteristic spectrum is superimposed upon continuous spectrum as shown in Fig. 31.4.

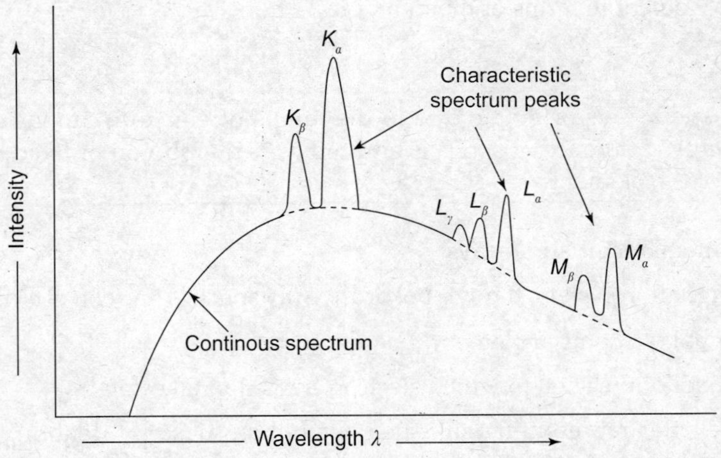

FIGURE 31.4 X-ray spectrum showing continuous and characteristic spectra from an X-ray tube

X-ray spectral lines are grouped in the series of *K*, *L*, *M*, … and so on which arise due to transitions from the higher levels to that particular level.

Origin of characteristic X-rays

The characteristic line spectrum of an element originates when electrons are ejected from the inner shells of its atoms and the electrons from higher levels fall into these vacancies in the inner shells. Each such transition causes energy loss equal to the energy difference of two shells which appears as an X-ray photon.

For example, the vacancy caused in *K*-shell is filled by transition from *L*-shell and a new vacancy in *L*-shell is filled by transition from *M*-shell as shown in Fig. 31.5. These transitions give rise to *K* X-rays and *L* X-rays, respectively, and are almost instantaneous, occurring

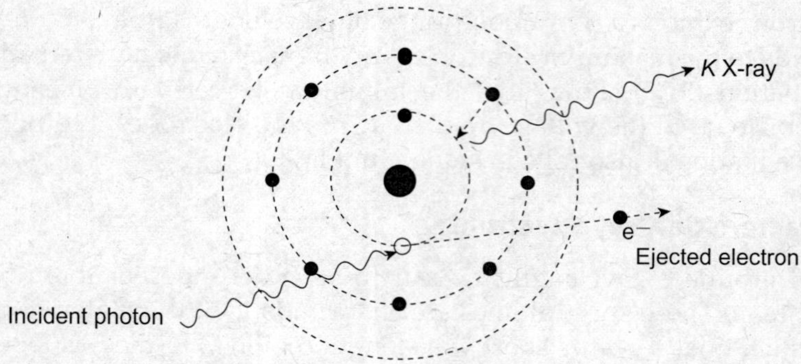

FIGURE 31.5 Creation of vacancy in K-shell by photo-ionization and emission of K X-ray

within 10^{-12} to 10^{-14} seconds of creation of vacancy. The difference of energy of respective shells in atoms of an element is a unique characteristic of that element. Thus, characteristic X-rays label the element in terms of their energies.

Note

The characteristic X-rays originate due to the jumps of electrons among atomic inner-shells. The energy difference of shells is emitted as X-ray photon of fixed energy that is characteristic of the element.

Nomenclature of characteristic X-rays

In Siegbahn designation system, the symbol of an X-ray spectral line consists of the following:

1. The symbol of chemical element,
2. The symbol of the shell to which electron transition takes place,
3. A Greek letter representing sub-shell symbol as subscript denoting the particular line in the series.

The K-shell has only one sub-shell in it, but other shells L, M and N have 3, 5 and 7 sub-shells, respectively. All these sub-shells in a shell do not have same energy. These different sub-shells are designated by a letter denoting shell along with a subscript in roman letters, for example, L_I L_{II} L_{III} for L-shell and M_I M_{II} M_{III} M_{IV} M_V for M-shell. The number of subscript increases with the increase in sub-shell energy, i.e. L_{III} sub-shell has higher energy than L_{II} and L_I. The transition from L-shell to K-shell results in line in K_α series where K_{α_1} line is due to transition from L_{III} to K-shell. Similarly K_{α_2} from L_{II} to K-shell. L_α series is due to transitions from M-shell to L-shell. K_β series results due to transition from M-shell to K-shell and so on. The energy levels with nomenclature are shown in Fig. 31.6.

Selection rules

Electron transitions from higher levels to lower ones do not occur arbitrarily. Rather, these take place with permitted changes in quantum numbers which are called selection rules. The selection rules for inner electron transitions are given as follows:

1. Change in principal quantum number n should not be zero, i.e. for $\Delta n = 0$, the transitions are not allowed. For transitions to take place, $\Delta n \neq 0$, where n can take only integral values.
2. Azimuthal or angular momentum quantum number l should change by unity, i.e. the transitions for which $\Delta n = \pm 1$ are allowed where $l = 0, 1, 2, \dots (n-1)$.
3. Total spin $J = l + s$ should change either by unity or by zero, i.e. for $\Delta J = \pm 1$ or 0 transitions are allowed.

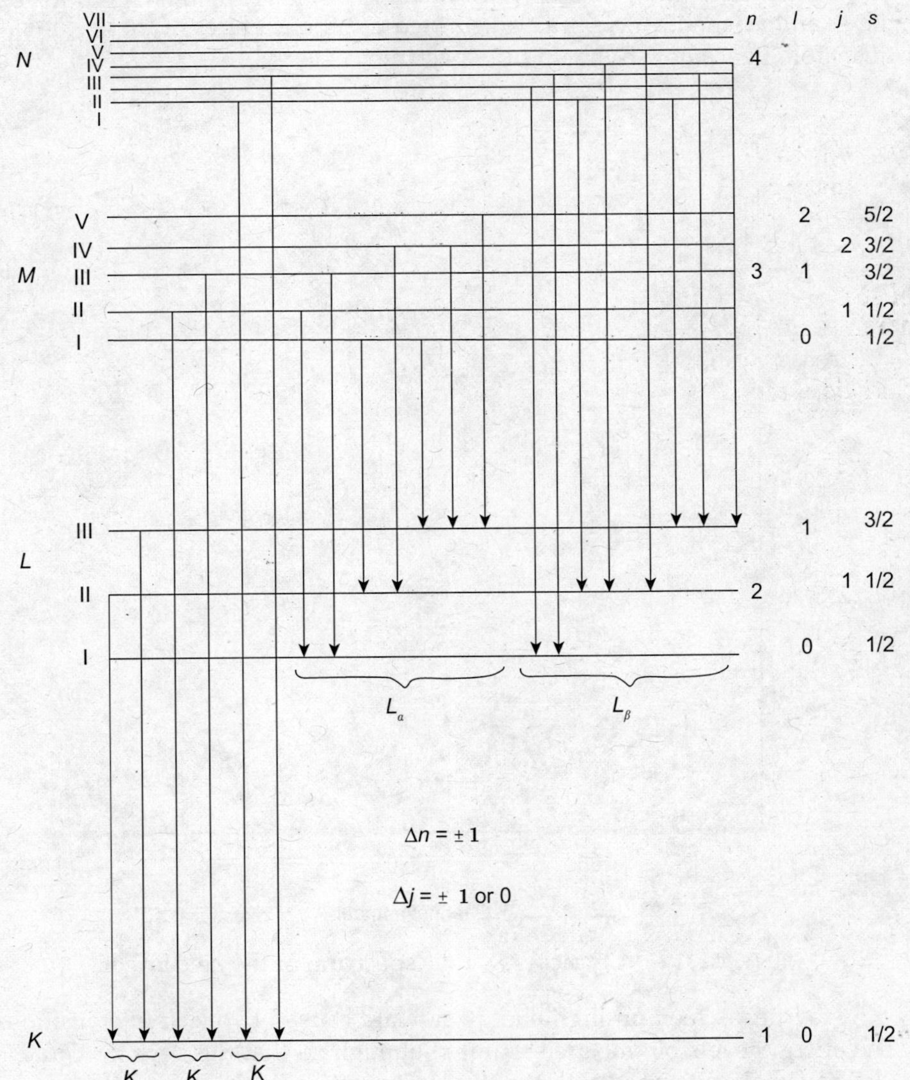

FIGURE 31.6 Nomenclature of X-ray transitions with selection rules of quantum numbers

The spectral lines that obey the selection rules are referred to as diagram lines and that violate them as non-diagram lines or forbidden lines.

X-ray spectra

A typical K-shell fluorescence X-ray spectrum (for zinc) is shown in Fig. 31.7. The two K X-ray energy peaks are corresponding to K_α and K_β of K-shell X-rays. Detector resolution is not enough to resolve the detailed line structure of different K_α and K_β X-rays. Therefore, these two peaks are composite of all K_α and K_β lines. For each element along with K-shell X-rays, the L- and M-shell X-rays are also emitted but the energies of these X-rays are too low for detection, hence not shown in this spectrum.

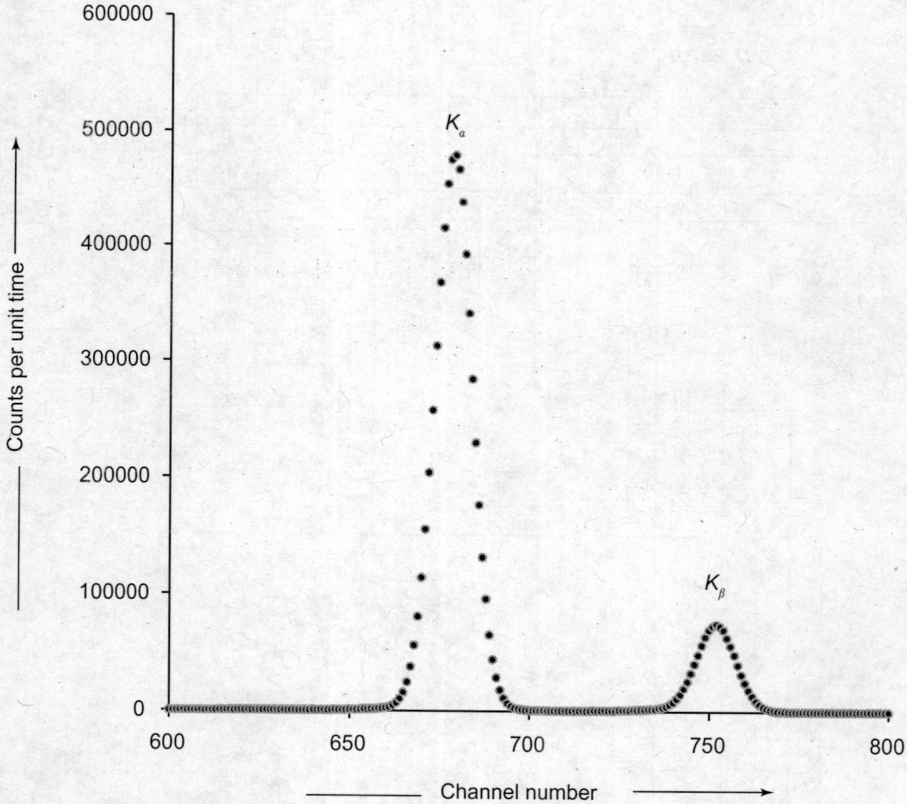

FIGURE 31.7 A typical K X-rays spectrum of the zinc target

Many K X-ray peaks seen on high energy side are caused by electron transitions to its K-shell. The energy levels of zinc are separated enough so that the detector could resolve K_α and K_β X-ray peaks as opposed to the two peaks under K_α that are labelled as $K_{\alpha 1}$ and $K_{\alpha 2}$ but not visible seperately. The separation of the energy levels within the L-shell is much

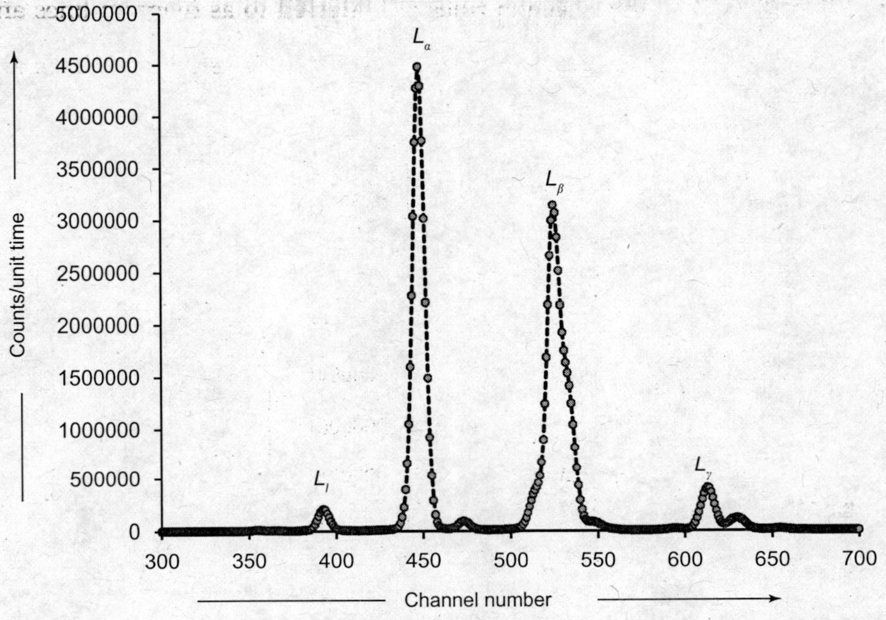

FIGURE 31.8 A typical L X-ray spectrum of the platinum target

smaller than the separation of the energy levels of the L-shell and those of the K-shell as shown in the L X-ray spectrum of platinum in Fig. 31.8.

Note

An X-ray spectrum is a plot of intensity of X-rays versus wavelength, energy or frequency.

31.4 ▶ Moseley's Law

Henry Moseley discovered that the wavelength (energy) of an X-ray depends on the nuclear charge of an atom. In 1913, he photographed the X-ray spectrum of ten elements that occupied consecutive places in the periodic table. He concluded that there was a fundamental quantity which increases by regular steps as we pass from one element to the next. In 1920, Rutherford identified this quantity as the atomic number (denoted by Z). The $K_{\alpha 1}$ lines shifted to higher energy with increased atomic number because the inner-shell electrons are more tightly bound by the higher number of protons in the nuclei (Fig. 31.9). The energy of transition is proportional to the number of shells changed; $K_{\alpha 1}$ radiation is less energetic than $K_{\beta 1}$.

Moseley's law describes the relationship between atomic number and wavelength of a spectral line as

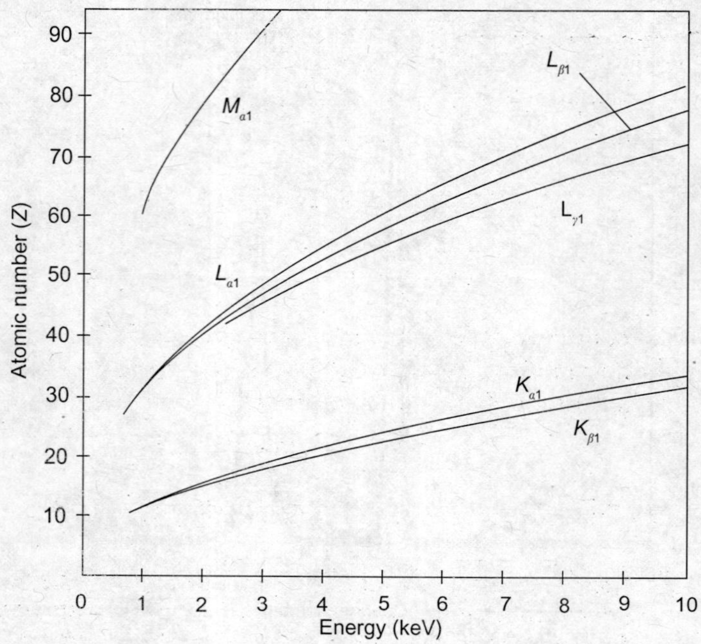

FIGURE 31.9 Energies of major X-ray emission lines below 10 keV of elements

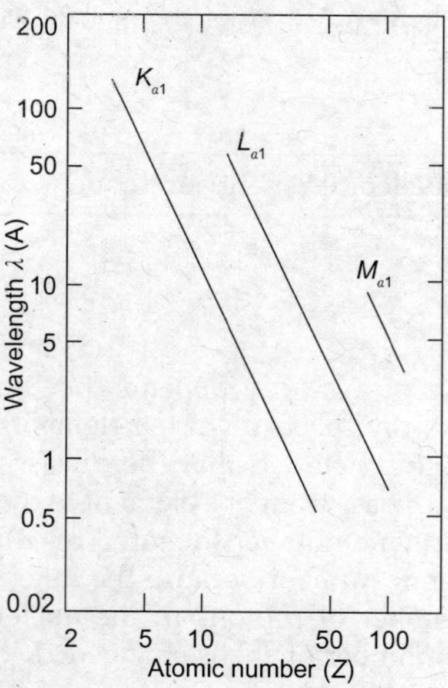

FIGURE 31.10 Moseley's relation between wavelength and atomic number for $M_{\alpha 1}$, $L_{\alpha 1}$ and $M_{\alpha 1}$ spectral lines

$$\boxed{\lambda = \frac{K}{(Z-\sigma)^2}} \tag{31.4}$$

where K and σ are constants for a given spectral line. The plot of λ versus atomic number Z is shown in Fig. 31.10.

The constant σ (sigma) is equal to 1 for the K-lines and 7.4 for the more shielded L-lines. For energy, this expression is approximately equivalent to

$$E\ (\text{keV}) = K(Z-1)^2$$

where Z = atomic number,

$K = 1.042 \times 10^{-2}$ for the K-shell

and $K = 3.446 \times 10^{-4}$ for the M-shell.

Moseley's determination of this relationship provided a simple test of the order of the elements according to Z. It showed where elements were missing from the periodic table and led to the discovery of some of these elements. For example, hafnium (Hf), which chemically is almost identical to zirconium (Zr), was identified by D. Coster and G. von Hevesy in 1923 from its X-ray spectrum.

31.5 ▶ Diffraction of X-rays – Bragg's Law

In 1912, M. Von Laue, W. Friedrich and E.P. Knipping first demonstrated X-ray diffraction by systematic crystals. They showed that a diffracting crystal intercepts X-rays of all wavelengths, but only those that undergo constructive interference are transmitted efficiently to the detector (Fig. 31.11).

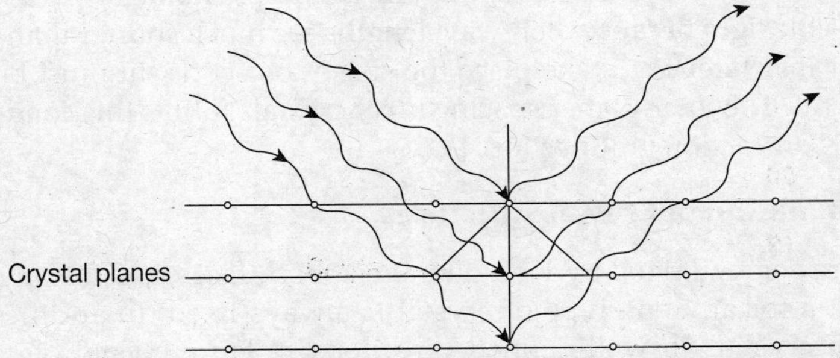

FIGURE 31.11 Representation of first-order X-ray diffraction

The path–length difference between X-rays diffracted from adjacent atomic layers corresponds to an integral number of wavelengths.

W.L. Bragg and W.H. Bragg built the first X-ray spectrometer based on this principle in 1913. The conditions necessary for constructive diffraction of X-rays (and light) by a crystal are described by Bragg's law (Fig. 31.12).

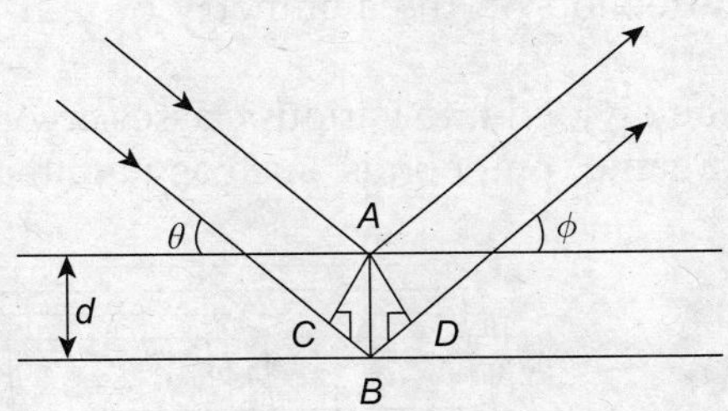

FIGURE 31.12 Diffraction of a coherent beam of X-rays from parallel planes of atoms, spaced a distance *d* apart

$$\boxed{n\lambda = 2d\sin\theta} \tag{31.5}$$

where n = an integer (1, 2, 3, …), λ = wavelength, d = spacing of the crystal and θ = incident angle (measured from crystal surface).

The diffraction condition is again that of coherence of the diffracted beam. The path lengths of the two rays shown differ by the distance *CBD*, which must therefore be equal to a whole (integral) number of wavelengths ($n\lambda$) for the rays scattered at the angle ϕ to be in phase with each other.

The X-rays that are in phase (those with wavelengths of $1\lambda, 2\lambda, 3\lambda, \ldots$) produce constructive interference, while out-of-phase waves (those with wavelengths like $1/2\lambda, 3/2\lambda, 5/2\lambda, \ldots$) produce destructive interference. Note that wavelengths of $\lambda/2, \lambda/3, \ldots$ are also diffracted constructively. These are termed as high-order wavelengths, because n must be a number higher than 1 for constructive diffraction to occur.

One may use monochromatic X-rays to determine the d-spacing of an unknown crystal as in X-ray diffraction (XRD), or conversely, use a crystal of known d-spacing to produce monochromatic X-rays and reject all except a specific wavelength. Since crystals are not perfect, the resulting X-rays are not perfectly monochromatic either.

It may be noted that X-rays do not show diffraction phenomena with normal slits and apertures like visible light because their wavelengths are much shorter than that of light and for diffraction or interference to take place the size of object or slit must be of the order of wavelength of light. The inter-planer spacing (d) of crystals fulfills this condition that is why the crystals act as diffraction gratings for X-rays

31.5.1 Determination of Crystal Structures

Large crystals are characterized by well-defined smooth faces and the angle between any pair of these faces for any one type of crystal is always fixed. In addition, it is possible to cleave certain crystals in well-defined directions. A large crystal of sodium chloride (table salt) may be cleaved many times, gradually reducing its size, and every time it is cleaved, a crystal is obtained that is physically similar in appearance to the one earlier. This phenomenon is what led to suggest that a crystal is built up of an infinite number of tiny unit cells, which are stacked together in a three-dimensional lattice by simple translation. A high proportion of all solid matter is built in this way on an atomic scale, even though the material may not show outward signs of crystallinity.

Each of the unit cells [Fig. 31.13(a)] is identical to all others and the symmetry and shape of the whole crystal depend on the symmetry of that cell to form whole lattice [Fig. 31.13(b)].

In general, materials can be divided into various classes according to the bond that hold the solid together. Typical organic compounds, such as naphthalene, crystallize into what

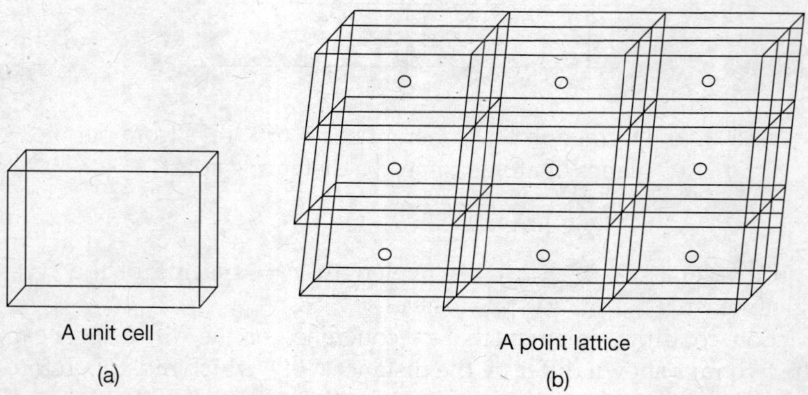

FIGURE 31.13 Unit cell (a) is building block of crystal lattice (b)

are called molecular or van der Waals solids. These are characterized by very low melting points because the melting point is a function of the weak attractions between the organic molecules. Very hard materials such as quartz and diamond are typical crystals, which are held together by a three-dimensional network of very strong covalent bonds. Sodium chloride is a typical example of an ionic solid.

The detailed discussion on crystal structure determination is given in Chapter 22.

31.6 ▶ COMPTON SCATTERING

Arthur H. Compton in 1922 observed the scattering of X-rays from electrons from carbon target and found that scattered X-rays have a longer wavelength than those incident upon the target. The shift or change of the wavelength increased with scattering angle according to the Compton formula:

$$\boxed{\lambda_f - \lambda_i = \Delta\lambda = \frac{h}{m_e c}(1-\cos\theta)} \tag{31.6}$$

Compton explained and modelled the experimental data by assuming a particle (photon) nature of X-rays and applying conservation of energy and conservation of momentum to the collision between the photon and the electron. The scattered photon has lower energy and therefore longer wavelength according to the Planck relationship (Fig. 31.14).

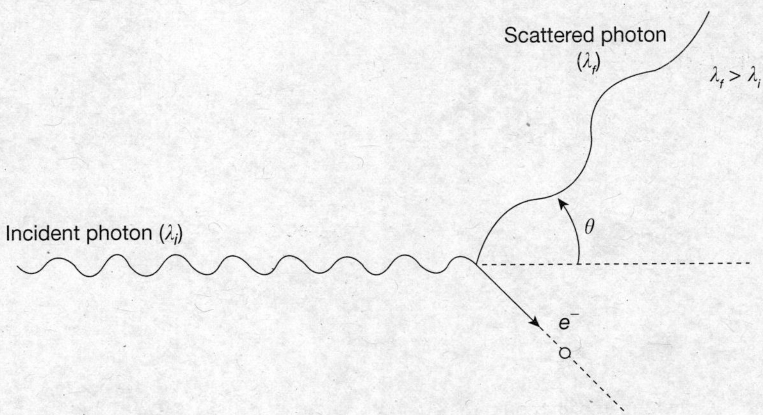

FIGURE 31.14 Compton scattering of photon of wavelength λ at an angle θ

At a time (in early 1920s) when the particle (photon) nature of light suggested by the photoelectric effect was still being debated, the Compton experiment gave clear and independent evidence of particle-like behaviour of photons. Arthur H. Compton was awarded the Nobel Prize in 1927 for the discovery of the scattering of photons from charged particles that is now known as Compton scattering. When the incoming photon gives a part of its energy to the electron, then the scattered photon has lower energy, and according to the Planck relationship has lower frequency and longer wavelength. The change in wavelength depends only upon the angle of scattering for a given target particle. The constant in the Compton formula above can be written as

$$\frac{h}{m_e c} = \frac{hc}{m_e c^2} = \frac{1240\,\text{eV nm}}{0.511\,\text{MeV}} = 0.00243\ \text{nm} \qquad \text{(Compton wavelength for the electron)}$$

It is called the Compton wavelength for the electron. The formula presumes that the scattering occurs in the rest frame of the electron.

31.6.1 Compton Scattering Experiment

In his original experiment (Fig. 31.15), Compton made use of molybdenum *K*-alpha X-rays, which have wavelength of 0.0709 nm. These were scattered from a carbon target and observed at different angles with a Bragg spectrometer. The spectrometer consists of a rotating frame with a calcite crystal to diffract the X-rays and an ionization chamber for detection of the X-rays. Since the spacing of the crystal planes in calcite is known, the angle of diffraction gives an accurate measure of the wavelength.

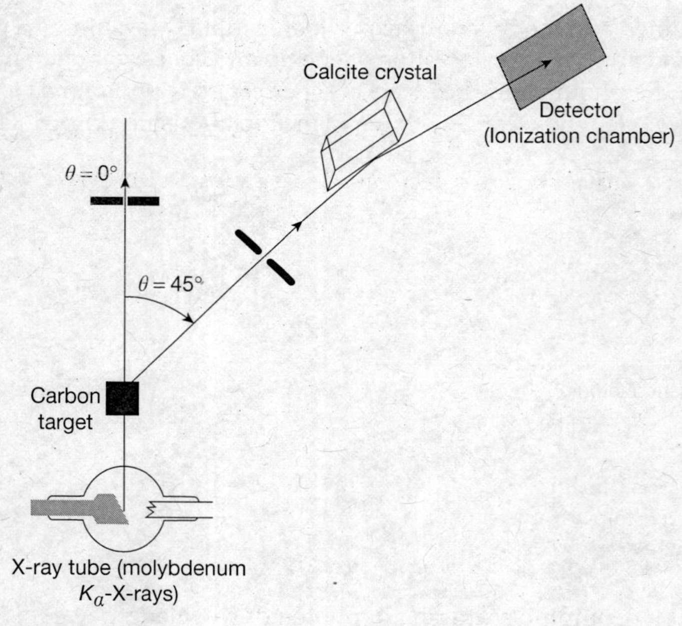

FIGURE 31.15 Compton's experiment of X-ray scattering from carbon target

Examination of the Compton scattering formula shows that the scattered wavelength depends upon the angle of scattering and also on the mass of the scatterer. At 90° scattering from stationary electrons, the formula gives a wavelength of 0.0733 nm. That is consistent with the right-hand peak in Fig. 31.16. The peak that is near the original X-ray wavelength is considered to be scattering of inner electrons in the carbon atoms, which are more tightly bound to the carbon nucleus. This causes the entire atom to recoil from the X-ray photon,

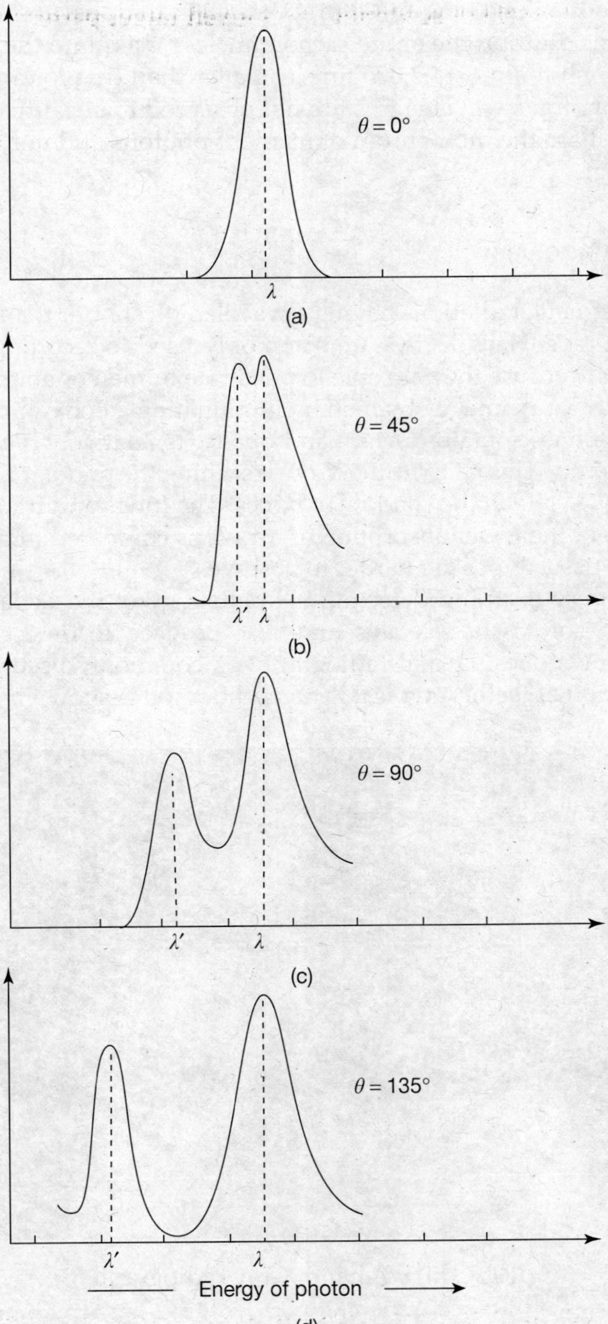

FIGURE 31.16 Compton scattering peaks at wavelengths λ' corresponding to scattering angles θ equal to 0° (a), 45° (b), 90° (c) and 135° (d)

and the larger effective scattering mass proportionally reduces the wavelength shift of the scattered photons. Putting the entire carbon nuclear mass into the scattering equation yields a wavelength shift almost 22,000 times smaller than that for an unbound electron, so those scattered photon's wavelengths are not observed to be shifted. Thus, confirming that only electrons share the momentum of incident photons and not the nucleus of atom of scattering material.

31.7 ▶ X-RAY RADIOGRAPHY

X-rays are electromagnetic radiations having a wavelength shorter than ultraviolet rays but longer than gamma rays. Hard X-rays are commonly used to record an image (known as radiographic images) because they are able to penetrate human or animal bodies.

A radiograph is an X-ray image obtained by illuminating a body part of the patient with a short X-ray pulse. Bones contain calcium and due to its relatively high atomic number, it absorbs X-rays efficiently. This restricts the X-rays reaching the photographic film or detector, making them clearly visible in the shadow of bones. The lungs and trapped gas can also be seen clearly because of the lower absorption of X-rays as compared to tissues; however, the differences between tissue types are harder to observe.

Radiographs are very useful in detecting the pathology of the skeletal system and also disease processes in soft tissues. X-rays may also be used to detect pathology such as gallstones and kidney stones. Dental radiography is commonly used in the diagnoses of cavities and dislocation of teeth. A typical X-ray radiograph is shown in Fig. 31.17.

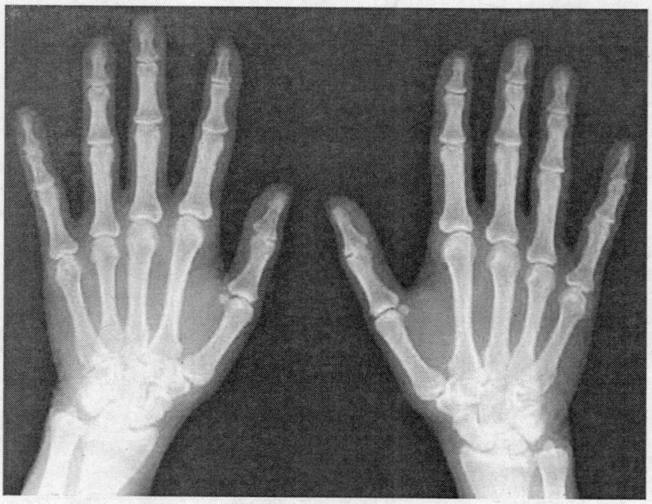

FIGURE 31.17 A hand X-ray radiograph

In medical diagnostic applications, soft X-rays having low energy are not very useful as they are totally absorbed by the body. They increase the radiation dose but do not contribute to the image. Thus, an X-ray filter, a thin metal sheet of aluminium, is placed over the window of the X-ray tube, absorbing the low energy part in the spectrum. It shifts the centre of the spectrum towards higher energy X-rays, that is why this process is called hardening the beam.

To generate an image of the arteries and veins (angiography), an initial image of the anatomical region of interest is taken. A second image of the same region after an iodinated contrast agent has been injected into the blood vessels is then taken. These two images are then digitally subtracted, which gives an image of the iodinated contrast outlining the blood vessels. The radiologist then compares the image obtained with the normal anatomical image to determine any damage or blockage of the vessel.

31.8 ▶ Advantages and Disadvantages of X-ray Radiography

The X-ray radiographs are commonly used for medical diagnostic purpose almost everywhere in the world. Besides having several advantages, the use of X-rays is also having some harmful effects that are the major disadvantages of the technique. Some advantages and disadvantages are listed below.

Advantages of X-rays

- X-rays are used to treat malignant tumours before these spread throughout the body.
- They help radiologists to identify infections, cracks, injury and abnormal bones.
- They also help in identifying bone cancer.
- X-rays help in locating alien objects like nail, bullets, glass, etc. inside the body or bones.

Disadvantages of X-rays

- X-rays make our blood cells to have higher level of hydrogen peroxide which could cause cell damage.
- There is a higher risk of getting cancer from X-rays in all parts of body exposed beyond a permissible limit.
- The X-rays are able to change the base of the DNA causing a mutation.
- Over exposure of dental exposure to eyes is harmful.

Recap Zone

Points to Remember

- The X-rays invented by Roentgen in 1895 were found to be electromagnetic waves of wavelength shorter than UV radiation.
- Immediately after their invention, the X-rays found versatile use in medical, civil and defence applications.

- The X-ray spectra are of three types, namely continuous, characteristics and band spectra.
- The exposure of high doses of X-rays is dangerous to living bodies and may lead to skin burns and cellular damage. Safety precautions must be followed while working with X-rays.
- Moseley in 1911 established atomic number Z as important atomic parameter of elements.
- The X-rays are not reflected but can undergo diffraction by crystals. The diffraction patterns reveal information about crystal structure.
- The X-ray photons share their momentum with the target electrons when scattered and the energy of scattered photons is less than the incident photon. The phenomenon is called Compton effect or scattering.
- Besides having several advantages, the X-ray radiography also has some harmful effects.

Key Terms with Definitions

- **X-rays** – X-rays are electromagnetic waves of wavelength shorter than ultraviolet light.
- **X-ray spectrum** – An X-ray spectrum is a plot of intensity of X-rays versus wavelength, energy or frequency.
- **Bremsstrahlung** – The process of loss of energy by charged particles radiation in a medium is called *Bremsstrahlung*, which in German means braking radiation.
- **Continuous spectrum of X-rays** – Some X-ray line peaks are superimposed on the continuous spectrum of X-rays. These lines are characteristic of the material used as a target in X-ray tube.
- **Moseley's law** – Moseley's law describes the relationship between atomic number and wavelength of a spectral line.
- **Bragg's law** – The conditions necessary for constructive diffraction of X-rays (and light) by a crystal are described by Bragg's law $n\lambda = 2d \sin \theta$.
- **Compton scattering** – The scattered X-ray photons from electrons have slightly longer wavelength than the incident one, the amount of change in the wavelength or energy increases with the scattering angle. This phenomenon is called Compton scattering.

Important Formulae and Equations

Equation Number	Equation	Remarks
(31.1)	$-\left.\frac{dE}{dX}\right\|_{\text{rad}} = \frac{nz_1^4 z_2^2}{m_0^2} Ef\left(\frac{E}{m_0 c^2}\right)$	Bethe-Heitler formula where $-\left.\frac{dE}{dX}\right\|_{\text{rad}}$ denotes the rate of loss of energy by the particle in the form of radiations n – Number of nuclei of the stopping medium per unit volume z_1 – Charge on the incident particle z_2 – Atomic number of atoms of the medium E – Initial total energy of the particle and $f\left(\frac{E}{m_0 c^2}\right)$ is a complex function of E, m_0 and c, the velocity of light m_0 is the rest mass of the incident particle.
(31.2)	$\lambda_{\min} = \frac{hc}{\text{eV}} = \frac{12396}{\text{V}} \text{Å}$	Minimum wavelength of X-ray in X-ray tube
(31.4)	$\lambda = \frac{K}{(Z-\sigma)^2}$	Moseley's law where K and σ are constants for a given spectral line
(31.5)	$n\lambda = 2d \sin\theta$	Bragg's law where n = an integer (1, 2, 3, ...), λ = wavelength, d = spacing of the crystal and θ = incident angle (measured from crystal surface)
(31.6)	$\lambda_f - \lambda_i = \Delta\lambda = \frac{h}{m_e c}(1-\cos\theta)$	Compton formula

Review Zone

Short Answer Questions

1. What are the properties of X-rays which distinguish these from visible light?
2. Discuss the origin of continuous spectra.
3. Describe the mechanism of production of characteristic X-rays.
4. What is Moseley's law? Discuss its significance.
5. Why X-rays are not reflected by ordinary mirrors?
6. What is Bragg's law?
7. Discuss the medical applications of X-rays.

Numerical Problems

1. An X-ray tube is operated with DC supply of 50 kV. What is the shortest wavelength of X-rays produced?
2. Calculate the heat generated per minute in the target of an X-ray tube if the voltage applied is 20 kV and the current is 10 mA. How much power is radiated if the efficiency is 0.2%?
3. Find the maximum speed of electrons striking the cathode under accelerating potential difference of 12 kV ($m = 9.1 \times 10^{-31}$kg, $e = 1.602 \times 10^{-19}$ C).
4. Find the wavelength for which the first-order Bragg reflection occurs at a 20° glancing angle from calcite crystal.
5. Find the shift in wavelength of an X-ray photon of energy of 5 keV when scattered at 45° from a stationary electron.

CHAPTER 32 Radioactivity

Learning Objectives

By the end of this chapter, the student will be able:

- To understand radioactivity, its types and decay processes
- To define half-life, mean life and decay constant
- To discuss the laws of radioactive decay, growth and equilibrium
- To explain radioactive series and application of radioactive radiation
- To describe radio carbon dating and clinical uses of radioactivity

32.1 ▶ Introduction

In 1896, Henri Becquerel used natural fluorescent minerals to study the properties of X-rays, which were discovered by Wilhelm Roentgen. He exposed potassium uranyl sulphate to sunlight and then placed it on photographic plates wrapped in black paper. He came to the conclusion that uranium absorbed Sun's energy and emitted it as X-rays. This hypothesis was proved wrong when his experiment was repeated. For some reason, Becquerel decided to develop his photographic plates. To his surprise, the images were strong and clear, proving that uranium emitted radiation without an external source of energy such as the Sun. Becquerel had discovered radioactivity that is a natural property of some elements.

Becquerel used an apparatus similar to the one shown in Fig. 32.1, to show that the radiation he discovered were not X-rays. X-rays are neutral and do not bend in a magnetic field, but new radiation were bent by the magnetic field indicating that the radiation must be charged and different from X-rays. When different radioactive substances were put in the magnetic field, they either deflected in different directions or showed no deflection at all. This predicted the existence of three types of radiations: negative, positive and electrically neutral.

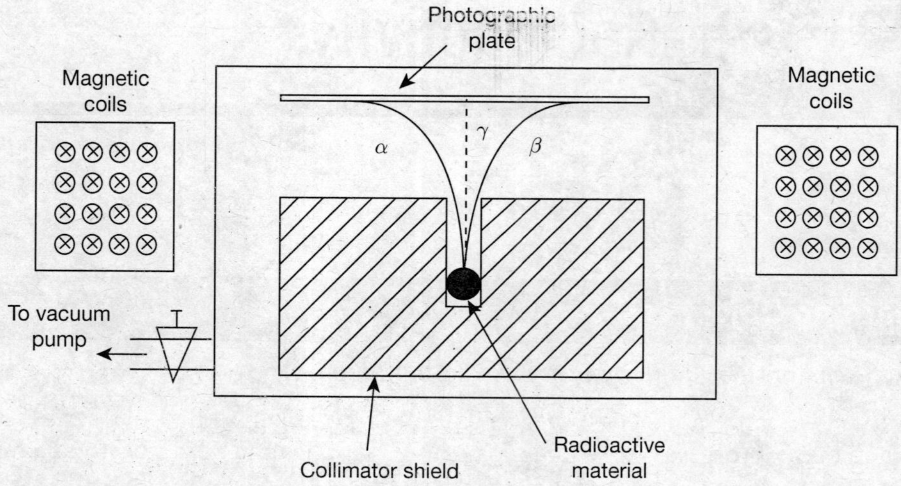

FIGURE 32.1 Experimental arrangement used to observe deflection of α- and β-particles from radioactive materials in magnetic field

The term radioactivity was later coined by a polish scientist Marie Curie, who together with her husband Pierre Curie, investigated the phenomenon of radioactivity. They extracted uranium from an ore and found that the residual ore showed more activity than the pure uranium. They concluded that the ore contained many other radioactive elements. This led to the discoveries of the elements, Polonium and Radium. It took four more years of processing tonnes of ore to isolate enough of each element to determine their chemical properties.

Ernest Rutherford, who conducted numerous experiments to study the properties of radioactive decay, named these radiations as alpha, beta particles and gamma rays, and classified them by their ability to penetrate matter. Rutherford used an apparatus similar to that depicted in Fig. 32.1. When air from the chamber was removed, the alpha source caused a spot on the photographic plate. When air was allowed to enter, the spot disappeared. Thus, it was confirmed that only a few centimetres of air was enough to stop the alpha radiation.

Since alpha particles carry more electric charge, are more massive, and move slowly compared to beta particles and gamma radiations, their interaction with matter is considerable. Beta particles are less massive and move faster, but are electrically charged. A sheet of one millimetre thick aluminium or several metres of air will stop these electrons and positrons. Because gamma rays carry no electric charge, they can penetrate large distances through materials before stopping. Several centimetres of lead or a metre of concrete is needed to stop gamma rays.

Note

The radioactivity is a spontaneous phenomenon, meaning that it occurs of its own without any external aid or stimulus.

32.2 ▶ Natural Radioactivity

Radioactivity is a natural part of our environment that arises due to instability of some nuclei that lie in the high Z range beyond $Z = 83$ as shown in the stability curve (Fig. 32.2). Present-day Earth contains all the stable chemical elements from the lowest mass $Z = 1$ (H) to the highest $Z = 92$ (U). Every element with atomic number higher than Bi ($Z = 83$) is radioactive. From Fig. 32.2, one can find that for the nuclei above $Z = 20$ the number of neutrons in nuclei keeps exceeding the number of protons. It helps the nucleus to remain stable by reducing the nuclear repulsion among its protons, but despite this instability of nuclei is not completely removed and that causes emission of excess energy in the form of radioactive radiation. The earth also contains several primordial long-lived radioisotopes that have survived to the present day in significant amounts. ^{40}K, with its 1.3 billion year half-life has the lowest mass of these isotopes and beta decays to both ^{40}Ar and ^{40}Ca.

32.2.1 Types of Radioactive Decay

It is found that nuclei with mass number greater than 100 can decay spontaneously into other types of nuclei. Such nuclei are said to be radioactive, and these decays are mainly of three types.

Alpha decay

Alpha (α) decay occurs by emission of an alpha particle $^4_2$He (helium nucleus). For example, the decay of uranium as:

$$\underset{\text{(parent)}}{{}^{238*}_{92}\text{U}} \rightarrow \underset{\text{(daughter)}}{{}^{234*}_{90}\text{Th}} + \underset{\text{(alpha-particle)}}{{}^{4}_{2}\text{He}}$$

In a decay such as that of $^{238*}_{92}$U, the nucleus (i.e., the parent nucleus) loses two protons and two neutrons (as seen by the atomic number (Z) and mass number (A) of the daughter nucleus, $^{234*}_{90}$Th). The total number of protons and the total number of nucleons are, however, the same on both sides of the reaction. Alpha particles are, in general, the least penetrating of the particles produced in a radioactive decay due to their relatively large size and charge ($+2e$). Alpha particles are usually stopped or absorbed by a few centimetres of air or a sheet of paper. * over U symbol indicates that it is highly unstable or in an excited state.

Beta decay

Beta (β) decay occurs by emission of a beta particle (either an electron or a positron). A positron is similar to an electron, except it has a positive charge ($+e$). An example of β-decay is the decay of carbon to nitrogen as:

$$^{14}_{6}\text{C} \rightarrow {}^{14}_{7}\text{N} + e^+ + \nu$$

ν is the neutrino, which has a negligible rest mass and is electrically neutral. The neutrino was proposed in 1930 by Pauli to account for the apparent loss of energy and momentum during β-decay. Neutrinos interact very weakly with matter, making them extremely difficult to detect.

Beta particles are more penetrating than alpha particles and can usually travel a few metres in air, or a few millimetres through aluminium sheet, before being absorbed or stopped.

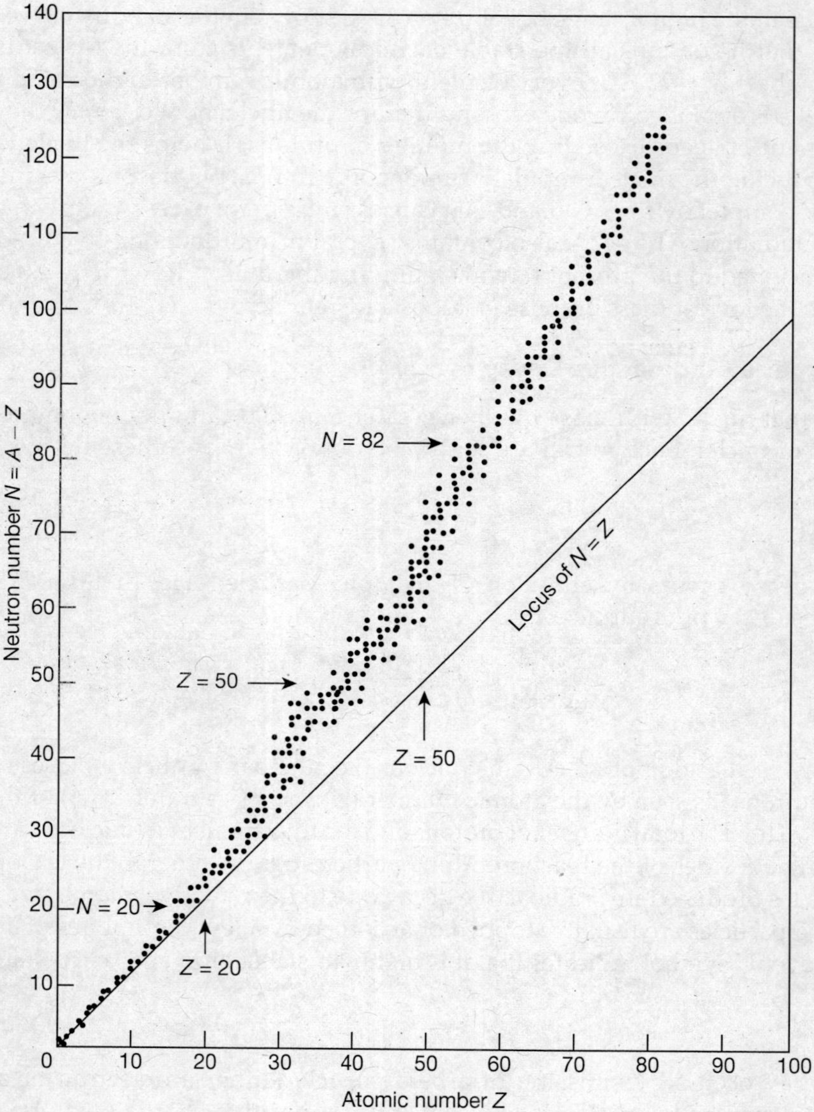

FIGURE 32.2 Stability curve of nuclides

Gamma decay

Gamma (γ) decay occurs by emission of gamma particles (photons or quanta of e.m. wave). It usually results after an alpha or beta decay, when nucleus is left in an excited state with excess energy. An example of this is the decay of Caesium to Barium in an excited state (Ba*) by β-decay, and then the excited barium (Ba*) decays by γ-decay:

$$\beta\text{-decay} \quad \left\{ {}^{137}_{55}\text{Cs} \rightarrow {}^{137}_{56}\text{Ba}^* + {}^{0}_{-1}e + \bar{\nu} \right\}$$

$$\gamma\text{-decay} \quad \left\{ {}^{137}_{56}\text{Ba}^* \rightarrow \underset{(\text{barium})}{{}^{137}_{56}\text{Ba}} + \underset{(\text{gamma ray})}{\gamma} \right.$$

The gamma decay does not change atomic number (Z) or mass number (A). Gamma rays are the most penetrating of the decay products, being capable of penetrating a few centimetres of lead if they have enough energy.

Apart from naturally occurring radioactive elements, the radioactivity can also be induced in other non-radioactive elements, which is called induced or artificial radioactivity. The details of such processes are discussed in the chapter on nuclear reactions.

Note

The origin of gamma rays is nucleus, but its energy can be in X-ray region (even soft X-ray). It is a mechanism by which an excited nucleus releases excess energy after β- or α-particle emission.

32.3 ▶ Radioactive Half-life

The radioactive half-life for a given nuclei is the time for the half of radioactive nuclei to undergo radioactive decay. After two half-lives, there will be one fourth of the originals left. The radioactive half-life gives a pattern of reduction to half in any successive half-life period.

For example, after three half-lives, one eighth of the original sample is left, and so on.

The half-life of a radioactive substance is a measure of how quickly the radioactive decay takes place. Suppose at some initial time we start with N_0 radioactive nuclei, and then measure the number N at a later time t, one finds N behaves as in the graph shown in Fig. 32.3.

Half-life, $T_{1/2}$, of a substance, defined as the time after which exactly one half of the originally active nuclei remains. In terms of this, the number of nuclei N remaining at time t, given N_0 at time $t = 0$, is given by

$$N = N_0 \left(\frac{1}{2}\right)^{t/T_{1/2}} \tag{32.1}$$

Thus, after a period of one half-life, half of a substance remains, after another half-life $\frac{1}{2} \times \frac{1}{2} = \frac{1}{4}$ of a substance remains, and so on. Half-lives of substances range from tiny fractions of a second to millions of years.

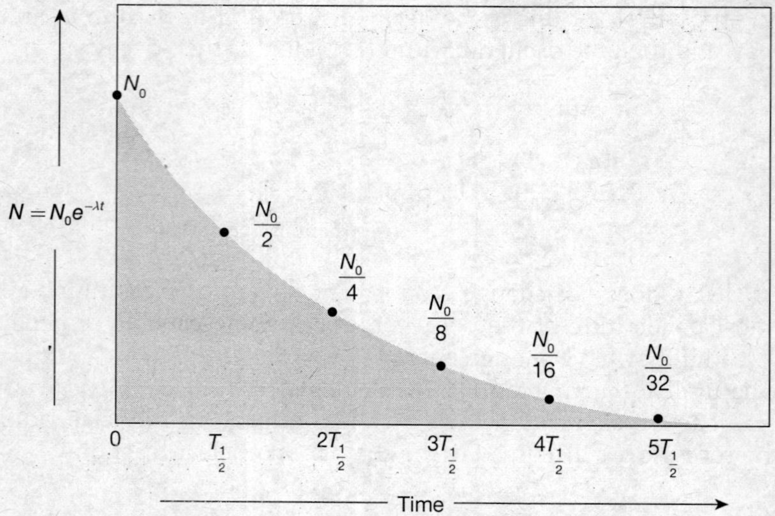

FIGURE 32.3 Exponential decay of radio nuclides with time

32.3.1 Radioactive Decay and Growth

Radioactive decay is a statistical process which depends on the instability of a particular radioisotope, but exact nucleus in any given sample is completely unpredictable. The decay process and the observed half-life dependence of radioactivity can be predicted by assuming that individual nuclear decays are purely random events. If there are N radioactive nuclei at some time t, then the number ΔN which would decay in any given time interval Δt would be proportional to N:

$$\Delta N = -\lambda N \, \Delta t \tag{32.2}$$

where λ is a constant of proportionality called decay constant.

Without any further assumptions, this leads to the exponential radioactive decay result:

$$\boxed{N = N_0 \, e^{-\lambda t}} \tag{32.3}$$

and also implies that the decay rate and the amount of emitted radiation also follow the same relationship as:

$$R = R_0 \, e^{-\lambda t} \tag{32.4}$$

With each decay of parent nuclei, the number of daughter nuclei increases. The decay and growth of these nuclei are shown in Fig. 32.4.

32.3.2 Decay Law

Although radioactive decay involves discrete reactions of nuclear disintegration, the number of these reactions is so large that it can be treated like a continuum and the

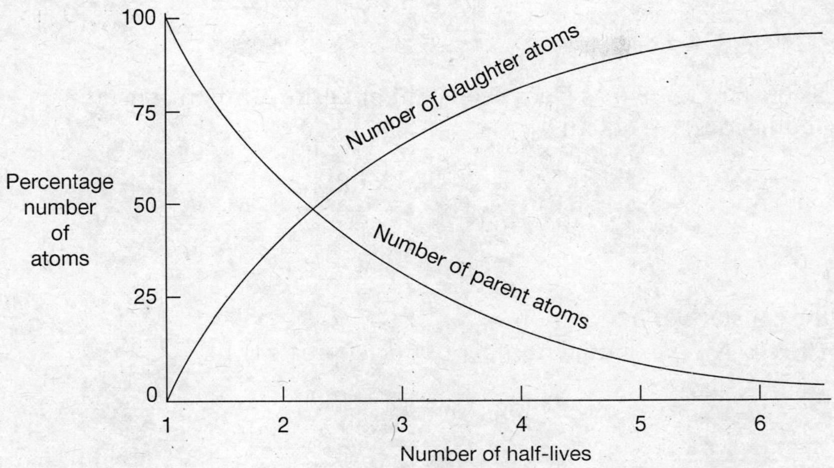

FIGURE 32.4 Growth and decay of number of radioactive atoms

methods of calculus can be employed to predict the behaviour. The decay probability can be written in the differential form as:

$$dN = -\lambda N\ dt \quad \text{or} \quad \frac{dN}{N} = -\lambda dt \tag{32.5}$$

This can be integrated directly to give ln $N = -\lambda t + C$, where C is the constant of integration. Taking the exponent on both sides gives

$$N = e^C - e^{-\lambda t} \text{ and } N = e^C = N_0 \text{ at } t = 0$$

so the standard form of the decay equation is

$$N = N_0\, e^{-\lambda t} \tag{32.6}$$

which also applies to all other quantities that are proportional to N.

32.3.3 Radioactive Decay Rates

The decay of radioactive nuclei takes place spontaneously and without any predictable chance. The process is governed by laws of probability. Let us consider that there are N number of radioactive nuclei present at time t. Out of these, dN disintegrates into daughter nuclei at an interval of dt. The rate of decay of the nuclei is directly proportional to the number of atoms present. Therefore,

$$-\frac{dN}{dt} \propto N \tag{32.7}$$

The negative sign indicates that with the increase in the decay, the value of N decreases. We can rewrite this equation as

$$\frac{dN}{N} = -\lambda dt$$

where λ is a constant known as Decay constant or Disintegration constant.
Integrating both sides, we obtain

$$\int \frac{dN}{N} = -\lambda \int dt$$

$$\log_e N = -\lambda t + k$$

where k is the constant of integration.
At time $t = 0, N = N_0$ (i.e., initial number of nuclei present).

$$\therefore \quad \log_e N_0 = k$$

$$\therefore \quad \log_e N = -\lambda t + \log_e N_0$$

$$\log_e N - \log_e N_0 = -\lambda t$$

$$\frac{N}{N_0} = e^{-\lambda t}$$

$$N = N_0 e^{-\lambda t}$$

This equation is known as Rutherford-Soddy equation and indicates that the number of nuclei present in the sample decreases exponentially with time. The decay curve of a radioactive material is shown in Fig. 32.4. The amount of daughter nuclei increases in the sample as the parent nuclei decays.

32.3.4 Half-Life Period ($T_{1/2}$) and Decay Constant (λ)

The time in which half of original nuclei decays into daughter nuclei is called half-life period. The half-life period is denoted by $T_{1/2}$.

$$N = \frac{N_0}{2} = N_0 e^{-\lambda T_{1/2}}$$

$$\Rightarrow \quad \frac{1}{2} = e^{-\lambda T_{1/2}}$$

$$\therefore \quad e^{\lambda T_{1/2}} = 2$$

$$\text{or} \quad \lambda T_{1/2} = \log_e 2$$

$$\therefore \quad \boxed{T_{1/2} = \frac{0.693}{\lambda}} \qquad (32.8)$$

The decay constant λ is also called disintegration constant.

This equation indicates that the nuclei with different disintegration/decay constant λ will have different half-life periods. Ideally, every material should take infinite time to decay completely. The actual life of a nucleus can be from negligibly small to infinity. Therefore, another parameter average life period is also used.

For example, free neutrons decay with a half-life of about 10.3 minutes. This corresponds to a decay constant of 0.067 and an average life-time of 14.8 minutes or 890 seconds.

32.3.5 Mean/Average Life Time (T_a)

The mean/average life of a radioactive nucleus is defined as the ratio of sum of lives of all atoms and the total number of atoms.

Let N_0 be the number of nuclei at $t = 0$ and N is the number of nuclei present at time t.

Let dN atoms disintegrate between t and $t + dt$.

The total life of dN atoms $= tdN$.

Because the life of total number N_0 nuclei varies from 0 to infinity,

$$\text{The life-time of } N_0 \text{ nuclei} = \int_0^\infty tdN$$

$$\therefore \quad \text{Mean/average life } T_a = \frac{\text{Total life time}}{\text{Number of nuclei}}$$

$$= \frac{\int_0^\infty tdN}{N_0}$$

$$\therefore \quad N = N_0 e^{-\lambda t}$$

$$\therefore \quad \frac{dN}{dt} = -\lambda N_0 e^{-\lambda t} \tag{32.9}$$

$$dN = \lambda N_0 e^{-\lambda t} dt \quad [-\text{ve sign has been ignored}]$$

$$\therefore \quad T_a = \frac{\int_0^\infty t\lambda N_0 e^{-\lambda t} dt}{N_0}$$

$$= \lambda \int_0^\infty te^{-\lambda t} dt$$

$$\therefore \quad T_a = \lambda \left[\frac{te^{-\lambda t}}{-\lambda} - \int \frac{e^{-\lambda t} dt}{-\lambda} \right]_0^\infty$$

$$= \lambda \left[\frac{te^{-\lambda t}}{-\lambda} - \frac{e^{-\lambda t}}{-\lambda^2} \right]_0^{\infty} = \frac{1}{\lambda}$$

$$\therefore \qquad T_a = \frac{1}{\lambda} \tag{32.10}$$

This relation relates average life-time T_a and decay constant λ.

32.4 ▶ Units of Radioactivity

The radioactivity of a substance is measured in terms of decays per unit time. Two popular units of activity are as follows:

Curie (Ci) defined as 1 Ci = 3.7×10^{10} decays/s

and Becquerel (Bq), defined as 1 Bq = 1 decay/s.

These units just measure the intensity of radioactive emission. These units do not measure the danger of a given amount of radiation to humans. For medical purposes/other units of radioactivity reflecting the health aspect of radioactive radiation are more popular in the medical world.

One of such units is Gray (Gy) that is defined as the amount of energy delivered to the tissue per unit mass. A related unit is rad, which is equal to 0.001 Gy:

$$1 \text{ rad} = 0.01 \text{ J/kg} = 0.01 \text{ Gy}$$

To quantitatively measure the biological damage done by the radiations, a factor called Relative Biological Effectiveness (RBE) has been introduced. X-rays of energy 200 keV are defined to have an RBE of 1 (Table 32.1), and all other radiation sources are compared to this standard.

TABLE 32.1 RBE values of various types of radiation

Types of radiations	RBE
Electrons	1–1.5
Slow neutrons	3–5
Protons	10
Alpha particles (Helium nuclei)	20
Heavy ions	20

The biologically equivalent dose, or simply equivalent dose, is measured in Sieverts (Sv) and is defined as equivalent dose (Sv) = RBE × Absorbed dose (Gy), which is a more accurate measurement of the biological danger posed by a given type of radiation.

Another common unit to measure this in is the rontgen equivalent man (rem), defined as Equivalent dose (rem) = RBE × Absorbed dose (rad).

32.5 ▶ Uses of Radioactive Radiation

Radiations from radioactive materials can destroy the cells of living bodies. Despite dangers of over exposures, the radioactive radiations in controlled settings have found a number of useful applications. There are two broad categories of this use:

1. Non-medical uses (Carbon dating, sterlization of equipment and food items)
2. Medical uses (Imaging, PET, tumour treatment, tracers)

32.5.1 Carbon Dating

Carbon-14 dating is a way of determining the age of certain archaeological artifacts of biological origin upto about 50,000 years old. It is used for dating bone, cloth, wood and plant fibres that existed in the relatively recent past.

Formation of carbon-14

Cosmic rays enter the Earth's atmosphere in large numbers every day. For example, every person is hit by about half a million cosmic rays every hour. It is common for a cosmic ray to collide with an atom in the atmosphere, creating a secondary cosmic ray in the form of an energetic neutron. These energetic neutrons collide with nitrogen atoms. When a neutron collides, nitrogen-14 (seven protons, seven neutrons) atoms turn into a carbon-14 atoms (six protons, eight neutrons) and a hydrogen atom (one proton, zero neutrons). Carbon-14 is radioactive, with a half-life of about 5730 years.

Carbon-14 in living things

The carbon-14 atoms that cosmic rays create combine with oxygen to form carbon dioxide, which plants absorb and incorporate into plant fibres by photosynthesis. Animals eat plants and take in carbon-14 as well (Fig. 32.5). The ratio of normal carbon (carbon-12) to carbon-14 in the air and in all living things at any given time is nearly constant. Maybe one in a trillion carbon atoms is carbon-14. The carbon-14 atoms are always decaying, but they are being replaced by new carbon-14 atoms at a constant rate. At this moment, our body has a certain percentage of carbon-14 atoms in it, and all living plants and animals have the same percentage.

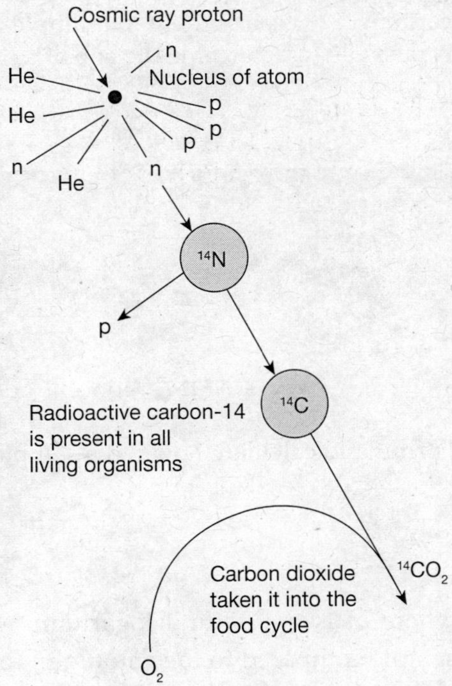

FIGURE 32.5 Absorption chain of ^{14}C in atmosphere

Dating a fossil

As soon as a living organism dies, it stops taking in new carbon (Fig. 32.6). The ratio of carbon-12

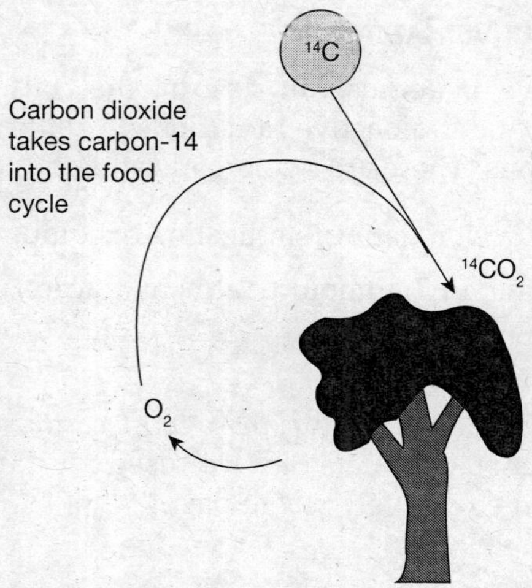

FIGURE 32.6 Carbon-14 is absorbed by plants from atmosphere during its life cycle. But the process of absorption stops when plant dies

to carbon-14 at the moment of death is the same as for every other living being, but the carbon-14 decays and is not replaced. The carbon-14 decays with its half-life of 5730 years, while the amount of carbon-12 remains constant in the sample. By looking at the ratio of carbon-12 to carbon-14 in the sample and comparing it to the ratio in a living organism, it is possible to determine the age of a living thing (Fig. 32.7).

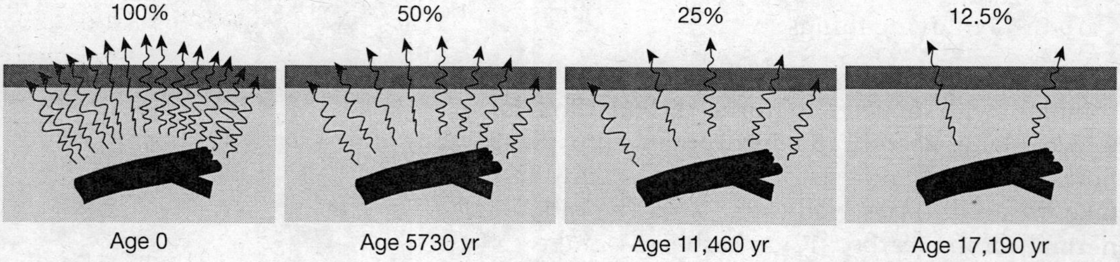

FIGURE 32.7 Concentration of carbon-14 is reduced with time

Formula to calculate how old a sample is by carbon-14 dating is

$$t = \left[\ln \frac{(N_f / N_0)}{-0.0693} \right] x t_{1/2}$$

where ln is the natural logarithm, N_f / N_0 is the per cent of carbon-14 in the fossil sample compared to the amount in the living tissue, and $t_{1/2}$ is the half-life of carbon-14 (5700 years).

So, if a fossil that has 10 per cent carbon-14 compared to a living sample, then the age of the fossil would be:

$$t = \left[\ln \frac{(0.10)}{(-0.693)}\right] \times 5700 \text{ years}$$

$$t = \left[\frac{(-2.303)}{(-0.693)}\right] \times 5700 \text{ years}$$

$$t = [3.323] \times 5700 \text{ years}$$

$$t = 18,940 \text{ years old}$$

Because the half-life of carbon-14 is 5700 years, it is only reliable for dating objects up to about 50–60 thousand years old. However, the principle of carbon-14 dating applies to other isotopes as well. Potassium-40 is another radioactive element naturally found in our body and has a half-life of 1.3 billion years. Other useful radioisotopes for radioactive dating include Uranium-235 (half-life = 704 million years), Uranium-238 (half-life = 4.5 billion years), Thorium-232 (half-life = 14 billion years) and Rubidium-87 (half-life = 49 billion years).

The use of various radioisotopes allows the dating of biological and geological samples with a high degree of accuracy. However, radioisotope dating may not work so well in the future. Anything that died after the 1940s, when nuclear bombs, nuclear reactors and open-air nuclear tests started changing things, will be harder to date precisely.

Sterilization of equipment and food

The fact that gamma radiations can destroy biological material/living organism is used in the sterilization of medical equipment. More controversial is its use on food, where it is used to destroy organisms that fasten spoilage.

Smoke detectors

In a smoke detector, a low-level radioactive material is present. This radiation can ionize particles contained in smoke, which are subsequently detected by their charge.

32.5.2 Medical Uses of Radioactivity

Radiation plays an increasingly important role in the diagnosis and treatment of various diseases.

Imaging in nuclear medicine

Looking inside human body is possible through a huge array of non-invasive techniques. Some of these nuclear medicine imaging techniques combine the use of computers, detectors and radioactive substances. These techniques include the following:

- Positron emission tomography (PET)
- Single photon emission computed tomography (SPECT)
- Cardiovascular imaging
- Bone scanning

All of these techniques use different properties of radioactive elements to create an image.

Nuclear medicine imaging is useful for detecting tumours, aneurysms (weak spots in blood vessel walls), irregular or inadequate blood flow to various tissues, blood cell disorders and inadequate functioning of organs, such as thyroid and pulmonary function deficiencies.

The use of any specific test, or combination of tests, depends on patient's symptoms and the disease being diagnosed.

Minimized tumour treatment

Radiation are used to selectively destroy tumours present in a body like cancereous ones. Damage to healthy cells is controlled by using radioactive samples with specific half-lives which subsequently are intended to decay very fast.

Tracer techniques

The body absorbs various elements and does not discriminate between different isotopes. In tracer technique, a radioactive isotope of such an element, namely, iodine, (for thyroid examination) is injected into the body. The signals coming from the ensuing radiation give a picture of the size and location of the area where the isotope has been absorbed. This helps to diagnose the metabolism or absorption of food or drugs. Usually isotopes of a relatively short or half-life, of the order of minutes or days, are used to minimize long-term radiation damages.

RECAP ZONE

POINTS TO REMEMBER

- Radioactivity was discovered accidentally by H. Becquerel in 1886 while handling natural minerals containing uranium and other heavy elements.
- Further studies revealed that these were spontaneously emitted by minerals and consisted of photons (Gamma rays), Alpha particles (He nuclei) or Beta particles (electrons).
- Later it was found that some sources also emit positrons (positively charged particles with mass equal to that of electron).
- These radiation have different penetrating powers in materials depending upon their energies.
- The activity or strength of a radioactive source is measured in terms of units like Becquerel and Curie. To account for the total biological damage or interaction, the units like Gray, RB Rad, REM, etc. are used. The standard biological unit of radiation dose in use these days is Sievert.
- The radioactive material is characterized by its half-life, i.e., the time taken by the material to decay half of its radioactivity.

- The radioactive decay is a spontaneous and statistical process and follows the exponential law $N = N_0 e^{-\lambda t}$.
- With the decay of parent nuclei, the number of daughter nuclei increases and a radioactive equilibrium is attained.

Key Terms with Definitions

- **Activity** – The activity of a piece of radioactive material (called a source) is a measure of how many radioactive decays taking place per second.
- **Alpha radiation** – Ionizing radiation consisting of alpha particles, emtitted by some substances undergoing radioactive decay.
- **Beta radiation** – Ionizing radiation resulting from the decay of radioisotopes where beta particle is emitted.
- **Gamma radiation** – Electromagnetic radiation emitted by atomic nuclei; the wavelength is generally in the range 1×10^{-10} to 2×10^{-13} m.
- **Half-life** – The time taken for the radioactivity of a specified isotope to fall to half its original value.

Important Formulae and Equations

Equation Number	Equation	Remarks
(32.3)	$N = N_0 e^{-\lambda t}$	**This exponential law is followed by nuclear decay process.**
(32.8)	$T_{1/2} = \dfrac{0.693}{\lambda}$	**Relationship between half-life period and decay constant,** **where** $T_{1/2}$ **= half-life period** λ **= decay constant**

Review Zone

Short Answer Questions

1. What is radioactivity? Why is it called a spontaneous phenomenon?
2. Discuss the properties of various particles emitted in radioactive processes.

3. What is half-life and average life of radioactive material? Derive an expression for the decay.
4. Describe various applications of radioactivity.
5. What is carbon dating? Discuss its limitations.
6. What are alpha particles? How these are emitted from radioactive elements?
7. What are medical applications of radioactivity?

Numerical Problems

1. If half-life of a radioactive material is 3, 8 days after what time (days) only 1/15 of original material will be left?
2. One gram of radium has 1 curie of activity. If half-life is 3.8 days, (a) in what time it will become 1 mCi, (b) what will be radioactivity of 1 mg of radioactivity?
3. Find the mass of 1 Ci of C^{14} of half-life of C^{14} is 5570 years.
4. If 1 g of sample takes 50 sec to decay to 0.1 g, find the half-life of sample.
5. The isotope ${}^{238}_{92}U$ successively has eight α-decays and six β-decays. What is the resulting isotope?
6. When a nucleus emits a gamma ray, what happens to its mass and atomic number?
7. Calculate the volume of radon gas that would be in equilibrium with 1 g of radium. Mean half-life of radium is 1600 years and of radium is 3.8 days.
8. In how many years 1 g of uranium will be reduced to one centigram? Half-life of uranium is 5000 million years.

CHAPTER 33

Interaction of Radiation with Matter

Learning Objectives

By the end of this chapter, the student will be able:

- To distinguish interactions of alpha particles, beta particles, gamma rays, neutrons and charged particles with matter
- To discuss the mechanisms of these interactions
- To describe the range and stopping power
- To describe the process of Compton scattering, photo-electric absorption and pair production
- To compute the attenuation of gamma rays in matter
- To explain elastic and inelastic interactions of electrons
- To discuss the interaction of neutrons with matter

33.1 ▶ Penetration of Radiation in Matter

Different types of radiation from radioactive sources interact differently with matter. Their penetration and processes of interaction are unique to them.

Though most massive and energetic of radioactive emissions, alpha particle is shortest in range because of its strong interaction with matter. The electromagnetic gamma ray is extremely penetrating, even to the several metre thicknesses of concrete. The electron of beta radioactivity strongly interacts with matter and has a short range.

Alpha and beta particles are high-energy charged particles. Neutrons are uncharged particles and gamma rays are electromagnetic radiations.

These four types of particles have different properties, the main difference being that they penetrate matter to different depths as shown in Fig. 33.1. This is an important consideration in protection of humans against radiation. The processes of interactions of each are discussed in the following sections.

Note

Different radiations from radioactive sources interact differently with matter.

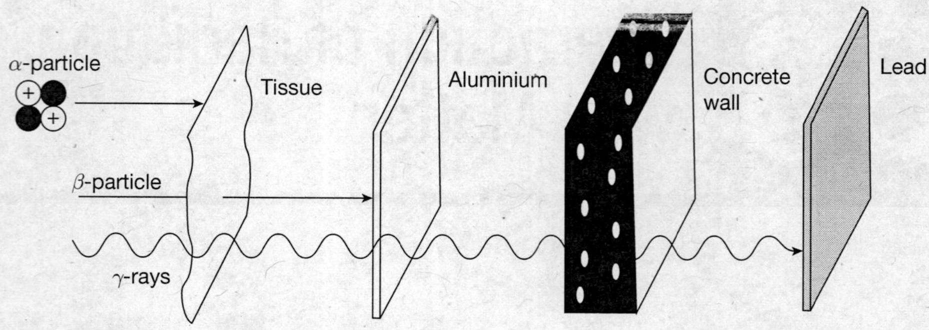

FIGURE 33.1 Relative penetration of radiation in matter

33.1.1 Alpha Particles

The alpha particle is a fast-moving packet of two protons and two neutrons, ejected as a single particle from a disintegrating nucleus. Since alpha particles produce heavy ionization, we expect them to travel much shorter distances before their energy is exhausted. That is why these have very little penetrating power. The alpha particles emitted from a radionuclide have a distinct energy. These energies, which are in the range of 4–8 MeV, are not sufficient to enable alpha particles to penetrate even the dead outer layer of the human skin. Hence, they do not present much of hazard externally. However, when alpha emitting radionuclides are consumed they cause serious internal radiation hazard. Figure 33.2 shows the range of alpha particles of different energies in air. At the end of their path, the alpha particles combine with two electrons from the free electron population around them to become ordinary helium atoms.

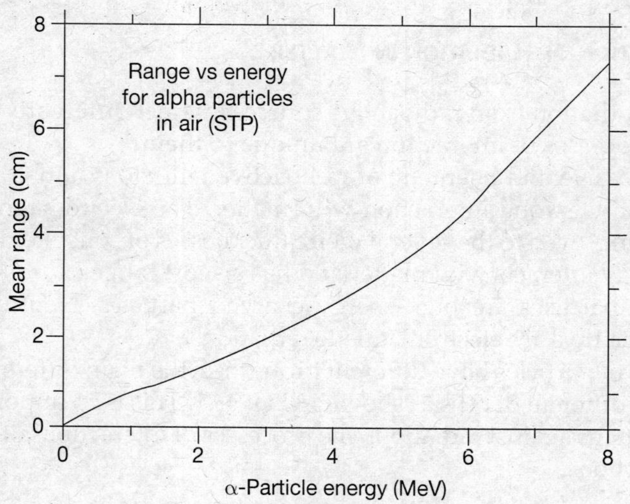

FIGURE 33.2 Range of alpha particles in air

33.1.2 Beta Particles

Beta particles are much more penetrating than alpha particles. Very high energy beta particles can penetrate to a depth of about a centimetre in tissue and can damage eye and skin if the source is strong. Therefore, beta sources present an external radiation hazard. They are, however, relatively easy to deal with by shielding. If the beta emitter is taken into the body, it presents a far greater hazard, though less than the alpha emitters. Beta particles emitted by a beta emitter have a range of energies with a maximum which is characteristic of the emitting nuclide. The beta energy spectrum emitted by strontium is shown in Fig. 33.3.

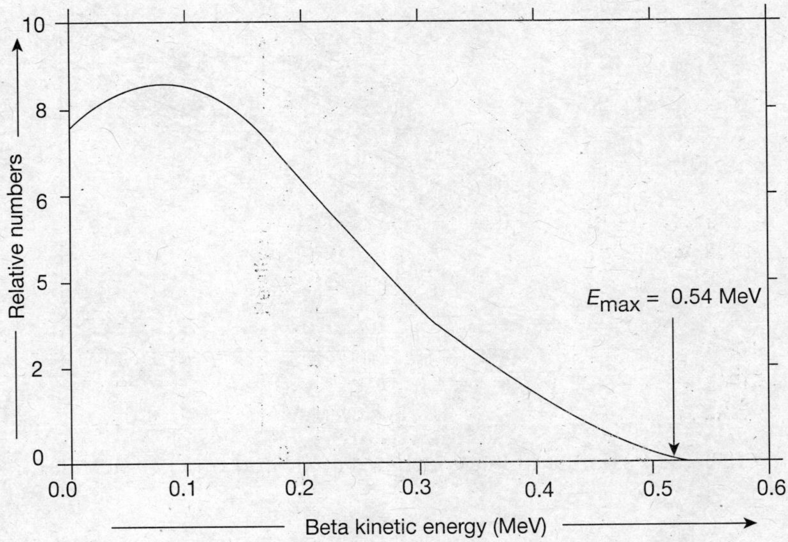

FIGURE 33.3 Beta energy spectrum of $_{90}$Sr

Note

Beta particles are much more penetrating than alpha particles. Very high energy beta particles can penetrate to a depth of about a centimetre in tissue and can damage eye and skin.

The maximum energy of beta radiation from the majority of nuclides lies in the range 0.5–3.5 MeV but some also have energy above and below this range. Materials with high atomic number, such as lead, are most effective in stopping beta particles. However, high energy beta particles produce X-rays known as Bremsstrahlung when absorbed, and this effect is more pronounced in heavy and dense absorbers. For this reason, lighter materials such as lucite or plywood are preferred as absorbers. The penetrating ability of beta particles (range) in air as a function of energy is shown in Fig. 33.4. When beta particles lose their energy, they join the free electron population as ordinary electrons.

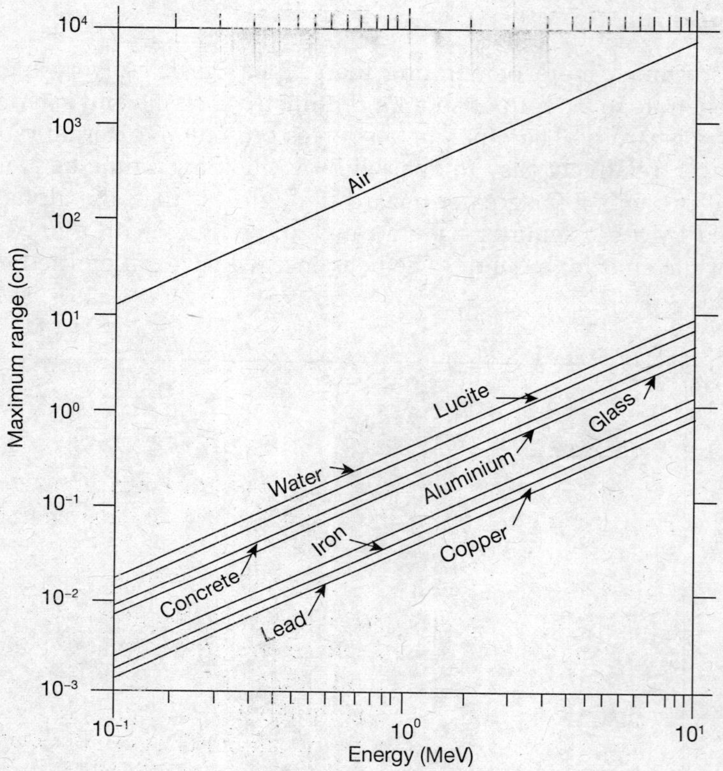

FIGURE 33.4 Range of beta particles in air and other materials

33.1.3 Gamma Rays

Gamma ray is the excess energy of an excited atomic nucleus, which is emitted as photon. Its energy range, although differs according to the level of excitement of the nucleus, covers range from keV to several MeV.

The interaction of gamma rays with matter is markedly different from that of charged particles. Gamma rays and X-rays, both being electromagnetic radiation, show a characteristic exponential absorption in matter and have no definite range such as that found for electrons or alpha particles. Three processes are mainly responsible for the absorption of gamma rays:

1. Photo-electric absorption
2. Compton scattering by the electrons in the atoms
3. Pair production, the creation of electron–positron pairs as a result of the interaction between gamma rays and the strong electric fields of atomic nuclei.

The process by which gamma ray/photons are absorbed in materials is shown in Fig. 33.5 with respective energy ranges.

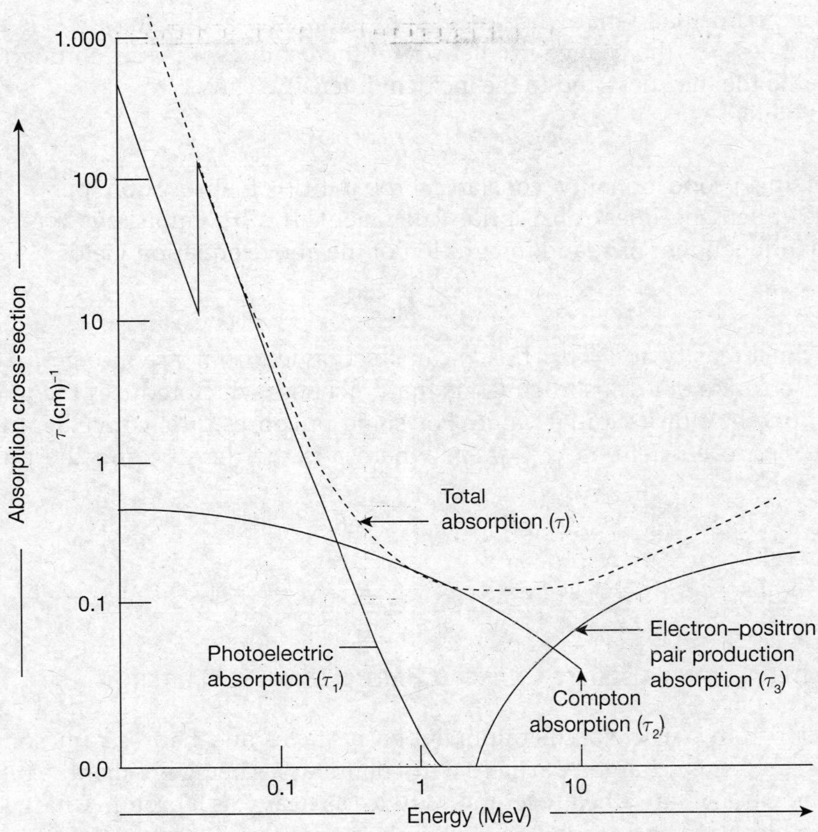

FIGURE 33.5 Energy regions of prominent modes of interaction of photons in matter

In the photo-electric effect, the photon imparts all its energy to an electron in an atom of the absorbing material, giving the electron kinetic energy equal to the original energy of the photon minus the binding energy of the electron. This type of interaction is most probable for a gamma ray with energy just in excess of the binding energy of the electron, the probability decreasing with increasing energy beyond this value. The photo-electric effect is the principal means of absorption at low gamma ray energies (below about 0.5–0.6 MeV), since electron binding energies range from a few thousand electron volts (keV) for light elements to about 100 keV for heavy elements. In the Compton effect, the photon imparts a portion of its energy to an electron and is deflected in the process. The relative probability of this type of interaction is quite high for gamma rays with energies ranging from about 0.6 to 2.5 MeV. The probability of this type of interaction with an atom is also proportional to the number of electrons, or to the atomic number, Z. In pair production, the entire energy of the photon is absorbed in the production of a positron and an electron. The minimum gamma ray energy necessary for this process is 1.02 MeV, equivalent to the sum of the rest masses of the positron and electron. Any gamma energy in excess of 1.02 MeV appears as kinetic energy of the particles involved. The probability of this interaction increases for increased gamma ray energy above the threshold, and also for increased atomic number of the absorbing material.

It is found experimentally that when a beam of gamma rays of intensity I_0 is incident on a slab of thickness dx, the change in intensity of the beam as it passes through the slab is proportional to the thickness and to the incident intensity. Thus,

$$\log(I_0 - I) \propto \mu dx$$

where μ is the proportionality constant. Probability of absorption per unit distance traversed is called the linear absorption coefficient. If all gamma rays are of the same energy, μ is independent of x, and integration of the above equation yields

$$I = I_0 e^{-\mu x}$$

where I_0 is the intensity at $x = 0$, Thus, μ can be obtained from a semi-log plot of I versus x. The half-thickness of an absorber $d_{1/2}$ is that thickness which reduces the photon beam intensity to one-half of its initial value. For some purposes, it is convenient to define a mass absorption coefficient as $\mu_m = \mu / \rho$, where ρ is the density, and the unit of μ_m is cm^2/g. Note that

$$I = I_0 e^{-\mu_m t} \tag{33.1}$$

where t is the absorber thickness in g/cm^2.

33.2 ▶ Interaction of Heavy Charged Particles with Matter

Atomic nuclei from particle accelerators, alpha particles and hadrons (mesons, baryons and protons) are heavy charged particles with their masses being much more than leptons. However, the mass of tau, a heavy lepton, is twice as heavy as a proton. Charged particles interact with electrons and nuclei via Coulomb interaction. These particles move at high speed. For example, an alpha particle with 1 MeV kinetic energy moves with a speed of 6.9×10^6 m/s, almost 2 per cent of the speed of light. At higher energies, Einstein's theory of relativity should be applied to estimate the speed, momentum and mass. Because of the high molecular density, the number of ion pairs per unit volume or per unit length on the path produced by an alpha particle is very high. Multiple ionization also takes place. Due to Coulomb interaction, alpha particles may excite electrons to a higher energy state, rather than knocking these off the molecule. Ionization and excitation break chemical bonds and generate reactive species that cause further chemical reactions.

The kinetic energy of the photoelectron is given by

$$T_e = E_r - B_e$$

where B_e = binding energy and E_r = incident photon energy.

33.2.1 Stopping Power and Ion Pair Density on the Path of Alpha Particles

The stopping power of a medium is the rate of energy loss per unit distance along the path. Born and Bethe have shown that the stopping power of a medium is proportional to the mass M, and to the square of atomic number, Z^2, of the atoms in the medium. Thus, a medium consisting of heavy atoms has high-stopping power. However, the stopping power

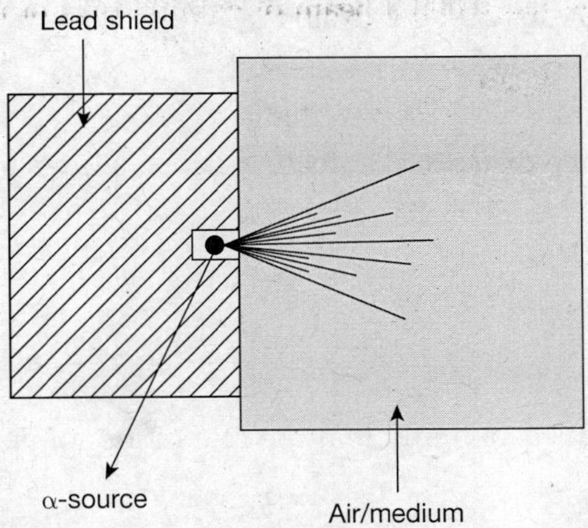

FIGURE 33.6 Observed paths of α-particles in the medium (the medium like nuclear emulsions, air in cloud chamber, bubble chamber, etc. are used to record these tracks)

is inversely proportional to the energy of the particle. A fast-moving particle deposits less energy per unit length on its track. High stopping power results in generating high ion pair density. As alpha particle losses its energy, the stopping power increases. At the end of its path, the stopping power is the highest. Thus, along the path, the ion-pair density is the highest at the path end. Heavy particles lose energy in a medium at a faster rate than light particles. They generate higher ion-pair densities. Thus, the tracks of alpha particles are different from those of fast-moving protons (Fig. 33.6).

Note

Alpha particles that have energy in the range of 4–8 MeV cannot penetrate even the dead outer layer of human skin.

33.2.2 Calculation of Stopping Power and Range of Heavy Particles

Heavy particles such as protons and alpha particles of certain energy lose all their energies at a definite distance in a medium is called the range. Range is the distance travelled by the particle. The calculations for stopping power can be made as follows.

To estimate the loss of energy of a particle of charge Z_e, let us consider an electron inside medium with which the particle interacts. Let the particle be moving with velocity v having distance of closest approach equal to a with respect to electron, at point A.

Momentum transferred to an electron is

$$= \frac{P^2}{2m}$$

where

$$p = \frac{2Ze^2}{av}$$

Therefore, kinetic energy loss per electron is

$$\frac{2Z^2e^4}{mv^2a^2}$$

The total energy loss for a distance $a_{\min}$ to $a_{\max}$ can be written as

$$\frac{dE}{dX} = \frac{4Z^2e^4n}{mv^2}\int_{a_{\min}}^{a_{\max}} \frac{1}{a}\,da$$

or

$$-\frac{dE}{dX} = \frac{4Z^2e^4n}{mv^2}\log_e\left(\frac{2mv^2}{\hbar\nu}\right)$$

The negative sign is added to imply that the rate of loss of energy decreases as energy increases.

This also implies that

$$-\frac{dE}{dX}\alpha\frac{Z^2}{v^2}$$

Meaning that the rate of loss of energy is independent of mass of incident charged particle. Range of radiation in the absorber medium can be given as

$$R = \int_0^{E_0} \frac{dx}{dE}.dE$$

$$R = \int_0^{E_0} \frac{dE}{S(E)}$$

where $S(E) = \frac{dE}{dx}$ is stopping power of the material/absorber.

For α-particles, $R = 0.318\ E^{3/2}$ where R is measured in cms and E in MeV.

EXAMPLE 33.1

On an average, 35 eV is required to produce an ion-pair. How many ion-pairs are produced by an alpha particle with a 5.0 MeV kinetic energy? If the range of the alpha particle is 10 cm what is the average ion-pair density?,

Solution: The number of pair and the ion-pair density is estimated as follows:

$$\frac{5.06\text{eV}}{35\text{eV}} = 142900 \text{ pairs}$$

$$\frac{142{,}900 \text{ pairs}}{10 \text{ cm}} = 14{,}290 \text{ ion-pairs/cm}$$

The number of ion pairs is very high, although the total charge is small.

33.3 ▶ Interaction of Gamma Rays with Matter

Gamma rays have energies that are characteristic of the radioactive decay that produced them. Measurement of these energies provides a useful method of identifying the type of radioactivity in a sample. Gamma rays penetrate to great depths in materials, and no amount of absorber can completely stop all of the gamma radiation. What is usually done in practice is to use sufficient thickness of an absorber to reduce the radiation level to an acceptable value.

The total attenuation coefficient for a gamma ray or X-ray travelling through a medium has contributions from scattering, photo-electric absorption and pair production processes. For this purposes, the photonuclear and other such small effects are neglected. These three main processes of interaction are discussed below.

33.3.1 Compton Scattering

This process of interaction of photons with matter is already discussed in Chapter 31 on X-rays. Figure 33.7 schematically shows the process of Compton scattering. In this case, an incident ray scatters from an outer shell electron in the absorber material at an angle θ, and some of the γ-ray energy is imparted to the electron. Conservation of energy and momentum leads us to the following expression for the energy of the scattered photon:

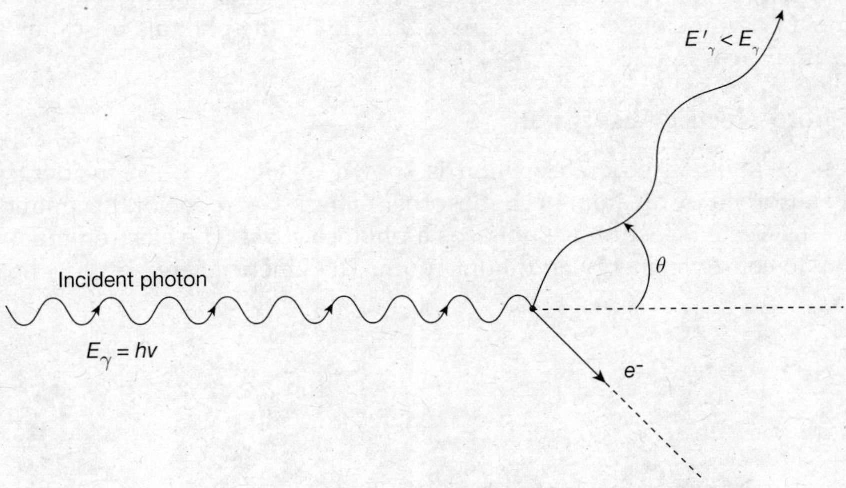

FIGURE 33.7 Process of Compton scattering

Energy of scattered photon $$E'_\gamma = \frac{E_\gamma}{1+(E_\gamma / m_0 c^2)(1-\cos\theta)} \tag{33.2}$$

where E_γ is the incident photon energy, E'_γ is the energy of the scattered photon, θ is the scattering angle and $m_0 c^2$ is the electron rest mass energy. The kinetic energy of the electron after the collision is given by

$$T_e = E_\gamma - E'_\gamma = \frac{E_\gamma^2 (1-\cos\theta)}{m_0 c^2 + E_\gamma (1-\cos\theta)} \tag{33.3}$$

It can be seen that, since all scattering angles are possible, the electron energy ranges from zero for $\theta = 0°$ to $2E_\gamma^2 / (m_0 c^2 + 2E_\gamma)$ for $\theta = 180°$, and that the photon never loses the whole of its energy in any one collision. The scattered photon can then continue through the absorber and interact again or scatter out of the absorber material completely. This process, where the scattered photon escapes, is very important for the γ-ray spectroscopist. If the full energy of the incident photon is not absorbed in the detector, then there is a continuous background in the energy spectrum, known as Compton continuum.

This continuum extends up to an energy corresponding to the maximum energy transfer, where there is a sharp cut-off point, known as the Compton edge.

Note

Gamma rays penetrate to great depths in materials, and no amount of absorber can completely stop all of the gamma radiation.

Compton scattering is the most probable process for photons in the intermediate energy range and the probability decreases rapidly with increasing energy. The probability is also dependent on the number of electrons available for the photons to scatter and hence increases with increasing Z.

33.3.2 Photo-electric Absorption

The process of photo-electric absorption is shown in Fig. 33.8. An incident photon is completely absorbed by an atom in the absorber material, and one of the atomic electrons is ejected. This ejected electron is known as a photoelectron. The electron must be bound to the atom, to conserve energy and momentum. The kinetic energy of the photoelectron is given by

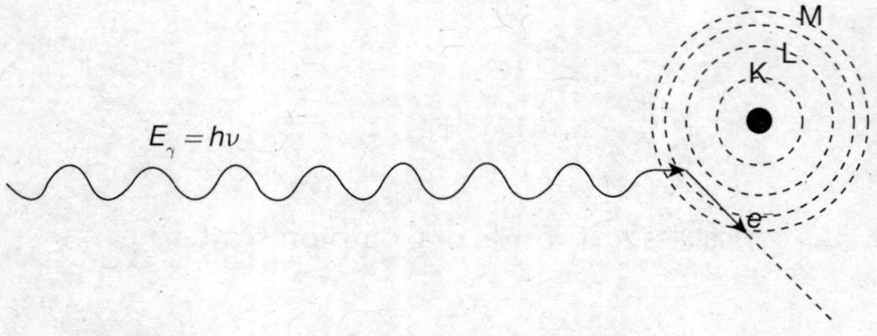

FIGURE 33.8 Process of photo-electric absorption

$$T_e = E_\gamma - B_e \text{ or } T_e = h\nu - B_e \tag{33.4}$$

where B_e is the binding energy of the atomic electron. The vacancy left in the atomic structure by the ejected electron is filled by one of the electrons from a higher shell. This transition is accompanied by an emission of X-rays. These X-rays are also absorbed by the detector.

Photo-electric absorption is the most favourable process for γ-ray spectroscopist, since the incident photon deposits all of its energy into the detector, but it is dominant only for low-energy photons (< 200 keV). The interaction is again dependent upon Z, and an approximate expression for the absorption probability τ is

$$\tau \propto \frac{Z^n}{E_\gamma^{3.5}} \tag{33.5}$$

Here, n is a constant that normally lies between 4 and 5 depending on the absorber material. This dependence on Z explains the choice of high Z materials such as lead for shielding purposes.

33.3.3 Pair Production

Soon after the discovery of positron by Anderson in 1932, Dirac's quantum theory was extended, which predicted that if a high-energy photon whose energy is of the order of 1.02 MeV or higher comes close enough to the nucleus of an atom, the electric field of the nucleus at such distance would be strong enough to annihilate the γ-ray and create in its place a pair of particles, an electron and a positron. These two particles according to the theory should have exactly same mass and equal but opposite charges. Blackett, Anderson and others looking for such pairs found these exactly as predicted by Dirac's theory. Gamma rays from a radioactive element like $_{81}Tl^{208}$ while passing through matter were observed to produce pairs of electron and positron.

When an electron–positron pair is created, conservation of energy and momentum requires the two particles to move in almost one straight direction, without magnetic field applied to the cloud chamber, the particles travel side by side in almost parallel paths, but with a magnetic field the path of positron bends to one side and of electron to the other (Fig. 33.9).

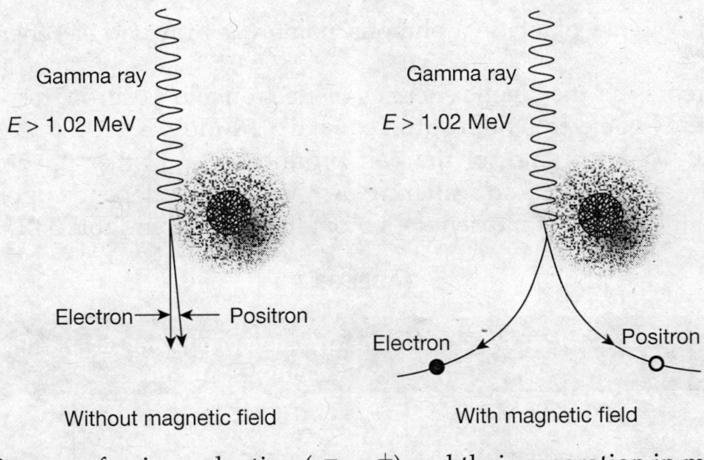

FIGURE 33.9 Process of pair production ($e^- - e^+$) and their separation in magnetic field

The pair of electron and positron is very unstable because as soon as a positron meets with an electron, the two get annihilated. Because a positron can annihilate an electron, thus it is called an antiparticle of electron.

Experiments indicate that all electrons and positrons spin around an axis through their centre of mass. There is a good evidence that when a positron and an electron come close together they frequently combine by revolving around each other like a double star, with their spin axes parallel to each other. Such a pair of electron and positron before annihilation is called positronium.

Positronium is very short lived because as soon as electron and positron combine they annihilate to give two γ-photons which are emitted in opposite directions to conserve the momentum (Fig. 33.10).

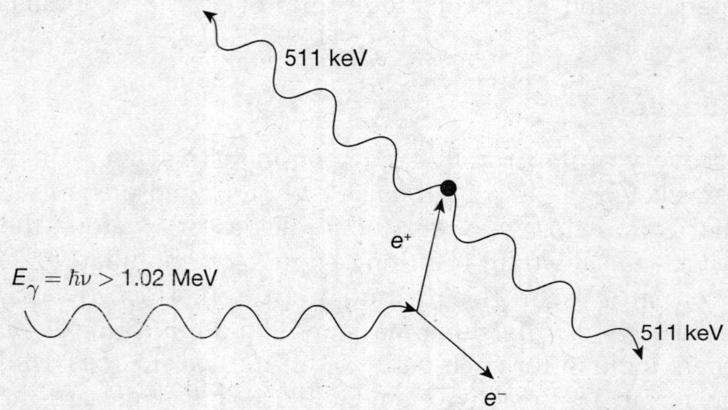

FIGURE 33.10 Process of pair production/annihilation

Energetics of pair production is as follows:

$$h\nu = (m_0 c^2 + T_{e^-}) + (m_0 c^2 + T_{e^+}) \tag{33.6}$$

where $h\nu$ is the frequency of incident photons, namely, a γ-ray and has energy greater than 1.02 MeV.

T_{e^-} and T_{e^+} represent the kinetic energies of electron and positron, respectively, m_0c^2 is the rest mass energy of electron or positron equal to 0.51 MeV.

The probability of occurrence of the pair production and other phenomenon, that is, photo-electric effect and Compton scattering by γ-ray photon depends upon the value of Z for absorber and the energy of incident γ-photon. It is clear from Table 33.1.

TABLE 33.1

Absorber	Photo-electric effect	Energy range of photons for Compton scattering	Pair production
$_{13}$Al	< 0.05 MeV	0.05–15 MeV	> 15 MeV
$_{82}$Pb	< 0.05 MeV	0.5–5 MeV	> 5 MeV

Figure 33.11 shows three major processes of interaction of photons with matter in their energy ranges.

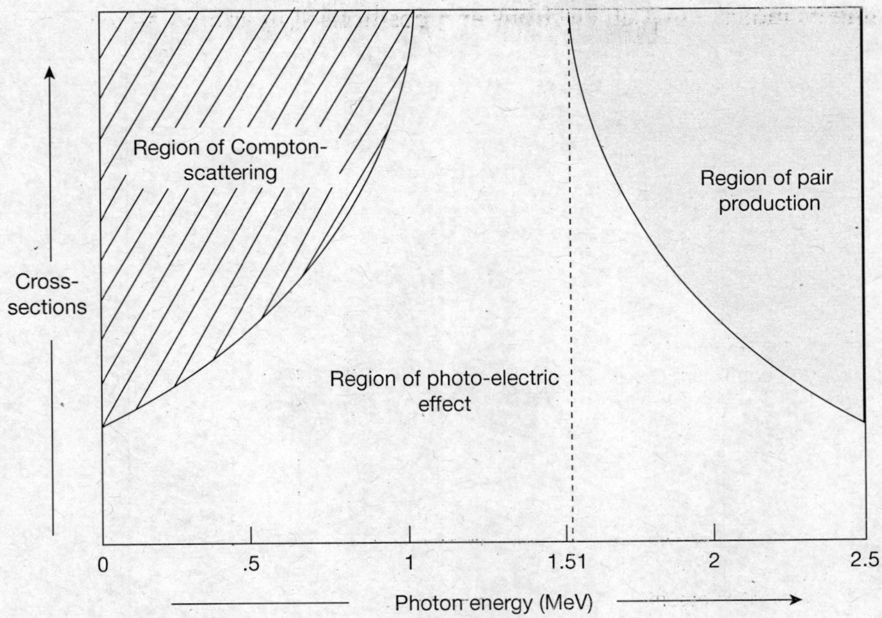

FIGURE 33.11 Comparison of cross-sections for Compton scattering, photo-electric effect and pair production

33.3.4 Attenuation Coefficients

As might be expected, the interaction of γ-rays varies depending on the matter in which the interactions take place. This variation can be described as a function of the atomic number of the material. Figure 33.11 shows linear attenuation coefficients for the three main interaction processes within germanium crystals. The total linear attenuation coefficient μ_{total} is given by

$$\mu_{\text{total}} = \mu_{pe} + \mu_{cs} + \mu_{pp} \tag{33.7}$$

where μ_{pe}, μ_{cs} and μ_{pp} are the attenuation coefficients for the photo-electric effect, Compton scattering and pair production, respectively. The intensity I of a γ-ray absorbed within the material then becomes

$$I = I_0 e^{-\mu_{\text{total}} x} \tag{33.8}$$

where I_0 is the incident γ-ray energy and x is the distance travelled within the absorber. From Fig. 33.12, it can be seen that Compton scattering is the most important process for the γ-ray energies involved in this (0.1–1 MeV).

Linear attenuation coefficients for each of the three main processes are plotted against γ-ray energy for germanium. The total attenuation coefficient is also shown μ_{cs}, μ_{pe}, μ_{pp} and μ_{total} are the attenuation coefficients for Compton scattering, photo-electric effect, pair production and the total, respectively.

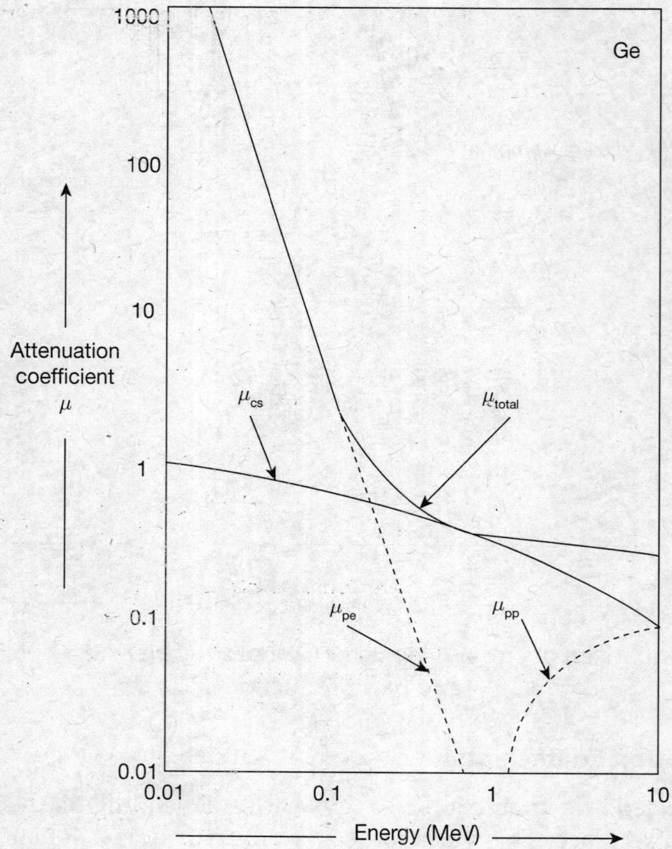

FIGURE 33.12 Attenuation coefficients μ for Ge as a function of photon energy in MeV. Various components are also shown

33.4 ▶ Interactions of Electrons with Matter

Two different types of interactions of electrons occur with matter.

33.4.1 Elastic Interactions

No energy is transferred from the electron to the sample. The electron either passes without any interaction (direct beam) or is scattered by the positive potential inside the electron cloud. These signals are mainly exploited in TEM and electron diffraction.

33.4.2 Inelastic Interactions

Energy is transferred from the incident electrons to the sample: secondary electrons, phonons, UV quanta or cathodo-luminescence is produced; shooting out inner shell electrons leads to the emission of X-rays or Auger electrons. These signals are used in analytical electron microscopy.

Inelastic interactions of electrons with matter

Ionization: The high-energy electrons of the incident beam can transfer critical amount of energy to inner shell electron of an atom, leading to the ejection of this electron. The ionization energy is provided by the incident electron, reducing its energy. This leads to an ionization edge in the electron energy loss spectrum (EELS). Subsequently, the hole in the inner shell is filled up by an electron with higher energy from an outer shell. This electron gives away a part of its energy, leading to the emission of characteristic X-rays or Auger electrons.

Secondary electrons: Electrons in the conduction or valence band do not need much energy (low work function) to be transferred into vacuum. Thus, the energy of secondary electrons (SE) is low (> 50 eV). The SEs are mainly exploited in SEM.

Phonons: Phonons are lattice vibrations, which are equal to heating the specimen. This effect may lead to damage of the sample.

Plasmons: Plasmons are longitudinal oscillations of free electrons, which decay either in photons or in phonons.

Cathodo-luminescence: In semiconductors, hit by high-energy electrons, electron–hole pairs are formed. Filling this hole with an electron from the conduction band leads to the emission of light with a frequency that corresponds to the band gap.

Note

Very high energy beta particles can penetrate to a depth of about a centimetre in tissue and can damage eye and skin if the source is strong.

33.4.3 Electron Interaction with Matter

Electron bombardment of a sample is unique to microprobe analysis where it produces a large number of effects from the target material (Fig. 33.13). The incident electrons interact with specimen atoms and are significantly scattered by them (rather than penetrating the sample in a linear fashion). Most of the energy of an electron beam will eventually end up heating the sample (phonon excitation of the atomic lattice); however, before the electrons come to rest, they undergo two types of scattering – elastic and inelastic.

In elastic scattering, the electron trajectory changes, but its kinetic energy and velocity remain essentially constant (due to large differences between the mass of the electron and nucleus). This process is known as electron backscattering (although later we will confine the term "backscattered electrons" to those which scatter out of the sample).

In inelastic scattering, the trajectory of the incident electron is only slightly perturbed, but energy is lost through interactions with the orbital electrons of the atoms in the specimen. Inelastic interactions produce diverse effects including phonon excitation (heating), cathodo-luminescence (visible light fluorescence), continuum radiation (bremsstrahlung or "braking" radiation), characteristic X-ray radiation, plasmon production (secondary electrons) and Auger electron production (ejection of outer shell electrons).

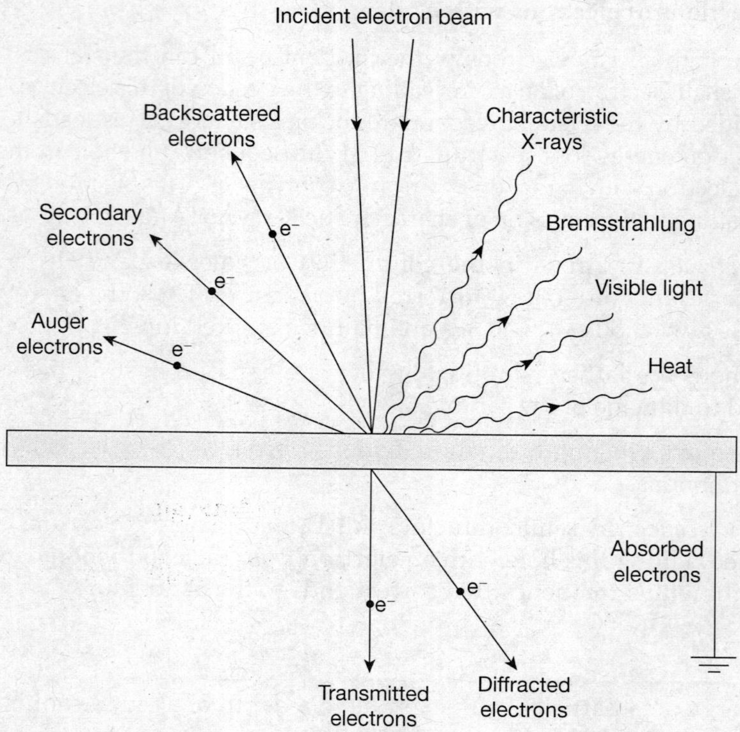

FIGURE 33.13 Effects produced by electron bombardment on a material

Two major factors with effect can be detected from the interaction volume. First, some effects are not produced from certain parts of the interaction volume (Fig. 33.14). Beam electrons lose energy as they traverse the sample due to interactions with it and if too much energy is required to produce an effect, it will not be possible to produce it from deeper portions of the volume. Second, the degree to which an effect, once produced, can be observed is controlled by how strongly it is diminished by absorption and scattering in the sample.

For example, although secondary and Auger electrons are produced throughout the interaction volume, they have very low energies and can only escape from a thin layer near the sample's surface. Similarly, soft X-rays, which are absorbed easier than hard X-rays, will escape more readily from the upper portions of the interaction volume. Absorption is an important phenomenon and is discussed in more detail.

Figure 33.14 gives generalized illustration of interaction volumes for various electron–specimen interactions. Auger electrons (not shown) emerge from a very thin region of the sample surface (maximum depth about 50Å) than secondary electrons (50–500 Å).

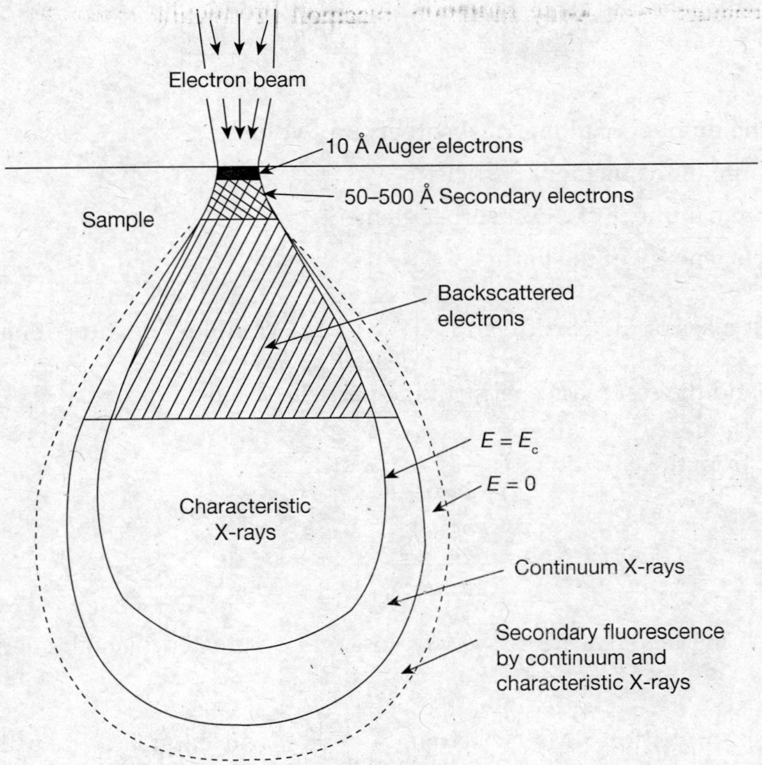

FIGURE 33.14 Relative interaction volumes for various processes for an electron beam in a thin sample

33.4.4 Bremsstrahlung Loss

Bremsstrahlung process or radiative collision of light particles is a nuclear phenomenon, in which fast-moving charged particle passing close to the nucleus losses its energy by radiation.

According to electrodynamics, whenever there is an acceleration or decceleration (retardation) of a charged particle, it loses energy in the form of radiation. Those radiation are called Braking radiation or *Bremsstrahlung*.

Bethe and Heitler in 1934 declared that when an energetic charged particle passes close to an atomic nucleus, its interaction with the strong field of the nucleus makes it suffer a large acceleration or retardation (depending upon the type of charge on the particle) which results in the transformation of its energy into electromagnetic radiation which is usually called the impulse radiation or *Bremsstrahlung* or braking radiations. The process is also called radiative collision.

Average loss of energy due to radiative collision of particle of mass m_0 is given after Bethe and Heitler as

$$-\left.\frac{dE}{dx}\right|_{\text{rad}} = \frac{nZ_1^4 Z_2^2}{m_0^2} E \cdot f\left(\frac{E}{m_0 c^2}\right) \tag{33.9}$$

where n is the number of atoms of absorber/unit volume,

Z_1 is the charge of the incident particle,

Z_2 is the atomic number of the absorber element,

E is the kinetic energy of the particle,

m_0 is the rest mass of the particle and $f\left(\frac{E}{m_0 c^2}\right)$ is a complex function of kinetic energy of the particle and the rest mass energy of the particle.

It is evident from the equation for $-\left.\frac{dE}{dX}\right|_{\text{rad}}$ that

(i)
$$-\left.\frac{dE}{dX}\right|_{\text{rad}} \propto E$$

which implies that *Bremsstrahlung* loss will be greater for high-energy particles as shown in Fig. 33.15.

For example, for 5 MeV electron, $-\left.\frac{dE}{dX}\right|_{\text{rad}}$ will be greater than that for 1 MeV electron.

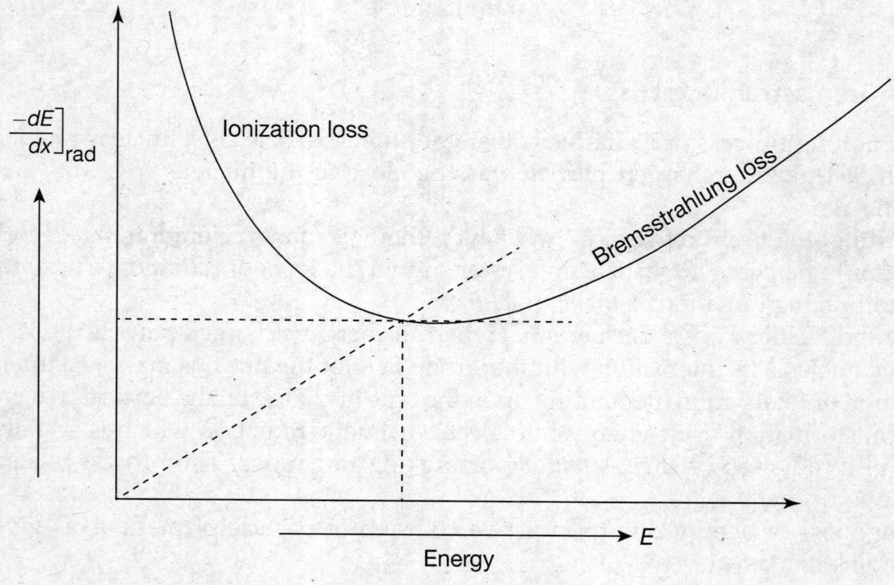

FIGURE 33.15 Radiative collision loss versus energy

(ii)
$$-\frac{dE}{dX}\bigg|_{\text{rad}} \propto Z_2^2,$$

i.e., *Bremsstrahlung loss* is directly proportional to the atomic number of absorber and therefore will be comparatively very large for lead $Z_2 = 82$ than for aluminium $Z_2 = 13$.

(iii)
$$-\frac{dE}{dX}\bigg|_{\text{rad}} \propto \frac{1}{m_0^2}$$

which means that *Bremsstrahlung loss* is greater for the particles of smaller mass. Thus it becomes very important in case of electrons, positrons; π mesons and so on, than in case of an α-particle. For positrons along with ionization and *Bremsstrahlung* loss, there is also a loss due to annihilation radiation.

Range energy relation for electrons

The range R of an electron can be given as

$$R = 412E_{\text{max}}^n \left(\text{mg/cm}^2\right)$$

where n is an index and is given as 0.01 MeV $< E_{\text{max}} <$ 2.5 MeV.

$$n = [1.265 - 0.09541\text{og } E_{\text{max}}]$$

$$n = 1.265 - 0.0954 \log E_{\text{max}}$$

For energies, 1 MeV $< E <$ 20 MeV

$$R\ (\text{mg/cm}^2) = 530E - 106$$

Note

High energy beta particles produce X-rays called Bremsstrahlung when absorbed, and this effect is more pronounced in heavy and dense absorbers.

33.4.5 Backscattered Electrons

Scattering of beam electrons produces backscattered and transmitted electrons as strong electrical field of the specimen's atomic nuclei deflects them; no additional electrons are produced from the sample. While transmitted electrons pass completely through the material after interacting with it, backscattered electrons are ejected from the top surface of the specimen at greater angles. Transmitted and backscattered electrons can have energies ranging from about 50 eV up to the accelerating voltage (E_0).

The number of backscattered electrons produced from a material may be quantified by its backscattering coefficient, n_b. This coefficient depends on a sample's average atomic number, Z.

Average Z is calculated using the weight fractions (w) of each element:

$$Z = w_1 z_1 + w_2 z_2 + w_3 z_3 + \cdots + w_n z_n$$

Thus, for SiO_2, with 0.4674 Si and 0.5326 O by weight:

$$Z_{SiO_2} = 0.4674 \times 14 + 0.5326 \times 8 = 10.8044$$

and $n_b = 0.142$. About 48 per cent of incident electrons are backscattered by a tungsten target ($Z = 74$), whereas only about 14 per cent are produced by sodium ($Z = 11$).

33.4.6 Secondary Electron Emission

When a sample is bombarded with electrons, the strongest region of the emitted electron spectrum is due to secondary electrons. The secondary electron yield depends on many factors, and is generally higher for high atomic number targets, and at greater angles of incidence. Secondary electrons are produced when an incident electron excites an electron in the sample and loses some of its energy in the process. The excited electron moves towards the surface of the sample undergoing elastic and inelastic collisions until it reaches the surface, where it can escape if it still has sufficient energy. Their energies are a function of E_0 and the surface work function, E_w, which defines the amount of energy needed to remove electrons from the surface of a material. One of the major reasons for coating a non-conductive specimen with a conductive material is to increase the number of secondary electrons that will be emitted from the sample (decrease E_w).

Secondary electrons, by convention, are those emitted with energies less than 50 eV. This is only a small fraction of the electrons emitted from the sample (Fig. 33.16).

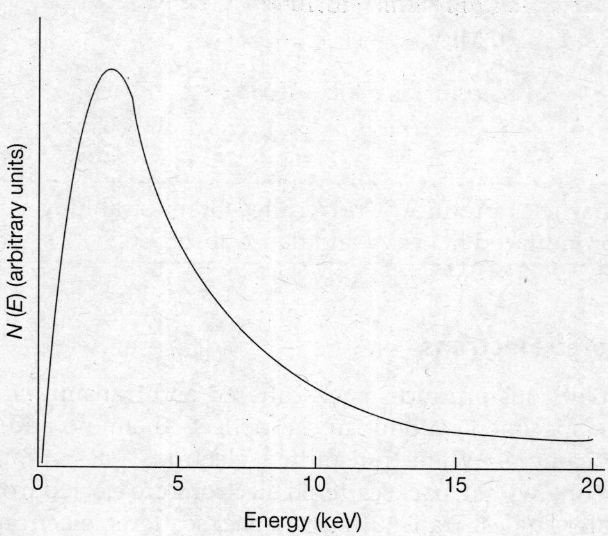

FIGURE 33.16 Energy distribution of secondary electrons

The mean free path length of secondary electrons in numerous materials is approximately 10 Å. Thus, although electrons are generated throughout the region excited by the incident beam, only those electrons that originate less than 10 Å deep in the sample escape to be detected as secondary. This volume of production is very small compared with back scattered electron (BSE) and X-rays. Therefore, the resolution using SE is better than either of these and is effectively the same as the electron beam size. The shallow depth of production of detected secondary electrons makes them very sensitive to topography and they are used for scanning electron microscopy (SEM).

33.4.7 Auger Electrons

The Auger process was discovered by Pierre Auger (pronounced as "O-jhay") in 1926, when he observed tracks of constant length in a cloud chamber, but Auger electrons were not used to study surfaces until late 1950s and 1960s. Auger electrons are electrons ejected by radiation less excitation of a target atom by the incident electron beam. When an electron from the *L*-shell drops to fill a vacancy formed by *K*-shell ionization, the resulting X-ray photon with energy $E_K - E_L$ may not be emitted from the atom. If this photon strikes a lower energy electron (e.g., an *M*-shell electron), this outer electron may be ejected as a low-energy Auger electron. Auger electrons are characteristic of the fine structure of the atom and have energies ranging between 280 eV (carbon) and 2.1 keV (sulphur). By discriminating between Auger electrons of various energies, a chemical analysis of the specimen surface can be made.

Auger electron energies are closely related to the corresponding X-ray energy and are usually described in X-ray notation. For example, the Si KL_1L_{23} transition, experimentally observed at 1620 eV, involves removal of an electron in *K*-shell allowing an electron from the L_1 shell to descend with the emission of energy of 1690 eV. This energy can either be emitted as a *Si-Ka* X-ray, or it can be transferred to the third electron, in this case in the $L_2 - L_3$ sub-shell, which has a binding energy difference of about 90 eV, ejecting it from the atom with an energy of around 1600 eV.

The probability of production of Auger electron increases as the difference between the energy states of the shells decreases. Light elements are more susceptible to the formation of Auger electrons by multiple ionizations. Thus, the proportion of radiation emitted at characteristic wavelengths is lower than for heavier elements. The proportion of Auger emission is greater than 0.5 up to about Z = 30 (zinc).

The Auger phenomenon is described by the fluorescence yield, ω, which for K-radiation is defined as:

$$\omega_k = n_k/N_k$$

where ω_k = fluorescence/yield,
n_k = number of X-ray photons emitted from the sample and
N_k = number of ionizations.

Fluorescence yields for the light elements are generally less than 0.2 for the *K*-lines. The X-ray yield increases sharply with increasing Z while Auger electron yield decreases (Fig. 33.17). Thus, Auger electrons provide a good basis for analysis of light atoms. One might expect that X-ray intensities would be lower at low Z because of increased Auger electron production, but lower fluorescence yield compensates for easier ionization.

Auger electrons are produced from depths of about a wavelength into the sample because their low energies make them easy to be reabsorbed. This makes Auger electrons particularly

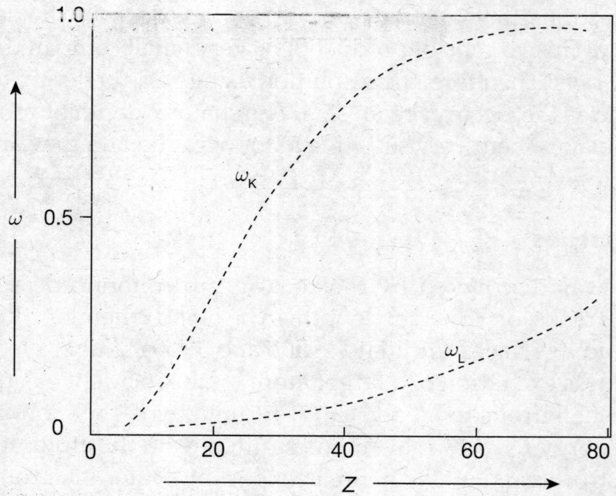

FIGURE 33.17 Schematic variation of fluorescence yields of the *K* and *L* spectral lines with atomic number (*Z*) of the emitting element

good for analysis of surface composition, but such analysis requires ultra-high vacuum to avoid absorption losses.

Figure 33.18 shows *K* shell fluorescence yield as a function of atomic number.

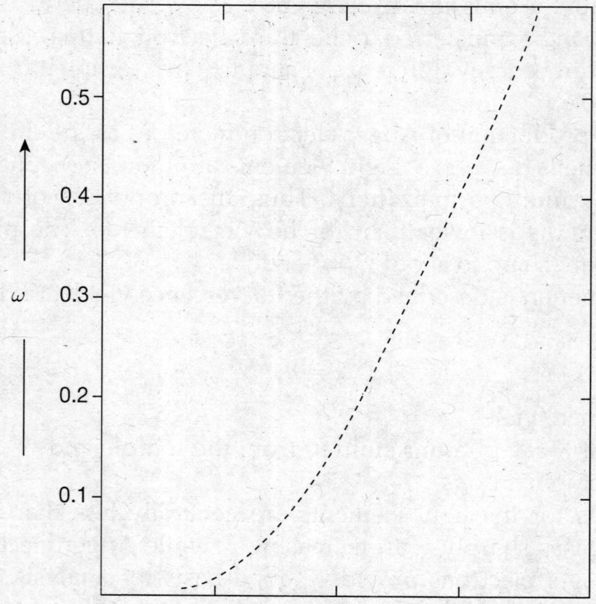

FIGURE 33.18 Experimentally determined *K*-fluorescence yield as a function of atomic number

33.5 ▶ Interaction of Neutrons with Matter

The interaction of neutrons with matter depends to a large extent on their energies, which can range from hundreds of MeV down to fractions of an eV. Neutrons are uncharged particles and do not interact with atomic electrons in the matter through which they pass but with the nuclei of these atoms. The nuclear force, which leads to these interactions, is very short ranged meaning that thereby neutrons have to pass close to the nucleus for the interaction to take place. Because of the small size of the nucleus in relation to the atom as a whole, the neutrons will have a low probability of interaction and could thus travel considerable distances in matter.

The most common neutron reactions are listed below:

- Spallation reactions
- Elastic scattering
- Inelastic scattering
- Transmutation
- Radiative capture

33.5.1 Spallation Reactions

At very high energies (above 150 MeV), neutrons may strike a nucleus producing a shower of secondary particles. These high energy secondary particles are harmful. They are in general not a problem, however, with the exception of neutrons and gamma-rays because:

- They disappear when the accelerator is not operating.
- Only neutrons penetrate the accelerator shielding when the accelerator is in operation.

33.5.2 Elastic Scattering

This is analogous to a billiard ball type of collision. The neutron collides with a nucleus and rebounds in a different direction. The energy lost by neutron is gained by the target nucleus, which moves away at an increased speed. If the neutron collides with a massive nucleus, it rebounds with almost the same speed and losses very little energy. Light nuclei, on the other hand, will gain a lot of energy from such a collision and will therefore be more effective for slowing down neutrons.

Elastic scattering, which is illustrated in Fig. 33.19, is not effective in slowing down neutrons with very high energy (above 150 MeV).

33.5.3 Inelastic Scattering of Neutrons

A neutron may strike a nucleus and form a compound nucleus instead of bouncing off as inelastic scattering. This nucleus is unstable and emits a neutron of lower energy together with a gamma photon which takes up the remaining energy. This process, called inelastic

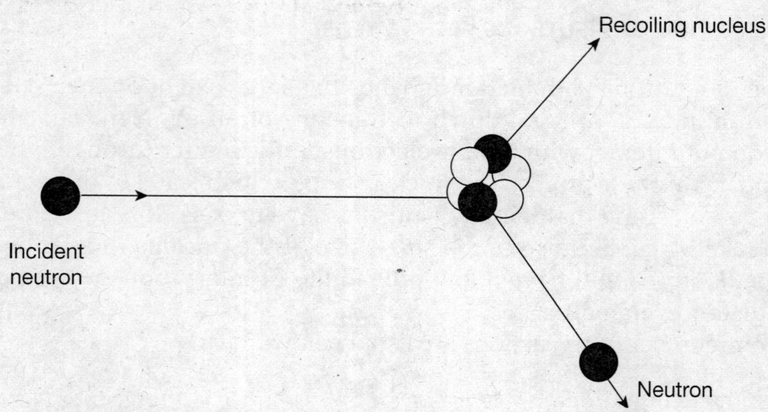

FIGURE 33.19 Elastic scattering of neutron

scattering, is most effective at high neutron energies in heavy materials, but at lower energies (a few MeV), elastic scattering becomes a more important reaction for energy loss provided there are light nuclei present. An example is shown in Fig. 33.20.

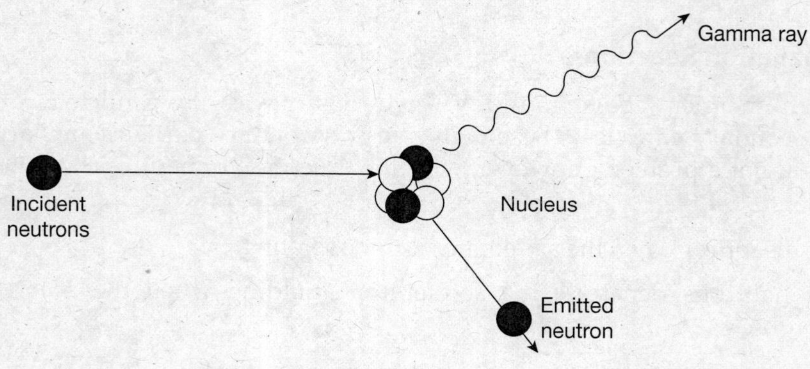

FIGURE 33.20 Inelastic scattering

33.5.4 Transmutation

Neutrons, protons or other secondary particles produced by spallation, strike a nucleus and form a compound nucleus. It ejects a different particle. A transmutation is said to have occurred. This is because the target nucleus is changed from one element to another. These nuclear reactions are most likely to occur when the energy of the incident particle is between a few MeV and several tens of MeV. An example is the neutron–proton (n, p) reaction with oxygen as described below:

$_{16}$O captures a high-energy neutron and emits a proton to form $_{16}N$. The product, $_{16}N$, is radioactive with a half-life of 7.2 seconds; it is a β-emitter and also emits very hard gamma radiations, which have energies of 6 7 MeV.

$$^{16}_{8}\text{O} + ^{1}_{0}n \Rightarrow ^{16}_{7}N + ^{1}_{1}p$$

This may be written in the shorter form as shown below:

$$^{16}_{8}\mathrm{O}(n,p)\,^{16}_{7}\mathrm{N}$$

In addition to (n, p) reactions, many other reactions are possible such as (p, n), $(p, 2n)$, $(n, 2n)$, (π, n) and (n, α). Most of these reactions result in nuclei which are deficient in neutrons and therefore decay by positron emission or electron capture.

The (n, α) reaction in boron (B) is an important reaction which is used for the detection of slow neutrons. Slow neutrons are captured by $_{10}B$ causing the following reaction:

$$^{10}_{5}\mathrm{B} + {}^{1}_{0}n \Rightarrow {}^{7}_{3}\mathrm{Li} + \alpha$$

or

$$^{10}_{5}\mathrm{B}(n,\alpha)\,^{7}_{3}\mathrm{Li}\,.$$

33.5.5 Radiative Capture

This is one of the most common neutron reactions. The neutron is again captured by a nucleus which emits only a gamma photon. This reaction occurs in most materials. It is important for neutrons with very low energy. The product nuclei of (n, γ) reactions are usually radioactive and are β- and γ-emitters.

Two of the neutron capture reactions which are important are the (n, γ) reaction in $_{59}$Co, which is normal stable cobalt metal and quite commonly occurs in steel, to produce $_{60}$Co, which is radioactive. The cobalt readily captures neutrons, and $_{50}$Co has a half-life of about 5 years. The other is the neutron capture in $_{23}$Na, which is normal, stable sodium. In this case, the product is the radioisotope $_{24}$Na. Traces of sodium are present in the concrete shielding. Figure 33.21 illustrates the latter reaction.

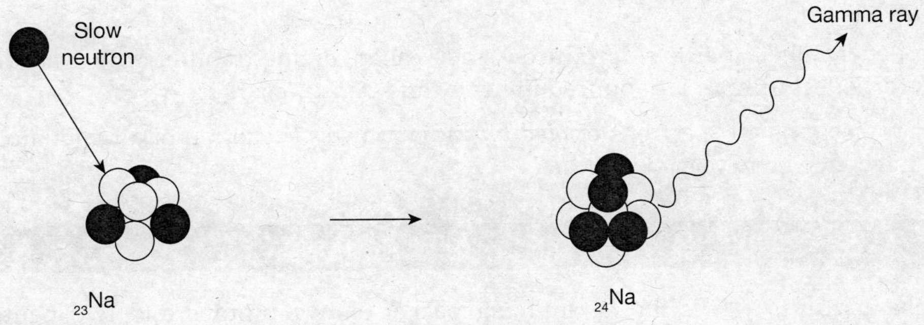

FIGURE 33.21 Neutron capture in $_{23}$Na

33.5.6 Interaction of Neutrons with Living Tissues

The human body is composed largely of water, about 60 per cent by weight, which contains many hydrogen nuclei. Elastic scattering of the neutrons with the hydrogen nuclei will cause the protons to recoil violently. Similarly, elastic collisions of neutrons with carbon, oxygen or other heavier nuclei will cause these to recoil. Because the mass of protons and the other recoiling nuclei is much greater than that of electrons, they generate a much denser ion path resulting in more damage to the tissue. Once neutrons have been slowed down by elastic collisions to thermal energy, 0.025 eV, they are readily captured by some of the reactions described earlier.

A very common reaction is the (n, γ) reaction, particularly with hydrogen. The gamma photon produced in this reaction always has energy of 2.2 MeV and will cause indirect ionization as described earlier. When neutrons are absorbed by an (n, γ) reaction in the body, the tissues will be further damaged by gamma radiation in addition to damage, which they receive in slowing down the neutrons.

Recap Zone

Points to Remember

- Alpha and beta particles are high-energy charged particles. Neutrons are uncharged particles and gamma rays are electromagnetic radiations.
- Three processes are mainly responsible for the absorption of gamma rays: photo-electric absorption, Compton scattering and pair production.
- In pair production, the entire energy of the photon is absorbed in the production of positron and electrons.
- The stopping power of a medium is the rate of energy loss per unit distance along the path.
- In inelastic scattering, the trajectory of the incident electron is slightly perturbed, but energy is lost through interactions with the orbital electrons of the atom in the specimen.
- High-energy beta particles produce X-rays called Bremsstrahlung when absorbed, and this effect is more pronounced in heavy and dense absorbers.
- Auger electrons are electrons ejected by radiation less excitation of a target atom by the incident electron beam.

Key Terms with Definitions

- **Alpha particles** – It is the fast-moving packet of two protons and two neutrons, expelled as a single particle from a disintegrating nucleus.
- **Beta particles** – They are much more penetrating than alpha particles. Very high energy beta particles can penetrate to a depth of about a centimetre in tissue and can damage eye and skin if the source is strong.
- **Gamma rays** – It is the excess energy of an excited nucleus which is emitted as photon. Its energy range, although differs according to the state of excitement of the nucleus, covers from keV to several MeV.

- **Stopping power** – Stopping power of a medium is the rate of energy loss per unit distance along the path.
- **Compton scattering** – It is the inelastic scattering of a photon by a charged particle, usually an electron. It results in a decrease in energy (increase in wavelength) of the photon, called the Compton effect.
- **Photo-electric absorption** – It is a form of interaction of X-ray or gamma photon with the matter. A low-energy photon interacts with the electron in the atom and removes it from its shell.
- **Pair production** – In pair production, the entire energy of the photon is absorbed in the production of positron and an electron. It is a creation of an elementary particle and its antiparticle.
- **Bremsstrahlung** – High-energy beta particles produce X-rays called Bremsstrahlung when absorbed and this effect is more pronounced in heavy and dense absorbers.
- **Auger electrons** – Auger electrons are electrons ejected by radiation less excitation of a target atom by the incident electron beam.
- **Elastic scattering** – In this process, the kinetic energy of a particle is conserved in the centre-of-mass frame, but its direction of propagation is modified.
- **Inelastic scattering** – It is a process in which the kinetic energy of an incident particle is not conserved and some of the energy of incident particle is lost or increased.

REVIEW ZONE

SHORT ANSWER QUESTIONS

1. What do you mean by alpha particle? Plot a graph between mean range and energy of alpha particles.
2. Differentiate between beta particles and alpha particles.
3. Briefly explain the three processes responsible for the absorption of gamma rays.
4. What are stopping power and range?
5. Write a short note on elastic and inelastic interactions.
6. What do you mean by Bremsstrahlung radiations? What is Bremsstrahlung loss, explain with the graph?
7. What are Auger electrons?
8. What is inelastic scattering of neutrons?
9. Explain the process of radioactive capture.

CHAPTER

34 Nuclear Physics

Learning Objectives

By the end of this chapter, the student will be able:

- To discuss the structure of the nuclei, size and constituents
- To estimate the nuclear mass, mass defect and binding energy
- To compare various models of the structure of nucleus
- To describe liquid drop model and semi-empirical formula for binding energy
- To introduce the nuclear magic numbers, stability of nuclei and shell model
- To discuss the collective model and the Fermi gas model of nucleus
- To compare various nuclear models

34.1 ▶ Introduction

With the discovery of neutron in 1932, the neutron acquired the status of a prime component of nucleus. But its bonding with proton and accumulation of large number of neutrons together in case of heavy nuclei always intrigued scientists. To explain the behaviour of atomic nuclei of atoms, various models of nucleus have been proposed from time to time but none could single handedly explain all features and details of various interactions involving atomic nucleus. The binding energy that comes in the form of mass defect of nuclei is a measure of nuclear force that binds the nucleons together. Several models like liquid drop model, Fermi gas model, shell model and collective models are in use to explain the detail of nuclear spectra and other properties.

34.1.1 Size of Nucleus

The relevant length scale for measuring the nuclear size is the femtometre, where 1 fm $= 10^{-15}$ metre. Physicists usually call this length one Fermi. The size of nuclei may vary from about one to a few Fermi in radius. Recall that the Bohr radius of Hydrogen is of the order of 10^{-10} metre; hence, the nucleus is far smaller than the atom. Nuclear size was first measured by Rutherford, by taking into account how close α-particles come to the nucleus before the scattering ceased to be pure Coulomb repulsion (at which point they were actually hitting the nuclear surface). Despite its small size, the nucleus has about 99.9 per cent of the mass of the atom.

34.2 ▶ Discovery of the Neutron

The neutron was discovered in 1932, when James Chadwick used scattering data to calculate the mass of this neutral particle. Since the time of Rutherford it had been known that the atomic mass number A of nuclei is slightly more than twice the atomic number Z for most atoms and that essentially mass of the atom is concentrated in the relatively tiny nucleus. In 1930, it was presumed that the fundamental particles were protons and electrons but the presence of electrons was a must in bound form in the nucleus to partially cancel the charge of protons. But by this time, it was known from the uncertainty principle and from "particle-in-a box" type confinement calculations that enough energy is not available to contain electrons inside the nucleus.

A rough scale of energy required for the confinement of a particle to a given dimension can be obtained by setting the de Broglie wavelength of the particle equal to the given dimension. For example, if we presume that the dimension of a hydrogen atom is about 0.2 nm, then the corresponding confinement energy is about 38 eV, the correct order of magnitude for atomic electrons. But to confine an electron to a nuclear dimension of about 5 Fermis requires energy of about 250 MeV. The maximum available confinement energy from the electrical attraction to the nucleus is given by

$$\boxed{\frac{Zke^2}{r} = \frac{79(1.44\ \text{MeV}\cdot\text{fm})}{5\ \text{fm}} = 23\ \text{MeV} << 250\ \text{MeV}} \tag{34.1}$$

So it is theoretically established that electrons cannot exist inside the nucleus.

An experimental breakthrough came in 1930 with the observation by Bothe and Becker that bombardment of Beryllium with alpha particles from a radioactive source produced neutral radiation which was penetrating, but non-ionizing. They presumed that it was the gamma rays, but Curie and Joliot showed that when a paraffin target is bombarded with this radiation, it ejects protons with energy of about 5.3 MeV (Fig. 34.1). This proved to be inconsistent with gamma rays, as can be shown from momentum and energy analysis below.

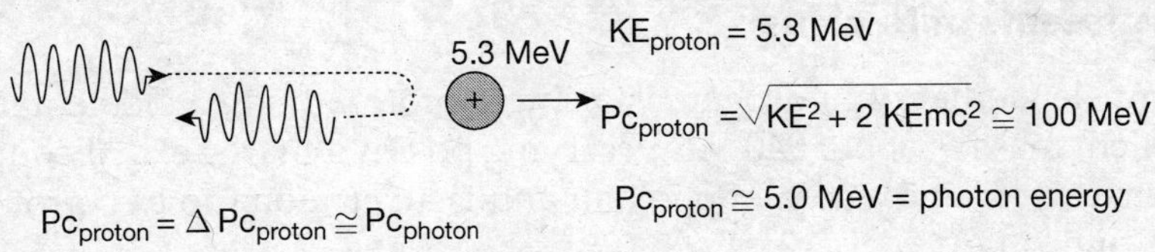

FIGURE 34.1

This analysis follows that for a head-on elastic collision where a small particle strikes a much more massive one. For the gamma ray, the necessary energy was much greater than any observed energy to be available from the nucleus; hence, the neutral radiation must be some kind of a neutral particle.

Note

Energy required for the confinement of a particle to a given dimension can be obtained by setting the de Broglie wavelength of the particle equal to the given dimension.

The 5.3 MeV energy of the ejected protons could be easily explained if the neutral particle had a mass comparable to that of a proton. For head-on collisions, this would require only 5.3 MeV from the neutral particle, a value within the range of observed nuclear particle emissions.

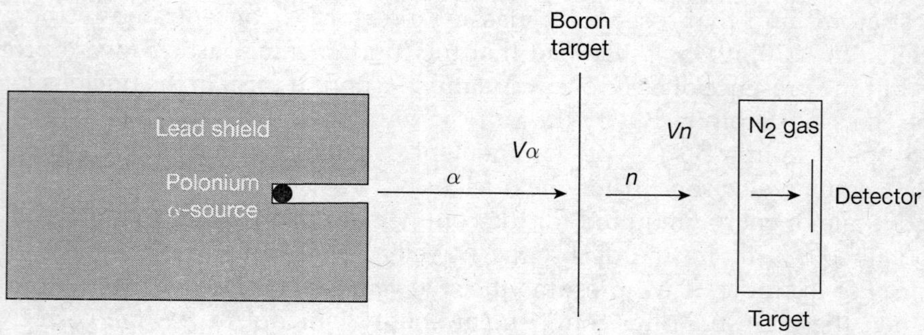

FIGURE 34.2 Chadwick's experiment for discovery of the neutron

Chadwick was able to prove that the neutral particle could not be a photon by bombarding targets other than hydrogen including nitrogen, oxygen, helium and argon in his experiment as shown in Fig. 34.2. Not only were these inconsistent with photon emission on energy grounds, the cross-section for the interactions was of the order of magnitude greater than that for Compton scattering by photons.

Assuming that the mass neutron is close to that of the proton, Chadwick bombarded hydrogen atoms with the produced neutrons in order to learn the speed of the protons after the collisions. Then setting the neutron speed equal to those of proton, he used the above energy expression to get a neutron mass of 938 ± 1.8 MeV. With a consistent set of experiments, Chadwick obtained the first value for the neutron mass, which compared well with the presently accepted value of 939.57 MeV.

34.2.1 Constituents of Nucleus

The simplest nucleus, that of hydrogen, is a single proton. It is an elementary particle of energy equivalent mass of about 940 MeV, carrying positive charge exactly opposite to that of electron's charge, having a spin of one-half and is a Fermion (no two protons can be in the same quantum state).

The next simplest nucleus, called the deuteron, is a bound state of a proton and neutron. The neutron, like the proton, is a spin one-half fermion, but it has no electric charge and is slightly heavier than the proton (by 1.3 MeV). The binding energy of the deuteron (analogous to 13.6 eV for hydrogen atom) is 2.2 MeV. A photon of this energy can "ionize" deuteron into a proton and neutron. However, it is not necessary to actually do this experiment to establish how tightly the deuteron is bound. One need to only weigh the deuteron accurately. It has a mass of

1875.61 MeV. The proton has a mass of 938.27 MeV, the neutron has 939.57 MeV, so together (but some distance apart!) they have a mass of 1877.93 MeV, 2.2 MeV more than the deuteron. Thus, when a proton and a neutron come together to form a deuteron, they must unload 2.2 MeV of energy, which they do so, by emitting a photon (called a γ-radiation at these energies).

Both protons and neutrons, being fermions, obey the exclusion principle, as two protons with spin up cannot be in the same state, although two with opposite spin directions could, and a proton and a neutron can, occupy the same spot at the same time.

Protons and neutrons are referred to as nucleons. The total number of nucleons in a nucleus is usually denoted by A, where $A = Z + N$ where Z is protons and N is neutrons. The chemical properties of an atom are determined by the number of electrons which is always equal to the number of protons Z. This is called the atomic number. Nuclei can have the same atomic number, but different numbers of neutrons. These nuclei are called isotopes (Greek "same place"), since they occupy same place in the periodic table. Nucleons attract each other within a short range, but by a very strong force, called the nuclear force. For electrons, the strong central force tends to dominate. In the nucleus, the nucleons are attracted mainly by their immediate neighbours. It is similar to an attractive force as being a potential well, as seen by an individual nucleon. In terms of the nuclei, as filling the lowest available quantum states in this well, similar to electrons in the atom. For example, the helium nucleus, $2p + 2n$ (the α-particle), is tightly bound four nucleons: a spin up proton, a spin down proton, a spin up neutron and a spin down neutron. All can occupy the lowest energy state in the well. However, some larger nuclei, like C, O, Fe, are actually a little more tightly bound than He (about 8.5 MeV per nucleon as against about 7.5 for He) because each nucleon is attracted to its close partners, and there are more close partners in these larger nuclei. It has also been argued that some of these higher nuclei strongly resemble bound states of α-particles. The total binding energy (usually expressed per nucleon) of any nucleus is easy to find – just as for the deuteron in the example above, the mass of the nucleus is subtracted from the sum of the masses of the separate nucleons. A relative scale model of an atom in comparison with our solar system is shown in Fig. 34.3.

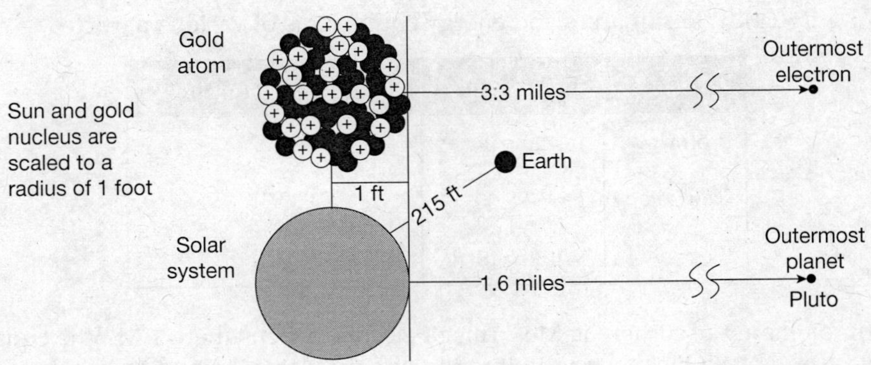

FIGURE 34.3 Relative scale model of an atom and the solar system

Note

Protons and neutrons are referred to as nucleons. The total number of nucleons in a nucleus is usually denoted by A, where $A = Z + N$ where Z is protons and N is neutrons.

34.2.2 Nuclear Notations

Atoms are electrically neutral, as they have an equal number of protons in the nucleus and electrons orbiting the nucleus. However, within the nucleus, there are other particles called neutrons, which are electrically neutral but have the same mass as protons. The nucleus is characterized by the following:
Z – atomic number, which equals the number of protons in the nucleus;
A – mass number, which is equal to the number of nucleons (protons plus neutrons) in the nucleus.

An element X is defined by the atomic number Z, while A denotes the particular isotope of that element. The usual notation for an element X is $^{A}X^{Z}$.

For example, there are four common isotopes of Carbon: $^{11}C^{6}$, $^{12}C^{6}$ and $^{14}C^{6}$, with $^{12}C^{6}$ being the most abundant (> 98 per cent). Notice that each isotope of a particular element has the same Z but a different A.

34.3 ▶ Nuclear Mass

The fundamental relation that is used to discuss properties of nuclei and nuclear reactions is Einstein formula:

$$\boxed{E = mc^2} \tag{34.2}$$

which relates energy E to mass m, where $c = 3.0 \times 10^8$ m/s being the speed of light in vacuum. Therefore, the amount of energy contained in 1 kg mass is enormous, of the order of 9 billion Joules. However, somewhat analogous to potential energy, this mass energy is useful when it gets converted to other forms of energy, namely, kinetic energy of the moving particles. This transformation of energy occurs mostly in nuclei, where the masses (and energy equivalents) involved are smaller, as shown in Table 34.1.

TABLE 34.1 Rest masses and energy equivalents of various particles

Particle	Mass (kg)	Equivalent energy (MeV)
Proton	1.6726×10^{-27}	938.27
Neutron	1.6750×10^{-27}	939.57
Electron	9.109×10^{-31}	0.51

The unit of energy used here is MeV (mega electron volt), where 1 MeV is equal to one million electron volts. One electron volt is the kinetic energy gained by an electron when accelerated through a potential difference of one volt (1.6×10^{-19} J).

Note

The simplest nucleus, that of hydrogen, is a single proton. It is an elementary particle of mass of about 938.27 MeV.

34.4 ▶ Mass Defect

Nuclei are made up of protons and neutrons, but the mass of a nucleus is always less than the sum of the individual masses of the protons and neutrons which constitute it. This is called mass defect. This difference is a measure of the nuclear binding energy which holds the nucleus together. This binding energy can be calculated from Einstein relationship as:

$$\text{Nuclear binding energy} = \Delta mc^2 \tag{34.3}$$

where Δm is the mass defect. The mass of a proton is take as 1 atomic mass unit (u), its actually a little heavier. The mass of a proton is 1.00728 atomic mass units (u), while neutrons weigh 1.00866 u.

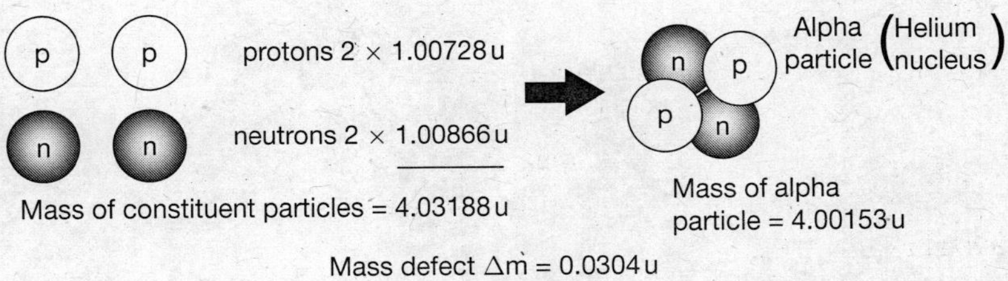

FIGURE 34.4 Difference of nucleon mass and nucleus is mass defect

Note

The mass of a nucleus is always less than the sum of the individual masses of the protons and neutrons which constitute it. This difference is called mass defect.

For alpha particle, $\Delta m = 0.0304\,u$, which is *mass defect* and the energy equivalent to it is calculated by Einstein relationship which comes to 28.3 MeV. This energy is the binding energy of alpha particle or helium nucleus.

34.5 ▶ Binding Energy of Nucleus

Binding energy of nucleus is defined as the energy equivalent to the mass defect of nucleus, i.e., when a nucleus is formed from nucleons, the mass of nucleus is not equal to the total masses of nucleons. The difference of these two is called mass defect.

If M_p is the mass of proton, M_n is the mass of neutron, Z is the number of protons in nucleus and A is the mass number of nucleus, then

Mass defect, $$\boxed{\Delta m = Z M_p + (A - Z) M_n - M(A \times Z)} \tag{34.4}$$

∴ Binding energy $$E_b = \Delta mc^2$$

If the binding energy is divided by the total number of nucleus A, then $\frac{E_b}{A}$ is called the average binding energy per nucleon.

∴ Average binding energy $$= \frac{\Delta mc^2}{A} \tag{34.5}$$

Average binding energy shows a characteristic variation in it with respect to the elements in periodic table. It is shown in Fig. 34.5.

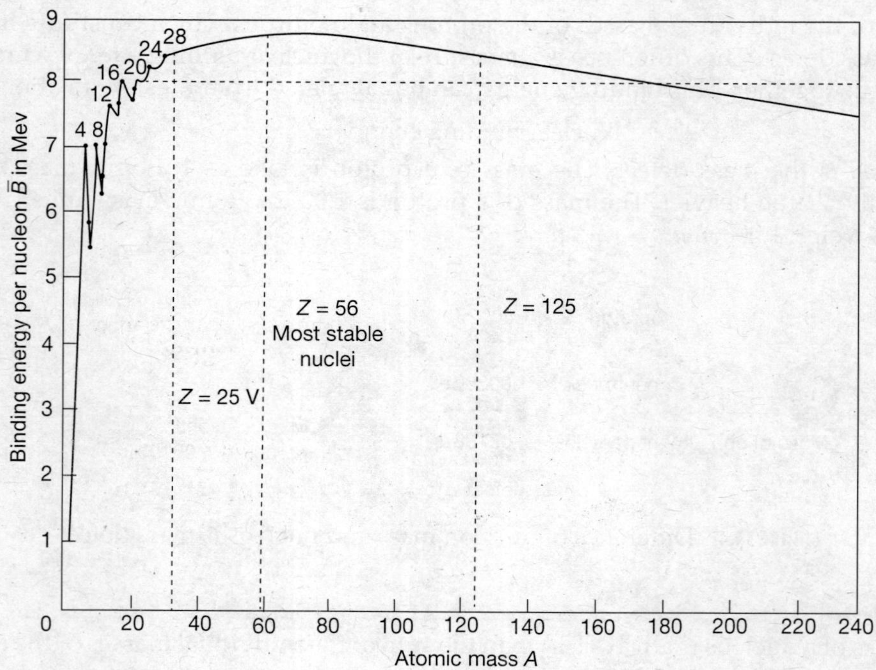

FIGURE 34.5 Binding energy per nucleon, showing broad maximum

The binding energy curve reveals:

1. Binding energy for all known stable nuclei is positive.
2. Peaks in binding energy at $A = 4n$ for light nuclei corresponds to He^4. Be^8, C^{12}, O^{16}, Ne^{20} and Mg^{24}.
3. In the medium mass region, i.e., between $A = 25$ and $A = 125$, the average binding energy per nucleon is constant.
4. There is a slight decrease in average binding energy for A greater than 125.

Basically, binding energy is a function of nature of binding forces of nucleon. From the binding energy curve, we can predict the following facts about the nature of nuclear forces.

Note

The Einstein mass–energy formula $E = mc^2$ relates energy E to mass m, where $c = 3.0 \times 10^8$ m/s being the speed of light in vacuum.

Binding energy of nucleus is defined as the energy equivalent to the mass defect of nucleus. When a nucleus is formed from nucleons, the mass of nucleus is not equal to the total masses of nucleons. The difference of these two is the mass defect of nucleus.

34.6 ▶ Isotopes, Isobars and Isotones

Isotopes

These are the elements having same atomic number but different mass number. They have the same atomic number because the number of protons inside their nuclei remains the same. The difference in their mass number is due to the difference in their number of neutrons. The hydrogen has three isotopes called hydrogen, deuterium and tritium ($^{1}H^{1}$, $^{1}H^{2}$, $^{1}H^{3}$). Similarly, $^{17}Cl^{37}$, $^{17}Cl^{35}$ are isotopes of chlorine. The isotope can occur either naturally or can be produced artificially in the laboratory.

Isobars

Isobars are atoms of different elements having the same atomic mass but different atomic number. Due to their different electron numbers, their chemical properties are different. The light nuclei have unstable isobars. Heavy nuclei have stable isobars and these occur in pairs. For example, the number of protons of one isobar matches with that of another; they are called mirror-nuclides of each other. Examples of isobars are $^{32}Ge^{76}$, $^{34}Se^{76}$ and $^{26}Fe^{58}$, $^{27}Ni^{58}$.

Isotones

Isotones are elements having the same number of neutrons. Examples of isotones are Chlorine - 37 and Potassium - 39. Both have 20 neutrons in their nuclei.

34.7 ▶ Nuclear Forces

The nuclear forces are very strong forces as compared to gravitation or electromagnetic forces as the binding energy per nucleon is quite high. Besides these forces are unique in nature due to the following features:

1. Nuclear forces are always attractive in nature except when the nucleons come very close. This is true as a result of positive binding energy for all nuclei. If the binding energy is negative, then the nuclei would not be stable and the repulsive forces would be existing in the nucleus. The nuclear forces are very strong in nature.
2. Nuclear forces become repulsive in nature at very short distances, i.e., shorter than 1 Fermi. This fact is supported by the fact that nucleus has a finite size. If at shorter distances the nuclear forces are not repulsive, then the nucleus will reduce to a point mass.

   Variation of nuclear force between two nucleons is shown in Fig. 34.6 with respect to distance between them.
3. α-particle structure has a peculiar stability, i.e., a state with Z protons and Z neutrons is a stable system. In this case, the spin of nucleus is zero. Because two protons align with opposite spins and similarly two neutrons also align with opposite spins, which

shows that the nuclear force between two nucleons depends upon their relative nuclear spin.

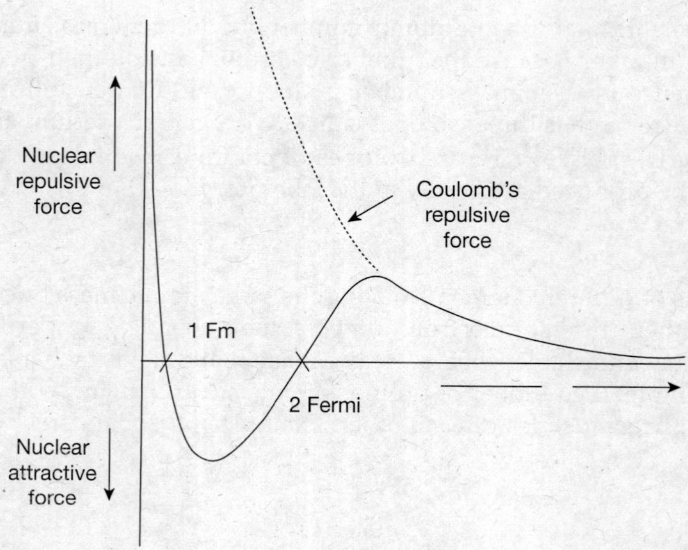

FIGURE 34.6 Variation of nuclear forces with distance

4. Binding energy of a nucleus is proportional to A, i.e. $B \propto A$ or $\frac{B}{A}$ is constant which shows that a nucleon in a nucleus does not interact with all the nucleons present in the nucleus.

   If the nucleon interacts with all its neighbouring particles, then the binding energy would be given as

$$B \propto \frac{A(A-1)}{2} \quad \text{if } A = 4$$

$$B \propto \frac{4.3}{2} = 6$$

   But if A would be very large, then $B \propto A^2$ which is not experimentally found. This predicts that the nucleons interact with only a limited number of adjacent nucleons, which shows that nuclear forces are short-range forces. This is also true of the forces in the formation of molecules from atoms which are electromagnetic in nature. Here nuclear forces do not come into play.

5. Nuclear forces are charge independent forces. It is verified by neutron–neutron or proton–proton scattering experiments that the forces between n-p, p-n, n-n or p-p are exactly same at short distances (of the order of Fermi) (Figs. 34.7 and 34.8).

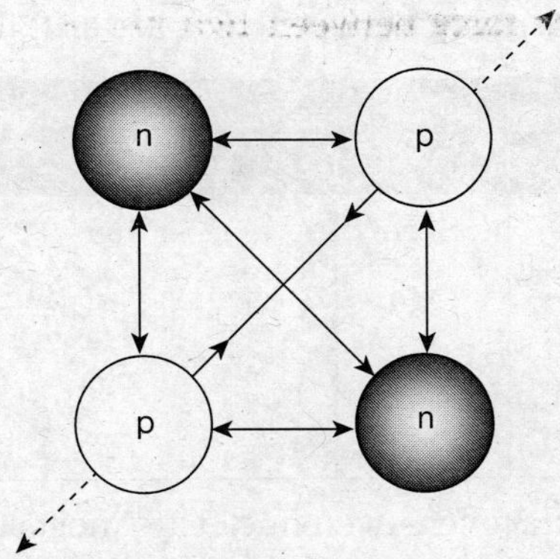

FIGURE 34.7 Various forces on nucleons in Helium nucleus

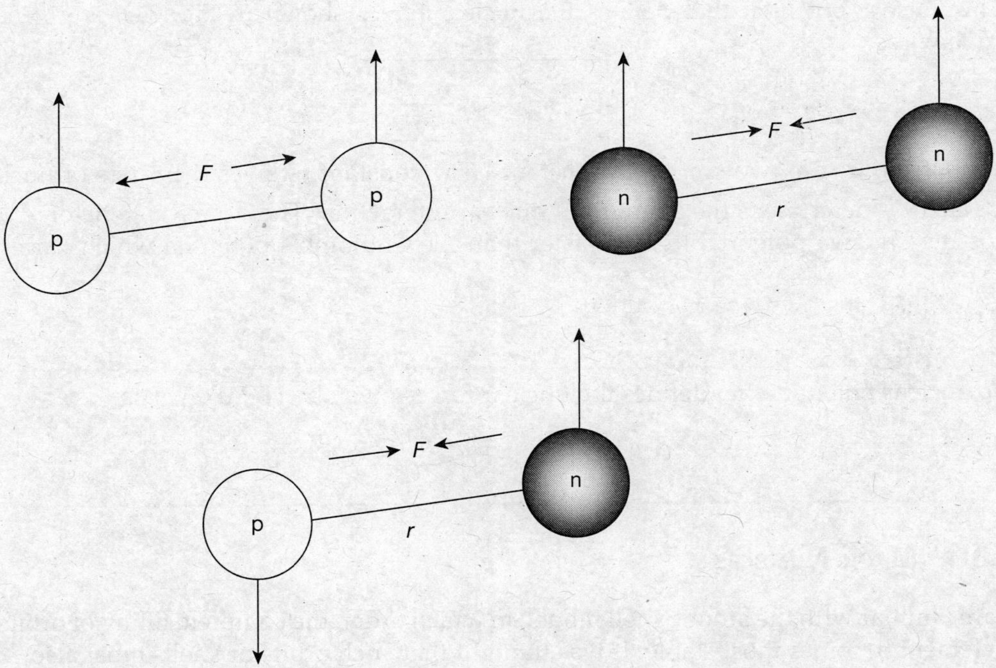

FIGURE 34.8 Charge independence of nuclear forces

6. Nuclear forces are saturated forces, i.e., they do not vary inside the nucleus and vanish as we move out of the nucleus, or they interact only in a very small region around a nucleon.
7. Nuclear forces are not completely central. But it depends upon the orientation of the spin of nucleons, relative to the line joining the centre of nucleons.

There are 274 naturally occurring isotopes, which are as follows:

Z	N	Number of isotopes
Even	Even	165
Odd	Odd	4 $\left({}_1H_1^2\ {}_3LI_3^6\ {}_5B_5^{10}\ {}_7N_7^{14}\right)$
Odd	Even	50
Even	Odd	55

The above data shows that the even–even nuclei are most stable and are found abundant in nature, odd–odd nuclei are unstable and hence are only four in nature. Odd–even or even–odd nuclei have almost same stability because their number is 50 or 55, respectively. This study shows that the nuclear forces are charge independent forces.

The nuclear potential that defines the nuclear forces shown in Fig. 34.6 is given by H. Yukawa as

$$\boxed{E(r) = -E_0 r_0 \frac{e^{-r/r_o}}{r}} \tag{34.6}$$

where E_0 and r_0 are two empirical constants. The constant r_0 gives the range of nuclear force and E_0 determines the strength of nuclear interaction. The decreasing factor e^{-r/r_0} drops the Yukawa potential to zero faster than the Coulomb's potential which varies as $1/r$.

Note

The nuclear potential that defines the nuclear forces given by H. Yukawa as

$$E(r) = -E_0 r_0 \frac{e^{-r/r_o}}{r}$$

34.8 ▶ MAGIC NUMBERS

We are familiar with the atomic shell model, in which order, the complete filling of orbits of the element becomes most stable. It was thought that such orbits or shells must also exist in the nucleus also. The binding energies for the nuclear Z values show sudden increase at $Z.= 2, 10, 18, 36, 54$ and so on. These atomic numbers correspond to the noble elements that are chemically most stable. Similar to this, it was observed by Mager and Jansen that the nuclei with proton number $Z = 2, 8, 20, 28, 50, 82$ and neutron number $N = 2, 8, 20, 28, 50, 82, 126$ show remarkable stability.

These numbers 2, 8, 20, 28, 50, 82, 126 are called magic numbers.

If we plot a graph of Z and N versus the number of stable isotopes, then we get graphs as shown in Fig. 34.9.

These curves show that there exists something special with these numbers and hence these numbers are called magic numbers.

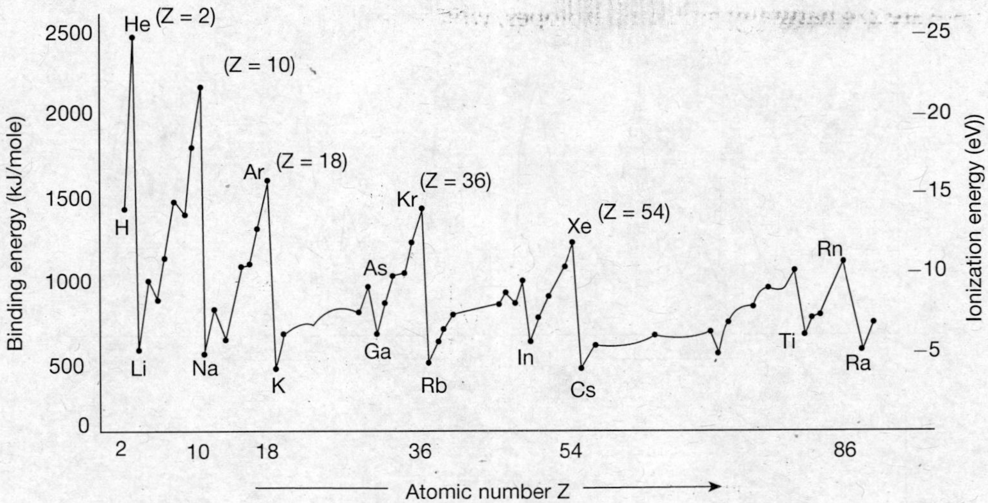

FIGURE 34.9 Nuclear binding energy vs Z

The existence of magic numbers is predicted by the following facts:

1. A nucleus having atomic number Z equal to one of the magic numbers are more stable isotopes than its neighbours.

   For example, ${}_{20}\mathrm{Ca}^{40}_{20}$ has 6-stable isotopes. Since its $Z = 20$ is a magic number ${}_{19}\mathrm{K}$ has 3-stable isotopes.

   ${}_{21}\mathrm{Sc}$ has 1-stable isotopes

   ${}_{50}\mathrm{Sn}$ has 10-stable isotopes whereas ${}_{49}\mathrm{In}$ has 2-stable isotopes

   ${}_{51}\mathrm{Sb}$ has 2-stable isotopes

   ${}_{82}\mathrm{Pb}$ has 4-stable isotopes and ${}_{81}\mathrm{Tl}$ has 2-stable isotopes

   ${}_{83}\mathrm{Bi}$ has 2-stable isotopes

2. Similarly, a nucleus which has its neutron number as one of the magic numbers has many more stable isotones than its neighbours.

   For example, for $N = 20$, there exist 5 isotones

   $N = 19$, there exists 1 isotone

   $N = 21$, there exists 1 isotone

   For $N = 50$, there exist six-stable isotones

   $N = 51$, there exists 1 stable isotone

   $N = 49$, there exists 1 stable isotone

3. From abundance of natural isotopes, the existence of these magic numbers is confirmed. For example,

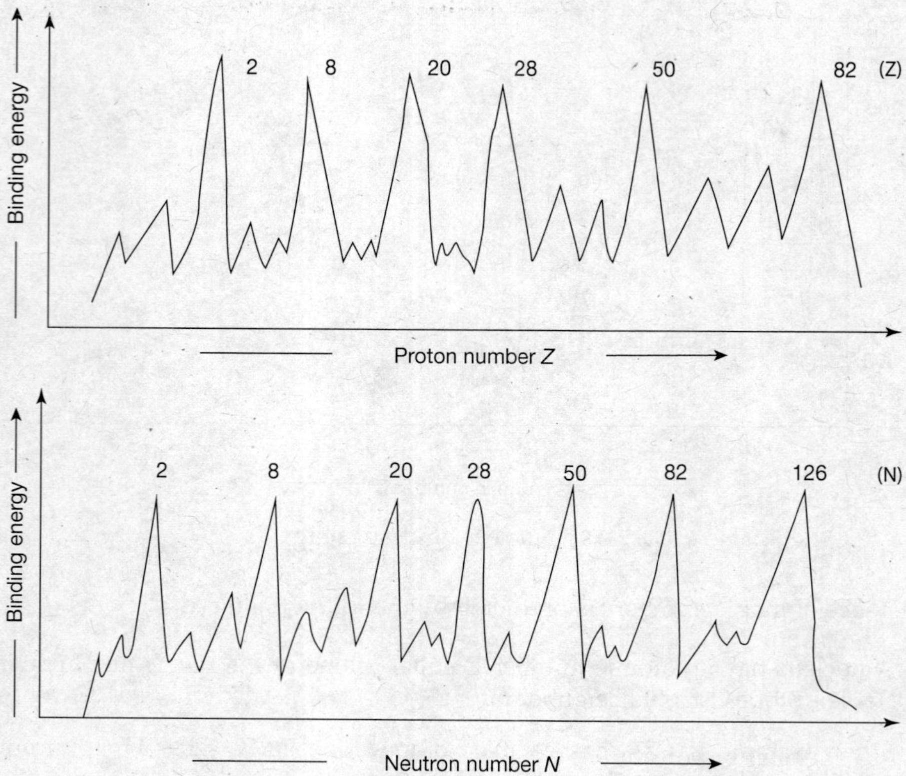

FIGURE 34.10 Curves showing the variation of binding energies with Z and N. Higher is the binding energy more stable in the nucleus

${}_{8}O_{8}^{16}$ occurs 99.8 per cent in nature and its Z and N both are magic numbers.

Similarly, $38Sr_{50}^{88}$ and ${}_{56}Ba_{82}^{138}$ are 82.6 per cent and 77.7 per cent.

Similarly, ${}_{58}Ce_{82}^{140}$ exist about 88.5 per cent.

4. The decay products of all radioactive series decay end up at lead, which has $Z = 82$, a magic number ${}_{82}Pb_{126}^{208}$ and $N = 126$ which is also a magic number.

   If N and Z both are magic numbers of a nucleus, then the nucleus is said to be doubly magic. Some examples of doubly magic nuclei are

$${}_{2}^{4}He,\ {}_{8}^{16}O,\ {}_{20}^{48}Ca,\ {}_{82}^{208}Pb$$

   Calcium provides a good example of being exceptionally stable as it possesses doubly magic nuclei. The existence of several stable isotopes of calcium corresponding to $Z = 20$, a magic number, can also be understood. The two highlighted isotopes of ${}_{20}Ca$ in Fig. 34.11 have neutrons equal to 20 and 28, which are also magic numbers.

Compared to the binding energy calculated from the Weizsacker formula, they both have more than the expected binding energy as shown in Fig. 34.11.

5. If we study the neutron capturing ability of various nuclei, then we find that ${}_{20}Ca^{40}_{20}$, ${}_{82}Pb^{208}_{126}$ and so on have very low neutron absorption cross-section, whereas ${}_{54}Xe^{135}_{81}$ and V^{50}_{27} have very high neutron absorption cross-section because by absorption of one neutron, their N can become a magic number.

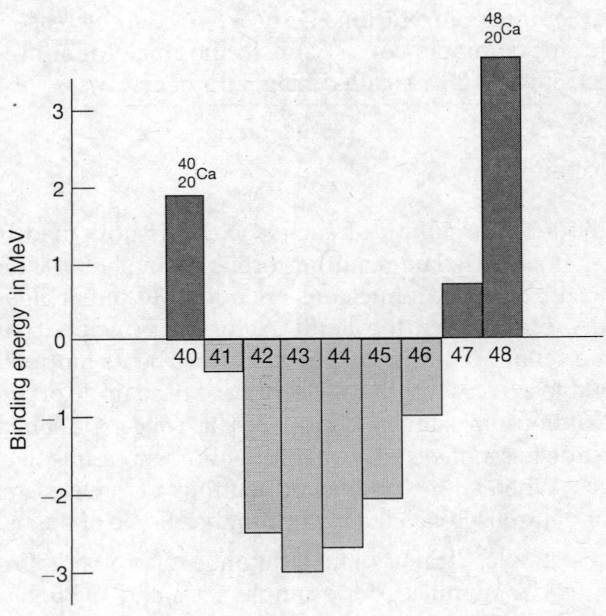

FIGURE 34.11 Binding energy of calcium isotopes from Weizsacker's semi-empirical values

6. We know only 4 isotopes which emit neutrons spontaneously, namely, ${}_{8}O^{17}_{9}$, ${}_{36}Kr^{87}_{51}$, ${}_{54}Xe^{137}_{83}$ as they can have N equal to a magic number by the loss of one neutron and ${}_{36}Kr^{89}_{53}$ can have its N equal to a magic number by the loss of 3 neutrons.
7. It is predicted that the nuclei having their Z and N as magic numbers should be spherical in shape, or they should have zero quadrupole moment or the nuclei, the shells of which are completely filled, should be spherical in shape. ${}_{71}Lu_{105}$ has the largest quadrupole moment, and hence, is not spherical in shape.

Energy level scheme of nucleus

The energy level diagrams of nucleus (Figs. 34.9 and 34.10) were first given by Mager and Jenson. They made the following assumptions while forming the energy levels of a nucleus to correspond with the magic numbers:

1. Each nucleon in a nucleus moves in the average field of other nucleons.
2. In addition to this average field, there is a strong coupling between spin and orbit, which is similar to *LS* coupling for atomic electrons and *J* is given as

$$J = L \pm S$$

3. $J = L + S$ lies lower than $J = L - S$. It is assumed that the forces that are responsible for the coupling are not of electromagnetic nature.

Note

The binding energy of nuclei for various *Z* shows sudden increase in value for some *Z* numbers. These atomic numbers correspond to the nuclei that are most stable. These numbers 2, 8, 20, 28, 50, 82, 126 are called magic numbers.

34.9 ▶ Shell Model

According to this model, the motion of each nucleon is governed by average attractive force of all the other nucleons. The resulting orbits form shells, which are similar to the orbits of electrons in the atoms. As nucleons are added to the nucleus, they drop into the lowest energy shells according to the Pauli exclusion principle, which states that each nucleon has a unique set of quantum numbers to describe its motion.

When a shell is complete (i.e., when the nucleons have used up all of possible sets of quantum numbers), a nucleus of unusual stability is formed. This concept is similar to the one found in an atom, where a complete set of electron quantum numbers results in an atom with unusual stability – an inert gas. When all the protons or neutrons in a nucleus are in filled shells, the number of protons or neutrons is called a magic number. Some of the magic numbers are 2, 8, 20, 50, 82 and 126. For example, ^{116}Sn has a magic number of protons (50) and ^{54}Fe has a magic number of neutrons (28). Some nuclei, for example, ^{40}Ca and ^{208}Pb, have magic numbers of both protons and neutrons; these nuclei have exceptional stability and are called doubly magic.

Filled shells have a total angular momentum *J* which is equal to zero. The next added nucleon (a valence nucleon) determines the *J* of the new ground state. When nucleons (single or in pairs) are excited, they change the angular momentum of the nucleus as well as its parity and isospin projection quantum numbers. The shell model describes how much energy is required to move nucleons from one shell to another and how the quantum numbers change. Figure 34.12 shows a energy diagram of two complete shells of the ground state of ^{12}C. Promotion of a nucleon or a pair of nucleons to an unfilled shell puts the nucleus into one of the excited states.

Excited nuclear states decay to more stable states, i.e. more stable nucleon shell. Measuring transition rates between nuclear energy levels require specialized alpha, beta and gamma detectors and associated electronic circuitry to determine the energy and half-life of the decay. Quantum mechanics and shell-model theory permit nuclear scientists to compute the transition probability (rate of decay) between nuclear states. For nuclei whose structure can be described by a small number of valence nucleons outside filled shells, the shell model calculations agree well with the measured values of spin and parity assignments and transition probabilities.

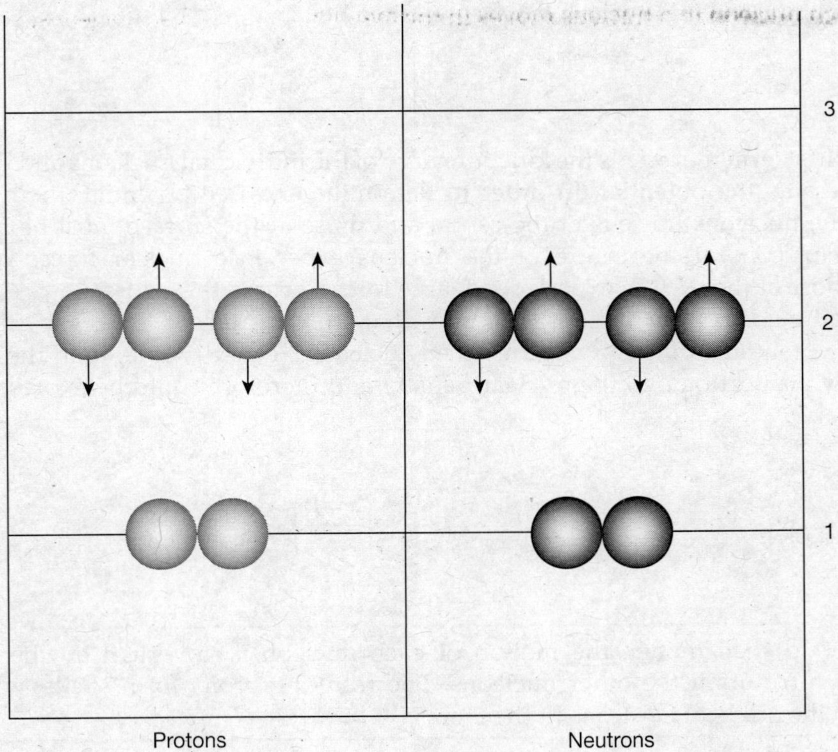

FIGURE 34.12 Energy diagram of protons and neutrons in ${}^{12}_{6}C$

34.9.1 Theory of Shell Model

In atomic physics, the shell model has proved successful in providing an explanation for the details of atomic structure. Atomic shells are filled with electrons in order of increasing energy, subject to the Pauli's exclusion principle. As a result of the success of the atomic shell model, this model has been extended to the realm of nuclei.

The shell model, also known as the independent particle model, treats nucleons individually as opposed to treating the nucleus as a whole. The long-range repulsive Coulomb force and the short-range strong attractive nuclear force interactions between nucleons are replaced by an average force. The shell model describes the nucleons in terms of groups of shells, with the position and energy of a nucleon dependent upon its quantum numbers.

However, there are several principles that needed reworking when compared with atomic structure. In atomic structure, the potential is provided by the coulomb field of the nucleus, whereas in the nuclear case, the motion of a single nucleon is governed by an average potential generated by all the other nucleons. If two nucleons collide, the energy required to excite one of them to a higher level is more than the nucleons are likely to transfer. Therefore, in the nuclei case, it is assumed that there are no nucleon–nucleon collisions and each nucleon moves in a unperturbed single particle shell within the nucleus.

The shell model Hamiltonian can be represented by

$$H = \sum_{i=1}^{A}\left[-\frac{h^2}{2m}\nabla_i^{\,2} + V(r_i)\right] \tag{34.7}$$

where the first term represents the kinetic energy of the individual nucleons and the second term represents the potential. In order to determine end-free potential, the differences between the nucleons nucleus at the centre and those at the surface must be taken into account. Nucleons at the centre of the nucleus experience nuclear forces uniformly, whereas those at the surface experience a large force towards the centre.

The shell model of the nucleus presumes that a given nucleon moves in an effective attractive potential formed by all the other nucleons. If that is true, then the potential is probably proportional to the nuclear density and therefore could be expressed in the form

$$V = \frac{-V_0}{1+\exp\left(\dfrac{r-R}{a}\right)} \tag{34.8}$$

Note

According to shell model, the motion of each nucleon is governed by the average attractive force of all the other nucleons. The resulting orbits form shells, which are similar to the orbits of electrons in the atoms.

34.10 ▶ Liquid Drop Model of Nucleus

Liquid drop model provides a reasonable explanation for a number of nuclear phenomena, which are not explained by shell model. These phenomena involve the masses and binding energy of nuclear ground levels, the energetics of β-decay, α-decay and nuclear reactions, the cross-section for resonance reactions and the energetics of nuclear fission. The liquid drop model is an antithesis of the independent particle model. The interactions between nucleons are assumed to be strong instead of weak. Nuclear levels are represented as quantized states of nuclear system as a whole, and not as states of single particle in an average field.

Liquid drop model originated was in Bohr's concept of the compound nucleus in nuclear reactions. When an incident particle is captured by a target nucleus, its energy appears to be shared among all the nucleons. The captured particle has a mean free path in nuclear matter, which is much smaller than nuclear radius. To account for such a behaviour, interactions between nucleons have to be strong and the particles cannot behave independently. The main quantitative aspects of liquid drop model are restricted to the explanation of nuclear masses, nuclear energy surface and the energetics of induced and spontaneous reactions. This involves the development of Weizsäcker's semi-empirical mass formula.

Semi-empirical mass formula

The mass of a neutral atom M whose nucleus contains Z protons and N neutrons is

$$M = ZM_H + NM_n - B$$

where B is the binding energy of the nucleus. It is made up of a number of terms each of which represents some general characteristics of nuclei. Thus,

$$\text{Binding energy } B = B_0 + B_1 + B_2 + B_3 + \cdots$$

The initial assumptions involved in the development of binding energy are as follows:

1. The nucleus is like a droplet of incompressible matter and all nuclei have the same density.
2. The distinction between the triplet (np) and singlet (np) force is neglected, i.e. the forces between nucleons are considered to be spin and charge independent. If the Coulomb force is turned off, then

$$(n-p) \equiv (n-n) \equiv (p-p)$$

3. These nuclear forces have short range and are effective only between the nearest neighbours.

Volume energy

The largest contribution to the binding energy of the nucleus is due to the number of nucleons

i.e., $$\text{Binding energy } B \propto A$$

$\Rightarrow$ $$B_0 = a_v A \tag{34.9}$$

where a_v is an arbitrary constant.

Surface energy

Those nucleons which lie at the surface of nucleus have fewer near neighbours than those which are deep in the nucleus and thus are more loosely bound to the nucleus than the inner ones. Due to this surface effect, the nucleons lying on the surface do not make the same contribution to the binding energy as is made by the inner nucleons.

Let R be the nuclear radius and d the nucleon diameter as shown in Fig. 34.13.

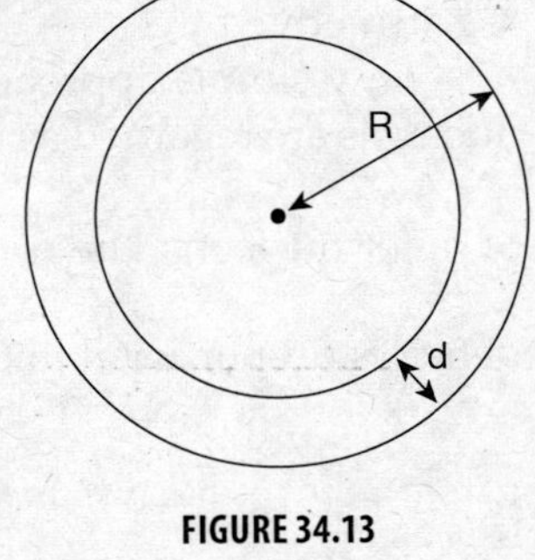

FIGURE 34.13

Density of nucleus $$\rho = \frac{A}{V} = \frac{A}{\frac{4\pi}{3} R_0^3 \cdot A}$$

$\rho = \dfrac{3}{4\pi R_0^3}$ is the number of particles/unit volume

Volume of a shell between the spheres of radii R and $R + d$

$$= 4\pi R^2 d$$

Number of particles in the shell

$$= 4\pi R^2 \frac{d.3}{4\pi R_0^3} = \frac{3d}{R_0^3} R^2$$

$$= \frac{3d}{R_0^3} \cdot A^{2/3} \cdot R_0^2 = \frac{3d}{R_0} \cdot A^{2/3}$$

$$\therefore \qquad B_1 = -\, a_s\, A^{2/3} \qquad (34.10)$$

Coulomb's energy

The only long range force in the nuclei is the Coulomb's repulsion between the protons. Assuming constant density nuclear radius, $R_0 A^{1/3}$, the loss of binding energy due to the disruptive coulomb energy is

$$B_2 = -\frac{3}{5} \frac{e^2 Z^2}{R_0 A^{1/3}} = -a_c \frac{Z^2}{A^{1/3}}$$

$$B_2 = -a_c \frac{Z^2}{A^{1/3}} \qquad (34.11)$$

Asymmetry energy

Another deficit in binding energy depends on the neutron excess (N – Z) and is proportional to $\frac{(N-Z)^2}{A}$. This asymmetry energy is a purely quantum mechanical effect in contrast with simple classical effects of surface and volume and Coulomb energies. Among the light elements, there is a clear tendency for the number of neutrons and protons to be equal, as in case of ${}_6C^{12}$, ${}_7N^{14}$, ${}_8O^{16}$ and so on. Heavy nuclei always contain more neutrons than the protons and are also found to be unstable because the Coulomb's large disruptive energy could not be overcome by the available *n*-*p*, *n*-*n* and *p*-*p* attractive forces.

A very simple approach to the form of the asymmetry energy is that if $|N - Z|$ excess nucleons are regarded as producing a deficit of binding energy, because they are out of reach of other nucleons, the fraction of the nuclear volume so affected is $\frac{(N-Z)^2}{A}$ and the total deficit is proportional to the product $\frac{(N-Z)^2}{A}$

i.e., $$B_3 = -a_a \frac{(N-Z)^2}{A} = -a_a \frac{(A-2Z)^2}{A}$$

$$\therefore \qquad B_3 = -a_a \frac{(A-2Z)^2}{A} \qquad (34.12)$$

Pairing energy

Pairing energy that δ corresponds to the even or oddness of nuclei is added.

$$\text{Pairing energy} \qquad B_4 = \begin{cases} +\delta & \text{even } Z \text{ even } N \\ 0 & \text{odd odd} \\ -\delta & \text{even } Z \text{ odd } N \text{ or odd } Z \text{ even } N \end{cases}$$

$$\delta = a\frac{1}{A}$$

Thus, semi-empirical mass formula can be written as $B = B_0 + B_1 + B_2 + B_3 + B_4$

$$B = a_v A - a_s A^{2/3} - a_c \frac{Z^2}{A^{1/3}} - a_a \frac{(A-2Z)^2}{A} \pm \delta$$

Thus, the mass of a nucleus is given as

$$\boxed{M = ZM_H + NM_n - a_v A + a_s A^{2/3} + a_c \frac{Z^2}{A^{1/3}} + a_a \frac{(A-2Z)^2}{A} \pm \delta} \qquad (34.13)$$

The experimentally observed values of various variables are as follows.

- $a_v = 14.1$ MeV,
- $a_s = 13$ MeV,
- $a_c = 0.595$ MeV and
- $a_a = 19$ MeV

$$\delta = \begin{cases} +33.5 \text{ MeV for even-even nuclei} \\ 0 \text{ MeV for odd-even nuclei} \\ 0 \text{ MeV for even-odd nuclei} \\ -33.5 \text{ MeV for odd-odd nuclei} \end{cases}$$

We apply the above formula to two mirror nuclei. Mirror nuclei are those nuclei in which Z and N get interchanged.

For example, ${}_7N_6^{13}$ and ${}_6C_7^{12}$ are mirror nuclei

$${}_7N_6^{13} \rightarrow {}_6C_7^{13} + e^+ + \nu$$

$\therefore$ Energy of disintegration of positron E_{dis} can be given as

$$E_{\text{dis}} = M(13,7) - \{M(13,6) + M_e\} \text{ is } -\text{ve}$$

$$E_{\text{dis}} = 7M_p + 6M_N - BE(13,7) - \left\{6M_p + 7M_n - BE(13,6) + m_e\right\}$$

$$E_{dis} = M_p - M_n + \{B(13,6) - B(13,7)\} - M_e$$

$$E_{dis} = M_p - M_n + \left\{ -\frac{3}{5}\frac{(6^2)}{R}e^2 + \frac{3}{5}\frac{(7)^2 e^2}{R} \right\} - m_e$$

$$E_{dis} = M_p - M_n + \frac{3}{5}\frac{e^2}{R}(49 - 36) - M_e$$

$$E_{dis} = \frac{3}{5}\frac{e^2}{R} \cdot 13 - \left[M_n(-M_p \right] - m_e$$

E_{dis} is the maximum energy of emitted positrons, which can be calculated experimentally. We know M_n, M_p, m_e, e and so on.

Thus, we can calculate R, that is, nuclear radius by

$$\boxed{R = R_0 A^{1/3}} \tag{34.14}$$

then the radius of nucleon R_0 can be calculated as

$$E_{dis} = \frac{39}{5}\frac{e^2}{R_0} A^{2/3} - (M_n - M_P) - m_e \tag{34.15}$$

Thus, the size of nucleus R can be determined experimentally.

Note

Liquid drop model provides explanation for a number of nuclear phenomena, which are not explained by shell model such as the energetics of β-decay, α-decay and nuclear reactions, the cross-section for resonance reactions and the energetics of nuclear fission.

34.11 ▶ Fermi Gas Model of Nucleus

In this model, the nucleus is regarded as a potential well like the one as shown in Fig. 34.14. Nucleons move in that potential. Nuclear forces between the nucleons are neglected. These forces are responsible for the existence of the potential well. In Fermi gas, we treat the nucleus statistically in a way the statistical treatment of a gas is done in the kinetic theory. This model does not give any information regarding the nuclear spins and parities, but is able to give the interpretation of the behaviour of nuclei at high levels of excitations. Starting at the bottom of the well, each energy level is regarded as being completely filled with as many protons or neutrons as Pauli's exclusion principle allows. There are no vacancies, and therefore, the nucleus is in its ground state. It, therefore, resembles a gas at absolute zero.

It can be explained on the basis of Fermi gas model, why the stable nuclei have more neutrons than protons. In the formulation of the Fermi gas model, we have neglected Coulomb's interaction between the protons. If V_0 is the electrostatic potential of the nucleus, then the bottom of the proton well will be raised by an amount $eV \approx 5$ volt as compared to the neutron well as shown in Fig. 34.14.

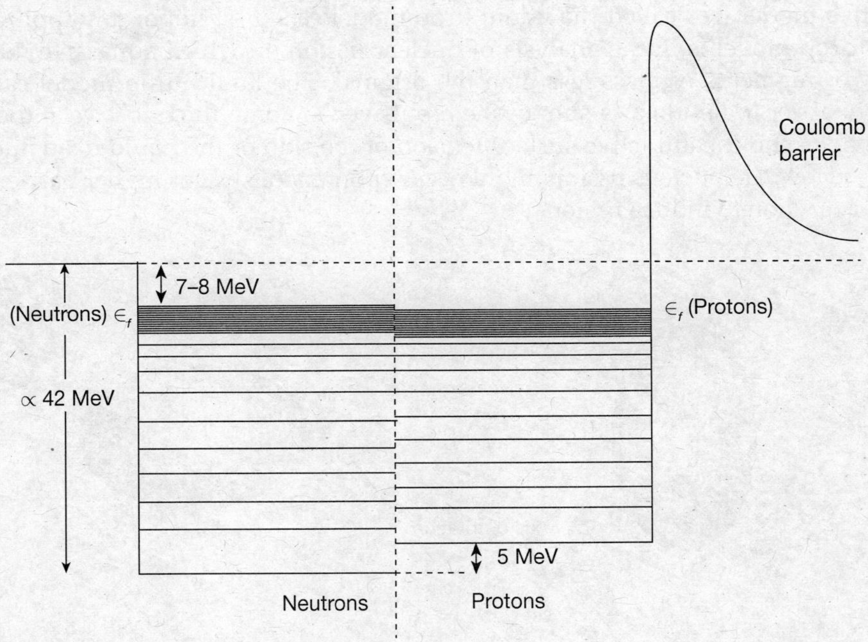

Energy levels of protons and neutrons as per Eermi model

FIGURE 34.14

For a given energy, a nucleus has more neutron levels than proton levels and, hence, more neutrons than protons in a stable state. The Fermi gas model also explains some terms in the empirical mass formula and the long mean free path of the nucleon scattering at extremely high energies.

Note

In Fermi gas model of nucleus, we treat the nucleus statistically in way the statistical treatment of a gas is done in the kinetic theory. This model does not give any information regarding the nuclear spins and parities, but is able to give the interpretation of the behaviour of nuclei at high levels of excitations.

34.12 ▶ COLLECTIVE MODEL OF NUCLEUS

In addition to the individual nucleons changing orbits to create excited states of the nucleus as described by the shell model, there are a number of nuclear transitions that involve many (if not all) of the nucleons. Since these nucleons are acting together, their properties are called collective, and their transitions are described by a collective model of nuclear structure. High-mass nuclei have low-lying excited states that are described as vibrations or rotations of non-spherical nuclei (Fig. 34.15). Many of these collective properties are similar to those of a rotating or vibrating drop of a liquid. In its early development, the collective model was called the liquid drop model. The first important application of the liquid drop model is in the analysis of nuclear fission, in which a massive nucleus splits into two fragments of mass less than the original. The liquid drop model calculates the energy barrier to fission as a sum of the repulsive Coulomb forces between the protons of the nucleus and the attractive surface tension of the skin of the liquid drop nucleus. If the barrier is low, the nucleus fission might occur spontaneously. For higher barriers, it takes a nuclear reaction to induce fission.

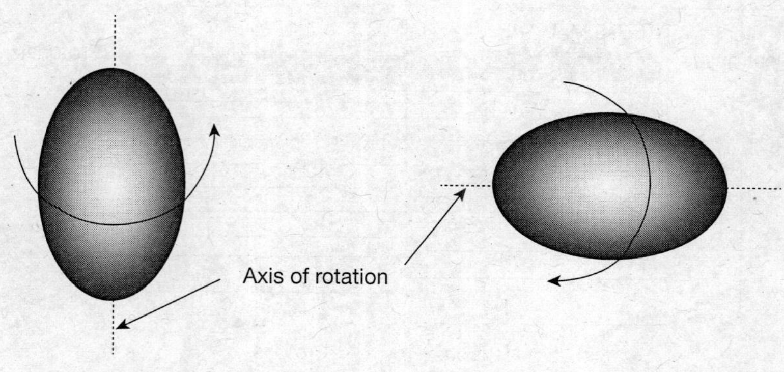

FIGURE 34.15 Non-spherical nucleus, rotation about axes

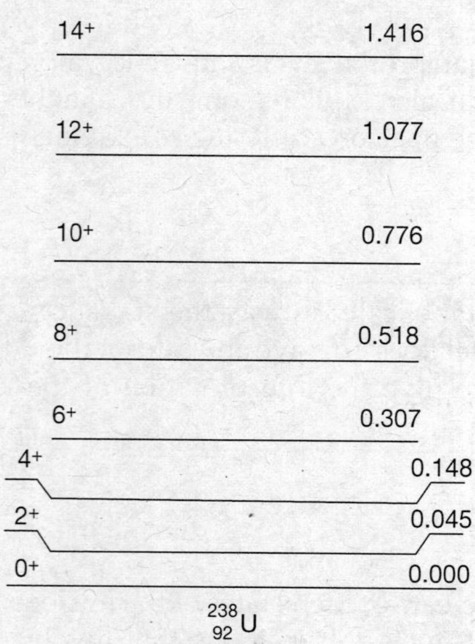

FIGURE 34.16 Energy levels of ^{238}U

Figure 34.16 shows energy levels of ^{238}U. The quantum numbers, level spacing and gamma radiation transition probabilities identify these levels as rotational states of a non-spherical nucleus. Nuclei showing collective properties are usually those with many valence nucleons, i.e., those with proton or neutron numbers that are far from filled shells. As with the shell model, the collective model permits the calculation of spin-parity assignments and transition probabilities that are in agreement with the measured properties of the collective nuclei.

Using accelerators, scientists can create nuclei having very high angular momentum. Nuclei respond to this rotation which can be as fast as a hundred billion revolutions per second. These nuclei loose some of their excitation energy and entire initial angular momentum by the emission of gamma radiation. The gamma radiation flash occurs only for less than 10^{-9} second during which 30 or more gamma radiations can be emitted.

A number of preferred pathways in the de-excitation process exist. They relate to favourable arrangements of protons and neutrons and can often be associated with specific symmetries or nuclear shapes. If a sufficient fraction of the decay flows down a particular quantized pathway or band, then the associated structure becomes observable and can be studied in detail.

34.13 ▶ Comparative Features of the Nuclear Models

- The shell model is based on the idea that the nuclear constituents move independently.
- The liquid drop model is opposite of shell model, in that the nuclear behaviour is based on collective motion.
- Each model only describes a limited set of phenomena and nuclear properties.
- The collective model of the nucleus combines the features of all the models.
- It assumes that nucleons in unfilled sub-shells move independently in a net nuclear potential produced by the core of filled sub-shells.
- The nuclear potential is not always spherically symmetrical potential of the shell model but undergoes deformations in shape.
- The deformations represent correlated (or collective) motion of the nucleons in the core that are associated with the liquid drop model.
- For a nucleus with one more than a magic number, the extra nucleon has a relatively high angular momentum.
- The nuclear interaction between the extra nucleon and the core distorts the core.
- A second extra nucleon will distort the core even more.
- The nucleus gets pulled along as under a tidal force.
- The moving deformation, which includes protons, constitutes a current with a magnetic dipole moment.
- It corrects the shell model value and brings the predictions in line with the experiments.
- It also produces an electric quadrupole moment of the nucleus to which hyperfine splitting is sensitive.
- Another result is that the core will acquire a moment of inertia with a distinct energy spectrum.
- This is observed in many even-even nuclei; we can derive energy spectrum semi classically

$$E = \frac{1}{2} I\omega^2$$

where I is the moment of inertia of the body about the axis or rotation and ω is the angular frequency of rotation.

- The discrepancies at higher excitations can be understood because at higher I values the nucleus will be slightly more deformed by the larger centrifugal force; hence, its moment of inertia becomes larger and the excitation energy decreases.
- The single particle shell model works best near closed shells.
- Further away from magic numbers cooperative effects appear which are best incorporated into collective models, in which rotational and vibrational motion can be built.

RECAP ZONE

POINTS TO REMEMBER

- The size of nucleus is estimated to be of the order of a few Femtometre (10^{-15} m).
- Energy required for the confinement of a particle to a given dimension can be obtained by setting the de Broglie wavelength of the particle equal to the given dimension.
- The constituent particles of the nucleus are protons and neutrons. The electrons cannot exist inside nucleus as explained on the basis of de Broglie wavelength.
- The mass of the nucleus is always less than the sum of mass of the total number of its nucleons, which is termed as mass defect.
- Isotones are the elements having the same number of neutrons.
- Isotopes are the elements having the same atomic number but different mass number.
- Isobars are atoms of different elements having the same atomic mass but different atomic numbers.
- The energy equivalent of this mass defect is the binding energy. Several models have been proposed to explain the properties of nuclei, the models have been proposed to explain the properties of nuclei, the models are Liquid Drop Model, Shell Model, Collective Model and the Fermi Gas Model.
- In Fermi gas model of nucleus, we treat the nucleus statistically like a gas in the kinetic theory. This model does not give any information regarding the nuclear spins and parities, but is able to give the interpretation of the behaviour of nuclei at high levels of excitations.
- Liquid drop model provides explanation for a number of nuclear phenomena which are not explained by shell model such as the energetics of β-decay, α-decay and nuclear reactions, the cross-section for resonance reactions and the energetics of nuclear fission.
- Protons and neutrons are referred to as nucleons. The total number of nucleons in a nucleus is usually denoted by A, where $A = Z + N$ where Z is protons and N is neutrons.
- The simplest nucleus, that of hydrogen, is a single proton. It is an elementary particle of mass of about 940 MeV.

- The mass of a nucleus is always less than the sum of the individual masses of the protons and neutrons, which constitute it. This difference is called mass defect.
- The Einstein formula $E = mc^2$ relates energy E to mass m, where $c = 3.0 \times 10^8$ m/s being the speed of light in vacuum.
- Binding energy of nucleus is defined as the energy equivalent to the mass defect of nucleus. When a nucleus is formed from nucleons, the mass of nucleus is not equal to the total masses of nucleons. The difference of these two is the mass defect of nucleus.
- The nuclear potential that varies the nuclear forces given by H. Yukawa as

$$E(r) = -E_0 r_0 \frac{e^{-r/r_o}}{r}.$$

- The binding energies for various Z show sudden increase in value for some Z numbers. These atomic numbers correspond to the nuclei that are most stable. These numbers 2, 8, 20, 28, 50, 82, 126 are called magic numbers.
- According to shell model, the motion of each nucleon is governed by average attractive force of all the other nucleons. The resulting orbits form shells, which are similar to the orbits of electrons in the atoms.
- Each model explains some property of the nuclei. The liquid drop model explains the fission process, shell model the magic numbers and collective model the fine structure and spin-orbit coupling.
- No single model is capable of explaining all features of the nuclei.

KEY TERMS WITH DEFINITIONS

- The protons and neutrons that reside inside the nuclei are called nucleons.
- The difference in nuclear mass and its constituent is called mass defect.
- The binding energy of a nucleus is equal to its mass defect (equivalent energy).
- The mass defect Δm can be converted to energy E as $E = \Delta mc^2$, where c is the velocity of light.
- Nuclear forces are always attractive except at a distance shorter than 1 Fermi.
- Nuclear potential is expressed as Yukawa potential $E(r) = -E_0 r_0 \frac{e^{-r/r_o}}{r}$
- The atomic numbers that show sudden increase in nuclear binding energy are called magic numbers (such as 2, 8, 20, 28, 50, 82 and 126).
- In Fermi gas model of nucleus, the nucleons are treated as gas molecules without spins.

Important Formulae and Equations

Equation Number	Equation	Remarks
(34.1)	$\frac{ZKe^2}{r}$	Maximum confinement energy due to electrostatic attraction of electron
(34.2)	$E = mc^2$	Einstein's mass energy relation
(34.4)	$\Delta m = ZM_p + (A - Z)M_n - M(A, Z)$	Mass defect
(34.6)	$E(r) = -E_o r_o \frac{e^{-r/r_o}}{r}$	Yukawa potential
(34.13)	$M = ZM_n + NM_n - a_v A + a_s A^{2/3} + a_c \frac{Z^2}{A^{1/3}} + a_a \frac{(A - 2Z)^2}{A} \pm \delta$	Semi-empirical mass formula
(34.14)	$R = R_o A^{1/3}$	Radius of nucleus

Review Zone

Short Answer Questions

1. Describe the method used by Rutherford to estimate the size of nucleus.
2. What is mass defect? How it is related to the binding energy of nucleus?
3. What are isotopes? Give examples.
4. What are isobars and isotones?
5. Explain with reasons, why electrons cannot stay inside the nucleus?
6. What is binding energy curve? Discuss its significance in determining the stability of nuclei.
7. Discuss various models to explain the properties of nucleus.
8. What is liquid drop model of nucleus? How it is used to explain the fission phenomenon?
9. What are magic numbers? How the magic numbers make the basis for nuclear shell model?
10. What is spin-orbit interaction in a harmonic oscillator potential model?
11. What is collective model of nucleus? Discuss its features.
12. Discuss the semi-empirical mass formula and its various terms.
13. Give a qualitative comparison of features of various nuclear models.

CHAPTER 35

Nuclear Fission and Fusion

Learning Objectives

By the end of this chapter, the student will be able:

- To understand the basics of nuclear reaction
- To recognize the key features of nuclear fission and fusion reactions
- To understand the mechanism of artificial nuclear transmutations
- To define and express nuclear cross-section
- To explain the mechanism of nuclear fission on the basis of liquid drop model
- To discuss nuclear fusion reactions
- To predict the controlled nuclear fission and various types of nuclear fuel
- To discuss spontaneous fission and working of nuclear bombs

35.1 ▶ Introduction

The reactions observed in radioactive materials occur spontaneously and involve a conversion of mass into (mainly) kinetic energy of the decay products. There is another class of nuclear reactions that can be artifically induced, an example of which is bombarding of a nitrogen nucleus with an alpha particle:

$$^{4}_{2}\text{He} + {}^{17}_{7}\text{N} \rightarrow {}^{17}_{8}\text{O} + {}^{1}_{1}\text{H}$$

In this case, the sum of the masses of the elements on the left-hand side of this equation is less than the sum of the masses of the decay products on the right-hand side. For the reaction to proceed, some initial kinetic energy must be supplied to the left-hand side reactants. Such energy, again through Einstein's mass energy, gets converted into mass of the products. Some examples will illustrate this later.

There are two general types of such nuclear reactions:

Fusion, where lighter elements fuse into heavier ones. The following are the examples of this reaction which is also shown in Fig. 35.1:

$$^{1}_{1}\text{H} + {}^{1}_{1}\text{H} \rightarrow {}^{2}_{1}\text{H} + e^{+} + \nu$$

and $$^{2}_{1}\text{H}+^{3}_{1}\text{H}\rightarrow\ ^{4}_{1}\text{H}+n+\text{Energy}$$

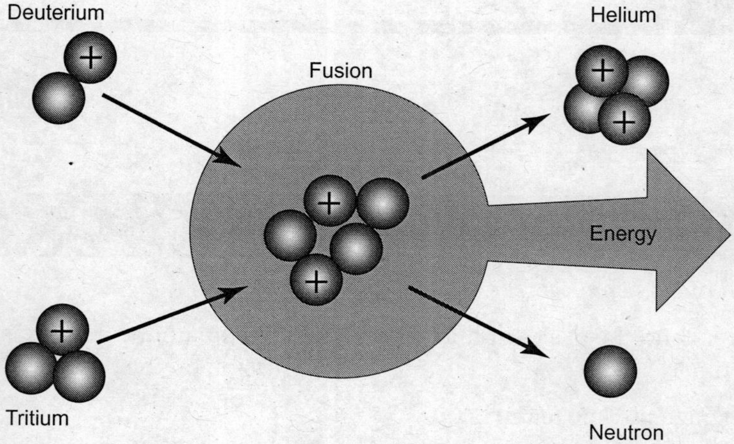

FIGURE 35.1 The process of nuclear fusion. Lighter nuclei ^{2}H and ^{3}H fuse together to form heavier nuclei ^{4}He thereby releasing large amount of energy

Fusion reactions are the main source of energy for the sun and are also used in certain types of nuclear weapons like hydrogen bomb.

Fission, where heavier elements are broken into lighter ones. An example of this is the following reaction:

$$n+\ ^{235}_{92}\text{U}\rightarrow\ ^{141}_{56}\text{Ba}\ +\ ^{92}_{36}\text{Kr}+3n+\text{Energy}$$

where n refers to neutrons. An interesting feature of this neutron induced reaction is the presence of neutrons in the products. The initial kinetic energy of the reacting neutrons is relatively small, and the released neutrons can induce reactions in nearby uranium atoms. Such a chain reaction (Fig. 35.2) is the basis of the design of nuclear reactors, as well as nuclear weapons.

The typical energies involved in nuclear reactions are of the order of MeV (10^{6} eV), compared to the eV scale of chemical reactions. Due to the relatively large kinetic energy needed for the initial reactants, fusion-based nuclear reactors are not economically feasible.

35.2 ▶ Nuclear Reactions

Nuclear reactions are completely independent of chemical reactions, in the sense that nuclear energies are of a much larger magnitude. Thus, an atom that has undergone a nuclear reaction, reacts chemically in a manner similar to atoms which have not. There are four major types of nuclear reaction:

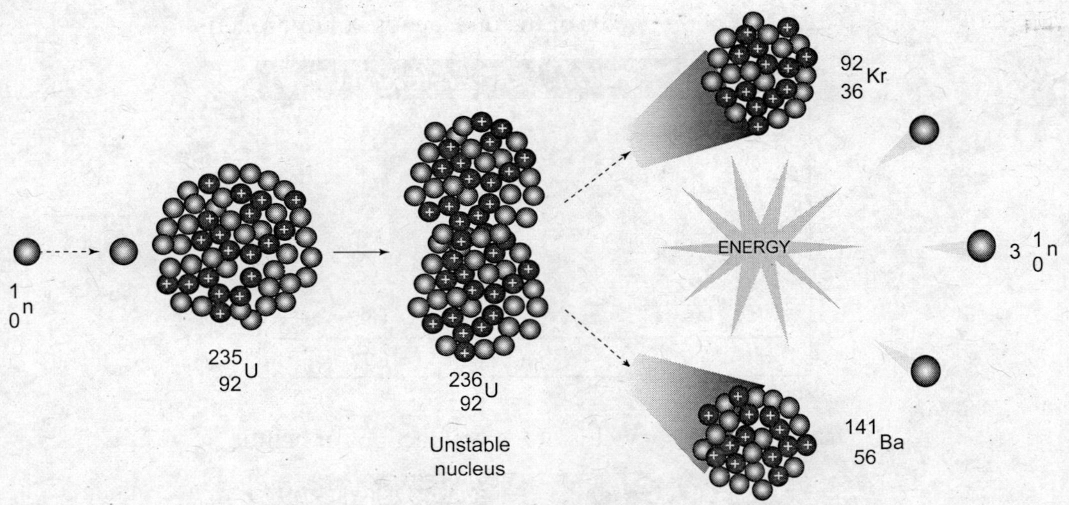

FIGURE 35.2 The process of nuclear fission. Heavier nuclei ^{235}U breaks or fragments into lighter nuclei ^{92}Kr and ^{141}Ba therby releasing large amount of energy

1. Fission is the splitting of a nucleus into two daughter nuclei:

$$\boxed{n + {}^{235}_{92}\mathrm{U} \rightarrow {}^{141}_{56}\mathbf{Ba} + {}^{92}_{36}\mathbf{Kr} + 3n + \mathbf{200\ MeV}} \tag{35.1}$$

2. Fusion of two parent nuclei into one daughter nucleus:

$$\boxed{p + p \rightarrow {}^{2}_{1}\mathrm{H} + e^{+} + \nu + \mathbf{0.42\ MeV}} \tag{35.2}$$

where ν stands for a neutrino. The energy yields of fusion reactions of hydrogen and helium are given in Tables 35.1 and 35.2.

3. Neutron capture is used to create radioactive isotopes in which the nuclear charge (Z, the atomic number) is unchanged, the nuclear mass (A = number of protons + neutrons is the atomic mass) increases by one, and the number of neutrons (N) increases by one.

Note

N is always equal to $A-Z$.

4. Other decay modes, in which nuclei spontaneously eject or capture one or more particles and loose energy to become nuclei of lighter atoms.

TABLE 35.1 Energy yields of fusion reactions of hydrogen

Reaction	Energy yield (MeV)
$p+p\rightarrow {}_1^2\text{H}+e^{+}+\nu$	0.42
$e^{+}+e^{-}\rightarrow\gamma$	1.02
${}_1^2\text{H}+p\rightarrow {}_2^3\text{He}+\gamma$	5.49
${}_2^3\text{He}+{}_2^3\text{He}\rightarrow {}_2^4\text{He}+p+p$	12.86
Total	**26.72**

TABLE 35.2 Energy yields of fusion reactions of helium

Reaction	Energy yield (MeV)
${}_2^3\text{He}+{}_2^4\text{He}\rightarrow {}_4^7\text{Be}+\gamma$	1.59
${}_4^7\text{Be}+p\rightarrow {}_5^8\text{B}+\gamma$	1.3
${}_5^8\text{B}\rightarrow {}_5^8\text{Be}+e^{+}+\nu$	10.78
${}_4^8\text{Be}\rightarrow {}_2^4\text{He}+{}^4\text{He}$	0.095
Total	**13.765**

Note that fission and neutron capture start initially as same but the difference is that in fission the excited nucleus (with the extra neutron) is too unstable to stay together. In neutron capture, the new nucleus is simply radioactive, i.e. it participates in the decay modes of the next series.

In this set of reactions, the first three steps occur twice for each occurrence of the fourth step.

This process takes place in all main reaction sequences in the stars. For the Sun, it will continue for about another 5 billion years, when hydrogen (p) supply will exhaust. It will then swell to encompass the orbits of the first four planets as a Red Giant. Then helium will be the main fuel.

After helium is exhausted, more massive nuclei will fuse until iron is produced. The fusion of iron is endothermic (requires more energy than it produces), so the star collapses under its own weight and each of them forms either a white dwarf (the remnant of a small star), or it will be a supernova (explode) and the core will become a neutron star (whose density is that of nuclei) or a black hole (from which nothing can escape even light!), depending on its initial mass. The Sun will probably end up as a dwarf, since its mass is below the Chandrasekhar mass required to become a black hole. It is interesting to note that everything in our solar system except primordial hydrogen and helium which were made during the creation of the universe, including all of us, is made from matter that was created by fusion processes in a star and dispersed into space in a supernova explosion. For this reason, the Sun is a second (or perhaps third, etc.) generation star.

35.3 ▶ Artificial Nuclear Transmutations

The nuclear structure can be investigated using nuclear reactions. It is possible to change nuclei by bombarding them with energetic particles a. Such reactions as below are called nuclear reactions.

$$a + X \rightarrow Y + b$$

where X is the initial nuclei and Y is the product nuclei.

Typical bombarding particles (a) might be protons, neutrons, deuterons, α-particles or other ions. The charged projectile must have enough kinetic energy to penetrate the region of Coulomb repulsion around the nucleus, otherwise Rutherford scattering might occur. The energy of the projectiles capable of causing reactions must be of the order of several MeV.

Such reactions obey conservation laws. Conservation of mass number, the total number of nucleons must be the same after the collision as before, conservation of charge and conservation of energy

$$M_X c^2 + T_a + M_a c^2 = M_Y c^2 + T_y + M_b c^2 + T_b$$

The total kinetic energy released or absorbed is called the reaction energy.

$$\boxed{Q = (T_Y + T_b) - T_a = (M_X + M_a - M_Y - M_b)c^2} \tag{35.3}$$

Conservation of angular momentum constrains the orbital and nuclear spin quantum numbers of the participating nuclides. The conservation laws lead to kinematical relations. For a reaction,

$$T_a + M_a^2 c^2 + M_X c^2 = M_Y c^2 + T_y + M_b c^2 + T_b$$

$$Q = (M_a + M_X - M_Y - M_b)c^2 = T_y + T_b - T_a$$

35.4 ▶ Nuclear Cross-sections

The cross-section σ is a measure of the probability of a particular nuclear reaction that will occur. The probability that an incident particle will interact with the target depends on the ratio of the effective area of the target nucleus to the area of the target. Each target nucleus X has an effective area σ called the cross-section. Cross-section will also depend on the energy of the incident target.

Consider a target foil of thickness x and area A with n target nuclei per unit volume. The total numbers of target nuclei are nAx. The total area exposed to the incident beam is σnAx (for a sufficiently thin target). The ratio of the rate of interactions R to the rate of incident particles R_0 is

$$\boxed{\frac{R}{R_0} = \frac{\sigma n A x}{A} = \sigma n x} \tag{35.4}$$

Likewise, the number of particles penetrating the target is given by (for dN particles interacting with the target, N is the number of incident particles)

$$-\frac{dN}{N}=\frac{nA\sigma dx}{A}=n\sigma dx$$

$$\int_{N_0}^{N}\frac{dN}{N}=-n\sigma\int_{0}^{x}dx$$

$$N=N_0\,e^{-n\sigma x}$$

So, for N_0 incident particles, N is the number of particles emerging out from the target.

Nuclear cross-sections have dimensions of area and units of the barn (b) defined as 1 barn $=10^{-28}$ m^2 are used.

Weak interaction cross-sections are much smaller than strong interaction cross-sections: e.g., the cross-section for the reaction $(\nu+p\rightarrow e^{+}+n)$ is of the order of 10^{-19}b in comparison to 4b for the reaction $(n+{}^{127}I\rightarrow{}^{127}I^{*}+n)$.

35.5 ▶ Nuclear Fission

Han and Strassman discovered in 1939 that the bombardment of uranium by neutrons produces elements that belong to the middle portion of the periodic table. A very large amount of binding energy is released in fission of a nucleus with a large atomic number into two nuclei of intermediate atomic number because of the reduction in the coulomb energy.

$$\boxed{Q\,(\text{prompt})=T_{y_1}+T_{y_2}} \qquad (35.5)$$

$$=[M({}^{235}\text{U})+M_n-(M_{Y_1}+M_{Y_2})]c^2$$

$$=B_{Tot}(Y_1)+B_{Tot}(Y_2)-B_{Tot}({}^{235}\text{U})$$

$$\sim 170\text{ MeV}$$

The total energy that is released including β decays and γ decays is found to be close to 200 MeV per fission, which is carried away largely by the two fission fragments along with 2 to 3 neutrons. These neutrons can be used to induce other uranium nuclei to fission, leading to a chain reaction. Representation of steps in the fission process with the liquid drop model of the nucleus is shown in Fig. 35.3.

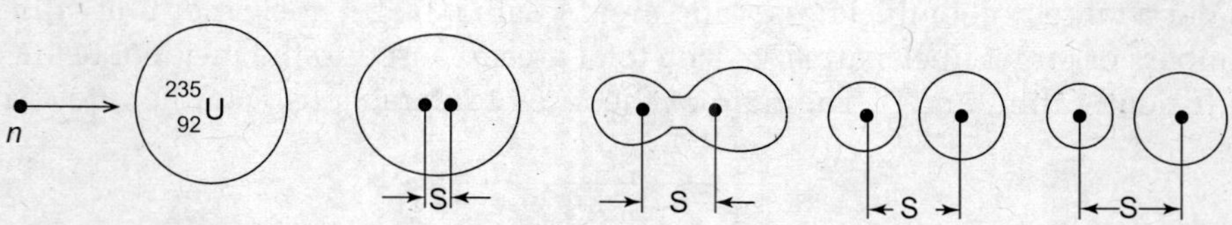

FIGURE 35.3 Representation of steps involved in the fission process in the liquid drop model

Consider a symmetric decay into two equal fragments. In the liquid drop model, the volume energy remains unchanged. Neglecting pairing energy, coulomb energy changes from

$$-a_c \frac{Z^2}{A^{1/3}} \quad \rightarrow \quad \frac{-2a_c (Z/2)^2}{(A/2)^{1/3}} - \frac{(Z/2)^2 e^2}{2R} + 0.9 a_c Z^2 / A^{1/3}$$

resulting in a change of surface energy by

$$-a_s A^{2/3} - 2a_s (A/2)^{2/3} \quad \text{or} \quad -0.26 a_s A^{2/3}$$

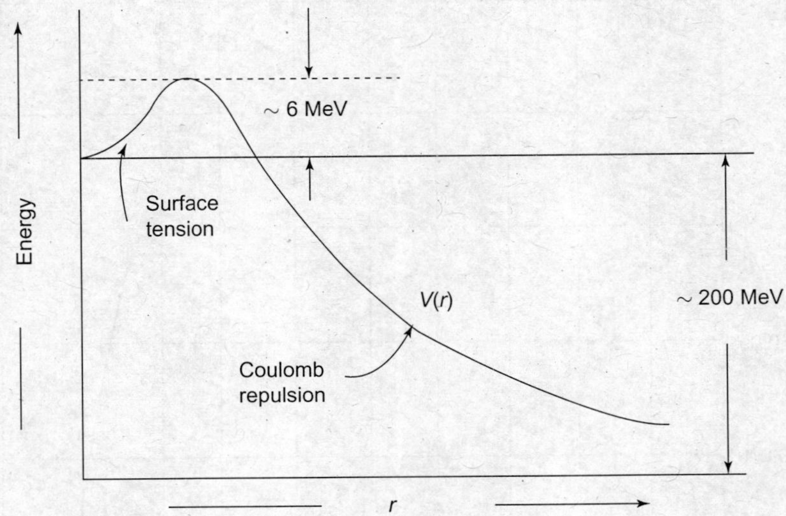

FIGURE 35.4 Energy diagram of a fissionable nucleus. The energy >6 MeV is required to overcome the fission barrier

In induced fission reaction, a nucleus captures a low-energy neutron. An incident particle enters the nucleus and transfers its energy to the nucleus, exciting it into a collective state where the energy is shared. The compound nucleus does not remember the way it was formed, so its decay should be independent of the way it was produced. As the binding energy of the last neutron in a high Z nucleus is about 6 MeV, the capturing nucleus receives enough energy to take it over to the top of the fission barrier (Fig. 35.4) and the nucleus splits in to two parts.

The excitation energy changes to collective vibrations in nucleus, which become sufficiently elongated to fission. The fission fragments are not usually symmetrical, with one of the fragments having Z and N values near the magic numbers 50 and 82 that $A = 132$ should be a prominent mass number on the final mass yield curve (Fig. 35.5).

But both fragments have the same Z/N ratio as the fissioning nucleus. These Z/N ratios are smaller than those of stable nuclei with these A values. Therefore, they undergo succession of β decays to bring them to stability curve. The fragments are very neutron-rich; hence, the prompt neutron emission is favoured. A single fission may yield 2 or 3 neutrons. Some reactions also have delayed neutron emission in which a daughter nucleus is insufficiently excited (above the neutron separation energy) to allow it to emit a neutron, leaving a stable nucleus. For example,

$$^{87}_{36}\text{Kr}^* \rightarrow {}^{87}_{36}\text{Kr}^* + e^- \rightarrow {}^{86}_{36}\text{Kr} + n$$

with characteristic $\tau_{1/2} = 55$ sec for $^{87}_{35}$ Br.

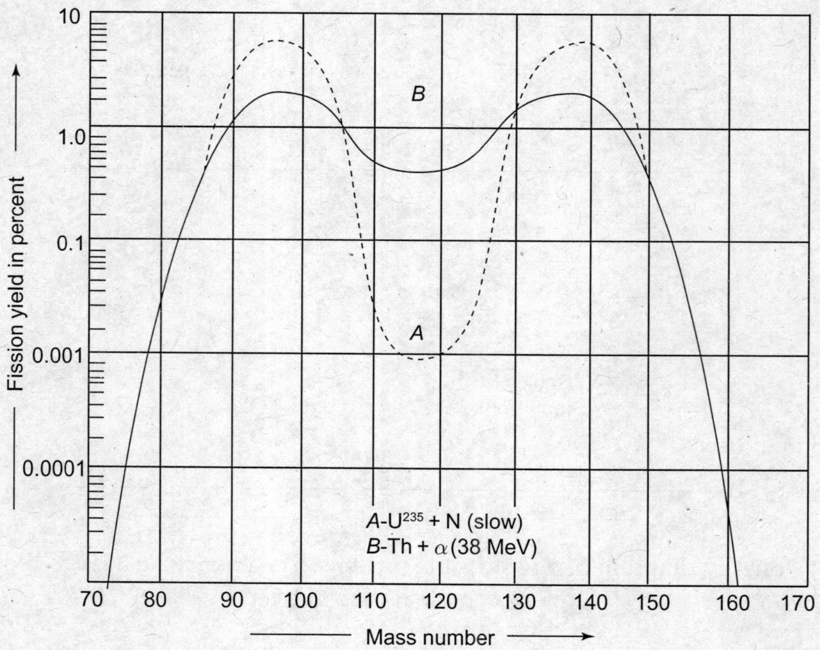

FIGURE 35.5 Mass yield curves for nuclear fission reaction of Uranium and Thorium. The atomic numbers of most fission fragments are centred around magic numbers 50 and 82

So, a typical fission event might be

$$n + {}^{235}\text{U} \rightarrow [{}^{236}\text{U}] \rightarrow {}^{95}\text{Sr} + {}^{141}\text{Xe} \rightarrow {}^{94}\text{Sr} + {}^{140}\text{Xe} + 2n$$

and through β decay:

$$\text{Sr}^{94} \rightarrow {}^{94}\text{Y} \rightarrow {}^{94}\text{Zr (stable)}$$

The total release of energy in U fission is roughly estimated as below

	MeV
Kinetic energy of fission products	167
Kinetic energy of prompt neutrons	5
Energy of instantaneous γ radiations	6
β particle energy including neutrinos	20
Secondary γ radiations	6
Total energy	**204**

^{238}U requires more energy to fission than ^{235}U because of the energetics of the pairing term. Hence, it will fission only with a fast neutron. So an odd-A nuclei can undergo fission with thermal neutrons while even–even nuclei have a high fission threshold.

Note

The fission fragments of all atoms in the reactor are not same. Some atoms disintegrate with different energies and products.

35.6 ▶ Nuclear Fusion

Nuclear fusion is the process in which two light nuclei ($A < 20$) combine to form a heavier nucleus. The mass of the final nucleus is less than the combined rest masses of the original nuclei; hence, there is a release of energy. Examples of fusion reactions that take place in the Sun are as follows:

$$ {}_1^1\mathrm{H} + {}_1^1\mathrm{H} \rightarrow {}_1^2\mathrm{H} + e^+ + \nu + 0.42\,\mathrm{MeV} $$

$$ {}_1^1\mathrm{H} + {}_1^2\mathrm{H} \rightarrow {}_2^3\mathrm{He} + \gamma + 5.49\,\mathrm{MeV} $$

$$ {}_1^1\mathrm{H} + {}_2^3\mathrm{He} \rightarrow {}_2^4\mathrm{He} + e^+ + \nu $$

$$ {}_2^3\mathrm{He} + {}_2^3\mathrm{He} \rightarrow {}_2^4\mathrm{He} + {}_1^1\mathrm{H} + {}_1^1\mathrm{H} + 12.86\,\mathrm{MeV} $$

These are the basic reactions in the proton–proton cycle, one of the basic cycles for energy generation in the Sun where about 25 MeV of energy is released. About 2 per cent of the energy is carried away by neutrinos which is not accounted for. When the density of helium becomes high, carbon is formed as follows:

$$ {}_2^4\mathrm{He} + {}_2^4\mathrm{He} \rightarrow {}_4^8\mathrm{Be} $$

$$ {}_2^4\mathrm{He} + {}_2^8\mathrm{Be} \rightarrow {}_6^{12}\mathrm{C} $$

When enough carbon is produced, carbon cycle is initiated as follows:

$$ {}_1^1\mathrm{H} + {}_6^{12}\mathrm{C} \rightarrow {}_7^{13}\mathrm{N} $$

$$^{13}_{7}\mathrm{N} \rightarrow ^{13}_{6}\mathrm{C} + e^{+} + \nu$$

$$^{1}_{1}\mathrm{H} + ^{13}_{6}\mathrm{C} \rightarrow ^{14}_{7}\mathrm{N}$$

$$^{1}_{1}\mathrm{H} + ^{14}_{7}\mathrm{C} \rightarrow ^{15}_{8}\mathrm{O}$$

$$^{15}_{8}\mathrm{O} \rightarrow ^{15}_{7}\mathrm{N} + e^{+} + \nu$$

$$^{1}_{1}\mathrm{H} + ^{17}_{7}\mathrm{N} \rightarrow ^{12}_{6}\mathrm{C} + ^{4}_{2}\mathrm{He}$$

Carbon is not consumed but participates as a catalyst. The quantity of carbon is negligible in the Sun in comparison with heavier stars. At higher temperatures, we get $^{16}_{8}\mathrm{O}$, $^{20}_{10}\mathrm{Ne}$, $^{24}_{12}\mathrm{Mg}$ and finally $^{56}_{26}\mathrm{Fe}$. Elements heavier than iron are not formed by fusion since at $A = 60$ the binding energy per nucleon maximizes.

Nuclei up to $^{209}_{83}\mathrm{Bi}$ are formed by neutron capture and β decay. When stars run out of hydrogen, they collapse under gravitational force and convert to supernova where heavier elements are produced.

Advantage of fusion reaction is that the fuel is inexpensive and a few radioactive byproducts are produced. The proton–proton interactions of the Sun require too high pressures and densities for a fusion reactor, instead if we use the following deuterium and tritium reactions:

$$\boxed{^{2}_{1}\mathrm{H} + ^{2}_{1}\mathrm{H} \rightarrow ^{3}_{2}\mathrm{He} + n + 3.2\,\mathrm{MeV}} \tag{35.6}$$

$$^{2}_{1}\mathrm{H} + ^{2}_{1}\mathrm{H} \rightarrow ^{3}_{1}\mathrm{H} + ^{1}_{1}\mathrm{H}$$

$$^{2}_{1}\mathrm{H} + ^{3}_{2}\mathrm{H} \rightarrow ^{4}_{2}\mathrm{He} + n$$

The main problem is that the coulomb repulsion between two charged nuclei must be overcome before they fuse. Thus, they need enough kinetic energy to overcome this barrier, which can be achieved by heating them to very high temperatures to the order of 10^8 K. Because particles have Boltzmann distribution, the average energy for a given temperature can be lower, since the tail of the distribution will have higher energy. The temperature at which the power generation exceeds the loss rate is called the critical ignition temperature.

The critical ignition temperature for deuterium–deuterium reaction is 4×10^3 K corresponding to 35 keV, while for deuterium–tritium reaction it is 4.5×10^7 K corresponding to 4 keV. There is energy loss due to bremsstrahlung radiation as a result of electron–ion collisions in the plasma. Two other critical parameters include the following:

1. Ion density n.
2. Confinement time τ, the time for which the ions are at a greater temperature than the ignition temperature.

The criterion for net energy output known as Lawson's criterion is given by

$$n\tau \geq 10^{14}\ \text{s/cm}^3 \quad \text{(D-T reaction)}$$

$$n\tau \geq 10^{16}\ \text{s/cm}^3 \quad \text{(D-D reaction)}$$

based on comparing the energy required to heat the plasma, $E_h = D_1 n$ to the energy generated by the fusion process $E_f = D_2 n^2 \tau$. D_1 and D_2 depend on the reaction.

Such fusion reactions are not feasible in any metallic chambers because such high temperatures cannot be withstood by any material. The magnetically compressed plasmas can create such high temperature, but confining these is difficult.

There are two methods to confine the plasmas, magnetic field confinement and inertial confinement. The concept of magnetic field confinement is shown in Fig. 35.6.

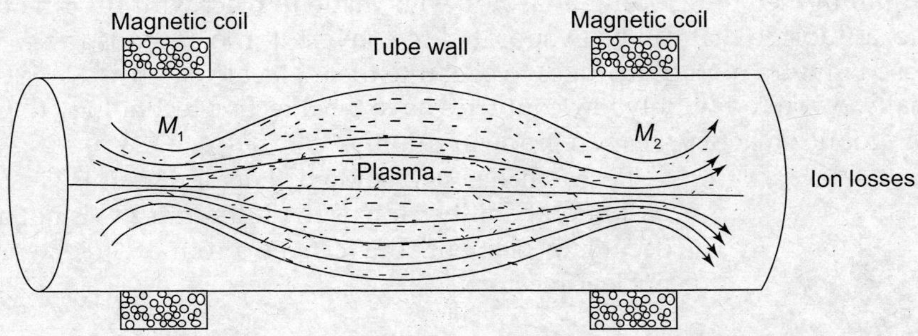

FIGURE 35.6 Magnetic confinement of plasma for controlled fusion reaction in laboratory

35.7 ▶ Controlled Nuclear Fission

To maintain a sustained controlled reaction, for every 2 or 3 neutrons released, a neutron must be allowed to strike another uranium nucleus. If this ratio is less than one, then the reaction will die out. If it is greater than one, it will go uncontrolled as an atomic explosion. Nuclear reactions are controlled by a neutron-absorbing material, such as graphite, cadmium and heavy water (Fig. 35.7).

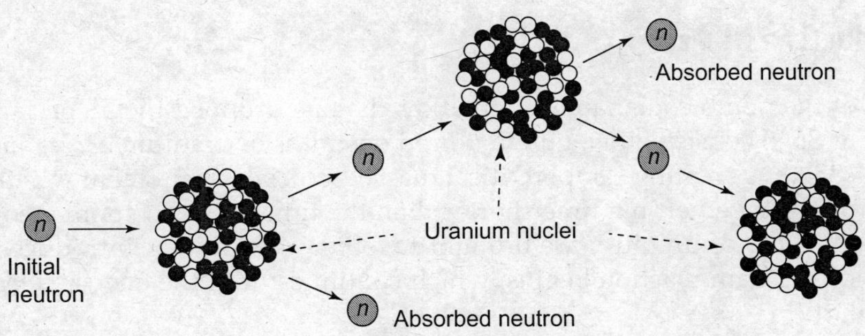

FIGURE 35.7 Controlled nuclear fission reaction of ^{235}U

35.7.1 Types of Nuclear Fuel

Scientists knew that the most common isotope, uranium 238, was not suitable for a nuclear weapon. There is a fairly high probability that an incident neutron can be captured to form uranium 239, instead of causing a fission. However, uranium 235 has a high fission probability.

Of natural uranium, only 0.7 per cent is uranium 235. This means that a large amount of uranium is needed to obtain the necessary quantities of uranium 235. Also, uranium 235 cannot be separated chemically from uranium 238, since the isotopes are chemically similar.

Alternative methods have been developed to separate the isotopes. This was the problem faced in the first atom bomb project. The process of increasing the contacts of U^{235} in U is called Uranium enrichment. The enrichment of uranium is done in gaseous form. First, the uranium metal (including all isotopes) is made to react with fluorine to produce Uranium hexa-fluoride that can be evaporated to convert it into gaseous form. This gas is then processed in ultra high centrifuges to separate the molecules with ^{239}U, ^{238}U and ^{235}U gravitationally. As on today, only few countries have the effective technology to process the uranium to produce reactor/weapon grade uranium.

Research has also shown that plutonium 239 has a high fission probability. However, plutonium 239 is not a naturally occurring element, but can be produced as byproducts in nuclear reactors.

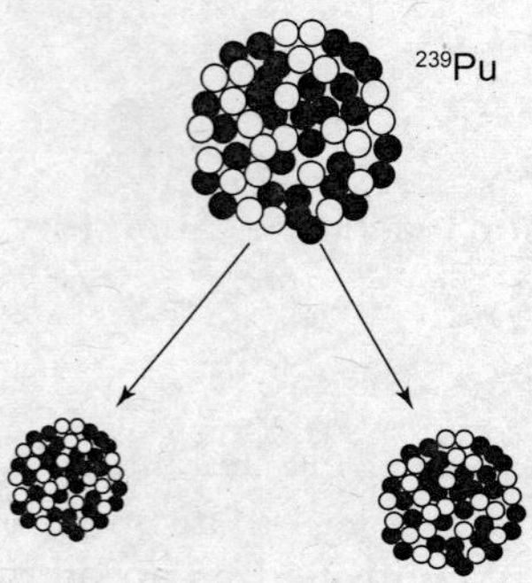

FIGURE 35.8 Spontaneous fission in ^{239}Pu

35.7.2 Spontaneous Fission

The spontaneous fission rate is the probability per second of a given atom to undergo fission spontaneously, that is, without any external intervention. If a spontaneous fission occurs before the bomb is ready, it could fizzle. Plutonium 239 has a very high spontaneous fission rate compared to the spontaneous fission rate of uranium 235 (Fig. 35.8). Scientists have to consider the spontaneous fission rate of each material when designing nuclear weapons.

35.8 ▶ Nuclear Bomb

The design of the bomb code named Little Boy, the one dropped by US on Japan in 1945 (shown in Fig. 35.9) consisted of a gun that fired one mass of uranium 235 at another mass of uranium 235, thus creating a supercritical mass. A crucial requirement was that the two pieces be brought together in a time shorter than the time between spontaneous fissions. Once the two pieces of uranium are brought together, the initiator introduces a burst of neutrons and the chain reaction begins, which continues until the energy released blows the bomb apart.

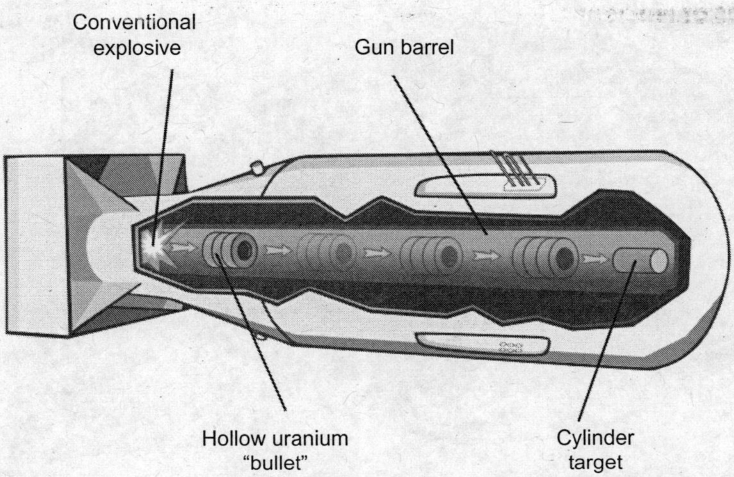

FIGURE 35.9 The design of fission bomb 'Little Boy' a Uranium-based fission bomb that was dropped at Hiroshima during World War-II by USA

Types of bombs

There are mainly three types of nuclear bombs, namely,

(i) Fission bomb (uranium, plutonium based)

(ii) Fusion bomb (hydrogen based, also called hydrogen bomb)

(iii) Neutron bomb (destroys only living creatures)

35.8.1 Fission Bomb

The fission bomb is a device that has fuel (fissile material) in two or more separate subcritical masses. These are made close to form a critical mass that is an essential condition for fission reaction to take place. This is achieved just before exploding the device either by mechanical or by some small chemical explosion mechanism. The neutron source placed at the centre of critical mass initiates the reaction that is uncontrolled one and hence highly explosive and destructive. It can raise the temperature of atmosphere to $\sim$ 6000°K and all living things can be completely destroyed by the enormous heat and the shock waves generated by the explosion. A typical construction of fission bomb (normally referred as atom bomb) code named as Fat Man is shown in Fig. 35.10.

The first nuclear bomb code named 'Little Boy' was dropped on Hiroshima, Japan on August 6, 1945. The second bomb code named 'Fat Man' was dropped over Nagasaki by USA during the Second World War. The little boy had an explosive power of $\sim$ 15 kilo TNT and weighed $\sim$ 4500 kg.

FIGURE 35.10 The Fat Man, a Uranium-based fission bomb that was dropped by USA at Nagasaki during World War-II

35.8.2 Fusion Bomb

The fusion bomb is also called hydrogen bomb that is more powerful than a fission bomb. The hydrogen or its another isotope deuterium is the fuel used in this type of bomb. In fact, it comprises a fission bomb along with hydrogen/deuterium as fuel. The fission bomb creates high temperature required for fusion reaction of hydrogen/deuterium to take place. As discussed in the previous sections, the nuclei of hydrogen, deuterium, tritium, etc. fuse together to release enormous amount of energy. The important features of fission and fusion bombs are compared in Table 35.3.

TABLE 35.3 Comparison of fission and fusion bombs

Fission (Atomic) bomb	Fusion (Hydrogen) bomb
(i) $U^{235} - Pu^{239}$	${}_1H^1$, ${}_1H^2$, ${}_1H^3$ is used as fuel
(ii) Neutron initiates chain neutron (uncontrolled)	Heat (from fission reaction) initiates fusion reaction
(iii) Explosion $\sim$ 1500–20,000 Tons of TNT and up to 500,000 Tons of TNT have been tested	About 1000 times powerful than fission bombs $\sim$ (10 Mega tons) of TNT. The most powerful bomb has a yield of 50 megatons of TNT
(iv) It is a single stage	Involves two stage explosion by fission that leads to fusion sufficient to build
(v) Easy to design and make	Difficult to build

35.9 ▶ CONSTRUCTION AND WORKING OF HYDROGEN BOMB

The first thermonuclear bomb was exploded in 1952 by the United States. Because extremely high temperatures are required in order to initiate fusion reactions, the hydrogen bomb is also known as a thermonuclear bomb. Hydrogen bomb derives a large portion of its energy from the nuclear fusion of hydrogen isotopes.

The construction of a typical hydrogen or thermonuclear bomb is shown in Fig. 35.11. At its centre is an conventional fission bomb that is surrounded by a layer of lithium deuteride (a compound of lithium and deuterium, the isotope of hydrogen with mass number 2) and around it is a thick outer layer, frequently of fissionable material, that holds the contents together in order to obtain a larger explosion. Neutrons from the atomic explosion cause the lithium to fission into helium, tritium (the isotope of hydrogen with mass number 3) and energy as below.

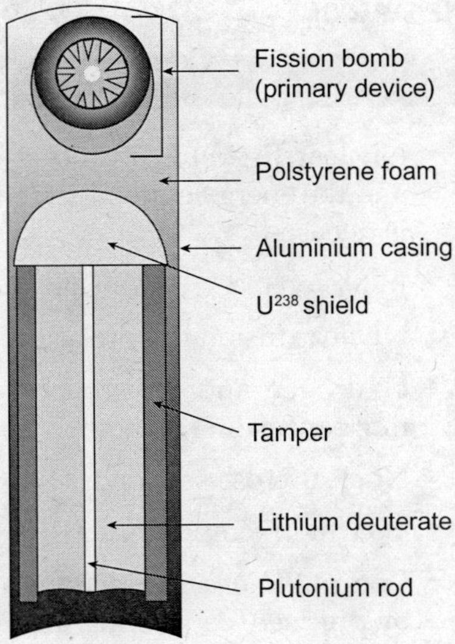

FIGURE 35.11 Construction of a typical hydrogen bomb

This reaction produces the tritium on the spot, so there is no need to include tritium in the bomb itself. In the extreme heat which exists in the bomb, the tritium fuses with the deuterium in the lithium deuteride. The atomic explosion also supplies the temperatures needed for the subsequent fusion of deuterium with tritium, and of tritium with tritium (50,000,000°C and 400,000,000°C, respectively). Enough neutrons are produced in the fusion reactions to produce further fission in the core and to initiate fission in the surrounding material.

Like other types of nuclear explosion, the explosion of a hydrogen bomb creates an extremely hot zone near its centre. In this zone of the high temperature, nearly all of the matter present is vaporized to form a gas at extremely high pressure. A sudden pressure far in excess of atmospheric pressure, propagates away from the centre of the explosion as a shock wave, containing most of the energy that is responsible for the major part of the destructive effects.

Since the fusion reaction produces mostly neutrons and very little radioactivity and therefore less radioactive fallout. Similarly the neutron bomb, developed in 1958 by the United States and successfully tested, produces blast effects and a hail of lethal neutrons but almost no radioactive fallout and little long-term contamination. This would cause minimal physical damage to buildings and equipment but kill most living things.

35.10 ▶ NEUTRON BOMB

The neutron bomb is also called as enhanced radiation weapon. It has less destruction power but kills all living beings by very high neutron radiation dose. The neutrons of high energy are released that can penetrate buildings and structures without physically damaging them but killing all living beings in the velocity. In principle, it works on the basis of a fission reaction that release high flux of energetic neutrons.

RECAP ZONE

POINTS TO REMEMBER

- Nuclear fission is a process where heavy nuclei are disintegrated into smaller ones and the energy equivalent to the difference of binding energy of atom and its products is released.
- In fission of U^{235} atom, an energy of $\sim$ 209 MeV per fission is released.
- Nuclear fusion is a process where lighter nuclei fuse together to form heavier ones.
- Hydrogen and its isotopes, deuterium, tritium and helium, are the main fuels in nuclear fusion reaction.
- Nuclear cross-section is the probability of interaction of incident particle with target.
- Very high temperatures $\sim 10^3$ to 10^8 K are required for fusion reactions to take place.
- Controlled fission reactions are the main processes that take place in nuclear reactors by precisely controlling the number of free neutrons available for reactions.
- Due to natural radioactivity, some atoms of radioactive isotopes may undergo spontaneous fission.
- In nuclear bombs (fission or fusion), the reactions are controlled by triggers but once triggered these are uncontrolled ones and cannot be stopped. That's why the use of nuclear weapons is highly destructive.

KEY TERMS WITH DEFINITIONS

- **Fission** – It is a process of disintegrating heavy nuclei into small fragments.
- **Fusion** – It is a process of fusing light nuclei to form heavier nuclei.
- **Nuclear reactions** – They are independent of chemical reactions
- **Nuclear reaction cross-section** – It is a measure of probability of interaction of a particle with the nucleus.

IMPORTANT FORMULAE AND EQUATIONS

Equation Number	Equation	Remarks
(35.1)	$n +_{92} U^{235} \rightarrow {}^{141}_{56}Kr + 3n + 200$ MeV	The fission of $_{92}U^{235}$
(35.2)	$p + p \rightarrow {}^{2}_{1}H + e^{+} + \nu + 0.42$ MeV	

(35.3) $$Q=(T_r+T_b)-T_a=(M_x+M_a-M_r-M_b)c^2$$

(35.4) $$\sigma=\frac{R}{R_o}\frac{A}{nA_x}=\frac{R}{R_o}\frac{1}{nx}$$

(35.5) $$Q\ (\text{Prompt})=T_1-T_2$$

(35.6) $${}_1^2\mathrm{H}+{}_1^2\mathrm{H}\rightarrow{}_2^3He+n+3.2\ \mathrm{MeV}$$

REVIEW ZONE

NUMERICAL PROBLEMS

1. Calculate the energy released in fission of 1 g of U^{235}. (Hint: Energy released per fission ~ 200 MeV)
2. Calculate the amount of consumption of U^{235} per year in a nuclear reactor which is operating at 250 Mega Watts. (Hint: Energy released per fission ~ 200 MeV)
3. Calculate the energy released if 1 g of Deuterium is fused to form Helium. (Hint: Energy released in two deuterium fusion ~ 3.2 MeV)
4. Calculate the energy in joules required to break 1 g of Helium nuclei into hydrogen nuclei. (Hint: The binding energies of $_1H^2$, $_1H^3$, Δ_2He^4)
5. Calculate the energy released in the following nuclear reaction:

$$ {}_3^6\mathrm{Li}+{}_0^1n\rightarrow{}_2^4\mathrm{He}+{}_1^3\mathrm{H}+\mathrm{Q}$$

(Use masses ${}_3^6\mathrm{Li}$ = 6.015 amu, ${}_2^4\mathrm{He}$ = 4.0 amu, ${}_0^1n$ = 1 amu, ${}_1^3\mathrm{H}$ = 3.01 amu)

APPENDIX I

Fundamental Physical Constants

Constant/Symbol	Value	Value in other forms
Planck constant h	$6.6260755.10^{-34}$ J.s	$\frac{h}{(2\pi)} = 1.05457266.^{10-}34$ J.s
Boltzmann constant k_B	$1.380658.\ 10^{-23}$ J/K	(= $8.617385.10^{-5}$ eV/K)
Elementary charge e	$1.60217733.10^{-19}$ C	
Avogadro number N_A	$6.0221367.10^{23}$ particles/mol	
Speed of light c	$2.99792458.10^{8}$ m/s	
Permeability of vacuum μ_0	$\mu_0 = 4\pi.10^{-7}\ T^2.m^3/J$	$12.566370614.10^{-7}.m^3J$
Permittivity of vacuum ε_0	$\varepsilon_0 = 1/(\mu_0 c^2)$	$8.854187817.10^{-12}\ C^2/J.m$
Fine structure constant α	1/137.0359895	
Electron rest mass m_e	$9.1093897.10^{-31}$ kg	
Proton rest mass m_p	$1.6726231.10^{-27}$ kg	
Neutron rest mass m_n	$1.6749286.10^{-27}$ kg	
Bohr magneton μ_B	$\mu_B = \frac{eh}{(4\pi m_e)}$	$9.2740154.10^{-24}$ J/T
Nuclear magneton μ_N	$\mu_N = \frac{eh}{(4\pi m_p)}$	$5.0507866.10^{-27}$ J/T
Free electron g factor g_e	2.002319304386	
Proton magnetic moment μ_p	$1.41060761.10^{-26}$ J/T	
Charge-to-mass ratio for the electron	e/m_e	$1.75880.10^{11}$ C/kg
Atomic mass unit amu	$1.66057.10^{-27}$ kg	
Bohr radius a_0	$5.29177.10^{-11}$ m	
Electron radius r_e	$2.81792.10^{-15}$ m	
Proton g factor (Lande factor) g_H	5.585	
Gravitational constant G	$(6.673 \pm 0.010).10\ m^3/kg.s^2$	
Acceleration due to gravity g	$9.80665\ m/s^2$	
Compton wavelength of the electron	$\lambda_c = h/(m_e c)$	$2.42631.10^{-12}$ m

APPENDIX II

Greek Alphabets

Α α alpha a	**Β β** beta b	**Γ γ** gamma g	**Δ δ** delta d	**Ε ε** epsilon e	**Ζ ζ** zeta z
Η η eta $\bar{e}$	**Θ θ** theta th	**Ι ι** iota i	**Κ κ** kappa k	**Λ λ** lambda 1	**Μ μ** mu m
Ν ν nu n	**Ξ ξ** xi ks, x	**Ο ο** omikron o	**Π π** pi p	**Ρ ρ** rho r, rh	**Σ σ** sigma s
Τ τ tau t	**Υ υ** upsilon u, y	**φ Φ** phi ph	**Χ χ** chi Kh, ch	**Ψ ψ** psi ps	**Ω ω** omega $\bar{o}$

APPENDIX III

Standard Notations of Units of Measure

ampere	A	electrostatic unit	esu
ampere-hour	A-h	erg	erg
angstrom	Å	farad	F
astronomical unit	AU	feet per second	ft/s, fps
atomic mass unit	atu	fermi (femtometre)	fm
atomic per cent	at.%	Foot	ft
atomic unit	a.u.	foot-candle	ft-c
atomic weight	at. wt.	foot-pound	ft.lb
barn	b	gallon	gal
baud	Bd	gauss	G
becquerel	Bq	giga (prefix)	G
billion electron volt	BeV	gram	g
British thermal unit	Btu	gravity	*g* (italic)
calorie	cal	gray	Gy
centimetre	cm	henry	H
coulomb	C	hertz (cycles per second)	Hz
counts per minute	counts/min	horsepower	hp
counts per second	counts/s	hour	h
cubic centimetre	cm^3	inch	in.
cubic foot	ft^3	inch-pound	in.lb
curie	Ci	inches per second	In./s, ips
cycles per second	cps	joule	J
decibel	dB	Kelvin	K (no °)
degrees Celsius (centigrade)	°C	kilogram	kg
degrees Fahrenheit	°F	kilovolt ampere	kVA
electromagnetic unit	emu	kilowatt-hour	kWh
electron unit	e.u.	lumen	lm
electron volt	e.V.	lux	lx

mega (prefix with other units)	M
metre	m
micro (prefix with other units)	μ
micrometre	μm
micron (micrometre preferred)	μ
miles per hour	mph
milli (prefix with other units)	m
molecular weight	mol wt
nano (prefix with others units)	n
nautical mile	n mile
newton	N
miles per hour	mph
oersted	Oe
ohm	Ω
parts per million	ppm
pascal	Pa
pico (prefix with other units)	P
pound	lb
radian	rad
revolutions per minute	rpm
second (time)	s
second (angle)	s
sievert	Sv
steradian	sr
tesla	T
volt	V
watt	W

APPENDIX

IV The SI Units

absorbed dose rate	gray per second	Gy/s
acceleration	metre per second squared	m/s^2
activity (of a radionuclide)	becquerel	Bq
angular acceleration	radian per second squared	rad/s^2
angular velocity	radian per second	rad/s
area	square metre	m^2
capacitance	farad	F (also C/V; $m^{-2}.kg^{-1}.s^4.A^2$)
Temperature	degree Celsius	°C
current density	ampere per square metre	A/m^2
dose equivalent	sievert	Sv (also J/kg; $m^2.s^2$)
electric charge density	coulomb per cubic metre	C/m^3
electric charge, quantity of electricity	coulomb	C (also s.A)
electric conductance	siemens	S (also A/V; $m^{-2}. kg^{-1}. s^{-3}.A^2$)
electric current	ampere	A
electric field strength	volt per metre	V/m
electric flux density	coulomb per square metre	C/m^2
electric potential difference,	volt	V (also W/A; $m^2.kg.s^{-3}.A^{-2}$)
electric resistance	ohm	W (also V/A; $m^2.kg.s^{-3}.A^{-2}$)
energy density	joule per cubic metre	J/m3
energy, work, quantity of heat	joule	J (also N.m; $m^2.kg.s^2$)
force	newton	N (also $m/kg.s^2$)
frequency	hertz	Hz (also s^{-1})
heat capacity, entropy	joule per Kelvin	J/K
illuminance	lux	lx (also lm/m^2)
inductance	henry	H (also Wb/A; $m^2.kg.s^{-2}.A^{-2}$)
length	metre	m

luminance	candela per square metre	cd/m2
luminous flux	lumen	lm (also cd.sr)t
luminous intensity	ampere per metre	A/m
magnetic field strength	testla	T (also Wb/m^2; $kg.s^{-2}.A^{-1}$)
magnetic flux	weber	Wb (also V.s; m^2 $kg.s^{-2}.A^{-1}$)
mass density	kilogram per cubic metre	kg/m^3
mass	kilogram	kg
moment of force	newtron metre	N.m
penneability	henry per metre	H/m
permitivity	farad per metre	F/m
plane angle	radian	rad
power, radian flux	watt	W (also J/s; $m^2.kg.s^{-3}$)
pressure, stress	pascal	Pa (also N/m^2; m–1.$kg.s^{-2}$)
radiance	watt per square metre steradian	$W/(m^2.sr)$
radiant intensity	watt per steradian	W/sr
solid angle	steraidan	sr
specific energy	joule per kilogram	J/kg
specific heat capacity, specific entropy	joule per kilogram Kelvin	J/(kg.K)
specific volume	cubic metre per kilogram	m^3/kg
speed, velocity	metre per seond	m/s
surface tension	newton per metre	N/m
thermal conductivity	watt per metre Kelvin	W/(m.K)
Temperature	K	Kelvin
time	second	s
volume	cubic metre	m^3
wave number	reciprocal metre	m^{-1}

APPENDIX

V Conversion Factors and Other Important Units

To convert	To	Multiply by
angstrom, Å	cm	10^{-8}
angstrom, Å	inn	0.1
astronomical unit, All	cm	1.496E13
feet (ft)	in	12
feet	m	0.3048
gauss	tesla	10^{-4}
gram (g)	pound	0.0022046
gram	kg	10^{-3}
horsepower	W	745.700
inch (in)	cm	2.54
inch (in)	mm	25.4
joule (J)	erg	10^{7}
joule	cal	0.239006
kilogram (kg)	g	10^{3}
kilogram	pound	2.20462
kilowatt	hp	1.34102
knot	mph	1.150779
litre	cm^3	10^{3}
litre	in^3	61.0237
meter	angstrom	1×10^{10}
meter	ft	3.28084
micron	cm	10^{-4}
mile	ft	5280
mile	km	1.60934
mm Hg	dyne cm^{-2}	1333.22

(Cont'd)

TABLE (Continued)

To convert	To	Multiply by
Newton	dyne	10^5
pound (Ipf)	newtron	4.4475
Torr	Millibar	1.333224
Torr	Millimeter of Hg	1
ton (Metric)	lb	2205
ton (Metric)	kg	1000
watt	$J\ s^{-1}$	1
yard	in	36
yard	m	0.9144
year	days	365.242198781
year	s	3.15576×10^7

Basic units

Quantity	Unit	Symbol
Length	meter	*m*
Time	second	*s*
Mass	kilogram	kg
Temperature	Kelvin	*K*
Electrical Current	ampere	*A*

Derived units

Quantity	Unit	Symbol	Dimensions
Force	Newton	*N*	$kg\ m\ s^{-2}$
Energy	joule	*J*	$kg\ m^2\ s^{-2}$
Power	watt	*W*	$kg\ m^2\ s^{-3}$
Frequency	hertz	*Hz*	s^{-1}
Charge	coulomb	*C*	$A\ s$
Capacitance	farad	*F*	$C^2\ s^2\ kg^{-1} m^{-2}$
Magnetic Induction	tesla	*T*	$kg\ A^{-1}\ s^{-2}$

SI prefixes

Y: 10^{-24} yotta	*Z*: 10^{-21} zetta
E: 10^{-18}exa	*P*: 10^{-15}pcta
T: 10^{-12} tcra	*G*: 10^{-9} giga
M: 10^{-6} mega	*K*: 10^{-3} kilo
h: 10^{-2} hecto	*da*: 10^{-1} deka
d: 10^{-1} deci	*c*: 10^{-2} centi
m: 10^{-13} milli	*μ*: 10^{-6} micro
n: 10^{-9} nano	*p*: 10^{-12} pico
f: 10^{-15} femto	*a*: 10^{-18} atto
z: 10^{-21} zepto	*y*: 10^{-24} yocto

Multiples of ten

	Unit	Symbol
10^{-12}	pico	*P*
10^{-9}	nano	*n*
10^{-6}	micro	*m*
10^{-3}	milli	*m*
10^{3}	kilo	*m*
10^{6}	mega	*M*
10^{9}	giga	*G*
10^{12}	tera	*T*

Fundamental Units

- *m* – meters (length)
- *kg* – kilogram (mass)
- *s* – seconds (time)
- *A* – ampere (electric current)
- *K* – kelvin (thermodynamic temperature)
- mol – mole (amount of substance)
- cd – candela (luminous intensity)

Derived Units

- *J* – joules (work, energy, heat) N . m, kg . m^2/s^2
- *N* – newton, (force) kg . $m/(s^2)$
- *Pa* – pascal, (pressure, stress) N/m^2
- *W* – watt, (power) J/s
- *C* – Coulomb, (electric charge) A . s
- *V* – volt (electric potential, emf) J/C, W/A
- Greek omega – ohm (resistance) V/A
- *S* – siemens (conductance) A/V
- *Wb* – weber (magnetic flux) V . s
- *H* – henry (inductance) Wb/A
- *F* – farad (capacitance) C/V
- *T* – Tesla (magnetic flux density) Wb/m^2, N/(A . m)
- *Bq* – becquerel (radioactivity) 1/s

Index

D

M

N